HEAT EXCHANGERS
Theory and Practice

HEAT EXCHANGERS
Theory and Practice

Edited by

J. Taborek
Heat Transfer Research, Inc.
Alhambra, California

G. F. Hewitt
Engineering Sciences Division and HTFS
U.K. Atomic Energy Authority
Harwell Laboratory, England

N. Afgan
International Centre for Heat and Mass Transfer
Belgrade, Yugoslavia

◉ HEMSIPHERE PUBLISHING CORPORATION
  Washington   New York   London

McGRAW-HILL BOOK COMPANY
  New York   St. Louis   San Francisco   Auckland   Bogotá
  Hamburg   Johannesburg   London   Madrid   Mexico
  Montreal   New Delhi   Panama   Paris   São Paulo
  Singapore   Sydney   Tokyo   Toronto

HEAT EXCHANGERS: Theory and Practice

Copyright © 1983 by Hemisphere Publishing Corporation. All rights reserved. Printed in the United States of America. Except as permitted under the United States Copyright Act of 1976, no part of this publication may be reproduced or distributed in any form or by any means, or stored in a data base or retrieval system, without the prior written permission of the publisher.

1 2 3 4 5 6 7 8 9 0   BCBC   8 9 8 7 6 5 4 3 2

Library of Congress Cataloging in Publication Data

ICHMT Symposium (14th : 1981 : Dubrovnik, Croatia)
   Heat exchangers—theory and practice.

   Includes bibliographies and index.
   1. Heat exchangers—Congresses. I. Taborek, J.
II. Hewitt, G. F. (Geoffrey Frederick) III. Afgan,
Naim. IV. International Center for Heat and Mass
Transfer. V. Title.
TJ263.I24 1981   621.402'5   82-6178
ISBN 0-07-062806-8   AACR2
ISSN 0272-880X

Contents

Preface xi

EVAPORATION AND CONDENSATION

INVITED LECTURE
Heat Exchangers with Phase Change
K. J. Bell 3

Flow Boiling Heat Transfer in Horizontal and Vertical Tubes
D. Steiner and M. Ozawa 19

An Assessment of Design Methods for Condensation of Vapors from a Noncondensing Gas
J. M. McNaught 35

Reflux Condensation Phenomena in a Vertical Tube Side Condenser
J.-S. Chang and R. Girard 55

Two-phase Pressure Drop for Condensation inside a Horizontal Tube
R. G. Owen, R. G. Sardesai, and D. Butterworth 67

Turbulent Falling Film Evaporators and Condensers for Open Cycle Ocean Thermal Energy Conversion
A. T. Wassell and A. F. Mills 83

Local Two-phase Flow Heat Transfer in Double-tube Heat Exchangers
K. Hashizume 95

Tube Submergence and Entrainment on the Shell Side of Heat Exchangers
I. D. R. Grant, C. D. Cotchin, and D. Chisholm 107

Heat and Mass Transfer in Counter Flow of Water and Air
S. Petelin, B. Gašperšič, and L. Fabjan 117

Experimental Investigation on a Full Scale Thermo-syphon Reboiler
P. von Böckh 131

Saturated Flow Boiling—Effect of Non-uniform Heat Transfer along the Tube Perimeter
W. Bonn 143

Film Condensation and Evaporation in Vertical Tubes with Superimposed Vapor Flow
F. Blangetti and R. Krebs 157

Design and Application Considerations for Heat Exchangers with Enhanced Boiling Surfaces
R. Antonelli and P. S. O'Neill 175

Comments to the Application of Enhanced Boiling Surfaces in Tube Bundles
	J. W. Palen, J. Taborek, and S. Yilmaz 193

Analysis of Local and Overall Heat Transfer Enhancement by Double Fluted Condenser/Evaporator Tubes
	G. Schnabel 205

Steam Heated Water Bath Evaporators for Liquefied Gases
	W. Süssmann and D. U. Ringer 221

Heat Transfer and Critical Heat Flux at Flow Boiling of Nitrogen and Argon within a Horizontal Tube
	H.-M. Müller, W. Bonn, and D. Steiner 233

Investigation of Heat Transfer and Hydrodynamics in the Helium Two-phase Flow in a Vertical Channel
	B. S. Petukhov, V. M. Zhukov, and V. M. Shieldcret 251

HEAT TRANSFER AND PRESSURE DROP IN TUBE BANKS

INVITED LECTURE
Problems of Heat Transfer Augmentation for Tube Banks in Cross Flow
	A. Zukauskas 265

Heat Transfer and Pressure Drop of Staggered and In-line Heat Exchangers at High Reynolds Numbers
	E. Achenbach 287

Heat Transfer and Flow Resistance of Yawed Tube Bundle Heat Exchangers
	H. G. Groehn 299

Surface Roughness as Means of Heat Transfer Augmentation for Banks of Tubes in Crossflow
	A. Žukauskas and R. Ulinskas 311

HEAT EXCHANGER TUBE VIBRATION

The Effect of Grid Generated Turbulence on the Fluidelastic Instability of Tube Bundles in Cross Flow
	B. M. H. Soper 325

Determination of Flow Induced Nonlinear Vibrations of Prestressed Heat Exchanger Tubes
	M. Wahle 339

AIR COOLED HEAT EXCHANGERS

INVITED LECTURE
State of Art for Design of Air Cooled Heat Exchangers with Noise Level Limitation
    P. Paikert and K. Ruff   **357**

Noise and Its Influence on Air Cooled Heat Exchanger Design
    P. E. Farrant   **383**

Heat Transfer and Flow Friction Characteristics of Louvred Heat Exchanger Surfaces
    C. J. Davenport   **397**

The Performance of a Counterflow Air to Air Heat Exchanger with Water Vapor Condensation and Frosting
    R. W. Besant and J. D. Bugg   **413**

COMPACT HEAT EXCHANGERS

INVITED LECTURE
Compact and Enhanced Heat Exchangers
    R. K. Shah and R. L. Webb   **425**

INVITED LECTURE
Heat Exchange Equipment for the Cryogenic Process Industry
    J. M. Robertson   **469**

Performance Calculation Methods for Multi-stream Plate-Fin Heat Exchangers
    L. E. Haseler   **495**

United States Ceramic Heat Exchanger Technology—Status and Opportunities
    V. J. Tennery   **507**

Study on Fluid Flow Distribution inside Plate Heat Exchangers by Thermographic Analysis
    G. Delle Cave, M. Giudici, E. Pedrocchi, and G. Pesce   **521**

FLUIDIZED BED SYSTEMS

Fluid Bed Heat Exchangers
    H. Martin   **535**

Status of Fluidized Bed Waste Heat Recovery
    M. I. Rudnicki, C. S. Mah, and H. W. Williams   549

REGENERATIVE HEAT EXCHANGERS

INVITED LECTURE
Thermal Energy Storage and Regeneration
    F. W. Schmidt   571

On the Thermal Characteristics and Response Behavior of Residential Rotary Regenerative Heat Exchangers
    M. H. Attia and N. S. D'Silva   599

HEAT EXCHANGER DESIGN

INVITED LECTURE
Heat Exchanger Design and Practices
    R. V. Macbeth   615

INVITED LECTURE
Heat Exchangers in Coal Conversion Processes
    G. Boruskho and W. R. Gambill   631

A New Energy-Cost Characteristic Diagram for Optimizing Heat Exchangers
    P. Le Goff and M. Giulietti   659

Complex Heat Exchangers Networks Simulation
    J. Castellanos Fernández   673

HEAT EXCHANGERS IN POWER GENERATION SYSTEMS

INVITED LECTURE
Nuclear Steam Generators
    G. Hetsroni   689

INVITED LECTURE
Some Aspects of Reliability and Operation Efficiency of Power Plant Heat Exchangers
    M. A. Styrikovich   707

Heat Transfer Apparatus of Nuclear Power Plants
    E. D. Fedorovich, B. L. Paskar, D. I. Gremilov, and I. K. Terentjev   721

On the Natural Circulating PWR U-Tube Steam Generators—Experiments and Analysis
  S. P. Kalra   733

A Model for Predicting the Steady-state Thermal-hydraulics of the Once-through Power Boilers
  G. Del Tin, M. Malandrone, B. Panella, and G. Pedrelli   747

Failures of the Turbines Condenser Tubes with the Direct Inflow of Make-up Water into Condenser
  N. Ružinski and M. Mustapić   757

Direct-Contact Heat Exchangers for Large Steam Turbine Installations
  G. I. Efimochkin   767

Calculation of Heat Transfer in Power Units with a Complex System of Boundary Surfaces
  O. G. Martynenko, O. V. Dikhtievsky, and N. V. Pavlyukevich   775

FOULING IN HEAT EXCHANGERS

INVITED LECTURE
Fouling of Heat Exchangers
  N. Epstein   795

Mechanisms of Furnace Fouling
  J. R. Wynnyckyj and E. Rhodes   817

Some Aspects of Heat Exchangers Surface Fouling Due to Suspended Particles Deposition
  M. A. Styricovich, O. I. Martynova, V. S. Protopopov, and M. G. Lyskov   833

Fouling in Crude Oil Preheat Trains
  G. A. Lambourn and M. Durrieu   841

Water Quality Effects on Fouling from Hard Waters
  A. P. Watkinson   853

Precipitation Fouling of Cooling Water
  L. Lahm, Jr. and J. G. Knudsen   863

Fouling in Plate Heat Exchangers and Its Reduction by Proper Design
  L. Novak   871

Fluid Bed Heat Exchanger: A Major Improvement in Severe Fouling Heat Transfer
  D. G. Klaren   885

PERFORMANCE OF ENHANCEMENT DEVICES

Performance Optimization of Internally Finned Tubes for Laminar Flow Heat Exchangers
    A. C. Trupp and H. M. Soliman   899

Two-dimensional Heat Flow through Fin Assemblies
    P. J. Heggs and P. R. Stones   917

Effect of Swirl Angle and Geometry on Heat Transfer in Turbulent Pipe Flow
    M. S. Abdel-Salam, M. M. Hilal, E. E. Khalil, and A. M. A. Mostafa   933

Spirally Fluted Tubing for Enhanced Heat Transfer
    J. S. Yampolsky   945

Heat Transfer Enhancement by Static Mixers for Very Viscous Fluids
    Th. H. van der Meer and C. J. Hoogendoorn   953

Index   965

Preface

Heat exchange equipment forms a vital part of many processes—chemical, petroleum, power generation, refrigeration, and so on. This equipment has always been of great industrial importance, but, in recent years, its significance has increased even more because of the energy crisis. It is imperative to make the best possible use of energy resources, and the conservation and reuse of thermal energy through the utilization of heat exchangers is critical in these endeavors. Thermal energy is more useful the higher its temperature. Thus, in recycling thermal energy, there is considerable emphasis on the minimization of temperature differences to reduce irreversibilities in the processes.

It was with this in mind that the subject of Advancement in Heat Exchangers was chosen for the 1981 Seminar of the International Centre for Heat and Mass Transfer. The seminars form an important focus for work on heat and mass transfer on a truly international basis. This latter point is illustrated by the fact that there were no fewer than 26 countries represented at the 1981 Seminar, reflecting the importance of the field.

In inviting lectures and soliciting papers for the 1981 seminar, the topics chosen represented a large spectrum of heat exchanger applications in industry. The seminar covered not only well established areas, but also new techniques and applications that are still in the early stages of development; for example, heat transfer equipment for coal conversion processes where the thermal energy recovery from such processes is a key to their economic acceptance. Among the new techniques discussed at the conference were the use of special types of ceramic heat exchangers, new forms of heat transfer enhancement, and the utilization of very small temperature difference processes as in cryogenic heat exchangers. A great deal of new information arose from the conference and selected material is presented in this volume. The material given is of two main types:

1. *State-of-the-art survey articles.* These were usually in the form of invited lectures and covered a wide range of topics in areas where current development is particularly dynamic. These survey articles will be of considerable use to those wishing an overview of the subject.
2. *General papers.* Around 70 papers were presented at the symposium based on original research. In selecting papers from this group for the present volume, emphasis has been given to those that are closest to practical application or have information on new developments that are likely to be significant in the field of heat exchangers in the future. A great deal of original research is reported in the present volume, which should serve as a major archival source in years to come.

In view of the closeness to practical application of the material in this volume, and the broadness of the subjects covered, we feel that the volume will be extremely valuable to all those involved in the design and operation of heat transfer equipment. However, as such is the breadth of the field, no volume of the present sort could claim to cover everything.

The editors would like to express their thanks to all those who helped to make the 1981 Seminar successful. First, we express our thanks to the staff of the International Centre for all the organizational work they carried out. Next, we would like to thank the members of the Seminar Committee and the Chairmen of the various technical sessions whose help in soliciting papers and evaluating the material was essential and invaluable. We also have appreciated very much the assistance from the publisher and his key staff in handling the manuscript material. Also, we would like to thank our own secretarial staff who have done a most competent job throughout the task of organizing the Seminar. Last but not least, we thank all the invited lecturers and authors without whose excellent material this Seminar, and hence this volume, would not have been possible.

J. Taborek

G. F. Hewitt

N. Afgan

EVAPORATION AND CONDENSATION

Heat Exchangers with Phase Change

KENNETH J. BELL
School of Chemical Engineering
Oklahoma State University
Stillwater, Oklahoma 74074, USA

ABSTRACT

After a brief survey of a few important general considerations of heat exchanger design, economics, and operation for the power and process industries, this paper addresses certain key problems in the design of vapor generators and condensers. Among the topics considered under Vapor Generators are the problems of vaporization with low temperature differences between surface and fluid, kettle reboiler bundle hydraulics, and vaporization of multicomponent mixtures. Under Condensers the topics covered include the problem of two-dimensional and three-dimensional flow calculations for large condensers, condensation in the presence of vapor shear, enhanced surfaces, design of condensers for multicomponent vapor-gas mixtures, reflux condensers, and direct contact condensers. These problems are all judged to be of industrial importance as well as providing suitable topics for research work at universities and research institutions.

1. INTRODUCTION

Any lecture with the title, "Heat Exchangers With Phase Change", must promise far more than it can deliver within realistic limits of time and human capabilities. Recognizing this, the author has elected to discuss in some small detail a few topics that seem to him to be of particular interest from the standpoint of industrial application of phase change heat exchangers. In order to justify the selection of topics, the author has allowed himself the liberty to set forth in a preliminary section those aspects of heat exchanger design and application which are important to the engineer engaged in industrial practice; not surprisingly, some of the considerations do not directly correspond to those of the scientist and researcher tending to select problems from the standpoint of their fundamental interest to him. However, it is the firm belief of the author that it is only as the researcher and academician has an insight into the industrial considerations that he can be assured of his results finding ready acceptance and understanding by the practicing engineer; pragmatically, in these days, there may also be some implications insofar as levels of research funding and support of research activities are concerned.

In Section 3, several topics in the general area of vapor generator design are introduced: vaporization with enhanced surfaces to improve low ΔT performance, kettle reboiler bundle thermal - hydraulics, and vaporization of multicomponent liquids.

In 4, problems in the design of condensers are presented: two and three dimensional flow analyses and the possibilities implicit in these analyses for improving condenser performance; condensation in the presence of vapor shear; enhanced surface treatments for condensation; design of condensers for multicomponent vapor-gas mixtures; the problems of two-phase countercurrent flow and modification of the heat transfer mechanisms and thermodynamic relationships in reflux condensers; and direct contact condensers.

It is hoped that this very brief presentation will encourage new approaches to the solution of these particular problems. But it is also hoped that the reader will be stimulated to think of other practical heat exchanger problems that require research and development work, and to devote new efforts in these areas as well.

2. CRITERIA FOR INDUSTRIALLY ACCEPTABLE HEAT EXCHANGERS

An important characteristic shared by medium and large size heat exchangers in the energy and process industries is that they are uniquely designed and constructed to meet the problem at hand. Further, there is no opportunity to test the thermal-hydraulic performance until the unit is installed and the entire plant brought on stream. Therefore, two critically important criteria for a successful heat exchanger are designability and operability. In this context, "designability" means that reliable procedures exist for calculating the thermal-hydraulic performance of the exchanger under any set of specified operating conditions, and "operability" means that the unit operates (or can be made to operate) as required in a stable, reliable fashion under the usual range of conditions encountered in the plant using regular plant personnel and with as little special attention as possible.

Those statements may appear so obvious as to be a waste of time and space here. But consider as an example the possible industrial application of dropwise condensation, which fails both criteria at present. Dropwise condensation, and therefore the possibly very high heat transfer coefficients obtainable with dropwise condensation, can not be counted upon (given our current knowledge and techniques) to actually occur in an industrial condenser. (It may or may not be important in a given case that the coefficient can not be very accurately predicted even if it does occur.) Therefore, the designer must design assuming filmwise condensation, resulting in a larger unit. In fact, the designer must take into account the possibility that dropwise condensation might occur, and this may be a distinct embarrassment: consider the case of a kettle reboiler operating near the peak heat flux in the design mode, but pushed into partial transition boiling by the higher condensing side coefficient.

The implications for the applied researcher are quite clear: if his work is to lead to useful application, he must strive for more than originality, cleverness, mathematical elegance, and spectacular results under laboratory conditions. He must show that his work leads to improved performance in real equipment operated by all-too-real people. Scientists may choose to live above and beyond these considerations, but they should not be too surprised if the rest of the world tends to ignore them.

3. PROBLEMS IN THE DESIGN OF VAPOR GENERATORS

3.1 Low ΔT Vaporization

A major consideration in cryogenic processing is the importance from both the capital investment and operating cost standpoint of minimizing the irreversibility of typical cryogenic cycles (for liquified natural gas production, helium liquefaction, hydrogen purification, etc.) by reducing the temperature differences between the hot and cold streams in the heat exchangers to the lowest feasible values. More recently, this consideration has become vital in energy conservation and alternative energy generation cycles as well. For example, the Ocean Thermal Energy Conversion (OTEC) concept operates with a total temperature differential between the source (surface ocean water) and sink (ocean water from below 800 m) of no more than 20 to 25°C, within which one must accommodate not only the temperature differences in the vapor generator and the condenser, but also the useful temperature difference from which energy may be extracted. Since ordinary engineering surfaces require surface superheats on the order of at least 3 to 5°C (and more generally 10 to 15°C) to insure nucleation of a stable boiling phase and the resulting heat fluxes at these levels are very low, there has been great interest in the last ten to fifteen years to develop surfaces which would allow nucleation at very much smaller superheats. The most widely known of the surfaces is by Union Carbide Linde, but more recently there have been commercial offerings from Hitachi, Wieland, and Trane (the latter apparently used mostly for their own air-conditioning and refrigeration equipment.)

Figure 1 shows typical curves from these surfaces as measured on single tubes in the laboratory, compared with a corresponding curve obtained experimentally from a plain tube (1). It should be noted that the surfaces do not offer remarkably improved peak heat fluxes, which in any case would not be expected because of the general understanding of the peak heat flux phenomenon as a hydrodynamic process rather than a surface-oriented process. However, these enhanced surfaces offer greatly improved heat fluxes at very low ΔT's and this is of course the preferred area of application.

It is not too hard to state the essential requirement for an enhanced boiling surface, namely, that it should have re-entrant cavities which trap vapor to initiate the nucleation of new bubbles in the boiling cycle, and/or that

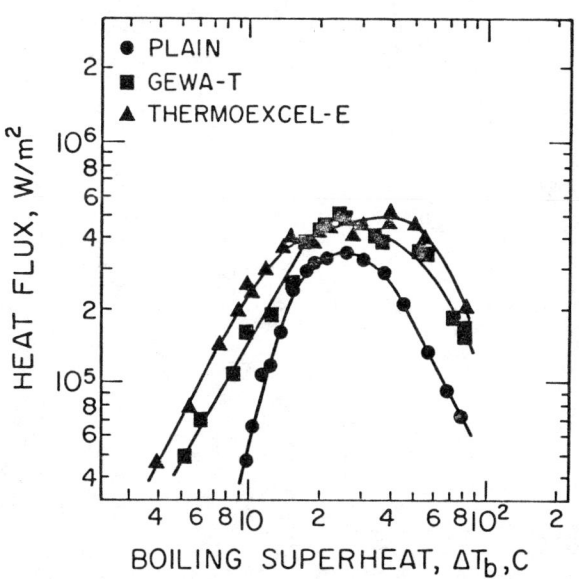

Figure 1. Single Tube Boiling Heat Transfer for Isopropyl Alcohol at 103 kPa (1).

a portion of the surface should be non-wetting. But if we are to know what geometry may be best for a specific fluid and in a specific application, we need to know a great deal more of the actual fluid mechanics and fluid/surface interaction.

Until that time comes, we may wish to focus upon more myopic design considerations. For example, reflect upon the following: If the only real advantage of these surfaces is to enhance nucleation of the vapor phase at very low ΔT's, then the proper utilization of this surface in kettle reboiler equipment (see below) is mainly in the bottom few rows of tubes in the bundle, where it is necessary to create a vapor phase. If, in fact, the surfaces offer no advantages - or even conceivably disadvantages (in addition to the substantially higher cost than plain tube) - in the convective flow region of the bundle, then their use is uneconomical in the upper regions of the bundle. The upper tubes in the bundle should be plain, and the heat transfer would be primarily by two-phase convective flow with a negligible nucleate boiling contribution.

To the author's knowledge, no studies have been made along these lines. However, some results have been published very recently on the average characteristics of a bundle of enhanced tubes, particularly showing the peak heat flux behavior. Typical results are shown in Figure 2 (2) and would seem to indicate that enhanced surface performance is not noticeably improved in a bundle, by contrast with plain tubes. This tends to support the argument made in the previous paragraph. In any event, we are presently working with a minimum amount of data, and it would be extremely useful to our better understanding of the application of these surfaces if we had a more complete mechanistic view of the heat transfer and fluid flow mechanisms operative on and near the surface of the tube. This is a study that might well be carried out on a relatively small scale, but one needs to compare the small scale studies with full bundle tests carried out under conditions more typical of actual process industry application.

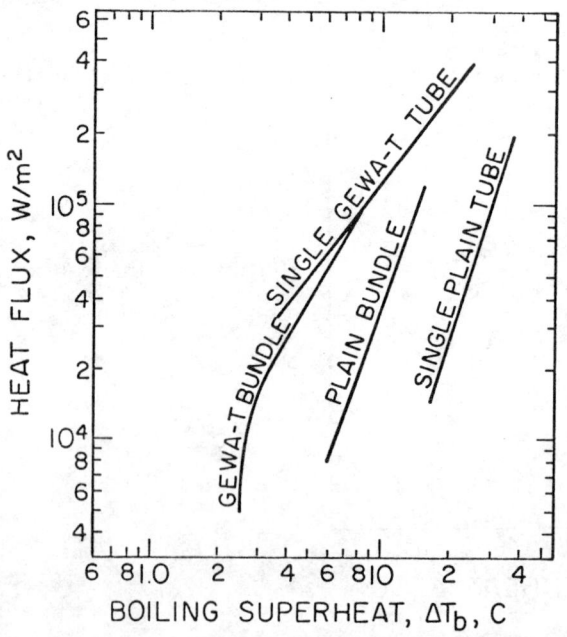

Figure 2. Comparison of Performance of Single Tubes to Bundles with p-Xylene at 103 kPa (2).

3.2 Kettle Reboiler Tube Bundle Hydraulics

For many years the kettle reboiler (or the flooded chiller, as it is referred to in some industries) was considered to be essentially a pool boiling device. That is, the heat transfer mechanisms were assumed to be those of a typical laboratory boiling experiment in which a single tube is immersed in a relatively quiescent pool of liquid at its saturation temperature; departure of the bundle from single tube boiling behavior was then interpreted as interference of the surrounding tubes with the vapor escaping from the tubes lower in the bundle. To the author's knowledge all presently published (3) (and at least some of the proprietary) design methods are based upon a single tube pool boiling correlation between heat flux and temperature difference, modified by a penalizing term for the presence of the other tubes. Originally, it was believed that the nucleate boiling correlation itself should be penalized because of bundle effects, but this was corrected when the first large scale full bundle boiling data were obtained by HTRI in the 1960's (4). The presence of the other tubes does significantly reduce the peak heat flux and modify the shape of the boiling curve in the range of temperatures characteristic of the peak heat flux and the subsequent transition boiling regime.

However, by the late 1960's, it was recognized by several workers that the picture of the kettle reboiler as a slightly modified pool boiling experiment was quite erroneous. In fact, the data that were being obtained at that time made it quite clear we were actually looking at a species of thermosiphon reboiler circulation, in which the clear liquid around the outside of the bundle furnished the driving head for the circulation of the two-phase flow inside the tube bank. The general picture of the circulation as we envisioned it at that time has subsequently been verified by movies taken at Heriot-Watt University. Figure 3 is a schematic version of this circulation. Under this interpretation,

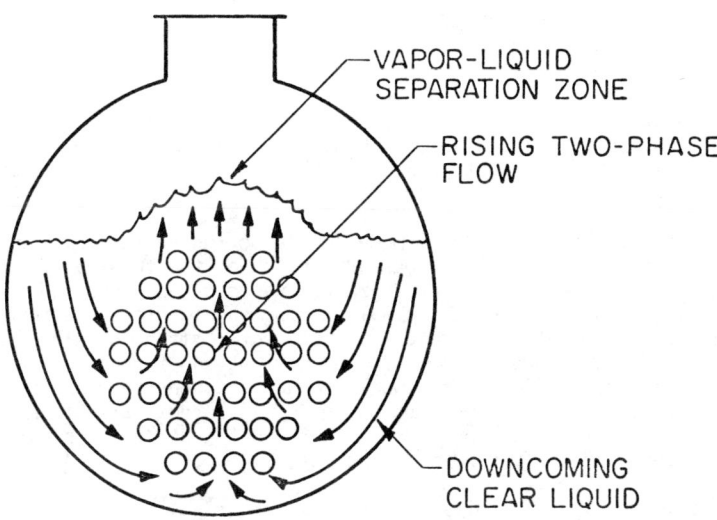

Figure 3. Schematic of Fluid Circulation in a Kettle Reboiler.

only the bottom few rows of tubes operate in a nucleate boiling regime, generating a low quality two-phase flow rising through the upper portions of the bundle. As this flow accelerates due to the growing vapor fraction, the heat transfer mechanism shifts to a convective two-phase flow, analogous to the annular climbing film regime already well-known and studied in connection with thermosiphon reboilers and boiling water nuclear reactors.

Until recently, there has been little interest in changing the design philosophy for kettle reboilers to the fundamentally and mechanistically more correct thermosiphon model. Partly, this is due to the fact that present design methods do a very good job of predicting satisfactory performance of kettle reboilers in practice. A further inhibiting factor in developing new design methods was that the thermosiphon model requires a good deal of information on the two-phase flow characteristics of tube bundles; until recently only two papers were available in this area. (5,6) This lack has been remedied over the last several years by work notably of Grant, Chisholm, and co-workers at the National Engineering Laboratory (e.g., 7, 8). Their extensive publications on the several aspects of two-phase flow in tube banks, including flow patterns and frictional pressure drop furnish at least the nucleus of information that is required to construct a thermosiphon model for the kettle reboiler. The heat transfer work is not in such good shape but might reasonably be expected to follow the general form of the Chen correlation (9) for combined nucleate and convective boiling inside tubes.

It is now know that the above pieces of information are being put together into a more sophisticated kettle reboiler design method by several organizations, most of them proprietary. It is expected that these new design methods will be available at least on a limited basis within the next six months to a year.

One might ask why, if the existing design methods are satisfactory, do we feel that it is necessary to go to a new basis? At least part of the answer is that new kettle reboilers are being designed to much large diameters than have been tested. (And faith in the existing design method rests heavily upon these tests.) One feels much more comfortable extrapolating into a relatively unexplored area (of bundle diameter, for example) a design method which is based on the actual mechanistic processes occurring in the equipment.

The present design methods predict a rapid deterioration of the peak heat flux as one goes to very large uniform bundles. This raises the question of whether we could not design better kettle reboilers if we had a better understanding of the flow patterns in them.

A solidly verified fundamental model could be run on the computer so as to indicate whether we should not increase the pitch of the tubes as one goes higher in the bundle, thereby providing a greater vapor escape space between the tubes. Or perhaps analysis would reveal that we should provide internal downcomers for clear liquid to penetrate into the center of the bundle. These downcomers presumably would need to be shrouded against vapor penetration into them from the surrounding bundle (which would reduce the density gradient and possibly reverse the flow). Perhaps computer studies would suggest that we should provide selected shrouding around the outside of the bundle in order to separate the downcoming liquid and the rising two-phase flow.

Another increasingly important consideration arises from the hydrostatic pressure gradient, which assumes significant proportions in deep (large diameter) bundles. The supression of the ΔT available at the bottom of the bundle is especially critical when one confronts the increasing tendency to design for smaller approach temperatures to minimize irreversibility losses in processing.

HEAT EXCHANGERS WITH PHASE CHANGE

3.3 Boiling of Multi-Component Liquids

The boiling of multi-component liquids is more complicated than for pure components because of the difference in composition between the vapor and the liquid phases dictated by phase equilibrium thermodynamics. The vapor phase will have a higher fraction of the more volatile components than the original liquid phase fed to the reboiler, and the remaining liquid phase therefore correspondingly less of the more volatile components. Every step in the vapor generation process is affected to some degree by the mass transfer phenomena and changes in physical properties thus introduced, and the bulk boiling fluid can no longer be treated as isothermal from start to finish in the equipment. Shock (10) has surveyed the extensive basic literature on multi-compoenent boiling.

From a design standpoint, the greatest effect occurs when a significant amount of the liquid has been vaporized; the liquid phase remaining in the immediate vicinity of the heat transfer surface has been depleted in the lighter components and therefore the effective boiling temperature is raised. While diffusional and turbulent mass transfer processes within the liquid phase will tend to move the more volatile components from the bulk liquid towards the interface, these processes tend to be relatively slow compared to the vaporization process itself. Therefore the effective boiling temperature of the liquid in the bundle is raised above that which would be predicted from a thermodynamic equilibrium anaylsis of the bulk average vapor-liquid phase compositions. Working with pool boiling data, if one defines the temperature difference effective for boiling in the usual fashion, i.e., surface temperature minus the saturation temperature of the bulk liquid phase, one overestimates the temperature difference actually available for heat transfer and correspondingly underestimates the heat transfer coefficient. Hence, much of the literature on binary liquid boiling heat transfer shows that the apparent heat transfer coefficient calculated by this means is less than for either of the two pure components.

Obviously, one would like to have a theoretical model of the boiling process which would properly incorporate these mass transfer phenomena, but this seems quite out of the question quantitatively at this time, because even the fluid mechanics and heat transfer mechanisms in boiling systems in practical geometries are only very poorly understood. The usual practice currently is either to incorporate a boiling range correction factor into the design procedure or to carry out the heat transfer calculations assuming a saturation temperature of the liquid phase after the required boil-up has been achieved in the equipment. These methods seem to work reasonably well for normal situations but lead to a high degree of uncertainty where one is vaporizing a very substantial fraction of the total flow or where one has a feed with large fractions of components at both ends of the volatility spectrum.

There is also considerable operational difficulty associated with the change in boiling characteristics from one part of the heat exchanger to another as the vapor disengages itself from the liquid and the remaining liquid becomes very difficult to boil. This severely limits the range of operational parameters that can be used in these pieces of equipment; a temperature difference that would achieve a reasonable heat transfer rate in the heavy liquid near the end of the vaporization process may be so high near the start of the vaporization as to lead to transition and even film boiling with attentent low heat transfer rates and possibly excessive fouling. This requires the designer to match temperature profiles in the equipment and may lead to limitations in the amount of heat transfer that can take place in any one reboiler. Of perhaps greater concern is that many designers do not recognize these potential problems and do not properly consider that local conditions may vary greatly from the average condition within the reboiler.

4. PROBLEMS IN THE DESIGN OF CONDENSERS

4.1 Multi-dimensional Flow Analysis for Vapor Condensers

It has been recognized for many years in the power industry (e.g., Ref. (11)) that it is necessary to consider the flow patterns within the tube fields of large condensers if the vapor is to be more or less uniformly distributed among all of the tubes without excessive pressure drop, and also if the flow of noncondensable gases is to be guided towards the vent system. Such two-dimensional flow calculations are now considered standard by major surface condenser manufacturers. (An excellent collection of papers primarily devoted to marine condensers, but equally valuable for large condensers of all kinds is given in Ref. (12)). Again, however, this is an area in which recognition of the problem has not necessarily implied a completely adequate procedure for solution. Many of the details of flow across the tube field are not well understood, especially since part of the vapor flow is continuously removed at the condenser tube surface. The condensate falls under the combined effects of gravity and the vapor flow field through the tube field in a way which is as yet very poorly understood. Large condenser tube fields have internal drain plates installed to withdraw the condensate formed on the tube rows above these plates so that it does not fall upon further heat transfer surface down towards the bottom of the shell.

There has been a great deal of concern about the deterioration of the heat transfer on the lower tubes in the tube field as a result of being "drenched" (or "inundated") by condensate falling from the upper tubes. At the same time, the heavy liquid flows tend to result in turbulent films with relatively quite high rates of heat transfer. Recent measurements tend to suggest that drenching is a less serious problem than had been feared, and that in fact much of the apparent deterioration of performance from this cause is due to one of two other causes: 1) The inability of vapor to penetrate into the lower reaches of the tube field without excessive pressure loss and therefore significant decrease in local condensing temperature, or (2) The accumulation of non-condensable gases in those reaches of the bundle where a great deal of the vapor has been condensed, leaving behind a particularly large fraction of non-condensable gases.

Obviously, it would be highly desirable to have a numerical procedure for calculating these flows taking into account all of the important terms in the momentum equations governing the flow through the tube field and also the interaction of the superimposed vapor flow on the heat transfer mechanisms in the film on the tube. More recently it has been recognized that a three dimensional analysis may very well be required because of the significant variation in local temperature difference from one end of the condenser to the other. A similar case exists among different passes where multiple tube side pass construction is employed. There still remains a good deal of work to be done in developing the complete set of pertinent equations in a computable form, as well as in determining the basic interaction parameters for two-phase flow in a complex geometry. With these tools in hand and properly integrated, the designer could then manipulate the layout of his bundle so as to secure optimal performance for a given investment and operating cost of the equipment. Such a tool would also offer the possibility of optimizing the on-line plant performance in response to daily and yearly changes in environmental conditions and load demands on the plant.

4.2 Effect of Vapor Shear on Heat Transfer in Condensing Systems

Even in his original papers (13), Nusselt recognized that the shear stress exerted by the generally more rapidly flowing vapor on the condensate film would influence the thickness of the film and therefore the rate of heat transfer. Later (14), it was recognized that an even more significant effect was the tendency of vapor shear to cause early turbulence in the condensate film and therefore to greatly increase the rate of heat transfer at lower Reynolds numbers, compared to the prediction of Nusselt theory. Since that time a great deal of mechanistic modeling has been done in order to develop sounder techniques for predicting heat transfer under conditions of vapor shear, (e.g., Ref. (15)). However, it is fair to say that the models that have been employed represent only a very idealized approximation of the actual flow structure. The recent work by Schlünder and co-workers (16) is providing a great deal of data for the refinement of these models for flow in tubes, and this general aspect of the problem can generally be considered to be fairly well in hand if present efforts are vigorously pursued.

However, two-phase condensing flows on the outside of the tube banks have been only very slightly explored to this point, and not all of what has been observed can be satisfactorily explained at this time. Early studies showed that vapor in crossflow in tube banks would significantly increase the heat transfer condensing coefficient for vapor Reynolds number above about 30,000 (compared to the Nusselt prediction for single horizontal tubes), rising to several times the value for Reynolds numbers above 50,000. However, other studies (17) since that time show that, if the vapor velocity is increased to very high values (corresponding to an appreciable fraction of sonic velocity), the heat transfer coefficient not only does not rise further but actually begins to decline. Various suggestions have been made to explain this phenomenon, including the reduction in the local saturation pressure at the low pressure zones where the vapor has been accelerated between two adjacent tubes. However, to the author's knowledge, these effects have not been verified nor even reconciled with the fact that the pressure losses implied by such a process are not consistent with reported overall pressure drops across the tube bank. A great deal more work remains to be done here before this problem can be said to be well understood and useful for design purposes. In this connection, it may be noted that some condenser designers regularly vary the baffle spacing on the shell side from a large value near the inlet to much closer spacing near the outlet; this is to keep the vapor velocity and therefore the local heat transfer coefficient high throughout the condensing process. The author feels that any pretension to precision and optimization is premature and not supported by the state of the knowledge.

4.3 Enhanced Surface Treatments for Condensation

Since surface subcooling required for condensation is only on the order of 0.01 to 0.1°C, the need for surface enhancement for condensation is perhaps not quite so great as for boiling . However, a number of techniques either are in common use or have been proposed for enhancing the condensing process, and it is not amiss at this point to briefly consider some of them. Low-finned tubes are in common use, and experience shows that the fin surface is almost 100% effective for condensation of liquid having low surface tension (below about 25 to 30 dynes/centimeter). Low-finned tubes are not effective with high surface tension substances, notably water, because the high surface tension causes the liquid phase to bridge the fins and effectively insulates the surface of the tube. So far as the author is aware, no one has looked at the problem of how far apart the fins would need to be in order to prevent the bridging action. Even though condensing steam coefficient are so high as to seldom constitute the limiting coefficient in a condenser, yet there might be some commercial advantage to be gained by the development of a tube with fins spaced far enough apart to avoid bridging.

Another enhancement which does possibly make sense for condensing steam is the fluted or Gregorig surface (18), Figure 4, in which the surface tension forces are deliberately used to pull the condensate film off of the convex surfaces of the tube and into the troughs where rapid drainage of the condensate occurs. This surface has been shown to give condensing heat transfer coefficients several times those that would be obtained on an equivalent plain tube (19, 20), Figure 5, and this enhancement works for water as well as other liquids. Several modifications of the surface have been proposed, such as the use of periodic skirts on the tube to cause the liquid draining in the troughs to be discharged, so that the process could start all over again with more bare surface exposed to condensation. As yet there is no substantial body of design procedure or large-scale operating experience that can be invoked to demonstrate the intrinsic superiority of these surfaces from a thermal-hydraulic point of view.

It is necessary at this point to mention the extensive amount of work that has been done on drop-wise condensation. By rendering a surface continuously nonwetting to the condensate, drop-wise condensation can be maintained resulting in heat transfer coefficients ten times and more the coefficients obtained by filmwise condensation with the same substances. The difficulty has always been how to render the surface nonwetting in the first place and to maintain the nonwetting characteristic of the surface under operational conditions. Techniques that are relatively easy to carry out under laboratory conditions are usually unacceptable for plant operation, and it is probably safe to say that nothing proposed to this time offers any interest for actual plant operation. Nonetheless, the knowledge that the phenomenon exists continues to be an inspiration to some workers to explore the problem and to seek new ideas for its possible commercialization.

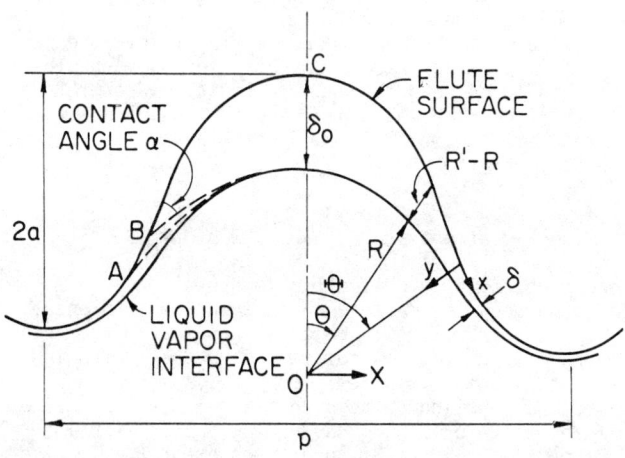

Figure 4. Schematic of a Gregorig Surface, with Condensation Occurring Primarily on the Convex Surfaces (20).

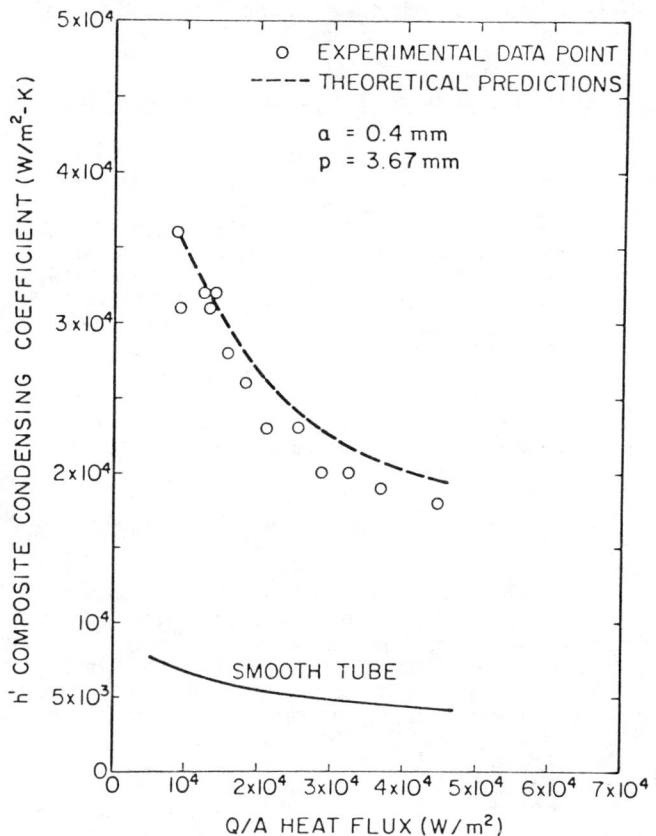

Figure 5. Condensation of Ammonia on a Gregorig Surface (19).

4.4 Design of Condensers for Multi-Component Vapor-Gas Mixtures

The remarks made above concerning the design of multi-component vapor generators can be applied analogously to the design of multi-component condensers with or without a non-condensable gas present. The thermodynamics of the situation result in a condensate phase having a different composition than either the original vapor or the remaining vapor, so that there are mass transfer effects which can constitute a significant portion of the heat load. Fairly fundamentally-oriented approaches are available on a practical design basis for the particular cases of the condensation of a vapor in the presence of a noncondensable gas (21) and for the total condensation of a binary vapor (22).

For the more general and commonly encountered case of many components approximation methods have had to be employed up to this time. Probably the most generally used (or copied) technique is by Silver (23, 24). This method does not explicitly incorporate the mass transfer terms that so greatly complicate the calculations, but it has been shown that there is an implicit consideration of mass transfer resistance. There is a deliberate over-estimation of the sensible heat transfer resistance required to cool the vapor. These techniques, while generally successful in producing a workable design for operating plants, do not give any real insight into either better design options or, more importantly, such operational problems as fog formation and optimal venting procedures.

Considerable work has been undertaken by workers at the University of Manchester in particular to develop reasonable computational procedures employing fundamental approaches to the problem of multicomponent diffusion in the vapor phase (25). The Stefan-Maxwell equations represent in a reasonably general way the diffusional mass transfer problem in the vapor phase. They however, result in the necessity of solving a set of simultaneous equations involving matrix operations. Usually, few of the elements in the matrix are known to an adequate degree of accuracy, giving rise to difficulties in getting suitable convergence of the inversion. Progress is being made in reducing the complexity of the calculations as well as generating better methods for predicting the diffusivities required for the calculations. However, two further difficulties must then be faced: First, the problem of diffusional resistance in the liquid phase is substantially more difficult to handle in fundamentally satisfactory way, and second, the problem at the interface is complicated by the fact that vapor-liquid interface phenomena are themselves very poorly understood from a purely fluid mechanical point of view. Another area of interest is the role of mist and droplet formation at the two phase interface, connected to the development of fog in the bulk vapor stream. Clearly there are enough problems in the multicomponent area to occupy many researchers for a very long period of time.

4.5 Reflux Condensers

Another kind of condenser that is quite widely used and yet not very well understood from a design standpoint is the so-called reflux condenser in which vapor enters at the bottom of the tubes and condenses on the side walls to form a condensate phase which drains downward (Figure 6). The more volatile components and the non-condensable components continue to rise upwards through the tube until they are vented from the top of the tube. Probably the single most critical problem in the design of such units is to insure that the vapor velocity upwards at the entrance of the tube does not exert so great a shear stress upon the falling condensate as to carry it up with the vapor, resulting in unstable operation of the equipment. Several correlations are available to predict the flooding velocity (26), and this problem, while not very well understood fundamentally, is at least on fairly sound ground from a design standpoint. However, there is relatively little in the literature to allow accurate heat transfer calculations to be made in the countercurrent two-phase flow regime further up the tube. One would like to have a correlation to predict the pressure drop in the vapor, along with the effect of vapor shear in thickening the draining film and modifying its heat transfer characteristics. The paper by Soliman, et al. (27) does begin to deal with this problem.

Another problem is a purely computational one, namely that of carrying out the necessary interface vapor-liquid equilibrium calculations and the overall material and heat balances for the two phases flowing in opposite directions. There are also problems associated with fog formation and the possibility of entrainment in the upper reaches of the tube where the fluid and vapor properties may be quite different than near the entrance to the tube. These considerations are primarily of concern in minimizing the loss of valuable condensate out the vent, but once again take us back to the question of mass transfer phenomena in the vapor phase and the hydrodynamics of the vapor-liquid interface.

HEAT EXCHANGERS WITH PHASE CHANGE

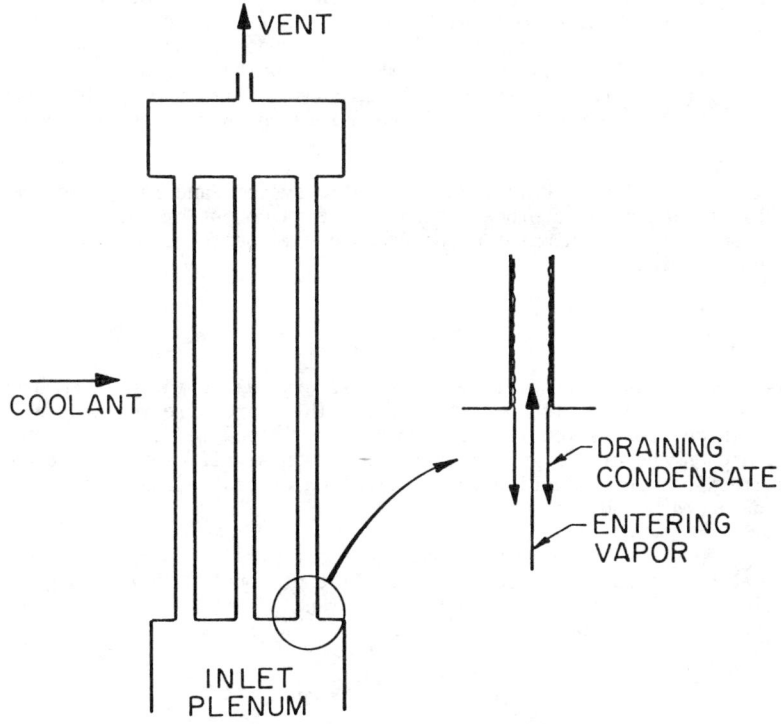

Figure 6. Schematic of a Reflux Condenser, Showing the Critical Design Point.

4.6 Direct Contact Condensers

The direct contact condenser, in which subcooled liquid is sprayed directly into the vapor space, is an old application receiving renewed attention. Most of the new effort, however, is inspired by the potential importance of direct contact condensation in heat removal and pressure suppression in nuclear reactor systems suffering a major loss-of-coolant accident. These results, then, are not likely to add very much to the data base and designability of direct contact condensers for the process and energy industries.

The attractiveness of direct contact is of course the absence of any heat transfer surface that can foul or corrode, and the presumed high volumetric heat transfer rates that can be achieved. The disadvantages include the inevitable mixing of condensate with coolant (miscible or immiscible) and the fact that the heat still must be rejected to the environment, now at a lower temperature.

Traditionally, the major application has been in barometric condensers, often the last element in an evaporator train, from which the effluent was generally worthless and often contaminated. Environmental restrictions are severely restricting the easy disposal or recycling (through a cooling tower) of the effluent. At the same time, however, the declining opportunity to use water as the ultimate heat sink in process or energy plants has given new incentive to the use of a direct contact condenser and secondary heat rejection from the coolant/condensate stream to air through a conventional air-cooled heat exchanger. The current main interest is the Heller-cycle power plant, and the most effort has been in Hungary in particular. Bakay and Jaszay (28) describe experiments with liquid sheets, and one can imagine some other configurations of contact surface (including various forms of wetted packing to retard drainage velocities and minimize the height of the apparatus and therefore the hydrostatic head loss which must be provided by pumps).

Still, all of this work is for steam-water systems and in parametric ranges uncharacteristic of organic process conditions. A modest amount of study in relatively small scale equipment might provide some useful insights into new areas of application.

5. SUMMARY

This lecture has very briefly surveyed several different problem areas in vaporization and condensation heat transfer. These areas represent, in the author's opinion, topics for research work on small and medium scales on the one hand, and on the other are likely to result in significant improvements and new directions in the design and application of heat exchangers in the process and energy industries.

This lecture has by no means exhausted the topics that come to mind; it would not be too hard to create another list of topics equally long and equally likely to be interesting in the long run. There is certainly no shortage of important problems to work on.

REFERENCES

1. Yilmaz, S., and Westwater, J. W., "Effect of Commercial Enhanced Surfaces on the Boiling Heat Transfer Curve," Paper presented at 20th National Heat Transfer Conference, Milwaukee, August, 1981.

2. Yilmaz, S., Palen, J. W., and Taborek, J., "Enhanced Boiling Surfaces as Single Tubes and Tube Bundles," Paper presented at 20th National Heat Transfer Conference, Milwaukee, August, 1981.

3. Palen, J. W., and Small, W. M., "A New Way to Design Kettle and Internal Reboilers," Hydrocarbon Proc. $\underline{43}$, No. 11, 199-208, (1964).

4. Palen, J. W., Yarden, A., and Taborek, J., "Characteristics of Boiling Outside Large-Scale Horizontal Multitube Bundles," AIChE Symp. Series, $\underline{68}$, No. 118, 50, (1972).

5. Diehl, J. E., "Calculate Condenser Pressure Drop," Pet. Ref. $\underline{36}$, No. 10, 147, (1957).

6. Diehl, J. E., and Unruh, C. H., ASME Paper No. 58-HT 20, (1958).

7. Grant, I. D. R., and Chisholm, D., "Two-phase Flow on the Shell-side of a Segmentally Baffled Shell-and-Tube Heat Exchanger," J. Heat Transfer $\underline{101}$(1), 38, (1979).

8. Grant, I. D. R., and Chisholm, D., "Horizontal Two-phase Flow Across Tube Banks," Int. J. Heat and Fluid Flow $\underline{2}$(2), 97, (1980).

9. Chen, J., ASME Paper No. 63-HT-34 (1963). Extensively cited in the literature, e.g., Collier, J. G., Convective Boiling and Condensation, McGraw-Hill (1973).

10. Shock, R. A. W., "Boiling in Multicomponent Fluids," Personal communication from the author, to appear in Multiphase Science and Technology.

11. Barsness, E. J., "Calculation of the Performance of Surface Condensers by Digital Computer," ASME Paper No. 63-PWR-2 (1963).

12. Marto, P. J., and Nunn, R. H., Editors, Proceedings of the Workshop on Modern Developments in Marine Condensers, March 26-28, 1980, Naval Postgraduate School, Monterey, California.

13. Nusselt, W., Zeits. VDI 60, 541, 569, (1916).

14. Carpenter, E. F., and Colburn, A. P., General Discussion on Heat Transfer London, 20-26, ASME, New York (1951).

15. Dukler, A. E., "Fluid Mechanics and Heat Transfer in Vertical Falling-Film Systems," Chem. Eng. Prog. Symp. Series 30 No. 56 "Heat Transfer-Storrs," 1-10 (1960).

16. Blangetti, F., and Schlunder, E. U., "Local Heat Transfer Coefficients on Condensation in a Vertical Tube," Proceedings 6th Int. Heat Transf. Conf., Toronto 2, 437-442 (1978).

17. Butterworth, D., "Developments in the Design of Shell-and-Tube Condensers," ASME Paper No. 77-WA/HT-24 (1977).

18. Gregorig, R., "Hautkondensation an feingewellten Oberflächen bei Berucksichtigung der Oberflächenspannung," Z. Angew. Math. Phys. 5, 36-46 (1954).

19. Combs, S. K., "Experimental Data for Ammonia Condensation on Vertical and Inclined Fluted Tubes," ORNL-5488, January, 1979. See also: Michel, J. W., "A Summary of Recent Experimental and Analytical OTEC Studies of ORNL," Proceedings of the 6th OTEC Conference, Washington, D.C., 19-22 June 1979, II, 11.7-1 - 11.7-6. (DOE Conf - 790631).

20. Panchal, C. B., and Bell, K. J., "Analysis of Nusselt-Type Condensation on a Vertical Fluted Surface," Num. Heat Transfer 3, 357-371(1980).

21. Colburn, A. P., and Hougen, O. A., "Design of Cooler - Condensers for Mixtures of Vapors with Non-Condensable Gases," Ind. Eng. Chem. 26, No. 11, 1178-1187(1934)

22. Colburn, A. P., and Drew, T. B., "The Condensation of Mixed Vapors," Trans. A.I.Ch.E., 33, 139-215(1937).

23. Silver, L., "Gas Cooling with Aqueous Condensation," Trans. Inst. Chem. Eng. 25, 30-42(1947).

24. Bell, K. J., and Ghaly, M. A., "An Approximate Generalized Design Method for Multicomponent/Partial Condensers," AIChE Symp. Ser. 69, No. 131 "Heat Transfer: Fundamentals and Industrial Applications," 72-79(1973).

25. Krishna, R., and Standart, G. L., "Mass and Energy Transfer in Multicomponent Systems," Chem. Eng. Comm. 3, 201-275(1979).

26. Diehl, J. E., and Koppany, C. R., "Flooding Velocity Correlation for Gas-Liquid Counterflow in Vertical Tubes," Chem. Eng. Prog. Symp. Series 65 No. 92 "Heat Transfer-Philadelphia," 77-83(1969).

27. Soliman, M., Schuster, J. R., and Berenson, P. J., "A General Heat Transfer Correlation for Annular Flow Condensation," J. Heat Transfer 90, 267-276 (1968).

28. Bakay, A., and Jaszay, T., "High Performance Jet Condensers for Steam Turbines," Proceedings 6th Int. Heat Transfer Conf., Toronto 2, 61-65 (1978).

Flow Boiling Heat Transfer in Horizontal and Vertical Tubes

DIETER STEINER and MAMORU OZAWA
Institut für Thermische Verfahrenstechnik
Universität Karlsruhe
Kaiserstrasse 12
D-7500 Karlsruhe 1, FRG

ABSTRACT

Although there are a lot of correlations about heat transfer at saturated flow boiling, a comparison between experimental and calculated heat transfer coefficients still shows significant deviations.

Considering the fact that forced convective boiling and nucleate boiling distinctly depend on system parameters, available experimental data were analysed under standardized conditions. While the heat transfer coefficient at forced convection increases with increasing mass velocity and/or flow quality and decreases with pressure and tube diameter, the heat transfer coefficient at nucleate boiling rises with heat flux and pressure and decreases with increasing tube diameter. It follows from the different pressure effect that forced convective boiling is not present for reduced pressures $p/p_c \gtrsim 0.2$ when the heat flux is not too low. Additional the perimeter averaged heat transfer coefficient at nucleate boiling is different for vertical and horizontal tubes. Whereas for forced convective boiling the experiments show no significant flow direction effect on the heat transfer coefficient.

1. INTRODUCTION

In the last three decades many studies have been made dealing with heat transfer at saturated flow boiling. Through the studies of Dengler /1/, Dengler and Addoms /2/, Mumm /3/, Mc Nelly /4/, Coulson and Mc Nelly /5/, Sterman and Styushin /6/, Sterman /7/, and Chawla /8/ the heat transfer during the progressive vaporization along a tube could be subdivided into regions with different mechanism. These regions are: forced convective boiling, nucleate boiling and the dryout or liquid deficient regime. The dependence of the heat transfer in these regions from the operating parameters such as heat flux q̇, mass velocity ṁ, flow quality ẋ, inner tube diameter d and pressure p were experimentally studied for different fluids, flow directions and geometries. For example, in forced convective boiling Dengler concluded 1952 that the heat transfer coefficient α does not depend on the driving temperatur difference ΔT or the heat flux. But in nucleate boiling the heat transfer coefficient is dominated by the heat flux. This is up to day the

Mamoru Ozawa's present address is Department of Production Engineering, Kobe University, Rokko-Dai-Cho, Nada, Kobe 657, Japan.

criterion for subdividing the experimental results in forced convective and nucleate boiling.

From his results in vertical tubes Sterman /7/ concluded that the heat transfer in nucleate boiling increases with $\dot{q}$ and pressure p. Also a rise in mass velocity increases slightly the heat transfer coefficient whereas an increase in volumetric quality shows no effect. Sterman also used a criterion in the form $f\{Bo, \rho_L/\rho_V, \Delta h_v/c_{p_L}T_s\}$ calculating the heat transfer in forced convective and nucleate boiling region with different equations.

In further experimental works up to 1962, e.g. of Guerry and Talty /9/, Tarasova, Armand et al. /10/, Schrock and Grossman /11, 12/, Bennett, Collier et al. /13/, Wright /14/ and Somerville /15/, the range of operating parameters was extended and different fluids have been tested. But the proposed correlations could not used beyond the range of the experimental data.

The mechanism at forced convective boiling was studied by Hsu /16/. He postulated that dependent on the thickness of the thermal boundary layer bubble nucleation can be suppressed. This was verified by experiments of Collier, Lacey et al. /17/.

Heat transfer in horizontal evaporator tubes was studied by Chawla /8/. He recommended separated equations for forced convective and nucleate boiling.

In the works mentioned above the heat transfer had been studied in vertical tubes, annular flow channels and in horizontal tubes. The effect of the flow direction on wetting had not been considered in the calculation of the perimeter averaged heat transfer coefficient. But the experiments of Styrikovich and Miropolskii /19/ in 1950 had shown that with an uniformly generated heat flux in horizontal tubes the separated flow patterns causes large temperature differences between top and bottom of the tube wall. These authors showed that the wetting does not only depend on the hydrodynamic conditions, but also on the heat flux, pressure and inclination of the tube.

In the following all these results will be summarized taking into account recently published data and own experimental results. The discussion will be limited to single-component fluids in the forced convective and nucleate boiling region.

2. PERIMETER AVERAGED HEAT TRANSFER COEFFICIENT

The heat flux to the fluid in vertical channels does only depend on the axial position z in the flow direction, except for time depend fluctuations. On the other hand the wall temperature in horizontal tubes shows a systematic variation from top to bottom, depending on various parameters as shown in the studies /19, 20, 21, 22, 23/. Due to this temperature distribution parts of the heat generated at the top flows to the bottom by heat conduction in the tube wall. Consequently, the radial heat flux transfered to the fluid is non-uniform. Knowing the local heat flux and the local inner wall temperature, a local heat transfer coefficient can be defined, see Bonn et al. /24/. For that purpose local measured data must be available, but at present this is not the case.

Consequently this paper deals only with perimeter averaged heat transfer coefficients defined by

$$\alpha(z) = \frac{\dot{q}(z)}{\bar{T}_W(z) - T_s} \qquad (1)$$

This heat transfer coefficient is calculated with the local saturation temperature T_s, perimeter averaged wall temperature $\bar{T}_W$ and heat flux $\dot{q}$ determined at the axial position z. The local flow quality $\dot{x}$ is the thermodynamic equilibrium quality evaluated from a heat balance

$$\dot{x} = \frac{h_{in} + 4\dot{q}z/(d\dot{m}) - h'}{\Delta h_v} \qquad (2)$$

where h_{in} is the inlet enthalpy of the total flow. The enthalpies h' and Δh_v are determined for the local pressure $p(z)$. All collected data are calculated by their authors in the manner stated above or are recalculated by us to keep consistency.

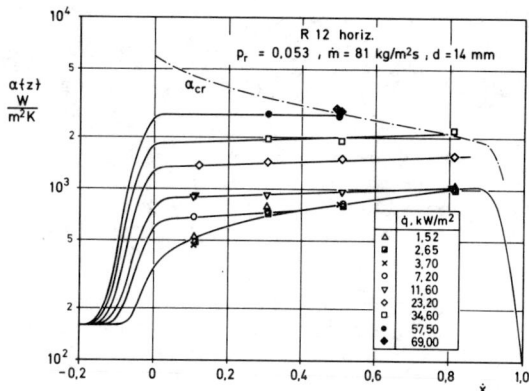

Fig. 1. Perimeter averaged heat transfer coefficient vs. flow quality

Heat transfer coefficients measured in a horizontal copper tube by Iwicki et al. /31/ are shown in Fig. 1. For heat fluxes $\dot{q} \leq 3.7$ kW/m² only forced convective boiling could be observed. In this region the increase of heat transfer coefficient with flow quality should be noted. However at heat fluxes $\dot{q} \geq 11.6$ kW/m² nucleate boiling was obtained for all flow qualities. The dotted line is the margin to burnout, i.e. the heat transfer coefficient calculated to the critical or burnout heat flux. The critical heat flux was calculated using equations given by Bonn et al. /25/. The experimental data at $\dot{x} = 0.5$, $\dot{q} = 69$ kW/m² and $\dot{x} = 0.81$, $\dot{q} = 34.6$ kW/m² was observed to be very close to the critical heat flux. The transition from single-phase heat transfer to boiling heat transfer, i.e. the region $\dot{x} < 0$ in Fig. 1, was calculated according to Levy /26/.

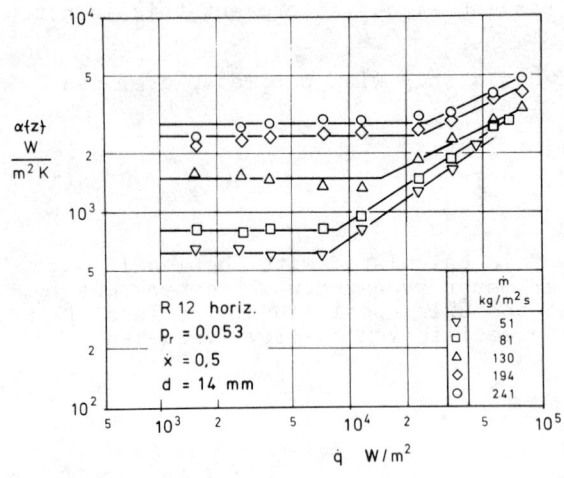

Fig. 2.
Perimeter averaged heat transfer coefficient vs. heat flux

For a vapour quality of 0.5 measured heat transfer coefficients for R 12 are plotted in Fig. 2. On the left hand side forced convective boiling canclearly be distinguished from nucleate boiling. While in the forced convective region an increasing mass velocity increases the heat transfer, this increase is reduced in the nucleate boiling region.

2.1 Heat Transfer Coefficient in the Forced Convective Boiling Region

Forced convective boiling exists when the thermal resistance of the boundary layer is less than the thermal resistance due to nucleate boiling. The heat flux across the boundary layer generates evaporation at the interface of the liquid to the vapour core.

A theoretical solution of the forced convective heat transfer is possible using Reynolds analogy and assuming an axissymmetrical thickness of the liquid film, e.g. at annular flow. Hewitt /27/ extended this method to calculate heat transfer coefficients for upward co-current flow and $Pr_L \neq 1$. Also Kunz and Yerazunis /28/ used the film model and derived

$$St^* = \frac{\alpha(z)_{con}}{c_{pL}\, \rho_L \sqrt{\tau_W/\rho_L}} = f(Re_L, Pr_L) \tag{3}$$

with

$$Re_L = \dot{m}\,(1-\dot{x})\,d/\eta_L \tag{4}$$

They compared their results to the empirical F-function of Chen /29/ and showed that their results are 30 % above the

F-function. Collier et al. /17/ made also a comparison between their experimental results and those calculated by Hewitt /28/. Again the calculated heat transfer coefficients are as much as 30 % above the measured ones. This shows that the agreement between experiment and theory is not sufficient.

The film model can also be used to derive the proportionality between heat transfer coefficient, mass velocity and diameter. For turbulent two-phase flow the effect of Re_L-number in equation (3) can be neglected. Using an approximation for wall friction, e.g. a Lockhart-Martinelli-type relationship

$$\tau_W = c_1 \, \dot{m}^{1.75} \, (1-\dot{x})^{1.75} \, d^{-0.25} \, \emptyset_{Ltt}^2 \qquad (5)$$

one obtains with equation (3)

$$\alpha\{z\}_{con} \sim \dot{m}^{0.88} \, d^{-0.13} \qquad (6)$$

However, the experimental data of Chawla /8/ with R 11 in tubes with 6, 14 and 25 mm inside diameter show

$$\alpha\{z\}_{con} \sim \dot{m}^{(0.9-1.3)} \, d^{-0.8} \qquad (7)$$

The effect of mass velocity on the heat transfer coefficient was further checked with measurements of R 12 in a horizontal tube with 14 mm inner diameter. Tests were performed in the mass velocity range of $50 \leq \dot{m} \leq 700$ kg/m²s and the results are plotted in Fig. 3. The dark symbols represent the results of Bandel /36/, the others are the results of Iwicki et al. /31/. From these data one obtains

$$\alpha\{z\}_{con} \sim \dot{m}^{1.05} \qquad (8)$$

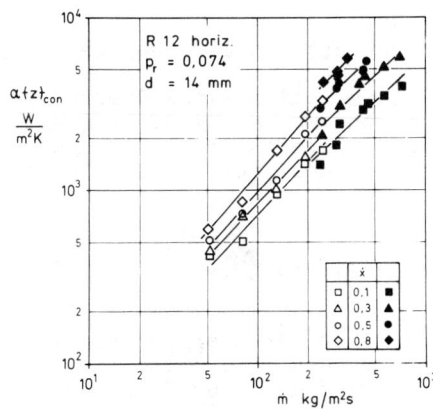 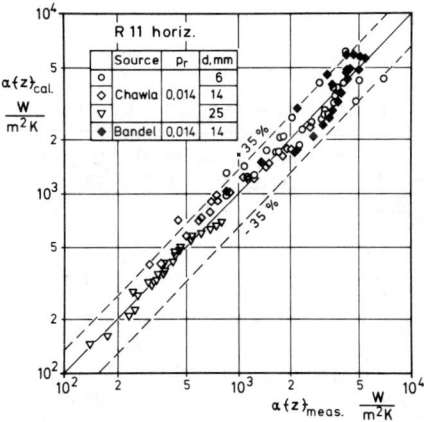

Fig. 3. Heat transfer coefficient vs. mass velocity

Fig. 4. Calculated vs. measured heat transfer coefficients

The effect of pressure on the heat transfer coefficient has been analysed. As a result $\alpha\{z\}_{con}$ decreases with increasing pressure. This is due to the reduction of wall friction with increasing pressure, see Steiner /23, Part 2/ and /30/. This has been experimentally confirmed for refrigerants /31, 37/ and water by Mumm /3/ and Sterman /7/ up to reduced pressures of 0.5. The experiments show also that there is no significant difference between heat transfer coefficients in the forced convective boiling region measured in vertical and horizontal tubes.

Consequently the following equation will be proposed for heat transfer at forced convective boiling in horizontal and vertical tubes /31/

$$St = \frac{\alpha\{z\}_{con}}{c_{pL} \dot{m}} = C_{con} \dot{x}^{0.45} (1-\dot{x})^{0.1} p_r^{-0.4} (d_o/d)^{0.8} \qquad (9)$$

where $d_o = 0.014$ m. p_c and C_{con} are given in Table 1.

Tab. 1. Critical pressures and values C_{con}

	Fluid		
	R 11	R 12	R 22
p_c M Pa	4.37	4.137	4.936
C_{con}	$4.76 \cdot 10^{-3}$	$6.64 \cdot 10^{-3}$	$7.03 \cdot 10^{-3}$

Calculated and measured heat transfer coefficients for R 11 studied in test units with three tube diameters are plotted in Fig. 4. The agreement is reasonable. The same can be shown for R 12 und R 22.

2.2 Heat Transfer Coefficient in the Nucleate Boiling Region

Heat transfer in the nucleate boiling region with superimposed forced convection is mainly governed by the mechanism of nucleation as it has been observed in pool boiling. Therefore commonly used equations for predicting the heat transfer coefficient are similar to equations at pool boiling. One may write

$$\frac{\alpha\{z\}_B}{\alpha_{0.3}} = C_B (\dot{q}/\dot{q}_o)^n F\{p\} F\{d\} F\{\dot{x}\} F\{\dot{m}\} F\{W\} \qquad (10)$$

where $\alpha_{0.3}$ and $\dot{q}_o$ are reference values. The heat transfer coefficient $\alpha_{0.3}$ is to calculate at the reduced pressure $p_r = 0.3$ and the reference heat flux $\dot{q}_o$ using the correlation given by Stephan and Abdelsalam /51/ for pool boiling

$$\frac{\alpha_{0.3} D_d}{\lambda_L} = 0.23 \left[\frac{\rho_V}{\rho_L}\right]^{0.297} \left[\frac{\rho_L}{\rho_L - \rho_V}\right]^{1.728} \left[\frac{a_L^2 \rho_L}{\sigma D_d}\right]^{0.35}$$

$$\cdot \left[\frac{\Delta h_v D_d^2}{a_L^2}\right]^{0.371} \left[\frac{\dot{q} D_d}{\lambda_L T_{so}}\right]^{0.674} \quad (11)$$

where

$$D_d = 0.0206 \cdot \beta^\circ \left[\frac{\sigma}{g(\rho_L - \rho_V)}\right]^{0.5} \quad (12)$$

The contact angle ß is shown in Table 2. For a better accuracy special equations are given for different fluids in /51/, which should then be used in (10).

As stated above heat transfer coefficients in the nucleate boiling region are different in vertical and horizontal tubes.

<u>Nucleate boiling in vertical tubes.</u> Heat transfer coefficients as a function of the heat flux for water at higher pressures are plotted in Fig. 5. For each pressure the heat transfer coefficient can be correlated by a relation of the form

$$\alpha\{z\}_B = C \dot{q}^n \quad (13)$$

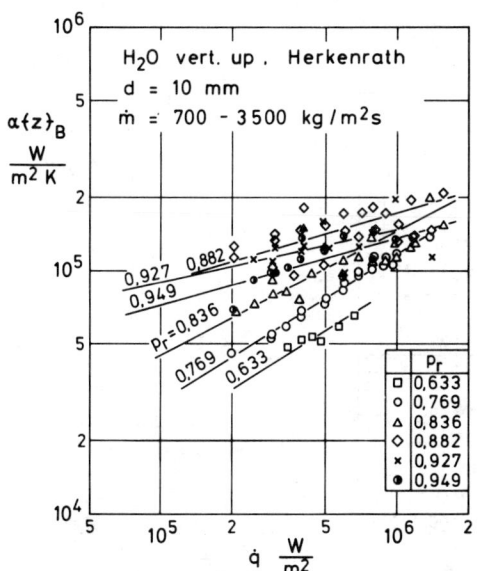

Fig. 5.
Heat transfer coefficient vs. heat flux at vertical upward flow

The exponent n in equation (13) varies with pressure. The same was found by Lukomskii and Madoskaya /32/ in 1951 for ethanol. They suggested

$$n = 0.73 - 0.01\,p \qquad (14)$$

with p in atmospheres.

The dependence of the exponent n as a function of the reduced pressure is shown in Fig. 6 for different fluids. Plotted are results at vertical upward flow and in horizontal channels with a heated bottom side, i.e. with a complete wetted surface.

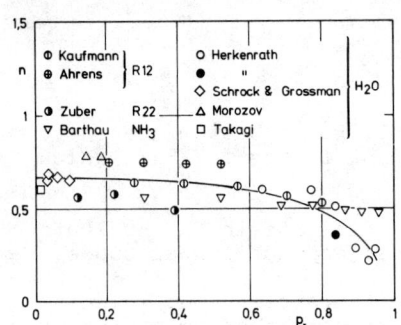

Fig. 6.
Exponent n as function of the reduced pressure

The result in Fig. 6 can be approximated by

$$n = 0.74\,(1-p_r)^{0.32} \qquad (15)$$

It should be mentioned that all experiments have been obtained in a wide range of heat flux (ratio about 1:5 up to 1:15) and that the critical heat flux was reached in most experiments.

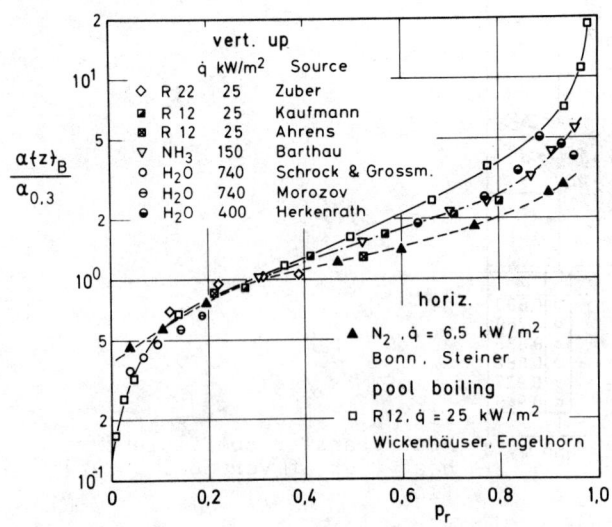

Fig. 7.
Reduced heat transfer coefficient vs. reduced pressure

The effect of pressure on the heat transfer coefficient is shown in Fig. 7 for different fluids and conditions. Pool boiling shows the steepest rise of heat transfer coefficient with pressure whereas flow boiling in horizontal tubes shows the lowest one. For nucleate boiling at vertical upward flow one may use

$$F(p) = p_r^{0.592} (2.14 + 5.42 \, p_r^{8.155}) \qquad (16)$$

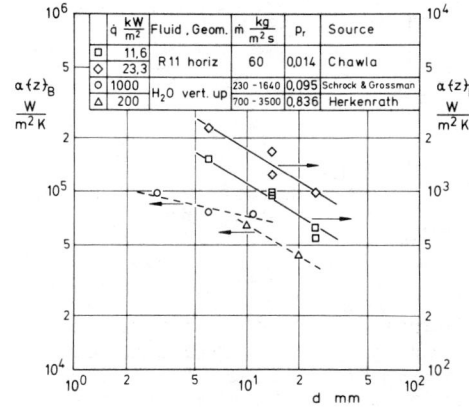

Fig. 8.
Heat transfer coefficient vs. inner tube diameter

No general correlation is known how the heat transfer in nucleate boiling depends on the inner tube diameter. Some authors stated no dependence of the heat transfer coefficient from tube diameter. Other authors however correlated $\alpha(z)_B \sim d^{-0.2}$, see for example /7, 11/. From their experiments Cumo et al. /33/ concluded $\alpha(z)_B \sim d^{-1}$. Fig. 8 shows the heat transfer coefficient plotted against the tube diameter. The experiments of Schrock and Grossman /11/ were made within seamless drawn stainless steel tubes AISI 347 so that equal roughness may be assumed. Their data might be correlated by $\alpha(z)_B \sim d^{-0.22}$. Herkenrath /46/ has measured in stainless steel tubes of material 1.4961 where equal roughness was anticipated. One finds $\alpha(z)_B \sim d^{-0.58}$. Also Purcupile, Riedle et al. /34/ measured nucleate boiling with R 11, R 12 and R 113 within vertical stainless steel tubes AISI 304 and correlated $\alpha(z)_B \sim d^{-0.6}$. As an average

$$F(d) \sim d^{-0.5} \qquad (17)$$

can be used.

Heat transfer coefficients measured at vertical upward flow are shown in Fig. 9 as a function of the flow quality. Nucleate boiling must be assumed since it is $\alpha(z) \sim \dot{q}^n$. Also plotted is the curve for forced convective boiling measured by Bandel /36/ at a reduced pressure of 0.074. Comparing these and other results one can conclude that at reduced pressures $p_r > 0.2$ normally nucleate boiling does exist as long as the heat flux is not too low. Further is to conclude from Fig. 9 and also from a lot of other experiments that the heat transfer coefficient for vertical flow is independent of flow quality, i.e.

$$F\{\dot{x}\} = 1 \tag{18}$$

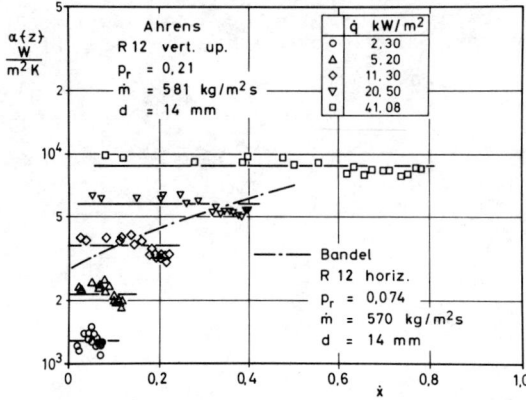

Fig. 9. Heat transfer coefficient vs. flow quality

Previous authors observed an increase of the heat transfer coefficient with increasing mass velocity, for example Cumo et al. /33/. Also Sterman /7/ correlated $\alpha\{z\}_B \sim \dot{m}^{0.1}$. No effect due to mass velocity was noted by other authors. Fig. 10 shows on the right hand side that for vertical flow of water and refrigerant R 12 the effect of mass velocity is negligible. As an approximation one may use

$$F\{\dot{m}\} = 1 \tag{19}$$

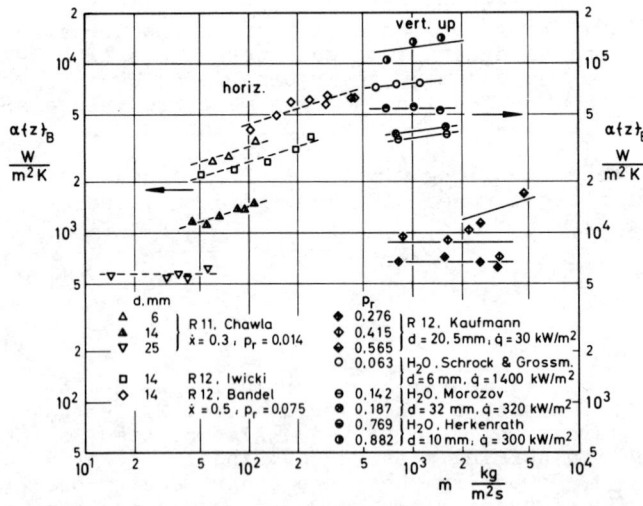

Fig. 10. Heat transfer coefficient vs. mass velocity

The effect of the wall material and the surface onto the heat transfer at nucleate boiling is not sufficient known. But the heterogeneous nucleation depends on the roughness. Stephan /18/ has used as characteristic dimension at pool boiling the mean smooth depth R_p (DIN 4762) and found $\alpha\{z\} \sim R_p^{0.133}$. Danilowa /35/ proposed at pool boiling $\alpha\{z\} \sim R_z^{0.2}$. At flow boiling Zuber et al. /43/ studied the effect of glass and stainless steel tubes on the heat transfer coefficient but no quantitative conclusion has been given. The roughness effect on the heat transfer was measured by Takagi /49/ in a rectangular horizontal channel. Because only the bottom side was heated (full wetted) the result can also be used for vertical conditions. His experiments show $\alpha\{z\}_B \sim R_p^{0.07}$ in the range $0.12 \leq R_p \leq 1.7$ µm. As an averaging result of the wall effect is therefore given

$$F\{W\} \sim R_p^{0.133} \tag{20}$$

Summarizing these facts the following equation is proposed for nucleate boiling in vertical tubes

$$\frac{\alpha\{z\}_B}{\alpha_{0.3}} = 0.75 \, (\dot{q}/\dot{q}_o)^{n\{p_r\}} \, p_r^{0.592} \, (2.14 + 5.42 \, p_r^{8.155}) \, (d_o/d)^{0.5} \cdot$$

$$\cdot (R_p/R_{po})^{0.133} \tag{21}$$

with $d_o = 0.014$ m, $R_{po} = 1 \cdot 10^{-6}$ m and n according to (15). $\alpha_{0.3}$ is to calculate at both the reduced pressure 0.3 and $\dot{q}_o$ as defined in (11). Reference conditions are taken from Table 2.

Tab. 2. Reference conditions and contact angle

		Water	Hydrocarbons + NH_3	Refrigerants	Cryogene
$\dot{q}_o$	W/m²	400.000	150.000	25.000	6.500
ß	Grad	45	35	35	1

<u>Nucleate boiling in horizontal tubes.</u> As already mentioned the temperature distribution along the tube circumference is not uniform at higher heat fluxes due to the non-uniformly wetted surface area and depends on the tangential heat conduction in the tube wall, see /24/.

This in particular affects the relation $\alpha\{z\}_B \sim \dot{q}^n$. In general the exponent n is smaller in horizontal tubes than in vertical tubes under the same conditions. Also the pressure effect is reduced, see Fig. 7. Mumm /3/ found at nucleate boiling of water in a horizontal stainless steel tube $\alpha\{z\}_B \sim \dot{q}^{0.46} \, \dot{m}^{0.34}$ (a+bx). Also the experiments with refrigerants R 11, R 12 and R 22 in copper tubes made by Bandel /36/, Chawla /8/, Iwicki /31/ and Naganagoudar /37/ gives the relationship $\alpha\{z\}_B \sim \dot{q}^{(0.5-0.67)} \, \dot{m}^{0.27}$. These results exhibit a remarkable effect of the mass velocity on nucleate boiling heat transfer, see also Fig. 10. Including the experimental data with nitrogen and argon /23, 38/ the heat transfer coefficient may be correlated by

$$\alpha(z)_B = C \, \dot{q}^{0.56} \, \dot{m}^{0.27} \, d^{-0.5} \, R_p^{0.133} \, F(p) \, F(\dot{x}) \tag{22}$$

where

$$F(p) = p_r^{0.35} \, (1 + 17.6 \, p_r^{30}) \tag{23}$$

and

$$F(\dot{x}) = \begin{cases} 0.2 \, (1+0.35 \, \dot{x}) & \text{if } \dot{q} \leq 500 \text{ kW/m}^2 \\ 0.2 & \text{if } \dot{q} > 500 \text{ kW/m}^2 \end{cases} \quad \text{for water} \tag{24}$$

$$F(\dot{x}) = \begin{cases} 1+0.35 \, \dot{x} & \text{if } \dot{q} \leq 50 \text{ kW/m}^2 \\ 1 & \text{if } \dot{q} > 50 \text{ kW/m}^2 \end{cases} \quad \text{for refrigerants} \tag{25}$$

$$F(\dot{x}) = 0.83 \, (1.01 - \dot{x}^{1.1})^{0.35} \, \dot{q}/\dot{q}_{cr} \quad \text{for } N_2 \text{ and Ar} \tag{26}$$

The critical heat flux $\dot{q}_{cr}$ is determined with the equations given in /25/. The constant C is to calculate with the physical properties and the saturation temperature T_{so} both taken at the reduced pressure 0.3.

$$C = \frac{0.015 \, \lambda_L^{0.74} \, \sigma^{0.225} \, (\Delta h_v \, \rho_V)^{0.133} \, \rho_L^{0.033} (\rho_L - \rho_V)^{0.072}}{\beta^{0.41} \, \eta_L^{1.03} \, T_{so}^{0.26}} \tag{27}$$

All quantities are in SI units. Contact angles ß are given in Table 3.

Tab. 3. Contact angle for use in equation (27)

ß Grad	Water	Hydrocarbons + NH_3	Refrigerants	Cryogene
	45	35	35	13.8

In order to predict the heat transfer coefficient for flow boiling inside vertical and horizontal tubes one has to apply both equation (9) for forced convective boiling as well as equation (21) or (22) for nucleate boiling. In any case the larger value is valid.

For comparison of the proposed correlations with experimental results about 5000 representative data for water at vertical upward flow from different sources /11, 33, 46, 47, 48/ were used. Fig.11 shows the reasonable agreement between calculated and measured heat transfer coefficients. It should be mentioned that already most of the experimental data show a large scatter of approximately ± 40 % overall deviation from the mean value.

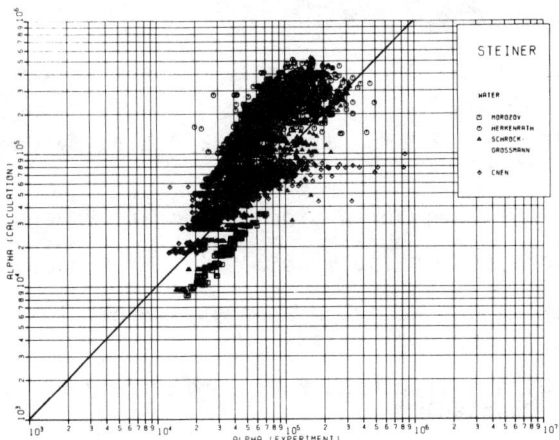

Fig. 11. Calculated vs. measured heat transfer coefficients

3. CONCLUSIONS

Today about 25.000 experimental data are available for saturated flow boiling heat transfer of single-component fluids. But the critical survey of the data shows that systematic data are few.

It has been determined that in forced convective boiling the increase of heat transfer coefficient is effected by an increase in mass velocity and flow quality and decreases with tube diameter and pressure. No significant effect between horizontal and vertical tubes was observed on the perimeter averaged heat transfer coefficient. Whilst in the nucleate boiling region (higher heat fluxes are to transfer) the flow stratification must be taken into account, the heat transfer coefficient is effected by tube orientation. In general the heat transfer coefficient at nucleate boiling is augmented with an increase of heat flux, mass velocity, pressure and surface roughness and decreases with tube diameter.

For the future the experiments should be carried out with more reliable measurement methods and by standarization of experimental devices. Thereby the effect of parameters should be systematically varied in a wide range.

ACKNOWLEDGEMENTS

The authors are indebted to Mr. G. Jöhl and Mr. P. Mauer for their assistance in computing the data. Appreciated is also the A. von Humboldt award for Dr. M. Ozawa.

REFERENCES

1. Dengler, C.E. 1952. Sc. D. Thesis in Chem. Eng. Mass. Inst. Techn.

2. Dengler, C.E. and Addoms, J.N. 1956. C.E.P. Sympos. Series 52, Vol. 18, pp. 95-103.

3. Mumm, J.F. 1954. Argonne National Laboratory Report ANL-5276.

4. McNelly, M.J. 1955. Ph.D. Thesis Univ. of London.

5. Coulson, J.M. and McNelly, M.J. 1956. Trans. Instn. Chem. Engrs. Vol. 34, pp. 247 - 257.

6. Sterman, L.S. and Styushin, N.G. 1951. Trudy UKTI. 21 Mashgiz (in Russian).

7. Sterman, L.S. 1954. Zh. Tech. Fiz. Vol. 24, No. 11, pp. 2046-53 (in Russian) A.E.R.E. Trans. No. 565.

8. Chawla, J.M. 1967. VDI-Forschungsheft 523.

9. Guerry, S.A. and Talty, R.D. 1956. C.E.P. Sympos. Series (52), Vol. 18, pp. 69-77.

10. Tarasova, N.V. Armand, A.A. et al. (1958). Teploenerg. Vol. 5, pp. 93 ff. (in Russian).

11. Schrock, V.E. and Grossman, L.M. 1959. University of California Institute of Engineering Research, Series No. 73308-UCX 2182, TID 14632.

12. Schrock, V.E. and Grossman, L.M. 1962. Nucl. Sci. Eng. Vol. 12, pp. 474-481.

13. Bennett, J.A.R., Collier, J.G. et al. 1959. Atomic Energy Research Establishment, Report AERE-R 3159

14. Wright, R.M. 1961. Lawrence Rad. Lab. UCRL - 9744.

15. Somerville, G.F. 1962. Lawrence Rad. Lab. UCRL - 10527.

16. Hsu, Y.Y. 1962. Trans. ASME, Series C, J. Heat Transfer Vol. 84, pp. 207-216.

17. Collier, J.G., Lacey, P.M.C. et al. 1964. Trans. Instn. Chem. Engrs. Vol. 42 pp. T 127 - T 139.

18. Stephan, K. 1964. Abh. des Deutschen Kältetechnischen Vereins, DKV Nr. 18, C.F. Müller, Karlsruhe

19. Styrikovich, M.A. and Miropolskii, Z.L. 1950. Dokl. Akad. Nauk. SSSR, Vol. 71 (2) (in Russian).

20. Styrikovich, M.A. and Miropolskii, Z.L. 1951. Dokl. Akad. Nauk. SSSR, Vol. 80 (1), pp. 57-60, (in Russian) Ind.Group Trans.: IGRL/T/R-4.

21. Chaddock, J.B. and Noerager, J.A. 1966. ASHRAE Transactions, Vol. 72, Part I.

22. Lis, J. and Strickland, J.A. 1970. Heat Transfer Paris-Versailles, Vol. V, Boiling B 4.6.

23. Steiner, D. and Schlünder, E.U. 1976. Cryogenics Vol. 16, pp. 387-398.Part 2, Pressure drop,Cryogenics Vol. 16, pp. 457-464.

24. Bonn, W., Steiner, D. et al. 1980. Wärme- und Stoffübertragung, Vol. 13, pp. 265-274.

25. Bonn, W., Steiner, D. et al. 1980. Wärme- und Stoffübertragung,Vol. 13, pp. 31-42.

26. Levy, S. 1967. Int.J.Heat Mass Transfer, Vol.10, pp.951-965.

27. Hewitt, G.F. 1961. U.K.A.E.A. Rep. No. AERE-R 3680.

28. Kunz, H.R. and Yerazunis, S. 1969. J. Heat Transfer, pp.413-420.

29. Chen, J.C. 1966. I & EC Process Design and Development, Vol. 5, pp. 322-329.

30. Steiner, D. 1980. Hochschulkurs Wärmeübertragung II, Teil III. Forschungs-Gesellschaft Verfahrens-Technik, Düsseldorf.

31. Steiner, D. and Iwicki, J. 1979. Interner Bericht Institut für Thermische Verfahrenstechnik der Universität Karlsruhe. AIF-Nr. 20:3531/3.

32. Lukomskii, S.M. and Madorskaya, S.M. 1951. Izvest. Akad. Nauk. SSSR Otdel. Tekn. Nauk. No. 9, pp. 1306 ff.

33. Cumo, M., Campolunghi, F. et al. 1977. CNEN-RT/ING (77) 10.

34. Purcupile, J.C., Riedle, K. et al. 1973. Fourteenth National Heat Transfer Conference, AIChE-A.S.M.E. Atlanta. AIChE Preprint 18.

35. Danilowa, G.N. 1965. Cholodilnaja Technika Vol. 42, No. 2, pp. 36/42 (in Russian).

36. Bandel, J. 1973. Dissertation, Universität Karlsruhe.

37. Steiner, D. and Naganagoudar, C.D. 1977. Interner Bericht Institut für Thermische Verfahrenstechnik der Universität Karlsruhe AIF-Nr. 20:3531/2.

38. Bonn, W. 1980. Dissertation Universität Karlsruhe.

39. Bier, K., Wickenhäuser, G. et al. 1977. In heat transfer in boiling. Academic Press, Hemisphere Publ. Corp., pp.137/158.

40. Engelhorn, R. 1977. Dissertation Universität Karlsruhe.

41. Ahrens, K.H. and Mayinger, F. 1979. In Two-Phase momentum, heat and mass transfer. McGraw-Hill Intern. Book Comp., Washington, pp. 591/602.

42. Kaufmann, W.-D. 1974. Dissertation Eidgenössische Technische Hochschule Zürich.

43. Zuber, N., Staub, F.W. et al. 1967. GEAP-5417, EURAEC 1949, Vol. 1 and 2.

44. Borishanskii, V.M., Paleev, I.I. et al. 1973. Int. J. Heat Mass Transfer, Vol. 16, pp. 1073/1085.

45. Shah, M.M. 1976. ASHRAE Trans. Vol. 82, pp. 66/86.

46. Herkenrath, H., Mörk-Mörkenstein, P. et al. 1967. Euratom, EUR 3658 d.

47. Campolunghi, F., Cumo, M. et al. 1974. CNEN-RT/ING (74) 17.

48. Morozov, V.G. 1969. In convective heat transfer in two-phase and one-phase flows. Israel Progr. for Scient. Transl., Jerusalem, pp. 106/114.

49. Takagi, T. 1967. Dr. Thesis, Osaka University.

50. Barthau, G. 1980. Private Mitteilung an das Institut für Thermische Verfahrenstechnik. Als Dissertation an Universität Stuttgart vorgesehen.

51. Stephan, K. and Abdelsalam, M.A. 1980. Int. J. Heat Mass Transfer, Vol. 23, pp. 73/87.

NOMENCLATURE

a	Thermal diffusivity	α	Heat transfer coefficient
c_p	Specific heat capacity	β	Contact angle
d	Inner tube diameter	η	Dynamic viscosity
D_d	Bubble departure diam.	λ	Thermal conductivity
h	Enthalpy	ρ	Density
Δh_v	Latent heat of vaporization	σ	Surface tension
$\dot{m}$	Mass velocity	τ	Shear stress
p	Pressure	φ	Perimeter angle
p_c	Critical pressure	$\emptyset_{Ltt}$	Frictional multiplier
p_r	Reduced pressure (p/p_c)		
$\dot{q}$	Heat flux	Bo	Boiling number ($\dot{q}/(\Delta h_v \dot{m})$)
R_p	Mean smooth depth	Pr	Prandtl number
T	Temperature	Re	Reynolds number
$\dot{x}$	Flow quality	St	Stanton number
z	Main flow direction	St*	Friction Stanton number

Sub- and Superscripts

B	Nucleate boiling	L	Liquid in satur. state
con	Convective boiling	s	Saturation state
cr	Critical heat flux	V	Vapour in satur. state
in	Inlet	W	Wall
		'	Saturated liquid state

An Assessment of Design Methods for Condensation of Vapors from a Noncondensing Gas

J. M. McNAUGHT
National Engineering Laboratory
East Kilbride
Glasgow, UK

ABSTRACT

Two types of design method, (a) Silver's method, based on a predetermined temperature/enthalpy relationship, and (b) film-theory mass-transfer models, for condensation of vapours from a non-condensing gas are discussed and compared. Particular attention is directed towards a comparison of the methods with some recently published data from condensation of vapours forming both miscible and immiscible liquid phases inside vertical tubes. The effect of including the film-theory mass-transfer correction term, described in a previous paper, in Silver's method is also considered.

N O T A T I O N

B	Parameter defined by equation (11)
c_p	Specific heat capacity
F	Parameter defined by equation (14)
h	Specific enthalpy
j_D	j-factor for mass transfer
j_H	j-factor for heat transfer
Le	Lewis number
$\bar{M}$	Molecular weight
N	Number of components
$\dot{N}$	Molar flowrate
$\dot{n}$	Molar flux
P	Total pressure
Pr	Prandtl number
p	Partial pressure

Q	Heat-transfer rate
q	Heat flux
Re	Reynolds number
S	Cross-sectional area
Sc	Schmidt number
T	Temperature
U	Local overall heat-transfer coefficient
x_g	Quality
$\tilde{y}$	Mole fraction
Z	Gradient dQ_G/dQ_T
α	Heat-transfer coefficient
β	Mass-transfer coefficient
ϕ_H	Heat-transfer rate factor

SUBSCRIPTS

d	Equilibrium
f	Condensate + wall + fouling + coolant
G	Bulk gas phase
g	Gas phase
i	Component identifier
O	Coolant
S	Condensate surface
SAT	Saturation condition
T	Total
V	Vapour
W	Wall

SUPERSCRIPTS

$\sim$	Molar quantity
$\cdot$	Corrected for mass transfer
$-$	Mean

1 INTRODUCTION

Condensers for partial or total condensation of mixtures of vapours are frequently designed using the method proposed by Silver[1] for the gas-phase resistance to heat and mass transfer. The basic method has been used and extended by others[2-5] and has been compared[5-7] computationally with more rigorous methods using design examples.

In Silver's method the mass-transfer rate from the gas phase to the condensate surface is evaluated implicitly by (a) calculating the sensible heat-transfer rate from the gas phase, and (b) assuming that the mixture follows some predetermined temperature/enthalpy relationship. The main advantages of the method are that it offers a rapid, simple calculation with the minimum of physical property information. The deficiencies are related to the use of a temperature/enthalpy relationship based on the assumption of equilibrium between the phases at the bulk gas-phase temperature. This assumption clearly contradicts the existence of a temperature difference between the bulk gas phase and the condensate surface, and there are other difficulties involved in, for example, a realistic simulation of wet-wall desuperheating.

A more precise method for condensation of a single vapour from a noncondensing gas is that of Colburn and Hougen[8], who use film theory to evaluate the mass-transfer rate explicitly from the concentration difference and the mass-transfer coefficient. This type of method can be extended to multicomponent systems using, for example, the matrix film model of Krishna and Standart[9].

Despite recent advances in the understanding of multi-component diffusion, and in the computational procedures, the Silver type of method will continue in use, particularly where component property data and mass diffusivity and vapour/liquid equilibrium information are absent or of poor quality, or where economic factors do not justify the more expensive calculation. This paper is therefore directed towards an improved understanding of the relative performance of Silver's method and the mass-transfer models. An analytical study is undertaken to quantify differences between the methods and particular attention is conferred on the capability of the methods to predict some recently published data from experiments.

2 DESIGN METHODS

The derivation of the Silver type of method is straightforward. The sensible heat flux q_G from the bulk gas phase is given by

$$q_G = \alpha_g (T_G - T_S) \tag{1}$$

where α_g is the gas-phase heat-transfer coefficient, which is generally evaluated as if the gas phase were flowing alone. The local overall coefficient, U, defined by

$$q_W = U(T_G - T_0) \tag{2}$$

is then given by

$$\frac{1}{U} = \frac{1}{\alpha_f} + \frac{q_G}{\alpha_g q_W}. \tag{3}$$

The predetermined temperature/enthalpy relationship is now employed to obtain approximations for q_G and q_W. Thus, where Q_G and Q_T represent the sensible heat-transfer rate from the gas phase and the total heat-transfer rate respectively according to the temperature/enthalpy relationship, the final relation for U is

$$\frac{1}{U} = \frac{1}{\alpha_f} + \frac{1}{\alpha_g} \frac{dQ_G}{dQ_T} \qquad (4)$$

If film theory is further used to correct the 'dry gas' coefficient α_g for the effect of mass transfer, U is given by[5]

$$\frac{1}{U} = \frac{1}{\alpha_f} + \frac{1}{\alpha_g^\bullet} \frac{dQ_G}{dQ_T} \qquad (5)$$

where

$$\alpha_g^\bullet = \alpha_g \frac{\phi_H}{\exp \phi_H - 1} \qquad (6)$$

and

$$\phi_H = \sum_{i=1}^{N} \frac{\dot{n}_i \tilde{c}_{p,i}}{\alpha_g} \qquad (7)$$

In the Colburn-Hougen method there is no need to introduce the predetermined temperature/enthalpy relationship. From film theory the vapour molar flux is given by

$$\dot{n}_V = \beta_g \ln \frac{1 - \tilde{y}_{VS}}{1 - \tilde{y}_{VG}} \qquad (8)$$

where the mass-transfer coefficient β_g is obtained from the analogy between heat and mass transfer, that is

$$\beta_g = \frac{\alpha_g}{\tilde{c}_{p,g}} \left(\frac{Pr}{Sc}\right)^{\frac{2}{3}}. \qquad (9)$$

From the relevant equations of downstream development, where film theory is used to evaluate local heat- and mass-transfer rates, it can be shown that for the Colburn-Hougen method the concentration and temperature gradients in the mainstream are related by

$$\frac{d\tilde{y}_{VG}}{dT_G} = \frac{\tilde{c}_{p,g}}{\tilde{c}_{p,V}} \frac{B^{(\tilde{c}_{p,V}/\tilde{c}_{p,g})(1/Le^{\frac{2}{3}})} - 1}{B - 1} \frac{\tilde{y}_{VG} - \tilde{y}_{VS}}{T_G - T_S} \qquad (10)$$

where

$$B = \frac{1 - \tilde{y}_{VS}}{1 - \tilde{y}_{VG}} \qquad (11)$$

and

$$Le = \frac{Sc}{Pr} \qquad (12)$$

Silver's method is tied to the equilibrium temperature/enthalpy relationship, so that in condensation of a single vapour from a non-condensing gas the mainstream concentration temperature gradient is determined by the vapour saturation curve, that is

$$\frac{d\tilde{y}_{VG}}{dT_G} = \frac{1}{P}\left(\frac{dp_{VG}}{dT}\right)_{SAT}. \tag{13}$$

Now for a given set of local bulk gas-phase conditions, the ratio F of the molar fluxes obtained by the Colburn-Hougen and Silver methods

$$F = (\dot{n}_V)_{CH}/(\dot{n}_V)_{Sil} \tag{14}$$

is given by the ratio of the concentration gradients dy_{VG}/dA predicted by the methods. If the film theory mass-transfer correction is applied to Silver's method then both methods use the same equation to obtain dT_G/dA. Hence from equations (10) and (13) F is given by

$$F = \frac{\tilde{C}_{p,g}}{\tilde{C}_{p,V}} \frac{B^{(\tilde{C}_{p,V}/\tilde{C}_{p,g})(1/Le^{\frac{2}{3}})} - 1}{B - 1} \frac{\tilde{y}_{VG} - \tilde{y}_{VS}}{T_G - T_S} \bigg/ \frac{1}{P}\left(\frac{dp_{VG}}{dT}\right)_{SAT}. \tag{15}$$

Values of F for four representative binary systems are plotted in Fig. 1. In the limiting case of low $\Delta T(= T_G - T_S)$ the term B in equation (15) is close to unity and the value of $(\tilde{y}_{VG} - \tilde{y}_{VS})/(T_G - T_S)$ is close to the slope of the saturation line. Hence, on expanding

$$B^m = (1 + \varepsilon)^m \simeq 1 + m\varepsilon$$

F is given by

$$F \simeq Le^{-\frac{2}{3}}. \tag{16}$$

It is apparent from Fig. 1 that this simplification no longer applies at moderate to high ΔTs. It should be emphasized that F represents local differences in prediction of mass flux for the same bulk gas-phase conditions. Over a typical condensing range these differences tend to decrease, since, for example, an over-prediction of heat flux by Silver's method is compensated downstream by the lower available temperature driving force at the next calculation point. In addition many of the curves in Fig. 1 straddle the line F = 1 and it might therefore be expected that the Colburn-Hougen and Silver methods yield similar overall results over a long condensing range in the systems depicted in Figs 1(a-c).

The discrepancies are more serious in the n-hexane/hydrogen system, where Silver's method yields much lower mass fluxes than the Colburn-Hougen method for $\tilde{y}_{VG} > 0.5$. This is mainly a consequence of the low value of the Lewis number, and use of the Colburn-Hougen method for this type of system must be recommended.

An alternative, possible only with the low ΔTs characteristic of vapour rich hydrogen systems, is to modify Silver's method using the approximation given by equation (16). The term dQ_G/dQ_T is given by

$$\frac{dQ_G}{dQ_T} = x_g \tilde{C}_{p,g} \frac{dT}{d\tilde{h}} \frac{\tilde{M}_g}{\tilde{M}_T} \tag{17}$$

and $d\tilde{h}/dT$ is given by

$$\frac{d\tilde{h}}{dT} = \frac{1}{\dot{N}_T}\left\{\dot{N}_t \tilde{C}_{p,\ell} + \frac{\dot{N}_g}{(1 - \tilde{y}_{VG})} \frac{d\tilde{y}_{VG}}{dT} \Delta\tilde{h}_V + \dot{N}_g \tilde{C}_{p,g}\right\}. \tag{18}$$

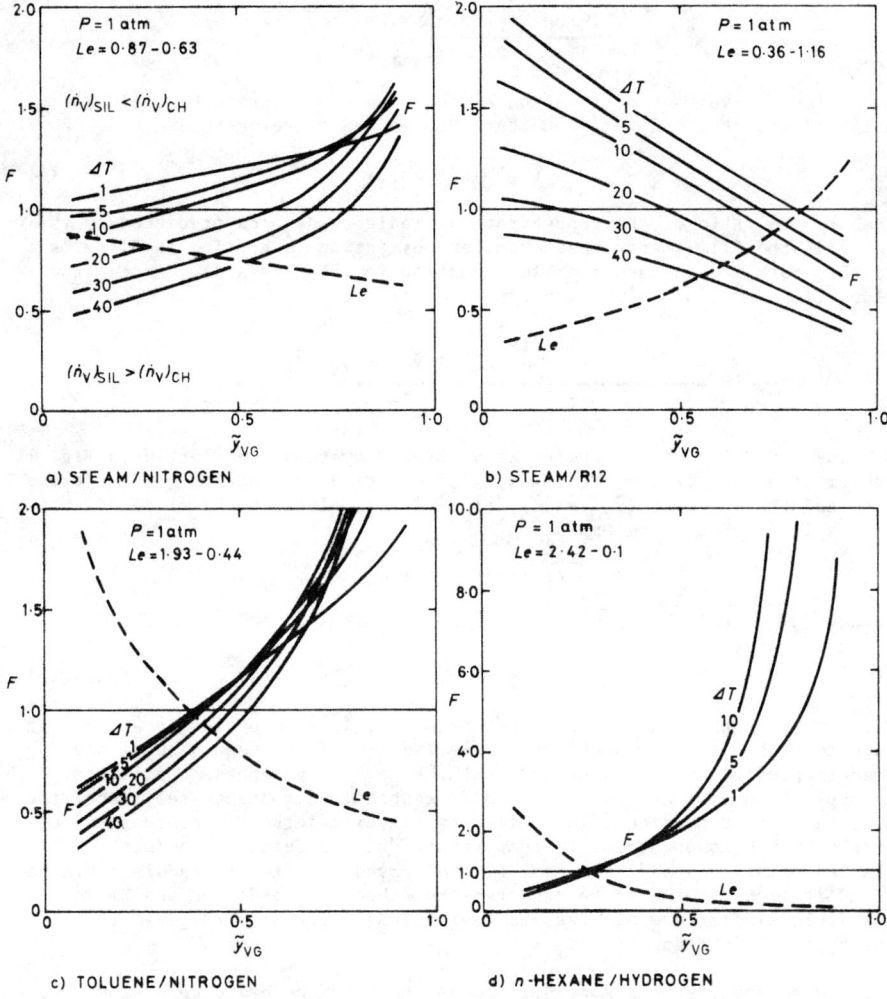

Fig.1. Parameter F for Vapour/Non-condensing Gas Systems

The necessary modification can be effected by replacing $d\tilde{y}_{VG}/dT$, which would normally be evaluated from the saturation curve, by the Colburn-Hougen value. At low values of ΔT, with the approximations inherent in equation (16), this is given by

$$\frac{d\tilde{y}_{VG}}{dT} \simeq Le^{-\frac{2}{3}} \frac{\tilde{y}_{VG} - \tilde{y}_{VS}}{T_G - T_S}. \qquad (19)$$

Again, at low ΔT, the term $(y_{VG} - y_{VS})/(T_G - T_S)$ can be replaced by the slope of the saturation curve, so that

$$\frac{d\tilde{h}}{dT} \simeq \frac{1}{\dot{N}_T} \left\{ \dot{N}_\ell \tilde{C}_{p,\ell} + \frac{\dot{N}_g}{(1 - \tilde{y}_{VG})} Le^{-\frac{2}{3}} \left(\frac{1}{P} \frac{dp_{VG}}{dT} \right)_{SAT} \Delta \tilde{h}_v + \dot{N}_g \tilde{C}_{p,g} \right\}. \qquad (20)$$

This value of $d\tilde{h}/dT$ is used only in the evaluation of dQ_G/dQ_T in equation (5). It represents a Lewis number correction to Silver's method, and is valid only at low ΔT.

In the next section measurements from experiments with a number of vapour/gas combinations are introduced. The measurements are used firstly to provide a bench-mark against which the accuracy of the methods can be compared and assessed, and secondly to provide a range of practical operating conditions which illustrate the effect of the parameters in equation (15).

A similar analysis is not possible in the general multicomponent case, where there is the additional complication of a greater degree of freedom for departure from the equilibrium temperature/enthalpy relationship. The fractional condensation curve is normally used in applying Silver's method, so that the vapour and accumulated condensate are envisaged to be in equilibrium at the bulk gas-phase temperature. However in the mass-transfer models it is assumed that the vapour and condensate are in equilibrium at the lower-interfacial temperature T_s. Thus it is possible for the more volatile component to be present in the liquid phase at a higher composition than would be predicted by the fractional condensation curve. Again the significance of this point is best examined relative to data from experiments.

3   EXPERIMENTAL RESULTS

Data from four separate studies, by Sardesai[10], Deo[11], Owen et al[12] and Lehr[13], are used here. The geometrical and process details of the tests are shown in Tables 1 and 2. In all cases condensation occurred in a single vertical tube surrounded by a cooling jacket, with co-current downwards flow of vapour and condensate.

In many of the tests analysed the vapour/gas mixture was superheated at entry by up to 35K. Practical experience, and the predictions of the Colburn-Hougen method, indicate that in all cases wet-wall desuperheating occurred at entry. Since the idea of Silver's method is to model mass-transfer rates and concentration differences using the corresponding heat-transfer parameters, it is appropriate to calculate the heat-transfer rate across the driving force $(T_d - T_s)$ instead of $(T_G - T_s)$, where T_d is the saturation temperature corresponding to the bulk vapour and liquid composition. Thus Silver's method is applied using Z which is given by the value of dQ_G/dQ_T at the temperature T_d. The local overall coefficient, U, now defined by

$$q_W = U(T_d - T_o) \tag{21}$$

is given by

$$\frac{1}{U} = \frac{1}{\alpha_f} + \frac{1}{\alpha_g} \cdot \frac{Z(T_d - T_s)}{\{T_d - T_s + Z(T_G - T_d)\}} \tag{22}$$

where the correction term $(T_d - T_s)/\{T_d - T_s + Z(T_G - T_d)\}$ accounts for the additional heat flux due to desuperheating.

TABLE 1

PROCESS CONDITIONS AND GEOMETRICAL DETAIL OF EXPERIMENTAL RESULTS – SINGLE VAPOUR WITH NON-CONDENSING GAS

Source	Fluids	No of runs	Inlet $Re_g \times 10^{-3}$	Inlet non-condensing gas mole fraction	Tube i.d. mm	Tube length m	Pressure bar	Symbol in Figs 1-4
Sardesai[10]	Isopropanol/nitrogen	15	6.0– 20.0	0.28–0.78	0.019	1.0	1.0	△
	Isopropanol/R-12	10						▽
Deo[11]	Isopropanol/nitrogen	17	10.0– 14.0	0.55–0.60	0.019	1.0	1.0	+
Owen et al[12]	Steam/nitrogen	13	40.0–100.0	0.14–0.88	0.025	3.0	1.0–1.6	□
Lehr[13]	Benzene/nitrogen	4	6.5– 26.0	0.18–0.60	0.040	5.0	1.0	×
	Toluene/nitrogen	4						◇

TABLE 2

PROCESS CONDITIONS AND GEOMETRICAL DETAILS OF EXPERIMENTS – TERNARY SYSTEMS

Source	Fluids	Number of runs	Inlet $Re_g \times 10^{-3}$	Inlet non-condensing gas mole fraction	Tube i.d. mm	Tube length m	Pressure bar
Sardesai[10]	iPA/water/N_2	8	5.6– 9.7	0.42–0.55	0.019	1.0	1.0
	iPA/water/R-12	7	10.7–20.1	0.27–0.48			
	n-heptane/water/R-12	3	24.2–29.4	0.14–0.32			
	toluene/water/N_2	5	13.1–15.2	0.30–0.44			
	toluene/water/R-12	5	17.6–23.8	0.19–0.29			
Lehr[13]	n-hexane/benzene/N_2	13	13.4–25.3	0.48–0.69	0.040	5.0	1.0
	Benzene/toluene/N_2	7	6.4–17.2	0.57–0.82			

The performance of the condensers was simulated by a numerical integration of the equations describing the downstream development of the stream temperatures and compositions. In all cases the condensate resistance was small compared to the gas-phase resistance, and in view of the relatively low liquid loadings, the resistances could be represented by Nusselt's[14] relations. The shared surface model of Bernhardt et al[15] was applied to systems forming immiscible liquid phases. The gas-phase heat-transfer coefficient α_g was evaluated from

$$\alpha_g = j_H \tilde{C}_{p,g} (\dot{N}_g/S)/Pr^{\frac{2}{3}} \qquad (23)$$

where
$$j_H = a_1 Re^{a_2}. \qquad (24)$$

A correlation derived by Sardesai[10], in which $a_1 = 0.037$ and $a_2 = -0.247$, for dry gas cooling in his apparatus was applied to the data of Sardesai and Deo. The values $a_1 = 0.023$ and $a_2 = -0.2$ were applied to the data of Owen et al and a correlation recommended by Lehr[13] was applied to his data.

The analogy $j_H = j_D$ was assumed in the evaluation of the mass-transfer coefficient β_g.

4 RESULTS AND DISCUSSION

Two important considerations in analysing data from condensation of vapours from a non-condensing gas are:

a   Local errors in prediction of condensation rate tend to smooth out over the condensing range. Thus a low prediction of local condensation rate is compensated downstream by the resulting higher available concentration driving force.

b   The local overall coefficient and temperature driving force can vary by an order of magnitude over the condensing range. A comparison of measured and predicted heat load does not therefore necessarily reflect the important design parameter of predicted/required surface area.

Ideally, then, overall measurements are valid if the condenser is long enough to prevent undue magnification of local errors in prediction, whilst avoiding a large relative variation in the local overall coefficient. It was found that the data of Sardesai, Deo and Owen et al satisfied this criterion. However it was necessary to restrict the analysis of Lehr's measurements to the top 2 m portion of the tube.

4.1   Single Vapour with Non-condensing Gas

Comparisons of the predictions of three methods:

a   Colburn-Hougen[8],

b   Silver[1], and

c   Silver's method with the film theory mass-transfer correction[5],

with the reported measurements of overall condensation rate and heat load are shown in Figs 2 and 3.

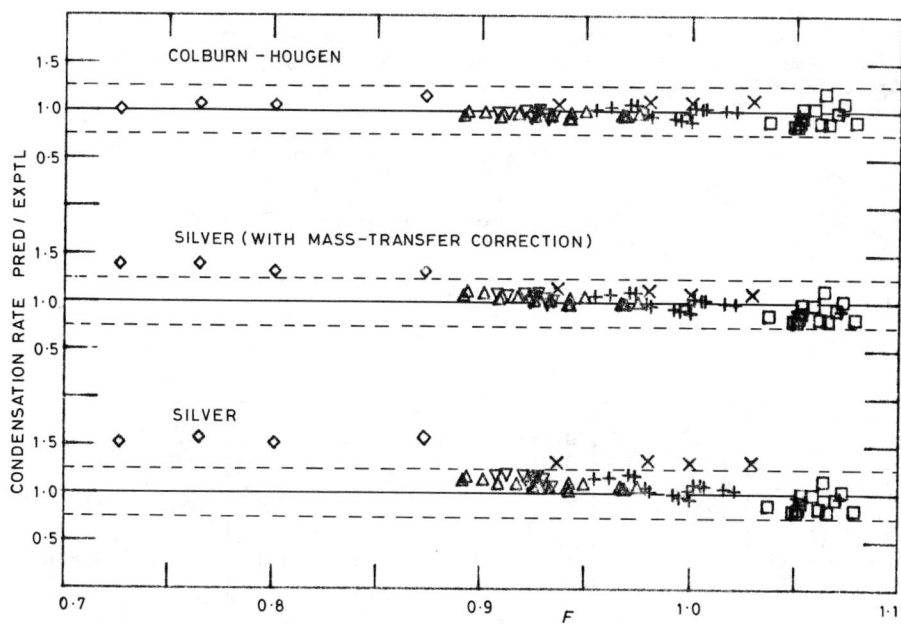

Fig.2. Comparison of Predicted and Measured Condensation Rates - Binary Systems

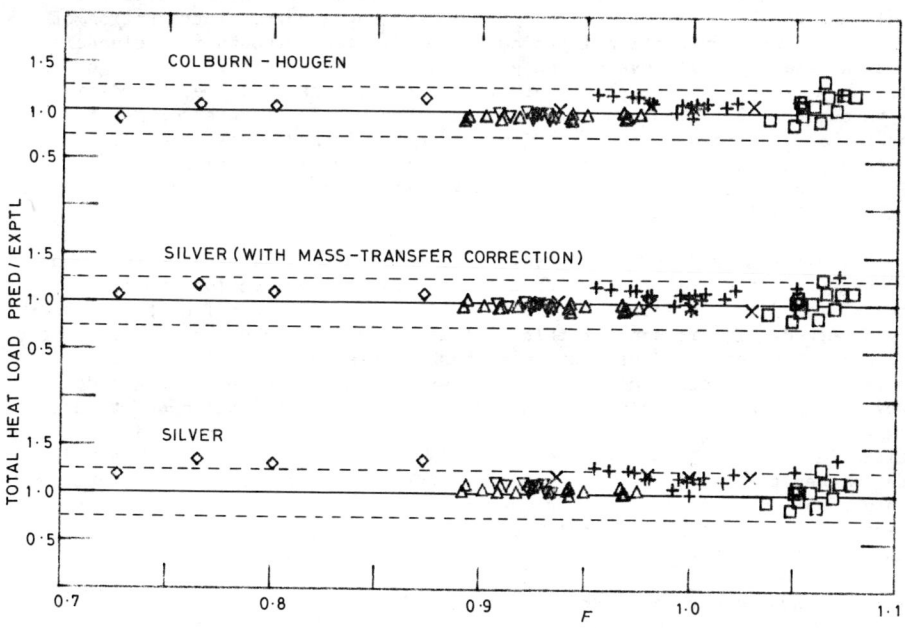

Fig.3. Comparison of Predicted and Measured Heat Load - Binary Systems

The abscissa is the parameter F, defined by equation (14). It was evaluated from equation (15) at the mean vapour flowrate using the concentration and temperature differences predicted by the Colburn-Hougen method. The mean relative errors for each data set are displayed in Table 3.

TABLE 3

MEAN RELATIVE ERRORS (%) IN PREDICTED CONDENSATION RATE AND HEAT LOAD - BINARY SYSTEMS

Source	Fluids	Colburn-Hougen ΔN_V	Colburn-Hougen Q	Silver (+ m.t.c.) ΔN_V	Silver (+ m.t.c.) Q	Silver ΔN_V	Silver Q
Sardesai	iPA/N$_2$	3.45	4.22	4.09	3.38	9.71	4.49
	iPA/R-12	3.31	3.13	5.30	2.34	14.3	8.09
Deo	iPA/N$_2$	4.30	11.1	5.54	12.0	8.08	19.3
Owen et al	Steam/N$_2$	9.89	12.4	11.9	9.8	11.12	9.89
Lehr	Benzene/N$_2$	9.75	5.73	11.4	2.64	33.4	12.0
	Toluene/N$_2$	7.30	8.19	35.7	11.04	54.9	30.0

It is apparent that the Colburn-Hougen method yields the best prediction of condensation rate. However, Silver's method gives satisfactory predictions of all but the benzene/nitrogen and toluene/nitrogen data. Introduction of the mass-transfer correction improves the predictions, though the toluene/nitrogen data are still overestimated, by around 36 per cent. The reason for this can be traced to the parameter F. The experiments were performed with low vapour concentrations and with a large temperature difference with the result that, as shown in Fig. 1c, the value of F is much less than unity. Silver's method therefore predicts a higher condensation rate than the Colburn-Hougen method, which is in turn within 10 per cent of the measurements in this particular system.

Silver's method tends to predict the total heat load data better than condensation rates. In fact, with the mass-transfer correction applied, the results from Silver's method are not significantly different from the Colburn-Hougen method. The reason for this is that Silver's method, being tied to the equilibrium temperature/enthalpy relationship does not allow for liquid cooling below the local saturation temperature. In the Colburn-Hougen method, however, the local heat flux calculation includes a contribution due to the formation of condensate at the surface temperature T_s. The effect is most noticeable in the toluene/nitrogen data, in which the mean temperature difference between the gas phase and condensate surface is relatively high (between 25 and 55K). Overprediction of condensation rate in Silver's method is compensated in the heat flux calculation by ignoring liquid cooling below the saturation temperature.

4.2 Two Vapours with a Non-condensing Gas

Data from experiments with condensation of two vapours from a non-condensing gas inside a vertical tube are available from two sources:

Sardesai[10] and Lehr[13]. The geometrical and process details of the tests are supplied in Table 2. The data include two systems, n-heptane/water and toluene/water, which form immiscible liquid phases. The measurements are compared with the predictions of

a   Silver's method with the film-theory mass-transfer correction applied[5], and

b   the matrix mass-transfer model of Krishna and Standart[9].

The matrix model represents a multicomponent generalization of the Colburn-Hougen method. Application of the model to condensation from ternary systems is described by Krishna and Panchal[16]. No mixing is assumed in the condensate film, that is, the local interfacial liquid concentration is given by the ratio of the local condensing molar fluxes.

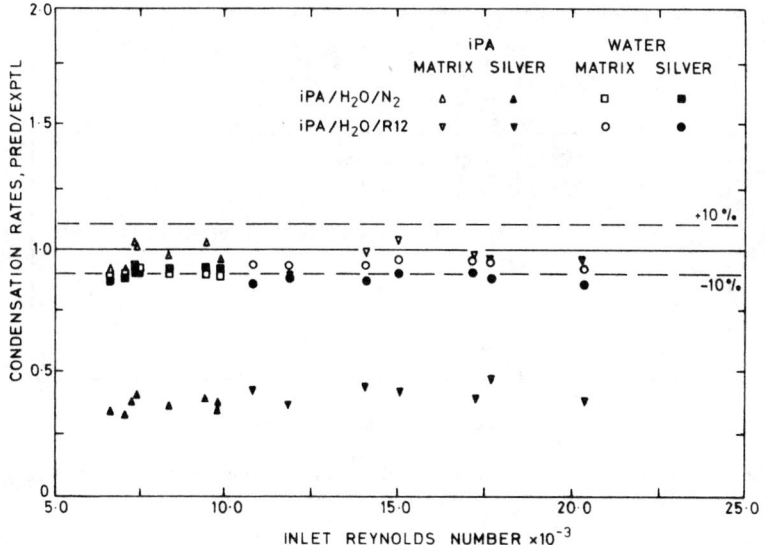

Fig.4. Comparison of Predicted and Measured Condensation Rates – Data of Sardesai[10] – Ternary Systems

Sardesai[10] has shown that his data for condensation of two vapours forming miscible liquid phases are well predicted by the Krishna-Standart matrix model. The power of this method is apparent in Fig. 4, which shows a comparison of Sardesai's data for condensation of steam and isopropanol from nitrogen and Refrigerant-12 with the predicted condensation rates. The individual condensation rates are predicted to within ±10 per cent by the matrix model. Silver's method is less successful. As anticipated, the condensation rate of the more volatile component, isopropanol, is less than that predicted by the mass-transfer model, and is quite poorly predicted. This is partly due, however, to the low condensation rate of isopropanol in the tests. Small differences in the predicted surface temperature imply large relative differences in the liquid concentration of isopropanol.

Fig. 5 shows a comparison of predicted and measured condensation rates for the top 2 m portion in Lehrs[13] data for n-hexane/benzene/nitrogen and

benzene/toluene/nitrogen systems. Though the agreement is not as good as in
Sardesai's data, the matrix method again gives the better prediction. Significantly the matrix method yields the better prediction of condensate composition. In these data Silver's method tends to overpredict the liquid concentration of the less volatile component compared to the matrix method. In the
benzene/toluene data this effect is exaggerated by a further over-prediction
of the toluene condensation rate for the same reason as in the toluene/
nitrogen data analysed above. However the overall error in total molar condensation rate is less serious than in the binary case, since there is another
component present which is fairly well predicted.

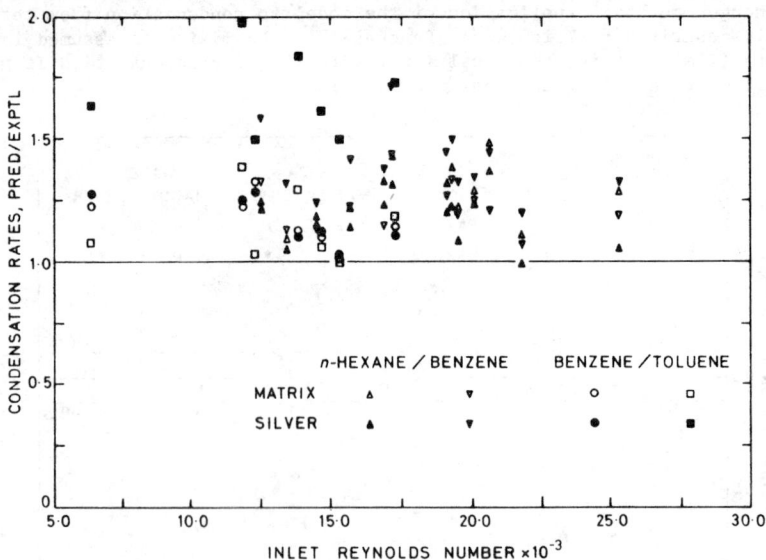

Fig.5. Comparison of Predicted and Measured Condensation Rates -
Data of Lehr[13] - Ternary Systems

Sardesai's data[10] also include three systems, n-heptane/water/
Refrigerant-12 and toluene/water with nitrogen and Refrigerant-12, which form
immiscible condensate phases. In the experiments, and in the matrix model,
the wall temperature at entry was below the transition point of the entering
stream, and both vapours condensed. However in all cases the dew-point of the
mixture at entry was above the transition point. As a result Silver's method
predicts that the component with the higher single-phase dew-point condenses
out first. In some examples the mixture dew-point remained above the transition point in the simulation, and the components with the lower single-phase
dew-point at entry remained in the vapour phase throughout. Consequently the
condensate oulet composition was generally poorly predicted by Silver's method
A comparison of both methods with experiment is shown in Fig. 6.

Silver's method is not really designed or intended for accurate
evaluation of local condensate compositions, and a much fairer test is provided by total heat load. A comparison, for all the available ternary data,
of measured and predicted heat load is shown in Fig. 7. Silver's method
clearly predicts heat load better than composition. This is partly due to the
compensating effect of ignoring liquid cooling below the equilibrium
temperature. In addition the molar latent heats of the components differ by

CONDENSATION OF VAPORS FROM A NONCONDENSING GAS 49

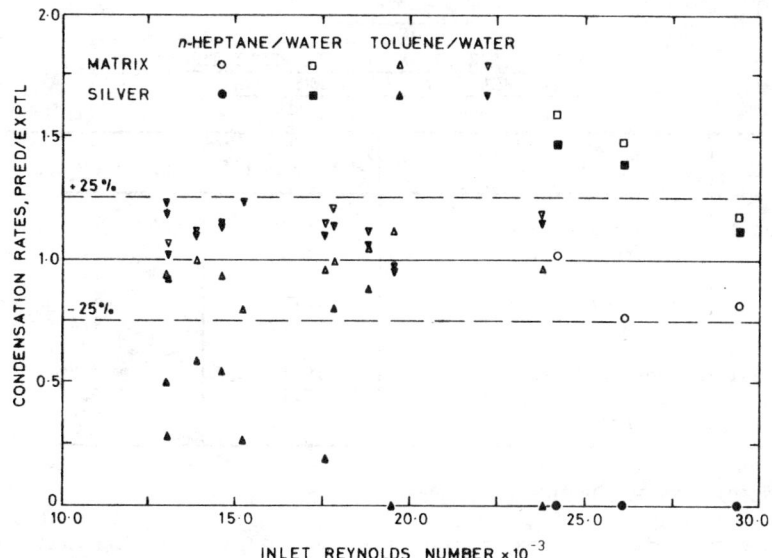

Fig.6. Comparison of Predicted and Measured Condensation Rates - Data of Sardesai[10] - Ternary Immiscible Systems

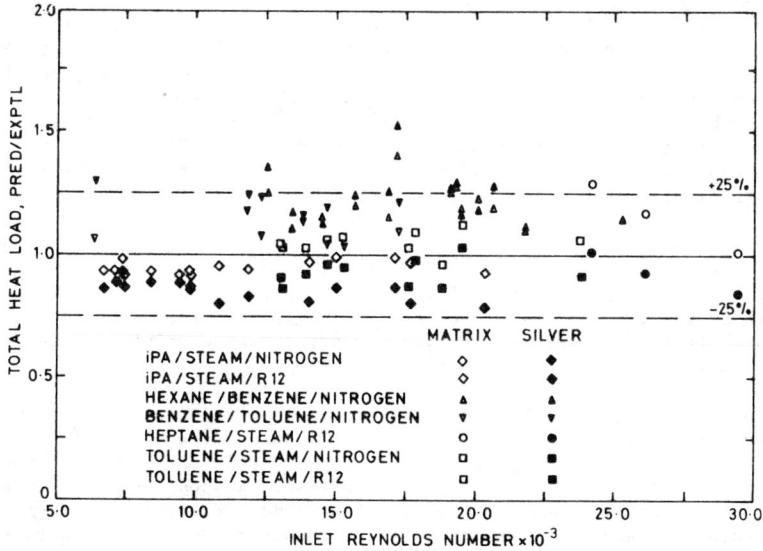

Fig.7. Comparison of Predicted and Measured Heat Load - Ternary Systems

no more than 30 per cent, so that errors in composition attenuate in the heat flux calculation. Generally, though the matrix model provides the more accurate estimation of heat load, the predictions of Silver's method are reasonable, with the majority of the data predicted to within ±25 per cent. Mean relative errors for each system are given in Table 4.

T A B L E   4

MEAN RELATIVE ERRORS (%) IN PREDICTED
TOTAL HEAT LOAD - TERNARY RUNS

Source	Fluids	Matrix model	Silver (+ m.t.c.)
Sardesai	iPA/steam/nitrogen	6.75	11.7
	iPA/steam/R-12	3.57	17.6
Lehr	n-hexane/benzene/nitrogen	20.9	24.0
	benzene/toluene/nitrogen	9.29	19.4
Sardesai	n-heptane/steam/R-12	16.1	7.70
	toluene/steam/nitrogen	5.14	7.88
	toluene/steam/R-12	7.10	7.90

5   IMPLICATIONS FOR DESIGN

The conditions of the experiments analysed in this paper are such that the gas-phase resistance to heat- and mass-transfer is the controlling one. The comparisons presented above between the design methods and experiment therefore give some indication of the margin of safety which should be applied to this part of the process in design.

The results for the single vapour/non-condensing gas systems indicate that the Colburn-Hougen method provides a reliable estimate of both condensation rate and heat flux. Its use is therefore to be recommended, particularly since the computations involved are not much more time-consuming than in Silver's method. Provided that the film-theory mass-transfer correction is applied and proper allowance is made for desuperheating, Silver's method yields an equally good estimate of the heat load data. However, theoretical considerations have identified at least one case, involving condensation from hydrogen, in which Silver's method tends to differ significantly from the Colburn-Hougen method. The general recommendation is that Silver's method should be treated with caution when applied to systems in which the Lewis number differs appreciably from unity or the temperature difference between the gas phase and the condensate surface is large, that is, where the temperature difference does not provide a representative driving force for mass-transfer.

If allowance is made for the difficulties involved in the measurement of condensate composition the comparisons with the data for condensation of two vapours from a non-condensing gas show that the matrix mass-transfer model successfully predicts both heat flux and condensate composition. It therefore appears to represent a sound model of the heat- and mass-transfer process in multicomponent condensation, and there is no reason to doubt its generality. In view of the simplicity of Silver's method relative to the multicomponent mass-transfer models, the predictions of heat load in the ternary systems appear remarkably good. Indeed, in the immiscible systems, which could be considered to present a severe test, Silver's method is not significantly poorer than the matrix model. However it is not possible to draw general conclusions with respect to other systems, and Silver's method must be subject to larger safety margins than the mass-transfer model. The data, and the comparisons, merely indicate that Silver's method can successfully predict heat load in multicomponent systems, and that there are factors inherent in the heat load calculation which compensate for errors in the condensate composition.

A number of caveats should be added to the interpretation of the results, since the experimental rigs are not completely representative of real heat exchangers. The condenser tubes are rather short and entry gas concentrations are comparatively high, with the result that the liquid loadings are low. Thus it has not been possible to examine the effects of condensate waves and entrainment, or to consider methods of handling the liquid cooling contribution to the heat load. A further point relates to the assumptions made regarding the liquid composition in contact with the vapour. Two extreme assumptions are possible in the application of Silver's method:

a   integral condensation, in which the accummulated condensate is perfectly mixed and remains in contact with the vapour phase, and

b   differential condensation, in which the condensate is considered either to be removed from the vapour phase or to be completely unmixed.

Corresponding assumptions in the multicomponent mass-transfer models are:

a   a perfectly mixed condensate remaining in contact with the vapour, and

b   a completely unmixed or removed condensate.

Differences between these assumptions are insignificant with the short condensing ranges and high non-condensing gas concentrations in the experiments, but are likely to be important in certain practical cases. It is common practice to apply the integral condensation curve in Silver's method, and it is well established[4] that this can lead to under-design if significant separation of the phases occurs. Safety margins applied to Silver's method in respect of this assumption should also be applied to the multicomponent mass-transfer models when the 'perfect mixing' model is used.

6   CONCLUSIONS

Data from condensation of one or two vapours from a non-condensing gas inside a vertical tube have been compared with the predictions of design methods based on:

a   film theory mass-transfer models, and

b   a predetermined temperature/enthalpy relationship such as Silver's method.

The mass-transfer models (Colburn-Hougen, Krishna and Standart) provide good estimates of both heat load and condensation rate. Provided that the film-theory mass-transfer correction is applied and that proper allowance is made for desuperheating, Silver's method yields heat load predictions of the data that are not significantly poorer than those of the mass-transfer models. However the mass-transfer models represent a consistent, and apparently sound, model of the heat- and mass-transfer processes, and can be applied in design with greater confidence than Silver's method.

Silver's method tends to differ locally from the Colburn-Hougen method when the temperature difference does not provide a representative driving force for mass-transfer, that is, when the temperature difference is large or the Lewis number differs significantly from unity. For binary systems local differences in predicted condensation rate can be quantified in terms of a parameter F, and a modification to Silver's method, which corrects for the Lewis number effect and is valid only at low temperature differences, has been suggested. Plots of F for specific examples have indicated that in many cases local differences in prediction tend to average out over long condensing ranges.

ACKNOWLEDGEMENTS

This paper is published by permission of the Director, National Engineering Laboratory, Department of Industry. It is British Crown Copyright. The work was carried out under the research programme of the Heat Transfer and Fluid Flow Service (HTFS) and was supported by the Chemical and Minerals Requirements Board of the Department of Industry, UK.

REFERENCES

1   SILVER, L. 1947. Gas cooling with aqueous condensation. Trans. Instn Chem. Engrs, Vol. 25, pp 30-42.

2   GILMOUR, C. H. 1954. Short cut to heat exchanger design - VI. Chemical Engng, Vol. 61, pp 209-213.

3   WARD, D. J. 1960. How to design a multi-component partial condenser. Petro./Chem. Engnr, Vol. 32, pp C42-48.

4   BELL, K. J. and GHALY, M. A. 1972. An approximate generlised design method for multi-component/partial condensers. AIChE Symposium Series, Vol. 69, (131), pp 72-79.

5   McNAUGHT, J. M. 1979. Mass-transfer correction terms in design methods for multi-component/partial condensers. Condensation heat transfer, 18th Nat. Heat-transfer Conf., San Diego, Calif., 6-8 August 1979. New York: ASME, pp 111-118.

6   PRICE, B. C. and BELL, K. J. 1974. Design of binary vapour condensers using the Colburn-Drew equation. AIChE Symposium Series, Vol. 70, (138), pp 163-171.

7   KRISHNA, R., PANCHAL, C. B., WEBB, D. R. and COWARD, I. 1976. An Ackermann-Colburn and Drew type analysis for condensation of multi-component mixtures. Letters in Heat and Mass Transfer, Vol. 3, pp 163-172.

8   COLBURN, A. P. and HOUGEN, O. A. 1934. Design of cooler condensers for mixtures of vapours with non-condensing gas. Ind. Engng Chem., Vol. 26, pp 1178-1182.

9   KRISHNA, R. and STANDART, G. L. 1976. A multi-component film model incorporating a general matrix method of solution to the Maxwell-Stefan equations. AIChEJ., Vol. 22, (2), pp 383-389.

10 SARDESAI, R. G. 1979. Studies in condensation. PhD Thesis. University of Manchester Institute of Science and Technology.

11 DEO, P. V. 1979. Condensation of mixed vapours. PhD Thesis. University of Manchester Institute of Science and Technology.

12 OWEN, R. G., SARDESAI, R. G. and PULLING, D. J. 1980. Heat- and mass-transfer coefficients for the condensation of binary mixtures in a vertical tube. 19th Nat. Heat Transfer Conference, Orlando, Florida.

13 LEHR, G. 1972. Heat and mass transfer in the condensation of two vapours out of a mixture with an inert gas (in German). Dr Ing. Thesis. Technological University of Hanover.

14 NUSSELT, W. 1911. The surface condensation of water vapour (in German). Z. Ver. Deutsch. Ing., Vol. 60, pp 541-546 and 569-575.

15 BERNHARDT, S. H., SHERIDAN, J. J. and WESTWATER, J. W. 1972. Condensation of immiscible mixtures. AIChE Symposium Series. Vol. 68, (118), pp 21-37.

16 KRISHNA, R. and PANCHAL, C. B. 1977. Condensation of a binary vapour mixture in the presence of an inert gas. Chem. Engng Sci., Vol. 32, pp 741-745.

Reflux Condensation Phenomena in a Vertical Tube Side Condenser

JEN-SHIH CHANG and R. GIRARD
Department of Engineering Physics and Institute
for Energy Studies
McMaster University
Hamilton, Ontario L8S 4M1, Canada

ABSTRACT

The basic phenomena occurring in reflux condensation in a vertical tube side condenser have been studied experimentally. The pyrex made annular single tube condenser (inner tube I.D. 1.76 cm and outer tube I.D. 10.16 cm) was used in present study. In order to study in detail the phenomena, the boundary conditions at both ends in this condenser are artificially controlled. The pressure, temperatures inside the inner tube are monitored along the axis together with secondary side heat flux distribution. The void fraction was measured by a capacitance method. The range of parameters of the present work cover total pressure drop across the condenser from 0 to 30 kPag at several system pressure levels, the plenum volume from 6 to 52.2 ℓ, and the total heat removal of 2 to 5 kw. A simple theoretical analysis by a lumped parameter model has been conducted.

1. INTRODUCTION

The use of reflux condensers is common industrial practice, particularly in conjunction with distillation equipment, however, the knowledge of basic mechanism of reflux condensation in a vertical tube is still at a primitive stage. One of the major reason is inadequate knowledge on the flooding condition, and this causes to limit optimum design of the condenser (Deakin 1977). Very few studies have been published specifically on this topic. A comparison of various flooding correlations has been reviewed by a Deakin (1977). He has shown that considerable discrepancies exist in the prediction of the flooding gas velocity. In this paper, the basic phenomena occurring in a vertical tube side condenser in reflux condensation mode have been studied experimentally. A simple analytical model is also conducted.

2. A REVIEW OF REFLUX CONDENSATION PHENOMENA IN VERTICAL TUBES

From the flooding studies in the adiabatic systems, the phenomena are expected to be influenced by the vapour entrance and exit geometries (Imura et al 1977), the inclination of the tube (Hewitt 1977) and tube diameters (Verschoon 1938). Those effects were examined by English et al (1963) for entrance geometry, Deakin et al (1978) for inclining tube, and Diehl and Koppany (1969) for tube diameter effect in reflux condenser. They presented several correlations from their experimental datas to predict flooding velocity under the reflux condensation phenomena. However, there are considerable discrepancies between the correlations in predicting the flooding velocity.

More recently, Tien (1977) presented simple analytical flooding correlation in the presence of vapour condensation, he started from Wallis (1969) correlation for adiabatic system

$$(j_g^*)^{\frac{1}{2}} + m \, (j_f^*)^{\frac{1}{2}} = C \tag{2.1}$$

and assumed vapour flow was reduced by condensation by the subcooled liquid during the flooding conditions, where m and c are empirical constant determined from experiment, and the dimensionless volumetric fluxes of gas and liquid are defined by

$$j_g^* = \left[\frac{\rho_g j_g^2}{gL(\rho_f - \rho_g)}\right]^{\frac{1}{2}}; \quad j_f^* = \left[\frac{\rho_f j_f^2}{gL(\rho_f - \rho_g)}\right]^{\frac{1}{2}} \tag{2.2}$$

where ρ is the density, j is the volumetric flux, L is a characteristic length, g is the gravitational acceleration and subscripts f and g refer to liquid and gas, respectively.

In most of experiments in adiabatic systems (Wallis 1969, Tien 1977) the sign of "m" was observed to be positive. However, in non-adiabatic systems such as reflux condensation the sign of "m" becomes negative in some conditions since j_g^* was reduced by condensation of increasing j_ℓ^* (Tien 1977). This type of correlation has been confirmed in experiments by Wallis et al (1980) and others in simple top flooding experiment.

Detailed mechanism of flooding in reflux condenser has been originally suggested by Deakin (1977) and Deakin et al (1978) from visual observation as follows: (1) At low vapour velocities a smooth falling liquid film was observed; (2) On increasing the vapour velocity small disturbance waves appear on the film, these were particularly significant at the vapour inlet; (3) A further increase in velocity causes the waves at the vapour inlet to bridge across the tube and an intermittent churn flow is established, however the reflux rate is still constant; (4) Eventually, a vapour velocity is reached which is sufficient to eject liquid from the top of the tube, this is accompanied by a dramatic rise in pressure drop across the tube and is commonly taken to be the flooding point; (5) A further increase in vapour flow resulted in climbing film annular flow. Relationship between pressure loss across a vertical tube and superficial vapour mass flux at inlet observed by

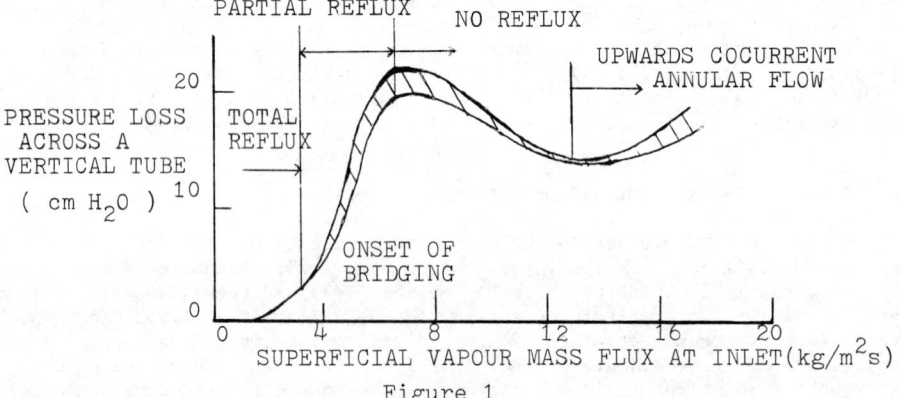

Figure 1

Deakin et al (1978) is shown in figure 1. More recently Banerjee et al (1980) observed the appearance of an oscillating single phase region (water column) on the top of a two-phase countercurrent churn and wavy annular flow region, in the case of very long vertical tube. This mode of operation seems to be halfway between the case where the vapour velocity leads to liquid ejection and the case where we have a climbing film. This mode was named (Banerjee et al 1980) complete reflux condensation, since all the inflow of steam is condensed.

Banerjee et al (1980) studied detail basic reflux condensation phenomena in a single tube, and the experimental results are summarized as follows: (1) The length of the single phase region increases with an increase in pressure difference between upper and lower plenum (total pressure drop), while the average length of the two phase region remains approximately constant; (2) The steam inlet flow rate for complete reflux condensation remains relatively constant with changes in pressure differences and not significantly affected by the heat flux to secondary side cooling water; (3) the length of the two-phase region decreases with increasing heat flux to secondary side, while the single-phase region length is not affected; (4) The temperature near the tube wall, generally falling liquid film temperature, increases slightly with increasing axial distance from inlet, and becomes maximum near interface between single-phase region; (5) The temperature of the centerline of the tube, generally two-phase mixture temperature, has a minimum near the upper edge of the two-phase region, and increase with increasing pressure differences; (6) The temperature profile inside the single phase region has a sharp gradient near the interface.

3. EXPERIMENTAL APPARATUS

The objective of present experiments was to study basic mechanism of reflux condensation, therefore, the boundary conditions on each end of the condenser tube were controlled to give essentially constant pressures. The experimental apparatus was made of 4.1 m long two concentric glass tubes, as shown schematically in Figure 2. The test section is a series of 3 consecutive double-pipe heat exchangers made of pyrex glass. The inner tube in which condensation of steam takes place is made of 3 sections of different lengths. The first section is .6056 m, the second .4572 m and the third 2.74 m long. The sections are linked together by Teflon spacers between cooling sections. The Teflon pieces allow measurement of pressure and

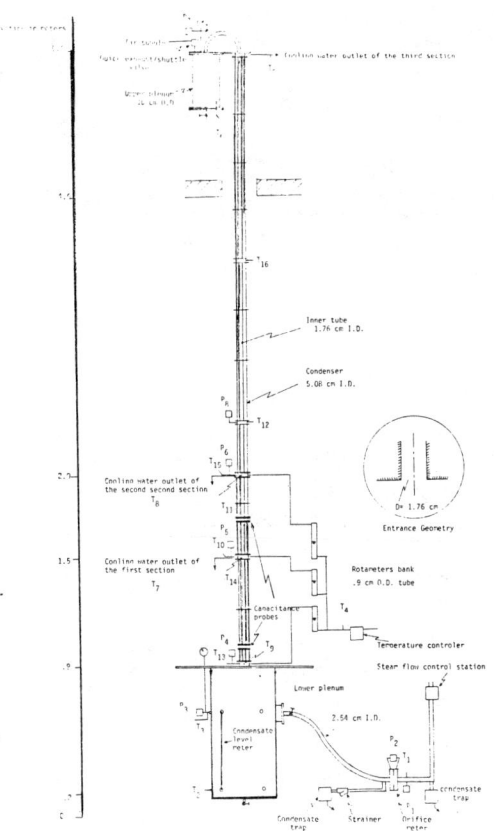

Figure 2

temperature in the inner tube. The inner tube is connected at the bottom to a steam inlet plenum and at the top to an outlet plenum, by a small section of stainless steel tubing in the form of an inverted U. The inlet (or bottom) plenum is constructed of mild steel. A capacitance probe for void fraction measurements (Wallis et al 1980, Banerjee et al 1980) was installed 22 and 84 cm above the plenum. Bonton capacitance meter, Validyne diaghram type pressure transducer and Omega Copper-Constantan thermocouple was used in the present study.

The steam flow rate is adjusted by the control station, then it goes through an orifice meter and flows to the plenum. Upstream of the orifice meter the thermodynamic state of the steam is evaluated. The steam flows through a 25.4 mm nominal diameter horizontal pipe on the side of the plenum with a baffle

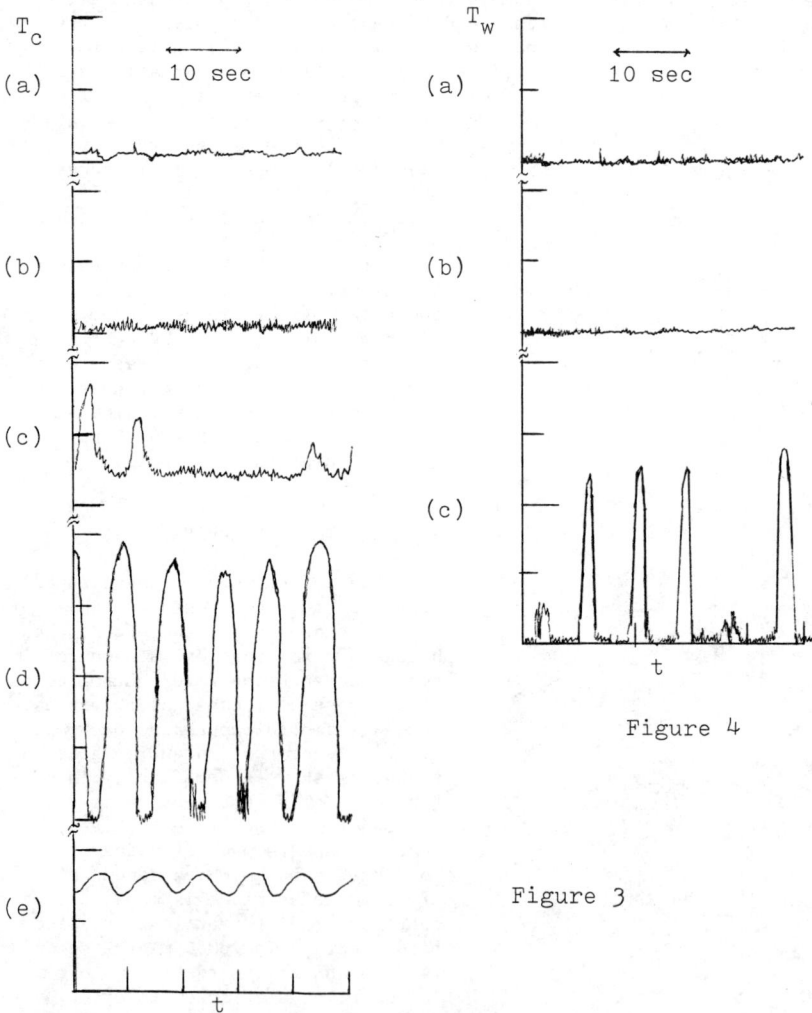

Figure 4

Figure 3

at the end before being discharged into the plenum. This is to allow a constant level of condensate to be maintained in the plenum and the condensate from going into the steam line.

4.1 Experimental Results and Discussions

Typical temperature fluctuations at the center line in the tube, T_c generally two-phase mixture temperature, along the axial direction from the tube inlet, Z, at the inlet cooling water temperature 26°C and total pressure difference drop 18.34 Kpag are shown in figure 3, where (a) Z=0; (b) Z=60 cm (c) Z=110 cm, near single phase/two phase interface, (d) Z=140 cm, inside water column; and (e) Z=400 cm, top edge of water column. Figure 3 shows that the temperature T_c fluctuation increases with increasing Z. Temperature fluctuations near the wall T_w along the axial direction from the tube inlet at same experimental condition are also shown in figure 4, where (a) Z=0; (b) Z=60 cm and (c) Z=110 cm. The temperature T_w fluctuations are relatively small compared with the center line of tube, except near single phase/two phase interface.

Pressure response along the axial direction from the tube inlet at same condition as shown in figure 5, where (a) Z=0; (b) Z=60 cm; (c) Z=110 cm and (d) Z = 140 cm. Figure 5 shows the pressure fluctuations increases with increasing Z. The increasing fluctuations in temperatures and pressure along the axial distance might mean that the local rate of condensation may increase with Z. This phenomena may be confirmed from the void fraction fluctuations as shown in figures 6 and 7, where figure 6 for location Z=22 cm at (a) total pressure drop of Δp_t=4 Kpag and (b) 26 Kpag, and figure 7 for location Z=84 cm at (a) Δp_t=4, (b) 11, (c) 19 Kpag and (d) for climbing film annular flow. Figures 6

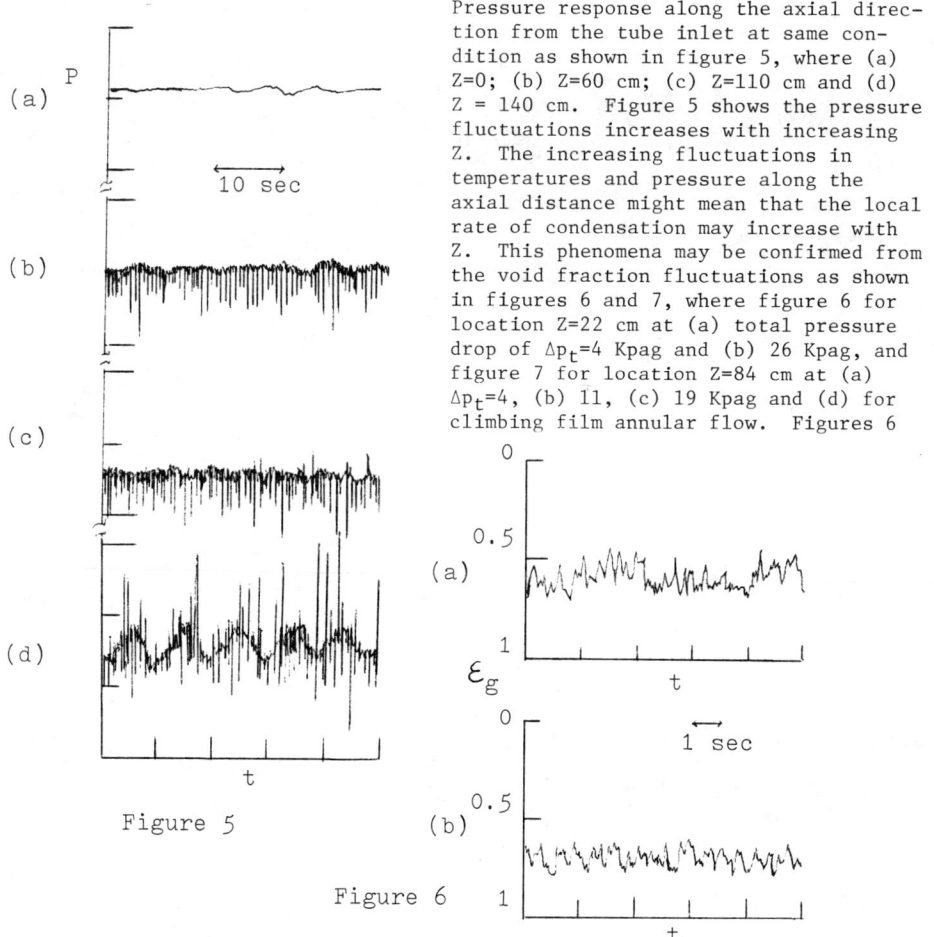

Figure 5

Figure 6

and 7 show that the void fraction fluctuation decreases with increasing pressure differences, and increases with increasing Z. By comparisons with the void fraction of climbing film annular flow region defined by Deakin region number 5 with complete reflux condensation phenomena in figure 7, we found that the liquid film thickness appears to be thicker due to reflux condensation.

The averaged length of single- and two-phase region in complete reflux condensation are shown as a function of system pressure difference in figure 8 at cooling water temperature T =26°C, where "averaged" used in here and later figures means "time averaged values". Figure 8 shows that the length of single phase region increases linearly with system pressure difference, while the two-phase region only very slightly increasing as observed by Banerjee et al (1980). The rate of condensation is also shown in figure 9 at same conditions. Figure 9 shows that the rate of condensation increases slightly with increasing pressure differences, where the rate of condensation used in the present figure is calculated from heat balance from measured temperature and agree with actual measurement of condensate in the plenum. The averaged temperature near the wall and center line of the tube are shown in figures 10 and 11, as

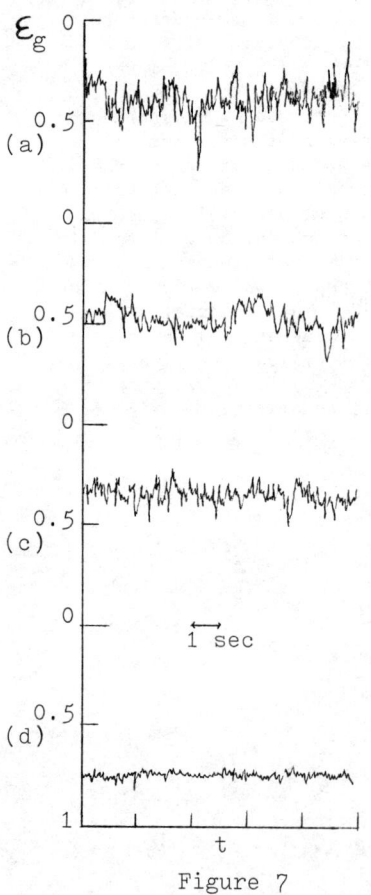

Figure 7

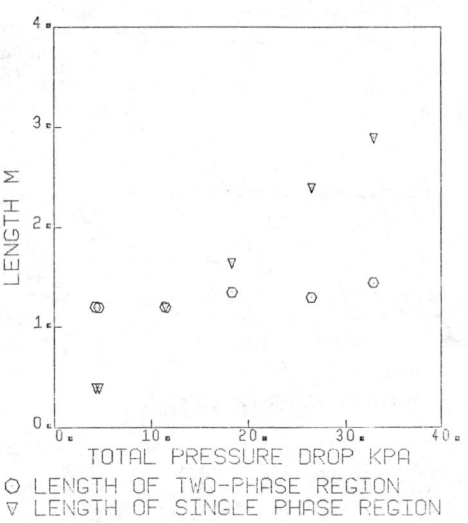

Figure 8

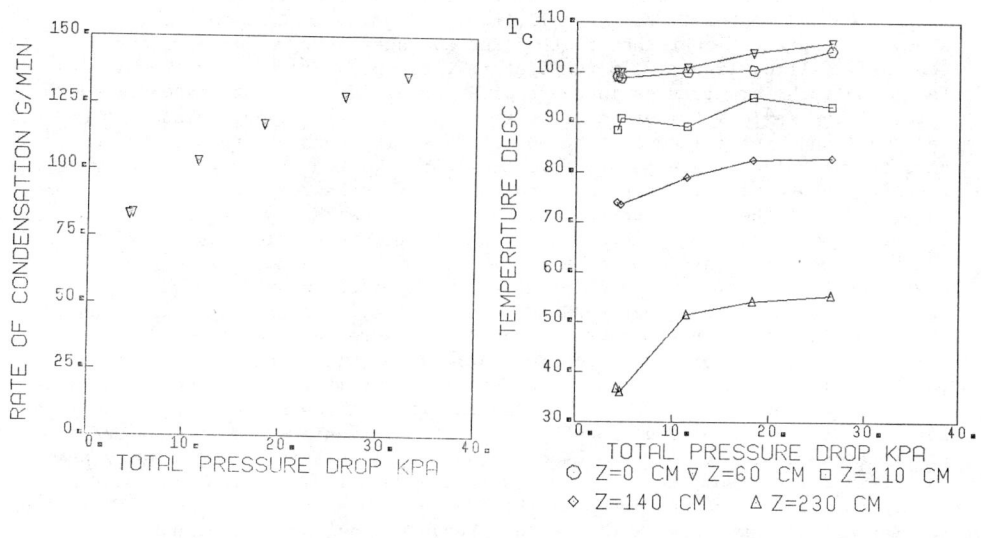

Figure 9

Figure 10

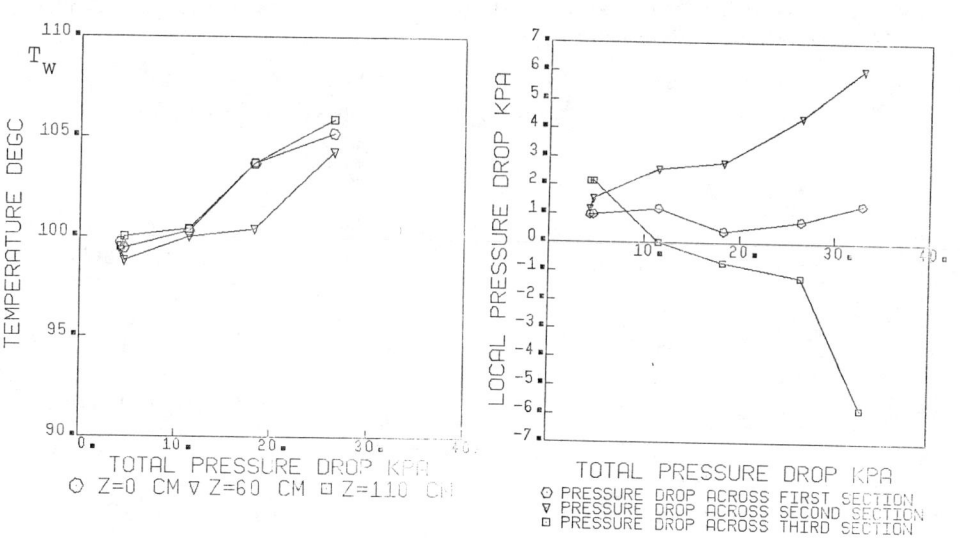

Figure 11

Figure 12

a function of total pressure drop, respectively, for various locations at
the same conditions. Figures 10 and 11 shows that the averaged temperatures
in the center line of tube, generally two-phase mixture temperatures, are
always few degrees lower than that of the averaged near wall temperatures,
generally falling liquid film temperatures, except near water column interface, and both temperatures increase with increasing total pressure drops.
However, the maximum temperature, as shown in figures 3a, 3b, 4a and 4b between
near wall and center line of tube do not differ by much, therefore, bottom
part of subcooled single-phase region may flow back   to reduced temperature
along the center line. This can also explain a large temperature fluctuation
in the edge of the two-phase region, as one can observe in figures 3c and 4c.
The other noticeable phenomena is strong temperature gradient inside the water
column as well as heated gas above water columns. This means that most of
heat loss for vapour, included latent heat, may occur near top edge of the
two-phase region, and direct condensation may play an important role in the
reflux condensation phenomena. The total   pressure drop dependence of
time averaged local pressure drop for various location long the tube are shown
in figure 12 at same conditions. For smaller total pressure drop, time
averaged the local pressure drop increases with increasing Z. However, for
larger   the time averaged local pressure drop takes a maximum and becomes
negative near the edge of two-phase region. This negative pressure drop again
can be explained from the flow reversal , and this flow reversal may happen
due to the rapid condensation near the edge of two-phase region. After flooding,
the rapid condensation will reduce the pressure and some part of subcooled single phase region will flow back. This flow reversal continues until
the new vapour-water churn flow coming from inlet will flood again the single
phase region. This might explain low frequency oscillations observed in
figure 3d and by Banerjee et al (1980) before. More discussion will be
presented in Sec. 6 from theoretical modeling. The averaged void fraction are
shown as a function of system pressure difference for various location at
same condition in figure 13. Figure 13 shows that the void fraction decrease
with increasing Z and total pressure drop Δp_t near interface, and increases
with increasing total pressure drop near entrance.

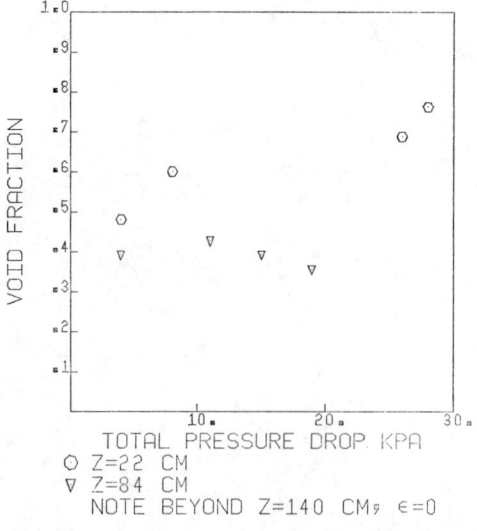

Figure 13

5. OPERATING CONDITIONS OBSERVED

As mentioned in section 4,
the oscillation of the single phase
region may be explained by the combination of cyclic flooding, rapid
condensation and flow reversal.
During this cycle, the counter-
current flow of liquid and vapour
will change, and it will fluctuate
around averaged values. The counter-
current flows of liquid and vapour
were based on a simple heat balance
and put in terms of Wallis parameter j_g^* and j_f^* defined as follows

$$j_g^* = W_t [gD\rho_g(\rho_f-\rho_g)]^{-\frac{1}{2}}A;$$

$$j_f^* = W_t [gD\rho_f(\rho_\delta-\rho_g)]^{-\frac{1}{2}}/A \quad (5.1)$$

where A is the tube cross-sectional
area, D is the tube diameter and

W_t is the condensation rate. Here we also assumed thermodynamic ratio R_T(=heat removed by secondary sides/total latent heat of steam) equals one (Wallis et al 1980). The relationship of nondimensional volmetric fluxes j_g^* and j_f^* are shown in figure 14 for various steam temperature in the plenum. Figure 14 shows that present average counter-current liquid and vapour flows are on the ascending branch of the flooding curve presented by Tien (1977), i.e. "m" is negative in equation (2.1). If we correlated, j_g^* and j_f^* by an equation (2.1), the averaged value of "c" and "m" do not change significantly with plenum steam temperature and equals nearly zero and -6.07, respectively, from figure 14, since we assumed R_T to be unity. Figure 15 shows that "c" decreases with increasing in steam enthalpy (i.e. in pressure and temperature). Finally, it should be noted that these operating conditions represent points of stable operation.

6. AN ANALYTICAL MODEL

In this section, a simple analytical model has been presented to investigate the order of magnitude of the frequencies of the sustained oscillations of the single-phase region (water 1 n.n)encounter in the present system. The schematics of the present reflux condenser used in the model is shown in figure 15 together with equivalent mechanical system, where L is the water column length, M is the mass of the water column, C is the related to the friction of the water column on the tube wall, and k_e is the spring constant related to the compression of the steam. In the present model, we assumed that the condenser is full of steam with no heat removed and steam added, although later we shall partially remove this restriction. We also assumed that the only restoring force in the present system is the compression of the steam. In this condition, the external constant pressure and the water column weight at equilibrium is compensated by the compression of the steam, therefore, the steam pressure at x = 0 is p_o, where V_o is the total volume of steam below the water column. From Newton's second law, we can express summation of the forces as follows:

$$\sum F_x = \dot{P}_{tot} = -F - A\Delta p \tag{6.1}$$

where $\dot{P}_{tot}$ is the time derivative of water column momentum, F is the net force of the fluid on the tube wall, $A\Delta p$ is the restoring force. By assuming laminar flow in water column, we have that

$$\dot{P}_{tot} = \rho L A \ddot{x} \text{ and } F = 8\pi\mu L \dot{x}$$

where ρ is the density of the water column, and μ is the absolute viscosity of the water column. In order to express Δp in terms of the coordinate x of the vibration of the water column, we assumed that the change of state to be polytropic (our first assumption was an adiabatic system) as follows:

$$pV^n = (p_o + \Delta p)(V_o + \Delta V)^n = p_o V_o^n = \text{const}$$

where n is the polytropic expornent. By assuming small values of Δp and ΔV, and using $\Delta V = -Ax$. We obtain $\Delta p = p_o nAx/V_o$. The momentum equation (6.1) then reduces to

$$\rho A L \ddot{x} + 8\pi\mu L \dot{x} + (p_o nA^2/V_o)x = 0 \tag{6.2}$$

If we compared this equation with damped harmonic oscillator motion equation

$$M\ddot{x} + C\dot{x} + k_e x = 0 \tag{6.3}$$

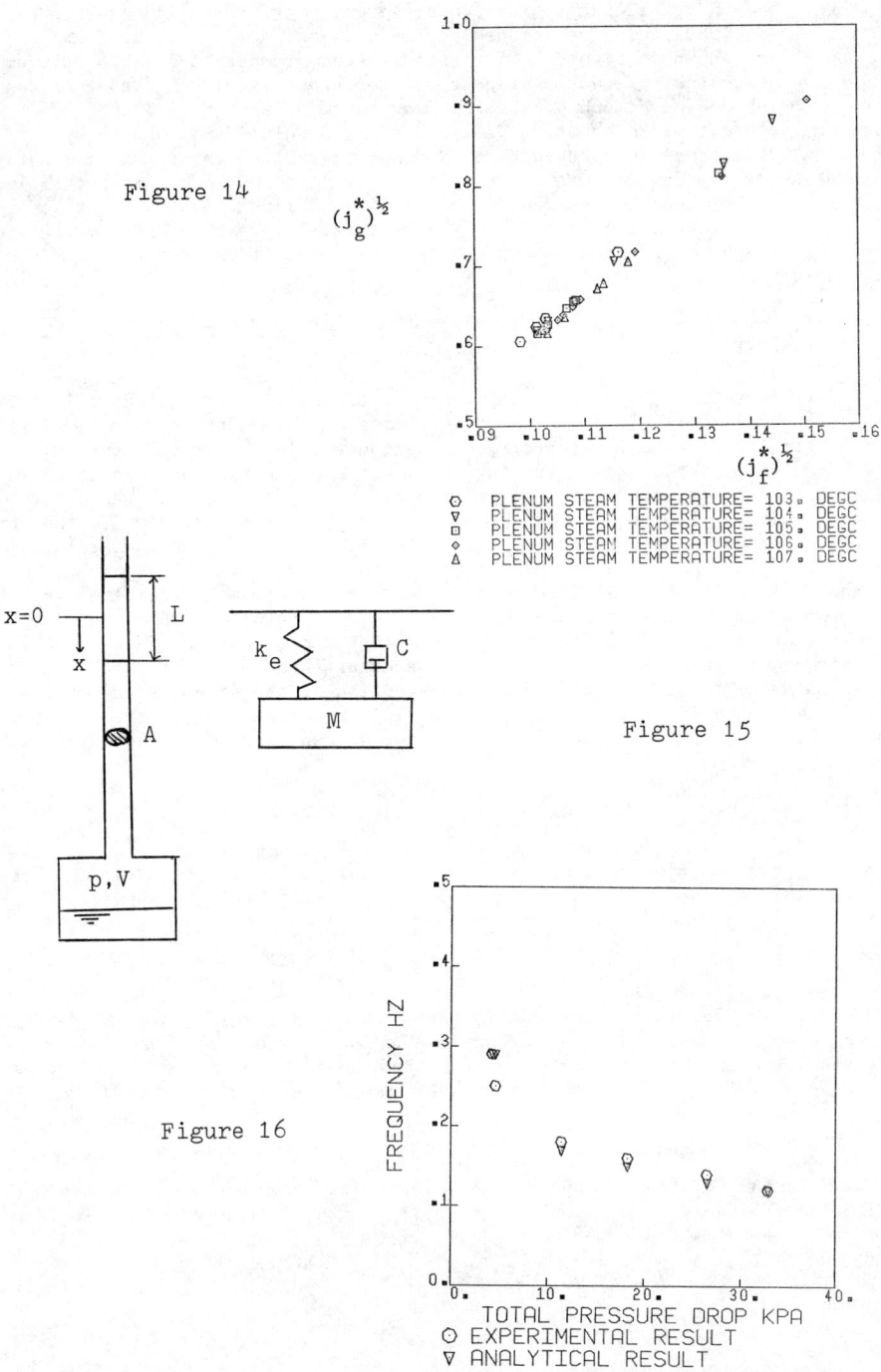

Figure 14

Figure 15

Figure 16

We obtain $M = \rho AL$, $C = 8\pi \mu L$ and $k_e = P_o nA^2/V_o$. If we neglect viscus dissipation effects, equation (6.2) reduces to the governing equation of the classical Helmholtz resonater (Rayleigh 1945).

Then, we obtain

$$f = \frac{1}{2\pi} \sqrt{\frac{P_o nA}{V_o \rho L} - 48 \left(\frac{\mu}{D^2 \rho}\right)}$$

$$\simeq \frac{1}{2\pi} \sqrt{\frac{P_o nA}{V_o \rho L}} \quad \text{for} \quad \left(\frac{32 \mu}{D^2 \rho} \ll 1\right) \quad (6.4)$$

For a forced system, the general motion equation becomes

$$M\ddot{x} + C\dot{x} + k_e x = F(t)$$

Experimental observations suggested that the external force $F(t)$ causing the cyclic flooding described in sections 4 and 5 is counter balance by the viscous dissipation, i.e.

$$\ddot{x} + k_e x = F(t) - C\dot{x} \approx 0$$

therefore, the frequency of the water column oscillations will be given by eq. (6.4b). Comparisons between present experimental values and equation (6.4b) are shown in figure 16 for a cooling water temperature at 26°C. Good qualitative and quantitative agreement between eq. (6.4b) and experiments has been obtained.

7. CONCLUDING REMARKS

The basic penomena occurring in reflux condensation in a vertical tube side condenser have been studied. The following concluding remarks are given:
(1) the appearance of a single phase region (water column) oscillating on the top of a two-phase countercurrent churn and wavy annular flow regions have been observed for a narrow range of inlet steam flow rates: In this condition inlet steam is completely condensed. The averaged length of the single-phase region and rate of condensation increases with increasing total pressure drop, while the averaged length of the two-phase region increases only slightly; (2) from the time dependent observation of temperature, pressure and void fraction profile along the axis, we condluded that the origin of the single-phase region oscillation may due to the combination of cyclic flooding, rapid condensation and flow reversal. A simple analytical model has been developed to predict oscillation frequency of the single-phase region and found out to be in good agreement with measured values both qualitatively and quantitatively;
(3) if we express the limiting non-dimensional volumetric vapour and liquid fluxes by j_g^* and j_f^* respectively, and the flooding correlation by $\sqrt{j_g^*} + m\sqrt{j_f^*} = c$, the negative value of "m" has been observed in present range of experiment.

ACKNOWLEDGEMENT

The authors wish to express their sincere thanks to Drs. S. Banerjee, V.S. Krishnan, J.S. Kirkaldy, C.S. Kim, W.I. Midvidy, S. Ogata and R.E. Pauls for valuable discussions and comments. This work supported partly by Natural Science and Engineering Council of Canada and Ontario Hydro of Canada.

References

1. Benerjee,S., Chang,Jen-Shih, Girard,R., Nijhawan,S., Krishnan, V.S. and Vandenbroek,M.A., 1980, Repot to Ontario Hydro on Transient Two-Phase Flow McMaster university (to be published)
2. Diehl,J.E. and Koppany,C.R.,1969, Flooding velocity correlations for gas-liquid countercurrent flow in vertical tubes. Chem.Eng.Prog.Symp.Ser. $\underline{65}$, 77-83.
3. Deakin,A.W. 1977, A review of flooding correlations for reflux condensers. HARWELL, UK Report No. AERE-M2923.
4. Deakin,A.W., Pulling,D.J. and Brogan, R. 1978, Flooding in reflux condensers HARWELL,UK Report No. HTFS RS 249.
5. English,K.G., Jones, W.T., Spillers,R.C. and Orr,V. 1963. Criteria of flooding and flooding correlation studies with a vertical updraft partial condenser Chem.Eng.Prog.Symp.Ser. $\underline{65}$,77-83.
6. Hewitt,G.E., 1977, Inference of end conditions,tube inclination and fluid physical properties on flooding in gas-liquid flows. HARWELL,UK Report No. HTFS-RS222.
7. Imura,H., Kusuda,H. and Funatsu,S. 1977. Flooding velocity in a countercurrent annular two-phase flow. Chem.Eng.Sci. $\underline{32}$,79-87.
8. Rayleigh,L. 1945. Theory of Sound. vol.II, Dover, New York,P.P. 170-172.
9. Tien,C.L. 1977. A simple analytical model for counter-current flow limiting phenomena with vapor condensation. Lett. Heat & Mass Trans.$\underline{4}$,231-238.
10. Wallis,G.B., deSieyes,D.C.,Rosselli,R.J. and Lacombe,J. 1980. Countercurrent annular flow regimes for steam and subcooled water in a vertical tube. EPRI Report NP-1336.
11. Wallis,G.B. 1969. One-dimensional Two-Phase Flow Modeling. McGraw-Hill Book Co.,New York.
12. Verschoor,H. 1938. Limiting vapour velocity in packed columns. Trans.Inst. Chem.Engrs.$\underline{16}$,66-76.

Two-Phase Pressure Drop for Condensation inside a Horizontal Tube

R. G. OWEN, R. G. SARDESAI, and D. BUTTERWORTH
Heat Transfer and Fluid Flow Service (HTFS)
Atomic Energy Research Establishment
Harwell, Oxfordshire OX11 0RA, England

ABSTRACT

Experimental data for the pressure drop of a condensing flow in a horizontal tube have been compared with the predictions of an annular flow model. The comparisons are made using data obtained by the authors and using further data from several other sources. The data span a mass flux range of 20 to 1250 kg/m^{2}s and are for tube diameters ranging from 0.007 m to 0.025 m. The experimental data are for the condensation of steam, propanol, methanol, refrigerant 113, refrigerant 12, refrigerant 22 and n-pentane.

It is found that the general trends of the predictions of the annular flow model, when used in its no entrainment limit, are in good agreement with the experimental data. However, the frictional component of the local pressure gradient predicted by the model is found to be sensitive to the correlation employed to describe interfacial roughness. Of two such correlations tested, one tended to overpredict and the other to underpredict the frictional pressure gradient. It is concluded that an annular flow model has the potential for accurate prediction of pressure drop for condensing flow in a horizontal tube. Particular care, however, must be taken in the choice of interfacial roughness correlation.

1. INTRODUCTION

Condensation inside horizontal tubes is important in chemical process and power industries. Tubeside condensation is often preferred to shellside condensation in shell and tube heat exchangers where the condensing stream is at high pressure, dirty, corrosive or where injection of glycol is required to prevent hydrate formation. Horizontal intube condensation is frequently employed in air-cooled heat exchangers.

Design engineers have to size condensers to perform the required heat transfer duty within specified constraints on pressure drop. They therefore need accurate methods for predicting pressure drop in horizontal tubeside condensers. In order to predict overall pressure drop, accurate prediction of local pressure gradient is required. Moreover, the local pressure gradient is frequently used to calculate the local shear stress which is employed in calculating the local heat transfer coefficients. Therefore, any inaccuracy in computing local pressure gradients will affect both the calculated overall pressure drop and the local heat transfer coefficients.

Two phase pressure drop in condensing flow is computed by integrating the local pressure gradient over the flow path. The local two-phase pressure gradient is usually evaluated from its frictional, gravitational and accelerational components. The gravitational and accelerational components may be simply calculated if there is some means available for estimating the local void fraction. The frictional component is traditionally evaluated from empirical correlations. Many of these correlations, such as that of Lockhart and Martinelli [1], are derived from adiabatic or boiling data. As illustrated in Figure 1, such correlations may not predict the correct distribution of pressure gradient for condensation inside a horizontal tube. Even correlations developed for condensing flows, such as that of Soliman, Schuster and Berenson [2], may not accurately predict the distribution of pressure gradient (refer again to Figure 1).

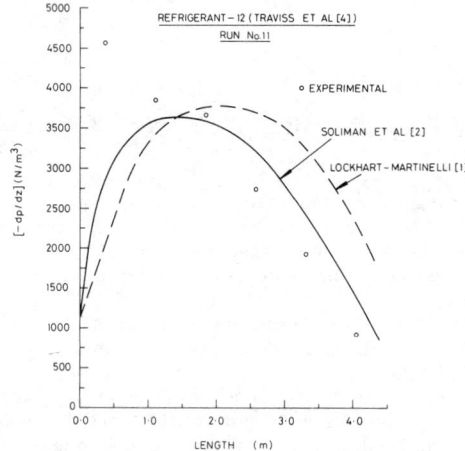

Fig.1. The Variation of Pressure Gradient Along the Condenser Tube

An alternative approach for calculation of the frictional component is to use a model which is based on a particular flow pattern. When condensation occurs in a horizontal tube, there is usually a region of annular flow and this may exist over a large fraction of the tube. In this region, the condensate flows as a sheared film on the tube walls and the uncondensed vapour flows in the core of the tube. Some of the condensate may be entrained into the gas phase as droplets. As the condensation proceeds, the liquid flowrate increases and the gas phase flowrate correspondingly decreases and a gradual transition may occur to a flow pattern such as stratified, stratified/wavy plug or slug flow.

The objective of this paper is to assess the performance of an annular flow model in predicting two-phase pressure drop for condensing flow inside a horizontal tube by comparing its predictions with experimental data. The annular flow model which has been used in this comparison is based on that of Whalley and Hewitt [3].

2. DESCRIPTION OF THE ANNULAR FLOW MODEL

For two-phase flow in a tube, a one-dimensional momentum balance based on the assumption of separated flow of the two phases yields [5]:

TWO-PHASE PRESSURE DROP FOR CONDENSATION

$$\frac{dp}{dz} = g\left\{\rho_g \alpha + (1-\alpha)\rho_\ell\right\} + \frac{d}{dz}\left\{\dot{m}^2\left[\frac{(1-x)^2}{\rho_\ell(1-\alpha)} + \frac{x^2}{\rho_g \alpha}\right]\right\} + \frac{S}{A}\tau_w \quad (1)$$

The three terms on the right hand side of equation (1) are respectively the gravitational, accelerational and frictional components of the local pressure gradient. Methods for evaluating each of these three components using an annular flow model are now discussed:

(a) **The gravitational component**: For flow in a horizontal tube, the gravitational component of pressure drop along the tube is equal to zero.

(b) **The accelerational component**: Equation (1) is based on the assumption of separated flow of the phases. The computation of the accelerational component of equation (1) requires knowledge of the void fraction, α. The void fraction has been simply calculated from the two phase frictional multiplier ϕ_ℓ^2 using a separate cylinders model [8] which gives;

$$\alpha = 1 - \frac{1}{\phi_\ell} \quad (2)$$

(c) **The frictional component**: There are three primary dependent variables in annular flow;

    (a) the interfacial shear stress, τ_i.
    (b) the average liquid film thickness*, m.
    (c) the flow rate of liquid in the film, W_{LF}.

To calculate these three unknown variables, three independent relationships between them must be formulated. These are listed below.

(i) The 'triangular' relationship. This is based on the assumption of some known velocity profile in liquid film, from which it follows that the interfacial shear stress and film thickness are related. The particular form of the triangular relationship used here was one due to Turner and Wallis [6] and Armand [7].

$$\frac{4m}{d} = \sqrt{\frac{(dp_F/dz)_\ell}{(dp_F/dz)}} = \frac{1}{\phi_\ell} \quad (3)$$

    where d = tube diameter.
    $(dp_F/dz)_\ell$ = the pressure gradient which would occur if liquid in the film occupied the whole cross-section of the tube and was flowing at the liquid film mass flowrate.

---

* For condensation inside a horizontal tube, the thickness of the liquid film will vary around the tube circumference because of gravity. The simplification is made here that the film thickness may be considered as uniform around the circumference but varying with axial distance. In this case, m can be considered as the average film thickness at any axial location.

(dp_F/dz) = the frictional pressure gradient in the two-phase flow.

$(dp_F/dz)_\ell$ is calculated from the liquid film friction factor which is plotted against Reynolds number in Hewitt and Hall Taylor [5] from the numerical data of Hewitt [9].

(ii) An interfacial roughness correlation. This is based on the assumption that the roughness presented by the liquid film is a function only of the film thickness. The correlation relates the interfacial shear stress, τ_i, to the shear stress at the wall for the gas core flowing alone in the tube with no liquid film, τ_o. The version employed here was proposed by Whalley and Hewitt [3]:

$$\tau_i = \tau_o \left[1 + \zeta \frac{m}{d} \right] \qquad (4)$$

where $\zeta = 24 \, (\rho_\ell / \rho_g)^{1/3}$

and where ρ_ℓ = liquid density
ρ_g = gas density

This is similar in form to the correlation proposed by Wallis [14] which may be expressed in the same form as equation (4) with $\zeta = 360$.

(iii) The limit of no entrainment for the annular flow model is employed. It is recognised that this is a substantial simplification and that some entrainment will usually occur. However, the degree of entrainment is expected to be lower in condensing flows than in corresponding adiabatic or boiling flows. This is illustrated in Figure 2 in which a comparison is given of the computed entrained liquid flowrate in condensing, adiabatic and boiling flows inside a vertical tube. Entrainment rates in Figure 2 were calculated using the entrainment correlation of Hutchinson and Whalley [10].

The frictional pressure gradient is calculated from the interfacial shear stress by assuming that the interfacial shear stress is uniform around the circumference of the tube, i.e.

$$\frac{dp_F}{dz} = \frac{4}{d} \tau_i \theta_F \qquad (5)$$

In equation (5), θ_F corrects the frictional pressure gradient for the effect of mass transfer to the vapour/liquid interface. Mickley et al [11] have shown that;

$$\theta_F = \frac{\phi_F}{1 - \exp(-\phi_F)} \qquad (6)$$

where $\phi_F = \dfrac{\dot{m}_c \, (U_g - U_i)}{\tau_i} \qquad (7)$

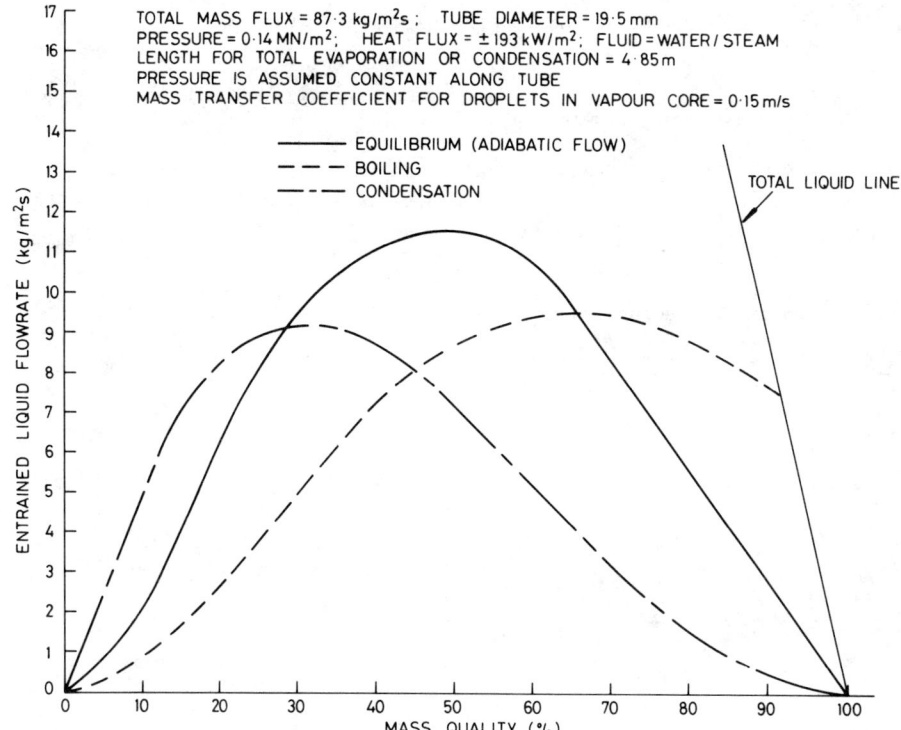

Fig.2. Computed Entrained Liquid Flowrate Variation for Condensation, Evaporation and Adiabatic Flow in a Vertical Tube

and $\dot{m}_c$ is the condensation mass flux
U_g is the gas phase velocity
U_i is the interface velocity (assumed to be small in comparison with U_g)

3. EXPLICIT SOLUTION OF THE ANNULAR FLOW MODEL IN THE NO ENTRAINMENT LIMIT

Equations (3) and (4) may be written in the form;

$$4 t = \frac{1}{\phi_\ell} \tag{8}$$

and $x^2 \phi_\ell^2 = 1 + \zeta t$ (9)

where $t = \frac{m}{d}$,

$\phi_\ell^2 = (dp_F/dz)_\ell / (dp_F/dz)$,

and X^2 = The Lockhart and Martinelli parameter = $(dp_F/dz)_\ell / (dp_F/dz)_g$

where $(dp_F/dz)_g$ is the pressure gradient which would occur if the gas phase alone was flowing in the tube.

Eliminating $\emptyset_\ell$ between (8) and (9) gives

$$X^2 = 16 t^2 \left[1 + \zeta t \right] \qquad (10)$$

There are two extreme solutions of equation (10).

Case A

$\zeta t \ll 1$

which gives $t_A = \dfrac{X}{4}$ \qquad (11)

Case B

$1 \ll \zeta t$

which gives $t_B = 0.397 \, \zeta^{-\frac{1}{3}} X^{\frac{2}{3}}$ \qquad (12)

A full solution to equation (10), accurate to within 2%, may be obtained by combining equations (11) and (12) in the following manner:

$$\frac{1}{t} = \left[\frac{1}{t_A^{2.5}} + \frac{1}{t_B^{2.5}} \right]^{0.4} \qquad (13)$$

Clearly, only solutions in which $t \leqslant 0.5$ (because $t = m/d$) are of interest. In fact, t must be restricted to values less than 0.25 to ensure, from equation (8), that $\emptyset_\ell > 1$. Hence, whenever equation (13) predicts a value of t greater than 0.25, a value equal to 0.25 is assumed.

Hence, combining equations (13) and (8), an explicit equation for $\emptyset_\ell$ is obtained

$$\emptyset_\ell = \frac{1}{4} \left[\frac{1}{t_A^{2.5}} + \frac{1}{t_B^{2.5}} \right]^{0.4} \qquad (14)$$

Now, from the definition of X and $\emptyset_\ell$, and including the correction factor for condensation, the frictional pressure gradient is given by;

$$\frac{dp_F}{dz} = \frac{4 f_g}{d} X^2 \phi_\ell^2 \frac{\rho_g U_g^2}{2} \theta_F$$

where f_g is the friction factor for the gas phase flowing alone in the tube. If the gas phase flowing alone in the tube is in turbulent flow then f_g may be evaluated from;

$$f_g = 0.079 \, Re_g^{-0.25} \qquad (15)$$

4. EXPERIMENTAL APPARATUS

The HTFS intube condensation facility was used for these tests* and a flow diagram is shown in Figure 3. Liquid was pumped from the storage tank via a calibrated rotameter through the evaporator where it was completely vaporized. The vapour was then passed through a heated tube to provide a controlled degree of superheat. To ensure that all the liquid droplets were evaporated in this section, a stainless steel twisted ribbon was inserted in the tube. The slightly superheated vapour was then passed to the test condenser. Any uncondensed vapour from the test condenser was condensed in a back-up condenser and passed to a storage tank via a cooler.

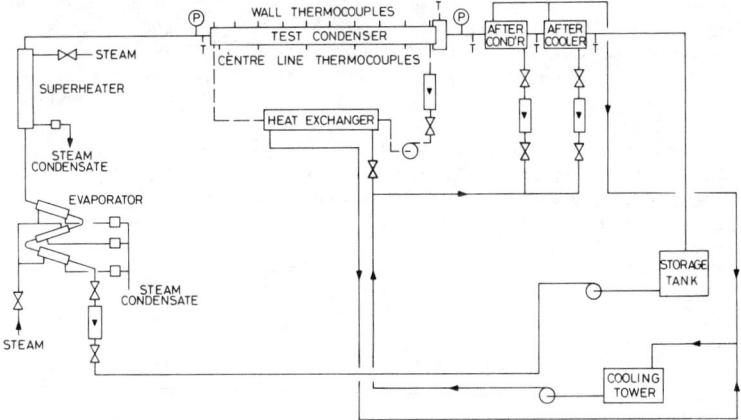

Fig.3. Flow Diagram of the HTFS Condensation Loop

The test condenser comprised a thick-walled stainless steel tube mounted horizontally and surrounded by a concentric cooling jacket. The effective cooled length was 2.92 m and the tube i.d. was 0.0244 m with a wall thickness

---

* The Heat Transfer and Fluid Flow Service (HTFS) intube condensation facility was used for the tests on water, methanol and propanol. The data for n-pentane and refrigerant 113 were taken on a similar experimental facility of Heat Transfer Research Incorporated (HTRI) and obtained by HTFS in a data exchange. The data for refrigerant 12 and refrigerant 22 were obtained from two Massachusetts Institute of Technology (MIT) reports [12], [4].

of 0.0045 m. The condenser tube was surrounded by a cooling-water jacket of 0.055 m i.d. with the cooling water flowing in counterflow to the condensing stream.

Mixing boxes were provided at the inlet and outlet of the cooling jacket in order to measure the inlet cooling water temperature and the rise in cooling-water temperature. The inlet and outlet pressures to the test condensers were measured by means of calibrated pressure transducers.

A mean condensation mass flux around the circumference of the tube was obtained at 10 equispaced axial locations 0.3048 m apart. The mean condensation flux was obtained by measuring the local heat flux at the top and bottom of the tube and taking an average. The local heat fluxes at the top and bottom of the tube were obtained using embedded pairs of thermocouples in the thick-walled stainless steel test section. The technique of using embedded thermocouple pairs for obtaining local heat fluxes and local inside wall temperatures is fully described by Butterworth [13]. Details are also given [13] of the methods of fitting and calibrating these thermocouple pairs.

5. EXPERIMENTAL PROCEDURE

The rig was fully degassed before starting the experimental procedure. In each experiment the procedure adopted was as follows. Fluid from the storage tank was pumped through the evaporator, its mass flowrate being measured by a variable area flowmeter. The steam supplied to the evaporator and to the superheater was adjusted so that the vapour at inlet to the condenser was very slightly superheated*. The cooling water flowrates to the test condenser jacket, to the after-condenser and to the after-cooler were adjusted to the desired values. The system was allowed to run until steady conditions were obtained, and then the data logging equipment was started to record the experimental readings automatically. Accurate measurements were also taken of the rise in test condenser cooling water temperature and of the condensing fluid and the cooling water flowrates. A heat balance for the test section was computed after each run. Where the heat balance was in error by more than 5%, the run was discarded and the experiment repeated.

6. RESULTS

The predictions of an annular flow model for two-phase pressure drop have been compared with experimental data for the condensation of vapour in a horizontal tube. The data is derived from three sources (i.e. HTFS, HTRI and MIT). The experimental systems and the parameter ranges are indicated in Table 1.

Figure 4 gives a comparison between predicted and experimental local frictional pressure gradients for the MIT data. The experimental frictional pressure gradient has been computed from the measured pressure gradient by subtracting the computed accelerational component. It may be observed from Figure 4 that the annular flow model is predicting a frictional pressure gradient which is about 30% too small. The low prediction of frictional pressure gradient leads to a correspondingly low prediction of overall pressure

---

* The purpose of providing the superheat was to ensure that the flow at entry to the test condenser was single-phase vapour.

TWO-PHASE PRESSURE DROP FOR CONDENSATION

drop as shown in Figure 5. For the MIT data (and also for the HTRI data) the mass flux is high and the total pressure drop is dominated by the frictional component. Any change in the model used for calculating the accelerational component of the local pressure gradient would therefore not significantly affect the results.

TABLE 1 - EXPERIMENTAL DATA EMPLOYED

Source	System	Tube diameter (m)	Mass flux range ($kg/m^2 s$)	Quality range
HTFS	Propanol	0.0244	25 - 100	0.01 - 0.98
	Steam	0.0244	25 - 100	0.05 - 0.98
	Methanol	0.0244	25 - 100	0.03 - 0.99
HTRI	n-pentane	0.0198	150 - 250	0.1 - 0.95
	Refrigerant 113	0.0198	1098	0.04 - 0.94
MIT	Refrigerant 12 [4]	0.008	150 - 1250	0.05 - 0.95
	Refrigerant 22 [4]	0.008	200 - 1000	0.02 - 0.95
	Refrigerant 22 [12]	0.0125	300 - 700	0.15 - 0.97

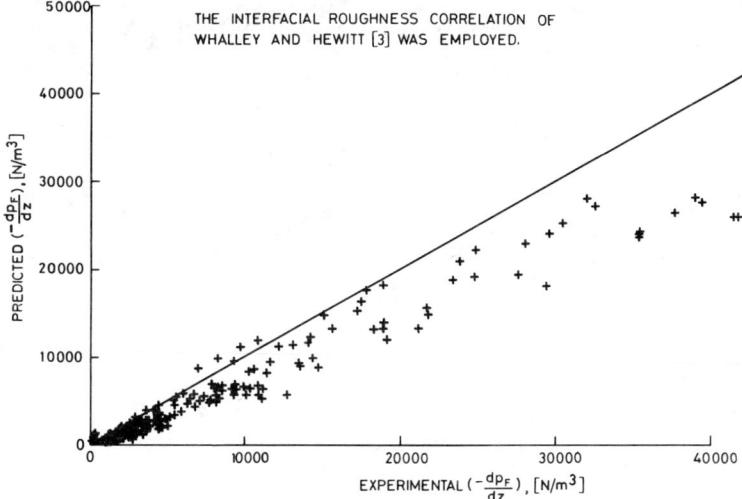

Fig.4. A Comparison of Predicted and Experimental Frictional Pressure Gradients Using MIT Experimental Data [4,12] for Refrigerants 12,22

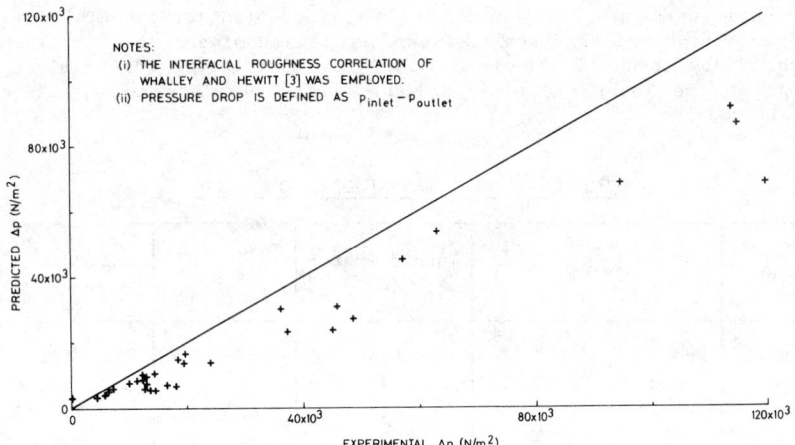

Fig.5. A Comparison of Predicted and Experimental Pressure Drops for the MIT Data

Figure 6 gives a comparison between the predicted and experimental <u>overall</u> pressure drops for the HTFS data*. As indicated in Table 1, the HTFS data is mainly for low values of mass flux and in this case the results are more sensitive to the model used for the accelerational component. A tendency for the low prediction of pressure drop is again evident in Figure 6.

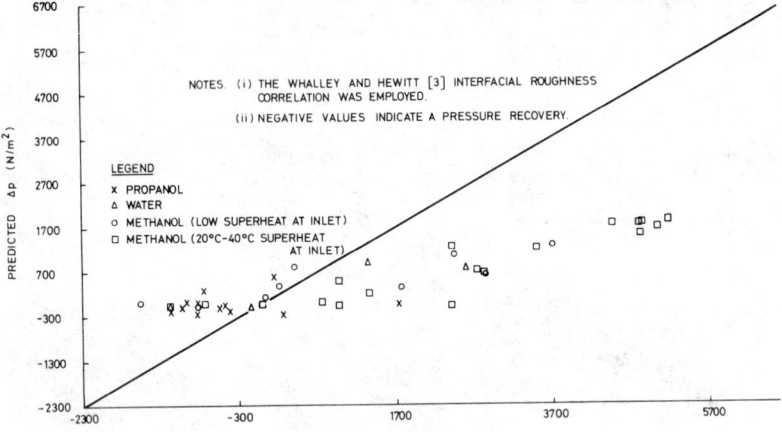

Fig.6. A Comparison of Predicted and Experimental Pressure Drops for the HTFS Data

In Figure 7 a comparison is given between the predicted and experimental <u>pressure variation</u> along the tube for one run of the HTRI data. Once again, the overall pressure drop predicted by the annular flow model is substantially

---

* Only overall pressure drop measurements are available for the HTFS data.

less than that measured. Further results for overall pressure drop using the HTRI data are given in Table 2.

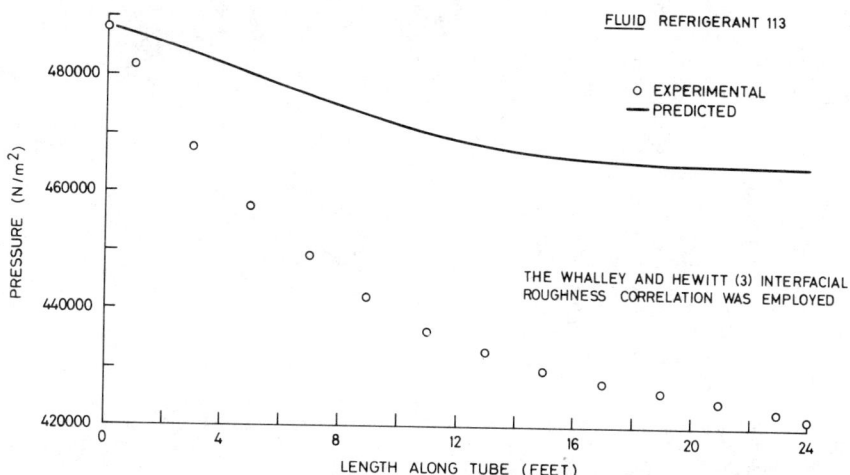

Fig.7. A Comparison of Predicted and Experimental Axial Pressure Variations for the HTRI Data

In order to test the sensitivity of the predictions of the annular flow model to the form interfacial roughness correlation employed, a corresponding set of results to those in Figures 4 to 7 were calculated using the Wallis [14] (see Section 2(c)) interfacial roughness correlation (but keeping all other aspects of the model unchanged). These results are given in Figures 8 to 11 and also in Table 2. In this case there is apparently a trend for the local frictional pressure gradient to be overestimated by up to 30%. The predictions, however, for the HTFS and HTRI data are closer to the measured values.

TABLE 2 - A COMPARISON OF MODEL PREDICTIONS AND THE HTRI DATA

Run No.	Fluid	Measured Δp (N/m^2)	Predicted Δp (N/m^2)	
			Using Whalley and Hewitt [3] interfacial roughness correlation	Using Wallis [14] interfacial roughness correlation
47	n-pentane	10000	4500	12643
48	n-pentane	9285	3392	9760
49	n-pentane	7678	2946	7777
92	Freon 113	72500	23437	83555

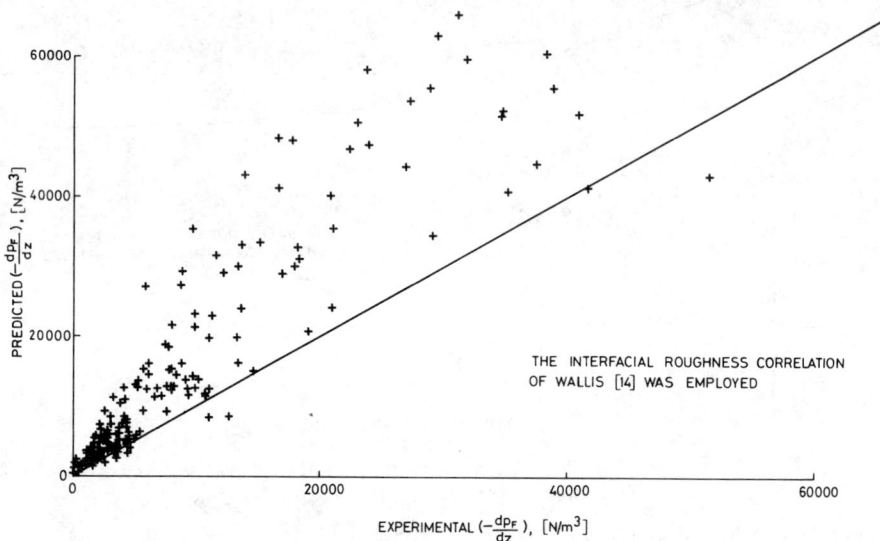

Fig.8. A Comparison of Predicted and Experimental Frictional Pressure Gradients for the MIT Experimental Data [4,12] for Refrigerants 12, 22.

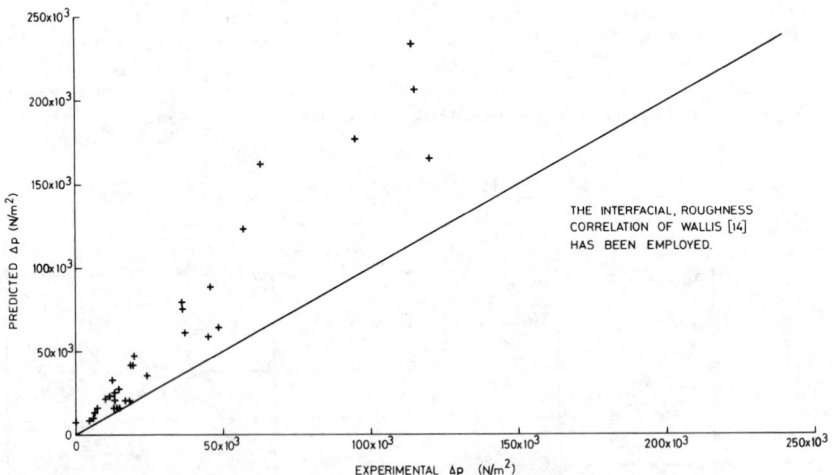

Fig.9. A Comparison of Predicted and Experimental Pressure Drops for the MIT Data

TWO-PHASE PRESSURE DROP FOR CONDENSATION

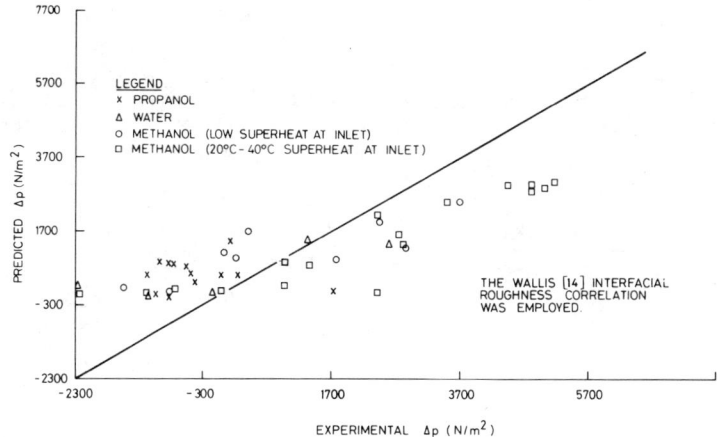

Fig.10. A Comparison of Predicted and Experimental Pressure Drops for the HTFS Data

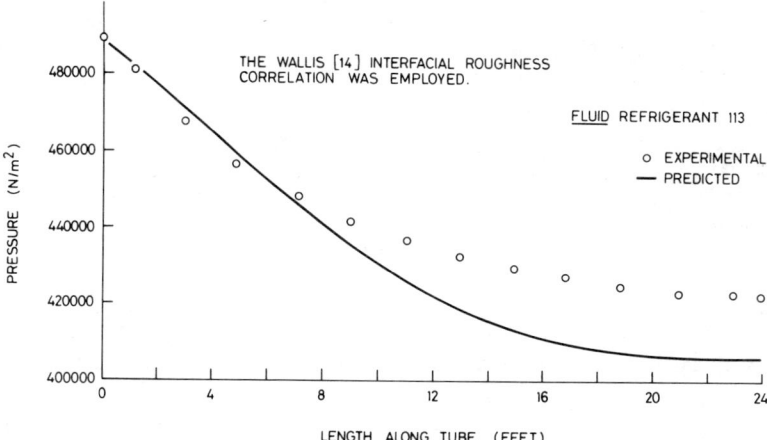

Fig.11. A Comparison of Predicted and Experimental Axial Pressure Variations for the HTRI Data

It is evident, therefore, that the frictional pressure gradient predicted by an annular flow model is sensitive to the form of interfacial roughness correlation used. The Wallis [14] correlation usually leads to overprediction of local pressure gradient whereas the Whalley and Hewitt [3] correlation usually results in underprediction. The extent of these under and over-predictions are illustrated in Figure 12 in which the variation of local pressure gradient along the tube is plotted.

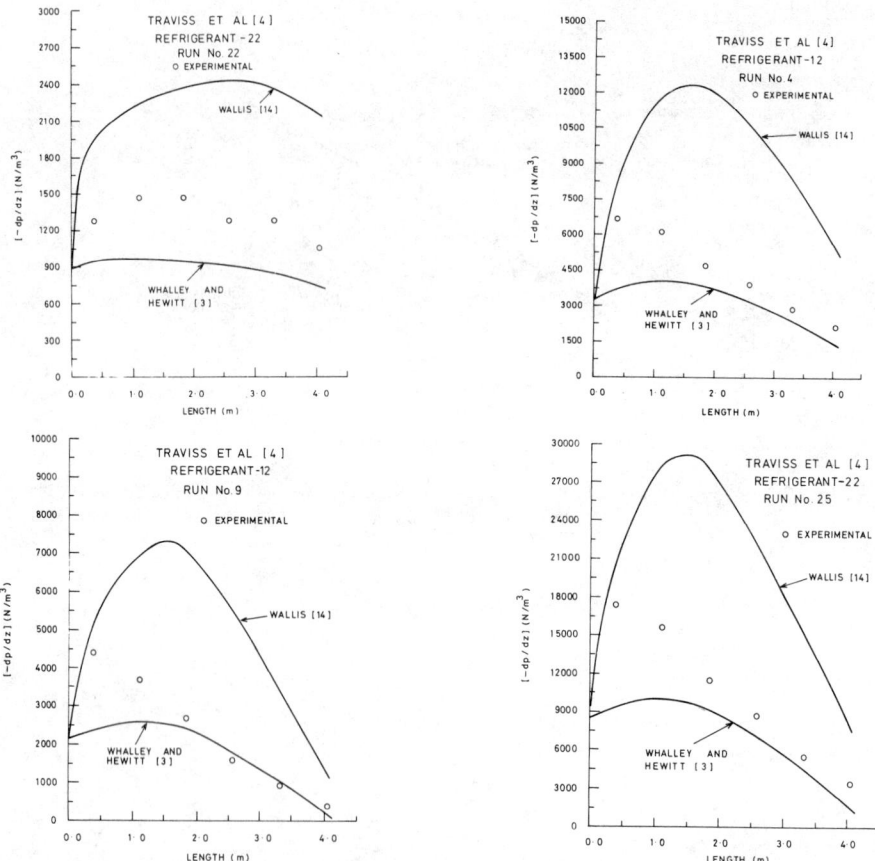

Fig.12. The Variation of Pressure Gradient Along the Condenser Tube

7. CONCLUSIONS

1. Experimental data for the pressure drop of a condensing flow in a horizontal tube have been compared with the predictions of an annular flow model in the no entrainment limit.

2. The predictions of the annular flow model are found to exhibit trends which are consistent with those of the experimental data.

3. The predictions of the annular flow model have been found to be sensitive to the interfacial roughness correlation used.

4. The interfacial roughness correlation of Whalley and Hewitt [3] leads to underprediction of the local frictional pressure gradient while that of Wallis [14] leads to overprediction. There is a need for the development of an improved and more general interfacial roughness correlation.

9. NOMENCLATURE

A	area (m^2)
d	tube diameter (m)
g	acceleration due to gravity (m/s^2)
$\dot{m}$	mass flux ($kg/m^2 s$)
$\dot{m}_c$	condensation mass flux to the vapour/liquid interface ($kg/m^2 s$)
m	thickness of condensate film (m)
p	pressure (N/m^2)
Re_g	gas phase Reynolds number (dimensionless)
S	tube perimeter (m) (= πd)
t	ratio of liquid film thickness to tube diameter (m/d)(dimensionless)
U	axial velocity (m/s)
W_{LF}	flowrate of liquid in the condensate film (kg/s)
x	mass quality (dimensionless)
X	Martinelli parameter (dimensionless)
z	axial coordinate (m)
α	void fraction (dimensionless)
ζ	parameter defined in equation (4) (dimensionless)
θ_F	correction factor for frictional pressure gradient caused by condensation occurring at the vapour/liquid interface (defined in equation (6)) (dimensionless)
μ	viscosity (Nm/s)
ρ	density (kg/m^3)
$\emptyset_F$	factor defined in equation (7) (dimensionless)
$\emptyset_\ell^2$	two-phase frictional multiplier (defined in equation (3)) (dimensionless)
$\emptyset_\ell$	square root of the two-phase multiplier (dimensionless)
τ_w	shear stress at the wall (N/m^2)
τ_o	shear stress at the wall for the gas core flowing alone in the tube with no liquid film (N/m^2)
τ_i	shear stress at the vapour/liquid interface (N/m^2)

Subscripts

a	accelerational
F	frictional
g	gas
i	interface
ℓ	liquid
w	wall

10. ACKNOWLEDGEMENTS

The authors wish to thank Dr. P. B. Whalley for preparing Figure 2 and also Mr. D. J. Pulling for conducting the experiments on the HTFS condensation loop. The work was carried out as part of the HTFS general research programme funded by the Chemicals and Minerals Requirements Board of the U.K. Department of Industry.

8. REFERENCES

1. LOCKHART, R. W. and MARTINELLI, R. C. (1949), 'Proposed correlation of data for isothermal two-phase, two-component flow in pipes', Chemical Engineering Progress, Vol. 45, No. 1, p.39.

2. SOLIMAN, M., SCHUSTER, J. R. and BERENSON, P. J. (1968), 'A general heat transfer correlation for annular flow condensation', Transactions of the ASME Journal of Heat Transfer, Series C, Vol. 90, No. 2, pp.267-276.

3. WHALLEY, P. B. and HEWITT, G. F. (1978), 'The correlation of liquid entrainment fraction and entrainment rate in annular two-phase flow', AERE-R 9187.

4. TRAVISS, D. P., BARON, A. B. and ROHSENOW, W. M. (1971), 'Forced convection condensation inside tubes', Report No. DSR 72591-74, Department of Mechanical Engineering, Massachusetts Institute of Technology, Cambridge, Mass., U.S.A.

5. HEWITT, G. F. and HALL-TAYLOR, N. S. (1970), 'Annular two-phase flow', Pergamon Press, Oxford.

6. TURNER, J. M. and WALLIS, G. B. (1965), 'An analysis of the liquid film in annular flow', USAEC (Dartmouth College) Report NYO-3114-13.

7. ARMAND, A. A. (1946), 'The resistance during the movement of a two-phase flow system in horizontal pipes', Isz Vseoyuz. Teplotekn. Inst., Vol. 1, pp.16-23.

8. TURNER, J. M. and WALLIS, G. B. (1965), 'The separate cylinders model of two-phase flow', Report No. NYO-3114-6, Thayer School of Engineering, Dartmouth College, Hanover, New Hampshire, U.S.A.

9. HEWITT, G. F. (1961), 'Analysis of annular two-phase flow; applications of the Dukler analysis to vertical upward flow in a tube', AERE-R 3680.

10. HUTCHINSON, P. and WHALLEY, P. B. (1973), 'A possible characterisation of entrainment in annular flow', Chem. Engng. Sci., Vol. 28, pp.974-975.

11. MICKLEY, H. S., ROSS, R. C., SQUYERS, A. L. and STEWART, W. E. (1954), 'Heat, mass and momentum transfer for flow over a flat plate with blowing or suction', National Advisory Committee for Aeronautics, Report No. NACA-TN-3208.

12. BAE, S., MAULBETSCH, J. S. and ROHSENOW, W. M. (1970), 'Refrigerant forced convection condensation inside horizontal tubes', Report No. DSR 72591-71, Department of Mechanical Engineering, Massachusetts Institute of Technology, Cambridge, Mass., U.S.A.

13. BUTTERWORTH, D., HAZELL, F. C. and PULLING, D. J. (1975), 'A technique for measuring local heat transfer coefficients for condensation in a horizontal tube', Transactions of the symposium on multi-phase flow systems', The Institution of Chemical Engineers Symposium Series, No. 38, paper D1.

14. WALLIS, G. B. (1970), 'Annular two-phase flow: Part 2, additional effects', J. Basic. Engng., Vol. 92, pp.73-82.

Turbulent Falling Film Evaporators and Condensers for Open Cycle Ocean Thermal Energy Conversion

A. T. WASSEL
Science Applications, Inc.
El Segundo, California 90245, USA

A. F. MILLS
University of California
Los Angeles, California 90024, USA

ABSTRACT

A design methodology has been developed for turbulent falling film evaporators and condensers of Claude open cycle ocean thermal energy conversion power plants. The mathematical model assumes one dimensional flow of liquid and vapor phases, and interphase transfer of momentum, energy and mass species are obtained from correlations which account for high mass transfer rates where appropriate. Parametric calculations have been made for a countercurrent evaporator and a co-current condenser suitable for a 100 MWe power plant. Parameters varied include geometric parameters such as exchanger cross-sectional area, plate spacing and film height, and flow parameters such as liquid and vapor stream flow rates, saturation temperatures, and liquid inlet temperatures.

NOMENCLATURE

A	area; exchanger free cross-sectional area	N_{tu}	number of transfer units
b	half channel width (m)	p	static pressure (N/m^2)
B_m	mass transfer driving force	Pr	Prandtl number, $C_p\mu/k$
C_{fg}	vapor phase skin friction coefficient	P	transfer surface perimeter
		Re	Reynolds number
C_p	specific heat (J/kg K)	Sc	Schmidt number, $\mu/\rho\mathcal{D}_{12}$
$\mathcal{D}_{12}$	binary diffusion coefficient (m^2/s)	T	temperature (K)
		u	velocity (m/s)
g	gravitational acceleration (m/s^2)	W	channel width (m)
h	heat transfer coefficient (W/m^{2}K)	x	mole fraction
$\hat{h}$	specific enthalpy (J/kg)	z	exchanger running length (m)
$\hat{h}_{fg}$	latent heat of vaporization (J/kg)		
H	total enthalpy (J/kg), exchanger height	Greek	
He	Henry number	ε	exchanger effectiveness
k	thermal conductivity (W/m K)	δ	film thickness (m)
K	mass transfer coefficient (kg/m^{2}s)	ρ	fluid density (kg/m^3)
Ka	Kapitsa number, $\nu^4\rho^3g/\sigma^3$	Γ	liquid film mass flow rate per unit width (kg/ms)
m	mass fraction	μ	dynamic viscosity (N s/m^2)
$\dot{m}''$	condensation/evaporation rate (kg/m^{2}s)	ν	kinematic viscosity (m^2/s)
M	molecular weight (kg/kmol)	σ	surface tension (N/m)
n	absolute mass flux (kg/m^{2}s)		
N	number of channels or plates		

Subscripts

b	bulk	s	s-surface
G	vapor phase	sat	saturation
H	heat transfer	u	u-surface
L	liquid phase, bulk liquid	1	water vapor
M	mass transfer	2	noncondensable gas

Superscript

"	per unit area	•	corrected for high mass transfer rate effects
^	specific propertry		

1. INTRODUCTION

The Claude open cycle thermal energy conversion system (OC-OTEC) utilizes sea water as the working fluid in a simple Rankine cycle. The heat source is warm sea water, which is partially flash evaporated in a barometric evaporator to generate low pressure steam: the steam is expanded through a turbine and condenses on cold sea water, which is the heat sink. The open cycle concept offers possible advantages over conventional closed cycle OTEC systems: surface exchangers can be eliminated, and environmental impact reduced, through elimination of the ammonia as the secondary working fluid. Conceptual studies, [1], have demonstrated the economic feasibility of OC-OTEC, and have indicated that design data are required for the heat exchangers, deaerators, and turbine. Various concepts are under consideration for the heat exchangers including falling turbulent films or planar jets, channels, and possibly a shell and tube condenser if fresh water production is also desired.

Since the heat transfer characteristics of the competing exchanger concepts differ substantially, it is necessary to develop near optimum designs for each concept before the relative merits of each can be evaluated. Presented here is a design methodology for turbulent falling film exchangers, and performance studies for a counter-current evaporator and co-current condenser. The mathematical model assumes one-dimensional flow of liquid and vapor phases, and interphase transfer of momentum, energy and mass species are obtained from empirical correlations extrapolated to the liquid and vapor phase Reynolds number regimes characteristic of OC-OTEC. A parallel experimental effort is underway to validate these extrapolations, and generate new correlations where needed.

Numerical results are presented for a 100 MWe plant over the range of pertinent geometric and flow parameters. The geometric parameters include exchanger cross-sectional area, plate spacing, and film height. The flow parameters include liquid and vapor flow rates, saturation pressures and warm and cold sea water inlet temperatures.

Most of the key technical issues are addressed in the modeling and calculations. For the condenser these include: (i) Water stream head loss. For a 100 MWe plant 1 m of head loss can correspond to as much as 3.6 MW parasitic power loss. (ii) Noncondensable gas effects. The effects of air desorbed in the evaporator and entering the condenser with the inlet steam, as well as air desorbed in the condenser are accounted for. (iii) Vapor phase pressure drop. The high vapor velocities associated with the characteristic low vapor densities lead to the possibility of vapor phase pressure drop through the condenser, rather than the usual pressure recovery. The corresponding change in saturation temperature must be accurately calculated since the available temperature difference is only of the order of 5 K. For the evaporator the key issues include: (i) Water stream head loss, as was the case for the condenser. (ii) Vapor phase pressure drop and associated decrease of saturation temperature up the evaporator. (iii) Film instability due to bubble nucleation and growth, and associated mist carry-over. This last mentioned issue is not addressed in this paper although it might well prove to be the critical issue. It is however being investigated in an accompanying experimental program.

2. ANALYSIS

2.1 Governing Equations

The mathematical model (Fig. 1) assumes a timewise steady state, one-dimensional flow of the vapor and liquid phases, and an ideal gas vapor-air mixture. The equations governing conservation of mass, momentum, total enthalpy and mass species in the vapor phase are

$$\frac{d}{dz}(\rho u_b) = (n_{1,s} + n_{2,s})/(b-\delta) \tag{1}$$

$$\frac{d}{dz}(\rho u_b^2) = -\frac{dP}{dz} + \rho g - \frac{1}{2} u_b^2 \dot{C}_{fG}/(b-\delta) \tag{2}$$

$$\frac{d}{dz}(\rho u_b H) = [\dot{h}_G(T_s - T_b) + u_s \dot{C}_{fG} \frac{1}{2} \rho u_b^2 + n_{1,s}\hat{h}_{1,s}]/(b-\delta) \tag{3}$$

$$\frac{d}{dz}(\rho u_b m_{2,b}) = n_{2,s}/(b-\delta) \quad \text{(air)} \tag{4}$$

The equations governing conservation of mass, energy and mass species in the liquid phase are

$$\frac{d\Gamma}{dz} = -(n_{1,s} + n_{2,s}) \tag{5}$$

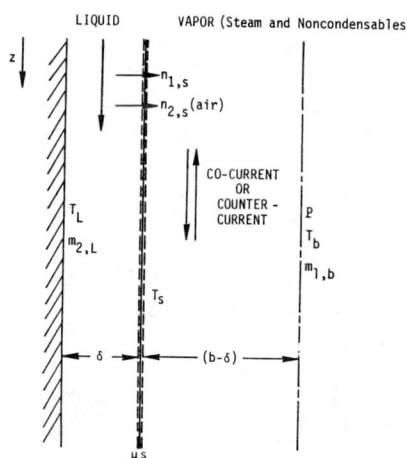

Fig.1 Model of the Liquid-Vapor System

$$\Gamma C_{pL} \frac{dT_L}{dz} = -\dot{h}_L(T_L - T_s) - \phi \frac{d\Gamma}{dz}$$

$$\times \int_{T_s}^{T_L} C_{pL} dT \tag{6}$$

$$\frac{d}{dz}(\Gamma m_{2,L}) = -n_{2,s} \tag{7}$$

The interface mass fluxes are expressed as follows. The mass flux of water vapor is approximated as

$$n_{1,s} \simeq \dot{m}'' = \dot{K}_G \mathcal{B}_m \; ;$$

$$\mathcal{B}_m = \frac{m_{1,b} - m_{1,s}}{m_{1,s} - 1} \tag{8}$$

which assumes $n_{2,s} \ll n_{1,s}$, i.e., the gas desorption rate is small compared to the evaporation or condensation rate. The desorption rate of dissolved air is given by

$$n_{2,s} \simeq K_L(m_{2,L} - m_{2,u}) \tag{9}$$

which assumes low mass transfer rates. Henry's law is used to relate air concentrations at the interface, $x_{2,u} = x_{2,s}/He$. The film surface velocity, u_s, in Eq. (3) is approximated by the film bulk velocity. The factor $\phi < 1$ in Eq. (6) accounts for subcooling in the liquid film.

The energy balance at the interface reflects a highly coupled heat and mass transfer process,

$$h_L(T_L - T_s) = \dot{h}_G(T_s - T_b) + n_{1,s}\hat{h}_{fg} \tag{10}$$

where $n_{1,s}$ is given by Eq. (8), and T_s is related to $m_{1,s}$ via the thermodynamic equilibrium relation

$$T_s = T_{sat}(P_{1,s}) = T_{sat}(m_{1,s}, P) \tag{11}$$

The separation of the conduction and convective components of the energy flux across the s-surface implicit in Eq. (10) is exact only for unity Lewis number, but since the conduction contribution is very small, the error incurred is negligible.

2.2 Transfer Coefficient Correlations

The transfer coefficients appearing in the governing equations are K_L, h_L, C_{fG}^*, $\dot{h}_G$ and K_G^*. The liquid side heat transfer coefficient h_L is of primary importance in both the evaporator and condenser, while K_G^* is also of primary importance in the condenser owing to the noncondensable gas effect. The liquid side mass transfer coefficient controls the amount of air desorbed in the exchangers, while C_{fG}^* determines whether or not there is pressure recovery in the condenser. Experimental data for these coefficients is very sparse, or non-existent, in the parameter range relevant to OC-OTEC, so that the following correlations are used in extrapolated form.

Based on gas absorption experiments of Won and Mills [2], and Chung and Mills [3], for $2000 < Re_L < 10,000$, the liquid side mass transfer coefficient is taken as

$$K_L/[\rho(\nu g)^{1/3}] = C\,Re_L^n Sc_L^m \tag{12}$$

where $C = 6.97 \times 10^{-9} Ka^{-1/2}$; $n = 3.49 Ka^{0.068}$; $m = -0.36 - 2.43\sigma$. The liquid side heat transfer coefficient is obtained by analogy from Eq. (12) with K_L replaced by h_L/C_p, and Sc_L replaced by Pr_L.

Due to lack of data at high enough Re_L, the correlation for the gas side mass transfer coefficient is taken from wavy laminar flow data as [4]

$$K_G/\rho u_b = 0.00814 Re_G^{-0.17} Re_L^{0.15} Sc_G^{-0.56} \tag{13}$$

and the heat transfer coefficient h_G is again found by analogy. The skin friction coefficient correlation used is that of Chien and Ibele [5]

$$C_{fG} = 0.92 \times 10^{-7} Re_G^{0.582} Re_L^{0.705} \tag{14}$$

The gas side transfer coefficients given above must be corrected for the effects of high mass transfer rates: the strong suction on the vapor phase in the condenser increases the transfer coefficients many-fold. Using stagnant film theory the following correction is used

$$\frac{\dot{\Phi}}{\Phi} = (n_{1,s}/\Phi)/[\exp(n_{1,s}/\Phi)-1] \qquad (15)$$

where $\Phi = \rho u_b C_{fG}/2$, h_G/C_{p1}, and K_G, respectively. No corrections were applied to the liquid side coefficients, since the effect was found to be relatively small.

2.3 Geometrical Relations and Boundary Conditions

The liquid film thickness δ is taken from the Brotz formula [6],

$$\delta = 0.0672(\nu^2/g)^{1/3} Re_L^{2/3} \qquad (16)$$

For both the evaporator and condenser the liquid inlet conditions are specified as $\Gamma_{in} = \dot{m}_{L,in}/2NW$, $T_L = T_{L,in}$ and $m_{2,L} = m_{2,L,in}$. In the condenser the inlet vapor velocity, temperature, pressure, and air concentration are specified, respectively, as $\rho u_b = \dot{m}_{G,in}/[NW2(b-\delta)]$, $T_b = T_{b,in}$, $P = P_{in}$, $m_{2,b} = m_{2,b,in}$. Since the evaporator is in counterflow vapor phase conditions are specified at the bottom as $u_b = 0$, with P, T_b and $m_{2,b}$ consistent with the exit liquid conditions.

2.4 Thermophysical Properties

For air dissolved in water the Schmidt number and Henry's number were based on data in [7],

$$Sc_L = 320(\mu^2/\rho T)/(\mu^2/\rho T)_{300 K} ; \qquad (17)$$

$$He = [310(T_s-300)(T_s-310)-740(T_s-290)(T_s-310)$$
$$+420(T_s-290)(T_s-300)]/P \text{ (atm)} \qquad (18)$$

The water surface tension σ was taken to be constant in the temperature range of interest at 74×10^{-3} N/m². Water properties were used in the form of curve-fits given by Kaups and Smith [8], while for steam-air mixtures, following mostly Harpole [9], curve-fits were used.

2.5 Exchanger Effectiveness

The heat transfer effectiveness is defined as the ratio of the actual to the maximum possible temperature change of the liquid stream,

$$\varepsilon_H = (T_{L,in}-T_{L,out})/(T_{L,in}-T_{sat}) \qquad (19)$$

For the evaporator T_{sat} corresponds to the pressure at the top of the evaporator in order to include the effect of gas phase pressure drop in evaluating exchanger performance.

Alternatively the performance of the evaporator can be represented by the "non-equilibrium" temperature difference, $\Delta T' = T_{L,out}-T_{sat}$, or

$$\Delta T' = (T_{L,in}-T_{sat})(1-\varepsilon_H) \qquad (20)$$

The mass transfer effectiveness of the exchangers is defined in the usual manner for a single stream exchanger,

$$\varepsilon_M = 1-(m_{2,L,out}/m_{2,L,in}) \tag{21}$$

and indicates the amount of air desorbed from the liquid stream.

2.6 Numerical Solution Procedure

The conservation equations were solved numerically using a fourth-order Adam-Moulton predictor-corrector method [10]. For the co-current condenser the boundary conditions are all specified at the top, and the problem is an initial value one, and is solved in a single sweep. For the counter-current evaporator the liquid phase boundary conditions are specified at the top, while the vapor phase is specified at the bottom. For these split boundary conditions iteration is required and typically 5 to 10 sweeps were required to obtain a converged solution. The method of bisection [11] was used to handle the non-linear energy balance at the interface, Eq. (10).

3. RESULTS AND DISCUSSION

3.1 Parameter Values

Design calculations of a 100 MWe plant were made over a range of geometric and flow parameters indicated by OC-OTEC conceptual studies [1]. The geometric parameters considered were plate spacing 2b (0.02-0.06 m); exchanger free cross-sectional area A = 2bNW (1200-2500 m^2) and exchanger height H (1-2 m). The flow parameters considered were sea water mass flow rate $\dot{m}_L$ (1.2-3.7x10^5 kg/s), steam mass flow rate at the condenser inlet $\dot{m}_G$ (1000-2000 kg/s), and available temperature difference at the exchanger inlet (3.5-5.5 K). The sea water was taken to be saturated with air at 1 atm pressure.

3.2 Results for the Evaporator

The performance of the evaporator approximates that of a single stream exchanger, the major deviation being caused by the decrease in T_{sat} up the evaporator due to a pressure decrease: a maximum decrease of $\Delta T_{sat}/(T_{L,in}-T_{sat})$ of 13% was noted in the results. Thus the essential features of the evaporator behavior can be deduced from the ε-N_{tu} relation.

$$\varepsilon = 1-e^{-N_{tu,L}} \tag{22}$$

where $N_{tu,L} = (h_L/C_p)PH/\dot{m}_L$ for heat transfer, and $K_L PH/\dot{m}_L$ for mass transfer. From Eq. (12), with P = 2 NW,

$$h_L, K_L \propto A^{-n}(2b)^n \dot{m}_L^n \tag{23}$$

and at 300 K, n ≃ 0.65; thus there is obtained

$$N_{tu,L} \propto A^{0.35}(2b)^{-0.35} \dot{m}_L^{-0.35} H \tag{24}$$

Figs. 2 and 3 show ε_H and ε_M versus A and 2b, respectively, for various $\dot{m}_L$; Fig. 4 shows ε_H, as well as $\dot{m}_G$, versus H for various $\dot{m}_L$. The trends shown are in line with the $N_{tu,L}$ dependence given by Eq. (24): in particular, for fixed $\dot{m}_L$, ε_H increases with increasing A and increases with decreasing (2b). Fig. 5 shows ε_H and $\dot{m}_G$ versus H for two values of T_{sat}, and two values of $\Delta T = T_{L,in} -$

TURBULENT FALLING FILM EVAPORATORS AND CONDENSERS

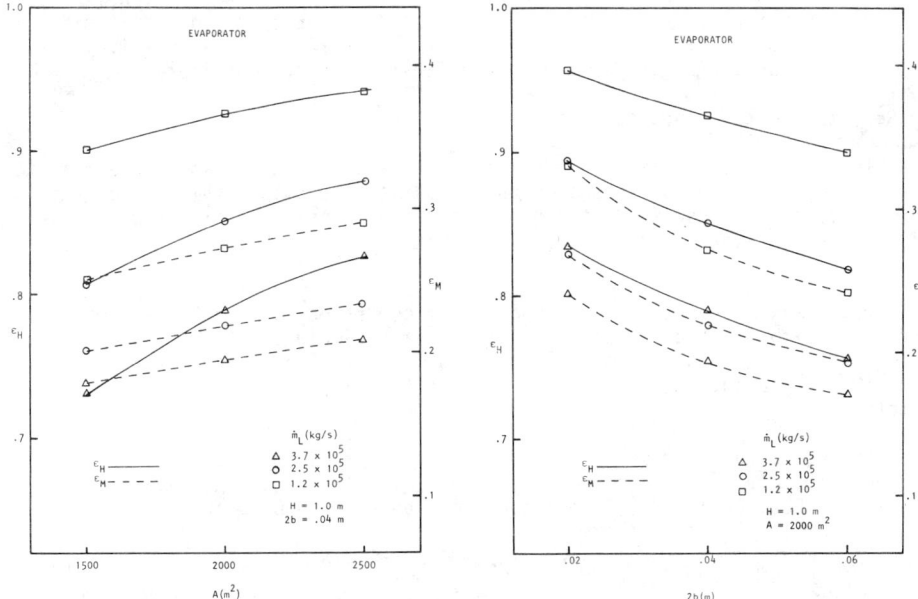

Fig.2 Heat and Mass Exchanger Effectiveness vs. Cross-Sectional Area for Different Water Mass Flow Rates

Fig.3 Heat and Mass Exchanger Effectiveness vs. Plate Spacing for Different Water Mass Flow Rates

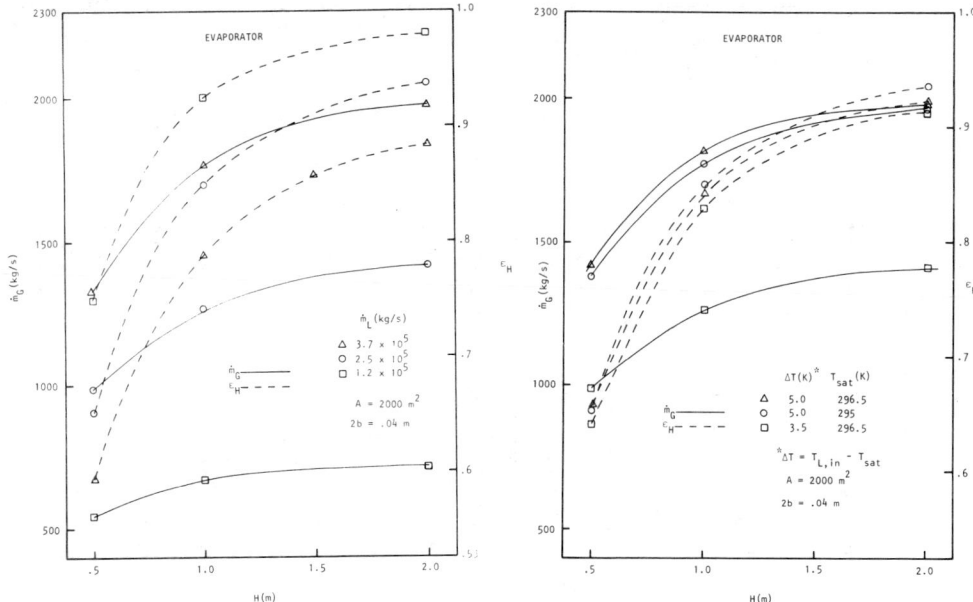

Fig.4 Steam Mass Flow and Heat Exchanger Effectiveness vs. Height for Different Water Mass Flow Rates

Fig.5 Steam Mass Flow Rate and Heat Exchanger Effectiveness vs. Height for Various Inlet Temperature Differences

T_{sat}: Fig. 6 shows the concentration of air in the outlet steam, which is also the inlet condition to the condenser: the mass fraction of dissolved air in the inlet water is 1.8×10^{-5}. It is seen that $m_{2,b}$ varies over a relatively small range, $0.69-1.05 \times 10^{-3}$ for the entire parameter range considered.

3.3 Results for the Condenser

Calculations were performed for a nominal value of inlet vapor stream air content of $m_{2,b,in} = 9 \times 10^{-4}$ based on the evaporator results: this value is an upper bound in the sense that it corresponds to no deaeration of the water entering the evaporator. Also $T_{sat,in}-T_{L,in}$ was fixed at 5.5 K, with $T_{sat,in}$ = 283 K.

Figs. 7 and 8 show the fraction of steam condensed for H = 2.0 m versus A and (2b), with $\dot{m}_L$ (Fig. 7) and $\dot{m}_G$ (Fig. 8) as parameters. Figs. 9 and 10 show ε_H and ε_M versus A with $\dot{m}_L$ (Fig. 9) and $\dot{m}_G$ (Fig. 10) as parameters. Figs. 11 and 12 show ε_H and ε_M versus (2b) with $\dot{m}_L$ (Fig. 11) and $\dot{m}_G$ (Fig. 12) as parameters.

With regard to air desorption, and the behavior of ε_M, the condenser is also approximately a single stream exchanger. The trends of ε_M shown are consistent with Eqs. (22) and (23) but now the exponent $n \simeq 0.73$ and

$$N_{tu,L} \propto A^{0.27}(2b)^{-0.27}\dot{m}_L^{-0.27}H \qquad (25)$$

The effect of $\dot{m}_G$ on ε_M seen in Fig. 10 is due to temperature dependent liquid properties. An increase in $\dot{m}_G$ increases $B_{\bar{m}}$, T_S and T_L: the resulting decrease in ν_L gives a higher Re_L and k_L, and hence a higher ε_M.

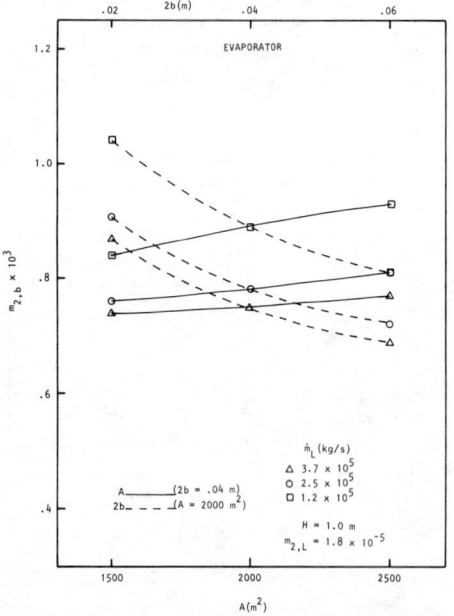

Fig.6 Fraction Air in Evaporator Exit vs. Cross-Sectional Area and Plate Spacing for Different Water Mass Flow Rates

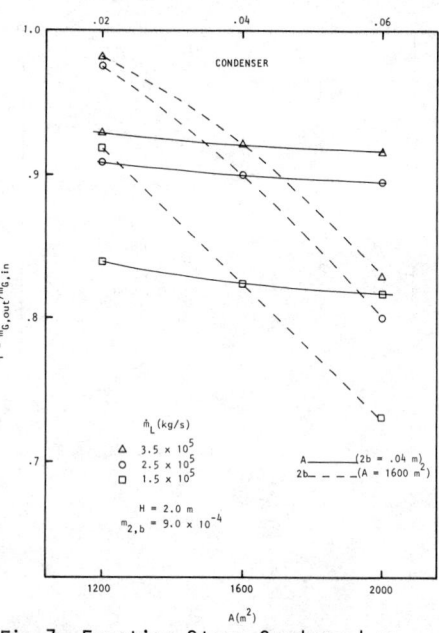

Fig.7 Fraction Steam Condensed vs. Cross-Sectional Area and Plate Spacing for Different Water Mass Flow Rates

TURBULENT FALLING FILM EVAPORATORS AND CONDENSERS

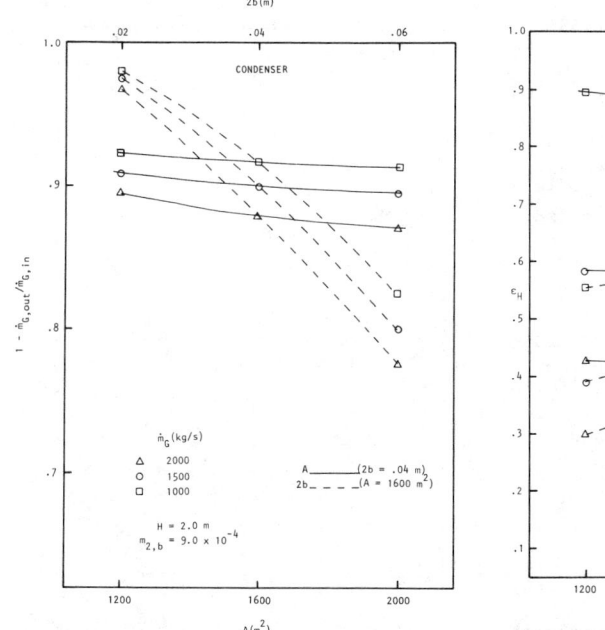

Fig.8 Fraction Steam Condensed vs. Cross-Sectional Area and Plate Spacing for Different Inlet Steam Mass Flow Rates

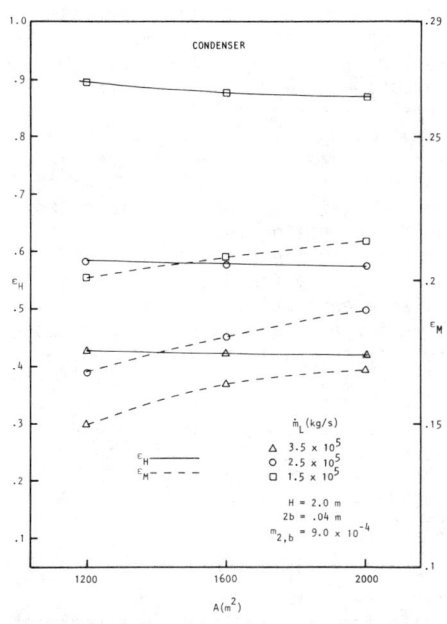

Fig.9 Heat and Mass Exchanger Effectiveness vs. Cross-Sectional Area for Different Water Mass Flow Rates

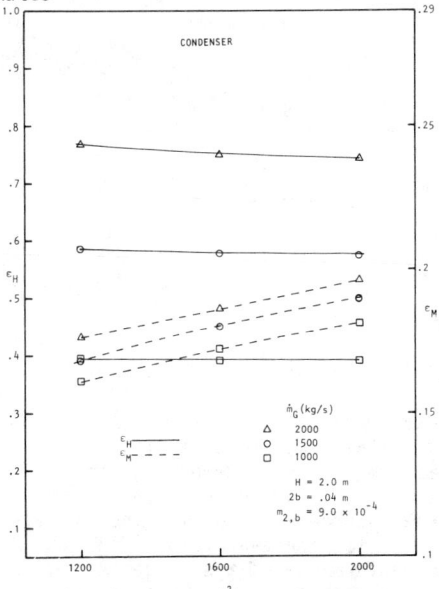

Fig.10 Heat and Mass Exchanger Effectiveness vs. Cross-Sectional Area for Different Inlet Steam Mass Flow Rates

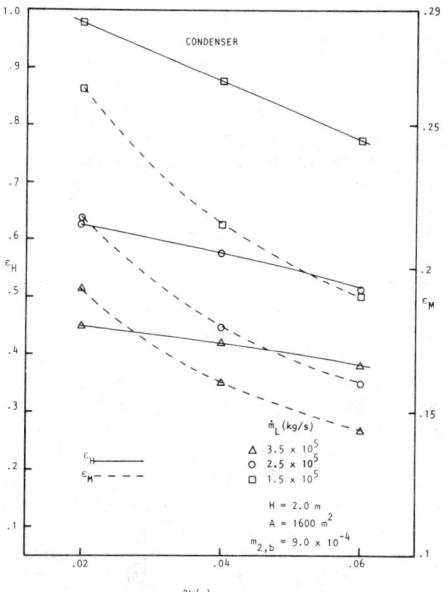

Fig.11 Heat and Mass Exchanger Effectiveness vs. Plate Spacing for Different Water Mass Flow Rates

The heat transfer behavior, and hence the amount of steam condensed, is rather complex since in this regard the condenser is a two-stream simultaneous heat and mass exchanger. The interface energy balance Eq. (10) plays a key role: it can be represented by an equivalent network as shown in Fig. (13). The network can be simplified to two resistances in series by noting that sensible heat transfer in the vapor phase, $h_G^*(T_s-T_b)$ is relatively small and can be ignored when examining the essential behavior of the condenser. The two resistances in series are $1/K_G^*\hat{h}_{fg}A_s$ in the vapor phase, and $1/h_L A_s$ in the liquid phase: these resistances are of comparable magnitude in the parameter range of concern, with the gas phase resistance tending to dominate. The dependence of h_L from Eq. (23) is

$$h_L \propto A^{-0.73}(2b)^{0.73}\dot{m}_L^{0.73} \qquad (26)$$

while from Eq. (13) the dependence of K_G is

$$K_G \propto \dot{m}_G^{0.83}\dot{m}_L^{0.15}A^{-0.98}(2b)^{-0.02} \qquad (27)$$

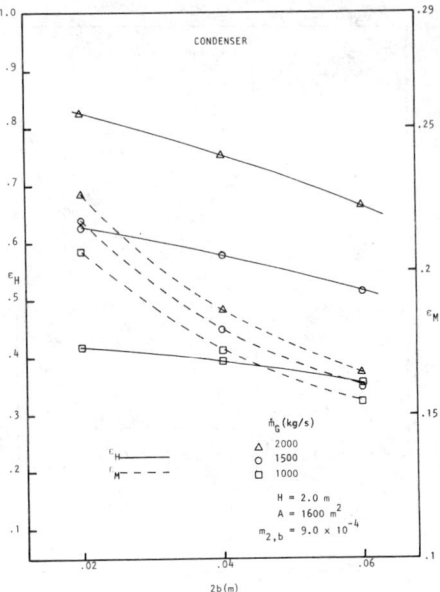

Fig.12 Heat and Mass Exchanger Effectiveness vs. Plate Spacing for Different Inlet Steam Mass Flow Rates

The effect of an increase in $\dot{m}_G$ increasing ε_H shown in Figs. 10 and 12 is partly due to the $K_G \propto \dot{m}_G^{0.83}$ dependence given by Eq. (27), but is also due to the increase in the capacity rate ratio $\dot{m}_G/\dot{m}_L$ which reduces the decrease in $m_{1,b}$ as the steam flows down the condenser, and thereby maintains a higher average mass transfer during force B_m. However, the fraction of steam condensed is shown in Fig. 8 to decrease slightly with increasing $\dot{m}_G$, in line with a mass transfer $N_{tu,G} = K_G^*PH/\dot{m}_G$ proportional to $\dot{m}_G^{-0.17}$. The effect of an increase in $\dot{m}_L$ increasing the fraction

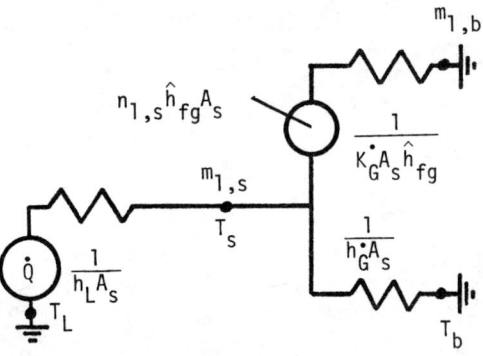

Fig.13 Equivalent Thermal Network

of steam condensed shown in Fig. 7 is due partly to the $h_L \propto \dot{m}_L^{0.73}$ and $K_G \propto \dot{m}_L^{0.15}$ given by Eqs. (26) and (27), but is also again due to the change in capacity rate ratio $\dot{m}_G/\dot{m}_L$. As $\dot{m}_L$ increases the rate of temperature rise of the water down the condenser is reduced, as is $m_{1,s}$, thereby maintaining a higher average value of B_m. The more marked decrease in ε_H with increasing $\dot{m}_L$ for the condenser (Fig. 11) compared to that for the evaporator (Fig. 2) confirms that, in this parameter range, the performance of the condenser is limited by the amount of steam available for condensation, rather than the heat sink capacity of the sea water stream.

Figs. 8, 9, and 10 show that the fraction of steam condensed and ε_H are nearly independent of area A; there is but a slight decrease with increasing A. This feature is mainly due to the gas side mass transfer $N_{tu,G}$ being essentially independent of A ($N_{tu,G} \propto A^{0.02}$), and as explained above, the performance of the condenser is vapor stream limited.

Fig. 8 shows that the fraction of steam condensed increases markedly as the plate spacing (2b) is decreased. Equation (27) shows K_G is almost independent of (2b), while $h_L \propto (2b)^{0.73}$: thus the overall transfer resistance actually decreases as (2b) is decreased. However $N_{tu,G} \propto b^{-1.02}$ so that the dominant effect is due to the increase in $N_{tu,G}$ with decreasing (2b). The increase in ε_H with decreasing (2b) shown in Figs. 11 and 12 reflects the increase in the fraction of steam condensed.

	H = 1 m		H = 2 m	
$\dot{m}_{2,b,in}$	ε_H	$1-(\dot{m}_G/\dot{m}_{G,in})$	ε_H	$1-(\dot{m}_G/\dot{m}_{G,in})$
9x10^{-4}	0.453	0.706	0.577	0.899
3x10^{-4}	0.487	0.758	0.592	0.921
0.0	0.638	0.990	0.645	1.00*

*all the steam is condensed in a little over a meter.

Table 1. Effect of noncondensables and condenser height on heat exchanger effectiveness and fraction of steam condensed. A = 1600 m^2, 2b = 0.04 m, $\dot{m}_L$ = 2.5x10^5 kg/s, $\dot{m}_G$ = 1500 kg/s.

Table 1 gives further insight into the performance of the condenser. The modest increase in ε_H as H increases from 1 to 2 m again illustrates that the performance is limited by the vapor stream capacity, and indeed the maximum possible ε_H for the parameter values shown is 0.645.

The foregoing results suggest that there is considerable benefit to be derived from reducing (2b) to a minimum value consistent with practical and cost contributions. In the condenser, where a spray type water distribution system might be feasible, small values of (2b) can be more easily realized than in the evaporator where mist carry-over considerations suggest that a slot type distributor is necessary. The results also indicate that the turbulent falling film concept promises to be competitive with other concepts since good performance is obtained at relative moderate values of exchanger height.

4. ACKNOWLEDGMENTS

The authors wish to thank Mr. John Farr for assistance in developing the computer program, and Mr. David Bugby for preparing the numerical results.

This work is sponsored by the Solar Energy Research Institute (SERI) under contract No. XJ-9-8190-2. The SERI project manager is Dr. T. Penney.

5. REFERENCES

1. Shelpuk, R., and Lewandowski, A. A. 1979. Alternate cycle applied to ocean thermal energy conversion. 11th Offshore Technology Conference, OTEC Paper No. 3589, Houston, Texas.

2. Won, Y. S., and Mills, A. F. 1981. Correlation of the effects of viscosity and surface tension on gas absorption rates into freely falling turbulent liquid films. To appear, Int. J. Heat Mass Transfer.

3. Chung, D. K., and Mills, A. F. 1976. Experimental study of gas absorption into turbulent falling films of water and ethylene glycol-water mixtures. Int. J. Heat Mass Transfer, Vol. 19, pp. 51-59.

4. Kafefgian, R., Plank, C. A., and Gerhard, E. R. 1961. Liquid flow and gas phase mass transfer in wetted wall towers. AIChE Journal, Vol. 7, pp. 463-469.

5. Chien, S. F., and Ibele, W. 1964. Pressure drop and liquid film thickness of two phase annular and annular-mist flows. J. Heat Transfer, Vol. 86, pp. 89-96.

6. Brotz, W. 1954. Uber die Vorausberechnung der Absorptions geschwindigwert von Gasen in Stramen der Fussigkeitschichten. Chem. Engrg. Tech., Vol. 26, pp. 470-478.

7. Edwards, D. K., Denny, V. E., and Mills, A. F. 1979. Transfer Processes, 2nd Ed., Hemisphere-McGraw Hill, New York, pp. 399-409.

8. Kaups, K., and Smith, A. M. O. 1967. The laminar boundary layer in water with variable properties. 9th ASME-AIChE National Heat Transfer Conference, Seattle, Washington.

9. Harpole, G. M. 1980. Droplet evaporation in a high temperature environment. Ph.D. Dissertation, School of Engineering and Applied Science, University of California, Los Angeles.

10. Carnahan, B., Luther, H., and Wilkes, J. O. 1969. Applied Numerical Methods, J. Wiley, New York.

11. Isaacson, E., and Keller, H. B. 1966. Analysis of Numerical Methods, John Wiley, New York.

Local Two-phase Flow Heat Transfer in Double-tube Heat Exchangers

KENICHI HASHIZUME
Energy Science and Technology Laboratory
Research and Development Center
Toshiba Corporation
Ukishimacho 4-1, Kawasaki, Japan

ABSTRACT

An investigation on local two-phase flow heat transfer in horizontal double-tube heat exchangers was conducted. Evaporation and condensation heat transfer coefficients for refrigerant R22 were measured both in tube and in annulus, and were compared with each other. Two-phase flow pattern also was observed in annulus.

1. INTRODUCTION

Double-tube heat exchangers have a simple configuration and are used frequently in refrigerators or chemical processes as evaporator or condenser. The flow channel for fluid to be evaporated or condensed can be chosen as either inside the inner tube or annulus between the outer and the inner tubes.

There have been a great deal of investigations conducted on two-phase flow heat transfer with evaporation or condensation in tube. In contrast to them, however, only limited information on in annulus flow is available. The purposes of this investigation are: (1) To obtain data on local two-phase flow heat transfer coefficients with evaporation and condensation in tube and also in annulus. (2) To compare them. (3) To examine the effect of tube corrugation on heat transfer enchancement. (4) To find some possibilities to improve double-tube heat exchangers.

The fluid used in this investigation was refrigerant R22. Evaporation temperature was 10°C and condensation temperatures were 35, 40 and 45°C.

2. EVAPORATION HEAT TRANSFER IN TUBE

Experimental equipment is shown in Figure 1. This is a conventional refrigeration cycle, but the evaporator section is modified. Refrigerant from the compressor is condensed in the condenser and subcooled in the subcooler. Then, its flow rate is measured by the flowmeter. Through the expansion valve, refrigerant flows into the pre-evaporator (horizontal double-tube), where the quality to the test section is settled, and into the

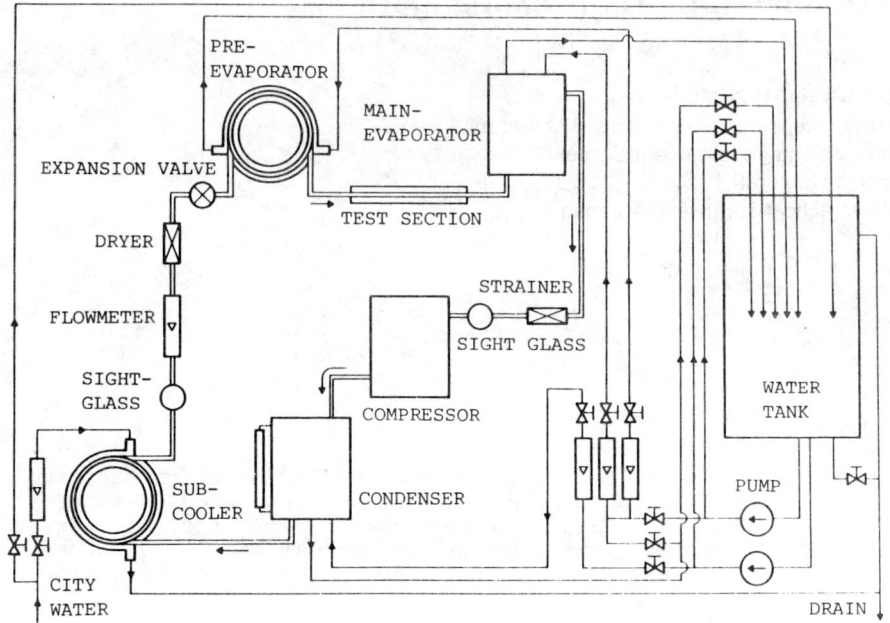

Fig. 1   Experimental Equipment for Evaporation Heat Transfer

horizontal test section. The main-evaporator (shell-tube) evaporates the remaining liquid phase refrigerant, and refrigerant in gas phase returns to the compressor. The condenser, pre-evaporator and main-evaporator are cooled or heated by water pumped from the water tank. The subcooler is cooled by city water.

The test section consists of a 5/8" Cu-tube (15.88 mm OD x 0.8 mm t) 400 mm long, around which a sheathed heater is wound, and thermally insulated with 50 mm glass wool. To measure wall temperature, two thermocouples are brazed on the upper side and two thermocouples are brazed on the bottom side of the tube. The average value for these four thermocouples was taken as the wall temperature. During measurements, the evaporation temperature was held at 10 ± 1°C, and refrigerant flow rate was from 240 to 300 kg/h.

Measured heat transfer coefficients are shown in Figure 2 in the form of α/α_{LO} versus Lockhart-Martinelli parameter X_{tt}, with Boiling number Bo as a parameter. α_{LO} was calculated by VDI-Wärmeatlas [1]. Here, also some empirical correlations [2] are shown. This experiment agrees well with Chaddock & Brunneman. However, in higher quality region, $X_{tt} \leq 0.025$, heat transfer coefficient decreases with an increase in quality. This is due to dryout on the heat transfer surface, with flow pattern transition to mist flow. In an experiment of water by Dengler & Addoms [3], this criterium was $X_{tt} \leq 0.014$. In case of horizontal flow of refrigerant, this is larger, i.e. dryout in horizontal flow can

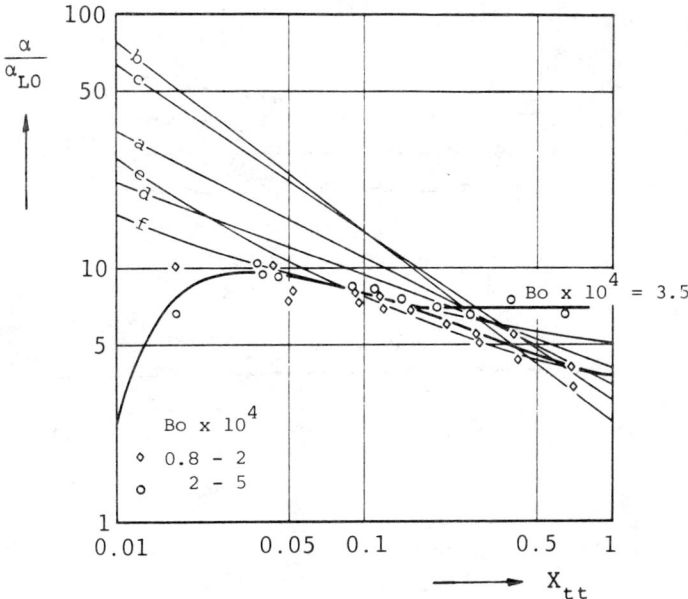

a Dengler & Addoms
b Schrock & Grossman
c Chaddock & Noerager
d Pujol & Stenning
e Schrock & Grossman   for Bo x 10^4 = 3.5
f Chaddock & Brunneman for Bo x 10^4 = 3.5
—— THIS EXPERIMENT

Fig. 2 Evaporation Heat Transfer in Smooth Tube

occur in lower quality than in upward flow.

Measured heat transfer coefficients in the corrugated tube, shown in Figure 3, are presented in Figure 4, where a marked effect can be seen on heat transfer enhancement versus data for a smooth tube.

3. EVAPORATION HEAT TRANSFER IN ANNULUS

Figure 5 shows the test section, which consists of a 5/8" Cu- inner tube (15.88 mm OD x 0.8 mm t) and a Cu- outer tube (25 mm ID). To observe flow pattern, a sight glass (Pyrex glass) is provided. Two thermocouples are brazed on the upper side and two thermocouples are brazed on the bottom side of the inner tube. Warm water at a controlled temperature was used for heating. The outside of this test section was thermally insulated with 50 mm glass wool except for the sight glass.

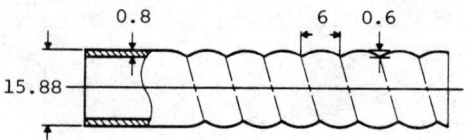

Fig. 3 Tested Corrugated Tube

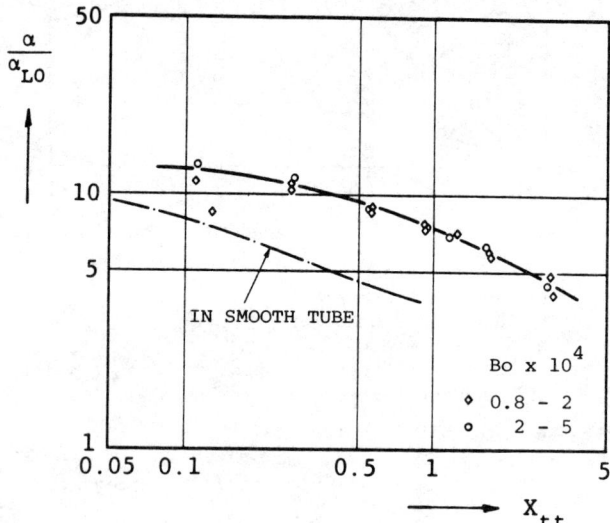

Fig. 4 Evaporation Heat Transfer in Corrugated Tube

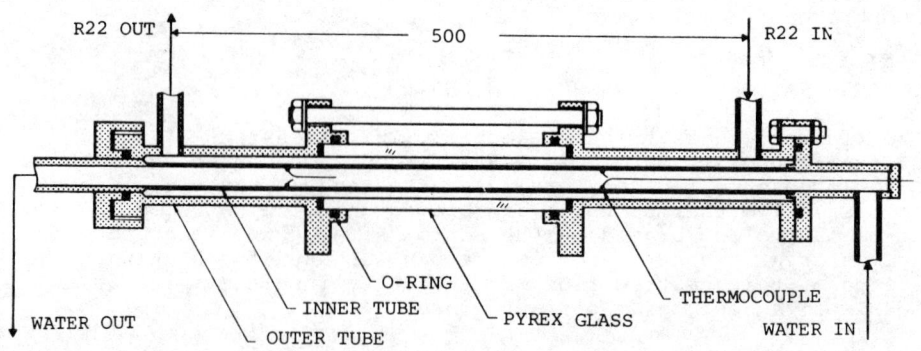

Fig. 5 Test Section for Heat Transfer in Annulus

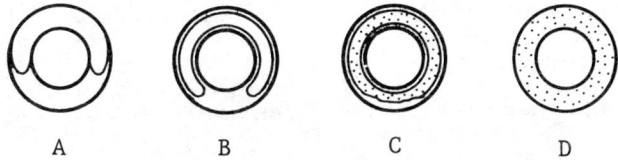

Fig. 6 Observed Flow Pattern During Evaporation in Annulus

Experimental conditions, i.e. evaporation temperature and flow rate, are the same as in tube evaporation.

Observed flow patterns are illustrated in Figure 6. They are stratified (A), semi-annular (B), annular-mist (C) and mist flow (D). Characteristic flow pattern in annulus, in contrast to in tube flow, is annular flow, where liquid film flow exists not only on the inner surface of the outer tube but also on the outer surface of the inner tube.

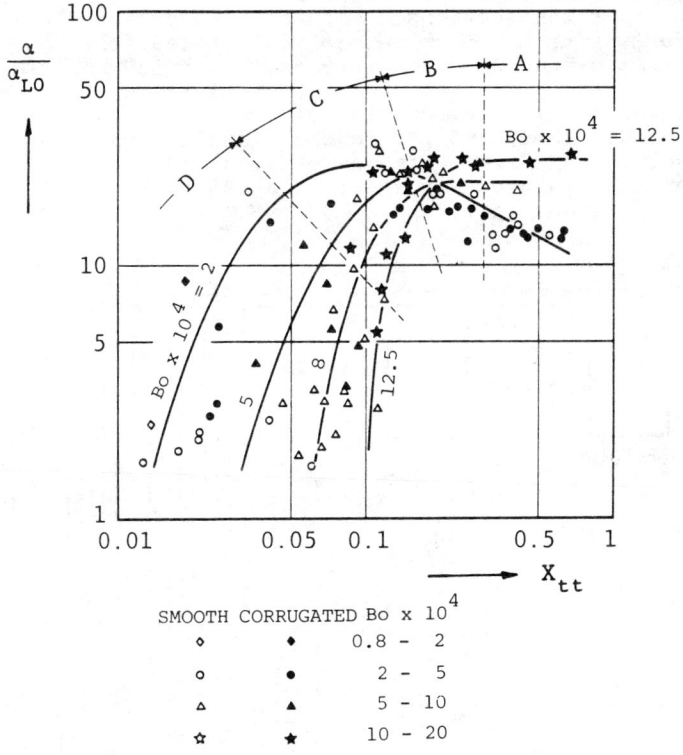

Fig. 7 Evaporation Heat Transfer in Annulus with Smooth and Corrugated Inner Tube

Measured heat transfer coefficients are shown in Figure 7 in the same manner as in tube evaporation. α_{LO} here was also calculated by VDI-Wärmeatlas. In this figure, approximate flow pattern region are also shown. In the stratified (A) to semi-annular flow (B), the variation in α/α_{LO}-value versus X_{tt} is similar to that in tube evaporation. In the annular-mist flow (C) region, however, heat transfer coefficient decreases rapidly with an increase in heat flux. Because the thin liquid film on the outer surface of the inner tube can easily diminish. In the mist flow (D) region, heat transfer coefficient decreases with an increase in quality and in heat flux.

Experimental results with the corrugated inner tube are also presented in Figure 7. However, here, practically no effect can be seen on heat transfer enhancement.

4. CONDENSATION HEAT TRANSFER IN TUBE

Experimental equipment to measure condensation heat transfer is illustrated in Figure 8. This is a modification of the equipment used in the experiment on evaporation. Here, the condenser is divided into three sections. They are the pre-condenser, where the quality to the test section is settled, the test section and the main-condenser. Horizontally settled test section is similar to that used in annulus evaporation, as shown in Figure 5. However, four thermocouples are brazed on the outer surface of the inner tube, and its leadings are through annular flow channel. Refrigerant flow rate was between 140 to 270 kg/h, and condensation temperatures were settled as 35, 40 and 45 ± 1°C.

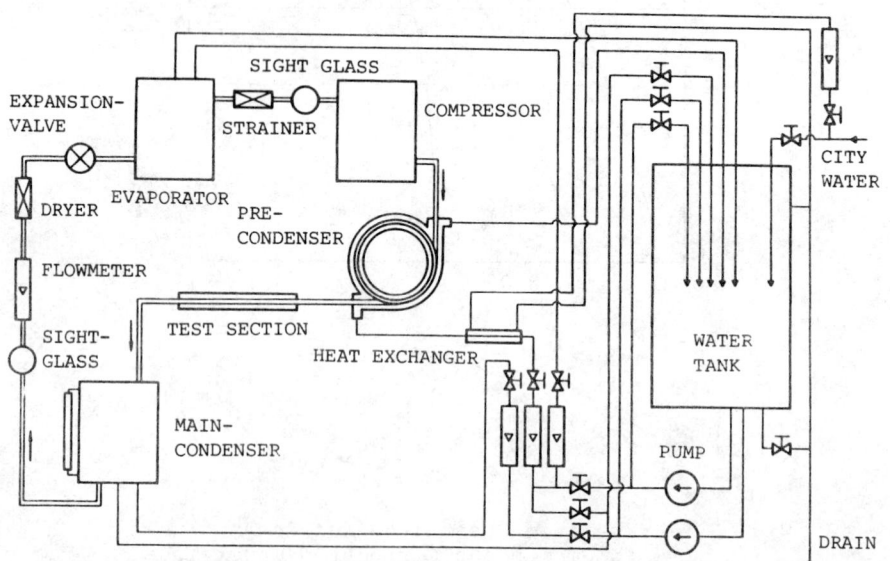

Fig. 8 Experimental Equipment for Condensation Heat Transfer

HEAT TRANSFER IN DOUBLE-TUBE HEAT EXCHANGERS 101

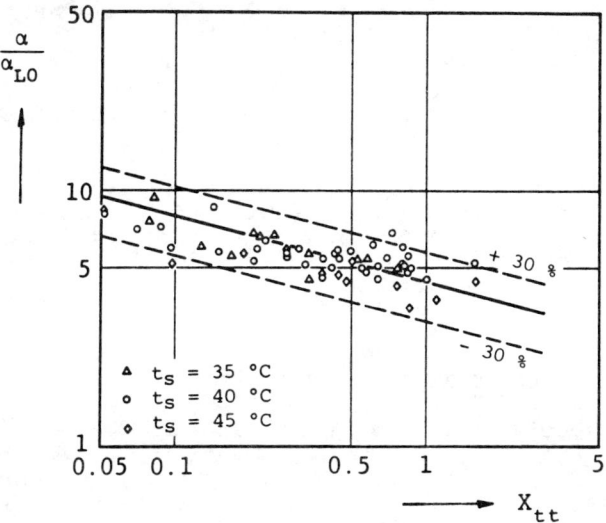

Fig. 9   Condensation Heat Transfer in Smooth Tube

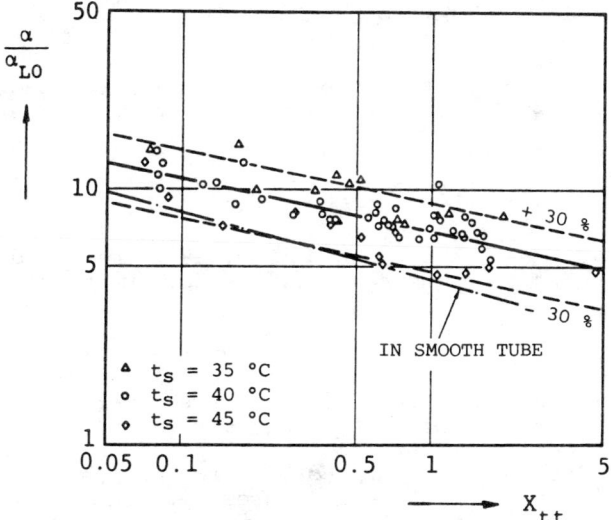

Fig. 10   Condensation Heat Transfer in Corrugated Tube

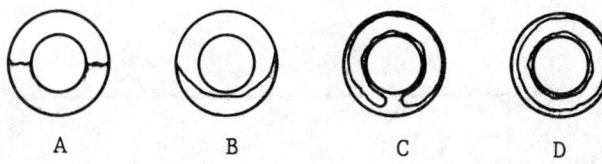

Fig. 11   Observed Flow Pattern During Condensation in Annulus

Figure 9 shows measured heat transfer coefficients in the same manner as in evaporation. Here, the effect of heat flux could not be recognized in contrast to evaporation. Measured heat transfer coefficients in the corrugated tube are presented in Figure 10, where a remarkable effect can be seen on heat transfer enhancement versus data for a smooth tube.

5. CONDENSATION HEAT TRANSFER IN ANNULUS

The test section is the same as used in the experiment on evaporation in annulus. Observed flow patterns are illustrated in Figure 11. They are stratified (A), wavy (B), semi-annular (C) and annular flow (D). In case of condensation, liquid film flow was always observed on the outer surface of the inner tube.

Figure 12 shows experimental results. Here, approximate flow pattern region after Figure 11 is shown, but no characteristic change in heat transfer coefficient with flow pattern transition could be recognized.

Experimental results with the corrugated inner tube are presented in Figure 13. In annulus condensation, the effect of inner tube corrugation on heat transfer enhancement is very small. Flow patterns with corrugated inner tube were almost the same as in Figure 11, and swirl flow of liquid film could not be observed.

6. COMPARISONS OF HEAT TRANSFER BETWEEN EVAPORATION AND CONDENSATION, AND BETWEEN IN TUBE AND IN ANNULUS

Experimental results described above are summarized in Figure 14. From comparisons among them, the following conclusions can be obtained.

(1) Evaporation and condensation heat transfer coefficients in tube are practically the same (cf. a and e). In lower quality region, however, the heat transfer coefficient for evaporation is lower than that for condensation, because the upper part of the inner tube does not contribute to heat transfer in stratified flow region. Further, in higher quality region, the heat transfer coefficient for evaporation decreases with an increase in quality, due to dryout with flow pattern transition to mist flow. This result confirms the qualitative consideration by Chawla [4].

HEAT TRANSFER IN DOUBLE-TUBE HEAT EXCHANGERS

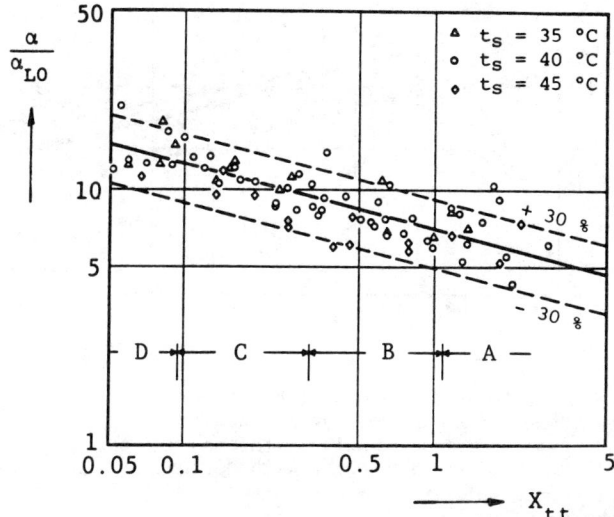

Fig. 12   Condensation Heat Transfer in Annulus with Smooth Inner Tube

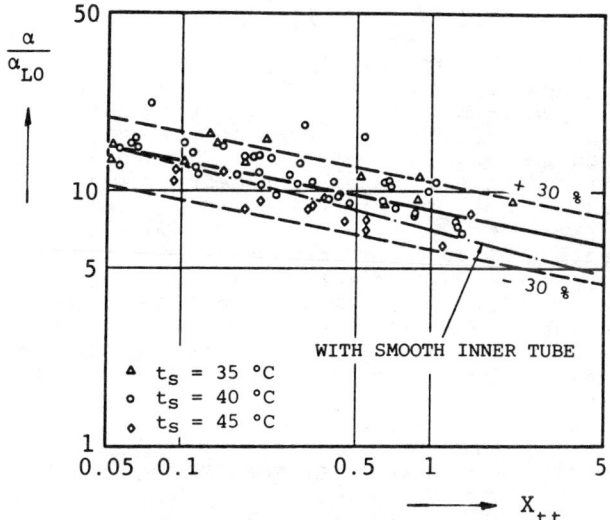

Fig. 13   Condensation Heat Transfer in Annulus with Corrugated Inner Tube

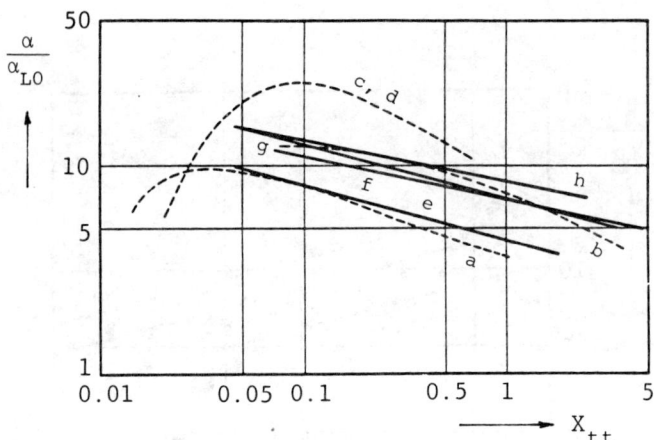

a   Evaporation in Smooth Tube
b   Evaporation in Corrugated Tube
c   Evaporation in Annulus with Smooth Inner Tube
d   Evaporation in Annulus with Corrugated Inner Tube
e   Condensation in Smooth Tube
f   Condensation in Corrugated Tube
g   Condensation in Annulus with Smooth Inner Tube
h   Condensation in Annulus with Corrugated Inner Tube

Evaporation Values are for $Bo \times 10^4 \leq 2$

Fig. 14   Comparisons of Heat Transfer Coefficients

(2) Heat transfer coefficients for evaporation and also for condensation in annulus are higher than in tube (cf. a and c, e and g). This is explained by thin liquid film on the outer surface of the inner tube, which, in case of evaporation, can easily diminish with an increase in heat flux. Therefore, heat transfer coefficient for in annulus evaporation decreases rapidly with an increase in heat flux in higher quality region.

(3) The effect of tube corrugation on heat transfer enhancement is marked in tube evaporation and in tube condensation (cf. a and b, e and f). Their heat transfer coefficients are practically the same (cf. b and f). However, in annulus evaporation or in annulus condensation, no effect of inner tube corrugation could be recognized (cf. c and d), or it was very small (cf. g and h).

7.   PROPOSED DOUBLE-TUBE EVAPORATOR FOR IMPROVING HEAT TRANSFER

Based on obtained results, a double-tube evaporator can be proposed for improving heat transfer. Lines I and II in Figure 15 represent heat transfer variation with quality for in annulus and in tube evaporation, respectively. In lower quality region, heat transfer coefficient for in annulus evaporation (I) is higher

HEAT TRANSFER IN DOUBLE-TUBE HEAT EXCHANGERS   105

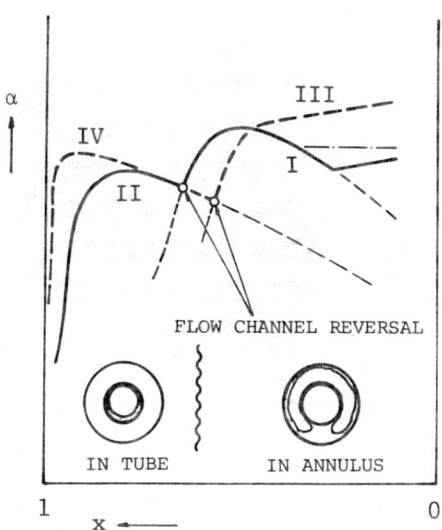

Fig. 15   Variation in Evaporation Heat Transfer Coefficient
          with Quality

than in tube evaporation (II). However, dryout in annulus evaporation occurs at lower quality than in tube evaporation. Therefore, in higher quality region, heat transfer coefficient for in tube evaporation becomes higher than in annulus evaporation. If the flow channel is reversed, higher heat transfer coefficient can be utilized in the whole quality region. This variation in heat transfer coefficient is shown in Figure 15 with a thick line.

That is, flow channel for the fluid to be evaporated is chosen in annulus in lower quality region, and then in tube in higher quality region. Figure 16 illustrates the configuration for the proposed double-tube evaporator. More heat transfer enhancement can be expected to provide some surface treatment, e.g. sintered layer, on the outer surface of the inner tube (shown by line III in Figure 15), and some mechanical manufacturing, e.g. fine groove, on the inner surface of the outer tube (shown by line IV in Figure 15).

8.   CONCLUSIONS

An investigation on local two-phase flow heat transfer in horizontal double-tube heat exchangers was conducted with refrigerant R22. Evaporation and condensation heat transfer coefficients were measured both in tube and in annulus. Also, the two-phase flow pattern was observed in annulus.

Experimental results have shown that the heat transfer coefficient for in annulus is higher than in tube. In higher

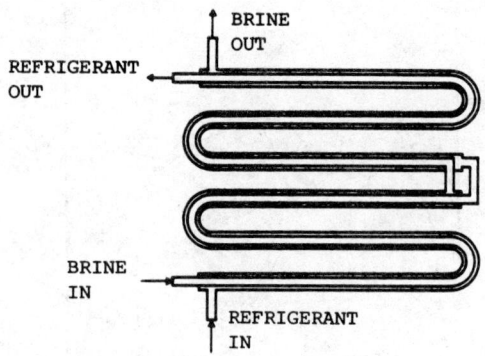

Fig. 16   Proposed Double-Tube Evaporator

quality region, however, the heat transfer coefficient for in annulus evaporation decreases rapidly with an increase in heat flux, because the liquid film on the outer surface of the inner tube can easily diminish, which was confirmed by flow pattern observation. The effect of tube corrugation on heat transfer enhancement could not be recognized or was very small for evaporation and condensation in annulus, but is remarkable for evaporation and condensation in tube. From these results, the author has proposed a configuration of double-tube evaporator, in which heat transfer characteristics can be improved.

NOMENCLATURE

$Bo = q/GH_v$: Boiling number
H_v : Latent heat, J/kg
t_s : Saturation temperature, °C
$X_{tt} = (1-x/x)^{0.9} (\rho_G/\rho_L)^{0.5} (\eta_L/\eta_G)^{0.1}$: Lockhart-Martinelli parameter
α : Heat transfer coefficient, W/m²K
α_{LO} : Heat transfer coefficient for $x = 0$, W/m²K
η : Viscosity, kg/ms
G : Mass flux, kg/m²s
q : Heat flux, W/m²
x : Quality
ρ : Density, kg/m³

REFERENCES

1. VDI-Wärmeatlas, 2. Auflage, 1974
2. Pujol,L. and Stenning,A.H., Effect of Flow Direction on the Boiling Heat Transfer Coefficient in Vertical Tubes, Symp. Ser. Can. Soc. Chem. Engng., 1969, pp.401-452
3. Mesler,R.B., An Alternate to the Dengler and Addoms Convection Concept of Forced Convection Boiling Heat Transfer, AIChE J., 23-4 (1977), pp.448-453
4. Chawla,J.M., Wärmeübergang in durchströmten Kondensator-Rohren, Kältetech.-Klimatisierung, 24 (1972), pp.233-240

Tube Submergence and Entrainment on the Shell Side of Heat Exchangers

I. D. R. GRANT, C. D. COTCHIN, and D. CHISHOLM
National Engineering Laboratory
East Kilbride
Glasgow, UK

ABSTRACT

Equations are developed for the liquid entrainment in the gas or vapour in two-phase horizontal crossflow over tube banks. Equations are also developed for the liquid cross-section, or tube submergence.

N O T A T I O N

B   Coefficient: equation (1)

B_H Coefficient: pseudo-homogeneous model, equation (8)

B_S Coefficient: separated flow model, equation (6)

D   Tube diameter

Dp   Pressure gradient due to friction and form drag

Dp_{LO} Pressure gradient if mixture flows as liquid

e_c Entrainment parameter, equation (15)

Fr_{LO} Froude number: total mixture flows as liquid

f_{GO} Friction factor: all gas flow

f_{LO} Friction factor: all liquid flow

G   Mass velocity

g   Gravitational acceleration

M_G Mass flowrate: of gas

M_L Mass flowrate: of liquid

M_{SL} Mass flowrate: of separated liquid

m   Exponent, equation (24)

N_T Number of tubes normal to flow

n   Blasius exponent, equation (2)

R   Term given by equation (13)

Re_{GO} Reynolds number: all gas flow

Re_{LO} Reynolds number: all liquid flow

v_G Specific volume: gas

v_L Specific volume: liquid

w   Proportion of liquid entrained in vapour, equation (18)

X   Lockhart-Martinelli parameter

x   Gas/total mass flowrate ratio

x_e Gas/total mass flowrate ratio in separated gas/liquid flow

α_{SL} Separated liquid fraction, Fig. 2

Γ^2 Physical property coefficient, equation (5)

μ_G Absolute viscosity of gas

μ_L Absolute viscosity of liquid

ρ_L Density of liquid

ϕ_L^2 Two-phase multiplier, equation (7)

ϕ_{LO}^2 Two-phase multiplier, equation (1)

ϕ_{LOe}^2 Two-phase multipler, equation (21)

ψ Parameter, equation (10)

ψ_H Parameter: corresponding to B_H, equation (15)

ψ_S Parameter: correspondinng to B_S, equation (15)

1. INTRODUCTION

Two-phase flow across tube banks occurs in condensers, boilers, mist flow coolers and similar types of equipment. A review of modern design methods for predicting the pressure gradients in these conditions is given in Reference [1].

This paper concerns the prediction of entrainment of the liquid in the gas or vapour, and the cross-section occupied by the un-entrained or separated liquid. Using a separated flow model an equation is developed for the entrainment in a form that can be readily used in practice.

2 PRESSURE DROP IN CROSSFLOW

The equation used in a number of papers [1-5] for predicting two-phase pressure gradients is

$$\frac{Dp}{Dp_{LO}} = \phi_{LO}^2 = 1 + (\Gamma^2 - 1)\{Bx^{2-n/2}(1-x)^{2-n/2} + x^{2-n}\} \quad (1)$$

where B is a coefficient which is either obtained from experiment or derived from theory. The all-liquid pressure gradient Dp_{LO} is used in preference to the liquid-only pressure gradient [4] Dp_L, since it does not vary along the flow path during a phase change.

The Blasius exponent, n, is defined by

$$\frac{f_{LO}}{f_{GO}} = \left(\frac{Re_{GO}}{Re_{LO}}\right)^n \quad (2)$$

where the Reynolds numbers are defined as

$$Re_{GO} = \frac{GD}{\mu_G} \quad (3)$$

$$Re_{LO} = \frac{GD}{\mu_L} \quad (4)$$

and f_{GO} and f_{LO} are the corresponding friction factors.

The physical property coefficient is

$$\Gamma^2 = \frac{f_{GO}}{f_{LO}} \times \frac{v_G}{v_L} \quad (5)$$

The zero interface shear model [6], associated with stratified flow, can be approximated [7] by

$$B_S = \frac{2^{2-n} - 2}{\Gamma + 1}. \quad (6)$$

The 'pseudo-homogeneous' equation given in Reference [8]

$$\phi_L^2 = 1 + \left\{\left(\frac{v_L}{v_G}\right)^{1/2} + \left(\frac{v_G}{v_L}\right)^{1/2}\right\}\frac{1}{x} + \frac{1}{x^2} \quad (7)$$

can also be approximated by [5]

$$B_H = \left(\frac{\mu_L}{\mu_G}\right)^{n/2}. \quad (8)$$

For the condition, n = 0, homogeneous theory is given by

$$B = 1.0 \quad (9)$$

in equation (1), which is consistent with equation (8).

For convenience we introduce the parameter [9]

$$\psi = \frac{\phi_{LO}^2 - 1}{\Gamma^2 - 1} \qquad (10)$$

and note that equation (1) can then be written

$$\psi = Bx^{2-n/2}(1 - x)^{2-n/2} + x^{2-n}. \qquad (11)$$

Analysis of data showed that over a considerable range of conditions for both tube [10] and crossflow [11]

$$\psi = Rx^{2-n} \qquad (12)$$

where, for crossflow,

$$R = 1.3 + 0.09 Fr_{LO} N_T^2 \left(\frac{\mu_L}{\mu_G}\right)^n. \qquad (13)$$

The Froude number Fr_{LO} for the total mixture flowing as liquid is given by

$$Fr_{LO} = \frac{G^2}{Dg\rho_L^2}. \qquad (14)$$

Fig. 1 represents diagrammatically the range of applicability of equations (6), (8) and (12). Equation (12) represents the transition, with increasing mass dryness fraction, from stratified flow to a relatively homogeneous flow distribution.

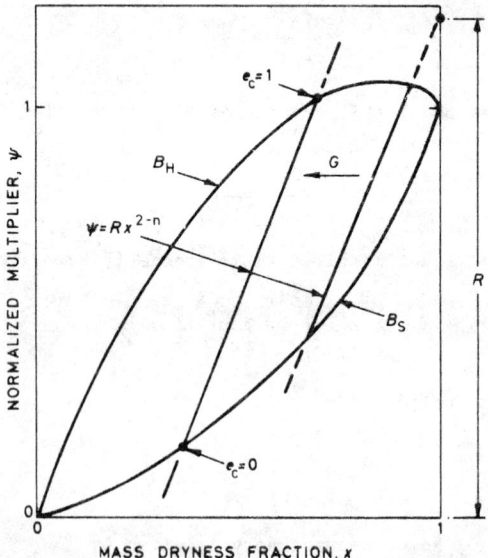

Fig.1. Normalized Two-phase Multiplier to a Base of Mass Dryness Fraction

3. THE ENTRAINMENT PARAMETER

An entrainment parameter e_c was defined [11] as

$$e_c = \frac{\psi - \psi_S}{\psi_H - \psi_S} \tag{15}$$

where ψ_S and ψ_H are evaluated using B_S and B_H respectively (equations (6) and (8)). In the range of applicability of equation (12) therefore

$$e_c = \frac{Rx^{2-n} - B_S\{x(1-x)\}^{2-n/2} - x^{2-n}}{(B_H - B_S)\{x(1-x)\}^{2-n/2}}$$

$$= \frac{(R-1)}{(B_H - B_S)}\left(\frac{x}{1-x}\right)^{2-n/2} - \frac{B_S}{B_H - B_S}. \tag{16}$$

Where this equation gives $e_c < 0$, e_c is made zero, and where $e_c > 1$, e_c is taken as unity. From equations (11) and (15)

$$B = B_S + (B_H - B_S)e_c. \tag{17}$$

The method of handling these equations when bubbly flow rather than separated flow occurs at low mass dryness fraction is discussed in Reference [11].

4. A SEPARATED FLOW MODEL

Fig. 2 shows a typical flow pattern in a horizontal two-phase crossflow and serves to define some of the notation used in the following development.

Define
$$w = \frac{M_L - M_{SL}}{M_L} \tag{18}$$

where M_L is the liquid mass flowrate, and M_{SL} is the liquid flowrate along the wall(s), the 'separated' liquid. The proportion of liquid contained in the vapour is therefore w.

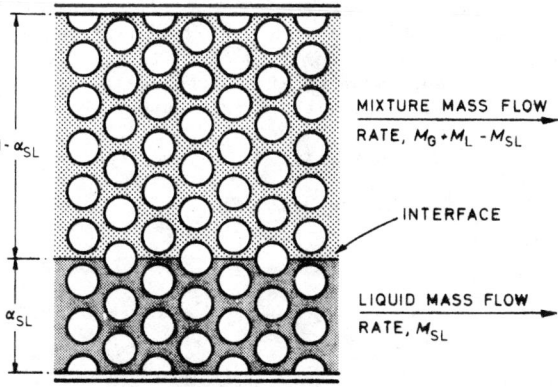

Fig.2. Flow Pattern in Crossflow

Considering the separated liquid flow, the pressure gradient is

$$Dp = Dp_{LO}\{(1 - w)(1 - x)\}^{2-n} \frac{1}{\alpha_{SL}^{2-n}} \qquad (19)$$

and for the gas/liquid mixture, assuming a common pressure gradient,

$$Dp = Dp_{LO}\{x + w(1 - x)\}^{2-n} \phi_{LOe}^2 \frac{1}{(1 - \alpha_{SL})^{2-n}}. \qquad (20)$$

The two-phase multiplier is

$$\phi_{LOe}^2 = 1 + (\Gamma^2 - 1)\{Bx_e^{2-n/2}(1 - x_e)^{2-n/2} + x_e^{2-n}\} \qquad (21)$$

where x_e is the proportion by mass of gas in the separated gas/liquid flow,

$$x_e = \frac{x}{x + w(1 - x)}. \qquad (22)$$

After some manipulation of the above equations the cross-section occupied by the separated liquid is obtained as

$$\alpha_{SL} = (1 - w)(1 - x)\Big/\Big[(1 - w)(1 - x) + [\{x + w(1 - x)\}^{2-n} + (\Gamma^2 - 1)\{B(wx(1 - x))^{2-n/2} + x^{2-n}\}]^{1/2-n}\Big]. \qquad (23)$$

Consequently from equations (19) and (23) the two-phase pressure drop can be evaluated if the entrainment w is known. It is presumed that in that calculation the relevant B-coefficient is given by equation (8).

5. THE APPARENT ENTRAINMENT IN CROSSFLOW

Equations (1), (17), (19) and (23) were solved numerically for the entrainment w. Examination suggested that the entrainment w and the entrainment parameter e_c were related

$$w = e_c^m \qquad (24)$$

where m is a constant to be obtained by numerical analysis.

As the original data were obtained with air/water mixtures at atmospheric pressure we will examine the equation for this case first. Fig. 3 shows a plot of the ratio of the pressure drop obtained from equations (1) and (17), to that from equations (19) and (23), with an entrainment

$$w = e_c^{1.35}. \qquad (25)$$

The predicted values agree to within ±6 per cent.

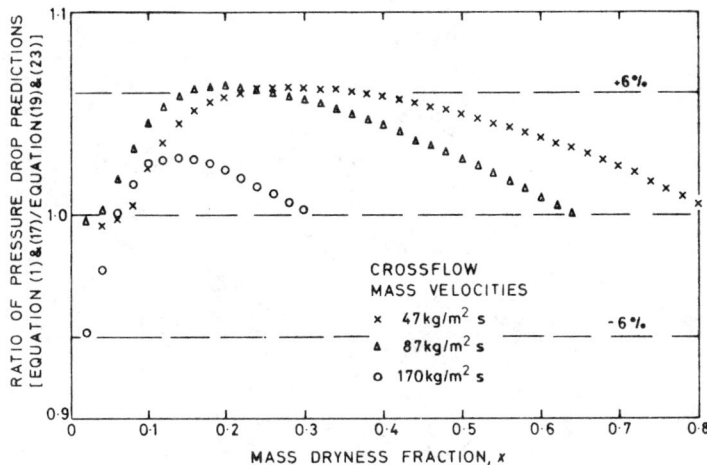

Fig.3. Ratio of Pressure Drop Prediction to a Base of Mass Dryness Fraction

The method has also been applied, so far to steam/water mixtures in the pressure range 0.1-100 bar. A value of m = 1.35 gives the agreement shown in Table 1, the maximum difference is -7 per cent to +13 per cent.

TABLE 1

COMPARISON OF PRESSURE DROP EQUATIONS FOR STEAM/WATER MIXTURES

(Entrainment $w = e_c^1 \cdot {}^{35}$)

Pressure, bar	0.1	1.0	10	100
Dp, eq. (1) & (17) / Dp, eq. (19) & (23)	0 to +13%	-2 to +8%	-5 to +4%	-7 to +1%

6. TUBE SUBMERGENCE

Experiments were carried out using air/water mixtures flowing through a tube nest containing 165 tubes, laid out as shown in Fig. 4. Tubes of 19 mm o.d. are arranged on an equilateral triangular layout of 1.25 tube pitch/diameter ratio. The model has one baffle giving two shell-side passes. The ratio of crossflow zone/window zone minimum flow area is 0.9. Half-tubes are located on the walls to prevent bypassing and tube/baffle leakage is eliminated to give ideal tube banks.

Glass walls allowed the depth of separated liquid to be measured. Experimental values of the separated liquid fraction α_{SL} obtained in this way are shown in Fig. 5. Also shown are curves, for various mass velocities, obtained using equation (23). The predicted values agree closely with experiment.

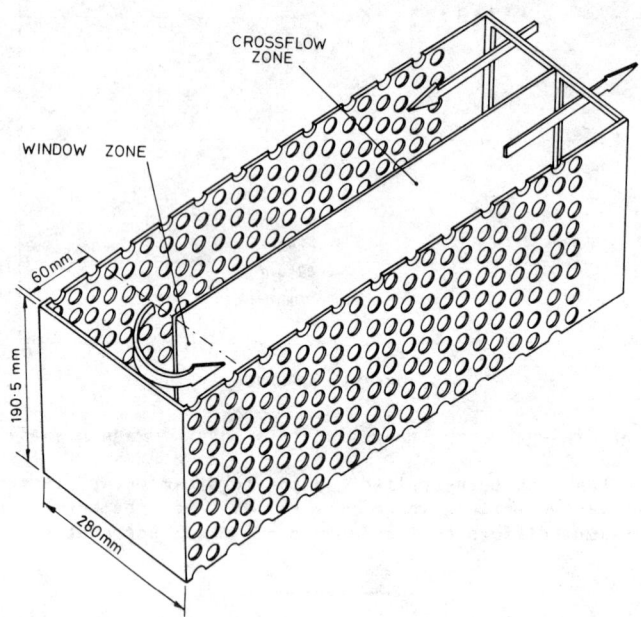

Fig.4. Experimental Model

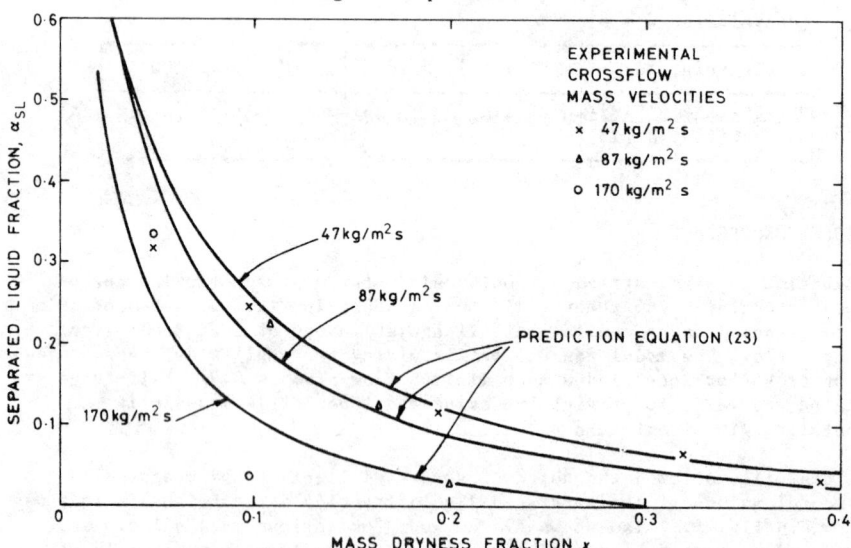

Fig.5. Separated Liquid Fraction to a Base of Mass Dryness Fraction

7. CONCLUSIONS

Using a separated flow model it has been shown that, in crossflow, liquid entrainment in the gas is given by

$$w = e_c^{1.35} \qquad (25)$$

where the entrainment parameter, e_c, is defined by equation (15).

ACKNOWLEDGEMENTS

This paper is presented with the permission of the Director, National Engineering Laboratory, UK Department of Industry. It is British Crown copyright. The work reported was supported by the Chemical and Minerals Requirements Board of the UK Department of Industry and was carried out under the research programme of the Heat Transfer and Fluid Flow Service (HTFS).

REFERENCES

1. Grant, I.D.R. and Chisholm, D. 1979. Two-phase flow on the shell side of a segmentally baffled shell-and-tube heat exchanger. Trans. ASME, J. Heat Transfer, 101(1), pp 38-42.

2. Grant, I.D.R. and Chisholm, D. 1980. Horizontal two-phase flow across tube banks. Int. J. Heat and Fluid Flow, 2(2), pp 97-100.

3. Grant, I.D.R. and Murray, I. 1974. Pressure drop on the shell side of a segmentally baffled shell-and-tube heat exchanger with horizontal two-phase flow. NEL Report No 560. East Kilbride, Glasgow: National Engineering Laboratory.

4. Chisholm, D. 1973. Pressure gradients due to friction during the flow of evaporating two-phase mixtures in smooth tubes and channels. Int. J. Heat Mass Transfer, 16, pp 347-358.

5. Chisholm, D. 1971. Prediction of pressure drop at pipe fittings during two-phase flow. Proc. 13th Int. Inst. Refrigeration Cong., Washington, DC, Vol. 2, pp 781-789.

6. Gloyer, W. 1970. Thermal design of mixed vapour condensers. Hydrocarbon Process, 49(6), pp 103-108.

7. Chisholm, D. 1973. Discussion of paper 'A theoretical solution of the Lockhart and Martinelli flow model for calculating two-phase pressure drop and hold-up, by T. Johannessen'. Int. J. Heat Mass Transfer, 16, pp 225-226.

8. Chisholm, D. 1967. Pressure gradients during flow of incompressible two-phase mixtures through pipes, venturis and orifice plates. Br. Chem. Engng, 12(9), pp 1368-1371.

9. Chisholm, D. 1970. Pressure gradients during the flow of evaporating two-phase mixtures. NEL Report No 470. East Kilbride, Glasgow: National Engineering Laboratory.

10. Chisholm, D. 1980. The turbulent flow of two-phase mixtures in horizontal pipes at low Reynolds nuumber. J. Mech. Engng Science, 22(4), pp 199-202.

11. Grant, I.D.R., Chisholm, D. and Cotchin, C.D. 1980. Shell-side flow in horizontal condensers. Joint ASME/AIChE National Heat Transfer Conference, Orlando, Florida, 27-30 July. Paper No 80-HT-56.

Heat and Mass Transfer in Counter Flow of Water and Air

S. PETELIN, B. GAŠPERŠIČ, and L. FABJAN
Jožef Stefan Institute
Ljubljana, Yugoslavia

ABSTRACT

A mathematical model was developed for two-dimensional laminar and turbolent flow of moist air in a channel between two vertical parallel plates with water film. In order to test the theoretical results obtained a corresponding experimental apparatus was built to simulate the counter flow of water and moist air.

1. INTRODUCTION

In the apparatures of process industries we often deal with counter flow and moist air. In these cases we have direct contact between water and air at the same time as heat, mass and momentum transfer. In direct contact of water and air different effects can be produced, such as: the water can be cooled, the air warmed and miostenred or the water warmed, the air cooled and dehumidified. Any kind of combination of these processes are possible. The local state at the interface of water and air determines what happens.

Because of the many possible processes which can take place with direct contact of water and air, we have limited theoretical and experimental work on the evaporative cooling of water. We hoped to obtain a realistic prediction of the physical process, using a digital computer.

2. GENERAL DESCRIPTION OF PHYSICAL MODEL

The packing of a cooling tower often consists of vertical plates on which the water film slides. Figure 1 shows a Section through two opposite plates. The physical process itself is shown quite simply. It is supposed that the whole quantity of water is in film flow.

The process of the heat and mass trasfer between the plates is adiabatic and stable. At the inlet in the channel uniform distribution of velocity, temperature and concentration fields is assumed. These change along the channel (coordinate z). We have a laminar flow of the air in the first part of the channel which at height $z=z_{cr}$ **starts** to change into turbulent flow.

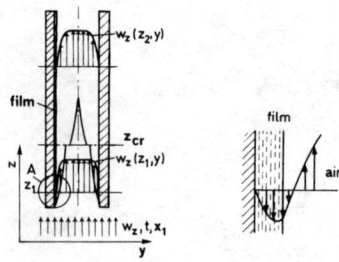

Fig. 1 Distribution of air velocity across the channel for laminar and turbulent flow.

The experimental device was built which is shown in figure 2.

Signs to figure 2

a - Asbestos-cement plate
b - Plexiglass plate
d - Wet and dry thermocouples
f2 - Thermocouples for inlet water temperatures
g2 - Thermocouples for outlet water temperatures
h - Water reservoir
i - Water reservoir
j - Electric heater for water
l - Regulation valve of water mass transfer rate
n - Water collector
o - Hot wire probe
o1 - Thermocouple
p - Orifice
r - Pressure tap
u - Air damper
v - Air filter
z - Electric heater for air
w - Fan

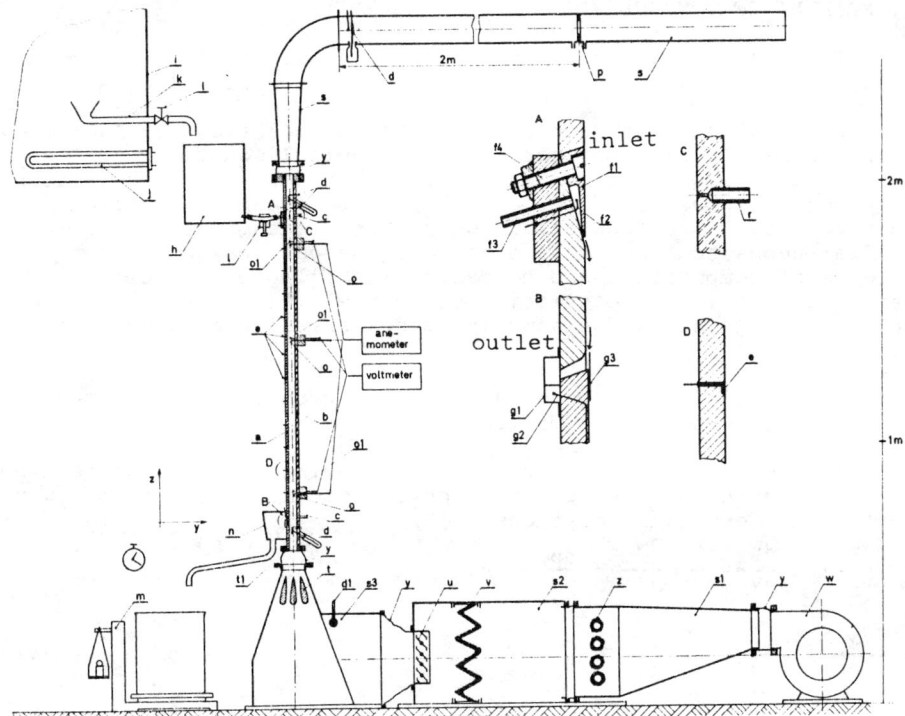

Fig. 2 Schematic arangement of experiment.

Details on figure 2:

A - Construction of water inlet and the manner of water distribution as film on the top of the channel

B - Construction of water outlet collection on the lower part of the channel

C - Construction of pressure tap

D - Method of placing termocouples for measuring temperatures along film

3. MATHEMATICAL MODEL

The distance between the plates is considerably smaller than their width, so we can a two dimensional mathematical model.

Recirculations of air flow in the channel are negligible. Such flow can be described by partial differential equations of the parabolic type which must be solved simultaneously. The equations for booth components of velocity field are the most important of them because forced convection contributes the greatest part at simulataneous heat and mass transfer. The conclusion is that influence of temperature and concentration filed on velocity one is negligible at forced convection in channels. In those cases the velocity field has a dominant influence on all processes in fluid flow.

The partial defferenetial equation were solved with the implict finite-diference method /1/.

At the entry part of them channel laminar flow is present and in this region the physical process is described by the following partial differential equations /2/.

$$\rho w_z \frac{\partial w_z}{\partial z} + \rho w_y \frac{\partial w_z}{\partial y} = \frac{\partial}{\partial y} \mu \frac{\partial w_z}{\partial z} - \frac{dp}{dz} - \rho g, \qquad (1)$$

$$\frac{\partial (\rho w_z)}{\partial z} + \frac{\partial (\rho w_y)}{\partial y} = 0, \qquad (2)$$

$$\rho c_p w_z \frac{\partial t_G}{\partial z} + \rho c_p w_y \frac{\partial t_G}{\partial y} = \frac{\partial}{\partial y}\left(\lambda \frac{\partial t_G}{\partial y}\right) + \mu \left(\frac{\partial w_z}{\partial y}\right)^2 + \rho D (c_{pws} - c_{pG}) \frac{\partial t_G}{\partial y} \frac{\partial \Psi_{ws}}{\partial y}, \qquad (3)$$

and

$$\rho w_z \frac{\partial \Psi_{ws}}{\partial z} + w_y \frac{\partial \Psi_{ws}}{\partial y} = \frac{\partial}{\partial y}\left(\rho D \frac{\partial \Psi_{ws}}{\partial y}\right). \qquad (4)$$

As laminar flow begins to change into turbulent /3/ with a vuscous sublayer at height

$$z_{cr} = 3 \cdot 10^4 \mu_G / (\bar{w}_{zr}). \qquad (5)$$

For turbulent flow the same equations (1), (2), (3) and (4) are applicable, provided all variables are taken time-averaged and effective turbulent flow properties used instead of fluid properties. Effective values are described by Prandtl's mixinglenght theory /4/.

$$\mu_{eff} = \mu + \rho \ell_m^2 \left|\frac{\partial w_z}{\partial y}\right| \qquad (6)$$

$$\ell_m/R = 0,09 + 0,20(1-y/R)^2 - 0,11(1-y/R)^4 \qquad (7)$$

For comparision the Nikuradse's formula for turbulent pipe flow is given below

$$\ell_{m/R} = 0,14 - 0,08(1-y/R)^2 - 0,06(1-y/R)^4 \tag{8}$$

In the viscous sublayer near the interface of water and air and near the dry wall of the channel the following equation, due to van Driest, was used:

$$\ell_m = 0,40y\{1-\exp(-y\sqrt{\tau\rho}/(26\mu))\} \tag{9}$$

The influence of cross flow velocity coused by evaporation was not taken into account in equation 9 because the experiments were limited to low interface mass transfer rate.

Turbulent Prandtl and Schimdt numbers were assumed both equal to 0,9.

$$Pr_T = \frac{(\mu_{eff}-\mu)c_p}{(\lambda_{eff}-\lambda)} = 0,9 \tag{10}$$

and

$$Sc_T = \frac{(\mu_{eff}-\mu)}{\rho(D_{eff}-D)} = 0,9. \tag{11}$$

4. BOUNDARY CONDITIONS

These partial deferential equations, which take into account laminar and turbulent air flow, can be solved with known boundary conditions.

At the channel entry we assume:

$$w_z(y,0) = w_{z1}, w_y(y,0) = 0, t_G(y,0) = t_{G1}, \Psi_{ws}(y,0) = \Psi_{ws1} , \tag{12}$$

and on dry wall:

$$w_z(2R,z) = w_y(2R,z) = \frac{\partial t_G(2R,z)}{\partial y} = \frac{\partial \Psi_{ws}(2R,z)}{\partial y} = 0. \tag{13}$$

The boundary conditions on the interface of water film and air are not known in advance like the conditions expressed by equations 13 and 12. They can be determined experimatelly in each case. This implies a deviation from the basic aim of our work in which we wish to simulate by a mathematical model (a physical process) without special experimental data. The boundary conditions are changing along the interface and are determined simultaneously during calculations.

The water slides down the vertical wall of the channel in the experimental apparatus. The velocity field and pressure drop in air flow influence the conditions in the water film, which we attempt to be take into account. The surface tension is neglected and laminar flow is supposed to take place at the whole range of Reynold's number.

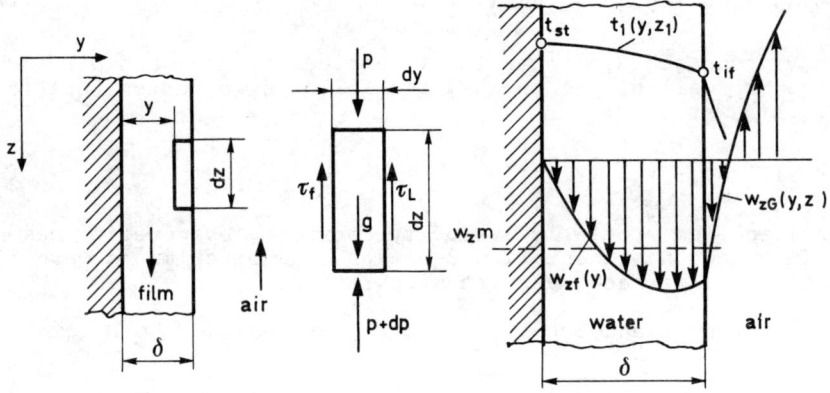

Fig. 3. Conditions in a falling water film.

If the equilibrium of forces on elementary matter is stated with an assumed developed velocity profile in water film we get the following expression for the velocity in the z direction at the interface.

$$w_z(\delta,z) = \frac{1}{\mu_f}\{(\rho g - \frac{dp}{dz})\frac{\delta^2}{2} - \tau_G \delta\}. \tag{14}$$

The velocity in y direction at the interface caused by interface mass transfer is determined by theory based on one way diffusion /2/:

$$w_y(\delta,z) = \frac{D}{1-\Psi_{ws,i}} \frac{\partial \Psi_{ws,i}}{\partial y} \tag{15}$$

The interface temperature is determined on the base of a one dimensional description of the process of air-water counter flow /5/. For this purpose energy and mass balance are considered for air and water. With the following simplications expresions are obtained

$$t_i = \frac{q_w + \dot{m}_G^0\left(c_s \dfrac{dt_{G,m}}{dA} + (c_{p,ws} t_G + r_o)\dfrac{dx_m}{dA}\right)}{\dot{m}_G^0 c_w \dfrac{dx_m}{dA}} \tag{16}$$

and

$$c_s = c_{pG} + x\, c_{pws}. \tag{17}$$

Derivatives $dt_{G,m}/dA$ and dx_m/dA in equation 16 are determined from the temperature and concentration field in the flow of moist air.

The heat transfer rate q_w in water from interface is determined by writing the energy balance for a loop which includes only air-water interface. This is shown on figure 4.

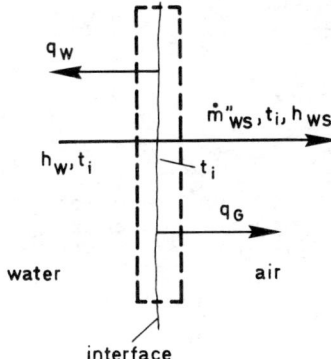

Fig. 4. Conditions on air-water interface.

$$q_w = -q_G + \dot{m}_G^0 \frac{dx_m}{dA}\left[(c_w - c_{p,ws})t_i - r_o\right]. \tag{18}$$

and

$$q_G = -\lambda_G \frac{dt_G}{dy}\bigg|_{y=\delta}. \tag{19}$$

For the concentration $\Psi_{w,s,i}$ of water vapor in the air on the interface, saturation is assumed corresponding to the interface temperature t_i.

5. RESULTS OF MEASUREMENTS AND CALCULATION

In order to test the theoretical results obtained an experimental apparatus was built to simulate the counter-flow of water and moist air. The test section can see in figure 1 and 2. The dimensions of test section are: width of the channel 463 mm, water film width 420 mm, distance between the plates is 29,4 mm and height 1150 mm. The parameters of the experiment were changed in these regions: mass flow water temperature 30-55 °C, entering air temperature 24-29 °C and humidity 4-9 g/kg.

All the **equantities** which define the heat and mass balance were measured. The distribution of velocity and temperature of the air flow, and pressure drop were also measured.

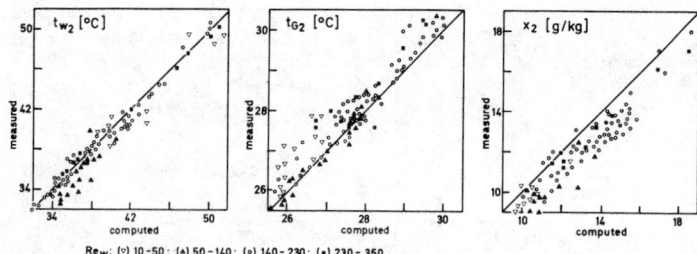

Fig. 5. Comparison of measurements and computated results for inlet water temperature, oulet temperature and moisture of air.

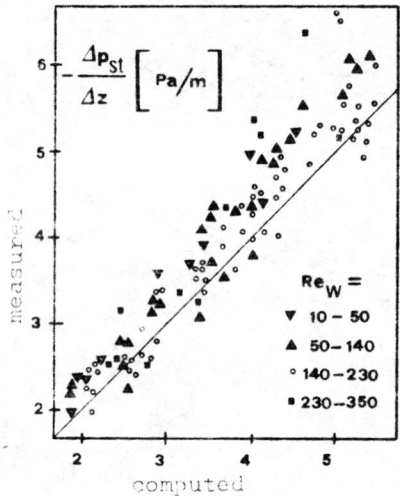

Fig. 6. Comparison of measurements and computed results for pressure drop in air flow.

The velocity distribution in the air flow was measured by a DISA hot wire anemometer. At measurements in the boundary layer with temperature gradient present, the temperature profile must be measured. This is done with thermocopuples. The hot wire probe has to be calibrated at vauries air velocities and temperatures /6/.

The anemomemter signals was analysed with the help of a Hewlett Packard Kolerator. The distribution of kinetic energy of turbulence was also determined. The results of those measurements can be seen on figure 7. Only two examples of measurements and computations are given. Similar results were obtained in other cases with different value of air and water flow.

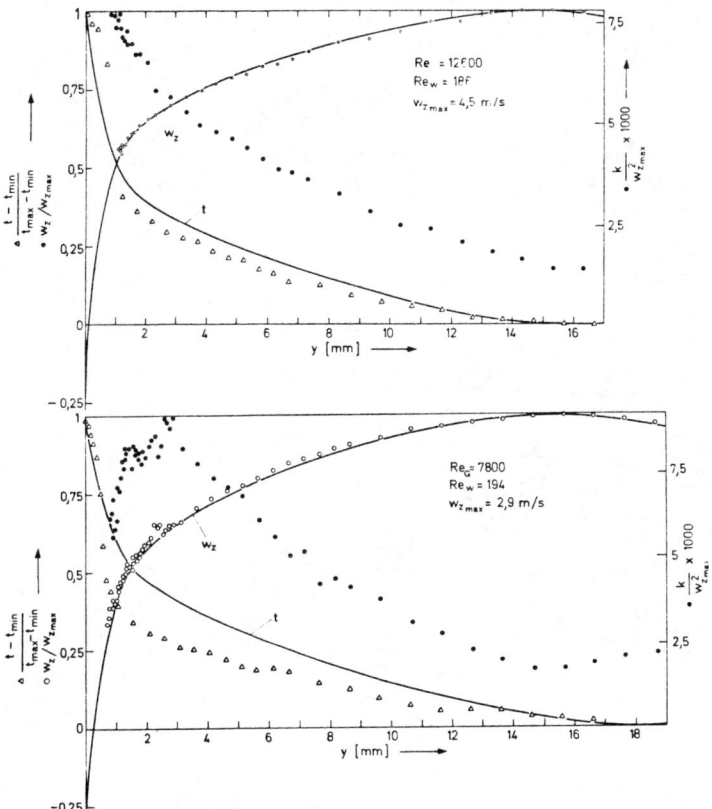

Fig. 7. Comparison of measurement and computation for velocity and temperature distribution at height of 740 mm, and measurement of kinetic energy of turbulence of the air flow.

The computer program determined the velocity, temperature and concentration field in the air flow. The local pressure drop, heat and mass transfer coefficient can be found with the help of these. For comparison with experimental results the overage values of these variables are needed which are determined with following equations:

$$\frac{dp}{dz} = \frac{1}{h_k} \int_0^{h_k} \frac{dp}{dz}\bigg|_z dz , \qquad (20)$$

$$\alpha_G = \frac{1}{h_k} \int_0^{h_k} \alpha_G(z) \, dz \qquad (21)$$

and

$$\beta = \frac{1}{h_k} \int_0^{h_k} \beta(z)\, dz \qquad (22)$$

The dependence of the friction coeficient on the water film Reynolds number is shown on figure 8. This is based on results of mathematical model.

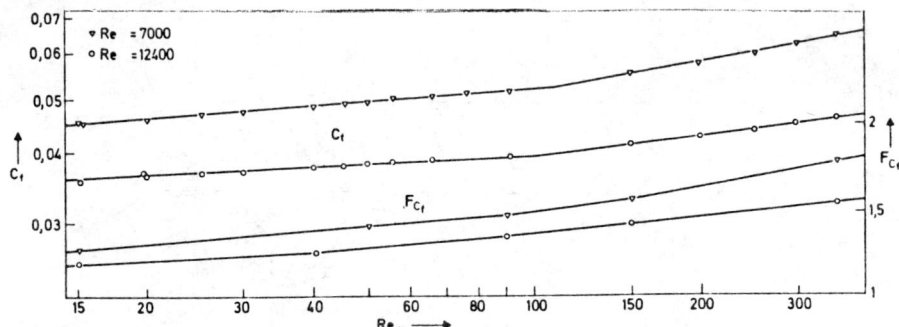

Fig. 8. Influence of Re_w on c_f from results of mathematical model. Fc_f represents increase of c_f with regard to dry channel.

Comparisons of experimental and mathematical model results for pressure drop, heat and mass transfer coeficient in the nondimensional form are shown on figures 9, 10 and 11.

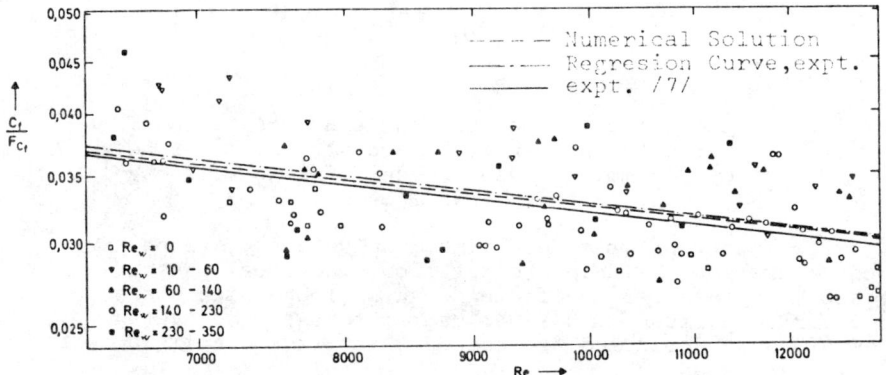

Fig. 9. Comparison of criction coeficient c_f.

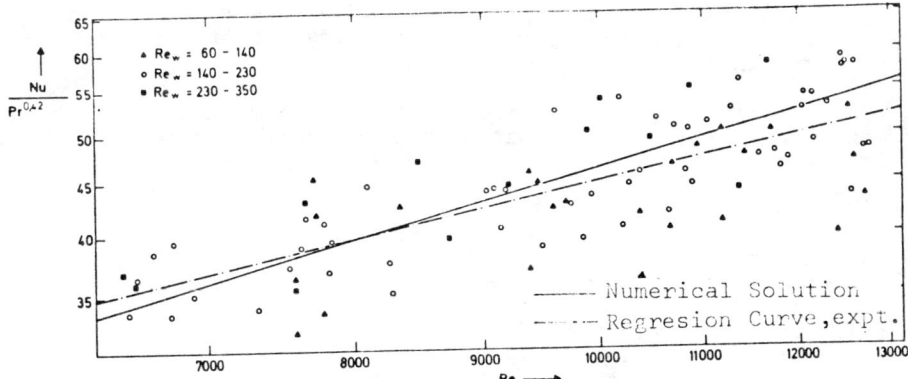

Fig. 10. Comparison of Nusselt number Nu.

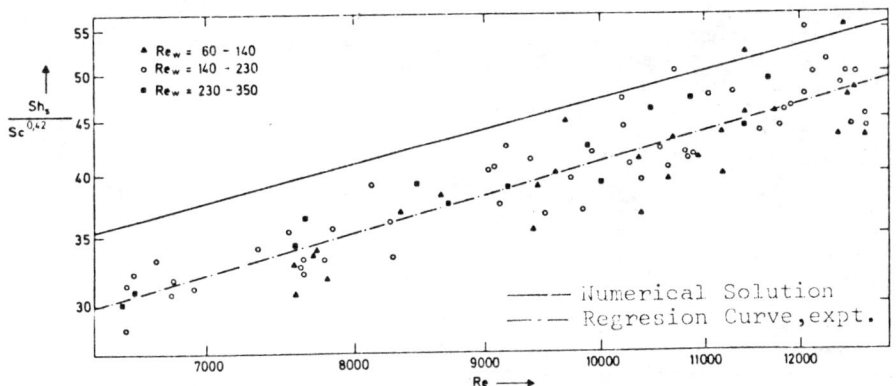

Fig. 11. Comparison of Sherwood number Sh.

The results on figures 9, 10 and 11 compare with the results shown on figures 5 and 6. The greatest differences are noticed in measurement of air humidity.

6. CONCLUSION

Complete agreement between measurements and computations is not obtained, but nevertheless it can be considered that the mathematical model is a realistic simulation of a counter flow of water film and moist air. The problem is asymmetric because the water film slides down only on one wall of the channel; the second wall is dry. There are no special limits in using the mathematical model for symmetric flow with both plates wet.

Although Prandtl's mixing-lenght theory is relatively simple it describes reasonably accurately the character of turbulence at air-water counter flow.

The experiments were carried out with rather narrow variations of water and air flow. The mathematical model is quite universal, so it could be used with confidence beyond the regions tested.

NONMENCLATURE

1   inlet of air and outlet of water
2   outlet of air and inlet of water
c   specific heat, J/kgK
D   diffusion coefficient, m^2/s
g   gravitational acceleration, m/s^2
h   height of the channel, m
k   turbulence kinetic energy, m^2/s^2
ℓ_m Prandtl's mixing lenght, m
$\dot{m}$ mass flow rate, kg/s
Nu   Nusselt number
p   pressure, Pa
q   heat transient rate, W/m^2
Pr   Prandtl number
R   channel half-width, m
Re   Reynolds number
r_o latent heat of vaporisation, J/kg
Sc   Schmidt number
Sh   Sherwood number
w   velocity, m/s
α_w heat-transfer coefficient, $W/m^2 K$
δ water film thickness, m
ρ density of moist air, kg/m^3
μ visconsity, Ns/m^2
τ shear stress N/m^2
Ψ mass concentration kg/kg
x   humadity of air, kg/kg
y   coordinate normal to the wall, m
z   streamvise coordinate, m

Subscripts

–   average
eff   effective
i   interface
cr   critical
m   middle
ws   water steam
G   gas (air)
f   film
y,z   direction
p   at constant pressure
w   water
r   relative
T   turbulent

REFERENCES

1. S.V.PATANKAR, D.B.SPALDING, Heat and Mass Transfer in Boundary Layers, Intertext Book, London, (1970).

2. R.B.BIRD, W.E.STEWART, E.N.LIGHTFOOT, Transport Phenomena, John Wiley & Sons, New York, London, (1969).

3. W.KAST, "The One Method for Dealing with Mass Transfer and Transmission of Heat by Convection", Verfahrenstechnik, vol. 6, n. 10 (1972), p. 346.

4. B.E.LAUNDER, D.B.SPALDING, Mathematical Models of Turbulence, Academic Press, London and New York (1972).

5. R.E.TREYBAL, Mass-Transfer Operation, McGraw-Hill Book Company, New York (1968).

6. A.M.KOPPIUS, G.R.M.TRINES: "The Dependence of Hot-Wire Calibration on Gas Temperatur at law Reynolds Numbers", Int. J. Heat Mass Transfer, vol. 9, pp. 967-974, Pergamon Press (1976).

7. G.S.Beavers, E.M.Sparov, J.R.Lloyd: Low Reynolds Number Turbolent Flow in Large Aspect Ratio Rectangular Ducts. Transaction of the ASME. Journal of Basic Engineering, June 1971, pp. 296-299.

Experimental Investigation on a Full Scale Thermo-syphon Reboiler

PETER VON BÖCKH
Heat Engineering and Planning Department (TGW)
Large Steam Turbo Set Division
BBC Brown Boveri and Company, Limited
CH-5401 Baden, Switzerland

ABSTRACT

On a thermo-syphon reboiler for turbine gland sealing steam generation in a BWR power plant, measurements were performed in a test loop before comissioning. At different heating and sealing steam pressures the steam production
was measured as a function of the virtual water level. The measurements provided detailed information on performance and the limits of stable operation.

The experimental data was used to compare different calculation methods for heat transfer and pressure drop. A computer code was developed for reboiler design. Computed performance curves agree well with the measured data, and also the limits of stable operation, such as dry-out and flooding, are also predictable.

1. INTRODUCTION

In BWR plants non-radioactive steam is needed as turbine gland sealing steam and for auxiliary purposes. This steam is generated economically in a reboiler heated by waste steam e.g. valve leakoff steam. BBC has developed a steam reboiler type for these purposes and has newly included these vessels in its product spectrum.

The reboiler is of the thermo-syphon type with vertical evaporator tubes. The evaporator is a tube-and-shell type heat exchanger. The heating steam condenses on the tubes on the shell side. Tube side water enters the evaporator tubes, and is partially evaporated while passing the tube.

The water carried along by the steam is separated in the space above the tube outlets and returned to the evaporator tube inlets through the downcomers. The circulation is maintained by the difference of densities in the downcomers and in the evaporator tubes. Integration of the downcomers in the shell is a special feature of the BBC reboiler.

In the field of two-phase flow the thermal and hydraulic design of a thermo-syphon reboiler poses an interesting problem. The saturated or subcooled water entering the evaporator tubes is evaporated. This leads to the various flow patterns such as bubble, plug, annular and spray flow. The calculation methods for design must take all types of flow encountered into account. A review of literature has revealed that results of the various proposed calculation methods have a rather wide range of scatter.

To obtain representative thermal and hydraulic design fundamentals and to be able to test the reboiler under extreme conditions measurement were performed on a sealing steam reboiler in a test loop prior to commissioning. The purpose of the measurements was to gain additional operating experience with the reboiler, and in this way to fulfill the stringent safety requirements in a BWR plant under all circumstances.

2. DESCRIPTION OF REBOILER AND TEST EQUIPMENT

2.1 Description of reboiler

Reboiler data:

	Heating steam	Sealing steam	
Pressure	0.6	0.2	MPa
Steam mass flow	5.0	4.5	kg/s
Steam temperature	180.0	120.2	°C
Evaporator tubes (carbon steel)	438 tubes with 24 mm OD and 1.5 mm wall thickness		
Downcomers	3 tubes with 200 mm ID		
Total weight		4200	kg
Thermal power		9.0	MW

The sealing steam reboiler is shown in Fig. 1. The evaporator tubes are arranged concentrically around the downcomers. A bellows expansion joint is installed to absorb the differential expansion between tubes and shell. The evaporator tubes and downcomers have below a common sump and above a common vapor space. In the vapor space the entrained water separates from the steam. In addition, the steam is dried by a moisture separator. The entrained water flows back to the sump through the downcomers. The evaporator tubes are heated from outside with condensing steam. By the evaporation of the water in the tubes the density of the two-phase mixture is substantially lower than that of the water in the downcomer. Since both tubes and downcomers constitute a communicating system the level in the evaporator tube will rise. The differences in density become so great that equalization can only be achieved by flow pressure loss. The mass flow rates in the evaporator and in the downcomers are approximately equal. Due to the considerably smaller diameter of the evaporator tubes and due to the evaporation the friction and

acceleration pressure drop is
higher. In the reboiler described
here the downcomers are heated as
well, but compared with the
evaporator tubes the ratio of
heating surface to mass flow is
smaller by magnitudes and the
evaporation rates are nearly
negligible. The steam leaving
the reboiler is replaced by
water. This subcooled water
is fed in the downcomer. The
water feed regulation is adjusted
such that the virtual liquid
level in the downcomer remains
constant. The level is denoted
virtual because this liquid
level is measured on the outside
of the vessel. However, it
corresponds to the sum of the
pressure losses in the downcomer,
comprising the static, friction
and acceleration pressure loss.
As a result of the evaporation
salts and impurities accumulate
in the sump, for which reason the
sump is blown down. The heating
condensate drain is regulated.
A constant condensate level is
adjusted in the heating steam
space. In order to prevent accumulation of inert gases in the
heating steam space, the space is vented continuously through a
perforated vent tube located in the center of the bundle. In
operation the reboiler is regulated such that with constant
heating steam pressure the sealing steam pressure required for
steam production is obtained. In the test it was possible to
adjust both pressures and hold them at a constant level.

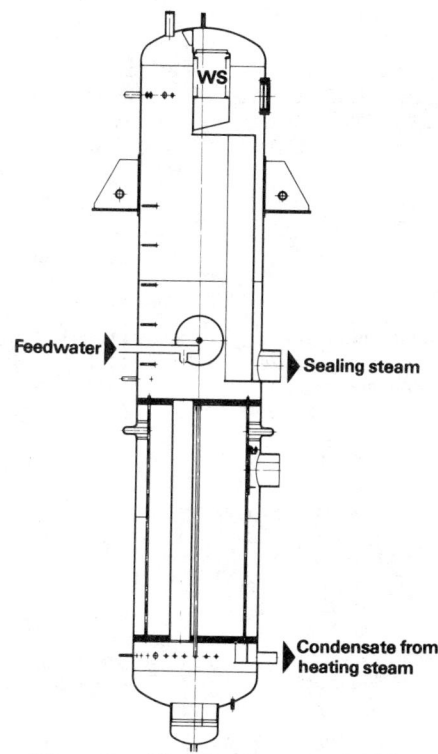

Fig. 1: Gland steam reboiler

The reboiler is a thermo-syphon evaporator with the
distinguishing feature that the circulation is maintained without
pumps solely by the difference in densities in the evaporator tubes
and downcomers.

When designing the reboiler and when the operating conditions
are, established care is to be taken that the recirculation is
sufficient. This prevents such high concentrations of impurities
as can pose a corrosion danger to the tubes. Flooding of the
reboiler vapor space, which can cause water to flow out of the
reboiler and endanger the downstream components, is also to be
avoided.

2.2 Test equipment

The reboiler arrangement in the test stand and the arrangement of the individual measuring devices is shown schematically in Fig. 2. The heating steam, live steam from the boiler house with a pressure of 1.6 MPa at 350 °C, is throttled down to the desired

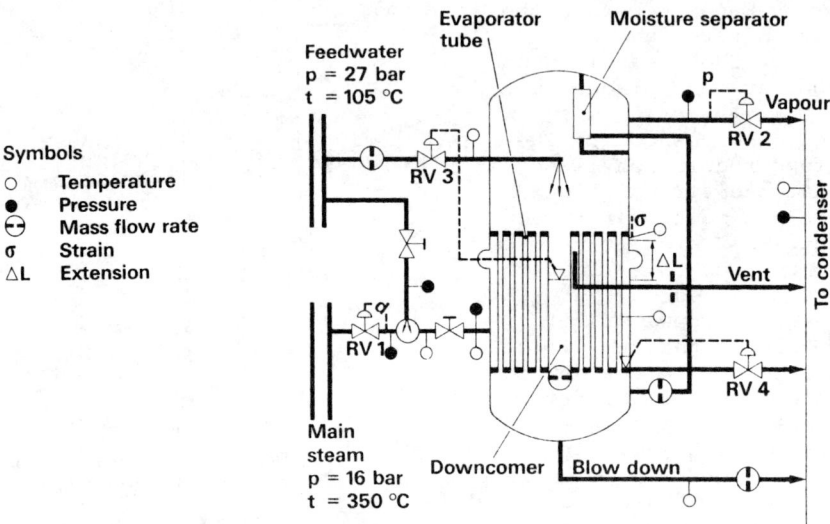

Symbols
- ○ Temperature
- ● Pressure
- ⊖ Mass flow rate
- σ Strain
- ΔL Extension

Fig. 2: Reboiler test loop

pressure with control valve RV 1, holding the adjusted pressure constant. Since the steam is highly superheated and subjecting the reboiler to excessive temperatures is to be avoided, the steam is sprayed down to obtain only slightly superheated steam. The condensate accumulating in the heating space is removed through control valve RV 4, which maintains a constant condensate level in the heating space.

Water is fed into the reboiler through control valve RV 3. The water supply is regulated with the reboiler virtual liquid level such that the level remains constant. To be able to determine the amount of steam production the water supplied and the water removed by blowdown were measured. For these flow measurements turbine flowmeters were used. The pressure of the produced steam is maintained constant with control valve RV 2. To check the moisture separator the quantity of separated water was measured. The steam moisture after the moisture separator was measured by tracer measurement. In order to measure the water recirculation orifices were installed in the inlets of a few evaporator tubes. These orifices were graduated into various measuring ranges. With the orifices it was possible to determine the mass flow of the entering water. After a pressure-loss correction it was then possible to calculate the recirculation rate. The points at which temperatures and pressures were measured are entered in Fig. 2. Chromel-Alumel jacket thermocouples were used for temperature measurement, Bell + Howell pressure transmitters for pressure measurement. The thermocouples were calibrated before and after the measurements. The pressure transmitters were recalibrated before each series of measurements.

In addition, the mechanical stresses in the shell in the area of the tubesheet and the bellows were measured. These measurements are not covered in this report.

The produced steam, the vent, the blowdown and all condensates were taken to a surface condenser.

Because of the great number of measurement data the measurement signals were registered and immediately evaluated with the aid of an automatic measurement signal processing system.

3. PERFORMANCE OF MEASUREMENTS

For the main part stationary tests were performed. The heating and sealing steam pressures were adjusted constant, and for various virtual liquid levels the steam production, overall heat transfer coefficient, water recirculation and amount of water separated in the moisture separator were measured. The adjustment ranges are tabulated below.

No. of test series	Heating steam pressure MPa	Sealing steam pressure MPa	Virtual water level %
100 and 200	0.47 - 0.68	0.21 - 0.40	20 - 70
300	0.26 - 0.46	0.18 - 0.26	20 - 57
400	0.40	0.21	38
500	0.65	0.36	37
600	0.45 - 0.85	0.19 - 0.70	17 - 99
700	0.094	0.054	38
800	0.67 - 0.85	0.26 - 0.35	32 - 40
900	0.45 - 0.85	0.25 - 0.70	23 - 52
1000	0.65 - 0.85	0.25 - 0.40	17 - 47

The test series 900 and 1000 were started after a 2-month outage period. Hence, it was possible to determine the effect of shutdown corrosion on heat transfer. A few tests to study the dynamic behaviour of the reboiler and the control were also performed. Since it was not possible to determine the steam wetness after the moisture separator, tracer measurements were made in the test series 800 to determine such wetness.

With the test series 1000 a few alterations were made to the reboiler. The downcomers were insulated and a few baffles were installed in the vapor space.

With the test series 600 the virtual water level was varied to determine the limits of the stable or allowed working range. With low water level the tubes dried out. With high water level the vapor space became flooded.

4. TEST RESULTS

The characteristic performance curves of the reboiler at 0.45, 0.65 and 0.85 MPa heating steam pressure are shown in Figs. 3 - 5. As can be seen from the diagrams the virtual water

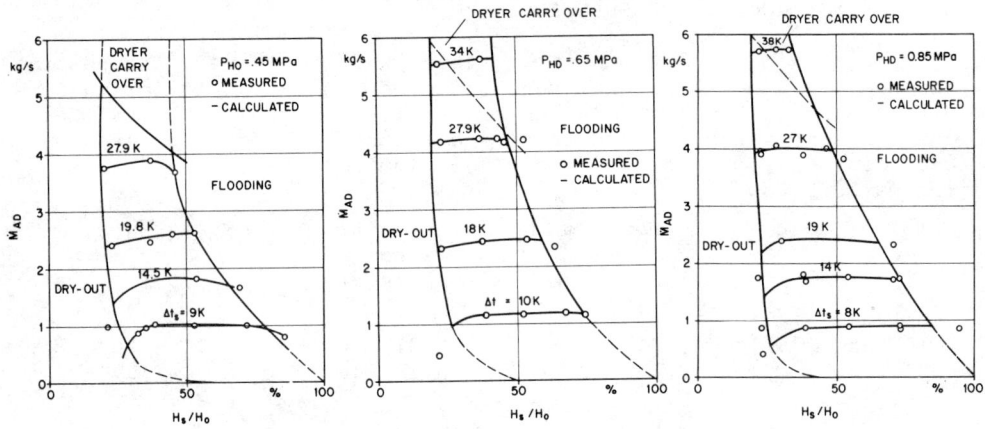

Fig. 3 - 5: Reboiler performance curves

level has only a slight effect on steam production. However, it is quite decisive for the limits of the allowed working range. As the curves show there are 3 phenomena which limit the working range.
1) With low virtual water level the upper sections of the tubes dry out, resulting in high concentration of impurities which can cause the appearance of corrosion on the tubes.
2) With excessive virtual water level the vapor space becomes flooded. In this situation water flows together with the steam through the moisture separator. It was possible to observe the vapor space through four sight glasses. On flooding a column of bubbles appeared in the vapor space.
3) With high steam production water starts to break through the moisture separator.

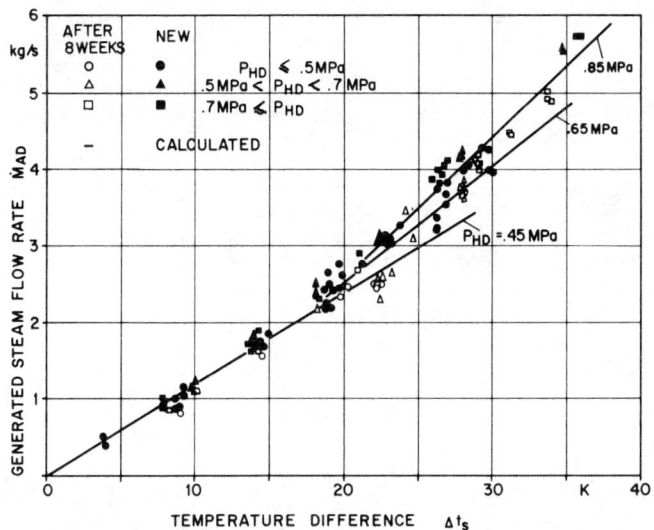

Fig. 6: Reboilers steam production vs. temperature difference

In Fig. 6 the steam production is plotted versus the difference in the saturation temperature of sealing and heating steam. The diagram shows that after 8 weeks of outage the steam production, and hence the overall heat transfer coefficient as well, has deteriorated by 12 - 16 %. An inspection shows that due to the entry of air the tubes were covered with a layer of rust.

5. CALCULATION OF A THERMO-SYPHON REBOILER

A thermo-syphon reboiler with one evaporator tube and one downcomer is shown in Fig. 7. In both tubes the pressure drop is equal because they are open to the same pressure space. The difference in the static pressures effected by the difference in density is equalized by the different flow pressure drop in the evaporator tube and the downcomer. The pressure drop in the evaporator tube and in the downcomer is calculated as follows:

$$\Delta p_E = \dot{m}_E^2 \{\frac{1.25}{\rho_{HE_1}} + (\frac{1}{\rho_{iE_2}} - \frac{1}{\rho_{iE_1}})\} + \int_1^2 \{(\frac{dp}{dl})_{R_E} + (\alpha_E \rho_g + (1-\alpha_E)\rho_1)g\}dl \quad (1)$$

$$\Delta p_D = \dot{m}_D^2 \{\frac{1.25}{\rho_1 \alpha_{D_2}^2} + (\frac{1}{\rho_{iD_1}} - \frac{1}{\rho_{iD_2}})\} + \int_2^1 \{(\frac{dp}{dl})_{R_D} + (\alpha_D \rho_g + (1-\alpha_D)\rho_1)g\}dl \quad (2)$$

The pressure drops comprise the inlet pressure drop, acceleration pressure drop and the static pressure drop. The inlet-pressure drops are calculated with the homogenious density ρ_H and acceleration-pressure drops are calculated with the impulse density ρ_i.

The downcomer is denoted D, and the evaporator tube E. 1 signifies the inlet and 2 the outlet level of evaporator tubes. $\dot{M}$ is the mass flow rate. The friction pressure drop $(dp/dl)_R$ and the void fraction α are a function of the steam quality $\dot{x}$. The steam quality along the tubes is governed by the transfered heat flow, and hence by the heat transfer coefficients. Since neither the heat transfer coefficients nor the pressure drop and the void fraction can be given as a simple function of the tube length, calculation must be done in steps along the tube. The pressure drop is calculated for an assumed mass flow and the mass flow is varied until the same pressure losses are established in the downcomer and evaporator tube.

Exact relationships are known for calculating pressure drop and heat transfer with single-phase flow. Many calculation methods and correlations are known in the field of two-phase flow. If the results of these calculations are compared, the differences are considerable. The calculation methods which have rendered the measurement results best are presented in the following sections.

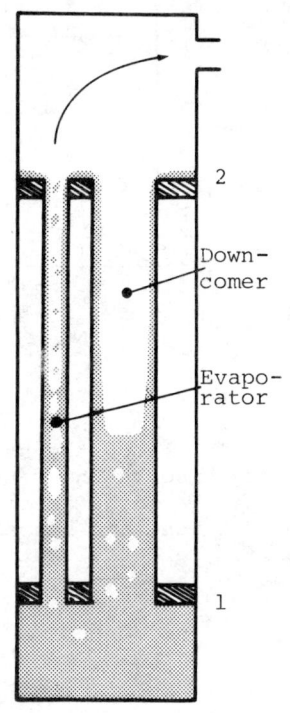

Fig. 7: Principle of reboiler

5.1 Calculation of friction pressure drop

For single-phase steam or water flow, the friction pressure drop coefficients can be determined from the diagrams by Colebrook and White [1].

For two-phase flow correlations proposed by Moussalli [2] and Kesper [3] showed the best agreement with the measured data. Moussalli's [2] correlation is developed for bubble and plug flow. Kesper's [3] correlation covers the area of slug annular and dispersed flow. Kesper's [3] correlation was slightly modified by adapt it to the present problem. The range of validity for both correlation was also modified.

5.2 Calculation of steam void fraction

For the void fraction calculation correlations proposed by Moussalli [2] and Chawla [5] showed the best agreement with the experimental data. Moussalli's [2] correlation is based on Bankoff's [4] variable density model, whereby the empirical constant is deduced theoretically. Chawla's [5] correlation is developed for slug and annular flow. The range of validity for both correlation was modified.

5.3 Shell-side heat transfer coefficient

In the lowest range where the tubes are flooded by condensate the heat transfer is calculated for free convection according to the VDI Wärmeatlas [6].

For the upper part of the tubes the Nusselt theory [7] for laminar film condensation can be used. For the turbulent condensate film there are only calculation methods which indicate the mean heat transfer coefficient over the tubes. These calculations already account for the laminar film. There is a sharp jump in the transition range. Since local heat transfer coefficients were required for the calculations, an equation has been developed for the local heat transfer in the turbulent film, the integral of which renders quite well the values of the mean turbulent heat transfer coefficient according to Grigull [8]. A comparison of the various calculations is shown in Fig. 8. The equation used here does not have a jump in the transition from laminar to turbulent.

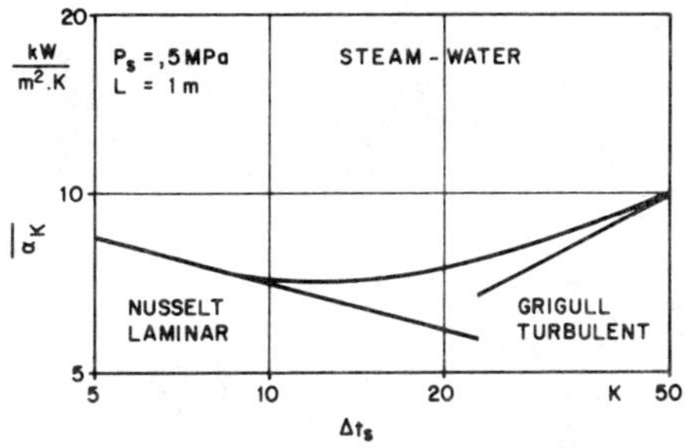

Fig. 8. Condensation heat transfer coefficient

5.4 Heat transfer in the tubes

In the field of single-phase flow the heat transfer in the tube is calculated according to Gnielinski [9].

For two-phase flow a calculation method according to Kesper [3] yields the best results. The Kesper method is based on the analogy between heat transfer and sheer stress. The transitions from bubble flow to plug and annular flow are accounted for by the analogy.

6. RESULTS OF THE CALCULATIONS

The above calculation methods were integrated in a computer program with which a thermo-syphon reboiler can be calculated. The program indicates when the tubes will dry out. With a level at which the vapor space is flooded equalization between the evaporator tubes and the downcomers is not possible, i.e. the iteration does not converge. The calculated steam production is compared with the test data already in Fig. 6. Agreement is very good. In Figs. 3 - 5 the calculated performance curves are compared with measured values. It can be seen that the limits of the allowable working range can also be calculated quite accurately.

With this computer program optimization of tube diameter and tube length are possible and feasible.

The effect of tube length on the heat transfer coefficient is shown in Fig. 9. With small temperature differences the shell-side condensate film is laminar. The overall heat transfer coefficients can be improved substantially by using shorter tubes, or by installing support plates which repeatedly interrupt the condensate film. With large temperature differences the condensate film is turbulent, i.e. here, long tubes improve the heat transfer.

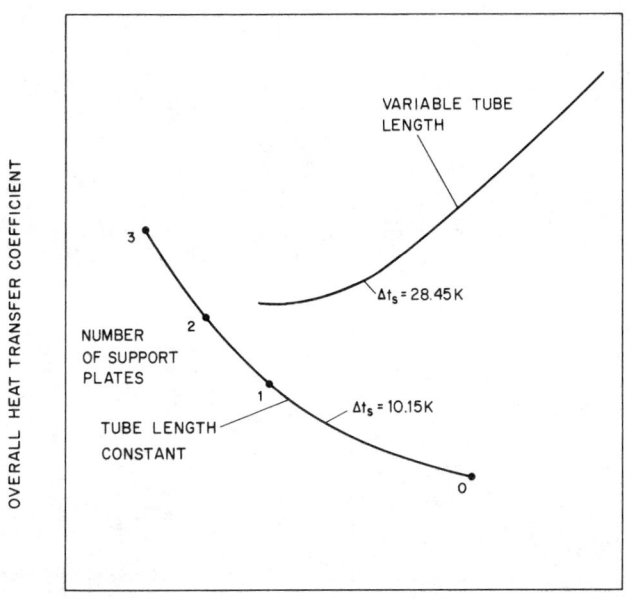

Fig. 9 : Overall heat transfer coefficient for different tube length

REFERENCES

[1] Colebrook and White: VDI Wärmeatlas (1974) Lb 2

[2] Moussalli, G.: Dampfvolumenanteil und Druckabfall in der Blasenströmung. Ph.D.-Thesis. University Karlsruhe, FRG (1975)

[3] Kesper, B.: Wandschubspannung und konvektiver Wärmeübergang bei Zweiphasen-Flüssigkeits-Dampfströmung hoher Geschwindigkeit. Ph.D.-Thesis. University Karlsruhe, FRG (1974)

[4] Bankoff, S.G.: Heat Transfer Trans. ASME Serie C 82 (1960) 11, 265-272

[5] Chawla, J.M.: CIT 41 (1969) 5 + 6, 328/330

[6] Konv. Wärmeübergang, VDI Wärmeatlas (1974) Fa 1-4

[7] Nusselt, W.: VDI-Z. 60 (1916) 541 and 569

[8] Grigull, U.: Forsch. Ing.-Wes. (1952) Nr. 10

[9] Gnielinski, V.: Forsch. Ing.-Wes. 41 (1975) Nr. 1

Saturated Flow Boiling—Effect of Non-uniform Heat Transfer along the Tube Perimeter

WERNER BONN
Institut für Thermische Verfahrenstechnik
Universität Karlsruhe
Kaiserstrasse 12
D-7500 Karlsruhe, FRG

ABSTRACT

In this work the heat transfer for forced convection boiling in a horizontal tube with constant heat flux is investigated. In the case of stratified and wavy flow the measured temperature distribution along the tube perimeter looks like a cosine function with a maximum at the top and a minimum at the bottom of the tube. Due to this temperature distribution the main part of the heat supplied at the top is conducted to the bottom.

A model which takes into account the tangential heat conduction was developed to calculate local heat transfer coefficients along the perimeter based on the measured temperature distribution. The results show two regions with different heat transfer behavior along the perimeter. In the upper part of the tube, which is in contact with the vapor, the heat transfer coefficient is very low and may be neclected. The heat transfer coefficient in the wet part can be described as a function of the local temperature difference. With this information it was possible to calculate the temperature distribution for tubes with different wall thickness.

The calculations predict an increase of the perimeter averaged heat transfer coefficient as well as of the critical heat flux with increasing heat conductivity and thickness of the tube wall. Therefore attention must be paid to the influence of tangential heat conduction when experimental data obtained with test sections of different materials, different wall thickness and different kinds of heat supply are compared.

1. INTRODUCTION

The heat transfer coefficient α is defined as the ratio of the heat flux to the temperature difference. If the heat transfer mechanism is uniform, the heat rate $\dot{Q}$ can be calculated with an area averaged heat transfer coefficient α:

$$\dot{Q} = \dot{q} \cdot A = \alpha \cdot A \cdot \Delta T \tag{1}$$

In flow boiling the heat transfer mechanism changes along the tube length due to increasing flow quality $\dot{x}$ and changing phase distribution. Therefore, long test tubes should be divided into short test sections with a small change of flow quality. Usually the obtained data for one of these test sections are used to calculate the so called "local" heat transfer coefficient $\alpha(z)$:

$$\alpha(z) = \frac{\dot{q}(z)}{T_w(z) - T_b} \qquad (2)$$

In this equation $\dot{q}(z)$ is the perimeter averaged heat flux $\dot{Q}/A_i$ and $T_w(z)$ is the perimeter averaged wall temperature. Therefore the heat transfer coefficient defined by equation 2 will be called perimeter averaged heat transfer coefficient.

An analysis of measured wall temperatures at several positions around the tube perimeter shows systematic variations. With the local temperature difference $T_w(z,\varphi) - T_b$ and the local heat flux $\dot{q}'(z,\varphi)$ from the inside surface to the fluid a local heat transfer coefficient can be defined as follows:

$$\alpha(z,\varphi) = \frac{\dot{q}'(z,\varphi)}{T_w(z,\varphi) - T_b} \qquad (3)$$

With the knowledge of single phase heat transfer the local heat transfer coefficient at the bottom of a horizontal tube can be assumed to be much higher than that at the top, if there is a separated flow as shown in Fig. 1.

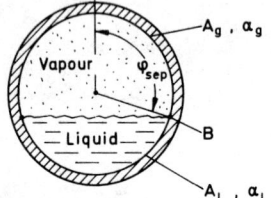

Fig. 1
Cross-section through a horizontal tube with separated two-phase-flow

Point B depicts the point of separation between the wet part and the dry part of the tube. The corresponding angle to point B is called φ_{sep}.

If the tube is heated with constant heat flux, the lower heat transfer coefficient α_g at the dry part the tube leads to a much higher wall temperature at the top of the tube than at the bottom. Due to this temperature distribution part of the heat supplied at the top will be conducted to the bottom and then transferred to the liquid phase (Fig. 2). This tangential heat conduction causes a non-uniform heat flux from the inner surface to the fluid, though the heat supply at the outer surface is uniform.

SATURATED FLOW BOILING

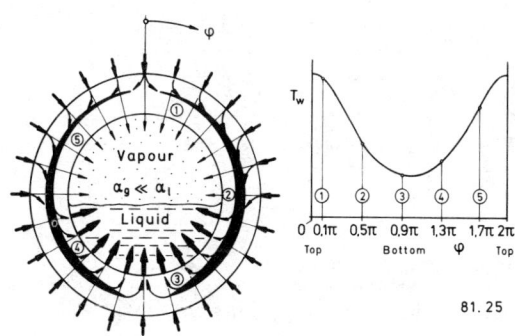

Fig. 2
Heat flux, phase - and wall temperature distribution in a horizontal evaporator tube

Assuming α_g and α_l to be constant the following two limiting cases can be considered:

1. If the tangential heat conduction of the tube wall is infinite, the wall temperature is constant along the perimeter and independend of the condition of heating (thermal boundary condition ΔT = const). The heat fluxes through the areas A_l and A_g will be $\dot{q}_l = \alpha_l \cdot \Delta T$ and $\dot{q}_g = \alpha_g \cdot \Delta T$. In this case the perimeter averaged heat transfer coefficient can be calculate by

$$\alpha(z)_{\Delta T=const} = \alpha_l \cdot (1-X) + \alpha_g X \qquad (4)$$

with $X = \varphi_{sep}/\pi$.

2. If the tangential heat conduction is zero and the heat flux at the outer surface is constant (thermal boundary condition $\dot{q}$ = const), the temperature differences for the two areas will be $\Delta T_l = \dot{q}/\alpha_l$ and $\Delta T_g = \dot{q}/\alpha_g$. The perimeter averaged heat transfer coefficient, obtained with an averaged temperature difference, can be then calculated from

$$\alpha(z)_{\dot{q}=const} = \frac{1}{\frac{X}{\alpha_g} + \frac{1-X}{\alpha_l}} \qquad (5)$$

Assuming, that the values of α_g, α_l and X are given for a certain heat transfer problem, the value of $\alpha(z)$ obtained from equ.(4) is higher than that of equ.(5). In reality the perimeter averaged heat transfer coefficient must be located between the two boundary values.

2. ALTERNATING EFFECTS BETWEEN THE WALL TEMPERATURE DISTRIBUTION AND THE LOCAL HEAT FLUX

The heat balance of a tube element, as shown in Fig. 3, yields the alternating effects between the wall temperature distribution and the local heat flux. The following assumptions are made:

1. Steady state heat conduction.
2. The axial component of heat conduction is zero.
3. The temperature distribution in the cross-section is only a function of the perimeter angle φ (one-dimensional heat conduction).
4. Thermodynamic equilibrium between the vapour and the liquid phase.

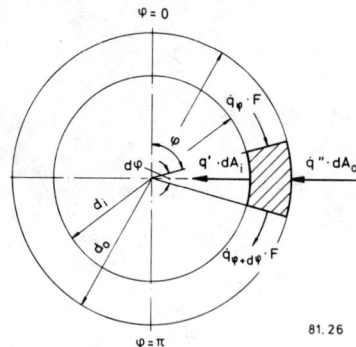

Fig. 3: Heat balance at a tube wall element

With these assumptions the heat balance can be expressed in the following way:

$$\dot{q}_\varphi \cdot F - \dot{q}_{\varphi+d\varphi} \cdot F + \dot{q}'' \cdot dA_o - \dot{q}'(z,\varphi) \cdot dA_i = 0 \qquad (6)$$

Substituting $\dot{q}'(z,\varphi)$ from equ. (3) into equ. (6), taking into account, that $\dot{q}'' = \dot{q}(z) \cdot dA_i/dA_o$ and using Fourier's law

$$\dot{q}_\varphi - \dot{q}_{\varphi+d\varphi} = \frac{2\lambda_w}{d_m} \cdot \frac{\partial^2 T_w(z,\varphi)}{\partial d\varphi^2} \, d\varphi,$$

the following equation is obtained:

$$\frac{\partial^2 T_w(z,\varphi)}{\partial \varphi^2} - \alpha(z,\varphi) \cdot (T_w(z,\varphi) - T_b) \cdot \frac{d_i \cdot d_m}{4\lambda_w \cdot s} + \frac{\dot{q}(z) \cdot d_i \cdot d_m}{4 \cdot \lambda_w \cdot s} = 0 \qquad (7)$$

3. CALCULATIONS WITH CONSTANT HEAT TRANSFER COEFFICIENTS α_g AND α_l FOR THE DRY AND WET REGION

If the heat transfer coefficients of the dry and the wet parts of the surface are assumed to be constant and if the angular position of the point of separation φ_{sep} is known, equ.(7) can be solved as described in /1/. The most interesting result is that the perimeter averaged heat transfer coefficients resulting from the two thermal boundary conditions can be compared directly:

$$\Psi = \frac{\alpha(z)_{\Delta T = const}}{\alpha(z)_{\dot{q} = const}} \qquad (8)$$

If the tangential heat conduction is infinite, Ψ will be unity and if the tangential heat conduction is zero, Ψ reachs maximum values, which can be calculated from equ.(4) and (5). These values are marked in Fig. 4 as a dotted line at the right axis. In all other cases Ψ is a function of the variables M, C and X, which are defined in Fig. 4.

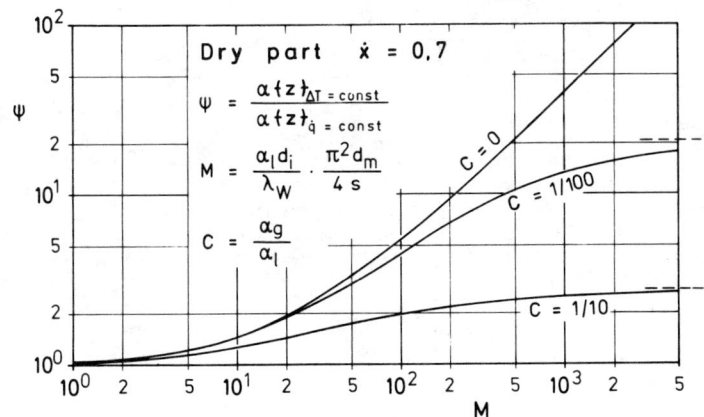

Fig. 4: Influence of the thermal boundary condition on the perimeter averaged heat transfer coefficient

In Fig. 4 the values of Ψ are plotted vs. M with C as parameter. The unwetted part $X = A_g/(A_g+A_l)$ is taken to be 0.7. M is calculated for different values of α_l and λ_w as seen from table 1.

Table 1: M for different values of λ_w and α_l with s = 3 mm and d_i = 14 mm

Material	$\lambda_w \frac{W}{mK}$	$\alpha_l \frac{W}{m^2 K}$			
		1000	3000	10000	20000
Copper	314	0,624	1,88	6,24	12,47
Steel, Nickel	60	3,262	9,79	32,62	65,25
Stainless steel	20	9,787	29,36	97,87	195,75

For a copper tube the values of M lie between 0.6 and 12.47. The values of Ψ, obtained from Fig. 4, are between 1 and 1.4. That implies, that for the chosen copper tube the influence of the thermal boundary conditions (ΔT = const, $\dot{q}$ = const) is insignificant on the perimeter averaged heat transfer coefficient. From Fig. 4, follows further, that in case of evaporation with $\dot{q}$ = const, the perimeter averaged heat transfer coefficient obtained in tubes of material having a low thermal conductivity, like stainless steel, can be much lower than those obtained with ΔT = const. That means also, that there will be a large temperature difference between the top and the bottom of the tube.

4. DETERMINATION OF LOCAL HEAT TRANSFER COEFFICIENTS FROM MEASURED WALL TEMPERATURES FOR EVAPORATION OF R12

Fig. 4 shows the solution of the differential equ.(7) for constant values of α_l and α_g. Now this equation will be used to determine the local heat transfer coefficient from measured wall temperatures. Rearanging of equ.(7) leads to:

$$\alpha(z,\varphi) = \frac{\dot{q}(z) + \frac{4 \cdot s \cdot \lambda_w}{d_i \cdot d_m} \frac{\partial^2 T_w(z,\varphi)}{\partial \varphi^2}}{T_w(z,\varphi) - T_b} \qquad (9)$$

The calculations are based on measured temperature distributions along the tube perimeter for evaporation of R12, which were carried out by Iwicki /2/. Fig. 5 shows the variation of the temperature difference vs. the perimeter angle φ for different heat fluxes. The curves are symmetrical and show horizontal tangents at angles φ = 0, π and 2π.

The second derivative of the wall temperature with respect to the angular position is obtained by graphical differentiation as shown in Fig. 6. (The upper curve of Fig. 5 is used in this example.)

The local heat transfer coefficient $\alpha(z,\varphi)$ at the bottom of the tube is much higher than the perimeter averaged heat transfer coefficient and even increases with increasing angle. At the point of inflection of the temperature distribution $\alpha(z,\varphi)$ rapidly decreases to very low values. Therefore it is assumed that the point of inflection is identical with the point of separation.

The utilization of other temperature distributions resulting from lower heat fluxes shows, that $\alpha(z,\varphi)$ is always very low in the upper part and will be set to zero for the following calculations. The local heat transfer coefficient in the wet proved to be a function of the local temperature difference $T_w(z,\varphi) - T_b$ only, as shown in Fig. 7.

SATURATED FLOW BOILING

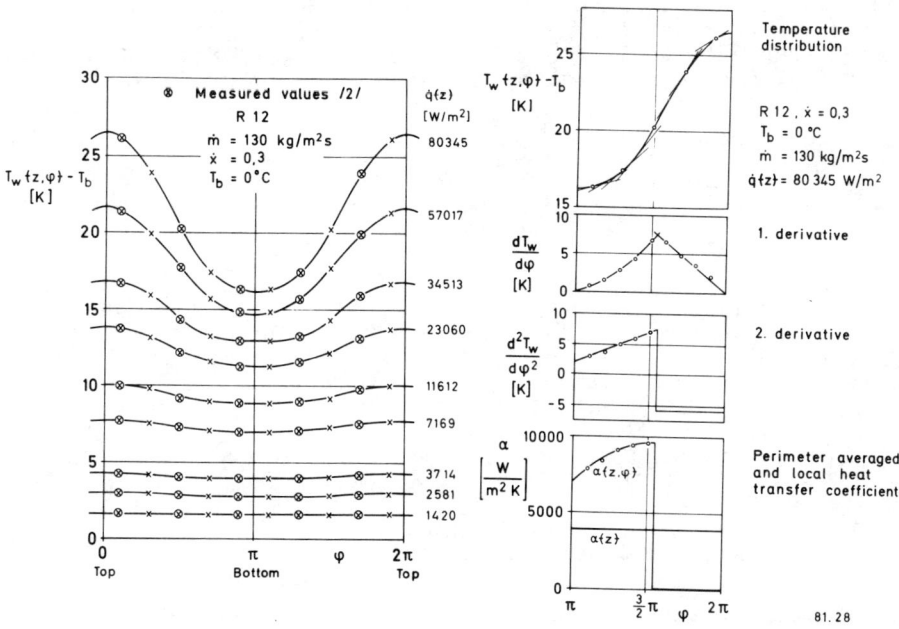

Fig. 5
Variation of temperature distribution vs. the perimeter angle for different heat fluxes

Fig. 6
Determination of local heat transfer coefficients from measured wall temperatures

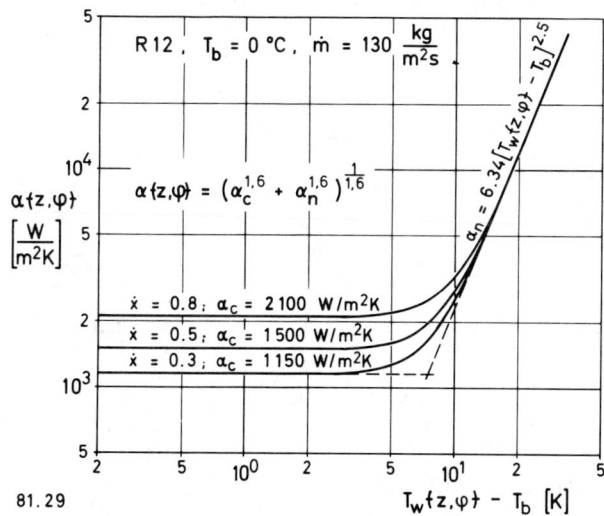

Fig. 7: Local heat transfer coefficient at the wet part as a function of the local temperature difference

For low values of $T_w(z,\varphi) - T_b$ (< 5K) $\alpha(z,\varphi)$ is constant (convective boiling c). At higher values of $T_w(z,\varphi) - T_b$ the local heat transfer coefficient increase with increasing temperature difference (nucleate boiling n). By superposition of the two regimes a description of $\alpha(z,\varphi)$ may be put in the following form:

$$\alpha(z,\varphi) = (\alpha_c^{1.6} + \alpha_n^{1.6})^{1/1.6} \qquad (10)$$

5. DETERMINATION OF THE WALL TEMPERATURE DISTRIBUTIONS BY USING THE LOCAL HEAT TRANSFER COEFFICIENTS

For the determination of the wall temperature distributions by using $\alpha(z,\varphi)$ a numerical method is applied. The local heat transfer coefficient in the wet part of the tube can be obtained from equ.(10), while that of the dry part is assumed to be zero, as described above. Adding one of the three following conditions, the temperature distribution can be determined:

a) Given point of separation (see chap. 6 and 7)

b) Given wall temperature at a certain position

c) Timing at a minimum deviation between measured and calculated wall temperatures.

With the foregoing assumptions and the last condition the wall temperature distribution will be calculated. Because of symmetry, consideration of one half of the perimeter is sufficient. This arc is devided into 100 segments. With a starting temperature at the bottom of the tube, the local heat transfer coefficient (equ. 10), the local heat rate and the derivative of the temperature distribution with respect to the angle at the boundary of the segment are calculated. Using the calculated temperature at the boundary of the first segment, the procedure will be repeated for the second one and so on.

As $\alpha(z,\varphi)$ is zero in the dry part of the tube, the whole electrical power must be transferred to the wet part. That means, that the sum of heat rates in the wet part must be equal to the total power input. With that conditions the end of heating zone can be predicted. The temperature distribution between the end of heating zone and the top of the tube (non heating zone) will be calculated following the same procedure, however, with $\alpha(z,\varphi) = 0$. After the first loop, the starting temperature at the bottom is to be changed until one of the conditions a), b) or c) is reached.

Testing the method the wall temperature distributions in Fig. 5 have been recalculated as shown in Fig. 8. The agreement between the measured wall temperature differences (symbol o) and the calculated values (full lines) is accurate. Also the calculated ends of heating zones (symbol Δ) agrees well with the points of separation, which are connected with a dotted line.

In Fig. 5 one observes, that the points of inflection and, in Fig.8, the end of the heating zones move towards the bottom of the tube when the heat flux increases. The commonly used flow pattern maps /3,4,5/ do not contain that effect. Therefore it is impossible to determine the point of separation in a heated tube by those flow charts.

SATURATED FLOW BOILING

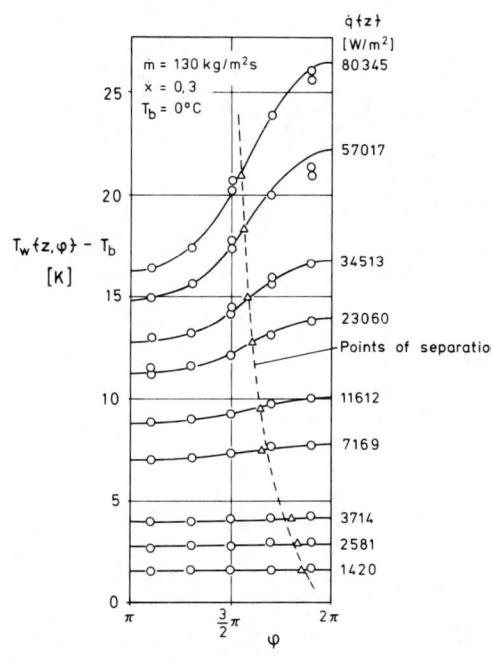

Fig. 8
Measured and calculated temperature differences along the tube perimeter

6. INFLUENCE OF THE WALL THICKNESS ON THE PERIMETER AVERAGED HEAT TRANSFER COEFFICIENT

In the literature no measured temperature distributions at constant parameters $\dot{m}$, $\dot{x}$, T_b, d and $\dot{q}$, however with various wall thicknesses or different thermal boundary conditions are to be found. So, the following calculations will lead to theoretical predictions. The experimental verification might be expected from future research.

For given parameters $\dot{m}$, T_b, $\dot{x}$ and d as in Fig. 8 the temperature distribution for a copper tube having a wall thickness of s = 5 mm will be calculated. Therefore, it is assumed, that the local heat transfer coefficient for the wet part can be calculated from equ. (10) and that of the dry part can be taken as zero. Furthermore it is assumed, that the point of separation depends on the local temperature difference as in Fig. 8. The heat flux is chosen to be $\dot{q}\{z\}$ = 80 345 W/m^2.

The results of the calculations using different starting temperatures are shown in Fig. 9. The true wall temperature distribution can be determined from the intersection between the curves through the point of separation and that through the end of heating zone.

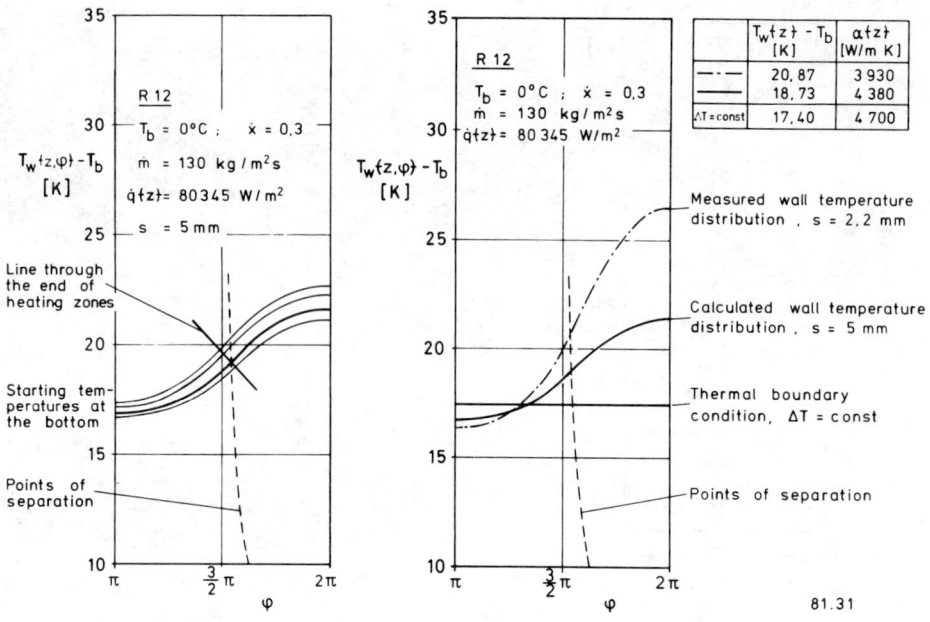

Fig. 9
Calculated temperature distribution in a copper tube.

Fig. 10
Influence of s on the perimeter averaged heat transfer coefficient

In Fig. 10 the calculated temperature distribution (s = 5 mm), the measured temperature distribution (s = 2.2 mm) and the case of heat transfer with the thermal boundary condition ΔT = const are presented. The temperature difference in the last case is calculated, so that the amount of heat transfer in the wet area with $\alpha(z,\varphi)$ according equ. (10) is equal to the total heat rate supplied to the tube.

It can be observed in Fig. 10, that the wall temperature distribution approaches a constant value as the wall thickness increases. The perimeter averaged temperatur differences and the corresponding $\alpha(z)$ are listed in Fig. 10. The same effect of s on $\alpha(z)$ can be observed in Fig. 4. For constant values of X and C the value of Ψ approaches unity if the parameter M becomes low by enlarging s. On the orther hand Ψ increases with increasing M caused by decreasing wall thickness.

7. INFLUENCE OF THE TUBE WALL THICKNESS ON THE CRITICAL HEAT FLUX

It can be seen from Fig. 10, that at the end of heating zone the wall temperature increases with decreasing wall thickness. Out of experiments it is known, that by overstepping the critical temperature difference, respectively the critical heat flux, the

SATURATED FLOW BOILING

heat transfer coefficient decreases suddenly. Therefore, the range of validity of equ. (10) must be bounded by a local critical temperature difference or by a local critical heat flux, resp. The critical heat flux in pool boiling can be calculated by using an equation of Kutateladze /6/.

$$\dot{q}_{cr} = K \cdot \Delta h_{g,l} \cdot \sqrt{\rho_g} \cdot \sqrt[4]{\sigma \cdot g (\rho_l - \rho_g)} \qquad (11)$$

With $K = 0.16$ for R12 /7/ and the physical properties according to the boiling temperature of 0 °C, one obtains $\dot{q}_{cr}$ = 370 000 W/m². Assuming, that $\dot{q}_{cr}$ according to equ. (11) is not affected by the flow velocity, this value should be true for the bottom of the tube.

In pool boiling $\dot{q}_{cr}$ for an inclined plane decreases with increasing angle of inclination ϕ. According to Vishnev /8/ this effect can be described by

$$\dot{q}_{cr}(\phi) = \dot{q}_{cr} \cdot \sqrt{\frac{190° - \phi}{190°}} \qquad (12)$$

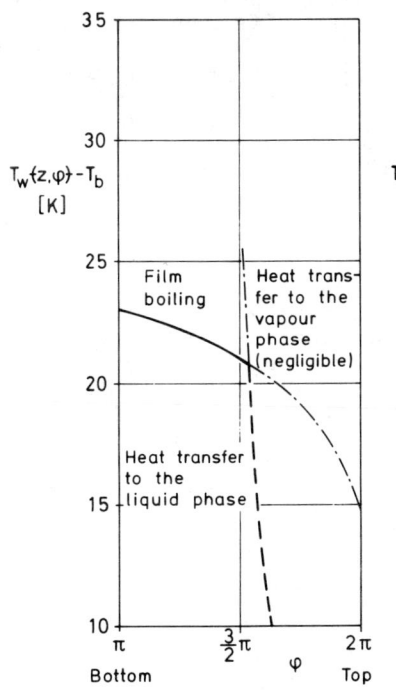

Fig. 11
Delimination of heat transfer to the liquid phase

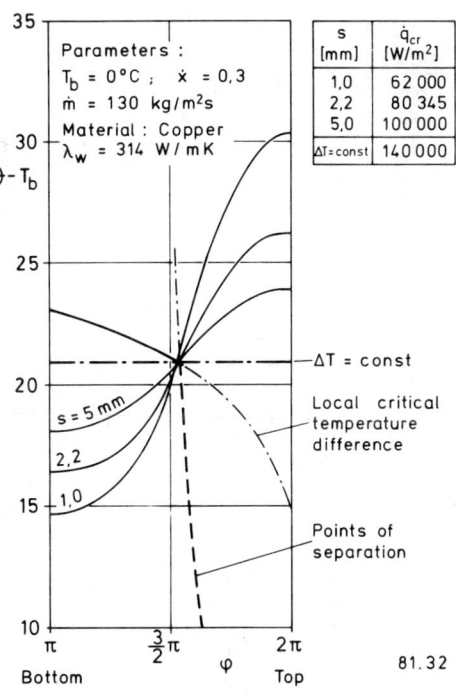

Fig. 12
Influence of tube wall thickness s on the critical heat flux

Assuming that the local critical heat flux at a tube wall element is as high as $\dot{q}_{cr}(\phi)$ of the corresponding inclined surface, it can be determined at any angular position. With equ.(10) one obtains the local critical temperature difference. The curves through the local critical temperature differences and the points of separation are plotted in Fig. 11. By that, the area of stable nucleate boiling in the wet part of the tube is entirely deliminated.

For a tube wall thickness of s = 5 mm the calculated temperature difference at the end of heating zone is approximately 2 degrees below the local critical temperature difference. Therefore a still higher heat flux seems to be applicable to this tube. The evaluation of the temperature distribution with a higher heat flux can be done by the same procedure as described above. The maximum heat flux is reached, when the end of the heating zone coincides with the point of intersection of the curves connecting the local critical temperature differences and the points of separation. The calculated temperature distributions of copper tubes with wall thicknesses of s = 1, 2.2 and 5 mm and the constant temperature difference for heat transfer with ΔT = const are plotted versus the perimeter angle in Fig. 12.

It can be concluded, that in horizontal evaporator tubes with constant heat flux the critical heat flux increases with the wall thickness. The maximum heat flux can be transferred under the thermal boundary condition of constant temperature difference.

To avoid misunderstandings: Section 7 of this paper is meant to demonstrate the effect of the wall temperature distribution due to tangential heat conduction in the wall on the critical heat flux. Though there are a large number of profound investigations of the critical heat flux phenomenon in the literature /9,10,11/, it seems to be worthwhile to point out this effect at some detail.

8. CONCLUSIONS

In this paper a local heat transfer coefficient as a function of the perimeter angle is defined. In horizontal flow boiling with constant heat flux this local heat transfer coefficient can be evaluated from measured wall temperature distributions. In the upper part of the tube, which seems to be -"thermally"- dry, the local heat transfer coefficient is negligible and assumed to be zero. At the bottom of the tube the local heat transfer coefficient is much higher than the perimeter averaged value. For constant parameters $\dot{m}$, x and T_b it is a function of the local temperature difference only!

The calculations for horizontal flow boiling yields a decreasing critical heat flux as well as decreasing perimeter averaged heat transfer coefficients with decreasing tube wall thickness. If the heat supply comes from condensing vapour at the outer surface, the thermal boundary condition ΔT = const seems to be achieved. Nevertheless the tangential heat conduction must be taken into account, if the maximum value of the local heat transfer coefficient at the evaporating side is at the same angular position as the minimum value of the heat transfer coefficient at the condensing side. For the lay out of an evaporator one should use local

SATURATED FLOW BOILING

heat transfer coefficients and take the tangential heat conduction into consideration.

REFERENCES

1. Bonn, W. 1980. Wärmeübergang und Druckverlust bei der Verdampfung von Stickstoff und Argon. Dissertation, Universität Karlsruhe.

2. Bonn, W., Iwicki, J., Krebs, R. e.a. 1980. Über die Auswirkung der Ungleichverteilung des Wärmeübergangs am Rohrumfang bei der Verdampfung im durchströmten waagerechten Rohr. Wärme- und Stoffübertragung 14, P. 31-42.

3. Baker, O. 1954. Multiphase flow in pipelines. Oil and Gas J. 53, Nr. 7, P. 185-195.

4. Mandhane, J.M., Gregory, G.A. and Aziz, K. 1974. A flow pattern map for gas-liquid flow in horizontal pipes. Int. J. Multiphase Flow Vol. 1, P. 537-553

5. Steiner, D. 1980. Zweiphasenströmung in Leitungen. Hochschulkurs Wärmeübertragung II, Teil III. Forschungsgesellschaft Verfahrenstechnik Düsseldorf.

6. Kutateladze, S.S. 1963. Fundamentals of heat transfer. Chap. 18.2, Edward Arnold Ltd. London.

7. Fink, D. 1968. Einfluß der Geometrie und der Materialeigenschaften der Heizfläche auf die Siedekrisis bei freier Konvektion. Bundesministerium für wiss. Forschung, Bericht K 68-05.

8. Vishnev, J.P. e.a. 1976. Study of heat transfer in boiling of Helium on surfaces with various orientation. Heat Transfer - Soviet Research 8, Nr. 4, S. 104-108.

9. Hewitt, G.F. e.a. Burnout and film flow in the evaporation of water in tube, AERE R 4864, Chemical Engineering Division, Harwell, March 1965.

10. Hewitt, G.F. 1964. A method of representing burnout data in two phase heat transfer for uniformly heated round tubes, AERE - R 4613.

11. Hewitt, G.F. Mechanism and prediction of burnout from S. Kakac, T.N. Veziroglu, Two phase flows and heat transfer, Vol. II, Hemisphere Publishing Corporation.

NOMENCLATURE

A	surface area	$\dot{q}''$	heat flux supplied to the outer surface
c	defined as α_g/α		
d	diameter	T	temperature
F	intersecting area	X	unwetted part of the surface
g	acceleration of gravity		
$\dot{m}$	mass velocity	$\dot{x}$	flow quality
$\dot{Q}$	heat rate	z	axial coordinate
$\dot{q}$	heat flux		
$\dot{q}'$	heat flux transferred to the fluid		

Greek letters

α	heat transfer coefficient
Δ	difference
$\Delta h_{g,l}$	phase change enthalpy
λ	thermal conductivity
ρ	density
σ	surface tension
ϕ	angle of inclination
φ	perimeter angle
Ψ	defined as $\alpha(z)_{\Delta T=const}/\alpha(z)_{\dot{q}=const}$

Subscripts

b	boiling
c	convective boiling
cr	critical
g	gas phase
i	inner side
l	liquid phase
m	mean value
n	nucleate boiling
o	outer side
w	tube wall

Film Condensation and Evaporation in Vertical Tubes with Superimposed Vapor Flow

FRANCISCO BLANGETTI and RAINER KREBS
Institut für Thermische Verfahrenstechnik
Universität Karlsruhe, FRG

ABSTRACT

Local heat transfer coefficients in film condensation inside a vertical tube are reported. A new model for heat transfer in turbulent film flow is presented, based on eddy viscosity profiles suggested by Levich for absorption of gases into films and by van Driest for one phase pipe flow. This model predicts the film condensation data and in particular the shift from laminar to turbulent heat transfer with the film Reynolds number very well. The model demonstrates the equivalence of film evaporation and film condensation in good agreement with experimental findings.

1. INTRODUCTION

Film condensation and evaporation inside vertical tubes is a field of very intensive theoretical and experimental investigation according to its practical importance. Laminar film condensation was analysed first by Nusselt /1/. For turbulent films several models were developed /2,3/. Most of these models are based on the assumption that momentum transfer in a turbulent film can be predicted by equations used sucessfully in turbulent one-phase pipe flow. Complementary to these so called "wall" models Levich /4/ supposed that the turbulent structure of two-phase film flow near the liquid/gas (L/G) interface cannot be identical with the structure of one-phase pipe flow. From theoretical considerations he concluded that the surface tension damps the turbulent motion near the free L/G interface. This basic idea was used /5/ to build up a Levich-type model for heat transfer in a falling liquid film in condensation and evaporation. Predictions by this model are compared to own and previous data both for film condensation and film evaporation.

To check their models the authors /2,3/ had to refer to experiments carried out with tube lengths between 0.5 and several meters /6 to 16/. In a long tube the vapour velocity varies from the maximum value at the inlet to zero at the outlet while the film mass flow rate varies accordingly in the opposite direction. Using such an experimental technique the complicated history of condensation along the tube is represented by the mean heat transfer coefficient only. Different local heat transfer coefficients

Francisco Blangetti's present address is Brown, Boveri and Cie, Baden, Switzerland.

due to different hydrodynamic conditions are levelled out. Therefore the information obtained from integral measurements are not adequate to analyse the transfer mechanisms in films particularly in the presence of significant shear stresses originated by the vapour flow. A more reliable starting point for such kind of analyses would be experimental data of local heat transfer coefficients obtained by independent variation of the external variables. Therefore an major objective of this contribution is to provide such class of information. Therefore the test section used in this investigation was a short segment of a long condenser tube. By independent variation of vapor and liquid flow any position in a long tube could be simulated.

2. TEST EQUIPMENT AND TEST PROCEDURE

A schematic diagram of the test facility is shown in fig. 1. The test unit (TU) was built up of a calming section of 3 m length and the test section (TS, pos.1) of 200 mm length. A precision metering pump (pos.9) recirculated the condensate. The recirculated liquid was preheated up to saturation with a PI-controlled heat exchanger (pos.11) before entering the top of the calming section (pos.4). Additionally the calming section was provided with a large heat exchange surface to equalize remaining small temperatures differences between vapour and liquid. By means of appropriate holes distributed on the perimeter of the tube the saturated liquid flowed inside the tube as a falling film cocurrent with the saturated vapour.

1 Test section
2 Total condenser
3 Condensate collector
4 Calming section
5 Boiler
6 Cyclohexan vapour cond.
7 Coolant collector
8 Overflow
9 Metering pump
10 Volumetric flow-meter
11 PI-controlled 3-way-valve
12 Circulation pump
13 Preheater

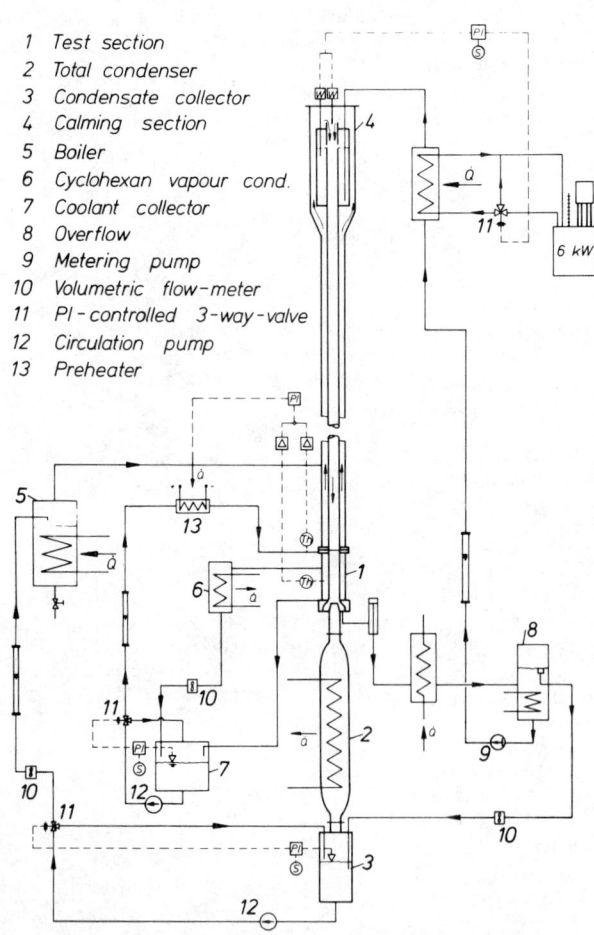

Fig.1: Schematic diagram of the test unit.

FILM CONDENSATION AND EVAPORATION IN VERTICAL TUBES

The vapour flow was generated in a glass boiler (pos.5) with 3 bar saturated steam. The vapour flowed upwards through the annulus formed by the jacket surrounding the calming section. This arrangement insured an isothermal tube wall. At the top of the calming section vapour entered into contact with the falling film.

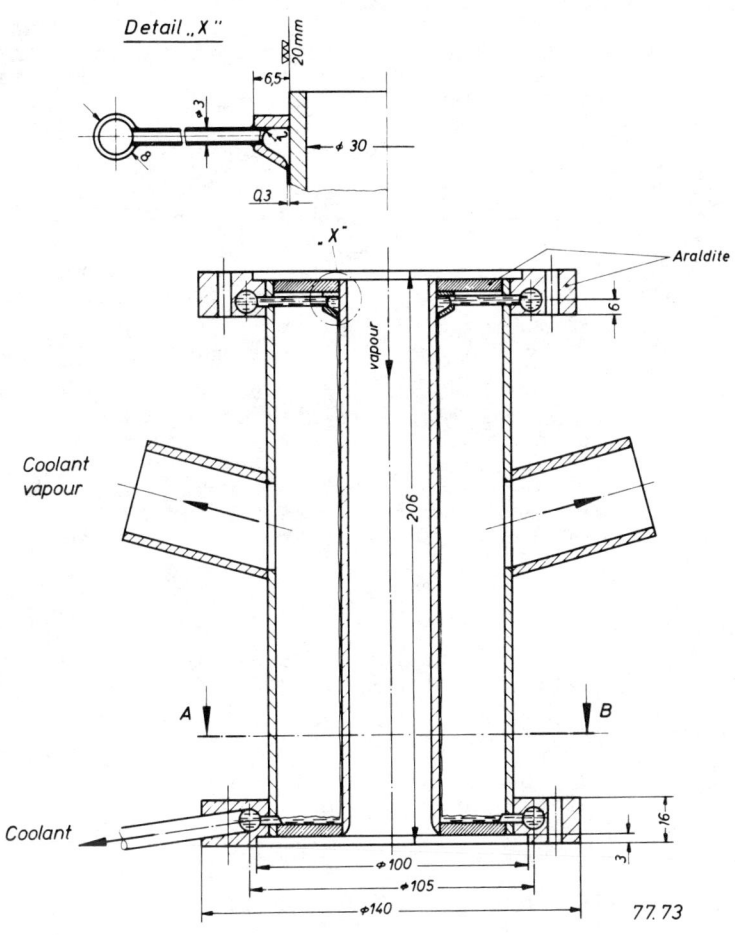

Fig. 2: Test-section

The TS consisted of a coaxial double tube (Fig.2) linked to epoxiresin (Araldite) flanges of 5 mm thickness. The flanges reinforced with several glassfiber mats was casted "in situ". As coolant saturated cyclohexan C_6H_{12} was pumped through a ring nozzle flowing down the outer side of the copper tube as a boiling film. With the vaporized mass of C_6H_{12} the heat exchanged with the test fluid inside the tube could be determined. The TS was provided with 7 thermocouples (TC) Ni/NiCr of 0.1 mm. The hot joints of five of them were located 0.15 mm from the inner wall. They were arranged helically, thus yielding an average wall temperature.

The C_6H_{12} vapour was fully condensed (pos.6). The liquid flows down by gravity to the container (pos.12). The mass flow rate of the coolant was measured with a rotameter and subsequently preheated nearly up to the saturation temperature by a controlled heat exchanger (pos.13).

Behind the TS the liquid was separated from the vapor and returned to the reciever (pos.8). The condensate in the TS was continuously removed by means of an overflow, thus keeping the liquid hold-up of the falling film loop constant at steady state conditions. From there the liquid returns to the receiver (pos.3). The vapour leaving the exit of the TS was fully condensed in a glass condenser. It also flows to the receiver at pos. 3.

A teflon pump circulated the condensate to the boiler. To keep the level of the liquid in the receiver at a constant height the evaporation loop was by-passed with of a pneumatic controlled 3-way valve. Before entering the boiler the condensate was measured with a precision flow meter. The vapour temperature was measured at the exit of the TS and the entry of vapour at the top of the TU by means of Ni/NiCr TC.

With 11 additional TCs, stream temperatures at different points of the TU were measured. The TC were connected to a data acquisition system (DAS) provided with a 0.1 μV resolution digital voltmeter with variable integration time (Solartron), printer and puncher. Additional 5TCs were connected to a multichannel recorder for visual observation of the state of the test unit. All TC were calibrated by means of a certificated Pt-resistance thermometer.

After 3 hours the TU reached steady state and the measurement routine at the DAS could be activated. The duration of a run was 40-45 min. Each channel was read out about once a minute. At the end of a run the information not recorded (barometric pressure, flow meters, runtime etc.) was typed on the keyboard of the DAS. Each run contained approximately 1000 samples. The tapes were processed in the computer centre of the university. Every run could be checked by means of the heat balance of the test section. For that purpose the mass flow rates of condensed vapour of the test-substance and of generated vapour in the cooling loop were compared.

3. ANALYSIS OF LAMINAR AND TURBULENT FILM FLOW

Using the turbulent viscosity concept the boundary layer equation for the conservation of momentum is written

$$u \frac{\partial u}{\partial x} + v \frac{\partial u}{\partial y} = \frac{\partial}{\partial y} (\nu_l + \varepsilon) \frac{\partial u}{\partial y} + (1 - \rho_G/\rho_l)g \qquad (1)$$

where ε, as introduced by Boussinesq, is the eddy viscosity. Neglecting the convective terms in eq.(1) /19/ and assuming a linear shear stress profile /20/, integration from the interface y=s to y yields the shear stress profile (see Fig.3)

$$\tau = g\rho_l (1 - \frac{\rho_G}{\rho_l})(s-y) + \tau_G \qquad (2)$$

FILM CONDENSATION AND EVAPORATION IN VERTICAL TUBES

Together with the Boussinesq equation

$$\tau = \rho_1 (\nu_1 + \varepsilon) \frac{du}{dy} \quad (3)$$

and the definition of the film Reynolds number

$$Re_F = \frac{1}{n} \int_0^s \rho_1 u(y) dy \quad (4)$$

integration and rearrangement yields an equation for calculating the film thickness:

$$Re_F = \int_0^s \int_0^y \frac{1}{\rho_1 \nu_1 (\nu_1 + \varepsilon)} \left[\tau_G + (1 - \frac{y}{s})(\tau_W - \tau_G) \right] dy\, dy \quad (5)$$

τ_G and τ_W are the shear stresses at the L/G interface (y=s) and at the wall (y=o) respectively. By definition of

$$y^+ = (y/\nu_1) u \quad (6); \qquad s^+ = (s/\nu_1) u \quad (7)$$

$$u_* = \tau_W/\rho \quad (8); \qquad \varepsilon^+ = 1 + \frac{\varepsilon}{\nu_1} \quad (9)$$

$$\tau_G^* = \frac{\tau_G}{g\rho_1 (1 - \frac{\rho_G}{\rho_1}) L} \quad (10); \qquad \tau_W^* = \frac{\tau_W}{g\rho_1 (1 - \frac{\rho_G}{\rho_1}) L} \quad (11)$$

$$\mathcal{L} = \sqrt[3]{\frac{\nu_1^2}{g}} \quad (12)$$

Eq.(5) can be normalized:

$$Re_F = \int_0^{s^+} \int_0^{y^+} \frac{1}{\varepsilon^+} \left[\frac{\tau_G^*}{\tau_W^*} + (1 - \frac{y^+}{s^+})(1 - \frac{\tau_G^*}{\tau_W^*}) \right] dy^+ dy^+ \quad (13)$$

Assuming thermodynamic equilibrium at the L/G interface /21/ and neglecting energy accumulation in the film the energy equation

$$\dot{q}_x = -\rho_1 c_{p1} (a_1 + \varepsilon_W) \frac{d\vartheta}{dy} \quad (14)$$

can be written in a similar way:

$$Nu_x = \frac{Pr\, s^{+1/3}}{F_2} \left\{ \frac{1 - \rho_G/\rho_1}{\frac{s^*}{s^* + \tau_G^*}} \right\}, \quad \text{with } F_2 = \int_0^{s^+} \frac{dy^+}{\frac{1}{Pr} + \frac{1}{Pr_t}(\varepsilon^+ - 1)} \quad (15)$$

The turbulent Prandtl number Pr_t is defined as:

$$Pr_t = \varepsilon/\varepsilon_w \qquad (16)$$

and the Nusselt number Nu_x as

$$Nu_x = \frac{\alpha_x L}{\lambda} \quad . \qquad (17)$$

The dimensionless film ticknesses s^+ and $s^* = s/\ell$ are interrelated by

$$s^{+2} = (1 - \rho_G/\rho_l)(s^{*3} + s^{*2} \tau_G^*) \qquad (18)$$

For $\varepsilon^+ = 1$, i.e. laminar flow eq.(13) and (15) yield the well known solution of the Nusselt theory (Fig.3):

$$Re_F = (1 - \rho_G/\rho_l)(\tfrac{1}{3} s^{*3} + \tfrac{1}{2} \tau_G^* s^{*2}) \qquad (19)$$

and

$$Nu_x = \frac{1}{s^*} \qquad (20)$$

For the complete specification of a steady state turbulent model now an appropriate ε^+-profil and a specified turbulent Prandtl-number have to be selected.

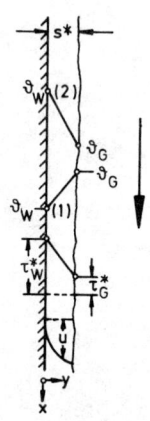

Fig. 3
Profiles in a falling film; temperature profile for the laminar case in (1) condensation and (2) evaporation.

A classical approximation of ε^+ and Pr_t is Rohsenow's model /3/. He assumes $Pr_t = 1.0$ and extrapolates the validity of the universal velocity profil of Prandtl-Nikuradse for turbulent pipe flow to turbulent film flow. This yields the proportionality

$$Nu_x \sim Re_F^{0.18} \quad . \qquad (21)$$

Levich /4/ pointed out that "wall"-models are not suitable to describe the hydrodynamic conditions at the liquid/gas interface. Levich, and later Davis /22/, suggested a model considering the damping of eddies at the interface by surface tension. Because of this a viscous sublayer is built up near the interface, restraining the transport of heat and mass. This model of Levich with the simple form

$$\varepsilon \sim \frac{(s-y)^2}{\sigma} \qquad (22)$$

was sucessfully applied to falling film absorption data /23,24/. The existence of a region with damped eddies at the interface was confirmed experimentally by interferometric measurements of CO_2-concentration profiles in a falling water film /18/.

FILM CONDENSATION AND EVAPORATION IN VERTICAL TUBES

Mills /24/ applied a van Driest ε^+-profile at the wall region and a Levich-type profile for the interface. He obtained good agreement with the experimental data of Chun and Seban /25/ for evaporation heat transfer without shear forces at the interface. In a later publication /30/ Mills desisted from Levich's approach. Borodin /26/ used such a profile and was able to describe the temperature measured on the free interface of a heated falling water film satisfactorily.

In this work, as in earlier papers /27,28/, an alternative Levich-type model is presented, derived from the experimental data of Lamourelle /23/ of absorption of gases into turbulent falling films:

$$\varepsilon^+ = 1 + 0.0161 \, Ka^{1/3} Re_F^{1.345} \left[\tau_G^* + s^* (1 - \frac{y^+}{s^+})\right] (s^+ - y^+) \qquad (23)$$

Eq.(23) includes both the damping of eddies by surface tension in form of the Kapitza number Ka similar to /24/

$$Ka^{1/3} = \frac{g \cdot \rho_1 \, L^2}{\sigma} \qquad (24)$$

and the effect of shear forces caused by the vapour flow. Eq.(23) is confined to small values of τ_G^*/τ_W^*, because appropriate experimental information on the gas absorption or desorption into turbulent films with significant gas velocities is lacking.

As corroborated by the electrolytic mass transfer experiments of Iribane /29/, the transport properties close to the wall can be described well by means of ε^+-profiles such as the van Driest profile. Patankar and Spalding /17/ recommend the use of this equation in the "damped" form:

$$\varepsilon^+ = \frac{1}{2} + \frac{1}{2} \sqrt{1 + 0.64 \, y^{+2} \tau^+ \left[1 - \exp(-\frac{y^+}{26} \, \tau^+)\right]^2} \qquad (25)$$

The range of validity of eq.(25) starts at the wall and for eq. (23) at the interface and ends for both equations at the intersection point (see Fig.4). To complete the model $Pr_t = 0.9$ is assumed following the recent practice in boundary layer theory.

In the calculation procedure first the film thickness s^+ has to be evaluated. For that purpose eq.(13) with eq.(23) and (25) was solved numerically by means of an integration procedure according to Romberg and accelerated by means of the Raphson-Newton technique. With s^+ the Nusselt number Nu_x could be determined by numerical integration of F_2 according to eq.(15).

As an example Fig. 4 shows the ε^+, u^+ and temperature profiles suitably normalized for a typical turbulent film. From the second viscous sublayer near the interface results a point of inflection in the temperature profile, which is a characteristic feature of this model.

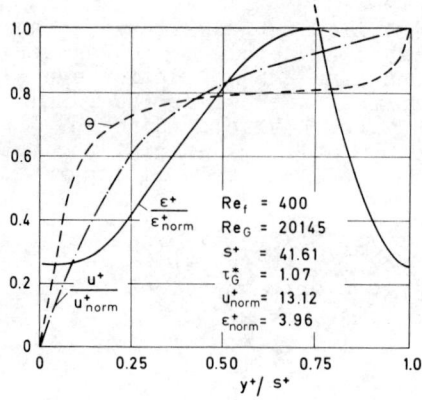

Fig. 4:
Viscosity, velocity and temperature profile in a condensation film (calculated).

With this model in fig. 5 Nu_x versus Re_F with τ_G^* as a parameter is plotted for water and MWA (Methoxi-isopropanol-Water aceotrope) at their normal boiling points. These substances differ mainly in the Prandtl and Kapitze numbers respectively. The Levich van Driest model includes the Nusselt theory for low Reynolds numbers, i.e. in the laminar regime (left hand assymptote) and predicts a strong increase of Nu_x

$$Nu_x \sim Re_F^{0.4} \qquad (26)$$

in the turbulent regime (right hand assymptote). The laminar and turbulent regimes are connected by a wide transition regime, characterized by a minimum in Nu_x. This minimum shifts to lower Re_F with increasing Ka, Pr, resp.

Since the model cannot be programmed on a computer very easily, the $Nu_x(Re_F, \tau_G^*)$-function was approximated in form of power laws:

$$Nu_x = (Nu_{x,lam}^4 + Nu_{x,t}^4)^{1/4} \qquad (27)$$

Nu_x results from the superposition of the laminar and turbulent asymptotes $Nu_{x,lam}$ and $Nu_{x,t}$. $Nu_{x,lam}$ can be calculated from eq.(19) and (20), $Nu_{x,t}$ from

$$Nu_{x,t} = a \, Re_F^b \cdot Pr^c (1+d \, \tau_G^{*e}) \qquad (28)$$

The coefficients a, b, c, d and e were evaluated by means of regression analysis of $Nu_x(Re_F, \tau_G^*)$ functions for 6 substances with Pr-numbers from 1 to 20. To obtain better fitting the shear stress values were devided into four ranges:

τ_G^*	a	b	c	d	e
0	$8.663 \cdot 10^{-3}$	0.3820	0.5689		
$0 < \tau_G^* < 5$	$8.663 \cdot 10^{-3}$	0.3820	0.5689	0.1450	0.5410
$5 < \tau_G^* < 10$	$2.700 \cdot 10^{-2}$	0.2071	0.5000	0.4070	0.4200
$10 < \tau_G^* < 40$	$4.294 \cdot 10^{-2}$	0.09617	0.4578	0.6469	0.4730

Tab.: Coefficients for calculation of $Nu_{x,t}$

FILM CONDENSATION AND EVAPORATION IN VERTICAL TUBES

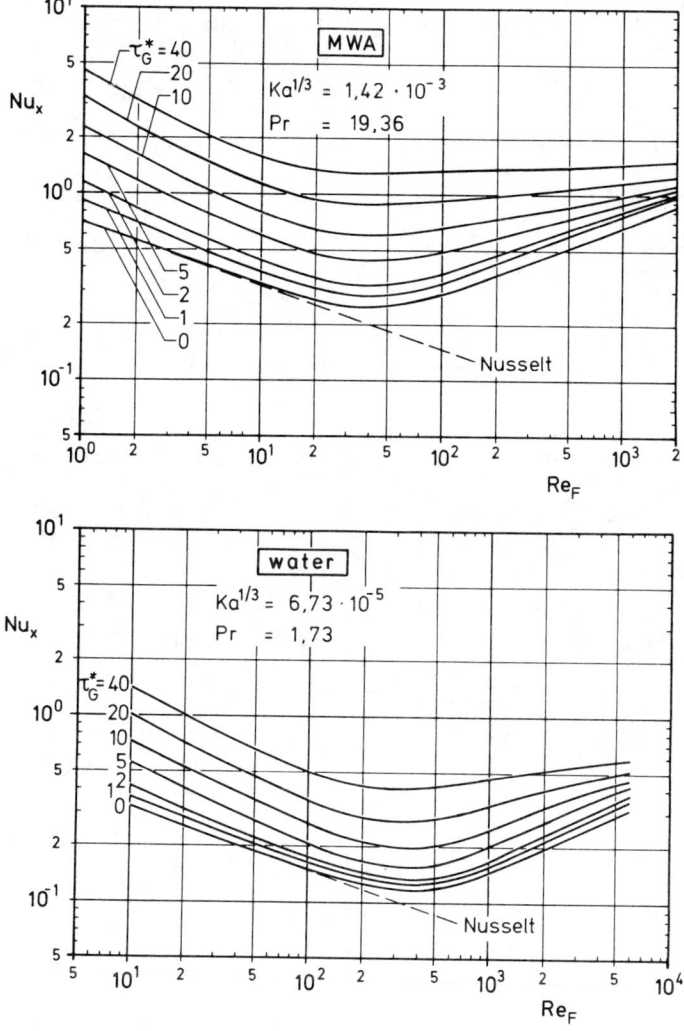

Fig. 5: Local Nusselt numbers calculated by means of the Levich type model.

In this approximation Ka could be omitted, although the Ka-number is a physically important parameter in Levich-type models. This is attributable to an "internal" correlation of Pr and Ka, representable by caused by an empirical correlation:

$$Ka^{1/3} = 5.34 \cdot 10^{-5} \cdot Pr^{1.17} \quad . \tag{29}$$

The shear stress τ_G^* can be evaluated from

$$\tau_G^* = \frac{\frac{f}{2} \rho_G (\bar{U} - u_o)^2}{g(\rho_1 - \rho_G) \mathcal{L}} \tag{30}$$

where $\bar{U}$ is the mean gas velocity and u_o the velocity of the film at the interface. As a rough approximation for calculating Fanning factor f the equation of Prandtl-Nikuradse for one-phase pipe flow may be used.

$$1/\sqrt{f} = 1.737 \ln(Re_G \sqrt{f}) - 0.40 \tag{31}$$

Whereas for more exact calculations the procedere recommended in /28/ should be used. Taking into account two phase flow effects and the distortion of the velocity profil by suction (in the case of condensation), an enhancement of Nu_X up to 50 % at constant Re_G can be expected.

In the comparison of our own experimental data to theoretical predictions, subcooling of the liquid film (thermal entry effect) was taken into account by an appropriate extension of the theory developed above. Therefore, the temperature profile was calculated from

$$u \frac{\partial \vartheta}{\partial x} = -\frac{\partial}{\partial y} - (a + \varepsilon_w) \frac{\partial \vartheta}{\partial y} \tag{32}$$

where ε_w was taken from the Levich-van Driest model. Then the derivative $\frac{\partial \vartheta}{\partial y}$ at the wall was used to calculate the Nusselt number Nu_X.

$$Nu_X = \frac{\mathcal{L}}{L(\vartheta_G - \vartheta_w)} \int_0^L \left.\frac{d\vartheta}{dy}\right|_{w,x} dx \tag{33}$$

$\tilde{Nu}_X$ is the Nusselt number to be expected, applying the test section described above.

4. EXPERIMENTAL RESULTS AND CONCLUSIONS

In fig. 6-11 heat transfer coefficients Nu_X and $\tilde{Nu}_X$ for condensation obtained from the test section described above and calculated values according to the models of Rohsenow and Levich-van Driest are plotted versus the film Reynolds number Re_F:

With water as test fluid 165 runs and with MWA 122 runs were made. Each diagram represents a series with constant gas Reynolds number Re_G. For water these Re_G-values were 7166, 15500 and 25431 (U = 4.8, 10.4 and 15.1 m/sec). Re_F was varied from 20 to 2300 for water and from 1 to 325 for MWA. The reported Re_F and Re_G are arithmetical means between inlet and outlet values of the short test section. The vapour temperature was nearly 100 °C for water and 97.5 °C for MWA, the temperature difference ΔT (wall-vapour) for water remained between 6.6 K and 4.1 K with a mean value of 5.0 K and for MWA between 7.8 K and 4.0 K with a mean value of 6.8 K.

FILM CONDENSATION AND EVAPORATION IN VERTICAL TUBES 167

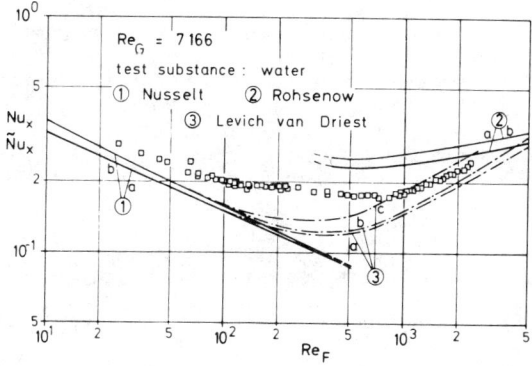

Curve a: $\tau_G^* = 0$; b: τ_G^* is variable; c: as b and thermal entry

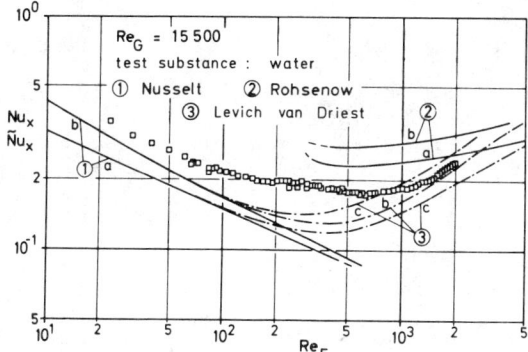

Curve a: $\tau_G^* = 0$; b: τ_G^* is variable; c: as b and thermal entry

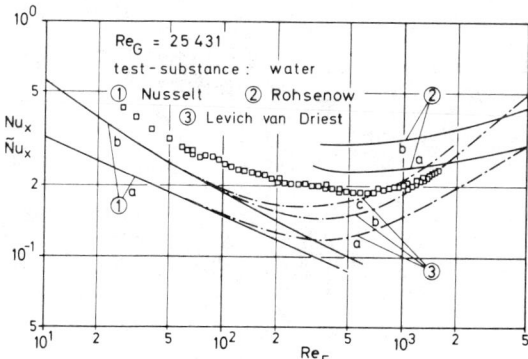

Curve a: $\tau_G^* = 0$; b: τ_G^* is variable; c: as b and thermal entry

Fig. 6-8: Local Nusselt numbers for condensation of water.

The theoretical curves in fig. 6 to 11 are shown according to the Nusselt-model (laminar regime, curve family no.1) to the Rohsenow-model (turbulent regime, curve family no.2) and to the suggested Levich van Driest model (laminar and turbulent regime, curve family no.3). Further differentiation are made: a) represents the basic curve for $\tau_G^* = 0$ ($U = u_o$), while b) represents Nusselt numbers Nu_x, neglecting entry effects for constant Re_G, but variable τ_G^* /28/. c) Represents Nusselt numbers $\tilde{Nu}_x$ taking into account the thermal entry effect in a short test section with τ_G^* according to b).

The following conclusions are drawn from the experimental results: A distinct transition film Reynolds number, which separates laminar heat transfer behavior (negative slope of Nu_x versus Re_F) from turbulent heat transfer behavior (positive slope of Nu_x versus Re_F), does not exist in condensation. The minimum Nusselt numbers are about 400-600 for water and 30-40 for MWA. The same is true for film evaporation. Data taken from /25/ are plotted in fig. 16 together with condensation data. It can be seen that there is no significant difference between film condensation and film evaporation heat transfer.

Comparing experimental and theoretical values one may conclude:

In the laminare type regime the Levich van Driest model and the Nusselt model yield identical values of Nu_x. Thermal entry effects can be neglected ($Nu_x = \tilde{Nu}_x$). The experimental $\tilde{Nu}_x$ values exceed the predictions within a range of 15-20 % at low vapour velocities. This agrees with data and simple correlations of authors /e.g.8/ who ascribe this fact to waves on the surface. Surprisingly, at higher vapour velocities (Re_G = 40 884) the deviation is increased up to 60-80 %.

In the turbulent-type regime thermal entry effects have to be considered. The experiments show a strong dependence of Nu_x on Re_F ($Nu_x \sim Re_F^{0.36-0.40}$). This is predicted by the Levich type model, adjusted for from entry effects, very well (curve 3b). The curves 3b, calculated with Rohsenow's model demonstrate that with wall models for the eddy viscosity the heat transfer in the turbulent regime cannot be predicted adequately. In particular the slope m of the curves $Nu_x \sim Re^m$ is underpredicted (m $\equiv$ 0.20).

The suggested Levich van Driest model is based on semi-empirical correlations for the eddy-viscosity distribution. It is combined from the van Driest equation for turbulent one-phase pipe flow and the Lewich model developed for absorption into turbulent films. Although obtained from physically quite different phenomena the combination of these two yields a theory which matches the experimental data in film condensation without additional fitting parameters. The Levich van Driest model also explains the similarity between film condensation and film evaporation (see fig. 12) and predicts quite accurately the location of the minimum Nusselt numbers.

FILM CONDENSATION AND EVAPORATION IN VERTICAL TUBES

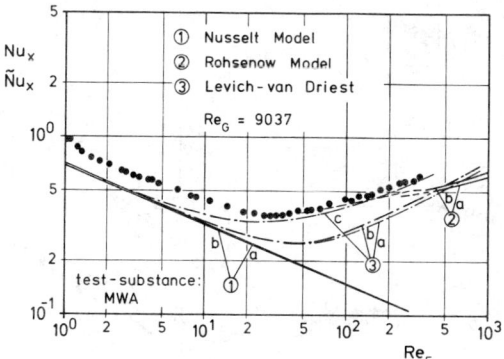

Curve a: $\tau_G^* = 0$; b: τ_G^* is variable; c: as b and thermal entry

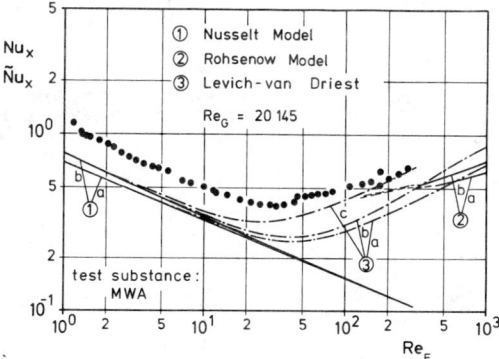

Curve a: $\tau_G^* = 0$; b: τ_G^* is variable; c: as b and thermal entry

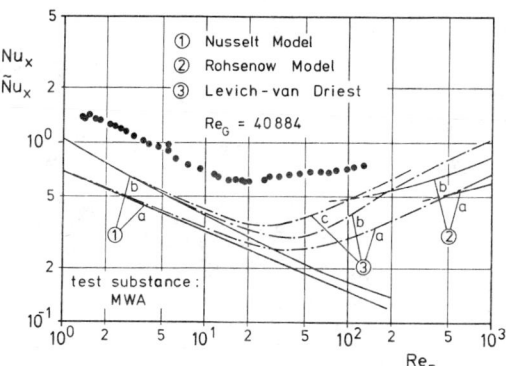

Curve a: $\tau_G^* = 0$; b: τ_G^* is variable; c: as b and thermal entry

Fig. 9-11: Local Nusselt numbers for condensation of MWA (Methoxi-isopropanol-water-azeotrope)

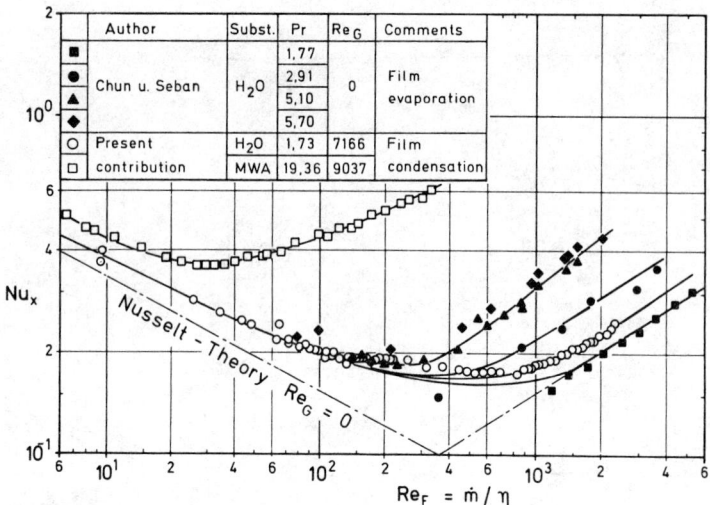

Fig. 12: Local Nusselt numbers for film condensation and film evaporation.

SYMBOLS
Latin letters

A	heat exchange area	m^2
a		m^2/sec
a,b,c,d,e	fitting parameter	
c_p	specific heat capacity	J/kg K
D	tube diameter	m
f	Fanning factor	–
F_2	defined in eq.(15)	–
g	acceleration of gravity	m/sec^2
$Ka = g^3 \rho^3 L^6 / \delta^3$	Kapitza number	–
L	length of test section	m
$\mathcal{L} = (\nu^2/g)^{1/3}$	characteristic length	m
$Nu_x = \alpha_x \cdot L / \lambda$	local Nusselt number	–
$Pr = \nu/a$	liquid Prandtl number	–
$Pr_t = \varepsilon/\varepsilon_w$	turbulent Prandtl number	–
$\dot{Q}$	heat flow rate	W
$\dot{q}$	heat flux	W/m^2

FILM CONDENSATION AND EVAPORATION IN VERTICAL TUBES

$Re_F = \dot{\Gamma}/\eta$	film Reynolds number	–
$Re_G = \bar{U}D/\eta$	gas Reynolds number	–
s	film thickness	m
$s^* = s/L$; $s^+ = (s/\nu)u_*$		–
u	velocity in the film in x-direction	m/sec
u_*	friction velocity	m/sec
$u^+ = u/u^*$		–
x	stream wise coordinate	m
y	coordinate normal to the x-direction	m
$y^+ = (y/\nu)u_*$		–

Greek letters

α	local heat transfer coefficient	$W/m^2 K$
$\dot{\Gamma}$	mass flow rate per unit perimeter	kg/msec
ε	eddy viscosity	m^2/sec
$\varepsilon^+ = 1+\varepsilon/\nu$	total normalized viscosity	–
ε_w		m^2/sec
η	dynamic viscosity	kg/msec
ν	kinematic viscosity	m^2/sec
ρ	mass density	kg/m^3
δ	surface tension	N/m
τ	shear force	N/m^2
$\tau^* = \tau/(g\rho_L(1-\rho_G/\rho_L)\mathcal{L}$; $\tau^+ = \tau/\tau_w$		–
ϑ	temperature	°C

Subscripts

G	gas
l	liquid
lam	laminar
t	turbulent
w	wall
x	local value
o	interface

Superscripts

∼	quasi local value
–	mean value

REFERENCES

1. Nusselt, W. (1916). Die Oberflächenkondensation des Wasserdampfes; Z. VDI, Vol.60, pp. 541/546 and 569/575.

2. Dukler, A.E. (1960). Fluid Mechanics and heat transfer in vertical falling film system. Chem.Engng.Prog., Symp.Series, Vol. 56, pp. 1/10.

3. Rohsenow, W.M., Weber, J.H. and Ling, A.T. (1956). Effect of vapor velocity on laminar and turbulent film condensation. Trans. Asme Vol. 78, pp. 1637/1643.

4. Levich, V.G. (1962). Physicochemical hydrodynamics, Chap.XII, Prentice-Hall Inc., Englewood Cliffs New Jersey.

5. Blangetti, F. (1979). Lokaler Wärmeübergang bei der Kondensation mit überlagerter Konvektion im vertikalen Rohr. Dissertation, Universität Karlsruhe, FRG.

6. McAdams, W.H. (1954). Heat transmission, 3. Ed, Chap. 13 McGraw-Hill, Inc., New York.

7. Grigull, U. (1952). Wärmeübergang bei der Filmkondensation. Forsch.Ing.Wes. Vol. 18, No.1, pp. 10/12.

8. Kutateladze, S.S. (1963). Fundamentals of heat transfer. Chap. 15, Edward Arnold LTD., London.

9. Kunz, H.R. and Yerazunis, S. (1969). An analysis of film condensation, film Evaporation and single-phase heat transfer for liquid Prandtl numbers from 1.E-03 to 1.E+03. Trans.Asme, J. Heat Transfer, Series C, Vol. 91, No. 3, pp. 413/420.

10. Jakob, M., Erk, S. and Eck. H. (1935). Verbesserte Messungen und Berechnungen des Wärmeübergangs beim Kondensieren strömenden Dampfes in einem vertikalen Rohr. Phys. Z., Vol. 36, No. 3, pp. 73/84.

11. Goodykoontz, J.H. and Dorsch, R.G. (1967). Local Heat transfer coefficients and static pressure for condensation of high velocity steam within a tube. NACA TN D-3953.

12. Goodykoontz, J.H. and Brown, W.F. (1967). Local heat transfer and pressure distribution for Freon-113 condensing in Downward flow in a vertical tube. NACA TN D-3952.

13. Krikbride, C.G. (1934). Heat transfer by condensing vapour on vertical tubes. Trans.Am.Inst.Chem.Engrs.,Vol.30,pp.170/193.

14. Hebbard, G.M. and Badger, W.L. (1934). Steam film heat transfer coefficients for vertical tubes. Trans.Am.Inst.Chem.Engrs. Vol. 30, pp, 194/216.

15. Badger, W.L., Monrad, C.C. and Diamond, H.W. (1930). Evaporation of caustic soda to high concentrations by means of diphenyl vapors. Ind.Engng.Chem.,Vol. 22, No. 7, pp. 700/707.

16. Carpenter, F.R. (1948). Heat transfer and pressure drop for condensing pure vapors inside vertical tubes at hich vapor velocities. Dissertation University of Deleware, USA.

17. Patankar, S.V., Spalding, D.B. (1970). Heat and mass transfer in boundary layers; Chap. 1, p.21; 2. edition; Intertext Books, London.

18. Jepsen,I.C., Crosser,O.K. and Perry,R.H.(1966). The effect of wave induced turbulence on the rate of absorption of gases in falling liquid film. AIChE J., Vol.12, No. 1, pp. 186/192.

19. Koh, J.C.Y., Sparrow, E.M. and Hartnett, J.P. (1961). The two phase boundary layer in laminar film condensation. Intern. J. Heat Mass Transfer, Vol. 2, pp. 69/82.

20. Hewitt, G.F. and Hall-Taylor, N.S. (1970). Annular two-phase flow. Pergamon Press, pp. 58/59, Oxford.

21. Rohsenow, W.M. and Hartnett, J.P. (1973). Handbook of Heat Transfer, Chap. 12, McGraw-Hill, Inc., New York

22. Davies, J.T. (1966). The effects of surface films in damping eddies at a free surface of a turbulent liquid. Proc.R.Soc., Vol. 290, No. 1423, pp. 515/526.

23. Lamourelle, A.P. and Sandall, O.C. (1972). Gas Absorption into a turbulent liquid. Chem.Engng.Sci.,Vol.27,pp.1035/1043.

24. Mills, A.F. and Chung, D.K. (1973). Heat transfer Across turbulent falling films. Internat. J. Heat Mass Transfer, Vol. 16, pp. 694/696.

25. Chun, K.R. and Seban, R.A. (1971). Heat transfer to evaporating liquid films. Trans. ASME, Series C, J. Heat Mass Transfer, Vol. 93, No. 4, pp. 391/396.

26. Borodin, A.S. and Picot, J.J.C. (1976). The measurement and prediction of temperatures at the free interface of water falling films under heat transfer. Can. J. Chem. Engng., Vol. 54, pp. 59/65.

27. Blangetti, F. and Schlünder, E.U. (1978). Local heat transfer coefficients on condensation in a vertical tube. 6th Int. Heat Transfer Conference, Toronto, General Papers, Vol. 2, pp. 437/442.

28. Blangetti, F. and Schlünder, E.U. (1979). Local Heat transfer coefficients in film condensation at high Prandtl numbers. 18th National Heat Transfer Conference, San Diego, Cal., Condensation Heat Transfer, ASME.

29. Iribane, A., Gosman, A.D. and Spalding, D.B. (1967). A theoretical and experimental investigation of diffusion controlled electrolytic mass transfer between a falling liquid film and a wall. Internat. J. Heat Mass Transfer, Vol. 10, pp. 1661/1676.

30. Hubbard,G.L., Mills,A.F. and Chung,D.K.(1976). Falling Film with Cocurrent Vapor Flow; J. of Heat Transfer, Vol. 98, pp. 319-320.

Design and Application Considerations for Heat Exchangers with Enhanced Boiling Surfaces

R. ANTONELLI
Union Carbide Benelux BV
Antwerp, Belgium

P. S. O'NEILL
Union Carbide Corporation
Tonawanda, New York 14150, USA

ABSTRACT

Design practice for bare shell-tube reboilers, evaporators, and refrigerant chilled condensers in hydrocarbon services for the Chemical Process Industries is well established, based on more than 50 years of experience. Application of enhanced evaporator tubing such as Union Carbide Corporation "High Flux" in the CPI is relatively more recent, but design procedures have evolved rapidly, for a wide range of enhanced exchanger types, orientations and fluid flow conditions.

This paper surveys and compares design procedures and heat transfer correlations developed for common High Flux exchanger services with those for bare tube. In horizontal shellside evaporators, the relative insensitivity of the enhanced boiling film coefficients to local mass velocity, quality, and tube spacing is discussed. Sizing procedures for large bundles operating at small temperature differences are reviewed. An example is given for an enhanced tube vertical thermosiphon reboiler, showing the individual resistances.

Enhanced surface application examples encountered over the past 10 years in refinery and light hydrocarbon services are given, and the more important design considerations and operation of each type are discussed.

1. INTRODUCTION

Design practice for commercial shell and tube heat exchangers in boiling service is well developed, and has grown in response to industry needs over the past 50 years. The rapid growth of the chemical process industries has led to increases in plant sizes of an order of magnitude or more over the period, thus resulting in larger heat exchangers with more complex design considerations. The needs of the petroleum refining and hydrocarbon processing industries have greatly increased the use of heat exchangers devoted exclusively to the vaporization of organics. Development of halocarbons ("freons") in the 1930's revolutionized the refrigeration industry, and created a specialized design practice for evaporators and condensers in large packaged chilling systems. Although boiling heat transfer is in general more complex than single phase convection, design procedures for bare tube reboilers and refrigerant evaporators rarely recognized the interaction of forced convection and nucleate boiling in determination of the boiling film coefficient. The mechanism of boiling in tube bundles, and the corresponding flow hydraulics, were for years poorly understood. As a result, certain areas of reboiler and vaporizer design practice tended to become conservative. The labor associated with hand calculation of functional design details discouraged the early development of more sophisticated thermal rating techniques (2). Today, thanks to the considerable body of accumulated heat transfer research, and the advent of high speed computation, the experienced design engineer can quickly consider or evaluate the numerous factors affecting thermal design. Exchanger performance has, as a result, become more reliable, and some of the arbitrary conservatism in reboiler design practice is beginning to disappear.

The application of improved or enhanced boiling surfaces in shell and tube heat exchangers has also grown in response to industry needs. One of the first nucleate boiling promoters applied commercially, which was of the enhanced type, was the Union Carbide "High Flux" surface. A porous metal film 0.25 to 0.35 mm thick was applied to the outside or inside of an evaporator tube. Large numbers of stable nucleation sites insured a 10-30 fold increase in the boiling coefficient. Several publications (5,6,7,8) have described the general application of these surfaces over the past 15 years. Application areas in the CPI where the surface has been found to be especially advantageous are in large packaged refrigeration systems (20,9), olefins plants (10,7), refinery/aromatics (11), and in reboilers for heat pumped distillation columns (12, 24). Originally this enhanced boiling surface was developed for air separation plant main condensers where large latent heat loads are encountered at overall temperature differences of only 1.5 - 2.0°C (21). Pilot plant qualification began in the mid 60's for ethylene plant services. Today, some of the world's largest ethylene plants have High Flux tubing in the critical refrigeration system services (25, 13, 7, 14).

Following successful use of the tubing in olefins plants, other plant services were qualified in the 1970's. Benzene/toluene/mixed xylene column reboilers (BTX) were especially attractive because lower pressure and less costly steam levels could be used to drive the High Flux exchangers without exceeding existing size limits. In very large heat pumped C_3 and C_4 splitters, considerable energy savings were realized with enhanced surface reboiler/condensers, designed for temperature differences of about 5°C. (Today more than half of the High Flux exchangers are designed for revamp applications where existing tube bundles are replaced with identical enhanced surface tubes in order to increase capacity, save energy, or utilize lower cost steam).

The remainder of this paper compares design procedures for High Flux and conventional horizontal kettle, horizontal thermosiphon, and vertical thermosiphon reboilers and chillers. These configurations are the three most common types of High Flux exchangers encountered in practice. Specific examples of confirmation of the design practice, based on laboratory and field performance, are also presented.

HEAT EXCHANGERS WITH ENHANCED BOILING SURFACES

2. EXCHANGER DESIGN CONSIDERATIONS

High Flux exchangers have been applied extensively in both the refrigeration and chemical process industries in a variety of reboiler, chiller, vaporizer and evaporator configurations. These exchangers may be horizontal, vertical, or kettle type, and have either once-through or circulating flows. For design purposes, the various combinations above can be collected into three groups: Figure 1 shows examples of these three common exchanger groupings or arrangements. The first and perhaps most important is the refrigerant chilled kettle

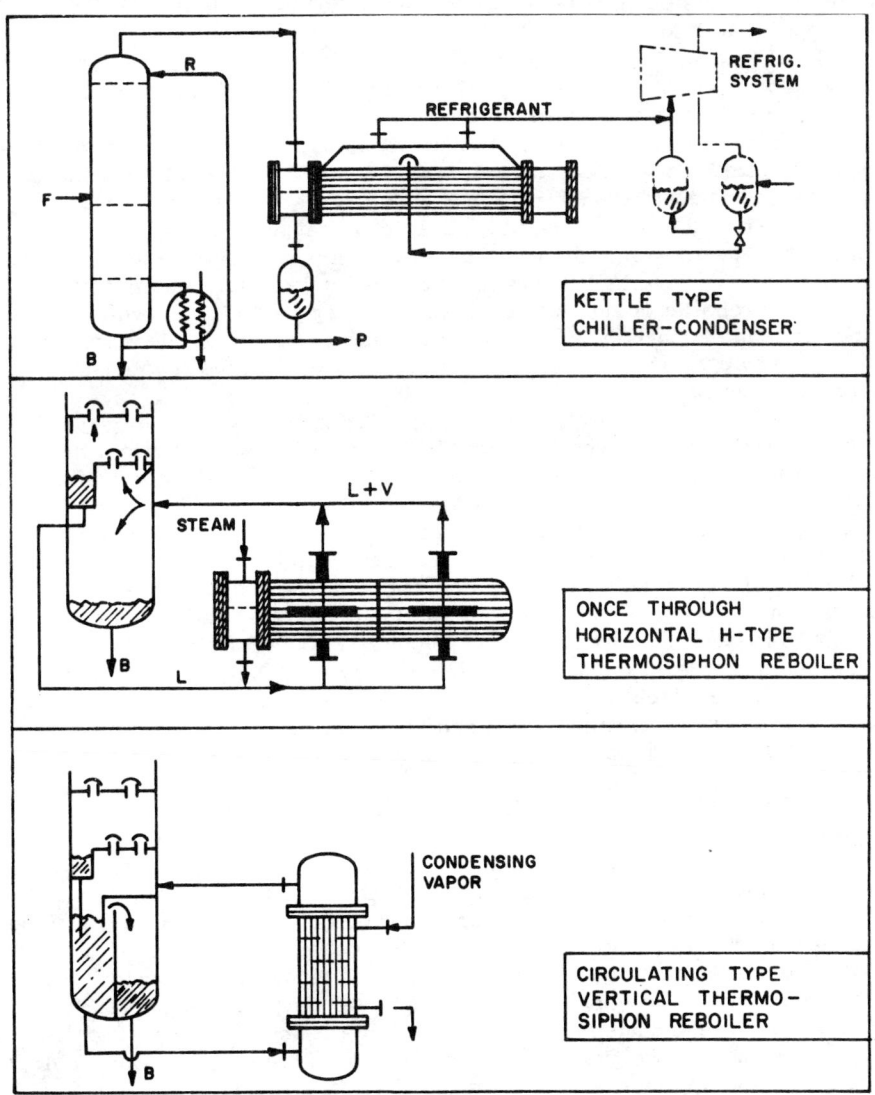

FIGURE 1   COMMON 'HIGH FLUX' HEAT EXCHANGER LAYOUTS

type condenser, typically used in refrigeration systems of large olefins and gas processing plants. An enlarged shell serves to remove entrained liquid from the vapor. Inlet liquid refrigerant usually enters the kettle after being throttled from a higher pressure. The boiling side is essentially isothermal except for a small hydrostatic elevation of boiling point due to liquid depth, while the process vapor usually condenses over a temperature range. These exchangers are often relatively large, with bundle diameters of 1.6 meters, shell diameters of 3 meters and lengths up to 12 meters. Design temperature levels may require the use of tubing material suitable for low temperatures; that is, fine grained or killed steel, and 3.5% nickel steel. Common exchanger TEMA designations are BKM, CKU, AKU, etc.

The second exchanger type in Figure 1 is the horizontal thermosiphon reboiler, often used when access to the tube bundle is required, or when tower elevations do not permit a vertical reboiler. These designs are more often found in petroleum refining or related services such as in BTX fractionator reboilers, naphtha stabilizer reboilers, depentanizer reboilers, etc. Easily removable bundles facilitate tube replacement for capacity increase or energy cost reductions with lower pressure steam. The elevation of the tower draw-off well and reboiler sets the available thermosiphon driving force (allowable pressure drop) and what reboiler design will be used. TEMA type "H" is shown. A circulating type, rather than once-through, may also be specified, depending on the tower control system. A pump is sometimes used when the reboiler is remote from the tower.

A final major design type for High Flux exchangers is the vertical thermosiphon reboiler, perhaps the most versatile and simple of all the configurations. Typical services for enhanced boiling surfaces are heat pumped light end fractionator reboilers, ethylene glycol evaporators, reboilers for ethylene and propylene splitters, and methanol refining column reboilers. When fluted tubes are used in conjunction with the porous boiling surface, higher overall coefficients can be achieved than with the horizontal units. The largest of these units have diameters over 2 meters, and lengths of about 6 meters. When revamping an existing exchanger to upgrade capacity or reduce ΔT, it is often most expedient to replace the existing "E section" shell and tube sheets as a whole with an identical enhanced tube section. The old heads and piping can then be reused, thus reducing downtime and related costs.

The configurations shown in Figure 1 are intended to serve as examples of some common arrangements. The application of enhanced tubing is by no means restricted to these designs. Often the existing tower details, allowable ΔP, or customer preference sets the kind of layout used. In general, any layout which would be suitable for a conventional (bare tube) exchanger would also be suitable for an enhanced surface as well. When reboiler mixtures with a narrow boiling range are encountered, the circulating type tends to be favored. The once-through types, however, usually give a better temperature difference profile through the exchanger when there is a large change in composition between the bottoms and the first tray. Services with a history of chronic fouling are avoided in the application of the enhanced surface tubing.

The remainder of the paper will discuss some of the major design considerations for the three most common types (kettle type, horizontal, and vertical thermosiphon). Emphasis will be placed on design criteria which is different or modified in some way when applied to High Flux tubing.

2.1 Kettle Type Reboilers Chillers and Vaporizers

The most important special considerations in the design of enhanced surface exchangers of this type are the prediction of the boiling heat transfer coefficient, the specification of the correct fouling allowance or design safety factor, optimization of the in-tube condensing coefficient, and the maximum allowable bundle heat flux which avoids vapor blanketing or dryout for a bundle of given size. Mechanical design criteria, prediction of tubeside pressure drop and nozzle sizes, etc. are for the most part similar to those of conventional exchangers and will not be discussed in detail. Tube supports or baffle holes are over-drilled about 0.4 mm to allow for the presence of the boiling heat transfer surface. Both U-tube and fixed tube units may be specified. The procedure for calculation of the overall heat transfer coefficient is the same for all the kettle type exchangers. Enhanced surface exchangers often have a higher heat flux, which means a greater boilup per unit of bundle volume. Larger shells, or alternately more shellside exit nozzles, may be required to maintain entrainment standards for chillers in refrigeration systems. Shell diameters are at least 600 mm greater than bundle diameters, and are on the order of 1.6-1.8 fold greater than the bundle diameter for the largest bundles. Typical tube pitch to tube diameter ratio is 1.25. Triangular pitch is preferred. Use of square, diamond, or rotated square pitch does not affect the heat transfer coefficient for High Flux kettle type exchangers.

Boiling heat transfer coefficients for the enhanced surface are 10-30 times larger than bare tube for pure fluids, and 5-10 fold greater for mixtures. Early research work on porous surfaces showed that when boiling inside vertical tubes or outside horizontal bundles, the boiling film coefficient is essentially equal to the pool boiling coefficient for a single tube or flat plate, for a wide range of mass velocities and local vapor qualities (23). An example of this phenomenon is shown in Figure 2. The insensitivity of the boiling coefficient to 2 phase forced convection in the bundle can be explained by considering that vaporization occurs within the thin porous matrix, rather than at the surface, and the mechanism of boiling is governed by surface tension forces within the matrix (15). This condition greatly simplifies design of not only kettle type exchangers but horizontal and vertical thermosiphons as well. Boiling heat transfer data measured on single tubes or flat plates can be used directly in the design of the reboiler bundle.

Figures 3 and 4 show typical measured boiling heat flux vs boiling film ΔT for a variety of pure fluids and mixtures encountered in some of the more important High Flux exchanger applications. The boiling coefficient ($h_b = (Q/A)/\Delta T_b$) may be calculated directly from the figures, or the entire curve can be specified as input to computer programs which account for the variation of boiling coefficient with heat flux. Variations in boiling pressure affect the coefficient. A conservative practice is to use the value at atmos-

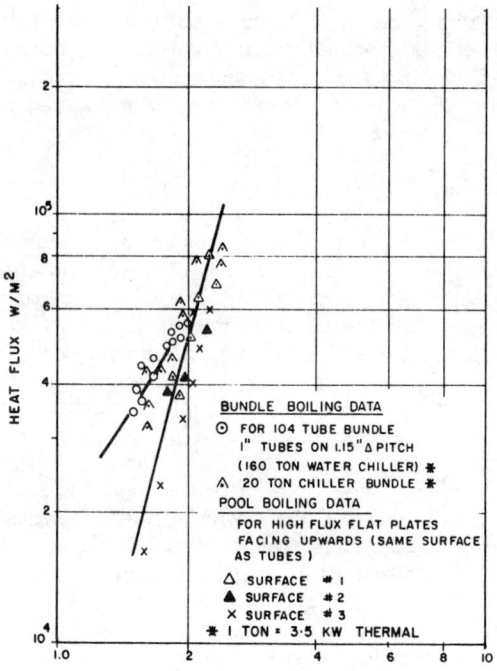

FIGURE 2 COMPARISON OF 'HIGH FLUX' POOL BOILING WITH BOILING PERFORMANCE OUTSIDE HORIZONTAL MULTI-TUBE BUNDLES IN R-11 AT 40 KPa

pheric pressure for services at elevated pressure. These boiling data were developed from measurements of pool boiling flux and film temperature difference using fluids typical of those found in chillers and in reboiler bottoms. In many cases, actual tower bottoms liquid was used in the tests. The design curves have allowances for normal fouling from oil, magnetite or pipe dust, etc. and need not be de-rated further. In practice, only a few boiling curves are needed for a number of exchangers of a given class. For example, more than 16 different chillers and condensers in an olefins plant refining section can be designed using only curve #4 in Figure 3 (7). In any chiller or reboiler, the curves in Figures 3 and 4 are used in that portion of the exchanger where saturated liquid exists, and fully developed nucleate boiling would occur. For the enhanced surface, boiling film temperature differences in excess of 0.5°C usually are great enough to promote boiling. (In some cases, temperature differences of 0.2°C are sufficient (21)).

Optimization of the non-boiling side heat transfer coefficient is another factor of importance in the design of enhanced surface exchangers. For kettle type chiller-condensers, the in-tube coefficient under condensing conditions is enhanced considerably at higher vapor velocities due to drag on the condensate film. High in-tube flow rates and Reynolds numbers tend to occur naturally in enhanced surface exchangers because of the lower tube counts involved. At

HEAT EXCHANGERS WITH ENHANCED BOILING SURFACES

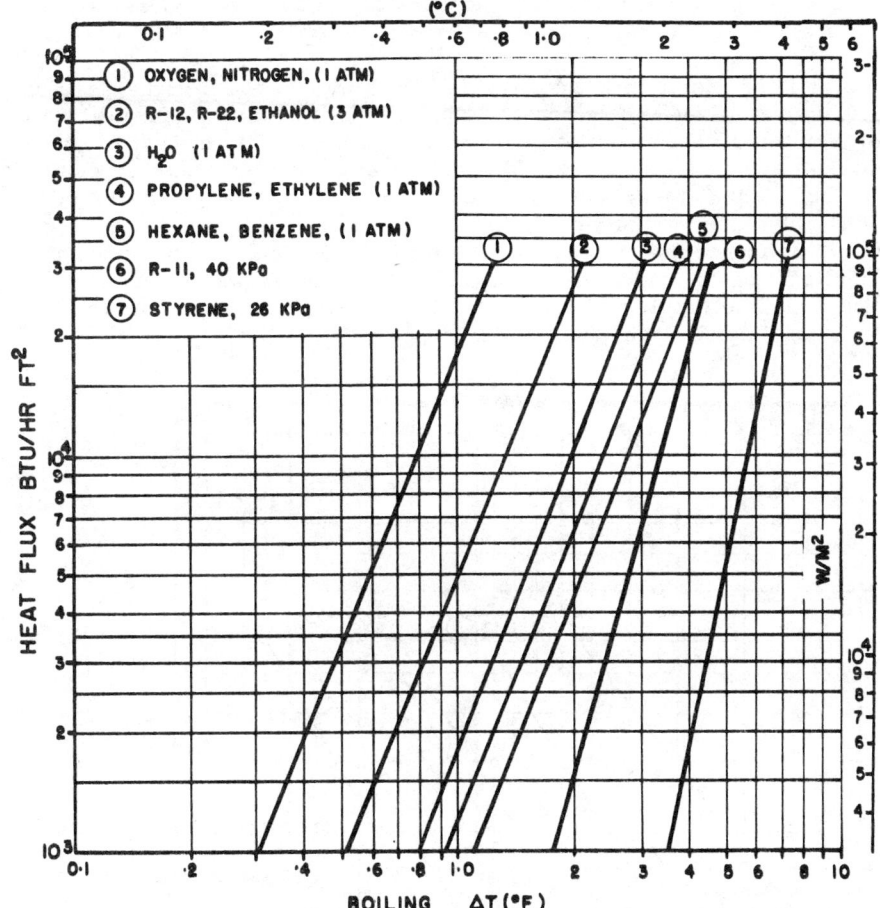

FIGURE 3 TYPICAL PURE FLUID BOILING HEAT TRANSFER DATA USED IN DESIGN OF 'HIGH FLUX' EXCHANGERS

very low flow rates the film coefficient tends to decrease with increasing condensate load. At higher flows, vapor drag assists in draining condensate, leading to a marked improvement in the condensing film coefficient. Akers et al (16) were among the first to present data for a variety of organics condensing at high Reynolds numbers. Proprietary in-tube condensing correlations have been developed from experimental measurements by Union Carbide Corporation, which are similar in form to those of Akers, but are specific to condensation of light hydrocarbons such as ethylene, propylene, cracked gas, etc. at pressures commonly encountered in practice. Reference (7) gives examples of condensing services for ethylene plants. When the boiling side is enhanced, the hotside then bears the dominant resistance to heat transfer. An exchanger design which "uses up" most of the allowable condensing side pressure drop will have the highest tubeside velocity, and as a result, the minimum area. Many High Flux

2 or 4 pass chiller-condensers used in olefins plants have total tubeside flow lengths in excess of 30 meters (7).

A final consideration in the design of kettle type chillers is the maximum permissible size of the bundle needed to prevent dryout. Since enhanced surface exchangers give higher overall heat transfer coefficients, larger heat fluxes would result at a given temperature difference, and maximum bundle sizes would be less than for bare tube. Several investigators have proposed criteria for calculation of dryout with shellside boiling in large multi-tube bundles (17,18). Gilmour (2) reasoned that a large bundle would exhibit declining performance when the local superficial vapor velocities in the bundle became equal to the velocities at the critical heat flux for a single tube. The critical heat flux $(Q/A)_{st}$ is calculated by the Kutateladze/Zuber equation (19).

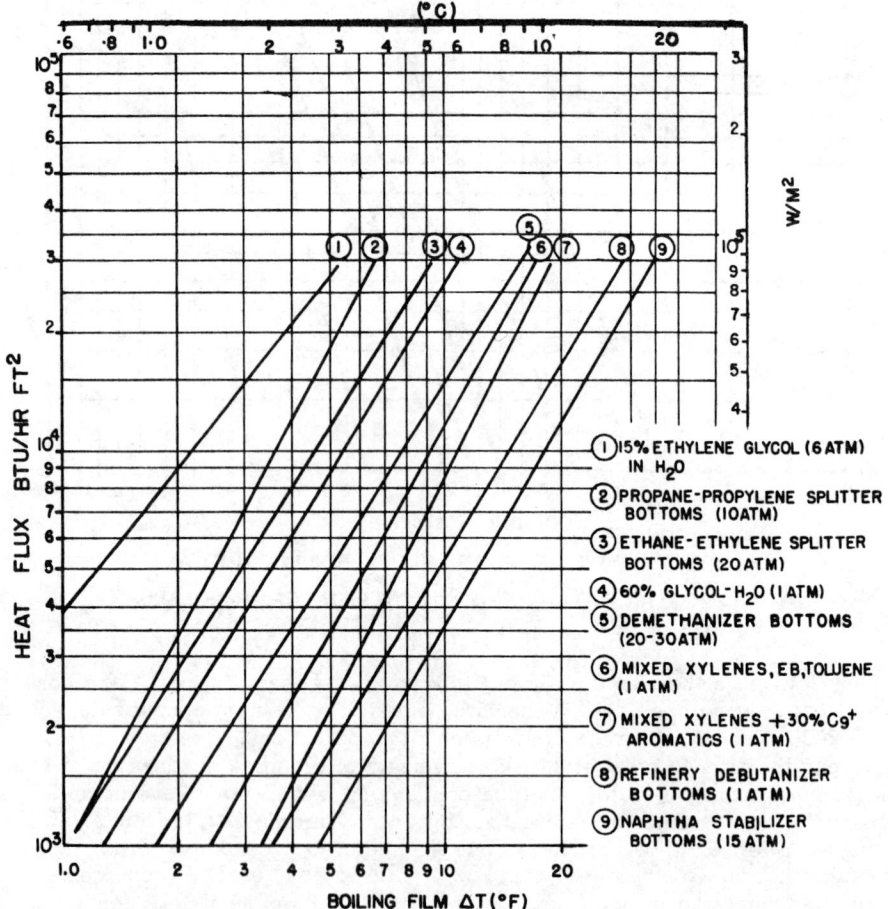

FIGURE 4 TYPICAL BOILING HEAT TRANSFER DATA FOR MIXTURES USED IN DESIGN OF 'HIGH FLUX' VERTICAL AND HORIZONTAL EXCHANGERS

$$(Q/A)_{st} = 0.18 \; \lambda \, \rho_v \left[\frac{(\rho_l - \rho_v) g g_o \sigma}{\rho_v^2} \right]^{1/4} \quad W/m^2 \qquad (1)$$

The velocity at the critical flux is

$$v_c = (Q/A)_{st} / \lambda \, \rho_v \approx v_p \qquad (2)$$

Because of recirculation in a bundle, the above criteria is very conservative.

It has been determined from tests on High Flux bundles (9) that the superficial velocity (based on projected bundle surface area) is at least six times the critical velocity of equation 2. That is:

$$v_p = 6 V_c = \frac{W_T}{\rho_v D_B L} \qquad (3)$$

The maximum bundle heat flux can in turn be expressed as

$$(Q/A)_B = \frac{W_T \lambda}{A_H} \qquad (4)$$

Eliminating W_T from equations 3 and 4 we get

$$(Q/A)_B = \left(\frac{6 D_B L}{A_H} \right) (Q/A)_{st} \qquad (5)$$

Where D_B is the bundle diameter, L is the length, and A_H is heat transfer area.

Equation 5 has been found to be a useful (although conservative) guideline for calculating High Flux maximum bundle size or maximum flux for a bundle of given size. It is interesting to note, however, that only in a few cases have enhanced surface exchangers been designed where the bundle flux equaled or exceeded the value in Equation 5. In the majority of applications, the temperature differences are small enough so that bundle heat fluxes, even for the enhanced tube, rarely exceeds the maximum flux. The current trend in the design of modern refrigeration systems, and in heat integration in chemical processes, is toward smaller temperature differences, where substantial energy savings can be realized. It is expected that in the foreseeable future, High Flux heat exchangers will be applied preferentially in low temperature difference services, where the likelihood of exceeding the limitations imposed by Equation 5 will become more remote.

Table 1 gives some examples of successfully operating units where the design flux has been exceeded in service, and the value is approximately that of Equation 5. These data indicate that the constant in Equation 5 is probably larger than 6.

TABLE 1

COMPARISON OF DESIGN AND MAXIMUM HEAT FLUX
FOR OPERATING HIGH FLUX EXCHANGERS

SERVICE	BUNDLE DIA. mm	BUNDLE LENGTH m	SURFACE AREA m^2	DESIGN HEAT FLUX KW/m^2	MAX BUNDLE FLUX KW/m^2 *
C_3H_6 VAPORIZER	355	1.5	8.02	164	242
C_9 AROMATICS REBOILER	914	6.1	269	76.0	63
NAPHTHA STABILIZER	1244	6.1	511	52.4	51
BENZENE COLUMN REBOILER	812	6.1	180	89.8	64
DEBENZENIZER REBOILER	1370	4.9	496	35.4	32

* Equation 5

TABLE 2

MEASURED EFFECT OF RECIRCULATION RATE ON PERFORMANCE
OF HIGH FLUX HORIZONTAL TOLUENE COLUMN REBOILER

	DESIGN CONDITION	HIGH RECIRC. RATE	REDUCED RECIRC. RATE
STEAM PRESSURE KPA	793	737	737
STEAM FLOW KG HR	36165	27918	36360
STEAM SAT. TEMP. °C	175	172	172
HEAT DUTY MW	20.3	15.7	20.5
SURFACE AREA M^2	767	767	767
BOILING TEMP. °C IN/OUT	163/165	154/158	160/165
REBOILER FLOW RATE KG HR	688,360	2,454,000	363,600
APPROX. % VAPORIZATION	33	7	62
MEAN TEMP. DIFFERENCE °C	9.9	15.5	9.4
OVERALL COEFFICIENT W/m^2C	2680	1442	2839 +

2.2 Horizontal Thermosiphon Reboilers

The previously discussed kettle type exchangers vaporized essentially all of the liquid fed to the boiling side. The remaining two types (as defined in Figure 1) vaporize only a part of the recirculating or once-through flow. Typically, High Flux horizontal thermosiphon reboilers vaporize 10-40% of the feed when of the circulating type and 30-70% of the feed when of the once-through type. The local boiling side coefficient, when the fluid is at the condition of saturation, is approximately the pool boiling coefficient given in Figures 3 and 4 and calculation of the overall coefficient for these exchangers is the same as for the kettle type discussed previously. Total pressure drop on the shellside is determined by considering individual 2 phase components through the nozzles, in cross flow, longitudinal flow, and in flow through the baffle windows, just as in the design of a bare tube (conventional) unit.

Common configurations are TEMA types BJU, BJM, BHM, CEN, or AES, with tube or bundle lengths of 4.8 and 6.1 meters and tube diameters of 19 and 25 mm.

Multi-component mixtures are frequently vaporized over a temperature range; therefore, the temperature-heat release curve must be taken into account in the design. In the most general case the fluid to be partly vaporized enters subcooled (at a temperature below the bubble point), due to the higher pressure at the reboiler inlet relative to that in the vapor space at the column bottom. Some heat transfer surface area near the reboiler inlet nozzle must be devoted to sensible warming of the feed to the bubble point. At the bubble point the liquid begins to vaporize with a continuously rising temperature until it reaches the outlet nozzle. Between the bubble point and outlet temperature the High Flux surface area is determined by the methods discussed for the kettle type exchangers. The surface area required for the vaporization section should ideally be a significant fraction of the total required surface. For optimum design of the enhanced tube thermosiphon reboiler, it is desirable to minimize the surface which must be devoted to sensible warming of the liquid. An enhanced tube containing a nucleate boiling promoted surface will usually not provide the high heat transfer coefficients in the sensible region. The differences between sensible and fully developed nucleate boiling coefficients is greater than for a smooth tube.

Table 2 shows performance data for a High Flux horizontal BHU type circulating thermosiphon reboiler at two widely varying reboiler recirculation rates. At the high liquid driving head, a large part of the bundle surface is involved in warming up the large recirculating flow to the boiling point (about 3°C higher than the inlet temperature), thus making the average coefficient only about half of the design value, which was based on a much smaller sensible heating load. When the measured reboiler recirculating flow was reduced to below the design value, however, the measured apparent overall coefficient became larger than design.

The example in Table 2 illustrates clearly that optimum performance can be best achieved with High Flux reboilers when liquid levels are low, and consequently the recirculation rates are restricted. Large tower liquid levels, and oversized piping can actually reduce the performance of the enhanced surface reboiler by creating an unnecessarily large sensible heating load. This is perhaps one of the most important design considerations in reboilers of this type.

2.3 Vertical Thermosiphon Reboilers

The final reboiler type where enhanced boiling surfaces can be used is the once-through or circulating vertical thermosiphon. Common application areas are ethylene refining section demethanizer, C_2 splitter, and propane-propylene splitter reboilers. Multi-effect evaporators and reboilers for concentration of aqueous solutions are also typically vertical. Reasons for selection of the vertical design are similar to those for conventional (bare tube); for example, availability of space around distillation columns, possibility of fouling, industry tradition, overall performance requirements, and designer preference usually dictate the selection. Higher overall coefficients can be achieved with this design when the axial fluted tube is used to promote condensing heat transfer. Typical tube lengths range from 3.6 to 6 meters, with tube diameters of 25 and 32 mm.

Design procedures for the vertical thermosiphon reboiler are in many respects similar to those of the horizontal type. Figures 3 and 4 are again used to determine the local boiling heat transfer coefficient inside the tube. When the fluted tube is used, the condensing film coefficient ranges from 8 to 12 fold greater than the corresponding value on a smooth tube, depending on whether steam or hydrocarbons, respectively, are condensing (6). It has been found that the condensing coefficient on vertical fluted tubes of selected geometries can be predicted from the equation

$$h_c = 0.925\, a \left[\frac{k^3 \rho^2 g}{\mu \Gamma}\right]^{1/3} \tag{6}$$

where the parameter "a" depends on fluted tube geometry and fluid class, and has the following experimentally determined values:

$a = 8.7$ steam condensing, Profile A

$a = 12.7$ light hydrocarbons condensing, Profile B

$a = 16.1$ nitrogen or methane condensing, Profile C

The equation has been found to be valid for tube lengths up to 6 meters, and at heat fluxes where the condensate film would be turbulent. The required profile geometry varies with fluid class, Profile A has a height of approximately 0.9 mm, while Profile B and C have a height of about 1.7 and 2 mm respectively.

Computer design for the High Flux vertical thermosiphon reboiler is carried out by first dividing the tube length into 20 calculation zones. Recirculation rates are calculated from the liquid level and the known piping and exchanger geometry. Fluid velocity or recirculation rate on the tubeside does not affect the overall coefficient in the boiling zone, but does affect the length of the zone. The liquid level can affect the available temperature difference across

HEAT EXCHANGERS WITH ENHANCED BOILING SURFACES 187

the reboiler. If excessive, both the ΔT and the length of the high performance boiling zone will be reduced. Typically, liquid levels are maintained between 50 and 100% of tube height, depending on column pressure and liquid density, with the top tube sheet (100%) being most common at elevated pressures. The optimum liquid level for a particular design maintains vaporization fractions less than about 40%, and the length of the sensible (non-boiling) zone less than about 10% of tube height. The roughness of the promoted boiling surface on the inside of the tube increases the friction factor about 2 fold over that for a smooth tube. This does not, however, lead to reduced recirculation rates in practice because the tube diameter is usually one size larger than the corresponding bare tube design would be. The rough surface also increases the tube-side coefficient in the sensible zone about 1.6 times over that of bare tube.

The magnitude of the various heat transfer coefficients and the design procedure can best be illustrated with a specific example. An ethylene-ethane splitter reboiler boils 97% ethane at a pressure of 2000 kPa, (-5°C) and condenses propylene refrigerant at approximately 10°C. The enhanced tube is externally fluted with Profile B, and has a length of 2.4 meters, and diameter of 25 mm. The tower has 2 vertical thermosiphon circulating type reboilers which are replaced with enhanced surface to increase capacity. The large temperature difference (15°C) is typical of older plants with only 2 stages of propylene refrigeration. (Newer plants with more efficient refrigeration systems would have design ΔT values of about 5°C for the service).

Figure 5 shows the individual film coefficients used in the design. The

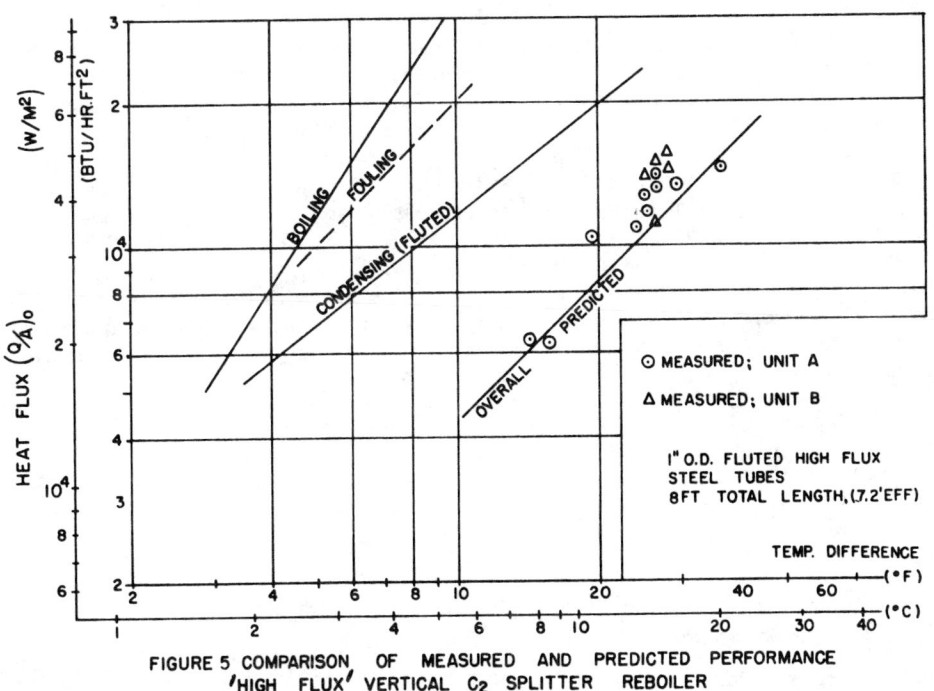

FIGURE 5 COMPARISON OF MEASURED AND PREDICTED PERFORMANCE 'HIGH FLUX' VERTICAL C₂ SPLITTER REBOILER

boiling coefficient is taken from curve 3 of Figure 4, while Equation 6 gives the condensing performance. Computer solution involves dividing the exchanger length into 20 calculation zones and determining the local overall coefficient from the individual values in Figure 5. The overall predicted curve contains allowances for the small sensible zone length at the bottom of the exchanger and some flooding of the tube surface due to condensate back up. Because of the high boiling pressure, the temperature difference profile through the exchanger is essentially constant. The measured performance data for the 2 reboilers is also shown, and was somewhat better than predicted, indicating that the fouling allowance had not been used. The reboilers were operating at essentially the "clean" design condition, as is typically the case with other High Flux reboilers of this type.

Piping and head design practice for the enhanced vertical thermosiphon reboilers is similar to that of bare tube. The most common heads have 90° nozzle entry and exit on both channels and bonnets. Mitred or long radius elbows are not often specified, since the High Flux reboilers are rarely used in vacuum service. On the shellside, an impingement plate is usually required at the nozzle inlet because of the higher velocities often encountered with the doubly enhanced tube in condensing service. A triple segmental or "window" baffle arrangement described in Reference (22) has been successfully used in numerous hydrocarbon reboiler/condensers to reduce pressure drop and help in non-condensible removal.

3. CONCLUDING REMARKS

In all of the heat exchanger designs previously discussed, fully developed nucleate boiling existed over a substantial fraction of the total surface area. This condition, along with relatively small temperature differences, and absence of excessive fouling are the necessary ones for ideal application of the High Flux surface. In all cases, the design boiling side heat transfer coefficient can be equated to the measured pool boiling coefficient, regardless of flow orientation, mass velocity, or quality, with little resulting error. This fact removes much of the uncertainty in the design of the enhanced surface reboilers. Also, certain types of reboiler flow instability are much less likely to occur when the boiling side heat transfer coefficient remains constant with flow rate or quality. Experience with more than a hundred units in various services confirms that the overall transfer rate can be predicted for the services with a high degree of confidence.

NOMENCLATURE

A   Surface Area m^2

D   Diameter m

g   Acceleration due to Gravity m/Sec^2

g_o Gravitational Constant

h   Heat Transfer Coefficient $W/m^2 K$

NOMENCLATURE

k	Thermal Conductivity	$W/m \cdot K$
L	Length	m
Q	Heat Transferred	W
U_o	Overall Heat Transfer Coefficient	$W/m^2 K$
V	Velocity	m/Sec
W	Flow Rate	Kg/Sec
λ	Latent Heat of Vaporization	Kj/Kg
ρ	Density	Kg/m^3
σ	Surface Tension	Kg/m
ΔP	Pressure Difference	kPa
ΔT	Temperature Difference	°C
μ	Viscosity	cp
Γ	Condensate Loading	Kg/Sec.m

SUBSCRIPTS

c	Critical
b	Bundle or Boiling
h	Heat Transfer
l	Liquid
o	Reference State
st	Single Tube
t	Total
v	Vapor
p	Projected

REFERENCES

1. Kern, D.Q. 1950. Process Heat Transfer. McGraw Hill, New York.

2. Gilmour, C.F. March 1954. Shortcut to Heat Exchanger Design (#6). Chem. Eng., 61, 209.

3. Myers, J.E. and Katz, D.L. January 1952. Boiling Coefficients Outside Horizontal, Plain and Finned Tubes. Refrigerating Engineering, p. 56.

4. Kun, L.C. and Czikk, A.M. 1969. Union Carbide Corporation. Surface for Boiling Liquids. U.S. Patent 3,454,081, issued July 8, 1969.

5. Gottzmann, C.F. and Milton, R.M. August 1972. High Efficiency Hydrocarbon Reboilers. CEP 68, #8, p. 56.

6. Gottzmann, C.F., O'Neill, P.S. and Minton, P.E. July 1973. High Efficiency Heat Exchangers. CEP 69, #7, p.69.

7. O'Neill, P.S., Ragi, E.G. and King, R.C. 1980. Application of High Performance Evaporator Tubing in Refrigeration Systems of Large Olefins Plants. AIChE Symp. Series, 76, #199, p. 289.

8. O'Neill, P.S., Gottzmann, C.F. and Terbot, J.W. 1972. Novel Heat Exchanger Increases Cascade Cycle Efficiency for Natural Gas Liquefaction. Advances in Cryogenic Eng., Vol. 17, pp. 420-437.

9. Czikk, A.M., Ragi, E.G. et al. 1969. Performance of Advanced Tubes in Flooded Water Chillers. ASHRAE Trans. 76, Part 1, p. 897.

10. Antonelli, R. and Chesnoy, A.B. December 1976. Applications of High Flux Heat Transfer Technology to the Chemical Process Industries. 6th National Convention of Industrial Chemistry, Madrid.

11. Kenney, W.F. March 1979. Reducing the Energy Demand of Separation Processes. CEP, pp. 68-71.

12. Wolf, C.W., Weiler, D.W. and Ragi, E.G. September 1, 1975. Energy Savings Prompt Improved Distillation. Oil & Gas Journal, pp. 85-88.

13. Wett, T. April 14, 1980. Flexible Olefin Plant Designed for Long Runs. Oil & Gas Journal, pp. 77-81.

14. Anon., September 1978. Union Carbide put 425 Million Dollars into Olefins Expansion Modification Project. Oil & Gas Journal, p. 83.

15. Czikk, A.M. and O'Neill, P.S. August 1979. Correlation of Nucleate Boiling from Porous Metal Films. Advances in Enhanced Heat Transfer, San Diego, p. 53.

16. Akers, W.W., Deans, H.A. and Crosser, O.K. 1959. Condensing Heat Transfer within Horizontal Tubes. CEP Symp. Series 55, #29, Heat Transfer-Chicago, p. 171.

17. Palen, J.W., Yarden, A. and Taborek, J. 1972. Characteristics of Boiling Outside Large-Scale Horizontal Multi-tube Bundles. AIChE Symp. Series, Vol. 68, #118, pp. 50-61.

18. Palen, J.W. and Small, W.M. November 1964. A New Way to Design Kettle and Internal Reboilers. Hydrocarbon Processing, p. 199.

19. Kutateladze, S.S. 1952. Heat Transfer in Condensation and Boiling. USAEC Rept. AECU-3770.

20. Starner, K.E. and Cromis, R.A. December 1977. Energy Savings using High Flux Evaporator Surface in Centrifugal Chillers. ASHRAE Journal Vol. 19, #12, pp. 24-27.

21. O'Neill, P.S. and Gottzmann, C.F. August 1980. Improved Air Plant Main Condenser, presented at ASME Century 2, Emerging Technology Conferences Cryogenic Processes & Equipment. San Francisco, California.

22. Lord, R.C. et al. March 1970. Design Parameters for Condensers and Reboilers. Chem. Eng., 77, #6, p. 127.

23. Czikk, A.M., O'Neill, P.S. and Gottzmann, C.F. August 1981. Nucleate Boiling from Porous Metal Films, Effect of Primary Variables. 20th National Heat Transfer Conference, Milwaukee.

24. Quadri, G.P. March 1981. Use Heat Pump for Propane/Propylene Splitter. Hydrocarbon Processing #3, p. 147.

25. Wett, T. April 6, 1981. Petrochemical Report, Oil & Gas Journal, 79, #14, p. 79.

Comments to the Application of Enhanced Boiling Surfaces in Tube Bundles

J. W. PALEN, J. TABOREK, and S. YILMAZ
Heat Transfer Research, Inc.
1000 South Fremont Avenue
Alhambra, California 91802, USA

ABSTRACT

Use of specially enhanced boiling surfaces can greatly increase heat transfer rates at low boiling side temperature differences. However, proper utilization of these surfaces in tube bundles requires an understanding of the tube bundle boiling mechanisms plus some basic data to define surface factors and nucleate boiling—convective boiling relationships. Single tube data cannot be used alone to design enhanced tube bundles. Adjustments must be made for bundle effects through the use of circulation boiling concepts. The amount of enhancement for bundles theoretically must be less than for single tubes, but can still be great enough to make such surfaces economically attractive in carefully selected applications.

1. INTRODUCTION

Because of the growing worldwide energy shortage it is becoming necessary to transfer heat in processes with less entropy gain than was previously considered acceptable. This usually means designing to lower temperature differences. In reboilers and vaporizers, the heat flux is very sensitive to the temperature difference, and at very low temperature differences it can be so low that some type of modification of the surface to enhance boiling can prove economical. The reentrant cavity enhanced boiling surface has been studied before by several workers, e.g., Refs. 1 — 4. These surfaces provide cavities which trap vapor and permit nucleation of vapor bubbles at much lower temperature differences than is possible with smaller naturally occurring surface nucleation sites.

One recent study (Ref. 5) for the first time compared data for boiling on single tubes with that for boiling outside tube bundles, using a common hydrocarbon and two types of commercial reentrant cavity tubes. The findings were significant since they showed that boiling data from enhanced single tubes do not necessarily indicate the amount of enhancement which can be obtained from an enhanced multitube reboiler or vaporizer. These data also showed that great care must be used in generalizing enhanced tube data to apply to conditions or fluids other than those used for the particular test.

In this paper some of the basic principles of boiling heat transfer for single tubes and bundles are reviewed and a system for application to enhanced surfaces is suggested. Although completely general design correlations cannot be recommended at this stage of development, certain limits to the range of applicability of enhanced surfaces for boiling are defined from basic principles, which should be useful in initial evaluations.

1.1. The Surface Factor Concept for Single Tubes

Before discussing design for boiling in bundles, the single tube nucleate boiling relationships, which form a basis for this work, should be reviewed. A large number of correlations are available in the literature. These have been investigated in several references, e.g., Refs. 6, 7, 8, and for the most part have been found to give unsatisfactory results over the wide range of physical properties which must be covered by the process industries. A recent comprehensive study by Stephan and Abdelsalam (Ref. 9) improved this situation considerably and produced separate correlations for the different fluid categories: water, hydrocarbons, cryogenics, and refrigerants. For hydrocarbons, the equation recommended by Stephan and Abdelsalam is

$$\alpha_{nb} = C_2 \, q^{0.67} \tag{1}$$

where C_2 is a function of pressure which varies slightly for different hydrocarbons.

A generalized form in terms of dimensionless groups is also given. A similar equation, which was previously recommended for design of hydrocarbon reboilers (Ref. 8), was presented by Mostinski (Ref. 10)

$$\alpha_{nb} = C_m \, q^{0.7} \, P_c^{0.69} \, F_p \tag{2}$$

The term F_p is a pressure correction function of reduced pressure which was curve-fit as a polynomial by Mostinski and simplified to a "safe" design equation in Ref. 11 as follows.

$$F_p \cong 1.8 \, (P/P_c)^{0.17} \tag{3}$$

The constant C_m for Eq. 2 in SI units was given as $3.75 \, (10)^{-5}$.

Equations 1 and 2 give similar results for typical hydrocarbons and are reasonable base equations for pure components and commercially available plain tube surfaces. For mixtures another correction (F_c) is necessary to account for composition gradient effects (Ref. 11).

For special surfaces, it is here suggested that a surface factor, F_s, be defined as the ratio of temperature differences (or heat transfer coefficients) at a given heat flux, as illustrated in Fig. 1. In terms of the nucleate boiling heat transfer correlation (Eq. 2) the surface factor is applied as

$$\alpha_{nb} = C_m \, q^{0.7} \, P_c^{0.69} \, F_p \, F_c \, F_s \tag{4}$$

Notice that the surface factor is a function of heat flux since the enhanced and plain tube boiling curves normally have different slopes. Usually F_s increases as q decreases.

An example of evaluation of F_s is obtained by referring to the data of Yilmaz et al. (Ref. 5) for a single plain tube and a single Gewa-T tube for boiling p-xylene at atmospheric pressure as shown in Table 1.

TABLE 1   Surface Factor (F_s) and Mostinski Nucleate Boiling Correlation Coefficient (C_m) For p-Xylene at Atm Pressure From Ref. 5 Data (SI units)

	Plain			Gewa-T		
q	ΔT_b	α	C_m (Eq. 2)	ΔT_b	α	F_s
40 000	22	1818	3.4 (10)$^{-5}$	4.3	9 302	5.12
100 000	30	3333	3.3 (10)$^{-5}$	8.6	11 628	3.49

Notice that the constant C_m derived from the measured data is very close to the generalized value from the Mostinski equation. The boiling curves for both the Gewa-T tube and the plain tube in the Yilmaz et al. data are almost straight

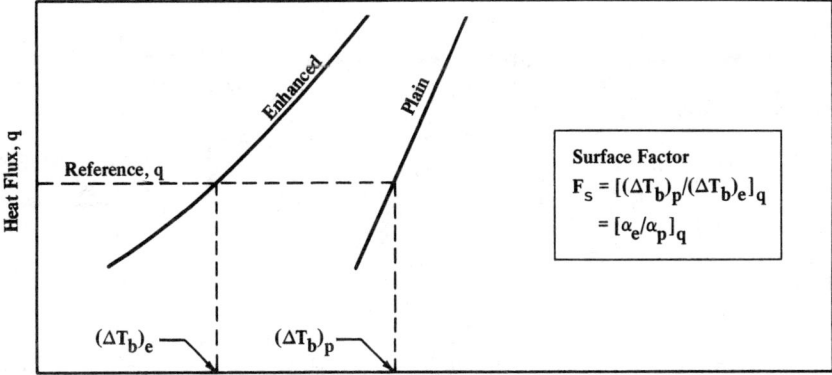

Fig. 1   Illustration of the Surface Factor Definition

lines on log-log paper, and the surface factor for this system (p-xylene at 1 atm pressure, Gewa-T tube) can be correlated as follows:

$$F_s = 410 \, q^{-0.414} \tag{5}$$

Equation 5 should not be extrapolated below a heat flux of about 10 000 W/m² (for which $F_s = 9.05$) and indicates that all advantage of the surface is lost ($F_s \to 1$) for high heat fluxes. Of course, enhanced surfaces are normally only used for designs at low heat fluxes for which the boiling side heat transfer coefficient is low and represents the controlling heat transfer resistance. It should be recognized that Eq. 5 is not general. Previous proprietary studies of a sintered surface tube are reported qualitatively in Ref. 8. It was found by HTRI that a more general form of the surface factor can be expressed as

$$F_s = C_s \, (q/q_r)^{m_1} \, (P/P_c)^{m_2} \, F_c^{m_3} \tag{6}$$

where the exponent, m_2 is negative (as is m_1) indicating a decrease in surface factor with increasing pressure. The exponent m_3 is positive, indicating a decrease in surface factor with an increasing mixture correction. The mixture correction F_c is shown in Ref. 10 to drop from the pure component value of 1.0 as the boiling range of the mixture increases to as low as 0.1 for fluids with very large boiling ranges. Speculated reasons for these effects are discussed in Ref. 8. In this paper attention will be turned to the effects on the surface factor of multiple tubes.

1.2   Extension of the Surface Factor Concept to Boiling in Tube Bundles

As suggested in Ref. 8, boiling heat transfer in tube bundles can be related to two types of heat transfer mechanisms, nucleate boiling and two-phase convection. A recent paper by Brisbane et al. (Ref. 12) describes the circulation effect in kettle reboilers. This effect was suggested earlier by Fair (Ref. 13) who also has recently initiated development of a mathematical model (Ref. 14). A further description of boiling in tube bundles is given by Collier (Ref. 15).

- **General.** Basically the local heat transfer coefficient in a kettle reboiler bundle can be defined in the same way as previously recommended for flow in tubes (Refs. 16, 17).

$$\alpha_b = s\,\alpha_{nb} + \alpha_{tp} \tag{7}$$

where s is a nucleate boiling suppression factor as will be discussed later. The nucleate boiling heat transfer coefficient, α_{nb}, can be obtained from Eq. 4 (or equivalent). The two-phase convective heat transfer coefficient, α_{tp} can be estimated as follows

$$\alpha_{tp} = \alpha_\ell \left(\frac{\Delta P_{tpf}}{\Delta P_\ell} \right)^{m_4} \tag{8}$$

The exponent, m_4, normally ranges between 0.4 and 0.5 depending on the efficiency of conversion of friction loss to heat transfer. The liquid-phase heat transfer coefficient (α_ℓ) and the liquid phase pressure drop (ΔP_ℓ) are obtained for crossflow of the liquid, as if flowing alone. Grant and Murray (Ref. 18) have published methods for the two-phase friction loss, ΔP_{tpf}, as have Diehl and Unruh (Ref. 19). Shellside two-phase pressure drop methods are also analyzed and evaluated by Ishihara et al. (Ref. 20). The greatest difficulty is in determining the actual flow rate in the bundle. This must be done by a pressure balance which is very sensitive to both the true driving head outside the bundle and the true two-phase density inside the bundle. The first of these depends on how much vapor escapes from the circumference of the bundle and the amount of disengaging area. The second depends on the amount of vapor-liquid slip in the bundle, which is a function of heat flux and vapor fraction, and for which no data presently exist in the open literature.

HTRI has recently developed a bundle circulation boiling model which was calibrated against overall performance of over 400 reboiler test runs. However, although the overall predictions by the method are very good, the circulation rates and true slip ratios could not be directly confirmed, and there is still a great need for basic data.

All circulation models show that there is a significant contribution of α_{tp} to the boiling process in bundles. In other words, the ratio of ($s\alpha_{nb} + \alpha_{tp}$) to α_{nb} is always significantly greater than 1. This ratio is termed F_b and has been developed empirically from data as described in Ref. 8. In Ref. 11, a conservative design value of $F_b = 1.5$ is suggested. However, values as high as 3 to 4 can be calculated depending on conditions. An empirical relationship for F_b has been developed from work described in Ref. 8 according to the following equation form

$$F_b = \frac{s\alpha_{nb} + \alpha_{tp}}{\alpha_{nb}} = 1 + \left[\frac{C_b}{(F_s)_1} \right]^{m_5} \left[\frac{1 - \psi_b}{\psi_b} \right]^{m_6} \tag{9}$$

The bundle geometry factor ψ_b is described in Ref. 11 and can be expressed as

$$\psi_b = \frac{\pi D_b L}{A} \tag{10}$$

As the number of tubes in the bundle approaches 1, ψ_b approaches 1 and F_b approaches 1. Therefore, α_{tp} becomes insignificant with respect to α_{nb}. Notice also that the surface factor $(F_s)_1$ is in the denominator of Eq. 9 so that as $(F_s)_1$ becomes large, F_b approaches 1, corresponding to α_{nb} becoming much larger than α_{tp}. However, the constants for Eq. 9 must be determined from actual bundle data, which are presently only in the proprietary domain.

The nucleate boiling suppression factor, s, in Eqs. 7 and 9 must also be determined from experimental data. This factor, as described by Fair (Ref. 16) and by Chen (Ref. 17), ostensibly accounts for the fact that with high velocity circulation, the thermal boundary layer in the vicinity of the nucleation sites is thinner, causing a slower bubble growth rate and deactivation of the larger nucleation sites. This phenomenon is especially significant for enhanced boiling sites which rely for their effectiveness on thick thermal boundary layers and relatively large bubbles.

ENHANCED BOILING SURFACES IN TUBE BUNDLES

Speculating on the results reported by Yilmaz et al. (Ref. 5) another theory of suppression with enhanced tubes is proposed by Bergles and Chyu (Ref. 21), who suggest that, with high flow velocity, surface tension forces (which feed liquid to the reentrant cavities) are overcome by shear forces. This starves the cavity of sufficient liquid for nucleation. This would suggest that nucleate boiling suppression would be greater for enhanced surfaces than for plain surfaces, since the boundary layer thinning and liquid starvation mechanisms would have an additive effect. Limited available data seem to indicate that this is true, as shown in a following section.

- **Effect of Bundle Boiling Mechanisms on Apparent Surface Factors for Bundles.** It is a logical assumption that enhancements, designed to increase the bubble nucleation capability of a surface, such as the reentrant cavity, would have the most effect on the nucleate boiling mechanism and much less on the convective boiling mechanism. Therefore, it can be deduced that the overall effect of surface enhancement on boiling heat transfer becomes less as the ratio of the convective boiling to the nucleate boiling component increases. This relationship may be expressed algebraically as follows.

We have defined a surface factor for a single tube above. An apparent surface factor for a bundle can be written in the same way. For the two cases we have

$$(\text{Single Tube}): \quad (F_s)_1 = \alpha_{e1}/\alpha_{p1} \qquad (11)$$

$$(\text{Bundle}): \quad (F_s)_b = \alpha_{eb}/\alpha_{pb} \qquad (12)$$

The heat transfer coefficients for the plain and enhanced bundles (assuming average values) are as follows:

$$\alpha_{pb} = s_p \alpha_{p1} + \alpha_{tp} \qquad (13)$$

$$\alpha_{eb} = s_e \alpha_{e1} + \alpha_{tp} \qquad (14)$$

The terms s_p and s_e are the nucleate boiling suppression factors for the plain and enhanced tubes, respectively. The nucleate boiling suppression factors are functions of α_{tp} and become equal to 1 when $\alpha_{tp} = 0$. It is assumed that α_{tp} is the same for both bundles and possible effects of surface roughness (which should be small) are ignored.

Equations 11 through 14 can be arranged to give a ratio of bundle surface factor to the single tube surface factor which provides some valuable insight into the effect of the convective component on the amount of enhancement possible in the tube bundles.

$$\frac{(F_s)_b}{(F_s)_1} = \frac{\left[\dfrac{s_e}{s_p}\right] + \left[\dfrac{\alpha_{tp}}{(F_s)_1 s_p \alpha_{p1}}\right]}{1 + \left[\dfrac{\alpha_{tp}}{s_p \alpha_{p1}}\right]} \qquad (15)$$

As a significant consequence, the following limits are seen:

A. If $\alpha_{tp} = 0$, $(F_s)_b = (F_s)_1$, i.e., as the two-phase convective component becomes insignificant, the bundle surface factor becomes equal to the single tube surface factor.

B. If $\alpha_{tp} \to$ large, $(F_s)_b \to 1.0$, i.e., as the two-phase convective component dominates, the performance of an enhanced tube bundle approaches that of a plain tube bundle.

C. If $\alpha_{tp} = \alpha_{p1}$, $\dfrac{(F_s)_b}{(F_s)_1} = \dfrac{\dfrac{s_e}{s_p} + \dfrac{1}{(F_s)_1 s_p}}{1 + \dfrac{1}{s_p}}$ \hfill (16)

D. If $\alpha_{tp} = \alpha_{e1}$, $\dfrac{(F_s)_b}{(F_s)_1} = \dfrac{\dfrac{s_e}{s_p} + \dfrac{1}{s_p}}{1 + \dfrac{(F_s)_1}{s_p}}$ \hfill (17)

Taking a more or less typical case with $F_b = 1.5$ from Ref. 11, we have

$$F_b = \frac{s_p \alpha_{p1} + \alpha_{tp}}{\alpha_{p1}} \tag{18}$$

$$\frac{\alpha_{tp}}{\alpha_{p1}} = 1.5 - s_p \tag{19}$$

Taking $s_p = 0.5$ (probably about a minimum value), $\alpha_{tp} = \alpha_{p1}$ and assuming $s_e = s_p$, then from Eq. 15,

$$\frac{(F_s)_b}{(F_s)_1} = \frac{1 + \dfrac{2}{(F_s)_1}}{3} \tag{20}$$

(For assumed case with $s_p = s_e = 0.5$, $F_b = 1.5$)

Therefore, if $(F_s)_1 = 5.12$ as in Table 1, at $q = 40\,000$ W/m^2, $(F_s)_b = 2.37$. In other words, if the ΔT_b would be reduced by a factor of 5.1 for a given heat flux for a single enhanced tube as compared to a single plain tube, the comparable reduction in ΔT_b for an enhanced bundle compared to a plain bundle would be only about 2.4. Although this is just one case with an assumed value of s_p, it is typical of the actual performance observed in Ref. 5. It is easy to show that, due to the influence of α_{tp}, the bundle surface factor can never be as great as the single tube surface factor and can potentially be much less.

Presently, more actual data on bundles are needed. However, as reliable circulation boiling models are developed, it should be possible to generalize the phenomena sufficiently to permit design of bundles based on measured single tube data. The HTRI circulation boiling model now does a reasonably good job of predicting overall enhanced bundle performance using single tube surface factors, suppression functions developed from plain tubes, and α_{tp} calculated from the two-phase pressure balance using empirically derived relations for the slip ratios.

One significant problem will be determination of the suppression factors. It is possible to estimate relationships for the apparent suppression factors for specific cases directly from data as follows. Taking as an example data from Ref. 5, the measured values are shown for single tubes and bundles of plain and Gewa-T enhanced tubes for boiling p-xylene at 1 atm pressure in Fig. 2.

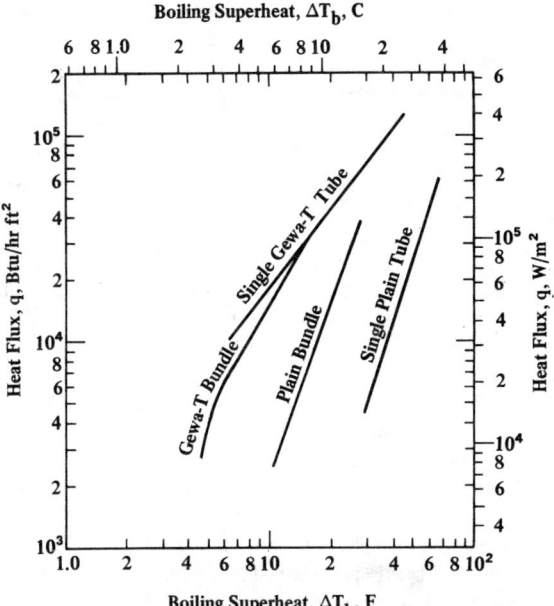

Fig. 2 Comparison of Performance of Single Enhanced and Plain Tubes to That of Enhanced and Plain Tube Bundles with p-Xylene at 15 psia (103 kPa) (Ref. 5)

At a heat flux of 40 000 W/m², $(F_s)_1$ = 5.12 from Table 1. From Fig. 2, $\alpha_{pb} \cong$ 3636 W/m² K, $\alpha_{eb} \cong$ 8000 W/m² K. Therefore, the apparent bundle surface factor $(F_s)_b$ = 2.2. Notice also that for this case, F_b = 3636/1818 = 2.0 for the plain tube bundle (greater than the value of 1.5 recommended in Ref. 11). Using Eq. 18, $\alpha_{tp}/\alpha_{p_1} = 2.0 - s_p$. If we assume s_p = 1.0 (the maximum value), α_{tp}/α_{p_1} = 1.0. From Eq. 15 or 16, it is possible to obtain a value for s_e/s_p.

$$\frac{2.20}{5.12} = \frac{\frac{s_e}{s_p} + \frac{1}{5.12}}{1 + 1} \tag{21}$$

$$\frac{s_e}{s_p} = 0.664 \tag{22}$$

Therefore, for this case it appears that the nucleate boiling suppression factor for the enhanced bundle can be less than that for the plain bundle. If this is true, it is consistent with the speculation of Bergles and Chyu (Ref. 21) which postulates an additional suppression mechanism for the reentrant cavity tubes.

Basic data are required before quantitative predictions can be made. However, the above simple algebraic relationships should be useful in evaluating new cases or interpreting data.

2. CONCLUSIONS

A review of some of the principles involved in enhanced boiling heat transfer has produced the following observations.

a. Definition of a surface factor can be very useful in describing and predicting enhanced boiling heat transfer.

b. A surface factor defined as the ratio of enhanced tube to plain tube heat transfer coefficient is shown to be an inverse function of heat flux for boiling a typical hydrocarbon at atmospheric pressure.

c. The surface factor concept can be extended to bundles with appropriate adjustments, and results in the conclusion that the improvement in heat transfer due to enhanced surface for bundles must always be less than that for single tubes.

d. In order to accurately predict the effect of enhanced surfaces in tube bundles from data obtained with single tubes, it would be necessary to have a reasonably reliable method of predicting the natural circulation induced flow in the tube bundle in order to obtain two-phase convective effects. However, if sufficient data are available, it may be possible to develop simplified empirical design equations of the form given by Eq. 9.

e. Nucleate boiling suppression due to thermal boundary layer thinning can have a significant effect on the performance of enhanced tube bundles. An additional postulated effect may cause suppression to be stronger for enhanced tubes than for plain tubes. Consequently, detailed design of enhanced tube geometry can be significant and will depend on the fluids to be boiled.

f. Additional data are required before acceptable design methods can be developed. It is especially important to investigate the effects of pressure and the effects of mixture composition. Based on present observations of the bundle, heat flux, and fluid property effects on boiling enhancement, the following tentative postulation is made. The amount of heat transfer improvement for a given reentrant cavity surface will *increase* as the following variables are *decreased*:

1) Number of tubes in a bundle. This decreases the α_{tp} so the relative effect of the nucleation enhancement is greater.
2) Heat flux. This makes the larger nucleation cavities relatively more effective.
3) System pressure. This also makes larger cavities more favorable nucleation sites.
4) Mixture boiling range. Pure components will tend to benefit more because the heavy components in mixtures may tend to concentrate in the reentrant nucleation cavities.

g. Even given all of the above potential problems and reservations, enhanced surfaces appear to offer great advantages in producing boiling heat transfer at low ΔT if properly understood and properly applied.

NOMENCLATURE

A	heat transfer area, m^2
C_b	constant, empirical relation for bundle boiling correction factor
C_2	constant, Stephan and Abdelsalam nucleation boiling heat transfer correlation
C_m	constant, Mostinski nucleate boiling heat transfer correlation
C_s	constant, HTRI suggested equation form for surface factor
D_b	bundle diameter, m
F_b	bundle boiling correction factor

ENHANCED BOILING SURFACES IN TUBE BUNDLES

F_c	mixture correction factor to nucleate boiling heat transfer coefficient
F_p	pressure function in Mostinski nucleate boiling heat transfer correlation
F_s	surface factor; ratio of enhanced surface heat transfer coefficient to plain surface heat transfer coefficient at a given heat flux
$(F_s)_1$	surface factor for single tubes
$(F_s)_b$	average effective surface factor for bundles
L	bundle length, m
$m_1, m_2, m_3, m_4, m_5, m_6$	constant exponents to be determined from data
P	operating pressure, Pa
P_c	critical pressure, Pa
q	heat flux, W/m²
q_r	reference heat flux, W/m²
s	nucleate boiling suppression factor
s_e	nucleate boiling suppression factor for enhanced surface
s_p	nucleate boiling suppression factor for plain surface
α	heat transfer coefficient, general term, W/m² K
α_b	boiling heat transfer coefficient for tube bundle, W/m² K
α_{e1}	nucleate boiling heat transfer coefficient for single enhanced tube, W/m² K
α_{eb}	boiling heat transfer coefficient for bundle with enhanced tubes, W/m² K
α_ℓ	convective heat transfer coefficient for liquid phase if flowing alone, W/m² K
α_{nb}	nucleate boiling heat transfer coefficient, W/m² K
α_{p1}	nucleate boiling heat transfer coefficient for single plain tube, W/m² K
α_{pb}	boiling heat transfer coefficient for bundle with plain tubes, W/m² K
α_{tp}	two-phase convective heat transfer coefficient, W/m² K
ΔP_ℓ	pressure drop for liquid phase if flowing alone, Pa

ΔP_{tpf} two-phase friction pressure drop, Pa

ΔT_b temperature difference between boiling fluid and tube wall, K

$(\Delta T_b)_e$ ΔT_b for enhanced tube, K

$(\Delta T_b)_p$ ΔT_b for plain tube, K

ψ_b bundle geometry factor; equal to bundle circumference divided by heat transfer area contained

REFERENCES

1. Czikk, A. M., Gottzmann, C. F., Ragi, E. G., Withers, J. G. and Habdas, E. P., "Performance of Advanced Heat Transfer Tubes in Refrigerant-Flooded Liquid Coolers," ASHRAE Trans., Vol. 76, Part I, No. 2132, 1970, pp. 96-109.

2. O'Neill, P. S., Gottzman, C. F., and Terbot, J. W., "Novel Heat Exchanger Increases Cascade Cycle Efficiency for Natural Gas Liquefaction," Advances in Cryogenic Engineering, Vol. 17, 1972, pp. 420-437.

3. Arai, N., Fukushima, T., Arai, A., Nakajima, T., Fujie, K., and Nakayama, Y., "Heat Transfer Tubes Enhancing Boiling and Condensation in Heat Exchangers of a Refrigerating Machine," ASHRAE Trans., Vol. 83, Part 2, 1977, pp. 58-69.

4. Nakayama, W., Daikoku, T., Kuwahara, H., and Nakajima, T., "Dynamic Model of Enhanced Boiling Heat Transfer on Porous Surfaces," Advances in Enhanced Heat Transfer, ASME, New York, 1979, pp. 31-43.

5. Yilmaz, S., Palen, J. W., and Taborek, J., "Enhanced Boiling Surfaces as Single Tubes and Tube Bundles," AIChE paper presented at the 20th National Heat Transfer Conference, Milwaukee, 1981.

6. Palen, J. W., and Taborek, J., "Refinery Kettle Reboilers," Chem. Eng. Prog., Vol. 58, No. 7, 1962, pp. 37-46.

7. Westwater, J. W., "Nucleate Pool Boiling," Part 2, Petro/Chem. Engr., Vol. 33, Sept. 1961, pp. 53-60.

8. Palen, J. W., Yarden, A., and Taborek, J., "Characteristics of Boiling Outside Large-Scale Horizontal Multitube Bundles," AIChE Symp. Ser., Vol. 68, No. 118, 1972, pp. 50-61.

9. Stephan, K. and Abdelsalam, M., "Heat-Transfer Correlations for Natural Convection Boiling," Int. J. Heat Mass Transfer, Vol. 23, 1980, pp. 73-87.

10. Mostinski, I. L., "Application of the Rule of Corresponding States for Calculation of Heat Transfer and Critical Heat Flux," Teploenergetika, Vol. 4, 1963, pp. 66-71, Engl. Abst., British Chem. Eng., Vol. 8, No. 8, 1963, p. 580.

11. Palen, J. W., "Shell and Tube Reboilers," Section 3.6, *Heat Exchanger Design Handbook*, E. U. Schlünder, ed., Hemisphere Publishing Co., Washington, D. C., (to be published 1981).

12. Brisbane, T. W. C., Grant, I. D. R., and Whalley, P. B., "A Predictive Method for Kettle Reboiler Performance," ASME Paper 80-HT-42, 19th National Heat Transfer Conference, Orlando, 1980.

13. Fair, J. R., "Vaporizer and Reboiler Design," Part I, Chem. Eng., Vol. 70, No. 14, July 8, 1963, pp. 119-224.

14. Moshinsky, A. K., "Analysis and Design of Process Reboilers," MS Thesis, Univ. of Texas, (J. R. Fair Advisor), 1981.

15. Collier, J. G., "Boiling Outside Tubes and Tube Bundles," Section 2.7.5, *Heat Exchanger Design Handbook*, E. U. Schlünder, ed., Hemisphere Publishing Co., Washington D. C. (to be published, 1981).

16. Fair, J. R., "What You Need to Design Thermosiphon Reboilers," Petroleum Refiner, Vol. 39, No. 2, 1960, pp. 105-123.

17. Chen, J. C., "Correlation for Boiling Heat Transfer to Saturated Fluids in Convective Flow," I & EC Process Design and Development, Vol. 5, No. 3, 1966, pp. 322-329.

18. Grant, I. D. R., and Murray, I., "Pressure Drop on the Shell-Side of a Segmentally Baffled Shell-and-Tube Heat Exchanger with Vertical Two-Phase Flow," NEL Report No. 500, February 1972.

19. Diehl, J. E. and Unruh, C. H., "Two-Phase Pressure Drop for Horizontal Crossflow Through Tube Banks," ASME Paper No. 58-HT-20, the ASME-AIChE Joint Heat Transfer Conference, Chicago, 1958.

20. Ishihara, K., Palen, J. W., and Taborek, J., "Critical Review of Correlations for Predicting Two-Phase Flow Pressure Drop Across Tube Banks," Heat Transfer Engineering, Vol. 1, No. 3, 1979, pp. 23-32.

21. Bergles, A. E. and Chyu, M. C., "Characteristics of Nucleate Pool Boiling from Porous Metallic Coatings," AIChE paper presented at the 20th Natl. Heat Transfer Conf., Milwaukee, 1981.

Analysis of Local and Overall Heat Transfer Enhancement by Double Fluted Condenser/Evaporator Tubes

GÜNTER SCHNABEL
Institut für Thermische Verfahrenstechnik
Universität Karlsruhe
Kaiserstrasse 12
D-7500 Karlsruhe, FRG

1. INTRODUCTION

In falling film evaporator plants condensing steam outside of tubes causes evaporation from thin liquid films flowing inside the tubes due to gravitational forces. For a given temperature difference $\Delta \vartheta = \vartheta_{s,c} - \vartheta_{s,ev}$ between the condensation and the evaporation side, the rate of heat transfer $\dot{q}$ can be improved by a factor of 2 or 3 if double fluted tubes as shown in Fig. 1 are used instead of smooth tubes /1,2/.

Fig. 1 Cross section of the double-fluted tube DF 1 /2/

Such high performance heat transfer surfaces are of great importance with regard to material savings and reduction of plant sizes. Moreover, the energy of steam available at small differences $\Delta\vartheta$ can be utilized.

The enhancement of overall heat transfer rates is due to improved heat transfer coefficients α_c and α_{ev} both for film condensation and evaporation on fluted surfaces.

Increased heat transfer coefficients for film condensation have first been reported by Gregorig /3/, who investigated experimentally and analytically the condensation on sinusoidal surfaces. However, no heat transfer coefficients were available for falling film evaporation on longitudinal fluted surfaces. Therefore the enhancement of overall heat transfer coefficients could not be explained nor predicted. This paper presents

- local heat transfer coefficients for the evaporation of falling liquid films on a fluted surface measured by a new technique

- overall heat transfer coefficients for double fluted tubes based on the experimental results for the evaporation side and on a modified Gregorig-model for the condensation side.

2. HEAT TRANSFER COEFFICIENTS FOR THE EVAPORATION OF FALLING FILMS ON A FLUTED SURFACE

2.1. Measuring Technique, Test Facility

The primary objective was to measure local boiling side heat transfer coefficients perpendicular to flutes of commercial /4/ sizes as shown in Fig. 2. They are defined as the local heat flux $\dot{q}$ divided by the difference between the wall temperature ϑ_w and the saturation temperature $\vartheta_s(p)$:

$$\alpha_{ev} = \frac{\dot{q}}{(\vartheta_w - \vartheta_s)_{ev}} \quad . \quad (1)$$

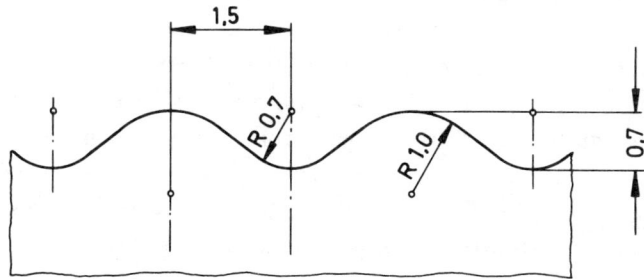

Fig. 2 Cross section of the fluted wall investigated in this work

According to eq. (1) it was required to measure $\dot{q}$ and ϑ_w locally along the flutes from the crests to the valleys with distances of a few 1/10 mm. This was not possible by use of conventional measuring techniques. Instead a new technique, which is shown schematically in Fig. 3, had to be developed.

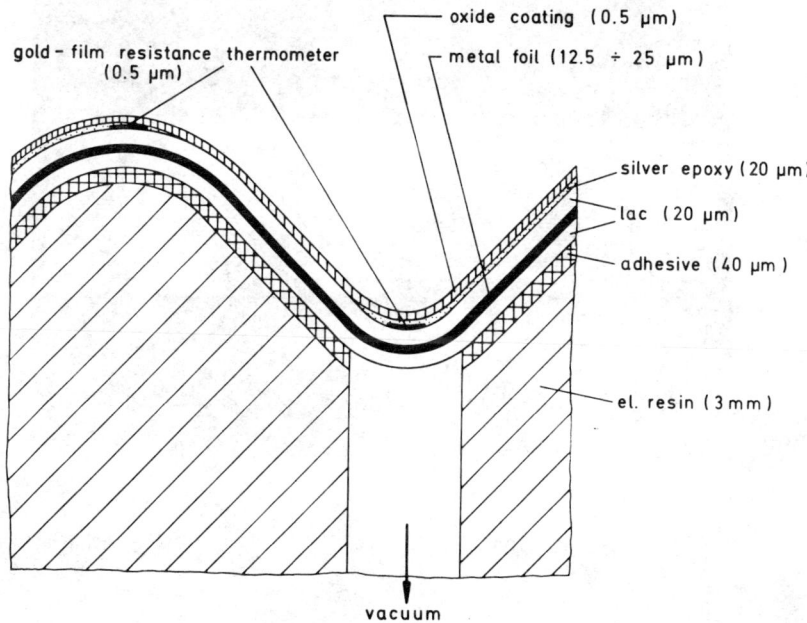

Fig. 3 Schematic of the measuring technique

The heated surface is a thin metal foil 12.7 or 25 µm thick. In this foil an
electric current generates a uniform heat flux density. The metallic surface
is insulated by a 25 µm thick coating of baking varnish. The insulated foil is
fixed on a base of electrical resin by use of adhesive and by vacuum in the
valleys. Because of the low thermal conductivity of the resin and the very
small thickness of the foil, no significant lateral heat flow occurs, even if
there are temperature gradients along the surface. Thus the distribution of
local heat transfer coefficients produces an analogeous temperature distribution.
To measure these local surface temperatures thin stripes of gold are deposited
along the flutes by vacuum evaporation using a metal mask with six
narrow 0.2 mm slits between the ridge and the valley. Each of the stripes is
used as a resistance thermometer. The connections are made of thin silver epoxy
layers, which do not disturb the film flow. At last, again by vacuum evaporation,
the whole test piece is coated with metal oxide thus causing the same
wetting behaviour as on a metal surface. Fig. 4 is a photograph taken of an
assembled test section being the core of the experimental setup schematically
shown in Fig. 5.

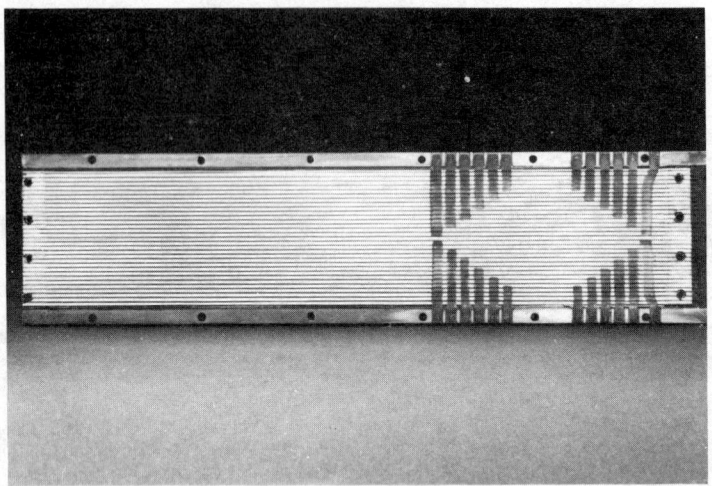

Fig. 4 View of a test section

DOUBLE FLUTED CONDENSER/EVAPORATOR TUBES

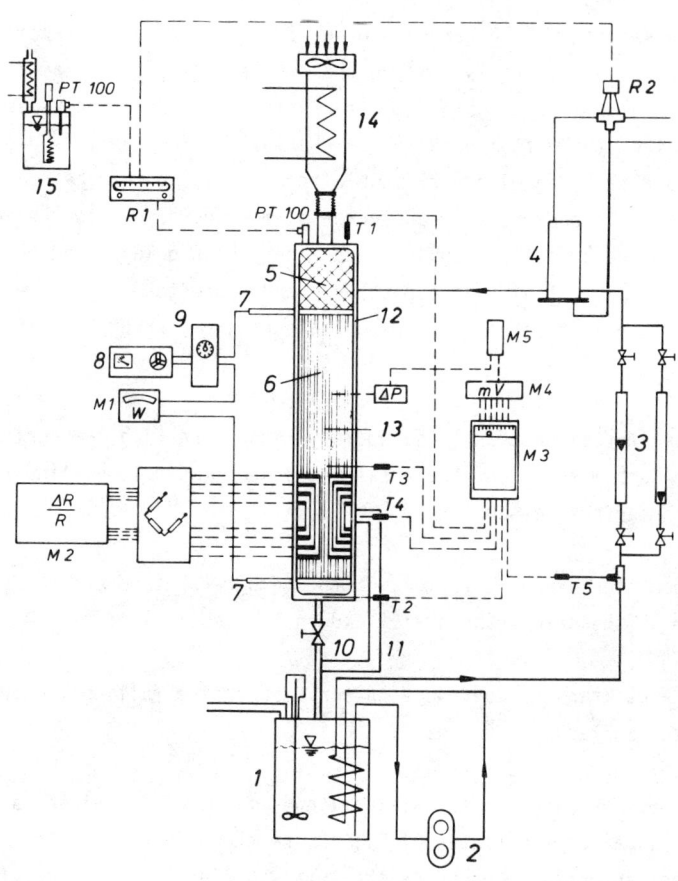

1 thermostat	10 liquid drain	M3 recorder
2 geared pump	11 vapor drain	M4 dig. voltmeter
3 rotameter	12 shell	M5 pressure diff. gauge
4 heat exchanger	13 double-paned window	
5 porous distributor	14 air heater	R1 temperature control
6 test section	15 boiler	R2 valve
7 connection	M1 wattmeter	T1-T5 thermocouples
8,9 transformers	M2 (strain)-gauge	

Fig. 5 Test assembly

The liquid circulates in a closed loop. It is pumped from a storage-thermostat through flow meters to a heat exchanger. There it is heated to the boiling point and subsequently penetrates a porous liquid distributor. From its front side a thin liquid film falls down on the test section. The steam is drawn off sideways and returns to the storage after being condensed. The remaining liquid drains from the bottom back to the thermostat. The shell is closed by a double-paned air-heated window to enable the observation of the boiling liquid film. The electric current is transferred from transformers to the metal foil of the test piece by gilden copper connections. Each gold stripe resistance thermometer is part of a Wheatstone bridge and connected to a gauge which supplies the bridges with a high frequency alternating current and records the resistances.

In seperate calibration runs, where - without heating the surface - the film temperature was varied, the resistances were found to be linear functions of the local surface temperatures.

Thermocouples are used to measure the temperatures of the vapor and the liquid film at the inlet, the outlet and on the test piece.

2.2. Local Heat Transfer Coefficients for Evaporating falling liquid Films on a fluted Surface

The experimental data for falling liquid films of deionized water evaporating at p=1 bar on the flutes of Fig. 2 are shown in Fig. 6. The local heat transfer coefficients α_{ev} based on the total boiling surface area are plotted vs. the projected ridge-valley spacing; z=o corresponds to the ridge, z=1.5 mm to the valley. For comparison the corresponding heat transfer coefficients on a smooth surface /5/ α_{sm} are noted and indicated by the dashed line. $\bar{\alpha}_{ev}$ is the average heat transfer coefficient of the measured α_{ev}-distribution. Parameters are the heat flux density $\dot{q}$ and the film Reynolds number Re=$\dot{m}/\eta$, where $\dot{m}$ is the total mass flow rate per unit of the projected width of the test section.
At the highest investigated Reynolds number (Re=1731) the local heat transfer coefficients do not greatly vary from the crest to the valley and also hardly differ from α_{sm}. With decreasing Reynolds numbers the liquid films on the ridges become thinner and consequently the heat transfer coefficients increase. This can be seen from the right to the left of Fig. 6. At Re=852 the measured

DOUBLE FLUTED CONDENSER/EVAPORATOR TUBES

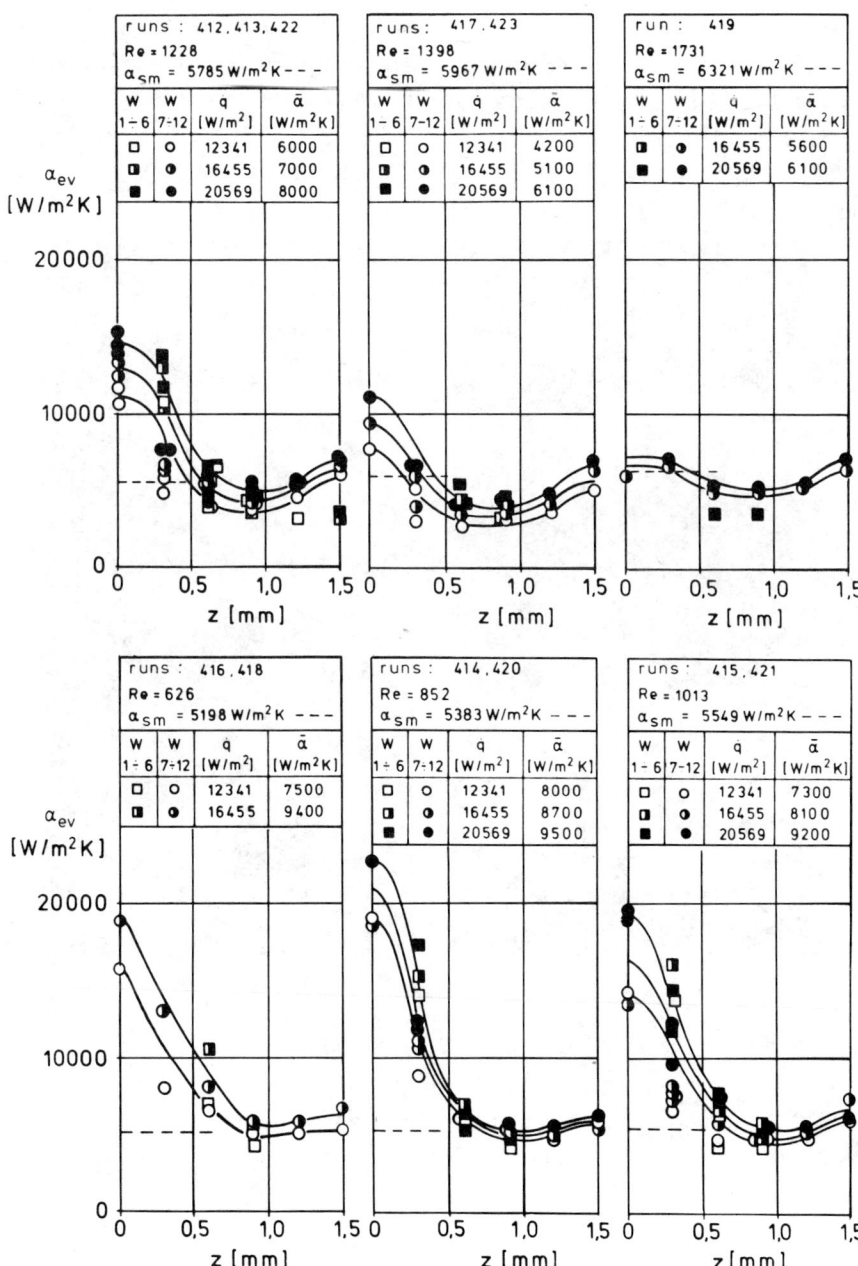

Fig. 6 Local heat transfer coefficients for evaporating water films at $p \simeq 1$ bar on a fluted surface as shown in Fig. 2.

α_{ev}-distribution is optimal. At $Re\tilde{=}600$ the crests become dry, causing minor heat transfer coefficients. At the best the local heat transfer coefficients on the ridges are 4 times higher than on smooth tubes. In the average the heat transfer is augmented by 50 to 80%. Further 20% must be added, allowing for the total to projected area ratio.

The heat transfer is slightly influenced by the heat flux density q. This becomes clear, looking at the boiling liquid films (Fig. 7).

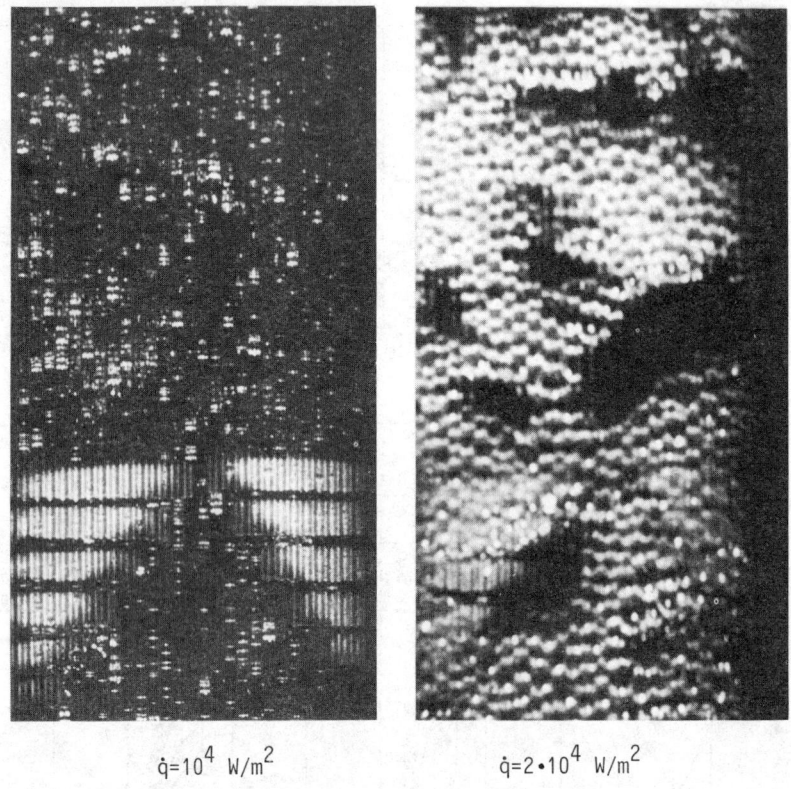

$\dot{q}=10^4$ W/m^2 $\dot{q}=2\cdot10^4$ W/m^2

Fig. 7 View of boiling water films at various heat fluxes

Fig. 8 shows the data for evaporating refrigerant R 113 films. The heat transfer was found to be quite similar to the water runs. Only the Reynolds numbers are lower.

DOUBLE FLUTED CONDENSER/EVAPORATOR TUBES

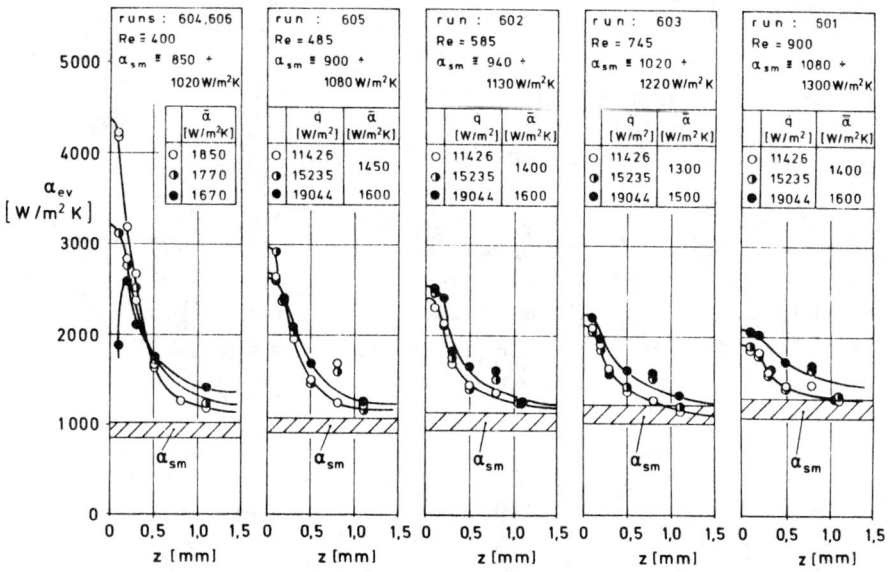

Fig. 8 Local heat transfer coefficients for falling liquid films of refrigerant R 113 on the fluted surface from Fig. 2.

If heat transfer coefficients are required for other flutes than investigated in this paper, it is proposed to use corresponding Reynolds numbers Re_{eq} and Re_{act} as schematically shown in Fig. 9. At both Reynolds numbers Re_{eq} and Re_{act} the different flutes are filled to the same length z_f.

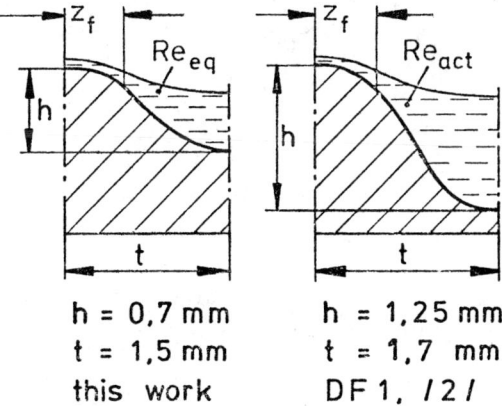

Fig. 9 Corresponding Reynolds numbers Re_{eq}, Re_{act} for various flutes. The liquid/vapor interfaces are parallel.

Then it is assumed, that - especially on the ridges - the distributions of the heat transfer coefficients are the same at Re_{eq} and Re_{act}. A method for calculating these Reynolds numbers is explained in /6/.

3. OVERALL HEAT TRANSFER ENHANCEMENT BY DOUBLE FLUTED TUBES

Now, from the measured local heat transfer coefficients α_{ev} on the evaporation side and from condensing film coefficients α_c, calculated by a modified Gregorig model /6/, overall heat transfer coefficients k can be computed. From Fig. 10 it is apparent, that the heat conduction in the tube wall significantly influences the heat transfer process.

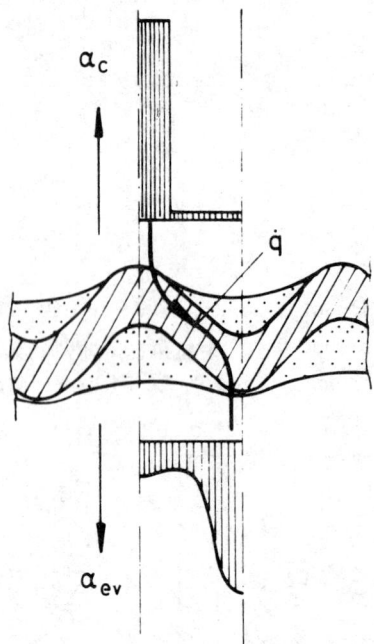

Fig. 10 Boundary conditions α_c and α_{ev} for heat transfer on double fluted tubes and main direction of heat flux $\dot{q}$.

Only if most of the heat is conducted from the ridge on the condensation side to the ridge on the evaporation side the overall heat transfer will be decisivly augmented. In order to compute overall heat transfer coefficients the two-dimensional heat conduction problem must be solved with the local distributions of α_c and α_{ev} as boundary conditions. A finite-difference method was used to calculate the distributions of the temperatures inside the tube wall and the local heat fluxes along both surfaces for the DF 1-tube. Fig. 11 shows the grid, which subdivides the DF 1-tube-wall, and sample boundary conditions α_{ev} and α_c.

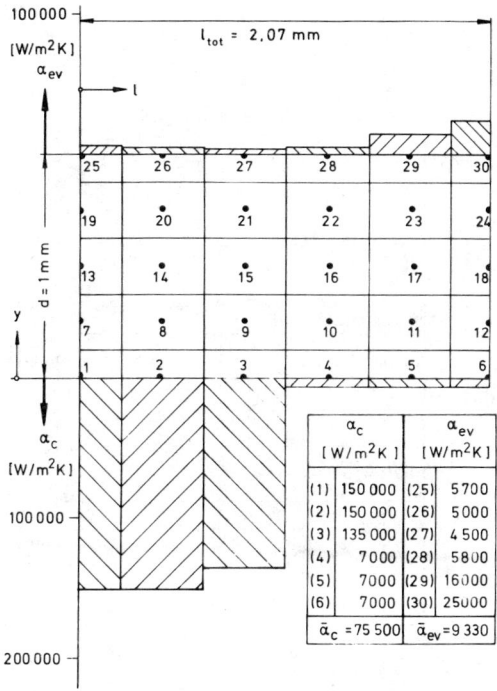

Fig. 11 Grid and boundary conditions for the numerical computation of temperature and heat flux distributions in fluted tube walls.

However, α_{ev} and α_c cannot be taken directly from Fig. 6 or the equations (in /6/), but must be determined by trial-and-error from the numerical solution, i.e. from the results for the surface temperature distributions $\Delta\vartheta_c$ and $\Delta\vartheta_{ev}$. The table of α_c and α_{ev} in Fig. 11 coincides with the final results, which have been obtained for condensing and evaporating water at an overall temperature difference $\Delta\vartheta = \vartheta_{s,c} - \vartheta_{s,ev} = 112.5 - 110.0 = 2.5\,K$ and a conductivity of the 1 mm CuNi10Fe-DF 1-tube of $\lambda = 46\,W/m^2K$. Fig. 12 shows the temperature distribution inside the tube wall and the local heat fluxes $\dot{q}$. From the average heat flux density $\dot{q} = 1.55 \cdot 10^4\,W/m^2$ an overall heat transfer coefficient k can be defined

as
$$k_1 = \dot{q}/\Delta\vartheta = 6200 \text{ W/m}^2\text{K}$$
based on the total area, or
$$k_z = k_1 \cdot l_{tot}/t = 6200 \cdot 2.07/1.7 = 7550 \text{ W/m}^2\text{K}$$
based on the projected area.

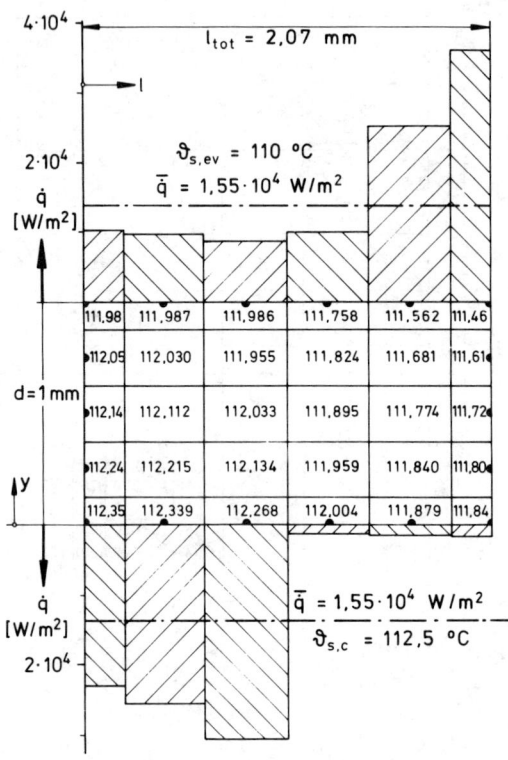

Fig. 12 Temperature distribution and local heat flux density.

These computations have also been carried out for other overall temperature differences and been compared with the experimental data from /2/. In Fig. 13 the numerical and the experimental overall heat transfer coefficients k_z are plotted vs. the overall temperature difference $\Delta\vartheta$. At differences $\Delta\vartheta > 2.5$ K the own experimental data for the evaporation film coefficients α_{ev} were evaluated by extrapolation. Because of the uncertainty, whether α_{ev} further increases with $\Delta\vartheta$ or not, the results of the computation lie within the cross-hatched field. The circles are the experimental data for the DF-1-tube. As comparison experimental and calculated overall heat transfer coefficients

DOUBLE FLUTED CONDENSER/EVAPORATOR TUBES

for smooth tubes are plotted. Fig. 13 proves, that fluted tubes can enhance heat transfer rates by a factor of 2 to 3. The experimental and theoretical results are in good agreement.

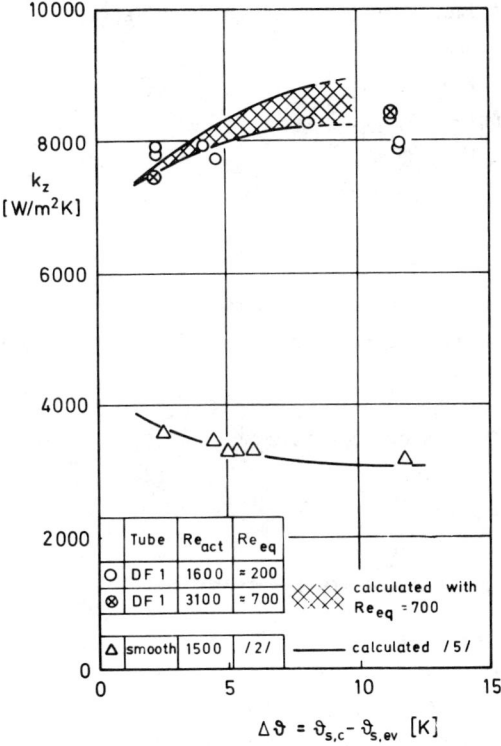

Fig. 13 Overall heat transfer coefficients for condensing and evaporating water on the double fluted tube DF 1 (based on the projected area).

Acknowledgment: The author wishes to thank the DFG (Deutsche Forschungsgemeinschaft) for the financiel support.

Symbols

h	m	height of a flute
k	W/m²K	overall heat transfer coefficient
l	m	coordinate along the flute surface
m	kg/ms	mass flow rate per unit width or length
p	N/m²	pressure
$\dot{q}$	W/m²	heat flux density
Re	-	Reynolds number: Re=m/η
s	m	film thickness
t	m	ridge-valley projected distance
u	m/s	velocity
x	m	vertical coordinate
y	m	coordinate across the film
z	m	horizontal coordinate

Greek letters

α	W/m²K	heat transfer coefficient
η	kg/ms	dynamic viscosity
ϑ	°C	temperature
λ	W/mK	thermal conductivity
ν	m²/s	kinematic viscosity

subscripts, superscripts

act	actual (see Fig. 9)
c	condensing side, condensate
eq	equivalent Reynolds number or wall thickness
ev	evaporating side
s	saturation temperature
sm	smooth
tot	total length along the flute surface
w	wall
z	based on the projected area

Literature

/1/ OSW Symposium on Enhanced Tubes for Distillation Plants, March 1969

/2/ Metallgesellschaft AG, Frankfurt: Entwicklung und Prüfung von Halbzeugen für Meerwasserentsalzungsanlagen. BMFT-Forschungsvorhaben CVEO 161 und 163

/3/ Gregorig, R.: Hautkondensation an fein gewellten Oberflächen. Zeitschrift für Angewandte Math. und Physik (ZAMP) 5 (1954), S. 36/49

/4/ Firma VDM, P.O.Box 100167, D-4100 Duisburg

/5/ Schnabel, G. und Schlünder, E.U.: Wärmeübergang von senkrechten, glatten Wänden an nichtsiedende und siedende Rieselfilme. Verfahrenstechnik 14 (1980), 1, S. 79/83

/6/ Schnabel, G.: Bestimmung des örtlichen Wärmeüberganges bei der Fallfilmverdampfung und Kondensation an gewellten Oberflächen zur Auslegung von Hochleistungsverdampfern. Dr. sci. thesis, Universität Karlsruhe, 1980

Steam Heated Water Bath Evaporators for Liquefied Gases

W. SÜSSMANN and D. U. RINGER
LINDE AG
Division TVT Munich
D-8023 Höllriegelskreuth, FRG

ABSTRACT

Steam heated water bath evaporators are frequently used for the evaporation of liquefied gases with low boiling points (e. g. O2, N2, CH4, C2H4). This paper presents investigations about increasing the efficiency of such evaporators as well as some design criteria to avoid severe problems during operation.

INTRODUCTION

Low temperature process plants frequently require heat exchangers where the liquefied gas is evaporated again. In many cases, the evaporator is not needed all the time and is operating then at stand by level. For these large changes in capacity, discontinuous operation or emergency purposes, the favoured solution for such an exchanger will often be a steam-heated water bath evaporator.

The main part of such evaporators is a pressureless vessel filled with water, in which a tube bundle is submerged. The water is kept at the required temperature by the injection of steam. The bundle may be similar to a TEMA type exchanger with U-tubes. If, however, tube length and heat transfer surface become large, a coiled bundle is preferred to reach a more compact design. LINDE AG manufactures both types of water bath evaporators. Figure 1a shows a schematic of such an apparatus with coiled tubes.

Typical conditions for the process fluid and the water are shown in figure 1b. The fluid flows inside the tubes and enters the heat exchanger in liquefied state. First, it is heated up to the boiling temperature. Then the evaporation takes place and afterwards the gas is heated up to the desired outlet temperature, which is in most cases close to the ambient one.

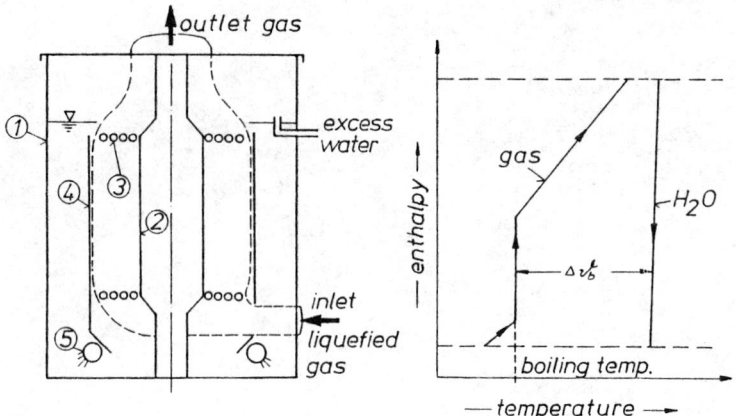

① pressureless vessel
② core-tube
③ coil
④ jacket
⑤ steam-injecting tubes

typical process conditions
for waterbath evaporators

1a) 1b)

Fig. 1: Schematic of a steam-heated water bath evaporator and typical process conditions

The installation of a water bath is recommended - instead of directly heated exchangers with steam on the shell side - if large and rapid changes in the gas flow rate occur and only small deviations in the gas outlet temperature are warranted. Water bath evaporators are also advantageous in emergency cases when the exchanger has to operate for a limited time (5 to 10 seconds) without sufficient supply of steam. The large heat capacity of the water volume acts as a buffer before regulations in steam supply responds to the changed process conditions. It is evident, that the required temperature of the water and its volume are basic aspects of design for such evaporators.

BASIC PHENOMENA

For the design of steam-heated water bath evaporators, three main points are of particular interest and will be discussed in more detail:

1   Heat transfer inside the tubes
2   Heat transfer across the bundle
3   Steam supply for large changes in heat duty

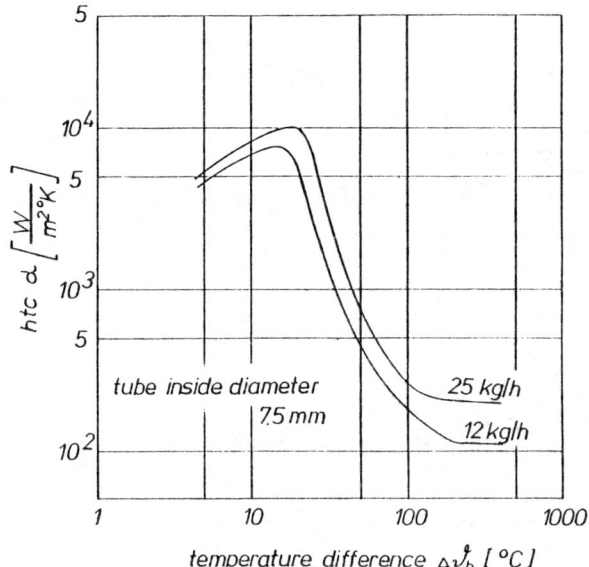

Fig. 2: Dependence of the heat transfer coefficient (htc) on the temperature difference $\Delta \vartheta_b$. Evaporation of N2 in a horizontal tube, according to Weishaupt /1/.

Heat Transfer Inside the Tubes (Gas Side)

The temperature difference $\Delta \vartheta_b$ (see fig. 1b) between the evaporating gas and the water is usually very high. In many cases the temperature of the water bath is about 60 °C. Considering e. g. nitrogen at a pressure of 10 bar, the boiling temperature is -168 °C. Hence we have a temperature difference of 238 °C. Weishaupt /1/ has made some investigations about the influence of the temperature difference on the heat transfer. His results for nitrogen are shown in figure 2. It can be seen, that for small values of $\Delta \vartheta_b$ (nucleate boiling), the heat transfer coefficient is about 50 times higher than for great $\Delta \vartheta_b$'s (film boiling, Leidenfrost Phenomenon).

According to figure 2, the point of operation for the above example lies in the film boiling region with low heat transfer rates. This is not a desirable condition, because it requires large heat transfer areas, but it is the common range of operation for water bath evaporators.

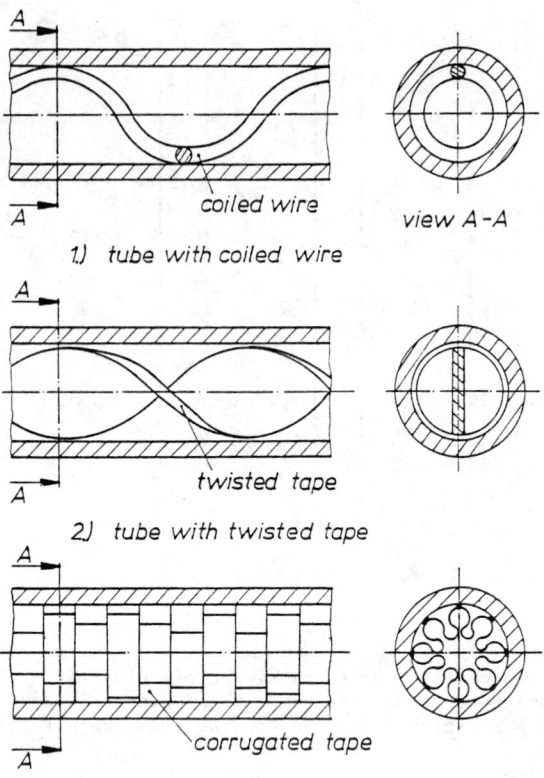

Fig. 3: Tested alternatives to improve heat transfer in the film boiling region

Therefore many investigations have been made to improve heat transfer rates in this range. The basic idea is to destroy or diminish the vapour film (Leidenfrost film) between the liquid and the wall. This can be achieved by higher mass fluxes - they give a turbulent and thinner film (see fig. 2) - or by inserting some special devices in the tube, which break up the film. The first alternative - increased mass flux - is not possible in most cases, because the flow rate is determined by process requirements. Furthermore, high liquid velocities would lead to very high gas speeds and thus high pressure drops because of the difference in density.

So the emphasis of research has been to investigate special devices, which could be inserted in the tube and would improve the heat transfer coefficient by destroying the Leidenfrost film. These devices should meet the following conditions:

- High efficiency in the film boiling region
- Low pressure drop
- Easy and economically to manufacture
- Possibility of coiling to produce large heat transfer areas within small volumes.

LINDE has tested three possible alternatives, which are shown in figure 3. The results for alternatives 1 and 3 are published by Weishaupt /1/, alternative 2 is discussed in this paper.

Alternative 1 is a simple coiled wire which fits at the inner tube wall. This gives an improvement of the heat transfer rates by 2 to 3, compared with a plain tube. But the improvement diminishes with higher flow rates and very high temperature differences.

A twisted tape - alternative 2, which was i. e. suggested by /2/ - gives about the same improvement in the heat transfer rates as the coiled wire without the limits in flow rate and Δv_b^2. But it has a higher pressure drop. Results for different tapes will be shown later. Both alternatives are easy and cheap to manufacture and they can be coiled.

The highest improvement of heat transfer rates was obtained with corrugated tapes in the tubes. The increase in pressure drop was moderate, comparable to the result for the twisted tapes. But these tubes cannot be coiled and furthermore they are difficult to manufacture and therefore very expensive.

Heat Transfer Across the Bundle (Water Side)

The heat transfer coefficient on the outer side of the tubes is determined by the velocity of the water which flows around them. For a given temperature of the water bath, the flow rate of the water has to be high enough to assure a wall temperature above $0\,^\circ C$, i. e. to avoid the formation of ice.

In the case of pure free convection, the water flow rate mainly depends on the temperature difference between water bath and tube wall. To maintain the temperature of the water, steam is injected. For a suitable design, this steam injection results in improved water flow conditions. If this amelioration is not sufficient, the water has to be revolved by a pump, which is not a preferred solution. For the moment, the interaction of water bath temperature, steam injection and water velocity can hardly be evaluated and extensive experience from operating plants is necessary for a designer.

Steam Supply System

As stated above, large differences in the flow rate of the liquefied gas, in some cases ranging from 0 - 100 %, are a common fact for water bath evaporators. Therefore the amount of steam, which covers the heat losses to the environment as well as to the gas, has to vary very much. Hence the steam injection system should be constructed in a suitable way.

Usually this system consists of simple tubes with some rows of holes. The steam is injected into the water bath through these holes. Both tubes and holes have to be designed in a way that during operation no water back flow into the tubes is possible. Otherwise the steam may condense in the tubes and cause dangerous pressure surges.

From the previous remarks it is evident, that more information is required for a safe and economic design of water bath evaporators. Thus LINDE has started an investigation program where different tubes, different twisted tapes and the influence of steam injection was tested.

EXPERIMENTAL SET-UP

For measurements of heat transfer coefficients inside the tubes and across a coiled bundle we used an experimental set-up which is shown in figure 4. The heart of the unit was a cylindrical vessel of 0.6 m in diameter and height. It contained 150 l of water.

In this water bath, three different coils with equal heat transfer surface but different tube diameters (48.3 x 2; 33.7 x 1.6; 25 x 1.5 mm) could be inserted. To obtain a definite circulation of the water, a second cylinder was installed in the centre of the vessel. This second cylinder could be covered at the top with a plate for tests with forced flow (see fig. 4). Otherwise, it was open at both ends. Steam was injected by a vertical pipe in the lower part of the inner cylinder. This construction assured downflow of the water across the bundle and upflow in the centre.

As test fluid we used technical pure nitrogen, which flew through the coiled tubes. Measured variables of the N_2 stream were inlet and outlet temperature, inlet and outlet pressure and flow rate.

The temperature of the bath was kept constant by a control loop for the steam. The amount of steam needed for this purpose was registrated. Thus, at known steam pressure, the heat duty could be calculated.

For calibration purpose of each coil, cold water was used. Here, flow rate and inlet and outlet temperature were recorded. The heat transfer coefficient for the water inside the tubes was calculated by means of published methods /3/, /4/. Then, together with the known heat flux from the needed amount of steam and the temperature of the water bath, an average value for the heat transfer coefficient across the bundle could be determined.

The obtained calibration curves are of the type:

$$\alpha_{out} = f(q) \tag{1}$$

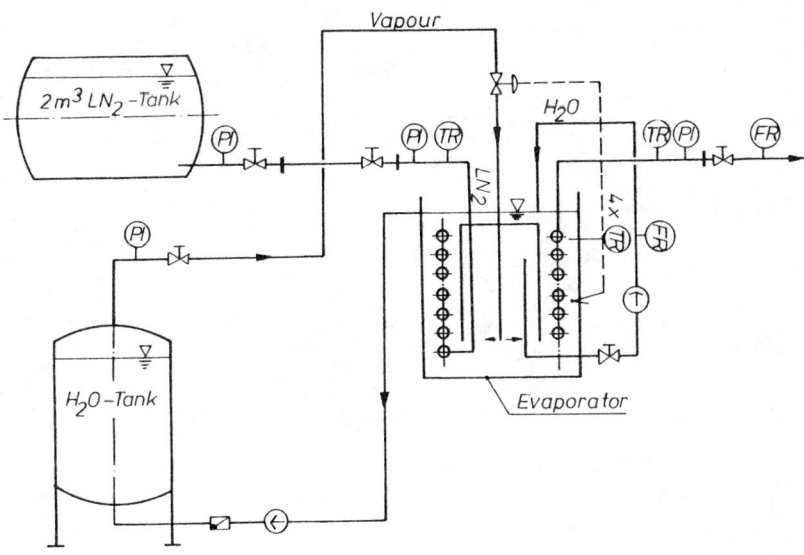

Fig. 4: Schematic of used experimental set-up

During the runs with nitrogen, the amount of steam for maintaining constant temperature of the water was recorded. This gave the required heat flux density q. From the calibration curves (see e. g. (1)), α_{out} for this case and thus an average heat transfer coefficient for the flow inside the tubes could be derived.

RESULTS

Heat Transfer in the Tubes

As stated above (see fig. 1b), heat transfer to the liquefied gas is combined with a change in state. Therefore one can distinguish several sections where different mechanisms of heat transfer occur.

First one has pure liquid which is heated up to the boiling point. Then film boiling occurs, where the heat transfer across the Leidenfrost film is the limiting parameter. As soon as the vapour content has reached a certain value, the remaining liquid is dispersed in the vapour and mist flow is present till the last drop is evaporated. The final section is heat transfer to pure gas.

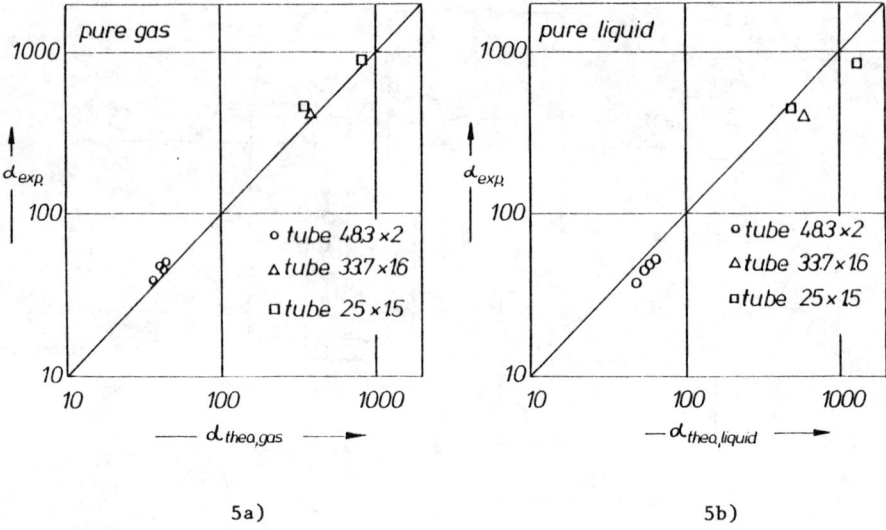

Fig. 5: Comparison of experimentally measured average heat transfer coefficients with calculated values for pure gas (5a) or pure liquid (5b) flow inside plain tubes

To calculate an average heat transfer coefficient which is equivalent to the measured values by combining all these steps is difficult. But if one mechanism is dominant, it could be sufficient to calculate this one and neglect the others. We supposed that the vapour phase is the dominant resistance along the tube. Therefore we calculated average heat transfer coefficients for pure gas flow in the tubes and compared them with measured values. For this calculation we used well known equations published in literature (see /3/, /4/). The result of this comparison is shown in figure 5. It can be seen, that the measured values are about 5 to 10 % higher than the calculated ones.

If we made the same calculations for pure liquid flow, we obtain figure 5b. Here, the theoretical values are 20 to 30 % too high. Talking into account the accuracy of the used equations and of the measured values, it seems to be sufficient to calculate the average heat transfer coefficient inside the tubes on the basis of pure gas flow. This is a reasonable and easy way on the safe side.

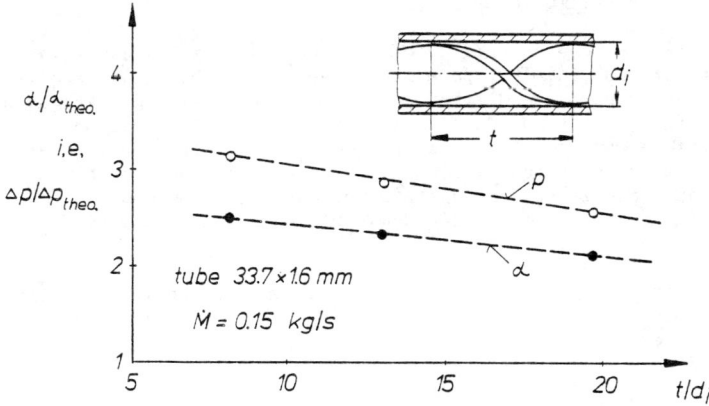

Fig. 6: Influence of twisted tapes with different drilling pitch on heat transfer coefficient and pressure drop. Theoretical values for pure gas flow in plain tubes.

To improve the heat transfer in the film boiling range, we tested three twisted tapes with different drilling pitch, which were inserted in the tubes. Average heat transfer coefficient and pressure drop were measured in the same way as for the plain tubes. As an example, the result for a tube diameter of 33.7 x 1.6 mm and a flow rate of 0.15 kg/s is given in figure 6. For better comparison, the experimental values are set into relation with theoretical values calculated for pure gas flow inside plain tubes according to /3/, /4/ (see fig. 5). It can be seen, that the increase in the heat transfer coefficient decreases with rising pitch. Also, most improvement is combined with highest rise in pressure drop. However, the rise inpressure drop is not acceptable in relation to the rise in heat transfer, so further tests with other tapes are on their way.

For the moment, a definite recommendation which device - coiled wire, twisted tape, corrugated tape - is best according to the above listed requirements is not possible. Each alternative has special advantages under special circumstances. The wire is sufficient for moderate flow rates and temperature differences, for higher ones a tape is better. For small heat transfer areas and no need for a coiled bundle, the corrugated tapes seem to be favourable. Though the best solution depends on the actual problem to solve.

Heat Transfer Across Coiled Tubes

The heat transfer on the outside of the tubes is determined by the water flow rate around them. For pure free convection, this flow rate is a function of the temperature difference between tube wall and water bath. The injection of steam causes additional water flow by its momentum, rising bubbles and differences in water density. The effect of this additional force on the heat transfer coefficient across the tubes depends on the geometrical design and is not evaluable numerically for the moment. However, it is evident, that steam-induced water flow should not conflict with convective flow.

Water bath evaporators have to operate even without steam supply for some time. Thus the basic design criterion for the required heat transfer area is free convection across the tubes. Therefore we used well known equations for free and additional forced convection (see /3/, /5/) to calculate theoretical heat transfer coefficients. We found, that for small amounts of steam, i. e. nearly pure free convection, theoretical and experimental values match quite well. For greater amounts of steam, a remarkable improvement up to a factor 2 to 3 was noted. The amelioration depends on the amount of steam injected and on the geometry of the coil and the steam injection system.

STEAM INJECTION SYSTEM

A proper design of the steam injection system is necessary for reliable operation of water bath evaporators. LINDE design criteria are based on experience from numerous types of exchangers operating under very different conditions.

One effect of poor design of the steam injection system is a reduced water flow across the tubes and therefore smaller heat transfer. But this is the minor problem. The worse case are very powerful steam surges, which can even destroy the whole apparatus. These surges are caused by water, which may flow through the injection holes into the steam pipes, if the system is operating at low steam rates. Then steam may condense just in the pipes and a pressure wave is built up. These problems do not occur at quick changes in steam supply, but at continuous operation of pipes, which were designed for high steam loads and are to operate at low ones. Therefore the problem is to design pipes and holes in a way, that no water may enter the steam injection system during operation.

A recent paper of Mersmann and al. /6/ deals quite well with this problem. These authors investigated the conditions of weeping and full operation of all holes on sieve trays. They found, that above certain values of the Weber- and Froude-Number, all holes of a sieve tray are operating and no weeping occurs. This basic idea has to be modified for a tube, which can be treated as a sort of bent sieve tray.

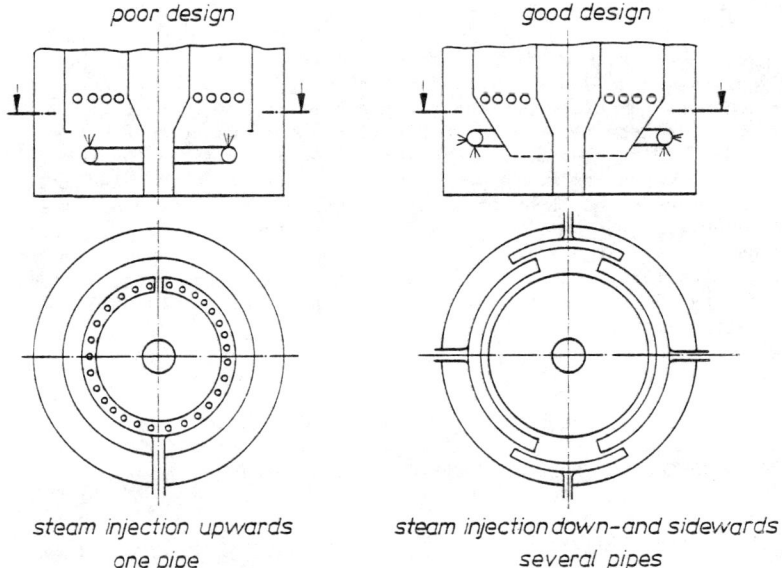

Fig. 7: Examples of poor and good design for the steam injection system

As a consequence of the ideas of Mersmann and al., large changes in heat duty, i. e. steam supply require not only one pipe for the whole range of operation, but different ones for different parts of the operating range. Because this splitting cannot be continued to infinity, a smallest pipe has to be set, where the occuring pressure surges are no longer dangerous. This smallest necessary diameter depends mainly on the total amount of steam required, e. g. for a total steam load of 40 t/h, a smallest tube of 50 mm diameter and 3 t/h steam load may be used.

Especially for stand-by services, small additional pipes are favourable to maintain the water temperature. In case of operation, they are switched off and the main pipe system takes over. An example for poor and good design of the steam injection system is given in figure 7.

CONCLUSION

The paper deals with basic design aspects for evaporation and heating-up of low boiling liquefied gas in water bath evaporators. As a result of the described measurements, that the average heat transfer coefficient inside plain tubes can be calculated with sufficient accuracy on the basis of pure gas flow. With special devices, which are inserted in the tube and break up the vapour film, the heat transfer rate can be doubled or tripled. Heat transfer on the water side can be estimated with the rules for free convection. The improvement of the heat transfer by the injected steam can be regarded as a safety factor. In designing the steam injection system it is important, that all holes are operating in the range of required heat duty. If necessary, several different pipes should be installed to assure safe operation.

REFERENCES

/1/ J. Weishaupt, Kältetechnik, 15(1963)8, p. 240-244

/2/ SUMITOMO PRECISION PRODUCTS CO. LTD., Osako, Japan
Company Publication, August 1977

/3/ VDI-WÄRMEATLAS, VDI-Verlag, Düsseldorf, West Germany, 1974

/4/ E. E. Abadzic, H. W. Scholz, Advances in Cryogenic Engineering, Vol. 18, Ed. K. D. Timmerhaus, Plenum Publishing Corp., N. Y.

/5/ W. Kast, O. Krischer, H. Reinécke, K. Wintermantel, Konvektive Wärme- und Stoffübertragung, Springer Verlag, Berlin Heidelberg New York, 1974

/6/ A. Mersmann, T. Pilhofer, K. Ruff, Chem.-Ing.-Techn., 48(1976)9, p. 759-764

Heat Transfer and Critical Heat Flux at Flow Boiling of Nitrogen and Argon within a Horizontal Tube

HANS-MICHAEL MÜLLER, W. BONN, and D. STEINER
Institut für Thermische Verfahrenstechnik
Universität (TH) Karlsruhe
D-7500 Karlsruhe, FRG

ABSTRACT

Local heat transfer and critical heat flux during flow boiling of nitrogen and argon have been studied experimentally in a smooth horizontal copper tube with 14 mm i.d. The main parameters such as heat flux $\dot{q}$, mass velocity $\dot{m}$, flow quality $\dot{x}$ and reduced pressure p_r have been varied in a wide range.

The influence of the main parameters on the heat transfer coefficients will be discussed. Using recommended equations, the calculated values do not agree satisfactory with experimental data. Therefore a modified correlation is proposed. The measurements show, that the critical heat flux $\dot{q}_{crit}$ depends considerably on the pressure and on the flow quality, while the influence of the mass velocity can be neglected in the observed range. Knowing this, a new correlation for the critical heat flux has been developed. The agreement between predicted and measured values is good.

During the measurements with argon a new phenomenon has been noticed. While the adjusted parameters remained constant, the temperature difference between wall and fluid increased slowly up to the burn out. This transient change of the heat transfer coefficient lasted up to 20 h and was probably caused by very small amounts of dissolved CO_2.

1. INTRODUCTION

Stimulated by the development of the nuclear technologie, the interest in flow boiling heat transfer has grown considerably during the last thirty years. The number of papers and reports dealing with this subject has reached 10 000 till 1970 and is still growing exponentially /9/. Yet, despite these efforts, the mechanism of boiling within a tube is still not understood. This is due to the large number of parameters, which influence the heat transfer. According to Chawla /7/ the heat transfer coefficient depends on the following 15 parameters: (Symbols see: Notation)

$$\alpha = f(\dot{m}, \dot{x}, d, l, k, g, \rho', \rho'', \eta', \eta'', c_p', c_p'', \lambda', \lambda'', \sigma) \qquad (1)$$

Bonn /3/ found, that the thickness and the heat conductivity of the tube wall have an influence on the heat transfer coefficients evaluated from experiments, too. To have a complete set of data not influenced by different test tubes and according to the increasing interest in experimental data for forced convection boiling at low temperatures, extensive measurements have been made by Steiner /14/ and Bonn /2/.

With an apparatus expecially constructed for this purpose, it was possible to vary the main parameters independently from each other and also to check seperately their influence on the heat transfer coefficient.

The parameters were varied in the following range:

heat flux $\quad\quad\quad 210 \text{ W/m}^2 < \dot{q} < 100\,000 \text{ W/m}^2$

mass velocity $\quad\quad 44 \text{ kg/m}^2\text{s} < \dot{m} < 470 \text{ kg/m}^2\text{s}$

flow quality $\quad\quad\quad\quad\quad 0 < \dot{x} < 1$

reduced pressure $\quad\quad 0.034 < p_r < 0.93 \quad$ for nitrogen

$\quad\quad\quad\quad\quad\quad\quad\quad\quad 0.034 < p_r < 0.4 \quad$ for argon

2. EXPERIMENTAL EQUIPMENT AND TEST PROCEDURE

2.1 Test assembly

Fig. 1 shows a schematic diagram of the test apparatus. To avoid pollution of the test gas, the flow system consists of two closed loops, the test loop and the cooling loop for recondensing of the test substance. The test substance is recondensed in the condenser (1), and then stored in the tank (2). The liquid is delivered by a pump (3) to the horizontal part of the test circuit. There the volumetric flow rate $\dot{V}$ is measured with a turbine flow meter. The flow rate is controlled by valves (7) and the pump bypass (4). Pressure p and temperature T of the liquid are measured behind the turbine, to make sure, that the liquid is still subcooled. Between the turbine meter and the test section, the desired vapour quality is produced by four pre-evaporators (8). In the following observation glass (9) the flow pattern can be observed. Then the two-phase flow enters the calming section (10) and passes through the guard zone (11) and the test sections (12). Finally the liquid-vapour mixture flows back to the condenser (1).

To keep the refrigeration losses small, test circuit and auxiliary circuit have high vacuum insulation systems (19)-(23).

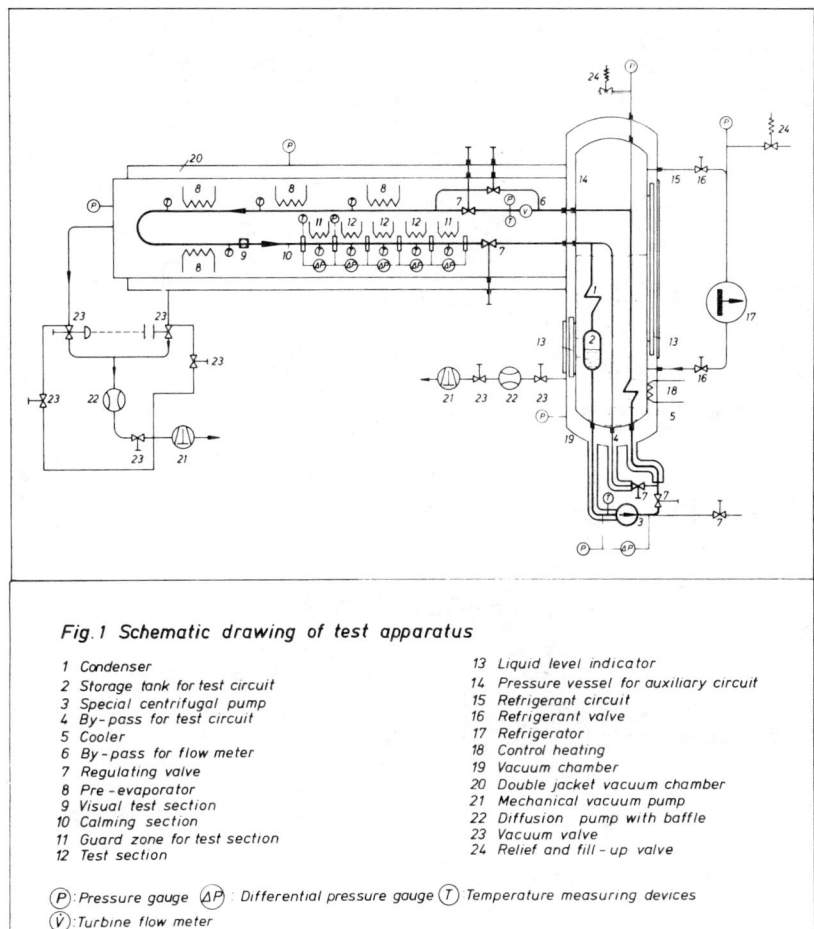

Fig.1 Schematic drawing of test apparatus

1. Condenser
2. Storage tank for test circuit
3. Special centrifugal pump
4. By-pass for test circuit
5. Cooler
6. By-pass for flow meter
7. Regulating valve
8. Pre-evaporator
9. Visual test section
10. Calming section
11. Guard zone for test section
12. Test section
13. Liquid level indicator
14. Pressure vessel for auxiliary circuit
15. Refrigerant circuit
16. Refrigerant valve
17. Refrigerator
18. Control heating
19. Vacuum chamber
20. Double jacket vacuum chamber
21. Mechanical vacuum pump
22. Diffusion pump with baffle
23. Vacuum valve
24. Relief and fill-up valve

(P): Pressure gauge (ΔP): Differential pressure gauge (T): Temperature measuring devices
(V): Turbine flow meter

2.2 Test section

The test sections, Fig. 2, and the guard zones were made from a smooth copper tube (1) with an inside diameter of 14 mm and a wall thickness of 3 mm. The three test sections, each 175 mm long, are identically equiped. Pressure and differential pressure are taken from pressure taps (2), at the inlet and at the exit of each unit. These taps consist of an annular chamber connected with the test tube by six holes (3). Five iron-constantan thermocouples (5) were embedded in grooves in each test section, uniformly distributed over length and circumference. The distance between thermocouples and the inner tube wall is 0.4 mm (6). Heat is supplied to the test sections by electrically heated thermocoax resistors, which are fitted bifilar into grooves at the outside of the tube. To obtain a homogeneous temperature distribution in the tube wall, all grooves and crevices are filled up with copper galvanically. More information about equipment and apparatus details are given in /14/.

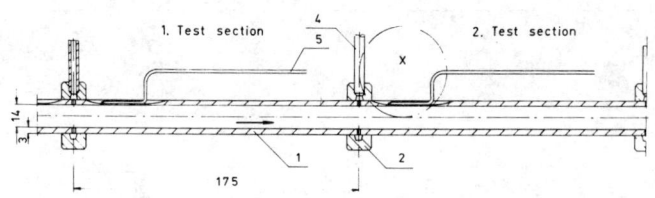

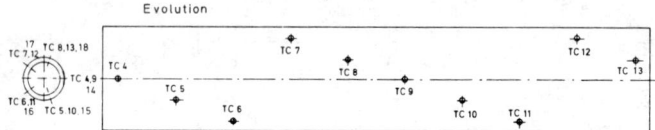

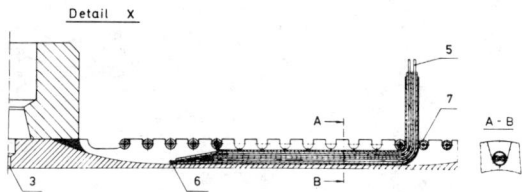

Fig. 2 Construction of the test section

2.3 Test procedure

Heat transfer coefficient

The local heat transfer coefficient is an average value over the tube circumference on a fixed axial position of the tube. It is calculated from

$$\alpha = \frac{\dot{q}_{eff}}{T_W - T_S} \qquad (2)$$

with

$$\dot{q}_{eff} = \frac{\dot{Q}_{eff}}{\pi \, d \, l} \qquad (3)$$

The total heat flow rate $\dot{Q}_{eff}$ is the sum of the electrical power $\dot{Q}_{el}$ and the heat transmission into the test section $\dot{Q}_{tr}$ which was measured daily.

$$\dot{Q}_{eff} = \dot{Q}_{el} + \dot{Q}_{tr} \qquad (4)$$

The average wall temperature is obtained by numerical integration of the five thermocouple readings around the tube circumference, applying Simpson's rule. This average value T_{meas} is corrected by the temperature drop between the location of the thermocouples and the inner tube wall.

$$\Delta T = \dot{q}_{eff} \frac{s}{\lambda_{Cu}} \qquad (5)$$

$$T_W = \overline{T}_{meas} - \Delta T \qquad (6)$$

Assuming thermodynamic equilibrium between liquid and vapour, the saturation temperature T_S can be calculated as a function of the pressure with an equation of state /10/. All test runs were performed after the test unit had reached the steady state at the requested pressure. Then, at a chosen mass velocity and heat flux, the flow quality was varied systematically. For all series, the measurements were carried out with a decreasing heat flux. To compensate statistic oscillations of temperature and pressure, the average value of 10-15 measurements is used for calculating the heat transfer coefficient.

Critical heat flux

To determine the critical heat flux $\dot{q}_{crit}$ the heat flux $\dot{q}$ was increased step by step, keeping the other parameters constant. At a certain heat flux, a rapid increase of the wall temperature caused by the oscillations of the mass velocity occures (Fig.17). The heat flux supplied at this moment is called the critical heat flux.

3. HEAT TRANSFER COEFFICIENT

3.1 Experimental results

In the following the relationship between the heat transfer coefficient of nitrogen and argon, and the main parameters heat flux $\dot{q}$, flow quality $\dot{x}$, reduced pressure p_r and mass velocity $\dot{m}$ is shown.

Influence of the heat flux $\dot{q}$ on the heat transfer coefficient α

Measured local heat transfer coefficients α of nitrogen and argon are plotted against the heat flux $\dot{q}$ in Fig. 3-6 for various reduced pressures p_r and flow qualities $\dot{x}$. In Fig. 3, two different boiling regions can be distinguished with respect to the influence of $\dot{q}$ on α. At lower heat fluxes, the heat transfer coefficient is independent of the heat flux. The mechanism in this region is convective boiling. The range of convective boiling is enlarged with increasing mass velocity and flow quality and decreasing pressure. The measurements with N_2 showed convective boiling only at $p_r = 0.034$ and $\dot{m} > 128$ kg/m^{2}s. For argon this range is larger. We even observed convective boiling at $p_r = 0.4$ and $\dot{m} = 470$ kg/m^{2}s for heat fluxes less than 35 000 W/m^2 and flow qualities greater than 10 %. In the nucleate boiling region, the heat transfer coefficient depends considerably on the heat flux. The experimental data for N_2 can be correlated by

$$\alpha \sim \dot{q}^{0.56} \quad . \tag{7}$$

Fig. 4 shows the heat transfer coefficient for nitrogen as a function of the heat flux for $p_r = 0.6$. At high pressure, the exponent in eq.(7) decreases with higher heat fluxes.

In Fig. 5 and 6, the heat transfer coefficient of argon is plotted versus the heat flux for $p_r = 0.4$ and $\dot{m} = 120$ kg/m^{2}s resp. 460 kg/m^{2}s. At the lower mass velocity we recognize nucleate boiling over the whole range (Fig.5). We find the proportionality

$$\alpha \sim \dot{q}^{0.4} \quad . \tag{8}$$

At higher mass velocity ($\dot{m}$ = 460 kg/m^{2}s) the slope of the curve $\alpha = f(\dot{q})$ varies unregularly with $\dot{q}$ (Fig.6). An explanation can not be given yet. Because most of the measurements were taken within the nucleate boiling regime, the following analysis of the influence of $\dot{x}$, p_r and $\dot{m}$ on α is only made for this region.

Influence of the flow quality $\dot{x}$

In Fig. 7 and 8 the heat transfer coefficient of N_2 and Ar is shown as a function of the flow quality $\dot{x}$ for various heat fluxes $\dot{q}$. The strong influence of the heat flux indicates nucleate boiling in both cases. At lower heat fluxes ($\dot{q}$ < 3500 W/m^2) the heat transfer coefficient α is almost independent of the flow quality $\dot{x}$. Due to the partial dryout of the heating surface, the heat transfer coefficient averaged over the tube circumference decreases with increasing flow quality at higher heat fluxes. The relation between heat transfer coefficient and flow quality can be expressed by

$$\alpha \sim 0.83 \, (1.01-\dot{x}^{1.1}) \, \overline{\dot{q}_{crit}}^{0.35\,\dot{q}} = F(x) \qquad (9)$$

$\dot{q}_{crit}$ is the critical heat flux, obtained from eq. 28, see below.

Influence of the reduced pressure p_r

With increasing reduced pressure p_r the heat transfer coefficient increases considerably (Fig. 9), because the needed superheat (T_W-T_S) to activate a given nucleation site diminishes /13/. Yet, the heat transfer coefficient in a tube is less affected by the pressure than in pool boiling. Bonn /2/ supposes this to be an effect of the temperature distribution along the tube perimeter, caused by the incomplete wetting of the heating surface. The realtionship between the reduced pressure and the heat transfer coefficient in flow boiling can be expressed by

$$\alpha \sim p_r^{0.35} \, (1+17.6 \, p_r^{30}) = F(p) \qquad (10)$$

Influence of the mass velocity $\dot{m}$

The influence of the mass velocity on the heat transfer coefficient of nitrogen and argon in the nucleate boiling region is shown in Fig. 10. An increase of the mass velocity always leads to an enhancement of the heat transfer coefficient. But, as a change of the mass velocity of a two phase flow may also affect the wetting of the heating surface (and by that the temperature distribution along the tube perimeter), the flow pattern and the boiling mechanism, a uniform relationship between mass velocity and heat transfer coefficient could not be found. Our measurements showed

$$\alpha \sim \dot{m}^{0\ldots0.4} \qquad (11)$$

For rough estimation we propose

$$\alpha \sim \dot{m}^{0.27} \qquad (12)$$

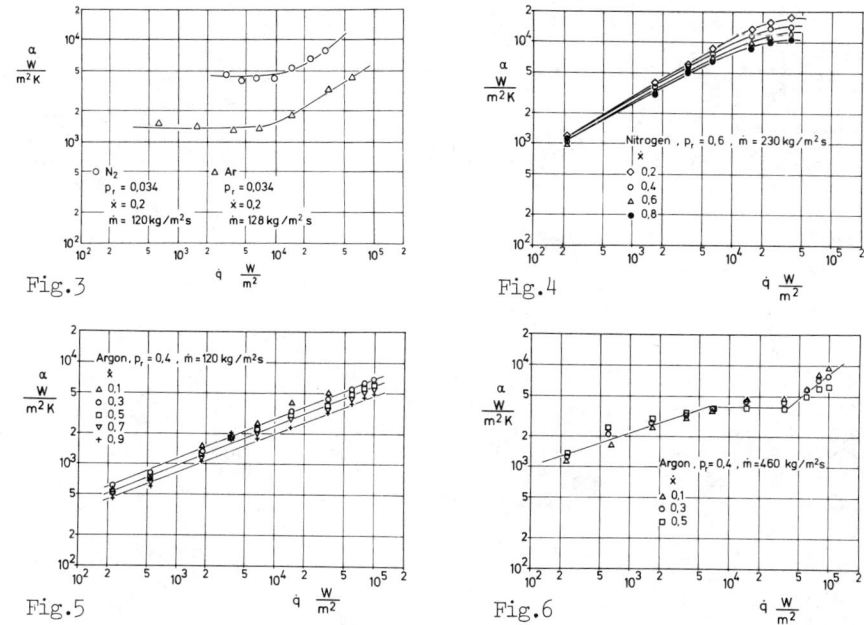

Fig.3 - Fig.6 Heat transfer coefficient of argon and nitrogen as a function of the heat flux

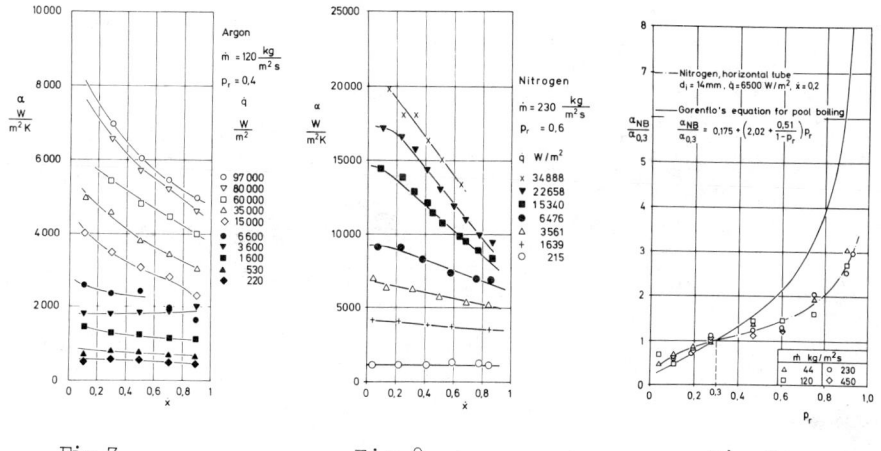

Fig.7   Fig.8   Fig.9
Influence of the flow quality on the heat transfer coefficient of argon and nitrogen   Heat transfer coefficient of N_2 as a function of the reduced pressure

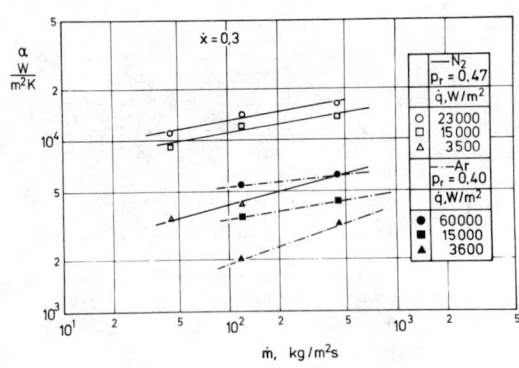

Fig. 10 Influence of the mass velocity on the heat transfer coefficient of Ar and N_2

3.2 Correlation of the heat transfer coefficient α

A great number of empirical and semi-empirical correlations for the heat transfer coefficient for forced convection two-phase flow are found in the literature. Four well known correlation methods are checked as to their applicability to nucleate boiling of cryogenic fluids in a horizontal tube. We tested the correlations of

1. Shah /13/
2. Borishansky /5/
3. Chen /8/
4. Chaddock-Brunemann /6/ .

Fig. 11-14 show a comparison between predicted and experimental heat transfer coefficients of nitrogen particularly with respect to the influence of the heat flux $\dot{q}$, the mass velocity $\dot{m}$, the flow quality $\dot{x}$ and the reduced pressure p_r. In general, the predictions according to the authors /13/,/5/,/8/ and /6/ fall far below the experimental data. Also the predicted change from nucleate boiling to convective boiling at higher pressures is incorrect. The mean deviation between calculated and measured values is more than 50% for all of the four equations.

To estimate heat transfer coefficients at low temperatures more precisely, we developed a new correlation. Starting from Stephan's equation for pool boiling /17/, this equation has been extended by terms which take into account the influences of mass velocity, flow quality and pressure:

$$Nu_{NB} = 1.05 \cdot 10^{-2} \cdot K_1^{0.3} \cdot K_2^{0.113} \cdot K_3^{0.56} \cdot K_4^{0.73} \cdot K_5^{0.5} \cdot F(p) \cdot F(x) \quad (13)$$

$$Nu_{NB} = \alpha_{NB} \, D_d / \lambda' \quad (14)$$

$$K_1 = D_d \cdot T_{so} \cdot \lambda' \cdot \rho' / (\eta' \sigma) \quad (15)$$

$$K_2 = R_p \, \rho'' \, \Delta h_v / (D_d^3 \, \rho' \, f^2) \quad (16)$$

$$K_3 = \dot{q} \, D_d / (\lambda' \, T_{so}) \quad (17)$$

$$K_4 = \sigma \rho' / (\dot{m} \, \eta') \quad (18)$$

$$K_5 = \dot{m}^2 / (g \, d \, \rho'^2) \quad (19)$$

The bubble departure diameter D_d can be calculated from

$$D_d = 0.0206 \cdot \beta \cdot \left[\sigma' / (g(\rho - \rho'')) \right]^{0.5} \quad (20)$$

HEAT TRANSFER AND CRITICAL HEAT FLUX OF NITROGEN AND ARGON 241

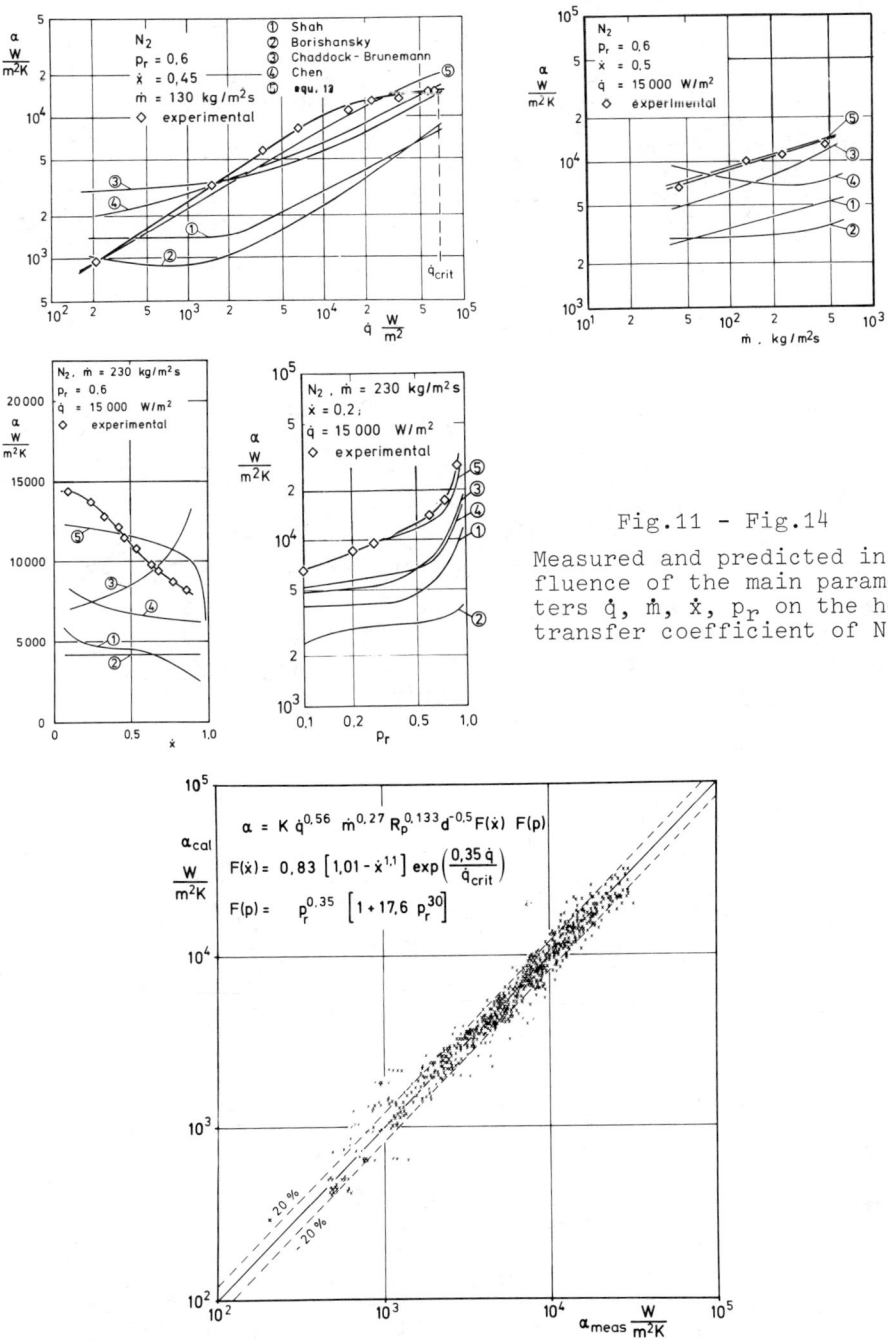

Fig.11 - Fig.14

Measured and predicted influence of the main parameters $\dot{q}$, $\dot{m}$, $\dot{x}$, p_r on the heat transfer coefficient of N_2.

Fig. 15 Comparison between predicted and measured heat transfer coefficients of N_2.

where β is the contact angle

$$\beta = 13.75° \text{ for } N_2 \text{ and } Ar \qquad (21)$$

The frequency of bubble departure is predicted by

$$f^2 = 0.314 \, g(\rho' - \rho'')/(D_d \, \rho') \, . \qquad (22)$$

The functions $F(p)$ and $F(x)$ are defined by equ.(9) and equ.(10). It is important to point out that the influence of the pressure is considered by a separate function $F(p)$ and not by the influence of the pressure on the physical properties. Therefore, the physical properties must be taken at $p_r = 0.03$. With these assumptions eq. (13) can be transformed into eq. (23)

$$\alpha_{NB} = C \cdot \dot{q}^{0.56} \, \dot{m}^{0.27} \, R_p^{0.133} \, d^{-0.5} \, F(p) \cdot F(x) \qquad (23)$$

The constant C only depends on the physical properties of the fluid.

$$C = 0.022 \, \frac{\lambda'^{0.74} \sigma^{0.225} \rho''^{0.133} \rho'^{0.033} (\rho' - \rho'')^{0.072} \Delta h_v^{0.133}}{\beta^{0.41} \, T_{so}^{0.26} \, \eta'^{1.03}} \qquad (24)$$

For nitrogen

$$C = 19.03 \, W^{0.17} \, \dot{m}^{0.567} \, K^{-1} \, s^{-0.54} \qquad (25)$$

In Fig. 15 the predicted heat transfer coefficients of N_2 are plotted vs. the measured data. The mean deviation of the calculated values is 13 %. 82 % of the predicted values are within a range of ± 20 % to the measured data. Equ.(13) applies for N_2 within a range of $\dot{q}$, $\dot{m}$, $\dot{x}$ and p_r as given on page 2.

4. BOILING CRISIS IN FORCED CONVECTIV FLOW

In convective boiling and nucleate boiling, the heat flux $\dot{q}$ increases with increasing wall superheat $T_w - T_s$ (Fig.16). If the critical heat flux $\dot{q}_{crit}$ (A) is exceeded, the transferable heat flux drops to the value of the Leidenfrost-point (C), because there is no more heat transfer to the fluid phase. A further increase of the wall temperature leads to heat transfer by film boiling, that means by convection, conduction and radiation through the vapour film.

If the critical heat flux $\dot{q}_{crit}$ is exceeded at a tube, heated with a constant heat flux, the new steady state in the film boiling region (B) lies at very high temperature differences. This may cause a rupture of the heating surface (Burnout).

Fig. 17 shows the rising of the thermocouple readings belonging to the three test sections during the boiling crisis. One can see, that the boiling crisis starts at the end of the heated zone, where the highest flow quality occures, and then goes upstream through all the three test sections. The flow quality at which the boiling crisis starts is called critical flow quality x_{crit}. On the way through the test sections, the boiling crisis also reaches zones, where $x < x_{crit}$, because the local heat flux increases by axial heat conduction from zones, where

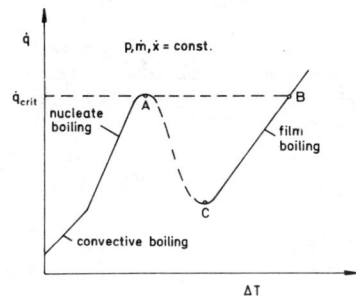
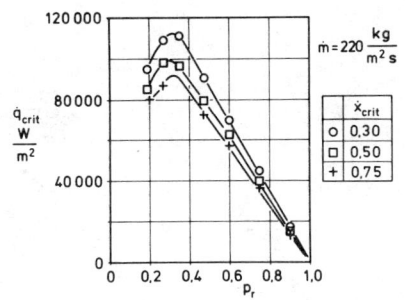

Fig. 16
Boiling curve

Fig. 18
Influence of the pressure on the critical heat flux.

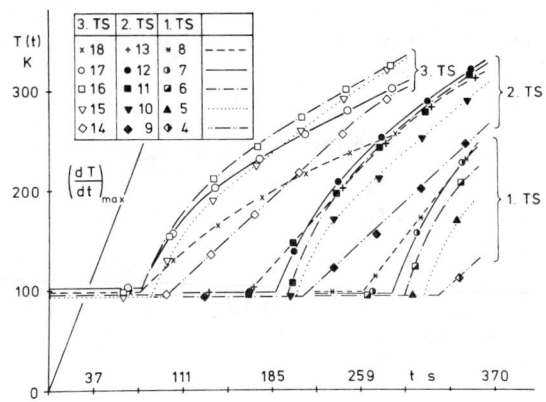

Fig. 17 Increase of wall temperatures during the boiling crisis. N_2, $p = 3.4$ bar, $T_s = 89.3$ K, $\dot{m} = 74$ kg/m^{2}s, $\dot{x} = 0.45$ $\dot{q}_{crit} = 50\,000$ W/m^2.

the boiling crisis has already occured. Additionally, the maximum temperature rising velocity $(dT/dt)_{max}$ of the tube is plotted in Fig. 17. $(dT/dt)_{max}$ is calculated with the assumption, that there is no heat conduction to the fluid and the whole heat flux is used to heat the test tube. The gradient of this line is similar to the gradients of the rising of the thermocouple readings at the beginning of the boiling crisis.

We could only make measurements of the critical heat flux for nitrogen, because argon showed a form of boiling crisis, which can not be described in the common way (see chapter 4.3).

4.1 Critical heat flux in forced convective boiling of nitrogen

The influence of the system pressure p, the flow quality $\dot{x}$ and the mass velocity $\dot{m}$ on the critical heat flux has been studied within the range of

$$44 \text{ kg/m}^2\text{s} < \dot{m} < 470 \text{ kg/m}^2\text{s}$$
$$1 \text{ bar} < p_r < 31 \text{ bar}$$
$$0 < \dot{x} < 1$$

Influence of the system pressure p on $\dot{q}_{crit}$

Fig. 18 shows the relationship between the reduced pressure p_r and the critical heat flux $\dot{q}_{crit}$. The critical heat flux passes through a maximum at $p_r = 0.35$. For the highest measured pressure, $p_r = 0.93$, only 10 % of the maximum value of $\dot{q}_{crit}$ remained. This behaviour is very similiar to pool boiling. The absolute value of $\dot{q}_{crit}$ however, is considerably lower in flow boiling than in pool boiling.

Influence of the flow quality $\dot{x}_{crit}$ on $\dot{q}_{crit}$

In Fig. 19 the measured critical heat fluxes $\dot{q}_{crit}$ are plotted vs. the flow quality $\dot{x}_{crit}$ at the end of the test section for various reduced pressures p_r. The mass velocity is 120 kg/m²s. Besides to the dominant influence of the system pressure, a considerable influence of the flow quality can be seen. $\dot{q}_{crit}$ decreases with increasing flow quality because of the diminution of the wetted surface area. This effect is remarkable especially at low flow qualities $\dot{x}_{crit}$.

Influence of the mass velocity $\dot{m}$ on $\dot{q}_{crit}$

To show the relation between the mass velocity $\dot{m}$ and the critical heat flux $\dot{q}_{crit}$, the full range of experimental data is presented in Fig. 20. The data with identical mass velocity are represented by fitted curves. It is interesting, that the critical heat flux shows a maximum at a mass velocity of 120 kg/m²s. The critical heat fluxes belonging to $\dot{m} = 44$ kg/m²s and $\dot{m} = 470$ kg/m²s are considerably lower. However, compared with the influence of system pressure and flow quality, the influence of the mass velocity on the <u>critical heat flux</u> is rather small. The effect of $\dot{m}$ on the <u>critical flow quality</u> at a given heat flux, however, can

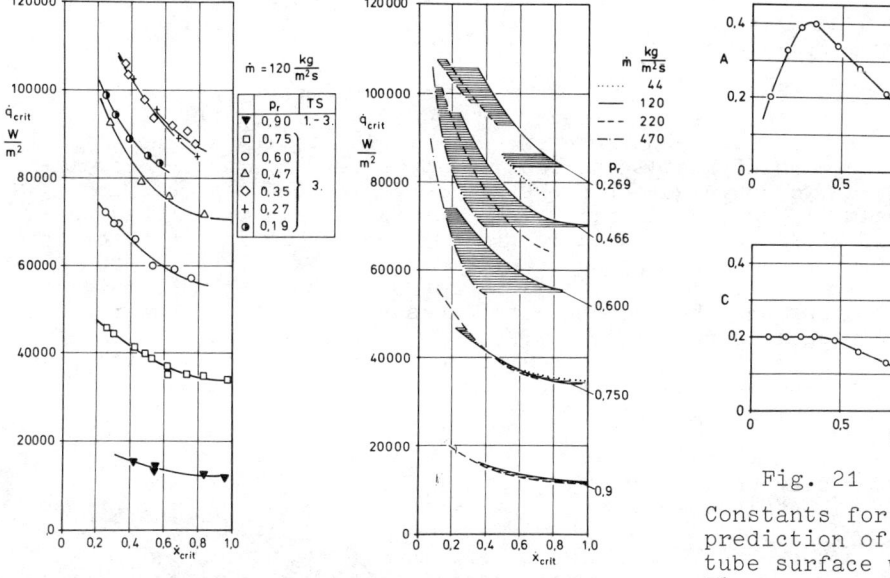

Fig. 21 Constants for the prediction of the tube surface wetting.

Fig.19 + Fig. 20 Influence of the flow quality and the mass velocity on the critical heat flux of nitrogen.

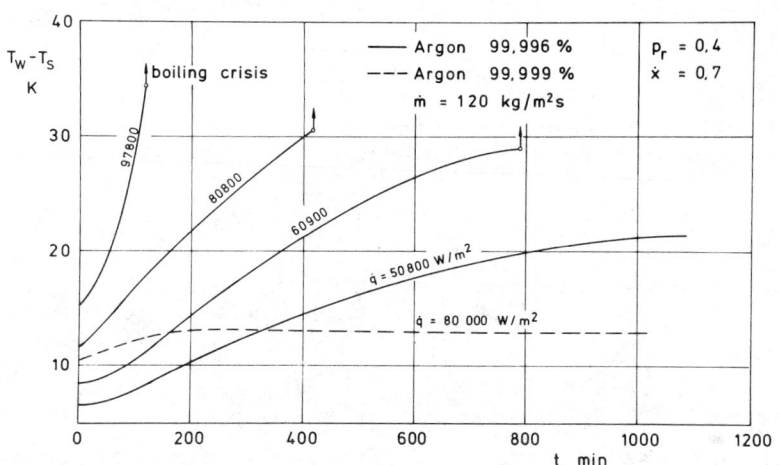

Fig. 22 Transient increase of the wall temperature during flow boiling of argon.

not be neglected at all. For example: an increase of the mass velocity from $\dot{m}$ = 120 kg/m²s to $\dot{m}$ = 470 kg/m²s at p_r = 0.6 and $\dot{q}$ = 60 000 W/m² leads to a diminution of the critical flow quality from $\dot{x}_{crit}$ = 60 % to $\dot{x}_{crit}$ = 25 %.

Furthermore, it should be mentioned, that there is still a number of second order effects on the critical heat flux. During experiments with Freon, Stevens et al. /18/ observed a relationship between critical heat flux and tube length. In our test apparatus, we could vary the overall length of the test section between 175 mm and 525 mm, using one, two or all three test sections. Within the observed range, no appreciable influence of the tube length could be noticed. Other influences, like the tube diameter, the surface finish, the method of heating, the material of the tube and the tube wall thickness could not be investigated. A theoretical study of Bonn /3/ shows, however, that in case of uniformly heated tubes there is a considerable effect of the tube material and of the tube wall thickness on the critical heat flux because of a non-uniform temperature distribution in the tube wall.

4.2 Correlation of the critical heat flux $\dot{q}_{crit}$ in a horizontal tube

In the observed parameter range $\dot{q}_{crit}$ was mainly influenced by the system pressure and the flow quality. Because the relation between the $\dot{q}_{crit}$ and the system pressure is similar to that in pool boiling, the suggested correlation is based on earlier equations proposed for pool boiling. The method developed by Bonn /4/ starts from the equation for the critical heat flux in pool boiling on a horizontal plate of Kutateladze /12/.

$$\dot{q}_{crit,PB} = K \, \Delta h_v \sqrt{\rho''} \sqrt[4]{\sigma g(\rho' - \rho'')} \qquad (26)$$

For cryogenic fluids K = 0.13 /19/. According to Vishnev /12/ $\dot{q}_{crit,PB}$ decreases with increasing angle of inclination ϕ

$$\dot{q}_{crit,PB}(\phi) = \dot{q}_{crit,PB}(\phi=0) \cdot \sqrt{\frac{190° - \phi}{190°}} \qquad (27)$$

Neglecting the heat transfer to the vapour, the perimeter averaged $\dot{q}_{crit}$ for forced convection boiling in a horizontal tube can be calculated from

$$\dot{q}_{crit} = \dot{q}_{crit,PB}(\phi=0) \cdot \frac{1}{180°} \cdot \int_0^{\phi_{crit}} \sqrt{\frac{190° - \phi}{190°}} \, d\phi \qquad (28)$$

ϕ_{crit} is the angle between the bottom of the tube and the end of the wetted zone. For ϕ_{crit} we found the correlation

$$\frac{\phi_{crit}}{180°} = 1 - \left[A \, \dot{x}^B + (1-A) \, \dot{x}^C \right] \qquad (29)$$

The value of B was found to be 30 for all measurements, the values of A and C can be taken from Fig. 21.

4.3 Critical heat flux of argon - the "drift" - phenomenon

During the measurements with argon, we observed a transient increase of the wall temperature, while all parameters were kept constant. This temperature drifting ended after several hours at a considerably higher wall temperature, or also with the burnout (Fig. 22).

The following observations were made:

1. The maximum heat flux at which we could make steady state measurements increased with increasing mass velocity and with decreasing system pressure and flow quality. For heat fluxes up to $\dot{q} = 100\,000\,W/m^2$, drifting was only observed for $p_r > 0.1$.

2. Keeping the heat flux constant, the drift velocity $d(T_w-T_s)/dt$ could be raised by increasing the pressure and the flow quality and reducing the mass velocity.

Naturally, all measurement devices were checked several times, to exclude errors. Later, the same measurments were repeated with nitrogen, but no drifting was observed. A similar effect had been observed by Bewilogua /1/ during the pool boiling of hydrogen. The author reports, that a deposit could be seen on the heating surface. Bewilogua supposes, that this deposit consisted of nitrogen or oxygen, which are sublimating at the observed temperature of 60 K.

The argon we used in our measurements had a purity of 99.996%. The remaining 40 ppm consisted of CO_2, O_2, N_2, H_2 and saturated hydrocarbons. A comparision of the solidification temperatures of these fluids lead to CO_2 as a possible cause of the drift-phenomenon. In a subsequent test series, we used argon with a purity of 99.999 %, which contains only 0.5 ppm of CO_2. This effected a drastical reduction of the drifting (Fig.22). Only at very high heat fluxes a small transient increase of the wall temperature could still be observed for high pressures and high flow qualities. Later, 500 ppm of CO_2 were injected into the test circuit, while $\dot{q} = 97\,800\,W/m^2$, $p_r = 0.4$, $\dot{x} = 0.9$ and $\dot{m} = 120\,kg/m^2s$ were adjusted in this run. After a delay of 2-3 minutes, a sudden burnout occured. Similar tests with O_2, N_2 and C_2H_2 did not cause any similar effect. Therefore, we assume, that the drifting is caused by the crystallization of CO_2 on the heating surface.

The fact, that impurities of such a small concentration can already produce a considerable reduction of the critical heat flux is very important for industrial applications. Therefore additional investigations are necessary, to get more informations about the mechanism of the drifting and about the relationship between CO_2 concentration and drift-velocity.

REFERENCES

1. Bewilogua, L., Görner, W., Knöner, R. e.a. 1974. Heat transfer in liquid hydrogen. Cryogenics 14, No.9, pp. 516-517.

2. Bonn, W. 1980. Wärmeübergang und Druckverlust bei der Verdampfung von Stickstoff und Argon. Dissertation, Universität Karlsruhe.

3. Bonn, W., Iwicki, J., Krebs. R. e.a. 1980. Über die Ungleichverteilung des Wärmeübergangs am Rohrumfang bei der Verdampfung im durchströmten waagerechten Rohr. Wärme- und Stoffübertragung, 14, pp. 31-42.

4. Bonn, W., Krebs, R. and Steiner, D. 1980. Kritische Wärmestromdichte bei der Zweiphasenströmung von Stickstoff im waagerechten Rohr bei Drücken bis zum kritischen Druck. Wärme- und Stoffübertragung, 13, pp. 265-274.

5. Borishanskij, W.W., Paleev, I.I., Agafonova, F.A. e.a. 1973. Some problems of heat transfer and hydraulics in two-phase-flows. Int. J. Heat Mass Transfer 16, pp. 1073-1085.

6. Chaddock, J.B. and Brunemann, H. 1967. Forced convection boiling of refrigerants in horizontal tubes. Phase 3: Report HL-113, School of Engineering, Duke University, Durham, N.Carolina.

7. Chawla, J.M. 1967. Wärmeübergang und Druckverlust in waagerechten Rohren bei der Strömung von verdampfenden Kältemitteln. VDI-Forschungsheft 523, Düsseldorf, VDI-Verlag.

8. Chen, J.C. 1963. A correlation for boiling heat transfer to saturated fluids in convective flow. ASME paper 63-HT-34.

9. Collier, J.G. 1981. Convective boiling and condensation. -2nd. ed. Mc Graw-Hill Inc.

10. Gosman, A.L., McCarthy, R.D. and Hust, J.G. 1969. Thermodynamic Properties of Argon. United States Departement of Commerce, NSRDS-NBS 27.

11. Gorenflo, D. 1977. Wärmeübergang bei Blasensieden, Filmsieden und einphasiger freier Konvektion in einem großen Druckbereich. Abhandlung des Deutschen Kälte- und Klimatechnischen Vereins, Nr. 22, Karlsruhe, Verlag C.F. Müller.

12. Kutateladze, S.S. 1963. Fundamentals of heat transfer. Chap. 18.2, Edward Arnold Ltd. London.

13. Shah, M. 1976. A new correlation for heat transfer during boiling flow through pipes. ASRHAE Trans., 82(2), 66-86.

14. Steiner, D. 1975. Wärmeübergang und Druckverlust von siedendem Stickstoff im waagerecht durchströmten Rohr. Dissertation, Universität Karlsruhe.

15. Steiner, D. and Schlünder, E.U. 1976. Heat transfer and pressure drop for boiling nitrogen flowing in a horizontal tube. Cryogenics 16, pp. 387-398 and 457-464.

16. Steiner, D. 1980. Zweiphasenströmung in Leitungen. Hochschulkurs Wärmeübertragung II, Teil III. Forschungsgesellschaft Verfahrenstechnik, Düsseldorf.

17. Stephan, K. 1964. Beitrag zur Thermodynamik des Wärmeübergangs beim Sieden. Abhandlung des Deutschen Kältetechnischen Vereins, Verlag C.F. Müller, Karlsruhe.

18. Stevens, G.F., Elliott, D.F. and Wood, R.F. 1964. An experimental investigation into forced convection burnout in freon, with reference to burnout in water. AEEW-R 321.

19. Verkin, B.I. and Kirichenko, Y.A. 1978. Soviet investigations on pool boiling of cryogenic liquids. ICEC London, lecture 11-61.

20. Vishnev, I.P., e.a. 1976. Study of heat transfer in boiling of helium on surfaces with various orientation. Heat Transfer Soviet Research, 8, No. 4, pp. 104-108.

NOMENCLATURE

A	internal tube surface area	m^2
c_p	specific heat	J/kg K
d	inside tube diameter	m
D_d	bubble departure diameter	m
f	frequency of bubble depature	Hz
F(p)	influence of pressure on α	-
F(x)	influence of flow quality on α	-
g	acceleration due to gravity	m/s^2
Δh_v	latent heat of vapourization	J/kg
k	surface finish of the tube	m
l	length of the tube	m
$\dot{m}$	mass velocity	$kg/m^2 s$
Nu	Nusselt number	-
p	pressure	bar
p_r	reduced pressure $p_r = p/p_{crit}$	-
$\dot{Q}$	heat flow rate	W
$\dot{q}$	heat flux	W/m^2
$\dot{q}_{eff}$	total heat flux	W/m^2
$\dot{q}_{el}$	heat flux due to electrical power	W/m^2
$\dot{q}_l$	heat flux due to refrigeration lows	W/m^2
R_p	surface roughness	m

s	distance between thermocouple and inner tube diameter	m
T	temperature	K
$\overline{T}$	mean temperature	K
T_{so}	boiling temperature at $p_r = 0.03$	K
t	time	s
$\dot{x}$	flow quality	–

Greek letters

α	heat transfer coefficient	$W/m^2 K$
β	contact angle	deg
η	viscosity	kg/ms
λ	heat conductivity	W/mK
ρ	density	kg/m^3
σ	surface tension	kg/s^2
φ	angle of inclination	deg

Subscripts

Ar	argon
crit	critical
Cu	copper
g	gas
meas	measured
l	liquid
NB	nucleate boiling
PB	pool boiling
s	at saturation
w	wall
'	liquid phase
"	gas phase

Investigation of Heat Transfer and Hydrodynamics in the Helium Two-phase Flow in a Vertical Channel

B.S. PETUKHOV, V.M. ZHUKOV, and V.M. SHIELDCRET
Institute of High Temperatures
USSR Academy of Sciences
Moscow, USSR

ABSTRACT

The experimental investigation results of heat transfer, critical heat flux and pressure drop in the helium two-phase upflow in vertical channel are presented. On the basis of analysis of our experimental data and those of number of other works, the correlation for calculation of heat transfer is suggested. It is found that in helium two-phase flow as well as in boiling ordinary liquids three types of boiling crisis which depend on the flow hydrodynamics may exist. Frictional pressure drop for both adiabatic and diabatic flows was measured. It was found that experimental data on pressure drop are not in agreement with calculation by homogeneous model. Correlation for calculation of pressure drop in diabatic helium two-phase flow is obtained.

1. INTRODUCTION

In connection with construction of superconducting devices for various applications, in which cooling of windings is performed by liquid helium forced flow, the studying of heat transfer and pressure drop in the helium two-phase flow over a wide range of regime parameters including post-dryout region is of great importance. The existing knowledge about heat transfer and hydrodynamics in the boiling helium flow is based now on comparatively limited amount of the experimental data |1-12|, some of them are preliminary because of methodical disadvantages and, therefore, these results should not be used for designing and construction of superconducting magnetic systems and other cryogenic devices. Alongside with this many problems of two-phase helium flow heat transfer and hydrodynamics are not investigated enough, e.g. mass velocity, quality, tube diameter and other factors influence on heat transfer; temperature regime of heat transfer surface in post-dryout region; mass velocity and quality influence on boiling crisis; behaviour of pressur drop against changes of regime parameters and a number of other problems. In this paper the results of investigation of heat transfer, boiling crisis and pressure drop in the helium two-phase upflow in a vertical stainless tube of 0.8 mm i.d. with thikness of the wall 0.1 mm at atmospheric pressure are presented within the following range of regime parameters: heat flux q = 20-7000 W/m^2, mass velocity $\bar{\rho w}$ = 50-330 kg/m^{2}s, inlet quality x_0 = 0-0.7, outlet quality x_{out} = 0-1, wall temperature difference ΔT = 0.03- 10°K. Experiments were carried out on the set up described in details in |13, 14|. Estimations showed that the root mean square rela-

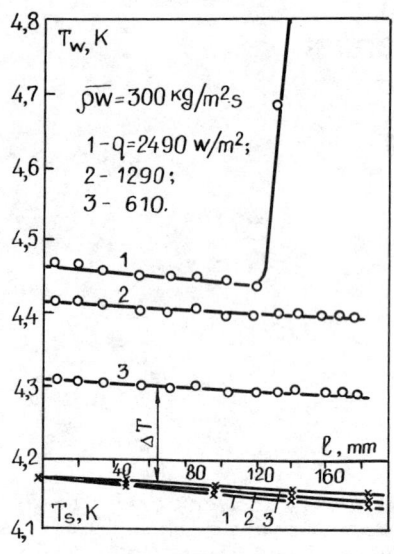

Fig.1. Typical wall and bulk temperature profiles for various heat fluxes.

tive error to be 1% for heat flux, 2.5%- for mass velocity, 5%- for quality, 15-20%- for temperature difference and 15-20% for frictional pressure drop.

2. HEAT TRANSFER

The heat transfer characteristics in the helium two-phase flow may be seen by analysing experimental temperature distribution of wall and flow along the channel. Fig.1 shows the typical curves of temperature of internal surface of the wall and saturation temperature of helium two-phase flow along the tube for different fixed values of q with $\overline{\rho w}$ = = 300 kg/m²s. Measurements showed that temperature difference defining heat transfer on the wall ($\Delta T = T_w - T_s$) remains constant for q = const and $\overline{\rho w}$ = const up to $x = x_{cr}$, at which boiling crisis in the flow occurs. In the boiling crisis contact of liquid with the heating surface breaks with sharp rise of the wall temperature.

Experiments showed that when $x < x_{cr}$ heat transfer did not depend on quality within the whole range of the operating conditions within the limits of the accurateness of experiment. This made possible to use average over the length of tube rather than local value of temperature difference. In the regims with sharp wall temperature rise in some cross-section, which corresponds to boiling crisis occurrence in this cross-section, in averaging temperature difference only a part of temperature profile till the rise was considered.

Experimental data on heat transfer when $x < x_{cr}$ with different mass velocities are presented on Fig.2. For the comparison the curve characterizing experimental data |15| on heat transfer in helium with pool boiling on the stainless steel surface under pressure 0.1 MPa is presented at the same figure. The comparison showed that forced flow essentially (3 or 4 times) enhances heat transfer in the helium boiling.

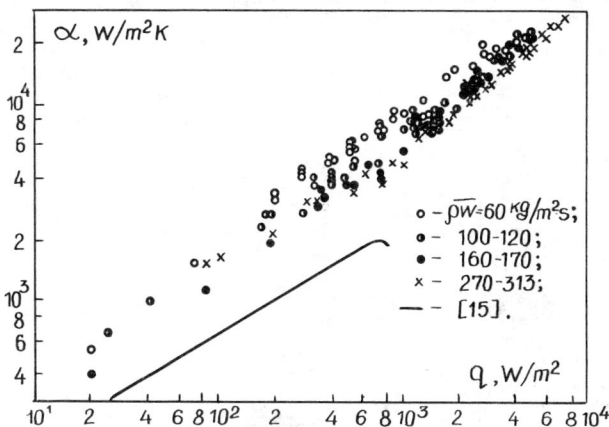

Fig.2. Heat transfer coefficient versus heat flux.

However, within the investigated range of mass velocities, increase of the helium flow rate at the inlet caused decrease of heat transfer coefficient α.

Experimental data analysis showed that heat transfer in forced helium two-phase flow changes with q analogously to dependence $\alpha \sim q^{0.7}$, which is typical for developed nucleate boiling of the ordinary liquids. This seems to mean that in forced flow heat transfer is performed mainly due to nucleate boiling at the tube wall independently on the flow regime. To describe the experimental data on heat transfer the following correlation is suggested

$$\alpha = 197 q^{0.7} (\overline{\rho w})^{-0.24} \qquad (1)$$

which generalizes the results of investigation within the whole range of regime parameters with the root mean square error $\pm 15\%$. Here α is heat transfer coefficient, $w/m^2 k$. Analysis and comparison of our experimental data with those of |1, 2, 4| for the region of developed nucleate boiling regime have shown that on including in correlation (1) terms allowing for pressure (P in MN/m^2) and tube diameter (d in m) influence on heat transfer it becomes

$$\alpha = 55.7 q^{0.7} (\overline{\rho w})^{-0.24} P d^{-0.5} \qquad (2)$$

and satisfactory ($\pm 30\%$) describes data in the range of heat fluxes from 20 to 7000 W/m^2, mass velocities from 20 to 313 $kg/m^2 s$, pressures from 0.1 to 0.15 MPa, and qualities from 0 to x_{cr}. Experimental data on heat transfer |5, 7, 9| with fixed regime parameters very differ from each other and therefore are not used in comparing with calculations by (2).

3. CRITICAL HEAT FLUX

Characteristic temperature difference profiles along the tube in the regimes with boiling crisis at different heat flux for two fixed values mass velocities are presented in Fig.3. With $\overline{\rho w}$=const increase of heat

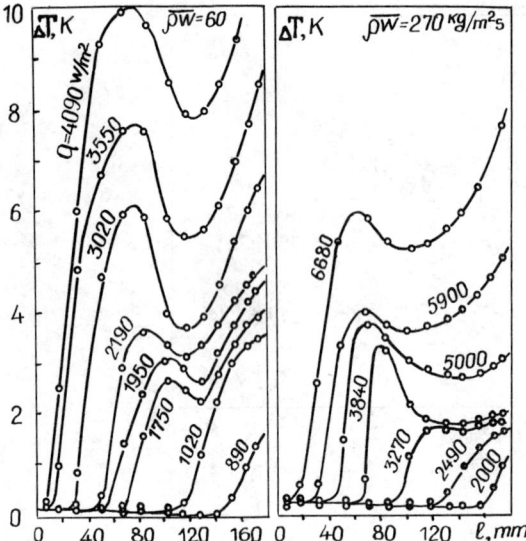

Fig.3. Typical wall temperature profiles for various heat fluxes and mass velocities.

flux causes temperature rise at the outlet of the test section due to boiling crisis occurrence. As heat flux increases boundary of boiling crisis occurrence moves to the inlet and rather sharp temperature rise with clear maximum on the wall temperature distribution curve occurs. Then wall temperature decreases reaches the minimum value and increases again to the outlet of the test section. It is seen from the figure that for the same q increase of $\overline{\rho w}$ causes decrease of temperature rise in the post-crisis region of the tube. Method of fixing boiling crisis used here made possible to determine coordinate of boiling crisis occurrence and critical quality x_{cr} with high accuracy on the base of measurements of wall temperature profiles along the tube. X_{cr} values in the cross-section where boiling crisis occurs (i.e., where wall temperature rise is observed) was calculated by heat balance equation. Experiments on investigation q_{cr} were made by three different procedures: 1. q=const, $\overline{\rho w}$=const, x_0=var; 2. x_0=const, q=const, $\overline{\rho w}$=var; 3. $\overline{\rho w}$=const, x_0=const, q=var. It should be noted that there were no discrepancies in experimental data regardless of used procedure.

Experimental data on helium boiling crisis for five fixed values of mass velocities are presented in Fig.4. It is seen that dependence q_{cr} on x_{cr} is analogous to dependence observed in investigating boiling crisis in vapour-water flows and shows reveals existence of three characteristic regions consistent with various crisis mechanisms. In the region $x_{cr} < x^°_{lim}$ q_{cr} decreases as quality increases which is in agreement with boiling crisis due to transition from nucleate boiling to film boiling. Region $x_{cr} = x^°_{lim}$ corresponds to dryout of the liquid film on the tube wall. The third region $x_{cr} > x^°_{lim}$ is performed when heat flux at the wall is small and qualities at the inlet are large ($x_0 \geqslant x_{\Delta p}$). In this case dispersed flow regime where boiling crisis is defined by deposition of droplets intensity is formed practically from the very inlet. Because values of q_{cr} are small and wall temperature is less than Leidenfrost point temperature ($\Delta T \leqslant 0.2K$), liquid droplets dispersed in the flow, deposit on the tube wall to form the liquid microfilm. In this case nucleate boiling in

HEAT TRANSFER AND HYDRODYNAMICS IN HELIUM TWO-PHASE FLOW

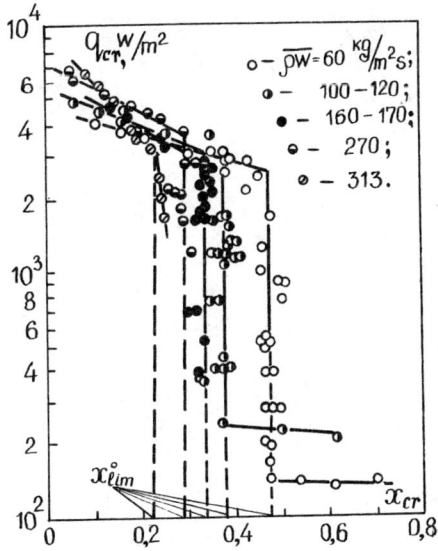

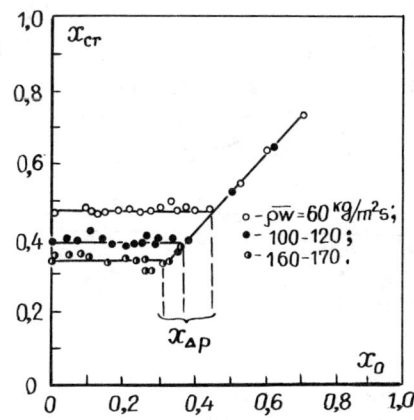

Fig.5. Critical quality x_{cr} as a function of inlet quality x_0.

Fig.4. Critical heat flux data.

the microfilm is absent and heat transfer from wall to flow is produced by the vaporization. As long as balance between amounts of liquid depositing from flow to the film and vaporizing from film is concerved, heat transfer surface remains wet. In moving downstream balance disturbs due to quality increase, film vaporizes absolutely and boiling crisis followed by the wall temperature rise occurs. This rise, however, is significantly less than that of in the case of dryout at $x°_{lim}$.

Analysis of the results of investigation of dryout in the region $x_{cr} = x°_{lim}$ have shown that quality change at the inlet with q_{cr} and $\overline{\rho w}$ constant did not affect quality $x°_{lim}$ as well as in the case of water boiling |16|. It is seen from the Fig.5 that in the wide range of q_{cr} for the fixed value of $\overline{\rho w}$ change of x_0 did not affect the value of $x°_{lim}$. However, in attaining at the inlet the quality correspondent to pressure drop crisis ($x_0 \geqslant x_{\Delta p}$) an increase of x_{cr} takes place. Such dependence $x°_{lim}$ on x_0 points out existence of deposition controlled burnout in the tube. As noted in the work of V.E.Doroschuk |16| such method of presenting the results of investigation of influence x_0 on x_{cr} allows to determine values $x_{\Delta p}$ which, as will be shown further, are in a good agreement with the experimental pressure drop data.

To define mass flow rate influence on $x°_{lim}$ experimental data are presented in Fig.6. It is seen that mass velocity increase couses progressive decrease of $x°_{lim}$. To describe this dependence an empirical correlation is suggested:

$$x°_{lim} = 1.82(\overline{\rho w})^{-1/3} \qquad (3)$$

which describes experimental data with maximum value of the root mean square error ±15%. Comparison between our data on q_{cr} with those of |2, 3, 8, 10| when $x_{cr} < x°_{lim}$ points out some reasonable discrepancy which can be explained by diameter influence on critical heat flux. This influence may be determined in treatment of data $q_{cr}=f(d)$ when x_{cr}=const; P= const; $\overline{\rho w}$= =const. Our data and those of |2, 3, 8, 10| presented in Fig.7 have shown

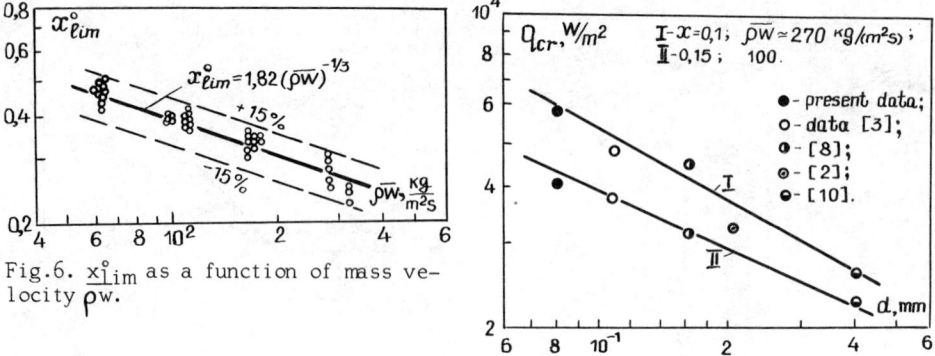

Fig.6. $x^°_{lim}$ as a function of mass velocity $\overline{\rho w}$.

Fig.7. Critical heat flux as a function of diameter channel.

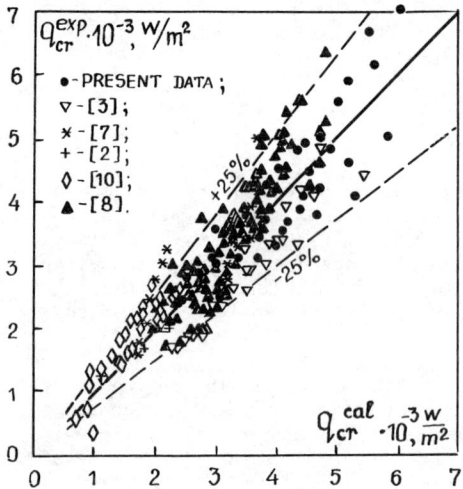

Fig.8. Predicted versus measured critical heat fluxes.

that connection between critical heat flux and diameter may be expressed as follows:

$$q_{cr} \sim d^{-0.5} \qquad (4)$$

as well as for the water boiling |16|. It is seen from the figure that small values q_{cr} |10| corresponds to the largest internal diameter of the tube (4 mm) and conversely large values of q_{cr} (this investigation data) corresponds to less diameter (0.8 mm).

Investigations of boiling crisis in the region $x_{cr} < x°_{lim}$ made in this paper and |8| have shown that mass velocity affects q_{cr} rather slightly. Data of |2, 3, 7, 8, 10| do not contain mass velocity influence. In the region of qualities near to 0, values of q_{cr} as seen from all known investigations on helium with allowance for the tube diameter are close to $q°_{cr}$, calculated by Kutateladze equation |17| for helium pool boiling. This feature of boiling crisis appropriate to transition from nucleate to film boiling made possible

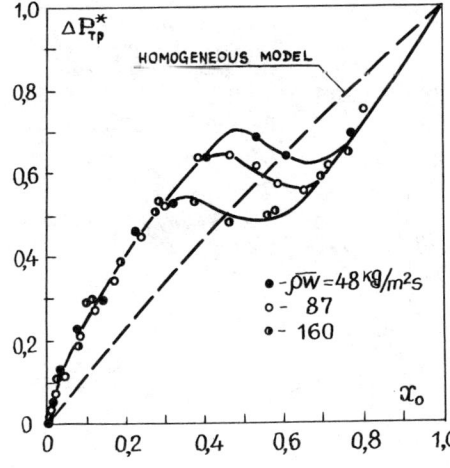

Fig.9. ΔP^*_{tp} as a function of inlet quality x_o for adiabatic flow.

to use Kutateladze equation for pool boiling. On the base of analysis of our data and those of |2, 3, 7, 8, 10| an empirical correlation for calculation of q_{cr} is suggested:

$$q_{cr} = (q°_{cr} - 1.37 \cdot 10^{-3} \frac{r\sigma\rho''}{\mu'} x \ e^{-x}) \sqrt{d_o/d} \qquad (5)$$

where $q°_{cr}$ - pool boiling critical heat flux calculated by Kutateladze equation, W/m²; r - latent heat, J/kr; σ - surface tension, N/m, ρ'' - vapour density, kg/m³; μ' - liquid vescosity, kg/m s; x - quality; $d_o = 8 \cdot 10^{-4}$ m - the tube diameter in our experiments; d - diameter for which calculation is made.

Numerical value of constant used in (5) was defined from treatment of all the experimental data of our work and those of |2, 3, 7, 8, 10|. Comparison of experimental data with calculation by this correlation (Fig.8) have shown that 90% of experimental values of q_{cr} within the range of pressures from 0.1 to 0.21 MPa, mass velocities from 6 to 630 kg/m² s, qualities from 0.24 to $x°_{lim}$ for tube diameter 0.8 mm (our work); 1.09 mm |3|; 1.63 mm |8|; 2.12 mm |2|; 2.13 mm |7|; 4 mm |10| are described with the root mean square error ±25% which is quite satisfactory. In the papers where $x°_{lim}$ are not pointed out |2, 3, 10| it was determined by (3). Correlation (3) suggested in this paper permits to calculate quality in the cross-section where boiling crisis due to film dryout on the wall occurs for given mass velocity and to define region of using correlation (2) for heat transfer calculation when $x < < x°_{lim}$.

4. FRICTIONAL PRESSURE DROP.

To define wall heating influence on pressure drop in the helium two-phase flow experiments were made without heating in changing inlet quality as well as with heating. Experiments with adiabatic flow were made with inlet quality changing from 0 to 0.8 at three fixed values of mass velocity - 48, 87, 160 kg/m² s. Results are presented in Fig.9 as a diagram $\Delta P^*_{tp} = f(x_o)$. Value of $\Delta P^*_{tp} = (\Delta P_{tp} - \Delta P')/(\Delta P'' - \Delta P')$ is dimensionless two-phase frictional pressure drop in helium two-phase flow, ΔP_{tp} is measured frictional pressure drop in helium two-phase flow; $\Delta P'$, $\Delta P''$ are frictional pressure drops in liquid and vapour flows, respectively. For the comparison the curve calculated

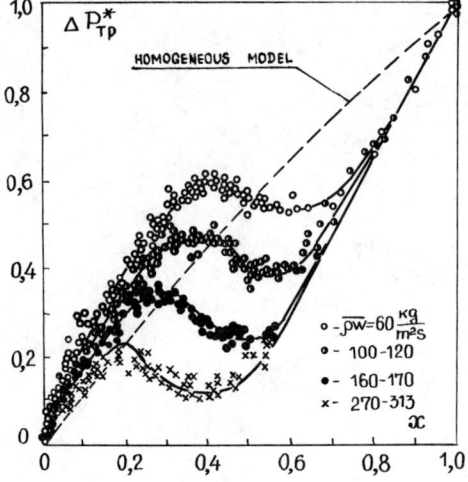

Fig.10. ΔP^*_{tp} as a function of quality x for diabatic flow.

Fig.11. Comparison between the predicted and experimental data.

by homogeneous model of helium two-phase flow is presented in the same figure. Fig.9 shows that experimental data within the investigated range of mass velocities and qualities differ considerably from calculated curve. When $x_0 < 0.3$ and $x_0 > 0.7$ data presented as $\Delta P^*_{tp} = f(x_0)$ are not affected by mass velocity and are described by one mean curve. In intermediate region $(0.3 < x_0 < 0.7)$ experimental curves exhibit not successive character. The peaks of these curves correspond to x_0 at which annular flow regime i.e. regime correspondent to pressure drop crisis forms immediately at the inlet. Similar character of ΔP^*_{tp} was mentioned in the papers on investigation of pressure drop in vapour water flow |18| as well as in helium two-phase flow |12, 19|. It should be noted that in the region of pressure drop crisis values of dimensionless frictional pressure drop and quality correspondent to $x_{\Delta p}$ essentially depend on mass velocity, with increase of mass velocity causing decrease of ΔP^*_{tp} as well as of $x_{\Delta p}$. E.G. at $\overline{\rho w} = 48$ kg/m² s $x_{\Delta p} \simeq 0.48$, and at $\overline{\rho w} = 160$ kg/m² s $x_{\Delta p}$ diminishes to $\simeq 0.35$, with ΔP^*_{tp} changing from 0.7 to 0.55, respectively. Experimental investigation of pressure drop in helium two-phase flow in diabatic tube was carried out within the range of heat fluxes densities at the wall from 20 to 7000 W/m², mass velocities from 60 to 313 kg/m²s and inlet qualities from 0 to 0.7. Experiments were carried out using two procedures: q = = const and inlet quality changes or x_0=const and heat flux changes.

Analysis the experimental pressure profiles measured in five taps have shown that at considerable heat fluxed on the curves P=f(l) points of bend occurs, i.e. change of pressure drop takes place, which corresponds to pressure drop crisis in this cross-section. Such peculiarities in the pressure drop characteristic can not be determined by measuring only two points, i.e. at the inlet and the outlet of the test section, while such procedures are often used in experiments. Experimental frictional pressure drop in helium two-phase flow in diabatic tube at four fixed values of $\overline{\rho w}$ and different q are presented in Fig.10. Analysis of experimental data has shown that for the case q=const, $\overline{\rho w}$=const, x_0= variable as well as for the case $\overline{\rho w}$=const, x_0 = const, q=variable discrepancy was not observed. Experimental data presented in the figure are qualitatively analogous to dependence $\Delta P^*_{tp} = f(x)$ for the fi-

HEAT TRANSFER AND HYDRODYNAMICS IN HELIUM TWO-PHASE FLOW 259

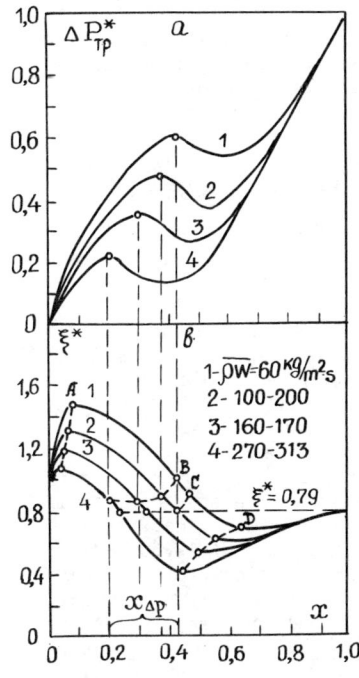

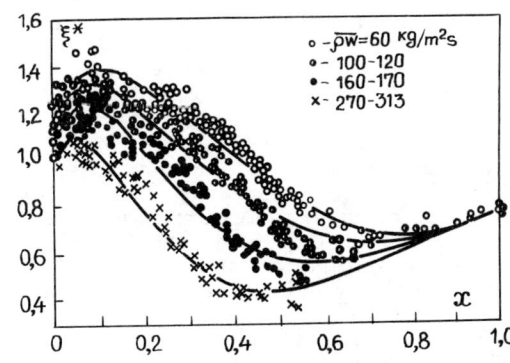

Fig.13. Predicted ξ^* and measured ξ^*_{exp}.

Fig.12. ΔP^*_{tp} and ξ^* as a function of quality x.

xed mass velocity as well as in the helium two-phase flow in adiabatic tube (Fig.9). The same behaviour of dependence of dimensionless frictional pressure drop on quality was observed in vapour water flows at diabatic conditions |18|.

As well as for adiabatic flow values of ΔP^*_{tp} for diabatic flow differs considerably from those of calculated by homogeneous flow model within all range of changing x and only at $\overline{\rho w}$=270-313 kg/m²s and $x < 0.2$ it may be used rather accurately. The figure shows that diffirently from adiabatic flow mass velocity influences on ΔP^*_{tp} in more wider range of qualities from 0 to 0.75 and only at $x > 0.75$ i.e. in the dispersed flow $\overline{\rho w}$ does not considerably affect. Mass velocity influences particularly strongly near the pressure drop crisis. For example, at $\overline{\rho w}$=60 kg/m²s ΔP^*_{tp} reaches 0.6 at x =0.4, while at $\overline{\rho w}$= =300 kg/m²s value of ΔP^*_{tp} increases essentially and reaches 0.2 at $x_{\Delta p}$ =0.2, i.e. dimensionless frictional pressure drop decreases approximately by the factor of 3. Heating influence on frictional pressure drop based on data of this paper are well illustrated in Fig.11 in comparing the curves, describing experimental data for adiabatic and diabatic conditions (curve 1 and 2). It is seen that $x_{\Delta p}$ for diabatic is slightly less than for adiabatic flow, which may be explained by additional entrainment of liquid from film by bubbles at nucleate boiling on the wall, which does not take place in adiabatic flow. In this connection film thickness diminishes more quickly and then pressure drop crisis occurs at smaller $x_{\Delta p}$. Radial flow of vapour occuring during liquid boiling at the wall causes diminishing of vapour axial velocity gradient at the vapour-liquid interfaces and vapour-wall in slug, annular and dispersed regimes of two-phase flow ($x > 0.1$). This seems to be the one of the reasons leading to decrease of frictional pressure drop in diabatic flow in comparison with adiabatic one.

Experimental data |19| for adiabatic flow in tube of 0.45 mm diameter and data |8, 12| for diabatic and adiabatic flows in tubes of 1.63 mm diameter at pressure 0.1 MPa and approximately equal mass velocities are presented in the same figure. Comparison between |8| and our data has shown that for adiabatic flow these data exhibit character qualitatively analogous to ΔP_{tp}^* = $= f(x)$. In |8, 12| mass velocity did not affect frictional pressure drop and $x_{\Delta p}$ both for adiabatic and diabatic flows. Curves describing experimental data |8, 12| show, that wall heating considerably affects ΔP_{tp}^* (curve 4, 5, 6). It should be noted that pressure drop was determined by readings of two tops, located at 250 mm from each other while heated length of tube was 180 mm. Therefore the assumption can be made that the whole complex pattern of pressure profile taking place at flow regime change is not observing.

For more detailed analysis of helium two-phase flow hydrodynamics mass velocity may be excluded from consideration and data may be presented as dependence of friction factor on quality. With this aim experimental data were treatied as $\xi^* = f(x)$, where $\xi^* = \xi_{exp}/\xi'$ - relative friction factor, $\xi_{exp} = 2\Delta P_{tp} \rho_{mix} / (\overline{\rho w})^2 (1/d)$ was calculated by measured values of ΔP_{tp}, $\rho_{mix} = \rho' \rho''/[x\rho' + (1-x)\rho'']$ -density of vapour-liquid mixture, $\xi' = 0.316/Re^{0.25}$ - friction factor calculated by Blasius formula in liquid helium flow in a round tube at atmospheric pressure. Averaged curves describing experimental points from Fig.10 are presented in Fig.12a,b. Let us consider behaviour of * vs quality for helium two-phase flow using curve 1 (Fig.12) at $\overline{\rho w} = 60$ kg/m²s. At x=0 ξ^* equals to its limit value 1, correspondent to single-phase liquid flow, i.e. $\xi_{exp} = \xi'$. As soon as even comparatively small amount of vapour buffles appears in flow their turbulent action causes sharp rise of pressure drop and then rise of ξ_{exp}. This rise is observed untill nucleate transits into slug regime (point A on curve 1) and flow rate does not seem to change considerably. Further as x rises increase of average mass velocity takes place which similarly to one-phase flow leads to decrease of ξ_{exp}. At ß=0.7- -0.75 which corresponds to x=0.25-0.3 flow regime transits to annular one which causes change of slope of curve $\xi^* = f(x)$, ie. rate of decrease of ξ^* slightly increases. It seems to be connected with the fact that decrease of ξ^* is affected not only by increase of average mass velocity, but also by film thickness change (i.e. change of disturbance wave amplitude at the phase interface with thinning of the film). At the complete disappearance of disturbance waves that is possible in the annular regime with micro -film on the wall of the tube, rate of decrease of ξ^* retards again and point of bend on the curve is observed (point B) which is in a good agreement with experimental $x_{\Delta p}$ from Fig.10. Occurrence of boiling crisis due to complete dryout of a liquid film on the wall should not considerably affect the friction factor because mechanism of interaction at the phase interface (smooth film-vapour and smooth wall-vapour) changes slightly. Therefore point C is not characteristic point of a curve and is defined by formula (3). Occurrence of dryout causes sharp rise of the tube wall temperature (crisis of transition from nucleate to film boiling and dryout are considered). In this case there are sufficiently high amount of liquid as small droplets in the flow. These droplets may be deflected from the direction of own movement along the axis of a tube because of action of turbulent oscillations and may come into superheated boundary layer of vapour on the wall where an intensive vaporization takes place. This results in the radial component of vapour velocity that comes decrease of axial velocity in boundary layer. If entrainment of liquid from film and droplets deposition from flow to boundary layer are absent dryout would take place at x close to 1 and decrase of ξ^* would seem bring observed only till point C, further ξ^* would equal to its second limit value $\xi'/\xi'' = 0.79$. Really because of entrainment from the core of flow and droplets vaporization in a boundary layer decrease of axial velocity gradient and ξ^* takes place near the wall, i.e. alongside with increase of average mass velocity one additional factor causing decrease of ξ^* appears due to x increase.

At x increasing in the region $x > x°_{lim}$ droplets concentration in the flow decreases and their velocity rises consequently droplets deposition in boundary layer region also decreases that ultimately leads to decrease of rate of diminishing ξ^*. At the point D ξ^* on reaching its minimum value increases till its second limit value. Analogous behaviour of $\xi^*=f(x)$ is observed within the whole range of mass velocities. As $\overline{\rho w}$ rises at x=const relative friction factor decreases. The reason of it may be the fact that larger values of $\overline{\rho w}$ at other equivalent conditions are correspondent with larger values of local void fraction φ. Therefore the less film thickness in the annular flow is the less amplitude of the desturbance waves on the surface of liquid film and then less pressure drop. This fact is rather well illustrated by experimental obtained at four fixed mass velocities and shown in Fig. 10.

The complex hydrodynamic behaviour of two-phase diabatic flow is characterized by complex dependence $\xi^*=f(x)$ which hardly may be described analytically. However taking into account sufficiently high reliability of experimental data on pressure drop obtained in this paper empirical correlation for $\xi^*=f(x)$ may be chosen. With the help of computer HP-9830 all data were treatied by method of least squares and polinome describing experimental data within the whole range of regime parameters with error $\pm 10\%$ was obtained.

$$\xi^* = 1 - 0.21x + (1-x)[A_1 x^{0.1\ln(\overline{\rho w}/21.6)} + A_2 x^{0.5/\ln(\overline{\rho w}/21.6)} + A_3 x^{1/\ln(\overline{\rho w}/21.6)} + A_4 x^3] \quad (6)$$

where: $A_1 = -95.4 + 64.7(\overline{\rho w}/21.6) - 12.92(\overline{\rho w}/21.6)^2 + 0.704(\overline{\rho w}/21.6)^3$;
$A_2 = 102.5 - 67.9(\overline{\rho w}/21.6) + 13.70(\overline{\rho w}/21.6)^2 - 0.736(\overline{\rho w}/21.6)^3$;
$A_3 = -9.3 + 3.9(\overline{\rho w}/21.6) - 0.92(\overline{\rho w}/21.6)^2 + 0.039(\overline{\rho w}/21.6)^3$;
$A_4 = 1.5 - 1.1(\overline{\rho w}/21.6) + 0.22(\overline{\rho w}/21.6)^2 - 0.010(\overline{\rho w}/21.6)^3$

Data of this investigation and curves calculated by (6) are presented in Fig. 13. The correlation (6) describes data in the following range of regime parameters: mass velocity from 60 to 313 kg/m²s, heat flux from 20 to 7000 Wt/m², quality from 0 to 1. The further calculation may be written as follows:

$$\left. \begin{array}{l} \xi_{tp} = \xi^* \cdot \xi' \ ; \\ \Delta P = \xi_{tp} \dfrac{(\overline{\rho w})^2}{2\rho'} [1 + x(\rho'/\rho'' - 1)] l/d \ ; \end{array} \right\} \quad (7)$$

where: $\xi' = 0.316/Re'^{0.25}$; $Re' = (\overline{\rho w}) d/\mu'$.

Thus the complex investigation of heat transfer and hydrodynamics in the helium two-phase flow was carried out which results may be used for designing and constructing the cryogenic heat exchangers and the cooling systems of the superconducting magnets.

REFERENCES

1. De La Harpe A. et al. 1969. Boiling heat transfer and pressure drop of liquid helium- I under forced circulation in a helicaly coiled tube. Adv. Cryog. Eng., Vol.14, p.170-177.
2. Johannes C. 1972. Studies of forced convection heat transfer to helium- I. Adv. Cryog. Eng. Vol.17, p.352-360.
3. Ogata H., Sato S. 1973. Critical heat flux for two-phase flow of helium- I. Cryogenics, Vol.13, No.10, p.610-611.
4. Keilin B.E. et al. 1975. Forced convection heat transfer to liquid helium- I in the nucleate boiling region. Cryogenics, Vol.15, No.3,

p.141-145.

5. Grigoriev V.A. et al. 1977. Experimental investigation of heat transfer in nitrogen and helium boiling in channels. Teploenergetika, No.4, p.11-14.

6. Hildebrandt G. 1972. Heat transfer to boiling helium- I under forced flow in vertical tube. Proc. ICEC-4, p.295-300.

7. Giarratano P.J. et al. 1974. Forced convection heat transfer to subcritical helium- I, Adv. Cryog. Eng., Vol.19, p.404-416.

8. Deev V.I. et al. 1979. Pressure drop and heat transfer crisis in helium boiling in tubes. Teploenergetika, No.1, p.60-63.

9. Ogata H., Sato S. 1974. Forced convection heat transfer to boiling helium in a tube. Gryogenics, Vol.14, No.7, p.375-380.

10. Belyakov V.P. et al. 1979. Helium boiling studies in crisis region under free and forced convection. Proc 21 Siberian Term. Sem., Novosibirsk, p.239-244.

11. Arkhipov V.V. et al. 1980. Investigation of boundary quality in helium boiling in tubes. Teploenergetika, No.4, p.19-21.

12. Deev V.I. et al. 1978. Pressure drop in a two-phase flow of helium under adiabatic conditions and with heat supply. Proc. 6th Int. Heat Transfer Conf., Vol.1, p.311-314.

13. Petukhov B.S., Zhukov V.M., Shieldcret V.M. 1980. Experimental investigation of pressure drop in helium two-phase flow in a vertical tube. Teplofizika Vysokih Temperatur, Vo.18, No.5, p.1040-1045.

14. Petukhov B.S., Zhukov V.M., Shieldcret V.M. 1980. Experimental investigation of pressure drop and critical heat flux in the helium two-phase flow in a vertical tube. Proc. ICEC-8, p.181-185.

15. Grigoriev V.A., Pavlov Yu.M., Ametistov E.V. 1977. Boiling of cryogenic liquids. "Energy", M., p.288 (in Russian).

16. Doroschuk V.E. 1970. Boiling crisis of water in tubes. "Energy", M., p.168 (in Russian).

17. Kutateladze S.S. 1979. Fundamental of heat transfer. "Atomizdat", M., p.416 (in Russian).

18. Miropolsky Z.L. et al. 1965. Effect of heat flux and velocity on pressure drop in the steam-water mixture flow in tubes. Teploenergetika, No.5, p.67-70.

19. Keilin V.e. et al. 1969. Device for measuring pressure drop and heat transfer in two-phase helium flow. Cryogenics, Vol.9, No.2, p.36-38.

HEAT TRANSFER AND PRESSURE DROP IN TUBE BANKS

Problems of Heat Transfer Augmentation for Tube Banks in Cross Flow

A. ŽUKAUSKAS
Academy of Sciences
Lithuanian SSR
Vilnius, USSR

ABSTRACT

The lecture considers problems of power efficiency increase and compactness of shell-and-tube heat exchangers in the aspect of the heat transfer augmentation.

NOMENCLATURE

a	—	relative transverse pitch, s_1/d
b	—	relative longitudinal pitch, s_2/d
c_f	—	friction factor, $2\tau_w/\rho u_1^2$
c_p	—	specific heat capacity, J/kg·K
d	—	tube outside, m
Eu	—	Euler number, $\Delta p/\rho w^2 z$
F	—	heat transfer surface area, m^2
k	—	height of the artificial surface roughness elements, m; overall heat transfer coefficient, $W/m^2 \cdot K$
k^+	—	dimensionless height of the artificial surface roughness elements, ku_*/ν
Nu	—	Nusselt number, $\overline{\alpha} d/\lambda$
Δp	—	pressure drop, Pa
Pr	—	Prandtl number, $c_p \mu/\lambda$
Re	—	Reynolds number, wd/ν
s_1, s_2	—	transverse and longitudinal pitches, respectively, m
St	—	Stanton number, $\overline{\alpha}/\rho c_p u_1$

Δt — surface-fluid temperature difference, $^\circ C$

u — main flow velocity, m/s

u_1 — velocity on the outer boundary layer, m/s

u_* — friction velocity, m/s

w — mean velocity in the minimum inter-tube space, (i.e. in the minimum free flow area), m/s

z — number of tube rows in a bank

α — local heat transfer coefficient, W/m^2.K

$\bar{\alpha}$ — average heat transfer coefficient, W/m^2.K

β — angle of attack, deg; compactness of exchanger

λ — thermal conductivity, W/m.K

μ — dynamic viscosity, Pa.s

ν — kinematic viscosity, m^2/s

ρ — density, kg/m^3

τ — shear stress, Pa

φ — angle measured from the front stagnation point, deg

Subscripts

f — main flow

w — wall

1. THE WAYS OF THE HEAT TRANSFER AUGMENTATION

Intensification of the heat transfer processes, and increase of power efficiency of heat exchangers constitute a leading tune in the whole history of shell-and-tube heat exchangers.

By augmentation of the heat transfer we imply size and volume restriction of heat exchangers, and decrease of the temperature difference involved. Whenever an increase of fluid velocity in the reasonable limits of pressure drop fails to produce either the necessary size of the apparatus or the prescribed wall temperature, ways must be found to augment the heat transfer in such a way, that its size could be decreased without any increase in the circulation power.

A bank of tubes in crossflow design is advantageous from the point of view of the heat transfer and exceeds that of apparatus containing yawed or longitudinal tubes. When reference is made to the fluid velocity in the minimum free space of a bank in crossflow, which lies at $\beta = 90^\circ$, the heat transfer is lower

HEAT TRANSFER AUGMENTATION FOR TUBE BANKS IN CROSS FLOW

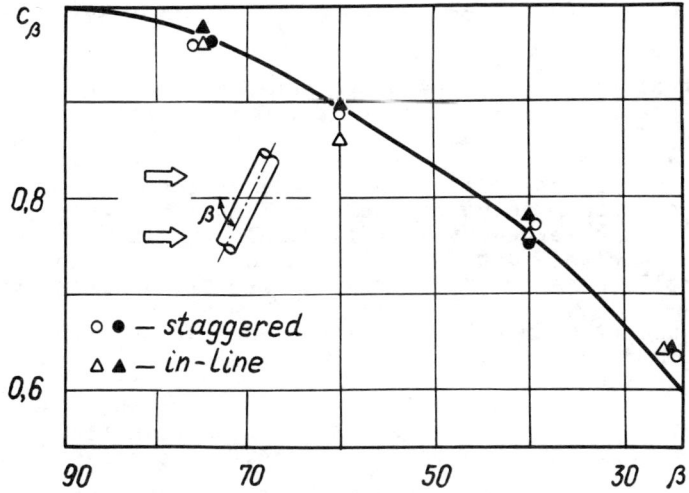

Fig. 1. Correction for the angle of attack in determining the heat transfer of a tube in a bank. White and black symbols- leading and inner rows, respectively

at lower angles of attack, be it a staggered or an in-line array (Fig. 1).

In Fig. 2 we present a comparison of thermal efficiency for banks of tubes in longitudinal and transversal flows of air. The value of $\bar{\alpha} = f(N/F)$ was determined under the assumption of $\Delta t = 1\,°C$ and circulation power (N) was related to unit area of the heated surface. For $Re \ll 10^5$ transverse banks are more efficient

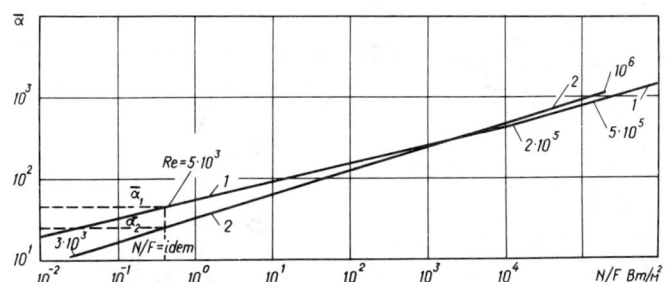

Fig. 2. Thermal efficiency of banks of smooth tubes in air: 1 - in-line bank, a = b = 1.25 in crossflow; 2 - square bank, s/d = 1.25 in longitudinal flow

than longitudinal ones, and ratio α_1/α_2 increases with a decrease of Re.

A most simple way of the heat transfer augmentation is through an increase of fluid velocity. Our studies with transverse flows of water on bank of smooth tubes in Fig. 3 showed a considerable augmentation of the heat transfer on the whole perimeter of an inner tube with higher fluid velocities, that is at higher Re. A maximum value of the heat transfer coefficient $\alpha = 1.5 \times 10^5$ was observed at Re = 1.52×10^6 in inner rows of a staggered bank with a x b = 2 x 2. But high fluid velocities involve naturally high power consumption for fluid circulation.

A feature of a one-phase flow on a solid non-permeable surface is a boundary layer, which constitutes the main source of resistance to the heat transfer. The thicker is the boundary layer and the lower heat conduction of the fluid - the lower heat transfer. From the point of view of the heat transfer, the most favourable conditions are formed in turbulent boundary layers.

In turbulent flows, efficient fluid viscosity and consequently, the rate of diffusion are considerably higher, than in laminar ones. The heat transfer is correspondingly higher: on a plate the heat transfer coefficient is $\bar{\alpha} \sim u^{0.5}$ and $\bar{\alpha} \sim u^{0.8}$ for laminar and turbulent boundary layers, respectively.

Naturally developed turbulence cannot provide for the maximum available heat transfer augmentation, because turbulence only occurs at very high velocities and is connected to high rates of

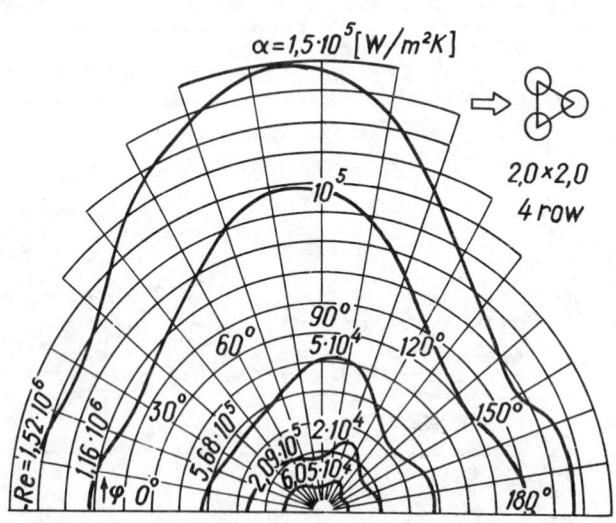

Fig. 3. Local heat transfer of a smooth tube in a bank

pressure drop. Augmentation of convective heat transfer can be accomplished either by artificial turbulence of the boundary layer, so that the processes of heat transfer are shifted to its turbulent region, or by decreasing the thickness of the boundary layer up to its complete elimination.

Specific surface elements introduced in the boundary layer cause more intensive circulation of the fluid, interchange of its volumes, and in many cases, particularly for $Pr > 1$, lead to augmentation of the heat transfer, which exceeds the accompanying increase of pressure drop. The greater this discrepancy, the more favourable is the heat transfer - pressure drop ratio, and the more advantageous the involved method of heat transfer augmentation.

Numerous ways of heat transfer augmentation can be classified as passive, active and combined ones.

Passive methods operate without any increase of the power consumption. They lie in the introduction of finned, rough or otherwise developed surfaces on the side of a fluid of a low heat transfer coefficient, as well as turbulizers, rods, swirlers a.o. devices of eliminating the bounday layer or decreasing its thickness.

Active methods of heat transfer augmentation employ supplementary external power. They employ mechanical devices to mix,or to scrape the heat carrier from the surface, as well as vibration and rotation of the surface itself, to decrease the thickness of the boundary layer and to increase the rate of wetting, which results in heat transfer augmentation. Sound vibrations may be introduced in the fluid, from 1 Hz to ultrasonic. An electric field may be introduced in the system, to increase convective motion of the fluid through specific forces in the dielectric fluid.The heated fluid may be sucked away through a porous layer.

Combined methods employ at least two of the above techniques simultaneously. Such are rough-surface tubes with swirlers. vibrating finned tubes a.o.

Transversal flow on a circular tube features normally a laminar boundary layer on the front part region. But our studies [1] and similar studies by E. Achenbach [2] showed that in specific conditions a turbulent boundary layer can be developed instead. The single-tube test presents a highly illustrative example. Our common studies with P. Daujotas and J. Žiugžda,performed in wide ranges of critical and supercritical values of Re, revealed that the initation of the laminar-turbulent transition in the boundary layer and formation of thermal turbulent boundary layer are dependent on the value of Re, Fig. 4, and free stream turbulence, Fig. 5. The thermal turbulent boundary layer was shifted upstream with an increase of the two factors, up to $\varphi = 30$ to $35°$, until it occupied a predominant part of the front region. Considerable augmentation of the heat transfer was observed.

Curves in Fig. 4 ($Re = 8.6 \times 10^5$) suggest a complex variation of the heat transfer in the critical flow regime. Laminar

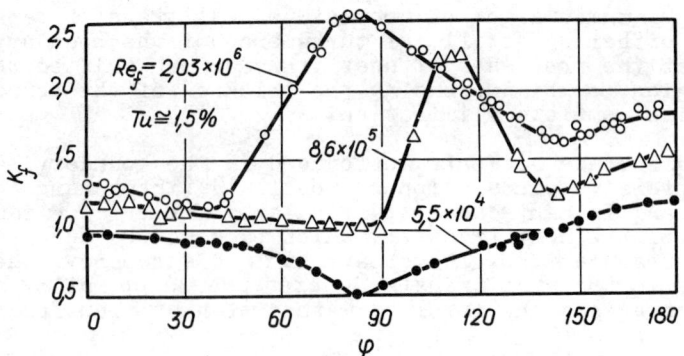

Fig. 4. The effect of Re on the local heat transfer of a cylinder in water at Tu = 1.5 %

boundary layer persists on the front region, similarly to the subcritical flow regime, and the heat transfer becomes lower with the increase of its thickness.

But the general tendencies of local heat transfer curves show some principal differences, as compared to those for the subcritical flow. Local heat transfer curves for air and water,

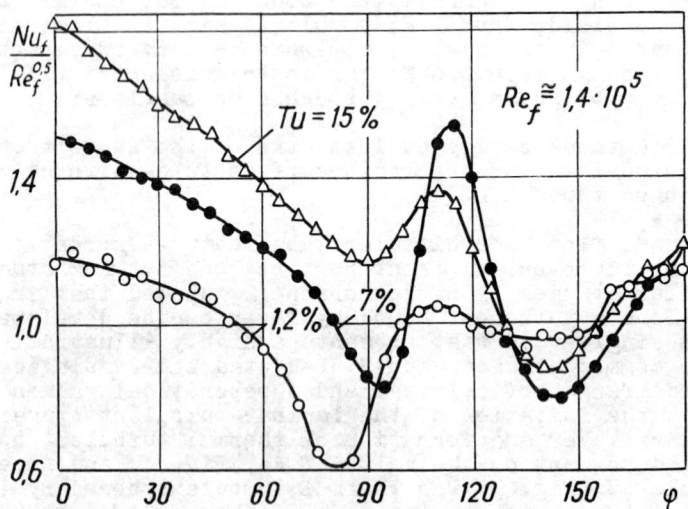

Fig. 5. The effect of turbulence on the local heat transfer of tubes in air

Fig. 4, 5, exhibit two minima, which reflect fluid dynamics on the cylinder surface.

With the critical flow regime, the first minimum at $\varphi = 80$ to $90°$ reflects the separation of the boundary layer and formation of the separation bubble, followed by its reattachment at $\varphi \approx 100°$ as a turbulent boundary layer. For the supercritical flow regime (Re = 2.03×10^6), the first minimum is interpreted as the initation of the laminar-turbulent transition in the boundary layer. The second minimum at $\varphi \approx 140°$ reflects separation of the turbulent boundary layer in any of the two regimes.

With a further increase of Re and Tu, the separation bubble dissapears, the transition point is shifted upstream and the laminar part of the boundary layer becomes shorter. A general conclusion is that even a low level of turbulence (Tu = 1 %) favours the onset of the critical flow regime. The effect of free stream turbulence is most pronounced in the front stagnation point, and decreases with angular distance. No regular relation between free stream turbulence and fluid dynamics can be observed in the rear. The highly vortical flow there must be insensitive to external disturbances.

Average heat transfer is augmented accordingly. Fig. 6 presents our studies of average heat transfer of circular cylinders diam 30.7 and 50 mm in flows of air and water at Re_f from 4×10^4 to 2×10^6. In the critical flow regime (Re > 2×10^5) average heat transfer is higher (dotted line). Results for Re > 3.5×10^5 form a single curve, with 0.8 tangence of the slope angle. Thus

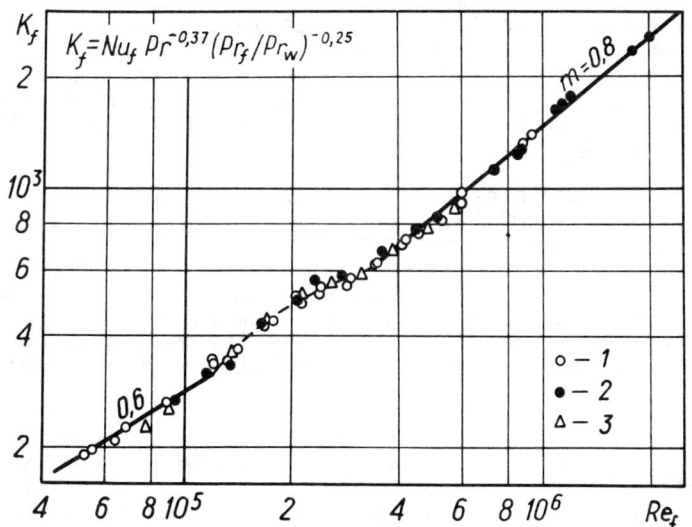

Fig. 6. Average heat transfer in water and air:
1, 2 - water, cylinder d = 30.7 and 50 mm;
3 - air, cylinder d = 32 mm

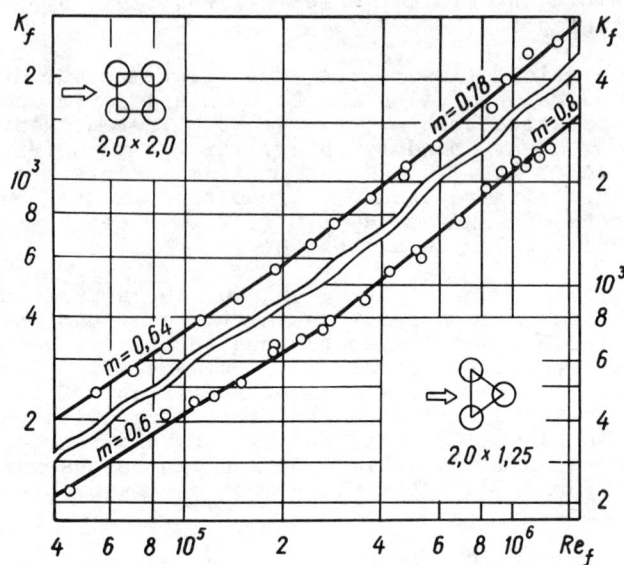

Fig. 7. Average heat transfer from tubes in inner rows of staggered and in-line banks in flows of water at high Re.
$K_f = Nu_f \, Pr_f^{-0.4} \, (Pr_f / Pr_w)^{-0.25}$

in the supercritical flow regime the heat transfer rate is augmented and can be described by a relation, in which Re is in the same power as for a turbulent flow on a plate. For Re from 1.5×10^5 to 3.5×10^5 the results deviate from the heat transfer curve, Fig. 6. This deviation reflects the critical flow regime—higher frequency of vortex shedding, separation bubble and transient flow dynamics.

Analogic changes are observed in banks of tubes in crossflow at high values of Re.

Fig. 7 presents our results [3] on the average heat transfer of an in-line bank 2 x 2 and a staggered bank 2 x 1.25 in water. For the staggered bank, the power index of Re is 0.6 at Re < 3×10^5, but increases to 0.8 for the supercritical flow regime. The corresponding values for the in-line bank are 0.64 and 0.78.

Heat transfer augmentation can be achieved by artificial flow turbulence. Fig. 8 presents our results [1] for average heat transfer of a single cylinder. Here, with an increase of turbulence from 1.2 to 15 %, average heat transfer is augmented approximately for 40 and 55 % for the critical and the supercritical flows, respectively.

A bank of tubes is in itself a sort of turbulizing device,

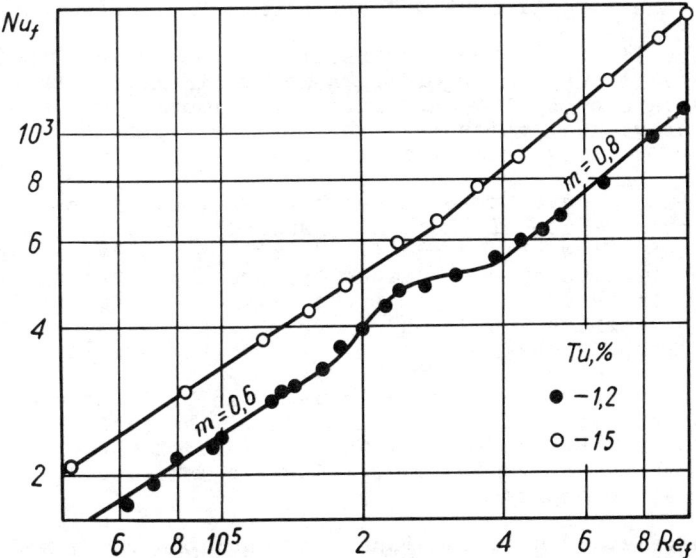

Fig. 8. The effect of turbulence on the average heat transfer from a cylinder in air

its internal turbulence and heat transfer depend on the type of tube arrangement [4].

Table 1   Heat transfer augmentation inside banks

pitches	Staggered							In-line			
	1.30 x 1.13	2.6 x 1.30	1.95 x 1.30	1.30 x 1.30	1.30 x 1.73	1.30 x 3.9	1.30 x 1.30	2.0 x 2.0	1.26 x 2.0	1.30 x 2.6	leading row
Heat transfer augmentation %	175	170	170	165	140	125	160	145	150	145	100

For a steady-state flow, a comparison of the heat transfer from the leading row and from the inner rows reflects the turbulizing effect of the bank. Such a comparison is shown in Table 1. Heat transfer inside a bank increases with a decrease of the longitudinal pitch. These results are in agreement with studies of a single tube at different distances from the turbulizing screen.

In the inner rows heat transfer is from 30 to 70 % higher as compared to the leading-row. Therefore in most bank geometries, the rate of heat transfer from the inner rows is a function of turbulence, which is higher at lower distances from the turbulizer, that is from preceding row.

Artificial turbulence can also be introduced in the boundary layer by rotation of the heated surface. Thus with cylinder rotating in a flow of viscous fluid, its particles are entrained in the rotation because of viscosity, and augmentation of the heat transfer occurs because of the higher relative fluid velocity on the surface. Rotation of the cylinder does not cause any increase of shape drag.

The amount of heat, transfered to the fluid at a constant termperature difference, is a function of the surface area and of the heat transfer coefficient. When a fluid of three and more times higher heat transfer coefficient is introduced, a finned surface must be employed.

We have at present numerous constructions of finned tubes with transverse and longitudinal fins, applied in transverse and longitudinal flows.

Fin configurations are also very different, Fig. 9. The most common are radial (Fig. 9b) and continuous fins, (Fig. 9d),

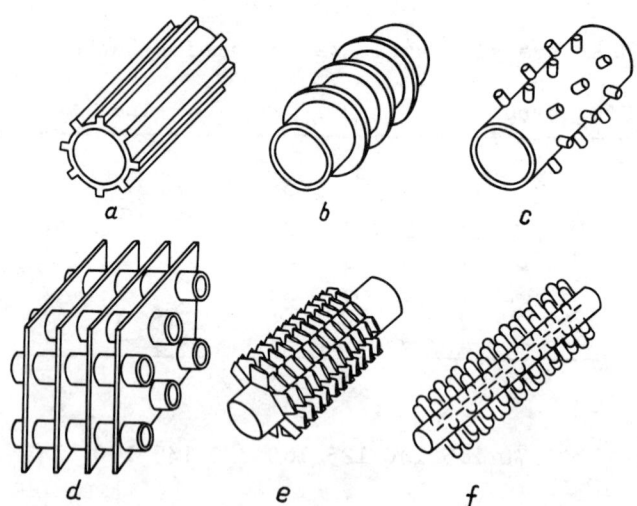

Fig. 9. Types of finned tubes: a - longitudinal fins, b - radial fins, c - spikes, d - continuous fins, e - elliptic profile fins, f - wire fins

but descriptions are known of experiments on flat, wavy, and grooved continuous fins.

Finned surfaces are introduced to improve performance and efficiency of recuperative heat exchangers, particularly in external gaseous flows. Fins of a limited height were found most effective in viscous fluids. Spiny tubes are most suitable as the mean of the heat transfer augmentation from banks in crossflow of gases. The spines may be cylindric, conical and semispherical. Cylindric spines are widely used because of their simple fabrication, low fouling rates and flame resistance.

Application of rough-surface tubes is also extending. Their heat transfer augmenting performance lies in the disintegration of the viscous sub-layer under turbulent flows, and in the loss of stability of the boundary layer. For other constant conditions the laminar-turbulent transition occurs at lower values of Re on rough surfaces.

Three different degrees of action can be distinguished on rough surfaces:

1. No action of roughness, when surface elements are low and lie within the viscous sub-layer.

2. Transient action of roughness, when surface elements protrude from the viscous sub-layer.

3. Full action of roughness, when the viscous sub-layer is completely destroyed.

Changes of fluid dynamics on rough surfaces lead to an earlier laminar-turbulent transition in the boundary layer. Surfaces with closely spaced surface elements have $S/d < 5$, where S - distance between surface elements. "Open" surfaces have widely spaced surface elements, and reattachment of the flow occurs after each separate element.

Inside staggered banks of smooth tubes, maximum heat transfer is observed around the front stagnation point, Fig. 10. Heat transfer decreases with the angular distance from the stagnation point because of the growing thickness of the laminar boundary layer, up to the separation point. Surface roughness introduces higher turbulence in the boundary layer. The effect depends on the ratio of height of the surface elements k, to the boundary layer thickness δ. For k considerably lower than δ, surface roughness are not reflected in the heat transfer. For similar values of k and δ heat transfer is augmented by the combined action of free stream turbulence and of velocity fluctuations. Partial desintegration of the laminar boundary layer by the surface elements causes a shift of the transition point from 90 to 60° at Re 2×10^5.

At Pr = 0.7, and surface roughness of k = 0.2 mm the effect on the heat transfer is not observed. It only appears at higher k.

Curves 1, 2, 3 in Fig. 11 illustrate the heat transfer from the inner rows of a 1.25 x 0.935 bank in transformer oil for

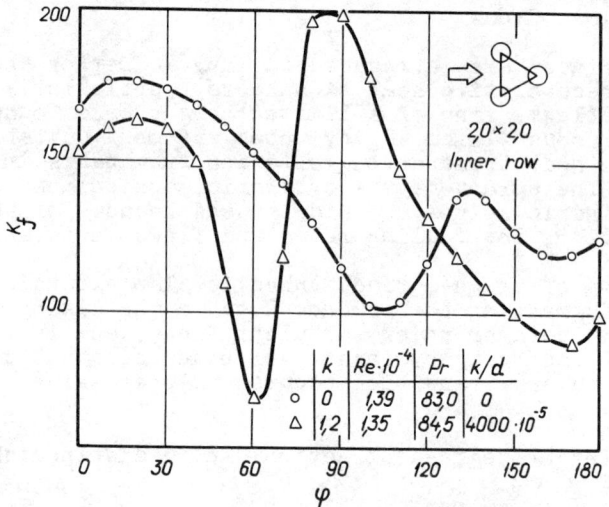

Fig. 10. The effect of rough surface on the local heat transfer of a tube in a bank

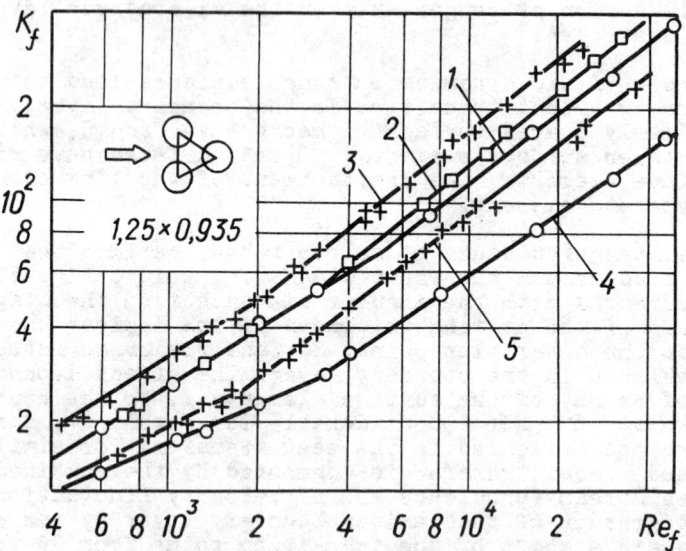

Fig. 11. Local heat transfer of tubes in different locations of a staggered bank 1.25 x 0.935 in transformer oil. Inner row: 1 - k/d = 0, 2 - k/d = 15 x 10^{-3}, 3 - k/d = 40 x 10^{-3}. Leading row: 4 - k/d = 0, 5 - k/d = 40 x 10^{-3}

$k/d = 15 \times 10^{-3}$ and 40×10^{-3}. Curves 4 and 5 correspond to the leading row for $k/d = 0$ and 40×10^{-3}. Average heat transfer increases with growth of k/d and at $Re = 10^4$ and optimal height of roughness, heat transfer augmentation is 44 and 53 % for the inner, and the leading rows, respectively.

Heat transfer augmentation is most pronounced, when surface roughness is introduced on the front regions of the tubes. We studied at the Institute of Physical and Technical Problems of Energetics, Lithuanian Academy of Sciences, a staggered bank ($a \times b = 1.25 \times 1.25$) of tubes, with surface roughness in the form of longitudinal rod-like fins on the front regions. Six to eight elements 0.5 diam were fabricated on each tube, to give the relative roughness $k/d = 0.01$ and the distance-to-height ratio of the elements about 20, so that reattachment of the separated shear flows occured on the tube surface, rather than on the nexts surface-element. In Fig. 12, similar curves of the heat transfer for this sort of tubes, are shown, but their average heat transfer as compared to smooth tubes continuously increases with Re, and reaches its maximum of 35 % at $Re = 2 \times 10^5$.

As to the pressure drop, it undergoes considerable changes because of this type of surface roughness. For a tube with eight surface elements on the front region, pressure drop is similar

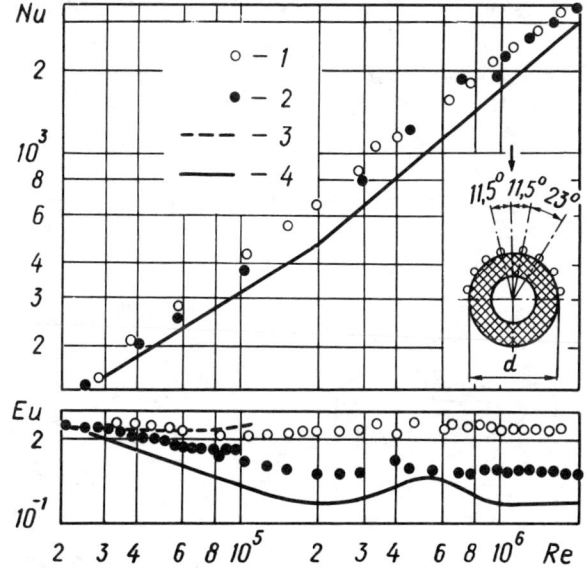

Fig. 12. Average heat transfer of a tube in the 5th row, and pressure drop of banks: 1, 2 - bank of finned tubes with eight and six fins, 3 - bank of rough tubes: $d = 50$ mm, $k/d = 0.0105$, 4 - bank of smooth tubes

to that on a tube with uniformly distributed surface elements of the same relative roughness. For a tube with six surface elements on the front, pressure drop is considerably lower. Consequently, significant heat transfer augmentation can be achieved by suitable surface roughness on the front, and consumption of circulation power may be even lower than with rough tubes.

Fig. 13 shows a tube with a partially developed surface, wich was constructed at the Institute of Physical and Technical Problems of Energetics. Here heat transfer augmentation was achieved by turbulizers (1) in the form of grooves (2), cylindrical in cross-section, to form a wavy surface. They give rise to fluid vortices, and level-out the distribution curve of the heat transfer coefficient by higher mixing and lower thicknes of the viscous sub-layer. Considerable augmentation of the heat transfer is observed along the sides of the tube.

To intensify forsed convection, mechanical scraping of the surface was suggested in [5]. A rotating blade was fixed on the surface, which removes periodically the boundary layer and decreases quite effectively the thermal boundary layer. Heat transfer augmentation is mainly determined by the velocity of scraping. The distance of the blade over the surface, which was varied from 0.7 to 3.8 mm had a minor effect. In this case heat transfer augmentation is most pronounced in laminar boundary layers. This technique can give as high as eleven times augmentation of the heat transfer, and is by far more efficient, than techniques based on the induction of the laminar-turbulent transition at similar Re.

As to different devices introdusing vibration of heated surfaces, their wide application in industrial heat exchangers is to be hardly expected, because their energy consumption exceeds the effect of the heat transfer augmentation.

We also studied [6] the effect of forced vibration on the

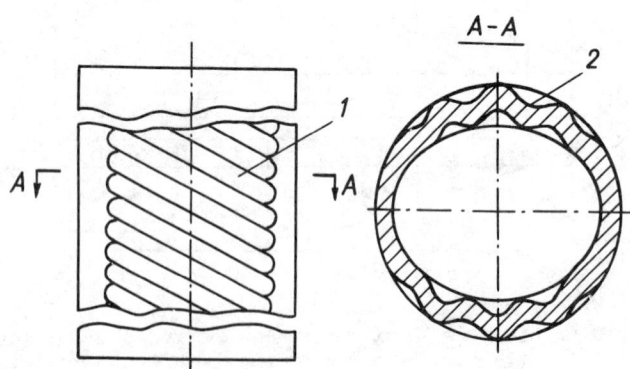

Fig. 13. Tube with partially wavy surface. 1 - turbulizers, 2 - inclined grooves

local and average heat transfer of a tube in crossflow of water at Re from 10^4 to 2×10^5. Vibration frequency was varried from 0 to 20 Hz, its relative amplitude A/d from 0.02 to 0.075, and the amplitudal velocity of vibration $u_a = 2\pi Af$ from 0 to 0.24 m/s (there A and f - vibration amplitude and frequency). The tube was caused to vibrate across the flow and vibration velocity was normal to the flow.

Velocity fluctuations induced in the flow were higher when directed across the boundary layer. They induce an earlier transition to the turbulent boundary layer. In Fig. 14, local heat transfer is mostly augmented at φ from 40 to $140°$, and reaches 30 % at $Re_a = 8 \cdot 10^3$ (Fig. 14 curves 1 and 3). At $Re > 10^5$ in the range of vibrational Re_a ($Re_a = u_a d/\nu$) covered no effect of vibration could be observed.

To describe average heat transfer of a vibrating tube in cross flow, we introduce a concept of resultant Reynolds number in the power of 0.72 ($Re_r = \sqrt{Re^2 + Re_a^2}$) to replace Re. The power index of Re was 0.6 for a stationary tube, and its increase is solely due to vibration.

Experiments with forced convection of a tube and of a plate [7] were also performed with ultrasonic vibration, from 27 to 697 kHz, perpendicular to the flow. Velocity of water and oil was varied from 0.07 to 1.0 m/s. Maximum augmentation of the heat transfer amounted to 80 % and was observed in a stationary wave on a thin plate. Our studies suggest an increase of the heat transfer with the intensity of vibration field. With a constant vibration field, a growth of velocity lead to a decrease of the effect. Results at 697 kHz suggest turbulization by micro streams on the surface, as a source of the heat transfer augmentation.

The decrease of the heat transfer at constant vibration and higher flow velocity can be interpreted as the predominant effect

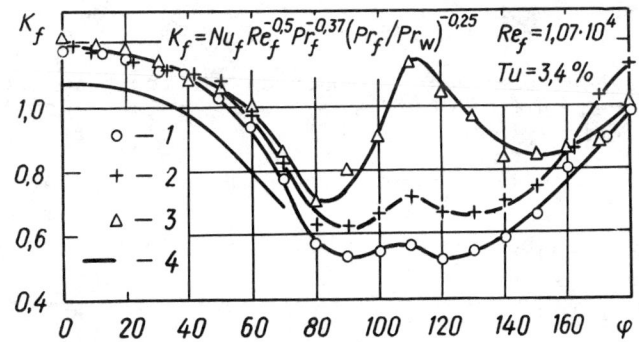

Fig. 14. The effect of Re_a on the local heat transfer of a tube in crossflow. Re_a: 1 - 0, 2 - 2.2×10^3, 3 - 8×10^3, 4 - analytical

of free stream turbulence. Ultrasonic vibration as a means of the heat transfer augmentation is only reasonable at free stream velocity under 1 m/s.

The list of solutions for the heat transfer augmentation is by no means exhausted. A more comprehensive study by A. Bergles [8] can be recommended.

By making the heat transfer processes more intensive, we opens the possibilities of constructing more compact heat exchangers.

2. COMPACTNESS OF HEAT EXCHANGERS

Compactness of heat exchangers β is described by the area of the heated surface for unit volume of the apparatus. For recuperators:

$$\beta = \frac{F_H + F_o}{V} \qquad (1)$$

where F_H - heated surface, F_o - cooled surface, V - total volume of exchanger.

For regenerators:

$$\beta = \frac{F_H}{V_H} = \frac{F_o}{V_o} \qquad (2)$$

where V_H - heat carryer volume, V_o - cooler volume.

There is no finite limit between compact and non-compact heat exchangers. Some time ago heat exchangers of $\beta > 200$ m²/m³ were considered compact enough, now we have apparatus with 700 m²/m³. In modern power generation, heat exchangers of $\beta = 150$ m²/m³ are common. Their tubes of 12 to 40 mm diam can operate at high pressures and high temperatures.

For pure heat carriers, compactness of the heat exchangers can be increased up to 250 m²/m³ by reducing tube diameters to 6 - or 12 mm. With plastic tube of 2.5 mm diam compactness of 650 m²/m³ can be reached. Modern automobile radiators of finned tubes can have β to 1100 m²/m³. Plate-and-membranne heat exchangers have $\beta = 5000$ m²/m³ and more.

Given equation

$$\frac{Q}{\Delta t} = k\beta V \qquad (3)$$

for equal k, the more compact a heat exchanger, the higher its heat transfer of unit volume. Compactness is in itself a condition of high efficiency, because of its limited free space, and the heat transfer coefficient is inversely proportional to the hydraulic diameter of the channel.

Higher compactness is ususally connected with lower amounts of metal used for unit heat transport. Compact heat exchangers are reasonable whenever one of the fluids is gaseous, and all the fluids are pure enough and exclude corrosion.

Ways of increasing compactness of heat exchangers are chosen in view of their application and construction. In banks of smooth circular tubes, compactness can be increased by reducing tube diameters. Heat transfer augmentation in banks of helical tubes is accomplished thanks to intermixing fluid flows directed from the core to the wall and vice-versa. The continuous transversal interchange of the fluid masses, and the higher degree of turbulence, as compared to non-compact are challenged by velocity fluctuations in the core.

Fig. 15 presents a shell-and-tube heat exchanger of increased efficiency. It is intended for chemical industries, where intensive heat transfer and high rates of mixing both on the tube and on the shell side are necessary. The apparatus consists of helical tubes of elliptic cross-section (1), fixed by direct circular ends in the tube holders (2). The tubes contact each other in their wider parts, so that the heat exchanger is highly resistant to vibration. Circulation of fluids in the tube side and in the shell side occurs in swirled flows, because of the helices. This heat exchanger has a 1.5 higher heat transfer and a

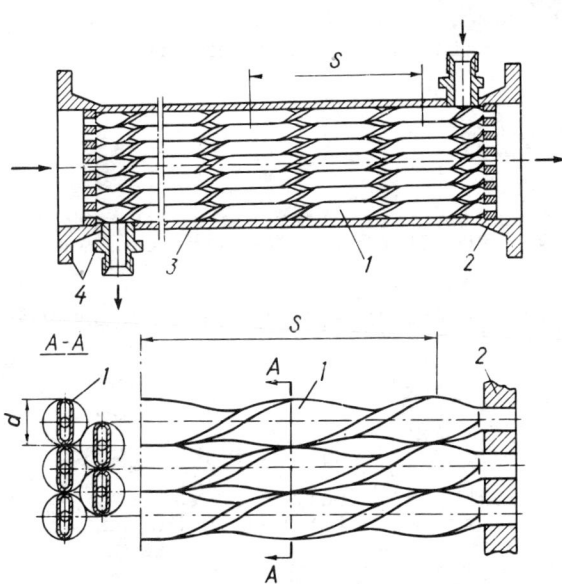

Fig. 15. Shell-and-tube heat exchanger of helical tubes. 1 - tube, 2 - tube holders, 3 - shell, 4 - supply collectors, d = = 6.15 mm - maximus diameter of the oval space; S = 12.2d - helix pitch

0.7 smaller volume, as compared to a similar plane-tube apparatus. Because of technological limitations, the method of increasing the heat transfer by increasing compactness is only applicable up to a certain limit. A solution was found in different constructions of baffles in shell-and-tube heat exchangers to monitor velocity and direction of the shell-side flow.

High efficiency of compact heat exchangers is also a result of specific geometries, which provide for higher heat transfer coefficients at similar flow rates. Indented fins and surfaces decrease the thickness of the boundary layer and induce its separation, or constitute conditions for the most favourable flow dynamics-laminar-turbulent transition.

Tube banks, curvilinear or wavy surfaces induce boundary layer separation and vorticity, and increase heat transfer. Thus helical tubes in longitudinal flows augment both external and internal heat transfer, so that the volume of a heat exchanger can be reduced to 1.5 times.

Efficiency increase of heat exchangers is achieved by their volume and weight reduction, economy of materials and maintainance. That is why heat exchangers are spreading in transport, cryogenics, electronics, chemical industries, and oil refineries.

3. EFFICIENCY OF HEAT EXCHANGERS

Power efficiency of a heated surface is described by the amount of heat transfered Q, for given surface area and circulation power N:

$$E = \frac{Q}{N} = \frac{\alpha F \Delta t}{V \Delta p} = \frac{\alpha \Delta t}{\frac{f}{F} w \Delta p} \tag{4}$$

where Q - amount of heat transfered from the surface, N - circulation power, V - fluid volume, f - minimum free space in the bank.

From Eq. (4), efficiency of a heat exchanger is a function of a construction factor f/F:

$$\frac{f}{F} = \frac{(s_1 - d)L \cdot z_1}{\pi d L z_1 z} = \frac{a-1}{\pi z} \tag{5}$$

where L - tube length, z_1 - number of tubes in a row. Thus f/F depends on the transverse pitch and on the number of rows. In a dimensionless form

$$E = \frac{\pi d^2}{a-1} \frac{\lambda_f \Delta t}{\rho_f \nu_f^3} Pr_f^n \frac{c}{k} Re^{m-r-3} \tag{6}$$

where $\Delta t = 1^\circ C$, c, k - geometry constants, m, r - power indices in equations

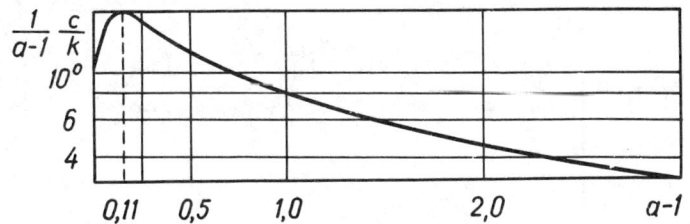

Fig. 16. Optimal relative pitch solution for symetric in-line banks. c and k - constant in $Nu = cRe_f^m Pr_f^n$ and $Eu = kRe_f^r$

$$Nu = cRe_f^m Pr_f^n \qquad (7)$$

$$Eu = kRe_f^r \qquad (8)$$

Eq. 6 contains an indication of the ways of increasing power efficiency. From optimization function $(c/k)/(a-1)$, Fig. 16, optimal relative pitches are a x b ≈ 1.1 x 1.1. A further decrease of the pitches is ineffective. In the presence of fouling, less efficient relative pitches must be used.

We consider now surface roughness of bodies in crossflow. The shape of surface elements is of secondary importance, but their height ought to be studied in detail.

A transverse flow on a blunt body features the point of attack, and a boundary layer, which becomes thickes with the angular distance up to the separation point—here its thickness is at the maximum. The thickness of the thermal boundary layer is a function of Pr, and is considerably lower at higher Pr.

For a known velocity distribution in the boundary layer, local friction on the tube surface is given by the integral momentum equation as

$$\tau_w dx = d \left[\int_0^\infty \rho u (u_1 - u) dy \right] \qquad (9)$$

Local values of the heat transfer coefficient are given by the integral energy equation as

$$q_w dx = d \left[\int_0^\infty \rho c_p u (\vartheta_1 - \vartheta) dy \right] \qquad (10)$$

where u - local velocity in the boundary layer, q_w - relative heat flux, ϑ, ϑ_1 - temperature on the surface and temperature in the outer boundary layer.

We express Eq. (9) and (10) through dimensionless variables u^+, ϑ^+, y^+ and arrive at the local friction coefficient

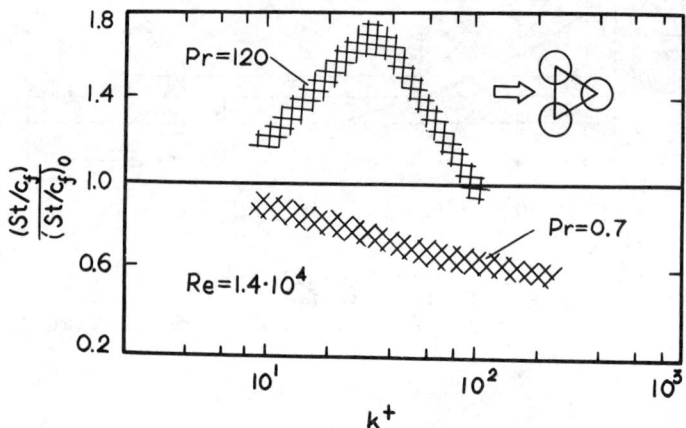

Fig. 17. The effect of surface elements in air and transformer oil

$$c_f = \frac{2\tau_w}{\rho u_1^2} = \frac{2}{u_1^{+2}} \qquad (11)$$

and the local Stanton number

$$St = \frac{q_w}{\rho c_p u_1 \vartheta_{max}} = \frac{1}{u_1^+ \vartheta_1^+} \qquad (12)$$

Fig. 17 presents experiments on the efficiency comparison of a rough surface (St/c_f) and a smooth surface $(St/c_f)_0$ in air and transformer oil (Pr = 0.7 and 120, respectively). Measurements were performed on tubes at $\varphi = 90°$. In air, the surface roughness is ineffective, and the heat transfer augmentation observed is proportional to the larger surface area. In transformer oil (Pr ≫ 1), an increase of k^+ leads to the heat transfer augmentation up to the limiting value k^+. With a further increase, the value of $(St/c_f)/(St/c_f)_0$ decreases and loses its effect. We conclude, that surface roughness of a limited height can be effective in fluids with Pr ≫ 1. At Pr = 120 and $\varphi = 90°$ local heat transfer increases 1.8 times at efficient roughness $k^+ \approx 40$.

After the boundary layer separation, in the recirculation region, velocity decreases, and the boundary layer develops from the rear to the separation point. Surface roughness is ineffective in this region, it only gives the heat transfer augmentation proportional to the larger surface area. Consequently, the effect of surface roughness is not uniform in the tube perimeter. The height of the surface elements ought to be variable, but this involves much higher cost of fabrication and is not reasonable.

REFERENCES

1. Žukauskas, A., Žiugžda, J. 1979. Teplootdacha Tsilindra v Poperechnom Potoke Zhidkosti. Heat Transfer of a Cylinder in Crossflow. Thermophysics 11. Mintis. Vilnius. 237 p. (In Russian).

2. Achenbach, E. 1975. Total and Local Heat Transfer from a Smooth Circular Cylinder in Cross-Flow at High Reynolds Number. International J. of Heat and Mass Transfer, Vol. 10, No 12, pp. 1387-1396.

3. Žukauskas, A.A., Ulinskas, R.V. 1978. Heat Transfer Efficiency of Tube Bundles in Crossflow at Critical Reynolds Numbers. Heat Transfer-Soviet Research, Vol. 10, No 5, pp. 9-15.

4. Žukauskas, A., Makarevičius, V., Šlančiauskas, A. 1968. Teplootdacha Puchkov Trub v Poperechnom Potoke Zhidkosti. Heat Transfer in Banks of Tubes in Crossflow of Fluid. Thermophysics 1. Mintis. Vilnius. 192 p. (In Russian).

5. Hagge, J.K., Junkhan, G.H. 1975. Mechanical Augmentation of Convective Heat Transfer in Air. Trans. ASME, J. Heat Transfer, Vol. 97, No 4, pp. 516-520.

6. Katinas, V., Šuksteris, V., Žukauskas, A. 1981. Teploobmen Tsilindra, koleblyushchegosya v Poperechnom Potoke Zhidkosti. Heat Transfer of Vibrating Cylinder in Cross-Flow of Fluid. Lietuvos TSR Mokslų Akademijos Darbai, Ser. B, Vol. 3(124), pp. 65-73 (In Russian).

7. Žukauskas, A.A., Šlančiauskas, A.A., Jaronis, E.P. 1961. Issledovanie Vliyaniya Ultrazvukovykh Voln na Teploobmen Tel v Zhidkostyakh. Investigations of Influence of Ultrasonic Waves on Heat Transfer of Bodies in Fluids. Inzhenerno-Fizicheskii Zhurnal, Vol. IV, No 1, pp. 58-62 (In Russian).

8. Bergles, A.E. 1973. Techniques to Augment Heat Transfer. Handbook of Heat Transfer. McGraw-Hill Book Co. New York, pp. 10.1-10.32.

Heat Transfer and Pressure Drop of Staggered and In-line Heat Exchangers at High Reynolds Numbers

E. ACHENBACH
Institut für Reaktorbauelemente
Kernforschungsanlage Jülich GmbH
Postfach 1913
5170 Jülich, FRG

ABSTRACT

Heat transfer and pressure drop of a staggered and in-line tube bundle have been measured in the range of Reynolds-numbers $4 \times 10^4 < Re < 7 \times 10^6$. It is shown that a critical Reynolds number exists beyond which heat transfer and pressure drop change their dependence upon the Reynolds number. The boundary layer phenomena associated with this effect are discussed referring to the local static pressure, skin friction and heat transfer distribution. Additionally, the effect of surface roughness on the pressure drop and heat transfer is considered.

1. INTRODUCTION

Cross flow heat exchangers applied to nuclear helium-cooled high temperature reactors operate at high Reynolds numbers due to the elevated pressure of 4o-7o bars in the primary circuit. Similar to a single circular cylinder in cross flow there exists a critical Reynolds number beyond which the heat transfer and pressure drop change their dependence upon the Reynolds number /1/, /2/, /3/. In a detailed fundamental study of the flow through staggered and in-line tube bundles /4/, /5/, /6/ it has been shown that the transition from laminar to turbulent flow the critical Reynolds number can be influenced by the surface roughness of the tubes. With increasing roughness parameter the critical Reynolds number decreases. The premature transition from laminar to turbulent boundary layer causes an enhancement of heat transfer, whereas the effect on the pressure drop depends on the geometrical conditions of the bundle. It has been demonstrated /6/ that the pressure drop coefficient of in-line arrangements may decrease with increasing surface roughness. Thus the efficiency of a heat exchanger can be improved. On a first view this phenomenon seems to be contradictory to the experiences made for the pressure drop of rough surfaced bodies. The pressure drop of a tube bundle, however, is predominantly due to the shape resistance whereas the friction drag only contributes to about five per cent /5/, /6/. Of course, with increasing surface roughness the shear stresses, which determine the heat transfer, also increase, but the effect on the total drag is minor.

The aim of the present work is to let the reader recognize the flow through tube bundles in cross flow as the flow past bluff bodies. The phenomena observed for the smooth and rough cylinder are assumed to be known /7/, /8/, /9/, /1o/. Therefore the description of what happens starts with the single row of tubes and proceeds to the whole array. The effect of surface roughness is pointed out, particularly by considering the total pressure drop as well as the total and local heat transfer. The experiments with rough bundles are still running. Nevertheless, the results already obtained reveal the phenomena occurring in a bundle.

The results are presented using the following dimensionless groups in which the tube diameter d is the length scale and in which the velocity u_c of the narrowest gap section is the characteristic velocity:

Reynolds number $\qquad Re = \dfrac{u_c\, d\, \varrho}{\eta}$

Pressure drop coefficient $\qquad \zeta = \Delta p / (\dfrac{\varrho}{2}\, u_c^2)$

Nusselt number $\qquad Nu = \dfrac{\alpha\, d}{\lambda}$

Prandtl number $\qquad Pr = \eta\, c_p / \lambda$

Roughness parameter $\qquad k_s/d$

Transversal pitch $\qquad a = s_t/d$

Longitudinal pitch $\qquad b = s_l/d$

A detailed description of the measurement techniques applied is given in the papers /4/ through /1o/. The method of electrically heated single cylinder in an unheated tube array has been chosen. The test cylinder was made from copper which nearly satisfied the boundary condition of constant wall temperature. The coolant was air. The high Reynolds numbers were produced in a high pressure wind tunnel using air up to a pressure of 4o bars.

2. SINGLE ROW OF TUBES

The dependence of the pressure drop coefficient of a single row of tubes on the Reynolds number and surface roughness parameter is evidently similar to that of a single circular cylinder in cross flow (Fig. 1). Four flow ranges can

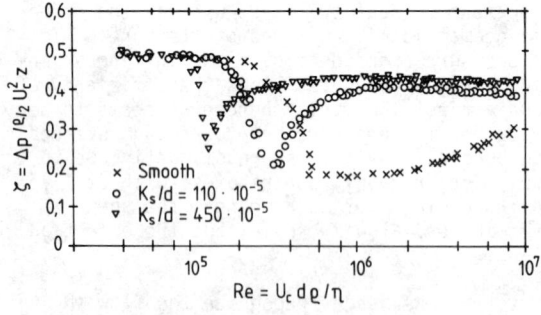

Fig. 1 Pressure drop of a single row of smooth and rough tubes.

be distinguished. At subcritical flow conditions ζ is independent of the Reynolds number. The succeeding critical flow regime is indicated by a sudden decrease of the pressure drop coefficient. The Reynolds number of ζ - minimum is defined to be the critical Reynolds number Re_{crit}. The range of increasing ζ

characterizes the supercritical one which leads to the transcritical range, where ξ is almost independent of Re again. Each of the four flow ranges is characterized by particular boundary layer phenomena as described for the single cylinder. Figure 2 which shows the local heat transfer of a single row of smooth

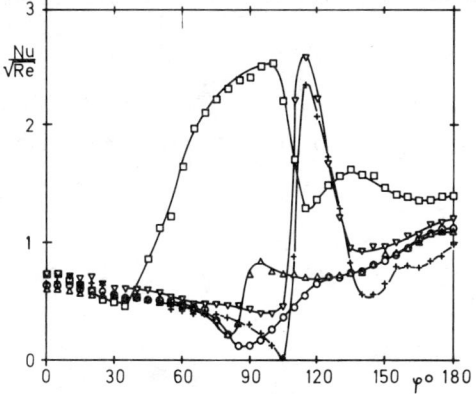

Fig. 2 Local heat transfer of a single row of smooth tubes, a = 2,04

$$Re \begin{cases} \circ & 1,8 \times 10^5 \\ \triangle & 4,4 \times 10^5 \\ + & 1,2 \times 10^6 \\ \square & 6,9 \times 10^6 \end{cases}$$

tubes gives an example that similar boundary layer phenomena occur: For all Reynolds numbers the value at the stagnation point starts from $Nu/\sqrt{Re} \approx 0.7$ which is equivalent to $Nu/\sqrt{Re} \sim 1$ for the single cylinder. At Re = 1.8 x 10^5 the flow is subcritical. The boundary layer is throughout laminar and separates immediately upstream the narrowest gap position. For Re = 4.4 x 10^5 the Reynolds number is close to the critical value. The laminar boundary layer does not finally separate at φ = 80° but reattaches to the wall. This phenomenon is completely established for Re = 1.2 x 10^6. The intermediate laminar separation at φ = 105°, the turbulent reattachment associated with a steep increase of local heat transfer and the far downstream shifting of the separation point to φ = 140° are characteristic for the critical flow regime. With further increase of Re an immediate transition from laminar to turbulent boundary layer occurs. In the supercritical range the transition point indicated by a strong increase of heat transfer shifts from about φ = 105° to the front position of the tube. For transcritical flow conditions the transition occurs in the vicinity of the front stagnation point.

The same effects are also observed for the single row of rough tubes. As demonstrated in Figure 3 the roughness causes transition to turbulence at

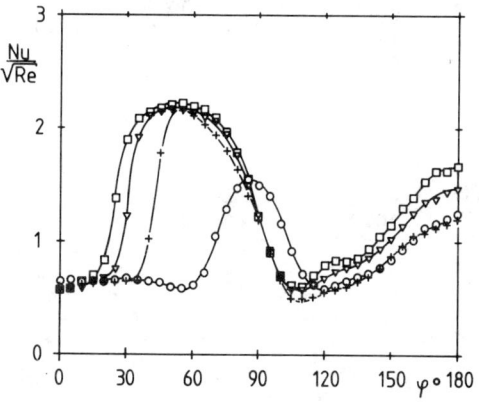

Fig. 3 Heat transfer from a single row of rough tubes.
a = 2,04   ks/d = 3oo x 10^5

$$Re \begin{cases} \circ & 1,8 \times 10^5 \\ + & 2,7 \times 10^5 \\ \triangledown & 3,4 \times 10^5 \\ \square & 4,0 \times 10^5 \end{cases}$$

lower Reynolds numbers compared to the smooth surfaces. For all Reynolds numbers the flow is supercritical. The upstream shifting of the transition point with increasing Reynolds number is evident.

Additionally Figure 4 reports on the local skin friction and static pressure

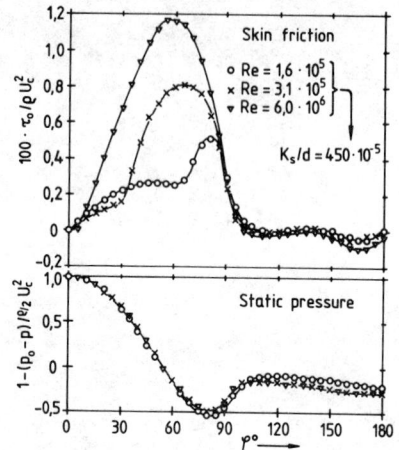

Fig. 4 Local skin friction and static pressure distribution of a single row of rough tubes. a = 2,04

distribution for a roughness parameter of $k_s/d = 450 \times 10^{-5}$ at variable Reynolds number. The comparison of the results in the Figures 3 and 4 yields that the increase of the shear-stresses at the transition point corresponds to an increase of local heat transfer. The position of boundary layer separation indicated by τ_o = o coincides with the position of minimum Nu number.

Finally Figure 5 represents the total heat transfer of a single row of tubes

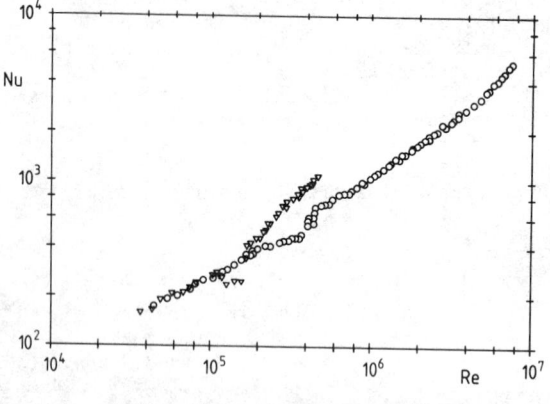

Fig. 5 Total heat transfer from a single row of tubes.
a = 2,04
O smooth
▽ $k_s/d = 300 \times 10^5$

depending on the Reynolds number and surface-roughness parameter. At subcritical flow conditions the results of the smooth and rough tubes collapse. The critical Reynolds number is indicated by a step in the curve which is caused by the instantaneous occurrence of the turbulent boundary layer in the rear of the tubes. Beyond Re_{crit} the heat transfer of rough surfaced tubes is higher since premature transition to the turbulent boundary layer improves the heat exchange normal to the wall. The curve for the rough tube row ends at Re = 5×10^5 because the high pressure wind tunnel was not available for further tests before preparation of this paper.

3. STAGGERED TUBE BUNDLE

The effect of surface roughness on the flow and heat transfer of staggered tube bundles has been studied for a tube arrangement with a transversal pitch of $s_t/d = 2.04$ and a longitudinal pitch of $s_l/d = 1.4$. The flow and heat transfer experiments have been conducted separately. The effect of two roughness heights on the flow has been investigated, while the results of only one roughness height are available for the heat transfer examination.

Figure 6 presenting the pressure drop coefficient depending on the Reynolds

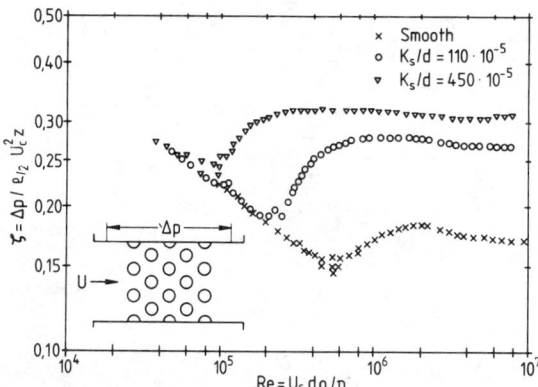

Fig. 6 Pressure drop coefficient of rough staggered tube bundles. a = 2,04, b = 1,4

number and roughness parameter demonstrates that for staggered tube arrays the character of bluff-body flow is existent. The minimum of the pressure drop coefficient occurring for each roughness parameter indicates the critical Reynolds number. Due to the high turbulence intensity in the bundle the decrease of ζ is not so steep as is observed for the single row of tubes. Nevertheless, the curves coincide at subcritical flow conditions. As expected the pressure drop coefficient increases with increasing roughness parameter for transcritical flow conditions. Details of the flow in this range can be recognized from Figure 7 which shows the skin friction and pressure distribution for different

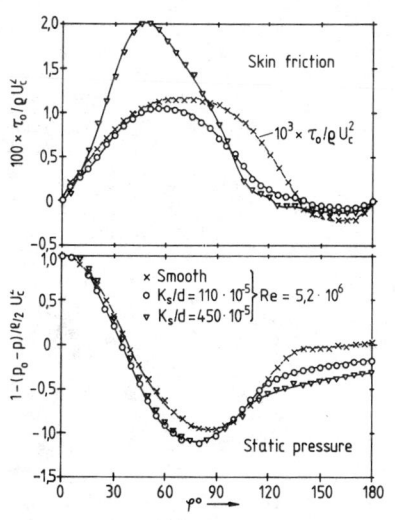

Fig. 7 Local skin friction and static pressure of a staggered rough tube bundle. a = 2,04; b = 1,4

roughness parameters. Particularly, the variation of the base pressure corresponds to the variation of the pressure drop coefficient at different surface roughness. Furthermore, the skin friction distribution indicates that high base pressures correspond to large angles of boundary layer separation and vice versa.

Additional information about boundary-layer effects comes from Figure 8

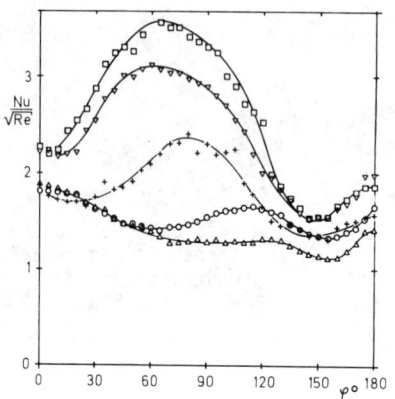

Fig. 8 Local heat transfer from a smooth staggered tube bundle. a = 2,04; b = 1,4

$$Re \begin{cases} \triangle & 2,0 \times 10^5 \\ \circ & 4,0 \times 10^5 \\ + & 1,0 \times 10^6 \\ \triangledown & 2,8 \times 10^6 \\ \square & 6,2 \times 10^6 \end{cases}$$

which represents the local heat transfer distribution for the smooth bundle at various Reynolds numbers. The augmentation of heat transfer in the stagnation point from unity to about double the amount may be due to the high turbulence intensity. The additional increase of heat transfer observed for the two highest Reynolds numbers is caused by a radial heat flux originated from the steep increase of heat transfer immediately downstream of the stagnation point. The position of transition from laminar to turbulent boundary layer shifts upstream with growing Reynolds number. As a demonstration that the boundary layer is turbulent in fact, the peak values of heat transfer increase with a power of the Reynolds number of 0.75 from Re = 10^6 to Re = 6.2 x 10^6.

The local heat transfer data reported in Figure 9 refer to a staggered rough

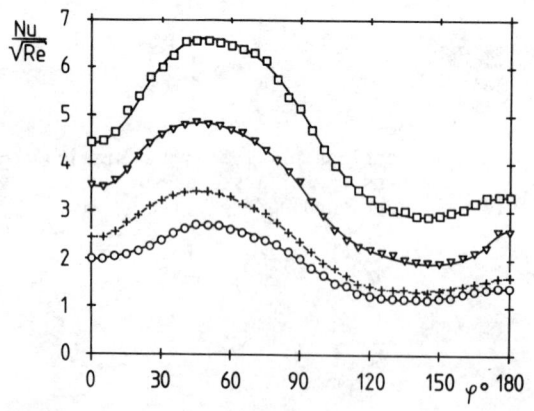

Fig. 9 Local heat transfer from a rough staggered tube bundle. a = 2,04; b = 1,4; k_s/d = 300 x 10^{-5}

$$Re \begin{cases} \circ & 1,6 \times 10^5 \\ + & 3,6 \times 10^5 \\ \triangledown & 1,3 \times 10^6 \\ \square & 7,2 \times 10^6 \end{cases}$$

bundle at transcritical flow conditions and k_s/d = 300 x 10^{-5}. For all Reynolds numbers examined the boundary layer transition occurs near the front stagnation point. Thus the peak values increase with a power of about 0.75 of the Reynolds number.

In Figure 1o the total Nusselt number of a smooth and a rough staggered tube

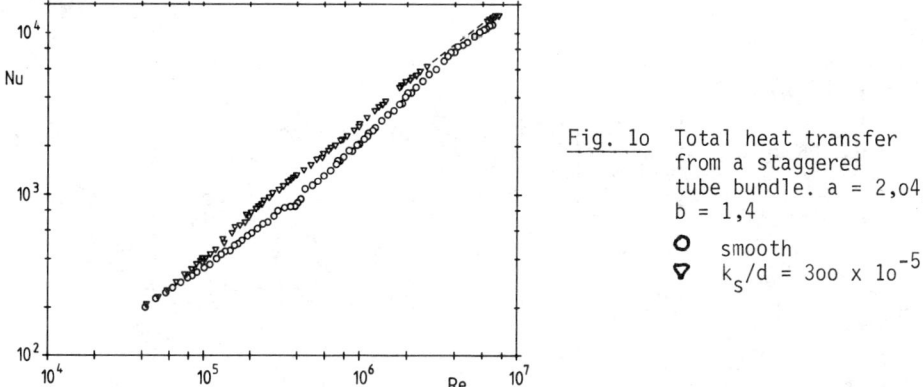

Fig. 1o Total heat transfer from a staggered tube bundle. a = 2,04; b = 1,4
○ smooth
▽ $k_s/d = 3oo \times 1o^{-5}$

bundle is plotted as function of the Reynolds number. For the subcritical flow range the data of the smooth and rough bundle coincide. The slope of the approximation curve is about o.6. Exceeding the critical Reynolds number which is Re = 6 x 1o^4 for the rough bundle and Re = 4.5 x 1o^5 for the smooth one, heat transfer increases nearly linear with Re, since with upstream shifting transition point the streaming length of the turbulent boundary layer grows. This phenomenon is typical for supercritical flow conditions. As this effect ends with the beginning of the transcritical flow regime the slope diminishes again and reaches a value of about o.75, which is characteristic for heat transfer through turbulent boundary layers. Since the mechanism described starts at lower Reynolds number for the rough bundle, the heat transfer curves diverge at lower Reynolds numbers and approach to each other at highest Reynolds numbers.

4. IN-LINE TUBE BUNDLE

For in-line arrangements succeeding tubes stand in the wake of each other. Thus not only the rear is cooled by recirculating flow, but this is also true for a part of the front portion. The point of impact of the flow separating from the preceding tube is somewhere between φ = 25° and φ = 45° depending on the Reynolds number and roughness parameter. From Figure 11 it is evident that its position determines the pressure distribution and that an interaction exists between the position of point of impact and point of boundary layer separation. The point of impact is indicated by vanishing of the shear-stresses at the front of the tubes. The smaller is the value of impact angle the larger are the pressure differences occurring around the tubes. Actually, a small value of impact angle is associated with a large angle of boundary layer separation also indicated by zero skin friction in the rear. Thus a large amount of the low pressure reached at φ = 9o^o is re-gained in the rear. However, the base pressure is still lower than for a flow with smaller angle of separation and larger angle of impact. Figure 11 shows that in the transcritical flow regime the separation angle is larger for smooth than for rough tubes. Thus it is expected that with increasing roughness the pressure drop coefficient decreases with increasing roughness parameter. Figure 12 demonstrates that this is true, indeed. While at subcritical flow conditions the pressure drop coefficients of all roughness parameters coincide it increases in the critical flow range where the separation point of the boundary layer shifts downstream. While

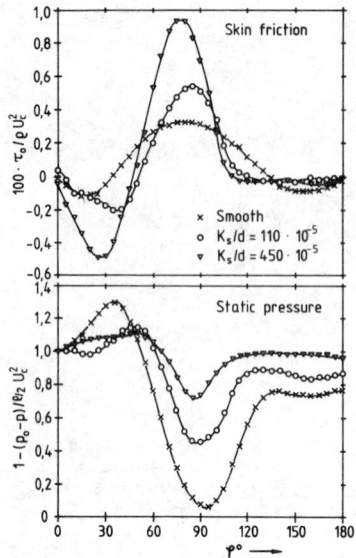

Fig. 11 Local skin friction and static pressure of an in-line tube bundle at Re = 4 x 10^6. a = 2,04, b = 1,4

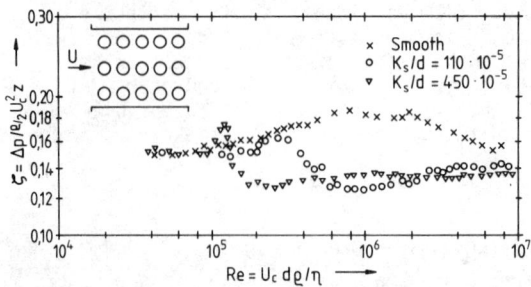

Fig. 12 Pressure drop coefficient of a rough in-line tube bundle. a = 2,04; b = 1,4

the separation point moves back again the pressure drop coefficient slightly diminishes in the supercritical regime and becomes nearly constant at transcritical flow conditions.

Figure 13 represents local heat transfer data for the smooth, Figure 14 for the rough in-line tube bundle. At subcritical flow conditions (Figure 13) the local heat transfer decreases downstream of the point of impact, which is indicated at $\varphi \sim 30°$ by a maximum. The boundary layer separates around $\varphi = 90°$ so that only about one third of the tube surface is covered by a laminar boundary layer. The increase of heat transfer at higher Reynolds numbers downstream of $\varphi = 30°$ indicates that the boundary layer undergoes transition to turbulent flow. This holds for the smooth bundle at Re $>$ 10^6 and for the rough bundle at all Reynolds numbers tested. This conclusion is confirmed substantiating that the peak values increase with about $Re^{0.75}$.

In a previous paper /11/ it has been stated that the total heat transfer is the same for smooth staggered or in-line tube bundles. Comparing Figure 1o with Figure 15 it is obvious that this also holds for rough surfaced tubes. As was observed for the staggered bundle the total heat transfer improves beyond

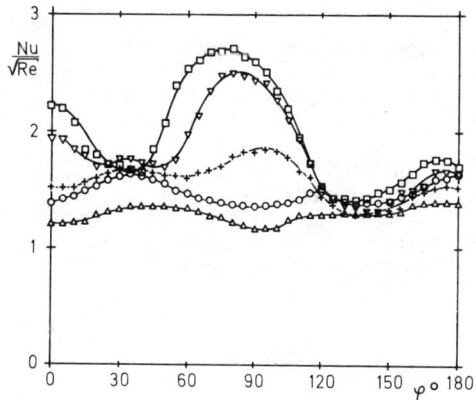

Fig. 13 Local heat transfer from a smooth in-line tube bundle. a = 2,04; b = 1,4

Re
- △ 2,3 x 10^5
- ○ 4,2 x 10^5
- + 1,1 x 10^6
- ▽ 2,8 x 10^6
- □ 6,0 x 10^6

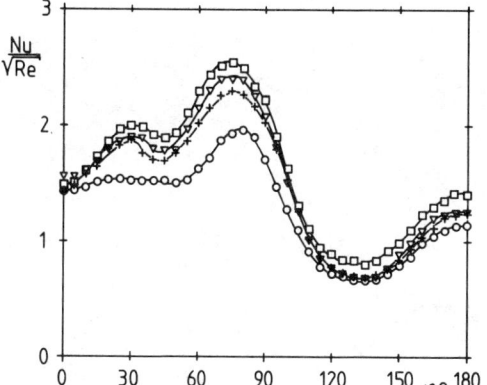

Fig. 14 Local heat transfer from a rough in-line tube bundle. a = 2,04; b = 1,4

Re
- ○ 1,6 x 10^5
- + 2,6 x 10^5
- ▽ 3,3 x 10^5
- □ 4,0 x 10^5

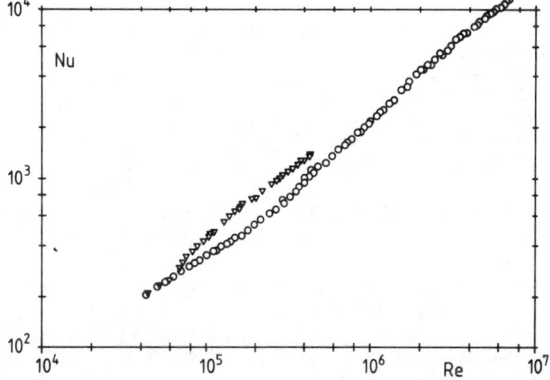

Fig. 15 Total heat transfer of an in-line tube bundle. a = 2,04; b = 1,4
- ○ smooth
- ▽ $k_s/d = 300 \times 10^{-5}$

the critical Reynolds number which has the same value for in-line and staggered arrangements. Remembering that the pressure drop coefficient of rough in-line bundles at transcritical flow conditions is lower than that of the smooth bundle the surface roughening of in-line arrays causes a remarkable improvement of the efficiency which may be quantified in terms of St^3/ξ /3/. This advantage, however, is impaired by the fact that in-line arrangements tend considerably more to tube vibration than staggered arrays /12/.

Comparing the present results of the smooth bundles with those reported by Hammeke et al. /1/ and Žukauskas /13/ an agreement within $\pm$ 15% is quoted in the overlapping range (Fig. 16). This is also true for the semi-empirial solution of Gnielinski /14/ up to $Re = 2 \times 10^6$. At higher Reynolds numbers Gnielinski's results for the in-line bundle are lower than the present ones by about 30%.

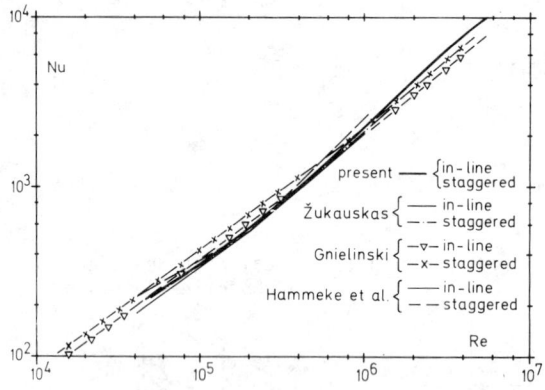

Fig. 16 Comparison of results of several authors

Figures 17 gives the comparison of own results with those of Groehn /2/ who

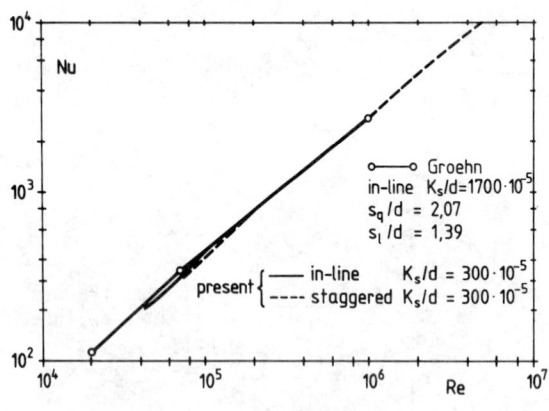

Fig. 17 Comparison of the present results with those of Groehn /2/.

investigated an in-line bundle of a similar roughness pattern as was used in the present tests. Again the own values of the in-line and staggered arrangement coincide. The agreement with Groehn's results is excellent though he applied a higher roughness parameter. It was shown, however, that for the single circular cylinder the heat transfer can not be increased using higher roughness parameters than $k_s/d = 300 \times 10^{-5}$ /10/. Probably this phenomenon also holds for tube bundles.

5. ENTRANCE EFFECTS ON HEAT TRANSFER

In a previous paper /10/ the entrance effect on heat transfer of smooth in-line and staggered heat exchangers has been investigated. The results indicated that a systematical dependence on the Reynolds number could not be elaborated. For staggered bundles the effect reaches up to the fourth row, whereas in-line arrays are affected predominantly in the first row. Figure 18 now gives a com-

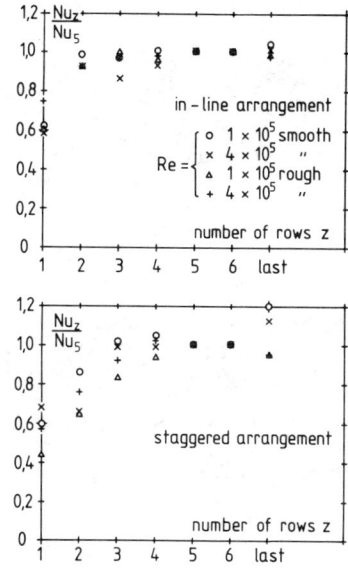

Fig. 18 Entrance effect on heat transfer of smooth and rough tube bundles.

parison of data obtained for smooth and rough tube bundles. The technique used was the same as has been described previously /10/. It can be concluded that the entrance effect seems to be independent of the roughness parameter, too. Again the effect is smaller for the in-line arrangement than for the staggered array.

6. CONCLUSION

A staggered and an in-line heat exchanger have been investigated up to Reynolds numbers of $Re = 6 \times 10^6$. At the critical Reynolds number the pressure drop and the heat transfer change their dependence on the Reynolds number. For staggered and in-line tube bundles the heat transfer improves. The pressure drop coefficient increases for the staggered array, similar to the drag coefficient of a single circular cylinder. This is also true for the smooth in-line bundle, but surface roughening causes a decrease of ξ at transcritical flow conditions.

The surface roughness of the tubes diminishes the critical Reynolds number. Thus the enhancement of heat transfer starts at a lower Reynolds number than is observed for the smooth bundle. The local heat transfer and skin friction distributions indicate that this effect is due to premature transition to the turbulent boundary layer.

The entrance effect on heat transfer is the same for smooth and rough surfaced bundles. For staggered arrangements the effect is greater than for in-line arrays.

REFERENCES

/1/ Hammeke, K.; Heinecke, E.; Scholz, F. Wärmeübergangs- und Druckverlustmessungen an querangeströmten Glattrohrbündeln, insbesondere bei hohen Reynolds-Zahlen, Int. J. Heat Mass Transfer 1o (1967), pp. 427-446

/2/ Groehn, H.G.; Scholz, F. Wärme- und strömungstechnische Untersuchungen an fluchtenden Rohrbündel-Wärmeaustauschern aus pyramidenförmig ausgerauhten Rohren. Jül-1437-(1977)

/3/ Groehn, H.G. Wärme- und strömungstechnische Untersuchungen an einem querdurchströmten Rohrbündel-Wärmeaustauscher mit niedrig berippten Rohren bei großen Reynolds-Zahlen. JÜL-1462-(1977)

/4/ Achenbach, E. Investigations on the flow through a staggered tube bundle at Reynolds numbers up to Re = 10^7. Wärme- und Stoffübertragung 2 (1969), pp. 47-52

/5/ Achenbach, E. Influence of surface roughness on the flow through a staggered tube bank. Wärme- und Stoffübertragung 4 (1971), pp. 12o-126

/6/ Achenbach, E. On the cross flow through in-line tube banks with regard to the effect of surface roughness. Wärme- und Stoffübertragung 4 (1971) pp. 152-155

/7/ Achenbach, E. Distribution of local pressure and skin friction around a circular cylinder in cross-flow up to Re = 5×10^6. J. Fluid Mech. 34 (1968) pp. 625-639

/8/ Achenbach, E. Influence of surface roughness on the cross-flow around a circular cylinder. J. Fluid Mech. 46 (1971) pp. 321-335

/9/ Achenbach, E. Total and local heat transfer from a smooth circular cylinder in cross-flow at high Reynolds number. Int. J. Heat Mass Transfer 19 (1975) pp. 1387-1396

/1o/ Achenbach, E. The effect of surface roughness on the heat transfer from a circular cylinder to the cross flow of air. Int. J. Heat Mass Transfer 2o (1977), 359-369

/11/ Achenbach, E. Total and local heat transfer and pressure drop of staggered an in-line tube bundles. Proc. Adv. Study Inst. in Heat Exchangers, Istanbul, Turkey, Aug. 4-15 (198o)

/12/ Mohr, K.-H. Messung instationärer Drücke bei Queranströmung von Kreiszylindern unter Berücksichtigung fluidelastischer Effekte. Diss. RWTH Aachen (1981)

/13/ Zukauskas, A. Heat transfer from tubes in cross-flow Advances in Heat Transfer 8 (1972) pp. 93-16o

/14/ Gnielinski, V. Wärmeübergang bei Querströmung durch einzelne Rohrreihen und Rohrbündel, VDI-Wärmeatlas, Abschnitt Ge, Düsseldorf (1977) VDI-Verlag.

Heat Transfer and Flow Resistance of Yawed Tube Bundle Heat Exchangers

H. G. GROEHN
Kernforschungsanlage Jülich GmbH
Jülich, FRG

ABSTRACT

Experimental results are reported which have been found out at an in-line tube bank in the Reynolds number range $2 \times 10^3 < Re < 10^5$. The yaw angle was varied between $15° \leq \varphi \leq 90°$ (cross flow). Entrance effects have been studied as well as the influence of an opposite inclination of neighboured tubes. The tests have shown that the principle of independence can be applied to predict the heat transfer coefficient of yawed tube banks. However, it is not suitable to describe the hydraulic resistance. Generally the hydraulic resistance coefficient increases with smaller yaw angles in the subcritical Reynolds number range. Besides, the dependence on the Reynolds number varies with the yaw angle.

1. INTRODUCTION

Oblique flow occurs in diverse types of heat exchangers. As an example Fig. 1 shows two different design of heat exchangers for High Temperature Gas Cooled Reactor (HTRG) plants. Right hand the recuperator of a closed gas turbine cycle is sketched. The mainly flow direction outside the tubes is parallel to the tube axis. In the exit and entrance areas, however, the angle between the tube axis and the flow direction changes to cross flow. At the left side of Fig. 1 a helical type heat exchanger is shown with a considerable lead of the tubes. In this case the yaw angle is constant along the whole tube bundle. Such an apparatus is favoured as an intermediate helium/helium heat exchanger in a nuclear process heat plant. The lead of the tubes is nearly $30°$. A special feature of this type of heat exchanger is a design whereby neighboured tube cylinders are coiled contarywise.

Besides specific thermal hydraulic problems which must be solved for each particular heat exchanger there are some general questions of oblique flow which are only insufficiently and partly inconsistently answered in the literature. For instance there is almost no information about the range of Reynolds number for which the data refered to are valid. On the other hand there is no doubt that the yaw angle influences the critical Reynolds number.

The flow pattern around a yawed cylinder is a three dimensional one. However, a theoretical solution is essentially simplified by the fact that the flow is independent of the axial extension of the cylinder. A consequence for the boundary layer equations is that the velocity components normal to the tube axis can be calculated independent of the axial velocity

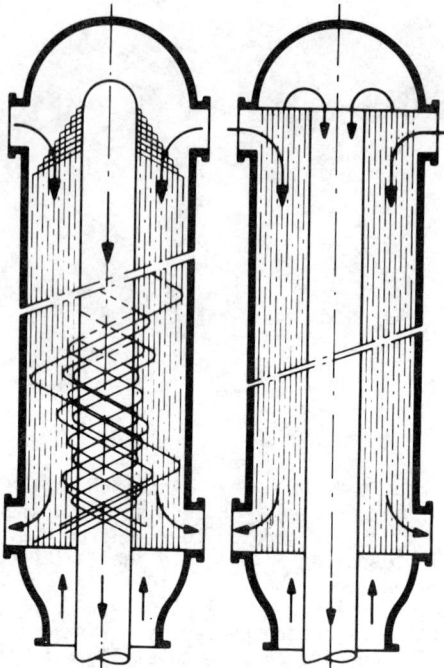

Fig. 1 Oblique flow in HTGR-heat exchangers

component. This fact is called the independence principle /1/. The velocity components normal to the tube axis are determined by the equations for the two-dimensional flow around a circular cylinder. Their solution is given, for instance, by Blasius /1/. His method has been extended by Sears /2/ and Görtler /3/ to find out the axial velocity component.

From the independence principle follows, for example, that the point of boundary layer separation of a yawed cylinder should be unaffected by the axial velocity component. In the same way the pressure distribution around the cylinder should not be influenced by the yaw angle. The boundary layer simplifications are not permissibled in the flow region downstream of the separation point. Therefore the validity of the indepence principle should be restricted in this flow region, too. Nevertheless measurements of Bursnal /4/ and also our own results /11/ have shown that the independence principle can be applied to describe the flow around a yawed cylinder down to a yaw angle of $\psi = 30°$ in the subcritical Reynolds number range. A main feature of this paper is to answer the question if the priciple of independence is appropriate to describe the thermalhydraulic behaviours of inclined tube bundles, too. Experiments should clarify down to which yaw angle the flow through an inclined tube bundle can be related to a bundle in pure cross flow.

2. APPARATUS AND TESTS

The experiments were designed as to represent the flow conditions of tube bundles in the subcritical Reynolds number range $2 \times 10^3 < Re < 10^5$. The Reynolds number is based on the velocity component U normal to the tube axis in the narrowest cross section. The relation between U and the velocity in the empty channel U_o is $U/U_o = S_T/(S_T-1)$. For S_T see Fig. 2. As the characteristic length in the Reynolds number as well as the Nusselt number Nu the tube diameter was chosen. Air at ambient conditions was used as fluid.

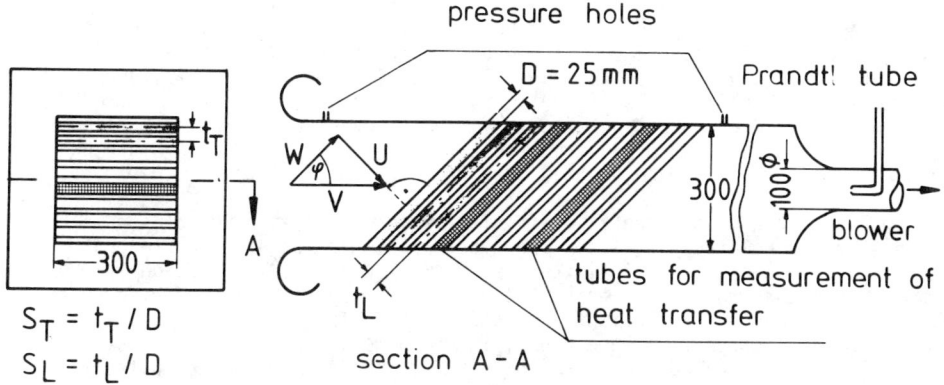

Fig. 2 Experimental setup

The dimensions of the test section were determined by the data of the blower available and by the necessity to arrange tube bundles with a representative number of tubes across and along the flow direction. The test set-up is sketched in Fig. 2. The cross section of the channel was 3oo times 3oo mm, the tube diameter 25 mm. The dimensions chosen allowed to arrange 8 tubes per row normal to the flow direction at a transversal tube pitch of $S_T = 1,5$ and 9,5 tubes at $S_T^* = 1,25$. In flow direction lo rows were used. The mass flow was determined means of a Prandtl tube. It was calibrated under different conditions of assembling. Pressure holes in front of and behind the tube bundle were used to determine the pressure loss over the bundle.

For heat transfer measurements electrically heated tubes were applied. A coax heater was inserted into a copper rod. It was pressed into one of the steel tubes of the bundle. Thus the test tube had the same surface conditions as the remainder dummy tubes. The temperature of the test tube was measured at the inner wall of the steel tube because the tube diameter was relatively small and the surface of the tube should not be hurt. The decrease in temperature through the tube wall was calculated. The temperature conditions were similar to those of tubes through which water runs. Three thermocouples were distributed over the active length of o,1 m of the heater and four over the circumference. The heaters were three divided. The outer parts acted as guard heater while the inner piece represented the active element. The particular sections were thermally insulated from each other and separately heated.

Tests were carried out several staggered and in-line tube banks. Because the results are similar with reference to the influence of the yaw angle /11/ the representation of the measured data can be restricted to those of one tube arrangement only. An in-line tube bank with the tube pitches $S_T = 1,5$, $S_L = 1,5$ was chosen.

3. RESULTS

3.1 Row to Row Measurements

Similarities or differences between the flow through oblique and cross flow tube bundles can be revealed among other things by carrying out row to row experiments of heat transfer and pressure drop. Initially it was not clear whether, for instance, the mass flow distribution is uniform along the tubes of a bundle consisting only of a few rows. Therefore the most of the tube arrays were investigated with variable row numbers. The heat transfer coefficient was determined from the first up to the eighth row. In each case at least two rows were mounted downstream of the test tube.

In Fig. 3 the heat transfer coefficient of each row at a Reynolds number of $Re = 10^4$ is related to that of the eighth row which stands for the heat transfer coefficient of an infinitely deep tube bundle. This relation is drawn over the row number in which the heated tube was inserted. The increase of the heat transfer coefficient from the first to the fourth row, well known from pure cross flow tube banks, is characteristic also for the smaller yaw angles, investigated. However, the increase seems to be weaker with smaller yaw angle.

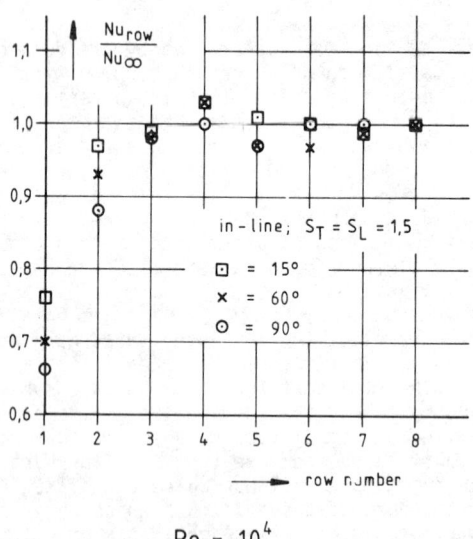

Fig. 3 Entrance effect on heat transfer of yawed tube banks

A similar coincidence between cross flow and oblique flow can be read from the flow resistance measurements with different row numbers, Fig. 4.

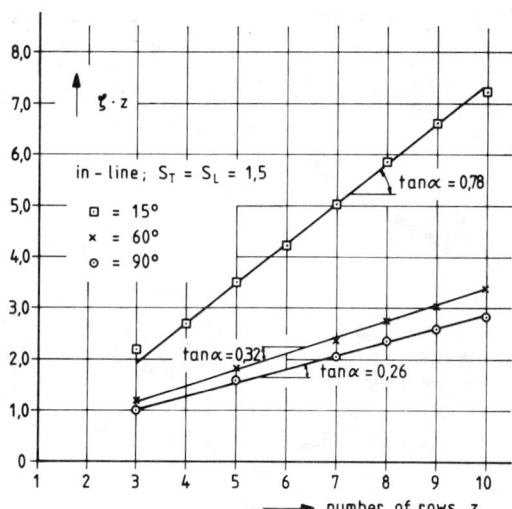

Fig. 4 Entrance effect on hydraulic resistance of yawed tube bundles

The flow resistance coefficient ζ is defined as usual $\zeta = \Delta p / (\rho/2 \cdot U^2 \cdot z)$ where Δp means the pressure loss over the tube bundle, z the number of tube rows in flow direction and $\rho/2\, U^2$ the dynamic pressure in the narrowest cross section normal to the tube axis. In the representation chosen an entrance effect on the hydraulic resistance should be apparent by a deviation from the linearity between $\zeta \cdot z$ and z. For the yawed tube banks as well as for the reference tube bank in cross flow the dimensionless pressure loss increases about linearly for more than three tube rows. From this evidence it can be concluded that the entrance and exit losses are negligable for $z > 4$ for all yaw angles. The slope of the curves can be related to the hydraulic resistance coefficient by $\zeta = \tan\alpha$. This equation is fairly satisfied if applied to a bundle consisting of ten rows. Summarizing it can be stated from the row to row measurements that the cross flow character of inclined tube bundles is kept down to yaw angles of $\varphi = 15°$.

3.2 Representative Data for Heat Transfer and Hydraulic Resistance

Besides the results discussed in the former paragraph it can be concluded from Fig. 3 and 4 that the heat transfer coefficient downstream of the fifth row as well as the hydraulic resistance coefficient of a tube bundle with more than four rows can be regarded as nearly independent of the number of tube rows. The data which are shown in the following diagrams were obtained from tube bundles with ten rows. The heated tube was always inserted in the eighth row. Therefore the results are representative for larger tube bundles, too. In Fig. 5 the measured data for the dimensionless heat transfer coefficient and the hydraulic resistance coefficient are given in dependence on the Reynolds number. The yaw angle is parameter in this representation. As can be seen from the upper part of Fig. 5 there is a good agreement between the experimental heat transfer data for all yawed tube banks and the tube arrangement with cross flow ($\varphi = 90°$). However, the

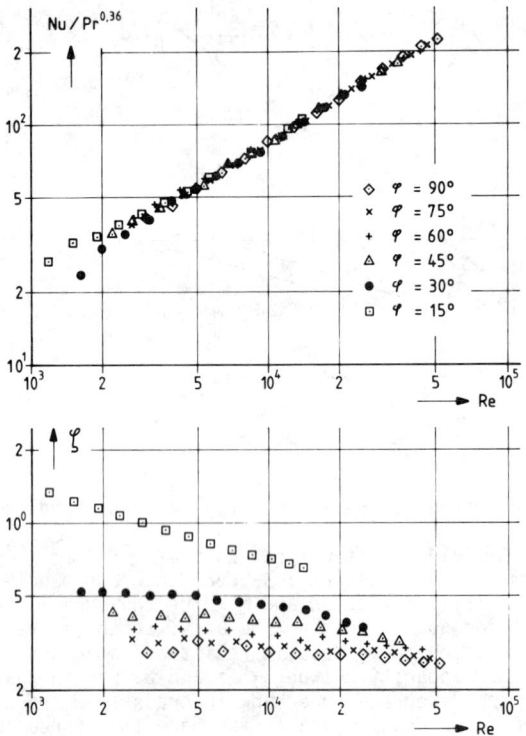

Fig. 5 Heat transfer and hydraulic resistance of yawed in-line tube banks; $S_T = 1,5$; $S_L = 1,5$

hydraulic resistance deviates systematically for different yaw angles. Generally the hydraulic resistance coefficient increases with smaller yaw angles in the subcritical Reynolds number range. The dependence on the Reynolds number becomes more evident with smaller yaw angle. The reason for that probably is the approach to the critical Reynolds number which decreases for smaller angles of inclination. Besides, the frictional part of the hydraulic resistance increases with smaller yaw angles because of the velocity component parallel to the tube axis becomes larger and the tube length increase. The slope of a curve through the data points for $\varphi = 15°$ is near to those of the relation for longitudinal flow.

YAWED TUBE BUNDLE HEAT EXCHANGERS

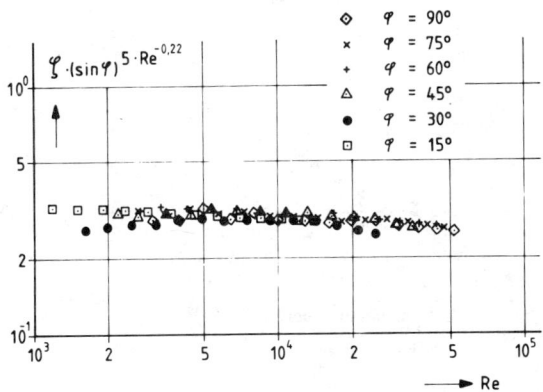

Fig. 6 Hydraulic resistance of yawed in-line tube banks, standardized; $S_T = 1,5$, $S_L = 1,5$

In Fig. 6 an attempt was made to describe the influence of the yaw angle on the hydraulic resistance in dependence on the Reynolds number. The factor

$$(\sin \varphi)^5 \, Re^{-0,22}$$

provides the desired standardization. The expression takes into consideration the prolongation of the tubes with smaller yaw angles which arise with $\sin \varphi$ as well as the relation between the friction factor and the Reynolds number for longitudinal flow.

3.3 Parallel and Crossed Yawed Tube Bundles

As stated in the introduction for a nuclear process heat application a helical type heat exchanger is planned neighboured tube cylinders of which are coiled contrarywise. For such heat exchanger type information on thermal hydraulic data is not yet available. To simulate the contrarywise coiling in this course of test series the neighboured tube layers of an in-line tube arrangement were crossed. Yaw angles of $\varphi = 60°$ and $\varphi = 75°$ were investigated.

The most important change in the flow when crossing the tubes results from the fact that the narrowest transverse cut of the parallel tube arrangement exists only at the crossing points of the tube layers. The same is true for the maximum velocity between the tubes which was chosen for the characteristic numbers. Theoretically the maximum velocity must be halved but this doesn't seem to be realistic. In the following diagrams the maximum velocity as defined for parallelly arranged tubes will be maintained for comparison in the dimensionless terms. Fig. 7 shows heat transfer and pressure drop of the two types of tube bundles with an angle of inclination of $\varphi = 60°$.

As expected the heat transfer and pressure drop are reduced for the crossed tube bundle due to the actual decreased velocity. The decrease in heat transfer is about 14% over the whole Reynolds number range examined, whereas the decrease in pressure drop runs up to amount to 5o% at $Re = 3 \times 10^3$ and to about 7o% at $Re = 5 \times 10^4$. That means that a rather small loss of

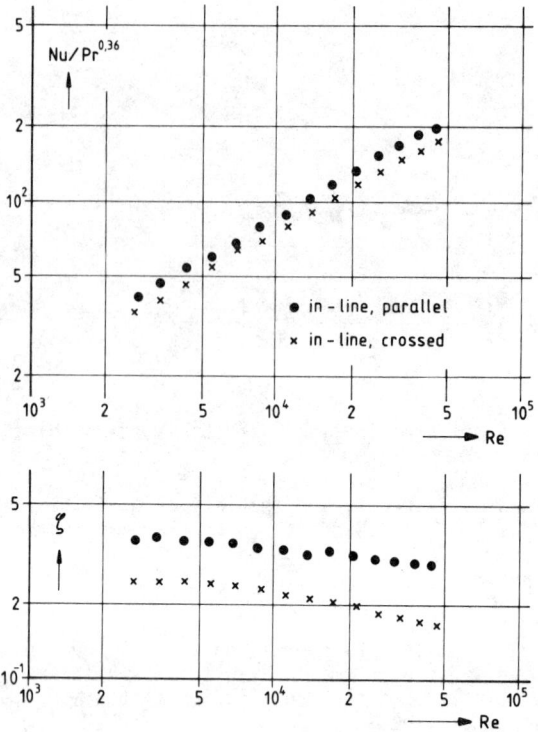

Fig. 7   Comparison between yawed tube banks with parallel and crossed tubes

heat transfer contrast with a considerable amount of pressure drop. An estimated calculation for heat transfer and hydraulic resistance coefficient of crossed tube bundles in relation to yawed banks with parallel flow can be obtained if the characteristic velocity is reduced by the factor $(\sin \varphi)^2$. In this case the data points for heat transfer of parallel and crossed tube bundles would coincide rather well. For the flow resistance this estimate is applicable only in the lower Reynolds number range. At higher Reynolds numbers the slope of the curves differs for parallel and crossed tubes and a deviation up to 20% occurs.

4. COMPARISON WITH LITERATURE

In the literature the influence of the yaw angle on heat transfer and pressure drop is accounted for by partition factors. The heat transfer coefficient α and the pressure drop Δp are related to the data of tube banks in cross flow. This representation is equivalent to that one where the Reynolds number and the hydraulic resistance coefficient are related to the velocity in the main flow direction V. The relation between V and the velocity component normal to the tube axis U is $V = U/\sin \varphi$ (see Fig. 2). To mark the changed reference velocity the subscript V is used for the Reynolds number and the hydraulic resistance coefficient (Re_V and ζ_V). Fig. 8 shows the experimental data of Fig. 5 in such a representation. The diagram reveals the decrease of heat transfer and pressure drop with smaller yaw angle.

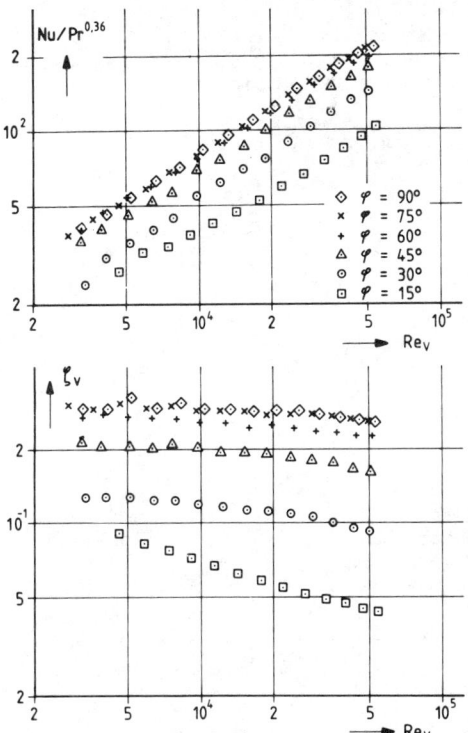

Fig. 8 Heat transfer and hydraulic resistance of yawed in-line tube banks; caracteristic numbers refered to velocity in main flow direction; $S_T = 1,5$; $S_L = 1,5$

As mentioned in the introduction no information is available in the literature about the Reynolds number range for which the partition factors are valid. Indeed, the experimental heat transfer data can be described by a partition factor which is valid over the whole Reynolds number range, investigated. That can be read from the slope of a curve through the heat transfer data which is nearly the same for all yaw angles with the exception of the data for $\varphi = 15°$ in the lower Reynolds number range. This changes definitely if the results for the hydraulic resistance coefficient are regarded.

The ratio $\Delta p_\varphi / \Delta p_{90°}$ differs by the factor 1,2 for a yaw angle of $\varphi = 30°$ and by the factor 1,7 for a yaw angle of $\varphi = 15°$ within the Reynolds number range, investigated. Nevertheless partition factors for heat transfer and pressure drop are shown in the following two diagrams in order to compare our results with data in the literature. Regarding the heat transfer, Fig. 9, there is a good agreement between the information given by Michejew /6/, VDI-Wärmeatlas /7/ and the measured data points for the two in-line tube banks. Additionally the heat transfer coefficient of a tube bundle with axial flow is given. The value was calculated with Pressers /1o/ prediction. The third curve in Fig. 9 was derived from the independence principle assuming a predominant cross flow influence on heat transfer. Using the fact that the dimensionless heat transfer coefficient is nearly independent of the yaw angle (see Fig. 5) and introducing the usual power relation for pure cross flow

$$Nu = C \cdot Re^d \qquad C, d \text{ constants}$$

as well as the relation between the velocity components normal to the tube axis U and in main flow direction V one obtains

$$\alpha_\varphi / \alpha_{90°} = (\sin \varphi)^d$$

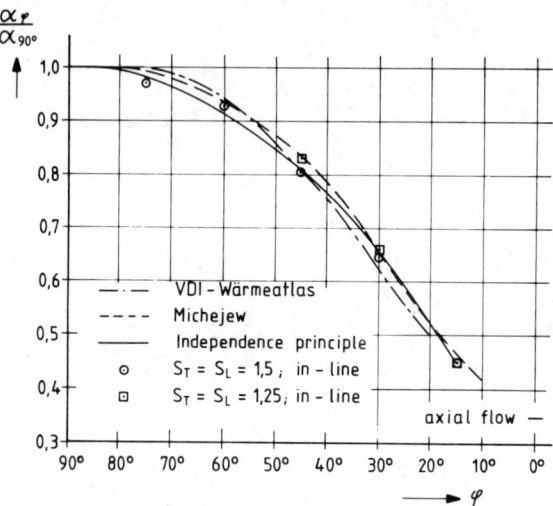

Fig. 9 Comparison of experimental heat transfer data with several predictions; $Re_V = 2 \times 10^4$

The equation is in excellent agreement with the experimental data down to $\varphi = 15°$. However, it doesn't describe the transition to longitudinal flow.

In a similar way an expression for the pressure drop of yawed tube banks can be deduced from the principle of independence. With the relationship for the hydraulic resistance of the tube bundle in cross flow

$$\xi = A \cdot Re^b \qquad A, b \text{ constants}$$

the partition factor for the pressure drop can be written

$$\Delta p_\varphi / \Delta p_{90°} = (\sin \varphi)^{(2 + b)}$$

This equation doesn't consider the frictional part of the hydraulic resistance which is going to be dominant at the smaller yaw angles due to the prolongation of the tubes. At a Reynolds number of $Re_V = 2 \times 10^4$ the best agreement to the experimental data is obtained with Michejews /6/ prediction. A wrong tendency of the influence of the yaw angle is given by the prediction in VDI-Wärmeatlas /7/.

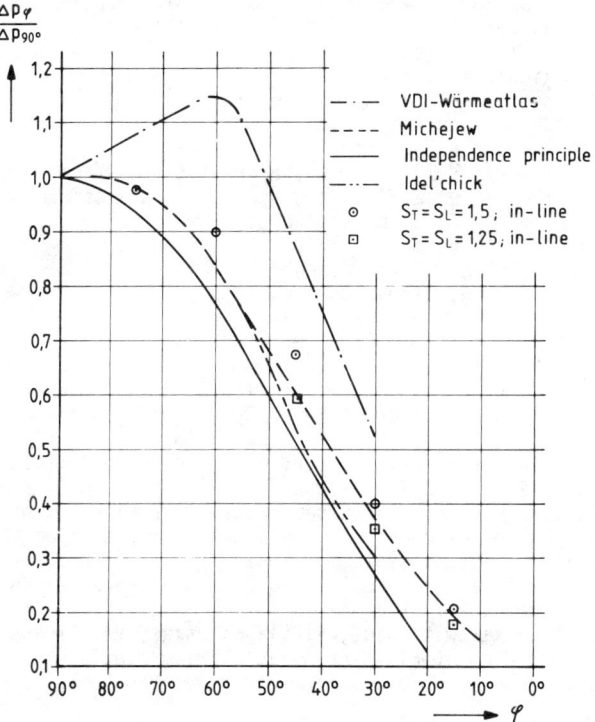

Fig. 1o Comparison of experimental hydraulic resistance data with several predictions; $Re_V = 2 \times 10^4$

5. CONCLUSION

Experiments at an in-line tube bank have shown that the independence principle is suitable to describe the heat transfer of yawed tube banks down to a yaw angle of $\varphi = 15°$ in the subcritical Reynolds number range. At the moment the hydraulic resistance is described most favourable by Michejews /7/ prediction. However, the dependence on the Reynolds number varies with the yaw angle. At $\varphi = 15°$ deviations up to 7o% occur.

First data for the calculation of crossed yawed tube bundles are given. In comparison to yawed tube bundles with parallel tubes a considerable decrease of the pressure loss is observed, whereas the decrease of the heat transfer coefficient is relatively small.

6. NOMENCLATURE

D	m	tube diameter
Δp	N/m^2	difference in static pressure over the tube bank
S_T, S_L		tube pitches, see Fig. 2
U_o	m/s	velocity component normal to the tube axis in the empty channel
U	m/s	$= U_o \cdot S_T/(S_T - 1)$, reference velocity in the characteristic numbers

V	m/s	$= U/\sin\varphi$, velocity in main flow direction
z		number of tube rows in flow direction
α	W/(m² K)	heat transfer coefficient
ν	m²/s	kinematic viscosity of the fluid
λ	W/(m K)	heat conductivity of the fluid
ρ	kg/m³	density of the fluid
φ	degree	yaw angle, see Fig. 2
ξ		$= \Delta p/(\rho/2 \cdot U^2 \cdot z)$, hydraulic resistance coefficient
ξ_v		$= \Delta p/(\rho/2 \cdot V^2 \cdot z)$, hydraulic resistance coefficient
Nu		$= \alpha \cdot D/\lambda$, Nusselt number
Pr		Prandtl number
Re		$= U \cdot D/\nu$, Reynolds number
Re$_v$		$= V \cdot D/\nu$, Reynolds number

7. REFERENCES

/1/ H. Schlichting, Grenzschichttheorie, Verlag G. Braun, Karlsruhe 1965
/2/ W.R. Sears, The Boundary Layer of Yawed Cylinders, Journal of the Aeronautical Sciences, Vol. 15, pp. 49-52, 1948
/3/ H. Görtler, Zur laminaren Grenzschicht am schiebenden Zylinder, Teil I Arch. Math. Bd. III, S. 216-231, 1952
/4/ W.J. Bursnal, L.K. Loftin, Jr., Experimental Investigation of the Pressure Distribution about a Yawed Circular Cylinder in the Critical Reynolds Number Range NACA TN-2463, September 1951
/5/ W.S. Chin, J.H. Lienhard, On Real Fluid Flow over Yawed Circular Cylinders, ASME Paper 67 - WA/FE-M., 1967
/6/ A. Zukauskas, V. Makarewitschius, A. Slanciauskas, Heat Transfer in Banks of Tubes on Cross Flow of Fluid (russian) Verlag Mintis, Vilnius, 1968
/7/ M.A. Michejew, Grundlagen der Wärmeübertragung, VEB Verlag Technik, Berlin, 1961
/8/ VDI-Wärmeatlas, 3. Auflage, VDI-Verlag, Düsseldorf 1977
/9/ I.E. Idel'chik, Handbook of Hydraulic Resistance, AEC-TR-663o, 1966
/1o/ K. Presser, Wärmeübergang und Druckverlust an Reaktorbrennelementen in Form längsdurchströmter Rohrbündel, Jül-486-RB, 1967
/11/ H.G. Groehn, Thermal Hydraulic Investigation of Yawed Tube Bundle Heat Exchangers, Advanced Study Institute on Heat Exchangers, pp. 97-1o9, Hemisphere Publishing Corporation, 1981

Surface Roughness as Means of Heat Transfer Augmentation for Banks of Tubes in Crossflow

A. ŽUKAUSKAS and R. ULINSKAS
Institute of Physical and Technical Problems of Energetics
Lithuanian Academy of Sciences
Kaunas, USSR

ABSTRACT

Experimental results are presented for the local parameter values in banks of rough - surface tubes in air and transformer oil. Local velocity and velocity fluctuations were measured by X - shaped sensors with a third wire introduced to measure temperature and its fluctuations. Calculated velocity and temperature profiles are given for the region of boundary layer separation.

Possibilities of augmenting the heat transfer of heat exchangers by introducing rough surfaces were explored. The study covered wide range of Re and Pr, relative transverse pitch a from 1.25 to 2.0, relative longitudinal pitch b from 0.935 to 2.0, and relative height of surface elements k/d from 6.10^{-3} to 40.10^{-3}. Calculation relations are suggested for the average heat transfer and pressure drop of banks of rough-surface tubes for Re from 10^2 to 2.10^6.

NOMENCLATURE

a	—	relative transverse pitch, s_1/d
b	—	relative longitudinal pitch, s_2/d
c_f	—	skin friction coefficient, $2\tau_w/\rho w^2$
c_p	—	specific heat capacity, J/kg.°C
d	—	tube diameter, m
Eu	—	Euler number, $\Delta p/\rho w^2 z$
k	—	height of the artificial surface roughness elements on the tube, m
k^+	—	dimensionless height of surface roughness, ku_x/ν
k	—	kinetic energy of turbulence, $1/2(u'^2+v'^2+w'^2)$

Nu	–	Nusselt number, $\bar{\alpha}d/\lambda$, $K = Nu_f/Pr_f^n (Pr_f/Pr_w)^{0.25}$
q_w	–	specific heat flux, W/m^2
Pr	–	Prandtl number, ν/a
Re	–	Reynolds number, wd/ν
St	–	Stanton number, $\alpha/\rho c_p u_1$
t	–	temperature, °C
Tu	–	turbulence level, $\sqrt{\overline{u'^2}}/u_o$
u	–	velocity component in streamwise direction
u_*	–	friction velocity, m/s
w	–	mean velocity in the minimum inter-tube space (i.e. in the minimum free flow area), m/s
x	–	streamwise coordinate, m
y	–	direction normal to the surface, m
y^+	–	dimensionless distance, yu_*/ν
z	–	number of tube rows in a bank
δ, δ^*, θ	–	thickness of the boundary layer, displacement thickness, momentum thickness, respectively, m
Θ	–	dimensionless temperature, $\vartheta/\vartheta_{max}$
ϑ	–	temperature difference between the wall and a point in the boundary layer, °C
ϑ_{max}	–	temperature difference, $t_w - t_f$ °C
ϑ_*	–	friction temperature, $q_w/\rho c_p u_*$ °C
$\alpha, \bar{\alpha}$	–	local and average heat transfer coefficients, respectively, W/m^{2}K
η	–	dimensionless ordinate, y/δ
λ	–	thermal conductivity, W/m·K
μ	–	dynamic viscosity, Pa·s
ν	–	kinematic viscosity, m^2/s
ρ	–	density, kg/m^3
τ	–	shear stress, Pa
Φ	–	pressure gradient factor, $(\delta/\tau_w)(dp/dx)$

φ — angular distance from the front stagnation point, deg

Subscripts

f — main flow

1 — outer boundary layer

w — wall

( )' — fluctuations

1. INTRODUCTION

Search for modern highly effective ways of heat transfer augmentation presents one of the actual industrial problems. Numerous efforts are directed towards size, weight and cost reduction of modern power apparatus, and simultaneous increase of their operation performance and economy of materials.

Surface roughness has some time ago drawn attention of researchers as a means transfer augmentation suitable for different fluids. A number of studies [1 through 5] on the average heat transfer and pressure drop on rough surfaces, in the flow of gas has been performed. Heat transfer augmentation under the effect of surface roughness and of tube arrangement was observed at Re about 10^4.

At the Institute of Physical and Technical Problems of Energetics, Lithuanian Academy of Sciences, we have performed studies on rough surface tubes with the object of optimizing the parameters of surface roughness, and of determining the effect of longitudinal and transverse pitches (a, b) of banks, their effect on the hydraulic drag in liquids of high viscosity. On the tube surface and in the free space inside the bank, local values of heat transfer, velocity components u, v, w and velocity fluctuation components u', v', w' were determined. Temperature and temperature fluctuations were measured.

2. EXPERIMENTAL TECHNIQUES

The studies were performed in the range of Re from 10^2 to 2.10^6 and of Pr from 0.7 to 300 with the most common staggered bank geometries 2.0 x 2.0, 1.25 x 1.25, 1.25 x 0.935 for tube diameters 30 and 150 mm.

Calorimetric test tubes were made of 0.2 mm thick constantan foil with surface roughness introduced by pressing. The pyramidal closely-space elements were of variable height k = 0.2, 0.45, 0.8, 1.2 mm.

Flows of transformer oil on tubes at Re from 10^2 to 2.10^5 were visualized, and pictures were taken on tubes of 150 mm diam. To cover a wider range of Re, smoke-coloured flows of air were observed.

Perimetric heat losses were measured and found insignificant and q_w = const was assumed at the wall. Surface temperature was measured by thermocouples welded inside the foil to the pyramid tops and foots at 10 deg intervals, so that temperature could be measured along the whole perimeter. Skin friction was measured with the small surface edge as in [1] and by the sensors of DISA-Electronics. Our results were close to [1]. Local velocity and velocity fluctuations in the inter-tube space were measured by X-shaped sensors. A third wire was introduced to measure temperature fluctuations. Near-wall measurements were performed with single-wire sensors. The sensors raised from special holes in the tube on microscrews.

Determinations of local values on rough surface are encumbered by the decrease of velocity on the surface elements, so that distance y can be determined either from a top or foot location, or in some intermediate location. The resultant flow is the same, but universal comparisons of the results are no longer possible. A rational technique of determining the y axis ought to give comparable results for the boundary layer parameters for smooth and rough surfaces. For this y = 0 must be located on a place of zero velocity. Such a location on a rough surface can only be found after a theoretical analysis supported by microphoto pictures. In our studies, the y distance was started from pyramidal tops.

Calculation of velocity profiles in deccelerating flows up to the separation point was decribed earlier [6]. In the present study, velocity and temperature profiles in the separation region were determined from differential equation of continuum and from time-averaged equations of Navier-Stokes. Turbulent viscosity μ_T was found experimentally and from the analytical model of turbulence "k-ε" after [7, 8], with the empirical constant $c_D \approx 0.09$. From the model

$$\mu_T = c_D \rho k^2 / \varepsilon \tag{1}$$

Kinetic energy of turbulence and its dissipation were found from

$$\frac{\partial}{\partial x}(\rho u k) + \frac{\partial}{\partial y}(\rho v k) = \frac{\partial}{\partial x}\left[\left(\frac{\mu_T}{\sigma_k}+\mu\right)\frac{\partial k}{\partial x}\right] + \frac{\partial}{\partial y}\left[\left(\frac{\mu_T}{\sigma_k}+\mu\right)\frac{\partial k}{\partial y}\right] + G - \rho \varepsilon \tag{2}$$

where: $\sigma_k = 1$, $G = \mu\left[2\left(\frac{\partial v}{\partial y}\right)^2 + 2\left(\frac{\partial u}{\partial x}\right)^2 + \left(\frac{\partial v}{\partial x}+\frac{\partial u}{\partial y}\right)^2\right]$.

$$\frac{\partial}{\partial x}(\rho u \varepsilon) + \frac{\partial}{\partial y}(\rho v \varepsilon) = \frac{\partial}{\partial x}\left[\left(\frac{\mu_T}{\sigma_\varepsilon}+\mu\right)\frac{\partial \varepsilon}{\partial x}\right] + \frac{\partial}{\partial y}\left[\left(\frac{\mu_T}{\sigma_\varepsilon}+\mu\right)\frac{\partial \varepsilon}{\partial y}\right] + c_1 \frac{\varepsilon}{k} G - c_2 \rho \frac{\varepsilon^2}{k} \tag{3}$$

where: $\sigma_\varepsilon = 1.3$, $c_1 = 1.45$, $c_2 = 2.0$.

3. EXPERIMENTAL LOCAL HEAT TRANSFER AND FRICTION

Highest practical interest lies in compact tube-banks. We present local parameters for a staggered bank 1.25 x 1.25 of

SURFACE ROUGHNESS AS MEANS OF HEAT TRANSFER AUGMENTATION

rough-surface tubes 150 mm diam. Relative height of the surface elements was $k/d = 3 \cdot 10^{-3}$ and $8 \cdot 10^{-3}$.

On the tube perimeter, the effect of surface roughness is observed up to the boundary layer separation. In the recirculation region the augmentation of the heat transfer insignificant, because of the low fluid velocity. Velocity distribution on the outer boundary layer $u_1 = f(x)$ is reflected in the changes of the boundary layer parameters: momentum thickness, shape factor, pressure gradient factor Φ and friction coefficient c_f (Fig. 1).

In accelerating regions (negative Φ) both θ and H increase gradually, but $\partial\theta/\partial x$ and $\partial H/\partial x$ increase considerably in the decelerating region (positive Φ). In the separation region, $H = 1.97$, Fig. 1. In other ranges of Re, the separation value of H varies from 1.8 to 3.5.

Friction coefficient c_f increases considerably on rough surfaces for similar Re, Fig. 1. The maximum value of c_f is observed in the acceleration region, instead of distance of maximal velocity $u_1(x)$. We conclude that the gradient ratio $\partial u/\partial y$ exerts its effect in the vicinity of the wall, its value being higher at $\Phi < 0$. After the boundary layer separation c_f is insignificant. The complex distribution of the bounday layer thickness on a tube, Fig. 1 leads to a non-uniform effect height of surface roughness on the heat transfer, so that it must be explored along the whole perimeter. To study the effect of different height surface roughness, we use function $St = f(k^+, \varphi, Re, Pr, a, b)$ and dis-

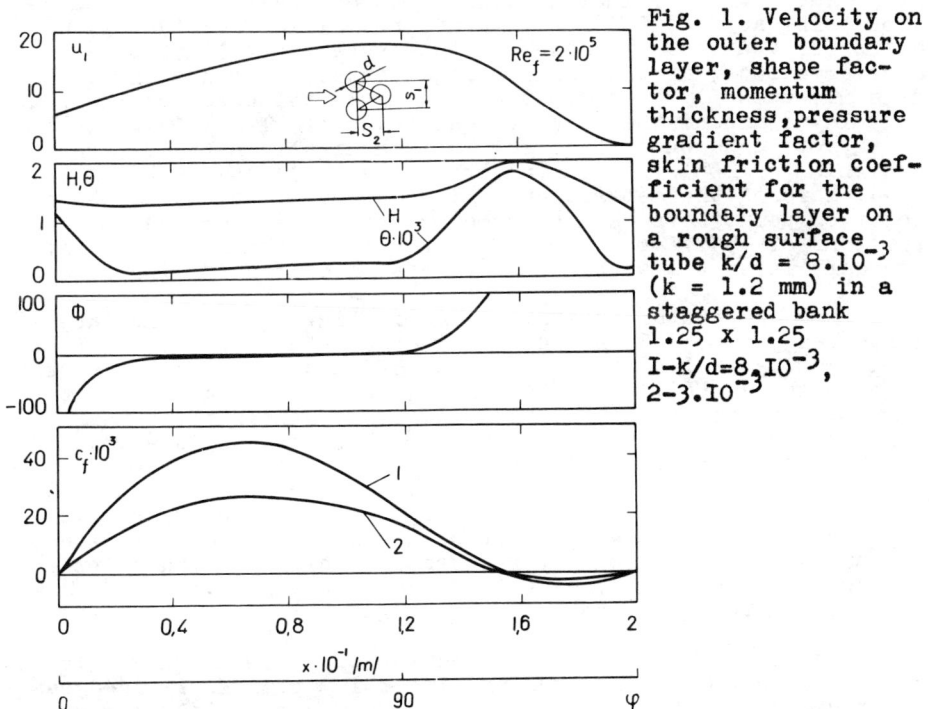

Fig. 1. Velocity on the outer boundary layer, shape factor, momentum thickness, pressure gradient factor, skin friction coefficient for the boundary layer on a rough surface tube $k/d = 8 \cdot 10^{-3}$ ($k = 1.2$ mm) in a staggered bank 1.25 x 1.25
$I - k/d = 8 \cdot 10^{-3}$, $2 - 3 \cdot 10^{-3}$

tinguish the acceleration and the deceleration regions on the tube. The region of deceleration comprises two different types of fluid dynamics, up to the separation point and defined after the separation point, where shear flow is opposite to the main flow direction u_o. At constant Re and a x b the distribution of $St = f(k^+, \varphi)$ for gases and liquids exibits an extreme value in the acceleration region. Its location is shifted to higher k^+ with an increase of angular distance. Judging from the distribution of $St = f(k^+, \varphi)$, optimal height of surface elements is $k^+ \approx 30-40$ at $Pr = 0.7$ and $k^+ \approx 70$ at $Pr_f = 80-300$ for interval φ from 0 to 135° along the perimeter. A detail analysis of $St = f(k^+, \varphi)$ for Re 10^4, 2.10^5 and 7.10^5 showed no significant effect of Re on the obove optimal values of k^+.

A theoretical interest presents the analysis of the momentum and heat transfer in the acceleration and deceleration regions of the boundary layer, particularly important is the deceleration part, where the boundary layer separation is followed by the recirculation region. We consider the distributions of velocity, temperature and kinetic energy of turbulence both in the boundary layer and in specific parts of free space inside. Fig.2 shows velocity distribution in the boundary layer expressed as $u/u_1 = f(\eta)$ where u local velocity component in Fig. 2a. In the acceleration region, velocity profile is effected by surface roughness k and by pressure gradient Φ. Function $u/u_1 = f(\eta)$ is equilibrium and can be described by relation $u/u_1 = \eta^n$, where the power index n of the ordinate value η is determined by pressure gradient at Re = idem. Deformation of the boundary layer (Fig. 2a, $\varphi = 120°$) increases with its approach to the separation point at $\varphi = 125°$. Velocity profile at $\varphi = 135°$ is specific for the separated flow, with a negative velocity, that is a reverse flow, at the wall up to a certain distance on η. The negative-velocity part occupies the whole thickness of the boundary layer around the back stagnation point ($\varphi = 170°$). Kinetic energy of turbulence (Fig. 2b) is much higher in the separation region ($\varphi = 120°$) than in the preceding locations on $k(\eta)$. We speak here of the "separation region", as oscillations of the separation point have been observed earlier and ascribed to fluctuations of $\pm 5°$. A maximum value of k is in the middle of the boundary layer - a specific feature of decelerating flows.

From temperature distributions in Fig. 2c over the tube at distances y = 0, 1.5, 4 and 8 mm in the inter-tube space of a stagered 1.25 x 1.25 bank of tubes 150 mm diam and relative surface roughness $k/d = 8.10^{-5}$, for liquids of high viscosity (Pr = 220), the temperature gradient $\partial t/\partial y$ is concentrated near the wall on the front part (y $\approx$ 1.5 mm). The distance is somewhat further in the rear part (y $\approx$ 2 mm).

Inspite of the intensive mixing inside the bank, heat transfer occurs in a thin wall layer of a viscous fluid. Inter-tube distances may be reasonably reduced to conserve circulation power consumed by the large part of the flow which does not participate in the heat transfer.

Fig. 2d presents the distribution of a dimensionless value of temperature Θ in the boundary layer in specific fluid-dynamical regions of flows of air and transformer oil. In viscous

SURFACE ROUGHNESS AS MEANS OF HEAT TRANSFER AUGMENTATION

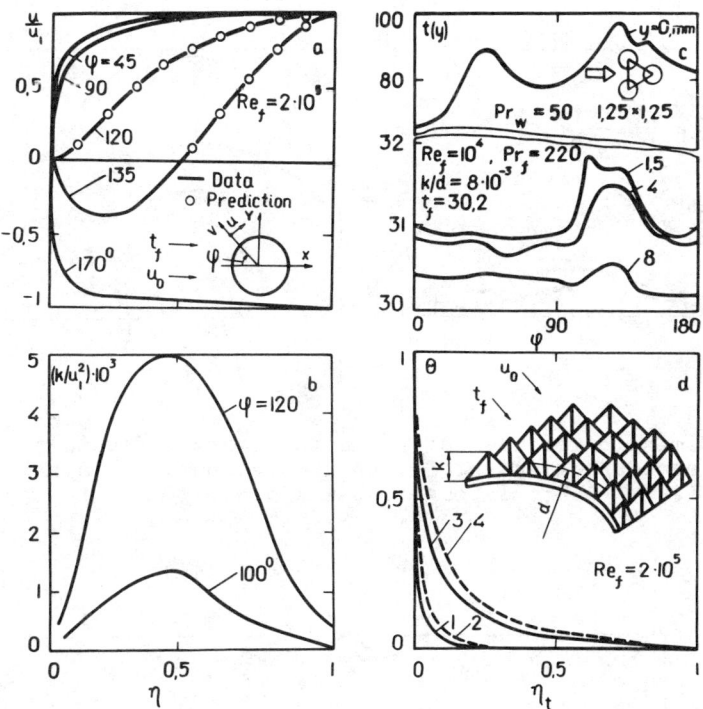

Fig. 2. Velocity distribution in the boundary layer in the acceleration, separation and recirculation regions a, kinetic energy of turbulence b, inter-tube space temperature c, dimensionless temperature d. Staggered bank 1.25 x 1.25 of rough tubes $k/d = 8 \cdot 10^{-3}$

fluids (curves 1, 2) the thermal boundary layer is considerably thinner, than in gases (curves 3, 4). Consequently, surface roughness of a limited height may be an effective means of the heat transfer augmentation in viscous fluids.

In gradient flows, velocity and temperature fluctuations are important features of momentum and heat transfer. Pressure gradient (wide ranges of Φ were covered) exerts both quantitative effect on the velocity distribution in the boundary layer, Fig. 3a. In the acceleration region the value of $\sqrt{u'^2}/u_*$ decreases under the effect of Φ, but increases in the deceleration region under positive Φ_+. The maximum value of $\sqrt{u'^2}/u_*$ is shifted to higher values of y^+, and momentum transfer deviates outside the logarithmic part of the velocity distribution.

Distribution analysis of the kinetic energy of turbulence on the basis of three components u', v', w' of the velocity fluctuation suggests a three-dimensional flow in the recirculation region. In the region of separation (for nearly zero values of c_f) vortical

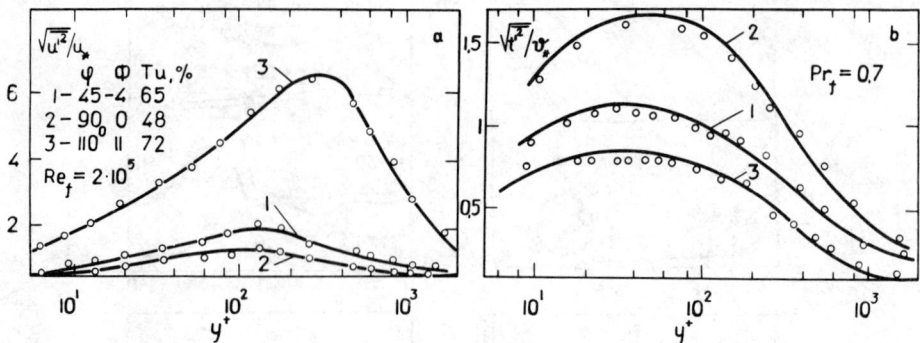

Fig. 3. Dimensionless velocity and temperature fluctuations for variable pressure gradient factor in the boundary layer. Staggered bank 1.25 x 1.25 of rough tubes $k/d = 8 \cdot 10^{-3}$

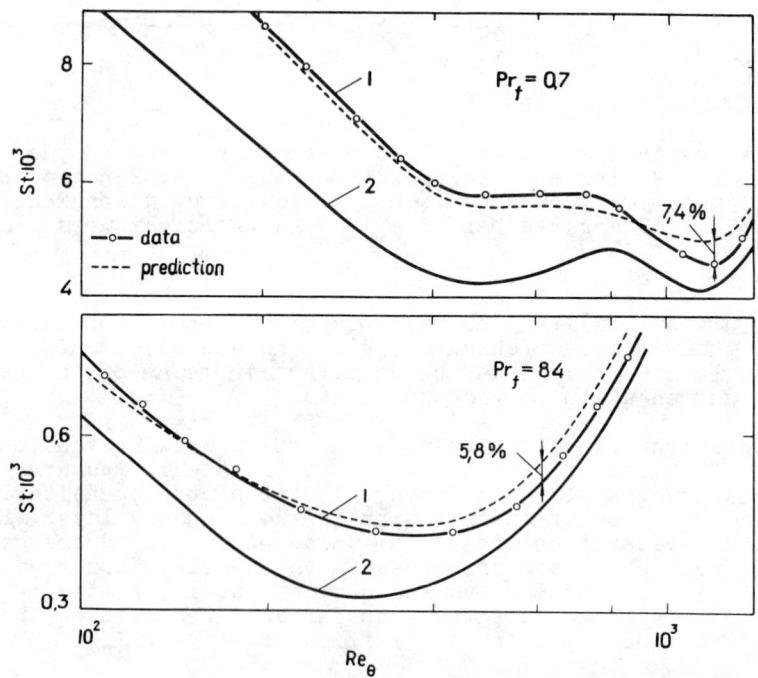

Fig. 4. Relation of St and Re in staggered bank 1.25 x 1.25: 1 — rough surface of $k/d = 8 \cdot 10^{-3}$, 2 — smooth surface, $Re_\theta = u_1 \theta / \nu$

fluctuations interact closely with the free stream, because at this angular distance (Fig. 2a, $\varphi = 135°$) velocity gradient $\partial u/\partial y$ is large at $\eta > 0.5$. The dimensionless distribution of temperature fluctuation varies in the boundary layer under the effect of Φ. The larger Φ, the larger absolute value of temperature fluctuation, Fig. 3b. But with the approach of the boundary layer separation, friction temperature ϑ_x increases considerably, and curve 3 is therefore lower than curve 2. A non-significant shift is observed in the maximum temperature fluctuation, under the effect of pressure gradient. It is now located at $y^+ \approx$ ≈ 30 to 50.

Fig. 4 presents calculated results of St and similar data for air and viscous fluids on rough surfaces. The agreement is better for Pr = 0.7. At higher Pr, the value of St is overestimated in the separation region. With the approach of the separation, the value of Φ approaches infinity, and the gradient $\partial\theta/\partial x$ becomes very high. In this range the prediction results differ from the experiment about 8-9 percent. One of the possible causes is an error in the boundary layer thickness, which increases significantly with the approach to the separation.

4. AVERAGE HEAT TRANSFER AND PRESSURE DROP OF BANKS OF ROUGH-SURFACE TUBES

Measured local value of drag and heat transfer were used to compile relations for the average pressure drop and heat transfer with the account of the transverse and longitudinal pitches and of relative roughness k/d, Fig. 5. From rough-surface tubes heat transfer was augmented at Re $(2-3)10^3$ in viscous fluids, but only at Re from $10^4-7.10^4$ in gases. At Re $\approx 10^5$ the power index of Re increases from 0.65 to 0.8.

On rough surfaces, the power index of Pr is 0.36 for laminar and transient flows in a wide range of Pr (from 0.7 to 5000). In turbulent flows, this value is 0.4 for Pr from 0.7 to 300.

For average heat transfer of staggered banks of rough-surface tube we suggest

$$Nu_f = 0.45(a/b)^{0.6} Re_f^{0.65} Pr_f^{0.36} (Pr_f/Pr_w)^{0.25} (k/d)^{0.1} \tag{4}$$

for Re from 10^2 to 10^5 and

$$Nu_f = 0.093(a/b)^{0.6} Re_f^{0.8} Pr_f^{0.4} (Pr_f/Pr_w)^{0.25} (k/d)^{0.15} \tag{5}$$

for Re from 10^5 to 2.10^6.

Average pressure drop is similarly

$$Eu = 7.2(a-1)^{-0.4} Re^{-0.3} (k/d)^{0.15} \tag{6}$$

for Re from 10^2 to 2.10^5 and

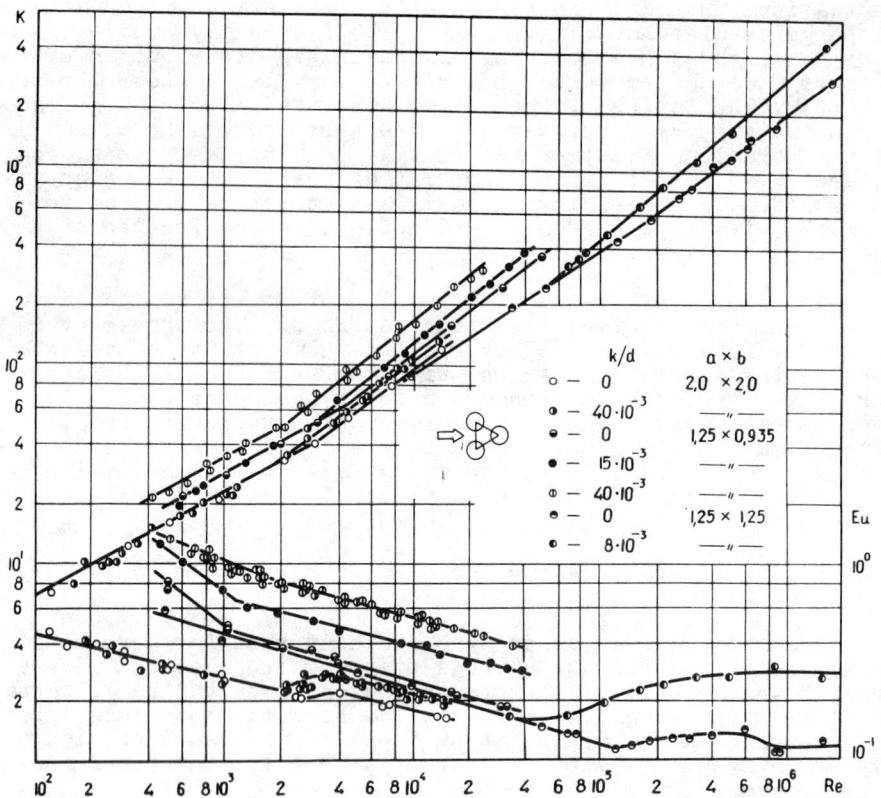

Fig. 5. Average heat transfer and pressure drop of rough surface tubes

$$Eu = 0.225(a-1)^{-0.45}(k/d)^{0.07} \qquad (7)$$

for Re from $2 \cdot 10^5$ to $2 \cdot 10^6$.

Relations (4 through 7) are applicable for relative roughness k/d from $6 \cdot 10^{-3}$ to $40 \cdot 10^{-3}$, relative transverse pitches from 1.25 to 2.0 and relative longitudinal pitches from 0.935 to 2.0. Relations (4 and 5) give average heat transfer within ±15 percent. Pressure drop is found from (6) within 10 percent, and from (7) - within 5 percent in the range of parameters covered.

Throughout the study, reference was made to base tube diameter, mean temperature inside the bank, mean surface temperature for all bank tubes, and mean velocity in the minimum inter-tube space. Average heat transfer coefficient was determined for the total surface area of a rough-surface tube.

REFERENCES

1. Achenbach, E. 1971. Influence of Surface Roughness on the Flow through a Staggered Tube Bank. Wärme-und Stoffübertragung, 4, S. p.p. 120-126.

2. Puchkov, P.I. 1948. Effect of Surface Roughness on the Heat Transfer of Tubes Banks in Crossflow. Kotloturbostroyenye, 4, p.p. 5-6.

3. Groehn, H.G., Scholz, F. 1970. Anderung von Wärmeübergang und Strömungswiderstand in Querangeströmten Rohrbündeln unter dem Einfluss verschiedener Rouhigkeiten sovie Anmerkungen zur Wahl der Stoffvertbezugstemperaturen. Heat transfer Conference, 3, paper FC. 7.10, p.p. 1-11.

4. Groehn, H.G., Scholz, F. 1976. Heat Transfer and Pressure Drop of Axial Tube Banks with Artificial Roughness in Crossflow. Teplomassoobmen V, Heat and Mass Transfer Conf. Minsk, Vol. 1, p. 2, p.p. 37-42.

5. Achenbach, E. 1977. On the Cross Flow through In-Line Tube Banks with Regard to the Effect of Surface Roughness. Wärme- und Stoffübertragung, 4, S. 152-155.

6. Žukauskas, A.A., Ulinskas, R.V. 1980. A parameter study of the turbulent boundary layer with adverse pressure gradient. Teplomassoobmen VI, Heat and Mass Transfer 6, Proc. Sixth Soviet Heat and Mass Transfer Conf. Minsk, Vol. 1, p. 2, pp. 91-100 (In Russian).

7. Gosman, A.D., Pun, W.M., Runchal, A.K., Spalding, D.B., Wolfshtein, M., 1972. Heat and Mass Transfer in Recirculating Flows. Academic Press, London-New York, 324 p.

8. Launder, B.E., Spalding, D.B., 1972, Matematical Models of Turbulence, Academic Press, London and New York, 169 p.

HEAT EXCHANGER TUBE VIBRATION

The Effect of Grid Generated Turbulence on the Fluidelastic Instability of Tube Bundles in Cross Flow

B. M. H. SOPER
Energy Technology Division
AERE
Harwell, Oxon, UK

ABSTRACT

Wind tunnel tests have been carried out to investigate the effect of grid generated turbulence on the fluidelastic instability of a tube bundle. A bundle of triangular tube layout, with a pitch to diameter ratio of 1.25, was used in every test and a range of grids were used in order to promote a range of turbulence scales and intensities.

Results are presented which show that, depending on the scale and intensity of the turbulence, the critical flow velocity for fluidelastic vibration of tubes in the first row of the bundle can be lowered by as much as 30% but can also be raised by up to 5%.

1. INTRODUCTION

Many tube failures in shell and tube heat exchangers can be attributed to the vibration mechanism of fluidelastic instability. The author has already demonstrated [1] that particularly severe vibration can result from this mechanism. The shell side flow velocity at which fluidelastic instability is initiated represents the upper limit to flow, and is therefore of particular significance to the designers of shell and tube heat exchangers. There is evidence to suggest that turbulence created upstream of the tube bundle, for example by an impingement plate or a partly closed valve, can lower significantly the flow velocity for the onset of fluidelastic vibration. This paper describes a series of tests in which a tube bundle of known vibration characteristics was subjected to various levels of free stream turbulence, produced by grids placed upstream of the bundle.

2. BACKGROUND

Franklin and Soper [2] showed that the critical flow velocity for the onset of fluidelastic instability could be reduced by as much as 30%, by the presence of turbulence in the fluid stream approaching the tube bundle. The results of other studies, however, have shown that this is not always the case.

Southworth and Zdravkovich [3] found from tests in air, on a two row tube array, that upstream turbulence supressed the formation of fluidelastic instability. They attributed this effect to the randomness imposed by the turbulence which inhibits the build up of fluidelastic excitation. Gross [4] also observed this effect, particularly for the tubes in the bundle, which, without the turbulence grid, vibrated at the lowest flow velocities. More recently, Gorman [5] has carried out tests in water on deep banks (in flow direction) of typical heat exchanger tubing. He found that the critical velocity was not significantly affected by the presence of an upstream turbulence grid.

Clearly, there is no simple answer to the question of whether turbulence is beneficial or not from the standpoint of avoiding fluidelastic instability. Therefore, it is essential that in any study of the effects due to turbulence that the turbulence itself should be characterised.

3. THEORY

3.1 Turbulence Properties

At a point in the flow downstream of the grid the instantaneous velocity, u, can be considered to comprise of a time averaged component, $\bar{u}$, plus a fluctuating component, u'. A turbulence intensity can then be defined as:

$$I = \frac{\hat{u}}{\bar{u}} \qquad (1)$$

where $\hat{u}$ is the RMS of the fluctuating component of velocity.

A measure of the size of eddies in the flow is given by the turbulence length scale. There are several definitions of length scale ranging from the macroscale, which is a measure of the larger sized eddies in the flow, to the microscale which gives a measure of the size of smaller eddies and depends on the viscosity of the fluid. It is considered that microscale is unlikely to affect significantly tube vibration and is not considered in this paper. Two methods of quantifying macroscale have been used, both of which are based on the time history of the flow velocity, measured by a single sensor at a point in the flow.

In the first method, the autocorrelation function, $R_{uu}(\tau)$, is computed where:

$$R_{uu}(\tau) = \lim_{T \to \infty} \frac{1}{T_o} \int_0^T u(t)\, u(t+\tau)\, dt \qquad (2)$$

The time delay, τ_o, at which the autocorrelation function just crosses the time delay axis is then used to compute the macroscale from the relation:

$$\Lambda = \tau_o\, \bar{u} \qquad (3)$$

Defined in this way the macroscale is a measure of the maximum size of eddy occuring in the flow.

The second approach, which is credited to Von Karman, requires computing the power spectral density (PSD) of the velocity time history. A macroscale can then be calculated from the value of the PSD at zero frequency, using the relationship:

$$\Lambda' = \frac{\bar{u}}{4(\hat{u})^2} \left[PSD \right]_{f=0} \qquad (4)$$

Defined in this way, Λ is an averaged value of the eddy size in the macroscale range.

3.2 Fluidelasticity Instability

Although fluidelastic instability is widely recognised as a cause of tube failures in shell and tube heat exchangers, the exact nature of this phenomenon is not well understood. Simple models of the phenomenom do, however, exist, for example Connors [6]. His stability analysis gives the critical velocity for the onset of fluidelastic instability, u_c, in terms of the properties of the flow and the structure:

$$\frac{u_c}{f_n D} = K \sqrt{\frac{m_e \delta}{\rho D^2}} \qquad (5)$$

where f_n is the tube natural frequency, D the tube outside diameter, K the instability factor, m_e the tube effective mass per unit length, δ the logarithmic decrement and ρ the density of the shell side fluid.

3.3 Effect of Turbulence on Critical Velocity

It has already been demonstrated [2], using an elementary argument, that free stream turbulence might be expected to lower the critical velocity. The velocity perturbations increase the rate at which energy is extracted from the fluid and therefore decrease the critical velocity. For the turbulent case, equation (5) can be written as:

$$\frac{\bar{u}_c}{f_n D} = \frac{1}{\sqrt{(1+I^2)}} K \sqrt{\frac{m_e \delta}{\rho D^2}} \qquad (6)$$

where I is the turbulence intensity as defined by equation (1).

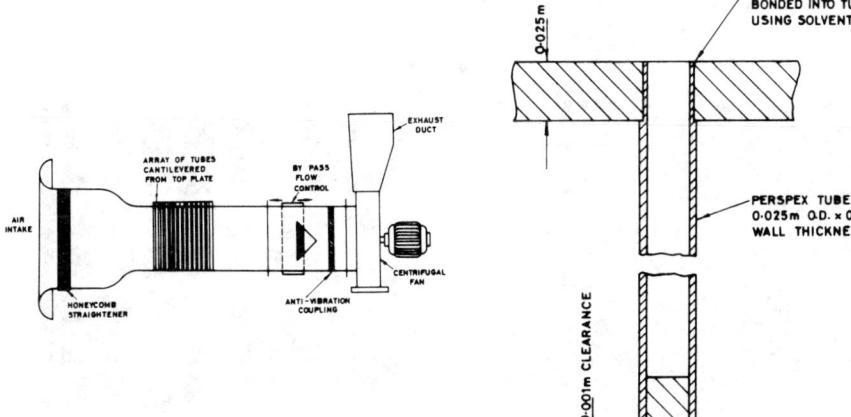

Fig 1. Wind Tunnel.　　　　Fig 2. Tube Mounting Arrangement.

4. TEST EQUIPMENT

4.1 Wind Tunnel

The wind tunnel is shown diagramatically in Fig 1. The working section measures 0.457m square and the contraction ratio is 3.2. Air is drawn through the wind tunnel by a centrifugal fan which is capable of giving a maximum empty tunnel velocity of 30m/s with a turbulence intensity of approximately 1%. The fan is run at constant speed and control of the flow through the test section is achieved by bleeding air into the tunnel downstream of the test section.

4.2 Test Section

All tests were conducted on a tube bundle of the normal triangle tube layout with a pitch to diameter ratio (P/D) of 1.25. The test section had

Fig 3a. Grid 1.

Fig 3b. Grid 2.

Fig 3c. Grid 3.

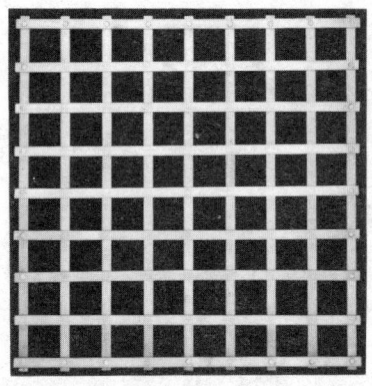

Fig 3d. Grid 4.

been designed for a study of the effect of tube layout on fluidelastic instability [1] and comprised ten rows of cantilevered 'Perspex' tubes as shown in Fig 2. The tube outside diameter was 25.4mm.

4.3 Turbulence Grids

Four turbulence grids were used in these tests, as shown in Fig 3. Grid 1 was fabricated from flat bar 25mm wide by 3mm thick and the mesh pitch was 50mm. Grid 2 was fabricated from the same section bar as Grid 1 but the mesh spacing was 54mm. Grid 3 was a piece of perforated sheet, 1mm thick, in which the perforations were 6.35mm square, separated by 1mm wide ligaments. Grid 4 was fabricated from flat bar 12.7mm wide and 3mm thick on a 54mm mesh pitch.

4.4 Instrumentation

The instantaneous velocity at a point in the flow downstream of a turbulence grid was measured using a single channel Prosser 6100 series constant temperature anemometer. The system comprised the anemometer bridge, linearizer, DC, RMS and turbulence intensity meters. A tungstem wire straight probe, type 6505/W05, was used throughout. Further processing of the velocity data was performed by a DEC PDP 11/60 computer 'on-line' to the experiment.

For the tube bundle vibration tests the superficial air flow velocity, u_s, was obtained from the pressure drop across the contraction in the wind tunnel. For the triangular layout tube bundle used in these tests, the

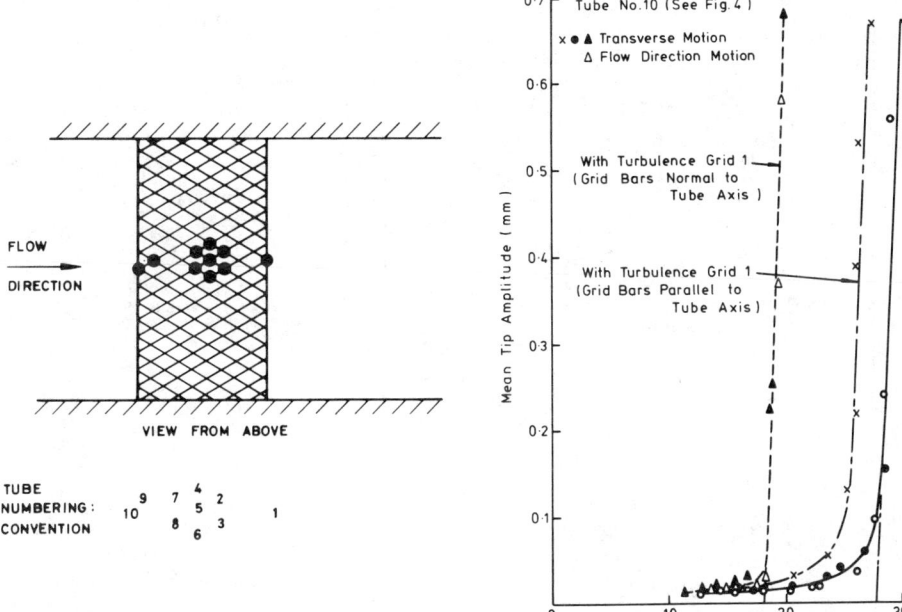

Fig 4. Position of Instrumented Tubes.   Fig 5a. Response of Tube in First Row.

maximum intertube velocity, u_g, (which is the velocity used in equation (5)) is related to the superficial velocity by:

$$u_g = \frac{P/D}{P/D-1} u_s = 5 u_s \qquad (7)$$

for the bundle under test.

The amplitudes of selected tubes, as shown in Fig 4, were measured using wire resistance strain guages, bonded to the tube bore at the position of maximum bending strain. The strain guage signal was amplified and then fed to a RMS meter with a 30s time constant or to an ultra violet recorder. For each instrumented tube, strain guages were provided to measure the components of the tube motion normal to and parallel with the flow direction. In later tests the strain guage signals, as well as an analogue signal proportional to the contraction pressure drop, were sent to the computer for processing.

5. TESTS CARRIED OUT AND RESULTS

5.1 Effect of Turbulence on Fluidelastic Instability

For the sake of completeness some of the results of an earlier study Franklin and Soper [2], using Grid 1, are included in this paper. Tests have been carried out to measure the vibration response of selected tubes within the test tube bundle as a function flow velocity, both with and without upstream generated turbulence.

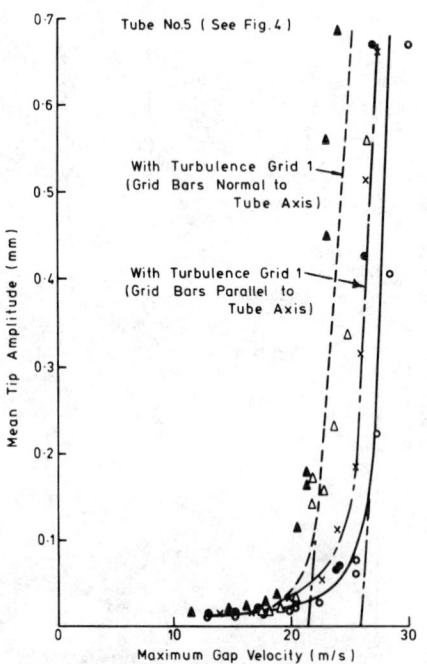

Fig 5b. Response of Tube in Centre Row.

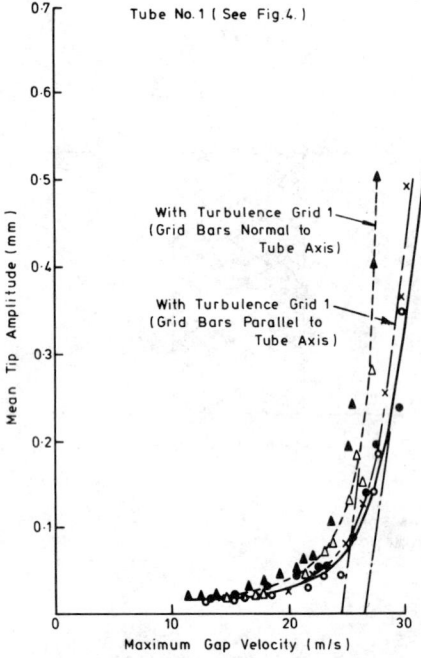

Fig 5c. Response of Tube in Last Row.

Using Grid 1, placed 235mm upstream of the bundle, the vibration response of tubes in the first, fifth and last rows was measured. The results of these tests are shown in Fig 5 from which it can be seen that the critical flow velocity was reduced by approximately 30% for the tube in the first row. This effect was less pronounced for the tube in the fifth row and was almost non-existent for the tube in the last row. This result suggests that the flow through a closely spaced tube bundle is so disorderly that the effect of upstream turbulence is rapidly lost as the flow penetrates the tube bundle. For these tests, the bars of the turbulence grid were arranged perpendicular to the longitudinal axis of tubes in the bundle. Further tests were undertaken with the grid bars parallel to the tube axis and these results are also shown in Fig 5. It can be seen that there was less effect with the grid in this orientation indicating that the turbulence was not homogeneous. Subsequent grids were manufactured with bars in two directions in an attempt to overcome this problem.

With Grid 2 in position, it was found again that the effect due to upstream turbulence diminished with distance into the bundle and so for the tests using Grids 3 and 4, vibration was monitored only for a tube in the first row, ie the most susceptible row. Fig 6 shows the effect of Grid 2 on tube vibration with the grid at different distances, x, upstream of the tube bundle. It can be seen that again the critical velocity was reduced by the

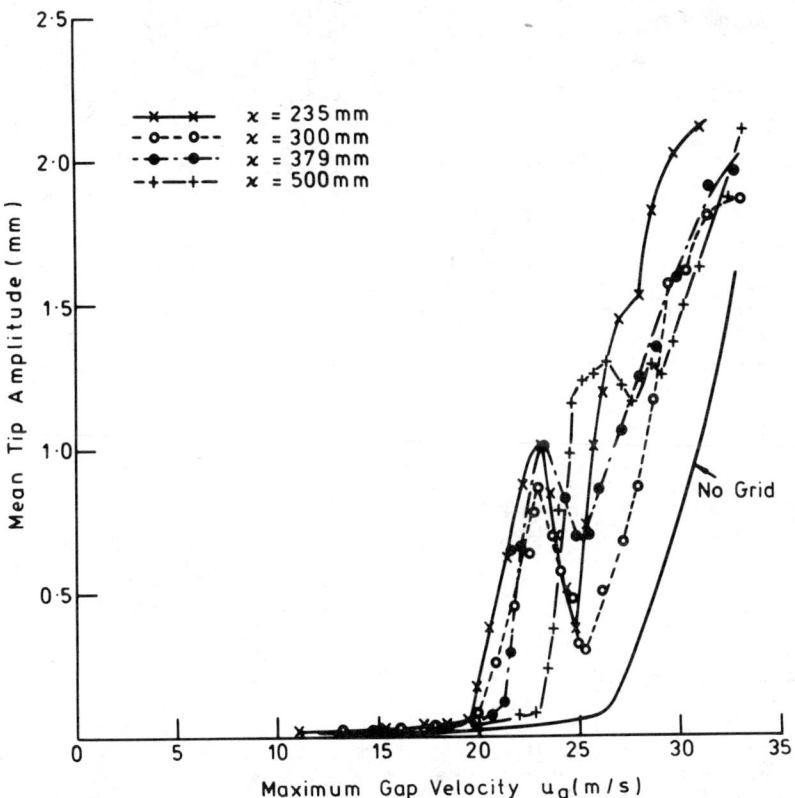

Fig 6. Effect of Grid 2 on Tube Response.

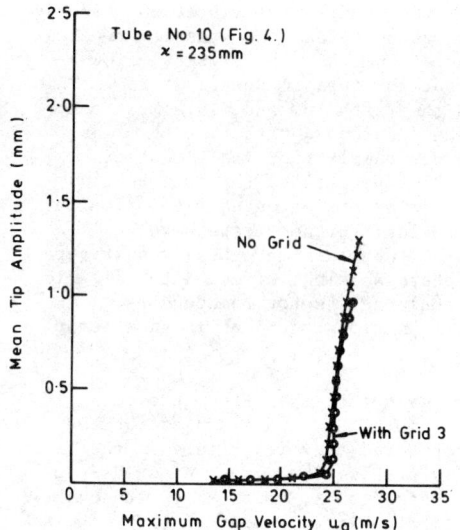

Fig 7. Effect of Grid 3 on Tube Response.

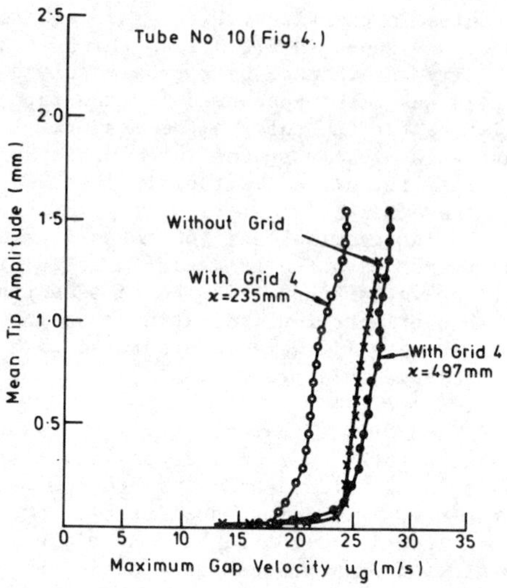

Fig 8. Effect of Grid 4 on Tube Response.

presence of the turbulence, the effect being greatest when the grid was nearest to the bundle and hence the turbulence intensity was at its greatest.

Grid 3 was used to generate low intensity turbulence and results are shown in Fig 7. The effect of this low intensity turbulence was to raise very slightly the critical flow velocity.

Grid 4 was aimed at producing mid range turbulence intensity and results are shown in Fig 8 for two positions of the grid. With the grid 235mm upstream of the bundle, the critical flow velocity was reduced by approximately 15% whereas with the grid 497mm upstream of the bundle the critical velocity was raised by approximately 5%.

5.2 Measurement of Turbulence Generated by Grids

For each of the four grids, turbulence measurements were made at downstream locations corresponding to the positions used in the vibration tests. For each measurement, 10^5 samples of the linearized anemometer output were acquired, at a rate of 1500 samples per second, by the PDP 11 computer. The mean velocity, turbulence intensity and length scale were then computed using the methods outlined in section 3.1. Data were collected for a range of flow velocities to enable the values pertinent to the critical flow velocity to be found by interpolation. A typical autocorrelation function and a power spectrum, for Grid 2, are shown in Figs 9 and 10. It can be seen that most of the energy is at low frequencies, implying that relatively large eddies are present. Results for each grid, at the critical velocity for the onset of fluidelastic instability, are given in Table 1. Where possible, the values of turbulence have been corrected for measurement errors arising from the high turbulence. The procedure for estimating these errors is summarized in Appendix 1.

Grid	x(m)	$\bar{u}_c$(m/s)	I_m	I	$\wedge$(m)	$\prec$(m)
1	0.235	18.2	0.225	0.24	–	0.045
2	0.235	19.5	0.51	–	0.53	0.125
"	0.300	20.0	0.54	–	0.45	0.100
"	0.379	21.5	0.41	–	0.61	0.110
"	0.497	23.0	0.38	0.51	0.57	0.075
3	0.235	24.4	0.032	0.032	0.028	0.006
4	0.235	20.4	0.216	0.228	–	0.021
"	0.497	25.1	0.104	0.105	–	0.029

Table 1. Properties of Grid Generated Turbulence

6. DISCUSSION

Because turbulence can either increase or decrease the critical velocity for fluidelastic instability, depending on its intensity and scale, there must be more than one mechanism involved. Southworth and Zdravkovich [3] have demonstrated that a grid of the same dimensions as Grid 1 was effective in drastically reducing the vibration of tubes in the first row of a two row array. Their results were, however, for a wide spaced array ($L_t/D = 2$, $L_1/D = 4$ compared with the present case of $L_t/D = 1.25$, $L_1/D = 1.08$ where L_t and L_1 are the transverse and longitudinal pitch respectively) and their results suggest that the effectiveness of grid generated turbulence in supressing fluidelastic instability diminishes as the tube spacing is reduced. Gross [4] tested closer space bundles and found that upstream turbulence raised significantly the critical flow velocity for the critical

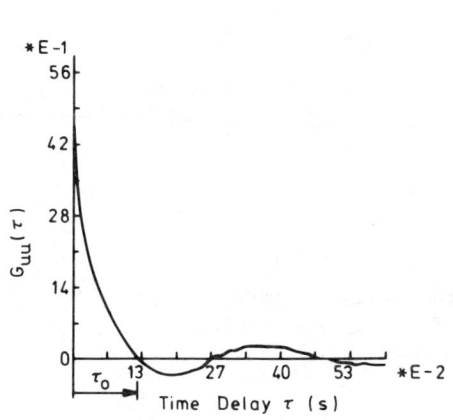

Fig 9. Typical Autocorrelation Function for Turbulence from Grid 2 (x = 300mm).

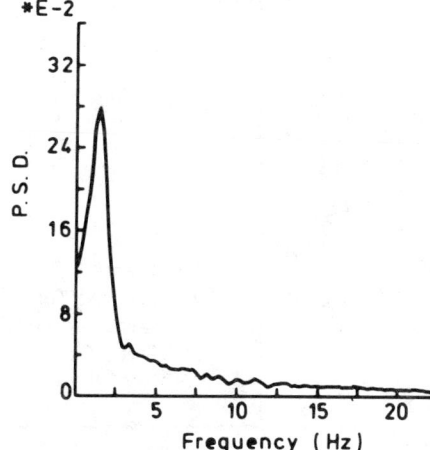

Fig 10. Typical Power Spectrum for Turbulence from Grid 2 (x = 300mm).

tube row (the first row to vibrate, normally the first or second row).
Although he tested a bundle of similar tube layout to the one in the present
study he did not, unfortunately, give details of the turbulence and so it is
not possible to make a close comparison of the results.

For the present results, Fig 11 shows a comparison of the observed change
in critical velocity with the simple prediction given by equation (6). The
circles represent the critical velocity defined as a tangent to the tube
response at large amplitude, this being the definition used by the author in
previous literature. The tube response with Grid 2 in place exhibited a peak
prior to the eventual large increase in tube amplitude, as seen in Fig 6,
particularly when the grid was nearest to the tube bundle. For this peak the
tube motion was still at the tube fundamental frequency of 36 Hz. Vortex
shedding resonance can be ruled out because a frequency match would occur at a
flow velocity of approximately 2m/s. The turbulence itself could cause tube
excitation if there was a predominant content of the turbulence power spectrum
at the appropriate frequency. This is not however the case as can be seen
from Fig 10. Goyder [7] has predicted that for a group of flexible cylinders,
there are many tube vibration patterns which can lead to instability, each
having a separate critical velocity. It is possible that the peak in the
vibration response curve arises as a result of a change of vibration pattern
brought about by an interaction between the incident turbulence and the tube
motion. If the initial increase in tube amplitude is taken as the onset of
instability, the corresponding reduction in critical velocity is as indicated
by the crosses in Fig 11 for the Grid 2 results.

Of the two methods used for obtaining the turbulence macroscale, the
method involving autocorrelation was not found to be very satisfactory. This
is because unless there is a strong harmonic content at low frequency the
required zero crossing of the autocorrelation function does not always occur.
The results from the power spectral density were found to be much more
consistent.

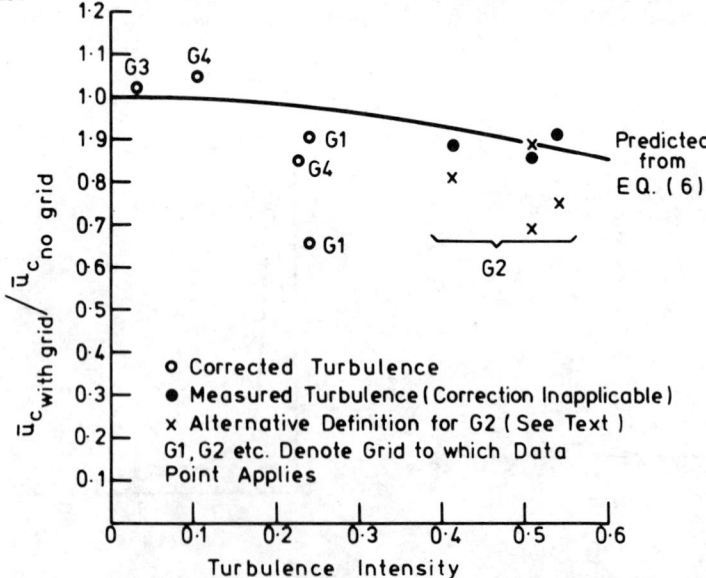

Fig 11. Effect of Turbulence on Critical Velocity
for Fluidelastic Instability.

7. CONCLUSIONS

The results of this work show that turbulence in the flow approaching a tube bundle may be either beneficial or detrimental to the susceptibility of the tubes to fluidelastic instability, depending on the nature of the turbulence.

Tubes in the first row are affected most by upstream turbulence and it is often these tubes in shell and tube heat exchangers which are the first to fail. The flow in the entrance region of a heat exchanger is known to be turbulent but there is not yet any information on the levels of turbulence. Measurements on real heat exchangers would give an insight into this phenomenon.

ACKNOWLEDGEMENTS

The author wishes to thank UKAEA and HTFS for permission to publish this paper. The HTFS work on vibrations research is supported by the Chemicals and Minerals Requirements Board of the UK Department of Energy.

REFERENCES

1. Soper, B.M.H. 1980. Effect of tube layout on the fluidelastic instability of tube bundles in cross flow. Proc. ASME Winter Annual Meeting. Flow-Induced Heat Exchanger Tube Vibration - 1980. HTD-Vol 9.

2. Franklin, R.E. and Soper B.M.H. 1977. An investigation of fluidelastic instabilities in tube bundles subjected to fluid cross flow. Proc. of 4th International Conference on Structural Mechanics in Reactor Technology, San Francisco, paper F 6/7.

3. Southworth, P.J. and Zdravkovich, M.M. 1975. Effect of grid turbulence on the fluidelastic vibrations of in-line tube banks in cross flow. J. Sound and Vibration, 39(4), 461-469.

4. Gross, H.G. 1975. Investigations in aeroelastic vibration mechanisms and their application in design of tubular heat exchangers. Dissertation for Technical University of Hannover.

5. Gorman, D.J. 1979. The effect of artificially induced up-stream turbulence on the liquid cross flow induced vibration of tube bundles. ASME Conference on Flow Induced Vibration of Power Plant Components, PVP 41.

6. Connors, H.J. 1970. Fluidelastic vibration of tube arrays excited in cross flow. Proc. ASME Symposium on Flow Induced Vibration in Heat Exchangers, pp 42-56.

7. Goyder, H.G.D. 1980. Unstable vibrations of a bundle of cylinders due to cross-flow. Proc. of Int. Conf. on Recent Advances in Structural Dynamics, ISVR, University of Southampton, UK.

NOMENCLATURE

D	Tube outside diameter, m
f	Frequency, Hz
f_n	Tube natural frequency, Hz
I	Turbulence intensity, dimensionless
I_m	Measured turbulence intensity, dimensionless
K	Instability factor, dimensionless
L_1	Longitudinal tube pitch, m
L_t	Transverse tube pitch, m
m_e	Tube effective mass per unit length, kg/m
P	Tube pitch, m
$R_{uu}(\tau)$	Autocorrelation function of velocity, m^2/s^2
T	Integration period, s
t	Time, s
u	Instantaneous flow velocity, m/s
$\bar{u}$	Mean flow velocity, m/s
u'	Fluctuating component of velocity, m/s
$\hat{u}$	RMS of fluctuating component of velocity, m/s
u_c	Critical flow velocity for onset of fluidelastic instability, m/s
u_g	Flow velocity through minimum gap between tubes, m/s
u_m	Measured component of instantaneous flow velocity, m/s
u_R	Resultant component of instantaneous flow velocity, m/s
u_s	Superficial flow velocity (approaching bundle), m/s
v', w'	Fluctuating velocity components perpendicular to main stream, m/s

Greek

δ	Logarithmic decrement of damping, dimensionless
Λ	Turbulence length scale based on eq. 3, dimensionless
Λ'	Turbulence length scale based on eq. 4, dimensionless
ρ	Density of fluid surrounding tubes, kg/m^3
τ	Time delay, s
τ_o	Time delay at first zero crossing of $R_{uu}(\tau)$, s

APPENDIX 1

Estimation of Errors in Hot Wire Anemometer Measurements

Consider a single wire probe as shown.

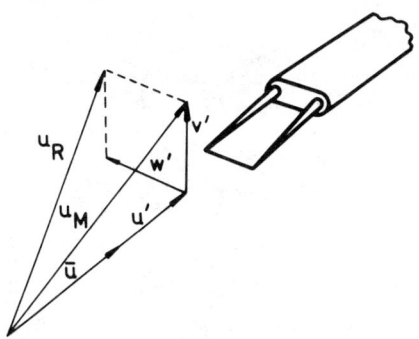

It is desired to measure $\bar{u}$ and $\acute{u}$ in the plane of the probe support but because of the fluctuating velocity components ν' and ω', the resultant instantaneous velocity is u_R and not $\bar{u} + \acute{u}$. The wire is relatively insensitive to the velocity component ω' and so for the present purposes, this term can be ignored. For a suitably linearized anemometer the instantaneous fluctuating velocity is given by:

$$u_M = \sqrt{(\bar{u} + \acute{u})^2 + \nu'^2} \tag{A1.1}$$

A series expansion yields:

$$u_M = \bar{u} \left[1 + \frac{\acute{u}}{\bar{u}} + \frac{\nu'^2}{2\bar{u}^2} - \frac{\acute{u}\,\nu'^2}{2\bar{u}^3} + \frac{\acute{u}^2\,\nu'^2}{2\bar{u}^4} - \frac{\nu'^4}{8\bar{u}^4} + \cdots \right] \tag{A1.2}$$

Taking the time average of equation (A1.2) and assuming that the turbulence is isotropic, in which $\acute{u}$ and ν' follow a normal Gaussian distribution and are also normally correlated, it can be shown that:

$$\frac{\bar{u}_M}{\bar{u}} = 1 + \frac{I^2}{2} + \frac{3}{8} I^4 + \cdots \tag{A1.3}$$

The measured fluctuating component of velocity, $\acute{u}_M$, is given by $u_M - \bar{u}_M$. The measured turbulence intensity, I, which is defined as

$$\sqrt{\bar{\acute{u}}_M^2} \;/\; \bar{u}_M$$

can be shown to be related to the actual turbulence intensity, I, by:

$$I_M = \frac{I\sqrt{1 - I^2}}{\left[1 + \frac{I^2}{2} + \frac{3}{8} I^4 + \cdots \right]} \tag{A1.4}$$

which is valid for values of I up to 0.39.

Determination of Flow Induced Nonlinear Vibrations of Prestressed Heat Exchanger Tubes

M. WAHLE
Institut für Leichtbau
RWTH Aachen
Postfach, 5100 Aachen, FRG

ABSTRACT

A step-by-step solution technique is presented for the evaluation of the nonlinear dynamic response of prestressed heat exchanger tubes. The computer-program requires the flow induced forces as input. Any kind of forces may be given. If only spectral information of the forces is available it is possible to generate time functions of the forces with a simulation procedure. The method calculates rms- and maximum-values of the response and determines probability distributions of the deflections and stresses of the tubes. It was shown that the procedure works quite well. Also in the case of random time dependent forces the calculations agree in an adequate manner with theoretical results in the linear and the nonlinear case.

1. INTRODUCTION

Flow induced vibrations of tubes very often have been observed in operating heat exchangers. The vibrations can be so severe that two tubes in a tube array are constantly impacting. There is the danger of structural fatigue due to high dynamic stresses. In some cases the vibrations broke the heat exchanger tubes and caused the shutdown of power plants with all the enormous economical consequences /1/.

Cross flow heat exchangers applied to nuclear helium-cooled high temperature reactors operate at high pressure of 40-70 bars in the primary circuit. Especially in this case it is very important that a designer should be able to predict the stress levels and displacements in the structure with adequate accuracy. He needs a method which describes the structure in a realistic manner and allows the consideration of various possible flow induced forces (harmonic, periodic and random) due to different excitation mechanisms.

2. REQUIREMENTS FOR A NUMERICAL METHOD

The special aim of the present work is to allow an accurate physical modelling of the vibrating structure and the boundary conditions of the tubes:

  i. The computer-program was developed for multi-degrees-of freedom systems. Straight tubes, curved tubes and even helically coiled tubes (Fig. 1) (three dimensional problem) of heat exchangers can be taken into account.

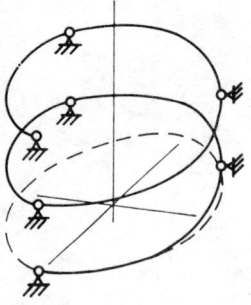

Fig. 1   Multi supported helically coiled tube of an heat exchanger

ii. The components of heat exchangers are subject to a wide band of temperature variation. There is an axial compression or tension force in the tubes, because the ends of the tube or the tube segment are usually not free at the supports but clamped, simply supported or elastically supported. The eigenvalues of such a prestressed tube are different to those of the tube without pre-stress and the dynamic response varies considerably.

iii. The fixed or nearly fixed ends of the tube or tube segment (axially restrained beam) produce a geometrically nonlinear vibration. The vibrations induce axial forces in the tube which depend on its displacement.

iv. In several cases we have to take nonlinear elastic constraints into consideration as in the case of loose support (Fig. 2).

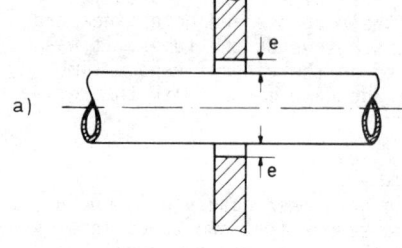

Fig. 2   a) Nonlinear elastic constraint as in the case of a loose support

b) Mechanical model of a loose support of a tube

(K: equivalent stiffness of the springs)

As input the computer-program requires the flow-induced forces acting on the tube in downstream and cross flow direction. The forces may be given in a time-history form or in the form of a Fourier-spectrum (amplitude and phase information) to describe harmonic or periodic excitation mechanisms. If only spectral information of the forces is available, as in the case of random excitation, it is possible to produce a time-history of the forces with a simulation procedure. The time-phase-angles of the different sinusoidal components of the force spectrum have to be calculated by a random generator and then are used to develop the time-history function required.

Random time dependent forces generally are the excitation mechanisms in the case of turbulent flow /2/.

Several research papers which deal with the problem of nonlinear random vibrations can be found in literature /3/, /4/, /5/, /6/. They all consider the axially restrained beam, partially with different boundary conditions. The classical methods reviewed by Iwan and Yang /7/ have been used to study the stochastically excited nonlinear structure. It will take many efforts to apply those methods in the case of more complicated structures (Fig. 3). Therefore a more practicable method of a step-by-step solution procedure together with the simulation process to determine the acting forces of the random excitation was developed.

a)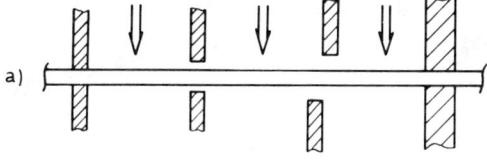

b)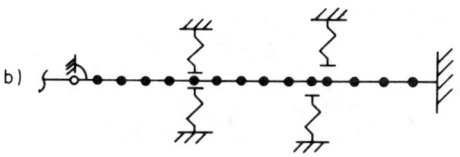

Fig. 3

a) Example for a tube segment with several nonlinear constraints and different boundary conditions

b) Mechanical model for the presented computer-program

3. THEORETICAL PROCEDURE

In the theoretical approach the determination of the response of the vibrating tube is carried out with a multi-degree-of-freedom-system. The dynamic equations can be written in the following matrix form

$$[M]\{\ddot{Y}\} + [C]\{\dot{Y}\} + [K]^*\{Y\} = \{P\} \qquad (1)$$

with $[M]$ = structural mass matrix. It is made up of the lumped masses consisting of half the weight of each member (including the dead weight of the inner tube) framing into a node. The rotary inertia contribution was computed by assuming that half of each member rotates as a rigid bar about the node while the torsional inertia is computed as the product of the polar moment of inertia and the mass density of the beam. Matrix $[C]$ = the damping matrix and $\{Y\}$ the displacement vector (u, v, w, θ_x, θ_y, θ_z at each node). The elements of $\{\ddot{Y}\}$ contain the accelerations. The vector $\{P\}$ = force vector represents concentrated forces acting at the nodes. The structural stiffness matrix $[K]^*$ depends on the deflections in the nonlinear case and considers the prestress of the tube. The prestressing force F_0 of the tube modifies the linear stiffness matrix $[K]$ after including F_0 at different positions in $[K]$. It is possible to consider the nonlinear effects by an induced force vector $\{P_i\}$ which has to be added to the right side of equation (1). In the case of geometrical nonlinearities $\{P_i\}$ only contains induced bending moments ΔM_{int} (Fig. 4b). These induced bending moments produce the internal bending moments M_{int} (Fig. 4c) in the tube due to the induced axial forces F_i (Fig. 4a). This method is useful for slender beam structures, because the potential energy of deformation will be represented mainly by the internal moments. A great advantage of this method

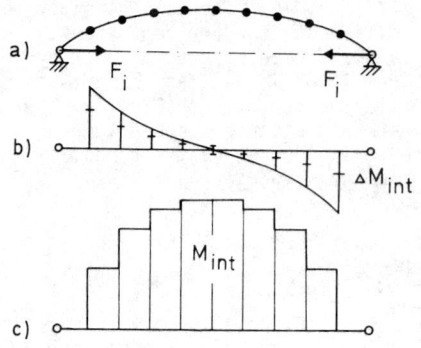

Fig. 4

a) Deformed lumped beam with induced axial forces F_i

b) Induced bending moments ΔM_{int} due to F_i

c) Internal bending moments M_{int} due to ΔM_{int}

is that it reduces computer time. This is a very important aspect, because a high number of integration time steps have to be performed to reach the steady-state-case of the response.

The dynamic equations were solved with the aid of a modified Wilson-θ-method /8/. This requires the assumption that the velocity and acceleration at a time level $t + \tau$ can be expressed as a function of the displacement at $t + \tau$, the time interval Δt, and the values at the previous state at t. To obtain the solution at $t + \Delta t$, we assume that the acceleration varies linearly over the time interval $\tau = \theta \cdot \Delta t$, where $\theta \geq 1.0$; we employed $\theta = 1.4$. Using the linear acceleration assumption it follows that

$$\{\dot{Y}\}_{t+\tau} = \{\dot{Y}\}_t + \frac{\tau}{2} (\{\ddot{Y}\}_{t+\tau} + \{\ddot{Y}\}_t) \qquad (2)$$

$$\{Y\}_{t+\tau} = \{Y\}_t + \tau \{\dot{Y}\}_t + \frac{\tau^2}{6} (\{\ddot{Y}\}_{t+\tau} + 2\{\ddot{Y}\}_t) \qquad (3)$$

which gives

$$\{\ddot{Y}\}_{t+\tau} = \frac{6}{\tau^2} (\{Y\}_{t+\tau} - \{Y\}_t) - \frac{6}{\tau} \{\dot{Y}\}_t - 2 \{\ddot{Y}\}_t \qquad (4)$$

and

$$\{\dot{Y}\}_{t+\tau} = \frac{3}{\tau} (\{Y\}_{t+\tau} - \{Y\}_t) - 2 \{\dot{Y}\}_t - \frac{\tau}{2} \{\ddot{Y}\}_t \qquad (5)$$

The equations of motion (1) should be satisfied at $t + \tau$. This leads to

$$[M] \{\ddot{Y}\}_{t+\tau} + [C] \{\dot{Y}\}_{t+\tau} + [K] \{Y\}_{t+\tau} = \{P\}_{t+\tau} + \{P_i\}_{t+\tau} \qquad (6)$$

As mentioned above the matrix $[K]$ considers the prestress of the tubes. $[K]$ is constructed for the equilibrium-position of the structure and therefore remains

constant. $\{P_{ij}\}_{t+\tau}$ represents the nonlinear effects due to geometrical nonlinearity.
$\{P\}_{t+\tau}$ is the 'projected' load equal to

$$\{P\}_{t+\tau} = \{P\}_t + \theta (\{P\}_{t+\Delta t} - \{P\}_t) \tag{7}$$

The $[C]$ matrix used in equations (1) and (6) is usually described by

$$[C] = \alpha [M] + \beta [K] \tag{8}$$

where α and β are constants. Equation (8) defines the assumption of proportional damping.
The damping factor D_i for the i-th mode is

$$D_i = \frac{\alpha}{4 \pi f_i} + \beta \pi f_i \tag{9}$$

with f_i = eigenfrequeny of the i-th mode.
Inserting (4), (5) and (8) into (6) after many manipulations leads to

$$[\tilde{K}] \{\tilde{Y}\} = \{\tilde{R}\} \tag{10}$$

in which

$$[\tilde{K}] = [M] \cdot b_2 + [K] \tag{11}$$

$$\{\tilde{Y}\} = b_0 \{Y\}_{t+\tau} - \frac{3\beta}{\tau} \{Y\}_t - 2\beta \{\dot{Y}\}_t - \frac{\beta \tau}{2} \{\ddot{Y}\}_t \tag{12}$$

$$\{\tilde{R}\} = \{P\}_{t+\tau} + \{P_i\}_{t+\tau} + [M] (b_4 \{Y\}_t + b_5 \{\dot{Y}\}_t + b_6 \{\ddot{Y}\}_t) \tag{13}$$

where b_0, b_2, b_4, b_5 and b_6 are constants in analogy to /8/. After solving equation (10) we receive $\{Y\}_{t+\tau}$. At the desired time $t + \Delta t$ the required accelerations, velocities and displacements are given by the linear acceleration assumption after using equation (4) to calculate $\{\ddot{Y}\}_{t+\tau}$:

$$\{\ddot{Y}\}_{t+\Delta t} = (1 - \frac{1}{\theta}) \{\ddot{Y}\}_t + \frac{1}{\theta} \{\ddot{Y}\}_{t+\tau} \tag{14}$$

$$\{\dot{Y}\}_{t+\Delta t} = \{\dot{Y}\}_t + \frac{\Delta t}{2} (\{\ddot{Y}\}_t + \{\ddot{Y}\}_{t+\Delta t}) \tag{15}$$

$$\{Y\}_{t+\Delta t} = \{Y\}_t + \Delta t \{\dot{Y}\}_t + \frac{\Delta t^2}{6} (\{\ddot{Y}\}_{t+\Delta t} + 2 \{\ddot{Y}\}_t) \tag{16}$$

The solution procedure works with an iteration process to determine the induced force vector $\{P_i\}_{t+\tau}$ accurately. Let the superscipt j denote the iteration step then

$$\{\Delta P_i\}^j_{t+\tau} = \{P_i\}^j_{t+\tau} - \{P_i\}^{j-1}_{t+\tau} \tag{17}$$

Equation (10) is solved iteratively until the value for $\{\Delta P_i\}^j_{t+\tau}$ approaches zero or j reaches a fixed number.

4. NUMERICAL RESULTS

The preceding theory was programmed in FORTRAN IV and run on the CYBER 175 computer. First the effectiveness of the program was tested by the example of a simply supported prestressed beam. Four different mechanical models of the beam were taken into consideration to determine the reduction of the lowest eigenvalue as the compression force increases (Fig. 5). Curve 1 of Fig. 5 describes the theoretical solution /9/:

$$(f/f_o)^2 = 1 - (F/F_{cr}) \tag{18}$$

with f = eigenfrequency of the prestressed beam, f_o = eigenfrequency of the beam without prestress, F = compression force and F_{cr} = critical load ($F_{cr} = (\pi^2 EI)/L^2$). (EI = flexural rigidity of beam, L = length of beam).

The curves 2 to 5 represent the results obtained with several mechanical models of the beam:

Curve 2: 10 equal beam segments (analogue to Fig. 12)
Curve 3: 20 equal beam segments
Curve 4: 10 un-equal beam segments (analogue to Fig. 7)
Curve 5: 20 un-equal beam segments.

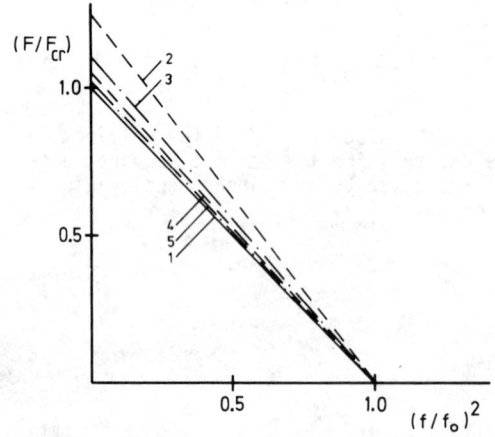

Fig. 5

Theoretical and calculated eigenfrequencies of a prestressed beam

The agreement between theoretical and calculated values is even quite good for the model with 10 un-equal beam segments, because suitable lengths of different segments were used.

Fig. 6 shows the increase of the vibration frequency according to the increase of the amplitude for the 1. mode of the beam with fixed supports due to geometrical nonlinearity. Curve 1 shows a theoretical result /10/ and the calculated results are described by curve 2 (10-element-model of the beam) and by curve 3 (20-element-model of the beam).

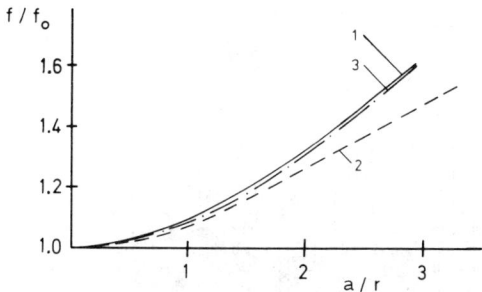

Fig. 6

Theoretical and calculated eigenfrequencies of a beam with fixed supports vibrating in the 1. mode as a function of the amplitude a

($r = \sqrt{I/A}$ = radius of gyration with A = cross-sectional area of the beam)

Now we have to consider forced vibrations. First we deal with the tube model shown in Fig. 7. The tube is

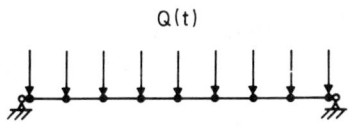

Fig. 7

Tube model with $EI = 5.10^6$ Nm², $A = 7.85 \cdot 10^{-6}$ m², $r = 1.777$ mm, $L = 1$ m, L/r = slenderness ratio $= 562.62$, β = damping factor (eq.8) $= 0.0005$

forced by a uniform pressure to compare the numerical calculations with the theoretical results of Seide /5/. Seide determined the linear and nonlinear response of a beam with fixed support excited by a random time dependent uniform pressure with a spectral input of 'white noise'. The random time dependent force needed as input for the program was developed by the simulation-procedure mentioned above. To reduce the numerical effort we considered a finite force spectrum from 2 to 48 Hz with a distance between the frequency lines of 2 Hz. The spectral components of the forces all have the same value of 0.1 N. The upper limit of 48 Hz is more than three times the fundamental frequency of the tube. Although Seide dealt with 'white-noise'-excitation the numerical calculations can be compared with his results, because he found out that the deflections can be obtained accurately with the use of only the first mode.

Fig. 8 to 10 show the time histories of the force and the deflection in the middle of the tube for three different cases of the simulation-procedure. Although the spectral input of the force is always the same the time-history functions of the force resulting are different. That is why the starting-point of the random-generator creating the phase-angles of the different spectral parts was varied intentionally. The damping factor for the first mode is about 2,1 %.

Although the maximum response values vary from 0.05569 m (Fig. 9a) up to 0.0665 m (Fig. 8a) the rms-values of the displacement remain constant up to 0.0317 m. The theoretical result received from Seide's formula is $Y_{rms}=0.0349$ m.

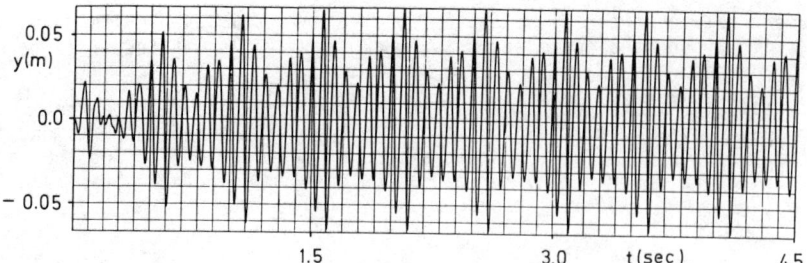

Fig. 8a   Displacement time-history of the middle of the tube in the linear case due to the force discribed in Fig. 8b

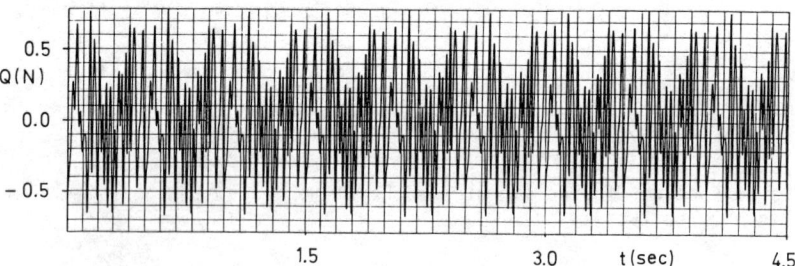

Fig. 8b   Force time-history as a result of the simulation procedure with the starting number 1 for the random generator

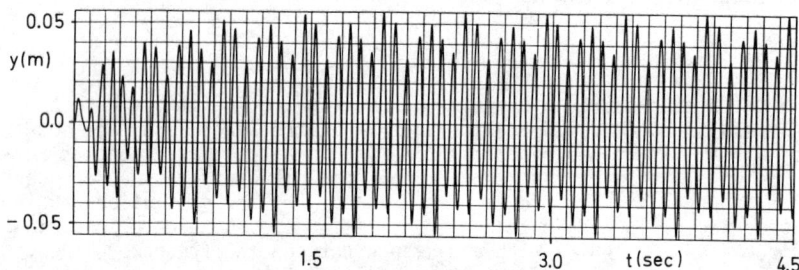

Fig. 9a   Displacement time-history of the middle of the tube in the linear case due to the force discribed in Fig. 9b

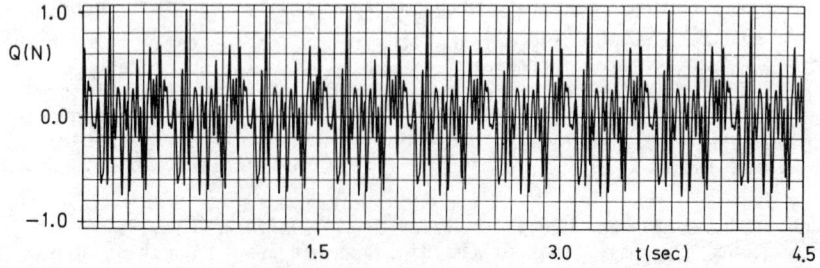

Fig. 9b   Force time-history as a result of the simulation procedure with the starting number 51 for the random generator

NONLINEAR VIBRATIONS OF PRESTRESSED HEAT EXCHANGER TUBES

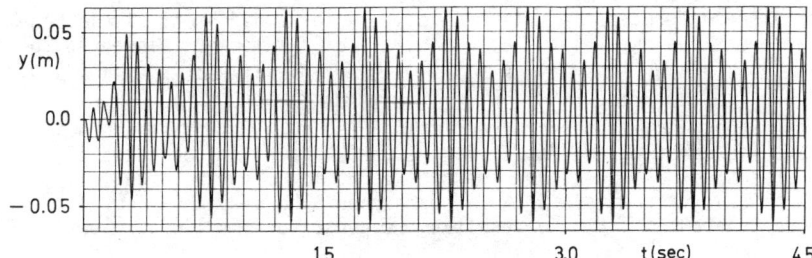

Fig. 10a Displacement time-history of the middle of the tube in the linear case due to the force discribed in Fig. 10b

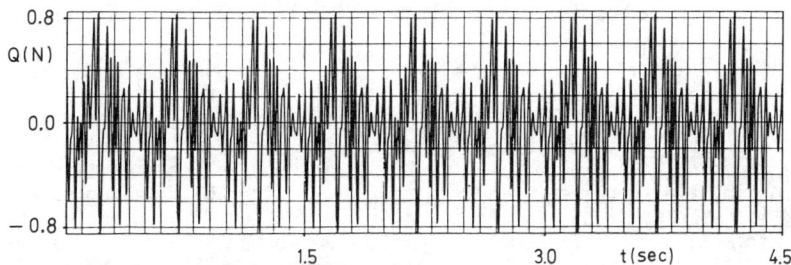

Fig. 10b Force time history as a result of the simulation procedure with the starting number 101 for the random generator

It seems that the agreement is quite good.
Next, the calculations were extended to the case of geometrical nonlinearity. Fig. 11 illustrates that an increasing ratio of the maximum frequency f_{max} of the force spectrum to the 1. eigenvalue of the tube f_1 leads to an increasing effective displacement Y_{rms} in the middle of the tube.

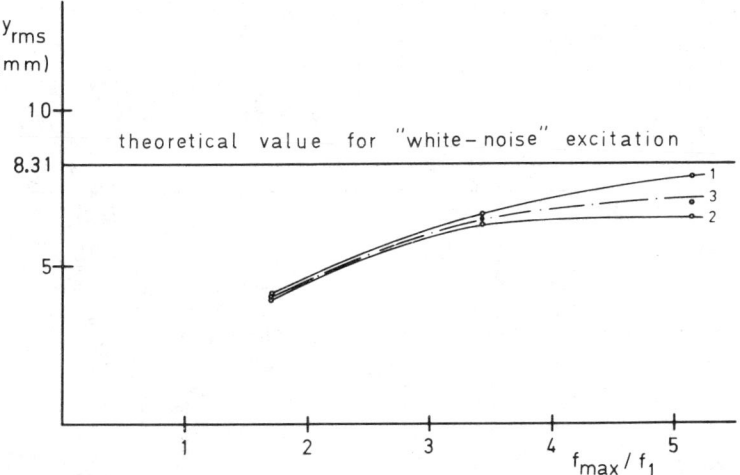

Fig. 11 Dependece of Y_{rms} in the middle of the tube to the band-wide of the exciting force spectrum for the nonlinear case. Upper boundary: theoretical result due to the 'white-noise' excitation for the example following Seide /5/

For three different ratios f_{max}/f_1 respectively three different simulation-procedures were carried out (Fig. 11). The numerical results get closer to the theoretical value for 'white-noise' excitation with increasing f_{max}/f_1 (/5/) as they are supposed to.

Furthermore wide-band excitations were calculated for the system shown in fig. 12. The force spectrum reaches from 3 Hz to 594 Hz with a distance of 3 Hz between the spectral lines. The constant value for the sinusoidal components of the force-spectrum is 1.0 N at each node. The damping factor for the first mode is about 1.4 %. The resulting response for the linear case is

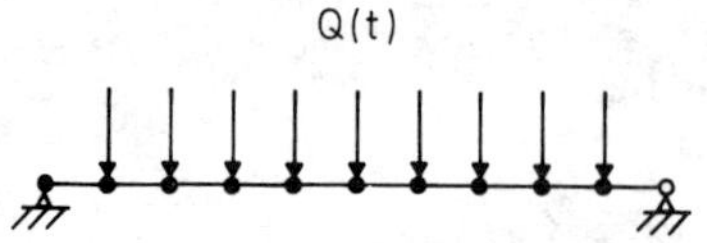

Fig. 12 Tube model with
$EJ = 955,2$ Nm², $A = 1,13 \cdot 10^{-4}$ m²
$r = 6.406$ mm, $L = 1.0$ m
$L/r = 156.1$
$\alpha = 8.814$, $\beta = 0$.

presented in Fig. 13a while Fig. 13b shows the corresponding displacement of the middle of the tube of the geometrical nonlinear system. Fig. 13c demonstrates the time function of the exciting forces which was created by the simulation process. It is obvious that the two response curves differ in their time history progress. The probability functions pictured in Fig. 14a (linear response) and Fig. 14b (nonlinear response) differ considerably although the nonlinearity is not very strong for this example. This fact is of great interest for the estimation whether structural fatigue may occur in a structure or not. The probability functions of the exciting force agree well with those of the Gaussian distribution (Fig. 14c).

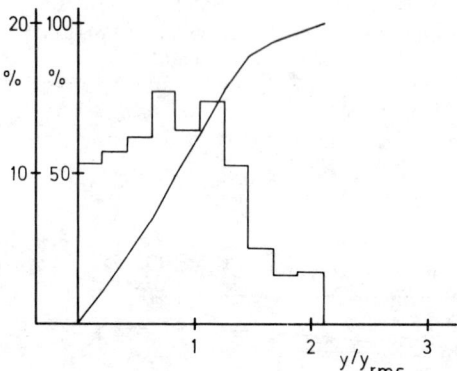
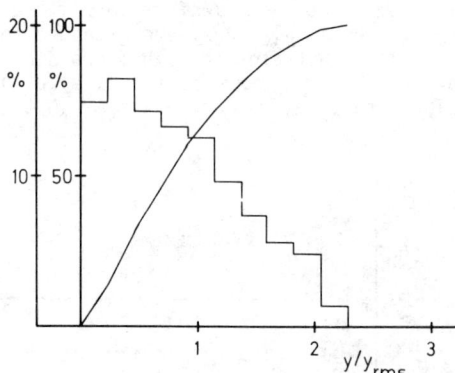

Fig. 14a Probability density and probability distribution of the linear response of Fig. 13a

Fig. 14b Probability density and probability distribution of the nonlinear response of Fig. 13b

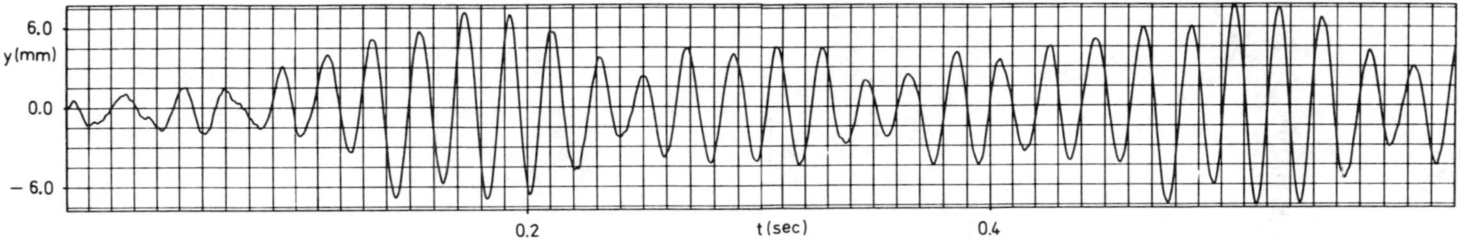

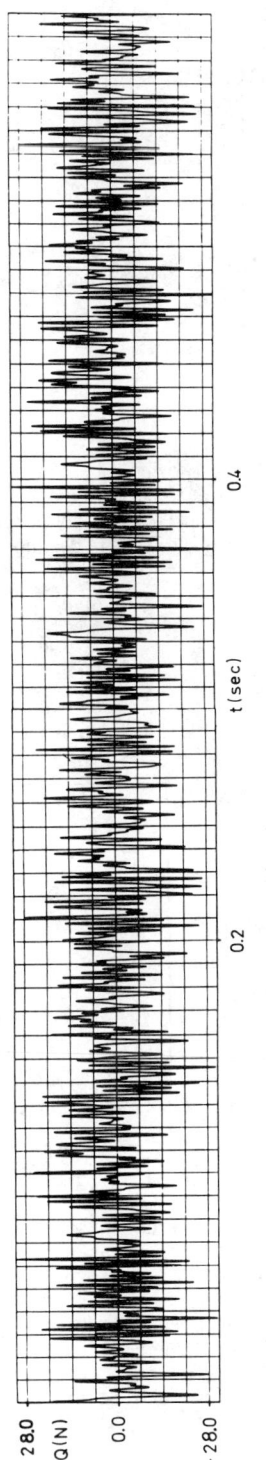

Fig. 13a Linear deflection time history in the middle of the tube due to the exciting force described in Fig. 13c

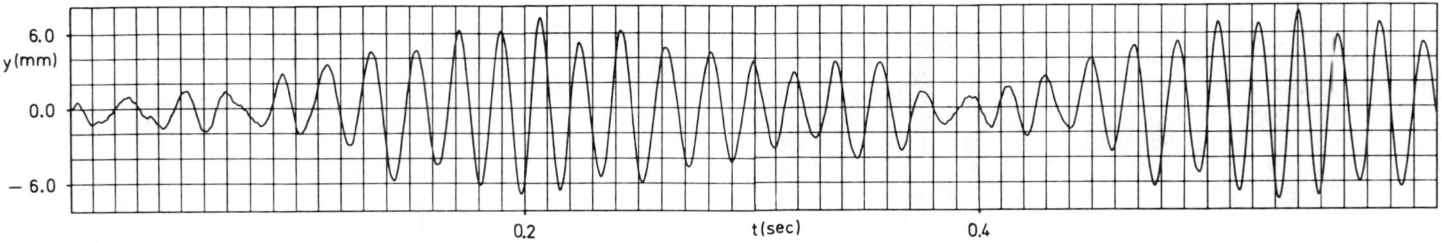

Fig. 13b Nonlinear deflection time history in the middle of the tube due to the exciting force described in Fig. 13c (Nonlinearity not very strong)

Fig. 13c Force time history as a result of the simulation procedure

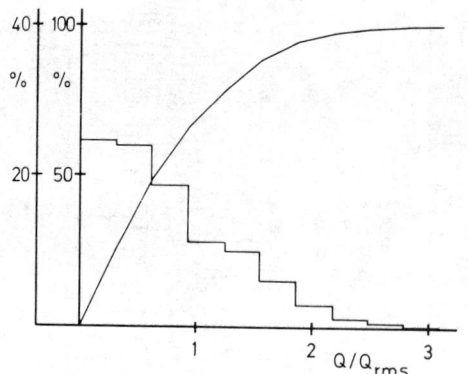

Fig. 14c
Probability density and probability distribution of the force time-history in Fig. 13c

Fig. 15 summarizes all the numerical calculations achieved with geometrical nonlinearity and compares them to theoretical results of Seide /5/. S describes the influence of the nonlinear effect to the response with

$$S = \frac{4}{\pi^5} \frac{S_0}{EA\,\gamma} \left(\frac{L}{r}\right)^4 \qquad (17)$$

where S_0 = power-spectral-density of the pressure field and γ = damping coefficient (definition see /5/).

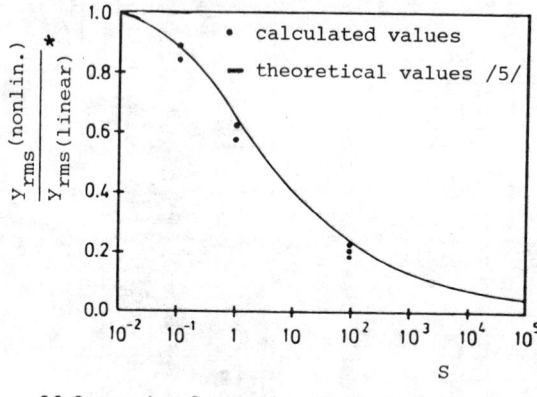

Fig. 15
Effectiveness of the presented method compared to theoretical results /5/ in the case of geometrical nonlinearity

* after Seide /5/

If S remains lower than 0.01 the effect of geometrical nonlinearity may be neglected.

Apart from this the effect of loose supports (Fig. 2a) on the behaviour of the structure was also studied. Using the mechanical model shown in Fig. 2b the induced force vector $\{P_i\}$ of equation (6) now consists of the forces on the tube due to the springs k. Fig. 16a reports the linear response of a tube excited by a random uniform pressure (Fig. 7) with a constant force spectrum up to 48 Hz. In Fig. 16b the corresponding nonlinear response of the tube with a loose support at the centre (K = 10 000 N/m, e = 1 mm) is shown. The two curves differ very much in maximum displacement value and the dominating vibration frequencies.

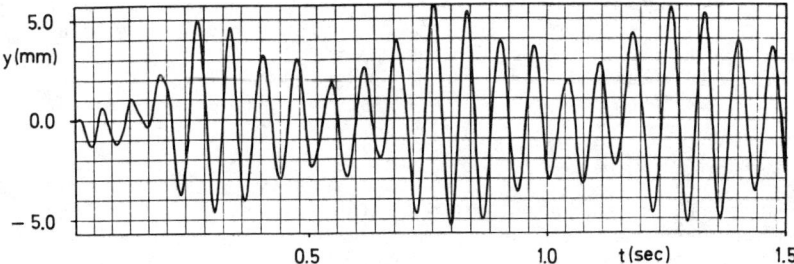

Fig. 16a Displacement time-history of the middle of the tube in the linear case due to the force described in Fig. 16c

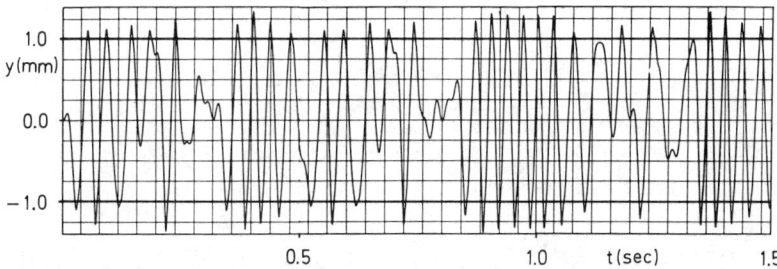

Fig. 16b Displacement time-history of the middle of the tube in case with loose support at the center

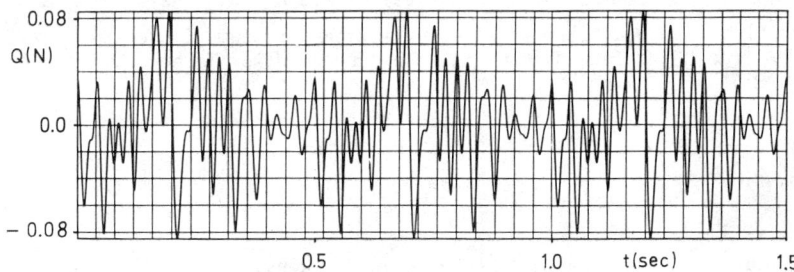

Fig. 16c Force time-history as a result of the simulation procedure

5. CONCLUSION

A computer program was developed which allows an accurate mechanical modelling of heat exchanger tubes. Three dimensional vibration problems with several boundary conditions of the vibrating structure can be tackled. It is possible to solve nonlinear problems due to geometrical nonlinearities as well as nonlinear constraints as in the case of loose supports.

Furthermore prestress of the tubes can be taken into account.[*] The program requires the flow induced forces as input. Any time functions of the forces given may be handled. An option exists to consider harmonic or periodic forces in a convenient manner. If only spectral information of the forces is available as in the case of random excitation it is possible to generate time functions of the forces with a simulation procedure. This way different

---

[*] Prestressing effects due to the flow in the tube as well as internal pressure may be considered.

possible flow induced excitation mechanisms (harmonic, periodic, random or any given function) may be considered. The method gives rms- and maximum values of the response and calculates probability distributions of the deflections and the stresses of the tubes which enables the designer to estimate the danger of structural fatigue. The method was proved for several problems in a satisfactory manner.

6. ACKNOWLEDGEMENTS

The research work was supported by the Kernforschungsanlage Jülich GmbH, Germany, and was performed in close connection with the Institut für Reaktorbauelemente of the institution mentioned above.

Furthermore the author would also like to thank Burkard Esser for his assistance in programming. The numerical calculations were done on the CYBER 175 computer of the computer-center of the RWTH Aachen.

REFERENCES

/1/ Hartlen, R.T., Recent Field Experience with Flow Induced Vibration of Heat Exchanger Tubes, in Proceedings of the International Symposium on Vibration Problems in Industry, Keswick, England, April 10-12, 1973, U.K. Atomic Energy Authority, England 1973.

/2/ Blevins, R.D., Flow-Induced Vibration, Van Nostrand Reinhold Company, New York 1977.

/3/ Herbert, R.E., Random Vibrations of a Nonlinear Elastic Beam, Journal of the Acoustical Society of America 36, Nov. 1964, No. 11, pp. 2090-2094.

/4/ Herbert, R.E., The Stresses in a Nonlinear Beam Subject to Random Excitation, Int. J. Solids and Structures 1, 1965, pp. 235-242.

/5/ Seide, P., Nonlinear Stresses and Deflections of Beams Subjected to Random Time Dependent Uniform Pressure, Israel J. of Technology 13, 1975, pp. 143-151

/6/ Martins, H., Stationäre stochastische Biegeschwingungen gerader und schwach vorverformter Balken, Dissertation Berlin 1976.

/7/ Iwan, W.D. - Yang, I-Min., Application of Statistical Linearization Techniques to Nonlinear Multidegree-of-Freedom Systems, J. of Applied Mechanics, June 1972, pp. 545-550.

/8/ Wilson, E.L. - Farhoomand, I. - Bathe, K.J., Nonlinear Dynamic Analysis of Complex Structures, Earthquake Engineering and Structural Dynamics 1, 1973, pp. 241-252.

/9/ Harris, C.M. - Crede, C.E., Shock and Vibration Handbook, Second Edition, New York McGraw Hill B.C. 1976, p. 7-18.

/10/ Woinowsky-Krieger, S., The Effect of an Axial Force on the Vibration of Hinged Bars, J. of Applied Mechanics, March 1950, pp. 35-36.

APPENDIX: NOMENCLATURE

A = cross-sectional area of the beam, m²

a = amplitude of the 1. mode of the beam, m

b_o, b_2, b_4, b_5, b_6 = constants in analogy to /8/

$[C]$ = damping matrix

D_i = damping factor of the i-th mode, dimensionless

E = Young's modulus, N/m²

NONLINEAR VIBRATIONS OF PRESTRESSED HEAT EXCHANGER TUBES

e	=	backlash of the loose support, m
F	=	prestressing force, N
F_{cr}	=	critical load for the beam, N
F_i	=	induced axial force due to geometrical nonlinearity, N
f	=	eigenfrequency of the prestressed beam, Hz
f_i	=	eigenfrequency of the i-th mode, Hz
f_{max}	=	maximum frequency of the force spectrum, Hz
f_o	=	eigenfrequency of a beam without prestress or nonlinear effects, Hz
I	=	moment of inertia, m^4
$[K]$	=	linear stiffness matrix including prestressing effects
$[K]^*$	=	nonlinear stiffness matrix including prestressing effects
$[\tilde{K}]$	=	effective dynamic stiffness matrix
k	=	equivalent stiffness of the springs of the loose support, N/m
L	=	length of the beam, m
$[M]$	=	structural mass matrix
M_{int}	=	internal bending moment due to ΔM_{int}, Nm
ΔM_{int}	=	induced bending moment due to geometrical nonlinearity, Nm
$\{P\}$	=	force vector
$\{P_i\}$	=	induced force vector which considers nonlinear effects
$Q(t)$	=	force acting at each node of the beam-model (uniform pressure), N
$\{\tilde{R}\}$	=	effective load vector
r	=	$\sqrt{I/A}$ = radius of gyration, m
t	=	time, sec
Δt	=	time interval, sec
u,v,w	=	displacements of a node within the displacement vector
x,y,z	=	cartesian coordinates, m
$\{y\}$	=	displacement vector
$\{\dot{y}\}$	=	velocity vector
$\{\ddot{y}\}$	=	acceleration vector
$\{\tilde{y}\}$	=	effective displacement vector
y_{rms}	=	rms-value of the displacement in the middle of the beam, m
α	=	damping factor, sec^{-1}
β	=	damping factor, sec
θ	=	factor in analogy to /8/, dimensionless
$\theta_x, \theta_y, \theta_z$	=	rotations of a node within the displacement vector
τ	=	$\theta \cdot \Delta t$ = time interval, sec

AIR COOLED HEAT EXCHANGERS

State of Art for Design of Air Cooled Heat Exchangers with Noise Level Limitation

P. PAIKERT and K. RUFF
GEA-GmbH
Bochum, FRG

ABSTRACT

The relations between aircooled heat exchangers and their sound propagation are shown. Primary and secondary measures are described to reduce the noise to a required level. Examples of units in service and the results of their noise reduction are shown.

1. INTRODUCTION

Aircooled heat exchangers require large airflow rates for heat evacuation.

In designing and building aircoolers, the air inlet and outlet must be unhampered, especially when planning complete plants.

For this very reason, aircoolers are often mounted on the top of process plants or factory buildings. These locations give unhampered cooling air inlet; however, they allow free propagation of noise created by the cooling air supply fans.

Considering the large cooling air flow rates, aircooled heat exchangers are generally driven by either forced or induces draft axial fans.

There are many ways to combine heat exchangers and axial fans.

Fig. 1 shows some basic arrangements of aircooled heat exchangers with forced and induced draft axial fans.

Fig. 2 represents an aircooler mounted on top of a twin-column for condensing and cooling of the column products (induced draft).

Fig. 3 shows aircooled condensers on top of a power station (roof type) with forced draft cooling air fans.

Instead of cooling air fans consuming about 3 % of the energy generated, natural draft cooling towers are used for major heat capacities above 100 MW.

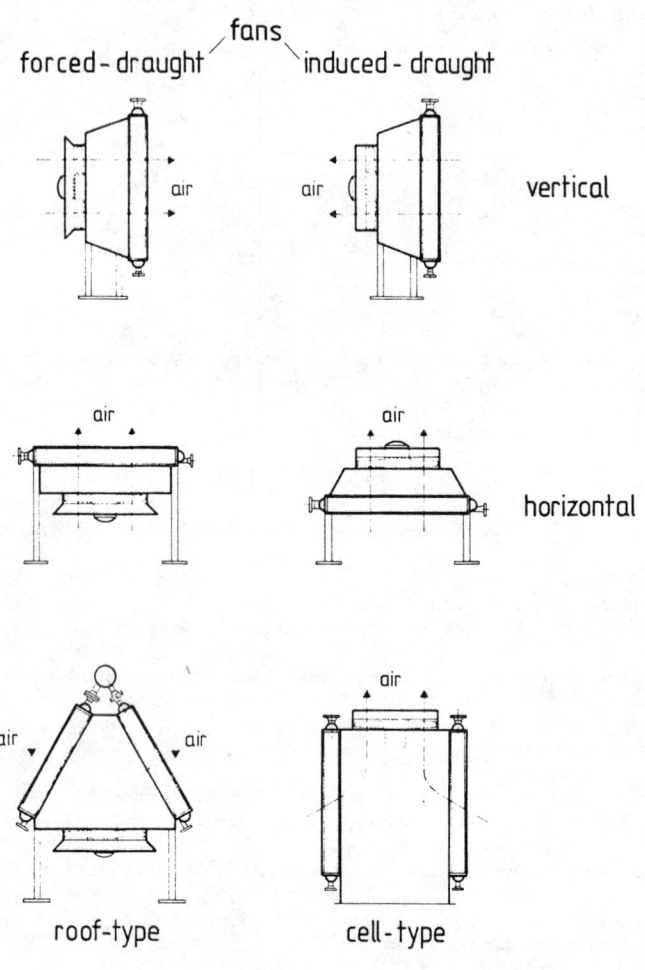

Fig.1 **Arrangement of Air-Cooled Heat Exchangers**

Fig. 2 Air Cooled Column Top Condenser

Fig. 3 Air Cooled Condenser for 160 MW Power Station

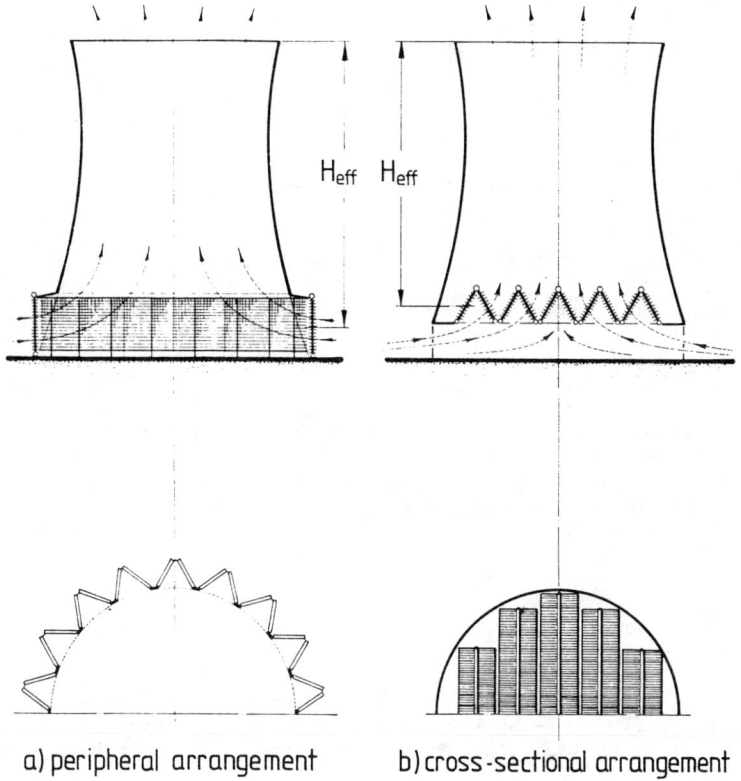

a) peripheral arrangement   b) cross-sectional arrangement

Fig. 4 <u>Arrangement of cooling bundles in natural draft dry cooling towers</u>

As almost any noise radiated by the aircooler is produced by the fan and its driving units, noise problems are generally unknown when a natural draft cooling tower is used. The required cooling tower height, however, is 100 m and above. For various reasons, cooling towers of such height are in many cases not permitted. The problem could possibly be solved by reducing the tower height and installing additional fans to compensate for the decreased natural draft. This alternative, however, results in a noise level increase.

Fig. 5 shows such a combination of natural and forced draft, ther pertinent cooling tower height being 80 m.

As unhampered air inlet requires an exposed installation, and in view of the aircooler size, a completely shielded installation with noise insulation cannot be provided. Sound level suppression is thus to be achieved by applying primary control means at the fan and its drive, cutting the sound level by up to 20 dB (A).

By applying correct secondary control means to an aircooled plant, it is possible to suppress the sound to any desired level.

2. PRIMARY NOISE CONTROL MEANS

Fan noise is produced by interrupted airflows at high velocities and by air turbulence. Its frequency is relatively low and is a result of blade number and rpm. The most efficient manner to attenuate the noise is by reducing the speed and thus the tip speed.

For speed reduction within the range of 30 to 40 kW, a V-belt drive is generally used. Proper design of the drive unit guarantees that any sound level increase will be avoided. Different types of reduction gears are used for the high hp-range. These gears are either installed at grade with a lengthened fan drive shafts, or mounted directly on a supporting structure at the fan.

Fig. 6 shows some standard fan drives. The top illustration represents a direct coupling between motor and fan, commonly used for both small fans up to 1.5 m dia. and large fans, if a special multi-pole motor is employed.

The lower illustration shows the well-known and proven V-belt drive for drive ranges up to 30 - 40 kW. The following two illustrations represent a mechanical reduction gear with direct coupling or with lengthened drive shaft.

Electric motors for direct driving of axial fans can be provided as extremely low-noise multi-pole unit up to 100 rpm. Reduction gears for driving of large fans using standard electric motors of 1000 or 1500 rpm are designed as low-noise units in form of spur wheel gears, planetary gears or worm gears. The workmanship is a decisive factor for the gear sound level. With poor workmanship, the sound level of the gear will be above that of the fan. A completely enclosed installation of the gear unit is often not very effective due to the sound radiated from solids over the large area fans. Isolating the gear unit and fan by way of insulation is hardly possible because of the required power transmission.

As noise level is produced in the fan, its level can be modified by choosing of another fan design, stressing that the main duty of the fan

AIR COOLED HEAT EXCHANGERS WITH NOISE LEVEL LIMITATION

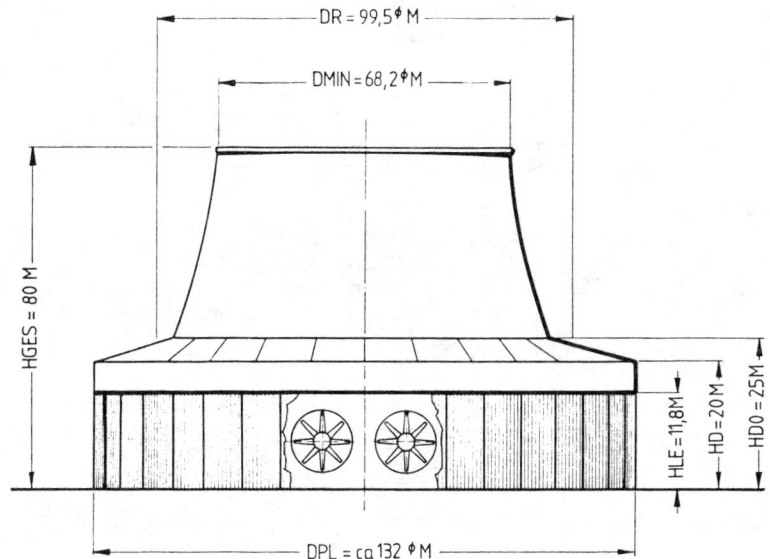

Fig. 5 M. 1 : 1000

Hybrid Cooling Tower with Forced and Natural Air Flow

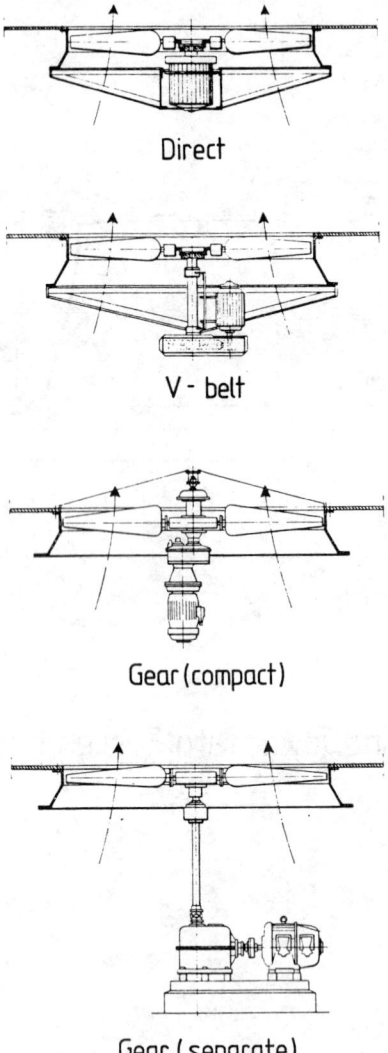

Fig.6 <u>**Normal Fan Drives**</u>

AIR COOLED HEAT EXCHANGERS WITH NOISE LEVEL LIMITATION

rests the supply of cooling air in ample volume and against the heat exchanger flow resistance. Measures taken to reduce the sound level necessitates appropriate countermeasures to safeguard the cooling air supply.

The tip speed is a decisive factor for noise generation and needs to be reduced to a minimum. To compensate for the reduced airflow, the number of blades and/or the blade chord must be increased.

The noise generation of a fan has a strong bearing on its efficiency. Thus, all measures taken to avoid shedding of vortices in the fan area favour both noise limitation and fan efficiency, especially as regards the inlet duct of the fan.

Sound power level is the measure for sound propagation of a fan, defined as fan sound power radiated to the environment. It is identical with the sound pressure referred to a reduced surface of 1 m^2.

The sound pressure level, on the contrary, is a subjective value of the measured sound pressure, referred to a random basic value. It is highly dependant upon the distance to the sound source.

Besides the tip speed of a fan, volume flow, static pressure difference ans fan diamter affect the sound power emitted by any type of fan, whereas both fan type and design are characterised by a constant value. The following correlation shows the bearing of the various parameters on the sound power level of an axial fan on the basis of empirical and theoretical laws:

$$L_w = c + 30 \log w + 10 \log \frac{\Delta p_{stat} \cdot \dot{V}}{1000} - 5 \log D \quad dB\ (A) \quad (1)$$

L_w	=	Sound power emitted	
c	=	Specific basic level 44 dB (A), standard design 37 dB (A), low noise design	
w	=	Tip velocity	m/s
Δp_{stat}	=	Static pressure rise	Pa
$\dot{V}$	=	Air quantity	m^3/s
D	=	Fan diameter	m

Fig. 7: Evaluation of Sound Power Level

It is shown that the tip velocity is the most important factor, its bearing being three times that of volume flow or static pressure. By increasing the fan diameter, the sound pressure level is slightly attenuated.

Fig. 8 is a diagrammatic representation, comparing different fan blade types designed for various sound level requirements. With reduced tip velocity, blade width and the hub ratio are increased to realize lower sound levels at unchanged volume flow.

The increase of the sound level due to several uniform or non-uniform sound sources is governed by the correlation stipulated in Fig. 9.

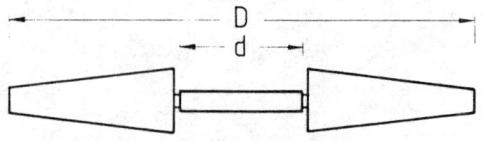

1. Standard sound level requirements
   w = 50-60 m/s
   n = 3/4/6 Blades
   d/D = 0,23 Hub ratio

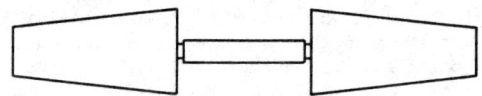

2. Mean sound level requirements
   w = 40-45 m/s
   n = 4/6 Blades
   d/D = 0,30 Hub ratio

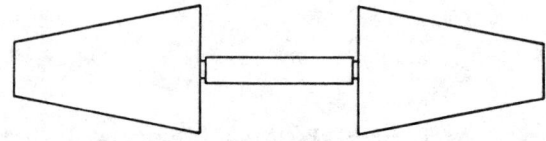

3. Stringent sound level requirements
   w = 28-32 m/s
   n = 8 Blades
   d/D = 0,36 Hub ratio

Fig.8 **Fan types used in GEA-recooling plants**

AIR COOLED HEAT EXCHANGERS WITH NOISE LEVEL LIMITATION

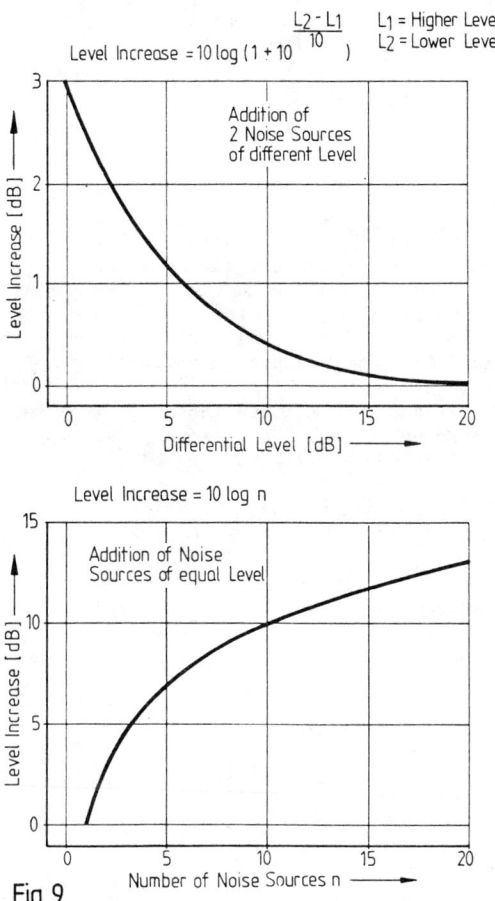

Fig. 9
Sound Power Level Determination

$$L_w = 10 \log \left(1 + 10^{\frac{L_2 - L_1}{10}}\right) \quad (2)$$

where L_1 = high level
L_2 = low level

This shows that for two uniform levels $L_2 = L_1$ the exponent in the second term of the expression is 0 and the term itself becomes 1. Sound increase for two uniform levels is consequently:

$$\Delta L_w = 10 \log 2 \quad (3)$$

The sound increase for n uniform levels:

$$\Delta L_w = 10 \log n \quad (4)$$

Determination of the sound power of an aircooler is made by measuring the sound pressure level at given distance from the sound source. The sound pressure level is calculated in relation to the reduced surface of the sound source according to the sound propagation law.

As the sound propagation is often asymmetrical, the measurement is to be repeated at several points around the sound source, and the arithmetic mean value is to be calculated from all readings. For final sound power level determination, the background sound level is to be taken into account. This is ascertained and recorded separately prior to the actual aircooler measurement.

Following to the law of sound source addition (Equation 2), the background sound level is to be deducted from the measured value.

From the corrected mean value of the sound pressure level and the size of the measuring surface A (m^2), the sound power level of the plant is calculated as follows:

$$L_w = L_p + 10 \cdot \log A \quad (5)$$

when L_p = Sound Pressure Level dB (A)

Normally, a tolerance of 2 dB is acceptable with respect to the reading.

3. SOUND PROPAGATION LAW

From the sound power level of an aircooler, the noise level at any distance can be determined by applying the sound propagation law. With unobstructed sound propagation the noise decreases with the square of the increased distance. If the source is mounted on the ground, a hemispherical propagation can be assumed in the ideal case.

Sound Propagation law:

$$L_p = L_w - 10 \log (2 \cdot \pi \cdot R^2) \quad (6)$$

where R = distance from source m

The relation shows that as the distance from sound source is doubled, the sound pressure level decreases by 6 dB.

Fig. 10 illustrates the law for sound attenuation with distance, and that this law is not strictly observed in practice. A minor decrease is observed near the aircooler (about 4 dB), and a major decrease at greater distances (about 7 dB).

For detail calculations, further factors must be taken into account, namely absorption due to ground screening, attenuation effect of the air, and influence of meteorological conditions, speaking of large-scale plants when sound propagation over a distance of several kilometers is to be considered.

4. SECONDARY NOISE CONTROL MEANS

Secondary noise control means are generally considerably more expensive than primary means. They should, as a consequence, only be considered if a further level decrease cannot be achieved by applying the described primary means or other methods, e.g. part-load operation during night hours. This possibility is often not made use of although the night temperatures are generally far below the design temperatures, and although part load operation (60 - 70 %) with corresponding level decrease of 8 to 10 dB could be realized.

A simple control means to shield the direct neighbourhood is the installation of sound absorbing screens of appropriate size and arrangement.

Fig. 11 shows an arrangement sketch and the attenuating effect of a screening (aircooler) to shield the residential building nearby.

Fig. 12 and Fig. 13 show the scheme and a photograph of a water recooling plant of 2 x 54 MW capacity provided with sound absorbing screens.

Fig. 14 is a view behind the attenuators towards the large number of cooling air fans. The attenuating effect of sound absorbing screens is shown in Fig. 15

5. SOUND ABSORPTION MEANS

If the sound power level has to be reduced considerably, the plant should best be equipped with sound attenuators. About 2/3 of the sound power is emitted directly by the fan of the aircooler, and about 1/3 by the cooling bundles. If a sound suppression of 5 to 7 dB (A) is required, it will generally suffice to equip the fan inlet area with sound baffles. For final attenuator calculation and design, the advice of expert companies should be asked for.

For preliminary estimates on the effect of sound damping baffles the aircooler designer may use the graphs and curves of Fig. 16.

Fig. 16 shows sound absorption per m flow length dependent upon baffle distance and sound frequency for two different baffle types.

Fig. 17 and Fig. 18 illustrates induced side, and induced/forced side sound absorption by means of baffles.

Fig. 19 shows the baffle design on induced and forced side of a hybrid forced draft cooling tower (using axial fans) in both the wet and the dry tower section

Fig. 20 is a comparative summary of the various means of attenuation.

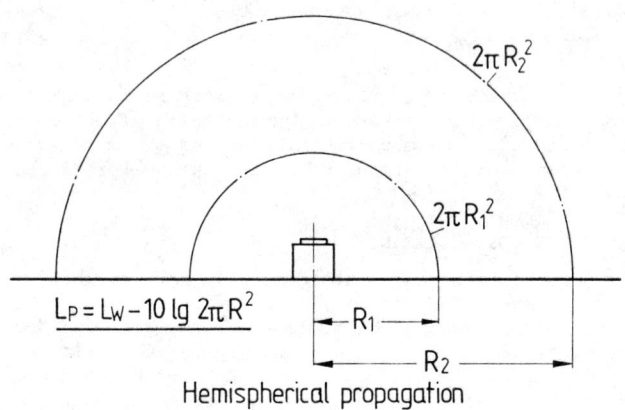

Hemispherical propagation

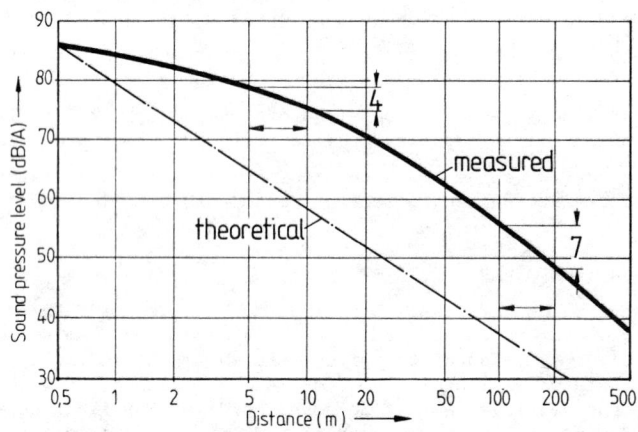

Noise level decrease measured on recooling plants

Fig. 10 **Sound Propagation**

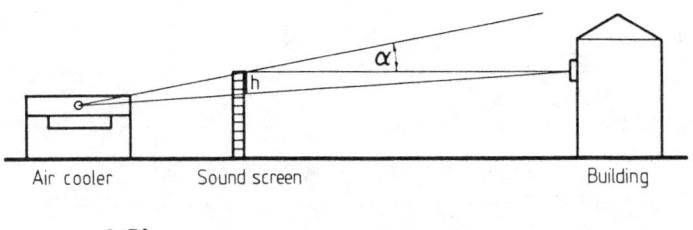

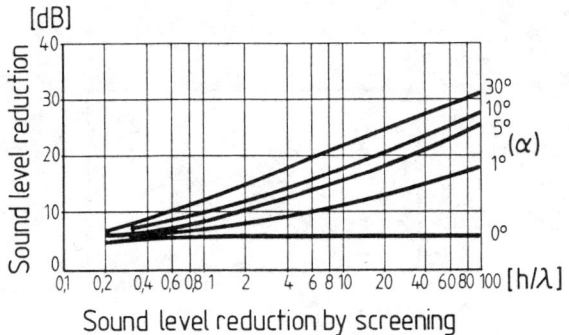

Sound level reduction by screening

λ = Wave lenght = $\frac{c}{f}$

The following values apply to ambient air of 18°C:

f	63	125	250	500	1000	2000	4000	8000 Hz
λ	5,3	2,7	1,33	0,67	0,34	0,17	0,08	0,04 m

Fig. 11
Sound screens and their sound level reduction

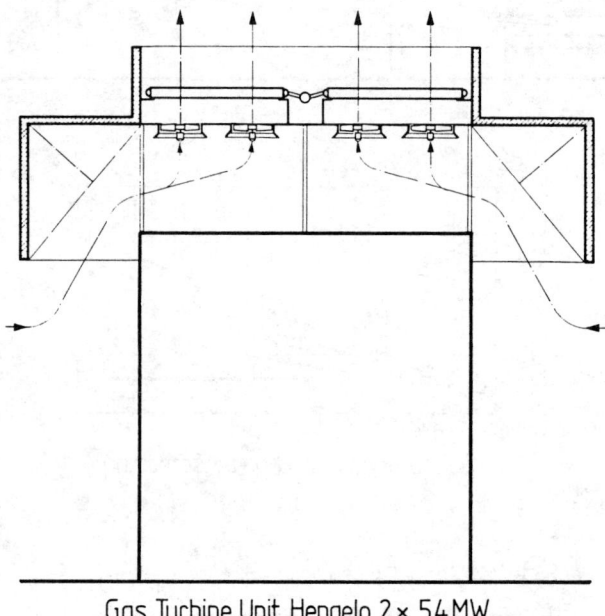

Gas Turbine Unit Hengelo 2 × 54 MW

Fig.12 <u>Recooling plant using screens</u>

Fig.13 View on Air Cooled Recooler with Sound Screens acc. to Fig. 12

Fig.14 View behind the Sound Screens of the Gas Turbine Unit acc. to Fig. 12, 13

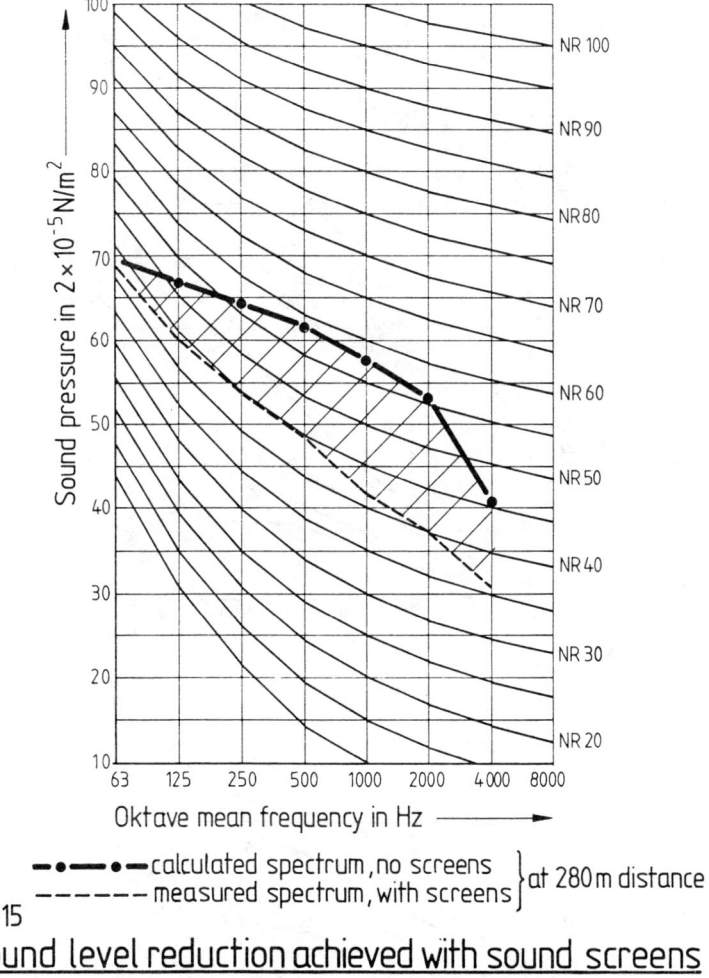

— • — • — calculated spectrum, no screens
– – – – – – measured spectrum, with screens } at 280 m distance

Fig. 15
Sound level reduction achieved with sound screens at Gas Turbine Unit Hengelo

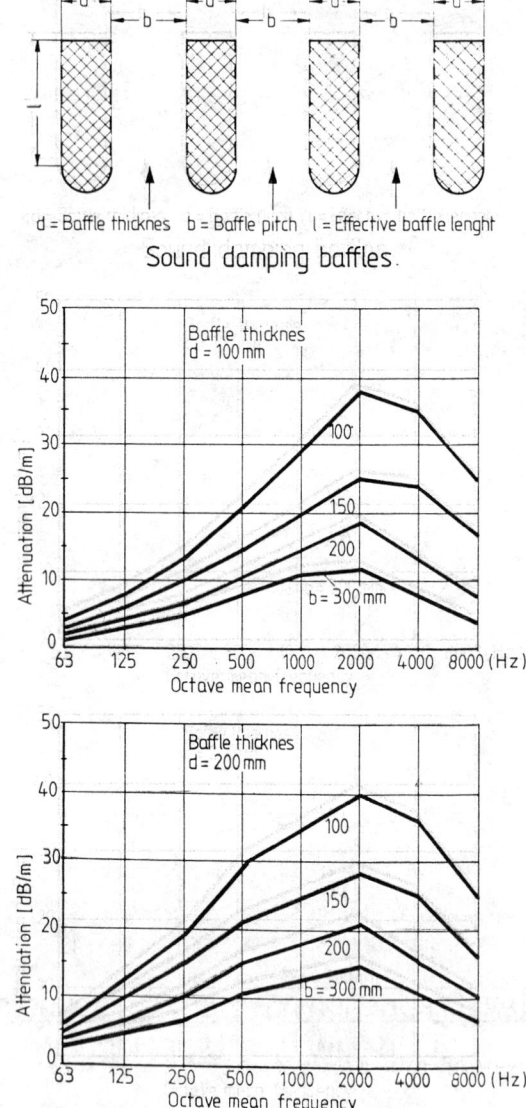

Fig.16 **Sound damping baffle design**

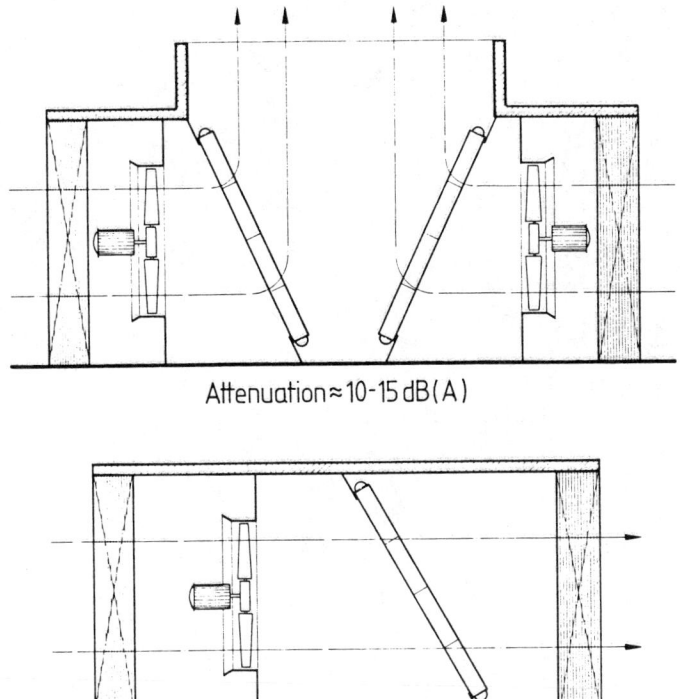

Fig. 17
GEA-Recooling Plants with sound damping baffles

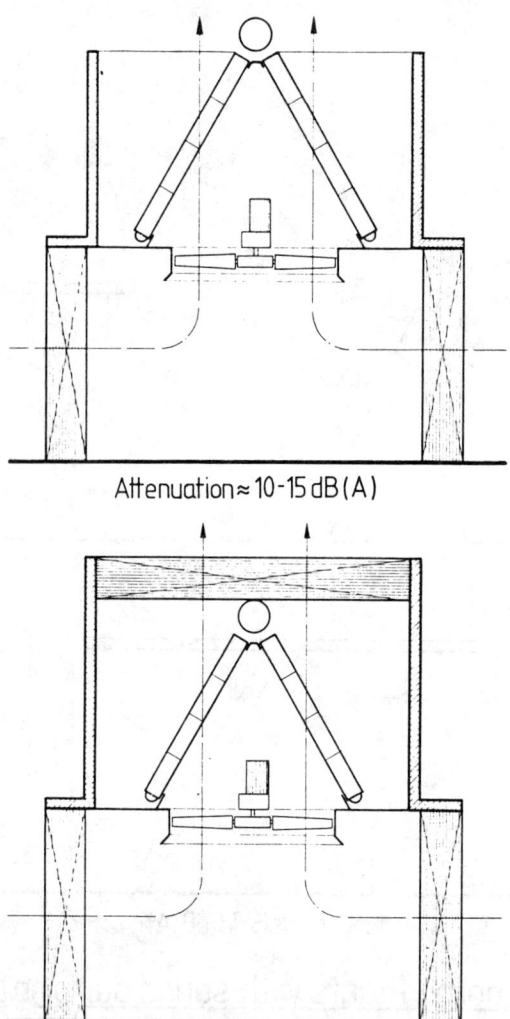

Fig. 18
Attenuation ≈ 15-40 dB(A)
Air Cooled Condensers with sound damping baffles

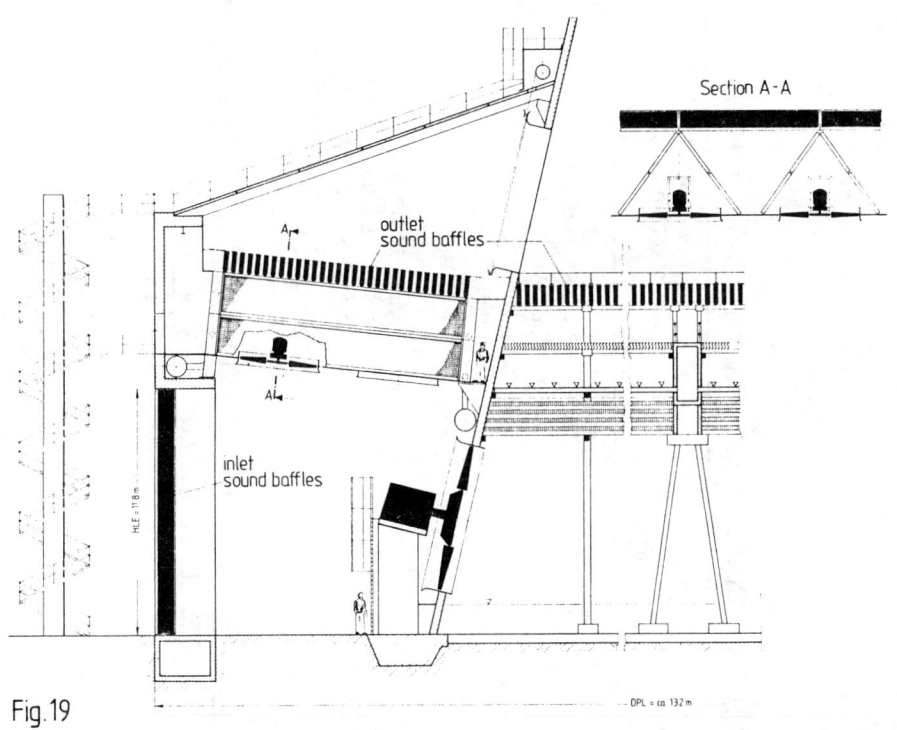

Fig. 19
Hybrid Cooling Tower with forced Air Flow and Sound Baffles

Example: One air cooled condenser with two 6,3m diam. fans

Arrangement	Sound Level 10m distance dB(A)	Measures	Power required %	Costs %
Normal assembly	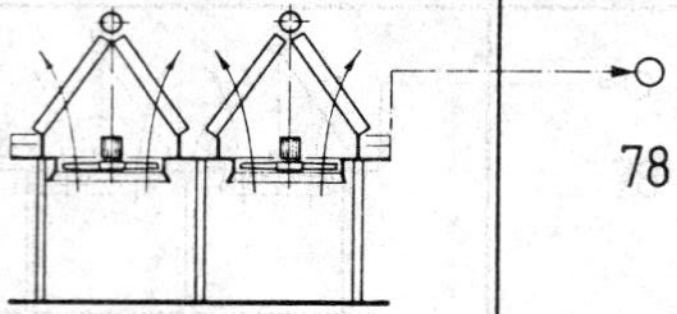 78	Optimum fan design and air flow will improve the level by 2 - 3 dB (A). Use screening effect of large buildings or earth walls, if any.	100	100
Increased cooled surface	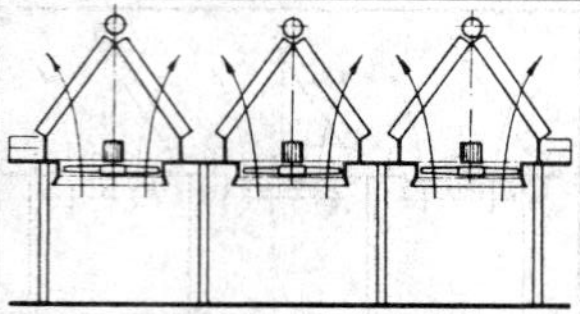 70	Reduced cooling air flow and discharge pressure at simultaneously increased cooling surface to compensate for deteriorated heat transfer coefficient.	40	130 ÷ 140

Fig.20a Measures to decrease the sound level of air cooled plants (equal cooling capacity provided)

Example: One air cooled condenser with two 6,3m diam. fans

	Arrangement	Sound Level 10m distance dB(A)	Measures	Power required %	Costs %
Low noise fans	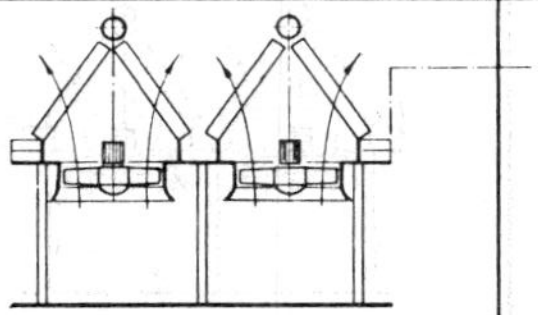	65	Fans with 30-40m/s tip speed, wide blades and large hub ratio, low noise motors and low noise reduction gears	100	110
Low noise fans and sound attenuators	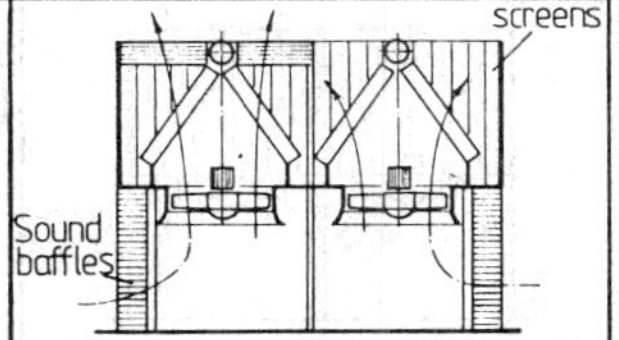	55 ÷ 40	Sound attenuation at suction and discharge sides. Costs and power as a funktion of required attenuation.	120 ÷ 150	130 ÷ 140

Fig. 20b **Measures to decrease the sound level of air cooled plants (equal cooling capacity provided)**

Noise and Its Influence on Air Cooled Heat Exchanger Design

P. E. FARRANT
National Engineering Laboratory
East Kilbride
Glasgow, UK

ABSTRACT

An equation is derived for air-cooled heat exchangers relating the exchanger air pressure drop, flowrate and density, the fan non-dimensional parameters specific speed and load coefficient and the fan total sound power level. An example is given where a reduction of the total sound power level in excess of 12 dB (re 10^{-12} W) could be expected for a noise optimized exchanger design in comparison to a lowest annual cost design. A method is presented for readily selecting and comparing different fans for an exchanger.

1 INTRODUCTION

The relationship between process plants and surrounding communities is becoming subject to more and more regulations. Part of these regulations frequently specify the maximum permissible noise levels at the plant perimeter fence. The source of a significant proportion of the noise reaching the surrounding community can be the fans associated with air-cooled heat exchangers (ACHEs). ACHEs tend to be placed in elevated banks to achieve unrestricted air flows. Consequently the fan noise is broadcast efficiently over wide areas and sound barriers such as walls and belts of trees have little effect. The reduction of ACHE fan noise emission has to be tackled at source, namely the ACHE fans, and as pointed out by Bijl and others[1] the best parameter for use in the control and specification of noise emission from ACHEs is the sound power level.

In this paper an equation is derived relating the fan total sound power level to significant fan and tube bundle parameters. Two ACHE designs for a particular duty are then outlined and compared. One design is optimized for lowest annual cost and the other on the basis of the derived sound power equation.

2 MINIMIZING FAN OVERALL SOUND POWER LEVELS

Deeprose, Bolton and Margetts[2] have proposed an equation for ACHE fan overall sound power level based on measurements for a wide range of axial-, mixed- and radial-flow fans.

The predicted total sound power level is

$$L_w = 10 \log(p_s \dot{V}) + 22 \log u_t + 26 \text{ dB (re } 10^{-12} \text{ W)}. \qquad (1)$$

This equation can also be expressed in the form

$$\hat{W}_{sound} = 4 \times 10^{-10} \times u_t^{2.2} p_s \hat{V}, \qquad (2)$$

where $\hat{W}_{sound}$ is the total sound power, W,
u_t is the fan tip speed, m/s,
p_s is the fan static pressure, Pa, and
$\hat{V}$ is the fan volumetric flowrate, m³/s.

Farrant[3] has shown that a useful coefficient for classifying the performance of axisymmetric fans and pumps is the load coefficient K_L which relates the useful power ($p\hat{V}$) produced by a machine at the best total efficiency point to the working fluid density ρ, impeller angular velocity $\dot{\phi}$, and the casing radius $D_t/2$. p is the total pressure rise through the machine. K_L is defined as

$$K_L = \frac{p\hat{V}}{\rho \dot{\phi}^3 (D_t/2)^5}. \qquad (3)$$

Most commercial fans tend to have load coefficients in the lower half of the range $0.04 < K_L < 0.08$. This tendency is independent of fan specific speed or type number N_s where

$$N_s = \frac{\dot{\phi} \hat{V}^{\frac{1}{2}}}{(p/\rho)^{\frac{3}{4}}} \qquad (4)$$

and p and $\hat{V}$ again correspond to the best efficiency point. For single-stage fans the three main fan types generally fall in the range

centrifugal $N_s < 2.0$,

mixed-flow $1.5 < N_s < 3.5$, and

axial $N_s > 3.0$.

In the ACHE industry it is usual to refer to fan performances in terms of fan static pressure rather than total pressure. The latter includes the velocity pressure associated with accelerating air from zero to the velocity at the fan outlet which in the usual ACHE amounts to an energy loss. Fan static pressure is predictably smaller than fan total pressure, and like the total pressure it obeys the fan laws relating geometrically similar fans of different diameters and speeds blowing differing density gases. Consequently it is valid to redefine load coefficient and specific speed as follows:

$$K_{Ls} = \frac{p_s \hat{V}}{\rho \dot{\phi}^3 (D_t/2)^5}, \qquad (5)$$

$$N_{ss} = \frac{\dot{\phi} \hat{V}^{\frac{1}{2}}}{(p_s/\rho)^{\frac{3}{4}}}. \qquad (6)$$

NOISE AND ITS INFLUENCE ON AIR COOLED HEAT EXCHANGER DESIGN

For given values of $\dot{V}$, ρ, ϕ and D_t, $K_{Ls} < K_L$ and $N_{ss} > N_s$ and K_{Ls} and N_{ss} can be expected to exhibit the same tendencies as K_L and N_s.

For axial and high performance mixed-flow fans the impeller diameter is almost identical to the casing diameter D_t thus by substituting $u_t = \phi D_t/2$ into equation (5) and rearranging

$$u_t = \frac{\phi^{0.4} p_s^{0.2} \dot{V}^{0.2}}{\rho^{0.2} K_{Ls}^{0.2}}. \tag{7}$$

By substituting equation (7) into equation (2)

$$\dot{W}_{sound} = \frac{4 \times 10^{-10} \times \phi^{0.88} p_s^{1.44} \dot{V}^{1.44}}{\rho^{0.44} K_{Ls}^{0.44}}. \tag{8}$$

From equation (6)

$$\phi = \frac{N_{ss} p_s^{0.75}}{\rho^{0.75} \dot{V}^{0.5}}. \tag{9}$$

Substituting equation (9) into equation (8)

$$\dot{W}_{sound} = \frac{4 \times 10^{-10} \times N_{ss}^{0.88}}{\rho^{1.1} K_{Ls}^{0.44}} p_s^{2.1} \dot{V}. \tag{10}$$

Alternatively the sound power level can be expressed as

$$L_w = 10 \log\left(\frac{\dot{W}_{sound}}{1 \times 10^{-12}}\right) \quad \text{dB(re } 10^{-12} \text{ W).} \tag{11}$$

Equation (10) shows that fan total sound power is minimized by:

a designing the tube bundle and plenum to reduce the product of the air pressure drop raised to the power of 2.1 and the air flowrate to a minimum;

b using fans with the lowest practical specific speed. That is using fans towards the centrifugal end of the centrifugal/mixed-flow/axial fan spectrum; and

c using fans with the highest practical load coefficient.

Fig. 1 shows a plot of sound power level against static air power for two specific speeds and two load coefficients. The left-hand ordinate corresponds to a static pressure drop of 150 Pa whereas the right-hand ordinate is for a pressure drop of 75 Pa. Thus for a fan static air power of 10 kW, a fan static pressue rise of 150 Pa, a static specific speed of 7, and a static load coefficient of 0.04 the predicted sound power level is 102.7 dB. This fan specific speed and load coefficient correspond to an 'average' axial fan. By contrast for the same static air power of 10 kW but with the tube bundle changed so that the fan static pressure rise need only be 75 Pa and with high performance mixed-flow fans fitted with a static specific speed of 2.5 and a static load coefficient of 0.15 the predicted sound power level is 92.9, a reduction of 10 dB.

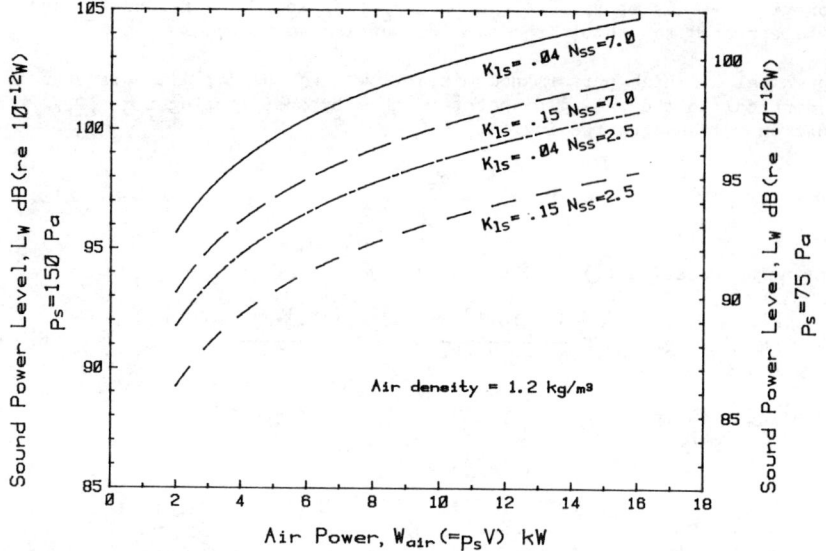

Fig 1 Fan Sound Power Levels

It should be noted that although a curve for $N_{ss} = 7$ and $K_{Ls} = 0.15$ has been included on Fig. 1 single-stage fans corresponding to these values would not be suitable for ACHEs as they would be highly unstable, that is there would be a very pronounced minimum at low flowrates in the pressure/flowrate characteristic.

3. TYPICAL CHARACTERISTICS OF AIR-COOLED HEAT-EXCHANGER FANS

Single-stage axial fans are used almost without exception on the larger ACHEs found on process plant. They are simple and effective, and by the use of variable pitch mechanisms or variable speed drives are readily controllable. They tend to be divided into two classes, one of lower price than the other. The former class tends to have narrow chord low solidity blading, which may be of extruded material. This class of fan is used where noise restrictions are not limiting. Where noise emission might be a problem the latter class of wider chord higher solidity blades, which may be constructed of glass-reinforced plastics, is used. This type of fan can generate a given pressure rise and flowrate at a lower running speed than the former and with the consequent reduction in fan blade tip speed will be correspondingly quieter. In terms of load coefficient and specific speed the former class has a lower load coefficient and a higher specific speed than the latter.

Fig. 2 shows the performance characteristics of three fans tested at NEL. 'Fan 1' is an example of the former class and 'fan 2' an example of the latter. Three pressure/flow and three static efficiency/flow curves for each fan are shown corresponding to three blade pitches. Table 1 lists the general features of the fans including sound power levels at typical operating points. More detailed noise data for these two fans are given by Deeprose[2]. 'Fan 3' is a mixed-flow cooling fan developed for a 'quiet' heavy goods vehicle and has been described by Farrant[4] and Bolton and others[5]. A cross-section of the fan fitted with a diffuser is shown in Fig. 3. While this fan has both a rotor and stator the blading is of a much simpler and more economical form than that for a high performance axial fan. Each blade is constructed from a single layer of thin gauge sheet material, rounded aerofoil shapes being unnecessary. The 'fan 3' characteristic curves on Fig. 2 exemplify the higher pressure and flow coefficients (and efficiency) that can be achieved by a mixed-flow fan in comparison to an axial. This is derived from the centrifugal action of mixed-flow blading which is additional to the aerofoil action of axial blading. The centrifugal action is particularly significant at low air flowrates when the blades are stalled as it can prevent a 'trough' in the fan pressure/flow characteristic. Consequently, while the high flow performance of an axial could be increased to higher pressure coefficients than for 'fan 2' it would be difficult to raise the low flow pressure coefficient significantly. The resulting 'trough' in the characteristic could cause the air flow to surge especially during gusts of wind across the fan inlet.

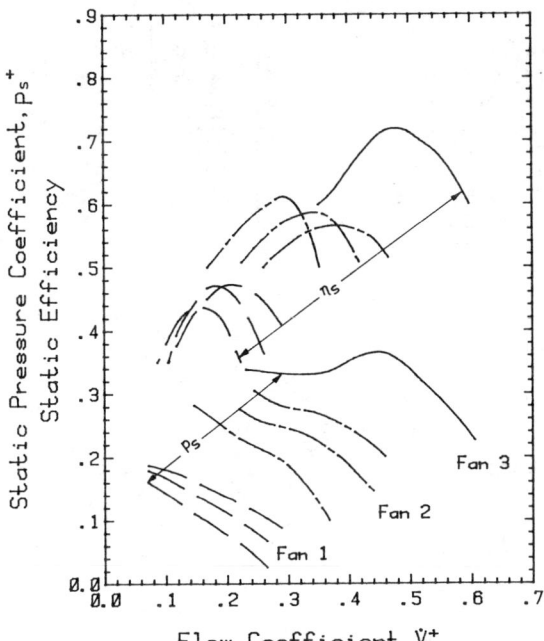

Fig 2 Fan Performance

The coefficients used in Fig. 2 are:

Static pressure coefficient

$$p_s^+ = \frac{p_s}{\rho \dot{\phi}^2 R^2}, \tag{12}$$

Flow coefficient

$$\dot{V}^+ = \frac{\dot{V}}{2\pi\phi R^2 B}, \qquad (13)$$

where R is the mean radius of the fan rotor blade at the trailing edge and B is the blade height at the trailing edge.

TABLE 1

GENERAL FEATURES OF THREE FANS

Fan No	1	2	3
Fan diameter, m	3.66	3.66	0.686
Hub-to-tip ratio (at rotor trailing edge)	0.40	0.45	0.74
Number of rotor blades	6	7	11
Rotor hub chord, m	0.38	0.64	0.24
Rotor tip chord, m	0.29	0.61	0.28
Number of stator blades	0	0	27
Test speed, r/min	28.5	178.5	1400
Flowrate, m^3/s	88.0	88.7	3.62
Static pressure, Pa	120	121	775
Measured total sound power level dB (re 10^{-12} W) (125 Hz to 4 kHz)	111.1	99.0	94.0
Predicted total sound power level dB (re 10^{-12} W) (equation (1))	105	100	98

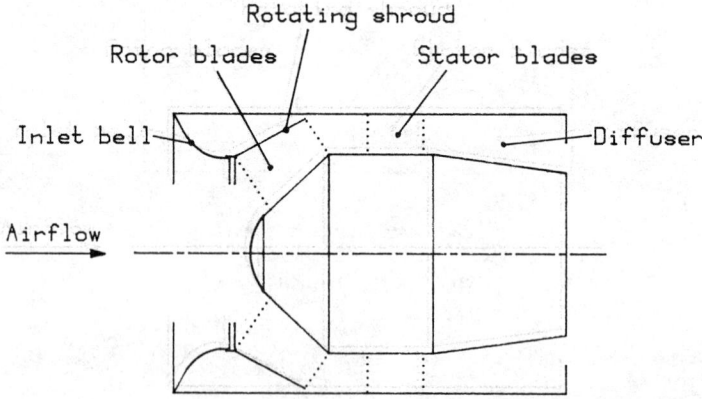

Fig 3 High performance mixed-flow fan

NOISE AND ITS INFLUENCE ON AIR COOLED HEAT EXCHANGER DESIGN

While these coefficients are often used by fan and pump designers for presenting performance data in non-dimensional form, this form is not suitable for matching fans to ACHEs as the coefficient denominators are design dependent. The static specific speed N_{ss} and static load coefficient K_{Ls} are not limited in this way and they provide a means whereby fan characteristics can be presented in a form independent of fan size, rotational speed and air density. Fig. 4 shows a transformation of the Fig. 2 characteristics to an N_{ss} basis. Static efficiency contours have been added.

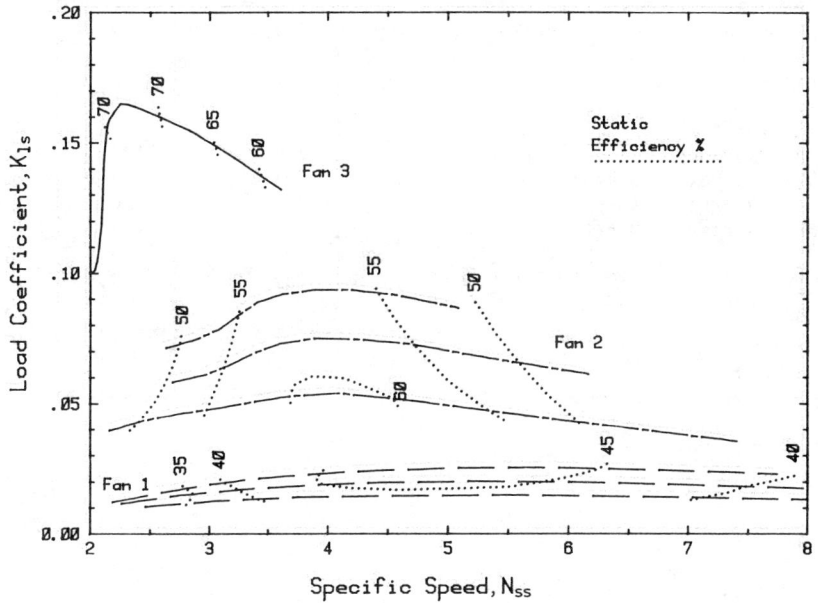

Fig 4 Fan K_{ls}/N_{ss} Characteristics

4. TUBE BUNDLE DESIGN

Equation (10) for sound power level includes the term $p_s^2 \cdot 1 \dot{V}$. The magnitude of this term is largely determined by the tube bundle design. The usual practice is to design an ACHE tube bundle for lowest annual cost or lowest capital cost, not for a minimum value of $p_s^2 \cdot 1 \dot{V}$. To illustrate the effect of designing for different criteria an example of a kerosene cooler is now considered.

A requirement is for an ACHE to cool 113 400 kg/h of kerosene from a temperature of 71°C to 52°C. The design dry bulb air temperature is 35°C. The Heat Transfer and Fluid Flow Service ACHE design program DACE was used to find two bundles to satisfy this requirement: one for lowest annual cost (LAC), the second for the lowest value of $p_s^2 \cdot 1 \dot{V}$ ($Lp_s^2 \cdot 1 \dot{V}$). The principal features of the designs are listed in Table 2 and illustrated in Fig. 9.

TABLE 2

GENERAL FEATURES OF TWO ACHEs FOR COOLING 113 400 kg/h OF KEROSENE FROM 71°C TO 52°C WITH AN AIR TEMPERATURE OF 35°C

ACHE	Lowest annual cost	Lowest $p_s^2 \cdot 1 \dot{V}$
Bundle length, m	9.00	8.00
Bay width, m	2.95	3.67
Number of tube rows	6	6
Number of passes	3	3
Base tube diameter, mm	25.4	25.4
Fin diameter, mm	50.8	57.2
Fin frequency, m^{-1}	276	433
Tube pitch (triangular), mm	53.8	60.2
Total number of tubes	312	336
Bare tube surface area, m^2	224	214
Extended surface area, m^2	2563	4996
Air static pressure drop (for bundle), Pa	153	86
Air flowrate (at density of 1.2 kg/m^3) m^3/s	68.6	57.3
Capital cost ratio	1	1.17
Power cost ratio	1	0.47

5. BUNDLE/FAN MATCHING

In matching fans to a given tube bundle there is usually scope to vary:

a   number of fans,

b   fan design,

c   fan blade pitch,

d   fan diameter with the limitation that the area covered by the fans is not less than 40 per cent of the bundle face area (see API 661[6]), and

e   fan speed, within limitations of maximum blade tip speed (see API 661[6]).

All these parameters can be brought together on one graph as follows.

Let the total area covered by the fans by A_f. Let the number of velocity heads lost in the fan entry, fan guards, plenum expansion losses etc be k. A value of k = 1.1 suitable for forced draught exchangers has been used in this paper. Then the required fan static pressure rise is

$$p_{sf} = p_{sb} + \frac{\rho}{2}\left(\frac{\dot{V}}{A_f}\right)^2 k$$

where p_{sb} is the tube bundle pressure drop.

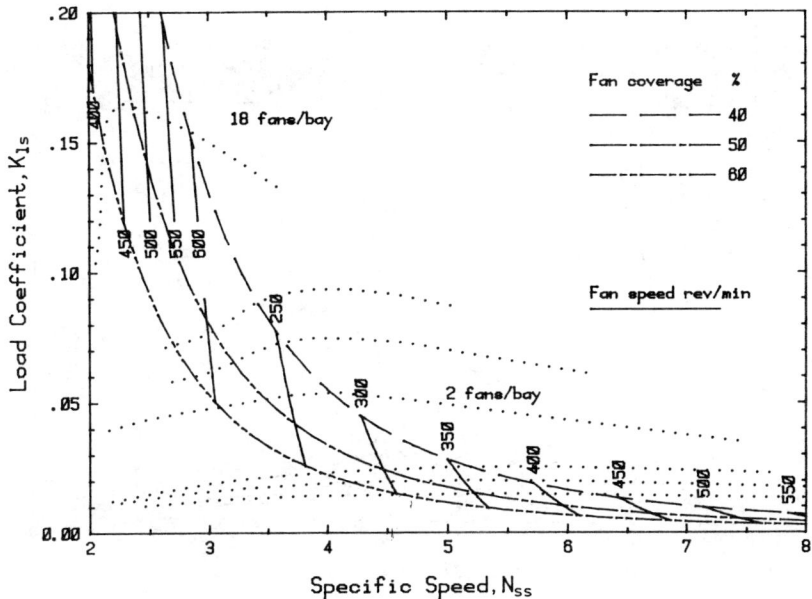

Fig 5 LAC ACHE fan selection

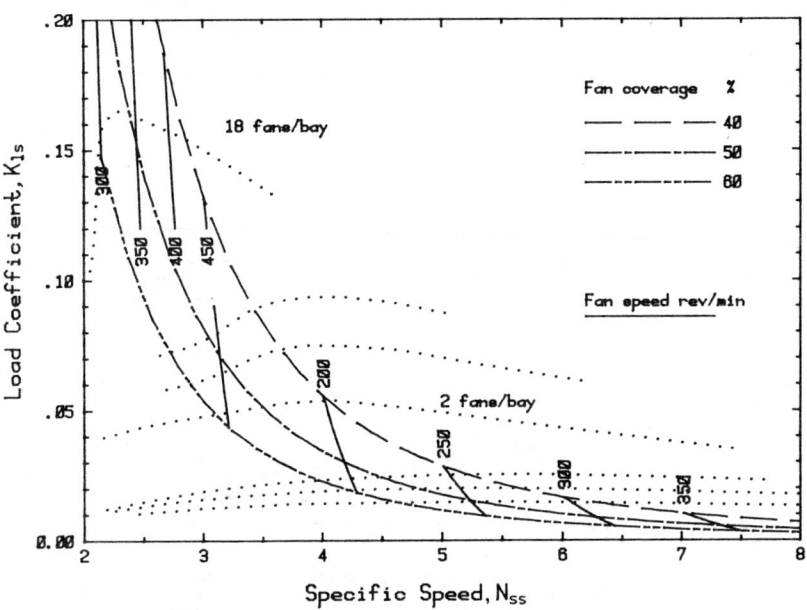

Fig 6 L.$p_s^{2.1}$V ACHE fan selection

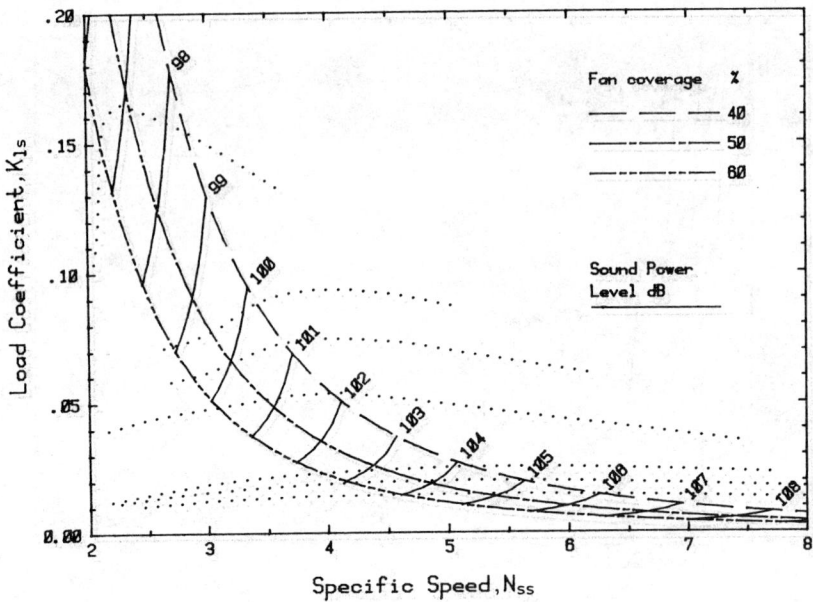

Fig 7 LAC ACHE fan noise levels

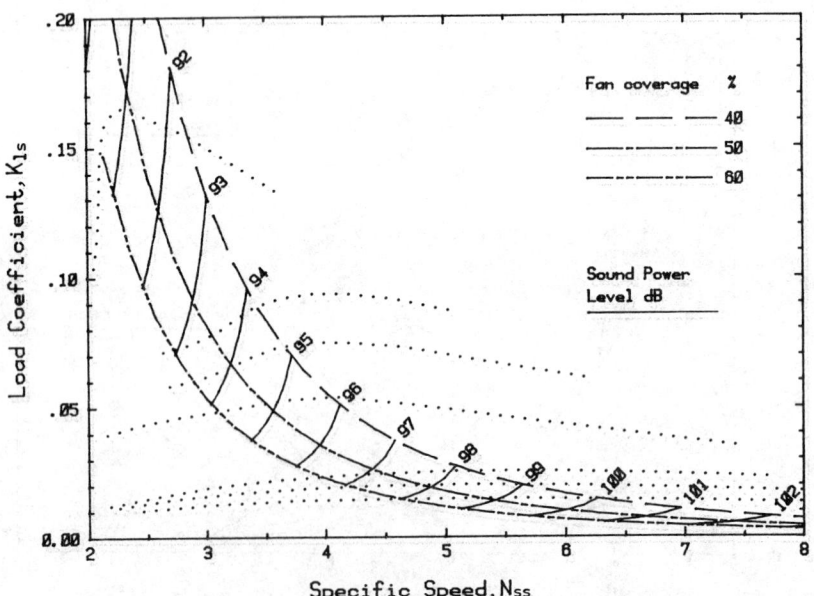

Fig 8 L.$p_s^{2.1}$V ACHE fan noise levels

By incrementing rotational speed and calculating the corresponding specific speed from equation (6) and load coefficient from equation (5) constant fan area curves can be plotted for the LAC ACHE design - Fig. 5, and the $Lp_s^2 \cdot 1\dot{V}$ ACHE design - Fig. 6. These curves are independent of the number of fans. After selecting possible fan numbers the fan area can be incremented and lines of constant speed plotted as shown on the figures. Also included on these figures are the fan characteristic curves which are identical to those in Fig. 4.

As a further aid to fan selection the sound power level equation (10) can be can be manipulated to enable sound power level contours, which are independent of the number of fans, to be plotted for the two ACHE designs - Figs 7 and 8. Table 3 summarizes possible operating points for the three fans. In this instance only the 40 per cent fan coverage has been considered as larger fan areas move the fan operating point too close to the onset of stall for fans 2 and 3 (cf. Figs 2, 4, 5 and 6). Fig. 9 shows views of the two extreme ACHE designs namely the LAC ACHE fitted with fan 1 and the $Lp_s^2 \cdot 1\dot{V}$ ACHE fitted with fan 3. The difference in sound power level according to equation (11) is 12.5 dB (re 10^{-12} W).

TABLE 3

SUMMARY OF SELECTED FANS AND PREDICTED NOISE LEVELS

ACHE Fan static pressure, Pa Total flowrate, m³/s	LAC 180 68.6			$Lp_s^2 \cdot 1\dot{V}$ 102 57.3		
Fan	1	2	3	1	2	3
Fans/ACHE	2	2	18	2	2	18
Fan diameter, m	2.60	2.60	0.87	2.73	2.73	0.91
Fan speed, r/min	392	255	596	280	178	429
Total sound power level, dB (re 10^{-12} W) equation (11)	105.0	100.8	98.5	99.0	94.8	92.5

6. DISCUSSION

Table 3 shows that for the ACHE duty considered in this paper a sound power level reduction of 6 dB can be achieved through changes in the tube bundle design and running the same design of fan at the appropriate lower speed. Reference to Table 2 shows that the capital cost penalty would be of the order of 15 per cent but the fan power would be halved. Additional noise reductions can be obtained by changing the fan design. Changing from the 'fan 1' axial fan type to the 'fan 2' type can bring a 4 dB improvement. A further reduction of 2 dB can be gained by using the 'fan 3' type.

Significant sound power level reductions are thus possible by the relatively simple means of suitably optimizing the tube bundle and by selecting the right class of fan. Other factors can also affect the noise level. One such factor is whether the exchanger is of the forced or induced draught type. A greater fan power is required for induced draught exchangers to handle the larger volumes of hot lower density air. The fan sound power level will therefore be correspondingly higher than for a forced draught

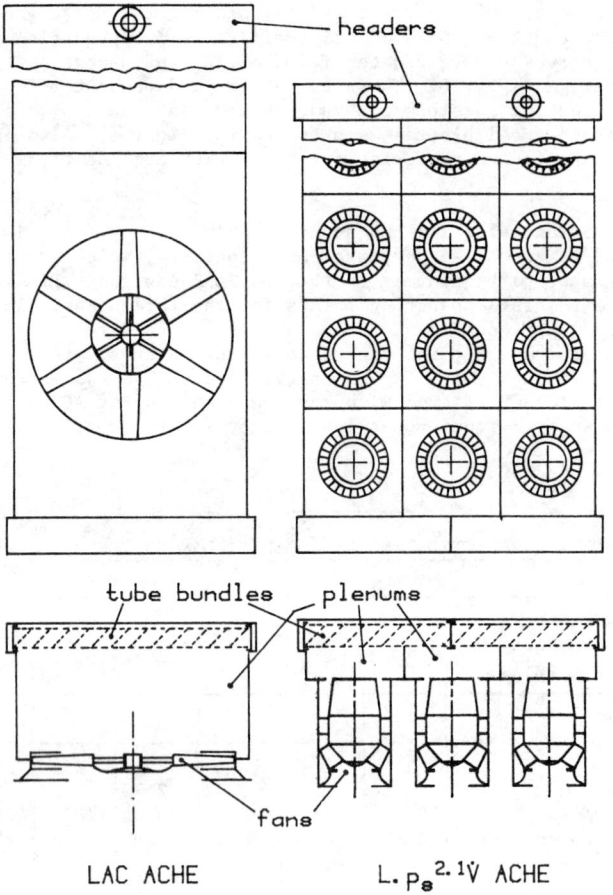

Fig 9 ACHE design comparison

exchanger. However with each fan situated above the tube bundles in a small chimney the sound energy leaving the fans has greater directionality. Consequently much of the sound energy goes directly upwards and that going downwards is attenuated by the tube bundle. At the plant perimeter fence the perceived sound can therefore be less for an induced draught exchanger than for a forced draught exchanger.

Fan blade number, aerodynamic efficiency, blade loading distribution and the manner in which the fan sound power is distributed through the frequency spectrum all have a part to play in the perceived sound. Wright[7] has give a full account of these and other factors. Table 1 indicates the effect fan detail design can have. Equation (1) has been fitted to data from many fans and its predictions for fans 1, 2 and 3 are given in Table 1. Fan 1 is much worse than 'average' whereas fan 3 is better. Thus, for the example illustrated in Fig. 9 the difference in sound power level between the two designs could be expected to be well in excess of 15 dB.

Reference to Figs 7 and 8 show that there are small gains in noise level to be had from using fan coverages in excess of 40 per cent of the tube bundle face area. This must be consistent with a suitable operating point on the fan pressure/flow characteristic.

Silencing is usually impractical for the conventional ACHE installation because of the large silencer volumes that would be necessary to handle the large fan sizes. Examination of Fig. 9 shows however that it would be relatively simple to fit silencers to each mixed-flow fan. The small diameter of the mixed-flow inlet bell also has the advantage that because of its small physical size it will attenuate low frequency sound.

The economics of the various fan configurations have not been examined. The use of the 'fan 3' type can be expected to be more expensive than the use of axial fans, but there are advantages with this type including the ready means of exchanger control by switching out rows of three fans when reduced cooling is required, excellent air flow distribution over the tube bundle, and high efficiency.

7. CONCLUSIONS

An equation has been derived relating the air pressure drop through an air-cooled heat exchanger, the air flowrate and density, the fan specific speed, and the fan load coefficient to the fan total sound power level.

This equation shows that the total sound power level is minimized by:

designing the tube bundle and plenum to reduce the product of the air pressure drop raised to the power of 2.1 and the air flowrate to a minimum;

using fans with the lowest practical specific speed, that is fans towards the centrifugal end of the centrifugal/mixed-flow/axial fan spectrum; and

using fans with the highest practical load coefficient.

An example is given where a reduction of the total sound power level in excess of 15 dB (re 10^{-12} W) could be expected. This figure is made up from 6 dB by optimizing the tube bundle, 6.5 dB by selecting a suitable fan type, and the remainer by optimizing the fan design.

A method is presented for matching fan type, speed, size, number and duty point to an air-cooled heat exchanger on one graph.

ACKNOWLEDGEMENTS

This paper is presented with the permission of the Director, National Engineering Laboratory, UK Department of Industry. It is British Crown copyright. The work reported was supported by the Chemical and Minerals Requirements Board of the UK Department of Industry and was carried out under the research programme of the Heat Transfer and Fluid Fow Service (HTFS).

REFERENCES

1. BIJL, L. A., MARSH, M., BERRYMAN, R. J. and RUSSELL, C. M. B. Comparison of methods of measuring sound power levels of air-cooler fans. Paper C219/78. Conf. Site Testing of Fans and Systems 1978. London: Institution of Mechanical Engineers.

2. DEEPROSE, W. M., BOLTON, A. N. and MARGETTS, E. J. Accurate noise measurement from a large fan installation. Paper C218/78. Conf. Site Testing of Fans and Systems 1978. London: Institution of Mechanical Engineers.

3. FARRANT, P. E. A method for selecting casings for mixed-flow pumps and fans. International Conference on Design and Operation of Pumps and Turbines. East Kilbride, Glasgow: National Engineering Laboratory, 1976.

4. FARRANT, P. E. An approach to mixed-flow fan design. Fluid Mechanics Silver Jubilee Conference. East Kilbride, Glasgow: National Engineering Laboratory, 1979.

5. BOLTON, A. N., FARRANT, P. E., McEWEN, D., MOORE, A. and WILSON, G. QHV (quiet heavy vehicle): the NEL contribution. Automotive Engineer, 1979, $\underline{4}$(3), pp 17-19.

6. AMERICAN PETROLEUM INSTITUTE. Air-cooled heat exchangers for general refinery services. API Standard 661, 2nd edition, Washington, 1978.

7. WRIGHT, S. E. The acoustic spectrum of axial-flow machines. J. of Sound and Vibration, 1976, $\underline{45}$(2), pp 165-223.

Heat Transfer and Flow Friction Characteristics of Louvred Heat Exchanger Surfaces

C. J. DAVENPORT
Industrial Heat Transfer Division
COVRAD
Sir Henry Parkes Road
Canley, Coventry, England

ABSTRACT

   This paper discusses the place of louvred surfaces in compact heat exchangers and presents heat transfer and flow friction data for eight louvred surfaces. The effect of the size of the louvre in the air flow direction is shown and used to obtain some correlation of the results. The correlation plots are compared with theoretical expressions for laminar boundary layers on flat plates and with previously published data on strip fins.

1. INTRODUCTION

   The demand for compact heat exchangers was created by the need to cool engines for transportation. Developments were intensified by the advent of the smaller gas turbines where regenerators required very compact surfaces. At the present time, compact liquid/gas and gas/gas exchangers are being increasingly employed for stationary duties, for example in diesel engine-based power generation plant and in other applications where the space available for cooling equipment is restricted.

   The overall heat transfer coefficients in a compact exchanger are usually limited by the gas-side coefficients. The first stage to improve overall coefficients is to use an extended surface on the gas-side. The gas-side coefficients will still be small, though operating on a larger surface area. The second stage is to improve gas-side coefficients by suitable surface features.

   These features may be divided into two classes: the first, in which the surface remains continuous and the second in which it is cut. Examples of enhancing features of the first class are dimples or waves in the fin, or corrugated fin formed with a herring-bone shape. They work by causing the gas to make sudden direction changes so that locally, the velocity and temperature gradients are increased. This causes local enhancement of both friction factor and heat transfer coefficient. There is a likelihood of separated flow on some parts of the surface and here, the heat transfer coefficients will be low and the pressure losses high. These pressure losses are associated with pressure drag of the surface which should be minimised to avoid excessive fluid pumping power requirement.

The second class of features is exemplified by strip and louvred fin illustrated in Fig. 1. In these, the surface is discontinuous but the air flow direction changes are kept to a minimum. They work by interrupting the boundary layer which would form on a flat continuous plate. Very close to a leading edge, friction factors and heat transfer coefficients are very high and both decrease downstream of the edge. Each cut in an interrupted surface restores the leading edge condition and maintains high values for the coefficients. Although high friction drag is produced, this has a counterpart in high heat transfer coefficient whereas pressure drag produced by separated flow has no such counterpart and represents wasted energy.

The enhancement obtained with interruptions is generally greater than that with continuous fin by about a factor of two. However, the tooling required to produce it is more costly and prone to wear and damage. There is also some resistance to the adoption of interrupted fins because of fears that it may be fouled by dust in the air stream although this danger is probably overestimated as shown by Cowell and Cross [1]. The pressure loss across an interrupted surface is much affected by the quality of the cut edge. If the edge is burred or bent then a pressure drag is introduced by the bluff shape encountered by the air stream. Interrupted surfaces consequently often have higher pressure losses than the ideal.

For these reasons, there remains a place for the lesser enhanced surfaces with dimples or waves. This is where great compactness is not required and where fluid pumping power is an important parameter. Where high compactness is a requirement, as in aircraft applications, automotive coolers and heaters, gas turbine regenerators and some air-conditioning equipment, interrupted surfaces are widely used. For non-mobile cooling equipment where compactness is less vital, continuous surface fins are still widely used. However, as the compactness requirement increases and as interrupted fin design and manufacturing techniques improve, we may expect an increasing use of interrupted fin in static equipment.

Strip fin (see Fig. 1) has been used in regenerator cores and is often used in plate heat exchangers for example in oil coolers and cryogenic equipment. For engine water cooling however, by far the most common surface enhancement in the last twenty years has been by louvres.

There are to date no published heat transfer and friction data on louvred surfaces of the type illustrated in Fig. 1. Kays and London [2] contains the most comprehensive collection of compact surface data including data on louvred fin. However, the louvres in [2] have a large distance between cuts and more complex shapes than those used on more recent surfaces.

LOUVRED HEAT EXCHANGER SURFACES

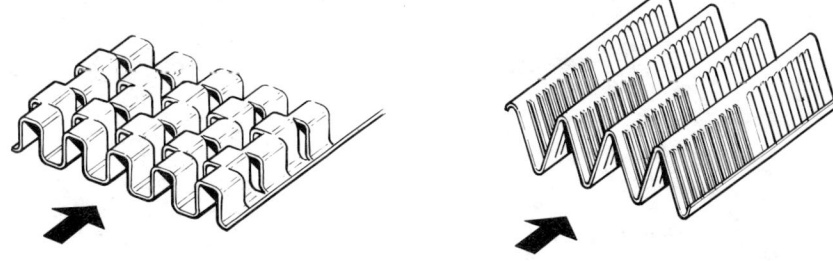

Fig. 1 Strip & Louvred Fin

An object of this paper is to present data for the type of louvred surface illustrated in Figs. 1 and 2 and to examine the effect of varying the uninterrupted flow length.

2. PREVIOUS WORK

Although there are no published design data there is work on the qualitive behaviour of air flows on louvred surfaces. For example Beauvais [3] working under the sponsorship of the Ford Motor Company used a ten-times scale model of a louvred surface with essentially triangular ducts. He used smoke flow visualisation to show how the louvres strongly directed the air flow. This disposed of the idea that louvres would simply act as a surface roughness with the main air flow direction being axially down the duct. The validity of the use of large scale models was established by Wong and Smith [4] who measured heat transfer on large scale models of louvred fin and compared the results with measurements on actual size louvres. The Nusselt Number/Reynolds Number plots had the same shape and the authors deduced that the same phenomena was occurring in both cases. [4] also contains some hot-wire velocity measurements on the model which tended to confirm the finding of Beauvais that the louvres directed the air flow.

Smith [5] has attempted to model the air pressure loss on louvred fin using "drag elements". The surfaces of the louvres were assumed to behave like flat plates with laminar boundary layers and the edges like bluff bodies. Smith also included the effects of turning the air in his model and obtained 15% agreement with pressure drops measured on a section of automotive radiator.

The advantages of louvres in air conditioning evaporator and condenser cores with circular tubes and plate fins is demonstrated in [6]. Here the authors state that heat transfer coefficients on a louvred surface were 1.6 times those on a conventional wavy surface. With the application to evaporators in mind Smith [7] attempted to model the heat transfer from louvred plate fins by dividing the fins into separate portions and adding together calculated coefficients for each portion. Some experiments with 3 times scale models were carried out and comparison with calculated results led the author to conclude that his model approach was promising.

Since the early seventies there has been very little published work on louvred surfaces, much of the data which has been obtained since then have remained proprietary.

3. THE TEST SAMPLES

Fig. 2 illustrates the type of samples used to measure the surface characteristics. They were formed from alternate sections of louvred fin and water tube. The test cores were 152 mm square and were fitted with headers to allow water to be pumped through the tubes.

The fin was .075 mm thick copper foil which was formed into the triangular section shown in Fig. 2 and soldered to the 0.2 mm thick brass tubes. The depth of all samples in the air flow direction was 40 mm.

The edges of the samples were terminated with water tubes so that the outermost section of fin was sandwiched between two water tubes. This was to maintain a similar temperature distribution in all the fin sections.

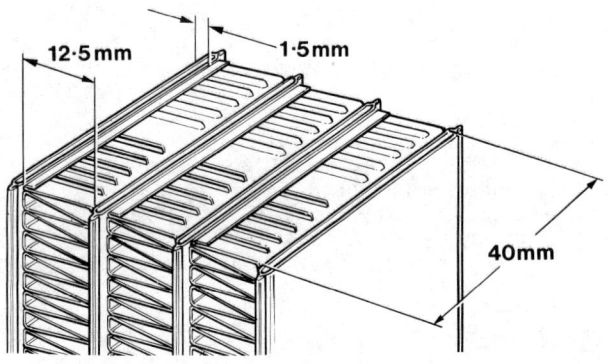

Fig. 2 Section of Test sample

4. FIN SURFACE GEOMETRY

Data are presented on eight different louvred surfaces whose geometries are summarised in Table 1. The parameters listed in Table 1 are defined in Fig. 3.

The experimental plan was designed to show the effect of varying the louvre pitch (i.e. uninterrupted flow length) with all other geometrical parameters held constant, this being done for two values of fin height. Four different louvre pitches were used and these are illustrated schematically in Fig. 4. The numbers of louvres for each pitch were chosen to give approximately equal louvred areas.

The most difficult parameter to control in practice was louvre height which varied by ± 3% for samples A, B, C and D and ± 10% for samples E, F, G and H. It will be noted that data for the latter four samples did not correlate so well as that for the former four and this is probably attributable to the greater fluctuation in louvre height. The choice of louvre height to be kept constant rather than louvre angle is mentioned in section 9.

Table 1 also gives area ratio data and hydraulic diameters. These are very similar for each group of four surfaces with the same fin height.

Table 1 - Surface Geometry Data

(Dimensions in mm)

Surface Ident.	Fin Height	Fin Pitch	Louvre Pitch	Louvre Length	Louvre Height	Min. free flow to frontal area	Ratio fin area to total area	Ratio total area to frontal area	Hydraulic Diameter
A	12.7	3.12	3.0	9.5	.29	.816	.891	49.9	2.80
B	12.7	3.12	2.25	9.5	.31	.816	.891	49.9	2.80
C	12.7	3.10	1.8	9.5	.29	.816	.892	50.2	2.78
D	12.7	3.17	1.5	9.5	.29	.817	.890	49.17	2.84
E	7.8	3.07	3.0	7.1	.36	.749	.837	49.55	2.61
F	7.8	3.02	2.25	7.1	.34	.748	.839	50.26	2.58
G	7.8	3.07	1.8	7.1	.31	.749	.837	49.55	2.61
H	7.8	3.1	1.5	7.1	.33	.749	.836	49.14	2.64

All samples 40 mm deep in flow direction

5. TEST EQUIPMENT

A schematic of the wind tunnel is shown in Fig. 5. It has a working section 152 mm square. Air is drawn through a gauze at the intake, passes through a flow straightener and then over a rotating vane anemometer. The anemometer was calibrated to give mean velocity at the working section by taking 36 pitot static tube measurements for a set airflow and counting pulses from the anemometer. A least squares straight line fit was then used to relate mean velocity at the working section to anemometer pulses.

The inlet air temperature used for calculating physical properties was sensed using a mercury-in-glass thermometer.

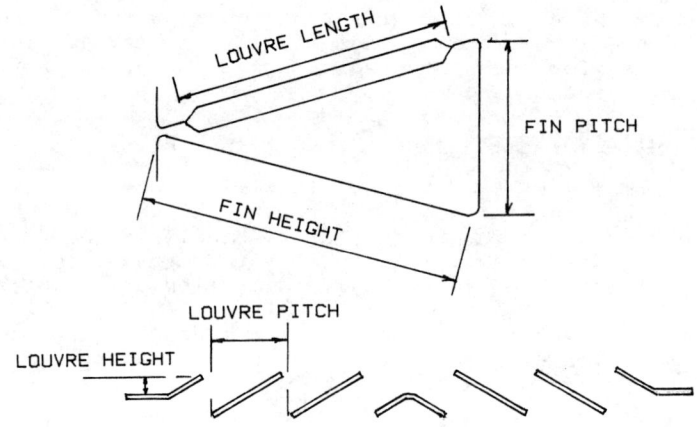

Fig. 3 Definitions of geometrical parameters

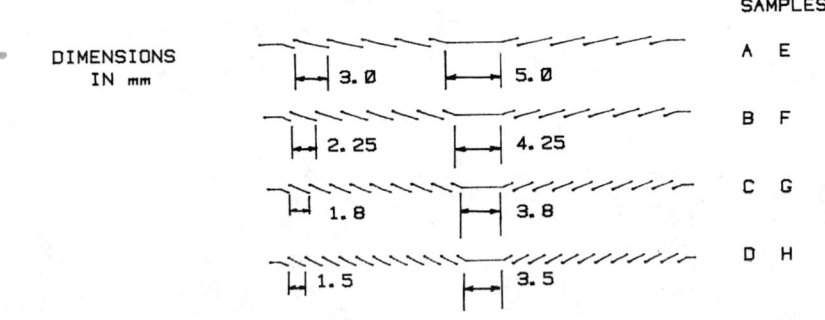

Fig. 4 The Four Louvre Bank Arrangements

The air temperature rise through the test sample was measured using nine copper/constantan thermocouples wired in series. The junctions were arranged in a square array upstream and downstream of the sample.

Static pressure loss across the sample was measured using an inclined manometer connected to piezo rings situated on either side of the sample. Pressure loss between sensing points was measured in the absence of a sample and the values obtained subtracted from the measured total.

Air inlet, water inlet temperature difference was measured using a chromel/alumel differential thermocouple.

Water was pumped through the samples from a heated reservoir through a variable area flow meter. The inlet water temperature (used for property calculations) was measured with a mercury-in-glass thermometer and water temperature drop across the sample by four chromel/alumel thermocouples in series. The water-side measurements were not used directly to calculate Stanton Number but as a check on the air-side. Heat balances of up to 6% were typical.

6. TEST PROCEDURE

The sample to be tested was mounted in the working section and about 100 ℓ/m of water at 85°C pumped through it. The e.m.f. across the water-side thermocouples was noted for nominally zero airflow. This gave a measure of the heat lost to the surroundings and any e.m.f. arising from differences in thermocouple materials. It was subtracted from subsequent water temperature difference readings.

The fan was then switched on to give approximately 15 m/s at the working section. Air velocity, air temperature rise, air/water inlet temperature difference, water temperature drop and air static pressure drop were noted when equilibrium was attained. This was repeated for about twenty air velocity settings.

The water and air inlet temperatures were taken once, half way through each test. Atmospheric pressure was read from a Fortin barometer and used for air property calculations.

Two tests were performed on each sample, one with hot water and the second without water, in isothermal conditions when only air flow rate and pressure loss were measured. These readings were used to calculate friction factor. The absence of air-heating removed the need to calculate pressure loss due to air acceleration, thus simplifying the derivation of friction factor and removing a source of error.

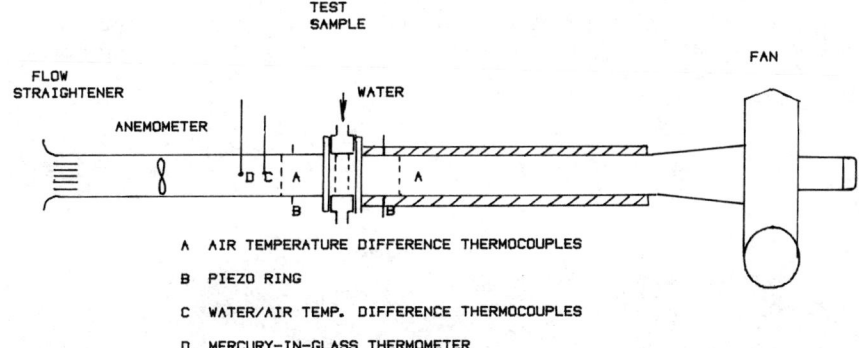

Fig. 5 Schematic of Wind Tunnel

7. METHOD OF DATA REDUCTION

It was required to express the heat transfer results as Stanton Number versus Reynolds Number plots. For a given air velocity, Stanton Number was calculated as follows:

1. Logarithmic mean temperature difference was calculated. This was used without correction for crossflow since the capacity rate ratio was .05 at the highest air flow which rendered the corrections insignificant.

2. The heat transfer rate per unit fin area was calculated using the air-side flow rate and temperature rise measurement.

3. The overall heat transfer coefficient, U, was calculated.

4. The water-side heat transfer coefficient was calculated from the Dittus Boelter relation valid for Reynolds Numbers in excess of 10,000. This condition was always satisfied.

5. The overall thermal resistance was resolvable into three parts, air-side, water-side and conduction term. The air-side was typically 95% of the total with the remaining 5% due mostly to the water-side. The conduction term was calculated from a measured mean solder thickness of .01 mm. Air-side thermal resistance was then calculated.

6. The fin efficiency was calculated using the expression given in [8] for infinite rectangular fins neglecting heat transfer from the edges. The fin effectiveness was then determined for the whole of the air-side heat transfer area including primary surface.

7. Stanton Number was then calculated from

$$St = \frac{h_a}{\rho_1 V_a C_{pa}}$$

Air-side Reynolds Number was calculated using air velocity and density evaluated at the front face of the sample and dynamic viscosity at the mean air temperature. It was formed using the hydraulic diameter of the triangular duct as the characteristic length.

$$Re = \frac{\rho_1 V_a D}{\mu_a}$$

The friction factors were derived from an expression in [2] (page 33) but simplified to the following for isothermal measurements.

$$\Delta P = \tfrac{1}{2} \rho_1 V_a^2 (k_{con} + k_{ex} + f A_t/A_c)$$

k_{con} and k_{ex}, the abrupt contraction and expansion coefficients were taken from [2] (page 96) for traingular ducts. Curves for infinite Reynolds Number were used as recommended by the

LOUVRED HEAT EXCHANGER SURFACES

authors for interrupted surfaces.

Experimental errors in the calculated parameters have been derived from estimates of errors in each measured quantity. The errors depend on the air velocity and are summarised below.

PERCENTAGE ERROR IN :		AIR VELOCITY (m/s)		
		.5	3	15
Re	±	5		5
St	±	4		3.6
f	±	12	7	6

The errors may be regarded as estimates of half the 95% confidence intervals.

Much of the error derived from the measurement of sample frontal area. The partial error in Stanton Number due to temperature measurements was between 1 and 2%.

8. TEST RESULTS

Figures 6 and 7 give the heat transfer and flow friction results for all eight samples. They are plotted against Reynolds Number based on hydraulic diameter but since this and air viscosity are almost constant for each group of samples, the plots are effectively against mass velocity.

At a given mass velocity the Stanton Number for 1.5 mm louvre pitch is about 33% higher than that for 3 mm pitch whereas the corresponding figure for friction factor is about 50%. The friction factor for sample D was exceptionally high and did not fit into the general pattern. This was attributed to some bending of the fin leading edges due to damage during manufacture.

Whereas the Stanton Number curves are almost straight lines, the friction factor curves show a tendancy to flatten at high Reynolds Number. This is characteristic of a mixture of friction and form drag, the latter becoming increasingly predominant at high Reynolds Numbers. The flattening is more marked on sample D which is consistent with bent leading edges which would contribute form drag.

9. CORRELATION

If the predominant mechanism on the louvred surfaces was due to laminar boundary layers then the Pohlhausen expression for heat transfer and the Blasius solution for friction should approximately apply: These are

$$St = .664 \, Pr^{-2/3} \left(\frac{G \, D}{\mu_a}\right)^{-1/2}$$

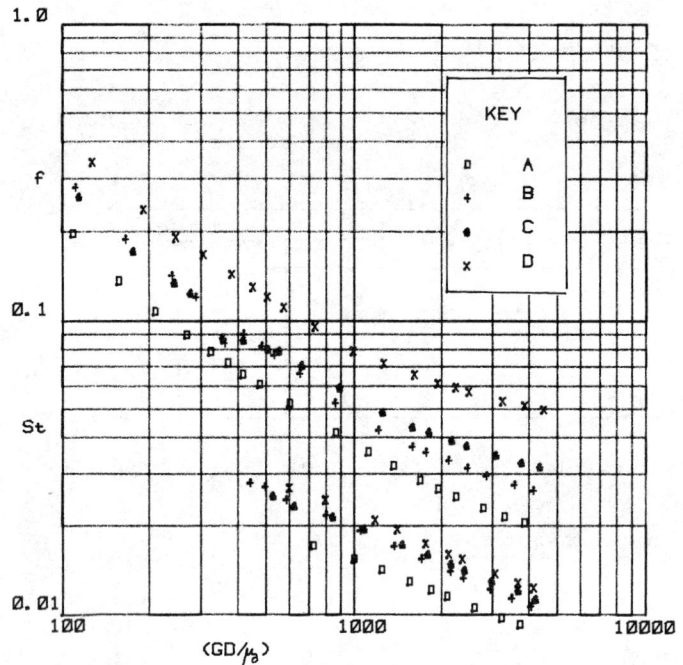

Fig. 6 Data for Samples A, B, C & D

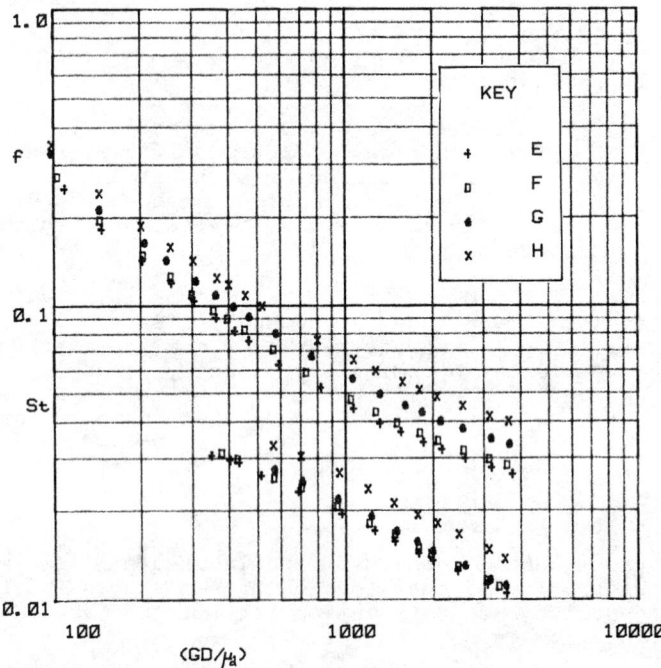

Fig. 7 Data for Samples E, F, G & H

and

$$f = 1.328 \left(\frac{G\,D}{\mu_a}\right)^{-\frac{1}{2}}$$

These give mean values on a flat plate with zero flow-angle of incidence, effectively zero plate thickness and zero pressure gradient. Since none of these conditions apply in the heat exchanger we would expect significant discrepancy between the predictions of these expressions and experimental results. However, the form of the equations using a Reynolds Number based on un-interrupted flow length prompts a method of correlation.

Figures 8 and 9 show the data re-plotted using the louvre-pitch-based Reynolds Number and significant compression of the results is achieved. The Stanton Numbers for samples A, B, C and D lie within about ± 6% and those for samples E, F, G and H within about ± 8%. The larger scatter in the results for the latter four samples is probably due to the larger range of louvre heights : sample E has the largest louvre height and plots highest in Fig. 9.

The friction factor curves are also brought together by the plot although sample D is clearly exceptional. The correlation for friction factors is clearly only partial and the curves are still left in the order they were before correlation. This is probably because reduction in louvre pitch increases the form drag due to bluff edges and also, since louvre angle is increased, the air turning losses. Neither of these two effects is allowed for in the correlation plot.

The compression of the data points supports the choice of louvre height rather than louvre angle as the parameter to hold constant.

The Pohlhausen and Blasius lines are shown in Fig. 8 for comparison. The slope of the Stanton Number curves is very similar to the Pohlhausen slope, suggesting a similar heat transfer mechanism, but the data points all lie about 40% below the theoretical. This may be expected, since only about 65% of the surface area on which the heat transfer coefficient is based has louvres on it. The remaining surface is unfeatured and would very probably have much lower local coefficients.

The friction factor data shows apparent agreement with the Blasius line. This is more likely to be due however to the cancelling effects of the form drag losses and the result of having louvres on only 65% of the surface. The slope of the friction curves is very close to the Blasius slope at low Reynolds Numbers, but flattens at high Reynolds Numbers. This comparison is consistent with the view that the mechanism on the louvres is predominantly due to laminar boundary layers with an addition to the skin friction arising from bluff edges and air-turning losses.

10. COMPARISON WITH STRIP FIN

Fig. 8 also shows curves for strip fin surfaces taken from [2]

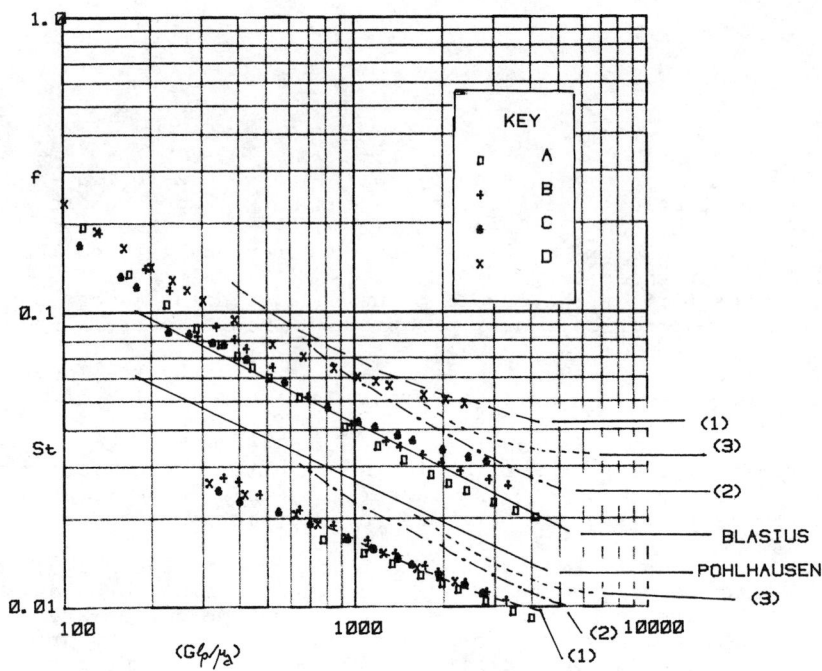

Fig. 8 Correlation Plot for Samples A, B, C & D

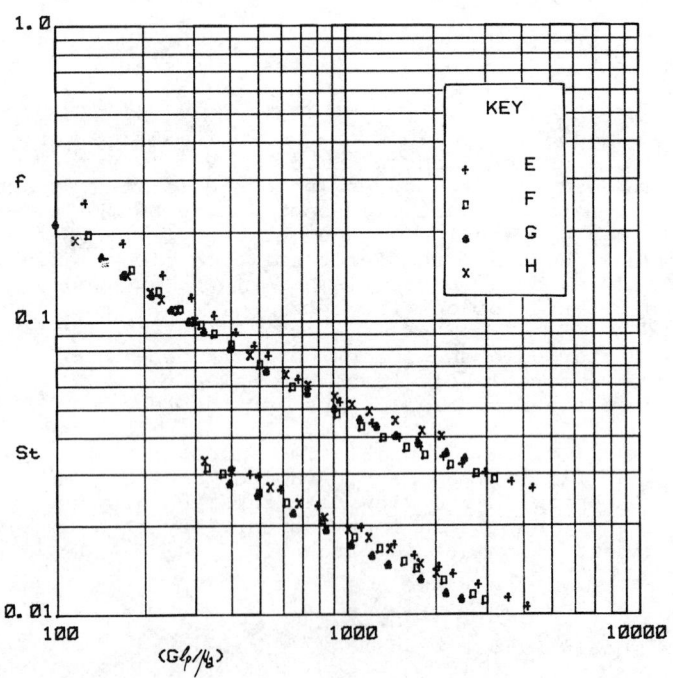

Fig. 9 Correlation Plot for Samples E, F, G & H

and [9] (curves marked (1) and (2) respectively). Curves (1) are taken from [2] Fig. 10-53. This gives data for a strip fin surface (Kays and London designation, 3/32-12.22) with 2.38 mm between cuts made from .1 mm thick copper foil with a 12.3 mm fin height. It is the surface with the shortest uninterrupted flow length given in [2]. The data has been replotted using the Reynolds Number based on this length to allow comparison with the louvred surfaces.

The Stanton Numbers fall very close to those for the louvred surfaces but the friction factors are much higher. This is probably due to burrs on the leading edges. Strip fin is very prone to this problem particularly near the base of the fin.

Subsequent strip fin data of Briggs and London [9] is much improved (marked (2) on Fig. 8). This surface (authors designation 15.76R-.153/.149-1/7(0)-.004(AR)) has a distance between cuts of 3.63 mm and was made from .1 mm thick aluminium.

Both the Stanton Numbers and friction factors are about 15% higher than those for the louvred surfaces. The ratio j/f often used to assess the overall efficiency of a surface is about .32 for surface (2) compared with 0.2 for the earlier surface (1).

Strip fin has the theoretical advantage that a higher proportion of the surface has leading edges since the cuts extend for the full height of the fin. It is not normally possible for louvres to occupy more than 90% of the fin height and even then the ends of the louvres taper down to the fin surface and do not present their full height to the air stream. This advantage probably accounts for the higher Stanton Numbers for surface (2).

Sparrow et al [10] have modelled the behaviour of strip fin by testing an array of staggered plates. They used a mass transfer method and the analogy between mass and heat transfer to determine local coefficients on individual plates. They found that the first plates to meet the air stream had lower coefficients on them than subsequent plates which they attributed to vortex shedding from the blunt trailing edges of the plates. They also measured pressure drops across the arrays and their results are plotted in Fig. 8 as curves (3). Curves (3) are for the lowest ratio of plate thickness to length (.04) which is the closest to practical fins. Sparrow et al used a constant plate length of 25.4 mm, so that the model was about ten times bigger than actual strip fin.

These coefficients are not averaged over an area only some of which is interrupted, but are the values on the fins themselves. It is to be expected therefore that they would be larger than those obtained on actual heat exchangers. This is indeed the case, but surface (2) comes quite close to meeting the Stanton Number values of (3).

The table below summarises the j/f values for all the surfaces considered in this paper including the model surfaces of Sparrow et al.

Surface	A	B	C	D	E	F	G	H	(1)	(2)	(3)
j/f	.36	.35	.29	.20	.32	.32	.28	.28	.20	.32	.33

The values have been taken at a pitch-based Reynolds Number of 2000. The samples with the lowest value both suffered from excess form drag which underlines the importance of keeping leading edges as clean as possible.

The surface with the highest ratio (A) also had the lowest Stanton Number for a given mass velocity. This points to the limitations of j/f as a comparison criterion since j/f tends to decrease as the inherent compactness increases. Generally surfaces with small louvre pitch are preferred despite their lower j/f value.

11. CONCLUSIONS

Heat Transfer and flow friction data have been presented for eight louvred surfaces with varied louvre pitches. Both Stanton Number and friction factor increased with decreasing louvre pitch. Plotting the data using a louvre pitch-based Reynolds Number gave significant compression of the data points suggesting the relevance of laminar boundary layers to the heat transfer and friction phenomena. The friction data points were only partially correlated and this was attributed to the presence of form drag on the bluff leading edges.

Published data on strip fin surfaces suggest a large variability of performance. Currently, at comparable fin areas per unit volume, louvred fin and strip fin offer similar performances.

ACKNOWLEDGEMENTS

The author wishes to thank the directors of his company, COVRAD, for giving permission to present this paper and Janet Evans for her patience in typing it.

NOTATION

A_t Total area on the air-side. (m^2)

A_c Minimum free flow area. (m^2)

C_{pa} Specific heat of air at constant pressure. (J/kgK)

D Hydraulic diameter of air passages $4\left\{\dfrac{A_c}{A_t}\right\}L$. (m)

f Friction factor

G Air mass velocity ($\rho_1 V_a$). (kg/m^2s)

h_a Air heat transfer coefficient. (W/m^2K)

j Colburns Modulus. $St\, Pr^{2/3}$

k_{con}, k_{ex} Abrupt contraction and expansion coefficients.

l_p Louvre pitch. (m)

L   Flow passage length. (mm)

Pr   Air Prandtl Number

Re   Air Reynolds Number

St   Air Stanton Number

U   Overall heat transfer coefficient. ($W/m^2 K$)

V_a Air velocity inside ducts. (m/s)

ρ Air density at inlet. (kg/m^3)

μ_a Air dynamic viscosity at mean air temperature. (kg/ms)

REFERENCES

1. Cowell T.A.T. and Cross D.A. 1980. Air-Side Fouling of Internal Combustion Engine Radiators. Society of Automotive Engineers, paper no. 801012.

2. Kays, W.M. and London A.L. Compact Heat Exchangers. McGraw Hill (2nd Edition) 1964.

3. Beauvais F.N. 1965. An Aerodynamic Look at Automotive Radiators. Society of Automotive Engineers, paper no. 650470.

4. Wong L.T. and Smith M.C. 1973. Airflow Phenomena in the Louvred Fin Heat Exchanger. Society of Automotive Engineers, paper no. 730237.

5. Smith M.C. 1968. Gas Pressure Drop of Louvred Fin Heat Exchangers. A.S.M.E. Journal of Heat Transfer, presented at AIChE conference August 1968.

6. Hosoda T., Uzuhashi H and Kobayashi N. 1978. Louvre Fin Type Heat Exchangers. Heat Transfer Jap. Res. Volume 6 part 2 pp 69 - 77.

7. Smith M.C. 1972. Performance Analysis and Model Experiments for Louvred Fin Evaporator Core Development. Society of Automotive Engineers. Paper no. 720078.

8. Kern D.Q. and Kraus A.D. Extended Surface Heat Transfer. McGraw Hill 1972.

9. Briggs D.C. and London A.L. The Heat Transfer and Flow Friction Characteristics of Five Offset Rectangular and Six Plain Triangular Plate Fin Heat Transfer Surfaces. Proc. of 1961-62 Heat Transfer conference. University of Colorado. A.S.M.E. paper no. 14 pp 122 - 134.

10. Sparrow E.M. and Hajiloo A. 1980. Measurements of Heat Transfer and Pressure Drop for an Array of Staggered Plates Aligned Parallel to an Air Flow. Trans. A.S.M.E. Vol. 102 pp 426 - 432.

The Performance of a Counterflow Air to Air Heat Exchanger with Water Vapor Condensation and Frosting

ROBERT W. BESANT and JAMES D. BUGG
Department of Mechanical Engineering
University of Saskatchewan
Saskatoon, Saskatchewan S7N 0W0, Canada

ABSTRACT

In this paper, a counterflow air to air heat exchanger is investigated as to its heat rate and temperature recovery when water vapour condenses, freezes, or sublimates on the heat exchanger surfaces. Experimental results were taken under a wide variety of operating conditions showing that temperature recovery ratio of 50 to 60 percent and heat rates of 100 to 300 w/m^2 of heat exchanger surface were measured under temperature differences from 23 to 46.5°C. A theoretical model of the heat exchanger was developed assuming steady, one dimensional flows with uniform properties. This theoretical model predicted heat rates and temperature recoveries within 15 percent of measured values for the case of water vapour condensation.

1. INTRODUCTION

With the advent of more expensive heating fuels for buildings, heat recovery from exhaust air can be attractive in cold climates where the largest single component of heat loss from buildings is often associated with ventilation air.[1] In the past such heat recovery, through the use of compact heat exchangers such as the regenerative wheel, was usually confined to ambient temperatures above freezing or to systems with air preheaters. A more efficient mode of operation is to permit freezing water on the surfaces of the heat exchanger accompanied with periodic deicing and defrosting. In this study, a counterflow plate type air to air heat exchanger is studied under conditions of condensation, icing, and frosting. The deicing and defrosting processes are not examined.

2. THEORETICAL CONSIDERATIONS

A counterflow heat exchanger, illustrated in Figure 1, was tested under various flow, temperature, and humidity conditions in a controlled environmental chamber. Under normal operating conditions, the heat exchanger may exhibit up to four distinct flow conditions on the warm side of air flow; namely, a dry region with no condensation, a condensation region with liquid water on the heat exchanger surface, an icing region with ice on the surface with a layer of liquid water on top and a frosting region with frost over a layer of ice on the surface. The cold air flow passage always remains dry. Not all the regions on the hot side need occur under a given set of operating conditions. For example, there may be no dry region or frosting region or there may be simply condensation. Generally, the size and location of each of these flow regions is determined by five inlet property conditions and the heat exchanger configuration.

In order to predict the performance of an air to air counterflow heat exchanger under operating conditions as described above, it is convenient to use a number of simplifying assumptions. These are
(i) inlet properties are steady,
(ii) the flow is one dimensional inside the heat exchanger,
(iii) properties in each flow region may be taken to be the average value,
(iv) properties inside the heat exchanger are steady or quasi-steady.

The second assumption implies that entrance effects are negligible while assumption (iii) implies somewhat uniform properties in each region. The last assumption implies that the rate of accumulation of ice and frost is sufficiently small that it does not change the heat rates and mass flow rates significantly.

For a dry air flow rate of $\dot{m}_h$ and $\dot{m}_c$ on the hot and cold side of the heat exchanger, the heat rate dQ across an element of length of heat exchanger dx is given by the equations

$$dQ = -\dot{m}_h dw_h \tag{1}$$

$$= -(\dot{m}di)_c \tag{2}$$

$$= U \theta \, dx \tag{3}$$

where i is the enthalpy, U is the overall heat transfer coefficient, and θ is the temperature difference between the hot and cold sides. The subscripts h and c refer to the hot and cold sides and f refers to liquid water condensate. The continuity equation for condensing liquid water is

$$d\dot{m}_f = -\dot{m}_h dw_h \tag{4}$$

where w is the humidity ratio. (In the case of icing and frosting of the heat exchanger surface, another basic equation must be added, that is, an unsteady equation to account for the accumulation of ice and frost.) Equations 1, 2, and 3 may be integrated in each region and combined with an interface equation which defines the property relationship between each region to give a set of algebraic equations which may be solved for the heat rates, temperature, and location of each flow region.

For example, in the case of hot air flow with dry region and a condensation region, the resulting equations are as follows for Region 1:

$$Q_1 = C_{h_1}(T_{h_i} - T_h^*) \tag{5}$$

$$= C_{c_1}(T_{c_o} - T_c^*) \tag{6}$$

and

$$\ln \frac{\theta^*}{\theta_i} = \lambda_1 U_1 x^* \tag{7}$$

where T is temperature, C is the mass flow heat capacity product, * denotes the interface between the dry and condensate region, o denotes outlet, and

$$\lambda_1 = -\frac{1}{C_{h_1}} + \frac{1}{C_{c_1}}$$

Similarly, in Region 2

$$Q = C_{h_2}(T_h^* - T_{h_o}) \tag{8}$$

$$= C_{c_2}(T_c^* - T_{ci}) \tag{9}$$

$$\ln\left(\frac{\theta_o}{\theta^*}\right) = \lambda_2 U_2 (1 - x^*) \tag{10}$$

where the heat exchanger is assumed to be of length 1.

The interface region between 1 and 2 is defined by a heat balance equation written in the form

$$T_h^* = [T_{dp_1} - T_c^*(U_1/h_{h_1})]/(1 - U_1/h_{h_1}) \tag{11}$$

where T_{dp} denotes dew point temperature and hh_1 is the heat convection coefficient in Region 1. Equations (5) to (11) can now be solved for the unknowns T_h^*, T_c^*, T_{c_o}, T_{h_o}, Q_1, Q_2, and x^*.

In the analysis, the heat transfer coefficients were taken as given by the correlation (2).

$$N_u = \frac{(f/8) R_e P_r}{1.07 + 12.7 \sqrt{(f/8)} (P_r^{2/3} - 1)}$$

except where condensation existed then

$$h_h = \left(\frac{di}{dT}\right) K_D$$

where the Lewis relationship was used (3)

$$K_D = h_h / [C_p(1 + w)]_h$$

All the thermodynamic relationships for air with water vapour were taken from emperical equations that have been shown to have a maximum error of about 1.0 percent (4).

3. EXPERIMENTAL APPARATUS

The heat exchanger used for the tests, shown in Figure 1, was constructed from galvanized sheet metal 0.381 mm thick. The exchange surfaces were 406 mm x 1473 mm and were spaced 12.7 mm apart. The hot side has seven flow passages while the cold side had six. The inlet and exhaust ports, each 152 mm x 1.27 mm, were located on the sides with guide vanes used on the inlets to more evenly distribute the air flows over the heat transfer surfaces. The heat exchanger was mounted vertically in such a manner that the hot air flowed down and the cold air flowed up through the heat exchanger.

Air and surface temperatures were measured using thermocouples calibrated to an accuracy of less than 0.5°C. Inlet and exhaust air temperatures were taken with the air well mixed. Air temperature profiles were taken using a thermocouple traversing device for each of the hot and cold flows. Air flows were taken using a pitot tube traversing method. The maxiumum uncertainty in

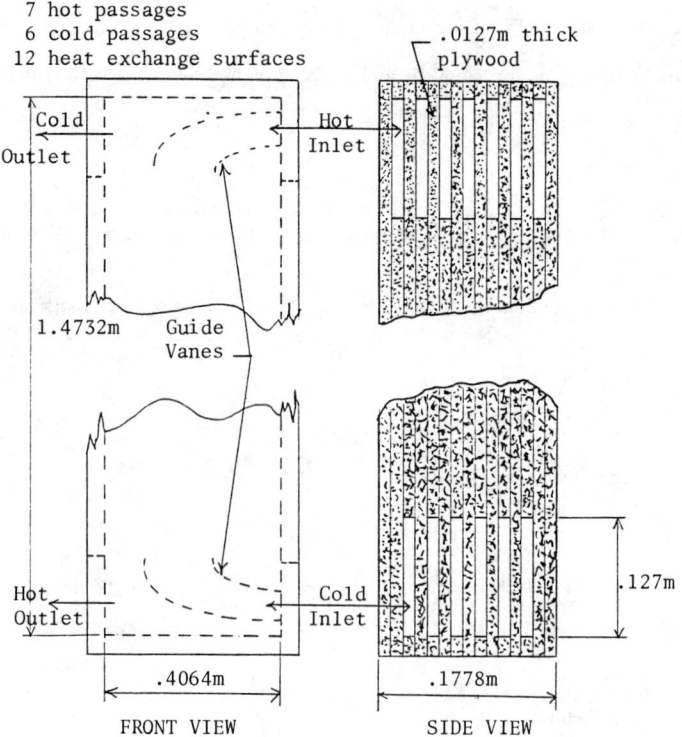

Figure 1. Flatplate Air to Air Heat Exchanger Configuration

mass flow rates was estimated to be less than 13 percent of the total mass flow rate. The inlet and outlet humidity measurements were made with a precision psychrometer which was capable of measuring temperatures to the nearest 0.2°C.

4. ANALYSIS OF TEST RESULTS

The heat exchanger was tested under a wide variety of temperature, humidity, and flow conditions. Air flow rates were nearly balanced for all test runs. The measured and computed results of these experiments are summarized in Table 1. All the experimental data was taken shortly after the heat exchanger reached steady state or quasi-steady state conditions. The computed results in this table include the uncertainties in the measurements but not the systematic or sampling errors. The magnitude of the systematic errors is implied by comparing the hot and cold heat rates. Since this small difference in Q's cannot be accounted for by heat loss to the surroundings (estimated to be less than 2%) due to axial conduction along the heat exchanger surfaces (estimated to be less than 1%) it is thought to be due to incorrect sampling. The property values were computed using the equations of (4). The Reynolds numbers were based mean value properties and twice the plate spacing. Typical temperature profiles within the device are presented in Figures 2 and 3. These temperature profiles can be used to infer the location of condensation, ice formation, and to a lesser extent frost formulation of the surfaces of the heat exchanger.

The temperature profile results such as those shown in Figures 2 and 3 show a distinct entrance region effect. Air entering the heat exchanger did so with about three times the speed and Reynolds number as the flow would have at the mid section to the heat exchanger. This would enhance the heat transfer coefficients of the hot and cold air flow at the inlets resulting in a heat exchanger plate temperature close to the respective inlet temperatures. It is estimated that for the test conditions in this particular heat exchanger that both the hot and cold for entrance regions occupied approximately 2/10 of the length of the heat exchanger each. Without the turning vanes at each entrance, the flow would be hightly nonsymmetrical, and the entrance regions would be larger.

5. THEORETICAL RESULTS

The theoretical model briefly described above was programmed on a computer for the case of air flow with condensation. Results were obtained by linearizing the equations and iteratively obtaining an algebraic solution for each test run. The results of these calculations are compared to a number of experimental test runs in Table 2.

This table suggests that the agreement between theoretical and experimental results is within the bound of the maximum experimental uncertainty. In spite of this agreement, it is noted that the theoretical model involves a number of simplifying assumptions which are not exactly similar to the experimental situation. The most significant descrepancy between the assumed model and the experimental test results is the entrance length effect discussed previously and the assumed one-dimensional flow with uniform properties. It is felt that this theoretical simplification could lead to large errors in some circumstances especially to the calcuation of the length of dry and wet regions. In Table 2 the predicted length of dry region compared to the measured length as determined by the dew point temperature on the heat exchanger surface can differ by as much as 24 percent.

The theoretical model of the heat exchanger with water vapour condensation

Table 1. Experimental Results For A Counterflow Air To Air Heat Exchanger With Water Vapour Condensation And Frosting

Run#	T_{hi} (C°)	T_{ci} (C°)	T_{ho} (C°)	T_{co} (C°)	T_{dp} (C°)	ϕ_{h_i} (percent)	ϕ_{h_o} (percent)	$\dot{m}_h$(dry) (kg/sec)	$\dot{m}_c$(dry) (kg/sec)	Re_h	Re_c	% Temp. Recovery
1	20	-2	9.5	10	5.2	36.7	--	.0663	.0698	2800	3650	52.3
2	20	-10	3.5	4.5	-.8	25.1	--	.0680	.0662	2910	3610	45.0
3	21.5	-25	-.5	-1.5	-7.3	13.4	--	.0729	.0659	3130	3870	50.5
4	21.5	-25	-.5	-1.5	-7.2	13.5	--	.0785	.0738	3380	4340	50.5
5	21.5	-2.5	12.5	11.5	12.9	57.1	--	.0729	.0705	3030	3650	58.3
6	21.5	-11.5	9.5	8.5	13.2	59.5	--	.0733	.0692	3070	3720	60.6
7	21.5	-11.5	10.5	8.5	14.0	63.6	100	.0763	.0692	3190	3720	60.6
8	20.5	-26	5.5	2.5	12.8	60.4	100	.0762	.0742	3230	4300	61.3
9	21.5	-2	10	11.5	6.5	36.6	69.4	.0489	.0545	2040	2810	57.5
10	21.5	-2	9.5	11	5.2	33.8	69.0	.0501	.0545	2090	2820	55.3
11	21	-11.5	4.5	6	.5	25.2	65.0	.0427	.0480	1810	2550	53.9
12	20.5	-25	-2.5	-.5	-3.9	17.5	74.4	.0500	.0534	2160	3110	53.9
13	20	-3	11	11.5	13.4	64.4	95.9	.0463	.0487	1930	2520	63.0
14	20	-12	7.5	8	12.7	61.4	100	.0503	.0526	2120	2330	62.53
15	20	-25	3	3.5	12.6	62.5	100	.0531	.0507	2260	2910	63.36
16	20	-25	4.5	4	12.5	63.7	100	.0529	.0529	2250	3030	64.49

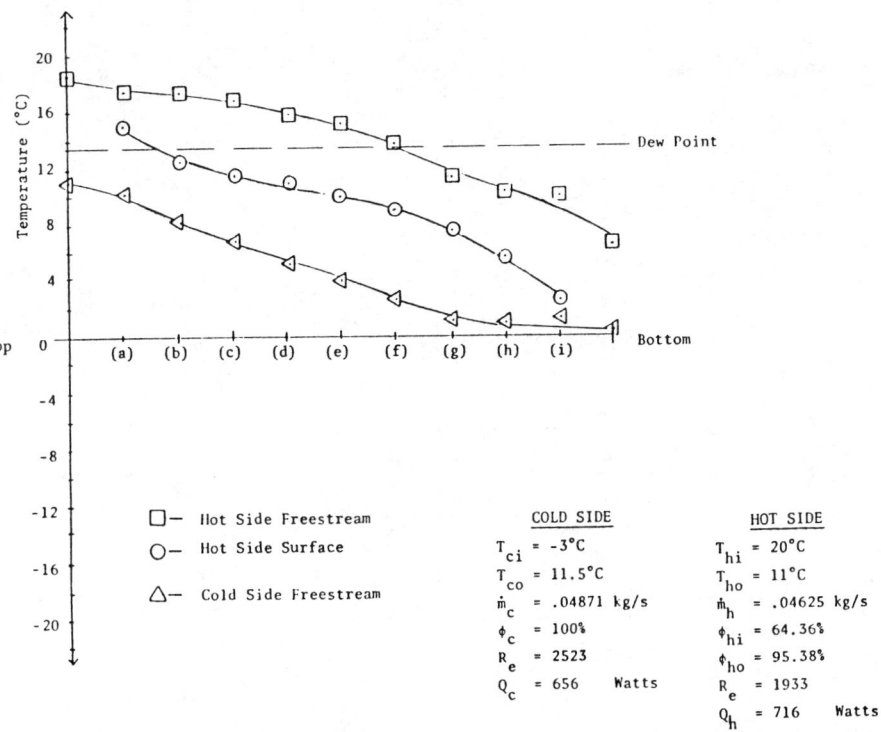

Figure 2. Temperature Profiles, Run 13

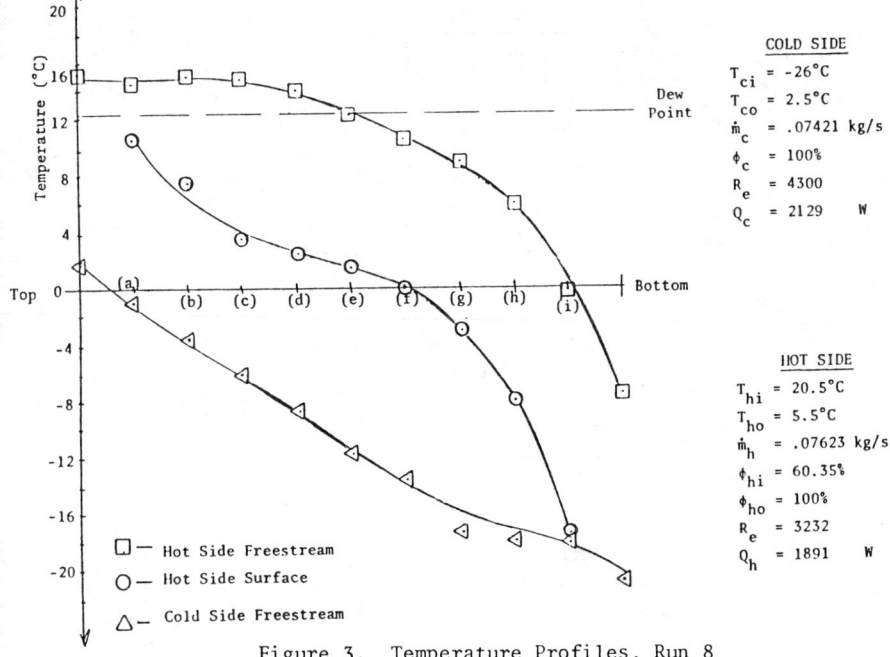

Figure 3. Temperature Profiles, Run 8

419

Table 2.
A Table Comparing Predicted And Measured Results
For A Counterflow Air To Air Heat Exchanger
Operating With Water Vapour Condensation

Run #	Air Flowrates (kg/s)	Hot Inlet R.H.(%)	Hot Outlet Temperature (°C) Predicted	Hot Outlet Temperature (°C) Experimental	Cold Outlet Temperature (°C) Predicted	Cold Outlet Temperature (°C) Experimental	Heat Rate (Watts) Predicted	Heat Rate (Watts) Experimental	Effectiveness Predicted	Effectiveness Experimental	% Dry Area Predicted	% Dry Area Experimental
1	≈.070	36.7	10.28	9.5	8.00	10.0	706.42	814	.45	.52	91.12	67
5	≈.070	57.1	14.47	12.5	11.22	11.5	978.31	1000	.57	.58	44.36	22
9	≈.049	36.6	10.80	10.0	9.25	11.5	620.62	705	.48	.57	82.65	62
13	≈.049	64.4	12.86	11.0	11.41	11.5	709.70	687	.63	.63	32.11	15

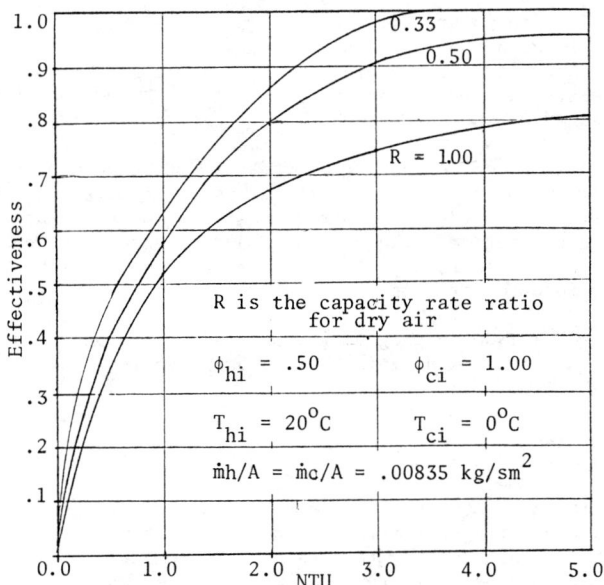

Figure 4. Effectiveness vs. NTU
For A Counterflow, Flat Plate Air To Air
Heat Exchanger

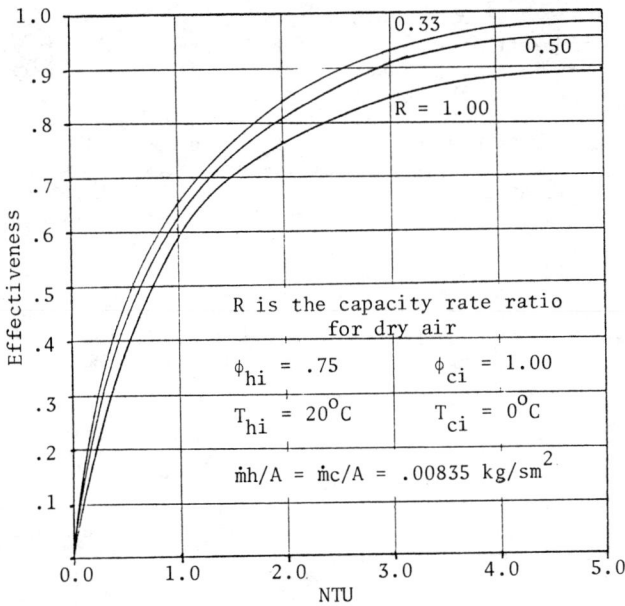

Figure 5. Effectiveness vs. NTU
For A Counterflow, Flat Plate Air To Air
Heat Exchanger

was subsequently used to predict the temperature recovery ratio or effectiveness as a function of the number of transfer units. These computed results are presented in Figures 4 and 5. These data, which are compared to the case of air flow with no water vapour condensation, show that the amount of water vapour in the inlet warm air increases the effectiveness subtantially.

The case of the heat exchanger performace with condensation, icing, and frosting is a much more complicated problem as it involves the simultaneous solution of fifteen nonlinear algebraic equations for any quasi-steady operating time interval. Furthermore, the unsteady mass balance equation must then be integrated for each time interval to predict the accumulation of water, ice, and frost as a function of time.

An attempt was made to program the unsteady icing and frosting conditions that occur in such heat exchangers, but comparative results are not available at this time.

6. DISCUSSION AND CONCLUSIONS

Test results of a counterflow air to air heat exchanger have been presented for a number of operating conditions. These results indicate that water vapour condensation, icing and frosting in such a heat exchanger can increase the heat rates and temperature recovery ratio by up to 40 percent for inlet relative humidities on the hot side of up to 64 percent with maximum temperature differences of 33°C between the hot and cold inlets. In general, the temperature recovery ratio and heat rate is increased as the inlet relative humidity is increased on the hot side and as the temperature difference between the hot and cold inlets is increased.

A simple theoretical model of a steady state one-dimensional flow counterflow heat exchanger has been presented for the case of water vapour condensation. Comparisions between predictions and experimental results suggest that such a model is accurate within the bounds of experimental uncertainty. More work is indicated to extend this model to the case of icing and frosting. The model when used to predict the heat rate and temperature recovery ratio indicates that high inlet relative humidities on the hot side with water vapour condensation will increase the performance of the heat exchanger by more than 30 percent.

REFERENCES

1. R.W. Besant, et al, "Design of a Low Cost Ventilation Heat Exchanger", 1977 International Seminar in Heat Transfer, Dubrovnik, Yugoslavia.

2. B.S. Petukhov, "Heat Transfer and Friction in Turbulent Pipe Flow with Variable Physical Properties", Advances in Heat Transfer, 6, 1970.

3. ASHRAE, Handbook of Fundamentals, 1977.

4. L.R. Wilhelm, "Numerical Calculation of Psychrometric Properties in SI Units", Transactions of the ASAE, 1976, pp. 318.

COMPACT HEAT EXCHANGERS

Compact and Enhanced Heat Exchangers

R. K. SHAH
Harrison Radiator Division
General Motors Corporation
Lockport, New York 14094, USA

R. L. WEBB
Mechanical Engineering Department
Pennsylvania State University
University Park, Pennsylvania 16802, USA

ABSTRACT

This paper starts with the definitions of compact and enhanced heat exchangers, and describes their unique features, subtle characteristics, surface geometries, and flow arrangements. For compact and enhanced heat exchangers, gas is the fluid on one side of the exchanger. Commonly used surfaces are the regenerative type and extended surface type. The latter includes plate-fin and tube-fin surfaces. The plate-fin surfaces considered are plain, wavy, offset strip, louver, perforated, and pin fins. There are two types of tube-fin surfaces: continuous fins on a tube array, and individually finned tubes. The continuous fins considered are the plain, wavy and louver fins on an array of circular or flat tubes. The fins on individual tubes considered are the circular (plain helical), segmented, studded, slotted, and wire loop fins. A comprehensive review of the literature has been made and the nondimensional single-phase heat transfer and flow friction characteristics are presented for all these surfaces. The information includes a summary of local flow structure and performance, analytical solutions, and experimental correlations.

1. INTRODUCTION

In this section, we define a compact and enhanced heat exchanger, compact heat transfer surface, and describe commonly used construction types, surface geometries, and flow arrangements for compact heat exchangers. In the following sections, we will present the local and average heat transfer and flow friction characteristics of compact surfaces. It will include the analytical solutions for simple flow passage geometries and experimental results and correlations for plate-fin and tube-fin surfaces.

1.1 Definitions and Subtle Characteristics

Loosely defined, a compact heat exchanger is one which incorporates a heat transfer surface having a high "area density," that is, a high ratio of heat transfer surface to volume. A compact heat exchanger is not necessarily of small bulk and mass. However, if it did not incorporate a surface of high area density, it would be much more bulky and massive.

Somewhat arbitrarily, we will specify that a <u>compact surface</u> has an area density β greater than $700 \text{ m}^2/\text{m}^3$. A spectrum of surface area density of heat exchanger surfaces has been shown by Shah [1]. A heat transfer surface is designated as a laminar flow surface when predominantly thermally developed or developing laminar boundary layers develop on the surface. Somewhat arbitrarily, we will define a laminar flow surface as a heat transfer surface having area density β greater than 3300 m^2/m^3.

The motivation for using compact surfaces is to gain specified heat exchanger performance, $q/\Delta t_m$, within acceptably low mass and box volume constraints.

$$\frac{q}{\Delta t_m} = U \beta V \qquad (1)$$

Clearly, a high β minimizes the exchanger volume. Moreover, compact surfaces generally result in a higher overall conductance U, again a contribution to a smaller volume. As compact surfaces can achieve structural stability and strength with a thinner section, the gain in lower exchanger mass is even more pronounced than the gain in lower volume. A heat exchanger of any construction is considered as compact if it employs a compact surface on either one or more sides of a two-fluid or a multifluid heat exchanger.[†]

An enhanced surface is the one which yields high heat transfer coefficients compared to an unenhanced plain surface. The high heat transfer coefficients result in three different ways: (1) by adding extended surfaces to the prime surface; (2) by interrupting the surface, for example, forming louvers or strips to plain triangular or rectangular plate-fin surfaces; and (3) by going to smaller and smaller hydraulic diameter flow passages. Both the first and third methods not only enhance the heat transfer coefficient, but also increase the surface area significantly. While not all of the extended or enhanced surfaces would be considered as compact according to the foregoing definition, many of the interrupted and small flow passage geometries fall into the category of compact surfaces. All of the compact surfaces would be classified as enhanced surfaces. Surface geometries having plain, wavy or interrupted fins between plates, finned tubes, or densely packed continuous or interrupted flow passages of various shapes fall into the category of compact and enhanced surfaces. Thus, for many surfaces, compactness and enhancement go hand in hand.

The convective heat transfer coefficient h for gases is generally one or two orders of magnitude lower than water, oil, and other liquids. Hence, for an equivalent hA, the heat transfer surface on the gas side needs to be much more compact and enhanced than can be practically realized with circular tubes. Thus, the major applications of compact and enhanced heat exchangers are the gas side of gas-to-gas, gas-to-liquid and gas-to-condensing or evaporating fluid heat exchangers. Compact and enhanced surfaces have also been used for two-phase applications or on the liquid side for some applications.

The uniquenesses of compact and enhanced exchangers are: (1) many surfaces available having different orders of magnitude of surface area density; (2) flexibility in distributing the area on the hot and cold sides as desired by design considerations; and (3) generally substantial cost, weight, or volume savings.

The subtle characteristics of compact exchangers are: (1) usually at least one of the fluids is a gas; (2) fluids must be clean and relatively noncorrosive; (3) the fluid pumping power (i.e., pressure drop) is often of equal importance to the heat transfer rate; (4) operating pressures and temperatures

---

[†] Note the distinction between the terminologies "compact heat exchanger" and "compact heat transfer surfaces".

are somewhat limited compared to shell-and-tube exchangers due to the brazing, mechanical expansion, etc., construction features; (5) with the use of highly compact surfaces, the resultant shape of the exchanger is one having a large frontal area and a short flow length; the header design of a compact heat exchanger is thus equally important for a uniform flow distribution; (6) the market potential must be large enough to warrant the sizeable manufacturing, research and development costs for highly compact enhanced surfaces.

A variety of methods have been employed to increase the heat transfer coefficient for the exchanger surfaces. An undesirable consequence of this enhancement is an increase in the friction factor. Since the friction power expenditure is equally very important for compact enhanced exchanger applications, only the passive techniques of heat transfer enhancement are employed in practice. These include interrupting the surface in the flow direction or making the flow passage as small as permissible by the design. The resultant heat transfer coefficients for high performing surfaces are 1.5 to 4 times higher than those for the equivalent unenhanced surfaces. Active mechanical devices and turbulator strips used for liquid side enhancement are not used on the gas side of compact enhanced exchangers due to the constraints on the required mechanical and frictional powers.

1.2 Construction Types, Surface Geometries, and Flow Arrangements

As mentioned earlier, a compact heat exchanger on the gas side requires a significantly greater amount of surface area for a specified heat transfer rate than for a liquid as a working fluid. The increase in surface area is achieved by employing surfaces that have high heat transfer surface area density β. The basic construction types employed in the design of a compact exchanger are: extended surface exchangers employing fins on one or more sides, regenerators employing small hydraulic diameter surfaces, and tubular exchangers employing small diameter tubes. Some of the basic criteria for selecting a particular construction type are the cost, operating pressure and temperature, fouling, fluid contamination, and manufacturing considerations. These are further discussed by Shah [2].

Two most common types of extended surface exchangers are the plate-fin and tube-fin types. In a plate-fin exchanger, fins or spacers are sandwiched between parallel plates (referred to as plates or parting sheets) as shown in Fig. 1; sometimes fins are incorporated in a formed tube [1]. Fins are attached

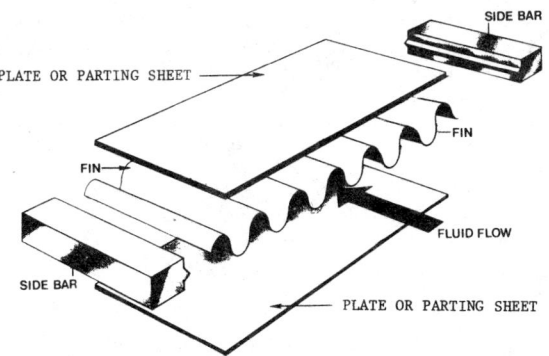

Fig. 1 An illustration of plate-fin assembly.

to the plates by brazing, soldering, gluing, welding, mechanical fit or extrusion. The plate fins are categorized as (1) plain (uncut surfaces) and straight fins such as plain triangular and rectangular fins; (2) plain but wavy fins; and (3) interrupted fins such as offset strip, louver, perforated and pin fins [1]. Examples of commonly used fins are shown in Fig. 2. Since the variations in the louver forms are not easily seen in the photographs, a picture of only the multilouver fin is shown in Fig. 2 along with a sketch of its louver form at Section AA; for all other louver forms, only the sketches of louver form (similar to the multilouver fin) are shown in Fig. 2. By varying the basic geometric variables for each type of plate-fin surface, it is possible to obtain a wide variety of specific surface geometries. Although typical fin densities are 120 to 700 fins/m, applications may exist for as many as 2100 fins/m. Fin thicknesses of 0.05 to 0.25 mm are common. Fin heights may range from 2 to 20 mm. A plate-fin exchanger with 600 fins/m provides about 1300 m^2 of heat transfer surface per cubic meter volume occupied by the fin.

In a tube-fin exchanger, round and rectangular tubes are the most common, although elliptical tubes are also being used. When fins are used, they are employed either outside, inside, or outside and inside of the tubes, depending upon the application. They are attached to the tubes by a tight mechanical fit, tension winding, soldering, brazing, welding, gluing or extrusion. Fins outside the tubes may be categorized as (1) continuous (plain, wavy or interrupted) external fins on an array of tubes;[†] (2) normal fins on individual tubes, referred to as individually finned tubes or simply as finned tubes; and (3) longitudinal fins on individual tubes. Fins inside the tubes are categorized as integral or attached fins [1]. Since an exchanger with longitudinally finned tubes or internally finned tubes is generally not a compact exchanger, we will not consider such finned tubes in the discussion of this paper. The typical fin densities for continuous fins vary from 300 to 600 fins/m, fin thickness from 0.1 to 0.25 mm, fin flow length from 25 to 250 mm. A tube-fin exchanger with 400 fins/m has about 725 m^2/m^3 surface area density. Typical examples of the tube-fin exchangers with continuous fins and finned tubes are shown in Figs. 3 and 4. More examples have been provided in [1].

In a regenerator, generally only the rotary regenerator is a compact exchanger; an exception is the fixed-matrix regenerator for the Stirling engine. Any of the plate-fin surface geometries can be used in the regenerator matrix. However, the interrupted surfaces are not used in a rotary regenerator because of a transverse flow leakage.

For an extended surface compact heat exchanger, crossflow is the most common flow arrangement. This is because it greatly simplifies the header design at the entrance and exit of each fluid. If the desired heat exchanger effectiveness is high (say greater than 75-80%), the size of a crossflow unit may become excessive. In such a case, an overall cross-counterflow multipass unit or a pure counterflow unit may be preferred. There are manufacturing difficulties associated with a true counterflow arrangement in a compact exchanger as it is necessary to separate the fluids at each end. Thus the header design is more complex. Some complex header configurations used in practice are presented in [1]. Multipassing retains the header and ducting advantages of the simple crossflow heat exchanger, while it is possible to approach the thermal performance for counterflow. A parallel flow arrangement having the lowest exchanger effectiveness for a given NTU is seldom used as a compact heat exchanger for single phase fluids.

[†]An exchanger having continuous fins on tubes is also referred to as a plate-fin and tube exchanger. In order to avoid confusion with plate-fin surfaces, we will refer to it as a tube-fin exchanger having continuous fins.

COMPACT AND ENHANCED HEAT EXCHANGERS

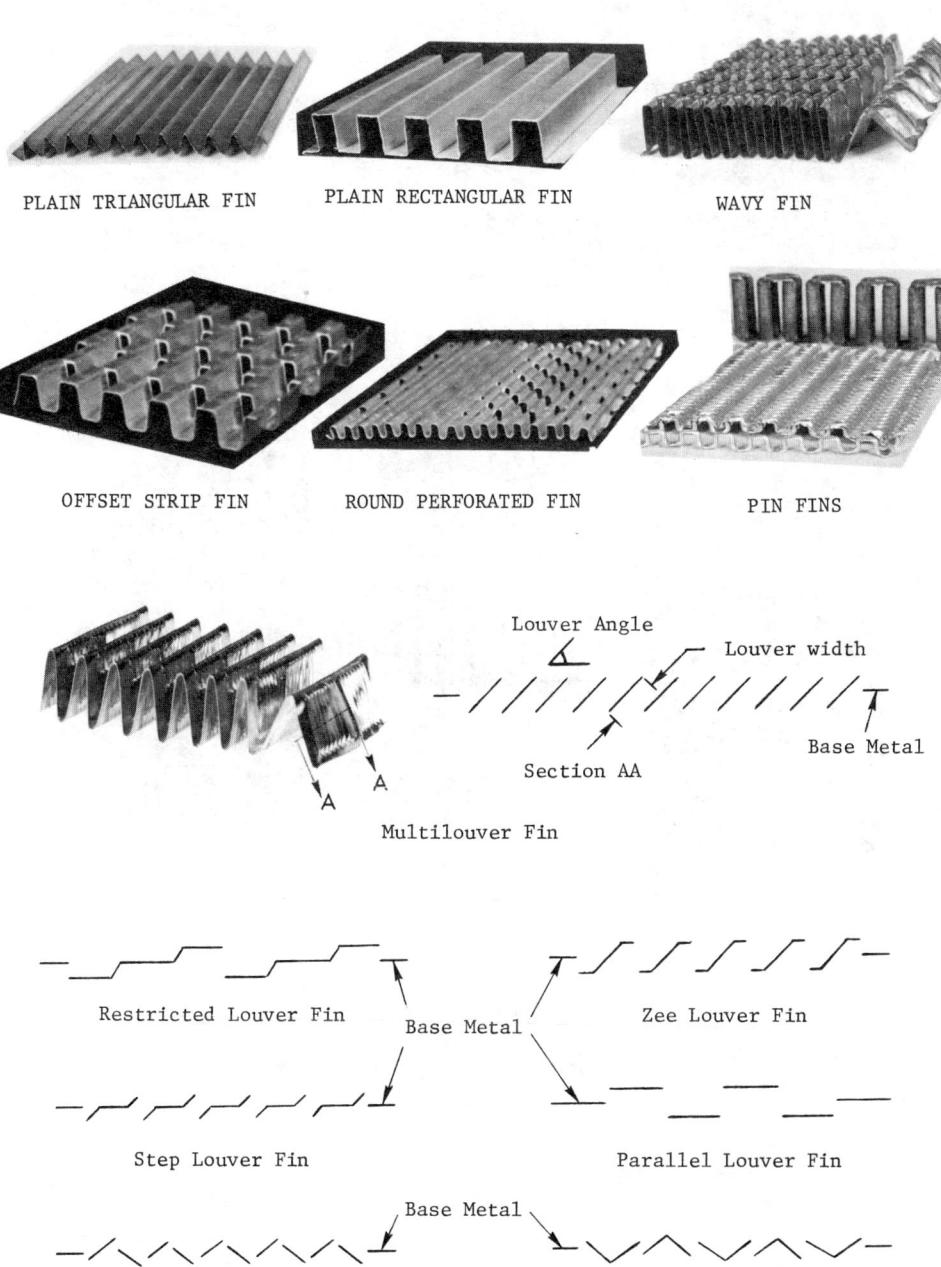

Fig. 2 Fin geometries for plate-fin heat exchangers

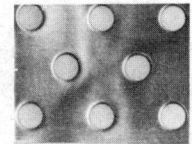
PLAIN FIN FOR STAG-
GERED ROUND TUBES

WAVY FIN FOR STAG-
GERED ROUND TUBES

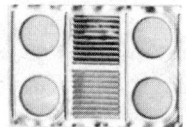
MULTILOUVER FIN FOR
INLINE ROUND TUBES

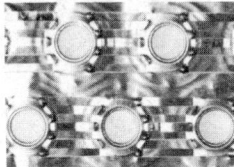
PARALLEL LOUVER FIN
FOR STAG. ROUND TUBES

PLAIN FIN FOR STAG-
GERED FLAT TUBES

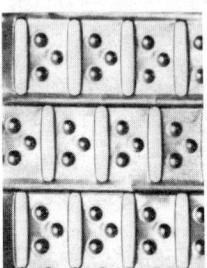
DIMPLE FIN FOR STAG-
GERED FLAT TUBES

WAVY FIN FOR STAG-
GERED FLAT TUBES

PARALLEL LOUVER FIN
FOR FLAT TUBES

MULTILOUVER FIN FOR
STAGGERED FLAT TUBES

Fig. 3 Continuous fins on an array of round or flat tubes.

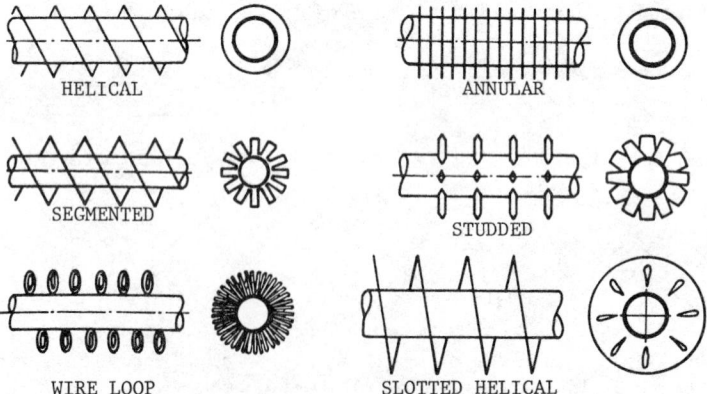

Fig. 4 Individually finned tubes

Single-pass and multipass crossflow arrangements and counterflow arrangements are used in plate-fin exchangers, while generally only the single-pass and multipass crossflow arrangements are employed in a tube-fin exchanger. The thermodynamically superior counterflow arrangement is usually employed in a regenerator.

2. SURFACE BASIC HEAT TRANSFER AND FLOW FRICTION CHARACTERISTICS

The nondimensional heat transfer and fluid flow friction (pressure drop) characteristics of a heat transfer surface are simply referred to as the surface basic characteristics, or surface basic data.[†] The nondimensional heat transfer characteristics are presented in terms of Colburn factor $j = St\, Pr^{2/3}$ vs. Reynolds number Re, Stanton number St vs. Re, or Nusselt number Nu vs. Re or $x^*(=x/D_h Re Pr)$. The nondimensional pressure drop characteristics are presented in terms of Fanning friction factor f vs. Re, or modified friction factor per tube row f' vs. Re_d.

Since the majority of basic data for compact surfaces are obtained experimentally, the nondimensional heat transfer and pressure drop characteristics of these surfaces are presented in terms of j and f vs. Re. Here the Reynolds number Re is based on the hydraulic diameter D_h. This approach is somewhat arbitrary since several variations of one basic type of surface geometry will not generally correlate on the j and f vs. Re basis. This is because geometric variables, other than the hydraulic diameter, may have a significant effect on surface performance. Because the values of j, f and Re are nondimensional, the test data are applicable to surfaces of any hydraulic diameter, provided a complete geometric similarity is maintained.

Correlation of heat transfer and friction data for compact heat exchanger surfaces is still an empirical art since there are many geometrical variables associated with the surface in addition to the operating and design variables. Hence, generally the j and f data are presented on a surface by surface basis. Some progress has been made in prediction or correlation of surface performance and will be described in the following sections.

Let us first discuss the typical design Reynolds number range for compact and enhanced exchangers. For a compact surface, the flow channel hydraulic diameter is small due to closely spaced fins or passages. Operation with low density gases will require excessive friction power, unless the gas velocity in the heat exchanger flow passage is low. These factors usually imply operational Reynolds numbers less than 10,000, the common range being 500-3000 for most compact surfaces. As the hydraulic diameter is further reduced, the Reynolds number is forced to yet smaller values. As a result, very compact surfaces may operate in the low laminar flow regime having the range of Re between 50 and 500.

Compact heat exchangers employ surfaces having either continuous flow passages or flow passages with frequent boundary layer interruptions. In compact continuous flow passages, the flow is generally fully developed over most of the flow length. This fully developed flow is laminar if the Reynolds number based on the passage hydraulic diameter is less than about 2300. This

---

[†] We will not use the terminology "surface performance data" since performance in industry means a <u>dimensional plot</u> of heat transfer rate and pressure drop as a function of the <u>fluid flow rate</u>.

critical Reynolds number is somewhat dependent upon the flow cross-sectional geometry and the entrance condition for noncircular passages. But as a first approximation, if Re $\leq$ 2300, we consider the flow as fully developed laminar for continuous flow passages. Since the heat transfer coefficients associated with plain (uncut) plate-fins (flow passages) are about 1/2 to 1/3 of those associated with interrupted plate-fins at the same hydraulic diameter, plain plate-fins are generally used in highly compact surfaces (very low D_h) or when the pressure drop is critical. So in many compact exchanger applications, when plain fins are used, the Reynolds number is usually well below 2300, and the idealization of fully developed laminar flow is a good one.

The flow is generally developing laminar for highly interrupted fins for Reynolds numbers up to about 10,000. To understand this, consider flow over a flat plate. A stable laminar boundary layer is observed over this plate for the length Reynolds number Re_x (= Gx/μ) $\leq$ 100,000 [3], although the main flow stream may be highly turbulent. The flow over an offset strip fin, multilouver fin and other interrupted fins may be treated as a flow over a flat plate of length x equal to the interruption length ℓ. Since ℓ/D_h for these surfaces is usually less than 10, the Reynolds number Re (= $Re_\ell D_h/\ell$) based on the hydraulic diameter will be at least 10,000 for the boundary layers to be developing laminar. In most compact surfaces, Re seldom exceeds 10,000 and $\ell/D_h \leq$ 10, hence the flow over each interruption is developing laminar flow, although the main flow stream may be turbulent. If there were no surface interruptions, the flow would have been fully developed turbulent at this high Re.

Also for offset strip fins, after the entrance region (5 to 10 strips), the velocity and nondimensional temperature profiles at the beginning of each strip are identical. This kind of flow is designated as periodic flow [4]; the velocity and nondimensional temperature profiles do vary along each strip, but they are invariant from strip to strip at the same axial location from the leading edge of the strip.

As a concluding remark, the design Reynolds number range for compact heat exchangers with gas flows is generally less than 10,000. In most cases, the flow is laminar, either developed, developing, or periodic. For non-compact enhanced exchangers, the flow will be laminar up to the range of Re $\sim$ 1000-2000 depending upon the type of enhancement employed, followed by a transition regime, and it will be fully developed turublent for the range of Re $\geq$ 3000-10,000 depending upon the type of surface.

Although mentioned earlier, it must be reemphasized that even in the laminar region, compact surfaces generally yield high heat transfer coefficients due to the small hydraulic diameter. For example, an 800 fins/m plain rectangular fin geometry in the laminar flow regime would provide the same heat transfer coefficient as a 19 mm diameter tube operated at Re = 25,000. If the plain fin geometry is replaced by an offset strip fin, the resulting heat transfer coefficients may be 2.5 times higher. Thus, the combination of small hydraulic diameters (large surface area per unit volume) and specially enhanced surface geometries can yield very high heat transfer rates per unit volume.

Now we will present the analytical solutions for laminar flow through compact exchanger passages of simple geometries, followed by experimental results and correlations for plate-fin and tube-fin surfaces. Results for non-compact enhanced finned tubes will also be included. Analytical solutions and correlations for turbulent flow are not included here due to space limitations, but the reader may refer to any recent heat transfer textbook [3].

COMPACT AND ENHANCED HEAT EXCHANGERS

3. ANALYTICAL SOLUTIONS AND CORRELATIONS FOR SIMPLE GEOMETRIES

3.1 Fully Developed Laminar Flow

The constant property fully developed laminar flow Nusselt numbers are constant, independent of Re and Pr, but dependent upon the flow passage geometry and thermal boundary conditions. The constant property product of Fanning friction factor and Reynolds number is also constant, independent of Re and Pr, but dependent upon the flow passage geometry, for fully developed laminar flow. Before summarizing the analytical results, three important thermal boundary conditions (T), (H1), and (H2) are defined first in Fig. 5. The (T) boundary condition refers to the constant wall temperature both axially and peripherally throughout the channel (or passage) length. For the (H1) boundary condition, the wall heat transfer rate is constant in the axial direction while the wall temperature at any cross section is constant in the peripheral direction. For the (H2) boundary condition, the wall heat transfer rate is constant in the axial direction as well as in the peripheral direction. Shah and London [5] describe these and other thermal boundary conditions in detail. The Nusselt numbers for these boundary conditions have subscripts T, H1 and H2, respectively. Nu_T is lower than Nu_{H1} for all passage geometries, and Nu_{H2} is lower than Nu_{H1} for noncircular flow passages [5].

Considerable information exists on analytical solutions for channel shapes commonly used in compact heat exchangers. Shah and London [5] present a compilation of analytical solutions for laminar heat transfer and friction of a total of 40 different channel shapes, out of which about 20 shapes could be applicable for compact heat exchanger design. The results include information on solutions for developed and developing velocity and temperature profiles, and the entry length required for fully developed flow. Solutions are given for a wide range of aspect ratios for each basic channel shape, e.g., triangular and rectangular channels. Webb [6] provides a good summary of these results.

The Nusselt numbers and fRe for some channel shapes of interest in compact heat exchanger design are summarized in Table 1. In addition to Nu_{H1},

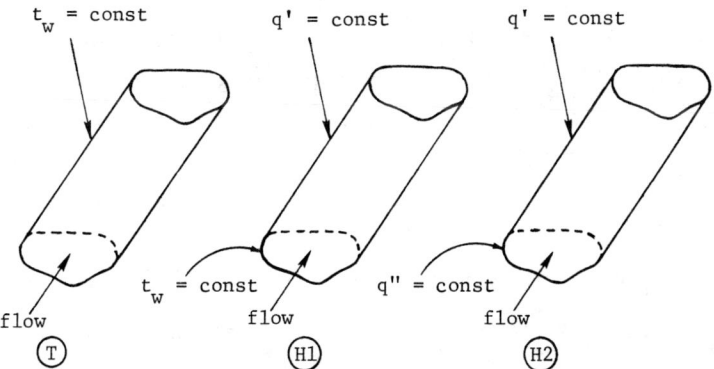

Fig. 5 Thermal boundary conditions for the channel flow.

Nu_{H2}, Nu_T and fRe, the area goodness factor j_{H1}/f is listed; j/f is proportional to the required exchanger frontal area for a specified hA, friction (fluid pumping) power, and flow rate [5]. Also included in Table 1 are the analytical (not experimental) values of L_{hy}^+ and $K(\infty)$. The hydrodynamic entrance length $L_{hy}^+ = x/D_h Re$ is the nondimensional channel length required to achieve a maximum channel section velocity of 99% of that for fully developed flow when the entering fluid velocity profile is uniform. Since

Table 1. Solutions for heat transfer and friction for fully developed laminar flow through specified ducts [5]

GEOMETRY ($L/D_h > 100$)	Nu_{H1}	Nu_{H2}	Nu_T	fRe	j_{H1}/f [†]	$K(\infty)$ [§]	L_{hy}^+ [¶]
Triangle, $2b/2a = \sqrt{3}/2$	3.014	1.474	2.39	12.630	0.269	1.739	0.04
Triangle 60°, $2b/2a = \sqrt{3}/2$	3.111	1.892	2.47	13.333	0.263	1.818	0.04
Square, $2b/2a = 1$	3.608	3.091	2.976	14.227	0.286	1.433	0.090
Hexagon	4.002	3.862	3.34	15.054	0.299	1.335	0.086
Rectangle, $2b/2a = 1/2$	4.123	3.017	3.391	15.548	0.299	1.281	0.085
Circle	4.364	4.364	3.657	16.000	0.307	1.25	0.056
Rectangle, $2b/2a = 1/4$	5.331	2.94	4.439	18.233	0.329	1.001	0.078
Rectangle, $2b/2a = 1/6$	6.049	2.93	5.137	19.702	0.346	0.885	0.070
Rectangle, $2b/2a = 1/8$	6.490	2.94	5.597	20.585	0.355	0.825	0.063
Parallel plates, $2b/2a = 0$	8.235	8.235	7.541	24.000	0.386	0.674	0.011

[†] $j_{H1}/f = Nu_{H1} Pr^{-1/3}/fRe$ with $Pr = 0.7$.

[¶] L_{hy}^+ for sine and equilateral triangular channels are too low [5] so use with caution. L_{hy}^+ for rectangular channels are based on the smoothened curve drawn through the recommended value in [5]. L_{hy}^+ for a hexagonal channel is an interpolated value.

[§] $K(\infty)$ for sine and equilateral triangular channel may be too high [5]; $K(\infty)$ for some rectangular and hexagonal channels are interpolated based on the recommended values in [5].

the flow development region precedes the fully developed region, the entrance region effects could be substantial even for channels having fully developed flow along a major portion of the channel. This increased friction in the entrance region and the change of momentum rate is taken into account by the incremental pressure drop number $K(\infty)$ defined by

$$\Delta p = \left[\frac{4f_{fd}L}{D_h} + K(\infty) \right] \frac{G^2}{2\rho} \qquad (2)$$

For most channel shapes, the mean Nusselt number and friction factor will be within 10% of the fully developed value if $L/D_h > 0.2$ Re (see Fig. 6). If the heat exchanger flow length is short ($L/D_h < 0.2$ Re), the fully developed analytical solutions may not be adequate, since Nu and f are greater in the developing flow region compared to the fully developed values. A heat exchanger consists of many flow passages in parallel. The actual flow passages are never ideal and uniform. Passage-to-passage nonuniformity discussed in [7,8] reduces heat transfer more than a gain by the thermal entrance effect, and hence the latter effect is generally neglected if $L/D_h \geq 100$ for gas flows. This passage-to-passage nonuniformity also reduces the friction factor (and Δp), although by a negligibly small amount, and hence it is generally neglected. The entrance length effect for pressure drop may not be neglected even for $L/D_h \simeq 100$ for gas flows, since it could be substantial.

The channel shapes are listed in Table 1 in increasing order of Nu_{H1}. High aspect ratio rectangular channels are superior to the triangular (corrugated or sine) channels; the latter are commonly used in compact exchangers due to the ease and cost advantage of manufacturing. High j/f is also associated with high Nu. Thus decreasing exchanger frontal area would be required for equal hA, friction power, and flow rate as one moves down the table. The high aspect ratio channels not only require less surface area to do the same job, but also require smaller frontal area. Notice in Table 1 that Nu_{H2} can be substantially lower than Nu_{H1} for noncircular ducts, and therefore one should be careful in designing compact exchangers that employ low thermal conductivity materials.

The following observations may be made from the solutions of Table 1 for laminar flow surfaces having fully developed flows: (1) There is a strong influence of thermal boundary conditions (T), (H1), and (H2) on the convective behavior. Depending on the flow cross-section geometry, the j factor for the (H1) boundary condition may be on the order of 26% greater than that for the (T) boundary condition. (2) As $Nu = hD_h/k$, a constant Nu implies a convective coefficient h that is independent of the flow velocity. (3) An increase in h is best achieved by reducing D_h, or by a change in the type of geometry, e.g., a change from a triangular to an 8:1 rectangular geometry. (4) The friction factor varies inversely with the velocity so that the fluid pressure drop tends to vary with the first power of the velocity.

The theoretical/analytical solutions presented in [5] have been verified by many researchers for flow and heat transfer in a single channel. Thus these results provide a valuable guideline for compact exchangers that employ many such channels in parallel. However, in such a case, passage-to-passage nonuniformity come into the picture that could affect the analytical solutions. Also the thermal boundary condition for heat transfer in an actual exchanger may not exactly correspond to any of the previously described boundary conditions. Hence, accurate (within about ±5%) j and f vs. Re data are generally determined experimentally even for simple geometries used in highly compact exchangers.

As an illustration, London et al. [9] present the following correlations for gas flow through triangular passages (40 < Re < 800):

$$j = 3.0/Re \qquad f = 14.0/Re \qquad (3)$$

London and Shah [10] present the following correlation for gas flow through hexagonal passages (80 < Re < 800):

$$j = 4.0/Re \qquad f = 17.0/Re \qquad (4)$$

<u>Developing Laminar Flow</u>. If the channel length is sufficiently short, it may be necessary to use the entrance length solutions to predict the mean Nu and f over the flow length. Shah and London [5] have summarized the thermal and hydrodynamic entry length solutions for a large number of practically important flow passage geometries.

The ratio Nu_m/Nu_{fd} is shown in Fig. 6 for several channel geometries having a constant wall temperature boundary condition. Here the abscissa $x^* = x/D_h RePr$ is a nondimensional length for the entrance region. Several important observations can be made from this figure: (1) The entrance region Nusselt numbers and hence the heat transfer coefficients can be two to three times higher over the fully developed values depending upon the value of nondimensional interrupted length $\ell^* = \ell/D_h RePr$. (2) At $x^* \simeq 0.1$, although the <u>local</u> Nusselt number approaches the fully developed value, the value of <u>mean</u> Nusselt number can be significantly higher for a channel of length $x^* = \overline{0.1}$. For example, for a

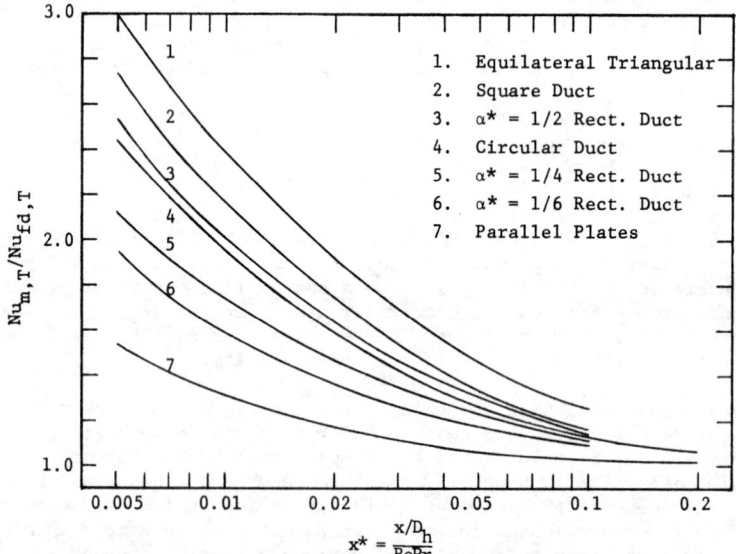

Fig. 6 The ratio of laminar thermally developing Nu to laminar thermally developed Nu for different channel shapes; the velocity profile developed for both Nu's.

triangular tube of length $x^* = 0.1$, while the local Nusselt number at $x^* = 0.1$ attains a value within 4% of the fully developed value [5], the mean Nusselt number Nu_m over the developing region is about 25% higher than Nu_{fd} as found in Fig. 6. Thus, the observations (1) and (2) imply that the entrance length effect could be substantial and advantageous in obtaining high heat transfer rates. (3) The order of increasing Nu_m/Nu_{fd} as a function of channel shape at a given x^* in Fig. 6 is just the opposite of Nu_{fd} in Table 1. For a highly interrupted surface, this means a basic inferior channel shape (such as triangular) will not penalize in terms of low Nu or low h in developing flow as it penalizes for the case of fully developed flow. (4) The higher the value of Nu_m/Nu_{fd} at $x^* = 0.1$, means that the flow channel will require a longer entrance region.

Shah and London [5] proposed the following correlations for thermal entrance solutions for both circular and noncircular channels having laminar developed velocity profiles and developing temperature profiles.

$$Nu_{x,T} = 0.427 \, (fRe)^{1/3} \, (x^*)^{-1/3} \tag{5}$$

$$Nu_{m,T} = 0.641 \, (fRe)^{1/3} \, (x^*)^{-1/3} \tag{6}$$

$$Nu_{x,H1} = 0.517 \, (fRe)^{1/3} \, (x^*)^{-1/3} \tag{7}$$

$$Nu_{m,H1} = 0.775 \, (fRe)^{1/3} \, (x^*)^{-1/3} \tag{8}$$

where f is Fanning friction factor for fully developed flow for a given channel, Re is the Reynolds number, and $x^* = x/(D_h Re Pr)$. For interrupted surfaces, $x = \ell$. The above equations are recommended for $x^* < 0.001$. The functional relationship of $Nu \propto (x^*)^{-1/3}$ of Eqs. (5)-(8) is quite useful for predicting the j data for a new surface if they are known for another surface of the same family and ℓ/D_h are known for both surfaces.

The following observations may be made from Eqs. (5)-(8) and the solutions in [5] for laminar flow surfaces having developing laminar flows: (1) The influence of thermal boundary conditions on the convective behavior appears to be of the same order as that for the fully developed based on the available solutions. (2) Since Nu is proportional to $(x^*)^{-1/3} = (x/D_h Re Pr)^{-1/3}$, Nu is proportional to $Re^{1/3}$ and hence, h varies as $u_m^{1/3}$. (3) The influence of the channel shape on thermally developing Nu is not as great as that for the fully developed Nu.

For simultaneously developing laminar velocity and temperature profiles, the entrance region Nusselt numbers are theoretically even higher than those of Fig. 6. However, these solutions are based on uniform velocity and temperature profiles at the entrance of a channel, and no wake effects or secondary flow effects (that are present in interrupted surfaces) are included in the analysis. Based on the experimental data, the compact interrupted surfaces do not achieve high heat transfer coefficients as predicted by the simultanously developing flows; the results of Fig. 6 or Eqs. (5)-(8) with developed velocity profiles are in better agreement with the experimental data and hence those are recommended for guidelines.

The analytical solutions for the friction factor in the developing laminar flow region have been summarized by Shah and London [5] for a circular tube, parallel plates, rectangular channels, isosceles triangular channels, and concentric and eccentric annular channels. Shah [11] combined the entrance region and fully developed region Δp^* to arrive at a correlation for $f_{app} Re$ factors given by the following equation.

$$f_{app} Re = 3.44(x^+)^{-0.5} + \frac{K(\infty)/(4x^+) + fRe - 3.44(x^+)^{-0.5}}{1 + C(x^+)^{-2}} \quad (9)$$

Here f_{app} is the apparent Fanning friction factor that takes into account both the skin friction and the change in momentum rate in the hydrodynamic entrance region. It is based on the static pressure drop from $x = 0$ to L. The definitions for f_{app} and Δp^* are

$$f_{app} \frac{4L}{D_h} = \frac{\Delta p}{G^2/2\rho} = \Delta p^* \quad (10)$$

$K(\infty)$, $f_{app} Re$ and C of Eq. (9) for rectangular, triangular and concentric annular channels are presented in Table 2. For the entrance region $x^+ < 0.001$, the first term on the right-hand side of Eq. (9) is dominant. Considering this term alone, and using the definitions of f_{app} from Eq. (10) and x^+, it can be shown that $\Delta p \propto u_m^{1.5}$ for the developing laminar boundary layer region.

Although Eq. (9) may provide some guidelines for friction factors for the interrupted surfaces, it includes only the effect of skin friction. The form drag associated with the blunt (smooth and burred) edges of the surface may contribute significantly to the pressure drop. Hence, analytical values of apparent friction factors are generally not used in designing interrupted surface compact exchangers. But, as a rule of thumb, $f = 4j$ or similar relationship is used to predict f factors for interrupted surfaces for which j factor is already known either from the theory or from experiments.

Table 2. $K(\infty)$, $f_{app} Re$ and C for use in Eq. (9), from Shah [11]

α^*	$K(\infty)$	fRe	C	r_i/r_o	$K(\infty)$	fRe	C
	Rectangular channels				Concentric annular channels		
1.00	1.43	14.227	0.00029	0	1.25	16.000	0.000212
0.50	1.28	15.548	0.00021	0.05	0.830	21.567	0.000050
0.20	0.931	19.071	0.000076	0.10	0.784	22.343	0.000043
0.00	0.674	24.000	0.000029	0.50	0.688	23.813	0.000032
				0.75	0.678	22.967	0.000030
2ϕ	Equilateral triangular channels			1.00	0.674	24.000	0.000029
60°	1.69	13.333	0.00053				

4. ANALYTICAL SOLUTIONS AND CORRELATIONS FOR PLATE-FIN SURFACES

A standard reference for the heat transfer and flow friction data of plate-fin heat exchanger surfaces is the monograph by Kays and London, Compact Heat Exchangers [12], published in 1964. Kays and London present j and f vs. Re plots for 56 different plate-fin surface geometries in a unified format. A partial list of additional surface basic data since 1964 is as follows: (1) Plain fins [13], (2) wavy fins [14-17], (3) offset strip fins [18-21], (4) louver fins [20,22-26], (5) perforated fins [27-33], and (6) pin fins [34,35]. We will now separately describe the average and local surface characteristics of plain, wavy, offset strip, louver, perforated and pin fin plate-fin geometries.

4.1 Plain Fins

Surface basic data for rectangular and triangular plain plate-fin geometries are available in [12]. Analytical solutions exist for most plain plate-fin geometries in the laminar flow region as discussed in the preceding section. These solutions provide valuable guidelines in laminar flow when no experimental results are available. In the turbulent flow region, the hydraulic diameter is the significant characteristic dimension. One may predict the turbulent region j and f data for plain fin geometries using the circular tube correlations with the hydraulic diameter as a characteristic dimension [3]. This procedure will generally yield good results, except for isosceles triangular channels having a very small apex angle. The transition region characteristics are unpredictable. No theories or correlations exist that would predict the transition region j and f data. In most cases, the experimental results are the only reliable sources.

Although the j factor for a plain plate-fin surface is lower than that for an interrupted surface (offset strip fin or louver fin) at a given Re, it will require a smaller flow frontal area for specified fixed values of heat transfer, pressure drop, and mass flow rate. Of course, the flow length will be longer for plain fins and so will have a higher overall volume. This would have been expected since the plain fin is not a high performing surface compared to the same D_h interrupted fin.

4.2 Wavy Fins

No specific correlations or prediction methods exist for the j and f data for wavy or herringbone fins. The only data available in the literature are for three wavy fin geometries [12]. Several studies of wavy channel geometries used in plate type heat exchangers [14,15] are applicable, using the principle of geometrical similarity. These studies relate specifically to very large aspect ratio channels, where the aspect ratio in a typical compact heat exchanger is on the order of 8:1. Goldstein and Sparrow [16,17] used a mass transfer technique to measure the local Sherwood number Sh[†] for a wavy fin having two complete waves, a wavy angle of 21°, the spacings between fins of 1.65 mm, and a total horizontal (projected) fin length in the flow direction of 18.5 mm. They measured local and average distributions of the Sherwood number and identified complex flow phenomena. They found that the enhancement

---

[†]Sherwood number for the convective mass transfer problem is analogous to the Nusselt number for the convective heat transfer problem.

in the transfer coefficients due to the wavyness of the wall was small at low Re (about 25% at Re = 1000), but was significant in the low Re turbulent regime (about 200% at Re = 6000-8000). The enhancement is due to Goertler vortices which form as the fluid passes over the concave wave surfaces. These are counter-rotating vortices, which produce a corkscrew-like flow pattern.

4.3 Offset Strip Fins

This is one of the most widely used enhanced fin geometries in compact heat exchangers. In addition to the fin spacing and fin height, the major variables are the fin thickness and fin strip length in the flow direction. The situation is complicated by burrs formed on the leading and trailing fin edges during the manufacturing operation. In addition to the surface basic data [18-21], considerable work has been reported in the open literature on a generalized correlation and analytical local performance. The information will be summarized now. Note that a typical strip length is 3.2 mm, and typical design Reynolds numbers, based on the strip length, are well within the laminar region.

Wieting [36] correlated available experimental heat transfer and flow friction data for 22 offset strip fin surfaces as follows.

For Re $\leq$ 1000,

$$f = 7.661 \, (\ell/D_h)^{-0.384} \, \alpha*^{-0.092} \, Re^{-0.712} \tag{11}$$

$$j = 0.483 \, (\ell/D_h)^{-0.162} \, \alpha*^{-0.184} \, Re^{-0.536} \tag{12}$$

For Re $\geq$ 2000,

$$f = 1.136 \, (\ell/D_h)^{-0.781} \, (\delta/D_h)^{0.534} \, Re^{-0.198} \tag{13}$$

$$j = 0.242 \, (\ell/D_h)^{-0.322} \, (\delta/D_h)^{0.089} \, Re^{-0.368} \tag{14}$$

Here ℓ is the strip length or uninterrupted fin flow length, δ is the fin thickness, D_h is the hydraulic diameter of the passages, and $\alpha*$ is the ratio of width to height of the passage. The following ranges were covered: $0.7 \leq \ell/D_h \leq 5.6$, $0.030 \leq \ell/D_h \leq 0.166$, $0.162 \leq \alpha* \leq 1.196$, and $0.65 \leq D_h \leq 3.41$ mm. Although 85% of all data are correlated within ±15% for f and ±10% for j, a few data have a maximum discrepancy as high as 40%. The deviation is probably influenced by burrs on the fin leading and trailing edges, whose effect is not considered in the correlation.

To obtain the f and j factors for a transitional Reynolds number, Wieting suggested the following procedure. Determine the reference Reynolds number for f and j from the following equations.

$$Re_f^* = 41 \, (\ell/D_h)^{0.772} \, \alpha*^{-0.179} \, (\delta/D_h)^{-1.04} \tag{15}$$

$$Re_j^* = 61.9 \, (\ell/D_h)^{0.952} \, \alpha*^{-1.1} \, (\delta/D_h)^{-0.53} \tag{16}$$

Here Re_f^* is the Reynolds number for an intersection point of the two f vs. Re curves, one for $Re \leq 1000$, and the other for $Re \geq 2000$. Similarly, Re_j^* is the Reynolds number for an intersection point of the two j vs. Re curves, one for $Re \leq 1000$, and the other for $Re \geq 2000$. If the Reynolds number of interest Re is lower than Re_f^*, use Eq. (11) for f, otherwise use Eq. (13). If $Re < Re_j^*$, use Eq. (12) for j, otherwise use Eq. (14). It should be emphasized that the Wieting correlation is strictly based on a limited amount of reported test data. No account is made of burrs on the leading and trailing edges in the correlation. Care must be exercized in extrapolating data for fin geometries that have geometrical parameters outside the range of those for the correlation.

In order to understand and appreciate the high performance of offset strip fin surfaces, comparisons of j and f data of offset strip fins and plain rectangular fins, having approximately the same D_h, have been made by Webb [6] and Shah [8]. Webb compared Surface 101 of [18] with scaled-down Surface 11.11(a) of [12]. He found $j_2/j_1 = 2.9$ and $f_2/f_1 = 3.7$ at $Re = 2000$. Here the subscripts 2 and 1 denote the offset strip fin and the plain fin respectively. Shah compared Surface 104 of [18] with actual Surface 16.96T of [12]. He found $j_2/j_1 = 4.1$ and $f_2/f_1 = 4.4$ at $Re = 2000$. Thus, the offset strip fin surface is a high performing surface having j factors 2 to 4 times higher and f factors 3 to 5 times higher than equivalent plain fin surface, although both surfaces are compact and have the same D_h and β.

A careful examination of all good data that are published has revealed the ratio $j/f \leq 0.25$ for strip fin, louver fin and other similar interrupted surfaces. This can be approximately justified as follows. The flow is developing along each interruption in such a surface. Based on the Reynolds analogy for flow over a flat plate, in the absence of form drag, j/f should be 0.5 for $Pr \simeq 1$. Since the contribution of form drag is of the same order of magnitude as the skin friction for such an interrupted surface, j/f will be about 0.25. Published data for strip and louver fins are questionable if $j/f > 0.3$ and such is the case for the results of Mochizuki and Yagi [21]. All the measurements and possible sources of flow leaks and heat losses must be checked thoroughly for all those basic data having $j/f > 0.3$ for strip and louver fins.

Several flow visualization and heat/mass transfer basic studies have been conducted on a few scaled-up strip fins. A simple model of heat transfer enhancement is the periodic growth and destruction of laminar boundary layers over each strip. However, the boundary layers are not completely destroyed within the wake region.

Several studies have been made dealing with the heat transfer from two colinear plates, with variable axial spacing s_x between the two plates [37-39]. Figure 7 shows the results of Zelenka and Loehrke [38] for two electrically heated plates. This figure shows j_2/j_1 (second plate j factor divided by first plate j factor) vs. s_x/ℓ (the wake length or streamwise spacing between plates divided by the plate length). For $s_x/\ell = 1$ (the actual offset strip-fin surface), Fig. 7 shows $j_2/j_1 \simeq 0.65$ at $Re_\ell = 592$ and $j_2/j_1 \simeq 0.8$ at $Re_\ell = 1627$. Since j_2/j_1 increases with Re_ℓ, it appears that the thermal wake is not dissipated totally. Cur and Sparrow [39] used a mass transfer technique to measure the average Sherwood number on eight colinear plates having the streamwise spacing between plates to the plate length

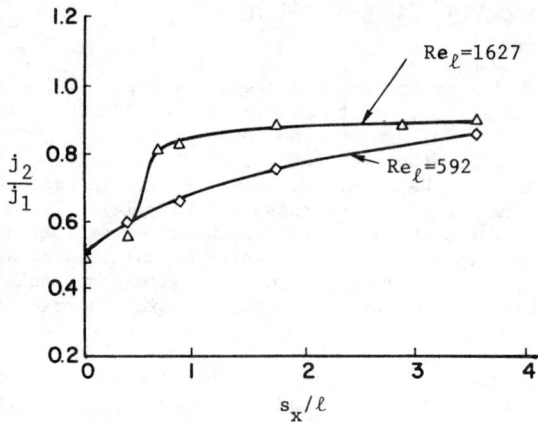

Fig. 7 Ratio of heat transfer from the second plate to that from the first plate of two colinear plates.

ratio as unity ($s_y/\ell = 1$), and the plate thickness to the plate length ratio (δ/ℓ) as 0.04, 0.08 and 0.12. Their findings for the first two plates are in reasonable qualitative agreement with those of Zelenka and Loehrke [38]. In the Re range of 1100 to 13,600, Cur and Sparrow found that fully developed periodic flow established at or prior to the eighth plate. The fully developed periodic flow Nusselt numbers increased with increasing plate thickness. The increase was most dramatic at the lowest Reynolds numbers, up about 65% when δ/ℓ was varied from 0.04 to 0.12. At high Reynolds numbers, the increase was constant at 40% for δ/ℓ varying from 0.04 to 0.12. They found $NuPr^{-0.4} = CRe^n$ with C is a constant dependent upon the δ/ℓ ratio and n varying from 0.72 to 0.74.

Sparrow and Hajiloo [40] extended the earlier mass transfer study to permit measurements in an array of offset strip fins (10 plates in the flow direction) with one strip coated with naphthalene to obtain the transfer coefficients. This plate was placed in successive rows of the array. The matrix geometry had $\ell/s = 3.8$ [plate length divided by fin (transverse) spacing] and $\delta/\ell = $ 0.04, 0.08 and 0.12. In contrast to the study of Zelenka and Loehrke [38], they found that their second row coefficients were higher than those of the first row. They also found that the fully developed periodic flow was observed for the second and all subsequent plates. In contrast to their earlier work [39], they did not find any increase in Nu with increasing δ/ℓ at Re ≃ 1200. At higher Reynolds numbers, Nu increased by about 60% for δ/ℓ varying from 0.04 to 0.12. The average j factors were higher than those of Wieting [36], about 25% for the $\delta/\ell = 0.04$ matrix. The increase was attributed to three reasons: (1) The high aspect ratio of passages ($1/\alpha^* \geq 20$) compared to small aspect ratios of actual exchanger passages; (2) 100% interrupted surfaces (actual exchanger also has continuous prime surface); and (3) Only one heated plate (wake effect exists in the actual exchanger) out of 30 plates. Sparrow and Hajiloo also measured the friction factors. They found that f is significantly affected by the plate thickness; f decreases gradually with Re for $\delta/\ell = 0.04$ and levels off for Re > 7000. In contrast, f is independent of Re for $\delta/\ell = 0.12$. This constancy of f with Re is indicative of the dominance of inertial losses associated with separation and mixing. The measured friction factors for $\delta/\ell = 0.12$ are substantially higher than those predicted by Eq. (13). This raises a question concerning the validity of the

COMPACT AND ENHANCED HEAT EXCHANGERS

data. This is because the scaled-up strip fins did not have burrs on the leading and trailing edges as would be found in an actual exchanger; one would have expected lower (and not higher) friction factors for the scaled-up strip fins compared to those by Eq. (13).

Several attempts have been made to predict analytically the heat transfer and flow friction characteristics of offset strip fins. One of the first attempts was made by Kays, first in his Ph.D. dissertation and then in [41]. He proposed to compute the heat transfer to or from an offset strip fin from the Pohlhausen laminar boundary layer solution for a flat plate. For friction, he modified the Blasius laminar boundary layer solution for a flat plate by the form drag associated with the leading blunt edge of the strip fin. He proposed

$$j = 0.664 \, Re_\ell^{-0.5} \tag{17}$$

$$f = \frac{C_D \delta}{2\ell} + 1.328 \, Re_\ell^{-0.5} \tag{18}$$

Kays suggested $C_D = 0.88$ based on the potential flow normal to a flat plate. In this model, it is idealized that the velocity and temperature boundary layers are destroyed in the wake region. Although this model is at best a first order approximation, it predicts the data with fair accuracy compared to the sophisticated numerical solutions; however, it does not provide the detailed insight into the mechanisms.

Sparrow, Patankar and coworkers [4,42] have attempted to predict the j and f factors numerically by a finite difference method for the offset strip fins considering the laminar boundary layers on each strip fin. The first study [4] was for strip fins of infinite height (two-dimensional problem) and zero thickness ($\delta/s = 0$) with ℓ/s varied from 0.2 to 5. Although the idealized model is simple, the numerical computation is complex. In [42], they extended the analysis for finite fin thicknesses, $\delta/s = 0.1$, 0.2 and 0.3. In this model, the impingement region on the leading edge and the recirculating region behind the trailing edge are considered in the analysis. At low Reynolds numbers or low δ/s, the recirculation zone behind the trailing edge is small as shown in Fig. 8. At high Re or high δ/s, this zone extends from the trailing edge of a strip to the leading edge of the succeeding strip. Thus at high Re or high δ/s, the throughflow confines to the minimum flow area (see Fig. 8) resulting in effective high velocity or a thinner boundary layer in the central portion of the strip fin. This results in higher j and f at high Re for a given δ/s, or at high δ/s for a given Re. The increase in the f factor is more pronounced with increasing δ/s because of both increased form drag due to the thicker δ and increased skin friction due to higher effective velocity over the strips at a specified Re. The comparison of the numerical results with a strip fin surface from Kays and London indicated a reasonable agreement with the f factors, but the predicted j factors are about 100% higher. The predicted slopes of j and f vs. Re curves are steeper than the test data.

Although a substantial amount of experimentation and numerical predictions have been performed, further improvements in the numerical model are needed to accurately predict and understand the performance of the strip fins. One of the major factors that cannot be taken into account accurately in the computation is the small burrs at the leading and trailing edges of the fin during its formation by a shearing operation. This burr increases the effective plate

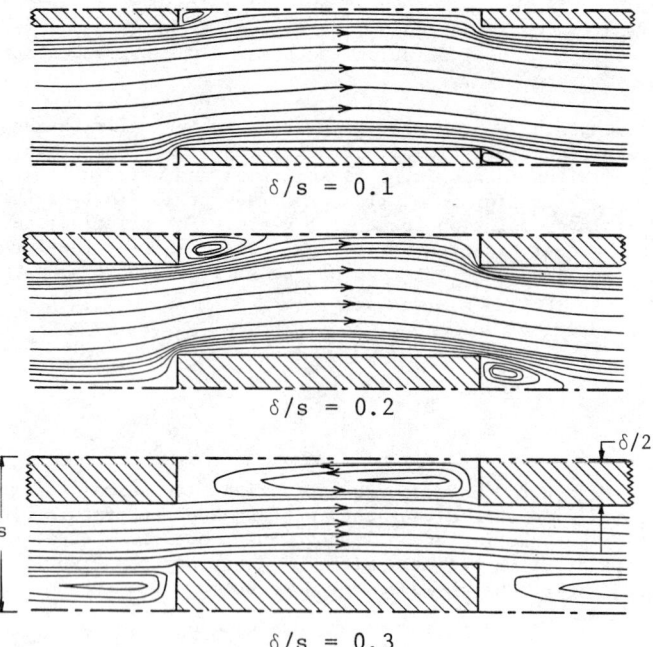

Fig. 8 Flow patterns over offset strip fins at different δ/s for $Re = \rho u_\infty D_h/\mu = 2000$, based on a numerical solution [42].

thickness, causing increased form drag. The possible existence of burrs causes uncertainty in the correlations or in the comparison of predicted values with data. Equation (18) may be used to predict the effect of burrs. For $Re_\ell = 2000$ and 3.18 mm strip lengths, the form drag contribution is 32% for 0.10 mm fin thickness, and increases to 54% with 0.25 mm fin thickness. Usually, the experimental friction factor exceeds that of Eq. (18).

4.4 Louver Fins

Louvers are formed by cutting the metal and either turning, bending, or pushing out the cut elements from the plane of the base metal. Louvers can be made in many different forms and shapes, some of which are shown in Fig. 2. Note that the parallel louver fin and offset strip fin both have small strips aligned parallel to the flow. The strips of parallel louvers are not pushed out as far as those of the offset strip fin, and they do not extend to the full height of the fin; the strips of offset strip fins do extend to the full height of the fin. The louver fin gage is generally thinner compared to the offset strip fin. The base metal has triangular (or corrugated) shape. And it is generally not as strong as the offset strip fin, since the latter has a "large" flat area for brazing, thus providing strength. The louver fins may have a slightly higher potential for fouling compared to the offset strip fins.

Louver fins are amenable to high speed mass production manufacturing technology, and as a result, are less expensive compared to the offset strip fins and other interrupted fins when produced in very large quantities. Desired fin spacing can be achieved by squeezing or stretching the fin, hence it allows some flexibility in fin spacing, without changes in the tools and dies. This flexibility is not possible with the offset strip fin. A wide range in performance can be achieved by changing the louver angle and louver width ℓ. These fins are extensively used in automotive heat exchangers.

Various louver fins have been tested extensively, and a considerable amount of proprietary data exists that is unavailable in the open literature. Some data and investigation have been reported by [12,20,22-26]. The multilouver fin has the highest heat transfer enhancement relative to the pressure drop in comparison with most other louver forms, and hence it is now extensively used by the industry. Some results for multilouver fins obtained by Fujikake [24] are presented in Fig. 9. Here the ratio of heat transfer enhancement is plotted against the frontal air velocity. The top curve shows the ratio of mean heat transfer coefficient for a multilouver fin of louver width $\ell = 1.7$ mm and fin flow length $L_f = 32$ mm to that for a plain unlouvered fin of the same length L_f, all other geometrical dimensions being kept the same. It is clear that the enhancement given by the multilouver fin is 2 to 2.5 times over the plain fins. When the fin pitch s is reduced from 2.85 mm to 2.5 mm, the enhancement ratio decreases from 5% to 25% as shown by the second curve from the top. When the louver width is reduced from 3.5 mm to 1.7 mm, the enhancement is about 1.4 times. As shown in the bottom curve, the periodic flow is fully developed at $L_f = 32$ mm ($L_f/\ell = 19$). Note that the louver angle for all of the multilouver fins of Fig. 9 was 28° except for the fin with $L_f = 38$ mm that had 14° louver angle.

Izumi and Yamashita [25] report that the heat transfer coefficient is proportional to $u_m^{0.6}$ and the pressure drop $u_m^{1.55}$ for multilouver fins. For simultaneously developing laminar flow in channels, Shah and London [5] show that $h \propto u_m^{0.5}$ and $\Delta p \propto u_m^{1.5}$. Thus the laminar boundary layer development over each thin louver is the major cause for heat transfer enhancement for multilouver fins.

Although slightly higher form drag may be associated with the louver fin compared to an offset strip fin, the performance of a well designed multilouver fin exchanger can approach that of an offset strip fin exchanger with possibly increased surface compactness and reduced manufacturing cost.

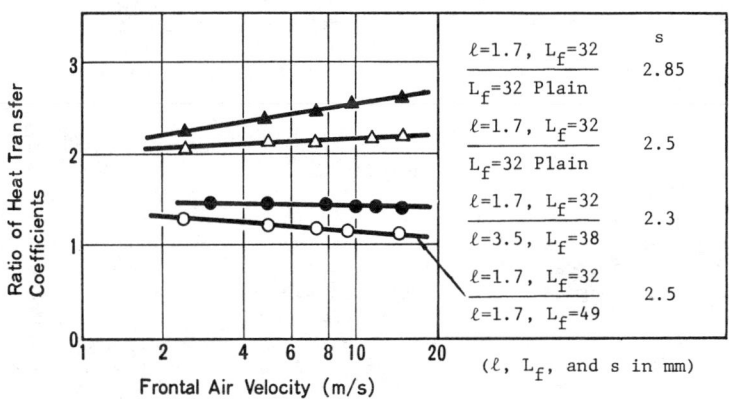

Fig. 9 Heat transfer performance comparison of multilouver fins [24,26].

4.5 Perforated Fins

After the pioneering study on a perforated fin, Kays [43] claimed that the perforated heat exchanger surfaces have a substantial increase in heat transfer performance without introducing a pronounced form drag due to frequent boundary layer interruptions. Since that study, in the next 15 years, seven groups of investigators tested 68 perforated cores for heat transfer, pressure drop, flow phenomena, noise and vibrations. All these results are summarized by Shah [30, 31]. Subsequently, some additional results have been reported [32,33]. Conclusions of Shah are: Microscopic performations ($d_{hole} \leq 0.8$ mm) may provide enhanced heat transfer in the laminar region, if plate open area due to performation is greater than 20%. Macroscopic performations ($d_{hole} > 1$ mm) do not enhance performance in the laminar region. Perforations induce early transition. However, the transition region j and f vs. Re characteristics are quite complex and unpredictable. Contrary to expectations, an increase in f is often higher than an increase in j. The turbulent flow j and f data are usually much higher than those of unperforated surfaces, and are usually associated with noise and vibration in transition and turbulent flow regimes if the fins are not rigidly attached (brazed) to the plates or parting sheets. The rectangular slot performations yield slightly better performance compared to the round perforations. Since the punched out material is wasted as well as considerable surface that may also be lost, the performance of the perforated surfaces is not superior as was originally claimed. So these surfaces are not used in compact gas-to-gas exchanger applications. The only perforated fin application that the present authors are aware of is in oil coolers as "turbulators". Perforated fins once used in two-phase cryogenics air separation exchangers are now replaced by the high performance offset strip fins.

4.6 Pin Fins

From the foregoing discussion, we know that the shorter the strip length, the higher will be the enhancement characteristics. From these viewpoints, a limiting strip fin geometry with good mechanical strength is a pin fin geometry. A large number of small pins are sandwiched between plates in either inline or staggered arrangement. Pins may have round, elliptical or rectangular cross-section. Pin fins can be manufactured at a very high speed continuously from a wire of proper diameter. After the wire is formed into rectangular passages (see the fin in the foregound Pin Fins in Fig. 2[†]), the top and bottom are flattened for brazing or soldering the pin fins with the prime surface (parting sheets).

Because of the potential for high enhancement characteristics of pin fins, they have been investigated extensively by industry. However, only limited amount of data are published. The j and f data for five round pin fins and one elliptical pin fin geometries (all inline) are presented by Kays and London [12]. Theoclitus [34] presented j and f data for nine pin-fin geometries with inline square pin spacing of $4 < b/d_o < 12$, where b is the pin length (plate spacing) and d_o is the pin diameter. Recently, VanFossen [35] presented heat transfer data for short pin fins having $b/d_o = 1.2$ and 2. He estimated that the heat transfer coefficient for pin fins is about 35% higher than that for the end walls (parting sheets). He found that the short

---

[†]After the pin fin in the foreground, in the middle of the picture, a number of fins are placed side by side with a braze sheet ready for brazing. In the background, shown are tall elliptical cross-sectional two pin fins standing one behind another.

pin fin heat transfer is about 100% higher than that for a channel of the same aspect ratio, but without fins. The heat transfer of long pin fins, $4 < b/d_o < 12$, of [12,34] is about 75-100% higher than that of the short fins.

Pin fins may be considered as tube banks having very small diameter tubes. As such, low Reynolds number correlations may provide a first order prediction of the j and f characteristics of the pin fin geometry. Indeed, the friction factor results of Theoclitus [34] "bracketed" the tube bank correlations. However, the heat transfer results were 20-35% lower for pin fins than predicted by the tube bank correlations. Possible reasons are: (1) There are no end walls (having low h) for tube banks; and (2) the tube walls are nearly at isothermal conditions due to fluid flow within the tube of a tube bank, whereas there is a significant temperature gradient along the pins due to the "fin" effect.

Pin fins have not become widely used in compact exchangers for the following reasons: (1) The compactness β achieved by the pin fin geometry is much less than that by the multilouver fins or offset strip fins; (2) The cost of wire is considerably higher than that of a thin strip per unit surface area; (3) From the manufacturing considerations, only the inline arrangement of the pins is desirable. This arrangement may not yield high heat transfer coefficients. Also, parasitic form drag is associated with the flow normal to the pins; (4) Due to vortex shedding behind the pins, noise and flow-induced vibrations are possible [34], which are generally not acceptable in most heat exchangers.

The potential application of pin fins is at very low flows (Re < 500), for which the pressure drop is of no major concern. Pin fins are used as electronic cooling devices with generally free convection flow on the pin fin side.

5. BASIC DATA AND CORRELATIONS FOR TUBE-FIN SURFACES

Heat transfer and flow friction data are summarized separately for two types of tube-fin surfaces: continuous fins on a tube array, and individually finned tubes. A more detailed review on the performance of tube-fin surfaces has been provided by Webb [44].

5.1 Continuous Fins on a Tube Array

This type of tube-fin geometry (shown in Figs. 3 and 4) is most commonly used in air conditioning and refrigeration exchangers[†] in which high pressure needs to be contained on the refrigerant side. As mentioned earlier, even though this type of tube-fin geometry is not as compact (having high β) as the plate-fin geometries, its use is becoming widespread in the current energy conservation era. This is because the bond between the fin and the tube is made by mechanically or hydraulically expanding the tube against the fin. Formation of this mechanical bond thus requires a very small energy consumption compared to the energy required to solder, braze or weld the fin to the tube. Because of the mechanical bond, the applications are restricted to those cases in which the differential thermal expansion between the tube and fin material is small, and preferably the tube expansion is greater than the fin expansion. Otherwise, the loosened bond will have a significant thermal resistance.

[†]These exchangers are simply referred to as "coils".

The flow structure within a finned tube bank is more complex than that in the previously discussed plate-fin channels. The presence of a circular tube causes flow acceleration over the fin surface and flow separation on the back side of the tube which may yield low velocity wake regions. Secondary flows may be associated with wavy fins, while flow separation and reattachment may be associated with the louver fins. The local and overall performance of plain, wavy and louver fins will be discussed separately next.

Plain Fins on a Round Tube Array.   Several investigators have measured the local mass transfer coefficient distribution on plain (flat) fins on circular tubes. Through the heat and mass transfer analogy, the nondimensional mass transfer coefficient, Sherwood number Sh, is analogous to the Nusselt number Nu in heat transfer. As a result, we will use the phrases "heat transfer coefficient" and "mass transfer coefficient" interchangeably in the following discussion. Krückels and Kittke [45] used a mass transfer technique to establish the local mass transfer coefficient distribution on plain fins on two row inline tubes, as shown in Fig. 10. The oncoming Re_d are 3700 and 19,000 for the upper and lower halves of the figure, respectively. For compact exchanger application, the design Reynolds numbers are in the 3700 range or lower. This figure shows that a low velocity wake exists in the region between the tubes, and that the heat transfer coefficients are low in this region. There is a 10:1 variation of the local mass transfer coefficient over the fin surface area. Saboya and Sparrow [46-48] have also used a mass transfer technique to establish the local mass transfer coefficient distribution on one, two and three row coils having a staggered tube arrangement. The results were interpreted in terms of heat transfer through heat and mass transfer analogy. They employed the following geometry: distance between fins 1.65 mm,[†] tube outside diameter 8.53 mm, transverse tube spacing 21.3 mm, and longitudinal tube spacing 18.5 mm. They covered the Reynolds number range of 160-1270. They found different transfer mechanisms operating in different portions of the fin. For the fin associated with the first row, two factors provide high heat transfer rates: the boundary layer on the forward part (leading edge) of the fin and a

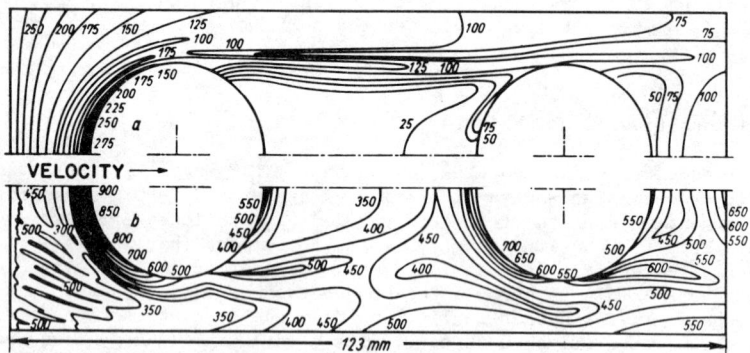

Fig. 10 Distribution of mass transfer coefficients (m^3/m^2 hr) on a two row plain continuous fin coil [45].

[†]For an assumed fin thickness of 0.15 mm, this translates into 556 fin/m.

vortex system on the fin portion in front and sides of the tube. Low heat transfer coefficients are found on the fins downstream of the minimum flow cross section at the first tube row. The boundary layer development is the most important factor for the first row fin with the vortex-induced transfer mechanism becoming an important factor at higher Reynolds numbers ($\sim$ 1000-1200). For the second and third row fins, there is no region of boundary layer development; the vortex system alone is responsible for high heat transfer. At low Reynolds numbers (Re $\simeq$ 200), the fin associated with the first row transfers about 50% of the total heat, the fin with the second row 28%, and the fin with the third row 22%. As the Reynolds number increases, stronger vortices are activated, and the relative contributions of the second and third row fins increase. At Re $\simeq$ 1100, heat transfer by fins of each row of a three row coil is nearly equal. The high heat transfer coefficients produced by the vortex may be seen in Fig. 10 at the front of the tube, and in the narrow zone behind the tube at approximately one tube radius from the centerline through the tubes.

Even with plain fins, there are many geometric variables (d_o, X_t, X_ℓ, s, L), and the flow structure is too complex to permit analytical predictions of the heat transfer and friction characteristics. Only limited progress has been made in the development of empirical correlations for continuous plain fin coils. Rich [49] obtained heat transfer and friction data for four rows deep plain fin coils having 13.3 mm outside diameter tubes equilaterally spaced on 31.8 mm centers. All fins were plain and 0.15 mm thick. The tubes and fins were made of copper and were carefully tin soldered to ensure good thermal bond and to minimize contact resistance. The geometries of all coils were identical, except for the fin density which was varied from 115 to 811 fins/m. Figure 11 shows the Fanning friction factor f and the Colburn factor j (smoothened data fit) as a function of the Reynolds number Re for the eight fin spacings tested. Entrance and exit losses have been lumped into the friction factors. The Reynolds number is based on the hydraulic diameter and the mass velocity in the minimum flow area. Figure 12 shows Rich's proposed correlation to account for the effect of the fin spacing. The Reynolds number in this correlation is $Re_\ell = GX_\ell/\mu$ where X_ℓ is the longitudinal spacing between tube rows. Rich proposed the following equations for the correlations of Fig. 12.

$$j = 0.195 \, Re_\ell^{-0.35} \tag{19}$$

$$f_f = 1.70 \, Re_\ell^{-0.5} \tag{20}$$

Figure 12 shows that the heat transfer coefficient is a function of the mass velocity G only, and is independent of the fin spacing. At the same mass velocity G, the bare tube bank heat transfer coefficient is about 40% larger than that of the finned tube bank. For the friction correlation, the total pressure drop is considered to be the sum of the pressure drop on a bare tube bank (geometry without the fins) and the pressure drop associated with the fins. Thus, f_f is the friction component resulting from the fins alone, defined as

$$f_f = \frac{\Delta p - \Delta p_t}{(G^2/2\rho)} \frac{A_c}{A_f} \tag{21}$$

The term Δp_t is that measured for a bare tube bank of the same geometry,

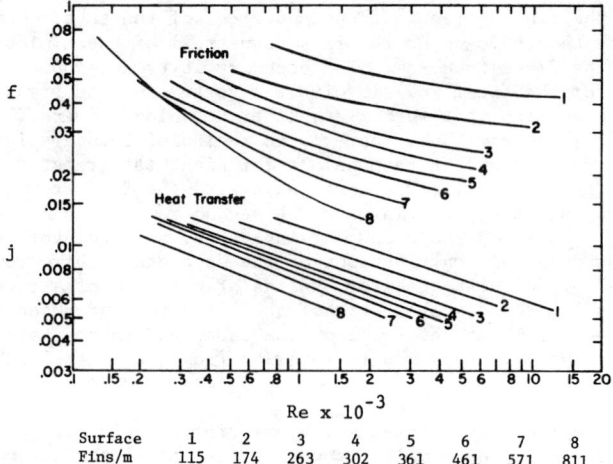

Fig. 11 Heat Transfer and flow friction characteristics of continuous plain fins on a round tube array [49].

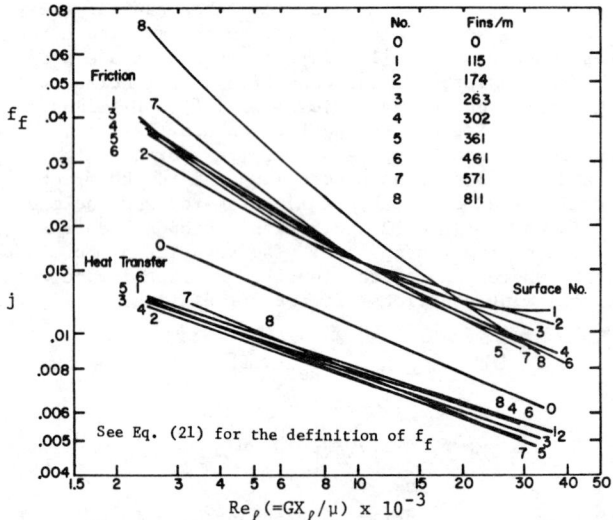

Fig. 12 Correlation of the results of Fig. 11 with f changed to f_f and Re changed to Re_ℓ [49].

but without fins. This means the fin blockage was ignored in the Δp_t computation, i.e., if the tube area exposed is 90% and the tube area under the fin is 10% say, for example, then Δp_t should have been 90% of Δp_t of the bare tube bank without the fins; however, Rich considered 100% Δp_t regardless of the fin density. Δp_t can be evaluated from the correlations for bare tube banks such as that of Grimison [50]. Both pressure drop contributions are evaluated at the same minimum flow area mass velocity. Figure 12 shows that the resulting friction correlation is reasonably good, except for the closest fin spacings. The present authors believe that the friction data on surfaces 7 and 8 may be somewhat questionable, since these surfaces show smaller j/f values than for

COMPACT AND ENHANCED HEAT EXCHANGERS   451

the other fin spacings at low Re_ℓ. This behavior is unexpected unless the fin spacing was nonuniform and some flow passages were totally blocked. Normally, the j/f ratio will increase as the fin spacing is reduced, since the fractional parasitic drag associated with the tube is reduced. Use of the length Reynolds number in Fig. 12 has no real significance, since all geometries tested had the same X_ℓ.

In a later study, Rich [51] used the same exchanger geometry with 571 fins/m to determine the effect of the number of tube rows on the j factor. Figure 13 shows the average j factor (smoothened data fit) for each coil as a function of Re_ℓ. The numbers on the figure indicate the number of tube rows in each coil. The row effect is greatest at low Reynolds numbers and becomes negligible at $Re_\ell > 15{,}000$. McQuiston [52] has provided a correlating equation for the Fig. 13 data.

The present authors believe that reduced data of Fig. 13 for three and more number of tube rows are in error, particularly for the values of Re_ℓ lower than those where j decreases with decreasing Re_ℓ. The reason is the observed "rollover" or "drop-off" in j with decreasing Reynolds numbers. Kays and London [53] and other investigators have shown that this rollover in j factors with decreasing Re occur due to the inaccuracies in the temperature and flow measurements and significant heat losses to the ambient at "low" air velocities when the core effectiveness is high. It requires extreme care and very accurate instrumentation to obtain the reliable j data at low air flows. An evaluation of the raw data of Rich [54] indicated significant heat losses to the surroundings or inaccuracies in the temperature measurements at low air flows, although he took extreme care in measurements. For example, a 11.4% difference was found between the measured airside and waterside heat transfer rate $(q_a - q_w)$ for the N = 3 coil at $Re_\ell = 2330$. These measured temperatures resulted in the exchanger effectiveness ε based on q_a and q_w as 82.0% and 96.3%, respectively. This amounted to a variation in j based on q_a vs. q_w of 213%. When the j factor is recomputed consistently on the average

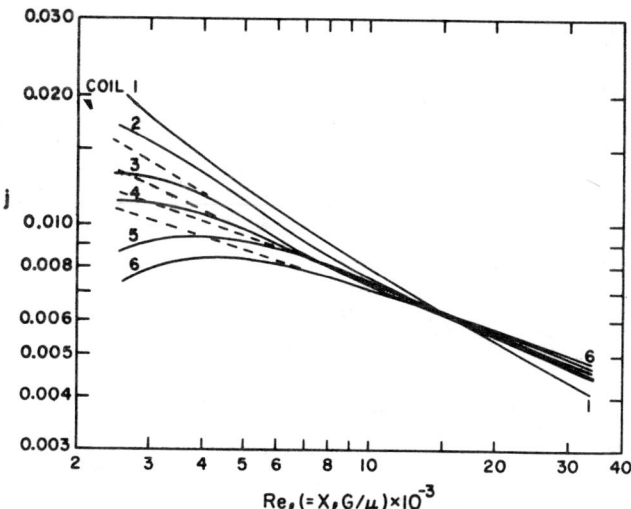

Fig. 13 Influence of number of tube rows on heat transfer of plain continuous fins on a circular tube array [51].

$q = (q_a+q_w)/2$, it falls on the dashed line as shown in Fig. 13. As another example, a 11.6% difference was found between q_a and q_w for the $N = 6$ coil at $Re_\ell = 2380$. The resultant ε based on q_a and q_w was 88.8% and 99.2%. This amounted to a variation in j based on q_a vs. q_w of 250%. Again, when the j factor is recomputed consistently on the average q, it falls on the dashed line as shown in Fig. 13. Hence, we recommend using the straight dashed line curves (thus eliminating the rollover in j) for the tube row effect on Rich's data. It should be reemphasized that it is very difficult to get accurate j data at low flows. Alternatively, a large change in j at low flows will have a small influence on the overall exchanger performance.

To further reconfirm the foregoing conclusion on Rich's data, let us review the mass transfer results of Saboya and Sparrow [48]. They did find decreasing j factors with increasing number of tube rows from one to three rows at low Re (< 1087), this decrease diminished with increasing number of tube rows. For example, for Re = 200, they found the decrease in j factor as about 14% from a one row to a two row coil, and about 7% from a two to a three row coil. At Re = 200 ($Re_\ell = Re \ X_\ell/D_h = Re \times 12.89 = 2578$),[†] Rich finds these decreases about 17% and 21%, respectively. Based on this comparison, the rate of decrease in j factors with an increasing number of tube rows at low Re should have been decreased and not increased as seen in Fig. 13. Incidentally, Rich did not report the tube row effect on the f factors in [51] since he did not find any significant tube row effect on f factors for $N \geq 2$ row coils within the experimental uncertainty [54].

McQuiston [52] developed the following simple empirical correlation for four row staggered banks of plain fins based on his data [55,56], Rich's data [49] and the Kays-London data [12].

$$j = 0.0014 + 0.2618 \ Re_d^{-0.4} \ (A/A_t)^{-0.15} \quad (22)$$

where $Re_d = Gd_o/\mu$ and d_o is the tube outside diameter, A is the total (fin + tube) airside surface area, and A_t is the outside area of the tubes without the fins. The coil geometries used by McQuiston in developing Eq. (22) are listed in Table 3. Ninety percent of data were correlated within ±10%. McQuiston also provided the following correlation for Fanning friction factor f based on D_h.

$$f = 0.004904 + 1.382 \left\{ Re_d^{-0.25} \left(\frac{r_o}{r^*}\right)^{0.25} \left[\frac{(X_t-d_o)\gamma}{4(1-\gamma\delta)}\right]^{-0.4} \left[\frac{X_t}{2r^*} - 1\right]^{-0.5} \right\}^2 \quad (23)$$

where

$$\frac{r^*}{r_o} = \frac{A/A_t}{(X_t-d_o)\gamma + 1} \quad (24)$$

and γ is fin density, fins/m, and δ is the fin thickness in m. Equation (23) correlated the data of the same coil geometries of Table 3 within ±35%.

[†] All Rich's coils [51] had $X_\ell/D_h = 12.89$. The core geometries of Saboya-Sparrow and Rich are not identical, but are similar and should provide the similar trends.

Table 3. Continuous plain fins and four row staggered tube coil geometries used by McQuiston [52] to derive Eqs. (22) and (23)

Tube dia. d_o, mm	X_t/d_o	X_ℓ/d_o	Fin Thickness, δ, mm	γ Fins/m	Number of coils	Ref.
13.3	2.38	2.06	0.15	115-811	8	[49]
10.2	2.49	2.15	0.33	315	1	[12]
17.2	2.22	2.59	0.41	305	1	[12]
9.96	2.55	2.21	0.15	157-551	5	[56]
10.3	1.97	1.70	0.17	315, 563	2	[55]

The present authors are unaware of any published surface basic data on multirow plain fin coils having an inline tube arrangement. This is because the flow bypasses between the tube rows in the flow direction (tube bypass effect), and substantially degrades the performance of an inline tube arrangement. This flow bypassing has been shown by Fukui and Sakamoto [57] through flow visualization. The degree of performance degradation has been established for circular fins, and is discussed in Section 5.4.

Wavy Fins on a Round Tube Array. This continuous fin geometry, also sometimes referred to as corrugated fin geometry, is a most commonly used design for air conditioning condensers and commercial heat exchangers. Because of its superior performance over plain fins and ruggedness, this fin is probably the most investigated by the industry. However, actual performance data on only one geometry have been published by Hosoda et al. [58] in dimensional form, and the performance comparison is made only with the parallel louver fin to be discussed in the next subsection.

Goldstein and Sparrow [59] used a mass transfer technique to measure the local mass transfer coefficients. Their test geometry simulated a one row wavy fin design having 556 fins/m (for an assumed fin thickness of 0.15 mm) on an 8.53 mm diameter tube. The wave configuration is the same as that tested by Goldstein and Sparrow [17] without the tube present; those test results were discussed in Section 4.2. Goldstein and Sparrow presented detailed local Sherwood number distribution on each upper and lower facet of the wavy fin. They identified the presence of several vortex systems and concluded that the windward facets of the wavy fins are primarily responsible for the enhancement of the transfer coefficients; the leeward facets have lower transfer coefficients and are strongly affected by the flow separation. A comparison of average transfer coefficients of this wavy fin with those of a similar plain fin showed that the enhancement due to the wavy fin surface increased with the Reynolds number. At Re = 1000, this one row design provided a 45% higher average transfer coefficient than a plain fin exchanger of the same basic geometry.

Louver Fins on a Round Tube Array. As discussed earlier, both strip fin and louver fin geometries have high performance and are most commonly used for plate-fin exchangers. One would think that the concepts of strips

---

†The dimensions s, X_t, X_ℓ and d_o are the same as those of plain fins tested by Saboya and Sparrow [46] and are listed earlier in the section on Plain Fins on a Round Tube Array in Section 5.1.

and louvers could be extended to continuous fins in order to obtain high performance compared with plain continuous fins. Many different forms of louvers on circular tubes have been investigated by the heat exchanger industry both with staggered and inline tube arrangements. Some of them are shown in Fig. 3. Hosoda et al. [58] showed that parallel louver fins on staggered tubes increased the heat transfer coefficient by about 50% over that by wavy fins on the same staggered tube array. Ito et al. [60] showed that a parallel louver fin geometry on staggered tubes increased the heat transfer coefficient by 35% over a plain fin geometry. Fukui and Sakamoto [57] showed that single cut louvers on continuous fins on inline tubes increased the heat transfer coefficient by about 50% over the flat fins.

Continuous Fins on a Flat Tube Array. Evaluation of Rich's data [49] for plain fins on circular tubes indicate that form drag on the tubes accounts for approximately 40 to 60% of the total pressure loss depending upon the fin density. In comparison, the use of flattened tubes (rectangular tubes with rounded or sharp corners) yields a lower pressure loss due to lower form drag, and avoids the low performance wake region behind the tubes. Also, the heat transfer coefficient is higher for flow inside flat tubes compared to circular tubes, particularly at low Re. The use of flat tubes is limited to low pressure applications, such as vehicular "radiators", unless the tubes are extruded with integral fins outside.

Kays and London [12] provide j and f data for one core having plain fins and inline flat tubes, and one core having plain fins and echelon[†] flat tubes. They also provide j and f data for one core having wavy fins and inline flat tubes, and two cores having wavy fins and echelon flat tubes. The wavy fins provided 30% higher heat transfer coefficient than the plain fins. Vlădea et al. [61] tested 0.1 mm thick flat fins on 2, 3 and 4 row inline and staggered flat tubes of 18.7 x 2.5 mm size. They showed that the heat transfer and friction factors were higher for the staggered tube arrangement compared to the inline tube arrangement, as expected.

Kovacs [62] investigated parallel louver steel fins on copper flat tubes and compared its performance with a similar flat fin geometry. The effective heat transfer coefficient $\eta_o h$ was about 50% larger than that for the flat fins.

Zozulya et al. [63] tested continuous fins with stamped rectangular holes (i.e., perforated fin) on flat tubes. Although the heat transfer coefficient increased by 38% and pressure drop by 33% over unperforated plain fins, the heat transfer surface area was significantly reduced and thus hA did not increase substantially. In a subsequent study, Zozulya et al. [64] investigated different continuous fin configurations on flat tubes. The fin geometries tested include parallel louvers, and three patterns of artificial roughness. They found that the parallel louver fin was the best performer and the fins with high and dense roughness elements offered no advantage over the plain fins.

[†] It means a staggered arrangement with the tube pattern repeating every fourth row; in commonly used "staggered" arrangement, the tube pattern repeats every third row.

COMPACT AND ENHANCED HEAT EXCHANGERS

5.2 Individually Finned Circular Tubes

Circular Fins. Substantial data have been published on the circular (helical or disk) fin shown in Fig. 4 in which circular plain fins are either wrapped on the tube in a helical fashion or mounted in a disk form. Most of the reported data have been taken with a staggered tube arrangement with six or more tube rows deep. Webb [44] presents a survey of the published data and correlations. Heat transfer and friction correlations are usually empirically based upon a multiple regression analysis of the basic nondimensional groups. Such correlations must account for five geometric parameters, which include the tube diameter d_o, the fin parameters δ, ℓ_f, and s, and the tube bundle geometry X_t and X_ℓ for inline or staggered arrangements. A number of correlations have been proposed, which differ in the choice of nondimensional groups, and of the characteristic dimension used in the Reynolds number.

For heat transfer, the recommended equation is that of Briggs and Young [65] as follows.

$$j = 0.134 \, Re_d^{-0.319} \left(\frac{s'}{\ell_f}\right)^{0.2} \left(\frac{s'}{\delta}\right)^{0.11} \tag{25}$$

This equation is based on the test data of air flow over 14 equilateral triangular tube banks. The following ranges were covered: $Re_d \sim$ 1100–18,000; $s'/\ell_f \sim$ 0.13–0.63; $s'/\delta \sim$ 1.01–6.62; $\ell_f/d_o \sim$ 0.09–0.69; $\delta/d_o \sim$ 0.011–0.15; $X_t/d_o \sim$ 1.54–8.23; fin root diameter $d_o \sim$ 11.1–40.9 mm; and fin density 246–768 fins/m. The standard deviation for Eq. (25) was 5.1%. Although Briggs and Young tried to correlate the friction factor data, they were not successful. The standard deviation for friction factor correlation was ±40%, and hence, a subsequent study [66] was conducted for the Δp correlation with a limited range of geometrical variables.

For isothermal pressure drops, the recommended equation is that of Robinson and Briggs [66] as follows.

$$f' = \frac{\Delta p}{N} \frac{2\rho}{4G^2} = 9.465 \, Re_d^{-0.316} \left(\frac{X_t}{d_o}\right)^{-0.937} \tag{26}$$

Here f' is a modified Fanning friction factor per tube row. This equation is based on the test data of air flow over 15 equilateral triangular and 2 isosceles triangular tube banks. The following ranges were covered: $Re_d \sim$ 2000–50,000; $s'/\ell_f \sim$ 0.15–0.19; $s'/\delta \sim$ 3.75–6.03; $\ell_f/d_o \sim$ 0.35–0.56; $\delta/d_o \sim$ 0.011–0.025; $X_t/d_o \sim$ 1.86–4.60; fin root diameter $d_o \sim$ 18.6–40.9 mm; and fin density 311–431 fins/m. The standard deviation for Eq. (26) was 7.8%.

A review of Eqs. (25) and (26) reveals that the tube pitch X_t (or $X_\ell = 0.866 \, X_t$) has no effect on the heat transfer coefficient, but the pressure drop is strongly affected by it. The pressure drop decreases with an increase in the tube pitch.

The Briggs and Young data included tubes having low fins, e.g., $\ell_f/d_o \approx$ 0.10 and high fin density (750 fins/m). Recently, Rabas et al. [67] have established an improved empirical correlation as follows for these high fin density low finned tubes using the data of five investigators including those of Briggs and Young and their own new data.

$$j = 0.292 \text{Re}^{[-0.415+0.0346\ln(d_e/s)]} \left(\frac{s}{d_o}\right)^{1.115} \left(\frac{s}{\ell}\right)^{0.257} \left(\frac{\delta}{s}\right)^{0.666} \left(\frac{d_e}{d_o}\right)^{0.473} \left(\frac{d_e}{\delta}\right)^{0.772} \tag{27}$$

$$f' = 3.805 \text{ Re}^{-0.234} \left(\frac{s}{d_e}\right)^{0.251} \left(\frac{\ell_f}{s}\right)^{0.759} \left(\frac{d_o}{d_e}\right)^{0.729} \left(\frac{d_o}{X_t}\right)^{0.709} \left(\frac{X_t}{X_\ell}\right)^{0.379} \tag{28}$$

These equations are valid for the following ranges: $\ell_f \leq 6.35$ mm, $1000 \leq \text{Re} \leq 25,000$, $X_\ell \leq X_t$, $N \geq 6$, fin root diameter $d_g \sim 4.76-31.75$ mm, fin density $\gamma \sim 246-1181$ fins/m, $X_t \sim 15.08-111.0$ mm, and $X_\ell \sim 10.32-96.11$ mm. These correlations predicted 94% of j data and 90% of f data within ±15%.

There is a considerably less choice of correlations for inline tube banks. This is because a strong bypass stream exists in the open and finned zones between tube rows; this bypass stream is dependent on the fin tip clearance. A bypass fluid stream is one which does not at all come in direct contact with the heat transfer surface. Schmidt [68] made an attempt to develop a heat transfer correlation based on data from 11 sources.

A staggered tube arrangement is preferred for multirow finned tube banks. The heat transfer coefficient of an inline finned tube bank is substantially lower than that for a staggered tube arrangement. The differences between staggered and inline tube arrangements will be discussed in a later section. Rosenman et al. [69] recommend as a rough guide the j and f' for inline circular finned tubes to be 67% and 60%, respectively, of those for the equivalent staggered finned tubes.

Figure 4 shows some of the enhanced surface geometries used on circular tubes. Webb [44] in his Table 2 provides references to information on the performance of these enhanced surface geometries for individually finned circular tubes. All of the concepts provide enhancement by the periodic development of thin boundary layers on small diameter wires or flat strips, followed by dissipation in the wake region between elements.

<u>Segmented and Spine Fins</u>. Segmented and spine fins have essentially the same basic geometry. They are used in a wide range of applications from boiler economizers to air conditioning. A segmented fin is generally rugged, has "heavy gage" metal, and is usually less compact compared to a spine fin. A segmented fin is also referred to as a serrated fin. A continuous strip of metal, after partially cut into narrow sections, is wound helically around the tubes. Upon winding, the narrow sections separate and form the narrow strips that are connected at the base. The steel segmented fin is attached to a tube by continuous welding of the base of the strip to the tube. The aluminum spine fin may be attached to the tube by epoxy.

A steel segmented fin geometry is used for boiler economizers and waste heat recovery boilers. The j and f' vs. Re_d characteristics for a four row staggered tube geometry is shown in Fig. 14 [70]. In this figure, also shown are the predicted j and f' factors for an equivalent plain fin tube bank using Eqs. (25) and (26). Weierman [71] presents design correlations for steel segmented fins and plain circular fins. The correlations, presented in graphical form, are for both staggered and inline banks. Based on the heat transfer correlation, the segmented fin heat transfer coefficient is 40% greater than that of the plain fin, when the fin height is 10 times its

COMPACT AND ENHANCED HEAT EXCHANGERS 457

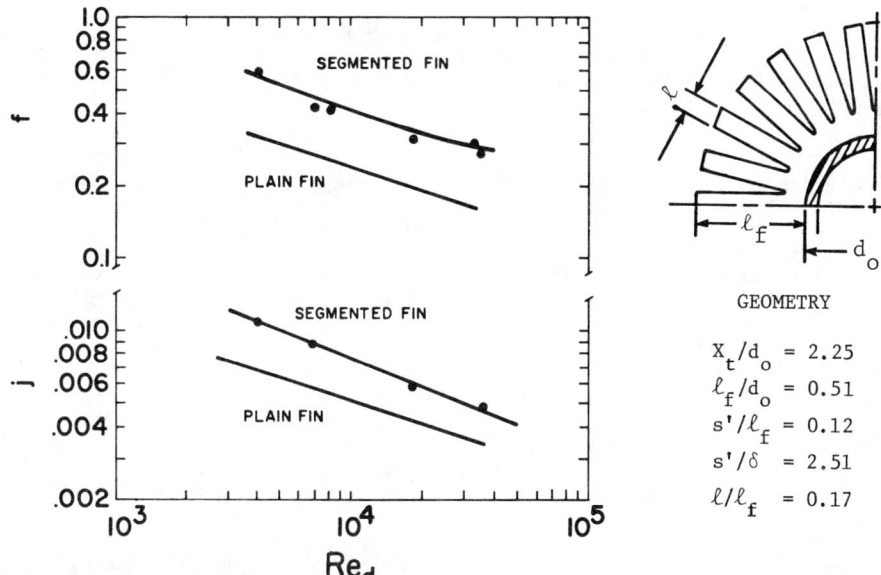

Fig. 14 Heat transfer and flow friction characteristics of staggered four row segmented fin [70].

spacing. The enhancement ratio decreases with reduced fin height-to-spacing. Rabas and Eckels [72] present additional data on steel segmented fins; they found that the inline arrangement yields heat transfer coefficients that are substantially lower than those for the staggered arrangement.

The spine fin is used extensively by General Electric in air conditioning condensers. The details of geometry, manufacturing and history are provided by Abbott et al. [73]. Moore [74] provides St and f vs. Re_d for one, two and three row staggered spine finned tube layouts.

Studded Fins. This fin geometry is made by welding individual "studs" around the base of the tube. The shape of the studs, and number of studs around the tube and along the tube are the variables. Because of ruggedness, this finned tube is used in steam/generator economizers. Ackerman and Brunsvold [75] presented $Nu_d Pr^{-1/3}$ vs. Re_d and $4f'$ vs. Re_d for five staggered and one inline tube bank arrangement. Since these plots were straight lines on a log-log paper, they provided correlations in closed form equations, the constants of which are dependent upon the specific geometry. They observed that the heat transfer performance was dependent upon the transverse pitch of the tube bank, the highest performance was given by the bank having the largest X_t. The inline tube arrangement was the worst performer.

Slotted Fins. When radially slitted fin material is wound on the tube, the slits open, forming slots whose width increases in the radial direction. This fin geometry offers an enhancement over tension wound plain fins. Preece et al. [76] found the heat transfer coefficient as much as 40% greater than that of a smooth fin.

Wire Loop Fins. This fin surface is formed by spirally wrapping a flattened helix of wire around the tube. The wire loops are held to the tube by a tensioned wire within the helix, or by soldering. Wall [77] describes the practical aspects of a commercially available wire loop design. He claims that the heat transfer coefficient is 2.5 times that of a plain fin, and that $h \propto d^{-0.4}$, where d is the wire diameter.

5.3 Individually Finned Flat Tubes

Oval and flat cross-sectional tube shapes are also applied to individually finned tubes. Figure 15 compares the performance of staggered banks of oval and circular finned tubes tested by Brauer [78]. Both banks have 313 fins/m, 10 mm high fins on approximately the same transverse and longitudinal pitches. The oval tubes gave 15% higher heat transfer coefficient and 25% less pressure drop than the circular tubes. The performance advantage of the oval tubes results from lower form drag on the tubes and the smaller wake region on the fin behind the tube. The use of oval tubes may not be practical unless the tubeside design pressure is low enough. Higher design pressures are possible using flattened aluminum tubes made by an extrusion process [44]. Such tubes can be made with internal full height ribs, which strengthen the tube and allow for a high tubeside design pressure. A variety of geometrical fin-and-tube shapes may be made by aluminum extrusions of different shapes.

A great deal of performance data have been published on the various types of enhanced fin geometries. A directory of published data is provided by Webb [44]. However, generalized correlations, which account for possible variation of the geometric parameters, do not exist. Rosenman et al. [69] have summarized the j and f vs. Reynolds number data for many of the enhanced fin geometries.

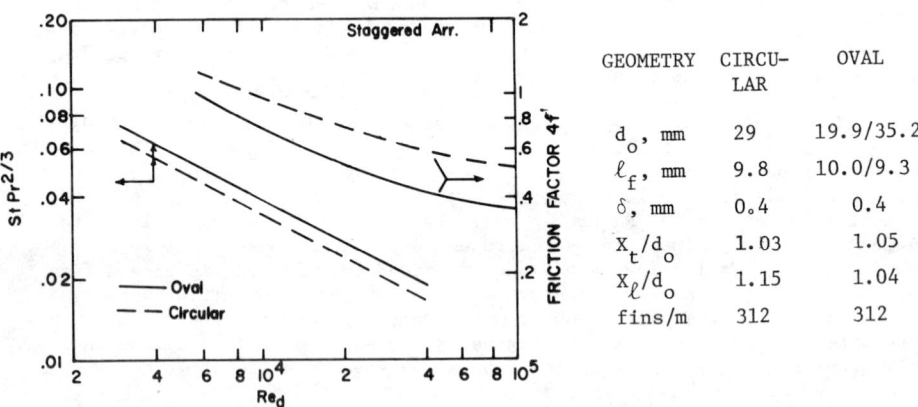

Fig. 15 Heat transfer and flow friction characteristics of finned tubes having circular and oval tubes in a staggered arrangement [78].

5.4 Staggered vs. Inline Tube Arrangements

There is a basic difference in the flow phenomena over staggered vs. inline finned tube banks. In the inline arrangement, the flow region upstream of the tubes of second and subsequent rows is the low velocity wake region. This results in low heat transfer through these tubes. Also there are straight throughflow channels (of width $X_t - d_o$) that offer the least flow resistance path, and as a result, a significant portion of the flow can bypass without touching the tubes (the heat transfer surfaces). Both these phenomena that yield low heat transfer and pressure drop are essentially not present in the staggered tube arrangement.

Figure 16 shows the sketches of the flow patterns of colored tracers in the liquid through staggered and inline tube arrangements [78]. The streamlines are shown by the dashed lines. The low velocity wake region or "dead spaces" are shown by the shaded area with fine dots. A much greater fraction of the fin surface area is contained in the low velocity wake region for the inline arrangement compared to the staggered arrangement. Consequently, the inline arrangement will have a lower surface average heat transfer coefficient. Outside the shaded zone, particularly between the fin tips, a strong bypass stream exists. Because of poor mixing between the wake stream and bypass stream, the weaker wake stream is quickly heated, and its mixed temperature is greater than that of the bypass stream. Thus, the actual temperature difference between the surface and the wake stream is much less than indicated by an overall LMTD based on the mixed outlet temperature. The staggered arrangement provides a good mixing of the wake and bypass streams after each tube row.

The performance of inline and staggered plain circular finned tubes, as measured by Bauer [78], is shown in Fig. 17. When the fins are short (low fins, $\ell_f/d_o = 0.07$), the staggered bank provided a 30% higher heat transfer coefficient. However, when the fin height is increased to $\ell_f/d_o = 0.54$, the staggered bank provided as much as a 100% higher heat transfer coefficient. The increase in friction factor is dependent upon the Reynolds number as found in Fig. 17(b). The staggered bank provides an increase in the friction factor over the inline bank, however, the increase is greater for the low finned tubes. Weierman et al. [70] found similar differences for a segmented fin; the staggered bank provided a 2.8 times higher Nusselt number with a friction increase of only 1.7. Rabas

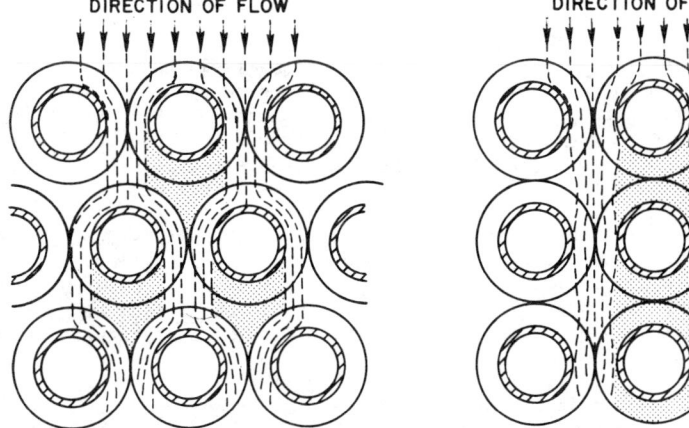

Fig. 16 Flow patterns observed by Brauer in staggered and inline finned tube arrangement [78].

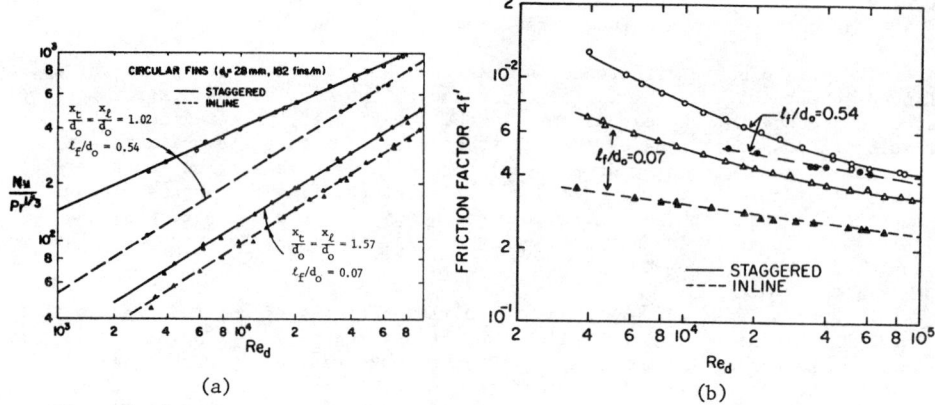

Fig. 17 (a) Heat transfer and (b) flow friction characteristics of staggered and inline plain circular finned tubes [78].

and Eckles [72] also found similar differences for segmented fins; the staggered bank provided a 1.7 higher j and 1.6 times higher f' over an inline bank of identical geometrical parameters at $Re_d = 10,000$.

No data are available for continuous fins on an inline tube array, and hence no quantitative performance comparison can be made between inline and staggered arrangements. However, based on the foregoing comparisons for individually finned tubes, the continuous fins with staggered tube arrangement will yield higher heat transfer coefficients and friction factors compared to the continuous fins with the inline tube arrangement.

5.5 Row Effect in Finned Tube Banks

The published correlations are generally for "deep" tube banks ($N \geq 4$), and do not account for row effects. The heat transfer coefficient will decrease with increasing number of tube rows in an inline bank due to the flow bypass effect. But, the coefficient will increase with rows in a staggered bank. This is because the turbulent eddies shed from the tube cause good mixing in the fin region of the downstream row.

<u>Staggered Banks of Circular Fins.</u> Ward and Young [79] measured the heat transfer coefficients in each row of a four row staggered tube bank (equilateral pitch) having 315 fins/m. The heat transfer coefficients of third and fourth rows were approximately equal, indicating that the row effect had been stabilized by the fourth row. Figure 18 shows the row effect as h_2/h_1 and h_4/h_1 vs. Reynolds number. From this figure, $h_4/h_1 = 1$ for $Re_d = 1900$, and $h_4/h_1 = 1.42$ at $Re_d = 10,000$. Therefore, the row effect is Reynolds number dependent. The increasing heat transfer coefficient with rows is due to the turbulent eddies shed from the tube which cause turbulent mixing in the fin region of downstream rows. The strength of these eddies increases with increasing Reynolds number. The bypass effect associated with the inline tube arrangement does not occur because of the good spanwise mixing of the staggered tube pattern.

Mirković [80] found that the stabilized coefficient occurred at approximately the sixth row independent of Re_d, and the asymptotic value was 30% larger than that for the first row. His tests were performed with larger tube

COMPACT AND ENHANCED HEAT EXCHANGERS

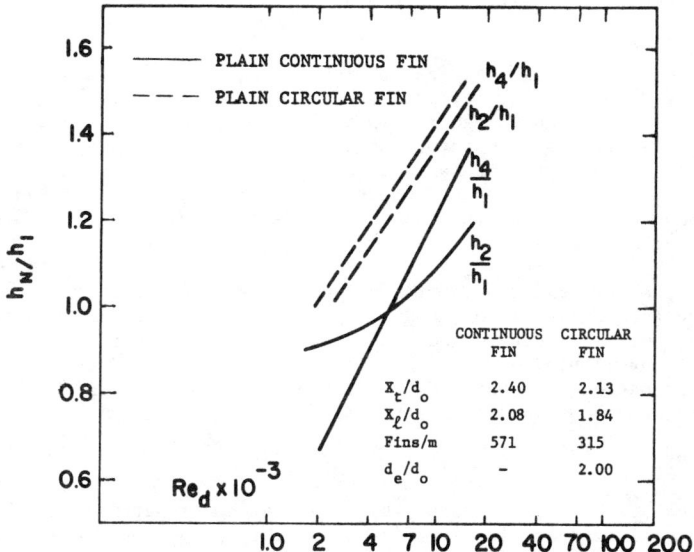

Fig. 18   The tube row effect on heat transfer for staggered finned tube banks: plain circular fins [79] and plain continuous fins [51].

pitches than used by Ward and Young. Neal and Hitchcock [81] found that the row-by-row performance of a staggered bank is highly dependent on the tube pitch layout. They measured heat transfer coefficients for the second and sixth row using three staggered tube layouts. They found that the heat transfer coefficient approaches an asymptotic value with a fewer number of tube rows by decreasing either the longitudinal pitch or the transverse pitch, although it is more sensitive to the transverse pitch.

Staggered Banks of Continuous Plain Fins. Rich [51] measured the individual row coefficients for the four row plain fin geometry, having 571 fins/m. As discussed earlier in Section 5.1, Rich's data were corrected by a straight line to eliminate the rollover in j for three and higher tube row coils (see Fig. 13). The subsequent row correction factors are shown in Fig. 18. The heat transfer coefficient of the fourth row does not exceed that of the first row until $Re_d > 5500$. For $Re_d < 5500$, the downstream rows have a smaller coefficient than that of the first row.

The row corrections factors, e.g., h_4/h_1 are different for the circular fin and plain continuous fin geometries. For example, at $Re_d = 11,000$, $h_4/h_1 = 1.42$ (circular fin) and $h_4/h_1 = 1.22$ (plain continuous fin). Apparently, there are no simple row correction factors for tube-fin surfaces. The correction factor depends upon the Reynolds number, tube pitches, and basic fin geometry (circular or continuous fin).

6. CONCLUDING REMARKS

In this paper, a comprehensive review is made of heat transfer and flow friction characteristics of plain passage, plate-fin, and tube-fin geometries. For plain passage geometries, the analytical solutions are summarized for developed and developing laminar flows. For plate-fin and tube-fin geometries, local flow phenomena, analytical solutions, and experimental correlations are discussed.

While the majority of the data published in the literature are either for constant fluid properties or for small changes in it, many of the heat exchanger applications involve either higher temperature differences or fluids other than the original test fluids. Therefore, the effect of the property variations must be included in the design calculations for a proper exchanger design. The correction factors are applied to modify the constant property j and f data or correlations presented in this paper. These correction factors have been summarized by Shah [82].

Although we have tried a comprehensive review of j and f data for plate-fin and tube-fin surfaces, no in-depth thorough scrutinization of all published data was made due to the time limitations. Since a vast amount of information is available in the open literature, we recommend to put all the published information on a unified format, weed out the inaccurate data, and provide firm design data and correlations.

ACKNOWLEDGEMENTS

The authors are grateful to Mr. E. H. Sanderson of Harrison Radiator for analyzing the raw data of Rich [51,54], and to Mr. R. S. Johnson and J. E. Farry, Jr., also of Harrison Radiator, for their critical review of this paper. The authors are very thankful to Dr. D. G. Rich of Carrier Corporation for providing the raw data of Fig. 13.

NOMENCLATURE

A	total heat transfer surface area (both primary and secondary, if any) on one side of a direct transfer type exchanger, total heat transfer surface area of a regenerator, m^2
A_c	minimum free flow (or open) area on one side of the exchanger, m^2
A_f	fin or extended surface area on one side of the exchanger, m^2
A_t	tube outside surface area considering it as if there were no fins, m^2
b	distance between two plates ("fin height") in a plate-fin exchanger, m
c_p	specific heat of fluid at constant pressure, J/kg°C
D_h	hydraulic diameter of flow passages, $4A_c L/A$, m
d_e	fin tip diameter of an individually finned tube (such as a helical fin), m
d_o	tube outside diameter, fin root diameter for an individually finned tube, m
f	Fanning friction factor, dimensionless
f'	modified Fanning friction factor per tube row, defined by Eq. (26), dimensionless
f_f	Fanning friction factor for fins alone excluding the tube effect, dimensionless
f_{fd}	Fanning friction factor for fully developed flow, dimensionless
G	mass velocity based on the minimum flow area, W/A_c, kg/m²s
h	heat transfer coefficient, W/m²°C

COMPACT AND ENHANCED HEAT EXCHANGERS

j	Colburn factor, $StPr^{2/3}$, dimensionless
$K(\infty)$	incremental pressure drop number for fully developed flow, defined by Eq. (2), dimensionless
k	fluid thermal conductivity, W/m°C
L	fluid flow (core) length on one side of the exchanger, m
ℓ	effective flow length between major boundary layer disturbances, distance between interruptions, m
ℓ_f	fin height [= $(d_e - d_o)/2$] for individually finned tubes, m
N	number of tube rows in the flow direction
Nu	Nusselt number, hD_h/k, dimensionless
NTU	number of heat transfer units [12], dimensionless
Pr	Prandtl number, $\mu c_p/k$, dimensionless
Δp	fluid static pressure drop on one side of a heat exchanger core, Pa
q	heat transfer rate in the exchanger, W
Re	Reynolds number based on the hydraulic diameter, GD_h/μ, $\rho u_m D_h/\mu$, dimensionless
Re_d	Reynolds number based on the tube outside diameter, Gd_o/μ, $\rho u_m d_o/\mu$, dimensionless
Re_ℓ	Reynolds number based on the interruption length, $G\ell/\mu$, $\rho u_m \ell/\mu$, GX_ℓ/μ, dimensionless
St	Stanton number, h/Gc_p, dimensionless
s	fin spacing, fin pitch, or transverse spacing for offset strip fins, m
s'	distance between two fins, $s - \delta$, m
s_x	distance between two colinear plates or between two offset strip fins in the flow direction, m
Δt_m	true mean temperature difference, defined by Eq. (1), °C
U	overall heat transfer coefficient, W/m²°C
u_m	fluid mean axial velocity occurring at the minimum free flow area, m/s
u_∞	free stream velocity, m/s
V	heat exchanger total volume, m³
W	fluid mass flow rate, $\rho u_m A_c$, kg/s
X_ℓ	longitudinal tube pitch, m
X_t	transverse tube pitch, m
x	Cartesian coordinate along the flow direction, m
x^+	axial distance, $x/D_h Re$, dimensionless
x^*	axial distance, $x/D_h RePr$, dimensionless
α^*	ratio of width to height of a rectangular duct, $(s-\delta)/(b-\delta)$, dimensionless
β	heat transfer surface area density, a ratio of the total transfer area on one side of a heat exchanger to the exchanger volume, m²/m³
γ	fin density, fins/m

δ   fin thickness, m

η_f fin efficiency or temperature effectiveness of the fin, dimensionless

η_o temperature effectiveness of total heat transfer area on one side of the extended surface heat exchanger, $\eta_o = 1 - (1-\eta_f)A_f/A$, dimensionless

μ   fluid dynamic viscosity coefficient, Pa·s

ρ   fluid density, kg/m^3

REFERENCES

1. R.K. Shah, Classification of heat exchangers, in Heat Exchangers: Thermal-hydraulic Fundamentals and Design, edited by S. Kakaç, A.E. Bergles, and F. Mayinger, pp. 9-46, Hemisphere Publishing Corp., Washington, D.C. (1981).
2. R.K. Shah, Compact heat exchanger surface selection, optimization and computed-aided design, in Low Reynolds Number Flow Heat Exchangers, edited by S. Kakaç, R.K. Shah, and A.E. Bergles, Hemisphere Publishing Corp., Washington, D.C. (1982).
3. W.M. Kays and M. Crawford, Convective Heat and Mass Transfer, Second Edition, McGraw-Hill, New York (1980).
4. E.M. Sparrow, R.R. Baliga, and S.V. Patankar, Heat transfer and fluid flow analysis of interrupted wall channels, with application to heat exchangers, Trans. ASME, Journal of Heat Transfer, Vol. 99, Series C, 4-11 (1977); discussion by R.K. Shah, Trans. ASME, Journal of Heat Transfer, Vol. 101, Series C, 188-189 (1979).
5. R.K. Shah and A.L. London, Laminar Flow Forced Convection in Ducts, Supplement 1 to Advances in Heat Transfer, Academic Press, New York (1978).
6. R.L. Webb, Matrix heat exchangers -- Thermal and hydraulic design, in Heat Exchanger Design Handbook, E.U. Schlünder, editor-in-chief, Vol. 3, Chapter 9, Hemisphere Publishing Corp., Washington, D.C. (1981).
7. R.K. Shah and A.L. London, Effects of nonuniform passages on compact heat exchanger performance, Trans. ASME, Journal of Engineering for Power, Vol. 102, Series A, 653-659 (1980).
8. R.K. Shah, Compact heat exchangers, in Heat Exchangers: Thermal-Hydraulic Fundamentals and Design, edited by S. Kakaç, A.E. Bergles, and F. Mayinger, pp. 111-151, Hemisphere Publishing Corp., Washington, D.C. (1981).
9. A.L. London, M.B.O. Young, and J.H. Stang, Glass ceramic surfaces, straight triangular passages -- heat transfer and flow friction characteristics, Trans. ASME, Journal of Engineering for Power, Vol. 92, Series A, 381-389, (1970).
10. A.L. London and R.K. Shah, Glass-ceramic hexagonal and circular passage surfaces -- heat transfer and flow friction design characteristics, SAE Trans., Vol. 82, Section 1, 425-434 (1973).
11. R.K. Shah, A correlation for laminar hydrodynamic entry length solutions for circular and noncircular ducts, Trans. ASME, Journal of Fluids Engineering, Vol. 100, Series I, 177-179 (1978).
12. W.M. Kays and A.L. London, Compact Heat Exchangers, Second Edition, McGraw-Hill Book Co., New York (1964).
13. Z.V. Tishchenko, V.N. Bondarenko, and L.I. Golechek, Heat transfer and pressure drop in gas-carrying ducts formed by smooth-finned plate-type heat-exchanger surfaces, Heat Transfer - Soviet Research, Vol. 11, No. 5, 117-124 (1979).
14. K. Okada, M. Ono, T. Tomimura, T. Okuma, H. Konno and S. Ohtani, Design and heat transfer characteristics of new plate type heat exchanger, Heat Transfer-Japanese Research, Vol. 1, No. 1, pp. 90-95 (1972).
15. G. Rosenblad and A. Kullendorf, Estimating heat transfer rates from mass transfer studies on plate heat exchanger surfaces, Wärme-und

Stoffübertragung, Vol. 8, 187-191 (1975).
16. L. Goldstein, Jr., and E.M. Sparrow, Mass-transfer experiments on secondary-flow vortices in a corrugated wall channel, International Journal of Heat and Mass Transfer, Vol. 19, 1337-1339 (1976).
17. L.J. Goldstein and E.M. Sparrow, Heat/mass transfer characteristics for flow in a corrugated wall channel, Trans. ASME, Journal of Heat Transfer, Vol. 99, Series C, 187-195 (1977).
18. A.L. London and R.K. Shah, Offset rectangular plate-fin surfaces -- heat transfer and flow friction characteristics, Trans. ASME, Journal of Engineering for Power, Vol. 90, Series A, 218-228 (1968).
19. R.K. Shah and A.L. London, Influence of brazing on very compact heat exchanger surfaces, ASME Paper No. 71-HT-29 (1971).
20. E.V. Dubrovskii and A.I. Fedotva, Investigation of heat-exchanger surfaces with plate fins, Heat Transfer - Soviet Research, Vol. 4, No. 6, 75-79 (1972).
21. S. Mochizuki and Y. Yagi, Heat transfer and friction characteristics of strip fins, Heat Transfer - Japanese Research, Vol. 6, No. 3, 36-59 (1977).
22. M.C. Smith, Performance analysis and model experiments for louvered fin evaporator core development, Society of Automotive Engineers, SAE Paper No. 720078 (1972).
23. L.T. Wong and M.C. Smith, Airflow phenomena in the louvered-fin heat exchanger, Society of Automotive Engineers, SAE Paper No. 730237 (1973).
24. K. Fujikake, Recent advances in automobile heat exchangers (in Japanese), Journal of JSME, Vol. 81, No. 714, 18-23 (1978).
25. R. Izumi, and H. Yamashita, Recent advances in heat exchangers and illustrative examples of their application (in Japanese), Kikai no Kenkyu (Study of Machines), Vol. 31, No. 1, 35-40 (1979).
26. Y. Mori, and W. Nakayama, Recent advances in compact heat exchangers in Japan, in Compact Heat Exchangers -- History, Technological Advancement and Mechanical Design Problems, edited by R.K. Shah, C.F. McDonald, and C.P. Howard, Book No. G00183, pp. 5-16, ASME, New York (1980).
27. P.F. Pucci, C.P. Howard and C.H. Piersall, Jr., The single blow transient testing technique for compact heat exchanger surfaces, Trans. ASME, Journal of Engineering for Power, Vol. 89, Series A, 29-40 (1967).
28. J.R. Mondt and D.C. Siegla, Performance of perforated heat exchanger surfaces, Trans. ASME, Journal of Engineering for Power, Vol. 96, Series A, 81-86 (1974).
29. H.L. Miller and C.A. Leeman, Heat transfer and pressure drop characteristics of several compact plate surfaces, Heat Transfer: Fundamentals and Industrial Application, AIChE Symp. Ser. 131, Vol. 39, 63-71 (1973).
30. R.K. Shah, Perforated heat exchanger surfaces. Part 1 - Flow phenomena, noise and vibration characteristics, ASME Paper No. 75-WA/HT-8 (1975).
31. R.K. Shah, Perforated heat exchanger surfaces. Part 2 - Heat transfer and flow friction characteristics, ASME Paper No. 75-WA/HT-9 (1975).
32. C.Y. Liang and W.J. Yang, Heat transfer and friction loss performance of perforated heat exchanger surfaces, Trans. ASME, Journal of Heat Transfer, Vol. 97, Series C, 9-15 (1975).
33. C.P. Lee and W.J. Yang, Augmentation of convective heat transfer from high-porosity perforated surfaces, Heat Transfer 1978, Vol. 2, 589-594, Hemisphere Publishing Corp., New York (1978).
34. G. Theoclitus, Heat transfer and flow friction characteristics of nine pin-fin surfaces, Trans. ASME, Journal of Heat Transfer, Vol. 88, Series C, 383-390 (1966).
35. G.J. VanFossen, Heat transfer coefficients for staggered arrays of short pin fins, ASME Paper No. 81-GT-75 (1981).
36. A.R. Wieting, Empirical correlations for heat transfer and flow friction characteristics of rectangular offset-fin plate-fin heat exchangers, Trans. ASME, Journal of Heat Transfer, Vol. 97, Series C, 488-490 (1975).

37. D.B. Adarkar, and W.M. Kays, Heat transfer in wakes, TR No. 55, Dept. of Mech. Eng., Stanford University, Stanford, CA (1963).
38. R.L. Zelenka, and R.I. Loehrke, Heat transfer from interrupted plates, in Advances in Enhanced Heat Transfer, edited by J.M. Chenoweth, J. Kaellis, J. Michel and S.M. Shenkman, Book No. I00122, pp. 115-122, ASME, New York (1979).
39. N. Cur and E.M. Sparrow, Measurements of developing and fully developed heat transfer coefficients along a periodically interrupted surface, Trans. ASME, Journal of Heat Transfer, Vol. 101, Series C, 211-216 (1979).
40. E.M. Sparrow and A. Hajiloo, Measurements of heat transfer and pressure drop for an array of staggered plates aligned parallel to air flow, Trans. ASME, Journal of Heat Transfer, Vol. 102, Series C, 426-432 (1980).
41. W.M. Kays, Compact heat exchangers, in AGARD Heat Exchangers, AGARD-LS-57, Advisory Group for Aerospace Research and Development, NATO, Paris (1972).
42. S. V. Patankar and C. Prakash, An analysis of the effect of plate thickness on laminar flow and heat transfer in interrupted-plate passages, Int. J. Heat Mass Transfer, Vol. 24, 1801-1810 (1981).
43. W.M. Kays, The heat transfer and flow friction characteristics of six high performance heat transfer surfaces, Trans. ASME, Journal of Engineering for Power, Vol. 82, Series A, 27-34 (1960).
44. R.L. Webb, Air-side heat transfer in finned tube heat exchangers, Heat Transfer Engineering, Vol. 1, No. 3, 33-49 (1980).
45. S.W. Krückels, and V. Kottke, Investigation of the distribution of heat transfer on fins and finned tube models (in German), Chemie-Ingenieur-Technik, Vol. 42, 355-362 (1970).
46. F.E.M. Saboya and E.M. Sparrow, Local and average transfer coefficients for one-row plate fin and tube heat exchanger configurations, Trans. ASME, Journal of Heat Transfer, Vol. 96, Series C, 265-272 (1974).
47. F.E.M. Saboya and E.M. Sparrow, Transfer characteristics of two-row plate fin and tube heat exchanger configurations, Int. J. Heat Mass Transfer, Vol. 19, 41-49 (1976).
48. F.E.M. Saboya and E.M. Sparrow, Experiments on a three-row fin and tube heat exchanger, Trans. ASME, Journal of Heat Transfer, Vol. 98, Series C, 520-522 (1976).
49. D.G. Rich, The effect of fin spacing on the heat transfer and friction performance of multi-row smooth plate fin-and-tube heat exchangers, ASHRAE Trans., Vol. 79, Part 2, 137-145 (1973).
50. E.D. Grimison, Correlation and utilization of new data on flow resistance and heat transfer for cross flow of gases over tube banks, Trans. ASME, Vol. 59, 583-594 (1937).
51. D.G. Rich, The effect of the number of tube rows on heat transfer performance of smooth plate-fin-tube heat exchangers, ASHRAE Trans., Vol. 81, Part 1, 307-317 (1975).
52. F.C. McQuiston, Correlation of heat, mass and momentum transport coefficients for plate-fin-tube heat transfer surfaces with staggered tubes, ASHRAE Trans., Vol. 84, Part 1, 294-309 (1978).
53. W.M. Kays and A.L. London, Heat transfer and flow friction characteristics of some compact heat exchangers surfaces - Part I: Test system and procedure, Trans. ASME, Vol. 72, 1075-1085 (1950); also Description of test equipment and method of analysis for basic heat transfer and flow friction tests of high rating heat exchanger surfaces, TR No. 2, Dept. Mech. Eng., Stanford University, Stanford, CA (1948).
54. D.G. Rich, Personal communication, Carrier Corporation, Research Division, Carrier Parkway, Syracuse, NY (1981).
55. F.C. McQuiston and D.R. Tree, Heat-transfer and flow-friction data for two fin-tube surfaces, Trans. ASME, Journal of Heat Transfer, Vol. 93, Series C, 249-250 (1971).

56. F.C. McQuiston, Heat, mass and momentum transfer data for five plate-fin-tube heat transfer surfaces, ASHRAE Trans., Vol. 84, Part 1, pp. 266-293 (1978).
57. S. Fukui, and M. Sakamoto, Some experimental results on heat transfer characteristic of air cooled heat exchangers for air conditioning devices, Bulletin of the JSME, Vol. 11, No. 44, 303-311 (1968).
58. T. Hosoda, H. Uzuhashi, and N. Kobayashi, Louver fin type heat exchangers, Heat Transfer - Japanese Research, Vol. 6, No. 2, 69-77 (1977).
59. L. Goldstein, Jr., and E.M. Sparrow, Experiments on the transfer characteristics of a corrugated fin and tube heat exchanger configuration, Trans. ASME, Journal of Heat Transfer, Vol. 98, 26-34 (1976).
60. M. Ito, H. Kimura, and T. Senshu, Development of high efficiency air-cooled heat exchangers, Hitachi Review, Vol. 26, No. 10, 323-326 (1977).
61. I. Vlădea, H. Theil, and F. Neiss, Wärmeaustauscher mit Flachrohren und Durchlaufenden Rippen, Heat Transfer 1970, Vol. 1, Paper HE 1.5, Elsevier, Amsterdam (1970).
62. G. Kovacs, Application of short-finned heat exchanger as air-cooled condenser (in French), La Revue Génerale du Froid, pp. 159-168 (Feb. 1963).
63. N.V. Zozulya, A.A. Khavin, and D.B. Kalinin, Effect of perforated transverse plate fins on oval pipes on the rate of heat transfer, Heat Transfer - Soviet Research, Vol. 2, No. 1, 77-79 (1970).
64. N.V. Zozulya, A.A. Khavin, and D.B. Kalinin, Effect of fin deformation on heat transfer and drag in bundles of oval tubes with transverse fins, Heat Transfer - Soviet Research, Vol. 7, No. 2, 95-98 (1975).
65. D.E. Briggs and E.H. Young, Convection heat transfer and pressure drop of air flowing across triangular pitch banks of finned tubes, Chemical Engineering Progress Symposium Series, Vol. 59, No. 41, 1-10 (1963).
66. K.K. Robinson and D.E. Briggs, Pressure drop of air flowing across triangular pitch banks of finned tubes, Chemical Engineering Progress Symposium Series, Vol. 62, No. 64, 177-184 (1966).
67. T.J. Rabas, P.W. Eckles, and R.A. Sabatino, The effect of fin density on the heat transfer and pressure drop performance of low-finned tube banks, ASME Paper No. 80-HT-97 (1980).
68. E. Schmidt, Heat transfer at finned tubes and computations of tube bank heat exchangers, Kältetechnik, Vol. 15, No. 4, p. 98 (1963); Kältetechnik, Vol. 15, No. 12, p. 370 (1963).
69. T. Rosenman, S.K. Momoh, and J.M. Pundyk, Heat transfer and pressure drop characteristics of dry tower extended surfaces. Part I: Heat transfer and pressure drop data; Part II: Data analysis and correlation, Report No. PFR7-100 and PFR7-102, Battelle Memorial Institute, Richland, Washington (1976).
70. C. Weierman, J. Taborek, and W.J. Marner, Comparison of the performance of inline and staggered banks of tubes with segmented fins, AIChE Paper No. 7 presented at the 15th National Heat Transfer Conference, San Francisco, CA (1975).
71. C. Weierman, Correlations ease the selection of finned tubes, The Oil and Gas Journal, pp. 94-100, September 6, 1976.
72. T.J. Rabas and P.W. Eckels, Heat transfer and pressure drop performance of segmented surface tube bundles, ASME Paper No. 75-HT-45 (1975).
73. R.W. Abbott, R.H. Norris and W.A. Spofford, Compact heat exchangers for General Electric products -- Sixty years of advances in design and in manufacturing technologies, in Compact Heat Exchangers -- History, Technological Advancement and Mechanical Design Problems, edited by R.K. Shah, C.F. McDonald, and C.P. Howard, Book No. G00183, ASME, pp.37-55(1980).
74. F.K. Moore, Analysis of large dry cooling towers with spine-fin heat exchanger elements, ASME Paper No. 75-WA/HT-46 (1975).
75. J.W. Ackerman and A.R. Brunsvold, Heat transfer and draft loss performance

of extended surface tube banks, Trans. ASME, J. Heat Transfer, Vol. 92, Series C, 215-220 (1970).
76. R.J. Preece, J. Lis and J.A. Hitchcock, Comparative performance characteristics of some extended surfaces for air-cooler applications, The Chemical Engineer (London), pp. 238-244 (June 1972).
77. A.J. Wall, Extended surface tubes, Chemical and Process Engineering (London), Vol. 48, No. 8, 136-138 (1967).
78. H. Brauer, Compact heat exchangers, Chemical and Process Engineering (London), Vol. 45, No. 8, 451-460 (1964).
79. D.J. Ward, and E.H. Young, Heat transfer and pressure drop of air in forced convection across triangular pitch banks of finned tubes, Chemical Engineering Progress Symposium Series, Vol. 55, No. 29, 37-44 (1959).
80. Z. Mirković, Heat transfer and flow resistance correlation of helically finned and staggered tube banks in crossflow, in Heat Exchangers: Design and Theory Sourcebook, edited by N.H. Afgan and E.U. Schlünder, pp. 559-584, McGraw-Hill, New York (1974).
81. S.B.H.C. Neal, and J.A. Hitchcock, A study of the heat transfer processes in banks of finned tubes in cross flow, using a large scale model technique, Proc. Third Int. Heat Transfer Conf., Vol. III, 290-298 (1966).
82. R.K. Shah, Compact heat exchanger design procedures, in Heat Exchangers -- Thermohydraulic Fundamentals and Design, edited by S. Kakaç, A.E. Bergles, and F. Mayinger, pp. 495-536, Hemisphere Publishing Corp., Washington, D.C. (1981).

Heat Exchange Equipment for the Cryogenic Process Industry

J. M. ROBERTSON
Heat Transfer and Fluid Flow Service (HTFS)
Atomic Energy Research Establishment
Harwell, Oxfordshire OX11 0RA, UK

ABSTRACT

Chemical processes carried out at low temperature include the separation and liquefaction of oxygen and nitrogen and the separation of olefin products from the cracking of heavy hydrocarbons in ethylene plants. Recovery and upgrading of gases such as carbon monoxide and hydrogen from waste-product streams is also undertaken at low temperature. Recently the liquefaction of natural gas for storage and for the separation of constituents, and also its vaporisation, has become of importance.

The economics of these processes are dominated to a certain extent by the energy requirements of compressors and thermodynamic considerations have emphasised the requirements for very small temperature differences between streams exchanging heat. This had lead to the development of heat exchangers which contain an enormous heat transfer surface area per unit volume, and this is associated with large numbers of flow passages for streams flowing at very low mass-fluxes.

To undertake these special duties, two main types of heat exchanger have evolved and are now widely used: the plate-fin or brazed-aluminium heat exchanger and the wound-coil or Hampson heat exchanger. These are constructed from materials such as aluminium, copper or stainless steel which are mechanically satisfactory at very low temperature. Both exchangers are frequently designed to contain several streams within the one unit. Inside a plate-fin heat exchanger, for example, heat may be exchanged among five or six streams.

There are, therefore, many features in these exchangers which are not normally encountered in industrial practice and these also influence the characteristics of boiling and condensing: the addition of secondary, or finned, heat transfer surfaces in the plate-fin heat exchanger; in the wound-coil exchanger, condensing may take place in upflow in the helically-wound tubes. The considerable number of passages or channels required in these heat exchangers is a result of the large surface area and highlights the importance of uniform distribution, particularly of two phases, of each stream so that thermal performance is not impaired.

In the paper, the heat transfer and flow characteristics of these exchangers, their thermal design and their structural details are described and the published information is reviewed.

1. GENERAL INTRODUCTION

There are two forms of heat-exchanger widely used in the cryogenic process industry: plate-fin and wound-tube exchangers. In cryogenic processes, heat-exchangers are sufficiently important that their design and operation can directly and considerably influence overall power requirements. Minimum power, for thermodynamic reasons, may require very small temperature differences between streams and a knowledge of the local heat transfer coefficients, as in boiling and condensing, becomes important. Combined or constant coefficients have been used where boiling and condensing streams are adjacent (1) but these cannot reflect the large effect of vapour quality and heat transfer along those streams. This variation in coefficient may be contrasted with single-phase operation.

The thermodynamic and transport properties of cryogenic fluids can also be expected to influence heat transfer in two-phase applications. For example, the low surface tension and latent heat commonly found with these fluids will affect the mechanics of boiling through wetting and vapour generation rates. The use of very accurate vapour-liquid equilibrium data becomes necessary with the very small temperature differences between streams.

In this paper published information about heat transfer characteristics of both forms of heat exchanger and thermal aspects of their design are examined. Certain aspects of this field have already been reviewed by the author (61).

2. PLATE-FIN HEAT EXCHANGERS

2.1 Introduction

During the thirty years since they were first introduced into air separation plants, brazed aluminium plate-fin heat exchangers have become widely used in the chemical process industry for single and two-phase applications over a range of temperatures from cryogenic to atmospheric. Several process streams can be accommodated in one exchanger block and heat may be exchanged among many different boiling, condensing or single-phase streams. Their assembly from corrugated sheet brazed to separating plates can be seen in Figure 1; fluid flows along the passages formed by the corrugations which act as fins. Many layers are used in one exchanger block which can be as large as 6.5 m long by 1.5 m square. Both counter-current and co-current arrangements of the process streams are used. Their construction and special features are fully described by Lenfesty (1) and more recently by Butt (40), among others.

They are extensively used in the separation of air to produce oxygen and nitrogen in gaseous or liquid form, and of olefin products from the cracking of hydrocarbons in ethylene plants. The recovery or upgrading of valuable gases such as hydrogen and carbon monoxide from waste-product streams is another important area of their use. For these processes, plate-fin heat exchangers are often utilised as thermosyphon reboilers for providing vapour for fractionation columns and also as product condensers. In the separation of air, two-stream exchangers, partly immersed in a pool of oxygen at the base of a fractionation column, are used to evaporate partially an oxygen thermosyphon stream against a condensing nitrogen stream (1). Thermosyphon reboilers are also widely used external to columns in the separation trains in ethylene plants and, in order to save energy, banks of these exchangers can be incorporated into a single cold box.

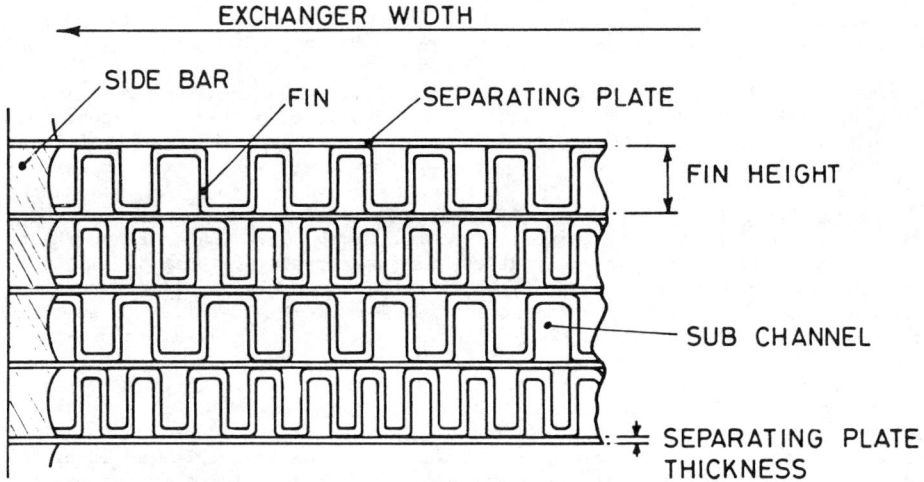

Fig.1. Cross-Section of Plate-Fin Heat Exchanger Showing Four Passages

More recently, the liquefaction of natural gas for the recovery of valuable heavy components and also for its transport and storage as a liquid, has lead to the development of plate-fin exchangers involving the boiling and condensing of multi-component refrigerants. These may incorporate many different hydrocarbon components as well as nitrogen with a boiling range as large as 175 K; in one process this can be accommodated within the length of one exchanger (2),(3).

In two-phase applications, the following special conditions of flow within the exchangers greatly influence the characteristics of boiling and condensing and also the two-phase flow patterns:

Channel shapes and dimensions. The various forms of finning used for single-phase operation are also used for two-phase: plain, perforated, serrated (lanced or multi-entry) and herringbone. However, the choice of the most suitable finning and its dimensions may not be made by only heat transfer or fluid flow considerations as the fins act also as stressed structural members. In two-phase flow, heat transfer coefficients and frictional pressure drops can be much higher. Higher coefficients tend to produce lower fin efficiencies and higher frictional pressure gradients will limit the choice of types of finning. For example, serrated finning would increase the low coefficients associated with post dry-out conditions but the pressure drop may be unacceptable at the high qualities preceding dry-out.

Flow rates. Mass fluxes lower than 150 kg/m^{2}s are normally used to keep pressure drops at reasonable levels and low Reynolds numbers are typical. While the flow of a vapour phase will generally be turbulent, the flow of liquid films may be laminar or turbulent.

Heat fluxes. The very small temperature differences between adjacent streams are accommodated by the enormous heat transfer surface area available but lead to low heat fluxes, typically around 1-4 kW/m^2. Panitsidis et al. (5) have examined the performance of these exchangers at very much higher heat flux levels in order to demonstrate their potential application in new fields.

Two-phase flow at inlet to any heat exchanger poses special problems in distributing each phase uniformly within it. This is of particular importance to plate-fin exchangers with their vast number of individual channels (often many thousands). Each phase of each stream requires to be uniformly distributed to all passages carrying that stream and also across the full width of each passage. Consequently, the phases are often separated and then recombined in a special form of two-phase mixer within or at inlet to the heat exchanger. This ensures that the mixture, in its correct proportions and in a suitable flow pattern, enters the inlet distributor uniformly (33). Integral inlet and outlet distributors are used in these exchangers to achieve a uniform distribution but with two-phase flow additional practical difficulties arise.

A special feature of two-phase operation is the possibility of flow instability. By this is meant the steady oscillatory flow of a boiling stream (with a condensing stream the phenomenon is unusual) or an excursive flow alteration, under certain operating conditions. This aspect of design involves checking the probability of the occurrence of flow instabilities during normal or abnormal plant operation. For this, a knowledge of the expected regions of stable and unstable flow behaviour is required. The large frictional pressure-drop occurring at the exit distributor of an evaporating stream can be of particular importance here.

An interesting two-phase application of plate-fin exchangers is their use as dephlegmators or non-adiabatic fractionation columns. While this is now of considerable interest in the context of energy saving, it is not new. One example of dephlegmator would have a stream of multi-component vapour entering its base and being partially or totally condensed within a channel in **counter-current** two-phase flow with the condensate flowing downwards. Heat would be removed from the condensing vapour by a cooling stream in adjacent passages. In plate-fin heat exchangers, **co-current** two-phase flow is standard practice (normally, but not always, vertically). A special type of finning suitable for the combined heat and mass transfer processes in plate-fin dephlegmators has been put forward by Haselden (4).

2.2 Single-Phase Operation

A widely used source-book for information on the heat transfer and pressure drop characteristics of the finned passages in plate-fin heat exchangers is that by Kays and London (37). Plots of Colburn heat transfer factor j and friction factor f are also given by several authors for various forms of finning, for example (1), and these may be used to predict heat transfer coefficients and pressure drop in the practical range of Reynolds number. Webb (59) provides many references to experimental work and discusses analytical approaches to the provision of suitable expressions for heat transfer and friction characteristics, for example (60), to rationalise the large amount of experimental data for the possible combinations of fin forms, thickness, height and pitch. However, it is recommended that in the assessment of heat exchanger performance it is desirable to use these figures with caution since, in the case of serrated-finning, they may refer to fins produced with new die-tools. In these cases a 15% or so possible increase in friction factor may require to be added to account for aging. Manufacturers of these fins also produce limited information in their advertising brochures and for accurate assessment of the performance of these exchangers, the manufacturers' data for a specific fin shape are preferable.

A typical plot of j factor against Reynolds numbers for a perforated fin passage is given in Figure 2. This plot, from (8), also shows the effect of Prandtl number on j in the laminar flow zone since this latter factor is more appropriate for correlating results in the turbulent zone. The plots for serrated finning demonstrate the improved heat transfer coefficients available, with no clear transition region, and obtained at the expense of increased friction.

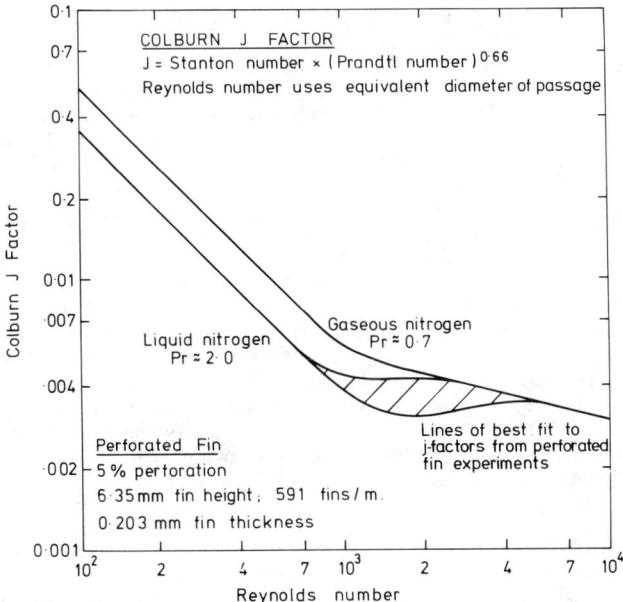

Fig.2. Plot of Colburn J Heat Transfer Factor Against Reynolds Number for Perforated-Fin Test Section, with Liquid and Gaseous Nitrogen (8)

2.3 Boiling

There are very few published papers on the boiling characteristics of plate-fin exchanger passages. Galezha et al (6) have undertaken laboratory thermosyphon experiments using Freon 12 and 22 (separately) with 200 mm long specimens of electrically-heated, serrated and plain finning. Wall temperature measurements were used to compute average boiling coefficients over the complete length without invoking a fin efficiency term. Coefficients range from approximately 1000 to 4500 W/m² K and show a steady increase with increasing pressure. Data are not provided on the quite separate effects of local vapour quality and mass flux on the boiling coefficients and the results must be regarded as indicative of the general magnitude of these coefficients.

Shorin et al (7) describe plate-fin test equipment on which they undertook thermosyphon tests using boiling oxygen and condensing nitrogen. It is however, difficult to extract details of the results since the tests appear to have been carried out to obtain the overall thermal performance of the apparatus.

Panitsidis et al. (5) have tested a very small plate-fin heat exchanger using steam to boil Freon 113 and isopropanol (separately) to obtain overall coefficients. Experimental results are compared with estimates from their predictive method developed for nucleate boiling from a single fin at high heat fluxes using pool-boiling data. They claim reasonable agreement between the predicted overall performance and the experimental results.

Robertson et al. (8,9) have reported results from tests using liquid nitrogen and Freon 11 on an electrically heated, serrated-fin test section 3.4 metres long. The finning was formed from 0.2 mm thick sheets with 519 corrugations per metre and 6.35 mm high. The results of these tests also give an indication of the general boiling characteristics of serrated and other forms of finning. These tests determined local boiling heat transfer coefficients at low heat-flux and required temperature differences often less than 1 K between wall and bulk. These very small temperature differences make testing and interpretation difficult and can be compared with the standard form of single-phase heat transfer test in which a short specimen of finning is heated by steam giving a temperature difference of approximately 80 K between the wall and the bulk of the test fluid, air.

The outstanding feature of the transition between the single-phase and boiling regions in a channel, with both liquid nitrogen and Freon 11, is the large wall and bulk liquid superheating which occurs before boiling commences with the production of the first vapour, (Figure 3, from (41)). The phenomenon has already been reported by Abdelmessih et al. (10) for large bore, circular tubes with Freon 11 and is quite separate from the onset of nucleate boiling. With liquid nitrogen a difference of up to 5 K has been measured with the electrically heated test section which has a constant heat flux imposed on its outer wall (41). Abdelmessih found this peak was not present on surfaces where boiling had already occurred (by lowering the heat inputs) and this has been partly confirmed in the liquid nitrogen tests. This feature points to the difficulty with cryogenic liquids, which are very good at wetting solid surfaces, in producing the first vapour from a nucleating point on a solid surface. Only very small cavities on an aluminium wall can retain a vapour core, by preventing the low contact-angle fluid from penetrating and condensing it, and a large wall superheat is necessary to create the first bubble. The Davis and Anderson correlation (11) has been used to analyse data from tests carried out with liquid nitrogen over a range of pressures and heat fluxes and it has been shown that the magnitudes of the wall and bulk peaks are highly dependent on heat flux (41). A tentative explanation of this dependency has also been put forward.

Although this phenomenon of high superheating has been found on electrically-heated test sections, it can be argued that the same superheating phenomenon may occur in industrial heat exchangers. There, the heat flux is governed mainly by the temperature of the primary wall and is thus a dependent variable. Frequently, the driving temperature difference between adjacent cryogenic process streams, available for creating the first vapour, may be less than the superheat which the testwork shows to be apparently necessary. This paradox has been partly explained in (41) by applying the results from these tests to the industrial case and by assuming that the heat flux must fall until it corresponds to the superheat of the wall of the exchanger passage. The tentative conclusion reached from this is that there may be a length of heat exchanger passage in which heat transfer is greatly reduced.

Once net vapour production has commenced in a channel it has been found (8) that convective boiling will be dominant with little nucleate boiling

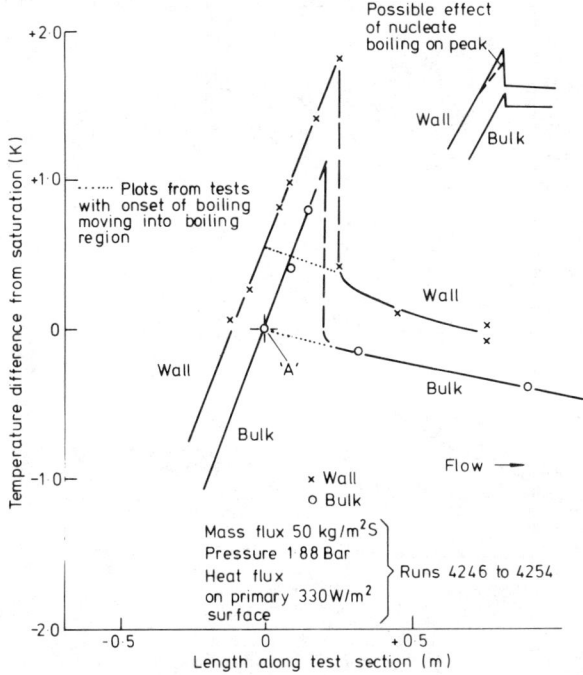

Fig.3. Temperature Profiles Along Serrated-Fin Test Section Showing Wall and Bulk Superheat Peaks at Onset of Evaporation (41)

present at the heat flux levels found in many industrial applications. If nucleate boiling in a flowing channel is closely associated with unrestricted pool boiling, then the temperature driving differences between wall and the saturated bulk will probably be too low for many cryogenic fluids for its inception. For example, the pool boiling curves given by Panitsidis et al. (5) for isopropanol show that about 10 K are required before fully-developed nucleate pool boiling commences.

Boiling coefficients have been obtained (8, 9) by making use of the concept of fin efficiency (standard in single-phase practice) which implies a uniform heat transfer coefficient round the perimeter of each small channel. At low mass fluxes the convective boiling coefficients obtained from a serrated-fin test section with Freon 11 and liquid nitrogen appear to be independent of mass flux; at higher mass fluxes the coefficients are dependent on mass flux to the power of 0.8 (Figure 4). These two regions with the serrated fin have, therefore, the heat transfer characteristics of laminar and turbulent flow and this is completely different from its single-phase performance where, in the Colburn j factor plot, a form of pseudo-turbulent flow appears to extend to Reynolds numbers as low as 100. An explanation for this behaviour has been advanced (8): the most prevalent two-phase flow pattern within these interrupted passages consists of a thin film of liquid flowing along each fin from leading to trailing edge where it is disrupted, at high vapour velocities, into the form of a spray and then reformed on the following fin. This film flow can be expected to be turbulent or laminar and it will not necessarily exhibit the same heat transfer characteristics of single-phase flow

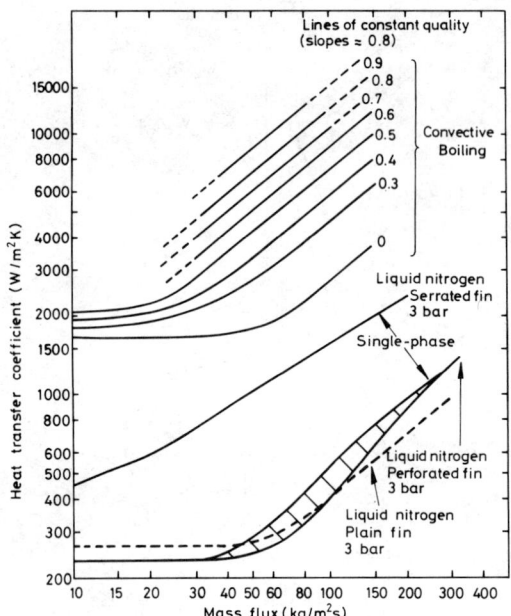

Fig.4. Plot of Convective Boiling Heat Transfer Coefficients on Serrated-Fin Test-Section Against Mass Flux with Nitrogen at 3 bar and Mixture Quality as Parameter (Single-Phase Coefficients Also Shown) (8)

in a serrated-fin passage.

If heat transfer is dominated by the evaporation of a thin liquid film, then it can be expected that the liquid film Reynolds number (which has the same value as if the liquid phase occupied the whole flow cross sectional area) might be an important correlating parameter. At low quality, this film Reynolds number tends to the value of the Reynolds number of the total flow regarded as a liquid. This latter Reynolds number has been used, tentatively, to correlate test results with Freon 11 and liquid nitrogen and a critical value of 1100 at the transition between the two regions holds for both fluids. With increasing vapour qualities, for the same mass flux, the film Reynolds number will decrease and eventually the film flow will become laminar. There is some evidence, particularly from the liquid nitrogen results, that the dependence of the boiling heat transfer coefficient on Reynolds number to the power 0.8 holds for values of **film** Reynolds number as low as 150. This may be a result of the interruptions to the film flow caused by this form of finning. At low qualities in the turbulent region it is to be expected that a form of slug flow may occur. This two-phase flow pattern is more likely to occur, at very low mass fluxes, over much of the quality range.

The effect of quality on the convective boiling coefficient is considerable and even greater than that of mass flux. For example, a range of 1.65 to 10 kW/m^2 K, for qualities between zero and 0.9, is apparent in Figure 3 for a doubling of mass flux. The much smaller effect of quality at low mass fluxes is also apparent. The high coefficients point to very thin films of liquid made thinner by the entrainment effect of the interrupted finning: much of the liquid, which acts as a thermal resistance if flowing on the fins, travels as droplets. (This effect should also be present during condensation and could be equally as important.) However, designers may not be able to use these high coefficients since they may be overshadowed by a low or controlling coefficient in an adjacent passage and, moreover, they will inevitably be accompanied by high, friction pressure drops.

A tentative correlation for the quality and mass flux ranges has been put forward in (9) based on the Reynolds number of the total flow as a liquid. For the turbulent flow region, a simple expression has been sufficient to correlate the available results from Freon 11 and nitrogen with the serrated-fin test section. A plot of this (Figure 5) shows that there appears to be a much greater effect of quality on the coefficients above a quality of 0.3. This has been explained as a critical point above which entrainment from the serrated fin, at high vapour phase flows, becomes important. This form of correlation has limited value and it will require to be replaced with either an empirical relationship of the type proposed by Chen (13), while ignoring the nucleate boiling component, or by a film flow model making use of local pressure gradients, viscosity, conductivity and the local Nusselt number of the liquid film. So far, the use of the Chen correlation has not been encouraging, possibly as it was developed for annular flow in round tubes.

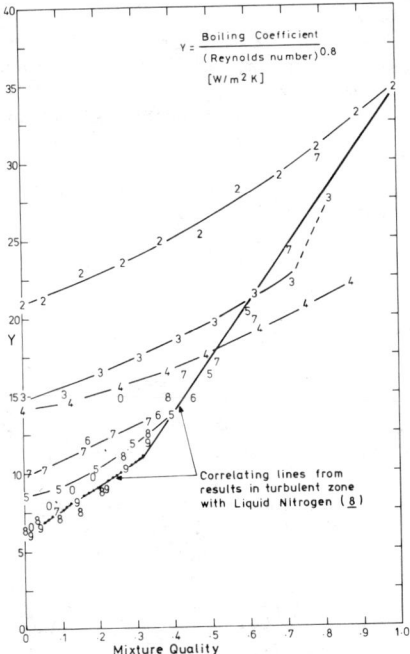

Fig.5. Correlation of Freon 11 Boiling Heat Transfer Coefficients Using Parameter Y Plotted Against Mixture Quality, For Serrated-Fin (9)

Some results have been obtained for the dry-out point in the electrically-heated, serrated-fin test section and these indicate that dry-out at the modest heat fluxes used will occur at qualities above 95%.

Yung et al. (42) have tested a plate-fin heat exchanger designed to evaporate (and condense) ammonia using water. Only overall thermal performance data were obtained but a theoretical analysis of the boiling streams' performance, using data from (8) for perforated finning, predicted the overall heat transfer satisfactorily. The theoretical analysis appeared to hold only for very low mass fluxes.

2.4 Condensing

Very little experimental information on condensation within plate-fin passages has been published. Using several refrigerants in a test section 0.2 m long, connected to a thermosyphoning vapour generator and cooled by water, Gopin et al. (15) obtained length-averaged condensing coefficients over a range of heat fluxes. Fin shapes tested included plain and serrated of various geometries. Mass fluxes used appeared to be very low at the maximum vapour inlet velocity 0.3 m/s. Condensing coefficients measured ranged from 1400 - 4000 W/m^{2}K up to the peak heat flux of 2×10^4 W/m^2. The results are somewhat obscured by the length averaging and the omission of quality as an analytical variable. The authors claim that the results are close to their predictions using a Nusselt form of correlation. No fin-efficiency term was used.

Several papers (17, 18, 19) deal with the condensation of cryogenic fluids, including oxygen and nitrogen, on vertical surfaces from a stationary vapour. The test results were well predicted by the Nusselt correlation. Others (20, 21, 22) deal with the enhanced condensation of partially-finned surfaces created by surface tension effects producing local thinning of condensate films. Although these effects may have a role in the high aspect-ratio channels of plate-fin exchangers, probably a more important effect has been studied by Patankar and Sparrow (23) who have analysed film condensation on a vertical fin or secondary surface, in a stagnant vapour. Their analysis has taken into account the variation of the film thickness from fin-root to fin-tip as well as the thickening of the film as it runs down the fin with increasing condensate present. The authors found that the calculated fin heat transfer coefficient is significantly less than that which would be predicted by an isothermal fin model. The specific application of this model to plate-fin exchanger passages has not yet been included.

Haseler (16) has reported results from tests with a plain-fin test section, 3.2 m long, for the downflow condensing of nitrogen over a range of mass flux and heat flux conditions commonly encountered in the cryogenic process industry. The test section was fed with two streams: a condensing stream of slightly superheated nitrogen vapour was supplied to the top of a plate-fin layer and two streams of subcooled liquid nitrogen at a lower pressure were supplied to the base of plate-fin layers on each side of the condensing stream layer. A thermal resistance of known characteristics, placed between the layers, permitted the estimation of local heat fluxes between boiling and condensing streams from a knowledge of the local measured temperature differences.

A typical bulk temperature profile for the condensing nitrogen is given in Figure 6 where measured and calculated stream temperatures and measured wall temperatures are plotted along the length of the test section. The desuperheating, condensing and subcooling regions are all apparent. The calculated bulk temperature of the stream has been obtained using the measured local heat flux, stream flowrates and pressure. The discrepancy between measured and calculated bulk temperatures in the condensing region arises because of errors in pressure measurement. In the subcooled region the temperature discrepancies reflect the small cumulative error in heat flux or flowrate measurements. In Figure 6 a marked decrease can be seen in the temperature difference between wall and bulk in the single-phase and condensing regions (for approximately the same heat flux). This reflects the considerably greater condensing coefficient. The almost flat bulk temperature profile is a consequence of the very small pressure gradients for this downflow condensing.

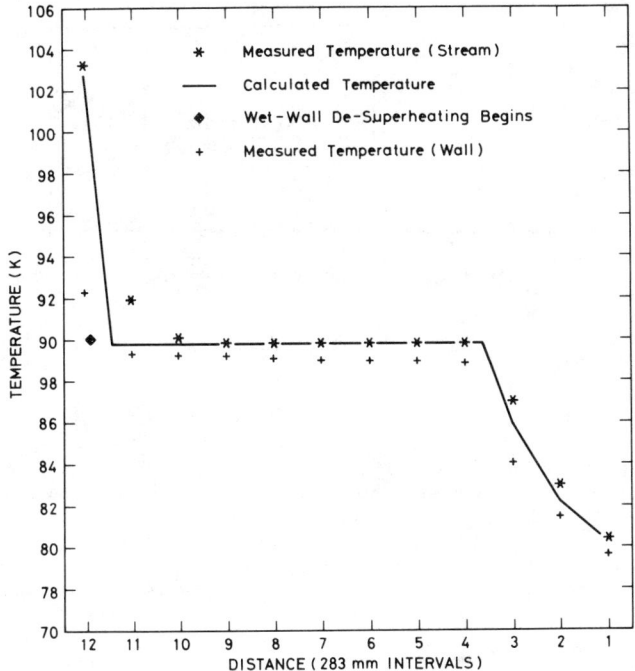

Fig.6. Temperature Profiles for Nitrogen Condensing at 3.6 bar and Mass Flux of 40.6 kg/m^{2}s

At the onset of condensation, wet wall desuperheating can be seen in Figure 6; the diagrams given in Figure 7 illustrate this phenomenon in more detail. At point A, wet wall desuperheating begins with the mixed bulk of the stream above the saturation temperature or dew point. After point A, a liquid film is created, essentially at saturation temperature, and the uncondensed vapour follows the temperature profile indicated by the dashed line. In the plate-fin test section, the liquid film will first occur on the primary heat transfer surface since the fins are at a higher temperature. The effects of wet wall superheating are particularly evident in condensation on heat flux controlled surfaces to which these experiments approximate. The extent to which it will affect local heat transfer will depend on the application but it will tend to reduce thermal performance.

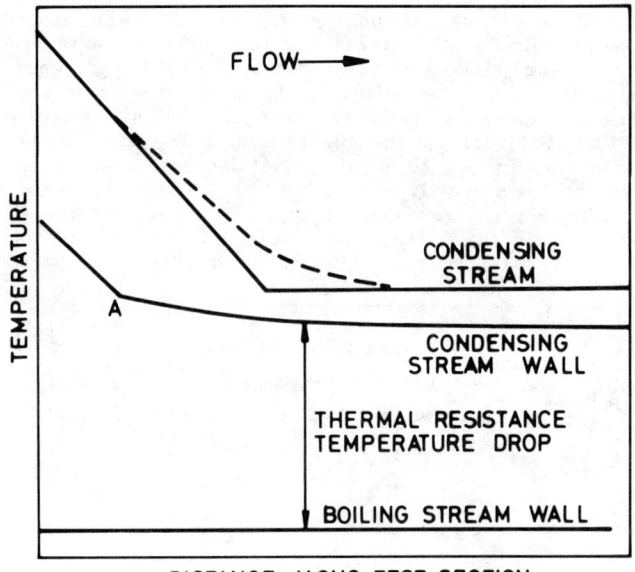

Fig.7. Diagram of Condensing Stream Temperature Profiles Showing Wet-Wall Desuperheating (16)

The effect of quality on the local condensing coefficient is shown on Figure 8 where the values have been derived from the analysis of the experimental results using a fin efficiency term. The curves are estimates from various theoretical condensing models: "S" has been obtained with Shah's correlation (24) and "B" from the correlation by Ananiev et al. (25). The experimental results are in good agreement with both correlations. The Nusselt model for condensation onto a falling film in a stagnant vapour produces much lower coefficients as expected. Results at lower mass fluxes indicate that all the correlations underpredict.

It is of interest to compare the condensing coefficients, as given in Figure 8 for downflow on a plain fin, with the results for upflow boiling given in Figure 3 for a serrated fin of approximately the same dimensions. With nitrogen at similar pressure, and the same mass fluxes at low qualities, the two coefficients have about the same magnitude.

2.5 Two-Phase Distribution

No published information from experiments on the distribution of a two-phase mixture to and within a plate-fin exchanger has been found. The topic is important since the performance of this type of heat exchanger is sensitive to flow maldistribution (34) particularly with a two-phase mixture at inlet. Flow maldistribution within the heated length of these exchangers is regarded as a possible reason for any thermal underperformance (33). For example, if the two phases of a multicomponent mixture, were distributed non-uniformly across the entrance to a passage in which evaporation was taking place, this would proceed under different vapour-liquid equilibrium conditions and at different rates across the passage, particularly if the temperature

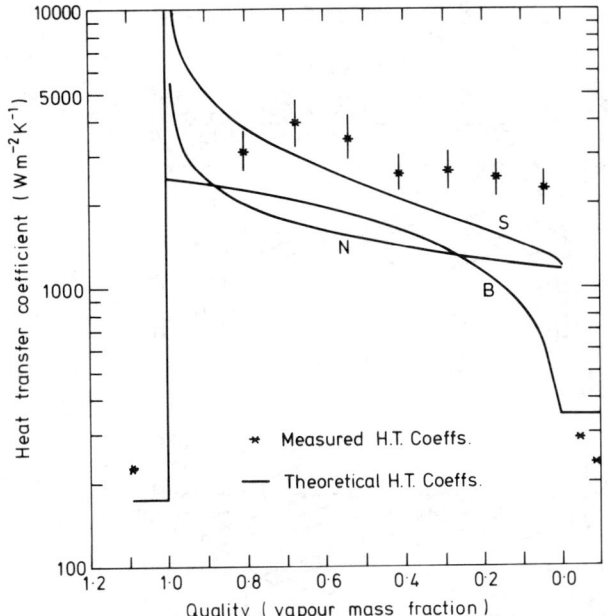

Fig.8. Experimental Condensing Coefficients for Nitrogen at 3.37 bar, and 40.6 kg/m^{2}s (16)

difference between this stream and an adjacent heating stream were very small.

Industrial experience with the introduction of two-phase flow into these exchangers has led to the deliberate separation of the two phases beforehand. Each phase is then pumped or gravity fed into a two-phase mixer. For a high quality mixture as in downflow condensing the mixer could be a liquid sprinkler system placed within a vapour plenum supplying many passages (Figure 9). The size of the droplets produced would be important in ensuring that any vapour cross flow did not produce drift. Careful positioning of the nozzles is necessary to ensure that each channel at entry to the inlet distributor receives its equal share of liquid. Other mixer systems are possible in principle, for example a device producing bubbles at the entrance to a low quality evaporating stream in upflow. The mixer, in principle, could also be placed at the entry to the thermal length and after the inlet distributor, which would be used here to distribute only the liquid phase across a passage. A back-to-back fin arrangement in which each phase has its own inlet distributor (within the thickness of one passage) ensures the phases are brought together at the entry to the thermal length. The mixer necessary for the distribution of a vapour phase in counter-current two-phase flow, as at the base of a condensing stream, in a dephlegmator, would require careful design since the possibility of flooding at the entry would need to be taken into account.

Inside the exchanger, the flow pattern across the thermal length is primarily controlled by the pressure drop characteristics of the inlet

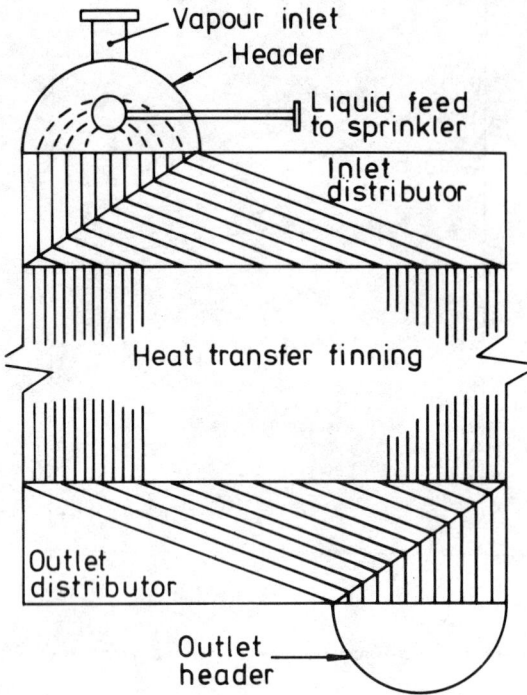

Fig.9. Diagram of One Passage in a Plate-Fin Exchanger Showing a Possible Flow Distribution System

distributor, the thermal length and the outlet distributor. The difference in pressure between the inlet and outlet plenums is common to all channels connecting them. Even if the two phases were introduced uniformly into the entrance of the inlet distributor, the resistance of all the channels can vary considerably and this will create an internal maldistribution.

There will also be a cross flow among all the channels since fins such as the perforated and serrated forms are permeable. Hence the flow pattern within a passage is two-dimensional. Cross-permeability can be enhanced locally by leakage paths at joins of finning and hence multi-region models are required for the determination of the flow patterns. These can be investigated realistically only in a suitable computer program. Two other factors are worth mentioning. In an evaporating stream the pressure gradients at exit are usually much greater than at inlet, hence the different path lengths in an exit distributor coupled with a large increase in mass flux (as the flow area decreases towards outlet), can influence maldistribution considerably. The other factor is concerned with the localised effect of heating or cooling from an adjacent passage. Local pressure gradients in two-phase flow are influenced greatly by local flow conditions. If these in turn are influenced by heat transfer to an adjacent stream (in which the flow may also be maldistributed) it can be seen that maldistribution can be transmitted from stream to stream by a thermal link.

Although flow instability is discussed in another chapter, it is worth adding here the peculiar difficulty which arises in two-phase co-current upflow with these broad passages encompassing many small channels. If the vapour velocity is not high enough, static legs of liquid can accumulate across the width of the passage. This separation of the two phases can usually be avoided by ensuring that the vapour phase velocity is sufficiently high so that the frictional drag may overcome the static pressure of the stagnant liquid.

2.6 Two-Phase Flow Instability

Cryogenic systems display many forms of flow instability such as geysering, rollover, thermal acoustic oscillations, etc. but equally important form from the aspect of process plant operation are the density-wave and Ledinegg forms in boiling streams. Both can occur in thermosyphon or once-through systems: the former is characterised by a stably oscillating flow with a period linked to the fluid residence time in the heated region; the latter by a flow excursion or bi-stable operation arising from a multi-valued flow resistance curve. The density-wave and other forms of instability, which occur with increasing levels of heat input into a thermosyphon stream, have been described by Chexel and Bergles (26). Bouré et al. have also reviewed this form of instability in once-through systems (27).

The only report dealing specifically with two-phase flow instability in plate-fin exchangers and its prediction under operating conditions is by Friedly et al. (28) who report progress in a programme of experiments to investigate systematically the effect of chosen system parameters. Developing Ishi's theoretical work (29), Friedly et al. uses ratios of two important system pressure drops, for example friction in the boiling section of a channel to the resistance at the channel inlet. He has obtained the boundary line between regions of stable and unstable flow for this and other ratios, with inlet subcooling and exit quality treated as parameters. In (28), experimental results from many sources have been compared with the predictions of this form of analysis with reasonable agreement. It has also been emphasised that the approach is particularly suitable to systematic experimentation, variable-by-variable, with the objective of producing a simple and understandable method for use by a designer to determine if unstable operation is probable and how it could be avoided.

The experiments were carried out with the same perforated and serrated-fin test sections used in (8), with once-through, vertical upflow boiling. The first series of experiments was designed to investigate the relationship between the exit-to-inlet pressure-drop ratio and the exit quality, with a fixed subcooling at inlet and when exit pressure-drop dominated. The results established the existence of the regions of stable and oscillating operation and also of excursive stability in accord with theoretical predictions. The second set of experiments established the effect of gradually increasing the importance of the friction of the heated section, at a fixed, high exit quality. The results from this series indicate that increasing friction of this region makes the system less stable by increasing the amplitude of flow oscillations and decreasing the operating exit-to-inlet pressure-drop ratio at the stability boundary. This latter effect has not yet been satisfactorily predicted by theory.

The application of this experimental information to the prediction of flow stability particularly of thermosyphons has been reported by Friedly and Hands (30) and also by Hands in two papers (31, 32). Hands has examined flow instabilities in a thermosyphon system with small-bore electrically-heated risers with liquid nitrogen. He has been able to produce a stability map on the heat-input/driving-head plane and has demonstrated that a "gravity pressure drop dominated form" of density wave flow instability may be the most likely for thermosyphon applications of plate fin exchangers. Further work in this area can be expected from these authors and this will be of importance in the systematic determination of flow instability in this form of exchanger. In industrial applications where many fractionation towers, each with a plate-fin thermosyphon reboiler, are interconnected, flow stability cannot be tolerated and research into the causes and prevention of instabilities will be of value to process plant designers.

2.7 Design Codes

Information obtained from experiments on plate-fin test sections will become more useful as its generality increases. However, its greatest impact on design will be to increase the applicability of and confidence in computer codes. While none has been published, several design evaluation methods, possibly more suitable for single-phase operation, are available (35, 36, 37). For the evaluation of the thermal performance of a given plate-fin exchanger with a completely known geometry and fluid flow conditions, codes can utilise the wealth of information that experimental work produces; hand methods would be inadequate.

The HTFS CEMP3 code, for example, can evaluate the heat exchanged among up to 6 streams in a fully-defined plate-fin exchanger with boiling, condensing or single-phase heat transfer in any stream and with single or multicomponent fluids (but not in cross-flow). Fluid conditions at either the cold end or warm end of an exchanger and physical property and vapour-liquid equilibrium data require to be supplied by the user. Geometric details of passages and finning and Colburn j and f factors are also required. Beginning with the given stream conditions at one end of the exchanger, including all flowrates, computation proceeds step-by-step along the length of the exchanger using local computed heat transfer coefficients. The program also computes local frictional and gravitational pressure gradients. Several hundred steps are normally used in the computation and the print-out provides enough information to construct detailed temperature and pressure drop profiles of each stream along the heat exchanger. This code also permits the assessment of the thermal performance of a reboiler with one thermosyphon boiling stream with up to five other streams. The thermosyphon reboiler may be immersed in a pool at the base of a fractionation column or connected to the column by pipework. The programme will also predict the flowrate for a given driving head of fluid within the column.

This computer program utilises a common wall temperature in the solution of the simultaneous equations governing heat transfer. However, this restriction could be removed by developing the program to deal with the heat transfer between adjacent passages with a passage-by-passage procedure. With multi-stream exchangers it is only possible to achieve an overall heat balance between all hot and cold streams grouped together; the individual heat balance on a stream can be an independent variable and this produces difficulties in performance evaluation codes which step from one end of the exchanger using both inlet and estimated outlet conditions for the various streams. This latter restriction has been removed by developing a program which can proceed from both ends simultaneously utilising only inlet fluid conditions and thereby producing more realistic estimations of performance.

In these codes it is possible to introduce features such as axial conduction along the exchanger, draw-off points to external fractionation systems, multiple finning, flow stability criteria, etc. A computer code which could evaluate the likely two-phase flow maldistribution within an exchanger would be a useful tool for any designer to complement initial design and selection procedures. Design codes can also be used to investigate the performance of plate-fin exchangers in process plants where unusual features in the plant's operation have been revealed. They can also be useful for incorporation in general codes for process plants to determine the optimum operating conditions and minimum power requirements.

3. WOUND-COIL HEAT EXCHANGERS

3.1 Introduction

In a wound-coil, or coiled-tube heat exchanger, tubes are coiled round a cylindrical core. Layers of tubes are helically wound alternately in opposite directions and separated by spacers, and a central mandrel on which the first coil is wound supports the coil winding. The completed winding is placed inside a vertical cylindrical containment vessel. Several streams of different process fluids can be arranged to flow through chosen tube layers. The fluid flowing on the shell side of the wound tubes will permeate all the coils and exchange heat with all other streams. (In a plate-fin heat exchanger heat is exchanged primarily between adjacent streams.) A cross-section of a typical wound-coil heat exchanger with three tube-side streams is shown in Figure 10 (from (43)). The mechanical design of these heat exchangers is dealt with in some detail by Gaumer et al. (39), Abadzic and Scholz (44) and Smith (45) among others. For cryogenic purposes, aluminium or copper is normally used for the construction of the tubes while stainless steel could be used for the tube sheets and shell (2). Tube bundles can be very large, for example, 3 m in diameter and 13 m long providing approximately 37,000 m^2 of heat transfer surface. Several bundles, comprising a complete cryogenic process exchanger train, may be combined in the one shell with considerable thermodynamic advantage as described by Crawford et al. (2). Since it utilises small-bore tubes, around 10 mm diameter, the wound-coil heat exchanger can, in principle, make use of higher pressures for the tube-side streams than is possible in a plate-fin exchanger.

These heat exchangers have been developed from the original Hampson design (47) to the enormous heat exchanger train contained within one shell in the liquefaction of natural gas described by Gaumer (39). However, they are also used in smaller sizes, for example as feed coolers and liquefiers for general cryogenic purposes. Wound-coil steam generators have also been used in gas-cooled nuclear reactors.

Normally shell-side and tube-side fluids are in counter-flow and boiling or condensing can take place within the tubes or on the shell side. In exchangers for liquefying natural gas, downflow boiling takes place on the shell side and is accompanied by upflow condensation of the refrigerant and feed streams on the tube side. As in plate-fin heat exchangers, mass flow rates are typically less than 150 kg/m^2s and heat fluxes are a few kW/m^2 reflecting the very small temperature differences available between shell-side and tube-side streams.

Flow on the shell side is particularly complex and the geometrical analysis of the coils' configuration to estimate the maximum, minimum and

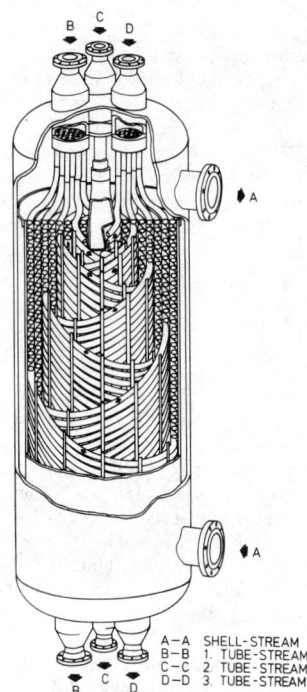

Fig.10. Illustration of a Four-Stream Wound-Coil Heat Exchanger (from (43))

average flow cross-sections between adjacent tube layers is given by Abadzic and Scholz (44). While flow across the coils will resemble cross-flow in a horizontal tube bank for single-phase flow, for two-phase flow the discontinuity in the film flow as it encounters and leaves each tube as the shell-side stream is progressibly evaporated in a downwards direction will produce boiling heat transfer coefficients which will differ from values obtained from results of experiments on boiling within tubes.

3.2 Single-Phase Operation

Abadzic (46) reviews experimental data from tests on shell-side heat transfer by several authors. The equations proposed by Grimison (49) and others for straight tubes can be used as a basis with tube diameter as a suitable characteristic dimension, but this leaves the difficulty of determining a suitable or characteristic cross-section for defining the local velocity. Glaser (48) investigated this aspect with air on coils of constant tube diameter with various tube pitches and using a characteristic mean cross-sectional area based on the Glaser and Hausen integrated mean area (48). Glaser correlated his results with the simple relationship between Nusselt and Reynolds numbers:-

$$Nu = 0.267 \, Re^{0.609} \qquad (1)$$

Dallmeyer (53), Scholz et al. (51) and Groehen et al. (52) also report results from similar tests. Abadzic gives results of tests on various coils and shows that these correlate well using the same form of relationship as Equ. (1) over the range 2×10^4 - 2×10^5 for one coil, with a change in slope at Re = 6.5×10^4. Abadzic reviews all the available data and by using the Glaser and Hausen mean area is able to correlate most of the results obtained with air by a set of three expressions, similar to Equ. (1), each being suitable for separate ranges of Reynolds number from 10^3 to 9×10^5. Abadzic compares the estimates of these expressions for Nusselt number against those from expressions given by Zukauskas (50) for banks of parallel tubes and concludes that provided the ratio of transverse to axial tube pitches is not too large, there is good agreement. By linking test results for wound-coils and straight-tube banks in this way, he deduces that the effect of Prandtl number on his expressions should be to the power 0.36.

For the shell-side, Gaumer et al. (39) suggest the use of a suitably modified Colburn correlation for heat transfer and the Grimison correlations for in-line tube banks (49) also suitably modified.

Abadzic and Scholz (44) suggest the use of the standard expressions for heat transfer and pressure drop inside coiled tubes. However, since much of the controlling thermal resistance is usually on the shell side, any increase in heat transfer coefficient as a result of the flow in the helically-wound tube will be negligible. The use of Schmidt's correlation for the increased pressure drop (54) is also recommended.

3.3 Two-Phase Flow Aspects

The reported experimental work in this area is very sparse and is confined to measurements of boiling and two-phase pressure drop on the shell-side of wound-coil models by Barbe et al. (55, 56). These authors used a model of a wound-coil exchanger consisting of three layers of coiled electrical cable of approximately 0.25 m coil diameter and 2.5 m high. Air and water were used as test fluids for the two-phase pressure drop measurements. For the boiling experiments, propane and a mixture of propane, ethane and methane were used. Heat fluxes used ranged from 2 to 5 kW/m^2 and mass fluxes from 45 to 115 kg/m^{2}s.

Heat transfer results were correlated by plotting the ratio of the boiling coefficient to the liquid-phase coefficient against the Martinelli parameter (the ratio of the pressure drop of each phase flowing turbulently alone). The results are well correlated in this way and show no dependence on mass flux or on heat flux since nucleate boiling would not be expected at these low levels (Figure 11). No information is given about the effect of the multi-component mixture on the boiling coefficients.

Results of two-phase pressure drop measured in upflow and in downflow are given in the form of plots of $\emptyset$, the square root of the ratio of the two-phase pressure drop to that of the single-phase pressure drop, with the total flow regarded as liquid, against the square root of the Martinelli parameter, X_{tt}. For the descending flow of air and water, at a pressure of 1 bar, the results were well correlated in this way by:-

$$\emptyset = \frac{1 + X_{tt}}{X_{tt}} \qquad (2)$$

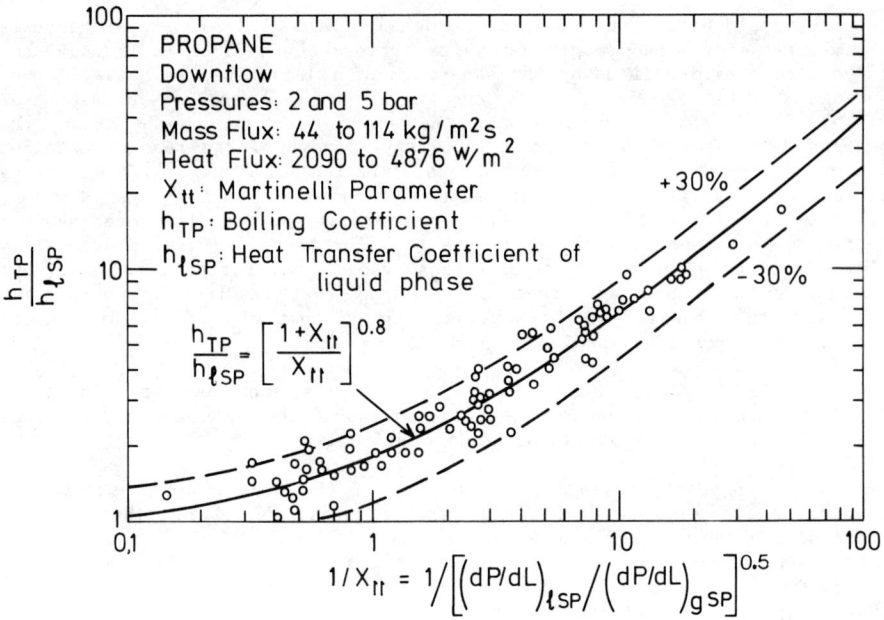

Fig.11. Plot of Ratio of Boiling Heat Transfer Coefficient to Liquid-Phase Coefficient Against Martinelli Parameter for Wound-Coil Test-Section (from (55))

For ascending flow with air and water, the results were correlated but only at high qualities. At lower qualities, the two-phase pressure drop tended to the value of the hydrostatic liquid head. The authors ascribe this to the low upward vapour velocities permitting stagnant columns of liquid to remain within the complex passageways of the coils. This may be compared with the similar conditions found in the multi-passage arrangements of plate-fin heat exchangers, and discussed in para. 2.6 above.

Gaumer et al. suggest the use of the Diehl (57) correlation for downward two-phase flow over parallel tube banks for the estimation of two-phase pressure drop, and a modified correlation of Davis and David (58) for convective boiling heat transfer. Here, the temperature difference between streams should be reduced, and the coefficients modified, for multi-component boiling. In the liquid-deficient region, where the tubes are no longer covered with a liquid film, heat transfer coefficients are calculated separately. Although these authors mention tests carried out to determine the dryout point, no results are reported.

For the tube-side, Gaumer et al. report that the Chenoweth-Martin correlation (38) is satisfactory for predicting two-phase pressure drops. The Davis and David correlation is also used to predict coefficients for condensing up-flow in the tubes.

3.4 Flow Distribution

In order to achieve uniform flow distribution on the tube side, wound-coil heat exchangers are usually designed with a constant tube length (and presumably tube diameter) (44). This results in a decreasing number of turns for each successive layer and this will affect the cross-sectional area for the shell-side flow across the layers of the winding. This variation will thus affect flow distribution on the shell side and could, in principle, be reduced by varying the spacer thickness from layer to layer. With a two-phase flow at inlet to the shell side, in downflow, the liquid and vapour are separated and introduced individually with the liquid-phase spray being graded radially to achieve a uniform throughout the windings (39). Care is taken to ensure that downflow velocities are high enough to prevent vapour at the bottom of the shell causing natural convection within the coils. Weimer and Hartzog (34) have estimated that the small effect of shell-side maldistribution has on overall performance can be attributed mainly to the ability of the two-phase flow to permeate the coil windings uniformly.

4. CONCLUSIONS

The published information on the heat transfer and fluid flow within plate-fin heat exchanger and wound-coil heat exchangers has been reviewed. These exchangers are widely used in the cryogenic process industry and their geometric features and applications are sufficiently unusual to make information obtained from test-work specifically aimed at revealing their characteristics, very valuable. However, only a limited amount of experimental information has been found, and much is in the form of design recommendations.

5. REFERENCES

1. Lenfesty, A.G. 1961. "Low Temperature Heat Exchangers," *Progress in Cryogenics*, Heywood & Co., London, pp.25-47.

2. Crawford, D.B., and Eschenbrenner, G.P. 1973. "Heat Transfer and Heat Exchanger Applications for LNG," *Chemical Engineering World*, Vol. VIII, No. 5, May, pp.77-94.

3. Stebbing, R. and O'Brien, J.O. 1975. "An up-dated report on the PRICO Process for LNG Plants," Gastech 75, Paper 15: International LNG/LPG Congress, Paris.

4. Haselden, G.G. 1957. "The Approach to Minimum Power Consumption in Low Temperature Gas Separation," Joint Symposium of Institute of Chemical Engineers, 26 November, pp.1-10.

5. Panitsidis, H., Gresham, R.D. and Westwater, J.W. 1975. "Boiling of Liquids in a Compact Plate-Fin Heat Exchanger," *International Journal of Heat Mass Transfer*, Vol. 18, pp.37-42.

6. Galezha, V.B., Usyukin, I.P. and Kan, K.D. 1976. "Boiling Heat Transfer with Freons in Finned-Plate Heat Exchangers," *Heat Transfer - Soviet Research*, Vol. 8, No. 3, May, pp.103-110.

7. Shorin, C.N., Sukhov, V.I., Shevyakova, S.A. and Orlov, V.K. 1974. "Experimental Investigation of Heat Transfer with the Boiling of Oxygen in Vertical Channels during Condensation Heating," *International Chemical Engineering*, Vol. 14, No. 3, July, pp.517-521.

8. Robertson, J.M. 1979. "Boiling Heat Transfer with Liquid Nitrogen in Brazed-Aluminium Plate-Fin Heat Exchangers," *AIChE Symposium Series*, No. 189, Vol. 75, pp.151-164.

9. Robertson, J.M. and Lovegrove, P.C. 1980. "Boiling Heat Transfer with Freon 11 in Brazed Aluminium Plate-Fin Heat Exchangers," ASME paper 80-HT-58, ASME/AIChE National Heat Transfer Conference, Orlando.

10. Abdelmessih, A.H., Yin, S.T. and Fakhri, A. 1973. "Hysteresis Effects and Hydrodynamic Oscillations in Incipient Boiling of Freon 11," *Proc. Intl. Meeting, Reactor Heat Transfer*, Karlsruhe, pp.331-350.

11. Davis, E.J. and Anderson, G.H. 1966. "The Incipience of Nucleate Boiling in Forced Convection Flow," *AIChE Journal*, Vol. 12, No. 4, pp.774-780.

12. Singh, A., Mikic, B.B. and Rohsenow, W.M. 1976. "Active Sites in Boiling," *Journal of Heat Transfer*, Vol. 98, No. 3, pp.401-406.

13. Chen, J.C. 1963. "A Correlation for Boiling Heat Transfer to Saturated Fluids in Convective Flow," ASME Paper No. 63-HT-34.

14. Thomé, J.R. 1978. "Bubble Growth and Nucleate Pool Boiling in Liquid Nitrogen, Argon and their Mixtures," D.Phil. Thesis, University of Oxford.

15. Gopin, S.R., Usyukin, I.P. and Aver'yanov, I.G. 1976. "Heat Transfer in Condensation of Freons on Finned Surfaces," *Heat Transfer - Soviet Research*, Vol. 8, No. 6, pp.114-119.

16. Haseler, L.E. 1980. "Condensation of Nitrogen in Brazed Aluminium Plate-Fin Heat Exchangers," ASME paper 80-HT-57, ASME/AIChE National Heat Transfer Conference, Orlando.

17. Haselden, G.G. and Prosad, S. 1947. "Heat Transfer from Condensing Oxygen and Nitrogen Vapours," Trans. Institution of Chemical Engineers, London, Vol. 27, p.195.

18. Ewald, R. and Perroud, P. 1970. "Measurement of Film Condensation Heat Transfer on Vertical Tubes for Nitrogen, Hydrogen and Deuterium," *Advances in Cryogenic Eng.*, Vol. 16, pp.475-481.

19. Leonard, R.J. and Timmerhaus, K.D. 1970. "Condensation Studies of Saturated Nitrogen Vapours," *Advances in Cryogenic Eng.*, Vol. 15, pp.308-315.

20. Rifert, V.G., Barabash, P.A. and Golubev, A.B. 1977. "Investigation of Film Condensation Enhanced by Surface Forces," *Heat Transfer - Soviet Research*, Vol. 9, No. 2, pp.23-27.

21. Rifert, V.G., Leont'yev, G.G. and Chaplinskiy, S.I. 1976. "Augmentation of Heat Transfer from Refrigerants Condensing on Vertical Tubes," *Heat Transfer - Soviet Research*, Vol. 8, No. 6, pp.120-127.

22. Zozulya, N.V., Karkhu, V.A. and Borovkov, V.P. 1977. "An Analytical and Experimental Study of Heat Transfer in Condensation of Vapour on Finned Surfaces," Heat Transfer - Soviet Research, Vol. 9, No. 2, pp.18-22.

23. Pantankar, S.V. and Sparrow, E.M. 1979. "Condensation on an Extended Fin," Journal of Heat Transfer, Vol. 101, No. 3, pp.434-440.

24. Shah, M.M. 1979. "A General Correlation for Heat Transfer during Film Condensation inside Pipes," Int. Journal Heat & Mass Transfer, Vol. 22, No. 4, pp.547-556.

25. Ananiev, E.P., Boyko, L.D. and Kruzhilin, G.N. 1961. "Heat Transfer in the Presence of Steam Condensation in a Horizontal Tube," Int. Dev. Heat Transfer, Part 2, pp.290-295.

26. Chexal, V.K. and Bergles, A.E. 1973. "Two-Phase Instabilities in a Low Pressure Natural Circulation Loop," AIChE Symposium Series, Vol. 69, No. 131, pp.37-45.

27. Bouré, J.A., Bergles, A.E. and Tong, L.S. 1971. "Review of Two-Phase Flow Instability," ASME Paper No. 71-HT-42.

28. Friedly, J.C., Akinjiola, P.O. and Robertson, J.M. 1979. "Flow Oscillations in Boiling Channels," AIChE Symposium Series, No. 189, National Heat Transfer Conference, San Diego, pp.204-217.

29. Ishii, M. 1976. "Study on Flow Instabilities in Two-Phase Mixtures," ANL Report No. ANL-76-23.

30. Friedly, J.C. and Hands, B.A. 1976. "Flow Oscillations in a Liquid Nitrogen Thermosyphon," 6th Intl. Cryogenic Engineering Conf., Grenoble, pp.319-324.

31. Hands, B.A. 1979. "Density Wave Oscillations: The Use of Vector Diagrams to Identify Some Different Types," AIChE Symposium Series, No. 189, National Heat Transfer Conference, San Diego, pp.165-176.

32. Hands, B.A. 1979. "The Flow Stability of a Liquid Nitrogen Thermosyphon with 8 mm Bore Riser," AIChE Symposium Series, No. 189, National Heat Transfer Conference, San Diego, pp.177-184.

33. The Trane Company, 1969-1971. "Method and Application for Two-Phase Heat Exchange Fluid Distribution in Plate-Type Heat Exchangers," US Patent 3559722.

34. Weimer, R.F. and Hartzog, D.G. 1972. "Effects of Maldistribution on the Performance of Multi-Stream, Multi-Passage Heat Exchangers," Advances in Cryogenic Engineering, Vol. 18, pp.52-64.

35. Fan, Y.N. 1966. "How to Design Plate-Fin Exchangers," Hydrocarbon Processing, Vol. 45, No. 11, pp.211-217.

36. Chato, J.C., Laverman, R.J. and Shah, J.M. 1971. "Analyses of Parallel Flow, Multi-Stream Heat Exchangers," Int. Journal Mass & Heat Transfer, Vol. 14, pp.1691-1703.

37. Kays, W.M. and London, A.L. 1964. "Compact Heat Exchangers," Second Edition, McGraw-Hill.

38. Chenoweth, J.M. and Martin, M.W. 1955. Pet. Refinev, 34, 151.

39. Gaumer, L.S., Geist, J.M., Hartnett, G.J. and Phannenstiel, L.L. 1972. "The design, fabrication and operation of large cryogenic heat exchangers," Session 11, Paper 15, LNG3 Conference.

40. Butt, A.G. 1980, "Mechanical design of cryogenic exchangers," ASME Winter Conference, Chicago.

41. Robertson, J.M. and Clarke, R.H. 1981. "The onset of boiling of liquid nitrogen in plate-fin heat exchangers," Paper to be presented at the ASME/AIChE National Heat Transfer Conference, Milwaukee, U.S.A.

42. Yung, D., Lorenz, J.J. and Panchal, C. 1980. "Convective vaporization and condensation in serrated-fin channels," ASME HTD - Vol. 12, pp.29-37.

43. Linde, A.G. Catalogue entitled "Rohrbündel-Wärmeaustaucher," Linde, A.G., Weskgruppe, TVT, Munich, Germany.

44. Abadzic, E.E. and Scholz, H.W. 1973. "Coiled tubular heat exchangers," Advances in Cryogenic Engineering, Vol. 18, Plenum Press, pp.42-51.

45. Smith, E.M. 1964. "Helical-tube heat exchangers," Engineering, Vol. 197, p.232.

46. Abadzic, E.E. 1974. "Heat transfer on coiled tubular matrix," ASME paper 74-WA/HT-64, Winter Meeting.

47. Hampson, W. British Patent No. 10156.

48. Glaser, H. 1938. "Heat transfer in regenerators," (Germ.), VDI Zeitschrift Beihefte Verfahrenstechnik, Vol. 82, No. 4.

49. Grimison, E.D. 1937. "Correlation and utilisation of new data on flow resistance and heat transfer for crossflow of gases over tube banks," Trans. Amer. Soc. Mech. Eng., Vol. 59, pp.583-594.

50. Zukauskas, A.A. 1969. "Heat transfer from single tubes and banks of tubes in cross-flow," Intnl. Centre for Mass and Heat Transfer, Conference.

51. Scholz, F., Meis, Th. and Groehen, H.G. 1970. "Final report on the tests on a helically wound heat exchanger for the steam generator of the THTR," Reports of the Nucl. Research Centre, Julich - 649 - RB.

52. Groehen, H.G. and Scholz, F. 1972. "Investigations on steam generator models of in-line tube arrangement and multistart helices design in pressurized air and helium," Conf. on Comp. Design in High Temp. Reactors, London 3/4 May.

53. Dallmeyer, H. 1964. "Investigations of heat transfer and pressure drop on coils," (in German), Diploma Thesis, T.H. Aachen.

54. Schmidt, E.F. 1965. Ph.D Dissertation, T.H. Braunschweig, Braunschweig, Germany.

55. Barbe, C., Grange, A. and Roger, D. 1971. "Echanges de chaleur et pertes de charges en écoulement diphasique dans la calandre des échangeurs bobinés," Proceedings of the XIII International Congress on Refrigeration, Vol. 2, pp.223-234.

56. Barbe, C., Grange, A. and Roger, D. 1972. "Two-phase flow heat exchangers and pressure losses in spool-wound exchanger shells," Pipeline and Gas Journal, Sept., pp.82-87.

57. Diehl, J.E. 1957. Pet. Refiner, 36, 147.

58. Davis, E.J. and David, M.M. 1964. I.E.C. Fundamentals 3, 111.

59. Webb, R.L. 1981. "Matrix heat exchangers - thermal and hydraulic design (compact heat exchangers)," Vol. 3, Chapter 8, Heat Transfer Design and Data Book, (Schlunder, E.U., Editor), Hemisphere Pub. Co.

60. Mochizuka, S. and Yagi, Y. 1977. "Heat transfer and friction characterisation of strip fins," Heat Transfer Japan Research, Vol. 6, No. 3, pp.36-59.

61. Robertson, J.M., "Review of Boiling, Condensing and Other Aspects of Two-Phase Flow in Plate-Fin Heat Exchangers, HTD-Vol. 10, pp 17-27, Compact Heat Exchangers, ASME Winter Annual Meeting, Chicago, 1980.

ACKNOWLEDGEMENT

The author wishes to acknowledge the kind permission of the American Society of Mechanical Engineers to use much of the material presented by the author at the Symposium on Compact Heat Exchangers at the Winter Annual Meeting, Chicago, 1980.

Performance Calculation Methods for Multi-stream Plate-Fin Heat Exchangers

L. E. HASELER
Heat Transfer and Fluid Flow Service (HTFS)
Atomic Energy Research Establishment
Harwell, Oxfordshire OX11 0RA, UK

ABSTRACT

In a plate-fin heat exchanger, there is heat transfer between streams in non-adjacent layers by conduction through the plate-fin matrix, as well as the usual heat transfer between streams in adjacent layers. It is shown that the matrix conduction heat transfer can be expressed in terms of a "by-pass efficiency" in each layer, which is a function of the same variables as the conventional fin efficiency. Methods of integrating the resulting heat transfer equations to calculate stream outlet conditions corresponding to given stream inlet conditions are suggested for co- and counter-current multistream exchangers.

1. PLATE-FIN HEAT EXCHANGERS

The main use of large multi-stream plate-fin heat exchangers is in the cryogenic process industries such as air separation plant, ethylene plant or in natural gas liquefaction. Smaller plate-fin units are also widely used for aero-space applications, because of their compactness. A plate-fin exchanger is built up of a large number of layers formed from a series of corrugated sheets or fins inter-leaved by flat 'parting sheets' or 'separating plates'.

A typical cross section through part of four layers of such an exchanger is shown in Figure 1. The various streams in the exchanger, sometimes up to 10, flow perpendicular to this cross-section, and are distributed among the layers, normally so as to form a pattern of alternate hot and cold streams. Each layer is typically between 5 and 15 mm high, and, for cryogenic process plant, the complete exchanger might consist of a hundred or more layers, to form a block up to a metre deep. The width of the exchanger might also be a metre, while its length along the flow direction could be several metres.

The fins are normally a fraction of a millimetre thick, there being several hundred fins in each layer per metre width of exchanger. The fins serve three purposes.

(1) They provide extra area for heat transfer.

(2) They enable the exchanger to withstand high pressures - up to 70 bar in certain cases.

(3) They permit heat to be transferred across the exchanger between non-adjacent streams.

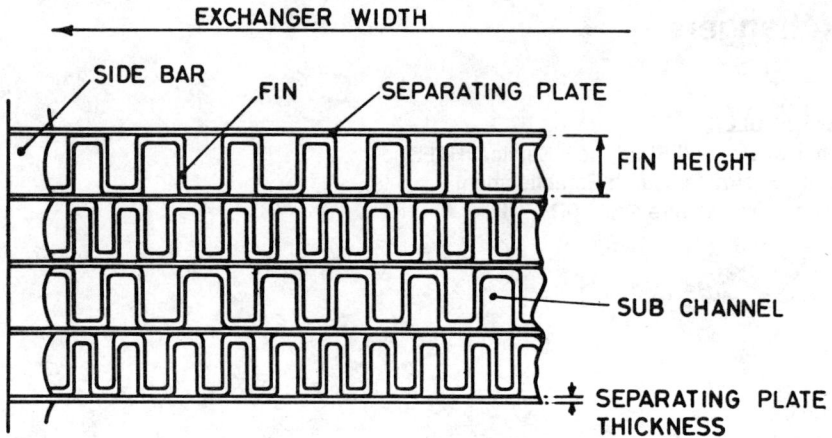

Figure 1. Cross Section of Part of a Plate-Fin Heat Exchanger Showing Four Layers

Once a particular process stream in a plate-fin exchanger - stream A, say - is divided up among a number of layers, say 25, the contents of the various individual layers are not mixed together again until they have passed completely through the exchanger. Although the designer will normally try to arrange for the heat gain (or loss) of all 25 layers of the A stream to be the same, there is no a priori reason why this must be so. In the most general analysis of a plate-fin heat exchanger, the contents of each layer must therefore be treated as a separate stream.

The "fin efficiency" describing heat transfer from fins with a free end, and defined in the standard textbooks (1,2), is readily adapted to cover fins between two plates at the same temperature. Cases involving plates at different temperatures have by comparison been examined by relatively few authors (e.g. 3,4).

Methods of integrating the heat transfer equation in multi-stream counterflow heat exchangers have been given for the simple case of constant coefficients (5), or where the coefficients are known as a function of position along the exchanger (3). Weimer and Hartzog (4) are the only authors known to have previously presented a more general solution method.

2. HEAT TRANSFER ACROSS THE EXCHANGER

2.1 Fin Temperature

Figure 2 shows the type of temperature profile T(x) that might be expected in the fins of three adjacent layers of a plate-fin heat exchanger. The temperatures of the streams in these layers are T_{i-1}, T_i and T_{i+1}, while the walls (separating plates) on either side of the ith layer are at temperatures Tw_i and Tw_{i+1}. In analysing the fin to stream heat transfer, the following assumptions are made:-

MULTI-STREAM PLATE-FIN HEAT EXCHANGERS

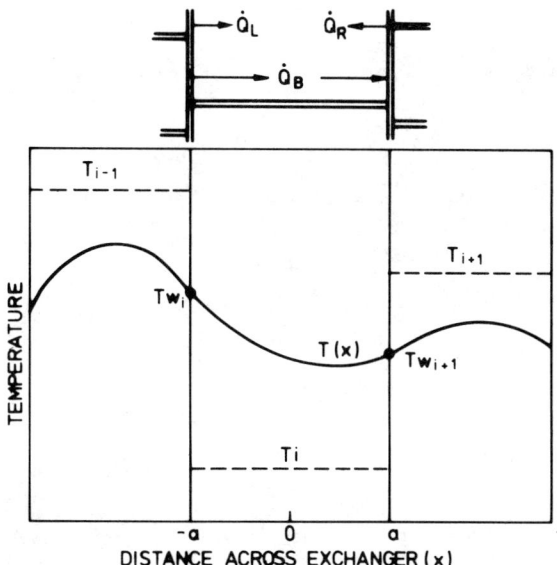

Figure 2. Temperature Profile across Three Plate-Fin Layers

(1) Each separating plate temperature does not vary in the x or y directions.

(2) The fluid in the channel between the two separating plates is at a uniform temperature T_i, and the heat transfer coefficient (α) is a constant, i.e. does not vary with position in the channel, or with temperature of the surface from which heat is transferred.

(3) The effect of any variation along the length of the channel may be ignored in calculating the variation across the channel.

The changing temperature gradient in the fin across the ith layer is due to heat being lost to the stream. The equation defining the temperature profile is:-

$$\frac{d^2 T}{dx^2} = \frac{2\alpha}{\lambda t} [T - T_i] \qquad (1)$$

α is the heat transfer coefficient of the ith layer
λ is the thermal conductivity of fin material (assumed constant)
t is the thickness of fin material in the ith layer

The solution is

$$T = T_i + \left[\frac{T_i + T_{i+1}}{2}\right] \frac{\cosh(\beta x)}{\cosh(\beta a)} + \left[\frac{T_{i+1} - T_i}{2}\right] \frac{\sinh(\beta x)}{\sinh(\beta a)} \qquad (2)$$

where $\beta^2 = \dfrac{2a}{\lambda t}$ (3)

It may be verified that this solution fits the necessary boundary conditions

$T(a) = Tw_{i+1}$ (4)

$T(-a) = Tw_i$ (5)

2.2 Heat Transfer

Heat flows away from the ith wall in the Tx direction both by direct heat transfer to the ith stream, and by conduction along the fin. This heat flow rate, per unit length along the exchanger, in a layer with N fins is thus

$$\dot{Q}_{LT} = \alpha A_1 (Tw_i - T_i) - N \lambda t \left.\dfrac{dT}{dx}\right|_{x=-a}$$ (6)

where A_1 is the primary surface area of the ith wall per unit length along the exchanger.

Differentiating Equation 2 and substituting in Equation 6 shows that $\dot{Q}_{LT}$ is the sum of two terms.

$$\dot{Q}_{LT} = \dot{Q}_L + \dot{Q}_B$$ (7)

$$\dot{Q}_L = \alpha (A_1 + \eta A_2)(Tw_i - T_i)$$ (8)

$$\dot{Q}_B = \eta' A_2 (Tw_i - Tw_{i+1})$$ (9)

where $A_2 = 2aN$ the secondary surface area per wall per unit length along the exchanger

and η is the fin efficiency as normally defined in the textbooks (1,2) for a fin, height a attached to one wall only,

$$\eta = \dfrac{\tanh(\beta a)}{\beta a}$$ (10)

and η' is the fin "by-pass" efficiency' defined by

$$\eta' = \dfrac{\coth(\beta a) - \tanh(\beta a)}{2\beta a} = \dfrac{\operatorname{cosech}(2\beta a)}{\beta a}$$ (11)

The heat flow in the fin in the $-x$ direction at $x = +a$ may be similarly calculated to be

MULTI-STREAM PLATE-FIN HEAT EXCHANGERS   499

$$\dot{Q}_{RT} = \dot{Q}_R - \dot{Q}_B \qquad (12)$$

$$\dot{Q}_R = \alpha \ (A_1 + \eta A_2) \ (Tw_{i+1} - T_i) \qquad (13)$$

The terms α, A_1, A_2, η and η' all refer to the ith layer but the subscript i is omitted for clarity.

The term Q_L is the usual expression for heat transfer in a plate fin layer with both walls at temperature Tw_i.

The term $\dot{Q}_B$ is equations (7) and (12) represents the heat which is not delivered to the streams, but is transferred by conduction along the fin directly from the ith wall to the $i+1^{th}$ wall. It can be seen that this term is directly proportional to the difference between the two wall temperatures, as might be expected. It also depends on the "by-pass efficiency" η' which is a function of exactly the same variables as the conventional fin efficiency.

η and η' are compared on Figure 3. When the fin conduction is very much better than the stream-to-wall heat transfer, i.e. high fin efficiency, the by-pass efficiency is also very high, with the corollary that the difference between Tw_i and Tw_{i+1} must be correspondingly low. At the other end of the scale, when the fin conduction is relatively poor, the by-pass efficiency is very low so almost all the heat from the walls is delivered to the stream, and very little heat is conducted directly from one wall to the other.

The above analysis and the definition of by-pass efficiency have not been given previously in this form. Equivalent formulations have been used by

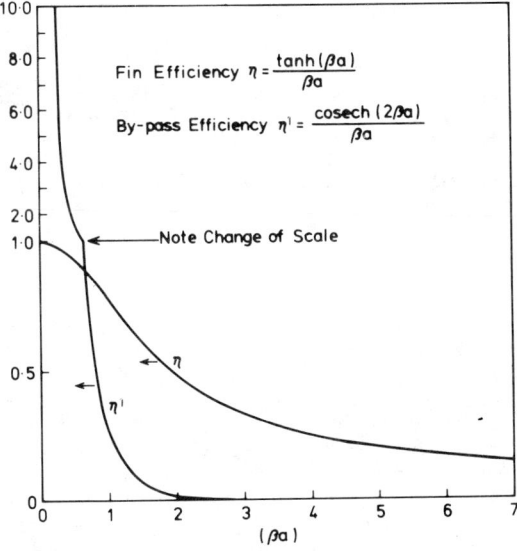

Figure 3. Fin Efficiency and By-pass Efficiency

other authors (3,4) who have analysed the problem, but they have left the solution in terms of complicated expressions involving integration constants and hyperbolic functions which obscure the simplicity of the processes involved.

2.3 Calculation of Wall Temperatures

When calculating the performance of a plate-fin heat exchanger, it is necessary to calculate the wall temperatures corresponding to a known set of stream temperatures. These may be found by consideration of the heat balance across the ith wall.

$$\dot{Q}_{R,i-1} - \dot{Q}_{B,i-1} + \dot{Q}_{L,i} + \dot{Q}_{B,i} = 0 \tag{14}$$

It is convenient to introduce the variables ϕ and ψ defined by

$$\phi = \alpha (A_1 + \eta A_2) \tag{15}$$

$$\psi = \alpha \eta' A_2 \tag{16}$$

then the heat balance equation is

$$\phi_{i-1} (Tw_i - T_{i-1}) - \psi_{i-1} (Tw_{i-1} - Tw_i) +$$

$$\phi_i (Tw_i - T_i) + \psi_i (Tw_i - Tw_{i+1}) = 0 \tag{17}$$

which may be re-arranged to give

$$Tw_i = \frac{\phi_i T_i + \phi_{i-1} T_{i-1} + \psi_i Tw_{i+1} + \psi_{i-1} Tw_{i-1}}{\phi_i + \phi_{i-1} + \psi_i + \psi_{i-1}} \tag{18}$$

For a given set of stream temperatures, T_i, this formula may be used iteratively to solve for the corresponding wall temperatures.

2.4 Simplifying Assumptions

The stream and wall temperatures in Equation 18 are of course functions of the distance along the exchanger. To integrate the heat transfer equations, with the help of Equation 18, separately for each layer is a formidable task in the general case of a plate-fin exchanger with a hundred or more layers.

In certain exchangers, where the "stacking pattern" in which the layers are arranged consists of a sub pattern of m layers, perhaps ten or fewer, repeated many times, Equation 18 may be used directly with T_{m+1} set equal to T_1. In exchangers where this does not apply, some other simplification is normally necessary, such as the common wall temperature assumption. This assumes that all Tw_i are equal and is defined by an overall enthalpy balance for all streams

$$\sum_i [\dot{Q}_{L,i} + \dot{Q}_{R,i}] = 0 \qquad (19)$$

which implies

$$T_w = \frac{\sum \phi_i T_i}{\sum \phi_i} \qquad (20)$$

The common wall-temperature assumption leads to a solution in which the enthalpy change of each stream is equal to the heat it gains from (or loses to) the wall. It is incorrect in that it does not deal correctly with the heat conduction through the fins from one wall to the other; it effectively assumes that this heat can flow unrestricted across the exchanger, and is thus equivalent to setting all the ψ_i to infinity.

The assumption provides a considerable simplification of the heat transfer problem, since it implies that all layers containing a given process stream behave identically. The number of streams to be considered is therefore equal to the number of process streams rather than the number of layers. The corresponding disadvantage is the preclusion of any distinction between various stacking patterns; the solution depends only on the total number of layers for each stream, not on the pattern in which they are arranged.

3. INTEGRATING THE HEAT TRANSFER EQUATIONS

The heat transfer per unit length from the walls to the ith stream is related to the rate of change of specific enthalpy, h, with respect to distance, z, along the exchanger.

$$Q_{LT} + Q_{RT} = Q_L + Q_R = \dot{M}\frac{dh}{dz} = \dot{M}c_p\frac{dT}{dz} \qquad (21)$$

where $\dot{M}$ is the mass flowrate of the stream

c_p is the specific heat of the stream

$\dot{M}c_p$ is referred to as the 'capacity rate' of the stream.

3.1 Variations of Coefficients

In a multi-stream exchanger in which the coefficients ϕ and $\dot{M}c_p$ are constant for each stream, it is possible to derive an analytical temperature profile expressed as a set of exponentials (5). In practice, however, the assumption of constant coefficients is rarely valid in a multi-stream plate-fin heat exchanger, given the duties for which these exchangers are used. Coefficients vary because of the large changes in temperature of the streams, and because there is often boiling or condensation of certain streams in all or part of the exchanger. Both heat transfer coefficient and capacity rate may thus vary significantly along the exchanger for each stream, and at the boundary between a single-phase and a two-phase region they may change discontinuously. Any practical calculation method for a multi-stream plate-fin heat exchanger must be able to deal with such effects.

3.2 Co-current Exchanger

In a co-current exchanger, the inlet states of all streams are known at one end of the exchange and integrating the equations for each stream, or in principle for each layer, along the exchanger is straightforward.

3.3 Counter-Current Exchangers

In the more usual case of counter current exchangers the inlet conditions of certain streams are known at one end of the exchanger, while those of the remainder are known at the other end. If at one end - say end B - only one stream enters, an iterative solution for the heat transfer may be found using the common wall temperature assumption. The outlet conditions of this stream at end A are estimated and the heat transfer equations for all streams integrated from this end.

If, however, more than one stream enters at each end of the exchanger, attempting to estimate the outlet conditions for more than one stream and then integrate all the heat transfer equations leads to difficulties.

In Figure 4, for a 4 stream exchanger, the solid lines represent the solution calculated by beginning at end A. Since this calculation is based on the outlet temperature of only one stream (stream 1), the calculation is stable, and a sensible set of temperature profiles is produced. The dashed lines show the result of using the calculated outlet temperatures at end B,

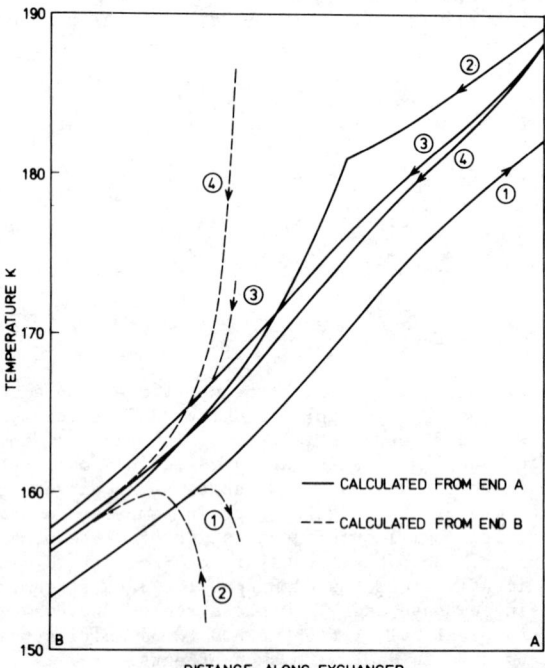

Figure 4. Temperature Profiles along a Four Stream Heat Exchanger

as the starting point for an integration from that end. The two streams with smaller flowrates, 2 and 4, rapidly deviate in temperature, while the two main streams, 1 and 3, soon follow them. Modifying the outlet temperature of just one stream by only 0.03 K and repeating the calculations leads to the sign of the divergences of streams 2 and 4 being reversed.

It must be stressed that this instability is not purely a mathematical artifice. If the exchanger to which figure relates were in fact to be fed with stream 1 as shown, with stream 2 at an extremely low temperature, and with streams 3 and 4 at extremely high temperatures, temperature profiles such as those indicated by the dashed lines would result.

3.4 Suggested Solution Method

In view of these instabilities it is desirable to integrate the heat transfer equations entering at end A (the "A" streams") in the A to B direction, and those entering at B in the reverse direction. This can be done, using the common wall temperature assumption and a step-wise integration procedure, if an estimate is available of $\dot{Q}_j$, the total rate of heat transfer per unit length from A streams to B streams is the jth step along the exchanger.

Using the estimated values of $\dot{Q}_j$, the enthalpy profiles along the exchanger, may be calculated, for streams entering at end A, as follows.

An effective wall temperature Tw_A is calculated from

$$Tw_{A,j} = \frac{\sum_A [\phi_{ij} T_{ij}] - \dot{Q}_j}{\sum_A \phi_{ij}} \qquad (22)$$

where the summation is over all streams entering at end A.

The corresponding heat loss by each stream is then

$$\dot{q}_{ij} = \phi_{ij} [T_{ij} - Tw_{A,j}] \qquad (23)$$

The enthalpy at the beginning of the next step is then calculated from

$$h_{i,j+1} = h_{ij} - \dot{q}_{ij} / \dot{M}_i \qquad (24)$$

After the calculation of the enthalpy and pressure profiles for the A streams, the response of the B streams is calculated using

$$h_{ij} = h_{i,j+1} + \dot{q}_{ij} / \dot{M}_i \qquad (25)$$

where $\dot{q}_{ij}$ is in this case determined by equations similar to 22 and 23 for B streams.

3.5 Convergence

When both A and B streams have been integrated along the exchanger, a set of wall temperatures Tw_j may be determined using Equation 20 in each interval, with summation over all streams. In general, if the estimated Q_j do not represent the true solution, these Tw_j will be different from both Tw_{Aj} and Tw_{Bj}.

A convergence criterion may be established by calculating $\dot{Q}'_j$ the heat transfer rate based on the stream temperatures and heat transfer coefficients, defined by

$$\dot{Q}'_j = \sum_A \phi_{ij} [T_{ij} - Tw_j] \qquad (26)$$

The first convergence check is whether the total heat transfer rates $\dot{Q}_{tot}$ and $\dot{Q}'_{tot}$ are equal to say 1%.

$$\dot{Q}_{tot} = \sum_j \dot{Q}_j \quad ; \quad \dot{Q}'_{tot} = \sum_j \dot{Q}'_j \qquad (27)$$

If agreement to 1% has been achieved for the totals a further check may be made that individual $\dot{Q}_j$ and $\dot{Q}'_j$ agree to say 10%. If so, a solution is deemed to have been found. It is useful to examine the ratio

$$r_j = \dot{Q}_j / \dot{Q}'_j \qquad (28)$$

A value of $r_j = 1.05$ may be interpreted as indicating that while the solution found is only approximate in the jth interval, it would have been correct had all the stream heat transfer coefficients there been 5% larger.

Such an interpretation is useful in that it permits an assessment of the solution in terms of a degree of confidence in the heat transfer coefficients used. The interpretation is slightly simplistic, since the different heat transfer coefficient in one step would affect the stream conditions at inlet to the next, so separate assessment of the solution in individual intervals is of limited validity.

3.6 Revision of Estimated Heat Transfer

A revised set of $\dot{Q}_j$ may be determined by dividing each by r_j and then applying a uniform scaling to give a revised total, $\dot{Q}_{tot}$. For revision of the total heat transfer, it is useful to have a value of Q_{max}, the maximum possible rate of heat transfer which may be determined from the enthalpy temperature profiles of the various streams without reference to exchanger geometry.

A successful revision procedure has been found to be

$$\dot{Q}_{tot}(new) = \dot{Q}_{max} \left[\frac{\dot{Q}_{tot}(old)}{\dot{Q}_{max}} \right]^c \qquad (29)$$

where $c = \dot{Q}_{tot} / \dot{Q}'_{tot}$ (30)

Equation (29) ensures that the new $\dot{Q}_{tot}$ is never greater than $\dot{Q}_{max}$ or less than zero. In many plate-fin heat exchangers $\dot{Q}_{tot}$ will be greater than 90% of $\dot{Q}_{max}$, and in this limit the revision formula approximates to

$$\dot{Q}_{tot}(\text{new}) = \dot{Q}_{tot}(\text{old}) + c \, [\dot{Q}_{max} - \dot{Q}_{tot}(\text{old})] \qquad (31)$$

The constant c is greater than unity if the estimated heat transfer is greater than that calculated from stream temperatures. The formula ensures that the new estimate of $\dot{Q}_{tot}$ will be lower than the old one. As convergence proceeds c becomes closer to unity, and the new estimate remains close to the previous one.

This method has been incorporated in the MUSE1 computer program for HTFS and used to calculate the performance of a range of plate-fin exchangers. Convergence has normally been rapid, fewer than 20 iterations being required in nearly all cases.

4. CONCLUSIONS

A by-pass efficiency has been defined to describe heat transfer between non-adjacent layers in a plate-fin heat exchanger.

Using the by-pass efficiency, together with the conventional fin efficiency, for each layer, the wall temperatures corresponding to a given set of stream temperatures in a plate-fin exchanger may be calculated.

Integrating the heat transfer equations along a co- or counter-current plate fin exchanger is a considerable task if each layer is treated separately. The "common-wall temperature assumption" simplifies the problem, since separate calculations are then only required for each stream rather than each layer.

For exchangers with two or more streams entering at each end, integration of the heat transfer equation is not straightforward. An iterative calculation procedure has been described in which the heat transfer equations of each stream are only integrated along the flow direction of that stream.

NOTATION

a	Fin half-height	(m)
A_1	Primary surface area per unit length per side of a plate-fin layer	(m)
$A_2 (=2Na)$	Secondary surface area per unit length per side of a plate-fin layer	(m)
c_p	Specific heat	(J kg^{-1} K^{-1})
h	Specific enthalpy	(J kg^{-1})
$\dot{M}$	Mass flowrate	(kg s^{-1})
N	Number of fins per layer	(-)

$\dot{Q}$	Heat flow/unit length along exchanger	(W m^{-1})
$\dot{Q}_B$	By-pass heat flow/unit length	(W m^{-1})
$\dot{Q}_L, (\dot{Q}_R)$	Heat flow from left (right) wall to or from stream/unit length	(W m^{-1})
$\dot{Q}_{LT}, (\dot{Q}_{RT})$	Total heat flow from left (right) wall per unit length	(W m^{-1})
$\dot{Q}_j$	Total heat flow/unit length from hot streams in j^{th} step	(W m^{-1})
t	Fin thickness	(m)
T	Temperature	(K)
Tw	Temperature of wall (separating plate)	(K)
x	Distance from centre of fin	(m)
z	Distance along exchanger	(m)
α	Heat transfer coefficient	(W m^{-2} K^{-1})
$\beta = \sqrt{2\alpha/\lambda t}$	Parameter in fin efficiency	(m^{-1})
η	Fin efficiency	(-)
η'	By-pass efficiency	(-)
λ	Thermal conductivity	(W m^{-1} K^{-1})
φ	Variable defined by equation (15)	
ψ	Variable defined by equation (16)	

Subscripts

i	Number of stream (or layer)	
j	Number of step along the exchanger	

REFERENCES

1. Kays, W.M. and London, A.L. (1966). Compact Heat Exchangers, McGraw Hill, second edition.

2. Kern, B.Q. and Kraus, A.D. (1972). Extended Surface Heat Transfer, McGraw Hill.

3. Kao, S. (1961). A Systematic design approach for a multi-stream exchanger with interconnected wall, ASME Paper 61 WA255.

4. Weimer, R.F. and Hartzog, D.G. (1972). Effect of Maldistribution on the Performance of multi-stream multi-passage heat exchangers, AIChE-ASME 13th National Heat Transfer Conference, Denver, Colorado.

5. Chato, J.C., Laverman, R.J. and Shah, J.M. (1971). Analyses of parallel flow multi-stream heat exchanger, Int. J. Heat Mass Transfer, Vol. 14, pp.1691-1705.

United States Ceramic Heat Exchanger Technology—Status and Opportunities

V. J. TENNERY
Oak Ridge National Laboratory
P. O. Box X
Oak Ridge, Tennessee 37830, USA

ABSTRACT

Results of ceramic heat exchanger (HX) design and materials studies in the United States are reviewed, including industrial waste heat recovery and advanced energy conversion systems. Both compact cellular (honeycomb) and tube and header HX designs have been analyzed and tested. Silicon carbide ceramics are currently leading candidates for tube and header configurations, while cordierite-based ceramics have received major attention for cellular geometry.

1. INTRODUCTION

This paper reviews the status of ceramic heat exchanger (HX) technology in the United States, including the basis for interest in designing, fabricating, and demonstrating these components for application in energy systems.

In this paper, the term "heat exchanger" or HX will be applied to devices having a fixed boundary between the primary and secondary sides. It will not review ceramic regenerative devices such as heat wheels or stationary mass regenerators whose operation requires cyclic gas flows. Until about 1975, there was little interest in ceramic HXs in the United States and the only units utilized on a significant commercial scale until that time were relatively large units used on steel soaking pit furnaces to heat combustion air. These multipass cross flow ceramic HXs are based upon designs developed in the 1930s, and typically operate at relatively low effectiveness values (see Eq. 1) of 0.3 to 0.4 to heat air to about 600°C with inlet flue gas temperatures to 1450°C. The low effectiveness values for these units have been attributed to gas leakage at the many joints between the ceramic tubes and the header tiles. Tubes made of silicon carbide and fireclay have been used.

Following the oil embargo of 1973, interest rapidly intensified in the United States for systems to reduce fuel consumption in industrial furnaces and in high efficiency energy conversion systems capable of burning dirty fuels such as coal, synthetic fuels, and residual oils. Recent work on ceramic HXs in the United States, reviewed in this paper, applies to engineering applications in two major areas, including (1) waste heat recuperation and (2) isolation of heat engines such as high pressure gas turbines from highly contaminated combustion gases. Waste heat recuperators typically operate at essentially atmospheric pressure on both the primary (flue gas) and secondary (air) sides and gas leakage up to a few percent from the secondary to the primary side can be tolerated without grossly reducing the effectiveness. On the other hand, ceramic HXs designed for use with high-pressure heat engines such as gas turbines can tolerate relatively little leakage without having a major effect upon system efficiency.

2. CERAMIC HEAT EXCHANGERS FOR WASTE HEAT RECOVERY

There are several conceptual applications in industrial processes using HXs to transfer heat from hot gas to another working fluid, such as air. The application which has been used extensively in the United States involves heating combustion air for use in the burners of fossil fuel-fired furnaces. In the United States, a HX used for this purpose is conventionally referred to as a recuperator. Both ceramic and metallic recuperators have been used by U.S. industries with the majority of these units being constructed of alloys, principally ferritic and austenitic steels. Due to the rapidly increasing fuel costs since 1973, the economic incentive for using recuperators to heat combustion air to high temperatures has increased due to the resultant fuel savings. Alloy recuperators have temperature limits for the partition separating primary (flue gas) and secondary (air) flows due to the mechanical and corrosion properties of available materials. This maximum temperature is currently considered to be in the range of 850 to 950°C for austenitic or ferritic steels and lower for low carbon steels [1]. This temperature limitation requires use of alloy recuperator designs in which the HX wall which separates the primary and secondary sides has a temperature considerably less than the flue gas temperature for many processes. This is achieved by limiting the thermal flux to the wall in the HX by radiation or the hot flue gas must be diluted with air at the recuperator inlet. Either of these alternatives results in a decrease in the efficiency of the heat recovery process in the system. Metallic radiation recuperators have been used extensively in U.S. industry for applications with flue gas temperatures entering the unit in the range 1400 to 1600°C. These radiation type units typically heat air in the 500 to 600°C range. The effectiveness, ε, of these recuperators is therefore in the range of 0.3 to 0.4 where is calculated using equation (1).

$$\varepsilon = (T_{oa}-T_{ia})/(T_{ig}-T_{ia}) \tag{1}$$

where

ε = effectiveness of the HX
T_{oa} = temperature of outlet air
T_{ia} = temperature of inlet air
T_{ig} = temperature of inlet flue gas.

Metal type recuperators are also widely used in which much of the heat transfer occurs by convection from the hot flue gas to the partition, and in this case, the inlet flue gas temperatures are usually limited via air dilution at the inlet.

The fuel savings possible in a high-temperature furnace due to preheating the combustion air can be calculated based upon thermochemical analyses for fuel combustion and assumed values for heat transfer to the objects being heated in the furnace. The dependence of fuel savings upon the preheat air temperature is shown in Fig. 1 for the case of fuel oil as a function of the furnace exhaust gas temperature.

It is apparent from Fig. 1 that air preheat temperatures higher than that achievable with metal radiation recuperators can increase fuel savings. Use of ceramic recuperators capable of heating air to 1000°C can reduce fuel consumption by 10 to 15% compared to a metal recuperator heating combustion air to about 550°C. It is obvious that only structural ceramic materials have the potential properties required for use in constructing such high-temperature recuperators.

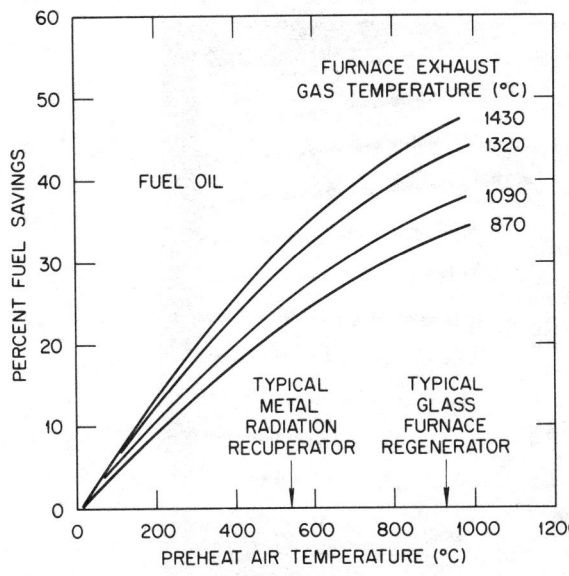

Fig. 1. Calculated Fuel Savings for Oil-Fired Furnace as Function of Air Preheat Temperature and Exhaust Gas Temperature

In the 1930s, a ceramic recuperator design was developed which employed ceramic tubes and tiles to define the gas flow paths. This unit is still used extensively on steel soaking pits in the United States. They typically have leakages of from 10 to 15% of the air from the secondary side into the flue gas and this leakage contributes significantly to the low effectiveness of these recuperators. No new units of this type have been installed in the United States for several years. Effort in the last five years or so in the United States on ceramic recuperators has been directed to development and demonstration of either compact honeycomb or advanced tube and header type structures. Some results of these efforts will now be reviewed.

2.1 Compact Cellular Ceramic Recuperators

Compact cellular (honeycomb) ceramic recuperators in which the passages for the hot flue gas and air are closely spaced, has the design advantage of having a small volume compared to more conventional tubular geometries because a large heat transfer surface area can be incorporated into a relatively small volume. Research and development in the United States was initiated in the early 1970s to identify ceramic materials having the required properties and fabrication characteristics for use in regenerator wheel heat recovery devices appropriate for use in heating combustion air for advanced vehicular gas turbines. This work led to identification of magnesium aluminosilicates and lithium aluminosilicates as having attractive properties. The former ceramics contain cordierite ($2MgO \cdot 2Al_2O_3 \cdot 5SiO_2$) (MAS) while the latter contain spodumene ($Li_2O \cdot Al_2O_3 \cdot 4SiO_2$) (LAS) as the principal phase. Due to their properties and economics, MAS ceramics have been subsequently developed for use in recuperator cores having flow passages up to 1.6 cm wide by 0.6 cm high. These passages alternate in a cross flow configuration and have been fabricated and performance tested in sizes up to 0.43 x 0.43 x 0.81 m [2]. A photo of a portion of a MAS ceramic recuperator core is shown in Fig. 2. Test results on a recuperator of this type having core dimensions 0.31 x 0.31 x 0.31 m on a bliss mill furnace heating Mo and W billets indicated an effectiveness of 0.55 while heating 5 x 10^{-2} m^3/s of air from 38 to 795°C. This recuperator, operating with a flue gas inlet temperature of 1430°C, transferred heat to

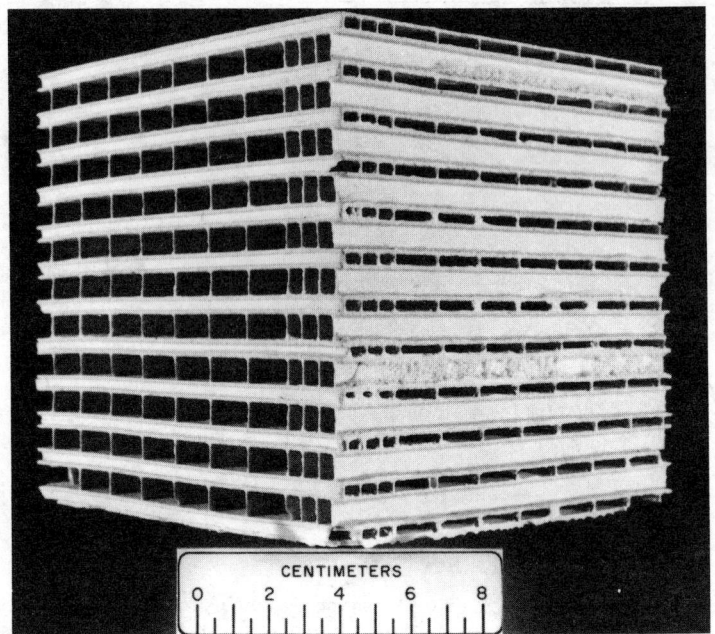

Fig. 2. Compact Cellular Ceramic Recuperator Core Fabricated of Cordierite (MAS)

the combustion air at a rate of 50.4 KW. The volume efficiency of these recuperators is illustrated by the high energy transfer rate per unit volume of core which was 1780 KW/m^3 for this example. Air leakage in these units has been reported to be about 5% of the air mass flow. Ceramic recuperators of this type have been installed on 100 industrial furnaces as of March 1980 [3]. The durability of these recuperators in highly contaminated hot process off-gas is still being evaluated. As will be noted later in this paper, there is some evidence to suggest that the thermal expansion of the cordierite-based ceramic core material may change during long term high-temperature exposure to some of the species in the combustion gas of residual oils and these changes could produce substantial internal stresses and possibly cracking in the MAS ceramic. This situation is still under investigation.

2.2 Tube and Header Ceramic Recuperators

Currently, there is one type of U.S. manufactured ceramic recuperator of relatively new design which utilizes a tube and header configuration. This CERHX recuperator consists of six rows of finned ceramic tubes, five rows deep along the flue gas flow which are made of a proprietary silicon carbide ceramic formulation. The tube to header seals in this unit are achieved by individually compression-loading the tubes on the air inlet header side. Sealing is achieved on the hot header via conical shaped ceramic seals on the tubes which mate to seals in the wall of the header. This unit has been tested extensively on a slot forge furnace having a slot 2.74 m wide [4]. The

UNITED STATES CERAMIC HEAT EXCHANGER TECHNOLOGY

recuperator was a single pass cross-flow type mounted on top of the furnace with the tubes oriented horizontally. Each tube had 0.93 m^2 of external area and 0.38 m^2 of internal area. The tubes which contained circumferential fins on the outside and longitudinal fins on the inside were 1.2 m long and 9.5 cm in diameter at the outer fin tips. The internal diameter was 3.4 cm to the internal fin tips and 3.9 cm diameter to the internal fin roots. The heated air entered an insulated plenum prior to passing to recuperated burners. The recuperator heated the combustion air for the burners to about 550°C at a mass flow of about 0.3 kg/s. The maximum flue gas mass flow rate was about 0.33 kg/s which entered the recuperator at about 1260°C at steady state conditions. The recuperator effectiveness under these full load conditions was 0.47, and energy input rate to the furnace from the fuel was 0.82 MW. The recuperator recovered 0.19 MW or 36% of the thermal energy in the furnace flue gas leaving the furnace.

A design and economic analysis of tube and header ceramic recuperator designs has been completed and recently published under U.S. Department of Energy sponsorship. This study [5] included a variety of design alternatives for ceramic recuperators sized to accept the flue gas and provide combustion air requirements for steel soaking pit, aluminum remelt, and glass melting furnaces. The design calculations and economic analyses for the recuperators were performed for Oak Ridge National Laboratory by Garrett-AiResearch Manufacturing Company (Torrance, California), and included determination of the tube configurations and recuperator mass as functions of multipassing the hot flue gas and air. Some conclusions of this study for the case of a recuperator sized for a steel soaking pit will be presented to illustrate the results obtained. For the purpose of estimating the cost of these recuperators and their simple payback times, silicon carbide was selected for the tubes because there are more data available on the oxidation and corrosion behavior of this ceramic in the relevant environments than other candidate materials. Silicon carbide is also the only candidate ceramic material for which reasonably realistic tubing cost projections could be obtained from commerical manufacturers at the time the work was performed.

For the steel soaking pit recuperator case, the flue gas and air mass flows were selected as 2.6 and 2.5 kg/s respectively, based on present soaking pit furnace parameters in the United States. A variety of tube surface geometries were included in the calculations to determine effects of external and internal tube fins and tube spacings on the recuperator size and effectiveness. The assumed flue gas inlet temperature to the recuperator was 1163°C and recuperator design calculations were performed for air outlet temperatures from 540 to 1093°C. Multipassing effects were analyzed for both configurations shown in Fig. 3, where L_i is the tube length and L_o is the tube bundle width. The results of calculations for these units as a function of multipassing the flue gas up to four times over the length of the tubes are illustrated in Table 1. Increasing the gas flow path above one pass significantly reduces the mass of the recuperator, but the advantages of more than two passes in reducing the HX mass are slight. The effect of increasing the air preheat temperature on the mass and number of tubes for the steel soaking pit recuperator is illustrated in Table 2 and in Fig. 4. These data show that the mass of the recuperator becomes prohibitively large as the air preheat temperature comes within about 100°C of the inlet flue gas temperature. The effectiveness of these recuperators can be quite high. An effective means for reducing the mass and therefore the cost of these recuperators employed the use of circumferential dimples on the inside of the tubes (air side), as the surface film coefficient at this surface was found to limit the heat transfer in these HXs. The simple payback period of these HXs was analyzed using projected costs for the silicon carbide tubing assuming that the ceramic costs could eventually be reduced from the present value of about $150/kg to $20/kg.

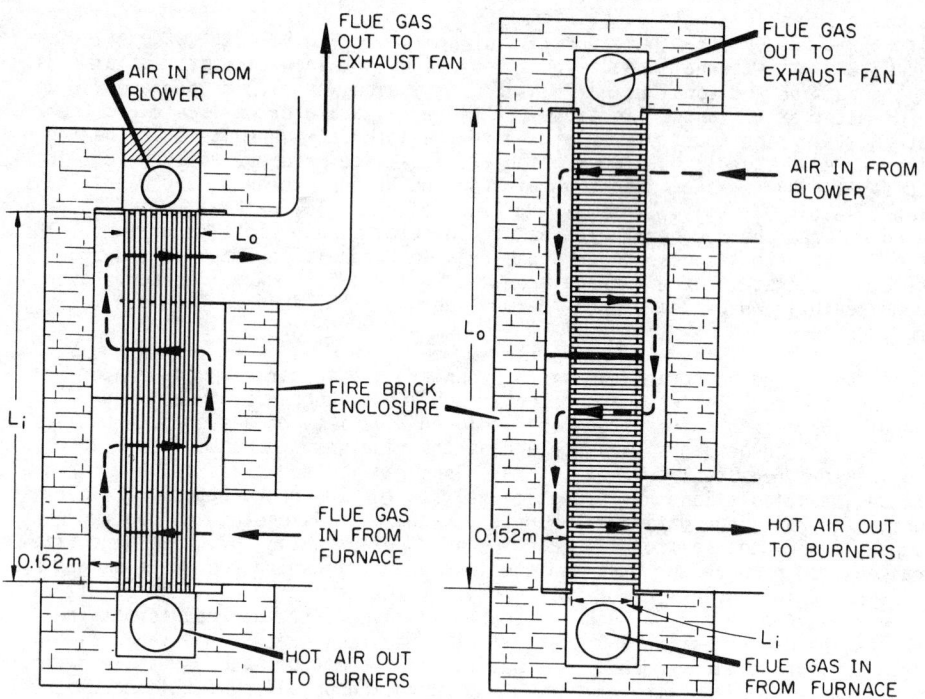

Fig. 3. Ceramic Recuperator Configurations Analyzed for Steel Soaking Pit Furnace (a) Multipass on Outside Tube Surface (b) Multipass on Inside Tube Surface

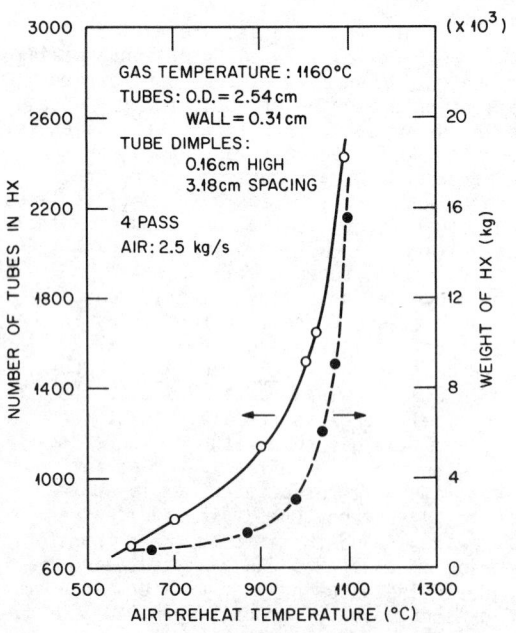

Fig. 4. Number of Tubes and Ceramic Recuperator Mass as Function of Air Preheat Temperature for Steel Soaking Pit Furnace

UNITED STATES CERAMIC HEAT EXCHANGER TECHNOLOGY

Table 1. Effect of Number of Passes upon Characteristics of Ceramic Heat Exchanger, Steel Soaking Pit, 870°C[a]

	Characteristic for Each Number of Passes			
	1	2	3	4
Number of tubes	1255	1124	1100	1088
Core dimensions, m				
Tube length (L_i)	2.44	1.95	1.86	1.83
Flow length across tubes (L_o)	2.87	1.04	0.732	0.518
No-flow length (L_n)	0.52	1.16	1.83	2.53
Core volume, m^3	3.6	2.6	2.4	2.4
Containment inner surface area, m^2	19.5	11.9	14.4	16.4
Mass, kg				
Tubes[b]	2136	1540	1419	1379
Ceramic header (hot)[c]	102	118	161	184
Steel header (cold)[d]	85	98	133	152
Ceramic baffles[e]	0	16	42	69
Cold air manifold[f]				
Total[g]	2323	1808	1799	1832

[a]The tube configuration for the HXs included X_t/D = 1.50, X_ℓ/D = 1.25 where X_t = tube centerline spacing perpendicular to the flue gas flow, X_ℓ = tube centerline spacing along the flue gas flow, and D = inner diameter of the tubes.
[b]All tubes 25.4 mm OD by 3.18 mm wall thickness.
[c]Ceramic plate, 38 mm thick.
[d]Steel plate, 12.7 mm thick.
[e]Ceramic plate, 6.4 mm thick.
[f]Steel, 1.6 mm thick
[g]Mass of refractory brick containment structure or ducting not included.

Table 2. Dependence of Ceramic Heat Exchanger Characteristics on Air Preheat Temperature with Gas Inlet Temperature of 1160°C, Steel Soaking Pit, Four-Pass Unit[a] (See Table 1 for footnotes.)

	Characteristic for Each Air Preheat Temperature, °C						
	540	650	870	980	1040	1066	1093
Number of tubes	670	775	1088	1416	1734	2015	2436
Heat exchanger effectiveness	0.44	0.54	0.74	0.84	0.89	0.91	0.94
Core dimensions, m							
Tube length (L_i)	0.67	0.88	1.83	3.10	4.60	6.04	8.75
Flow length across tubes (L_o)	0.18	0.24	0.52	0.91	1.34	1.80	2.59
No-flow length (L_n)	4.72	3.87	2.53	1.89	1.55	1.34	1.13
Core volume, m^3	0.54	0.82	2.38	5.24	9.57	14.6	25.5
Containment inner surface area, m^2	11.6	12.0	16.4	23.8	34.6	47.2	76.9
Mass, kg							
Tubes[b]	309	483	1379	3052	5567	8493	14,857
Ceramic header (hot)[c]	236	208	184	191	201	216	244
Steel header (cold)[d]	196	172	152	158	167	179	202
Ceramic baffles[e]	75	68	68	78	87	96	111
Cold air manifold[f]	66	58	49	50	54	59	69
Total[g]	882	898	1832	3529	6076	9042	15,482

This cost reduction will require a major effort by ceramic manufacturers. Other costs associated with the recuperator construction and installation were estimated based upon present U.S. engineering experience. Simple payback times ranged from 1.1 to 1.4 years for air preheat values of 870 and 980°C, respectively, where the fuel cost was assumed to be $3.16/GJ. These payback times are quite attractive at present to U.S. industrial companies.

3. CERAMIC HEAT EXCHANGERS FOR ADVANCED ENERGY SYSTEMS

The recuperator ceramic HXs discussed in the previous section operate with pressure differentials across the wall or partition of perhaps up to 10 KPa. The physical requirements on the joints and seals to minimize leakage from the higher pressure secondary side into the hot flue gas are relatively modest, although leakage values below 2 to 3% of the air mass flow are still very difficult to achieve. In the past five years, there has been substantial work on ceramic HXs in which large pressure differentials will be required and this section of the paper will review these applications and the status of results.

The potential applications of ceramic HXs in advanced energy systems which have been considered to date include gas turbines in various configurations and coal gasifiers. The gas turbine system configurations for which ceramic HXs have been considered for electric power generation include those in which a dirty fuel such as coal is the heat source to drive the turbine or solar energy is focused by heliostats to a central receiver to heat the gas working fluid in the HX to operate the turbine. In the case of coal fired configurations, the ceramic HX would serve to isolate the turbine from the combustion gases as shown in Fig. 5. In this case, the ceramic HX must be capable of operating with the combustor firing directly on the primary side [6]. In order to achieve acceptable cycle economics, this type of system will require a steam bottoming turbine as shown in Fig. 5. System calculations indicate these systems to be capable of thermal efficiencies in the range of 40 to 50%. Both open and closed cycle turbine systems of this type have been considered and the closed system using helium as the turbine working fluid potentially offers cycle efficiency advantages, but the leakage constraints for the ceramic HX are more severe due to the economic penalty associated with helium loss.

Conceptual designs of large ceramic HXs rated at 586 MW$_t$ and 400 KW$_t$ have been completed for Brayton turbine systems [7]. The former are the largest studied to date in the United States. The 586 MW$_t$ HX design was part of a 300 MWe advanced coal fired system study in which the initial HX design employing bare ceramic tubes included 12 modules, each with a nominal module rating of 49 MW$_t$. Parametric analyses were performed for a two-pass cross-counterflow U-tube configuration with helium at 3.5 MPa inside the tubes and coal combustion gas at 0.1 MPa outside the tubes. The HX was to heat helium from 690°C at the inlet to 980°C at the outlet with a mass flow of 386 kg/s. The combustion gas entered the HX at 1650°C and exited at 760°C. The ceramic HX having the minimum tube mass of 10^5 kg contained 62,590 tubes. This design using bare tubes always had some undesirable characteristic, such as very long tubes, large mass, or a very small dimension in the direction of the coal combustion gas flow. The ceramic tubes in some cases had lengths of about 6.5 m. Subsequent design studies were done to determine effects of using finned tubes (outer surface) on the HX configuration. Substantial mass and tube length reductions were achieved using the finned tube configuration shown in Fig. 6. The HX mass was reduced about 40% compared to the bare tube case for 3-m long tubes about 3 cm in diameter with fin lengths of about 0.6 cm. The tube mass for the HX design shown in Fig. 6 using 2.54-cm-diam tubes was 1.55×10^5 kg and the design required 49,000 tubes.

UNITED STATES CERAMIC HEAT EXCHANGER TECHNOLOGY

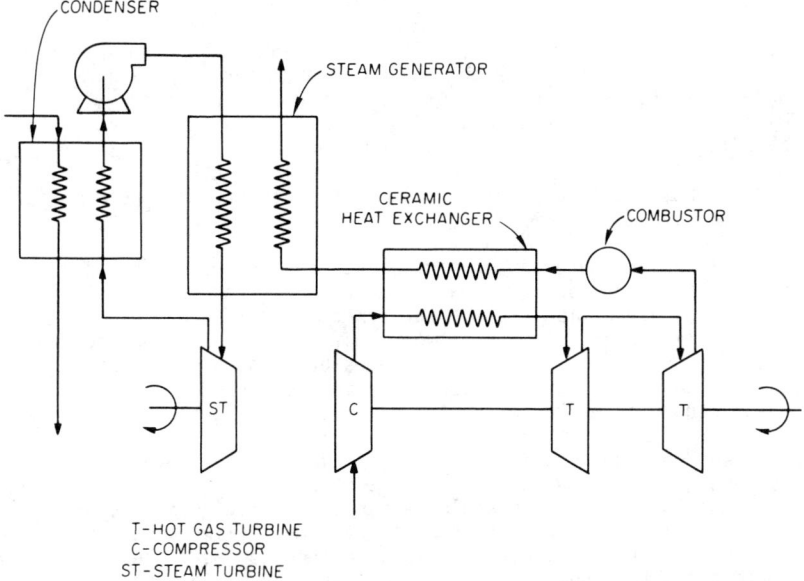

Fig. 5. Indirectly Fired (Exhaust Heated) Gas Turbine Combined Cycle Diagram Including Major Components

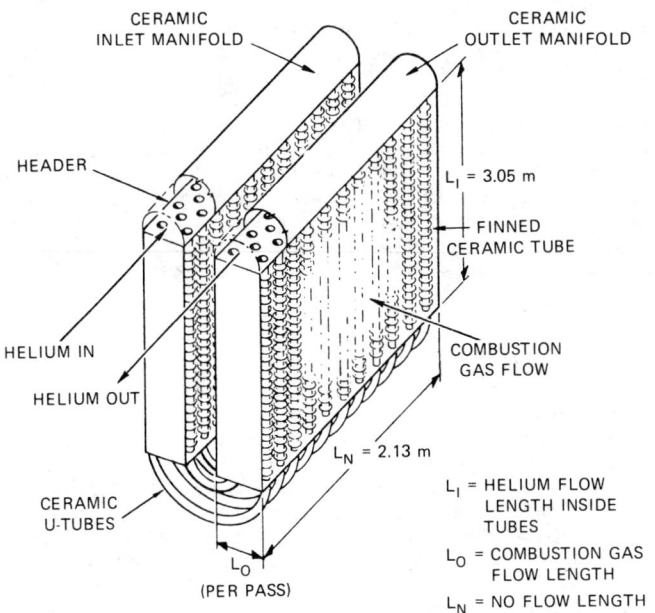

Fig. 6. Finned Tube Ceramic HX Module Sized for Utility Gas Turbine

Other analyses of this type for ceramic HXs capable of transferring 0.4 MW$_t$ and suitable for interfacing with a commercial 0.15 MW closed-loop Brayton-cycle engine have been performed [8]. The HX configuration also employed finned ceramic U-tubes either 1.9 or 2.54 cm in diameter with various fin lengths and spacings. Argon was the working fluid on the secondary side of this HX with a mass flow of 2.5 kg/s. In this case, the HX mass could be reduced by using external fins compared to using smooth tubes only for fin thicknesses less than about 0.4 mm regardless of fin length. The final HX configuration consisted of 6 modules of 30 tubes each having finned surface lengths of 0.98 m.

Stress analyses and material properties studies indicated that siliconized silicon carbide is an attractive ceramic material for fabrication of these pressurized HXs. Fabrication studies performed by manufacturers of silicon carbide ceramics indicated that prototype manifolds and heat transfer tubes could be manufactured using present techniques but that significant improvements would be required to produce the configurations required by the HX designs described in order to reduce the ceramic costs to acceptable levels.

Other recent work on ceramic HXs for potential use to isolate gas turbines from contaminated combustion gases has identified an axial counter flow configuration as having several desirable features [9]. This configuration is shown in Fig. 7. In this case, the hot gas at 1204°C flows along the tubes in which air flows in counter current fashion. In this concept, the heated air would operate a gas turbine. The axial flow design has the advantage of better temperature uniformity among the ceramic tubes compared to the cross counterflow designs discussed previously. A ceramic HX suitable for interfacing with a 5 MW industrial gas turbine was analyzed. Each header would contain 40 tube positions, be about 0.6 m long, and 11.4 cm in diameter. The air is heated in tubes about 1.9 cm in diameter and 6.1 m long. The improved tube temperature uniformity of this design is significant in reducing thermal expansion stresses in the ceramic components of the HX. The tubes are joined to the headers using mechanical compression joints loaded from the air inlet header. A ceramic HX module containing 28 tubes and based upon this design was constructed and successfully operated at 689 KPa on the air side with a hot gas inlet temperature of 1370°C and an air outlet temperature of about 1150°C using No. 2 diesel fuel oil combustion on the primary side. The inlet air temperature was about 430°C. The silicon carbide tubes in this module were 4.6 m long and included a solid joint at their midlength. The tubes had a wall thickness of 0.64 cm, and had a surface clearance of 0.64 cm between adjacent tubes. Individual Inconel 718 bellows were used for each ceramic tube at the cold header, and spherical compression joints were employed on both ends of the tubes. No data are available for this ceramic HX configuration under direct coal firing conditions.

Design analyses have been reported for a ceramic HX to heat air in an open cycle Brayton turbine system in which solar energy is focused into an appropriate receiver containing the HX [10]. The design included four independent cavities containing ceramic HXs which would face 90° sectors of the heliostat field. The combined solar flux into these HXs would heat air from 480 to 980°C, the latter being the turbine inlet temperature. The ceramic HX design chosen included 10-cm-diam silicon carbide tubes having straight sections 12.2 m long and U-tubes to join straight tube pairs at the upper end of the HX. Each cavity would contain 70 U-tubes. These HXs would nominally deliver 100 to 200 MW$_t$ by heating air at about 1 MPa to temperatures of about 980 to 1100°C. The nominal electrical rating of this system was 60 MWe. Extensive testing of various tube to header joint concepts during this work indicated that a mechanical compressive loaded joint at the headers had promise of satisfying the low leakage requirements for the secondary side of this type of system

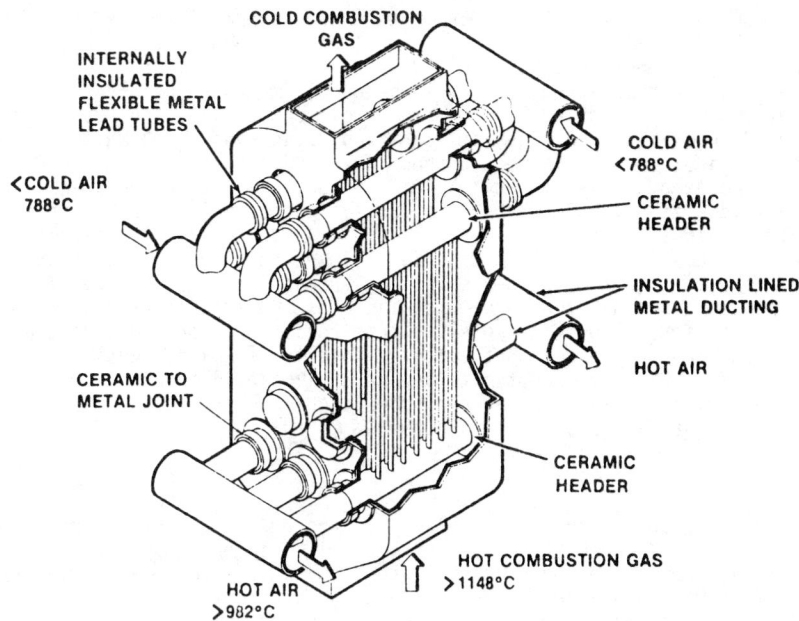

Fig. 7. Axial Counter Flow Ceramic HX Configuration

(∿1%) and simultaneously being able to maintain thermal expansion stresses at acceptable levels in the ceramics.

4. CERAMIC MATERIALS FOR ADVANCED SYSTEM HEAT EXCHANGERS

The U.S. data base for structural ceramic materials presently considered to be serious candidates for use in pressurized HXs is presently limited primarily to siliconized silicon carbide and sintered alpha silicon carbide. Most of the data pertaining to mechanical properties include fracture strength and fracture strength distributions for relatively simple flexure bar or short tubular specimens of a given material. Little is known about these properties when measured for specimen configurations where the stress geometries are more appropriate to those for HX designs using long ceramic tubes and tube to header joints. There is considerable evidence that the present manufacturing processes used to produce silicon carbide tubes do not produce material having consistent and well controlled microflaw concentrations or sizes. Since these flaws usually control fracture in ceramic materials, it will be necessary in the future for advances to be made in the processing required to produce such tubes. There is essentially no data base pertaining to subcritical crack growth behavior of silicon carbide or similar ceramics in the presence of environmental species anticipated in the hot combustion gases and at the temperatures of interest appropriate to these HXs. If subcritical crack growth is significant in these materials under the stress and environmental conditions in the HX, the measured fast fracture strength will have to be replaced in the design basis by appropriate crack growth criteria. Since the fast fracture strength of commercial silicon carbide ceramics varies from about 276 to 400 MPa, and the HX designs discussed previously generally assume maximum working stresses in the ceramics of perhaps one half the fast fracture strength, effects of environmental species such as coal slag could be quite significant.

Recent work at ORNL [11,12] has shown that when these ceramics are exposed to residual oil or coal combustion for several hundred hours, there are corrosion reactions for both siliconized and sintered alpha silicon carbide which change the fast fracture strength when subsequently measured at 25°C. During 500 h of exposure at about 1200°C to residual oil combustion, the fast fracture strength at 25°C of sintered alpha silicon carbide decreased 30% while that of siliconized silicon carbide increased about 40% [10]. During a similar 500-h exposure of these materials at about 1240°C to coal combustion and severe slagging, siliconized silicon carbide had an 11% increase in fracture strength while sintered alpha silicon carbide exhibited a 4% decrease. The causes of these changes are still under investigation. There is also evidence that under some conditions of high-temperature exposure, the helium permeability of ceramics such as sintered alpha silicon carbide can increase by at least three orders of magnitude with ΔP = 520 KPa across the tube walls. An improved understanding of the chemical and mechanical behavior of candidate ceramics, particularly for use in pressurized HXs, is expected to become available in the next two to four years and will greatly improve our understanding of how to use these materials in high efficiency energy systems.

5. SUMMARY

Significant advances have been achieved in the United States in design and materials evaluations for ceramic HXs for several engineering applications. Work presently underway includes joint systems to achieve acceptably low gas leakage, and materials studies to demonstrate satisfactory stability of candidate ceramic materials as well as efforts to reduce fabrication costs. Fabrication and testing of relevant size modules of these HXs for advanced conversion systems remain to be accomplished. Success in these efforts will provide HXs having unique capabilities not presently available and which can have a significant impact on the energy conversion efficiencies of many processes.

REFERENCES

1. Tennery, V. J. and Wei, G. C. 1978. Recuperator materials technology assessment. Report ORNL/TM-6227.
2. Kohnken, K. H., Cleveland, J. J., and Gonzalez, J. M. 1980. Ceramic heat recuperators for industrial heat recovery. Report DOE/EC/02162.
3. Kohnken, K. H. 1981. GTE Products Corporation, Towanda, Pennsylvania, Private Communication.
4. Bjerklie, J. W. and Curtis, R. H. 1978. Demonstration of fuel conservation in high temperature industrial furnaces. ASME Paper 78-WA/Ener-8, presented at the ASME Winter Annual Meeting, San Francisco, California, Dec. 10–15.
5. Tennery, V. J. 1981. Economic application, design, analysis, and material availability for ceramic heat exchangers. Report ORNL/TM-7580.
6. McDonald, C. F. 1980. The role of the ceramic heat exchanger in energy and resource conservation. Journal of Engineering for Power, Vol. 102, pp. 303-315.
7. Pietsch, A. 1978. Coal fired prototype high temperature continuous flow heat exchanger. Report EPRI AF-684, by Garrett AiResearch Manufacturing Company of Arizona.
8. Coombs, M., Kotchick, D., and Warren, H., High-temperature ceramic heat exchanger development. Report EPRI-FP-1127, by Garrett AiResearch Manufacturing Company of California.
9. Ward, M. E., Solomon, N. G., Gulden, M. E., and Smeltzer, C. E. 1980. Development of a ceramic tube heat exchanger with relaxing joint. Report FE-2556-30, U.S. Department of Energy, by Solar Turbines International, San Diego, California.
10. Grosskreutz, J. C. 1978. Solar-thermal conversion to electricity utilizing a central receiver, open-cycle gas turbine design. Report EPRI ER-652, by Black and Veatch Consulting Engineers, Kansas City, Missouri.

11. Wei, G. C. and Tennery, V. J. 1981. Evaluation of Tubular Ceramic Heat Exchanger Materials in Residual oil combustion environment. Report ORNL/IM-/578.
12. Ferber, M. K. and Tennery, V. J. 1981. Evaluation of tubular ceramic heat exchanger materials in acidic coal ash from coal-oil-mixture combustion. ORNL report in preparation.

Study on Fluid Flow Distribution inside Plate Heat Exchangers by Thermographic Analysis

G. DELLE CAVE, M. GIUDICI, E. PEDROCCHI, and G. PESCE
Institute of Fisica Tecnica
Polytechnic of Milan
Milan, Italy

ABSTRACT

A method is proposed for studying the distribution of fluid flow inside modular channels of plate heat exchangers by using infrared thermography. Advantages and drawbacks of the method compared to the ones heretofore employed are reported. Theoretical and practical artifices employed to attain good precision and to overcome drawbacks of the method are shown. Tests on commercial models of plates have been carried out as illustration of the proposed method.

1. INTRODUCTION

Plate heat exchangers offer advantages compared to tubular ones, especially for applications of alimentary and pharmaceutical industry. Among them we recall modularity of exchanging unit, easy maintenance and high heat transfer coefficients.   (1)
A right distribution of fluids on the whole surface of plate can prevent the formation of unobstructed channels and stagnation areas. This is especially fundamental when, due to properties of the process fluid and technology of treatment, all particles should spend the same time at different temperatures of the cycle.
It is impossible to determine the real distribution of the fluid inside the modular channel by a theoretical study of the velocity and temperature profiles, due to complexity of a turbulent and tridimensional flow. So we tried an experimental approach.
From available literature we learnt the existence of two experimental methods:

1) the one due to Dikerson et al.   (2)
2) the one due to Watson et al.   (3)

The former, that we will call "colorimetric", resorts to a direct insight of the flow inside the modular channel by chromatic contrast of the fluid particles To allow observations of the inside, modular channels are made by assembling a commercial metallic plate and an equal plexiglas plate. The authors inject a fixed amount of dye upstream the modular channel and record the transient of the coloured front by a cinecamera: the film is then examined with a moviola. The method enables to show flow patterns of inside fluid, and to measure residence times.   The latter, that we will call "conductometric", resorts to the injection of a step of salt solution inside the modular channel, usually fed with water, and to the recording of the output perturbation with conductometric pick-up. This method enables to detect only residence times but not flow patterns.

G. Delle Cave's present address is Instituto de Cibernetica CNR, Naples, Italy.
M. Giudici's present address is Foster Wheeler Italiana of Milan, Milan, Italy.
G. Pesce's present address is University of S. Maria, Valparaiso, Chile.

2. THE THERMOGRAPHIC METHOD

The method we propose resorts to the detection, by infrared thermography, of the time evolution of the plate surface temperature when a modular channel is fed with steps of cold and warm water. We, as to say, colour thermally the fluid.   (4)
We suppose that the temperature front of the fluid propagates istantaneously on the plate surfaces. We have verified it is true at first approximation owing to the small plate thickness and its good thermal diffusivity. Then we can detect, in real time, the propagation of the fluid inside the channel. If, at the same time, we record the input and output average temperature of the fluid, we can check the residence times by a method similar to the conductometric one. Our method offers the following advantages:

1) it is possible to follow the distribution of the fluid inside a modular channel made with normal metal plates, avoiding expensive equipment to realize plexiglas plates, and to analyse many plates in a short time;

2) it is possible to collect experimental data similar to the ones provided by combination of both the previous methods.

We have to take into account that the experimental equipment (the two plates and connecting and supporting elements in thermal contact with the first ones) has a thermal capacity that is not negligible compared with capacity of the internal fluid and perturbs the investigated phenomenon; colorimetric and cond= uctometric methods do not share this draw-back because their "marking" of fluids is not affected by the experimental apparatus. So even if we had a piston flow, due to equipment thermal capacity there would be a progressive decrease of the initial temperature step between cold and warm water. To estimate how these perturbations affect our experimental method, we have worked out a mathematical model of the modular channel.

3. THE MATHEMATICAL MODEL

The mathematical model contains the following simplifications:

a) heat dissipated toward air during transient is negligible because the water convective coefficient is much greater than the air convective coefficient

b) flow inside the channel is "piston-type" (flat velocity profile)

c) fluid temperature is constant at any time on all sections orthogonal to the flow (flat temperature profile)

d) plates temperature is constant at any time on all sections orthogonal to the flow

e) in the plate thermal balance the conductive contribution is negligible on flow direction

The system of differential equations describing the phenomenon is:

$$\frac{\partial T_l}{\partial t} + \bar{v}\frac{\partial T_l}{\partial z} + \frac{2h_l}{\rho_l C_l s}(T_l - T_p) = \alpha^{(t)}\frac{\partial^2 T_l}{\partial z^2} \qquad (1)$$

$$\frac{\partial T_p}{\partial t} + \frac{h_l}{\rho_p c_p l}(T_p - T_l) = 0 \qquad (2)$$

where :
T_l, T_p = water and plate temperatures

$\bar{v}$ = average fluid velocity

h_1 = water convective coefficient

$\alpha^{(t)}$ = coefficient of turbulent thermal diffusion

ρ_1, ρ_p = water and metal densities

C_1, C_p = water and metal specific heat

z = spatial coordinate

t = time

s, l = modular channel and plate thicknesses

Initial and boundary conditions are:

a) $T_1(0,z) = T_1$

b) $T_1(t,0) = T_2$

c) $T_1(t,\infty) = T_1$

d) $T_p(0,z) = T_1$

To estimate the diffusive term, we supposed an adiabatic modular channel; in this case equation (1) becomes:

$$\frac{\partial T_1}{\partial t} + \bar{v}\frac{\partial T_1}{\partial z} = \alpha^{(t)}\frac{\partial^2 T_1}{\partial z^2} \tag{3}$$

with conditions a), b), c),

The solution is

$$T_1(t,z) = T_1 - \frac{T_1-T_2}{2}\left[\text{erf}\frac{z-\bar{v}t}{2\sqrt{\alpha^{(t)}t}} + e^{\frac{z\bar{v}}{\alpha^{(t)}}}\text{erf}\frac{z+\bar{v}t}{2\sqrt{\alpha^{(t)}t}}\right] \tag{4}$$

We introduced in equation (4) the values of parameters valid for our experiments and values of $\alpha^{(t)}$ reasonable for the case under examination. We could verify that the diffusive term $\alpha^{(t)} \cdot (\partial^2 T_1/\partial z^2)$ is very negligible as regards flow term ($\bar{v} \cdot \partial T_1/\partial z$).

Therefore we take into account the system of equation (1) and (2):

$$\frac{\partial T_1}{\partial t} + \bar{v}\frac{\partial T_1}{\partial z} + \frac{2h_1}{\rho_1 C_1 s}(T_1-T_p) = 0 \tag{5}$$

$$\frac{\partial T_p}{\partial t} + \frac{h_1}{\rho_1 \cdot C_1 \cdot l}\cdot(T_p-T_1) = 0 \tag{2}$$

whose solution obtained with Laplace transform method with conditions a), b), d), is:

$$T_1(t,z) = T_i - (T_i - T_2) H(t - \tfrac{z}{v}) e^{-\tfrac{a}{v} z} \left[e^{-b(t-\tfrac{z}{v})} I_o\left(2\sqrt{\tfrac{ab}{v} z (t-\tfrac{z}{v})}\right) + b \int_o^{t-\tfrac{z}{v}} e^{-b\tau} I_o\left(2\sqrt{\tfrac{ab}{v} z \tau}\right) d\tau \right] \quad (6)$$

$$T_p(t,z) = T_i - (T_i - T_2) H(t - \tfrac{z}{v}) b e^{-\tfrac{a}{v} z} \int_o^{t-\tfrac{z}{v}} e^{-b\tau} I_o\left(2\sqrt{\tfrac{ab}{v} z \tau}\right) d\tau \quad (7)$$

where

1) $a = \dfrac{2h_1}{\rho_1 C_1 s}$

2) $b = \dfrac{h_1}{\rho_p C_p l}$

3) H = is the Heaviside function

4) I_o is the modified Bessel function

Numerical solution valid for commercial stainless steel plates of surface $0,3 m^2$, length 1,15 m, thickness 1 mm, $T_1 = 60^0 C$, $T_2 = 20^0 C$, average fluid velocity 0,18 m/s equal to a flow rate of 600 l/h, corresponding to the worst experimental conditions, i.e. when system thermal capacity is greatest and flow rate is low= est, is plotted in fig.1.
The values of $20^0 C$ and $60^0 C$ for temperatures are a compromise between good ac= curacy and set up power.

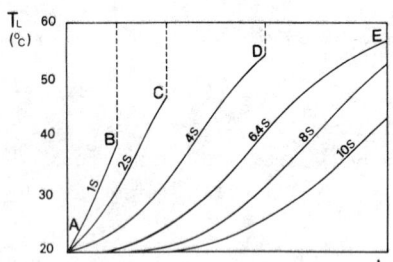
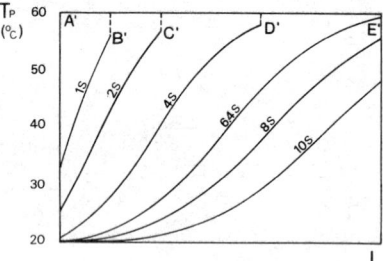

Fig.Ia- Fluid temperature distrib= ution versus channel length, with time as parameter.

Fig.Ib- Same as in fig.Ia but for plate temperature.

FLUID FLOW DISTRIBUTION INSIDE PLATE HEAT EXCHANGERS

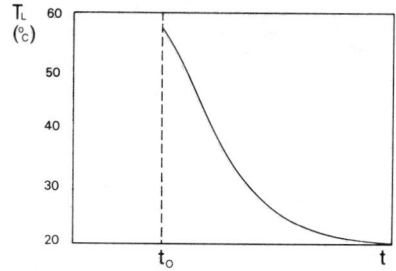

Fig.Ic- Temperature of fluid at output of modular channel versus time.

In fig.Ia points A,B,C,D,E represent temperatures increasing during transient of first fluid front incoming at 20^0C. It is possible to see that fluid becomes progressively warmer through the channel and reaches at output, apart from few degrees, the initial plate temperature (60^0C), owing to the only system thermal capacity.
In fig. Ib points A', B', C', D' represent plates temperatures corresponding to the advancing fluid front. At time of arrival of cold front, plate temperature decreases just few degrees compared to the initial one (60^0C).
In fig. Ic we can see that, even if we suppose a piston flow, the input temperature step is dumped at output because of the system thermal capacity; t_o represents the time needed by the fluid to get out of the channel in the piston flow case.

4. THE EXPERIMENTAL APPARATUS

The experimental apparatus is made of a hydraulic circuit that can feed with different flow rate the modular channel with steps of cold and warm water by commutating a three way valve (Fig.2).
Water is contained in two vessels at fixed temperatures, kept at the same pressure by a compressor. Water flow rate in the modular channel is controlled by regulation valves. After the channel, water flows through a flowmeter and reaches a centrifugal pump that maintains circulation.
The thermographic system we have employed is AGA thermovision 680 -(5). It is composed by a telecamera and a display. The first one collects infrared radiation emitted by body undertest and concentrates it, by means of lenses and collimators, on an AsIn detector kept at -196^0C to minimize noise. Two revolving prisms explore the whole framed surface. The detector converts e.m. radiation in an electrical signal and sends it to the display. The last one exibits the different temperatures of the framed surface by a continuous scale of the tonality between white (high temperatures) and black (low temperatures). Due to high scanning frequency and very short propagation time of signal, image is produced in real time without perturbing temperature of body under test. Attainable precision is $\pm 0,2^0C$.
Thermographic system can also work with a "thermal cut": i.e. it is possible to have a picture of framed surface resulting black for temperatures lower and white for temperatures higher than a certain threshold value.
From the mathematical model solution (see for example fig.Ib) it is possible to infer that, to follow the fluid front incoming at 20^0C, the level of the "thermal cut" should be just below 60^0C, otherwise there would be a delay, between the position of the front and the change in colour of thermograms. It would be possible to study either a cooling or a warming process; we preferred the first one to avoid all the noise from environment that would have been present with a "thermal cut" at 20^0C.

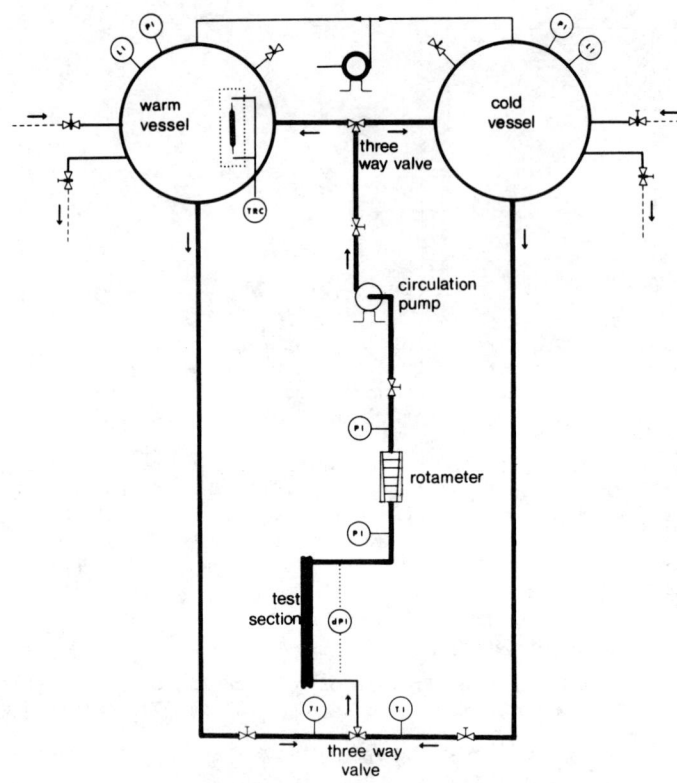

Fig. 2 - Experimental apparatus

Tests have been carried out on five modular channels. Tested stainless-steel plates are made by an Italian company and are normally available on the market for the milk heat treatment (fig.3). Useful area of each plate is about 0.3 m^3, with a lenght of 1.15 m and a thickness of 1 mm. Plates are provided with the usual assembling gaskets in special rubber.
Experienced modular channels are obtained by assembling for different plates in the following way:

channel A : two plates type 1 (fig.3)
channel B : two plates type 2
channel C : one plate type 3 and one plate type 4
channel D : two plates type 3
channel E : two plates type 4

FLUID FLOW DISTRIBUTION INSIDE PLATE HEAT EXCHANGERS 527

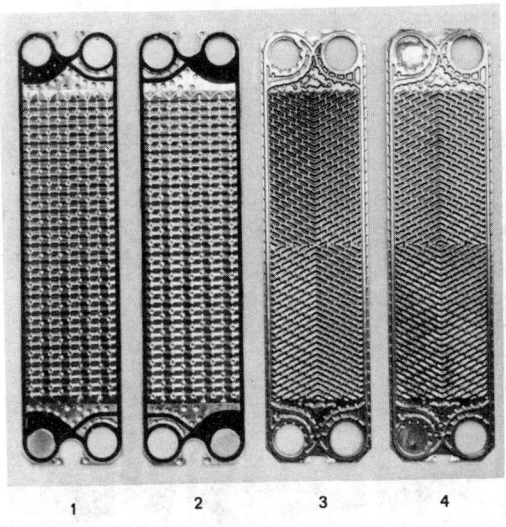

Fig. 3 - Tested plates

To obtain a constant volume, relative pressure inside the modular channel has been kept about zero by means of regulation valves, in any case the deformat= ion of the modular channels is prevented by suitable rigid reinforcements.

5. EXPERIMENTAL RESULTS

In this paper we report results relevant to flow rates of 600,1200 and 1800 l/h.
Thermal transient detected by AGA is recorded on film by means of a 16 mm cine= camera with speed of 8 frames per second. Films have been examined by a moviola; in fig. 4 we show a series of pictures separated by a second. In fig. 5 we show flow patterns for the examined modular channels.
Patterns represent position occupied by fluid front at following times. From their inspection it is possible to check out unobstructed channels and/or stagn ation areas, if any.
The best patterns, from a fluidodynamic point of view, is the flattest. Film scanning allows the determination of maximum and minimum residence time. If we consider as initial picture of transient the one showing first a black area at input, then minimum residence time corresponds to picture showing first a black

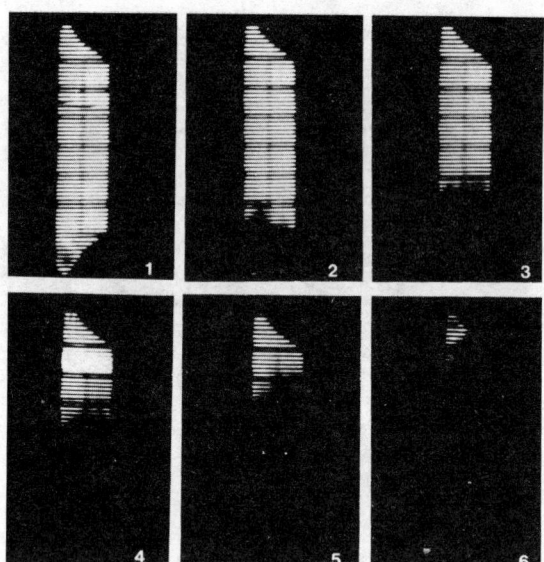

Fig. 4 - Series of pictures referring to a mod= dular channel for a flow of 1200 l/h

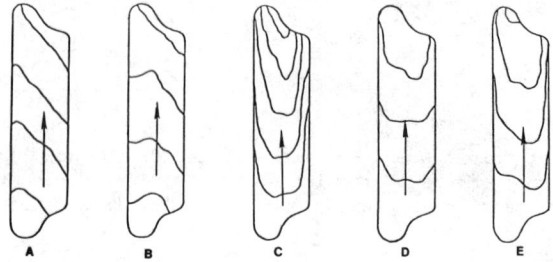

Fig.5-Flow patterns for modular channels at a flow rate of 1200 l/h

area at output and maximum residence time to the first completely black picture.
In Table I we report residence times after normalization to average theoretical time $\bar{t} = V/\Gamma$ where V is the volume and Γ the flow rate. The better $\Delta t/\bar{t} = (t_{max} - t_{min})/\bar{t}$ tends to zero, the more uniform is the distribution inside the channel. Normalization allows to compare modular channels with the same external dimens= ions of plates, but with different morphology and volumes.
Table I shows that no plate attains best results at all flow rates. On average, model D ranks best, model C worst. These conclusions agree with the ones att= ainable from flow patterns analysis. (Fig.5). Input and output fluid temperat= ures have been measured by thermocouples connected to a galvanometric recorder: in fig.6 we show a sample of recordings.
From temperature trends it is possible to determine residence times. Referring

TABLE I

Flow rate (l/h)	Residence time (s)	Mod. A	Mod. B	Mod. C	Mod. D	Mod. E
600	$\bar{t}$	6.57	6.57	6.86	6.86	6.86
600	t_{max}	8.25	7.88	11.57	7.25	8.88
600	t_{min}	5.38	5.63	4.13	4.50	5.13
600	Δt	2.88	2.25	7.63	2.75	3.75
600	$\Delta t/\bar{t}$	0.44	0.34	1.11	0.40	0.55
1200	$\bar{t}$	3.29	3.29	3.43	3.43	3.43
1200	t_{max}	5.00	4.38	5.38	4.00	4.25
1200	t_{min}	3.00	2.88	2.63	2.65	2.63
1200	Δt	2.00	1.50	2.75	1.25	1.62
1200	$\Delta t/\bar{t}$	0.61	0.46	0.80	0.36	0.47
1800	$\bar{t}$	2.23	2.23	2.32	2.32	2.32
1800	t_{max}	3.63	3.38	3.88	2.75	3.00
1800	t_{min}	2.13	2.00	1.63	2.00	2.00
1800	Δt	1.50	1.38	2.25	0.75	1.00
1800	$\Delta t/\bar{t}$	0.67	0.62	0.97	0.32	0.43

to graph of fig. 7, we have considered as minimum time the difference between points A and C, representative of the arrival of first cold front at input and output thermocouples. We had more difficulties in defining the maximum residence time.
Owing to the asymptotic trend of traces at 20^0C, we have considered times referred to 22^0C, i.e. 95% of total thermal step. In this way we obtain points B and E, representative of the arrival of cold water to thermocouples. Difference between E and B cannot be accepted as maximum time because the trace of output temperature is affected by thermal capacity of the system and by presence of unobstructed channels. We had to take into account only the influence of unobstructed channels. In order to eliminate the influence of system thermal capacity, we retained approximately valid to substract the delay time calculated with our simplified mathematical model. So the calculated curves of fig.1c type have been plotted on experimental diagrams starting from point C for all tests(fig.6) Therefore, as the difference between D and C represents the contribution of system thermal capacity, we can define $(E-B)-(D-C)$ as the maximum residence time.

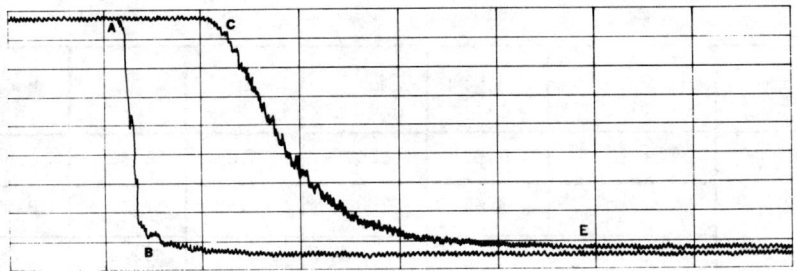

Fig.6-Recordings of water temperature at input (trace ABE) and at output (trace ACE) of a typical modular channel during a cooling transient for a flow rate of 1200 l/h.

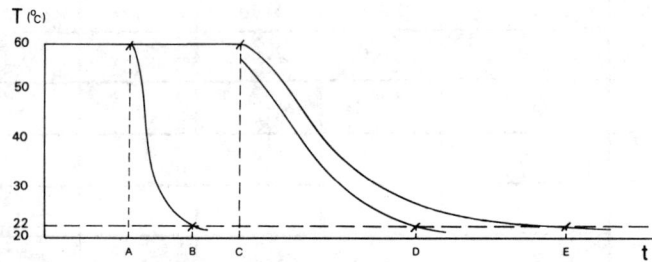

Fig.7-Graphic definition of residence times

Also these data allowed us to evaluate $(t_{max}-t_{min})/\bar{t}$ for all considered modular channels.The values of this ratio is shown as example for 1200 l/h in fig.8 ; in this figure the analogous ratio relevant to thermographic determination from Table 1 is also shown. Even if a systematic difference between the two values is evident, we can remark how both our determination well agree with the analysis of undertest channel performances.

6. CONCLUSIONS

We believe that experimental results confirm the validity of the method we propose. It offers: 1) the possibility of studying in a short time a large number of commercial plates; 2) good precision and reproducibility; 3) easy operation. Moreover, recording at the same time input and output temperature allows to combine advantages of "colorimetric" and "conductometric" methods. Using our method it is possile to obtain very useful information to design new models of plates, expecially as concerns their extremities. The main draw= back of our method is the not negligible heat capacity of the experimental equipment compared with the capacity of the internal fluid; a careful analysis of experimental data is therefore necessary.
We think this method can be applied, with suitable care, to other kinds of heat exchangers.

FLUID FLOW DISTRIBUTION INSIDE PLATE HEAT EXCHANGERS

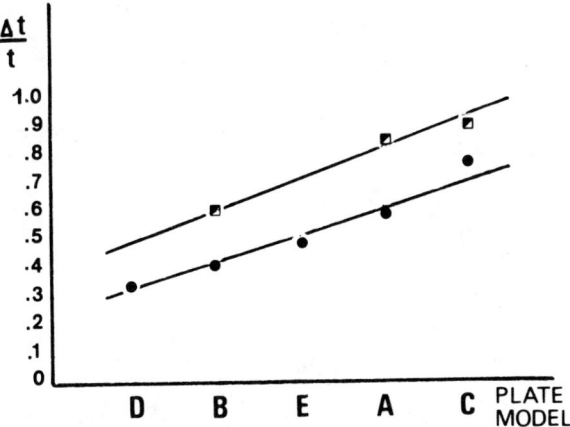

Fig. 8-Comparison between the performances of several channels, obtained with the thermography (●) and with the thermocouples (■) at 1200 l/h

REFERENCES

1. Troupe,R.A.and Morgan,J.C.and Prifti,J.The plate heater,versatil chemical engineering tool. Chem.Eng.Progr.56, 1,Jan.(1960).

2. Dickerson,W.R.and Read,R.B. Flow profiled in the plates of a milk pasteur= izer. Jour.of Dairy Sc.,57,1 June (1973).

3. Watson,E.L.and Killop,A.A.Mc.and Dunkley,W.L. - PHE flow characteristics. Ind.andEng.chemical,52,9,Sept.(1960).

4. Giudici,M. Studio della distribuzione del fluido nel canale modulare di un PHE con la termografia all'infrarosso. Graduation thesis,(1978)

5. AGA Thermovision 680. Operating manual.

6. Skoczylas,A.and Kaplan,J. Characteristics of flow and heat transport in plate heat exchangers". I.A.Ch. 16,2,1,(1977).

FLUIDIZED BED SYSTEMS

Fluid Bed Heat Exchangers

HOLGER MARTIN
Institut für Thermische
Verfahrenstechnik der
Universität (T.H.) Fridericiana
Karlsruhe, FRG

ABSTRACT

Recently a new model has been developed for the wall-to-bed heat transfer from solid surfaces immersed in fluidized beds /1/. This model makes use of some basic ideas adopted from the kinetic theory of gases in order to describe the mechanism of energy transfer through the moving particles ("particle convection"). The parallel - and in most cases less important - gas convective and radiative contributions are calculated from well known equations from the literature. In this paper the predictions of the new model are compared with a number of experimental data from the literature especially to show the influence of particle diameter, properties of particles, properties of fluidizing gas, bed expansion, temperature, and pressure on heat transfer coefficients in fluid bed heat exchangers.

1. INTRODUCTION

A fluid bed heat exchanger is shown schematically in Fig. 1. A gas stream flowing through a bed of solid particles (sand, coal or catalyst for example) can "fluidize" this bed within a certain range of flowrates or gas velocities. In the fluidized state the gas-solid system is very well mixed (with a uniform temperature ϑ_B all over the bed) due to the free mobility of the solid particles.

In many technical applications (combustion, coal gasification, or other chemical reactions) heat has to be removed from - or added to - a fluidized bed. This can be done by using immersed heating or cooling elements like single tubes, tube bundles, coils (see Fig. 1) etc.

Heat transfer between fluidized beds and the surfaces of immersed heating or cooling elements has been a subject of research since more than 30 years. A huge amount of literature has been published in this field and a lot of experimental results as well as theoretical models can be found in the textbooks on fluidization /2-13/.

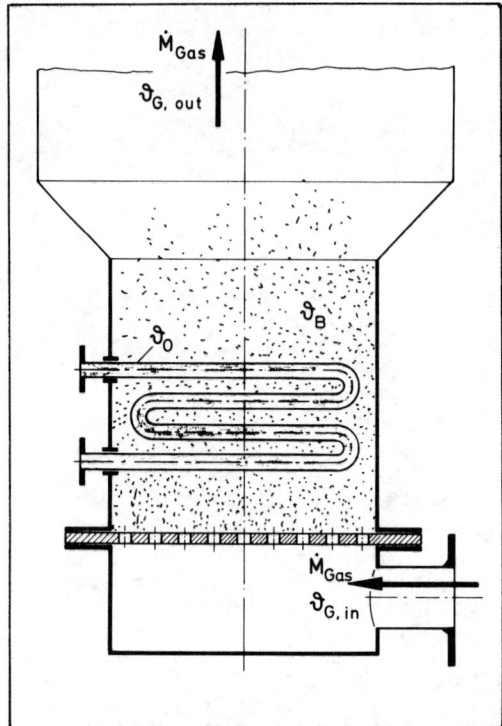

Fig.1 Fluid bed heat exchanger

Despite the abundance of experimental data and theoretical work, there seems to be no generally accepted calculation method for heat transfer between a gas fluidized bed of particles and the surface of an immersed solid body.

Most experimental data have been obtained with single cylindrical heaters immersed vertically in the center of the fluidized bed. One of these experimental arrangements is shown schematically on the right hand side of Fig. 2, which is replotted from the recently published thesis by R. Wunder /14/.

The heat rate $\dot{Q}$ measured via the electric power input is divided by the heater surface area A and the temperature difference between that surface and the bed ($\vartheta_0 - \vartheta_B$) to get the heat transfer coefficient α.

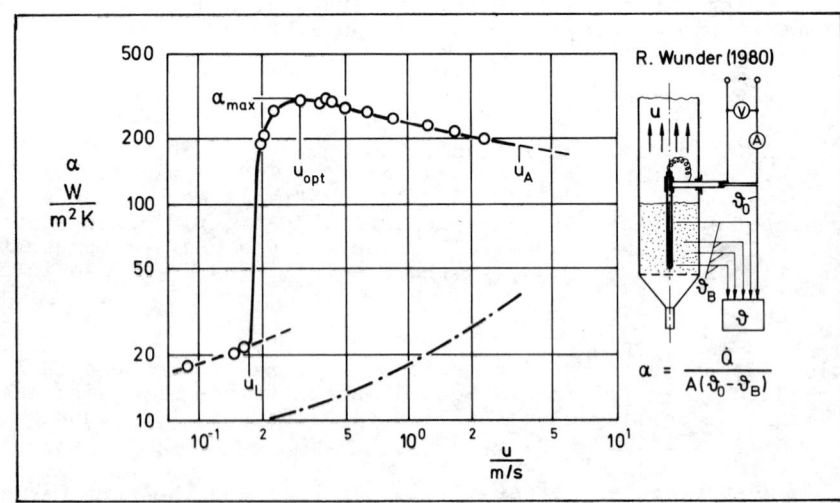

Fig. 2 Experimental setup /14/, definition of heat transfer coefficient α, and α vs. u (gas velocity) plot for fluid bed of 400 µm glass beads with 25°C, 1 bar air. Dotted line: single phase gas convection.

On the left hand side of Fig. 2 is a plot of heat transfer coefficients α vs. gas velocity u. The data points are also taken from Wunders thesis /14, p. 132/. The dotted lower line shows the values which are measured without particles - the heat transfer coefficients rise monotoneously with gas velocity. With particles they show the same behaviour only in the fixed bed region, where gas velocity is lower than the minimum fluidization velocity u_L. If u is increased only slightly above u_L the heat transfer coefficients rise by about a factor of ten, reach a maximum value α_{max} (at a certain gas velocity u_{opt}) and slightly tend downward for higher gas velocities. If the gas velocity passes u_A, the particles are carried away with the gas flow.

With increasing gas velocity the bed expands from its initial bed height L_L = 285 mm (for $u \leq u_L$) to L ≅ 600 mm for the last data point at u = 2.3 m/s. The void fraction varies from ψ_L = 0.4 to $\psi(2.3 \text{ m/s})$ = 0.72 and tends to unity if u tends to $u_A (\psi(u_A) = 1)$. Experimental results similar to those shown in Fig. 2 have been obtained by numerous other authors too.

Obviously, the steep rise of heat transfer coefficients has to be attributed to the free mobility of the solid particles in a fluidized bed.

2. MECHANISMS OF ENERGY TRANSFER

In a fluidized bed, heat transfer from an immersed solid object to the gas-particle system can be imagined to be composed of three parallel mechanisms: Particle Convection, Gas Convection, and Radiation.

$$\alpha = \alpha_P + \alpha_G + \alpha_R \qquad (1)$$

2.1 Particle Convection

A particle from the bulk of the bed (which is regarded to be completely mixed, for example by strong bubble induced solids convection) comes into contact with the heater surface by a random component of its motion. During the contact, heat is transferred mainly by conduction through the gaseous gap in the vicinity of the contact point. The particle stores a certain amount of thermal energy (limited by its heat capacity and the fact that it cannot exceed the temperature of the heater wall) and carries that surplus energy back into the bulk where it is transferred almost instantaneously to the gas and to the other particles.

This particle convective mechanism of energy transfer has been modelized by making use of some basic ideas adopted from the kinetic theory of gases (applied to the solid particles) in order to obtain expressions for the rate of exchange of particles between bulk and surface and for the contact time. Details may be found elsewhere /1/. The equations resulting from this theoretical approach give the particle convective heat transfer coefficient α_P as a function of the particle diameter d, the void fraction at minimum fluidization ψ_L, the volumetric heat capacity of the particles $(\rho c)_P$, the heat conductivity λ_g, a modified mean free path of the gas molecules σ /15/, and the bed expansion, or void fraction ψ:

$$\alpha_P = \alpha_P(d, \psi_L, (\rho c)_P, \lambda_g, \sigma, \psi) \qquad (2)$$

In terms of a particle convective Nusselt number $Nu_P = \alpha_P d/\lambda_g$ the model equations write:

$$\boxed{\begin{array}{l} Nu_P = (1-\psi) \, Z \, (1 - e^{-N}) \\[6pt] Z = \dfrac{1}{6} \dfrac{(\rho c)_P}{\lambda_g} \sqrt{\dfrac{gd^3(\psi - \psi_L)}{5(1-\psi_L)(1-\psi)}} \; ; \quad N = \dfrac{Nu_{WP(max)}}{C \cdot Z} \\[10pt] Nu_{WP(max)} = 4\left[(1+\dfrac{2\sigma}{d}) \, \ln \, (1+ \dfrac{d}{2\sigma}) - 1\right] \end{array}} \qquad (3)$$

The constant C in the model equations relates the contact time to the rate of exchange of particles between wall and bulk (or particle velocity). From a comparison with the experimental data of Wunder and Mersmann /16/ this constant has been determined to be

$$C = 3 \, . \qquad (3a)$$

This numerical value means that the average contact time t_c is in the order of magnitude of the time to move a particle over a path of one particle diameter on its way from bulk to surface or vice versa /1/.

The quantity Z in eq.(3) may be interpreted as a dimensionless particle capacity exchange rate $(Z \sim (\rho c w)_P)$. The term $(1-\psi)Z$ becomes zero for $\psi=\psi_L$ (no particle motion) and tends to zero again for $\psi \to 1$ (the number of particles per volume tends to zero). The argument (N) of the exponential function in eq.(3) is a nondimensional contact time (or a "Number of Transfer Units"). $Nu_{WP(max)}$ is the Nusselt number defined with the maximum heat transfer coefficient between a plane wall and a spherical particle during (point) contact. The formula for $Nu_{WP(max)}$ was obtained /15/ by integrating the local conduction heat flux across the gaseous gap between a sphere and a plane surface over the projection area of the sphere on the plane. In deriving this equation, one has to account for the fact, that the gap width always becomes less than the mean free path of the gas molecules in the vicinity of the contact point (Knudsen range).

Schlünders formula /15/ for $Nu_{WP(max)}$ gives a simple and physically reasonable explanation for the fact that there is a limiting heat transfer resistance for short contact times, which only depends on the physical properties of the gas and on the particle diameter: $Nu_{WP(max)} = Nu_{WP(max)}(2\sigma/d)$.

The group $(2\sigma/d)$ is a modified Knudsen number with

$$\sigma = 2\Lambda\left(\frac{2}{\gamma} - 1\right), \qquad (3b)$$

where Λ is the mean free path of the gas molecules and γ is the accommodation coefficient ($\gamma < 1$) accounting for the incompleteness of energy transfer during the molecule wall collisions. For practical application, the mean free path Λ can be calculated from the viscosity of the gas:

$$\Lambda = \frac{16}{5} \cdot \sqrt{\frac{\tilde{R}T}{2\pi\tilde{M}}} \cdot \frac{\eta}{p} \qquad (3c)$$

For air at $T = 298$ K and $p = 10^5$ Pa, Λ becomes 67.9 nm. The accommodation coefficient γ depends on the kind of gas molecules involved and on temperature. An empirical equation for γ based on the measurements of Reiter, Camposilvan and Nehren /17/ was given in /1/:

$$\lg\left(\frac{1}{\gamma} - 1\right) = 0.6 - \frac{1000 \text{ K}/T + 1}{C_A} \qquad (3d)$$

The constant C_A can be found from one measured value of $\gamma(T)$. The following values of $\gamma(298$ K$)$ are recommended:

Gas	H_2	He	Ne	H_2O	Ar	air,CO_2	Kr	Xe
γ(298K)	0.20	0.235	0.573	0.80	0.876	0.90	0.933	0.956
C_A	∞	50	6	3.62	3	2.80	2.5	2.25

2.2 Gas Convection

The gas convective contribution to the total heat transfer coefficient is calculated from Baskakovs formula /18/:

$$Nu_G = 0.009 \, Pr^{1/3} \cdot Ar^{1/2} \qquad (4)$$

$$\left(Pr = \frac{\eta_g c_{p\,g}}{\lambda_g}; \quad Ar = \frac{gd^3}{\eta_g^2} \cdot \rho_g(\rho_P - \rho_g)\right)$$

2.3 Radiation

The radiative contribution, which becomes important for higher temperatures may be estimated from

$$\alpha_R \cong 4\,\varepsilon\,c_s\,\bar{T}^3 \qquad (5)$$

In eq.(5) ε is an effective emissivity ($\varepsilon \leq 1$) depending on the surface properties of the heater and on those of the bed.

3. COMPARISON OF MODEL PREDICTIONS WITH DATA FROM THE LITERATURE

From the model equations (1-5) one can find, that the heat transfer coefficient α depends on

particle properties (d, ψ_L, $(\rho c)_P$, ρ_P),
gas properties (λ_g, σ, η_g, ρ_g, c_{pg}) and
operation parameters($\psi\{u\}$, T, p).

3.1 Particle properties

The influence of particle diameters on maximum heat transfer coefficients is shown in Fig. 3. The data points are taken from Wunder's thesis /14/ (Run No. 1-15, glass beads O, ● and Run No. 20-21, aluminum particles △ with narrow size distributions). The curves are calculated from equations (1-4) (radiation was neglected) with the properties of air at 25 °C and $(\rho c)_P$ = 1.875 MJ/m³K) ρ_P = 2500 kg/m³, ψ_L = 0.4 (glass beads).

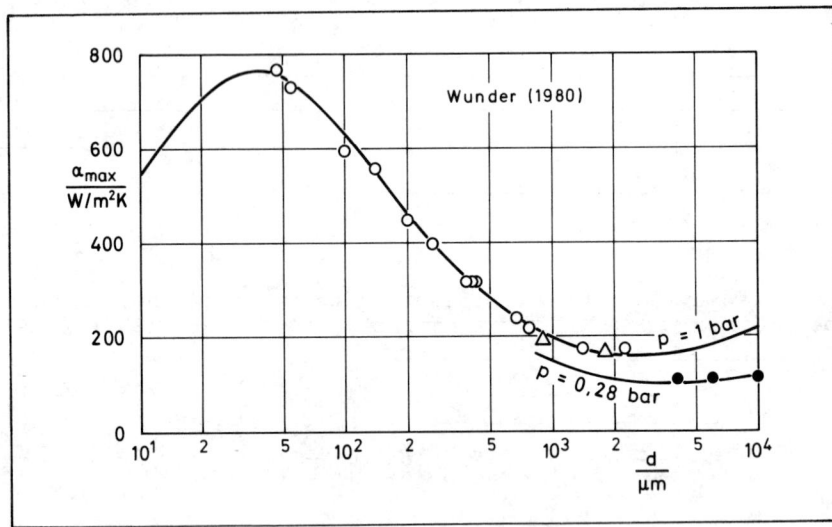

Fig. 3 Maximum heat transfer coefficient vs. particle diameter. Data points: Wunder (1980). Curves calculated from eqns. (1-4).

Since Wunder operated his fluid bed apparatus with a suction pump, the high gas flowrates which are necessary to fluidize large particles reduced the average pressure in the bed to values below atmospheric pressure. While this effect was small for particle sizes less than ∿3 mm (average pressure in the bed close to 1 bar), it became significant for the particles with 4, 6 and 10 mm (full symbols). In these runs (Run No. 13-15) the average pressure in the bed was 0.28 bars. Lower pressure increases the mean free path of the molecules (see eq.(3c)) which in turn (slightly) decreases $Nu_{WP(max)}$ and thus α_p. The lower pressure also decreases the density of the gas ρ_g, leading to a lower gas convective contribution (see eq.(4)). Density and heat capacity of aluminum is not very much different from the value for glass. The hundredfold higher heat conductivity of aluminum obviously has no influence (the conductivity of the particle material does not appear in the model equations!). The range of very fine powders, where the model predicts decreasing heat transfer coefficients with decreasing particle size is shown in Fig. 4.

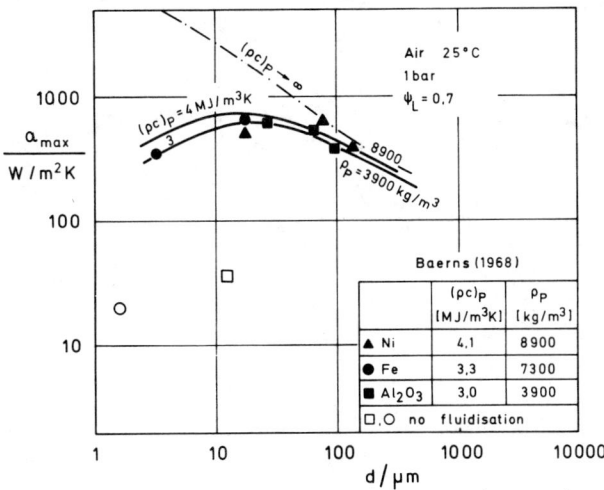

Fig. 4 Max. heat transfer coefficients vs. particle diameter. Data points: Baerns (1968). Curves calcd. from eqns. (1-4).

The experimental data of Baerns /19/ with fine powders of nickel, iron, and alumina seem to confirm that prediction. It is very difficult, however, to measure in this range, since fine powders form agglomerates and do not fluidize easily (the open symbols in Fig. 4 show fixed bed values of heat transfer, fluidization was not possible for these powders).

3.2 Gas properties

The influence of the physical properties of the gas on the maximum heat transfer coefficients is shown in Fig. 5, where α_{max} is plotted vs. the heat conductivity of the gas for a particle diameter of 315 μm.

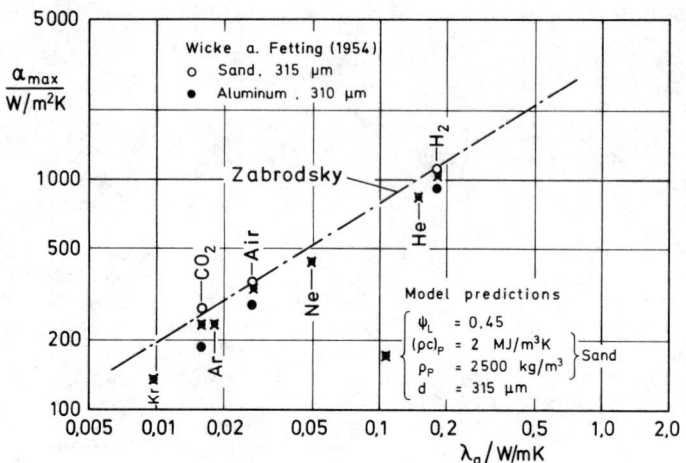

Fig. 5 Max. heat transfer coefficients vs. heat conductivity of fluidizing gas.

The open circles are data for sand, the full ones those for aluminum measured by Wicke and Fetting /20/ with CO_2, air, and hydrogen as fluidizing gases. The crossed symbols show values calculated from the model equations. The dotted line represents the empirical relationship

$$\alpha_{max} = 35.8 \, \lambda_g^{0.6} \, \rho_P^{0.2} \, d^{-0.36} \qquad (6)$$

(all quantities in basic S.I.-units)

which has been given by Zabrodsky /8/. Again the model predictions are in very good agreement with experimental results.

3.3 Operation parameters

Most authors give their measured heat transfer coefficients as a function of superficial gas velocity (or gas flowrate) rather than bed expansion or void fraction. In order to compare those data with the model predictions (- eq.(3) contains only ψ, not u -) one has to use a bed expansion relation $\psi = \psi(u)$. In Fig. 6 Wicke and Fetting's data /20/ for 315 µm sand particles fluidized by air and hydrogen are compared with curves calculated in that way from the model. The bed expansion was calculated from the simple Richardson Zaki Equation (see /1/).

Wunder's data /14/ can be compared more directly with the model, since he also gives the measured bed expansion L/L_L for each data point so that the void fraction ψ can be calculated easily:

$$\psi = 1-(1 - \psi_L) \frac{L_L}{L} \qquad (7)$$

FLUID BED HEAT EXCHANGERS

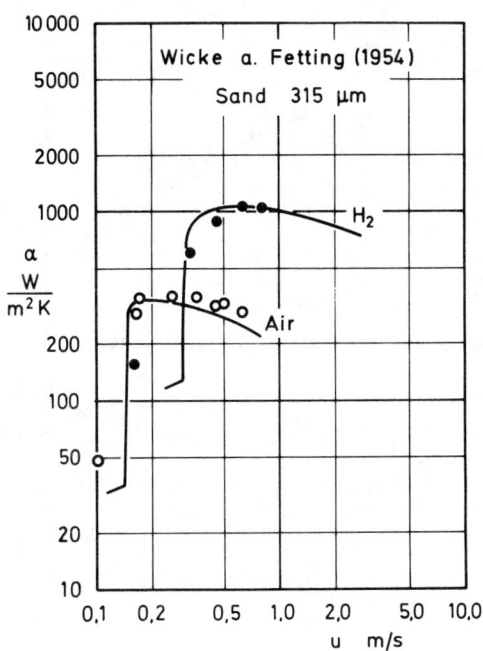

Fig. 6
Heat transfer coefficients vs. gas velocity. Curves calcd. from eqns.(1-4) with Richardson Zaki bed expansion equation.

Fig. 7 shows his data α vs. ψ for four different particle diameters in comparison with the calculated curves (for glass beads). Similar results have been obtained more than 30 years earlier by Mickley and Trilling /21/. These authors measured heat transfer coefficients from a vertical cylindrical rod in an air fluidized bed of glass beads. They used a conical gas inlet section in place of a distributer plate and they did not start from the fixed bed side (minimum fluidization) but from the other end of the range of fluidization ($\psi \rightarrow 1$). So their data cover the range of void fractions $\psi_{opt} < \psi < 1$.

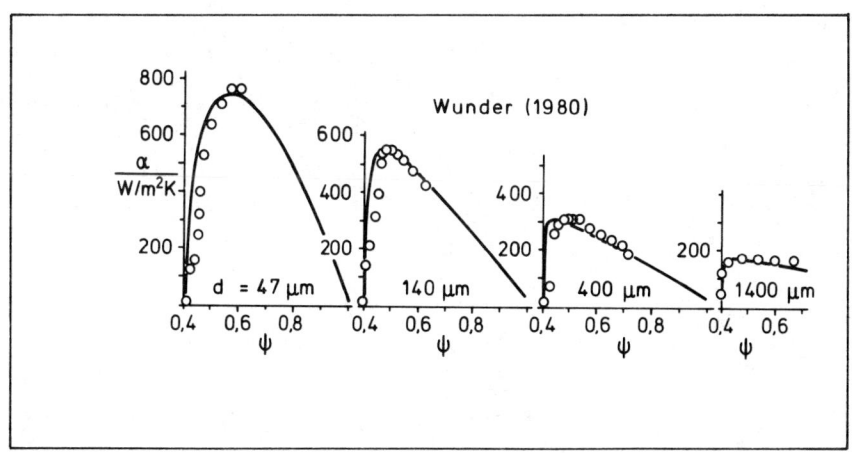

Fig. 7 Heat transfer coefficients vs. void fraction. Data: Glass beads, air 25°C, 1 bar, Wunder (1980). Curves calculated from eqns.(1-4).

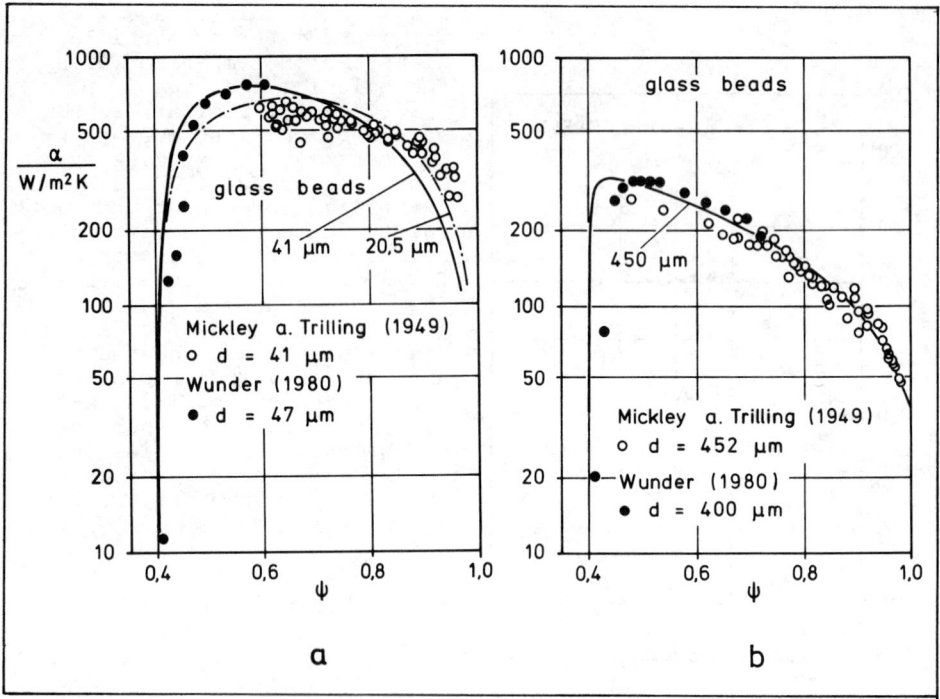

Fig. 8 Heat transfer coefficients vs. void fraction. Data: Glass beads, Mickley a. Trilling (1949) and Wunder (1980). Curves calculated from eqns.(1-4).

Figures 8a and 8b show the data for their smallest (41 μm) and their largest particles (452 μm) in comparison with the curves calculated from the model equations (air 1 bar, 120°C, glass beads). Additionally Wunder's data /14/ for particle diameters close to those used by Mickley and Trilling are shown in these figures.

The larger deviations for the smaller particles might be caused by a broader particle size distribution (the authors /21/ give only the average particle size). The dotted line calculated for 20.5 μm shows a somewhat better agreement with the data. Mickley and Trilling measured bed expansion via the pressure drop, they tabulated solids volume fractions $(1-\psi)$ (in their paper denoted by α) together with flowrates, heat fluxes, temperatures heat transfer coefficients a.s.o. for all their data.

The contribution of radiation to the total heat transfer coefficient is shown in Fig. 9. Experimental data from Janssen's thesis /22/ for maximum heat transfer coefficients from a horizontal tube in an air fluidized bed of sand particles with an average diameter of 717 μm are plotted vs. the mean temperature $\bar{\vartheta}$ ($\bar{\vartheta} = (\vartheta_0 + \vartheta_B)/2$). The dotted line is calculated from eqs.(1-5) with $\varepsilon=0$ (no radiation, i.e. this line gives $(\alpha_P + \alpha_G)_{max}$ as a

FLUID BED HEAT EXCHANGERS

function of temperature. The full line was calculated with $\varepsilon=0.5$ which seems to be a reasonable value for the effective emissivity of a metallic heater surface in a bed of sand particles. The broken line with $\varepsilon=1$ gives the maximum possible radiative contribution. One may see from this figure that radiation can usually be neglected for temperatures less than about 200 °C.

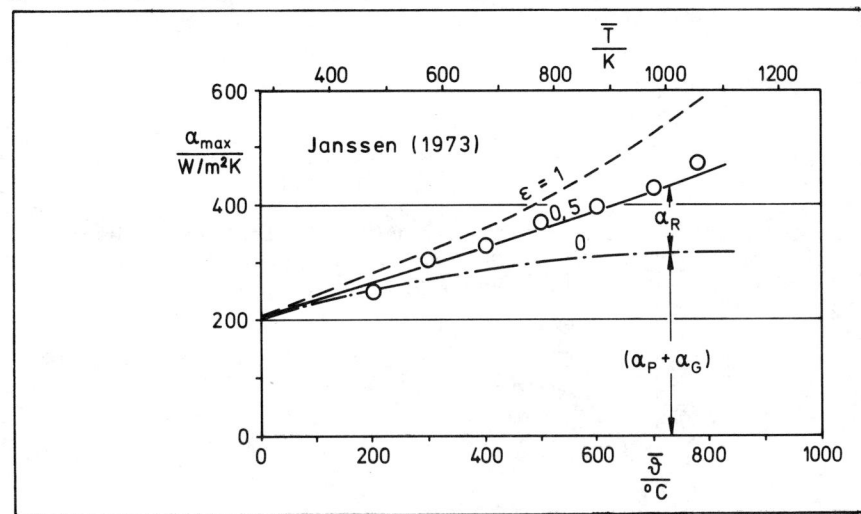

Fig. 9 Max. heat transfer coefficients vs. bed temperature. Influence of radiation. Data: Sand 717 μm, air 1 bar, horizontal cylindrical heater. Janssen (1973).

Finally in Fig. 10 the influence of higher pressure together with high temperatures is demonstrated. The data are taken from publications coming from Bergbauforschung, Essen /23, 24/.

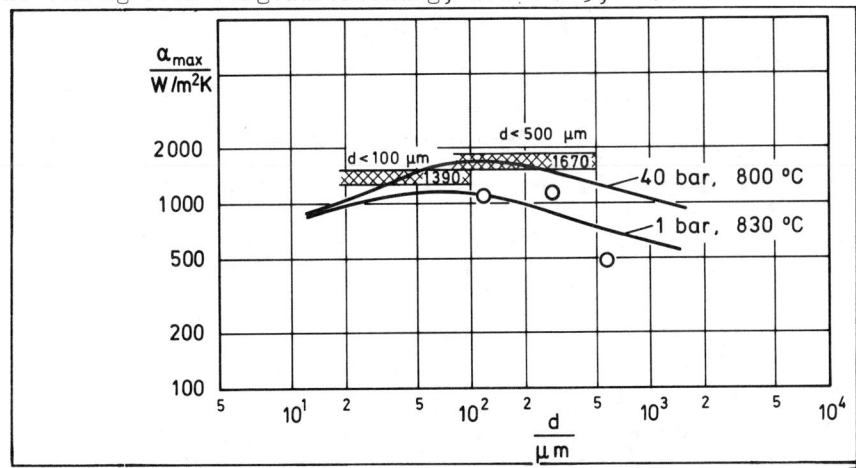

Fig. 10 Max. heat transfer coefficients vs. particle diameter. Data: Allothermal gasification of coal with steam in fluid bed gasifier, Bergbauforschung, Essen. Curves calculated from eqns.(1-5).

The open circles (O) are data from a laboratory fluid bed coke gasifier with pure steam (1 bar, 830°C) as the fluidizing gas and three sieve fractions of coke. The crosshatched rectangular areas are data from a pilot scale allothermal coal gasification plant with 40 bar, 800 °C steam and two sieve fractions of coke with d < 500 µm and d < 100 µm respectively. Again the curves are calculated from eqns.(1-5).

4. CONCLUSIONS

From the comparisons carried out so far one can find, that the model is able to represent the observed phenomena in wall-to-bed heat transfer between heat exchanger elements and fluidized beds with a reasonable degree of approximation. Even the nonmonotoneous variations of heat transfer coefficients with bed expansion (or void fraction) (Figs. 6,7,8) and with particle diameter (Figs. 3 and 4) are calculated from eqns.(1-5) in good agreement with the data.

Since the model equations are easily applied - there is no need for big computer programs - they might become a valuable tool for the design of fluid bed heat exchangers.

5. NOMENCLATURE

A	m^2	surface area
C	-	constant in eq.(3)(C=3)
c	J/kg K	heat capacity per mass
c_s	$W/m^2 K^4$	Stefan-Boltzmann's constant
d	m	(average) particle diameter
g	m/s^2	gravitational acceleration
L	m	height of fluidized bed
N	-	"number of transfer units" in eq.(3)
p	Pa	pressure
$\dot{Q}$	W	heat rate
$\bar{R}$	J/kmol K	gas constant (universal)
T	K	temperature (absolute)
t	s	time
u	m/s	gas velocity
V	m^3	volume
w	m/s	particle velocity
α	$W/m^2 K$	heat transfer coefficient
γ	-	accommodation coefficient
ε	-	effective emissivity

FLUID BED HEAT EXCHANGERS

η	Pas	viscosity
ϑ	°C	temperature
λ	W/mK	heat conductivity
Λ	m	mean free path of gas molecules
ρ	kg/m³	density
σ	m	modified mean free path
ψ	-	void fraction

Dimensionless numbers

$$Ar = \frac{gd^3}{\eta_g^2} \cdot \rho_g(\rho_P - \rho)$$

$$Nu = \frac{\alpha d}{\lambda_g}$$

$$Pr = \eta_g c_{pg}/\lambda_g$$

Subscripts

A	upper limit of fluidization ("airlift point")
L	lower limit of fluidization (minimum fluidization)
g	gas
G	gas convective
P	particle convective, particle
R	radiative
B	bed, bulk
WP	wall-to-particle

REFERENCES

1. Martin, H. 1980. Wärme- und Stoffübertragung in der Wirbelschicht, Chemie-Ing.-Techn., Vol. 52, pp.199/209

2. Othmer, D.F. 1956. Fluidization, Reinhold Publ. Co., New York.

3. Leva, M. 1959. Fluidization, McGraw Hill, N.Y.

4. Zenz, A.F. and Othmer, D.F. 1960. Fluidization and Fluid Particle Systems, Reinhold Publ. N.Y.

5. Schytil, F. 1961. Wirbelschichttechnik, Springer, Berlin.

6. Davidson, J.F. and Harrison, D. 1963(and 1971). Fluidized Particles, Cambridge University Press.

7. Beranek, J., Sokol, D. and Winterstein, G. 1964. Wirbelschichttechnik, VEB Deutscher Verlag für Grundstoffindustrie, Leipzig.

8. Zabrodsky, S.S. 1966. Hydrodynamics and Heat Transfer in Fluidized Beds, MIT Press, Cambridge, Mass.

9. Vanecek, V., Markvart, M. and Drbohlav, R. 1966. Fluidized Bed Drying, Leonard Hill Books, London.

10. Drinkenburg, A.A.H.(Editor) 1967. Proceedings of the International Symposium on Fluidization, Eindhoven 1967, Netherlands University Press, Amsterdam.

11. Kunii, D. and Levenspiel, O. 1969. Fluidization Engineering, Wiley, N.Y.

12. Botterill, J.S.M. 1975. Fluid Bed Heat Transfer, Academic Press, N.Y.

13. Beranek, J., Rose, K. and Winterstein, G. 1975. Grundlagen der Wirbelschichttechnik, Krausskopf-Verlag, Mainz.

14. Wunder, R. 1980. Wärmeübergang an vertikalen Wärmetauscherflächen in Gaswirbelschichten. Dr.-Ing.-Thesis, T.U.München.

15. Schlünder, E.U. 1971. Wärmeübergang an bewegte Kugelschüttungen bei kurzfristigem Kontakt, Chemie-Ing.-Techn. Vol.43, pp. 651/654.

16. Wunder, R. and Mersmann, A. 1979. Wärmeübergang zwischen Gaswirbelschichten und senkrechten Austauschflächen, Chemie-Ing.-Techn. Vol. 51, p. 241.

17. Reiter, T.W., Camposilvan, J. and Nehren, R. 1972. Akkommodationskoeffizienten von Edelgasen an Pt im Temperaturbereich von 80 bis 450 K, Wärme- und Stoffübertragung, Vol. 5, pp. 116/120

18. Baskakov, A.P. et.al. 1973. Heat Transfer to Objects Immersed in Fluidized Beds, Powder Technology, Vol.8, pp.273/282.

19. Baerns, M. 1968. Verfahrenstechnische Eigenschaften von Wirbelschichten aus staubförmigen Feststoffmaterialien, Chemie-Ing.-Techn. Vol 40, pp. 737/

20. Wicke, E. and Fetting, F. 1954. Wärmeübertragung in Gaswirbelschichten, Chemie-Ing.-Techn., Vol. 26, pp. 301/356

21. Mickley, H.S. and Trilling, Ch.A. 1949. Heat Transfer Characteristics of Fludized Beds, Ind.Eng.Chem. Vol. 41, pp. 1135/2247.

22. Janssen, K. 1973. Beitrag zur Berechnung von Wärmeübergangszahlen zwischen Fluidatbetten und darin eintauchenden Wärmetauschflächen in Abhängigkeit von den Strömungsbedingungen inhomogener Fluidisierungszustände, Dr.-Ing.Thesis,TH Aachen

23. Petrovic, V. 1978. Messung und Berechnung des Wärmeübergangs von einem Heizrohr an eine Kohlewirbelschicht in Wasserdampf und Inertgas, Dr.-Ing. Thesis, TH Aachen.

24. Jüntgen, H. van Heek, K.H. 1977. A Technical Scale Gas Generator for Steam Gasification of Coal Using Nuclear Heat. Nuclear Technology. Vol. 35, pp. 581/590.

Status of Fluidized Bed Waste Heat Recovery

M. I. RUDNICKI, C. S. MAH, and H. W. WILLIAMS
Aerojet Energy Conversion Company
Sacramento, California, USA

ABSTRACT

Fluidized bed waste heat recovery is developing rapidly as an advancement in heat exchange. Recent years have seen a substantial acceleration in research, development, and industrial applications of fluidized bed technology. The work spread from Great Britain and the U.S.A. to Japan and the rest of Europe.

The paper describes the general understanding of fluidized bed characteristics, performance, problems, and it explains how the knowledge is rapidly evolving. Status of research, development, demonstrations, and field-operations is covered.

Advanced equipment installations in the United States, Great Britain and in Japan are described. This work will substantially advance the state of the art in heat transfer co-efficients, self-cleaning, pollution control, and other features of fluidized bed technology.

INTRODUCTION

Waste heat recovery equipment can significantly improve process efficiency in reactors, furnaces, ovens and process heaters. Fluidized bed heat exchange has important characteristics for waste heat recovery.

Shallow fluidized bed heat exchangers have been intensively studied in recent years for various waste heat applications and found to be competitive with conventional heat exchangers.

In the design of heat exchangers for waste heat recovery, the designer is faced with the limitations of conventional, gas-to-surface, heat transfer coefficients of 30 to 100 W/m^2K. This heat transfer limitation results in the requirement for large heat exchange surfaces, and the typical waste heat recovery boiler will have approximately four times more heat exchange surface than the typical liquid-to-liquid heat exchanger for the same heat duty.

Fluidized beds enhance the gas-to-surface heat transfer. Heat transfer coefficients as high as 700 W/m^2K have been measured.

Fluidized beds also have other potential advantages when used as heat exchangers. These advantages include:

- Self cleaning. The solid particles forming the fluid bed provide gentle scrubbing to keep the heat-exchange surfaces clean.
- Uniform temperature. The even temperature in the fluid bed minimizes temperature gradients in the heat exchanger, thereby simplifying the structural design.
- High thermal capacitance. The bed material in the fluid bed provide a high capacitance to absorb large temperature change transients in the fluidizing gas, thus allowing fast startups. The bed material can also be used for heat storage for systems with intermittent operations.
- Pollution control. The fluid bed can be used as a filter for undesirable particulates in the waste gas stream. The bed material can also be selected to absorb gaseous pollutants such as SO_2.
- Inherent damper. The fluid bed slumps on the distributor plate when not fluidized and thereby acts as a built-in damper.

STATUS OF FLUIDIZED BED WASTE HEAT RECOVERY

Because of these possibilities, investigators have been exploring the technology and application of fluidized beds for waste heat recovery, and commercial companies are beginning to bring out equipment and actively market in the area. This paper describes some of the current activities which will advance the state-of-the-art.

BACKGROUND

Fluidized Bed Heat Exchanger

A fluidized bed consists of a bed of fine particles continuously being buoyed up by a fluid stream. This buoyed-up bed, generally referred to as a dense-phase, bubbling fluidized bed, looks like a boiling liquid (Figure 1) and will exhibit the same buoyancy and height/pressure characteristics as a liquid.

A fluidized bed heat exchanger will have heat exchange surfaces within the bed to take advantage of the high heat transfer rates. In a conventional design for waste heat recovery (Figure 2) the hot gas is first admitted into a plenum before it is evenly introduced into the particulate bed by way of many small openings in the distributor plate. The gas, while percolating through the particulate bed, gives up its available heat to the bed, which in turn transfers the heat to the coolant tubes.

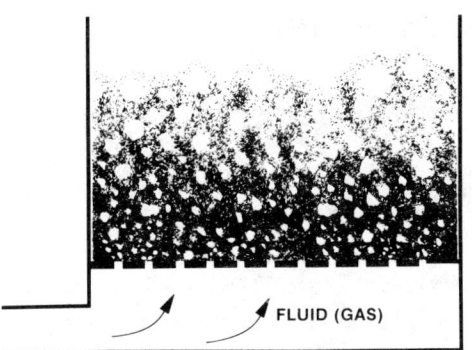

FIGURE 1. BUBBLING FLUIDIZATION

Heat Transfer in Fluidized Beds

The physical process of heat transfer to a surface immersed in a fluidized bed consists of three interrelated phenomena. These include gas-phase convection, solid phase convection/conduction, and radiation from both the particles and the gas. Various physical models have been proposed to explain the heat transfer mechanism in fluidized beds. These models interpret the physical process of heat transfer as shown in Table 1.

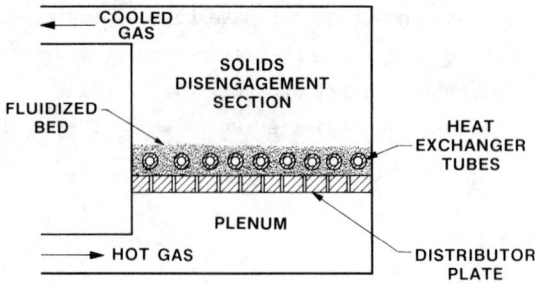

FIGURE 2. WASTE HEAT RECOVERY - FLUIDIZED BED HEAT EXCHANGER

Table 1. MODELS FOR FLUIDIZED BED HEAT TRANSFER

INVESTIGATOR	MODEL
Leva (Reference 1)	Heat transfer is limited by conduction through a thin film.
Zabrodsky (Reference 2)	Moving particles dominate the heat transfer process.
Mickley and Fairbanks (Reference 3)	Heat transfer is through transient conduction to "packets" of solids periodically displaced by bubbles.

Basically, the thin film model assumes that the scouring action of the solid particles decreases the effective film thickness and thereby increases the heat transfer rate. The concept assumes a steady state heat transfer process and results in the conclusion that heat transfer can be increased by increasing the fluidizing gas flow. The conclusion is generally not consistent with test results.

The moving particle model assumes that the particles are bathed by fluid at the wall temperature, absorb heat by transient conduction, then give up the heat to the gas. The model results in a conclusion that suggests particle size has very little effect on heat transfer. This conclusion is also generally not consistent with the test results.

The transient conduction model is the most comprehensive of the physical models of heat transfer. In this model, "packets" of particles are brought onto the heat transfer surface, and heat is transferred via conduction from each of

STATUS OF FLUIDIZED BED WASTE HEAT RECOVERY

the packets to the surface. The packets are then exchanged at the heat exchanger surface through the bubbling action of the fluidized bed. This model has been the focus of many of the more recent studies (References 4-7).

In a series of experiments with a 0.6m diameter fluidized bed heat exchanger, the Aerojet Energy Conversion Company defined the effects of the bed materials properties on heat transfer. Test set up is shown on Figure 3.

Properties varied were size, thermal conductivity, density and specific heat.

The approach taken by AECC to effectively define the impact of bed material on heat transfer was to isolate each material parameter in direct pair comparison testing. In this approach, matched pairs of materials were selected

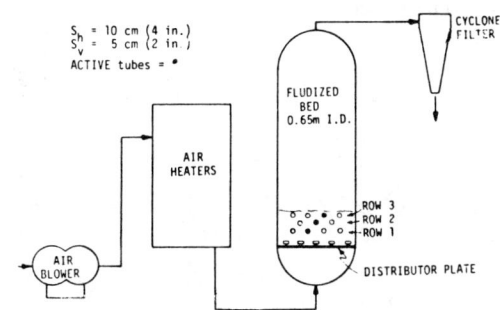

FIGURE 3. TEST CONFIGURATION

which would ideally have the same thermophysical properties except for the variable to be isolated and tested. The effect of each material property on the fluidized bed heat transfer coefficient could then be uniquely determined without the influence of extraneous factors.

In the selection of materials for testing, a general screening of materials and their properties were first made based on information available in the literature. Those selected for use in the test program are listed in Table 2. The materials are grouped according to their thermochemical properties used in the direct pair comparison test.

Table 2
FLUIDIZED BED HEAT TRANSFER TEST MATERIALS

TEST SERIES	MATERIAL	MEAN PARTICLE SIZE DIAMETER (μm)	SPECIFIC GRAVITY	SPECIFIC HEAT J/Kqt (BTU/lb°F)	THERMAL CONDUCTIVITY W/M-K (BTU/hr ft°F)
THERMAL CONDUCTIVITY	SAND	172	2.3	816 (.195)	1.73 (1)
	GRAPHITE	137	2.25	846 (.202)	120-160(70-94)
DENSITY	316 STAINLESS STEEL	106	7.9	490 (.117)	14 (8)
	ZIRCON SAND	147	4.5	594 (.142)	2 (1.11)
	ALUMINA	170	3.9	1042 (.249)	20 (11.69)
	EXPANDED ALUMINA	163	1.6	1042 (.249)	21 (12)
SPECIFIC HEAT	NICKEL	91	8.9	460 (.11)	92 (53)
	SOLDER	88	8.9	188 (.045)	47 (27)
PARTICLE SIZE	SAND	172	2.3	816 (.195)	1.73 (1)
	SAND	279			
	SAND	439			
	BALLOTINI	164	2.5	1000 (.24)	1.73 (1)

It was found that a heat transfer coefficient of 400 W/m^{2}K is representative of small (<500μm) particle fluidized beds (Reference 8) and the empirical relation developed from the data from tests on various materials with widely varying thermophysical properties was as follows:

$$h = C \cdot d_e^{-0.5} \cdot \rho^{0.2} \cdot c_p^{0.2}$$

where h = gas-side heat transfer coefficient; W/m^2-k
C = equation constant
d_e = equivalent particle diameter, μm
ρ = specific gravity of particle
c_p = specific heat of particle, J/kg-K

Bed material thermal conductivity had little effect on heat transfer rate, but size, density and specific heat effects can be significant.

STATUS OF FLUIDIZED BED WASTE HEAT RECOVERY

For comparison, the data from the AECC tests have been plotted along with that of a number of other researchers. These researchers used a variety of bed materials and widely differing experimental conditions. The bed materials used included sand, glass beads, copper, iron, alumina, nickel, solder and magnesite. Experimental fluidized beds used in the testing varied from 3.5cm diameter round beds to 0.60 m x 1.50 m rectangular beds. Heat transfer surfaces used included spherical probes, horizontal tubes and vertical tubes with bed temperatures from room temperature to 548K(9527°F).

The AECC data falls in the middle of the data from these other nine researchers (see Figure 4). Since the consensus results agree with the AECC results, it can be concluded that the AECC results has a high probability of being representative of the real effects of the thermophysical parameters of the bed material.

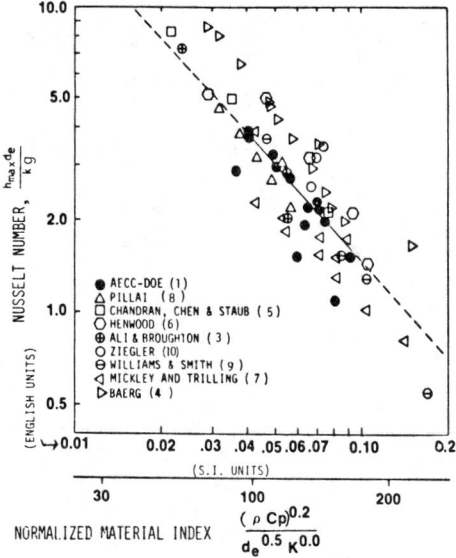

FIGURE 4. HEAT TRANSFER DATA COMPARISON

Pollution Control with Fluidized Beds

A fluidized bed provides excellent contact between the fluidizing gas and the bed particles. This was the very quality which led to the original development of the fluidized bed for the Fluid Catalytic Process first put into operation in 1942. This good gas-solids contact results in the possiblity of using the fluidized bed for gas pollutant cleanup. Processes using simple oxides such as CaO as absorbents for SO_2 have been evaluated with success, especially in conjunction with coal combustion. The good gas-solid contact can also be used to promote the adsorption of NO_x with materials such as activated carbon.

The fluidized bed can also be used as a particulate filter. Such particulates (sized <5μm) have been found to agglomerate onto the bed material, and high cleanup rates have been observed in experiments. Success has also been achieved with electrically charging the particles to enhance collection.

The subject of using a fluidized bed heat exchanger for pollution control was discussed at length in Reference 9. Significant findings by various investigators are shown in Table 3.

Table 3. POLLUTANT CONTROL CAPABILITY OF FLUIDIZED BED

Pollutant	Bed Material	Typical Removal Effectiveness, %	Data Source
SO_2	CaO	90	Ref. 10
NO_x	Carbon	40-90	Ref. 11
Particulates	Sand	83-92	Ref. 12
Alkalis	Bauxite	98	Ref. 13

Physical Parameters for Fluidized Bed Heat Exchanger Design

The fluidized bed expands significantly when it changes from a static bed to bubbling fluidization. The degree of expansion depends on many parameters (Reference 14). In general, the ratio of the expanded bed height to the static bed height is in the order of 1.5 to 1.7 at superficial fluidizing velocities of 1-3 m/sec.

The fluidizing gas will suffer a pressure drop as it passes through the fluidized bed. The rule of thumb is that the pressure drop is equal to the weight of the bed per unit area. With common bed materials, this roughly translates into 100 pascals per cm of bed height.

There is a maximum packing of heat exchange tubes in the fluidized bed without a significant degradation in heat transfer. The tube spacing used in design is usually in the order of 2-4 tube diameters (Reference 15). Finned tubes are also used. Fin spacing is usually greater than 10 particle diameters and up to 5 times the bare-tube heat transfer rates can be achieved (Reference 16).

Heat Exchanger Thermodynamics

The fluidized bed has very good mixing and can maintain a very even uniform temperature throughout the bed. The consequence of this good mixing is that the bed temperature is equal to the gas temperature at the heat exchanger outlet for the waste heat boiler application. This condition lowers the temperature potential between the hot gas and the water/steam, and, at the same time, the amount of heat that can be transferred is limited to the same degree as a parallel flow heat exchanger.

This limitation can be illustrated with a typical example. For a boiler generating steam at 3.45×10^5 Pa(420K) with a 590K waste heat source, the difference between a conventional, counterflow boiler and a fluidized bed boiler would be as shown in Table 4.

Table 4. Boiler Comparison

	Conventional Counterflow Boiler	Fluidized Bed Boiler
Overall Heat Transfer Coefficient, W/m²K	50	200
Log Mean Temp Difference °K	100	47
Heat Transfer Area	A	0.53A

The ideal temperature profiles for the two cases are shown in Figure 5.

CONVENTIONAL BOILER

FLUIDIZED BED WASTE HEAT BOILER

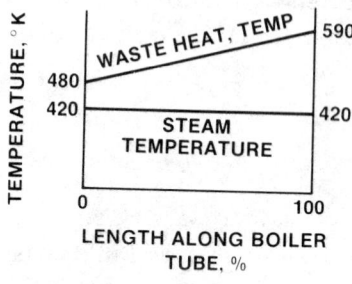
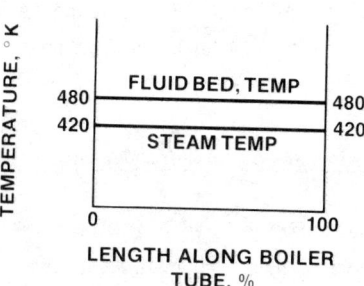

FIGURE 5. TYPICAL TEMPERATURE PROFILES

Fluidized Bed Heat Exchanger Applications

The classic fluidized bed, waste heat recovery heat exchanger has a shallow (10-15 cm) bed to minimize the gas-side pressure drop and uses finned tubes to minimize the tubing requirements. The bed material is usually common silica sand or aluminum oxide.

STATUS OF FLUIDIZED BED WASTE HEAT RECOVERY

One good example of this genre is the waste heat boiler provided by Stone-Platt Fluidfire Ltd., UK, for service in the Shell International Marine tanker Fjordshell, a 32,000 ton ship (Reference 16) (see Figure 6). The boiler is installed on the exhaust of the 900KW diesel engine used to drive the ship. This unit, the largest industrial fluidized bed heat exchanger in operation, has overall dimensions of 2.7m x 3.7m x 5.5m high (Figure 7). Internally, the boiler has three shallow beds in parallel with a total bed area of almost 30 m^2. Chain driven steel brushes are used to periodically clean up the soot that would otherwise accumulate on the distributor plate.

Table 5. Boiler Specifications, Waste Heat Recovery Boiler

Gas Flow	74,460 Kg/hr.
Gas Inlet Temperature	320°C
Gas Outlet Temperature	238°C
Steam Pressure	8×10^5 Pa
Steam Flow	2800 Kg/hr.
Heat Recovered	2000 KW

FIGURE 6. STONE-PLATT FLUIDFIRE WASTE HEAT BOILER

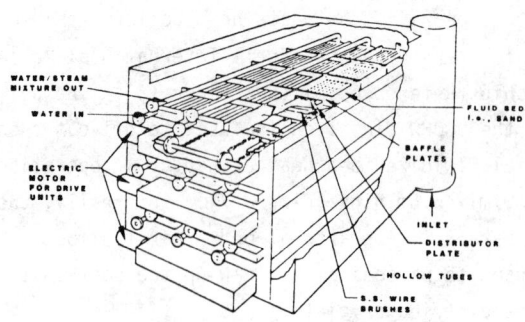

FIGURE 7. TRIPLE-BED, FLUIDIZED-BED WASTE HEAT RECOVERY BOILER

The potential for waste heat recovery using fluidized bed technology was studied by the Argonne National Laboratory under the sponsorship of the United States Department of Energy (Reference 17). In this effort, a 300KW fluidized bed hot water heater was designed and evaluated. It was concluded that no operating or technical difficulties exist that would hinder the use of fluidized bed systems in waste heat recovery applications.

Industrially, Stone-Platt Fluidfire, Ltd. is leading the way in the applications of their Fluidfire fluidized bed waste heat recovery systems with about 20 installations (Reference 18). These units can handle gas temperatures of up to 1000°C and have ratings of 75-2000 Kw. Several of the more recent installations are shown in Table 6 (Reference 19).

Table 6. Some Recent Installations of Fluidized Bed Waste Heat Recovery Systems, Stone-Platt

Site of Installation	Rating	Heat Source
The Carborundun Co. St. Helens Merseyside, Manchester	115 KW	Tunnel Kiln
Union Carbide (UK Ltd. Hardley Southampton)	300 KW	Boiler (economiser)
Simms Steel Industries Ltd., Co., Antrim Northern Ireland	400 KW	Galvanising Bath (zinc plating)
TF & JH Braime Limited, Leeds, 10	115 KW	Furnace-heat treat
Howmet Turbine Components Corp. Exeter, Devon	235 KW	Furnace-heat treat
Templeborough Rolling Mills, Ltd. Rotherham	500 KW	Furnace-steel wlk.bm.

STATUS OF FLUIDIZED BED WASTE HEAT RECOVERY

Gadeluis/Fluidfire is also beginning to make installations in Japan. About a dozen installations have been made as of June 1980. Several of the more recent installations are shown in Table 7.

Table 7. Some Recent Installations of Fluidized Bed Waste Heat Recovery Systems-Gadelius/Fluidfire

Site of Installation	Rating	Heat Source
Mitsui Toatsu Ohimuta	725 KW	Boiler
Daihachi Nayagawa	250 KW	Boiler
Ajinomoto Kawasaki	25 KW	Boiler
Tsuyakin Kogyo Tsushima	450 KW	Diesel
Daiichi Koyyo Seiyaku Kyoto	100 KW	Boiler
Honshu Seishi Shiga	350 KW	Boiler

Developments in Fluidized-Bed Waste Heat Recovery

Research needs in fluidized bed systems were subjects of a symposium held in October 1979 (Reference 21). The programs necessary to improve the fluidized bed exchanger and help accelerate the implementation of fluidized bed waste heat recovery was discussed in Reference 22. In general, improvements in the heat exchanger were considered to be desirable and the demonstration of the systems in severe industrial environments was considered necessary. In the new technology area, the focus has been on means and methods of getting around the problems of low gas velocities through the heat exchanger (which means large size), low heat exchanger effectiveness, and pollutant-laden waste streams.

U. S. Department of Energy (DOE) Programs

The U. S. DOE funded programs for improving conventional fluidized bed waste heat recovery heat exchangers. In one program, the Aerojet Energy Conversion Company, Sacramento, California, extended the bed materials study into the Fluid bed cost optimization: it was found that the bed material can make a significant impact on the fluidized bed system's economics (Reference 8).

Approximately 100 materials from each of the eight standard Dana mineral classifications as well as man-made materials were evaluated. The general trend showed that heat exchangers with oxides and minerals as bed materials were the most cost effective at the lower bed temperatures and that heat exchangers with metallic bed materials were the most cost effective at the higher bed temperatures. The specific results, shown in Figure 8, indicate the relationship of density and heat exchanger costs for bed temperature of 700K (800°F). In another program, the United Technologies Research Center, Hartford, Connecticut, extensively tested and evaluated a commercial fluidized bed heat exchanger; it was concluded that the anti-fouling quality of the fluidized bed heat exchanger was its most advantageous feature (Reference 23).

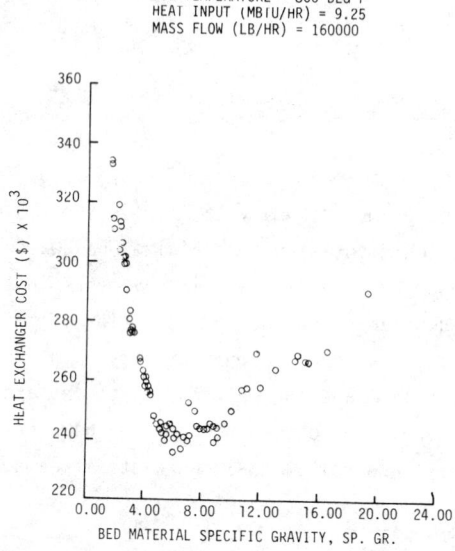

FIGURE 8. EFFECT OF BED MATERIAL ON HEAT EXCHANGER COSTS AT 800°F

In the demonstration of the capability of conventional fluidized bed heat exchangers in a severe industrial environment, the DOE and Aerojet are planning a joint program to install a fluidized bed boiler in the exhaust of an aluminum melting furnace. This application features a waste gas stream with a considerable amount of particulates that are known to cause heat exchanger fouling without fluidized bed cleaning. This application will also be made somewhat difficult because intermittent operation on a three hour cycle is required. Some important characteristics of the application are shown in Table 8.

Table 8. Specifications for Aerojet Fluidized Bed Waste Heat Recovery Demonstration

Waste Gas Conditions
    Flow, M^3/sec   8.5
    Temperature, °C   600
    Particulate loading, g/m^3 0.5
    Cl_2 (max.), Vol %   0.035
    Operational Cycle, hr.   3
Steam Conditions
    Pressure, MPa   1.3
    Flow, Kg/hr   3500

In the demonstration of a moving bed type heat exchanger (Figure 9), the DOE and the Thermo-Electron Corporation, Waltham, Massachusetts, are conducting a joint program to install a system also on the exhaust of an aluminum furnace. This system, similar to that of the Econco-Therm Corporation of Englewood, New Jersey, has a solids stream flowing countercurrent to the waste gas stream. The heated solids are then flowed countercurrent to the cold combustion gas stream to preheat it. Heat exchange effectiveness as high as 0.80 can be achieved. After cooling, the solid particles are recycled. The specification for the system is shown in Table 9.

Table 9. Specification for Thermo-Electron
Fluidized Bed Heat Recovery Demonstration

Upper Bed Conditions	
Inlet Gas Temperature, °C	1100
Outlet Gas Temperature, °C	650
Inlet Particle Temperature, °C	340
Lower Bed Conditions	
Inlet Air Temperature, °C	38
Outlet Air Temperature, °C	540
Inlet Particle Temperature, °C	870
Net Heat Transfer, KW	500

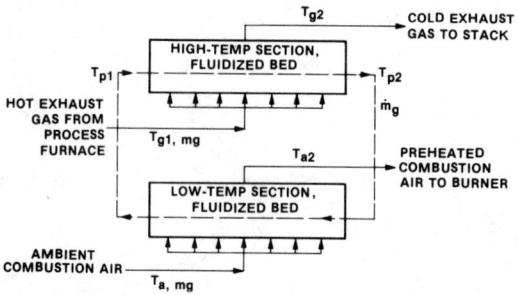

FIGURE 9. FLOW SCHEMATIC OF MOVING BED HEAT EXCHANGER

Other Developments

Interest in the falling cloud type heat exchanger is extensive. The Universite de Compiegne is investigating a similar design which they call a raining packed bed (Reference 18), and Stone-Platt is doing considerable development work in this area (Reference 24).

Fluidization of magnetic particles in the presence of a magnetic field is being studied (Reference 25). The magnetic field, oriented along the gas flow path, smooths the gas flow so that the solids/gas mixture remains an emulsion as the bed expands. This type of system can offer true counter-current contacting between the gas and solids. It was also found that the bed was an effective particulate filter.

High velocity, turbulent fluidized beds are being investigated extensively, especially for coal combustion systems. This type of beds have high velocity gas flow, but do not have large bubbles and instability (Reference 26). The high velocity results in low bed cross-sectional areas.

In another development, the bed is pulsed periodically at frequencies up to 4 Hertz. The pulsing improved heat transfer, in some cases up to 80% (Reference 27).

Summary

Fluidized bed waste heat recovery heat exchangers are being applied in industry. These heat exchangers, while found to be very successful in applications in many countries, represent a first generation effort. There are many avenues along which the system can be improved. Research and development are now being carried out along many of these avenues, and prospects for improving the existing systems are good.

REFERENCES

1. M. Leva, <u>Fluidization</u>, McGraw Hill, New York 1959

2. S. S. Zabrodsky, <u>Hydrodynamics and Heat Transfer in Fluidized Beds</u>, MIT Press, Cambridge, 1966.

3. H. S. Mickley, D. F. Fairbanks, "Mechanism of Heat Transfer to Fluidized Beds," <u>AIChE Journal</u>, Vol. 1, 1955

4. T. F. Ozkaynak, J. C. Chen, "Emulsion Phase Residence Time and Its Use in Heat Transfer Models in Fluidized Beds," <u>AIChE Journal</u> Vol, 26, No. 4, July 1980.

5. N. S. Grewal, <u>Experimental and Theoretical Investigations of Heat Transfer Between a Gas Solid Fluidized Bed and Immersed Tubes</u>, PhD Thesis University of Chicago, 1979.

6. W. B. Krause, <u>An Investigation of the Heat Transfer Mechanisms Around Horizontal Bare and Finned Heat Exchange Tubes Submerged in an Air Fluidized Bed of Uniformly Sized Particles</u>, PhD Thesis, University of Nebraska-Lincoln 1978.

7. A. M. Xavier, <u>Heat Transfer Between a Fluidized Bed and a Surface</u>, PhD Thesis, University of Cambridge, UK 1978.

8. H. W. Williams, R. Hernandez, C. S. Mah, "Choosing the Optimum Bed Material for a Fluidized Bed Heat Exchanger," Paper 819302, <u>Proceedings of the 16th Intersociety Energy Conversion Engineering Conference</u>, Atlanta, Georgia. August 10-14, 1981.

9. G. J. Vogel, P. J. Grogan, "A Description of Emission Control Using Fluidized-Bed, Heat Exchange Technology," Argonne National Laboratory Report ANL/CNSV-TM-64, June 1980.

10. K. S. Murthi, D. Harrison, R. K. Chan, "Reaction of Sulfur Dioxide with Calcined Limestone and Dolomite," <u>Environmental Science and Technology,</u> Vol 5, Sept. 1971.

11. H. L. Faucett, J. D. Maxwell, T. A. Burnett, "Technical Assessment of NO_x Removal Processes for Utility Application," U. S. Environmental Protection Agency Report EPA-600/7-77-127, November 1977.

12. J. P. Pilney, E. E. Erickson, "Fluidized Bed Fly Ash Filter," Journal of the Air Pollution Control Association, Vol. 18, November 1968.

13. S. H. D. Lee, W. M. Swift, I. Johnson, "Screening of Granular Sorbents for the Removal of Gaseous Alkali Metal Compounds from Hot Flue Gas," Argonne National Laboratory Report ANL/CEN/FE-79-17, November 1979.

14. C. Y. Shen, H. F. Johnstone, "Gas-Solid Contact in Fluidized Beds," AIChE Journal Vol. 1, 1955.

15. N. K. Gelperin, V. G. Ainshtein, L. A. Korotyanskaya, "Heat Transfer Between a Fluidized Bed and Staggered Bundles of Horizontal Tubes," International Chemical Engineering, Vol. 9, January 1969.

16. D. R. Cusdin, M. J. Virr, "A Marine Fluidized Bed Waste Heat Boiler-Design and Operating Experience," Transactions of the Institue of Marine Engineers, Vol. 91, 1979.

17. G. J. Vogel, P. J. Grogan, A. R. Evan, "Application of Fluidized Bed Technology to the Recovery of Waste Heat," Argonne National Laboratory Report ANL/CNSV-TM-34, August 1979.

18. J. H. Mannon, "Fluidized Solids Reach for Heat Recovery Uses," Chemical Engineering, March 23, 1981.

19. "Heat Recovery Equipment," Sone-Platt Crawley Circular, 1980.

20. "Gadeluis/Fluidfire Systems & Products," Gadelius Circular, 1980.

21. H. Littman, ed., Fluidization and Fluid-Particle Systems Research Needs and Priorities, proceedings of the NSF Workshop held at Renselaer Polytechnic Institute, October 17-19, 1979, NTIS PB80207640.

22. W. Thielbahr, M. Perlsweig, "Origin and Status of the DOE Heat Exchanger Technology Program to Improve Energy Conversion Efficiency," Paper 790250 presented at the Society of Automotive Engineers Congress and Exposition, Detroit, Michigan February 26 - March 2, 1979.

23. W. E. Cole, M. Suo, "Waste Heat Recovery with Fluidized Beds," Chemical Engineering Progress, December 1979.

24. M. S. Sagoo, "The Development of a Falling Closed Heat Exchanger - Air and Particle Flow and Heat Transfer," Heat Recovery Vol. 1, No. 2, 1981.

25. R. E. Rosenweig, "Fluidization: Hydrodyanmic Stabilization with a Magnetic Field," Science, Vol. 24, 6 April 1979.

26. F. W. Staub, "Solids Circulation in Turbulent Fluidized Beds and Heat Transfer to Immersed Tube Banks," ASME Journal of Heat Transfer, Vol. 101, August 1979.

27. S. C. Bhattacharya, D. Harrison, "Heat Transfer in a Pulsed Fluidized Bed," Transactions of the Institution of Chemical Engineers, Vol. 54, 1976.

REGENERATIVE HEAT EXCHANGERS

Thermal Energy Storage and Regeneration

F. W. SCHMIDT
Mechanical Engineering Department
The Pennsylvania State University
University Park, Pennsylvania 16802, USA

ABSTRACT

There are many types of thermal energy storage devices in use today. These have been classified in three groups: regenerators, heat storage units and heat storage exchangers. Areas of application, methods available for prediction of performance and general observations are presented for each of these groups.

1. INTRODUCTION

1.1 Basic Concepts of Energy Storage

The importance of energy to the economic, social, and political well-being of the countries of the world has been well documented. It has had a profound effect on all segments of society including the individual, as well as, large energy consuming industry. A number of industrial and commercial organizations have developed energy management plans in order to reduce the cost of the energy which must be purchased. Energy conservation including waste heat recovery and the development of energy efficient machines and processes have resulted in considerable savings. Alternative sources of energy, most notably solar, are also being developed.

In many instances the available of an energy source does not coincide timewise with the demands for energy. The development of the capability to store and retrieve energy may thus be a critical component of an energy management system or a system developed to use an alternative source of energy. The energy storage devices may take many different forms.

The National Research Council of the United States undertook a study on the potential of advanced energy storage system. The results of this study (1) indicated that the ultimate decision on the installation of an energy storage system would be based on the following operating characteristics of the storage device:

1. Storage capacity
2. Storage/retrieval rate
3. Replacement lifetime
4. Weight, volume and other physical limits

5. Critical safety parameters
6. Environmental standards
7. Acceptable capital and operating costs.

The committee also noted that the areas where storage devices could be used were so varied that no specific type of device could be expected to have proper operating characteristics for all possible applications. In this paper only thermal energy storage devices will be discussed.

1.2 Classification of Thermal Energy Storage Devices

The physical locations of sources of energy and the device or process requiring energy seldom coincide and one must be concerned with the transport of energy from one location to another as well as the storage of energy. The ideal situation occurs when the energy transporting fluid also served as the storage medium. There are many examples of these types of systems. Hot water storage is used in most residential and commercial units for domestic water service. A large number of solar units use hot water for storage. Units using a fluid as both the transport and storage media are, however, restricted to low temperature applications.

Thermal energy storage devices employing steam or hot water accumulators have been used in industry in Europe for many years. A complete discussion of these systems has been presented by Goldstern (2). The first type of these units developed were essentially high pressure steam or water tanks from which the fluid was expanded as required to a lower pressure for the generation of power or for use in an industrial process. Reay (3) describes new units which can be used for both steam generation and storage. These are called "thermal storage boilers".

The thermal energy storage devices to be discussed in this paper are those in which the storage medium and the energy transporting fluids are not the same. The storage units are thus of the indirect type in which heat is transferred from the hot fluid stream to the storage material and then from the storage material to the cold fluid stream.

The following classification system for thermal energy storage devices will be used in this paper.

<u>Regenerator</u> - A thermal energy storage device in which the heat storage and heat retrieval processes are repeated in a periodic fashion. Only one of the fluid streams is in contact with the storage material at any instant of time so both the hot and cold fluids can use the same flow channels. When both the hot and cold fluid streams are flowing continuously two or more regenerators can be arranged to transfer the heat continuously from the hot fluid stream to the cold fluid stream. The system of regenerators can thus be used in place of a recuperator.

<u>Heat Storage Unit</u> - A thermal energy storage device in which there is no periodic pattern to the heat storage and heat retrieval processes. The mass flow rates, inlet fluid temperatures and duration of flow for the hot and cold fluids vary in an arbitrary timewise fashion. The same flow channels are used for both the hot and cold fluid streams.

<u>Heat Storage Exchanger</u> - A thermal energy storage device in which two or more fluid streams may be simultaneously in contact with the storage material. Each fluid stream used different flow channels. One might think of these units as thick wall exchangers. A portion of the heat removed from the hot fluid stream is transferred directly to the cold fluid stream while the

THERMAL ENERGY STORAGE AND REGENERATION

remainder is stored in the walls of the unit.

In classifying thermal energy storage devices in the manner described above, consideration was given to the design operating mode of the unit. Thus a heat storage device designed for periodic operation is called a regenerator even though during certain periods of operation, when changes in the inlet fluid conditions occur, the unit displays non-periodic operating characteristics while it adjusts to the new conditions.

1.3 Thermal Storage Materials

Most thermal energy storage devices store the energy in solids and use two or more fluid streams to transport the energy between the source, the storage material, and the energy using device. There are many factors that must be taken into consideration when selecting the storage material. A set of desirable characteristics would include the following:

1. High specific heat
2. High density
3. High thermal diffusivity
4. Reversible heating and cooling
5. Chemical and geometrical stability
6. Noncombustible, noncorrosive and nontoxic
7. Low vapor pressure to reduce the cost of containment
8. Low cost for material and containment
9. Sufficient mechanical strength to support the stacking of the storage core.
10. Proper operating temperature range.
11. Long operating life

At the present time there are three major groups of heat storage media utilized for thermal energy storage. These are nonmetals, metals and phase change materials. Firebricks formed from clay, olivine, chrome, magnesite and various mixtures of these represent an important group of nonmetals used in storage devices operating at high temperatures. In an effort to obtain a material having a high volume heat capacity Feolite composed of ferric oxide, Fe_2O_3, was developed. The commercial material, called Tenemax, utilizes low cost enriched iron ores and uses virtually standard brick making manufacturing processes.

Concrete is another nonmetal which is attracting considerable attention as a sensible heat storage material. Although its thermal characteristics are marginal, its cost and the fact that the unit can be fabricated on site are attractive factors. Architects are using concrete and other construction material in the design of buildings for passive storage of solar energy. The last nonmetal to be noted is gravel. This is used primarily in packed beds for low temperature solar energy storage.

Castable metals including gray cast irons and cast irons containing alloying ingredients such as silicon and aluminum have been used in storage devices in the past. Their advantage offered by having a very high heat storage capacity per unit volume is more than offset by there high cost.

As a result, most storage devices using these materials are no longer economically practical. The only exceptions are rotary regenerators which are fabricated using metallic matrices. The matrices are constructed using very thin metal strips and are designed to give a very large surface to volume ratio. The material which is used depends to a large extent on the application. Stainless steel is used where there is a corrosive atmosphere present.

Storage devices utilizing the phase change properties of the storage material (PCM) are receiving considerable attention at the present time. These units are of interest because they have a small temperature swing as one cycles from storage to retrieval since the major portion of the energy is stored or removed while the material is at a nearly uniform temperature. The major advantage of these units, however, is that they utilize the high heat of fusion of the storage material and thus have a very high heat capacity per unit volume or weight. The mean storage temperature can be controlled to a large extent by the selection of the storage material. A set of criteria for phase change storage materials has been presented by Lorsch (4).

Salt hydrates have very high heats of fusion. A list of some hydrates which have been used as PCM storage materials are:

	Chemical Compound	Melting Point °C	Heat of Fusion kJ/kg	Density Mg/m^3
Calcium chloride hexahydrate	$CaCl_2 \cdot 6H_2O$	29-39	175	1.63
Sodium carbonate decahydrate	$Na_2CO_3 \cdot 10H_2O$	32-36	247	1.44
Disodium phosphate dodecahydrate	$Na_2HPO_4 \cdot 12H_2O$	36	265	1.52
Sodium sulfate decahydrate	$Na_2SO_4 \cdot 10H_2O$	31	251	1.46
Sodium thiosulfate pentahydrate	$Na_2S_2O \cdot 5H_2O$	48	209	1.67

The most promising material tested todate is sodium sulfate decahydrate ($Na_2SO_4 \cdot 10H_2O$) commonly known as Glauber's salt.

Parafin waxes such as Fulfoax 33 which melts between 50-55°C and has an apparent heat of fusion of 205 kJ/kg are also attractive for certain applications. Eutectic mixtures of $NaNO_3$ or KNO_3, which have a fusion temperature of 220°C; sodium hydroxide with a fusion temperature of 315°C; lithium hydroxide, lithium hyride and lithium flouride with fusion temperature in the 1000-1700°C range are under consideration for high temperature PMC storage systems.

1.4 Design Consideration

It has previously been noted that thermal energy storage devices can be used in a large variety of areas to reduce the overall energy consumption of a system; to allow for the use of an alternative source of energy; or to allow existing systems to operate continuously at or near their design point regardless of the variation of the load placed upon it. Economic justification of the use of the storage devices is necessary. One is thus concerned with obtaining an accurate assessment of the potential cost savings which would result from the use of a storage device before making the final decision. This can only be accomplished if the performance of the storage device and perhaps the complete system can be predicted accurately. Since the performance of the storage device depends on its design it becomes necessary to conduct the preliminary design of the device and the economical feasibility study together. The various items which must be considered in the design of a thermal energy storage device will now be discussed. They will, to a large extent, be dependent upon the specific application that the storage device is to be used for.

Storage material - The importance of this item in the design of a thermal energy storage device has been properly emphasized by the discussion presented in the previous section. This item is listed only for completeness of factors which must be considered in the design of storage devices.

Pressure drop - The allowable pressure drop across the storage device will be strongly dependent on the application for which the device is intended. In many cases this is not a critical item in the design of the storage device while in other applications it represents a severe design limitation. The pressure drop is related to the properties of the fluid, the mass velocity of the fluids flowing through the channels in the storage device, the geometry of the fluid flow passages and the length of the flow channels. The magnitude of the convective heat transfer coefficient is also dependent on many of these same factors. A compromise is required since the designer ideally desires to obtain a high film coefficient and a low pressure drop. To assist him in making a decision an understanding of the influence of the convective heat transfer coefficient on the performance of the thermal storage device is required.

Configuration - The need to design a device with realistic dimensions is obvious. Special restrictions, however, are frequently imposed because of the physical location of the device or the particular system in which the storage device is to be used. The restrains usually involves the length, breath, weight or volume of the device.

Amount of available energy stored - In many installations, particularly those involving heat storage units, the hot stream is discharged to the surroundings immediately after it leaves the storage unit. Design consideration can include the maximization of the amount of energy removed from the hot stream, however, care must be taken or an extremely long unit will result.

Utilization of storage material - This is another item which may be of considerable concern to the designer of a heat storage device. A criteria which might be used for the assessment of the performance of a storage unit could be the total amount of heat stored per unit volume of storage material. The maximum amount of heat is stored when the mean temperature of the storage material is equal to the temperature of the fluid entering the storage unit. Basing the design only on obtaining the maximum energy storage per unit volume is not recommended since during the latter part of the storage process little energy is removed from the hot fluid stream. The amount of the energy available in the hot streams which is stored is thus very small.

Operating characteristics - Although there are many operating characteristics which may effect the design of the storage device, most of these are contained within the parameters used to calculate the performance of the device. They in a sense act as independent variables. An operating characteristics which is of the dependent variable type is the temperature of the cold stream leaving the storage device. A minimum acceptable temperature is usually established by the device or process which uses the retrieved energy. In many systems where heat storage devices are used, it is necessary to maintain the heated fluid at a uniform temperature. Since the temperature of the cold fluid leaving a storage device during the retrieval process will decrease as the duration of retrieval increases, a combination of several storage devices or a bypass arrangement is necessary to ensure the availability of a constant temperature fluid stream.

In an earlier section, thermal energy storage devices were classified into three types; regenerators, energy storage units and energy storage exchangers. A detailed look into each type of storage device will now be made. Particular items to be discussed include the areas of applications and methods for predicting performance.

2. REGENERATORS

2.1 Applications

A heat storage device in which the heat storage and retrieval processes are repeated in a period fashion has been classified as a regenerator. Identical fluid passages are used during the heat storage and retrieval processes. The hot and cold fluids usually flow in opposite directions (counter flow operation). Two types of regenerators are in common use. A rotary regenerator is a unit in which the storage material moves physically from one fluid stream to the other in a periodic fashion. A regenerator in which the storage material is stationary and a series of valves and ducts used to alternately direct the hot and cold streams through the storage material is called a fixed bed or fixed matrix regenerator. In nearly all the current applications using regenerators the hot and cold fluids are gases.

The primary function of the regenerator is to transfer heat from the hot fluid stream to the cold fluid stream in a periodic or continuous fashion. If the fluid inlet temperature and the flow rates of both fluid streams are relatively time invariant a recuperator could also be used to exchange heat between the two fluids. The major advantages for using a regenerator under these operating conditions include:

- A high heat transfer surface area per unit volume is obtainable.
- Only one set of flow channels are needed since the fluids flow through the same channels alternatively.
- An even distribution of pressure within the regenerator
- Counterflow of the two fluid streams can give a cleaning action thereby reducing fouling.
- Vapors which are condensed during the heat retrievel process may be vaporized during the heating process and carried away with the hot fluid stream.

When the stream passing through the flow passages of the regenerator is changed, some residual fluid is left in the passages and is mixed with the incoming fluid stream. A slight decrease in the operating efficiency of the regenerator may result. If the two streams react or if the contamination of the streams is undesirable, regenerators should not be used.

Rotary regenerators are used extensively in the electric power generation industry for air preheating. A diagram of a typical rotary regenerator for

THERMAL ENERGY STORAGE AND REGENERATION

such an application is shown in Fig. 1. The hot exhaust gases leaving the furnace are used to preheat the air supplied to the furnace for combustion. Heat transfer surface area to regenerator volume ratios of over 6000 m^2/m^3 are attainable. More limited applications of rotary regenerators are for vehicular gas turbine power plants and in cryogenic refrigeration units. A major problem encountered in the design of a rotary regenerator is the sealing of the hot and cold streams from each other.

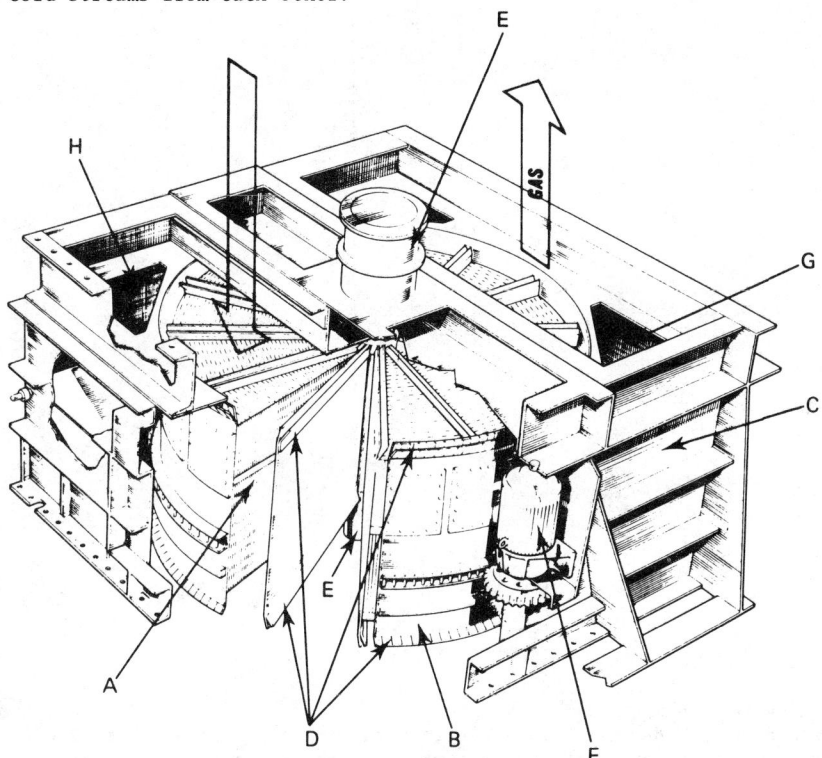

Fig. 1 Rotary regenerator. A, Heating surface elements. B, Rotor in which the elements are packed. C, Housing in which the rotor rotates. D, Seals and sealing surfaces. E, Support and guide bearing assemblies. F, Drive mechanism. G, Gas by-gas. H, Air by-pass. (Courtesy of James Howden & Co., Glasgow, Scotland.)

Fixed regenerators are used in the metallurgical, glassmaking, and chemical industries. A sketch of a fixed regenerator, Cowper stove, used for a blast furnace in the steel making industry is shown in Fig. 2. The air is preheated to a temperature of 800 - 1200°C. Regenerators used in the glass industry must be designed to withstand entrance gas temperatures of the order of 1600°C. The storage media, called checkerworks, is made of ceramic brick material.

If the hot and cold fluid streams are continuous at least two fixed bed regenerators must be used so that hot gases can pass through one regenerator while the cold gases are passing through the other regenerator. The fluid streams are then periodically switched from one regenerator to the other.

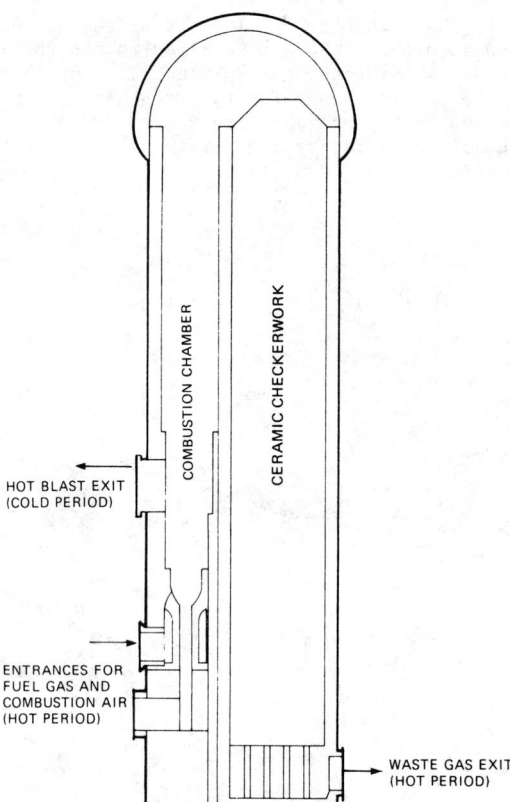

Fig. 2 Cowper stove regenerator

In many applications the temperature of the cold fluid leaving the fixed bed regenerator must be maintained within a rather narrow temperature range. There are two arrangements of two or more regenerators that can be used to satisfy this operating condition. During the retrieval process in the "series-parallel" arrangement, a portion of the cold fluid is passed around the regenerator and recombined with the portion of the fluid which went through the regenerator. The amount of cold fluid which bypasses the regenerator is controlled so that the temperature of the recombined fluid stream is held at a constant value.

In the "staggered-parallel" arrangement two regenerators are connected in parallel. One operates at a "high temperature" while the other is operating at a "low temperature". The cold stream is divided with one portion of it passing through the "low temperature" regenerator and the remainder travels through the "high temperature" unit. The temperature of the recombined stream is held within the desired temperature limit by varying the flow rates of the fluid passing through the two regenerators. Once the "low temperature" regenerator has given up most of its stored energy it is removed from the system, the "high temperature" regenerator is switched to become the "low temperature" regenerator, and a fully charged regenerator is installed as a new "high temperature" regenerator. This arrangement is similar to the

THERMAL ENERGY STORAGE AND REGENERATION

"series-parallel" arrangement except that the bypassed stream is heated by sending it through a partially discharged regenerator. A detailed discussion of these arrangements and their control is presented in Schmidt and Willmott (5).

When the hot and cold streams are not flowing at the same time a single fixed regenerator can be used. The temperature of the cold air stream can be regulated by using a bypass arrangement during the retrieval period of operation.

2.2 Mathematical Model

The performance of a regenerator is obtained by solving the mathematical model which describes the operating characteristics of the unit. This model consists of the conservation of energy equation for the fluid streams and storage material, the appropriate initial and boundary conditions, and the pressure drop equation. In most applications using regenerators the pressure drop is not a controlling design factor and since it does not directly enter into the determination of the performance characteristics of the regenerator it will not be discussed in this section.

In the derivation of the energy equation for the two fluids and the storage material the following assumptions will be made:

- The mass flow rates of the two fluids are time invariant
- The velocity and temperature of the fluids entering the regenerator are uniform over the entire flow section.
- The properties of the fluid and material are constant
- The heat transfer coefficients are uniform
- The regenerator is completely insulated from the surroundings
- The heat conduction in the fluids and storage material in the direction of fluid flow is negligible
- The thermal conductivity of the storage material transverse to the flow direction is infinite.
- Counterflow operation of regenerator

The energy equations in differential form are:

Heat Storage

Fluid stream
$$\bar{h}'A(t_m - t_f') = \dot{m}_f' c_f' L \frac{\partial t_f'}{\partial x} + m_f' c_f' \frac{\partial t_f'}{\partial \tau} \tag{1}$$

Storage material
$$\bar{h}'A(t_f' - t_m) = M_m c_m \frac{\partial t_m}{\partial \tau} \tag{2}$$

Heat Retrieval

Fluid stream

$$\bar{h}''A(t_m - t_f'') = -\dot{m}_f'' c_f'' L \frac{\partial t_f''}{\partial x} + m_f'' c_f'' \frac{\partial t_f''}{\partial \tau} \qquad (3)$$

Storage material

$$\bar{h}''A(t_f'' - t_m) = M_m c_m \frac{\partial t_m}{\partial \tau} \qquad (4)$$

The single prime indicates hot stream and the double prime indicates the cold stream. At reversal the temperature distribution in the storage material at the start of the new period is assumed to be identical to that at the end of the old period. A very complete description of the various methods which can be used to solve the set of equations has been presented by Willmott (5) and Hausen (6). In addition, several survey papers which deal with the prediction of regenerator performance have been recently published by Hausen (7), Razelos (8) and Shah (9).

The results of the solution to the above set of equations are expressed in terms of dimensionless parameters. Two different sets of variables have been used. The most common group was that proposed by Hausen and will be referred to as the $\lambda - \pi$ group. A less commonly used grouping involves the NTU and effectiveness. This is used primarily when working with rotary regenerators. The $\lambda - \pi$ group will be used in this discussion although the two groupings can be shown to be equilivant (7).

The dimensionless parameters to be used are:

	hot period	cold period
Reduced length	$\Lambda' = \bar{h}'A/\dot{m}_f' c_f'$	$\Lambda'' = \bar{h}''A/\dot{m}_f'' c_f''$
Reduced time	$\pi' = \bar{h}'A(P' - m_f'/\dot{m}_f')/M_m c_m$	$\pi'' = \bar{h}''A(P'' - m_f''/\dot{m}_f'')/M_m c_m$
Temperature	$T_f = \dfrac{t_f - t_{fi}''}{t_{fi}' - t_{fi}''}$	$T_m = \dfrac{t_m - t_{fi}''}{t_{fi}' - t_{fi}''}$

The thermal ratio or effectiveness is defined as the ratio of the actual amount of heat stored (or retrieved) during the period to the thermodynamically maximum amount of heat which could be stored (or retrieved) during the period. If the time average exit temperatures are defined as $\bar{t}_{fo}'$ and $\bar{t}_{fo}''$ the expressions for the thermal ratio reduce to

$$\eta'_{REG} = \frac{t_{fi}' - \bar{t}_{fo}'}{t_{fi}' - t_{fi}''} \quad \text{and} \quad \eta''_{REG} = \frac{\bar{t}_{fo}'' - t_{fi}''}{t_{fi}' - t_{fi}''}$$

The heat stored will equal the heat retrieved in a regenerator operating in a periodic fashion. If the heat ratios of the hot and cold period are equal, $\eta'_{REG} = \eta''_{REG}$, the regenerator is said to be balanced. It can be shown that for a balanced regenerator,

THERMAL ENERGY STORAGE AND REGENERATION

$$\frac{\pi'}{\pi''} = \frac{\lambda'}{\lambda''} = k$$

If the reduced lengths and periods are equal, k = 1, the regenerator is further classified as being symmetric. The relationship between the reduced lengths period and thermal ratio for a symmetric counterflow regenerator is given in Fig. 3.

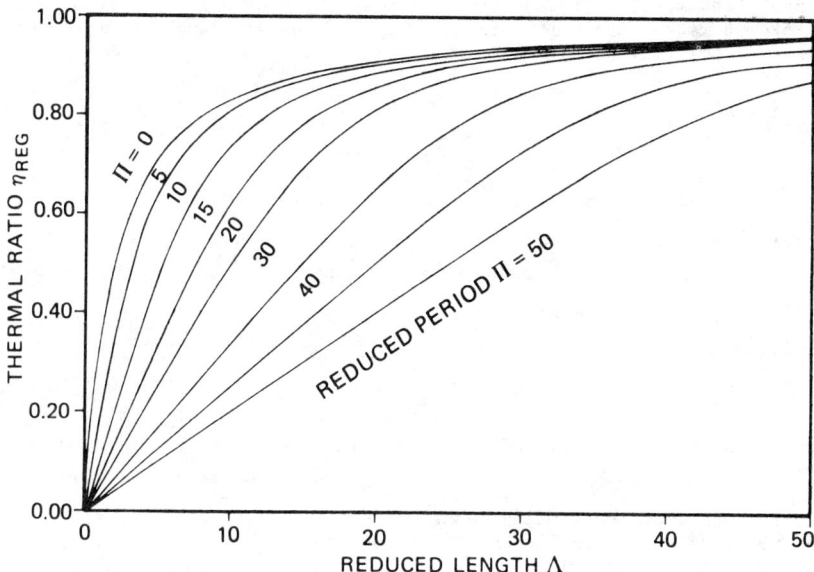

Fig. 3 Graph of thermal ratio in relation to reduced length and reduced period for symmetric counterflow regenerators.

If the regenerator is balanced but unsymmetric, Hausen has suggested that the performance of the regenerator can be approximated under certain conditions ($3 < \lambda'' < 18$ and $3 < \pi'' < 18$) by an equivalent symmetric regenerator using the harmonic mean length and period defined as

$$\frac{1}{\lambda_H} = \frac{1}{2\pi_H} (\frac{\pi'}{\lambda'} + \frac{\pi''}{\lambda''})$$

and

$$\frac{1}{\pi_H} = \frac{1}{2} (\frac{1}{\pi'} + \frac{1}{\pi''})$$

If the regenerator is unbalanced or balanced and unsymmetric, but falls outside the variable limits previously stated, the reader should consult references (5-9) for a detailed description of technique which may be used to obtain the operating characteristics of the regenerator. Some results are also given in these references.

Several restrictive assumptions were made in the formulation of the mathematical model described by eqs. 1-4. The most questionable assumption may be that which considers the thermal conductivity of the storage material to be infinite in the direction perpendicular to the flow. The resistance to the transfer of heat within the storage material may be of the same order as that offered at the fluid boundary for combinations of a high convection heat transfer coefficient, thick regenerator wall and low storage material thermal conductivity. The computer program developed by Willmott (10) can accommodate these conditions. In many instances, however, the concept of a lumped heat transfer coefficient first proposed by Hausen gives a reasonable result. For a regenerator with storage walls of semi-thickness w, the lumped heat transfer coefficient is

$$\frac{1}{\bar{h}} = \frac{1}{h} + \frac{w}{3k_m} \phi$$

where

$$\frac{w^2}{\alpha_m}\left(\frac{1}{P'} + \frac{1}{P''}\right) < 5 \qquad \phi = 1 - \frac{w^2}{15\alpha_m}\left(\frac{1}{P'} + \frac{1}{P''}\right)$$

$$\frac{w^2}{\alpha_m}\left(\frac{1}{P'} + \frac{1}{P''}\right) > 5 \qquad \phi = \frac{2.142}{\sqrt{0.3 + 4w^2\,(1/P' + 1/P'')/2\alpha_m}}$$

The lumped heat transfer coefficient is used in determining the reduced length and period.

The influence of longitudinal heat conduction, carry-over leakage, variable physical properties, and radiation effects on the performance of regenerators have been reported (5-9). The transient performance of these units have also been reported in these references.

2.3 Observation

To predict the performance of a regenerator the following information is needed:

- A correlation for the calculation of the convective heat transfer coefficients. The factors which must be taken into consideration include the mass flow rates and properties of the fluid streams and the geometrical cross-section, surface roughness and length of the flow channels. Radiation heat transfer effects are frequently reflected in the value used for the heat transfer coefficient. If necessary the lumped heat transfer coefficients may be calculated to account for the internal thermal resistance to the transfer of heat offered by the storage material.

- The relationship between the regenerator's operating parameters, reduced length and period for the hot and cold streams, and the thermal performance of the regenerator. These relationships were discussed in the previous section.

The pressure drop across the regenerator can be calculated by taking into account the same items considered in the evaluation of the heat transfer coefficient plus the entrance and discharge nozzles of the regenerator.

When a number of regenerators are connected in "series-parallel" or "staggered-parallel" arrangements the analysis must consider the complete system. The mass flow rate of the cold fluid stream will be time dependent. In the "staggered-parallel" arrangement both the mass flow rate and inlet fluid temperatures of the cold streams will be time dependent. The performance of the complete arrangement is best obtained by the simultaneous simulation of each individual regenerator using numerical techniques.

The thermal performance curve, Fig. 3, allows one to calculate the amount of heat removed from the hot fluid and that picked up by the cold fluid. With this information it is possible to determine the time averaged temperatures of the hot and cold streams leaving the regenerators. Although the amount of heat retrieved is important in many applications, the minimum temperature of the hot stream leaving the regenerator is also of considerable interest. This information has been determined for a balance symmetric regenerator and is shown in Fig. 4.

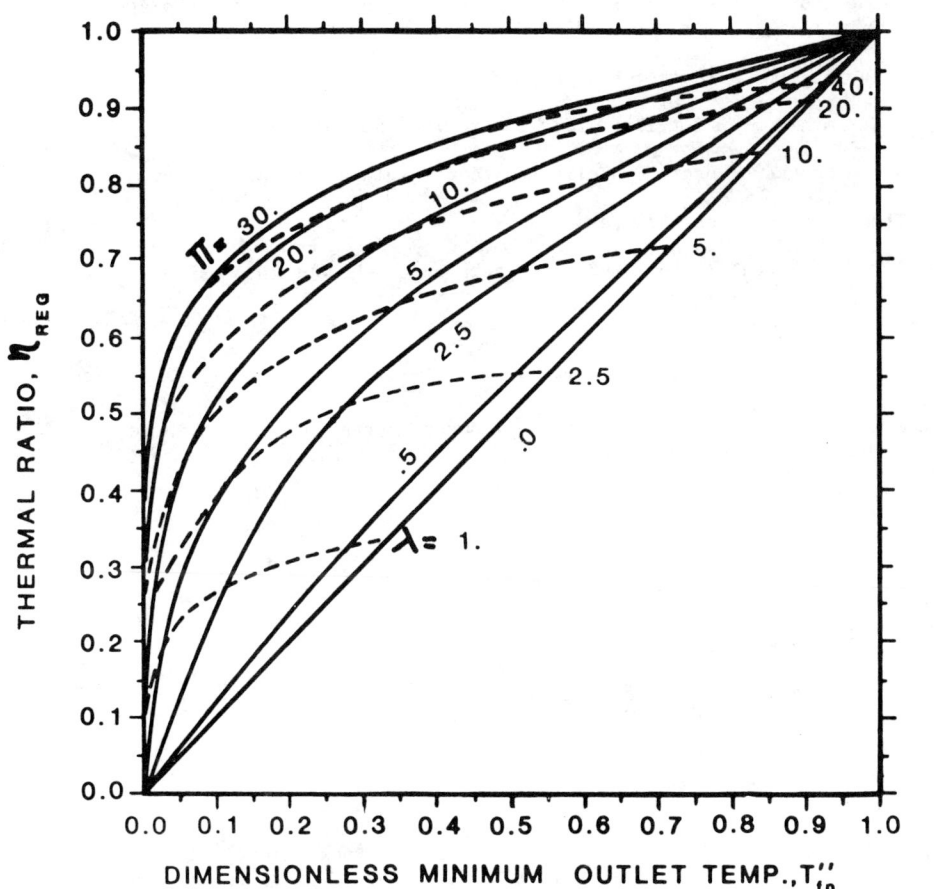

Fig. 4 Graph of thermal ratio in relation to reduced period, reduced length and minimum cold fluid outlet temperature for symmetric counterflow regenerators.

Although for a given value of the reduced length and period double interpolation is necessary, the possibility of obtaining both the thermal ratio and the dimensionless outlet fluid temperature at the end of the cold period is attractive.

An explicit design procedure can be used which will allow a regenerator for a particular application to be sized directly. To illustrate the procedure, a balanced-symmetric regenerator is to be designed. The following operating characteristics of the unit are specified:

- Inlet fluid temperatures and mass flow rates for the hot and cold fluid streams,

- The duration of the hot and cold periods,

- The physical properties of the fluids and the thermal storage material,

- The geometrical cross-section of the flow channels in the storage material (type of checkers),

- The maximum allowable pressure drop,

- The minimum outlet fluid temperature for the cold fluid stream,

- The amount of heat, Q, delivered to the cold stream per cycle.

The following steps are taken:

Step 1 - The thermal ratio for the regenerator can be obtained using

$$\eta_{REG} = \frac{Q}{\dot{m}'_f \cdot c'_f (t'_{fi} - t''_{fi}) P'}$$

Step 2 - The appropriate values of the reduced length and period can be determined using Fig. 4.

Step 3 - The mass of the storage material is obtained

$$M_m = \frac{\lambda}{\pi} \frac{Q}{c_m (t'_{fi} - t''_{fi}) \eta_{REG}}$$

Step 4 - The regenerator porosity, ε_R, which is defined as the cross-sectional area of the flow channels per unit total frontal area of the checker pattern can be calculated. The heat transfer coefficient can be found directly from

$$h = \frac{1 - \varepsilon_R}{4\varepsilon_R} \pi' \frac{\rho_m c_m D_h}{P'}$$

Step 5 - The correlation between the heat transfer coefficient and the mass velocity of the fluid is used to determine the mass velocity of the flow through the regenerator.

Step 6 - The expression for the pressure drop is used to determine the length of the regenerator.

THERMAL ENERGY STORAGE AND REGENERATION

Step 7 - The total frontal area of the regenerator is determined using

$$S_{fr} = \frac{M_m}{\rho_m (1 - \varepsilon_R)}$$

3. HEAT STORAGE UNITS

3.1 Applications

In many applications only one fixed thermal energy storage device is used and the hot and cold fluids pass alternately through the unit. The mass flow rates and the inlet temperatures of the two fluids may vary timewise in an arbitrary manner. There may even be certain periods of time when neither fluid is flowing through the storage unit. A thermal energy storage device operating under these conditions has been classified as a heat storage unit.

A heat storage unit is a necessary component in an active solar energy system. The energy is collected during the daytime and is transported by the hot fluid to the storage unit. If the transporting fluid is water, it most probably will be used as the energy storage material. A sensible heat or phase change materials will be used to store the energy when the transporting fluid is air. A schematic sketch of a solar energy system is shown in Fig. 5. The most common type of a storage unit found in such systems is a packed rock bed although other bed configuration are also used. In order to make solar systems more economically feasible the possibility of utilizing the structural elements of the building for storage purposes has been advocated (12).

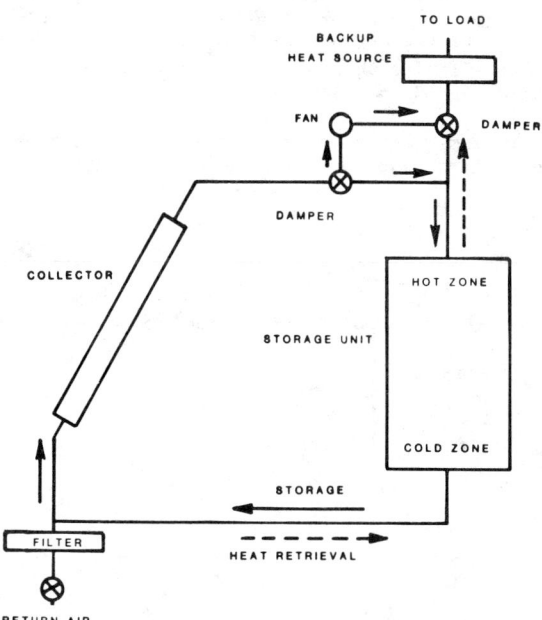

Fig. 5 Typical air based solar space-heating system.

Waste heat recovery systems for industries which utilize the heat supplied by a batch type process is another application area where heat storage units can be used. The energy transporting fluid can be either a gas or a liquid. The transporting fluid depends to a large extent on the operating range of the process supplying the heat. The storage units are usually constructed in a manner quite similar to that used for fixed bed regenerators. Either parallel or counter flow arrangements are used for the fluid streams.

Another application of a heat storage unit is in the design of a "blow down" facility for aerodynamic research. A domestic hot air heating system was recently proposed which uses a heat storage unit to allow the burner to operate continuously for a longer duration at its most efficient condition (13). The unneeded energy is stored for later use. A bypass arrangement was used to maintain the air used for space heating at a uniform temperature. Considerable savings in operating and maintenance cost can be realized by this system because it minimizes the frequent start-stop mode of operation normally used.

3.2 Mathematical Model

More difficulties are encountered in predicting the performance of heat storage units because of the large number of combinations of mass flow rate and inlet fluid temperature which exist for each fluid stream. Two approaches can be used to obtain the desired results. The energy equations for the fluid streams and storage material are linear. A solution to the fundamental problem, to be noted as the "single blow" problem, can be obtained. Superposition techniques can then be used to account for arbitrary timewise variations in inlet fluid temperature and mass flow rates.

The second approach involves the numerical simulation of the storage unit. Willmott (5) has described a method for the simulation of a regenerator which can be used for the heat storage unit. A numerical method which considers the storage unit to have a finite thermal conductivity in all directions has been described by Schmidt and Szego (14).

The fundamental problem, namely the "single blow" problem, will now be discussed. A storage unit composed of a series of flat slabs, Fig. 6, will be considered.

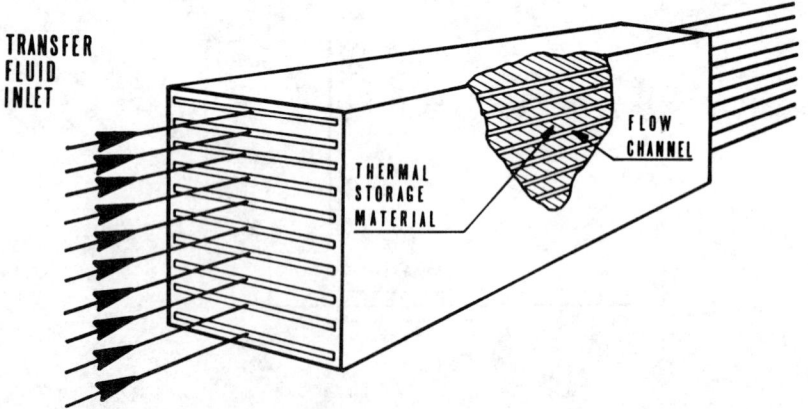

Fig. 6 Flat slab heat storage unit

THERMAL ENERGY STORAGE AND REGENERATION

The following assumptions will be made:

- constant fluid and material properties
- uniform heat transfer coefficient
- step change in inlet liquid temperature
- uniform initial temperature distribution in the storage material
- uniform fluid velocity
- negligible heat loss to the surroundings.

Four mathematical models can be considered for the solution of this problem.

Infinite fluid heat capacity - It is assumed that the temperature of the fluid remains nearly constant as it passes through the storage unit. This will be strictly true only if the fluid undergoes a change in phase as it passes through the unit. There are, however, occasions when the heat capacity of the fluid is sufficiently high to allow this model to be used. If in addition the temperature gradients within the storage material can be neglected, the model is referred to as the lumped parameter model.

Simplified - The thermal conductivity of the storage material in the direction of flow is zero while it is assumed to be infinite in the transverse direction. These assumptions are identical to those used in the development of the regenerator model.

Finite Conductivity - The thermal conductivity of the storage material is considered to be finite in all directions. This is the most accurate model of the storage unit.

A comparison of the three methods for the flat slab heat storage unit under "single blow" operating conditions has been presented by Szego and Schmidt (15). The regions where the simplest possible model that could be used without introducing appreciable errors are shown in Fig. 7. The Biot number is defined in terms of the semi-thickness of the flat slab, w,

$$Bi = \frac{hw}{k_m}$$

The dimensionless length and time are defined in a manner similar to that used for regenerators

$$\lambda \equiv \frac{hA}{\dot{m}_f c_f} \quad \text{and} \quad \eta \equiv \frac{hA(\tau - x/v)}{S_m L \rho_m c_m}$$

The solutions for the infinite fluid heat capacity model are:

$$Bi \leq 0.1 \qquad\qquad Bi > 0.1$$

storage material temperature

$$T_m = 1. - \exp(-\eta) \qquad T_m = 1. - \sum_{j=1}^{\infty} 2\frac{\sin M_j}{\sin M_j \cos M_j + M_j} \exp[-(M_j^2)\frac{\eta}{Bi}] \cos(M_j Y)$$

$$\text{where } M_j \tan M_j = Bi$$

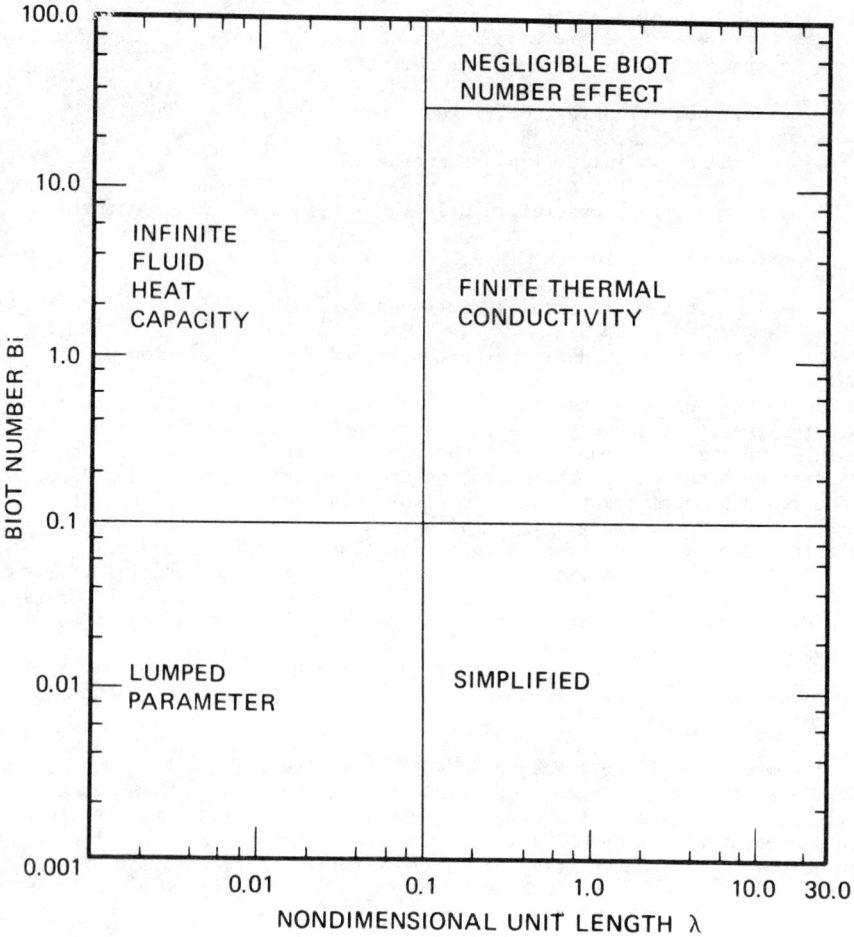

Fig. 7 Applicable solution regions for flat slab heat storage unit under "single blow" operating conditions.

heat storage

$$Q^+ = 1. - \exp(-\eta) \qquad Q^+ = \sum_{j=1}^{\infty} 2\left[\frac{\sin^2 M_j}{M_j \sin M_j \cos M_j + (M_j^2)}\right]\left[1. - \exp\left[-(M_j^2)\frac{\eta}{Bi}\right]\right]$$

The dimensionless fluid outlet temperature is one for both cases.

In Fig. 8 and 9, the results for the simplified model are presented. The dimensionless heat storage is defined as

$$Q^+ \equiv \frac{Q}{Q_{max}} = \frac{t_m - t_o}{t_{fi} - t_o}$$

where t_o is the initial temperature distribution in the storage material and

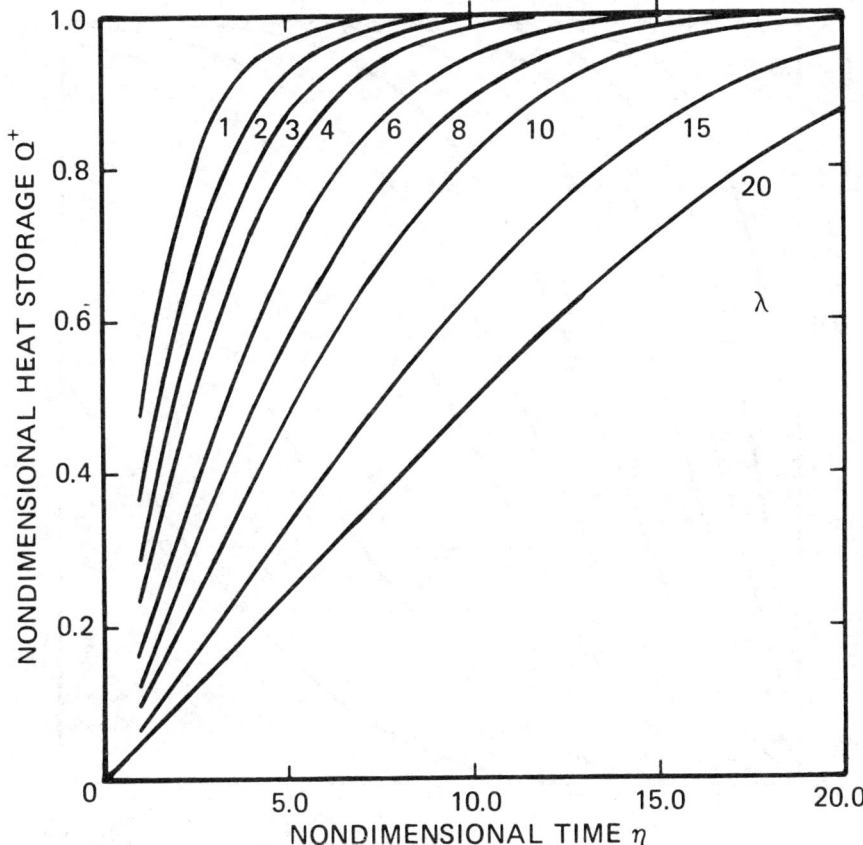

Fig. 8 Nondimensional heat storage using simplified model.

T_m is the dimensionless temperature,

$$T_m \equiv \frac{t_m - t_o}{t_{fi} - t_o}$$

The solutions for the finite conductivity model are not easily presented but they may be found in reference 14. Please note that the geometrical configuration of the heat storage unit directly effect the calculations only when using the finite conductivity model of the heat storage unit.

The simplified model may be used to predict the performance of a packed bed. The dimensionless time is defined as,

$$\eta = \frac{hA(\tau - x/v)}{S_{fr}L(1-\epsilon)\rho_m c_m}$$

while the same definition is used for the dimensionless length.

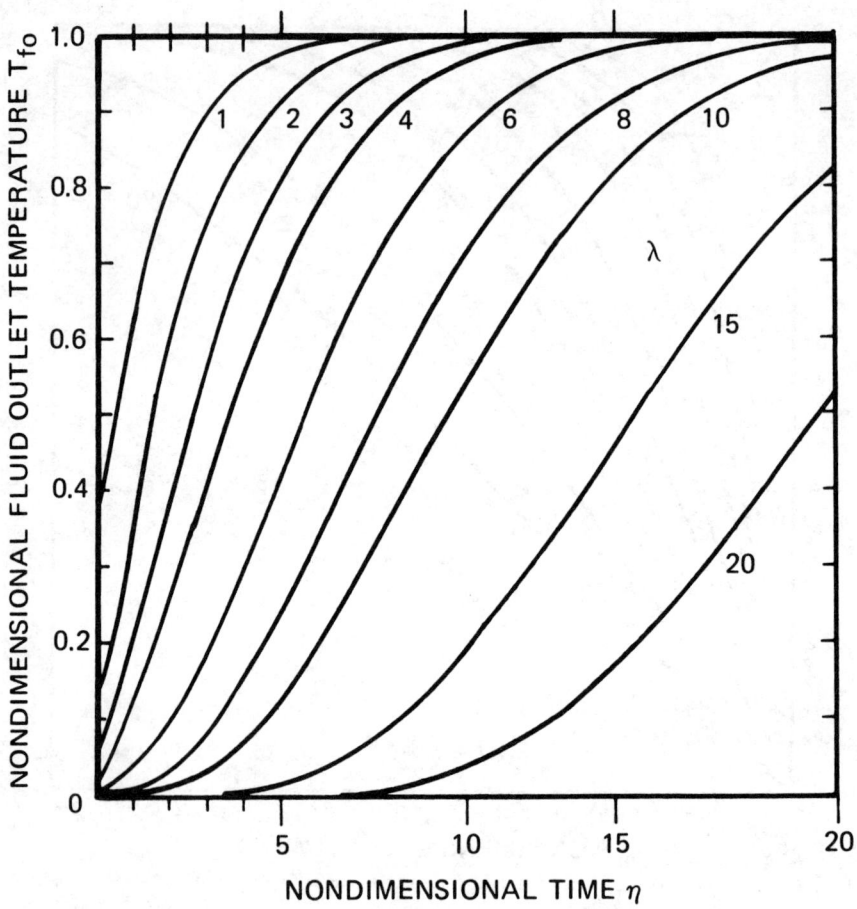

Fig. 9 Nondimensional outlet fluid temperature using simplified model.

The results presented above can be applied for either heat storage or retrieval. The only restriction is that the storage material must be at a uniform temperature and experience a step change in the inlet fluid temperature.

Superposition techniques are used when the inlet fluid temperature experience timewise arbitrary variations in fluid mass flow rate or inlet fluid temperature. They also are used when the initial temperature distribution within the storage material is not uniform. These techniques are described in detail in reference 5.

A very limited amount of work has been published which deals with the prediction of the performance of thermal energy storage units using phase change materials. Three of the most recent papers are those by Green and Vliet (16), Smith, Ebersole and Griffin (17) and Shamsunder and Srinivasain (18).

THERMAL ENERGY STORAGE AND REGENERATION

3.3 Observation

The design of a thermal storage unit is very difficult because the unit is required to perform under a great variety of operating conditions. As an example consider a unit for an active solar energy system. At the present time the person designing a storage unit for a solar energy system is give very little guidance. One of the most popular method for sizing a solar system in the United States is the "f" chart method (19). The size of the storage unit is related to the collector surface area with a storage volume of 0.25 m^3/m^2 of collector surface area recommended. The mass flowrate of the air used to transport heat from the collector to the storage unit is 0.01$m^3/s/m^2$ of collector surface area. No guidelines are given for length and frontal area. Even less information is given in a design and installation manual developed specifically for thermal energy storage units by Cole et al. (20). Kulakowski and Schmidt (21) recently reviewed the design procedure for a storage unit for a solar system. They have shown that a considerable reduction in the required bed volume can be obtained using a closed fluid system for collector-storage and a bypass system for the heat retrieval process.

It is very clear that much more work is needed to generate simple yet accurate information to assist those constructing solar energy systems which utilize thermal storage units.

4. HEAT STORAGE EXCHANGERS

4.1 Application

In energy management system there may be many occasions when it is necessary to work with several fluid streams which may, on an intermittent basis, result in heat being added and withdrawn from storage at a given instant of time. If the fluid are different, liquid and gas, the design of the storage system may get quite complicated with heat exchangers, as well as, storage units used. An example, the waste heat available in the exhaust gases of an industrial process may be used at a later time for domestic hot water and space heating. This could be accomplished by using a heat storage exchanger. The major advantage of a heat storage exchanger is that a number of liquid and gas streams can be used simultaneously for intermittently. If the fluids are flowing simultaneously the heat can be transferred directly between the streams. Any excess heat can be stored for future use. Examples of thermal storage exchangers are shown in Fig. 10.

4.2 Mathematical Model

The mathematical model for the storage unit shown in Fig. 10a has been developed by Szego and Schmidt (22). The unit has two fluid streams. The finite thermal conductivity model is used. The results for particular sets of operating condition have been presented in ref. 5 or 22. In presenting the results, a large number of dimensionless parameters were required. Superposition techniques can be used for arbitrary timewise variations in inlet fluid temperature.

If the two fluid streams are flowing continuously but have time varying fluid inlet conditions the techniques used for the transient analysis of heat exchangers can be used. These have been summarized by Schmidt (23).

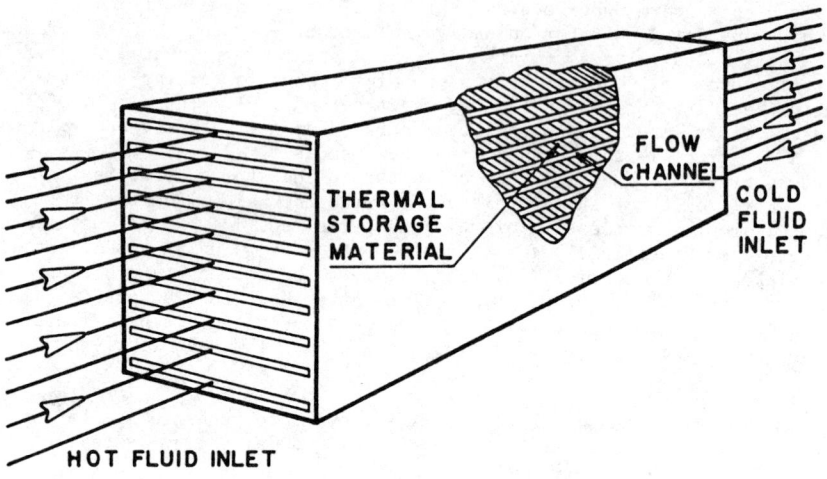

(a) Flat slab

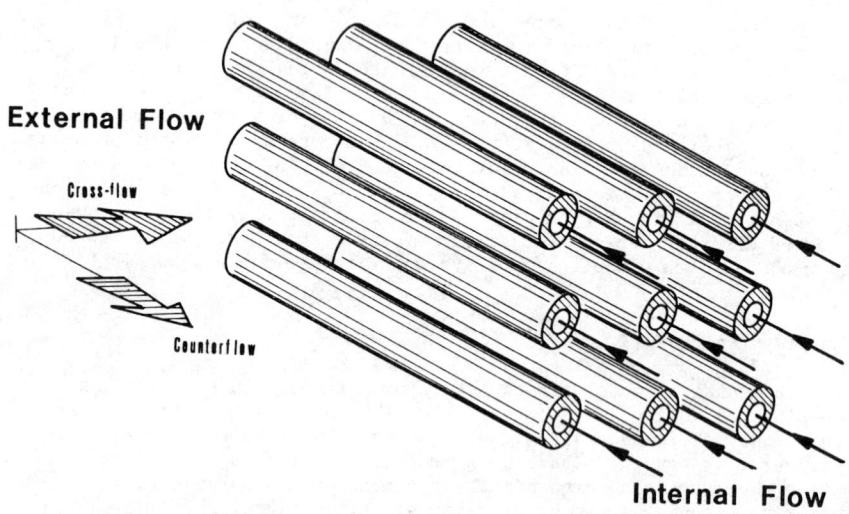

(b) Shell and tube

Fig. 10 Heat storage exchangers

4.3 Observation:

It is difficult to make any definition observations other than to note that these types of storage devices are not in widespread use today. The potential advantages to be gained by there use has been demonstrated and can be appreciable. One must wait to see if the value of potential savings obtained using these units will make them more popular. While some solutions are available in graphical form, numerical simulations may prove to be the best approach for predicting the performance of these units because of the varying operating conditions they are exposed to.

5. CONCLUSIONS

A great deal of information has been written about the use of thermal energy storage devices to assist in the more efficient use of existing energy sources and to enable us to develop new energy systems using replenishable energy sources. In spite of all these articles and reports, the increased usage of thermal energy storage devices by industry has been almost negligible.

The major industrial applications of thermal energy storage devices, regenerators, is as an alternative to recuperators. The regenerators have a high heat transfer surface area to unit volume, relative even internal pressure distribution and more favorable fouling characteristics due to the cleaning action of the gas which alternate in their flow direction. Regenerators are used primarily in the steel making and glass industries. The techniques for predicting the performance of these units was developed in Germany in the late 1920's. They have been refined and modernized to take advantage of the computation capabilities offered by high speed digital computers. The techniques for the design, construction and control regenerators are well documented. Since the world's industrial growth in the industries which use regenerators has slowed somewhat during the past few years, there has not been a noticable increase in the industrial use of regenerators.

The major activity in the past several years which has created an increase in the awareness of the importance of thermal energy storage has been associated with the rapid growth of solar energy systems. While a great amount of research has been devoted to understanding and improving the performance of solar collectors, relatively little significant work has been done in optimizing the design and control of the storage units. This is not an easy task since the complete solar system must be considered and many factors which directly influence the performance of the storage unit are time dependent with different time scales which extends from minutes to months.

Little adequate design information is currently available. The techniques for the predicting of the performance of these units are well known and accurate results are obtained when the operating conditions of the unit are specified. The primary user of the design information is the small local contractor who constructs these units for residential and commercial building. Generally they do not have large computers available and must rely on charts or small programable calculators to design the storage unit. The development of simple yet accurate design procedures for the storage units is a very necessary step if economically feasible solar systems are to be constructed.

A very small realization of the potential offered by energy storage has been achieved in energy and waste heat management. The reason for our failure to take advantage of the possible savings that are possible is not clear. It may in part be associated with our unawareness of the potential and thus our failure to seriously consider energy storage in the preliminary design and economic feasibility studies of new energy consuming and producing systems. We must realize that in many cases the saving that are possible through the use of an energy storage device do not justify the expense involved in designing and constructing the storage unit.

In nearly all thermal energy storage devices in use today, the fluid transporting the energy to and from the storage device is a gas. The heat storage exchangers described in this paper are ideal storage devices for handling two or more streams which may be liquids as well as gases. We can predict their performance accurately enough to allow a careful economic evaluation of a system containing these types of units to be made. The current interest in the optiminization of large chemical processing plants involving large numbers of heat exchangers might realize greater flexibility and improved performance through the utilization of energy storage exchangers.

In summary although it is an accepted fact that thermal energy storage devices have potential for enabling one to achieve a reduction in the energy consumption of a large number of industries, relative little use of these devices can be found. The capability for predicting the performance of these devices is available. The engineering community must become more concious of the possible benefits to be achieved through the use of storage devices and actually perform feasibilities studies to see if systems employing these devices are economically justified.

6. ACKNOWLEDGEMENTS

The author wishes to acknowledge the increased awareness of the potential uses for regenerator and their performance predictions obtained through close work with Dr. A. J. Willmott, University of York. His graditude is also extended to Dr. B. Kulakowski for his review of this manuscript and the generation of the data needed for Fig. 4.

7. NOMENCLATURE

A	Heat transfer surface area, m^2
Bi	Biot number, hw/k_m
c	Specific heat at constant pressure, kJ/kg K
D_h	Hydraulic diameter, m
h	Convective film coefficient, $W/m^2 °C$
$\bar{h}$	Overall heat transfer coefficient, $W/m^2 K$
k	Thermal conductivity, $W/m°C$
L	Length of unit, m
M	Total mass of storage material, kg
m_f	Mass of fluid in storage channels, kg
$\dot{m}_f$	Mass rate of flow, kg/s
P	Duration of hot or cold period for regenerator, s
Q	Total heat stored, kJ
Q_{max}	Maximum heat storage, kJ

THERMAL ENERGY STORAGE AND REGENERATION

Q^+	Nondimensional heat storage
S	Cross-sectional area, m^2
T	Nondimensional temperature
t	Temperature, °C, K
v	Fluid velocity, m/s
w	Semithickness of storage material for heat storage units
x	Axial coordinate, m
Y	Nondimensional transverse distance, y/w

Greek

α	Thermal diffusivity, m^2/s
ε	Porosity of packed bed
ε_R	Porosity of regenerator
η	Nondimensional time of heat storage unit
η_{REG}	Thermal ratio, dimensionless
Λ	Reduced length $hA/\dot{m}_f c_f$
Λ_H	Harmonic mean reduced length
λ	Nondimensional length of heat storage unit
π	Reduced period, $\overline{hA}(P-m_f/\dot{m}_f)/M_m c_m$, dimensionless
π_H	Harmonic mean reduced period
ϕ	Overall heat transfer correction factor
τ	Time, s

Subscripts

f	Fluid
fi	Fluid at entrance to unit
fo	Fluid at exit
fr	Frontal cross-sectional area of packed bed
m	Storage material
o	Initial condition

Superscripts

'	Hot period for regenerator
''	Cold period for regenerator
-	Time averaged

REFERENCES

1. Committee on Advanced Storage Systems, <u>Criteria for Energy Storage R & D</u>, National Academy of Science, Washington, D.C., 1976.

2. W. Goldstern, Steam Storage Installations, Pergamon Press, Oxford, 1970.

3. D. A. Reay, Industrial Energy Conservation, Peragamon Press, Oxford, 1977.

4. H. G. Lorsch, K. W. Kauffman and J. C. Denton, Thermal Energy Storage for Heating and Air Conditioning, Future Energy Production Systems, Vol. 1, Academic Press, New York, 1976, P. 69.

5. F. W. Schmidt and A. J. Willmott, Thermal Energy Storage and Regeneration, McGraw-Hill, NY, 1981.

6. H. Hausen, Warmeübertragung in Gegenstrom, Gleichstom und Kreuzstrom, Springer, Berlin, 2nd ed., 1976.

7. H. Hausen, Developments of Theories on Heat Transfer in Regenerators, Compact Heat Exchangers - History, Technological Advancement and Mechanical Design Problems, HTD-Vol. 10, ASME, 1980.

8. P. Razelos, History and Advancement of Regenerator Thermal Design Theory, Compact Heat Exchanger-History, Technological Advancement and Mechanical Design Problems, HTD-Vol. 10, ASME, 1980.

9. R. K. Shah, Thermal Design Theory for Regenerators, Heat Exchangers-Thermohydraulic Fundamentals and Design, edited S. Kakac, A. E. Bergles and F. Maginger, Hemisphere Publishing Corp., New York, 1981.

10. A. J. Willmott, The Regenerative Heat Exchanger Computer Representation, Int. J. Heat Mass Transfer, Vol. 12, 1969, pp. 997-1014.

11. H. Hausen, Vervollstandigte Berechnung des Warmeaustausches in Regeneratoren, Z. Ver. Deutsch. Ing., Beiheft Verftk No. 2, 1942, pp. 31-43.

12. F. W. Schmidt, Prediction of the Transient Response of Concrete Structural Building Elements used for Thermal Energy Storage, Solar Energy Storage Options, NFIF Conf.-790328-P2, pp. 391-400, 1979.

13. B. T. Kulakowski and F. W. Schmidt, Discrete Control Algorithm for a Heat Storage System, J. Dynamic Systems, Measurement and Control, Trans. ASME, Vol. 102, pp. 226-232, 1980.

14. F. W. Schmidt and J. Szego, Transient Response of Solid Sensible Heat Thermal Storage Units-Single Fluid, J. Heat Transfer, Trans. ASME, Vol. 98 p. 471, 1976.

15. J. Szego and F. W. Schmidt, Analysis of the Effects of Finite Conductivity in the Single Blow Heat Storage Unit, J. Heat Transfer, Trans. ASME, Vol. 100, p. 740, 1978.

16. T. F. Green and G. C. Vliet, Transient Response of a Latent Heat Storage Unit: An Analytical and Experimental Investigation. ASME paper # 79-HT-36, 1979.

17. R. N. Smith, T. E. Ebersole and F. P. Griffin, Heat Exchanger Performance in Latent Heat Thermal Energy, J. Solar Energy Engr., Trans. ASME, Vol. 102, pp. 112-118, 1980.

18. N. Shamsunder and R. Srinivasan, Effectiveness NTU Charts for Heat Recovery from Latent Heat Storage Units, J. Solar Energy Engr., Trans. ASME, Vo. 102, p. 263-271, 1980.

19. W. A. Beckman, S. A. Klein and J. A. Duffie, Solar Heating Design, Wiley Interscience, New York, 1977.

20. R. L. Cole, K. J. Nield, R. R. Rohde and R. M. Wolosewicz, Design and Installation Manual for Thermal Energy Storage, Argonne National Laboratory, ANL-79-15, Feb. 1979 (2nd ed. Jan. 1980.)

21. B. Kulakowski and F. W. Schmidt, Design of a Packed Bed Thermal Storage Unit for a Solar System, submitted to J. Solar Energy Engineering, Trans. ASME, 1981.

22. J. Szego and F. W. Schmidt, Transient Behavior of a Solid Sensible Heat Thermal Storage Exchangers, J. Heat Transfer, Trans. ASME, Vol. 100, p. 148, 1978.

23. F. W. Schmidt, Numerical Simulation of the Thermal Behavior of Convective Heat Transfer Equipment, Heat Exchanger: Design and Theory Sourcebook, N. Afgan and E. U. Schlunder (eds), Hemisphere, Washington, D.C., p. 491, 1974.

On the Thermal Characteristics and Response Behavior of Residential Rotary Regenerative Heat Exchangers

M. H. ATTIA and N. S. D'SILVA
Tribology and Mechanical Processes Group
Mechanical Research Department
Research Division, Ontario Hydro
Ontario, Canada

ABSTRACT

A mathematical model, based on the finite difference approximation has been developed for computer simulation of the thermal response behaviour of an air-to-air rotary heat exchanger. This model recognizes the nonlinearity of the system associated with the mutual coupling between the heat and mass transfer processes. The effect of the circumferential heat conduction in the solid matrix is introduced. This model is capable, therefore, of portraying the two-dimensional quasi-steady state temperature field in the matrix as well as the temperature profile of air streams. Frost formation, which constitutes a serious operational concern, can also be predicted. The results indicate the significance of the effect of both the system nonlinearity and the circumferential heat conduction on the thermal characteristics of the regenerator.

1. INTRODUCTION

The rotary air-to-air heat exchanger has potential advantages as an energy conservation device, which alternately stores and releases energy from and to two different air streams. It consists of a heat storing solid honeycomb matrix which comprises a very large number of air flow passages and is, thus, distinguished by its large heat transfer area per a unit volume. Through its continuous rotation, the matrix is exposed periodically to a hot and a cold air stream which flow countercurrently and sequentially through the same passages.

Recently, this device has been adopted in Ontario Hydro, Canada, for applications related to heat recovery in residential ventilation system designed for air-tight homes [1]. In such as system, heat is exchanged between the air exhausted at room temperature and the supply air at outdoor temperature.

Under certain weather conditions, experimental evidence has indicated that condensation and frosting may occur in the solid matrix. The mechanism of frosting or icing is of a complex nature as it is governed by numerous interacting variables. Obviously, frosting may reduce the flow and in general terms, the performance of the heat exchanger. It is the aim of the present paper to develop a mathematical model to investigate the thermal response behaviour of the axial flow type, rotary air-to-air heat exchanger for counterflow arrangement and to predict frost formation.

Previously published studies under steady state [2-6] or transient [7,8] conditions are, to a certain degree, of limited application as they do not offer necessary information to predict the temperature distribution and frost formation within the solid matrix. Lambertson et al [2], and Schalkwijk [3] have studied the idealistic case of zero thermal conductivity and finite periodicity. On the other hand, Hahnemann's solution, cited in [4], covered the case the case of finite thermal conductivity and infinite rotational speed. Later in 1964, Bahnke et al [5] introduced the effect of longitudinal heat conduction on the performance of the heat exchanger. In the investigation carried out by Holmberg [6], both the heat and mass recovery have been accounted for. As a general commentary, it can be concluded that the effect of the circumferential heat conduction has never been considered in any of the previous solutions found in the literature. The mathematical model developed in this paper takes into account the nonlinearity associated with the process of mass transfer and the effect of the circumferential heat conduction.

In addition, the analysis presented allows us to evaluate more accurately the effectiveness of the heat regenerator and to provide the designer with basic data to calculate the thermal distortion of this device which may aggravates leakage.

2. NATURE OF THE PROBLEM

2.1 Nonlinear Behaviour of the Heat Regenerator

The operational cycle of the rotary heat regenerator is composed, as shown in Figure 1, of two parts; the heating and cooling periods. In the heating period, heat is transferred from the hot stream and stored in the honeycomb matrix. As this elment (control volume) of the matrix enters the cold air compartment, the stored heat is released to the cold stream. The simultaneous process of moisture transfer from, or to the rotating matrix is of a complex nature, due to the nonlinearity associated with the time-dependent heat-mass transfer closed loop interaction. This concept is presented schematically in Figure 2. It indicates that the thermal behaviour of the regenerator is controlled by the conventional design parameters; namely: the flow streams capacity ratio C_h/C_c, the matrix-to-compressed air capacity ratio C_r/C_c, NTU, the convective conductance ratio $(\overline{hA})$, the conduction area ratio $\overline{A}_k$, and the total conduction parameter λ. The consideration

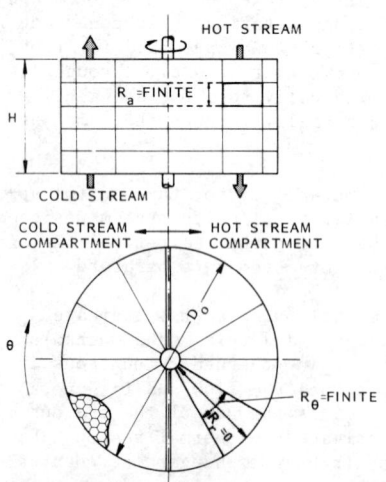

FIG. 1. Schematic representation of a typical element (control volume) in a rotary heat exchanger-counter flow arrangement.

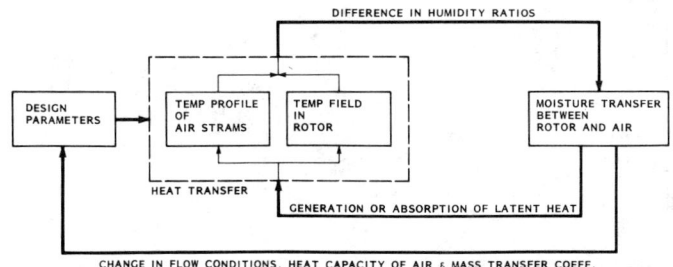

FIG. 2. Schematic representation of the nonlinear behaviour of the heat exchanger due to the heat and mass transfer coupling.

of the circumferential heat conduction requires the ratio between the thermal resistances in the axial and circumferential directions R_a/R_θ to be introduced as an additional dimensionless variable. Under a certain combination of design parameters, a definite quasi-steady temperature field is set up within the solid matrix and along the direction of flow of air streams. Depending on the temperature of the matrix surface T and the dew point temperature t_d of oncoming air, moisture transports from air to solid matrix or vise versa. As a result, latent heat will either be generated or absorbed within the matrix. In other words, mass transfer results in a distributed heat source or sink which causes, in turn, the redistribution of the temperature field within the matrix and air streams, and so on. This cycle of coupling proceeds with time till reaching the quasi-steady state conditions.

In this perspective, the process of mass transfer, which causes frosting, is both controlling and being controlled by the mechanism of heat transfer. The fact that the thermal loading is unknown in advance defines the nonlinearity in the behaviour of the heat regenerator, due to this mutual interaction.

Another aspect of nonlinearity arises from the possible effect of frosting on changing the flow conditions of air streams as well as the heat capacity of flowing air and the coefficient of mass transfer which depend on the moisture content of the air.

2.2 Mechanism of Mass Transfer and Phase Change

Convective mass transfer takes place due to the gradient in the concentration of water vapour of oncoming air stream and that of the wetted surface of the rotor. Expressed in terms of the difference between the humidity ratio of air stream v and the saturation humidity ratio of rotor surface v_s the rate of moisture transfer is given by:

$$\dot{m}\, dv = K_D \cdot A_c \cdot (v - v_s) \tag{1}$$

where $\dot{m}$ = mass flow rate of air, kg dry air/s

K_D = coefficient of mass transfer, kg dry air/m^2·s
A_c = convective surface area contained within the matrix control volume, m^2

The direction of mass transport is defined by the sign of $(v-v_s)$ or equivalently by the ratio (v/v_s). For a given flow pressure, the humidity ratios v and v_s are merely defined by the dew point of the oncoming air t_d

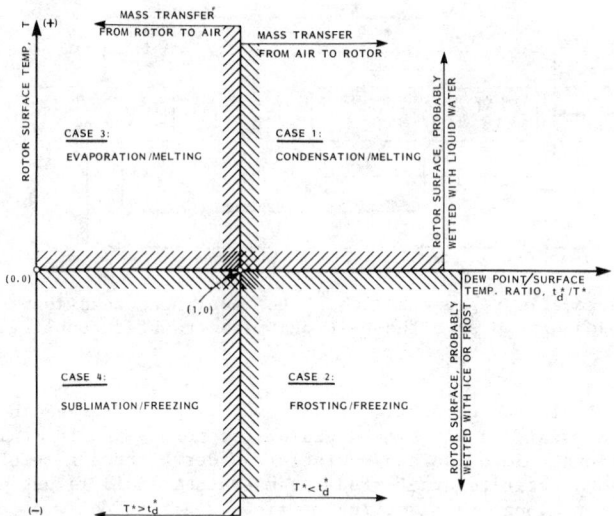

FIG. 3. Effect of the rotor surface temperature and the dew point temperature on the mechanism of mass transfer and change of phase.

and the rotor surface temperature T [9]. Therefore the ratio (v/v_s) can be substituted by (t_d^*/T^*); where the superscript * denotes absolute temperatures. Illustration of the effect of T and the ratio (t_d^*/T^*) on this mechanism is given in Figure 3. The two lines T=0 and t_d^*/T^*=1 divide the $(T - t_d^*/T^*)$ plane into four quarters correspond to four distinctive cases that could be encountered during one operational cycle. For $t_d^*/T^* > 1$, the quantity $(v-v_s)$ is positive indicating that moisture transfers from air to the rotor (case 1 or 2). It is of paramount importance to notice that in the two cases 3 and 4, the convective moisture transfer is limited by the amount of water or frost existing on the surface of the control volume and by the amount which can be absorbed by the air to reach the saturation point. This will be referred to as "Transfer Constraint".

The rate of latent heat generation $\dot{G}$ within the matrix control volume depends in its sign and magnitude on the case to be encountered:

<u>case 1</u>; where T>0 and $(t_d^*/T^*) > 1$:

$$\dot{G} = \dot{m} L_c \cdot dv - L_f \cdot \dot{F} \qquad (2)$$

where L_c, L_f = latent heat of condensation and melting respectively, J/kg.

$\dot{F}$ = rate of melting of frost existing on the surface of the matrix element, kg/s

<u>case 2</u>; where t_d^*/T^* still >1 but T is allowed to drop below zero:

$$\dot{G} = \dot{m} (L_c + L_f) \cdot dv + L_f \cdot \dot{W} \qquad (3)$$

where $\dot{W}$ = rate of freezing of water existing on the surface of the rotor, kg/s

case 3; where $t_d^*/T^* < 1$ and $T < 0$:

$$\dot{G} = -(\dot{m} \cdot L_c \cdot dv + L_f \cdot \dot{F}) \tag{4}$$

case 4; t_d^*/T^* is kept <1 but T drops below zero:

$$\dot{G} = -\dot{m}(L_c + L_f) \, dv + L_f \cdot \dot{W} \tag{5}$$

where, as in case 3, the amount $(\dot{m}dv)$ is limited by the 'Transfer Constraint'.

3. FORMULATION OF THE PROBLEM

The presented mathematical model simulates the nonlinear behaviour of the nonhygroscopic heat regenerator and accounts for the effect of the axial and circumferential heat conduction in the solid matrix. The thermal resistance of the rotor in the radial direction is neglected. The assumptions on which this model is based are:

1. The system operates under quasi-steady state conditions and counter-flow arrangement.

2. The inlet temperature and humidity ratio of the air streams are constant and uniformly distributed over the inlet cross sections.

3. The effect of frost formation on the flow of airstreams is ignored.

4. No cross contamination; ie, no leakage or carryover.

5. Excessive water condensed on the rotor surface is retained at the same cross-section.

6. Physical and transport properties are constant.

7. Thermal storage of air streams is neglected.

Based on these assumptions, the differential equations governing this process can be expressed, for the hot-air side, as follows:

a. Performing a heat balance on the air contained within a typical element (control volume) of the rotor:

$$\dot{m}_h \cdot c_{p_h} \cdot \frac{\partial t_h}{\partial y} = - \frac{h_h A_c}{\ell} (t_h - T) \tag{6}$$

where $\dot{m}_h$ = mass flow rate of hot air through the matrix control volume, kg/s

c_{p_h} = specific heat of hot air, at constant pressure,
     = $c_{p_a} + v c_{p_v}$, J/kg moist air °C

c_{p_a}, c_{p_v} = specific heat of dry air and water vapour respectively

A_c = area of convective heat transfer within the control volume, m^2

ℓ = height of the matrix control volume, m

t_h, T = temperature of hot air and solid matrix, respeciively, °C

b. Similarly, and based on the heat and mass transfer analogy, the convective mass transfer on the air contained within the control volume is governed by:

$$\dot{m}_h \cdot \frac{\partial v_h}{\partial y} = \frac{-K_D \cdot A_c}{\ell}(v_h - v_s) \qquad (7)$$

c. Energy balance of the control volume of the rotor (shown in finite difference form in Figure 5) yields:

$$\left(\frac{c_{p_r} \cdot M_r}{\ell}\right) \cdot \frac{\partial T}{\partial \tau} = (\dot{m}_h \cdot c_{p_h}) \cdot \frac{\partial t_h}{\partial y} + A_a \left(k_a \cdot \frac{\partial^2 T}{\partial y^2} + k_\theta \cdot \frac{1}{r^2} \cdot \frac{\partial^2 T}{\partial \theta^2}\right) + \frac{\partial \dot{G}}{\partial y} \qquad (8)$$

where
- M_r = mass of matrix control volume, kg
- c_{p_r} = specific heat of rotor matrix, J/kg·°C
- A_a = cross sectional area of rotor control volume, m^2
- k_a, k_θ = apparent (effective) thermal conductivity in axial and circumferential directions, respectively, W/m-°C
- T, t_h = temperature of rotor and air, respectively, °C
- $\dot{G}$ = rate of internal heat generation, W

Similarly, a set of differential equations governing the process, for the cold stream side, can be obtained.

The boundary conditions which define, uniquely, the solution of this problem are:

i. At the inlet of the hot stream side: $t_h = t_{hi}$, $v_h = v_{hi}$

ii. At the inlet of the cold stream side: $t_c = t_{ci}$, $v_c = v_{ci}$

iii. Longitudinal heat conduction is zero at the upper and lower edge of the rotor:

$\frac{\partial T}{\partial y} = 0$, at y=0 and y=L

iv. For satisfaction of the quasi-steady state condition, the temperature profile along the y-axis at any angular position θ, repeats itself after a complete turn: $T(\theta, y) = T(\theta + 2\pi, y)$

4. SOLUTION OF THE PROBLEM

The computer program developed to solve the governing differential equations is based on the finite difference method, represented by Euler's Explicit scheme. With respect to the space frame (y,θ), the central difference is used. This routine of solution allows us to obtain the sustained (quasi-steady) solution. The time interval Δτ, which satisfies the stability criterion, corresponds to a rotational angle of π/6.

To analyze the heat conduction in the solid matrix, the impractical situation of modelling individual honeycomb cells was surmounted by estimating the equivalent thermal conductances in the axial and circumferential direction for the whole finite difference lump (coefficients H_a and H_θ in Eq. 11).

The two dimensional finite difference approximation of the heat regenerator is shown in Figure 4. Energy balance of a typical element is provided schematically in Figure 5. It shows that the rate of change of the enthalpy of the element (ΔĖ) which results in the change in its temperature from T(i,j)

RESIDENTIAL ROTARY REGENERATIVE HEAT EXCHANGERS

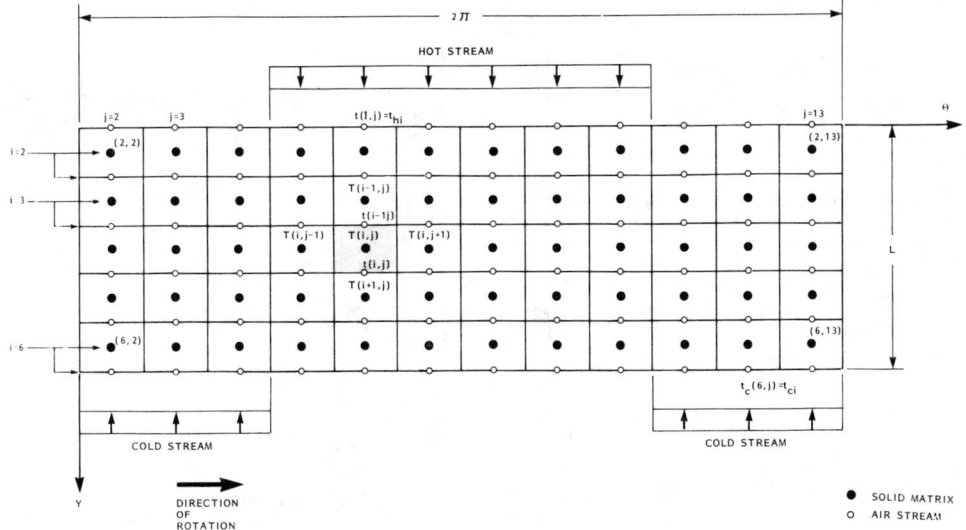

FIG. 4. Nodal representation of the rotary heat regenerator-finite difference approximation for instantaneous two-dimentional heat transfer.

at $\tau=\tau$ to $T'(i,j)$ at $\tau=\tau+\Delta\tau$ is due to: the net rate of heat conduction Q_K into this element, the change in rate of heat content of flowing air $\Delta \dot{E}_s$ and the internal heat generation $\dot{G}$. For the element (i,j), on the hot steam side, the finite difference form of the three interrelated differential equations can be represented as follows:

a. Equation of convective heat transfer

$$t(i-1,j) - t(i,j) = \frac{h_h \cdot A_c}{\dot{m}_h c_{p_h}} \cdot \left[\frac{t(i-1,j) + t(i,j)}{2} - T(i,j) \right]$$

which leads to the following expression

$$t(i,j) = F_1 \cdot t(i-1,j) + F_2 \cdot T(i,j) \qquad (9)$$

where $F_1 = (2 \dot{m}_h \cdot c_{p_h} - h_h \cdot A_c)/(2\dot{m}_h \cdot c_{p_h} + h_h \cdot A_c)$

$F_2 = (2 h_h A_c \cdot)/(2\dot{m}_h \cdot c_{p_h} + h_h A_c \cdot)$

b. Equation of mass transfer:

$$v(i-1,j) - v(i,j) = \frac{K_D \cdot A_c}{\dot{m}_h} \left[\frac{v(i-1,j) + v(i,j)}{2} - v_s(i,j) \right]$$

yields the following:

$$v(i,j) = F_3 \cdot v(i-1,j) + F_4 \cdot v_s(i,j) \qquad (10)$$

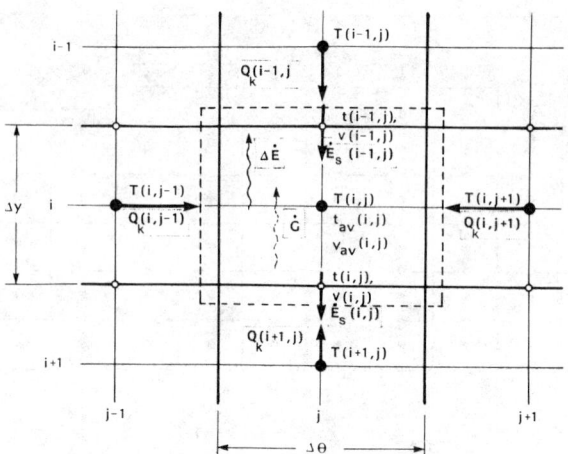

FIG. 5. Energy balance of a typical control volume

where $F_3 = (2\dot{m}_h - K_D \cdot A_c)/(2\dot{m}_h + K_D \cdot A_c)$
$F_4 = (2K_D \cdot A_c)/(2\dot{m}_h + K_D \cdot A_c)$
v_s = the saturation humidity ratio at temperature $T(i,j)$, kg vap/kg dry air

c. Equation of energy balance of rotor element:

$$T'(i,j) - T(i,j) = \frac{\Delta\tau}{c_{p_r} \cdot M_r}\left[\{H_a \cdot (T(i-1,j) + T(i+1,j)) + H_\theta \cdot (T(i,j-1) + T(i,j+1)) - H_n \cdot T(i,j)\} + \{c_{p_h} \cdot \dot{m}_h (t(i-1,j) - t(i,j))\} + \{\dot{G}(i,j)\}\right]$$

Thus, the rotor temperature $T'(i,j)$, at the end of the time interval, resulting from the change in the enthalpy of the matrix lump $\Delta\dot{E}$ (Figure 5) is expressed as:

$$T'(i,j) = F_5 \cdot T(i,j) + F_6\left[\{H_a(T(i-1,j) + T(i+1,j)) + H_\theta(T(i,j-1) + T(i,j+1))\} + \{\dot{m}_h \cdot c_{p_h} \cdot (t(i-1,j) - t(i,j))\} + \{\dot{G}(i,j)\}\right] \quad (12)$$

where $F_5 = 1 - \frac{\Delta\tau \cdot H_n}{c_{p_r} \cdot M_r}$, $F_6 = \frac{\Delta\tau}{c_{p_r} \cdot M_r}$

$H_n = 2H_a + 2H_\theta$; in which H_a and H_θ denote the finite difference heat conductance coefficient, W/°C

The term $\dot{G}(i,j)$ which represents the rate of heat generation within the lump (i,j) due to mass transfer and/or change of phase is defined by one of Eqs. 2 to 5.

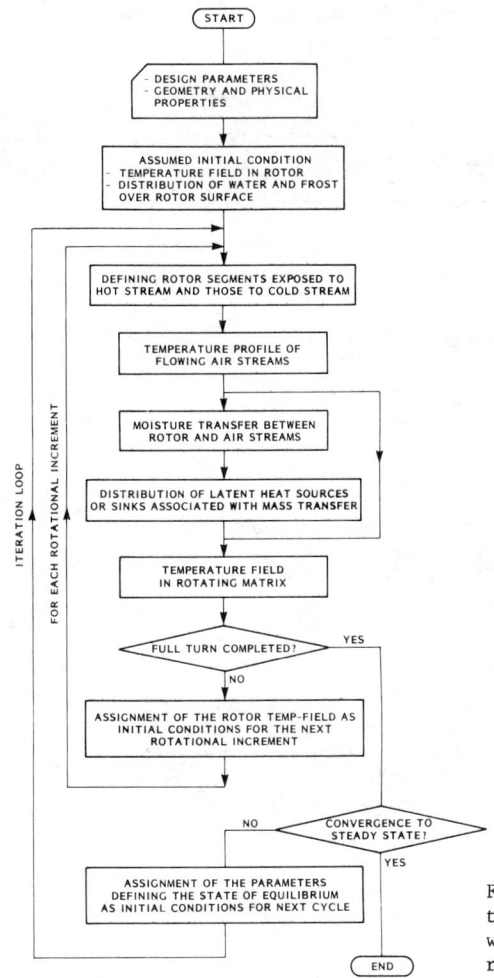

FIG. 6. Flow chart for predicting the temperature field and frost formation, with the nonlinear behaviour of the regenerator taken into account.

In a similar manner, finite difference equations for the element (i-1,j) on the cold stream side can easily be obtained. To incorporate the boundary conditions i to iv, fictitious elements have been added as rows i=1, 7 and columns j=1,14. Solutions of finite difference equations proceeds according to the flow chart given in Figure 6, in which the state of equilibrium (correct solution) is sought for through iterations. Convergence has been accepted when the absolute difference, in rotor temperatures, between two successive iterations is <0.0005; i.e., till converging to 3 significant digits.

The approach of solution presented in Figure 6 can be extended to a 'consecutive-iteration' routine to cover the transient portion of the solution. In this case, the iteration scheme is applied to search for the correct solution at the end of a complete cycle. With this as an initial condition, the iteration scheme is repeated again for the next cycle. Variations in the flow, or inlet conditions can also be considered.

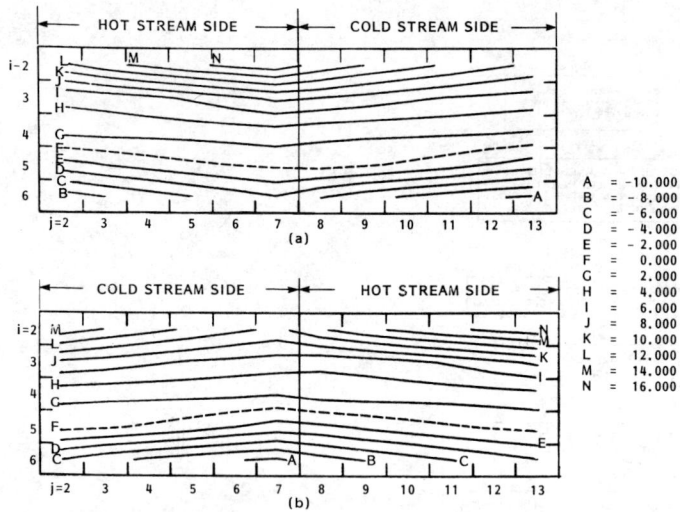

FIG. 7. Quasi-steady state temperature field in the solid matrix, without accounting for the mass transfer:
a. at the beginning of the cycle; b. after a half cycle.

5. RESULTS AND DISCUSSION

As a case study, the effect of the closed loop interaction, on the thermal response of the air-to-air heat regenerator specified in [10] has been investigated. The following conditions have been assumed:

$\dot{m}_h/\dot{m}_c$ = 1.0, C_r/C_c = 7.5, $(\overline{hA})$ = 1.0, $\overline{A}_k$ = 1.0, NTU_o = 19 λ=0.065 and R_a/R_θ = 0.3. The inlet conditions of air streams are taken as:

t_{hi} = 23°C, V_{hi} = 0.009 kg Vap./kg dry air

t_{ci} = -15°C, v_{ci} =0.0008 kg vap/kg dry air.

The quasi-steady state temperature fields, for the case of no mass transfer, at the beginning of the cycle and after a half complete revolution are shown in Figures 7a and 7b respectively. It is seen that the hotest spot is always located on the inlet side of the hot stream at that portion of the rotor which has been in the hot compartment for the whole hot period. The reverse is true for the coldest spot. It can also be noted that temperature gradient which affects significantly the thermal distortion of the heat regenerator structure, is relatively steep at these two spots. The temperature variations in the θ-direction is more pronounced at the inlet cross sections of both air-streams The effect of the mass transfer on the temperature field can be concluded from the comparison of Figures 7 and 8. While the location of the hot and cold spots are persistent, the pattern of the temperature field is significantly changed. The location of the isothermal line corresponds to 0°C (shown as dotted line on both Figures), which motivates the frost formation in the existance of moisture on the matrix surface, has moved noticeably towards the inlet cross-section of the cold stream, due to internal heat generation.

Results indicated that only the temperatures at the inlet side of the cold stream, i=6 remain, below zero during the whole cycle. Thus, it can be

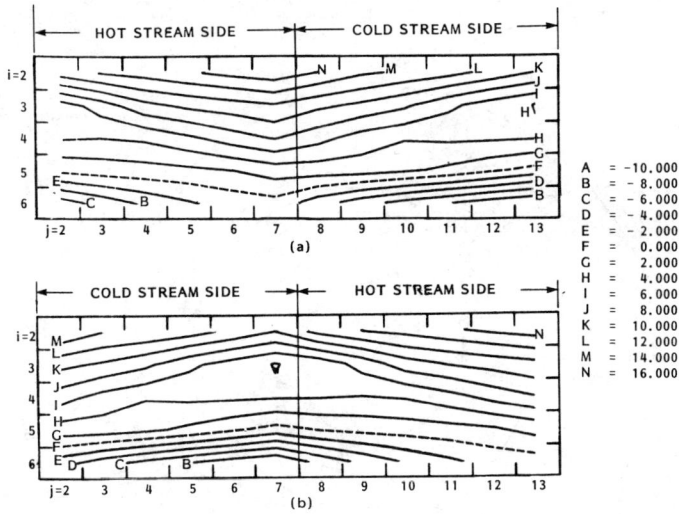

FIG. 8. Quasi-steady state temperature field in the solid matrix, with the mass transfer taken into account:
a. at the beginning of the cycle; b. after a half cycle.

concluded that the 'Frosting Limit', at which frost will begin to build up is reached. The transient nature of the frost build up process is currently under study.

The periodic time variation of the quasi-steady state temperatures is presented in Figure 9, for points located at different levels i=2, 4 and 6. The same results are obtained for various points located at the same level but with a phase difference ranges from 0 to 2π.

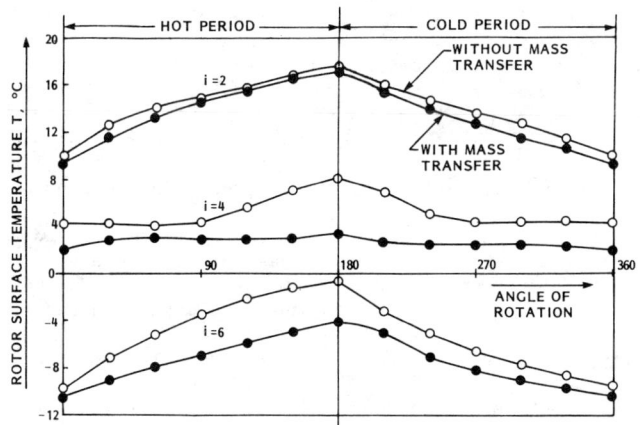

FIG. 9. Time variation of the temperature, at different levels; i=2, 4 and 6 under the quasi-steady state conditions.

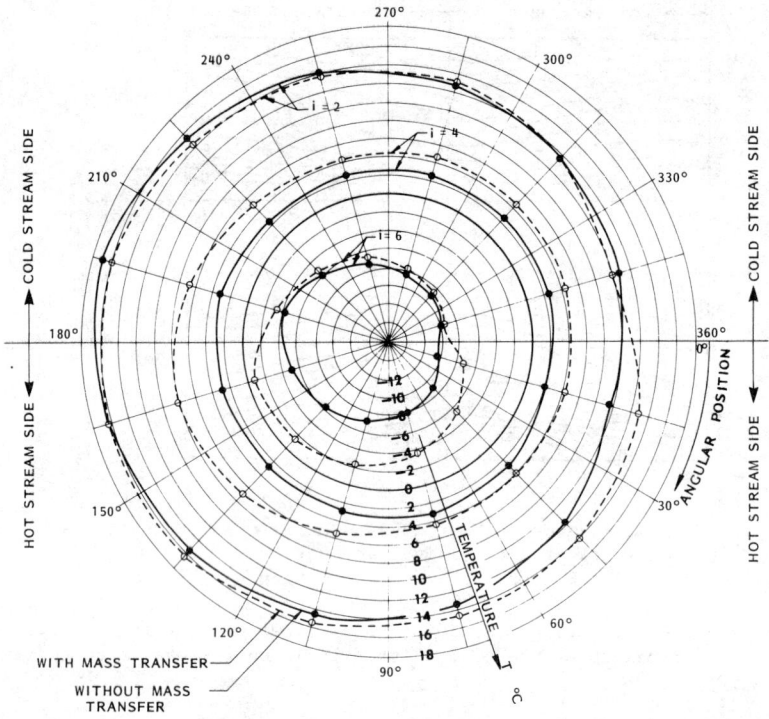

FIG. 10. Circumferential temperature variations at different levels i=2, 4 and 6, at the beginning of the quasi-steady state cycle.

The circumferential temperature variation of the momentary temperature field can best be appreciated in Figure 10. It emphasizes the importance of not ignoring the circumferential heat conduction in the rotor, as it was the case in all available solutions found in literature. The significance of the effect of mass transfer on the temperature field is also evident from this graph.

CONCLUSIONS:

From the analysis presented in this paper, the following conclusions can be drawn:

1. A solution scheme is developed to simulate the thermal response behaviour of the rotary heat regenerator under quasi-steady state conditions. It recognizes the effects of circumferential heat conduction and the inherent nonlinearity associated with the heat and mass transfer processes.

2. Frost formation and build up can be predicted by this method of solution.

3. Characteristics of the two dimensional, quasi-steady state-temerature field in the rotor matrix have been underlined. It is concluded that the circumferential heat conduction and mass transfer have significant effect on the thermal response of the regenerator.

REFERENCES

1. D'Silva, N.S. "Operating experience with a residential mechanical ventilation system with heat recovery." Ontario Hydro Res. Div. 81-51-K, 1981.

2. Lambertson, T.J. "Performance Factors of a periodic flow heat exchanger." Trans ASME, vol 80, 1958, pp 586-592.

3. Schalkwijk, W.F. "A Simplified regenerator theory." J. Engrg for Power, Trans. ASME, Vol 81, 1959, pp 142-150.

4. Mondt, J.R. "Vehicular gas-turbine periodic-flow heat exchanger solid and fluid temperature distributions." J. Engrg for Power, Trans. ASME, vol 86, 1964, pp 121-126.

5. Bahnke, G.D. and Howard, C.P. "The Effect of longitudinal heat conduction on periodic-flow heat exchanger performance." J. Engrg for Power, Trans. ASME, vol 86, 1964, pp 105-120.

6. Holmberg, R.B. "Heat and mass transfer in rotary heat exchanger with nonhygroscopic rotor materials." J. Heat Transfer, Trans. AME, vol 99, 1977, pp 196-202.

7. London, A.L., Biancardi, F.R., Mitchell, J.W. "The Transient response of gas-turbine-plant heat exchangers - regenerators, intercoolers, precoolers, and ducting." J. Engrg for Power, Trans. ASME, vol 81, 1959, pp 433-448.

8. London, A.L., Sampsell, D.F., McGowan, J.G. "The Transient response of gas turbine plant heat exchangers - additional solutions for regenerators of the periodic-flow and direct-transfer types." J. Engrg for Power, Trans. ASME, vol 86, 1964, pp 127-135.

9. ASHRAE Handbook of Fundamentals, American Society of Heating, Refrigerating, and Air Conditioning Engineers, New York, N.Y., 1972.

10. Shoukri, M. "The Use of a regenerative air-to-air rotary heat exchanger for heat recovery in residential ventilation systems." Presented at the Winter Annual Meeting. ASME, 1979, Paper No 79-WA/HT-32.

HEAT EXCHANGER DESIGN

Heat Exchanger Design and Practices

ROBERT V. MACBETH
UKAEA, Atomic Energy Establishment
Winfrith, Dorset, England

ABSTRACT

In the thermal design of heat exchangers we have to deal with something like 12 to 15 independent variables. We deal with most of them simply by doing what is customary. There is usually no alternative. It is the consequence of insufficient experimental data, although the underlying cause is the large number of variables. We do not have to examine every permutation, but we are talking about many hundreds of systematically performed experiments.

Professor Bergelin and Co-workers at the University of Delaware made an onslaught on the problem in the 1950's, but heat exchanger testing is very time consuming and they barely scratched the surface. What was needed was a drastic speeding up of the process of data acquisition. The break-through came in 1967 when Gay and Co-workers at the University of Aston in Birmingham showed that a particular mass transfer technique produced results, comparatively quickly and cheaply, that were practically identical to the Delaware data. In 1971 the Harwell based consultancy and testing service (HTFS) set in operation a scheme to produce large quantities of heat exchanger design data using the mass transfer technique. This scheme has been in operation now for 10 years and a vast amount of data has been produced. In this lecture the author describes some of the important findings that have emerged from the work and their likely effect on heat exchanger design and practices.

1. INTRODUCTION

This lecture gives an apportunity to say something about experiments on shell-and-tube heat exchangers that have been going on steadily for the past ten years at the UK Atomic Energy Establishment, Winfrith. It is proposed therefore, to present some of the interesting shell-side performance data obtained and discuss their implications. The work was sponsored by the Chemicals and Minerals Requirements Board and was carried out under the direction of HTFS who have given permission to publish some of the findings.

Some inkling of the work was given (1) to the Toronto Heat Transfer Conference, 1978, but this mainly described the experimental method used, an electrochemical mass transfer modelling technique, and indicated the scope of the experimental programme. It is not proposed to go over the experimental method again here. Some doubts about its validity have occasionally been expressed, but the facts show that it does give accurate heat transfer data, using the Chilton and Colburn analogue (2). At Winfrith, exact facsimilies have been tested of heat exchangers used in the extensive work done by

Delaware University (3) and agreement has been obtained within the estimated experimental accuracy of 8½ per cent. (The Delaware work involved direct heat transfer measurements). The University of Aston in Birmingham, where the electrochemical technique was first applied to shell-and-tube heat exchangers, has also obtained good agreement with Delaware data and with other sources (4) (5) (6).

The big advantage of the electrochemical method is its speed. At Winfrith, for example, the turn-round time is about five working days for assembling, testing, data processing and disassembling a complete heat exchanger bundle and the whole operation can be performed by one person. This turn-round time is for a 300-tube bundle with a facility for measuring transfer coefficients on 80 individual areas, practically simultaneously, anywhere within the bundle. The tests are usually done at 12 different electrolyte flowrates which cover a wide range of Reynolds numbers. Thus, nearly 1000 individual heat transfer coefficients are recorded for each heat exchanger bundle tested. It would be very difficult and would involve many months of work to obtain the same amount of information by conventional, direct heat transfer measurements.

The considerable speeding up of data acquisition that the electrochemical method achieves opens the way to unravelling the large number of independent variables that characterise the shell side of shell-and-tube heat exchangers. Even so, the amount of testing required is extensive and although about 200 tube bundle arrangements have so far been tested and many special studies have been made, such as measuring the effect of inlet nozzle impingement plates, there is still much to be done.

2. SCOPE OF EXPERIMENTAL PROGRAMME

Experimental costs are reduced and handling made easier by reducing bundle size and therefore, a fairly small outside tube diameter of 9.53mm (0.375 in) has been used in nearly all the experiments done so far, with tube pitch/ diameter ratios of 1.25 and 1.33. Since a more typical tube size for shell--and-tube heat exchangers is 19.06mm, the Winfrith programme can be said to use typically half-scale models. This does not affect the usefulness of the results however, as they are easily scaled up to larger tube sizes, or scaled down, using the principles of dimensional similarity.

Tests have been performed mostly with 80-tube and 300-tube bundles with corresponding inside shell diameters of 133.5mm and 248.8mm respectively. Shell/baffle and tube/baffle clearances have been varied from zero up to values typical of manufacturing practice, and different tube layouts have been explored.

Pressure drop measurements have proved very important in assessing the affect of geometry variables and two tubes in every bundle tested have been specially constructed to measure the static pressure in the bundle at 3 or 6 positions. Figure 1 shows the location of these positions in a variety of baffle arrangements that have been tested; baffle cuts used have varied from 18.4 to 37.5 per cent, besides zero cut or orifice baffles. An investigation has also been made of rod baffles (7)(8) and rules for their design have been established. The effect of baffle spacing has been examined and so too has baffle plate thickness. Bundle orientation within the shell has also been investigated.

The practice has been to measure transfer coefficients over full lengths of tube contained within one baffle spacing. Measurements are made practically

HEAT EXCHANGER DESIGN AND PRACTICES

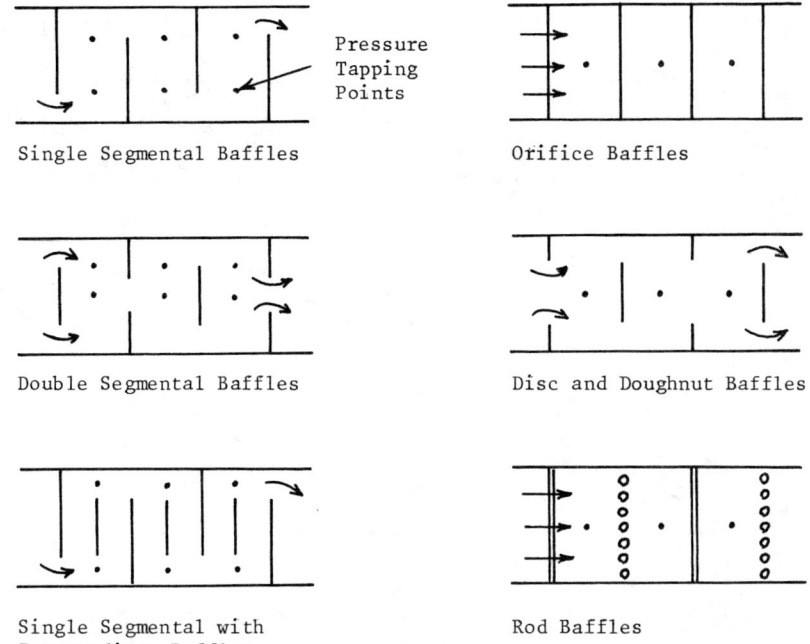

Fig. 1 Location of Pressure Drop Tappings in the Baffle Arrangements Tested.

simultaneously on 78 of the tubes in the 80-tube bundle and on 80 tubes in the 300-tube bundle, thus giving a detailed picture of the variation of heat transfer coefficient within a baffle compartment. Most tests have been done on a mid-bundle compartment, but the relative performance of other compartments has been investigated, including tests on inlet nozzle compartments with and without an impingement plate. It has been found that most of the data analysis can be done satisfactorily using average compartment heat transfer coefficients. It is these averages that are quoted in the following examples of experimental results, unless otherwise stated.

3. BAFFLE CLEARANCES

Tube/baffle and shell/baffle clearances are a practical necessity and they impose on any shell-and-tube heat exchanger a substantial degree of axial flow. Thus, for a given shell-side flowrate, as clearances increase, heat transfer coefficients can be expected to decrease because axial flow tends to produce a thicker boundary layer than cross-flow. Experiments show that the effect is always substantial and Figure 2 is an example from tests on the mid-bundle compartment of an 80-tube bundle using a staggered square tube layout (1.25 tube pitch/diameter ratio) with single segmental baffles (baffle cut 18.4 per cent and baffle spacing 36.4 per cent of shell diameter). The middle curve shown in Figure 2 has clearances typical of manufacturing practice and in this test the total clearance or leakage area is nearly half that of the baffle window area, allowing for tube and tie rod obstructions. The leakage is of course larger than the above relative areas would indicate, since the pressure

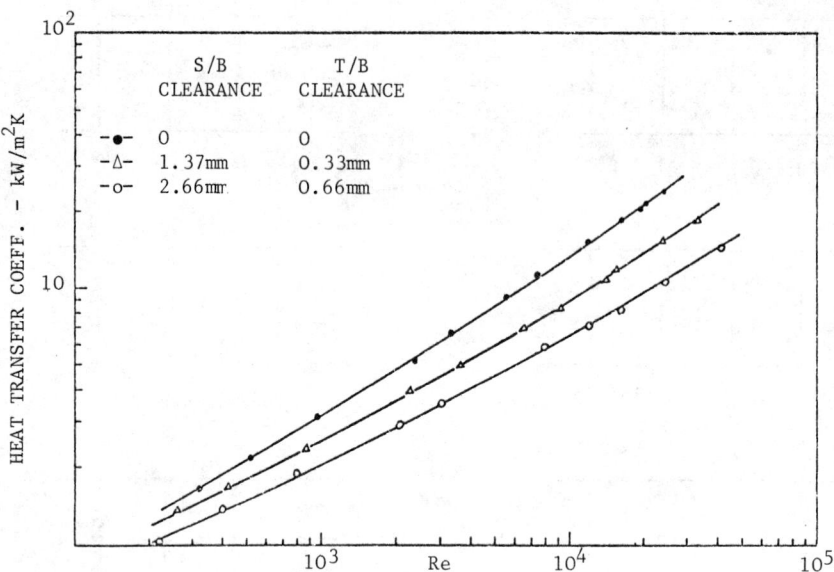

Fig. 2 Effect of Baffle Clearances on Heat Transfer

drops across the baffle clearances are mostly bigger than across the window zone. On the other hand, measurements indicate that the clearances behave like sharp edged orifices and have discharge coefficients in the region of 0.65.

Figure 2 illustrates the very large increase in heat transfer performance that is possible by eliminating leakages. For example, at $R_e = 10^4$ the potential increase is 50 per cent, comparing the top curve with the middle curve. (R_e is here based on the conventional minimum flow area in the region of the mid-plane of the heat exchanger bundle compartments). It must not be assumed however, that the elimination of clearances is desirable because there is the penalty of pressure drop. Figure 3 shows, for example, that the above 50 per cent gain in heat transfer coefficient must be paid for by more than a factor of 2 on pressure drop. The decision would normally rest with the economics of the situation, the cost of pressure drop tending to be much lower with liquid flow than with gas or vapour flow. There is more to this issue of heat transfer versus pressure drop however, and we shall return to it later on.

It is worthwhile noting here that clearances have the effect of making the heat transfer coefficient variation among the individual tubes in a bundle remarkably small. This applies to all the segmental baffle arrangements tested, both single and double as well as disc and doughnut. No significant distinction has been found between the so-called window zones and the cross-flow zone. As a general rule, the coefficients for individual tubes in a baffle compartment differ by less than ± 10 per cent from the average. The situation is different however, when the baffle clearances are zero. In these circumstances, the test results show consistently that heat transfer coefficients in a cross-flow zone

HEAT EXCHANGER DESIGN AND PRACTICES 619

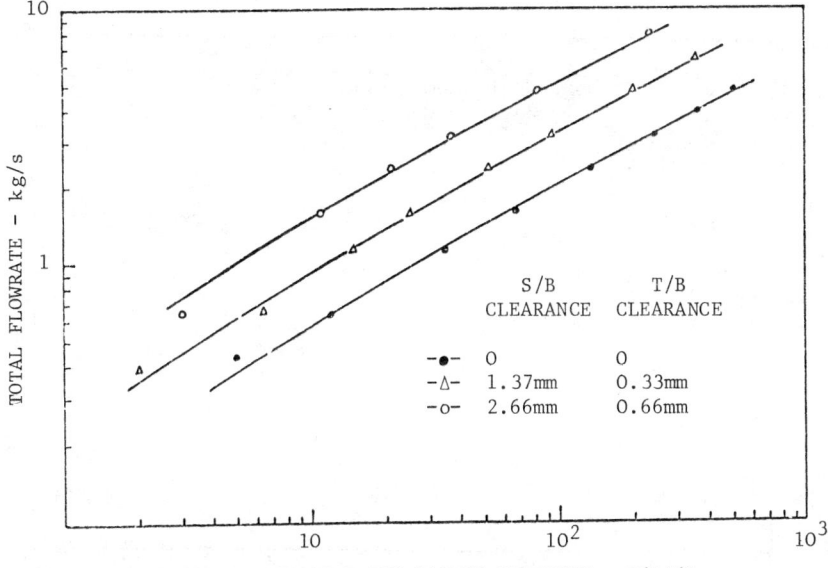

Fig. 3 Pressure Drop of Tube Bundles shown in Figure 2

are about 25 per cent higher than in the adjoining window zones.

The comments made in this section on baffle clearances have been made with reference to Figures 2 and 3, which are typical examples of results obtained from mid-bundle compartment tests. Experiments on other compartments however, show that the same comments apply, except for the inlet and outlet nozzle compartments of a shell-and-tube exchanger.

4 SINGLE VERSUS DOUBLE SEGMENTAL BAFFLE ARRANGEMENTS

Figures 4 and 5 are presented to illustrate a general observation that if we take any bundle arrangement with double segmental baffles, then irrespective of the flowrate, exactly the same heat transfer coefficients and pressure drops can be achieved by substituting single segmental baffles with a larger baffle cut. Disc and doughnut baffles can be similarly replaced.

The bundle used in Figures 4 and 5 was an 80-tube staggered square layout with a 1.25 tube pitch/diameter ratio. Baffle spacing was fixed at 36.4 per cent of shell diameter and the tube/baffle and shell/baffle diametrical clearances were also fixed at 0.66mm and 2.66mm respectively. The double segmental baffle cut of 18.4 per cent refers to the top and bottom cut (both are equal) of the one-piece baffle, see Figure 1, and as customary the two-piece baffle has the same window area as the one-piece baffle. To ensure that the comparison is correctly made, the Reynolds number used in Figure 4 and 5 for the double segmental baffle arrangement is calculated on the same basis as the single segmental baffle, viz, the conventional minimum cross-flow area in the region of the mid-plane of a bundle compartment.

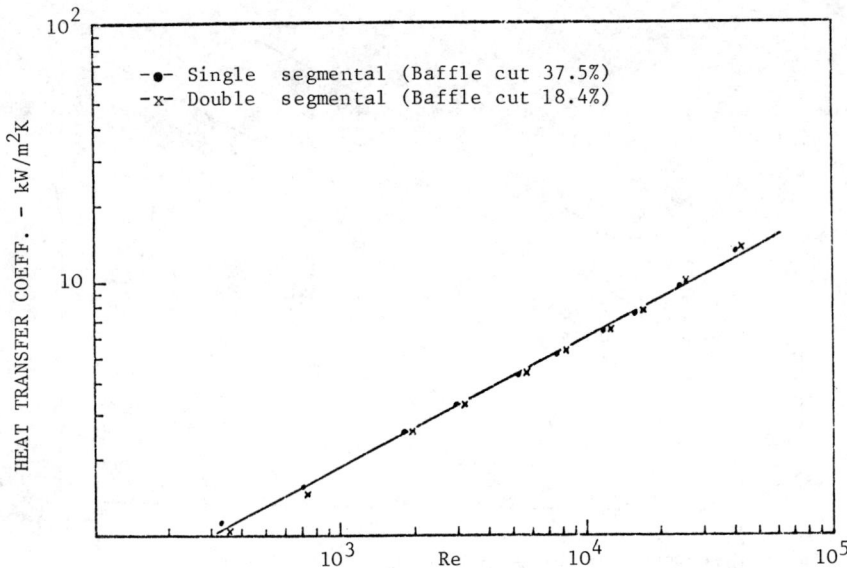

Fig. 4   Effect of Baffle Type on Heat Transfer

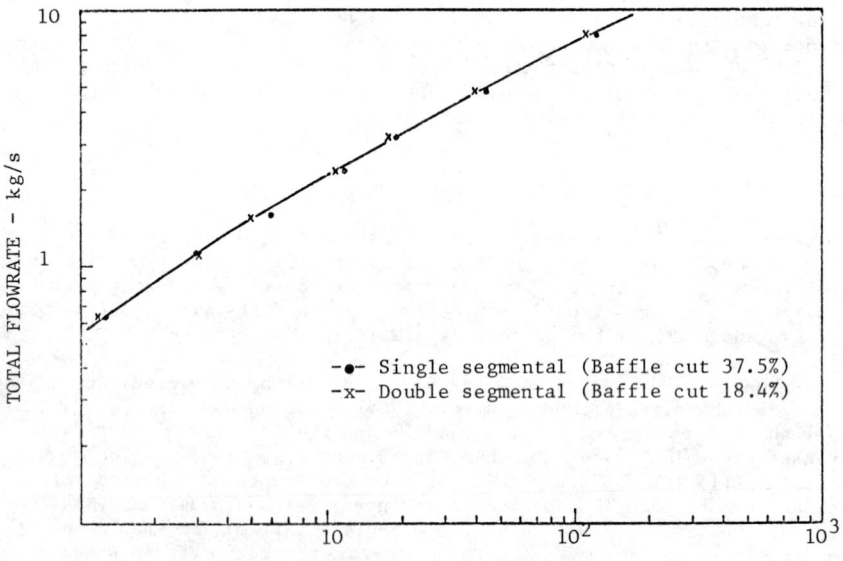

PRESSURE DROP OVER TWO BAFFLE SPACINGS - Millibar

Fig. 5   Effect of Baffle Type on Pressure Drop

HEAT EXCHANGER DESIGN AND PRACTICES

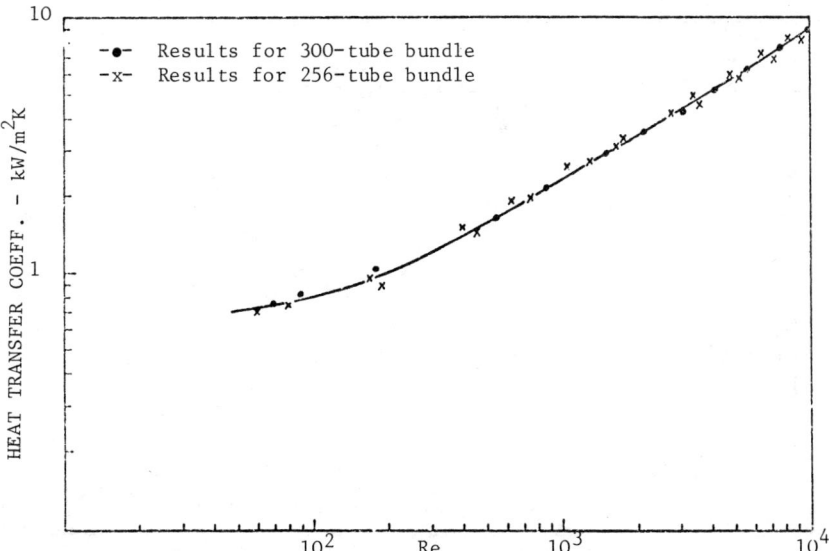

Fig. 6 Effect on Heat Transfer of Removing 44 Tubes from the Periphery of a 300-Tube Bundle

Figures 4 and 5 offer no surprise in so far as they show that heat transfer or pressure drop, looked at individually, are unaffected by changing from double to single segmental baffles, provided the baffle cut is suitably adjusted. What is a little surprising is that the adjustment necessary is the same for both heat transfer and pressure drop. This result is an example of a close link that has been found to exist between the two and we shall say more about it later.

5 THE EFFECT OF REMOVING TUBES FROM A BUNDLE

A curious result found is that heat transfer coefficients remain unaffected by the removal of tubes from anywhere in a bundle, even in large numbers. Whenever this has been done, the baffle holes with missing tubes have been closed off with thin rubber buttons. Figure 6 shows an example in which 44 tubes were removed, more or less uniformly, from around the periphery of a 300-tube bundle. The tube layout was staggered square (1.25 tube pitch/ diameter ratio) with single segmental baffles (cut 37.5 per cent and spacing 39% of shell diameter) and diametrical clearances 0.15mm shell/baffle and 0.20mm tube/baffle. It must be remembered that although the heat transfer coefficients are hardly affected, the heat transfer area is reduced by 15 per cent. The corresponding reduction observed in the pressure drop was about 25 per cent.

Another example is shown in Figure 7 which was obtained from tests on a 300-tube bundle using single segmental baffles (cut 25 per cent and spacing 39 per cent of shell diameter) fitted with additional intermediate baffles as shown in Figure 1. The tube layout was staggered square (1.25 tube pitch/ diameter ratio) with diametrical clearances 1.22mm shell/baffle and 0.20mm tube/baffle. Experimental results were obtained first with the full compliment of 300 tubes and then with all the window zone tubes removed (top and bottom) leaving the bundle with only 194 tubes. Baffle holes with missing tubes were

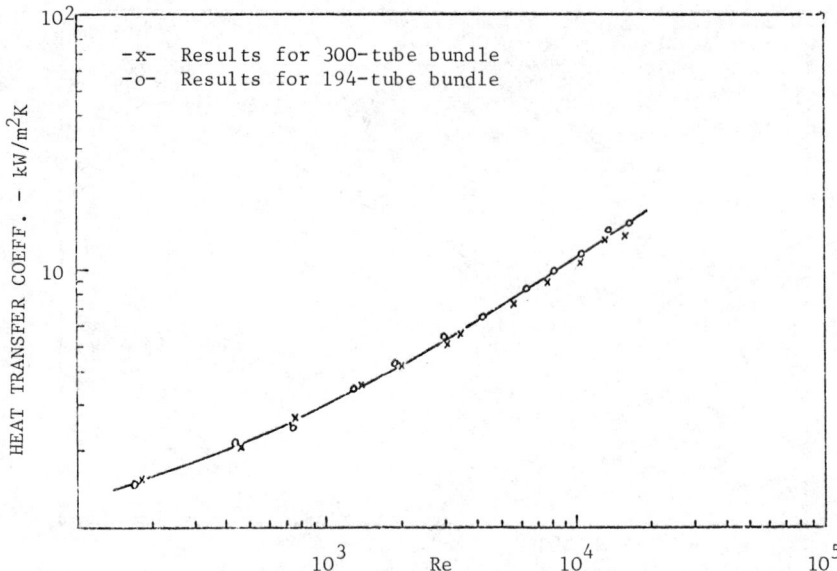

Fig. 7 Effect on Heat Transfer of removing 106 Window Zone Tubes from a 300-Tube Bundle

plugged as in the previous example. Figure 7 shows that the heat transfer coefficients are hardly affected, although the heat transfer area is reduced 35 per cent by removing the window tubes. The pressure drop falls by about 30 per cent.

Some additional measurements that have been made in these intermediate baffle tests show that the heat transfer coefficients on the upstream side of an intermediate baffle are generally about two-thirds of those on the downstream side; the measurements being made on tubes in the cross-flow zone between the intermediate baffle and the adjoining single segmental baffles. This axial variation in the heat transfer coefficients is something that has often been noted, whether intermediate baffles are present or not.

It is worth noting that from a heat transfer/pressure drop point of view the above intermediate baffle arrangement with all window zone tubes removed is the best arrangement that has been tested. Its drawback is that it makes poor use of the capital cost invested in the shell.

In Figures 6 and 7, Reynolds number is based on the conventional minimum flow area in the region of the mid-plane of the heat exchanger bundle compartments. In Figure 7 therefore, Reynolds number is unaffected by removing the window zone tubes. In Figure 6 however, removing tubes from the bundle periphery increased the flow area, as defined above, by 14.7 per cent. Thus, for any given Reynolds number in Figure 6, the 256-tube bundle has a 14.7 per cent higher flowrate passing through it. Interpretation of Figure 6 therefore, requires some care.

It is considered unlikely that Figures 6 and 7 indicate any general rule regarding the removal of tubes from a bundle, but obviously under some conditions the heat transfer coefficients are unaffected by substantial tube removals. This is considered to be an area well worth investigation.

6. A SIMILARITY RULE FOR TUBE BUNDLES

In the Winfrith experimental programme two bundle sizes have mostly been used, viz, an 80-tube bundle (shell diameter 133.5mm) and a 300-tube bundle (shell diameter 248.8mm). Consideration has been given to combining heat transfer results obtained with these bundles. It is not straightforward because tube diameter has been kept constant in all the tests, whereas strict modelling rules require dimensional similarity. Nevertheless, it is still possible to specify a set of conditions that ensures that all flow areas in the two bundles are in the same proportion, whether they be tube/baffle or shell/baffle leakage areas, window zone or cross-flow zone areas, however defined. The conditions are that the following must be the same in both bundles: (a) tube diameter, tube layout and pitch; (b) shell diameter/baffle spacing ratio; (c) baffle cut; (d) tube/baffle clearance; (e) baffle/shell clearance to shell diameter ratio.

It would now normally be simply a matter of ensuring that flowrates through the two bundles were in the same ratio as the above flow areas to have complete similarity. Using the same fluid and the same fluid temperature, the heat transfer coefficients would be identical. However, the situation is distorted by the tube diameter being fixed. Axial flow components are not affected, but in the smaller of the two bundles the cross-flow resistance will be less than it should. Nevertheless, a shell-and-tube heat exchanger bundle with practical clearances has great powers of accommodation and can be expected to easily absorb a fair degree of distortion. The test lies in trying it and Figure 8 is an example of a wide range of 80-tube and 300-tube bundle data conforming quite closely with the set of modelling conditions specified above. This example has been chosen since it also indicates the general observation that baffle cut has little effect on the heat transfer coefficients.

It is hoped sometime to obtain performance data from a very large heat exchanger bundle and then test a small model of it, using the above modelling conditions, to see how far they can be taken. Potentially, the method could be useful in the design of large heat exchangers where there is a cost incentive to obtain accurate design data.

A similar modelling procedure to that above has been found to work with the pressure drop data, but there are some improvements to be made and these are currently being investigated.

7 ORIFICE BAFFLES

Although the orifice or zero cut baffle is rarely used today, tests have shown that with the right size of orifice, determined by the tube/baffle diametrical clearance, its heat transfer/pressure drop performance is usually better than conventional baffles. Orifice baffles were fairly common among heat exchangers prior to the early 1930's and it is not clear why they disappeared. They may have been misjudged on the grounds that cross-flow heat transfer is generally superior to axial flow heat transfer, a fact which emerged in the early 1930's as experiments got under way.
Alternatively, they may have been misjudged because it was not realised that large orifices are essential for good performance. They are sensitive to geometric proportions, making them quite different to conventional baffles which are singularly unresponsive to changes as some of the examples in this paper have shown.

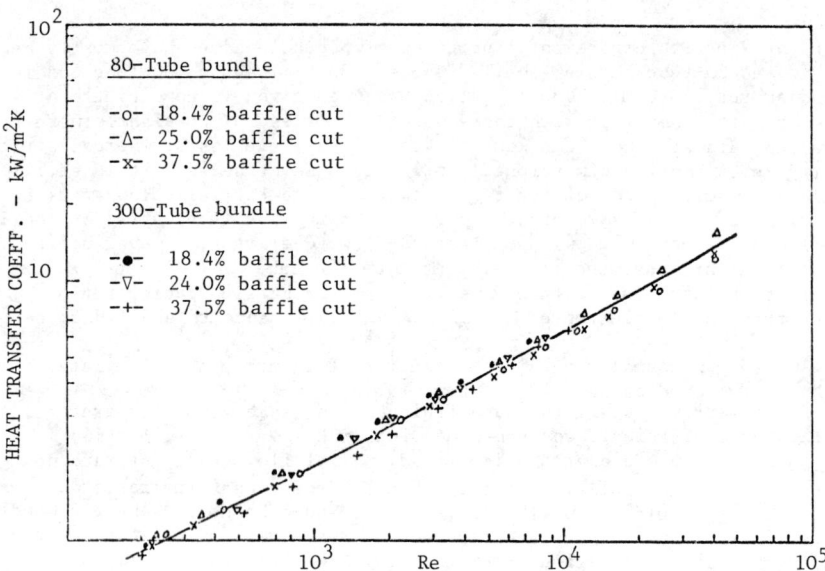

Fig. 8 Comparison of Heat Transfer Performance of Geometrically Similar 80-Tube and 300-Tube Bundles

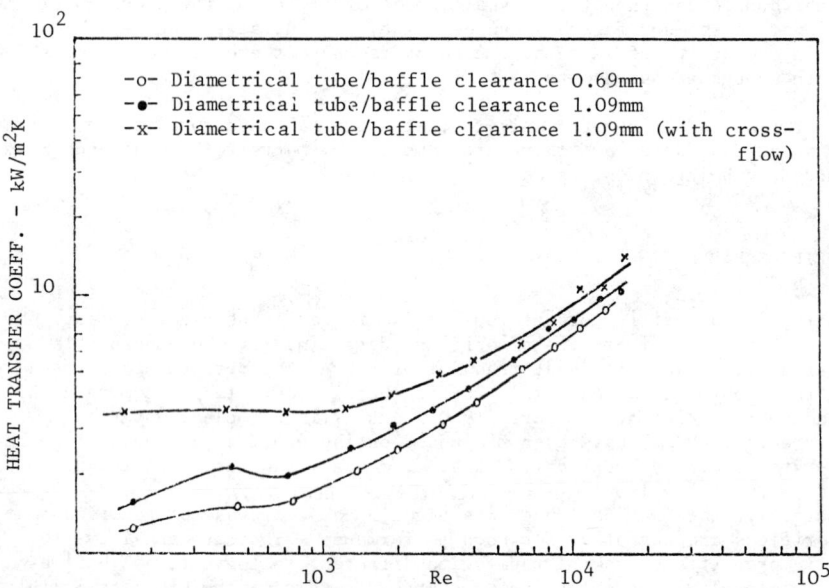

Fig. 9 Heat Transfer Performance of Orifice Baffles

HEAT EXCHANGER DESIGN AND PRACTICES

Figure 9 is an example of three variations to a 300-tube bundle with an orifice baffle spacing of 39 per cent of shell diameter using a staggered square tube layout with tube pitch/diameter ratio of 1.25. The Reynolds number used in Figure 9 is for convenience based on the convention described earlier for segmental baffles. The first variation to note is the increase in the heat transfer coefficients as a result of enlarging the tube/baffle diametrical clearance from 0.69mm to 1.09mm; the corresponding decrease in pressure drop is about 30 per cent at $Re = 10^4$ and 50 per cent at $Re = 10^3$. This increase in the heat transfer coefficients is surprising as one would expect a decrease. The phenomenon is something to do with the tube/pitch/diameter ratio as experiments show that it does not occur at a ratio of 1.6. It is believed to be associated with the momentum exchange distance after fluid passes through the orifices.

The 1.09mm clearance in Figure 9 was the largest possible in these particular tests without making the baffle ligaments too thin. It has been discovered however, that a further gain in heat transfer is obtainable by introducing a small amount of cross-flow. This has been done in the upper curve shown in Figure 9 and was achieved by removing 11 of the top-most and bottom-most tubes in the bundle and alternately sealing off, top and bottom, the corresponding baffle holes. The same effect is achievable by giving the orifice baffles say a 5 per cent cut. The gain in heat transfer is accompanied by a slight reduction in pressure drop of a few per cent. A most interesting feature of the top curve in Figure 9 is the way the heat transfer coefficients remains practically constant whilst a change in Reynolds number by a factor of nearly 10 occurs. This feature has been observed regularly in other orifice baffle tests whenever a small amount of cross-flow is introduced. So far there is no explanation as to its cause.

In order to try and re-create an interest in orifice baffle exchangers, one was designed and installed at Winfrith about 3 years ago on a shell-side water heating duty of 2.5 MW rating. A feature introduced into this design was to offset the holes in each alternate baffle plate so that when the bundle was inserted into the shell, the tubes were firmly supported and any risk of tube vibration damage avoided. This exchanger has performed well and has satisfactorily passed a recent inspection. It is probably the only one of its kind in Britain.

8. HEAT TRANSFER AND PRESSURE DROP CONNECTION

As a last example of the kind of experimental results coming out of the extensive Winfrith programme, we shall look at a remarkably close relationship existing between heat transfer coefficients and pressure drop. There was reason to look for such a relationship initially because, in process plant equipment a close link often occurs between some index of performance of the equipment and the cost of achieving this performance. A good example is to be found in gas dedusting equipment. Stairmand (9) showed that for a given gas flowrate and dust loading, plotting a range of dust collection effeciencies against the corresponding total annual costs for various forms of cyclones, wet scrubbers, bag filters and electrostatic precipitators, produced results lying fairly close to a common curve. In plain language, Stairmand was demonstrating that one gets what one pays for, at least some of the time.

In applying the above principle to heat exchangers we can liken average heat transfer coefficients to a performance index and the pressure drop per unit heat transfer area to the effort required to achieve the performance. It was expected that some restriction might apply and in fact it was found that the 80-tube and the 300-tube bundle data must be dealt with separately. It is

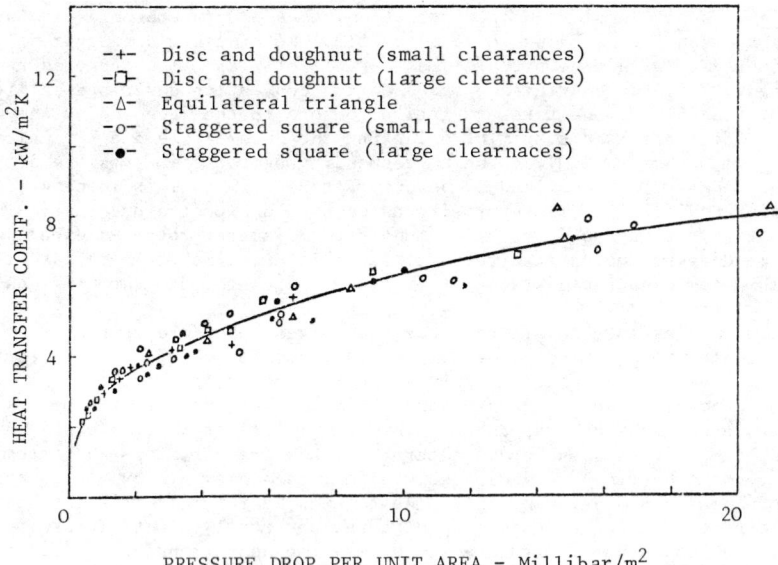

Fig. 10 Heat Transfer/Pressure Drop Relationship for Various Tube Layouts and Baffle Arrangements at all Flowrates Tested

now apparent however, that the similarity rule described earlier, once the pressure drop issue has been cleared up, will enable the tube bundle size restriction to be removed. This is something we hope to do soon.

Figure 10 shows an example of data from 20 different 300-tube bundle arrangements including disc and doughnut, equilateral triangle and staggered square tube layouts; baffle cuts varying from 18.4 to 37.5 per cent with different baffle spacings and a wide range of baffle clearances. Figure 10 includes all flowrates, although the higher flows produce pressure drops of over 100 millibar/m^2 and they are not shown in Figure 10 to avoid crowding of points at the lower pressure drop end. These higher values nevertheless lie close to an extension of the best fit curve shown which, as can been seen, fits practically all the heat transfer coefficients to within $\pm$ 10 per cent. The best fit curve (including the extension to higher pressure drops) conforms precisely with the relationship:-

Pressure drop per unit area $\propto$ (average heat transfer coefficient)^3

Exactly the same relationship applies to the 80-tube bundle data.

Some experiments have been done at Winfrith on 300-tube bundles with grossly malproportioned geometries. Their purpose is to provide data for testing more severely than usual computer design programmes being developed by HTFS. For example, experiments have been done on a disc and doughnut baffle arrangement with proportions as shown to scale in Figure 11.

HEAT EXCHANGER DESIGN AND PRACTICES 627

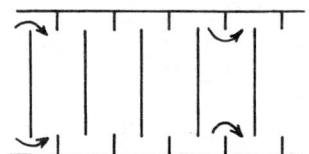

Fig. 11 Malproportioned Disc and Doughnut Baffle Arrangement Tested

The experienced eye will judge right away that the baffle spacing is much too small and will tend to cause flow stagnation in the zone between adjoining discs. This diagnosis is supported by the experimental results which show that tubes in the area outside the disc have a mean heat transfer coefficient about 30 per cent higher than the mean coefficient for the tubes within the doughnut hole. Such malproportioned geometries are not included in Figure 10 since they would normally be regarded as unacceptable. It is worth noting however, that they still plot within -20 per cent of the best fit curve in Figure 10.

In referring above to malproportioned geometries tested at Winfrith it is necessary to make the qualifying remark that there may well be some geometries which depart significantly from the cubic relationship shown by Figure 10, which is based primarily on tube bundles conforming to what may be regarded as the conventional or normal range of baffle spacing etc. For example, a bundle with extremely wide baffle spacing, although probably impractical, would tend to approach purely longitudinal flow conditions. Data in the Handbook of Heat Transfer (10) show that with purely longitudinal flow in a bundle, the cubic relationship shown in Figure 10 gives way to something slightly greater than a square law.

Orifice baffle and rod baffle data have not been included in Figure 10. There is a reason for this. The orifice and rod baffle arrangements are axial-flow systems and therefore, capable of producing a pressure drop per unit area as low as one likes, the heat transfer coefficient being fixed. It is merely a matter of increasing the number of tubes in the bundle. This is not true of the arrangements shown in Figure 10 which are essentially mixed systems, ie, their performance depends on the interaction between cross-flow and axial flow components.

Some idea of the performance of orifice and rod baffles is given by Figure 12. This shows results of tests on a 21-tube bundle, the tube outside diameter being 19mm, layout staggered square and tube pitch/diameter ratio 1.33 (A small bundle is quite adequate for testing axial flow systems, the measurements being taken from the centre tube in the bundle). The rod baffle results shown in Figure 12 include Phillips (segmental and alternate) and other types, all with rods of 6.35mm diameter. The orifice baffle results refer to a 3.2mm diametrical tube/baffle clearance.

Figure 12 includes data for different baffle spacings and flowrates; it is similar to Figure 10 except that the pressure drop is given per unit tube length, which is more appropriate. It can be seen that a close link between heat transfer coefficient and pressure drop is again confirmed, the best fit curve showing a cubic relationship as before. Also shown in Figure 12 is the curve representing axial flow through the bundle without any baffles, calculated for different flowrates from data in the Handbook of Heat Transfer (10). The curve follows a relationship slightly greater than a square law, as mentioned above. It shows the penalty that is paid for putting baffles into heat exchangers. They are necessary of course to support the tubes and stop them from vibrating, but an interesting thought with which to conclude this lecture is, should rod baffles be aerodynamically streamlined?

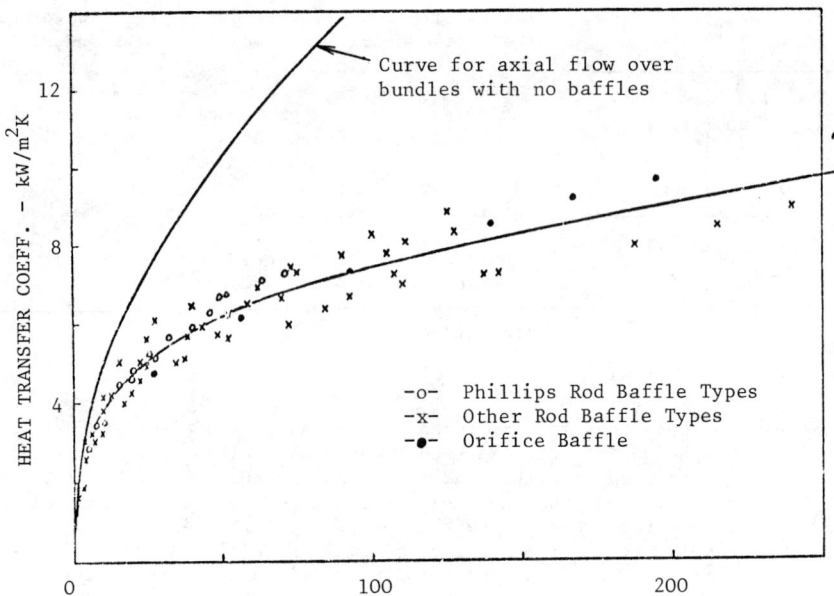

Fig. 12 Heat Transfer/Pressure Drop Relationship for Orifice and Rod Baffle arrangements at all Flowrates Tested

REFERENCES

1. Macbeth, R. V. 1978. Shell-and-tube heat exchanger data produced by an electrochemical mass transfer modelling technique. Paper No HX7, 6th International Heat Transfer Conference, Toronto.

2. Chilton, T. H. and Colburn, A. P. 1934. Mass transfer coefficients. Ind. Engineering Chem., Vol 26, pp 1183-1187.

3. Bergelin, O. P., Leighton, M. D. and Lafferty, W. L. and Pigford, R. L. 1958. Heat transfer and pressure drop during viscous and turbulent flow across baffled and unbaffled tube banks. University of Delaware Engineering Experimental Station, Bulletin No 4.

4. Gay, B. and Williams, T. A. 1968. Heat transfer on the shell-side of a cylindrical shell-and-tube heat exchanger fitted with segmental baffles. Bundle and zonal average heat transfer coefficients. Trans. Instn. Chem. Engrs, Vol 46, T 95- T 100.

5. Gay, B. and Roberts, P. C. O. 1970. Heat transfer on the shell-side of a cylindrical shell-and-tube heat exchanger fitted with segmental baffles. Flow patterns and local velocities derived from the individual tube coefficients. Trans. Instn. Chem. Engrs, Vol 48, T 3 - T 6.

6. Gay, B. and Mackley, N. V. and Jenkins, J. D. 1976. Shell-side heat transfer in baffled cylindrical shell-and-tube heat exchangers. An electrochemical mass-transfer modelling technique. International Journal of Heat and Mass Transfer, Vol 19 (9).

7. Small, W. M. and Young, R. K. 1977. Exchanger design cuts tube vibration failures. The Oil and Gas Journal, Sept 5th.

8. Rod design solves a baffling problem. Process Engineering, November 1977.

9. Stairmand, C. J. 1955. The design and performance of modern gas cleaning equipment. Paper presented to the Inst. of Fuel, 23 November.

10. Rohsenow, W. H. and Hartnett, J. P. 1973. Handbook of Heat Transfer, Pub. McGraw-Hill Book Company.

Heat Exchangers in Coal Conversion Processes

G. BORUSHKO
Exxon Research and Engineering Company
Florham Park, New Jersey 07932, USA

W. R. GAMBILL
Oak Ridge National Laboratory
Oak Ridge, Tennessee 37830, USA

ABSTRACT

A survey of the capabilities of commercial heat recovery equipment and related processes was conducted to determine their potential in coal coversion complexes. Major categories of heat exchangers investigated, in addition to shell and tube, included regenerators and direct contact units; Rankine cycle applications were also reviewed. Primary applications included feed-effluent and other process stream interchangers, combustion air preheaters, and steam generators (waste heat boilers).

The importance of heat recovery and utilization in coal conversion plants is recognized, as are the general problems associated with the design of heat exchangers. It is agreed that at present the rating (process design) of heat exchangers for coal conversion facilities is more difficult than the fabrication. These designs are based on the process conditions, physical property data, and materials of construction specified by the purchaser. No warranties are made concerning the effects of corrosion, erosion or fouling on performance.

There is a definite need for a research, development, and testing program to provide more physical property and heat transfer data for multi-phase streams and a better understanding of fouling, corrosion, and erosion.

1. INTRODUCTION

The specific investment costs (dollars per million Btu of input coal) for converting coal or lignite to electricity, synthesis gas, SNG, methanol, and liquid fuels are, broadly speaking, of the same order of magnitude; however, the product costs will vary significantly with the process thermal efficiencies. One factor obviously affecting these efficiencies is the degree of heat recovery and reuse that can be attained reliably and economically in the plants. For this reason and because of the degree of hostility encountered in coal-conversion facilities (e.g., dense slurries, dirty and corrosive streams, coke formation, and cyclic operation), a review of the heat transfer equipment and related processes that are potentially applicable to coal conversion was desirable.

A survey was conducted by reviewing the general literature and the current catalogs and brochures of a number of companies specializing in the design and fabrication of large and/or severe-service heat exchange equipment. Discussions were also held with a number of individuals conversant with the present state of the process heat transfer field.

The problems of designing and fabricating reliable heat exchange equipment for coal conversion plants should be recognized as not fundamentally different from those affecting exchangers for other plants. However, there are many uncertainties in this new field. These include stream compositions, thermophysical properties, rates of fouling, corrosion and erosion, and the method of designing for heat interchange between process streams that may include unusually large amounts of solids and/or highly viscous liquids. These problems can be solved only by focused research and development, field testing, and documented operating experience.

This paper addresses heat recovery, transport, and utilization. Operations such as primary heat extraction from methanation catalyst beds and rejection of heat to the atmosphere or surface water are mentioned only in passing, if at all.

2. SCOPE

2.1 Process Streams

Heat recovery and reuse are possible in two general types of equipment. Process vessels such as chemical reactors involve heat interchange because they contain exothermic reactions or because heat is used to sustain endothermic reactions. It may be necessary to protect the vessel or a catalyst from excessively high temperatures. In such cases, quench fluids are introduced or heat-removal surfaces may be installed. Jackets and coils have been used for this type of service; heat pipes are currently being considered. The size and geometry of a process vessel often determines whether internal heat recovery surface or an external exchanger is preferable. Heat extraction within a process vessel itself is excluded from the scope of this paper.

The second general type of equipment used for heat recovery is the heat exchanger or waste heat boiler using the flows between process vessels. This equipment type is the subject of this paper.

Coal gasification plants include at least some of the following operations:

o coal treatment

o gasification (in one or more stages), 900 to 3300°F

o catalytic shift conversion (in one or two steps), 700 to 800°F

o gas purification, -60 to 260°F, and

o methanation, 600 to 900°F.

Coal liquefaction plants vary considerably, depending on the liquids and associated gases produced. Such plants may include the following operations:

o feed preparation

o dissolution or pyrolysis

o gasification

o distillation

o filtration and drying

- o hydrotreating
- o ammonia removal, and
- o auxiliary units for fuel gas production and for provision of utilities.

The temperature drops between operations provide opportunities for heat recovery for all but coal feed preparation.

Types of streams from which heat may be recovered include:

- o solids - char, ash, and solidified slag, in some cases slurried with water or oil
- o liquids - distillation product streams and bottoms
- o gases - air, synthesis gas, SNG, and steam
- o multiphase mixtures - gasifier or carbonizer off-gas with suspended solid char and solid ash and with varying amounts of liquid oil and tar; slurries of ash in water; and dissolver liquid/solid products.

Streams being heated may include:

- o solids - moist coal feed
- o liquids - feed streams to reactors and distillation columns
- o gases - air, oxygen, and steam-feed streams
- o multiphase mixtures - reactor slurry feed composed of coal in oil with hydrogen gas, water for steam generation, and filter cakes.

Several auxiliary processes serve coal conversion plants, some of which are partially or wholly proprietary in nature, They include the power, sulfur recovery, and cryogenic air separation (O_2) plants. These are not discussed in detail in this paper.

2.2 Heat-Recovery Processes

This paper summarizes the methods used to remove heat from a process stream and

- o exchange it with another process stream
- o transfer it to combustion air
- o exchange it with an intermediate transfer fluid;
- o increase its value by converting it to electrical or mechanical energy by a Rankine cycle.

Also included in the paper is a discussion of several processes that have been applied to the cooling of hot solids.

2.3 Selecting Heat-Recovery Equipment

The selection of heat-recovery equipment for optimal efficiency in coal-conversion plants encounters several obstacles. Some of these problems are discussed.

Evaluating fully the life cycle costs of all promising alternative combinations of process conditions and heat-recovery equipment can be costly. The number of alternatives increases significantly if one is dealing with other than a specific plant site and is therefore obliged to consider a range of unit costs for electric power and water. Accurate trade-offs are impossible and simplifying assumptions may not be justified.

Assuming that the plant site is known, the cost of specialized equipment is not easily determined. Trade-offs may be made in equipment design by manufacturers to yield differing designs for the same equipment item. Particularly vulnerable are high-temperature devices. Given the requirement to handle a hot process fluid of known composition, various manufacturers may differ as to the best way to compensate for:

o lower metal strength and higher corrosion rates at high temperatures,

o thermal stresses caused by startup and shutdown or related to the exposure of various parts to different temperatures, and

o high-heat fluxes often associated with recovery of large heat loads at high temperatures.

Should cooling surfaces become fouled in a high-temperature service, metal temperatures could rise considerably. Fouling rates may vary with fluid temperature, and hotter surfaces generally are more susceptible to corrosion. Selection of the process operating temperature itself can thus involve trade-offs between higher product yields and increased equipment maintenance costs (or increased costs for maintainability of equipment).

One impediment to the optimal design of heat recovery equipment is the handling of coal-conversion projects. Ideally, this would progress from pilot plant though demonstration plant to the commercial design. The pilot unit is scaled up to a size smaller than a commercial plant but large enough to demonstrate reliable integrated operation in a commercial design. The experience gained in operating the demonstration plant should permit a more refined rating and mechanical design of the heat recovery equipment. This could result in a better design of the same type or a different, improved approach to heat recovery.

However, many demonstration plants, when they are built, stress process development with little thought to the operation of equipment such as heat exchangers. This is particulary true if the developer of the process does not plan to be the owner/operator of a commercial facility.

Relying on previous knowledge, the usual practice of soliciting several proposals and comparing them against such other serves to give a good selection in petrochemical operations. Unfortunately, the extension of this practice to many coal conversion streams cannot be done at this time with sufficient accuracy to justify significant expenditures. Both the engineering and the manufacturing firms may be unexperienced concerning fluid properties, fouling, corrosion, operating uncertainties and other factors of such streams. A recent study[1] in which severe service heat exchangers were submitted to vendors for proposals shows an extraordinary range in exchanger surface, type and arrangement. Table 1 shows a comparison of the offerings of four manufacturers for a flue gas/combustion air preheater service with a heat duty of 1074 million Btu/h. It would be a major effort to make a comprehensive study of the installed cost of these proposed offerings.

Table 1

Manufacturers' Proposals for a Flue Gas/Combustion Air Preheater

Equipment Type	Flow Arrangement	Heat Transfer Area per Unit ft^2	No. Units	Total Area ft^2
Heat wheel	Horizontal disc, counterflow	274,000	6	1,644,000
Rectangular shell w/10 vertically mounted rectangular tube bundles	Crossflow, air shellside	654,000	1	654,000
Shell & tube, fixed tubesheet	Air shell side, 100 parallel sets of 2 in series	14,820	200	2,964,000
Shell & tube, U-tube	Air shell side, shell double-divided flow	75,900	10	759,000

Although attracted by the potential market in heat recovery equipment for coal conversion plants, it is likely that few manufacturers will begin to test their own designs for such services. Most fabricators have neither the facilities nor the capital resources to embark on any extensive testing programs. Instead, the majority of advances by the industry will come from the more aggressive and innovative fabricators who will extend the limits of their equipment based upon feedback from the operation of previous designs coupled with general improvements in materials.

It is not likely that manufacturers will alter significantly the present policy of warranting equipment only for mechanical design and thermal performance when operated at the specified design conditions and fouling factors. Also, they will assume no responsibility for excessive fouling due to any cause.

The situation is similar for engineering contractors. Essentially all heat exchanger development and testing will be limited to engineering firms who are also engaged in coal conversion process research. Such information is invaluable in enhancing the marketability of the process. However, as in the process, the details of any out-of-the-ordinary heat exchange equipment are kept proprietary.

It is the licensee owner of the coal conversion facility who faces the most difficult decisions. Although he is the one who may make the largest monetary gains, he is also vulnerable for the greatest losses. Often he is in no position to evaluate within the allotted time frame the heat exchange equipment specified by the licensor. Obviously, he cannot embark on a program of development and testing. Reliance on excessive conservatism to offset lack of solid information on heat transfer coefficients may aggravate problems such as corrosion and fouling. This also has an adverse effect on rates of return in an industry whose financial picture could be better.

3. REVIEW OF HEAT-RECOVERY PROCESSES

3.1 Heat Interchange Between Process Fluids

Heat recovered from a process stream may be used to heat another process stream. When both streams are associated with one equipment item or process, the heat recovery is accomplished in a feed/effluent exchanger. These will usually be more attractive than alternative indirect third-fluid systems when:

o inadvertent contacts of feed and effluent do not create a hazardous condition

o distances and elevations are such that the available pressures can maintain flow.

It is important that process-fluid temperatures and properties be known with sufficient exactness that an economical equipment size can be determined. Attempts to size equipment conservatively often cannot compensate for inadequate data.

This is particularly true for temperature data of streams undergoing a change of phase. Non-linear heating/cooling curves can result in a gross reduction of temperature difference (temperature pinch) which even massive increases in heat transfer surface cannot offset.

<u>Shell and tube exchangers</u>. The shell and tube exchanger is the most common type and has been applied in the process industry at high pressures and temperatures in large sizes. A serious deficiency of the shell and tube exchanger stems from the velocity changes of the baffled shellside flow and of the tubeside flow at the tubesheets. The consequences of velocity changes are not so significant with single phase gases or liquids; however, multiphase flow can be a problem because phase separation can occur at turning points and at changes in cross-sectional area. When solids are present, the particles can accumulate in low velocity spaces and reduce the available heat transfer area. In high velocity areas, this can cause erosion. Erosion has been countered in some shell and tube exchangers by using impingement plates or solid rod shields as well as special ferrules or projection of tubes beyond the tube-sheet at the inlet end.

A survey of costs for large shell and tube heat exchangers compiled from manufacturers' estimates and quotations has been published.[2] This survey was limited to shell-and-tube units with the following specifications: carbon steel construction, surface areas of 20,000 to 30,000 ft^2, tubeside pressures of 200 to 2000 psia, shellside pressures of 100 to 4000 psia, 3/4- and 1-in. OD tubes 30 to 40 ft long, 1.25 tube pitch, single pass tube and shell sides. In current U.S. dollars, the average costs ranged from $10 to $90 per square foot.

In another article,[3] unit costs were compared for shell and tube heat exchangers that were identical except for the materials for the tubes and the cladding of the tubesheets. Figure 1 shows the relative costs for a range of surface areas for heat exchangers fabricated with 3/4 in. OD, 20 ft long tubes. The heat exchangers involved all have carbon steel shells and are TEMA type BEM (bonnet, fixed tubesheet) design.

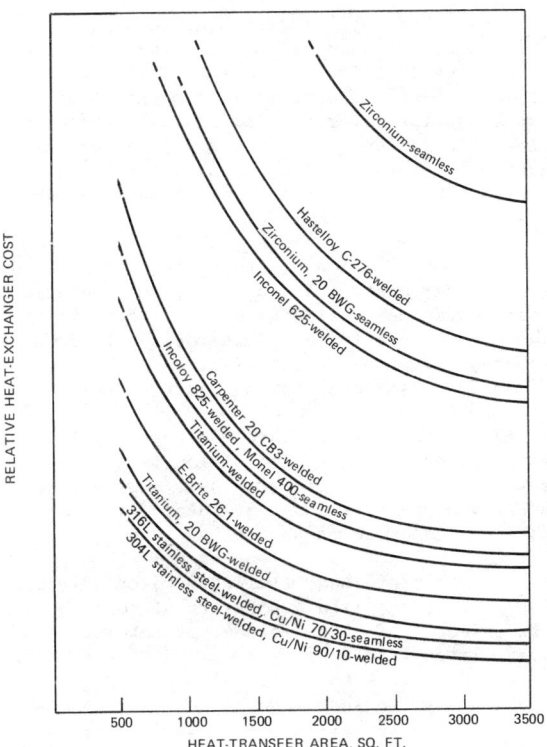

Fig. 1 Cost comparison for S&T heat exchangers, mild steel shells, tubing 16 BWG except as shown otherwise.

The above cost data may be of some use to the designer to give him a feel for the impact of various parameters in the pricing of shell and tube heat exchangers. However, other than for product finishing operations, it is almost impossible to generalize on the costs of the special design shell and tube units required for the severe services in coal processing complexes. Believable cost figures require reasonably detailed mechanical design and pricing information developed with a fabricator having experience working with that or a reasonably similar design.

The Phillips Petroleum Company licenses a variant of the shell and tube unit called the rod-baffle exchanger.[4] Practical considerations limit this configuration to the square tube pattern. In this concept, a matrix of rods in a flat plane is arranged perpendicular to the tubes. Rods equal in diameter to the spacing between adjacent tubes are inserted between every other tube at each baffle matrix. Rods at successive baffles are inserted between alternate tubes; each pair of baffles is at 90° to the preceding pair. Thus a baffle set of four constrains each of the tubes from four directions.

A virtue of the rod-baffle exchanger of particular interest to coal conversion operations is its possible use in processing slurries on the shellside. Such a unit, arranged vertically, with a U-tube or bayonet tube bundle, may be able to handle even slurries prone to saltation without appreciable solids buildup.

Spiral plate exchangers. This type of exchanger consists of an assembly of two long strips of metal plate, spaced apart and wound around an open, split center to form a pair of concentric spiral passages. The distance between the metal surfaces in both spiral channels is maintained by means of distance pins or studs welded to the metal strips prior to winding. Alternate edges of the passages are closed so that the fluids flow through continuous channels with no possibility of inter-leakage. Covers are fitted on each side of the spiral assembly to complete the unit. The usual counterflow arrangement is of interest in coal conversion plants.

The manufacturers quote good experience with spiral plates handling slurries with up to 30% solids. They claim that the turbulence induced by the single curving passages tends to reduce or eliminate solids deposition by reentrainment. This turbulence is also said to reduce fouling tendencies. Another advantage for processing erosive/corrosive coal fluids would be the relatively large thickness of plate used to fabricate spiral plate units.

A distinct disadvantage of the spiral plate exchanger is its size limitation of 2000 ft^2 and design pressure limitation of 150 psig. The full-faced gasket on the closure may also be subject to erosion. Complete welding of one of these passages can raise the pressure and temperature limits inherent with gaskets but then that passage cannot be mechanically cleaned.

Plate-type exchangers. The plate heat exchanger reportedly is efficient for highly viscous fluids such as the liquid products of coal conversion plants. These units consist of an assembly of thin rectangular embossed plates clamped together between two cover plates into a compact unit. Gaskets serve to seal the perimeter of the plates as well as the internal ports which conduct the fluids to and from the passages. Pressure loads transferred to the cover plates by multipoint contact of the embossed plates are carried by a plurality of bolts connecting the cover plates at their periphery.

The plate exchanger is limited to low pressure (about 300 psig) in smaller sizes and to about 150 psig in the largest sizes by the design of the end plates. Elastomeric gaskets limit the plate units to hot fluid inlet temperatures of about 300°F; this limit can be raised to about 450°F with the use of asbestos composition gaskets at some sacrifice of sealing effectiveness.

The maximum size of units suitable for process stream interchange is about 4000 ft^2. Units up to 13000 ft^2 with plate sizes 9 ft 8 in by 4 ft 6 in are available for water to water services.

Services with relatively long heating/cooling curves and close approach temperatures which result in a severe temperature cross favor the normally countercurrent plate heat exchangers. Claims are made for designs capable of a 2°F approach and an attainable heat recovery of 90% or more. The highly tortuous flow path between the plates results in turbulent flow in all but the most viscous streams, giving coefficients 3 to 5 times those possible in shell and tube units for the same pressure drop. This high degree of turbulence also results in less fouling than in comparable shell and tube units.

Plate heat exchangers can tolerate moderate quantities of suspended granular solids. Particles lodging in the crevices at corrugation contact points can be removed by backflushing. Backflushing is mandatory when solids size exceeds the spacing between plates since the ports act as efficient edge-strainers.

Safety considerations preclude the use of plate heat exchangers in low flash or toxic fluid services. The combustible peripheral gasket material which forms the major portion of the exterior surface is vulnerable to fire. Also the multiplicity of external gasketed joints increases the probability of leakage. External shielding has been proposed to mitigate these hazards; the efficiency of such shielding is highly dependent on its design. A truly efficient shield will negate the ease of inspection feature of the standard plate exchanger.

3.2 Heat Exchange Between Process Fluid and Combustion Air

The industrial fired heater market currently uses both recuperators and regenerators for heat recovery. Thermal regenerators (described later) use internal thermal storage matrices that are alternately heated by hot flue gas and then cooled by atmospheric combustion air directed through them for preheating. Recuperators, on the other hand, use a metallic surface between the two continuously flowing streams. The surfaces may be constructed to maximize radiant or convective heat transfer. Two sections may be installed in series, a radiant section at the hot gas inlet end and a convective section at the cooler end.

Recuperators. The radiant section of a stack-type recuperator consists of two concentric cylinders. When lateral pressure differentials are significant, the high-pressure stream is sent through the inner cylinder and the low-pressure stream through the annulus to minimize the stresses in the intervening wall. Two radiant sections can be arranged in series, the cooler-gas end in counter-flow and the hot-gas end in parallel flow. This arrangement can be used to protect the recuperator wall from excessively high temperatures at the hot gas inlet. The basic radiant type is generally used for hot stream temperatures above 1600 to 1800°F, at which temperatures thermal radiation from CO_2 and steam is the principal transfer mode. Below about 1300°F, convection sections are used. Convective tubular exchangers, either vertical or horizontal and with either gas or air through the tubes, are installed to receive flue gas already cooled in the radiant zone. Generally, the radiant-type recuperator is used for low-pressure service (< 50 inches H_2O), but higher pressure units (2 psig) can also be built. Both field-fabricated and shop-fabricated units are available. Manufacturers' literature suggests that radiant-type recuperators may be used with gas entering at temperatures as high as 2500 to 2600°F. Convective recuperators with flue gas through the tubes can handle dirty gas continuously at temperatures as high as 2000°F. Tubes of 2.5- to 4-inch diameter are used, and each is fitted with an expansion joint. The shell also may have an expansion joint to allow for thermal growth. For a 950°F preheat temperature, such devices reportedly are designed for flue-gas velocities of about 10 ft/sec and for air velocities of about 35 ft/sec at standard conditions.

In addition to the tubular recuperators described, modular canal type units are made for low-pressure service. The tubular modules are suspended in large brickwork "canals" or flues bounding a horizontal flow of flue gas. The stream to be heated passes through vertical tube bundles. Flow is generally 2 to 4 passes through the tubes with flexible connections between transfer boxes at the base of two adjacent bundles. Square or rectangular bundles are used. Heating fluid temperatures as high as 2200°F are used to preheat air to 1200°F.

Another type of recuperator is the cage type, so called because the tube bundle resembles a cylindrical cage with an open center. This type is used for the higher temperatures encountered, for example, in the glass industry (2500°F maximum gas inlet temperature). The cage is housed in a refractory-lined vessel, and heat transfer is primarily by radiation. Air may be handled at high pressure (250 psig) in these units.

Recuperators may be designed for preheating media other than air (e.g., hydrogen). If the heating stream is hot and dirty, radiant-type recuperators are attractive because there are fewer obstacles to trap dirt.

Regenerators. Batch-type thermal regenerators (i.e., heat exchangers with thermal storage) are so identified because they are operated in an alternating batch mode. Cold-and hot-gas streams are alternately valved to direct a given stream to one of two, or sometimes more, regenerator chambers so that the streams may alternately exchange heat with a high-heat-capacity matrix. The matrix is stationary and is usually made of a ceramic material (either crude pebbles or rubble) or, for lower pressure drop, of preformed checker-work elements of simple or intricate shapes. For gas flows, the containment vessels are necessarily large. Only a small fraction of the heat stored in the matrix is removed during a cooling cycle because of the low thermal conductivity of the matrix.

Batch-type regenerators have been used extensively in the steel and glass industries for preheating air to 1650 to 2200°F, usually using a checkerwork matrix. One summary[5] describes blast-furnace stoves using 9 by 4.5 by 2.5 inch firebrick assembled to form flues 3.25 in. sqare. An air flow of 100,000 scfm is preheated from ambient to 900 to 1200°F (1000°F time average) using four stoves. One unit heats air while three are being heated by hot gas. The shells for the stoves are 24 feet in diameter and 100 feet high; they contain 124,000 ft^2 of surface area. Bypassing air flow is used for control to maintain a reasonably constant preheat temperature. Open-hearth-furnace regenerators require larger flue openings because of particulates. A less expensive pebble-bed regenerator has been used successfully[5] to preheat air to 2800°F in a 25 ton-per-day blast furnace that produces pig iron and ferroalloys.

Disadvantages of the batch regenerator include:

o the need for large switching valves

o a possible requirement for control valves if reasonably constant preheat temperatures are needed

o difficulty in cleaning

o cross contamination of streams unless purging between cycles is performed

o sluggishness in bringing the regenerator to temperature

Nevertheless, such regenerators are relatively attractive for use with high-temperature gas flows bearing particulates that might plug the smaller passages in equipment such as rotary regenerators. Batch regenerators are most attractive for streams having large temperature differences and moderate pressures. Although shop-fabricated units are available, the sizes most economical for a coal conversion plant would likely require field fabrication.

Equations for the thermal design of batch regenerators have been summarized by Mueller[6] and by Schack[7].

Combustion air for a fired heater or a boiler can be heated in a continuous mode by means of a "heat wheel" type of preheater. In these units, a thermal matrix rotates and cyclically exposes a sector of the matrix, first to a hot flue-gas stream and then, as it turns further, to a cooler inlet-air stream. The heat-wheel matrix is usually made of ceramic coated material in those sections that are to operate above 600 to 700°F. It may be made of two layers, one being a more corrosion-resistant material to resist acidic cold-end condensation. If the layer shows corrosion, the corroded matrix may be replaced; in some units it may be removed, reversed, and reinstalled to present a fresh surface to the cold inlet air. Leakage rates typically increase with corrosion of seals and matrix and often reach 15 to 20% before corrective measures are taken. Other options for the heat wheel are soot blowers and washing jets for keeping the surface clean to maintain low-pressure drops, though the added deposits can help thermally because of their heat capacity. Heat wheels are normally used in moderate temperature environments and at low pressures.

A device resembling the heat wheel is the "Rothemuhle air heater," which uses a heavy stationary matrix for transiently storing heat. A sector of the frontal area of the disk-shaped matrix is subjected to a flow of cool air diverted through the disk by slots in two rotating hoods, one on either side of the disk. The hoods direct hot air through that portion of the disk not exposed to the slotted openings. The two currents of air are swept in a circular motion.

Flow-stream orientations are reversed in some installations. The advantage of the Rothemuhle air heater is said to be the lighter duty involved in rotating the less massive hoods, which constitute about 20% of the device weight. In carbon steel it is used for temperatures as high as 800°F.

In both types of heat wheels, there is considerable area to be sealed if hot and cool gas streams are to be kept separate. Heat wheels in the utility industry normally operate with pressure differences of less than 40 inches of water. Heat wheels have been reported[8] with thermal efficiencies of as much as 85% and diameters of 70 feet. The same source also reports that purging has been used to limit contamination in a laminar heat wheel to as little as 0.04% and particle contamination to 0.2%. Single heat wheels have been built to preheat as much as 9 to 10 million pounds per hour of air.

Equations and design curves for rotary air preheaters are available in several references.[6,9] Smaller power plants rely on tubular air preheaters in addition to the rotary regenerative type.

Costs for rotary preheaters vary considerably, depending on pressure drop limitations and moisture loadings in the plant to be served. Estimated prices for large units currently range near 70 cents per pound per hour of 750°F flue gas ($3/scfm of flue gas).

A recent inquiry concerning several heat exchangers in typical coal-conversion-plant evaluations resulted in a reply from one vendor which suggested studying the use of a rotary heat wheel for conditions involving high-pressure differentials between the two flow streams. Similar work was reportedly undertaken in the 1950's in a project for gas-turbine use. Conditions were to have been 1000°F and 4 atm pressure differential at the wheel. More recently (1974 to 1976), testing has been performed on ceramic heat wheels of small size for gas-turbine and Stirling engines.[10] Equipment from several suppliers made of two materials (lithium aluminum silicate and aluminum silicate) was tested. Matrix geometry was also investigated. The tests were based on operation of a 28 inch diameter unit at 1475 and 1830°F.

The problems of using heat wheels are aggravated by high temperature (warping of many matrix materials makes sealing difficult), high pressure differentials (high sealing forces require the matrix to be sufficiently strong to support the torque that the drive must produce), and large heat loads (large size and large resultant deflections can cause leakage). In addition, the pressure drop across the heat wheel may require a fan to maintain flow. Controls and a bypass will be required in some cases[11] to ensure safety should the heat wheel stop. Standard industrial heat wheels accommodate high temperatures and low pressure differentials in small sizes[12] or moderate temperatures and low pressure differentials in large sizes.[13]

Heat pipe arrays. Heat pipes are axial high-conductance, heat transport devices that have undergone considerable improvement since their initial development about 15 years ago. Such a pipe is a closed structure containing a working fluid that transports heat from one end (evaporator) to the other end (condenser) by liquid vaporization in the evaporator, axial vapor flow in the central core region, vapor condensation in the condenser, and condensate return by gravity and/or capillary forces to the evaporator, as illustrated in Figure 2.

The relevant literature is abundant; a good overall treatment may be found in Chi's book.[14] A lucid summary of the several operating limits imposed by fluid dynamics considerations (limits relating to boiling, capillary pumping, sonic flow, and entrainment) is given in Tien's article.[15] The potential of the heat pipe in coal-conversion processes has been addressed by Ranken in his overview.[16] Design concepts include the incorporation of heat pipes into indirectly heated coal-steam gasifiers, into methanators for heat removal, and for energy recuperation in general, especially in first-generation plants where the shift reactor effluent gas is cooled to a near-ambient temperature for sulfur removal and then reheated for methanation. Also included are descriptions of Biery's heat-pipe recuperator concept,[17] devised to increase the efficiency of heat recovery from methanator product gas.

Strimbeck and co-workers[18] reported earlier results from an experimental test program, which used three heat pipes to couple thermally a fluid-bed, coal-air combustor at 2200°F with a fluid-bed, coal-steam gasifier at 1800°F. In the subject program, 14-inch long, 0.84 inch-OD heat pipes made of Inconel 601 and lined with a wick structure of 316 stainless steel screen were operated

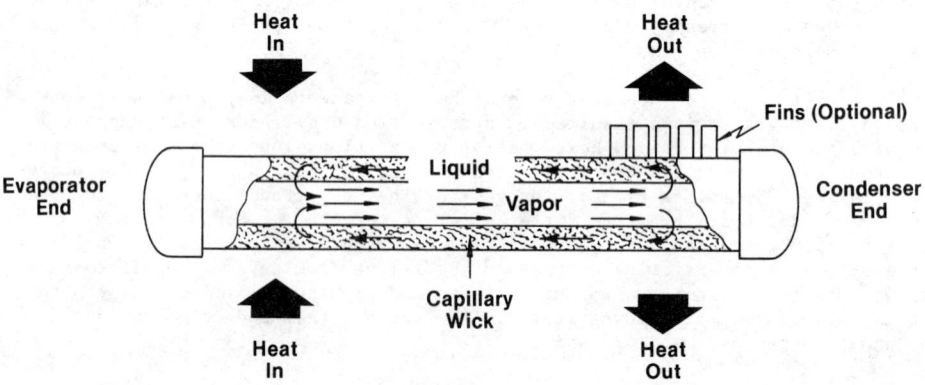

Fig. 2 Schematic of heat pipe

at about 2000°F with pure sodium fill fluid. Based on the measured decrease in condenser-end wall thickness (0.109 inches initially) caused by erosion/corrosion in the hostile gasifier environment, complete consumption of the pipe wall would have occurred at approximately 3500 hours (4.8 months). To extend the lifetime, proposals have been made to use protective coatings, sleeves, or composite wall construction.

Exchangers have been fabricated from arrays of laterally finned heat pipes.[19] Standard units cover an operating (external fluid) temperature range from -60 to 900°F. The central partition plates used to separate the hot and cold gas streams can be made of carbon steel or some other suitable metal, insulated if required. Fins, ranging from 4 to 14 per inch, and pipes are also available in a variety of metals as required by the service. The pipes are usually inclined slightly, with the evaporator end low, to augment condensate return by gravity. Face velocities and depth of array are determined by the available pressure drop. Custom units of stainless or aluminized carbon steel pipes and fins can handle flows up to 1×10^6 scfm and temperatures up to 1500°F. Design considerations and correlations applicable to heat pipe heat exchangers have been published.[20,21] Figure 3 shows a proposed ceramic recuperator utilizing heat pipes.

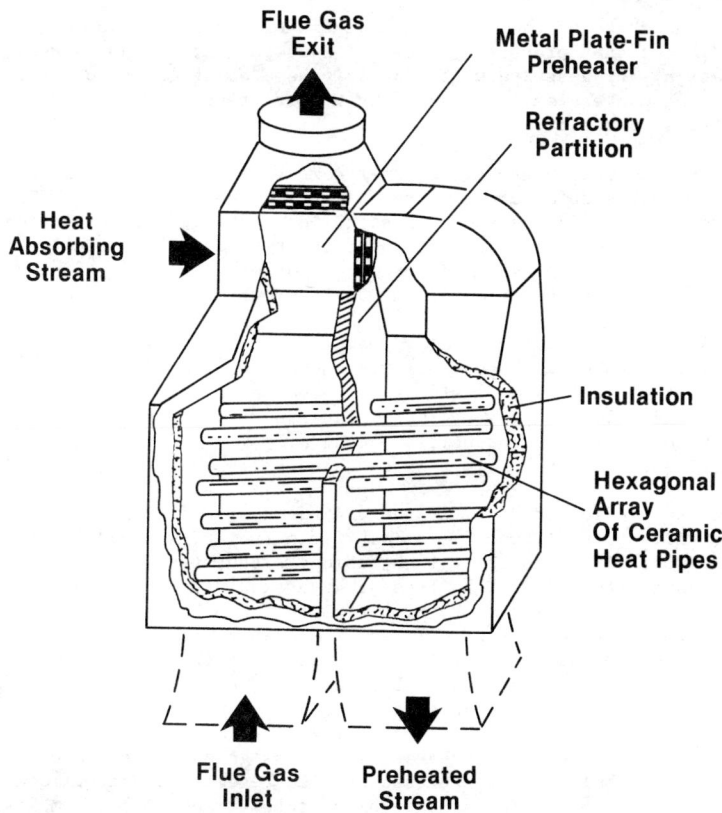

Fig. 3 Ceramic Heat Pipe Recuperator

3.3 Heat Interchange Using Intermediate Heat Transfer Fluids

Waste heat boilers. Waste heat or heat recovery boilers may be used to generate steam for utility use (e.g., to drive turbines) or for process use. They may also be used to vaporize other liquids such as organic heat transfer fluids. Excessive thermal degradation of the organics limits their use to a temperature of 750°F. Waste heat boilers have been built for flue gas flow rates as high as 1 million scfm.

Coal conversion plants using waste heat boilers have been built overseas, and recent U.S. plant designs incorporate such exchangers.[22] High temperature waste heat (2400 - 3000°F) from slagging entrained coal gasifiers can be recovered effectively in a series of radiant and convection waste heat boilers. Heat exchange is done in stages to minimize the size of the costly pressure shell required by the hot unit. Medium level waste heat leaving fixed bed type gasifiers can be recovered more economically by convection type boilers or shell and tube heat exchangers.

The use of waste heat boilers is attractive for more reasons than just increased thermal efficiency. It permits some degree of corrective action for a poorly performing unit. Use of a waste heat boiler may also allow the designer to deal with fewer unknowns in a given heat exchanger, provided the water is treated to a high qualtiy.

The current state of the art in the design of waste-heat boilers as described[23] indicates that pressures of 650 to 2400 psig are commonly used for power generation. Process steam is usually generated at 125 to 650 psig. In the case of coal conversion processes, the higher pressure steam may be used as one of the feeds to the process.

Designs for heat-recovery steam generators have been developed for several industries. Boiler manufacturers, fired heater companies, vendors of shell and tube heat exchangers, suppliers of transfer line exchangers (for ethylene furnace product quenching), and fabricators of reactor vessels have all shown originality in their approaches. Articles on waste heat boilers[23-26] summarize the pros and cons of various traditional designs, which are discussed below.

Flue gas boilers normally handle flue gas at near-atmospheric pressure (typically below 2 psig). The usual arrangement is steam tubeside; this allows containment of the flue gas in a simple box-type casing. With higher flue gas pressures (over 5 - 10 psig), a cylindrical casing is required. Inlet gas temperatures for an all-convection unit are generally limited to about 1800°F, although some manufacturers claim they can handle flue gas up to 2400°F.

One author[24] indicated that "if the unit is designed by a boiler manufacturer, he will typically use 1-1/2 inch plain tubing, whereas a fired heater contractor is more likely to use 4 inch tubing, frequently with extended surface heat transfer fins." Such boilers of a vertical tube configuration typically use natural circulation; but, if horizontal, the boilers normally are provided with forced circulation to lessen the chance of dryout in the tubes. Selection of type depends on space limitations. In either case, a mixture of water and steam cools the tubes and tubesheet or tube header before being directed to a steam drum for separation of the steam.

The sensible heat in high temperature process gases is recovered by radiant-type waste heat boilers. Low pressure units resemble a conventional boiler radiant firebox. The pressure radiant-boiler tube arrangement resembles cylindrical fired heaters where cage shaped tube panels are used. While the mechanical construction of the waste heat boiler is simple, heat transfer calculations are rather difficult. Reliable gas emissivity information for gasifier product components at elevated pressure is proprietary and not available from literature information sources. This problem is further complicated by gas emissivity spectral overlap caused by the opaqueness of one gas to another within a certain wavelength range. However, boiler designers should have no problem in scaling up pilot plant heat transfer data for a commercial unit.

The same convection heat recovery equipment used for dirty gases in zinc oxide plants and industrial kiln exhausts is also suitable for gasifier applications provided the gas if free of hot slag particles. Convection boilers basically fall into two categories: firetube boilers and water tube boilers.

If the process gas is at high pressure and dirty, it is common to route the gas on the tube side (firetube) where velocities are better managed. Also, the larger diameter tubes are more resistant to total stoppage of flow. Some firetube boiler designs have used an internal by-pass pipe with a valve for control of gas outlet temperature. Refinery operators have reported trouble with such valves.

Thermal stresses in the tubesheet of the firetube boiler can be a problem, particularly if thick tubesheets are used. One manufacturer also points out[25] that firetube boilers are somewhat different from conventional shell and tube exchangers because the former are often fixed tubesheet designs with the tubes, tubesheets, and shell kept at relatively the same temperature by the water boiling in the shell. This factor permits some reduction in tubesheet thickness in a boiler as compared with a sensible-heat shell and tube exchanger.

The inlet tubesheet in a firetube boiler is particularly vulnerable, especially in vertical units. Protection may be provided by insulating the hot tubesheet outer face with a refractory material (also the inlet channel and perhaps the outlet channel) and effectively insulating the tubesheet at each tube penetration by using an insulating layer between a tube ferrule and the tubesheet or tube wall. Tube ferrules are used to direct the hot gas from the inlet channel through the refractory lining and the tubesheet into the tube at a safe distance from the tubesheet. The ferrule also protects the tube from erosion at the highly turbulent inlet end. Insulating the tubesheet permits it to be thinner because the material is stronger at lower temperatures. Insulation also permits the tubesheet to operate nearer the temperature of the shell and thus reduces thermal stresses. Proper circulation of water is important to carry off steam, prevent dry-out, and disperse particulates to reduce fouling. Thick tubesheets are unavoidable in some cases (such as a U-tube high pressure design), for which shell diameters may be minimized to reduce the total load on the tubesheets.

An interesting double-tube steam generator is used for rapidly quenching low pressure effluent gases from ethylene furnaces. One can think of the unit as a firetube boiler with a small concentric shell around each tube. The exchanger consists of several water-jacketed firetube elements. The central pipe of each element extends at both ends through oval cross-section headers and is welded to the headers. The jacket does not extend through the headers but is so attached that each oval header feeds the annular jacketed space. Several elements are installed in parallel between the two oval headers and form a register. Several registers are combined to form a vertical tube bundle. The exterior surfaces of oval headers in adjacent registers are welded to each other to form a leak-tight "tubesheet," somewhat similar in concept to a membrane wall.

Each oval header connects outside the exchanger to a piping manifold. The net result is that the exterior of each double-tube element can be at atmospheric pressure, surrounded only by insulation and a weatherproof jacket. The concept is illustrated in Figure 4.

The double-tube boiler seems to have considerable merit in permitting the use of simplified shop construction (small welds might be automated), and it eliminates the requirements for thick tubesheets. With some double-tube designs, the firetubes are straight and easily decoked. A possible application of these exchangers in coal conversion is the rapid cooling of the stream from a flash hydropyrolysis reactor. In this process, pulverized coal is mixed with hydrogen at elevated pressures and heated rapidly. A mixture is formed that, when quenched rapidly, yields valuable hydrocarbon liquids and gases. Typically, the reactor effluent stream would be cooled from 1600 - 1800°F down to 600 - 800°F, a temperature level low enough to freeze the chemical reactions but high enough that the light oils remain in the vapor phase. Currently, direct-contact quenching seems to be the concept of choice, using a water spray, low-temperature steam, higher-temperature liquid metals, or molten salts. The liquid metal or molten salt could be recovered and used to generate steam or for process heating.

As stated earlier, process gas boilers may be either of the fire-tube or watertube variety. Watertube boilers for waste heat recovery are often based on a patented or proprietary design. Generally, a vertical tube orientation is preferred to minimize the potential for dry-out in the tube. Any impurities in the feedwater concentrate at points of dry-out, and accelerated corrosion can occur in such zones. A recent article summarizes operating problems with waste heat boilers of the most common designs.[24] Failure mechanisms in watertube boilers have included:

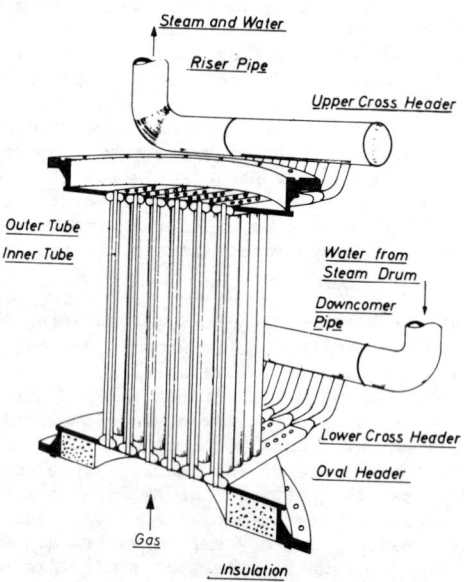

Fig. 4 Double-Tube Steam Generator

- Creep rupture of individual tubes caused by overheating following complete or partial flow blockage. Debris and magnetite scale dislodged during plant upsets have caused such blockages.

- Rapid tube corrosion following massive acid breakthrough from the water-treatment plant.

- Accelerated corrosion beneath deposits in regions of high heat flux.

- Creep rupture of the shell because of failure of the refractory caused by a combination of poor design, workmanship, and quality control.

- Extended tubewall dry-out.

Using conventional sootblowers, watertube boilers are easier to clean than firetube units. Stationary sootblowers are preferred over a retractable type for high pressure applications where penetrations in the pressure shell must be sealed off. To minimize the quantity of high pressure cooled process gas required, sootblower nozzle design for sonic exit gas velocity is highly desirable. If convergent-divergent nozzles are used, supersonic exit velocity can be obtained with lower sootblower gas pressure.

Variations of watertube boilers include bayonet-tube arrangements and the U-tube design. These two types require especially clean water. Circulation can be either natural or forced.

The reported experience with heat-recovery steam generators used to cool raw gasifier gas is limited. Franzen, in responding to several questions concerning the waste heat boilers used with Koppers-Totzek coal gasifiers, stated[27] that the steam generation pressure is from 40 to above 100 atmospheres, that the gas space velocity is "high," and that tube life with respect to erosion is "satisfactory." In his review of ammonia production from coal,[28] Waitzman cites high pressure steam generation "in a commercially-proven waste heat boiler" as a benefit of the Koppers-Totzek process.

A review of the first 18 months of operating experience with the coal-based ammonia plant at Modderfontein, South Africa,[29] however, indicates that ash-impingement erosion did occur in the waste heat boilers serving the Koppers-Totzek gasifiers, which operate at about 2900°F. Quench water is injected into the hot gas and molten ash leaving the gasifier top outlet, resolidifying the ash, and cooling the gas to about 1650°F. The gas then passes sequentially through a radiant boiler and two parellel tubular (convective) boilers which are connected by three tangent-tube boilers, all of which generate steam at 800 psia. The gas changes direction in passing through the tangent-tube boilers, and the abrasive high silica and alumina ash particles impinge on the inner surface of the boiler tubes. The worst-affected areas were initially eroded at a rate of about 3/16 inch per 1000 operating hours, corresponding to a maximum local wall thinning rate of 1.644 inches per year. The localized erosion was arrested by installing protection plates made of 310 stainless steel. As an experiment, all tangent-tube inner surfaces on one set of boilers were coated with an abrasion-resistant alumina material.

A waste heat boiler design for cooling process gas from partial oxidation gasifiers has been reported.[30] Gas, oil, and coal feedstocks are mentioned, as are inlet gas temperatures of 2400 to 2725°F and pressures up to 1150 psia. The design consists of a vertically suspended bundle of bayonet tubes cooled internally by boiling water in either natural or forced circulation. Though

"ensured good control" over the extreme operating conditions - which include inlet heat fluxes as high as 317,000 Btu/h ft^2, soot and erosive ash entrainment, and corrosion by H_2S - is mentioned, no specific operating experience is cited.

During the 1950's, a coal-based ammonia plant was operated at Belle, West Virginia, using two gasifiers, a smaller unit of semiplant scale, and a later larger one of commercial scale. The operating experience with the heat-exchange equipment in this plant is probably reflected in statements made in a preliminary evaluation of the technical and economic aspects of producing fuel-grade methanol from coal. Prior to methanol synthesis, the 330 psia gas leaving the gasifier was cooled to 1200°F by extracting heat in steam superheater banks. Following this, particulates were removed in multiclones and process steam generated in waste heat boilers. The gas then flowed at 300°F to a venturi scrubber and to a water-scrub tower. As stated in the evaluation, prior to water scrubbing of the synthesis gas, "the high temperature sour reducing environment within the gasifier represents a high potential corrosion problem for all heat exchange surfaces, and the ability to maintain high metal temperatures in a corrosive gasifier environment must be demostrated."

There is other successful experience involving waste heat boilers with inlet gas temperatures of 2700°F. In one case, 12 units were installed on a nickel smelter to cool very dusty off-gas at approximately 2 psig to 1800°F. Gas enters the 3 inch diameter tubes at 500 to 600 fps, generating steam at pressures as high as 150 psig. The units reportedly have worked well since their installation about 5 years ago. In the second application, a boiler was installed on an oil-fed partial-oxidation unit about 20 years ago. Gas enters at about 2700°F and 400 psig, laden with soot and bearing H_2S. Steam at 600 psig is generated as the gas is cooled to about 500°F. Initial corrosion problems with hot sulfidation of the tube walls presumably have been solved.

In summary, waste heat boilers that cool raw coal-gasifier off-gas are subjected to extreme conditions that can be handled by special designs, though apparently not without problems. A recent program recommendations[31] stated that development work and plant experience are needed to complete and confirm designs of gasifier waste heat boilers that are exposed simultaneously to high temperature and high pressure. It was suggested that such boilers be designed and installed in selected pilot plants.

In-plant thermal energy transport. Thermal transport systems are designed to transfer thermal energy from a central source such as a fired heater to a series of sometimes distant energy users by means of a heat transfer fluid. The American Schack Company, Inc., Pittsburgh, Pennsylvania, for example, offers a series of such "Heat-Energy Systems" ranging in capacity from 2 x 10^6 up to about 150 x 10^6 Btu/hr.[32] These systems encompass liquid- and vapor-transport fluids in both atmospheric and pressurized loops.

A 1.5-mile-long closed waste heat loop is being installed at a large refinery in Louisiana.[33] Refined kerosene will be used to absorb heat from 11 sources (mainly 800°F flue gas) and transfer it in heat exchangers to preheat boiler feedwater and to provide reboiler heat to a butylene fractionating unit. The kerosene will be circulated at a rate of 3940 gpm between high and low temperatures of 417 and 210°F. Heat savings will total about 4.8 billion Btu/day.

Fried has summarized the characteristics, physical properties, and heat transfer efficiency factors of 14 high-temperature heat transfer fluids, with emphasis on the temperature range from 350 to 1000°F.[34] Both liquid-phase and vapor-phase systems are treated. The fluids include water, heat-transfer salt, and a variety of heat transfer oils.

The use of thermal transport can be extended conceptually to include thermal energy storage, which enables retention of available surplus or waste heat for recovery and use later when it is needed. A significantly variable capacity factor in a coal-conversion plant might make the use of thermal storage attractive. Primary emphasis to date has been on electric-power systems.[35,36]

Rankine cycles for converting waste heat to electrical or mechanical energy. In a low-temperature cycle, a waste heat stream is used to generate steam, which condenses in a shell and tube exchanger and generates Freon vapor to drive a rotary, positive-displacement expansion engine. The expander isentropic efficiency with Freon 12, at a rated P of 315 psi, is 80%. The Freon vapor is condensed in a water cooled exchanger and recirculated to the vaporizer. Units producing 10 kW to 10 MW, or alternatively, mechanical power for pumping or compression, are said to be available. Cycles have been operated on Freons 11, 12, and 114 at onstream availabilities exceeding 99%. The system is presumably applicable to any waste heat source temperature of 150°F or higher; with a difference between heat-source and heat-sink temperatures of 100°F or above, typical payout times of 3 years or less are claimed.

High temperature cycles that use steam as the working fluid are also available. These packaged, closed-cycle systems use steam from a heat recovery boiler to drive a turbine generator, following which the steam is condensed and recirculated as water to the boiler. With clean (low-particle loading) exhaust gases, the boiler can be made more compact by using externally extended surfaces and forced-water circulation. For dirty hot gases, bare tubes and natural water circulation are used, and steam or air soot blowers and collection hoppers are provided. Units are offered that generate 500 kW to 15 MW and that operate as either condensing or backpressure systems. Turbine inlet steam conditions can vary from 15 psig at 250°F (saturated) to 600 psig at 700°F. For a nominal 1500-kW system with turbine inlet steam at 450 psig/700°F and a condenser temperature of 115°F, the steam rate is 13 pounds of steam per kilowatt hour. Typically, these units cool fairly clean waste gases from temperatures of 1000 to 1500°F down to about 350°F, or to a temperature slightly above the dew point. The most recurrent problem is boiler surface fouling by particles in the hot gas stream.

3.4 Solids Coolers

Coal-conversion process equipment must deal with hot solids such as ash, char, catalysts, and thermal transport media or with mixtures of these materials.

For waste materials posing no contamination problem, a fluid having good heat transfer properties (water, e.g.) can be mixed with the solid. One method for recovering heat from the ash or slag is to quench it in a water bath, flashing steam from the water in a special vessel. In a proposed gasifier, for example, 50 psig steam is generated in a quench tank, leaving behind a slag slurry (23 wt % ash at 298°F). Considerable heat and water escape with the waste slurry. In the agitated quench-bath water, the ash-containing slag is converted to a frit or sand-like material that can be conveyed easily by the water. Variations in heat recovery from the water include the use of jacketed vessels or coils from steam generation or for process heating. If pressure differences between fluids could cause leakage through the heattransfer walls, the process fluids should be mutually compatible.

For cases in which contamination of the solid by a bath liquid would seriously decrease its value, a fluidized-bed heat exchanger could be used. The fluidized bed requires a gaseous fluidizing medium. A compatible fluid can be separated from the carry-over solids downstream of the cooler. An ERDA-sponsored study[37] proposed a fluidized-bed char cooler to receive char at 555 psia and 1600°F. With the recovered heat, 600 psia steam was generated in a tube bundle immersed in a bed of char fluidized with steam. Although such fluidized-bed coolers are not standard commercial equipment, they have been built for special applications, such as for cooling alumina.

Several configurations of air-cooled clinker heat-recovery units are used in the cement industry.[38] These units require large flows of air, which are used to heat incoming feed material. The air flow is cleaned of carry-over solids between stages. The equipment is currently limited to operation at near atmospheric pressure.

Heat recovery from solids in coal conversion plants does not appear attractive with the metallic conveyor-belt coolers used, for example, in the food industry. The belts may be made of thin stainless steel but require cooling to prevent annealing. Although hot solids may be handled, the cooling medium, generally water, is allowed only a small temperature rise of about 2°F. The coolant is not useful as a heat source in other parts of the plant, although it is uncontaminated. Metallic conveyor belts are not currently made for operation at pressures significantly above atmospheric.

3.5 Direct-Contact Coolers and Condensers

In cooling and/or condensing hot gases and vapors such as pyrolysis gases, considerations relating to pressure drop, fouling, and economics may render indirect heat exchange through metal walls undesirable or infeasible. An important application for which direct-contact quenching of a gas with a liquid has been used is the partial cooling of an ethylene furnace effluent process stream. Such quenching cools the stream rapidly through a critical temperature range and prevents an undesired continuation of the reaction, i.e., freezing of the chemical equilibrium. Direct-contact cooling is also employed while entrained dust is simultaneously scrubbed from dirty hot gases.

Fair's review article[39] summarizes briefly design rules of thumb and equations for direct-contact cooler/condensers. The design relations are quite approximate, many are derived by analogy from mass transfer data or correlations, and none are applicable directly to tower diamters exceeding 6 feet (usually restricted to $D \leq 2$ feet). At best, the equations are based on data obtained with air/water, air/oil, and H_2/light hydrocarbon/oil. To account better for all the influential variables, more comprehensive correlations are needed. Areas addressed should include the degree of back-mixing and of entrainment in large diameter spray chambers and the axial drop-size distribution in pipeline contactors with turbulent mist flow. Data obtained in studies of water-injection line desuperheaters for steam and with other gas/liquid combinations indicate that the uniformity of initial phase dispersion is quite important because thermal effectiveness decreases rapidly with nonuniformity of dispersal.

The Ozaki quench cooler (patented in Japan) is soon to undergo prototype testing.[40] Employed in a light-crude-oil cracking process, the cooler is claimed to prevent coking on the tube side of firetube waste heat boilers by causing quenching oil to flow concurrently with the gas on the inner wall surfaces. Gas from the cracking process typically might be cooled (tubeside) from 1650 to 750°F while generating steam (shell side) at 600 to 1400 psia. A thin oil film is maintained by high hot-gas velocities (65 to 165 fps). Other advantages claimed include low quench-oil degradation and minimal wall corrosion by gases such as hydrogen sulfide because of the protective quench-oil coating.

Since direct-contact coolers are custom designed in the larger sizes, adequate generalizations concerning their costs are not available.

4. BACKGROUND DATA AND PROBLEM AREAS

4.1 Heat Exchanger Thermal Experience

Even in services typically cleaner than those characteristic of coal conversion processes, the difference between calculated and actual heat transfer rates can be significant. Illustrative of this difference is the summary paper by Smith,[41] who tested several hundred heaters, coolers, and condensers in a petrochemical complex. Streams included pure and mixed liquids and gases at pressures from 1 to 360 atm abs. The ratio of the calculated clean to measured overall heat-transfer coefficients derived for each heat exchanger varied from 1.1 to 4.1. The deviation between calculated and observed values was attributed primarily to heat transfer surface fouling and secondarily to stagnant fluid regions in the shells and to flow maldistribution among parallel tubes.

Many tubeside liquid and slurry process flows in coal conversion plant heat exchangers have been and are expected to be in the laminar or transition flow regimes. The corresponding low heat transfer rates have been documented in a report[42] which indicates that overall heat transfer coefficients for operating coal liquefaction plant heat exchangers have varied over the following typical ranges:

Exchanger type	Coefficient range ($Btu/hr \cdot ft^2 \cdot F$)
Recycle gas coolers	30 to 100
Heavy vapor condensers	40 to 80
Reactor effluent coolers, preheating coal-paste feed	10 (dirty) to 50 (clean)

Film coefficients for process streams in German coal hydrogenation plants were reported by Laughrey et al.[43] For heating coal pastes at a velocity of 8 fps, h varied from 80 to 120 $Btu/hr \cdot ft^2 \cdot F$; for heating mixtures of middle oil and hydrogen (15 to 30 ft^3 of H_2/lb of oil) at 6 to 14 fps, h ranged from 130 to 160 $Btu/hr \cdot ft^2 \cdot F$; and for cooling heavy oil letdown streams, the average coefficient was 90 $Btu/hr \cdot ft^2 \cdot F$ at bulk temperatures above 400°F and at velocities of 9 to 15 fps.

Similar data obtained for various types of heat exchange equipment in six continuous tar distillation plants in England also indicate quite low heat transfer rates.[44] For about 30 heat exchangers (vapor/tar, oil/tar, pitch/tar, condensers, fired heater convection tubes, and batch heaters), the range of surface heat flux was from 160 to 3000 $Btu/hr \cdot ft^2$ and that of overall heat transfer coefficient from 0.4 to 26 $Btu/hr \cdot ft^2 \cdot F$.

The shell and tube exchanger used to preheat the coal/water slurry before it enters the centrifuges at the Mohave generating station operates at a mean Reynolds number of about 9800. Though no fouling has occurred, tube-inlet erosion has been severe. Stainless steel ferrules at the tube inlets had worn out within 3 months, and hard rubber inserts were being tried.

The salient points are that fouling of various types can cause a wide divergence between calculated and operating heat transfer rates, that many streams are quite viscous and in laminar or transition flow, and that overall heat transfer coefficients are relatively low (< 100 Btu/hr·ft^2·F).

4.2 Laminar-Flow Heat Transfer

The Reynolds moduli for liquid and slurry streams in heat exchangers used in coal liquefaction pilot plants were estimated from available data.[42] The minimum moduli, based on <u>maximum</u> viscosities,[42, 43] varied from 6 to 1600, indicating stable laminar flow. Even when superimposed free convection is substantial, heat transfer coefficients are generally small for laminar flow in tubes of standard diameters. A comprehensive review of industrial laminar-flow heat transfer by Porter[45] provides an informative summary, including 263 references.

Augmentation of laminar-flow heat transfer in tubes would obviously be desirable. Bergles has reviewed past studies in which curved tubes, rotating tubes, displaced promoters, surface roughness, internal extended surfaces, and swirl flow have been used. In the same report[46] and in a later one,[47] he described the results of tests using twisted-tape inserts, internally finned tubing, and static-mixer inserts. Large increases in heat transfer rate were obtained with modest increases in pressure drop with the internally finned tubing. The relative merit of the inserts, on the basis of the ratio of heat transfer to pressure drop, followed the sequence (decreasing order) of twisted tape   static mixers. With the simple twisted-tape configuration, the heat transfer coefficient was increased by a factor of 2 to 3.

An extension of these efforts to enhance laminar-flow heat transfer rates with heavy oils and coal-oil slurries should be worthwhile. In addition, erosion effects in handling coal slurries need defining. There is also the possibility that the increased turbulence will minimize and eliminate some types of fouling.

4.3 Property Data and Coefficient Correlations

Relevant physical-property data or methods for estimating the properties are generally available for pure and mixed gases and for pure light liquids.[48,49] For heavy oils, many liquid mixtures, slurries, slags, refractories, and some construction materials, however, it is clear that many more data are needed.[50-52] For heat exchanger design purposes, it is important that vapor-liquid equilibrium (VLE) data and viscosities and thermal conductivities be known with reasonable accuracy. Assuming adequate VLE data to establish the relative flow and composition of vapor and liquid streams, the thermal design problem reduces largely to calculating the heat-transfer coefficient. As an illustration of the relative importance of the various physical properties, consider the heat transfer coefficient for turbulent, single-phase tube flow, for which the coefficient h is proportional to $k^{2/3} \cdot c_p^{1/3} \cdot $ 0.8/ 0.467. Densities ( ) and heat capacities (Cp) are usually available or can be reasonably estimated, but a threefold error in the thermal conductivity (k) results in a two fold error in h; likewise, an error in viscosity ( ) of 4.4-fold leads to a twofold error in h.

Knowledge of heat transfer coefficients is the primary requirement in designing heat transfer equipment. Adequate correlations for predicting coefficients are available for some coal conversion applications, but are not for others. The need for reliable correlations for direct-contact coolers and condensers, for coal slurry heaters and slurry-to-slurry exchangers, and for surfaces in fluidized-bed combustors and gasifiers has been addressed by Kasper et al.[51]

4.4 Fouling and Other Process-Fluid Effects

A recent Westinghouse report[53] is explicit about the ways in which fouling, corrosion, and erosion render current heat-recovery system design for hot coal gas both difficult and uncertain. Problems addressed include tar and slag fouling, residual hydrocarbon deposition, carbon laydown, hydrogen and H_2S attack, nitriding, and aqueous and chloride-salt corrosion. Statements in the report that "only actual test operation can provide data for guaranteeing exchanger cleanliness" and that "it remains for actual results in a real system with coal gas to determine the severity of the problem and the efficacy of the solution" are realistic.

In general, the problems of fouling in coal conversion process exchangers are unique and are anticipated to be more severe than those experienced currently in most other industries. Such fouling is a costly and largely unresolved problem throughout refineries and some petrochemical plants. The deposits decrease heat transfer rates, increase pressure drops, and impede process flows. Provisions must be made for periodic mechanical, chemical, or hydrojet cleaning of such exchanger surfaces. Fouling-rate data and associated velocity dependencies, if such exist, are needed for a large variety of process streams and conditions if the onstream time of dirty-service heat exchangers is to be reasonable. There are theoretical treatisis on some types of fouling phenomena, such as thermophoresis, the movement of particles to a lower temperature surface under the influence of a temperature gradient.[54] However, only specific testing and/or accumulated experience in pilot plants can provide the needed fouling factors as well as the corrosion data needed for material selection and the erosion data for application of specific mechanical design techniques.

REFERENCES

1. W. R. Gambill and W. R. Reed, "Survey of Industrial Coal Conversion Equipment Capabilities: Heat Recovery and Utilization," ORNL-TM-6073, Oak Ridge National Laboratory, Oak Ridge, Tenn. (July 1978).

2. S. L. Milora and J. W. Tester, Geothermal Energy as a Source of Electric Power, MIT Press, Cambridge, Mass., 1976, p. 84.

3. J. R. Pudlock, "Corrosion-resistant Shell and Tube Exchanger Costs Compared," Oil Gas J. 75(37): 101 (1977).

4. W. M. Small and R. K. Young, "Exchanger Design Cuts Tube Vibration Failures," Oil Gas J. 75(37): 77 (1977).

5. R. H. Perry and C. H. Chilton, Eds., Chemical Engineers' Handbook, 5th ed., McGraw-Hill, New York, 1973, p. 9-48.

6. A. C. Mueller, "Heat Exchangers," p. 18-67 in Handbook of Heat Transfer, 1st ed., W. M. Rohsenow and J. P. Hartnett, eds., McGraw-Hill, New York, 1973.

7. A. Schack, Industrial Heat Transfer, translated from 6th German ed., John Wiley & Sons, New York, 1965, p. 293.

8. National Bureau of Standards, Waste Heat Management Guidebook, prepared for Federal Energy Administration, Washington, D.C., February 1977, p. 146.

9. A. P. Fraas and M. N. Ozisik, Heat Exchanger Design, 1st ed., John Wiley & Sons, New York, 1965, pp. 166 and 169.

10. J. A. Cook et al., Automotive Gas Turbine Ceramic Regenerator Design and Reliability Program, Final Annual Report - July 1, 1975, to Sept. 30, 1976, COO-2630-18, Energy Research and Development Administration, Washington, D.C. (October 1976).

11. C. D. Spangler, "Air Preheater System Safely Controls Heater Fuel/Air Ratio," Oil Gas J. 74(1): 63 (1976).

12. C. B. Gentry, "Refractories and Insulation - Ceramic Heat Exchanger Properties and Uses in Heat Recovery," Ind. Heat. 43(6): 54 (1976).

13. Combustion Engineering, C-E Air Preheater - Ljungstrom, Wellsville, N.Y., 1974.

14. S. W. Chi, Heat Pipe Theory and Practice, A Sourcebook, Hemisphere Publishing Co., Washington, D.C., 1976.

15. C. L. Tien, "Fluid Mechanics of Heat Pipes," p. 167 in Annual Review of Fluid Mechanics, vol. 7, Annual Reviews, Inc., Palo Alto, Calif., 1975.

16. W. A. Ranken, The Potential of the Heat Pipe in Coal Gasification Processes, LA-UR-76-954, Los Alamos Scientific Laboratory, Los Alamos, N. Mex. (September 1976).

17. J. C. Biery, Methanation - With High Thermodynamic Efficiency Energy Recovery, LA-6656-MS, Los Alamos Scientific Laboratory, Los Alamos, N. Mex. (January 1977).

18. D. C. Strimbeck, D. C. Sherren, and E. S. Keddy, "Process Environment Effects on Heat Pipes for Fluid-Bed Gasification of Coal," p. 1015 in Proc. 9th Intersociety Energy Conversion Eng. Conf., San Fransisco, Calif., August 1974.

19. "Heat Exchanger Exploits Heat-Pipe Concept," Chem. Eng. 82(7): 70 (1975).

20. K. T. Feldman and D. C. Lu, "Preliminary Design Study of Heat Pipe Heat Exchangers," p. 451 in Proc. Second Internat. Heat Pipe Conf., Bologna, Italy, March 31-April 2, 1976, N76-32374-432, National Technical Information Service, Springfield, Va., 1976.

21. L. L. Vasiliev et al., "Study of Heat and Mass Transfer in Heat Pipe-Based Exchangers," p.463 in Proc. Second Internat. Heat Pipe Conf., Bologna, Italy, March 31-April 2, 1976, N76-32374-432, National Technical Information Service, Springfield, Va., 1976.

22. S. C. Lou and Harvey Wen, "Heat Recovery from Coal Gasifiers," presented at the Third Annual Industrial Energy Conservation Technology Conference, Houston, Texas, April 26-29, 1981.

23. J. P. Fanaritis and H. J. Streich, "Heat Recovery in Process Plants," Chem. Eng. 80(12): 80 (1973).

24. P. Hinchley, "Waste Heat Boilers: Problems and Solutions," Chem. Eng. Prog. 73(3): 90 (1977).

25. D. Csathy, "Latest Practice in Industrial Heat Recovery," presented at the Energy-Source Technology Conference and Exhibition, New Orleans, Louisiana, February 4, 1980.

26. E. Seher, "Waste Heat Boilers for the Chemical and Metallurgical Industries," Br. Chem. Eng. 10(5): 314 (1965).

27. J. E. Franzen and E. K. Goeke, "SNG Production Based on Koppers-Totzek Coal Gasification," p. 217 in Proc. Sixth Synthetic Pipeling Gas Symp., Chicago, Ill., October 1974, American Gas Assocation, Arlington, Va., 1974.

28. D. A. Waitzman, "Ammonia from Coal: A Technical/Economic Review," Chem. Eng. 85(3): 69 (1978).

29. L. J. Partridge, "Coal-Based Ammonia Plant Operation," Chem. Eng. Prog. 72(8): 57 (1976).

30. Schmidt'sch Heissdampf FmbH. Waste Heat Boilers for Chemical, Metallurgical, and Other Processes, Kassel-Bettenhausen, W. Germany (n.d.).

31. A. L. Wilson, Mechanical Development Recommendations, Interim Report, ERDA FE-2240-44, C. F. Braun and Company, Alhambra, Calif., June 1977, p. 12.

32. The Directory of Industrial Heating and Combustion Equipment, U.S. Manfacturers, 1977-1978, Bermont Books, Washington, D.C., 1977, p. 99.

33. E. Crawford, "Refinery `Loop´ Shuttles Waste Heat," Energy User News, July 25, 1977, p. 7.

34. J. R. Fried, "Heat-Transfer Agents for High-Temperature Systems," Chem. Eng. 80(12): 89 (1973).

35. E. G. Kovach, Ed., "Heat Transfer and Thermal Energy Transport," p. 35 inThermal Energy Storage, report of a NATO Science Committee Conference, Turnberry, Scotland, March 1976.

36. M. D. Silverman and J. R. Engel, Survey of Technology for Storage of Thermal Energy in Heat Transfer Salt, ORNL/TM-5682, (January 1977).

37. C. L. Crawford, Fluid-Bed Char Gasifier: Process Design, FE-2213-5, TRW Systems Group, McLean, Va. (Dec. 2, 1976), pp. 3, 33, and 35.

38. Federal Energy Administration, Proceedings of the FEA-PCA Seminar on Energy Management in the Cement Industry, F.E.A. Conservation Paper 47, Washington, D.C., October-November 1975.

39. J. R. Fair, "Designing Direct-Contact Coolers/Condensers", Chem. Eng. 79(13): 91 (1972).

40. "Chementator" section, Chem. Eng. 84(20): 41 (1977).

41. R. A. Smith, "Performance Factors in the Design of Heaters and Coolers for the Heavy Chemical Industry", p. 52 in Proc. Gen. Discuss. Heat Transfer, The Institution of Mechanical Engineers, London, 1951.

42. S. Kimmel et al., Coal Liquefaction Design Practices Manual, Final Report AF-199 from Fluor Engineers and Constructors, Inc., to Electric Power Research Institute, July 1976, pp. IV-B-1 through IV-C-21.

43. P. W. Laughrey et al., "Design of Preheaters and Heat Exchangers for Coal-Hydrogenation Plants", Trans. ASME 72: 385 (May 1950).

44. The Coal Tar Data Book, 2nd ed., Section C, The Coal Tar Research Association, Leeds, England, 1965, p. 12.

45. J. E. Porter, "Heat Transfer at Low Reynolds Number (Highly Viscous Liquids in Laminar Flow)", Trans. Inst. Chem. Eng. 49: 1 (1971).

46. S. W. Hong and A. E. Bergles, "Augmentation of Laminar Flow Heat Transfer in Tubes by Means of Twisted-Tape Inserts", Am. Soc. Mech. Eng. Pap 76-HT-QQ, November 1975; also Iowa State University Heat Transfer Laboratory Report HTL-5, December 1974.

47. A. E. Bergles, Laminar Flow Heat Transfer in Horizontal Tubes Under Normal and Augmented Conditions, HTL-11, Iowa State University Heat Transfer Laboratory, Ames, December 1975.

48. J. C. Janka and R. Malhotra, Estimation of Coal and Gas Properties for Gasification Design Calculations, Office of Coal Research, Department of the Interior, R&D Report No. 22, Interim Report No. 7, Washington, D.C., January 1971.

49. R. C. Reid, J. M. Prausnitz, and T. K. Sherwood, The Properties of Gases and Liquids, 3rd ed., McGraw-Hill, New York, 1977.

50. L. C. Yen et al., "Data Deficiency Hampers Coal-Gasification Plant Design", Chem. Eng. 84(10): 127 (May 9, 1977); "Fundamental Data Needs for Process Design of Coal Gasification Plants", p. 87 in Proc. 55th Annu. Conv. Gas Process. Assoc., Gas Processors Association, San Antonio, TX, March 1976.

51. S. Kasper, D. Henzel, and R. Nene, "Heat Transfer Problems in Coal Conversion", AIChE Paper No. 4a, AIChE 83rd Nat. Meeting., Houston, TX, March 1977.

52. H. G. Hipkin, T. C. Lin, and C. K. Lu, "Survey of Design Data Needs for Refining Synthetic Crudes", Preliminary Report prepared for the Refining Department of the American Petroleum Institute, Washington, DC, May 1981.

53. Westinghouse Electric Corp., High Temperature Turbine Technology Program: Phase I - Program and System Definition, Section 10.5, Heat Recovery System Design, Report FE-2290-27, Lester, PA., February 1977.

54. G. Nishio, S. Kitani, and K. Takahashi, "Thermophoretic Deposition of Aerosol Particles in a Heat-Exchanger Pipe," Ind. Eng. Chem. 13(4) 1408 (1974).

A New Energy-Cost Characteristic Diagram for Optimizing Heat Exchangers

PIERRE LE GOFF
Laboratoire des Sciences du Génie Chimique
CNRS-ENSIC
Institut National Polytechnique de Lorraine
Lorraine, France

MARCO GIULIETTI
Instituto de Pesquisas Tecnologias do Estado de São Paulo
São Paulo, Brasil

ABSTRACT

We present a new "energy-cost-characteristic" diagram (CAREC) for optimising heat exchangers. It is based on the monetary cost and the total consumption of primary energy (energetic cost). This diagram gives the characteristics of the exchanger which corresponds either to the minimum monetary cost or the minimum energetic cost and also the extra-monetary cost required to pass from the first optimum to the second.

The use of the CAREC is demonstrated for the case of a boiler water feed pre-heater and a standard shell and tube heat exchanger.

1. GENERAL STATEMENT OF THE PROBLEM

A great number of industrial systems can be considered as a "black box" which converts low value inputs into outputs with a greater value (fig. 1). This transformation requires that a certain quantity of energy be furnished to the system. The equipment used in such an industrial system can be classified in three categories. Those which *supply* energy to the system, those in which energy is *degraded* and those which are *independent* of the transformation of energy.

We have proposed (ref. 1,2) calling the first category

Pro-energetic equipment. This includes motors, pumps, compressors boilers... The cost of buying and installing these is an *increasing* function of their nominal capacity and is practically always an *increasing* function of the flux of primary energy consumed in the system (for example expressed in tonnes of petrol equivalent).

In the same way the second category is called

Anti-energetic equipment. These, for example, include the pipelines in which mechanical energy of pressure is degraded into heat by friction, the lagging round heated reservoirs through which heat is lost to the surroundings. The cost of this sort of equipment is practically always a *decreasing* function of the flux of primary energy consumed. However there are a certain number of exceptions to this rule (ref. 3,4).

For a given technical project (such as those dealt with later on) the total investment cost can be expressed as a function of the flux of operating energy consumed $\dot{E}_O$ in the system (for example expressed in TPE/year). This cost is the sum of 3 terms : an increasing term (the pro-energetic investment), a decreasing term (anti-energetic investment) and a constant term.

Therefore there exists an optimum value E_i of the energy consumed in the system which corresponds to the minimum in the total investment (see fig. 2).

But suppose that, in addition, we decide to include the operating cost of the energy consumed during the whole period over which the plant is to operate. We must therefore add a fourth term to the preceeding 3. In this case, we call the resulting total cost the "Integral" cost of the industrial operation because it results from the *integration* of the operating cost over the whole life of the plant. This should include a procedure to actualise the costs so that future expenditure is expressed in money terms at the date at which the equipment is bought (that is the "zero" year).

The minimum of this total integral cost corresponds to a value $\dot{E}_t$ of the flux of energy consumed which is often very different from the preceeding value E_i for example 4 times smaller (ref. 2).

2. EXAMPLE OF A HALF-EXCHANGER AND ITS ASSOCIATED PUMP

Consider a given mass flowrate $\dot{M}$ of a fluid which is to be heated (or cooled) from T_e to T_s in a heat exchanger using a heat exchange fluid at temperature T_e'. Therefore here the heat flux in the heat exchanger is fixed (see fig. 3) and the unknowns which remain to be determined are :

- the type of heat exchanger
- for the type of heat exchanger chosen, the total exchange surface A and various complementary geometrical factors must be determined. For example in the case of a shell and tube heat exchanger : the sizes of the tubes (length, diameter, wall thickness), their number and spacing.
- the flowrate $\dot{M}'$ of the heat exchange fluid.

Experience shows (ref. 5) that as a first approximation it is often possible to optimise each half exchanger *separately* and then readjust by iteration over the sum of the two partial optimums to find the overall optimum.

Let us therefore consider the mechanical energy degraded by friction in one of the half exchangers. This is the quantity $\dot{E}_O$ given as abscissa on figure 2. In this case the pro-energetic equipment is the pump unit which supplies fluid to the exchanger at the pressure $(P_e - P_s)$. The anti-energetic equipment is composed of the half exchanger itself with

$$\dot{E}_o = \frac{\dot{M}}{\rho} (P_e - P_s) \tag{1}$$

3. THE ENERGY COST CHARACTERISTIC (CAREC)

Figure 2 seems to show that any displacement from point I towards lower values of $\dot{E}_O$ is beneficial both from the point of view of the personal interest of a private decision-MAKER (since by an initial extra-investment it gives a reduction in the total integral cost) and also from the point of view of the overall good as it leads to a reduction in the consumption of the operating energy. But this economy in the operating energy has been obtained by extra-equipment... which itself requires energy to make. Therefore if the *energy*

ENERGY-COST CHARACTERISTIC DIAGRAM FOR OPTIMIZING HEAT EXCHANGERS 661

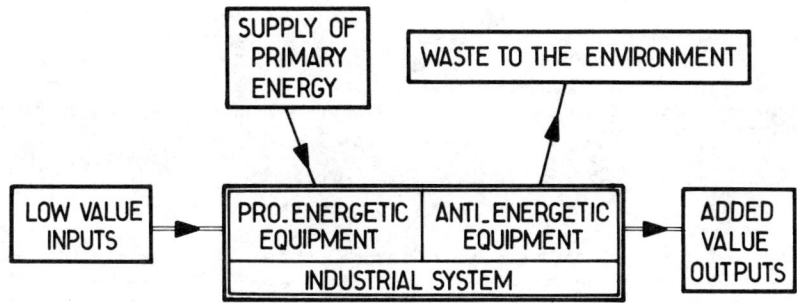

Figure 1

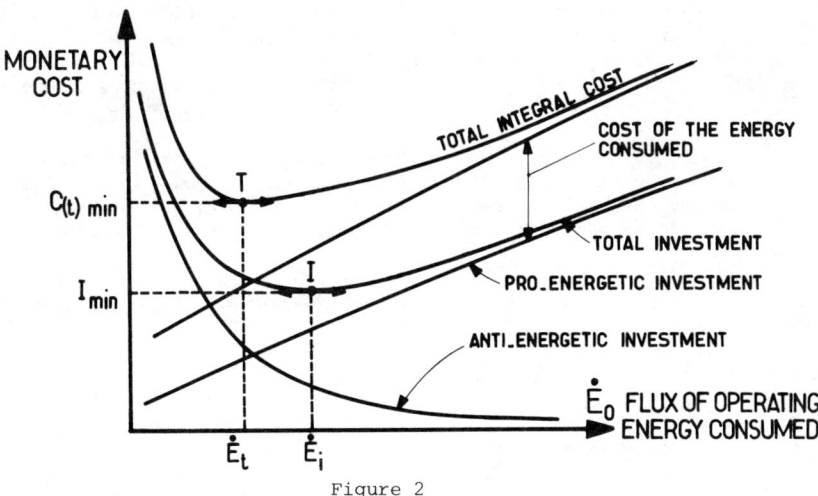

Figure 2

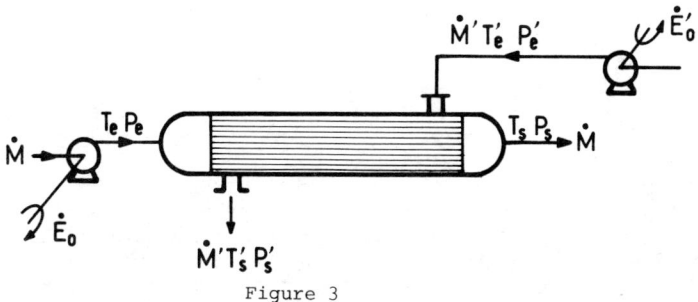

Figure 3

content of this extra-equipment is included it is not necessarily true that there is an overall saving in energy.

To solve this problem it is necessary to make a complete energy balance. Thus not only should we include the operating energy E_ON consumed over the N years of operating the process, but also the energy content CE_m of the pump unit (pro-energetic equipment) and the CE_c of the heat exchanger (anti-energetic equipment).

Let us suppose, to a first approximation, that the energy contents CE_m and CE_c are both proportional to the monetary cost CM_m and CM_c of the equipment. In this case the pump unit on one hand and the heat exchanger on the other hand can be represented by two straight lines through the origin (see fig. 4) and the curve of the sum of the investments will lie between these two straight lines. This is constructed by adding the two component vectors (CE_m, CM_m) and (CE_c, CM_c) as shown in figure 4. Therefore on this curve there is a point I' which corresponds to the minimum monetary cost and a point I" corresponding to the minimum total consumption of primary energy.

Now to these vectors let us add a 3rd component vector corresponding to (CE_c, $\dot{E}_ON$) the operating energy. This leads to (cf. fig. 4) a curve giving the total quantity of primary energy consumed. On this curve there is a point T' corresponding to the monetary minimum and a point T" corresponding to the energetic minimum.

Thus, for a given project or piece of equipment, we plot the total energy consumption in exergy or primary energy ($\equiv$ petrol equivalents) as abscissa and the total monetary cost, expressed in national money or foreign exchange, as ordinate. These graphs we propose calling Energy Cost Characteristic Diagrams (CAREC).

In previous publications we have applied this method to various specific examples such as : transporting a given liquid flowrate over a given distance using a pipeline (ref. 2), storing heat from a solar panels (ref. 6), preheating the water feed to a boiler (ref. 7).

To obtain a CAREC of more general interest which will allow the comparison of various types of heat exchanger we propose using normalised curves. That is with quantities expressed with reference to the energy transferred from one phase to the other. These quantities X and Y are defined as :

$$X \equiv \frac{\begin{pmatrix}\text{Total consumption}\\ \text{of primary energy}\end{pmatrix}}{\begin{pmatrix}\text{Total thermal}\\ \text{energy transferred}\end{pmatrix}} = \frac{CE_t}{\dot{Q}N} = \frac{\text{joule}}{\text{joule}} \qquad (2)$$

$$Y \equiv \frac{\begin{pmatrix}\text{Total monetary cost}\end{pmatrix}}{\begin{pmatrix}\text{Total thermal}\\ \text{energy transferred}\end{pmatrix}} = \frac{CM_t}{\dot{Q}N} = \frac{\text{francs}}{\text{joule}} \qquad (3)$$

In the general case where the energy contents are not proportional to the monetary costs, the vector construction of figure 4 is not valid. However the general shape of the curve is the same. The CAREC is always in the form of a parabola with its two arms asymptotic to the two curves which represent the pro-energetic and the anti-energetic components as shown in figure 5.

ENERGY-COST CHARACTERISTIC DIAGRAM FOR OPTIMIZING HEAT EXCHANGERS

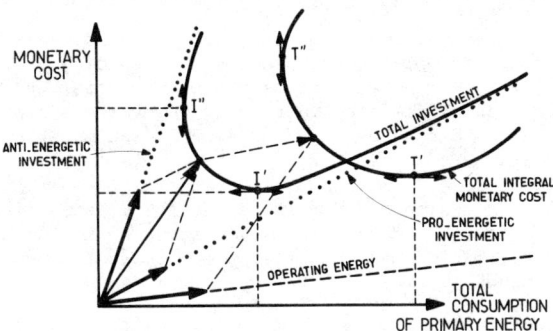

Figure 4

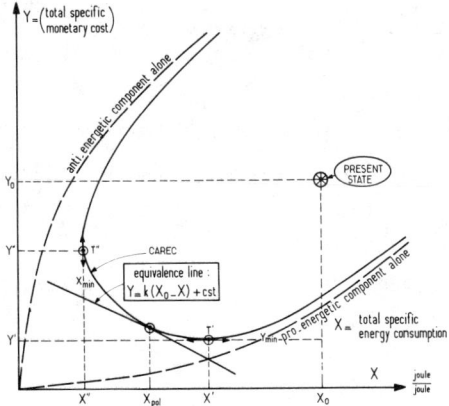

ENERGY COST CHARACTERISTIC CURVE (= CAREC)
OF A HEAT EXCHANGER

Figure 5

4. POLITICAL CHOICE FUNCTION

A technical decision maker in a private company normally has the objective of minimising the total monetary cost of the operation. He will therefore chose a point on the CAREC close to T'.

On the other hand, the planetary objective of reducing primary energy consumption would lead to choosing a point as close as possible to T" even though this will bring about an extra-investment cost. However the capital available for investment can also be considered to be a rare "raw material" and be subject to market laws.

Thus a political decision maker must define the relative weight to be given to the energetic "cost" and the monetary cost. He will therefore form a certain function $F(X,Y)$ which may be called the *political choice function*, which is to be minimised.

- A first example of such a function is where a relative weight is given to each optimum objective by writing

$$Y_{pol} = \omega Y_{min} + (1-\omega)(Y)_{X_{min}} \qquad (4)$$

If for example, we chose $\omega = 0.80$ this means that we give a weighting of 80 % to the objective of minimising the monetary cost and 20 % to the objective of minimising the "energetic cost".

- Another example of a different form of this sort of function is where, starting from any given state, we say that any saving in monetary cost of ΔY is equivalent to a saving in energy of $k\Delta X$. Thus

$$\Delta Y = k\Delta X$$

Here k is the *coefficient of political equivalence* between the quantity on the ordinate and that on the abscissa. For example the k may be taken to be the international market price for fuel oil that is 1000 francs/tonne.

Let us call $Y = C(X)$ the CAREC equation and let (Y_o, X_o) be the coordinates of the existing process from which we measure the energy saving to be obtained.

The political monetary cost of the operation is then

$$Y_{pol} = C(X) - k(X_o - X) \qquad (5)$$

This is the function which has to be minimised. Its derivative is

$$C'(X) + k = 0$$

that is $\boxed{C'(X) = -k} \qquad (6)$

The optimum point is that on the CAREC which has a tangent of -k (see fig. 5).

5. A PRE-HEATER FOR WATER FED TO A BOILER

The first example is of projet to add a pre-heater heat exchanger for the feed water to a boiler. This is to increase the feed water temperature from 20 °C to 119 °C. The feed rate is 5 kg/sec flowing in an exchanger composed of a pipe of diameter d and length L in contact with the hot flue gases from the boiler and which are present in excess. We assume that the tube wall is held uniformly at a temperature of 120 °C (cf. fig. 6). In ref. 7 where we dealt with such a mono-tubular heat exchanger we established a general relationship between the mechanical power degraded by the pressure drop $\dot{E}_o$ and the area of the transfer surface A. This is

$$\dot{E}_o = K_1 A^{\frac{4}{n-2}} \tag{7}$$

where K_1 is a constant and n an exponent which for commercial steel tubes is 1 in the laminar regime and 0.164 in the turbulent regime.

The whole of the costs can therefore be expressed either as a function of only $\dot{E}_o$ or as a function of only A. And the dimensions of the pipe d and L are fixed when a value of A is chosen. *There is only one degree of freedom therefore a single infinity of solutions.*

The final result of these calculations can therefore be presented in the form of a curve (a CAREC) on the diagram of total monetary cost as a function of the total energetic cost. This CAREC is given in figure 7 on log log coordinates and the extremity is given on a larger scale in figure 8.

51. The total "cost" minimums

The points on the curve at the horizontal and vertical tangents represent respectively the minimum in the total cost in francs and the minimum in the total "cost" in primary energy.

The dimensions and characteristics of the corresponding heat exchangers are given in columns (2) and (5) of Table I. It can be seen that the minimum cost heat exchanger T' is a tube 58 m long and 5.9 cm diameter (area A = 10.9 m^2). At the same time the heat exchanger which gives the minimum total consumption of primary energy is a tube of 11.2 cm diameter and 100 m length (A = 35.2 m^2).

52. Comparison with the original installation

In the existing installation water is heated from 20 °C to 119 °C by using an excess fuel consumption in the boiler. This corresponds to the point (X_o, Y_o) on the CAREC, fig. 7. Now let us compare this point with the point (X'Y') corresponding to the minimum total monetary cost. This gives

$$\frac{Y_o}{Y'} = 300 \quad \text{and} \quad \frac{X_o}{X'} = 4460$$

It can be seen that adding such a heat exchanger (with an amortizment period of 10 years) brings about a reduction of 300 to 1 in the total monetary cost and a reduction in the total consumption of primary energy of 4460 to 1 !

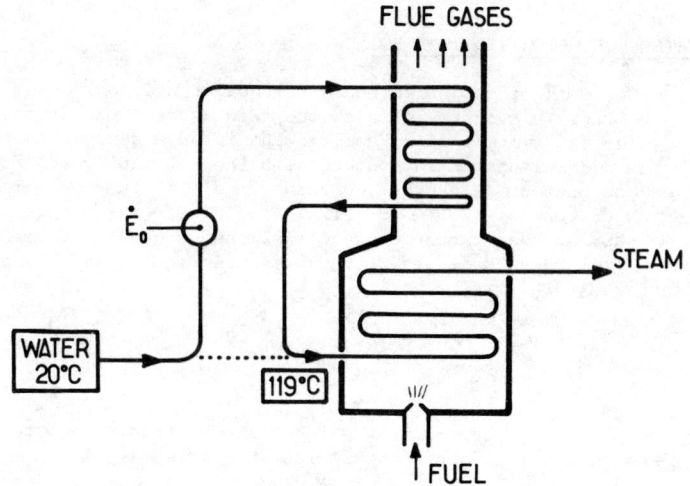

Figure 6

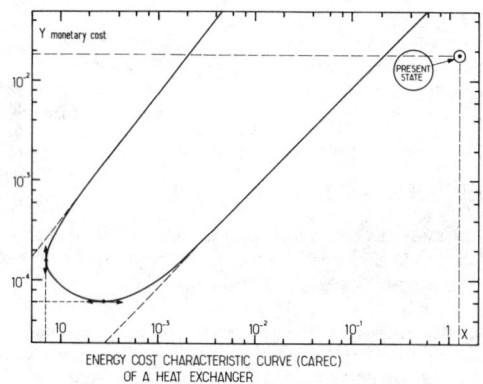

ENERGY COST CHARACTERISTIC CURVE (CAREC)
OF A HEAT EXCHANGER

Figure 7

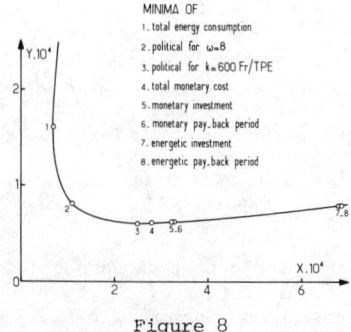

Figure 8

Column 1	2	3	4	5	6	7	8	9
	MONETARY MINIMUMS			ENERGETIC MINIMUMS			POLITICAL MINIMUMS	
PARAMETERS	total cost	Investment	pay-back period	total consumption	investment	pay-back period	with $\omega = 0.80$	with k=600 F/TPE
Diameter d (m)	5.9×10^{-2}	5.7×10^{-2}	5.7×10^{-2}	11.2×10^{-2}	4.7×10^{-2}	4.7×10^{-2}	7.9×10^{-2}	6.1×10^{-2}
Length L (m)	58.5	56.6	56.7	100.0	48.1	48.2	74.6	59.5
Transfer surface A (m²)	10.9	10.1	10.1	35.2	7.06	7.10	18.5	11.4
Flow velocity u_m (m s⁻¹)	1.86	2.00	2.01	0.51	2.97	2.95	1.04	1.72
Re	2.63×10^5	2.74×10^5	2.75×10^5	1.35×10^5	3.34×10^5	3.33×10^5	1.98×10^5	2.53×10^5
Mechanical power degraded Ė (watts)	185.6	217.3	217.2	14.4	477.6	472.0	58.3	167.2
Total energy consumption CE_t (MJ)	125 300	144 740	144 700	$\boxed{31200}$ min	306 300	302 850	49 500	113 870
Total cost C_t (Frs)	$\boxed{27340}$ min	27 530	27 530	72 200	35 400	35 180	36 300	27 400
Specific energy consumption X (J/J)	28×10^{-5}	32.3×10^{-5}	32.3×10^{-5}	$\boxed{6.97 \times 10^{-5}}$ min	68.4×10^{-5}	67.6×10^{-5}	11.1×10^{-5}	5.4×10^{-5}
Specific monetary cost Y (Fr/J)	$\boxed{6.11 \times 10^{-5}}$ min	6.15×10^{-5}	6.15×10^{-5}	16.1×10^{-5}	7.91×10^{-5}	7.86×10^{-5}	8.11×10^{-5}	6.12×10^{-5}
Pay-back period (hours) energetic T_{re}	5	5.03	5.03	12	4.72	$\boxed{4.70}$ min	7.0	5.1
monetary T_{rm}	189	187.5	$\boxed{187.4}$ min	538	224	223	266	191

Table I : Comparison of the monetary, energetic and political optimums.

In reality this exceptionally beneficial result is obtained because we have considered that the energy used to pre-heat the water is available *free* from the combustion flue gases at present rejected to the atmosphere. We should therefore complete the calculation by adding to the energy balance a term we have neglected namely the pressure loss by the flue gas across the heat exchanger. The full calculation has been presented in another publication (ref. 7).

53. The minimum investments

On the basis of the excellent profitability of this project, it seems obvious that a financial manager will decide to go ahead with it immediately. However many industrial decision makers consider that capital is even more scare than energy. They prefer to choose the heat exchanger which minimises the initial monetary investment and accept an extra-operating cost.

The solution corresponding to this case is given in column (3) of Table I and on figure 8. It can be seen that it differs little from that of the minimum total monetary cost solution. The reason being that in all this domain the integral operating cost (even over a period as long as 10 years) is practically negligible in comparison with the monetary investment (less than 6 %).

For comparison it is interesting to calculate the dimensions of the heat exchanger which requires the minimum energetic investment. This is shown in column 6 of Table I. It can be seen that, contrary to the previous case, this heat exchanger is very different from that which minimises the total energy consumption. This one only has an area of 7 m^2 instead of 35 m^2 for the latter case. This is due to the fact that the energy required for operation is not negligible in comparison with the energy content of the equipment and is in fact 8.4 times greater.

54. The pay-back period for the investments

A standard criterion for evaluating the profitability of an investment consists in comparing the annual profit which it provides. In this case this is given by the difference between the saving in fuel consumption in the boiler and the extra-consumption of electricity by the pump. This gives the time required to repay the investment more commonly called the pay-back period.

A financial decision maker may therefore have the objective of minimising the monetary pay-back period. This calculation gives a minimum of 187 hours. The dimensions of the corresponding heat exchanger are given in column 4 of Table I. It can be seen that this exchanger is very similar to that of column 2.

In the same way we have calculated the energetic pay-back period and determined its minimum. The result is 4.7 hours (see column 7 of the Table I).

Clearly both in monetary accounting and in energy accounting, the minimum investment is very close to the minimum pay-back period.

55. The political minimums

To take into account both the monetary and energetic aspects at the same time, we have calculated the dimensions of the heat exchanger which corresponds to the political minimum. First we have used a value of $\omega = 0.8$ in equation 4 and have calculated the dimensions of the resulting heat exchanger corresponding to a value of k = 600 Fr/TPE in equation 6 and given its characteristics in column 9 of Table I.

It can be seen that with k = 600 Fr/TPE there is no notable difference from that corresponding to the monetary minimum. However using ω = 0.8 can be seen to lead to a very different heat exchanger. This heat exchanger requires an extra-monetary cost (investment-operation) of 33 % and gives a saving in primary energy of 250 % with respect to that of column 2 (minimum total cost).

Figure 8 gives a good comparison of the 8 optimums identified here :

- the three monetary optimums (total cost, investment and pay-back period)
- the three energetic optimums analogous to the above
- the two political optimums.

6. A SHELL AND TUBE LIQUID/LIQUID HEAT EXCHANGER

The second example considered here is that of a standard shell and multi-tube fixed head heat exchanger with a single pass tube-side and shell-side.

61. The optimisation method

In the first example of a mono-tube half-exchanger there was *only one* degree of freedom. In this case, there is at least *four* degrees of freedom : the tube outside diameter, the heat exchange area, the number of tubes and the Reynolds number for the shell side fluid (see fig. 3).

In addition there are also certain technological constraints imposed by the characteristics of the fluid flow notably those set by heat exchanger manufacturers (for example the tube pitch should be between 1.25 and 2.5).

The total amount of mechanical energy degraded in the exchanger by one or the other fluids is therefore a function of these 4 parameters

$$\dot{E}_o + \dot{E}'_o = f(d_o, A, n, Re') \qquad (8)$$

This function has been developed in another publication (ref. 8). We have also deduced expressions for the total monetary cost and the total energetic cost and have established the corresponding CAREC diagrams.

When one of the parameters in (8) is varied and the other 3 parameters are held constant the point representing the system describes the classic CAREC parabola between the pro-energetic and the anti-energetic asymptotes. If a technological constraint is encountered, the parabola is cut and only one of the branches is left.

If a second parameter is then varied, whilst the 2 others held constant, a new parabola which is the envelope of the preceeding CAREC is obtained as is shown on figure (9).

This procedure can then be repeated for a 3rd and then a 4th parameter (obviously this is done numerically by computer and not graphically). The final result is also a parabolic curve, a super-CAREC, which is the envelope of all the possible CAREC corresponding to all possible values of the 4 parameters and the technological constraints.

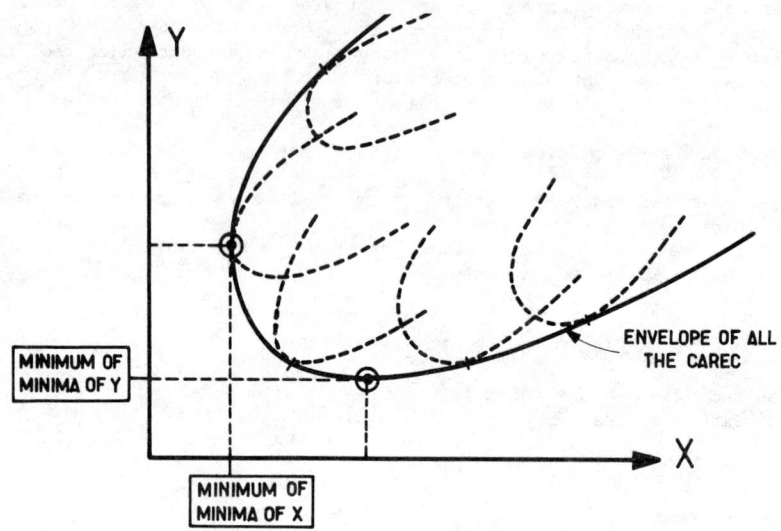

Figure 9

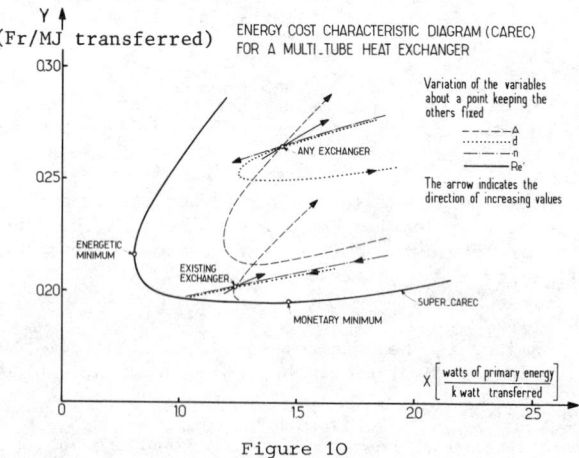

Figure 10

This curve will have 2 characteristic points here corresponding to the minimum of minima in the total monetary cost and the minimum of minima in the total energetic cost.

62. Example of an industrial heat exchanger

We have applied the method described above to a heat exchanger which has already been designed and constructed by a french engineering company.

The heat exchanger has to cool 58 kg s^{-1} of an aqueous solution of caustic soda from 53 °C to 47 °C (in the tubes) with water available at 12 °C. The other specifications are detailed elsewhere (ref. 8).

Figure 10 and Table II give the results of the calculations. Figure 10 shows the super-CAREC with the monetary and energetic minima as obtained by a Rosenbrock search together with the points corresponding to the heat exchanger in question and one corresponding to a non-specified exchanger. This diagram also shows the arms of the CAREC which are obtained by varying variables at each point whilst keeping the others fixed. The dimensions of the following exchangers are given in columns (2) (3) and (4) of Table II

- the heat exchanger which was actually built by the engineering company using their normal optimisation procedure
- the heat exchanger corresponding to the minimum of the total *monetary* cost minima
- the heat exchanger corresponding to the minimum of the total *energetic* cost minima

It should be noted that the optimal heat exchangers are not normalised but the results would not be very different even if they were. It can be seen that the exchanger which was built is close to the heat exchanger corresponding to the monetary minimum (it is less than 7.5 % more expensive). The heat exchanger has therefore been well designed from the point of view of a private industrial decision maker.

However a decision-maker with a planetary view trying to minimise the consumption of primary energy would require a heat exchanger at the energetic minimum. In this case, the extra-monetary cost would be 10 % more with respect to the monetary minimum and only 2 % more than the heat exchanger which was really built. This would nevertheless give an overall economy in primary energy of 45 % with respect to that designed to the monetary minimum and 42 % with respect to the existing heat exchanger.

The political choice obviously dictates using the heat exchanger corresponding to the energetic minimum since it only requires a small extra-monetary cost and will provide a large overall saving in energy.

7. CONCLUSION

The new Energy-Cost Characteristic diagram (called CAREC) is a powerful tool for the technico-economic analysis of heat exchangers. It allows the profitability of a project to be judged with respect to a reference state and to make the best choice as a function of a pre-determined optimisation criterion. It has also been shown to be capable of taking the energy content of the equipment into account and to allow the design of heat exchangers which are a little more expensive but which economise a lot of primary energy.

Table II : Characteristic parameters of the existing heat exchanger and those corresponding to the two minima

Column 1	2	3	4
Parameters	Existing exchanger	Monetary minimum	Energetic minimum
Exchange surface (m^2)	70,0	64,0	85,0
Number of tubes	184	323	495
Tube diameter (m) 10^{-2}	2,54	1,94	2,34
Shell side Reynolds number	5 513	5 578	3 029
Tube side Reynolds number	12 630	10 000	5 191
Tube length (m)	4,877	3,240	2,338
Shell diameter (m)	0,584	0,503	0,748
Number of baffles	23	18	9
Tube pitch	1,30	1,25	1,25
Mechanical energy degraded (watt)	169,0	267,3	32,9
Total monetary cost (Fr)	176 670	164 290	180 600
% capital investment	95 %	91,5 %	99 %
% operating	5 %	8,5 %	1 %
Total energetic cost (MJ)	255 550	269 000	147 860
% capital investment	59 %	38 %	86 %
% operating	41 %	62 %	14 %
Specific monetary cost (Fr/MJ transfered).10^{-3}	9,77	9,07	1,00
Specific energetic cost (watts of primary energy / kwatt transfered)	1,41	1,46	0,818

REFERENCES

1. Le Goff, P. 1978. Rev. Gen. Therm. France, n° 193, p. 68-98 et n° 194, p. 89-103.
2. Le Goff, P. 1980. "Energétique Industrielle", Ed. Technique et Documentation Lavoisier, Vol. 1 (1979), Vol. II (1980).
3. Le Goff, P. and Midoux, N. (nov. 1978). Compte-rendu du 1er Congreso Mediterraneeo de Ingeniera Quimica, Barcelone.
4. Le Goff, P. 1980. Chem. Eng. Science, Vol. 35, pp. 2029-2063.
5. Le Goff, P. 1980. The Chem. Eng. Journal, 20, p. 197-209.
6. Le Goff, P., Bordet, J. and Giulietti, M. Janvier 1981. Compte-rendu du colloque "Le stockage de l'énergie solaire appliquée au bâtiment", Lyon, Villeurbanne.
7. Le Goff, P. and Giulietti, M. 1980. Entropie n° 93, p. 58 à 73.
8. Le Goff, P. and Giulietti, M. Entropie (à paraître).

Complex Heat Exchangers Networks Simulation

JULIÁN CASTELLANOS FERNÁNDEZ
División de Proceso
Instituto Mexicano del Petróleo
México 14, D.F.

ABSTRACT

A mathematical tool for the automatized analysis of complex heat exchangers arrangements is summarized in this paper. Special attention to the actual temperature gradients is given, where a strict mathematical model is solved to produce temperature profiles for the shell and each tube pass. The model is represented by a set of $\sum(N_i+1)$, $i=1,m$ differential equations (N_i=number of passes for the ith equipment, m=number of equipments) plus the corresponding continuity equations for each stream joint or split. A numerical integration method is applied for the solution, coupled with an iterative technique imposed by the class of the system boundaries.

1. INTRODUCTION

Often, the most frequent pieces of equipment in process plants are heat exchangers; also very frequently these equipments are present not only as individuals, but as networks accomplishing the task of heat exchanging between hot and cold process streams together with services using steam, heating oil, refrigerants and cooling air or water. Typical examples are the cooling systems in cryogenic units for ethane and LPG recovery from natural gas, and the heating systems of thermally integrated atmospheric and vacuum units in petroleum refineries (See fig. 2). The importance of having adequate tools for the analysis of these systems of equipments becomes readily apparent.

The behaviour of heat exchangers is basically governed by kinetic and thermodynamic laws; the former deals with the heat (and some times mass) transfer resistance, normally handled in terms of film coefficients, wall thermal conductivity, fouling factors, etc. The latter deals with the transfer gradients, in this case temperature differences (some times partial pressure differences) that allow the heat flow (mass flow). The resistance problem has been tackled by empirical and semiempirical expressions derived from dimensional analysis, theoretical concepts and experimental determinations. Several equations and methodologies are presently available for the estimation of the resistances, covering the different processes experimented by the exchanging fluids (single phase heating or cooling, condensation, evaporation, etc.). The gradient problem has been solved by applying the Zeroth Law of thermal equilibrium together with energy balances, including enthalpy changes and heat flow, that allow estimation of average fluid temperature differences.

Along with the heat transfer phenomena, special attention to the fluid flow aspect is required, and pressure drop correlations are the pragmatic approach to this problem. In all this context it is necessary to mention that the equipment geometry and its general characteristics are of prime importance for the behaviour definition, and for heat exchangers two equipment classes may be distinguished: shell and tube heat exchangers and compact heat exchangers.

The purpose of this paper is to present the approach given to the development of a mathematical tool oriented to the simulation of shell and tube heat exchangers networks. Particular attention is given to the gradient problem where a numerical iterative solution is proposed in order to avoid the simplifications normally considered for the mathematical model. Of special interest is the simultaneous solution of the complex set of differential equations representing the network. A general presentation of the generated computer program will be given, describing the modular approach, mentioning the heat transfer and pressure drop correlations employed. Finally, typical results for a case of study will be presented.

2. PHYSICAL MODEL SCOPE

Physically, the model consists of a set of process streams interconnecting heat exchangers, furnaces, reactors, filters and some other pieces of equipment; the streams may also mix or split, experimenting flow, temperature and pressure variations, according to the following causes:

a) Streams mixing, where two process streams join without pressure change to generate a third stream with a flow equivalent to the addition of the mixing streams flows, and with a temperature such that the resulting stream enthalpy corresponds to the addition of the mixing streams enthalpies.

b) Streams splitting, where one process stream is splitted to generate two new streams with equal temperature and pressure and with flows such that the original stream flow corresponds to the addition of the two new streams flows.

c) Heat elimination or addition to a process stream, generating a new stream with equal flow and with a temperature such that the resulting stream enthalpy corresponds to the original stream enthalpy plus the positive or negative added heat. A pressure variation may also be experienced. This represents a model simplification for a fired heater, or any other (not shell and tube) heat exchanger, where a heat duty and pressure drop is specified.

d) Temperature and/or pressure change of a process stream, generating a new stream with equal flow and with a new temperature and/or pressure, according to a fixed ΔT and/or ΔP. This represents a model simplification for any piece of equipment where a temperature and/or pressure change may take place, such as a reactor, a filter, etc.

e) Heat exchange of two process streams, producing a pressure and temperature change on both streams; the exchange takes place through the walls of the tube bundle of a shell and tube heat exchanger. The relative directions of the fluids may vary depending on the internals arrangement, and for the purpose of this work, the following alternatives are contemplated according to the TEMA classification: E arrangement with one or any even number of tube passes up to 12; the fluid in even passes may be parallel or counterflow with respect to the fluid in the shell by changing the relative positions of the shell and tube inlet nozzles; J arrangement with a central inlet nozzle for the shell and divided flow to produce two independent outlets and with tube side trajectories similar to E; I arrangement, complement to J, with two independent inlet nozzles for the shell and flow converging to a central outlet; G arrangement, with split flow for the fluid inside the shell, produced by a longitudinal baffle, a central inlet and two diverging flows in the upper part that converge to a central outlet in the lower part.

The heat exchange referred to in point (e) may produce not only pressure and temperature changes in the process fluids, but also condensation or vaporization depending on the conditions and fluid characteristics. The hot fluid may flow inside the shell or the tube bundle, and the equipments may be in horizontal or vertical position.

3. MATHEMATICAL MODEL

According to the physical model described in point 2, five mathematical models have to be developed for representing: streams mixing, streams splitting, heat addition, temperature and pressure change and heat exchange. In general, the physical model may be considered as a set of modules where the five described actions take place. These modules have input and exit streams and the modeling is oriented to produce relationship laws between the streams properties and certain modules parameters. For the following development refer to nomenclature.

3.1 Streams Mixing

This module has two input and one exit streams. The modeling consists of simple mass and energy balances and a pressure equality assignment.

$$Wm_{i,1} + Wm_{i,2} = Wm_{i,3} \qquad i=1,M \qquad (1)$$

$$Wm_{i,1} H(Tm_{i,1}) + Wm_{i,2} H(Tm_{i,2}) = Wm_{i,3} H(Tm_{i,3}) \qquad i=1,M \qquad (2)$$

$$Pm_{i,3} = Pm_{i,2} = Pm_{i,1} = MIN (Pm_{i,1}, Pm_{i,2}) \qquad i=1,M \qquad (3)$$

3.2 Streams Splitting

This module has one input and two exit streams. The modeling consists of simple equality assignments.

$$Ws_{j,1} = Ws_{j,2} + Ws_{j,3} \qquad j=1,S \qquad (4)$$

$$Ws_{j,2} = Fs\ Ws_{j,1} \qquad j=1,S \qquad (5)$$

$$Ts_{j,2} = Ts_{j,1} \qquad j=1,S \qquad (6)$$

$$Ts_{j,3} = Ts_{j,1} \qquad j=1,S \qquad (7)$$

$$Ps_{j,2} = Ps_{j,1} \qquad j=1,S \qquad (8)$$

$$Ps_{j,3} = Ps_{j,1} \qquad j=1,S \qquad (9)$$

3.3 Heat Addition

This module has one input and one exit stream. The modeling consists of simple mass and energy balances and a pressure equality assignment.

$$Wq_{k,2} = Wq_{k,1} \qquad k=1,r \qquad (10)$$

$$Wq_{k,1} H(Tq_{k,1}) + Q_k = Wq_{k,2} H(Tq_{k,2}) \qquad k=1,r \qquad (11)$$

$$Pq_{k,2} = Pq_{k,1} \qquad k=1,r \qquad (12)$$

3.4 Pressure and/or Temperature Change

This module has one input and one exit stream. The modeling consists of simple equality assignments:

$$Wc_{l,2} = Wc_{l,1} \qquad l=1,c \qquad (13)$$

$$Tc_{l,2} = Tc_{l,1} + \Delta Tl \qquad l=1,c \qquad (14)$$

$$Pc_{l,2} = Pc_{l,1} - \Delta Pl \qquad l=1,c \qquad (15)$$

3.5 Heat Exchange

This module has two input and two exit streams. The modeling is based upon the following considerations:

a) The shell fluid temperature is an average isothermal temperature at any cross section.

b) The heating surface in each tube pass may be different.

c) The individual heat transfer resistances along the exchanger may vary.

d) The rate of flow of each fluid is constant.

e) The specific heat of each fluid along the exchanger may vary.

f) Phase changes (evaporation and/or condensation) may occur in a part of the exchanger.

Considering a differential element of the exchanger, the exchanged heat rate per unit area at any point may be represented by:

$$\frac{dq}{dA} = U(T_s - T) \qquad (16)$$

Where

$$\frac{1}{U} = \frac{1}{h_t} + \frac{1}{h_s} + R_m + R_{dt} + R_{ds} \qquad (17)$$

On the other hand, the exchanged heat rate in the differential element, can be related to the fluids enthalpy changes:

$$dq = W d_H(T) \qquad (18)$$

The elemental area can be expressed in terms of the area per unit length:

$$dA = A_u dx \qquad (19)$$

COMPLEX HEAT EXCHANGERS NETWORKS SIMULATION

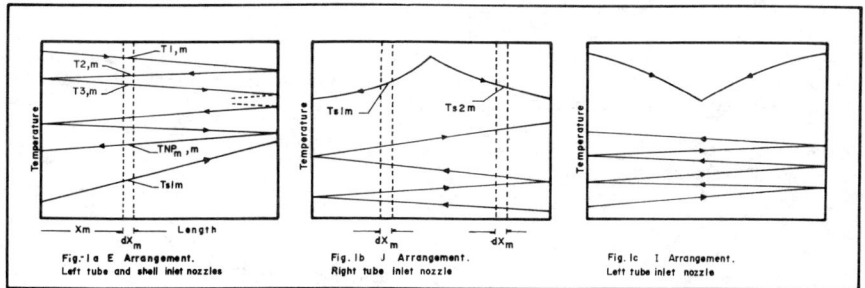

Fig. 1a E Arrangement. Left tube and shell inlet nozzles

Fig. 1b J Arrangement. Right tube inlet nozzle

Fig. 1c I Arrangement. Left tube inlet nozzle

The enthalpy, being a function of the fluid temperature, can be expressed as:

$$dH(T) = H'(T)dT \tag{20}$$

The principles of expressions (16), (18), (19) and (20) can be applied to each one of the tube passes of all heat exchangers: (See Fig. 1).

$$(-1)^{j+f} W_{tm} H'(T_{jm}) dT_{jm} = Au_{jm}(T_{sim} - T_{jm}) dX_m \quad j=1, N_{pm}, \quad m=1, N_h \tag{21}$$

Rearranging expression (21)

$$\frac{dT_{jm}}{dX_m} = (-1)^{j+f} \frac{Au_{jm} U_{jim}}{W_{tm} H'(T_{jm})} (T_{sim} - T_{jm}) \quad j=1, N_{pm} \quad m=1, N_h \tag{22}$$

The positive or negative sign of the set of expressions (22) depends upon the nozzles positions for the fluid inside the tubes. When the inlet is located at the left-hand side of the exchanger, a sign given by a factor of f=1 will correspond; when the inlet is at the right-hand side, a factor of f=0 will apply.

The temperature variation of the fluid inside the shell can be related to the temperature variations of the tube fluid by an energy balance at any differential element of the exchanger:

$$W_{sim} dH(T_{sim}) = W_{tm} \sum_{j=ti(1)}^{ti(2)} (-1)^{j+f+fs} dH(T_{jm}) \quad i=1, N_{sm}, \quad m=1, N_h \tag{23}$$

The differential enthalpy changes may be expressed according to expression (20) and with certain rearrangement, equation (23) can be transformed to:

$$\frac{dT_{sim}}{dX_m} = \frac{W_{tm}}{W_{sim}} \frac{1}{H'(T_{sim})} \sum_{j=ti(1)}^{ti(2)} (-1)^{j+f+fs} H'(T_{jm}) \frac{dT_{jm}}{dX_m} \quad i=1, N_{sm}, \quad m=1, N_h \tag{24}$$

The positive or negative sign of the set of expressions (24) depends upon the relative flow direction of the fluid inside the shell with respect to that inside the tubes. For a right-to-left direction in the shell, a sign given by a factor of fs=0 will correspond; for a left-to-right direction, a sign given by a factor of fs=1 will apply.

For an E, J or I arrangement, $ti(1)=1$ and $ti(2)=N_{pm}$, as all the tube passes are in thermal contact with every fluid trajectory in the shell. For a G arrangement, the situation differs; the model scope contemplates 1, 2 or 4 tube passes for this arrangement. For one pass, each shell trajectory will be in contact with the pass; for two passes, the top shell trajectories will be in contact with one pass and the bottom shell trajectories will be in contact with the other pass. Finally, for four passes, the top shell trajectories will exchange heat with two tube passes, and the bottom shell trajectories with the other two. An identification (ti) of which tube passes exchange heat with the ith shell trajectory is then included in the model.

Another important factor for completing the model definition is the boundary limits establishment:

For $X_m = 0$

$T_{1,m} = T_{in\ m}$	(Left tube inlet nozzle)
$T_{2j+1,m} = T_{2j,m}$ $j=1, N_{pm}/2-1$	(Left tube inlet nozzle and $N_{pm} > 2$)
$T_{2j,m} = T_{2j-1,m}$ $j=1, N_{pm}/2$	(Right tube inlet nozzle and $N_{pm} > 1$)
$T_{Npm,m} = T_{out\ m}$	(Left tube inlet nozzle and $N_{pm} > 1$)
$T_{Npm,m} = T_{out\ m}$	(Right tube inlet nozzle and $N_{pm} = 1$)
$T_{s1m} = T_{s\ in\ 1m}$	(E arrangement with left shell inlet nozzle and I arrangement)
$T_{s1m} = T_{s\ out\ 1m}$	(E arrangement with right shell inlet nozzle and J arrangement)
$T_{s1m} = T_{s2m}$	(G arrangement)

For $X_m = L_m/2$

$T_{s1m} = T_{s\ in\ 1m}$	(J and G arrangement)
$T_{s1m} = T_{s2m}$	(J arrangement)
$T_{s1m} = T_{s\ out\ 1m}$	(I arrangement)
$T_{s1m} = T_{s3m}$	(G arrangement)
$T_{s2m} = T_{s\ out\ 1m}$	(G arrangement)
$T_{s2m} = T_{s\ out\ 2m}$	(I arrangement)
$T_{s4m} = T_{s\ out\ 2m}$	(G arrangement)

For $X_m = L_m$

$T_{1,m} = T_{in\ m}$		(Right tube inlet nozzle)
$T_{2j+1,m} = T_{2j,m}$	$j=1, N_{pm}/2-1$	(Right tube inlet nozzle and $N_{pm} > 2$)
$T_{2j,m} = T_{2j-1,m}$	$j=1, N_{pm}/2$	(Left tube inlet nozzle and $N_{pm} > 1$)
$T_{Npm,m} = T_{out\ m}$		(Right tube inlet nozzle and $N_{pm} > 1$)
$T_{Npm,m} = T_{out\ m}$		(Left tube inlet nozzle and $N_{pm} = 1$)
$T_{s1m} = T_{s\ in\ 1m}$		(E arrangement with right shell inlet nozzle)
$T_{s1m} = T_{s\ out\ 1m}$		(E arrangement with left shell inlet nozzle)
$T_{s2m} = T_{s\ in\ 2m}$		(I arrangement)
$T_{s2m} - T_{s\ out\ 2m}$		(J arrangement)
$T_{s3m} = T_{s4m}$		(G arrangement)

For $0 \leq X_m < L_m/2$

i = 1 for E, J and I arrangements
i = 1 for G arrangement if ($t_1(1)$ or $t_1(2)$) = jth tube pass
i = 2 for G arrangement if ($t_2(1)$ or $t_2(2)$) = jth tube pass

For $L_m/2 \leq X_m \leq L_m$

i = 1 for E arrangement
i = 2 for J and I arrangement
i = 3 for G arrangement if ($t_3(1)$ or $t_3(2)$) = jth tube pass
i = 4 for G arrangement if ($t_4(1)$ or $t_4(2)$) = jth tube pass

It is important to mention here that the following conventions have been adopted:

a) The single inlet for E, J and G arrangements is referred to as first inlet, and the single shell trajectory for E arrangement as first trajectory (i=1)

b) The left and right shell inlets for I arrangement are referred to as first and second inlets.

c) For J arrangement, the center-to-left shell trajectory is referred to as first trajectory (i=1), the left outlet as first outlet; on the contrary, the center-to-right shell trajectory is referred to as the second trajectory (i=2), and the right exit as the second exit.

d) For I arrangement, the left-to-center shell trajectory is referred to as first trajectory (i=1) and the right-to-center shell trajectory as second trajectory (i=2); two outlets are considered, 1 from first trajectory and 2 from second trajectory.

e) For G arrangement, trajectory (i=1) is center-to-left, trajectory (i=2) is left-to-center, trajectory (i=3) is center-to-right and trajectory (i=4) is right-to-center. Two outlets are considered: 1 from second trajectory and 2 from 4th trajectory.

In order to relate the two exits considered in the model for I and G arrangements with the single physical outlets for these equipments mass and energy balances have to be applied:

$$Ws1m + Ws2m = Wsm \tag{25}$$

$$Ws1m\, H(Ts_{out\ 1m}) + Ws2m\, H(Ts_{out\ 2m}) = Wsm\, H(Ts_{out\ m}) \tag{26}$$

$$Ws2m + Ws4m = Wsm \tag{27}$$

$$Ws1m\, H(Ts_{out\ 1m}) + Ws2m\, H(Ts_{out\ 2m}) = Wsm\, H(Ts_{out\ m}) \tag{28}$$

Equations (25) and (26) apply for I arrangement and (27) and (28) for G arrangements.

Finally, the heat exchanger model is completed with inlet and outlet pressure relationships according to the following expresions:

$$P_{out\ m} = P_{in\ m} - \Delta P_m \tag{29}$$

$$Ps_{out\ m} = Ps_{in\ m} - \Delta Ps_m \tag{30}$$

Certainly, the model includes different correlations for the estimation of heat transfer coefficients and pressure drops. Table 1 shows the correlations employed.

3.6 Integral Mathematical Model

The individual models developed in sections 3.1 to 3.5 can be combined and organized in order to represent any complex arrangement constituted by the described modules. This can be done by simply establishing the interconnecting fashion that will be the basis for setting the variables relationship between the different particular modules. One adequate way for this relationship setting

		Tube Side		Shell Side	
		Horizontal	Vertical	Horizontal	Vertical
Single Phase	Heat Transfer	Sieder & Tate(16)	Sieder & Tate(16)	Bell(2)	Bell(2)
	Pressure Drop	Sieder & Tate(17)	Sieder & Tate(17)	Bell(2)	Bell(2)
Condensation	Heat Transfer	Kern(12)	Colburn(7)	Kern(13)	Kern(13)
	Pressure Drop	Dukler(10)	Dukler(10)	Kern(14)	Kern(14)
Evaporation	Heat Transfer	Altman(1)	Chen(9)	X	X
	Pressure Drop	Dukler(10)	Okiszewski(15)	X	X

Table 1. Pressure Drops and Film Coefficients Correlations.

is the use of a topology matrix, where the inlet and outlet streams for each module are defined. The topology will incorporate some new equality assignments to the model of the following form:

$$W_{in_{j,k}} = W_{out_{i,n}} \quad j = 1, N_{in_k} \quad k = 1, N_M \tag{31}$$

$$T_{in_{j,k}} = T_{out_{i,n}} \quad j = 1, N_{in_k} \quad k = 1, N_M \tag{32}$$

$$P_{in_{j,k}} = P_{out_{i,n}} \quad j = 1, N_{in_k} \quad k = 1, N_M \tag{33}$$

The corresponding i,n indexes associated to each j,k values are precisely defined by the topology matrix.

This leaves a consistent integral mathematical model constituted by a set of differential, linear and non linear algebraic equations.

4. MODEL SOLUTION

The solution of the integral model presents four basic problems, that will be described in the following sections: The differential equations handling, the required iterative calculations, the initial guess and the solution sequence. For the scope of this paper, the solution is oriented for those cases where all the external process streams and modules parameters are totally defined, being the unknowns the interconnecting and exit process streams.

4.1 Differential Equations.

Equations (22) and (24) have been analytically solved for single equipments and particular arrangements when several simplifications are assumed, leading to logarithmic mean temperature differences and correction factors (3,4,8,11,18). However, for a strict solution a numerical technique is required and the Runge-Kutta approach has been found adequate for this system.

The direct application of these techniques requires the definition of boundary limits of the type $Y_i(X) = C_i$ at $X=K$. Although the model includes this type of boundary limits, it also includes limits of the type $Y_i(X) = Y_j(X)$ at $X=K_i$. This implies that an iterative solution will be mandatory; the first type of boundary limits have to be assumed, and after the numerical integration, the actual limits will be checked. The particular number and type of assumptions depends upon the shell arrangement and the relative shell and tube nozzles positions. In general the iterative solution may be formulated as a set of non linear algebraic equations to be solved:

$$f_i(Y_i(K) \quad i=1,n) = 0 \quad i=1,n \tag{34}$$

Where $Y_i(K)$ are the assumed boundary limits for the differential equations $dY_i(X)/dX$ and f_i are error functions that represent the deviations of the integrated equations with respect to the actual boundary limits:

$$f_i (Y_i(K) \quad i=1,n) = Y_i(K_i) - Y_j (K_i) \quad i=1,n \tag{35}$$

4.2 Iterative Calculations

The solution of equation (2) for the streams mixing module and of equation (11) for the heat addition module requires an iterative calculation with a single variable, the exit temperature, $T_{mi,3}$ for mixing and $T_{qk,2}$ for heat addition. The Zepeda and Cano (19) technique has been found adequate for this problem.

The more complex problem of the heat exchanger model where, as described in the previous section, an iterative solution is imposed by the class of the system boundaries, has been solved by a combination of a generalized algorithm for multivariable non linear algebraic equations solution proposed by. J. Castellanos and S. Ortiz de Montellano (5) and a particular approach to this system, proposed by J. Castellanos, D. Salazar and E. Mercado (6).

4.3 Initial Guess

Of special importance for the solution of the integral model is the set of the assumed variables that has to be made in order to initiate the iterative calculations. Due to the extensive computing effort required for each iteration, it is highly recommended to pay particular attention to the initial guess. One adequate approach for this problem is an initial simplified solution of the mo-

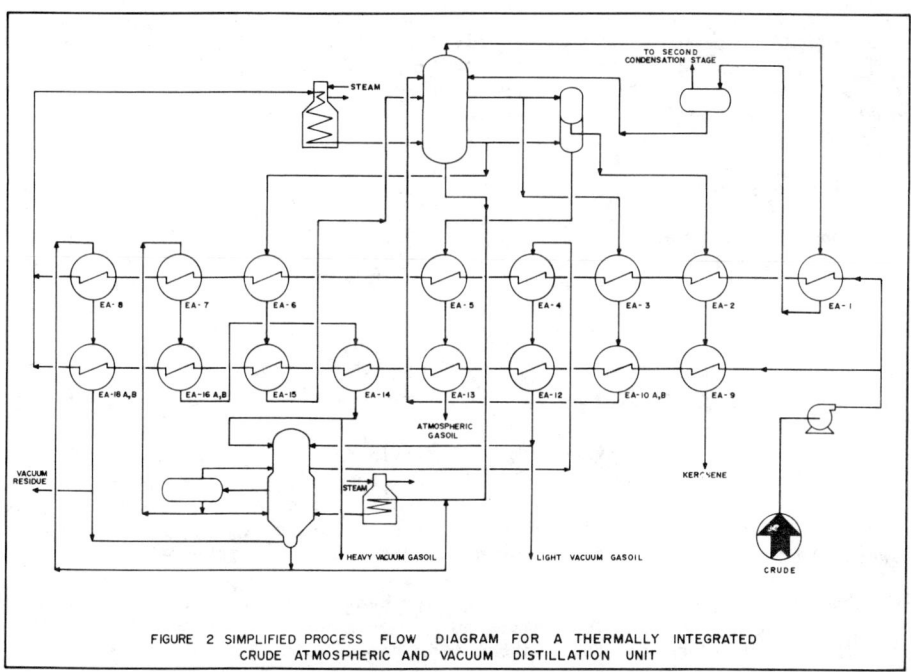

FIGURE 2 SIMPLIFIED PROCESS FLOW DIAGRAM FOR A THERMALLY INTEGRATED CRUDE ATMOSPHERIC AND VACUUM DISTILLATION UNIT

HEAT EXCHANGER DESIGN

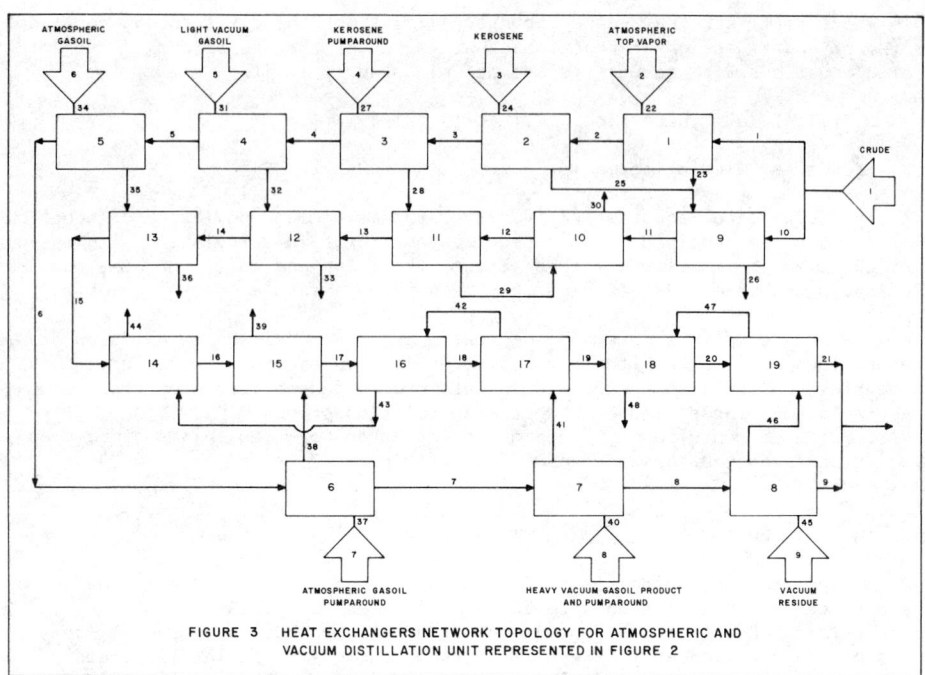

FIGURE 3 HEAT EXCHANGERS NETWORK TOPOLOGY FOR ATMOSPHERIC AND VACUUM DISTILLATION UNIT REPRESENTED IN FIGURE 2

del, using the logarithmic mean temperature difference definition for each heat exchanger, for which average heat transfer coefficients and specific heats may be considered. A Newton-Raphson numerical technique is adequate for this solution, which will generate the interconnecting process streams temperatures. The individual shell trajectories and tube passes temperatures may be estimated from theoretical considerations used in the LMTD derivations.

4.4 Solution Sequence

Although the model is solved by a modular approach, the individual heat exchangers modules are not converged at each iteration; after a one-by-one numerical integration, a guess resetting using the whole boundary limits error functions is accomplished. The sequencial solution is based on a minimization of guessed variables and results a function of the system topology.

5. COMPUTER PROGRAM

A computer program, SIMPROTERM, written in Fortran V has been prepared for the solution of the described model. The program allows the analysis of complex heat exchangers networks, calculating the pressure and temperature of the interconnecting and exit process streams, the heat duties for each equipment and the temperature profiles for the fluids in the shell and each tube pass.

The program can generate the individual heat transfer coefficients when a complete definition of the equipments and process streams are provided; optionally, the user may provide this information for all or some of the equipments.

COMPLEX HEAT EXCHANGERS NETWORKS SIMULATION

The process streams definition can be made in terms of their compositions of chemical species or petroleum fractions pseudocomponents. In this case, an internal generation of properties takes place, using the IMP Thermophysical Properties Package and its associated IMP Components Data Bank. Alternatively, the properties may be provided by the user as tabulated values of properties vs. temperature. When the heat transfer coefficients are externally defined, the only required streams information is the enthalpy, provided as a polynomial function of temperature.

The data to be provided to the program may be classified in the following categories: a) General information, including number of simulations, and desired messages to be printed in the results. b) System topology, provided as an interconnection matrix. To prepare this information, an arrangement scheme like that shown in Fig. 3 is useful; each module and process stream is numbered, as these numbers are used as the matrix elements. c) Calculation controls, including maximum number of iterations, integration step, calculation options definition. d) Streams definitions, temperature, pressure, flow, composition, thermophysical properties correlations selection. e) Heat exchangers internal construction arrangement and dimensions.

Figure 2 represents a simplified process scheme for a thermally integrated atmospheric and vacuum petroleum distillation unit. The computer program was successfully applied to a case of study for the network represented in that figure; the run required 50 seconds in a 1100/80 A UNIVAC Computer. The calculation included internal properties and heat transfer coefficients generation, calculated at each integration step. A typical result for exchanger No. 6 is ilustrated in figure 4.

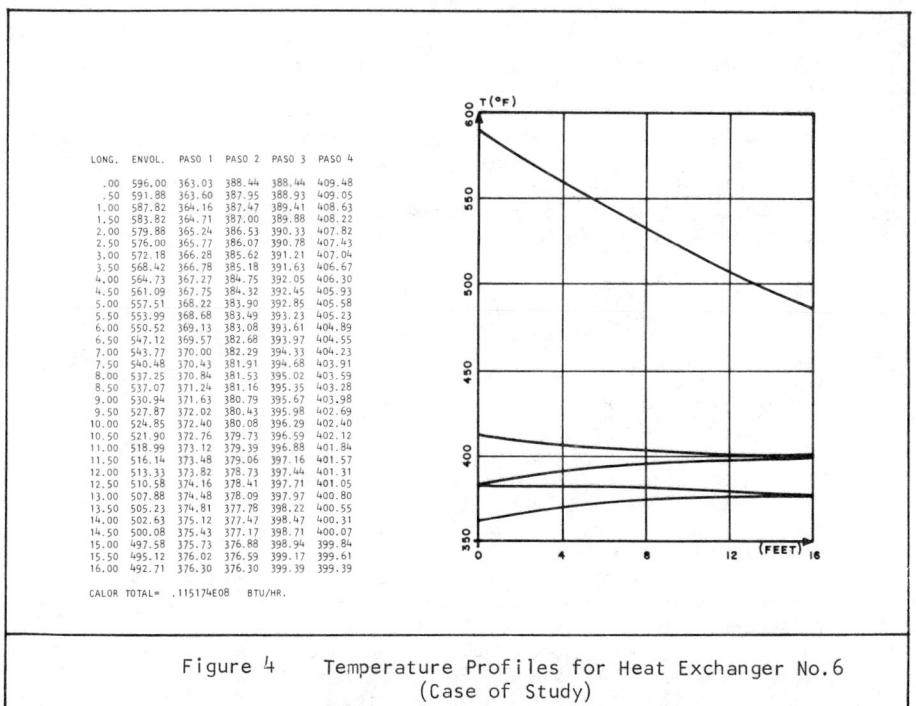

Figure 4   Temperature Profiles for Heat Exchanger No.6
(Case of Study)

NOMENCLATURE

A	Heat transfer area
A_u	Heat transfer area per unit lenght
A_{ujm}	Heat transfer area per unit lenght for ith tube pass and mth heat exchanger.
c	Number of Pressure/Temperature changing modules
d	Differential operator
f	Sign definer, according to inlet tube nozzle position
f_s	Sign definer, according to inlet shell nozzle position
F_s	Split fraction
$H(T)$	Enthalpy at temperature T
$H'(T)$	dH/dT at temperature T
h_t	Tube side film transfer coefficient
L_m	Tube length of mth heat exchanger
M	Number of streams mixing modules
MIN	Minimum function operator
N_h	Number of heat exchangers modules
$N_{in\,k}$	Number of stream inlets of kth module
N_M	Total number of modules
N_{pm}	Number of tube passes of mth exchanger
N_{sm}	Number of shell trajectories of mth exchanger
$Pc_{l,1}$	Pressure of lth changing module inlet stream
$Pc_{l,2}$	Pressure of lth changing module exit stream
$Pin_{j,k}$	Pressure of kth module jth inlet stream
Pin_m	Inlet pressure of mth heat exchanger, tube side
$Pm_{i,1}$	Pressure of ith mixing module first inlet stream
$Pm_{i,2}$	Pressure of ith mixing module second inlet stream
$Pm_{i,3}$	Pressure of ith mixing module exit stream
$Pout_{i,n}$	Pressure of nth module ith outlet stream
$Pout_m$	Outlet pressure of mth heat exchanger, tube side
$Pq_{k,1}$	Pressure of kth heat addition module inlet stream
$Pq_{k,2}$	Pressure of kth heat addition module exit stream
$Ps\,in\,m$	Inlet pressure of mth heat exchanger, shell side
$Ps_{j,1}$	Pressure of jth splitting module inlet stream
$Ps_{j,2}$	Pressure of jth splitting module first exit stream
$Ps_{j,3}$	Pressure of jth splitting module second exit stream
$Ps\,out\,m$	Outlet pressure of mth heat exchanger, shell side
q	Heat rate.
Q_k	Heat rate of kth heat addition module
r	Number of heat addition modules
R_{ds}	Shell side fouling factor
R_{dt}	Tube side fouling factor
R_m	Tube wall thermal resistance
S	Number of splitting modules
T	Temperature
$ti(1)$	First tube pass in thermal contact with ith shell trajectory
$ti(2)$	Second tube pass in thermal contact with ith shell trajectory
$Tc_{l,1}$	Inlet temperature of lth changing module
$Tc_{l,2}$	Outlet temperature of lth changing module
$Tin_{j,k}$	Temperature of kth module ith inlet stream
Tin_m	Inlet temperature of mth heat exchanger, tube side
Tj_m	Temperature of jth tube pass, mth heat exchanger
$Tm_{i,1}$	Temperature of ith mixing module first inlet stream
$Tm_{i,2}$	Temperature of ith mixing module second inlet stream
$Tm_{i,3}$	Temperature of ith mixing module exit stream
$Tout_{i,n}$	Temperature of nth module ith outlet stream
$Tout_m$	Outlet temperature of mth heat exchanger, tube side

$T_{qk,1}$	Temperature of kth heat addition module inlet stream
$T_{qk,2}$	Temperature of kth heat addition module exit stream
T_s	Shell side temperature
$T_{sj,1}$	Temperature of jth splitting module inlet stream
$T_{sj,2}$	Temperature of jth splitting module first exit stream
$T_{sj,3}$	Temperature of jth splitting module second exit stream
T_{sim}	Temperature of ith shell trajectory, mth heat exchanger
Ts in 1m	Inlet temperature of first shell trajectory, mth heat exchanger
Ts in 2m	Inlet temperature of second shell trajectory, mth heat exchanger
Ts out m	Temperature of combined shell outlets of mth heat exchanger
Ts out 1m	Outlet temperature of first shell trajectory, mth heat exchanger
Ts out 2m	Outlet temperature of second shell trajectory, mth heat exchanger
U	Overall heat transfer coefficient
U_{jim}	Overall heat transfer coefficient at jth tube pass, ith shell trajectory, mth heat exchanger
W	Mass flow rate
$W_{cl,1}$	Mass flow rate of lth changing module inlet stream
$W_{cl,2}$	Mass flow rate of lth changing module exit stream
$W_{inj,k}$	Mass flow rate of kth module jth inlet stream
$W_{mi,1}$	Mass flow rate of ith mixing module first inlet stream
$W_{mi,2}$	Mass flow rate of ith mixing module second inlet stream
$W_{mi,3}$	Mass flow rate of ith mixing module exit stream
$W_{outi,n}$	Mass flow rate of nth module ith outlet stream
$W_{qk,1}$	Mass flow rate of kth heat addition module inlet stream
$W_{qk,2}$	Mass flow rate of kth heat addition module exit stream
$W_{sj,1}$	Mass flow rate of jth splitting module inlet stream
$W_{sj,2}$	Mass flow rate of jth splitting module first exit stream
$W_{sj,3}$	Mass flow rate of jth splitting module second exit stream
W_{sim}	Mass flow rate at ith shell trajectory, mth heat exchanger
W_{sm}	Total mass flow rate of mth heat exchanger shell side
W_{tm}	Mass flow rate of mth heat exchanger, tube side
X_m	Length meassured from left-hand-side of the mth exchanger
ΔP_l	Pressure change of lth changing module
ΔP_m	Tube side pressure drop of mth heat exchanger
ΔP_{sm}	Shell side pressure drop of mth heat exchanger
ΔT_l	Temperature change of lth changing module.

ACKNOWLEDGEMENTS

The initial development of this work was prepared with the contributions of Enrique Aguilar, Elías Mercado and Máximo Téllez. The programming effort was the responsibility of Daniel Salazar who also provided many useful ideas for the mathematical model solution and computer program structure. Presently, several improvements are being made, under responsibility of José Antonio Amozurrutia and Julio Landgrave, under the direction of José Luis Cano. The author also wishes to express his gratitude to María Esther Prado for patiently typing and retyping the manuscript of this paper, accomplishing the time objectives.

REFERENCES

1. Altman M., Straub F.W. and Norris R.H. Chem. Eng. Prog. Symp Ser. 56, 30 p. 151 (1960).

2. Bell K. J. Petro. Chem. Eng. p. C26-C40c Oct. (1960).

3. Bowman R.A. Ind. Eng. Chem. 28 p. 541-544 May (1936).

4. Bowman R.A., Mueller A.C. and Nagle W.M. Trans. Am. Soc. Mech. Engrs. 62 p. 283-294, May (1940).

5. Castellanos J. and Ortiz de Montellano S. Rev. IMIQ XVIII, 7-8, p. 86-96, July-Aug (1977).

6. Castellanos J., Salazar D. and Mercado E. Rev. IMIQ XVIII, 5-6 p. 64-80, May-June (1977).

7. Colburn A.P. Trans. AIChE 30, p. 187-193 (1934).

8. Correa A., Acosta R., Rosales M.A. and Garcés P.A. Rev. del Inst. Mex. del Petr. IV, 4, p. 42-51 (Oct. 1972).

9. Chen. N.H. Chem. Eng. p. 141-146 March 9 (1959).

10. Dukler A.E. Chem. Eng. Prog. Symp. Series 56, 30 p. 110 (1960)

11. Jarzebski A.B. Lachowski A. I. Szponarski and Gasior S. Canad. Jour. Chem. Eng. 55 p. 741-743. Dec. (1977).

12. Kern D.Q. Process Heat Transfer, Mc. Graw Hill-Kogakusha p. 269 (1950)

13. Ibid, p. 266-267.

14. Ibid, p. 273.

15. Okiszewski J. Jour. Petro. Tech. June (1967).

16. Sieder E.N. and Tate G.E. Ind. Eng. Chem. 28, p. 1429-1436 (1936).

17. Standards of Tubular Exchanger Manufactures Association.

18. Underwood A.J.V. Jour. Inst. Petro. Tech. 20, p. 145-158 (1934).

19. Zepeda R. and Cano J.L. Rev. IMIQ XVII 9-10, p. 38-42 Sept.Oct. (1976).

HEAT EXCHANGERS IN POWER GENERATION SYSTEMS

Nuclear Steam Generators

G. HETSRONI
Danciger Professor of Engineering
Technion, Israel Institute of Technology
Haifa, Israel

1. INTRODUCTION

Most of the new electric capacity installed these days, conventional or nuclear, utilizes steam. In conventional boilers steam pressure reaches 250 bars and temperatures up to 540°C. In nuclear steam generators the steam is about 48 to 75 bars and 265°C to 315°C.

Modern conventional steam boilers operate safely and dependably, and remain in service for many years with cleaning and repairs usually required only at scheduled outage periods. These boilers owe their dependability to more than a quarter of a millenium of experience in the design, fabrication and operation of such units. The first recorded commercial steam generator was the Haycok boiler (1720) which was merely a kettle filled with water and heated by fire.

Since then steam generators evolved to more sophisticated designs and the properties of water and steam have been thoroughly investigated and accurately tabulated. Also, a wealth of information was developed on heat transfer and flow of single phase and two-phase fluids.

The NSSS (nuclear steam supply system) is a relatively recent development, and has been in use for about twenty years. During this time there were constructed and put in operation 107 Pressurized Water Reactors (PWR), 47 of which are in the U.S.; 61 Boiling Water Reactors (BWR), 26 of which are in the U.S.; 11 light-water cooled graphite-moderated reactors (LGR) and 17 pressurized heavy water moderated and cooled reactors (PHWR) - all over 30 MW. In addition, to the end of the 80s, there were expected to be in operation 163 more PWRs, 56 BRWs, 12 LGRs and 18 PHWRs. However, this growth depends largely on the political climate and various economical, ecological and emotional issues.

The NSSS provided in the past service which was mostly safe and trouble free. Recently, however, there is an increasing number of PWR nuclear steam generators which develop technical problems, such as denting, intergranular attack (IGA), vibration of tubes which cause wear and fatigue, wastage of tubes, pitting, erosion-corrosion, water hammer, etc. Any of these can lead to a breach of the integrity of the tubes and to leakage of primary (contaminated) coolant into the secondary fluid. Here primary fluid is defined as the fluid transferring heat from the reactor core to the steam generator. Secondary fluid is the one boiling in the steam generator, leaving the containment vessel to the turbines. The secondary fluid must not be radioactive, and must not be contaminated by primary fluid. Therefore, when leaking tubes are detected (with leakage at a rate of <u>about</u> 400 liters per day), the plant must be shut down for repairs, at great costs and loss of revenue.

Today, there is no assurance that present designs and practices will result in the 40-year steam generator expected lifetime. Moreover, experience to date warns against such an expectation. Since these concerns are relatively recent, and there is little literature on the subject, it is discussed here in detail.

2. CLASSIFICATION OF PWR NUCLEAR STEAM GENERATORS

For reasons associated with the construction of the containment vessel, all PWR steam generators in the U.S. are vertical.

The horizontal and vertical steam generators have certain advantages and suffer from certain drawbacks. Here, only vertical steam generators are discussed: the Once-Through Steam Generator (OTSG) and the Recirculation Steam Generators or the U-Tube Steam Generators (UTSG).

2.1 The Once-Through Steam Generator (OTSG)

The OTSG is typically associated with a NSSS which has the following general characteristics:

	Oconee 1	Max Unit*
Rated power, MWe	860	1,300
Rated heat output, MWth	2,568	3,760
System pressure, bar	149	153
Primary coolant flow rate, 10^6 kg/hr	29.5	35.6
Primary coolant temperature, °C	317.8	329.7
Number of loops	2	2
Number of pumps	4	4
Number of steam generators	2	2
Shell side design pressure, bar	71.4	84.0
Steam flow at full load, 10^6 kg/hr	5.0	7.4
Steam temp. at full load, °C	299	308
Steam superheat at full load, °C	19.4	19.4
Feedwater temperature, °C	235	241
Number of tubes	15,530	16,000
Diameter of tubes, mm	15.9	15.9

A general view of the system is depicted in figure 1, and the general arrangement of the OTSG is given in figure 2.

Reactor coolant water enters the steam generator at the top, flows downward through the tubes and out at the bottom. The high pressure parts of the unit are the hemisphere heads, the tubesheets, the straight tubes between the tubesheets. The tube material is Inconel Alloy 600. Tube support plates, with trefoil holes, hold the tubes in a uniform pattern along their length. The unit is supported by a skirt attached to the bottom tubesheet.

---

* Conforming to current Nuclear Regulatory Commission (NCR) size limitation of 3800 MWth.

NUCLEAR STEAM GENERATORS

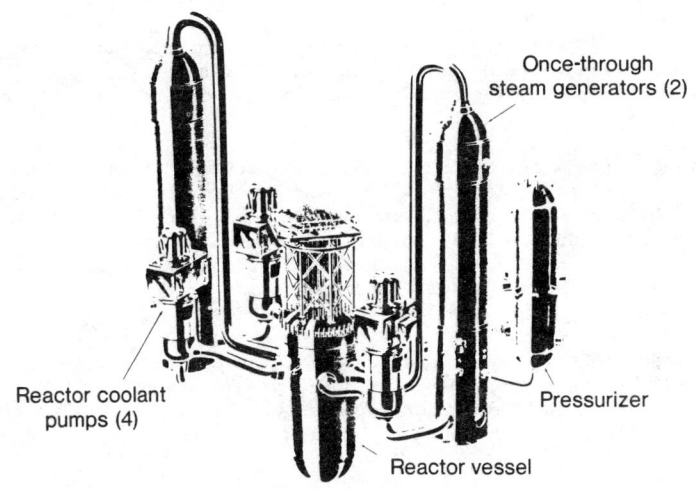

Figure 1. B&W nuclear steam system (Steam, B&W).

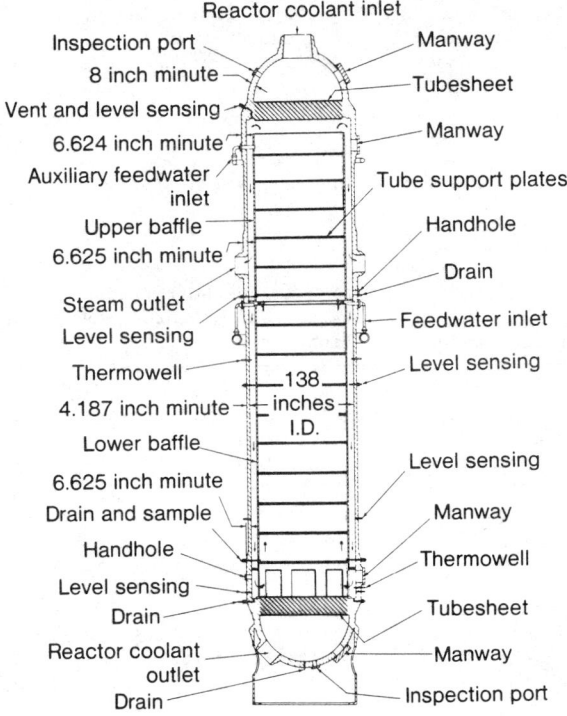

Figure 2. Once-through steam generator (Steam, B&W).

Figure 2 indicates the flow paths on the steam side of the unit. Feedwater enters the side of the shell. It flows down through an annulus just inside the shell where it is brought to saturation temperature by mixing with steam. The saturated water enters the heating surface at the bottom and is converted to steam and superheated in a single pass upward through the generator.

The shell, outside of the tubes, and the tubesheets form the boundaries of the steam producing section of the vessel. Within the shell, the tube bundle is surrounded by a shroud, which is in two sections with the upper section the larger of the two indiameter. The upper part of the annulus between the shell and the baffle is the superheater outlet, while the lower part is the feedwater inlet heating zone. Vents, drains, instrumentation nozzles and inspection openings are provided on the shell side of the units.

Superheated steam is produced at a constant pressure over the power range. At full power, the steam temperature of 300°C provides about 19°C of superheat. As load is reduced, steam temperature approaches the reactor outlet temperature, thus increasing the superheat slightly. Below 15% load, steam temperature decreases to saturation.

The ability to maintain a constant (secondary) steam pressure is achieved by varying the boiling length with the load, i.e., the mass flow rate of the feedwater is controlled (above 15% load) by the pressure. When the outlet steam pressure decreases, the mass flow of the feedwater increases, thus increasing the boiling length and the steam flow rate. This can best be illustrated as follows: the total heat transferred in the steam generator Q is proportional to the overall heat transfer coefficient U, to the total transfer area A, and to the temperature difference between the primary and secondary fluids ΔT, i.e.,

$$Q = UA\ \Delta T$$

This equation can be written in an equivalent form

$$W_s \propto U\ L_B\ (T_p - T_{sat})$$

where W_s is the steam flow rate, U is the heat transfer coefficient which is mainly in the boiling region (since the amount of heat transferred in the superheated steam region is relatively small), L_B is the boiling length, T_p is the primary temperature, and T_{sat} is the saturation temperature.

Note that the above equations represent average quantities -- a temporal averaging and averaging over the cross section of the generator. Actually, heat transfer coefficients and local temperatures are complex functions of space and time and depend on local flow rates, quality etc. In order to make more precise evaluation of the heat transferred in the steam generators, one needs two phase, three dimensional, thermal-hydraulic computer codes which are now becoming available in the industry.

Another feature of the OTSG is its smaller size (compared with UTSG of comparable capacity) - mainly because it does not have moisture separators. The smaller size of the OTSG results, of course, in low secondary water inventory, which is about a sixth of the water inventory of a UTSG.

Since it was always expected that corrosion may be a problem in OTSG, water chemistry was emphasized. ALL plants have almost full flow condensate polishing (powdered resin or deep bed system), and an all volatile chemical treatment of the feedwater. This treatment involves the addition of ammonia for pH control, and hydrazine for oxygen control. The specification for OTSG feedwater during normal operation and layup is given in table 1.

Table 1. OTSG Feedwater-Specifications
Normal Power Operation

Total solids	50 ppb max.
Cation conductivity	0.5μmhos/cm max.
Dissolved oxygen as O_2	7 ppb max.
Hydrazine as N_2H_4	20-100 ppb
Silica as SiO_2	20 ppb max.
Total iron as Fe	10 ppb max.
Total copper as Cu	2 ppb max.
pH @ 77F	9.3-9.5 or 8.5-9.3
Lead as Pb	1 ppb max.

Startup

Total iron as Fe	100 ppb max.
Cation conductivity	1.0μmhos/cm max.
Dissolved oxygen as O_2	100 ppb max.
Ammonia as NH_3	10 ppm nom.
	2 ppm-20ppm range
pH @ 77F	9.5-10.5
Hydrazine	200 ppm initial
	50 ppm min.
Sodium	1.0 ppm max.

2.2 The U-Tube Steam Generator (UTSG)

The U-Tube Steam Generator is installed in a wide variety of commercial NSSS, from the early (1957) 90-MWe Shippingport 4 loops system, to new 1300-MWe plants. Two typical systems are characterized in the following table:

	Surry	Model 414
Rated power, MWe	822	1,300
Rated heat output, MWth	2,441	3,819
System pressure, bar	153	153
Primary coolant flow rate, 10^6 Kg/hr	45.3	67.9
Primary coolant temperature, °C	318.7	332.0
Number at lumps	3	4
Number at pumps	3	4
Steam generators		
Number	3	4
Shell side design pressure, bar	53.4	77.0
Steam flow at full load, 10^6 Kg/hr	4.8	7.9
Steam temperature, °C	268.9	294.1
Feedwater temperature, °C	225.4	239.4
Number of tubes	3,388	6,970
Diameter of tubes, mm	22.2	17.4
Heat transfer area, m^2	4,784	7,665

Figure 3. Combustion Engineering system 80 two-loop design.

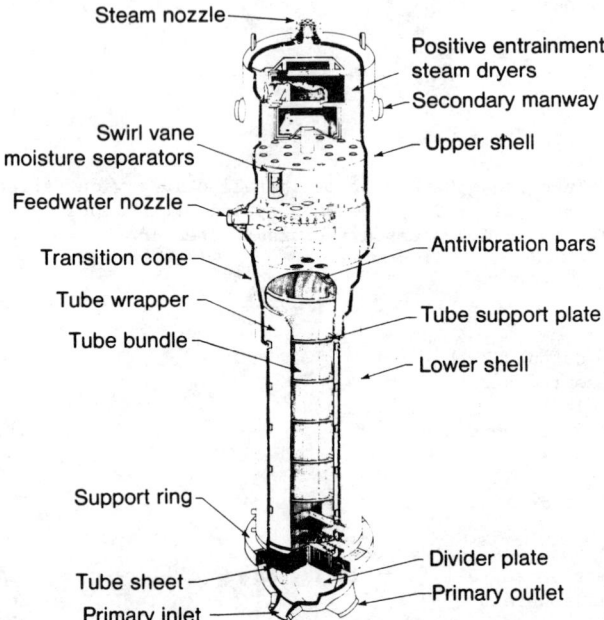

Figure 4. Westinghouse steam generator

A general view of a system is depicted in figure 3 and the general arrangement of the UTSG is given in figure 4.

Reactor coolant enters the steam generator at the inlet nozzle (on the bottom left-hand side of figure 4) and enters the tube bundle in the hot leg, completing the U-bend through the cold leg to the primary outlet.

Feedwater enters the side of the shell. It flows down through an annulus just inside the shell (downcomer), where it mixes with water coming from the separator deck. The water enters the heating surface (tube bundle) at the bottom and is increasing in quality as it rises through the steam generator. The steam-water mixture enters the steam water separators - where steam is passed through to the driers and then to the steam nozzle and the water is recirculated, through the downcomer, to the bottom of the heating surface.

The shell, outside of the tubes and the tubesheet form the steam production boundaries. Within the shell, the tube bundle is surrounded by a wrapper (or shroud). The tube material is usually Inconel Alloy 600. Tube support plates, with quatrefoil holes or egg crate supports, hold the tubes in uniform pattern along their length. The U-tube region of the tube is additionally supported by antivibration bars. Vents, drains, instrumentation nozzles, and inspection openings are provided on the shell side of the units.

The UTSGs are characterized by a widening of the shell about 2/3 of the height of the steam generator. This is done in order to increase the area available for the separators at the separator deck.

In the UTSG there is always a water level in the downcomer, in order to balance the pressure losses of steam-water mixture as it flows through the tube bundle. The pressure drop in the tube bundle is due to friction along the tubes, pressure drop at the tube support plate and separators, acceleration of the flow and hydrostatic head. Quality in the downcomer, e.g. steam bubbles, reduces the density and thus the available head substantially. This can be a result of imperfect separation of steam water at the separators and is termed carryunder.

The ratio of flow rate of the steam water mixture which flows through the steam generator tube bundle, to the flow rate of steam out of the steam nozzle, is called the circulation ratio. It is desirable to maintain a high circulation ratio (say, over 7) to reduce concentration of chemicals, debris, etc., in various places in the steam generator. Current steam generators have frequently circulation ratios of around 3 - which is undoubtedly one of the causes of the recently mounting difficulties with these units.

The steam pressure variation with power generated is depicted in figure 5.

3. DESIGN OF STEAM GENERATORS

3.1 General

The design of steam generators is a complicated procedure, which involves many steps, iterations, and interaction with other components of the system.

The NSSS, as part of the power station, is designed to minimize the power cost. This consideration is subject to many constraints -- the primary one being safety. Other constraints are imposed by availability of major equipment (e.g. primary coolant pumps), manufacturing capabilities (e.g. can vessel be fabricated), shipping considerations, etc.

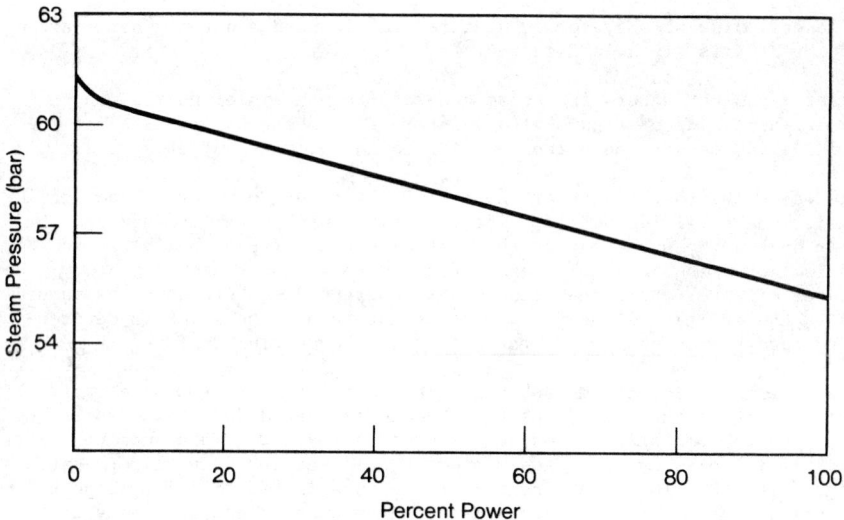

Figure 5. Steam pressure variation with power generation for a U-tube steam generator.

Thus, the design of the steam generators is subject to many outside constraints and requirements. For example, the primary fluid conditions (i.e., temperatures, allowable pressure drop in the steam generator) is determined mostly by the reactor design and availability of pumps. The performance requirement, i.e., steam pressure, temperature, and flow rates, are determined mostly by the turbine design and as part of the system performance. It follows, therefore, that the tube bundle size (namely the required heat transfer area, as well as allowable prssure drop of the primary fluid) is determined by system consideration. Typically, steam generator vendors will get these specifications (termed F̲unctional D̲esign D̲escription) from the group dealing with the design of the system and with optimization of the power cost.

The first step in a design of a steam generator would generally be the preparation of layout drawings which include the basic configuration and sizing of the steam generator. These drawings are prepared after preliminary structural, mechanical and thermal-hydraulic designs.

The preliminary structural design is performed in accordance with ASME's Boilers and Pressure Vessel Code, Section III. Also, steam generators, as all power plant components, are required to be designed to withstand various accident situations. For steam generators, these consist of the following conditions:

Small steam line break, loss of feedwater, turbine trip, etc.
LOCA (Loss of Coolant Accident)

MSLB (Main Steam Line Break)
SSE (Safe Shutdown Earthquake)

The preliminary design may be concluded by a design review where it must be shown that the proposed (preliminary) design responds to the functional design description and that it is in compliance with plant control and safety criteria.

Detailed design includes detailed structural and thermal-hydraulic analyses and studies, and is later used to prove to the customer and regulatory agencies that the steam generator design is in compliance with ASME code, NRC regulations, etc.

3.2 Thermal-Hydraulic Design of Steam Generators

Boiling heat transfer and two-phase flow are complex subjects even for idealized conditions. For conditions such as encountered in steam generators, modeling is very difficult. As a result, empirical correlations are used, and mostly only average quantities are calculated. A typical design procedure for OSTG is given in a ORNL design guide (1975).

The U-tube steam generator has four main regions. The first includes the U-tube zone, where heat is added to the secondary fluid and the boiling occurs. The second is the riser (see figure 4), the zone between the tube bundle and the the separator deck. The riser increases the natural circulation driving force. Above the riser is the moisture separation region where saturated water is removed from the steam and returned to the top of the downcomer where it mixes with cold feedwater. The subcooled fluid then flows to the bottom of the downcomer to complete the circuit. The natural circulation driving force is provided by the difference between the density of the water in the downcomer and that of the steam-water mixture in the heating zone and riser.

There is a corresponding circuit in the OTSG, less the separators.

The driving force for the natural circulation flow is resisted by the pressure losses. There are pressure losses in all regions described above. Of these, the single phase pressure drop is fairly straight forward to calculate using acceptable correlation. The pressure drop along the tube is usually calculated by using a well known two-phase multiplier such as Martinelli-Nelson (1948), Baroczy (1965) or Friedel (1977).

The pressure drop in the tube support plates and U-tube region depends largely on the geometry and is therefore specific to any particular design. Mostly, the calculations are based on coefficients which are determined by the various vendors experimentally. The same is true for the steam separators.

One of the more important problems associated with the thermal-hydraulic design is the stability of water level in the downcomer. A temporary change in the feedwater flow causes a change in the void fraction in the tube bundle, and and the water seeks a new level. As a result of feedback in this sequence of events, the water level may start to oscillate, and it is possible for these oscillations to become unstable. This is a problem which most vendors of PWR steam generators have encountered, usually at a power level between 40 to 60%.

4. PROBLEMS WITH STEAM GENERATORS

A schematic outline and problem areas of a UTSG is depicted in figure 6. The problem areas are associated with tube denting, intergranular attack (IGA), wastage and corrosion, flow induced vibrations (FIV), carry over and carry under. These are discussed in the following sections.

4.1 Problems with UTSG

Denting. The most widespread and serious problem observed in UTSG at this time is "denting" (figure 7). Denting is a reduction of steam generator tube diameter which occurs when a magnetite corrosion product forms in the annulus between the carbon steel tube support plate and the Inconel tube. The volume that the corrosion product occupies is larger than the volume of the original material which was attacked. Eventually, enough magnetite is produced to cause an inward displacement of the tube wall which an eddy current probe indicates as a dent. In extreme cases of denting, the buildup of forces can be sufficient to distort the tube support plate and cause cracking.

It is now obvious that denting occurs due to inadequate chemistry control of the water in the steam generator and of the feedwater. The most common source of ingress of impurities is from leaks in the condenser. Therefore, it is also obvious that cooling water leakage control is the key to steam generator chemistry control.

Denting type corrosion has been found in seawater and fresh water cooled units: in the U.S. 11 out of 13 are dented, 4 out of 14 units outside the U.S. are dented. In Japan 10 units are not dented. Freshwater and cooling tower plants: in the U.S. 10 out of 20 units are dented, 3 out of 16 units outside the U.S. are dented. It was recently found that denting with drilled hole carbon steel TSP occurs in a short time, even when chloride control is excellent and that aggravating factors, other than acid chloride, must also be controlled. Denting increases linearly with the total chloride exposure (i.e., chloride concentration in ppm times days) once denting has been initiated. The implication is that it may be difficult to avoid denting in plants with drilled hole carbon steel TSP. However, with careful chemistry control denting may be greatly retarded, if not prevented.

Some of the parameters affecting denting are as follows:

a. Chlorides - sample intersections of tubes and TSP removed from dented steam generators have shown local chloride concentration of over 4000 ppm in the dented region. A contributing factor to this high local chloride concentration is inadequate local thermal-hydraulic conditions in the crevice between the tube and the TSP, causing a concentration of the aggressive species. In particular, it is hypothesized that alternate wetting and drying of the crevice, due to some instability of the two-phase fluid, can cause the observed increase in concentration.

Denting has been successfully reproduced in the laboratory indicating that non-protective magnetite (NPM) cannot be produced by neutral chlorides alone, but that the presence of reducible metal cations introduce NPM formation and the presence of oxygen induces NPM in neutral chloride solutions. It is also apparent that other factors affect production of NPM - but their effect has not been quantified yet.

b. Oxygen, Copper and Nickel - the roles of oxygen, copper and nickel in denting is not fully understood. In some laboratory tests they seem to have

NUCLEAR STEAM GENERATORS

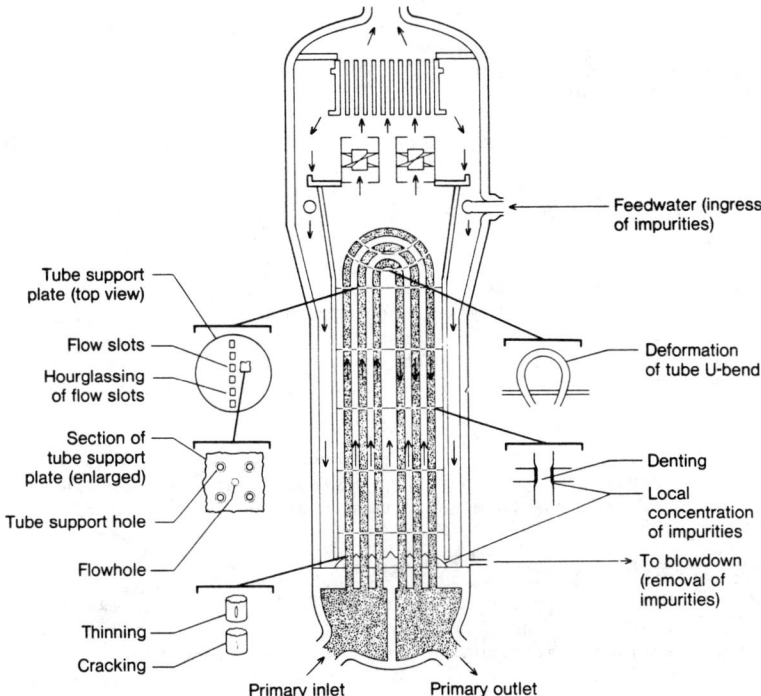

Figure 6.   Corrosion Problems in Steam Generators

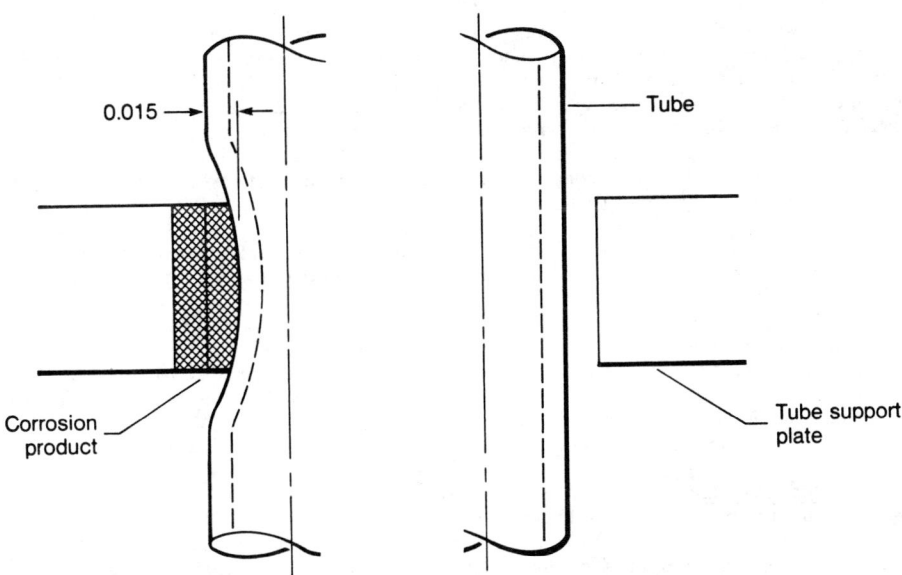

Figure 7.   Denting at Tube/Tube Support Plate Intersection.

accelerated the denting - but these results are yet inconclusive. Oxygen in laboratory tests has been shown to produce an acid environment. If a large amoung of neutral chloride is present, a small amount of oxygen can produce an acid chloride environment.

c. Sulfates - the role of sulfates as a cause of denting must be evaluated since they are used to treat cooling tower water and in demineralizers. Preliminary tests are yet inconclusive, but it is possible that sulfate behavior would be similar to the behavior of chloride, thereby causing denting type corrosion.

d. Thermal-hydraulic - power level and local thermal-hydraulic conditions seem to play a significant part in the denting corrosion process. As indicated above, local instability of the two-phase flow can cause alternate wetting and drying in the crevice - a process which leads to local concentration of chemicals. Concentration of at least 20,000 times had been calculated for some laboratory experiments.

Initial results from laboratory tests show that the tube wall superheat in the crevice are significantly higher than the superheat existing away from the crevice. Values of 22°C superheat have been observed, with heat flux of about 315 kw/m^2.

e. Denting control - experience indicates that steam generator corrosion has been limited with rigorously controlled all-volative-treatment (AVT) chemistry. In addition, limitation of the ingress of cooling water impurities, especially chlorides and dissolved oxygen, helps minimize local concentration of aggressive chemicals. Addition of various inhibitors is also being contemplated, but this is still being studied.

The major methods of inhibitor application are the soak, the on-line application, or a combination of the two. The soak application is defined as a short and intermittent application of the inhibitor to the steam generator, while at low power conditions. Another mode of soaking considers off-line, hot shutdown conditions for the steam generators. While the exact sort of inhibitor, and its mode of application has not been determined yet, it is clear that frequent soaks (e.g., boric acid soak) are beneficial and help in returning chemicals from hideout. There have been very significant returns observed. For example, a unit operating at full load with chloride level at ∿0.05 ppm, had an increase of chloride concentration to 5 ppm (a factor of 100!) when the unit was shut dowr to zero load.

Since the major success in controlling denting was thus far achieved by maintaining proper chemistry, it is worthwhile to quote some of the specifications maintained in Japan:

"The goal is to maintain feedwater to the steam generator as pure as practically possible, to maintain dissolved oxygen as low as practically possible and to keep the pH at the proper value. In particular, feedwater has oxygen less than 100 ppb, chloride less than 50 ppb and sodium less than 100 ppb. The feedwater pH is controlled between 8.8 and 9.3 by adding hydrazine and ammonia".

<u>Denting near the tubesheet</u>. There are several plants which have identified tube degradation at or below the top of the tubesheet. This damage has been identified as intergranular attack (IGA) and circumferential stress corrosion cracking, and appears to be caustic related. This tubesheet denting has so far been reported only in plants with open crevices (figure 8) i.e., where the tubes were not expanded to close the full length of the crevices. While

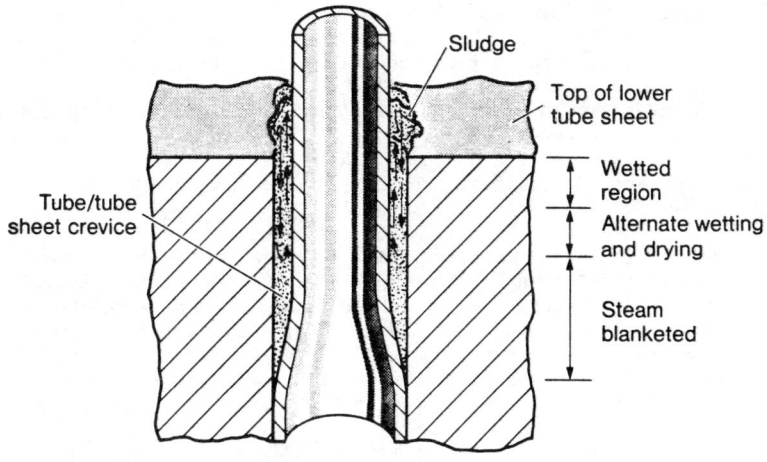

Figure 8. Tube/Tubesheet Crevice

all new plants have fully expanded tubes, some older plants have crevices between the tubes and the tubesheet. Here, again, it is postulated that concentration of chemicals occurs in the region where alternate wetting and drying takes place. The presence of sludge on the tubesheet undoubtedly accelerates this process.

Another mechanism of chemical concentration that has been postulated to operate involves the ability of any solute to raise the boiling point of water. The water is postulated to migrate into the crevice until it is heated to its boiling point, as determined by its, say, sodium hydroxide concentration. At this point vaporization takes place, and the caustic begins to concentrate in the remaining water, which raises the boiling point even further and allows further migration of the water. Through this process the temperature of the solution can ultimately reach that of the hot primary fluid (∼325°C) and the concentration of sodium hydroxide in the solution can reach a maximum value of about 660,000 ppm. Thus, it is possible for highly concentrated solution of sodium hydroxide to penetrate the entire length of the tubesheet crevice and cause substantial corrosion.

Because the crevice cannot be cleaned directly, other means were sought to alleviate the problem. One process, developed in Japan, is crevice flushing. This involves filling the crevice with water and then depressurizing the steam generator to cause bubble nucleation in the crevices which forces the liquid containing dissolved chemicals out of the crevice top.

In one case the steam generator was filled with water up to 60 cm deep and the system was brought to a temperature of 120°C, with the main steam isolation valves closed. The pressure in the steam generator reached the saturation pressure of about 0.2 MPa. The reactor coolant pumps were stopped and the unit was allowed to remain in this condition for a short time to allow for dissolution of deposits in the crevices. The water on top of the tubesheet was then subcooled by adding cold water and the relief valves were opened and the generator was rapidly depressurized for about 8-15 minutes. The tubesheet was then cooled and the generator was drained. The cycle which took about 6 hours was then repeated. Results of these procedures were very encouraging.

Wastage. Sludge (a porous material which settles on the lower tubesheet) has been observed to build to a height of 30 cm at some spots on the lower tubesheet. Since 1 cm of sludge height is equivalent to approximately 180 kg of sludge on the tubesheet, this would indicate huge quantities of sludge inside the steam generator. It is not unusual to remove several hundreds of kg of sludge from the lower tubesheet, if removal is possible. Frequently the sludge hardens and is impossible to remove by the conventional lancing methods.

The presence of large quantities of sludge indicates clearly that blowdown is inefficient for corrosion product removal in current design and at current rates. It is also expected that demineralizers, and proper treatment of feedwater can reduce sludge buildup by a factor of two to ten. Demineralizers also reduce ionic leverls during condenser in-leakage. It has also been demonstrated that there is a relationship between air in-leakage and sludge accumulation. The air in-leakage occurs at subatmospheric locations in the secondary plant including the low pressure turbines, the condensers, and the inlet end of the condensate pump.

Tube wall thinning has been observed previously in the region above the tubesheet when phosphate chemistry has been used. Since the industry shifted to AVT chemistry, the progression of tube wall thinning appears to have diminished to the point that it is not a serious industry-wide problem.

Vibration. Tube fretting and wear is caused by flow induced vibrations, which cause impact and tangential sliding against the tube support plate, adjacent tubes or antivibration devices. The tube vibration can be induced by fluid flow perpendicular to the tubes or parallel flow along the tubes. Because the movement between the rubbing surfaces, i.e., the tube and support plate, is oscillatory and usually small in amplitude, the rubbing process taking place is termed "fretting".

In addition to the detrimental effects of the tube metal loss, fretted regions are highly sensitive to fatigue cracks. Under fretting conditions, fatigue cracks can be initiated close to the material surface at very low stresses, well below the fatigue limit of nonfretted tube test material. If the nominal cyclic stresses on the tube produced by the induced vibration is sufficient to cause these fretting-induced cracks to grow and propagate, early failure of the tube can occur. Furthermore, if there is impacting of the tube on the tube support plate, then it is possible to get local grain boundary slippage, which is then more sensitive to corrosion attack.

Carry-over and carry-under. The moisture, usually in the form of small water droplets which is carried by the steam out of the steam outlet nozzle, is termed carry-over. The steam, usually in the form of small bubbles, which is carried by the separated water into the downcomer, is termed carry-under. The separation of the water from the steam in UTSG is done by the steam separators.

NUCLEAR STEAM GENERATORS

Two systems of separators are used in a steam generator to remove water from the steam: the primary steam separators, which are usually of an upflow centrifugal design, and the secondary separators, or dryers, which are usually of an inertia type, namely the Chevron dryers. The cylindrical primary separators are contained in a deck which seals the lower portion of the generator containing the low quality steam from the upper portion containing the high quality steam. The primary separators swirl the steam-mixture so that the higher density water is thrown to the sides of the separators cylinder as a rising, rotating layer which is skimmed off. The mass carry-over of water at the primary separator exit varies from 1% to 30%, again depending primarily on separator design. The water skimmed off is piped onto the deck where it flows into an annular downcomer and is recirculated to the lower portion of the steam generator. The secondary separators remove most of the remainder of the moisture from the steam. Some gravity separation also occurs in the riser between the primary and secondary separators.

Modern turbine design requires that the steam produced by the steam generator be of high quality. Current steam separator technology delivers steam with a quality of 99.75% by mass, however, it is desirable to reduce carry over further and thereby produce qualities higher than 99.9%. Two particular turbine problems are due to steam wetness. First, severe erosion occurs in the blades of the low pressure turbines. Saturated steam of 0.2% wetness and 50-70 bar pressure entering the first high pressure stage of the turbine will produce a wetness of 12% or more at the turbine exit. In order to avoid grave erosion, a limitation of turbine exit wetness is necessary. The second is that the thermal efficiency of turbine stages operating in the wet steam region is considerably lower than that of dry stages. One percent of wetness present in a stage is likely to cause about a 1% decrease in efficiency.

Steam separators allow some steam bubbles to be drawn downward with the recirculating water into the downcomer annulus. This lowers the effective density of the water and, therefore, lowers the circulation ratio. Only 1% carryunder reduces the density of the circulation mixture by approximately 20%. Lower circulation ratio means a lower flow rate through the tube bundle and higher quality of the steam in the two-phase mixture surrounding the U-tube bundle. Both factors increase the tendency for chemical hideout, adversely affect local thermal-hydraulics in the tube bundle, and thereby increases the tendency for the localized corrosion that is experienced in operating units.

Naturally, vendors seek steam separators which have minimum pressure drop, and allow minimal carry-over and carryunder.

<u>Water Hammer</u>. Water hammer has occurred in feedwater piping of feed ring steam generators and can potentially occur in the preheat section of economizer type steam generator designs which are now in construction. This is true for UTSG as well as OTSG. Such events could occur following a loss of normal feedwater flow which allows the susceptible area (feed ring and associated piping or preheat section) to fill with steam. Initiation of cold auxiliary feedwater flow can then lead to entrapment of a volume of steam. The subsequent rapid collapse of the steam valume (by heat transfer to the cold water) can lead to a pressure differential which can accelerate a slug of water through the susceptible region resulting in a water hammer event when this slug contacts a barrier. Because of the complex nature of the formation of the steam pockets, the lack of data on direct contact condensation rates, and the uncertainty regarding the liquid slug length and interface geometry, the forces resulting from such an event cannot be accurately predicted. Therefore, current industry practice involves a combination of design modifications and administrative con-

trols to prevent formation of the initiating steam volume, and testing to verify that the event either will not occur or that if it does occur it will not result in damage.

Specifically, the modifications to provide top-discharge feed rings (J or T tubes) and short, downward sloping piping runs with elbows immediately upstream of the steam generator per the NSSS vendor balance-of-plant design criteria appear to have eliminated the problem in feed-ring type steam generators.

4.2 Problems with OTSG

The main problems encountered with the OTSG are associated with droplet flow in the tube free lane, flow-induced vibration, suspended solids, and possibility of water hammer in the feed-ring.

The tube free lane, depicted in figure 9 was thought to be helpful for inspection purposes. It does, however, provide a low resistance path and two-phase flow at relatively high velocities flows through it. Due to the high velocities, water droplets are entrained in the steam flow and carried upwards. There is needed about 25°C superheat to evaporate the droplets, and if this is not available they reach the upper tubesheet (UTS), evaporate and deposit their salt content on the UTS. The chemicals thus accumulated cause stress corrosion, which may result in cracking.

Those tubes having cracks (depicted in figure 9 for a particular plant), as well as other tubes which have significant eddy current signal distortions, must be stabilized by insertion of a solid rod and must be plugged at the ends.

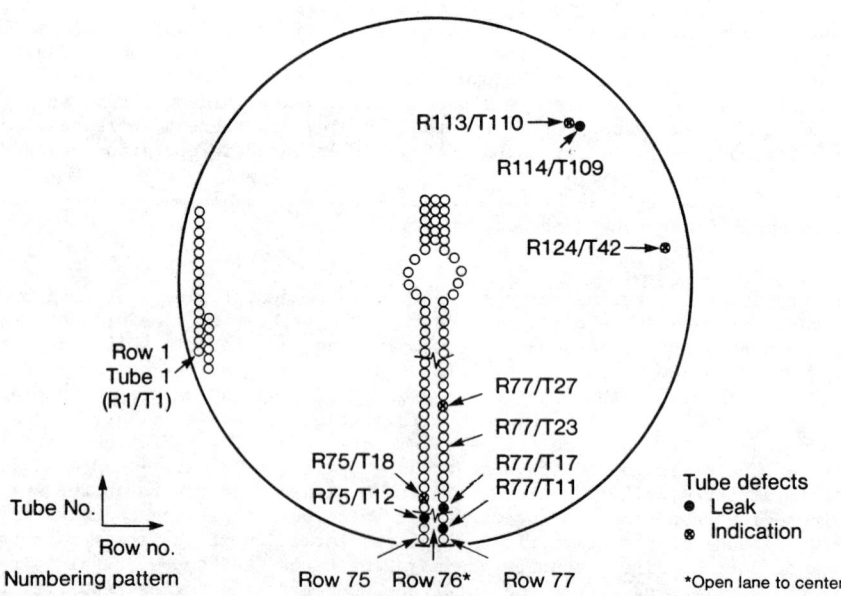

Figure 9. OTSG Tube Identification Plan View

Another problem associated with OTSG is that of pitting. Although pitting is not a serious industry-wide problem at the present time, it has been observed as a problem in laboratory tests under faulted conditions with AVT (all-volatile treatment). Pitting has been observed on a tube sample removed from a once-through steam generator. Improper layup of steam generators is generally thought to be the most likely condition to cause pitting.

Flow-induced vibrations are primarily due to cross flow between the upper tube support plate (TSP) and the upper tubesheet - where the flow is deflected from axial flow to cross flow. There is also cross flow between the ninth and tenth TSPs, where steam is aspired by the cold feedwater and used to heat it to about saturation temperature.

Recently, fretting damage has been postulated to have occurred in the vicinity of the 15th support plate of an OTSG. An affected tube evidence rectangular wear patterns corresponding to the shape of the interfacing land areas of the plate. Metal loss varied but in some cases reached 0.3 mm loss of tube wall.

The problem of flow induced vibrations is being investigated theoretically and experimentally. An interesting experimental program undertaken by the vendor and utility and supported by EPRI, included instrumenting an operating steam generator with accelerometers. The data analyzed thus far did show flow induced vibration with peak-to-peak magnitude at midspan of up to 1.3 mm. But even these anomalous vibrations result in stresses of less than 400bars at the UTS. These stresses are not thought to be enough to cause tube cracking by themselves.

REFERENCES

1. Baroczy, C.J. 1965 A Systematic Correlation for Two-Phase Pressure Drop, AICHE preprint 37; 8th National Heat Transfer Conference, Los Angeles.

2. Design Guide for Heat Transfer Equipment in Water-Cooled Nuclear Reactor Systems; July 1975; ORNL-TM-3578; Oak Ridge Nat. Lab., Oak Ridge, Tn. 37380.

3. Friedel, L. 1977 Momentum Exchange and Pressure Drop in Two-Phase Flow. Two-phase Flow and Heat Transfer (Ed. Kakac, S. and Mayinger, F.), Hemisphere Pub. Co., Washington, D.C.

4. Martinelli, R.C. & Nelson D.B. 1948 Prediction of Pressure Drop During Forced Circulation Boiling Water, Trans. ASME 70, 695-702.

Some Aspects of Reliability and Operation Efficiency of Plant Heat Exchangers

M. A. STYRIKOVICH
Institute of High Temperature
Korovinskoye Road
127412 Moscow 1-412, USSR

ABSTRACT

The mechanisms by which salt concentration can occur in power plant heat exchangers are discussed. High concentration of salt can occur and can cause corrosion and failure. The phenomena are illustrated by data from the High Temperature Institute and elsewhere. The evidence is reviewed for the pre-dryout and post-dryout regions respectively. In the pre-dryout region, effects of tube orientation and non-uniformity in heat flux have been investigated. In the dryout and post-dryout regions, very high concentration factors are possible and the experimental evidence supports their existence.

1. INTRODUCTION

Power plants heat exchangers determine to a large extent both the reliability and economics of energy producing units. In recent years, due to the increase in the capacity of individual units, the wide use of nuclear energy, together with the increasing demands in environmental protection, there has been a sharp increase in costs brought about by forced and planned outage of units and, consequently, increases in demands for their reliability.

The development of nuclear energy has lead to the appearance of new types of heat exchanges (in particular, steam generators) with two phase flow in channels of very complicated geometry.

Calculations as well as experimental investigations of heat- and especially mass transfer for such systems are very difficult to carry out. As a result, numerous cases have occurred of damage, connected with corrosion under conditions of high local augmentation, by several orders of magnitude, of the concentration of the corrosion-active species.

For some heat-exchangers, operating under conditions of complete evaporation of the water, especially at moderate (one-through steam generator) and low (reheaters) pressures, it turned out to be necessary to investigate the possibility of some contact between the remaining water droplets (highly concentrated in active species) with heat exchange surfaces, resulting in stress corrosion and corrosion fatigue. These processes too, lend themselves with difficulty to any experimental investigation and existing calculation models do not reflect the physical picture of the processes well enough.

In particular, it has to be noted that, as a rule, the use of electric heating (q=const) in experiments can more or less satisfactorily simulate only those heat exchangers with nuclear or radiation heating (reactor cores, or heating surfaces on the wall of the furnaces of conventional steam generators). This method is not suitable for simulation of apparatus heated by a heat carrier (steam generators, heated by liquid metals and, even more, by pressurized water), where heat flux decreases sharply in the zone of nucleate boiling deterioration associated with a reduction of the wall temperature rise.

It is important to note that the presence, in the water film near the heated surface, of various impurities, whose concentrations differ from bulk by several orders of magnitude, can take place also in steam condensation in preheaters and main turbine condensers. Here high concentrations of salts and alkalies slightly soluble in steam can occur in the first condensate droplets; high concentrations of volatile species and gases like ammonia, carbon dioxide and oxygen can occur in the condensation of the last part of the steam.

In addition to their role in large nuclear and fossil fuel units, heat exchangers are of great importance in the development of other energy-producing equipment. In such equipment the role of the heat exchanger is often crucial, determining the prospects for wide use of the equipment. For example, in the utilization, in high-temperature technological processes, of heat produced in helium cooled nuclear reactors, perhaps the most difficult task is the creation of a helium-tight heat exchanger for working temperatures of 800-900°C. The role of heat exchangers in developing geothermal installations is also large, especially in cases when chemical composition excludes discharge of waste waters into surface basins. In the design of power plants using the temperature difference between the surface and depth of tropical seas, there is a need for the development of giant heat exchangers which are cheap, corrosion- and biofouling-resistant.

Many other examples could be cited, but even a sketchy enumeration is beyond the possible volume of this lecture. That is why I will restrict myself to two small examples, connected with the problem of the concentration of slightly volatile species in the nucleate boiling zone and post dry-out region.

2. MASS CONCENTRATION STUDIES IN THE PRE-DRYOUT REGION

In the High Temperatures Institute of the Academy of Sciences (Mass Exchange Department), during the last decade, a complete investigation has been carried out of mass exchange phenomena in steam generating systems (1-6). Extensive experimental data has been accumulated on mass exchange in steam-generating channels with non-permeable surface (Table 1). As it can be seen from the table, a sufficiently broad range of parameters has been investigated to reveal the general features of the process.

Let us consider the influence of individual factors on the concentration degree n (n is a parameter representing the ratio of water impurity concentration in the boundary layer and bulk flow). Clarification of the influence of heat flux q, pressure P and mass-velocity ρW on the parameter "n" in coordinates n = fn (x), where x is the steam quality, is a very difficult task because at a given value of x the conditions of mass exchange at various P,q and ρW are different. This is connected, in particular, with the fact, that at the same x values, the difference between x and the critical steam quality x_{cr} and between x and the quality at the beginning of surface boiling x_{bsb}, are different. Because of this, for data systemization a generalized parameter

$$X = \frac{x - x_{bsb}}{x_{cr} - x_{sb}}$$

TABLE 1 - RANGE OF PARAMETERS INVESTIGATED INFLUENCING MASS EXCHANGE IN STEAM GENERATING CHANNELS WITHOUT DEPOSITS(1-6)

Series No.	P bar	q kW/m^2	ρW kg/m^{2}s
1	1.18	500	1130
2	1.18	500	1430
3	18.6	350	1620
4	69	250	1000
5	98	580	1000
6	137	291	1000
7	137	580	1000
8	137	870	1000
9	137	640	2000
10	137	580	3500
11	137	580	1000

can be used, whose magnitude changes in the range from 0 to 1.0. This parameter is remarkable, because it allows to show in convenient coordinates results of experiments, carried out in significantly different conditions.

On figure 1, are given results of such a generalization. Analysing the influence of individual factors (pressure, heat flux and mass velocity) it can be noted that concentration degree increases with increasing pressure and heat flux but decreases with increasing mass velocity. This is in good accordance with physical concepts of the nature of exchange processes in steam generating channels (1-2).

The experiments were carried out mainly in vertical tubes with uniform heating. However, studies in horizontal channels with uniform and nonuniform heating around the perimeter are also of considerable scientific and practical interest. Experiments in horizontal channels with uniform heating were carried out with a tube 6 mm internal diameter > 2 mm wall thickness and 1100 mm heated length. Another section of the tube had also a 6 mm inner diameter, but the wall thickness varied from 2 mm on the upper generatrix to 0.6 mm on the lower, thus providing nonuniform heat release around the perimeter when electric heating of the section was employed. Maximum heat release was on the upper part of perimeter. The heat flux profile over the tube perimeter varied approximately by a cosine law. The heated length in this section was 415 mm.

Investigations of mass exchange were carried out, using the "salt" method, developed by the author more than 10 years ago (2). With the uniformly heated tube experimental conditions were: P = 13.7 MPa, ρW = 1000 kg/m^{2}sec., q = 297 kW/m^2 and q = 580 kW/m^2. In the case of nonuniform heating, experiments were carried out at P = 13.7 MPa, ρW = 1000 kg/m^3 sec., q = 568 kW/m^2, and ρW = 2000 kg/m^2 sec and $\bar{q}$ = 870 kW/m^2.

Results for mass exchange in channels with uniform heating are given in figures 2 and 3. Also shown are data for vertical channels under comparable conditions. It can be seen that for $x = x_{bsb}$ deposit-formation starts at a

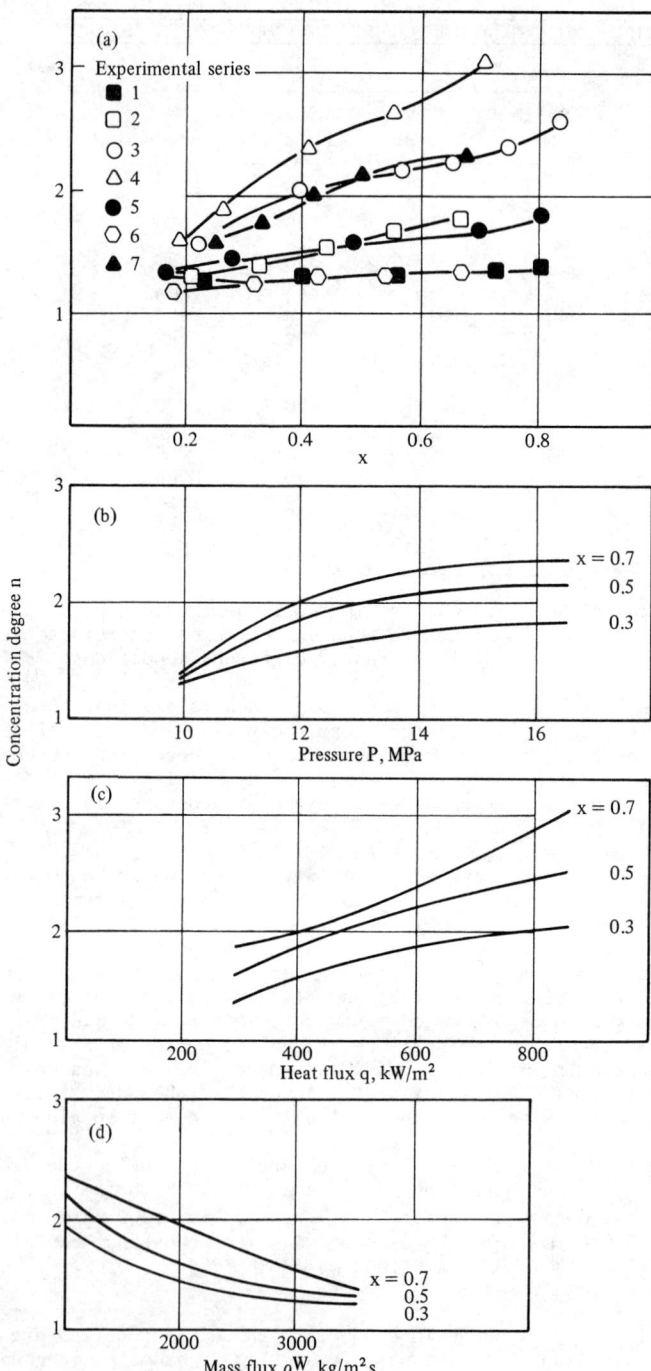

Fig. 1. Data for concentration degree in vertical tubes obtained at the High Temperature Institute.

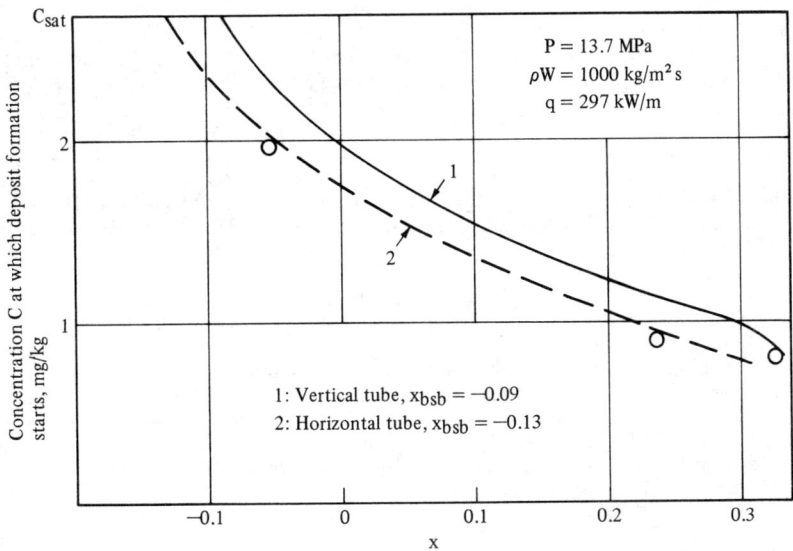

Fig. 2. Concentration at which deposition of salt starts for uniformly heated vertical and horizontal tubes.

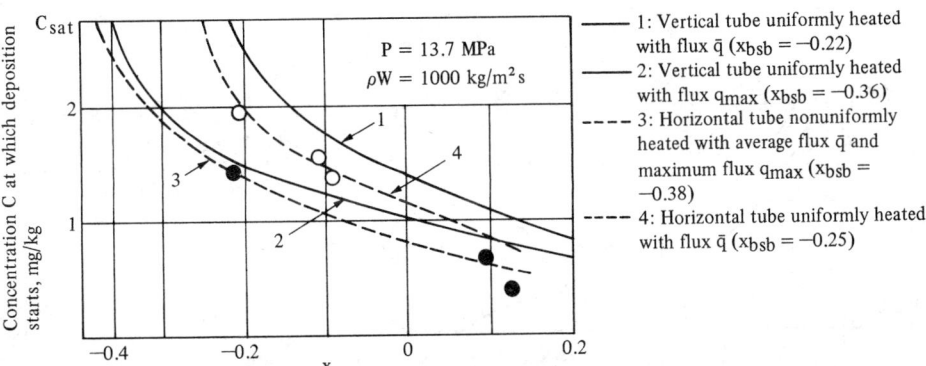

Fig. 3. Comparison of salt deposition characteristics of uniform and non-uniformly heated horizontal tubes with results for vertical channels.

concentration equal to the saturation concentration, i.e. $C = C_{sat}$. With increase of steam quality x, the limiting inlet concentration decreases. Comparison of data for horizontal and vertical channels shows that, in horizontal channels, deposit formation begins at lower flow enthalpy, the difference being higher, the higher the heat flux. Obviously this difference results from the fact, that at equal average steam quality values of the flow, the local steam quality in the upper part of the horizontal channel is higher. With increased mass velocity, the difference between the limiting concentrations in vertical and horizontal tubes is likely to fall.

The influence of non-uniform circumferential heating on the limiting concentration is shown on figs. 3 and 4. Comparison of curves 3 and 4 (fig. 3) shows that, with equal average heat fluxes in the horizontal tube with non-uniform heating, the boundary of the onset of salt deposition is shifted to lower quality. The difference is especially large in the subcooled boiling region.

With increasing steam quality, the influence of the non-uniformity of heating falls. Comparison of curves 3 and 2 shows that the boundary concentration in the non-uniformly heated horizontal channel is somewhat lower than that for a uniformly heated vertical channel whose heat flux is equal to the maximum flux in the horizontal tube. This is additional evidence of flow stratification. A similar result has been obtained at $\rho W = 2000$ Kg/m^2 sec (figure 4). The decline of limiting concentration in this case is higher in comparison with the previous case, (figure 3), probably reflecting the greater difference in heat flux.

On the whole, it has to be noted that there are considerable differences in C between horizontal and vertical channels, especially in the case of circumferentially non-uniform heating, the maximum heat flux coinciding with the channels upper generatrix. Under these conditions divergence in C values can reach 80%. It is necessary to consider this fact when developing apparatuses of boiling type with horizontal steam generating tubes.

3. MASS CONCENTRATION IN THE POST-DRYOUT REGION

In some equipment, heat transfer is accomplished under post dry-out

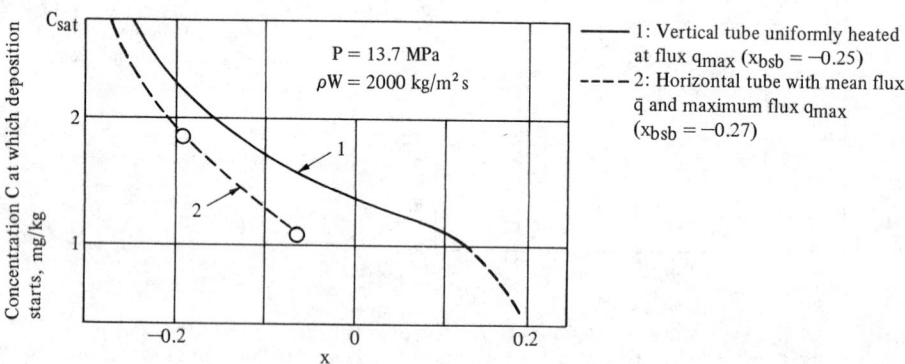

Fig. 4. Comparison of salt deposition characteristics of a uniformly heated vertical tube with those of a non-uniformly heated horizontal tube with the same maximum heat flux.

conditions. With steam quality approaching its critical value, the
concentration of some species in the liquid layer can rise considerably. In
case of corrosion-active species, corrosion rates increase in this region.

When dryout occurs at low steam qualities and high heat fluxes in cases
where the flux is not reduced due to the wall temperature rise (nuclear or
radiation heat input), the increase of the metal's temperature in post dry-out
region evidently exceeds the tolerable value and a slowly developing corrosion
problem is not the limiting factor. In those cases when the post-dryout rise
does not exceed the maximum safe temperature (taking into account of material
strength), long term operation in the post dry-out area is possible and
concentration of corrosion-active impurities has to be considered. These
latter conditions usually occur at low heat flux or where there is a sharp
decrease in heat flux when the post-dryout region is entered (e.g. in the
case of heating by a fluid at moderate temperature).

It has to be emphasized that highly soluble compounds with positive
solubility coefficients, in particular sodium chloride and sodium hydroxide,
are the most dangerous. Under the operational conditions of thermo-technical
equipment in nuclear plants, these species can in principle form very highly
concentrated solutions, causing active corrosion of even very resistant
materials of construction (fig. 5). In particular, at pressure and temperatures
typical for nuclear plants with light water-cooled reactors (P = 6. 8-7.0 MPa,
T_{wall} - 310-325°C) solutions of sodium chloride with a concentration up to
40% can occur under reheater conditions (P = 0.5-1.2MPa, T_{wall} up to 260°C).

It is even worse with sodium hydroxide, as can be seen in fig. 5; in

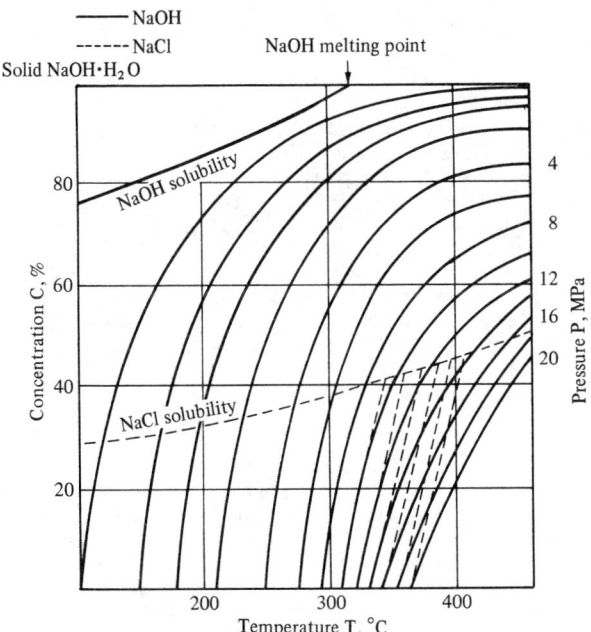

Fig. 5. Solubility-pressure-temperature relationship for aqueous solution of
NaCl and NaOH.

conditions typical of a PWR steam generator, concentrations of 50% can occur. In the intermediate reheater concentrations up to 60% are possible. Concentrations up to 90% can occur in the superheater of liquid metal cooled nuclear plants, at 15.0 MPa and wall temperature up to 520-530°C. But the possible achievement in the boundary layer of such thermodynamically ultimate concentrations depends on many factors.

The liquid phase concentration near the wall is determined by the carry-over of sodium chloride and sodium hydroxide into the steam, by their initial concentration in the flow and by the conditions of mass exchange between the two phase bulk flow and liquid boundary layer. In nucleate boiling, mass exchange is very intensive and the concentration degree does not exceed n > 5-10, this being quite acceptable from the corrosion point of view since the initial concentrations in nuclear plants run in the worst case up to tenths of parts per billion. Nevertheless, in the case of the wall liquid layer drying up, when its flow rate and its thickness tends to zero, the degree of concentration can be very high.

Such physical situations can occur, as has been mentioned already, in post-dry out areas, i.e. in once-through steam generators and steam super-heaters. For nuclear plants with water cooled reactors and once-through steam generators, of for example those supplied by the Babcock and Wilcox Company, crisis conditions are characterized by low mass velocities (ρW = 200-300 kg/m^2 sec) moderate pressures (P = 5,0-8,0 MPa) and heat fluxes before film disappearance of about q = 100 KW/m^2.

It has to be emphasized that after practically complete water evaporation, the heat flux drops sharply since heat exchange on the low pressure side decreases strongly while the heating media temperature remains almost constant. For intermediate reheaters of nuclear plants with typical LWR's the characteristic conditions are low pressures, small initial water content in the steam (< 1%) and very low heat flux density.

In the usual arrangement for two-step reheat, the heating in the first step has a temperature about 220-230°C, with the boiling temperature varying from 150 to 180°C. There are not enough data for the initial zone heat transfer coefficient (from wall to media) which is intensified by deposition droplets on the heating surface, particularly for conditions usually found in reheaters (transverse pass-around of the condensing steam over tube bundles). Nevertheless one can assume the average wall temperature to be only slightly below the heating steam saturation temperature. I am sorry to say that under such conditions no experiments on mass exchange have been performed. So one has to be guided by experiments carried out under other conditions, in particular inside circular straight tubes at constant heat flux density along their length and, as a rule, at higher pressures. Therefore, looking through the existing experimental data, one has to evaluate critically their applicability for constructions of nuclear plant equipment.

The predictive, as well as the experimental aspect, of mass exchange in the post dry-out region was studied in (7). Let us consider this work a little closer. The following physical model of the process was proposed (fig. 6); the liquid layer (film) on the wall is thinned step by step along the channel. In some cross-section (z = 0) dry-out takes place and the wall temperature starts to rise. However, the liquid film does not disappear instantaneously because, due to water evaporation and the consequent increase in concentration of impurities in the wall region. Simultaneously, boiling temperature of the solution increases. The concentration rise in the film is defined by the distribution coefficient, diffusion and droplet break-away.

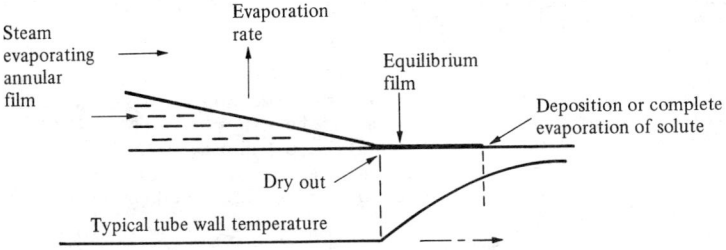

Fig. 6. Mechanism of salt concentration under conditions of film dryout.

The length of the concentrated film ($z = L$) is governed by the impurity properties. In the case of sodium chloride it is limited by the beginning of depositing of solid phase salt on the wall. The existence of a concentrated sodium hydroxide solution film in steam generators ($P > 0.1$ MPa) is limited only by the length of the steam generating tubes.

In the proposed model it is assumed that the film's concentration depends only on temperature on the wall in the post dry-out area and the extent of the film depends on concentration and flow parameters defining steam solubility.

In the work a case was studied where the critical steam quality is near unity ($x_{cr} = 1.0$) and the wall temperature increases linearly from the point of departure from the critical cross section.

With regard to analytical predictions of the final concentration, I shall mention only some comparisons of calculated data with those obtained from experiments carried out on a once-through steam generator model and aimed at verification of the calculation method used.

Experiments were carried out, using the slightly modified "salt" method under conditions when the concentration excess on the wall above equilibrium should not be expected to be very high. The heat flux was not large ($0.11 MW/m^2$) but because electric heating was used ($q = const$), the wall temperature rose rapidly in the post dry-out area. The mass-velocities were very low (approximately 50 $kg/m^2 \cdot sec$); under these conditions, bubble formation in the film ceases long before complete film dryout occurs. This results in droplet carry-over (not considered in the model) having only a small role near the dry-out cross-section.

It is important to note that calculated as well as experimental values of the concentration degree were considerably higher than those for equilibrium conditions; deposits started to form at feed water concentrations lower than the sodium chloride steam solubility. The divergence between calculation and experimental data was only about 30% at 17.0 MPa, but rose up to a ratio of 12 at 10.0 MPa. The calculated values were considerably close to the equilibrium ones than to the experimental ones and increased, though more slightly, with decreasing pressure. Even under these experimental conditions the kinetic factor led to considerable concentration degree increase in comparison with simple calculations considering only complete equilibrium conditions, especially at low pressure. This means that for feed water with specifications such as those for once-through PWR steam-generators, an approximately 10-fold rise of maximal film concentration occurs, compared with that defined by thermodynamic equilibrium values. Of course, this conclusion has to be considered as a preliminary one, taking into account the difference between experimental and real steam-generator operating conditions, though in terms of main parameters, this difference in the study (7) was less than that of other investigators.

The significant point is that for PWR steam-generator's conditions (where the heat carrier's temperature is fixed and not the heat flux density) q-values in the post-dryout zone at conditions similar to experimental ones (7), are decreasing after complete moisture evaporation by approximately 10 times. That is why the last film residues will evaporate much slower. Besides, the evaporation cross-section should not be considered as fixed neither in time nor at the whole tube perimeter.

Cohen (9) has considered further the model given in reference 7 and has also made an approximate evaluation of concentration degree for a "rivulet" model. The possible role of droplet carry-over was also considered. In the first variant on Mann's approach, the case when critical steam quality differs from unity was more exactly defined. Calculations (9) carried out for a liquid-metal cooled steam-generator have shown that the calculated concentration factor is of the same order as in (7). However, for this case the applicability Mann's model is controversial because at a critical steam quality much lower than unity, the zone of the film dryout will co-exist with a flow core with many droplets. These, depositing on the heating surface, will considerably delay the wall temperature rise and, because of this, extend the drying region, thus sharply reducing the role of kinetic restrictions.

The evaluation of concentration degree in the "rivulet" model is carried out in (9) extremely arbitrarily (the zone of drying was assumed to be about 25.4 mm long and to have about $\frac{1}{3}$ of its area wetted). Under these conditions the concentration factor was calculated as $n = 6.55 \times 10^6$. For the mass exchange scheme (7), the concentration factor was $n = 4.2 \times 10^5$, and under conditions of droplet exchange the calculated value was $n = 5.24 \times 10^4$. Estimates made for n were compared with experimental data on steel corrosion in the post dry-out area. The experiments showed that the concentration factor was less than $n = 5.24 \times 10^4$.

In an experimental study by English scientists (10), the phenomena of deposition and corrosion were investigated. In addition, data on the concentration factor in post-dryout area were obtained. Radioactive sodium chloride and sodium hydroxide solutions were used. However, data on radiation rates at various tube cross sections should not be considered as numerical values of concentration increase in the boundary layer since in the counts received included those from the droplets in the core of the flow. An indirect indication of deposit formation was given by the higher than usual count rate which persisted for some time after the discontinuance of the radioactive compound dosage of the inlet flow.

CONCLUSION

On the whole, in spite of some contradictions in the interpretation of mass exchange, it is undoubtedly true that high concentration occurs in the post-dryout, especially at low pressures when carry-over of the impurities into the steam is difficult because of their very low steam solubility. It seems also probable that concentrations in the film can significantly exceed the values, corresponding to equilibrium between the liquid and main steam flow. This is especially true at high heat fluxes in the evaporation zone, but with the film being thin enough to exclude bubble formation and liquid carry-over in the form of droplets. At moderate (steam generators) and low (reheaters) pressures, occurance of high concentrated solutions has to be considered, even in the case

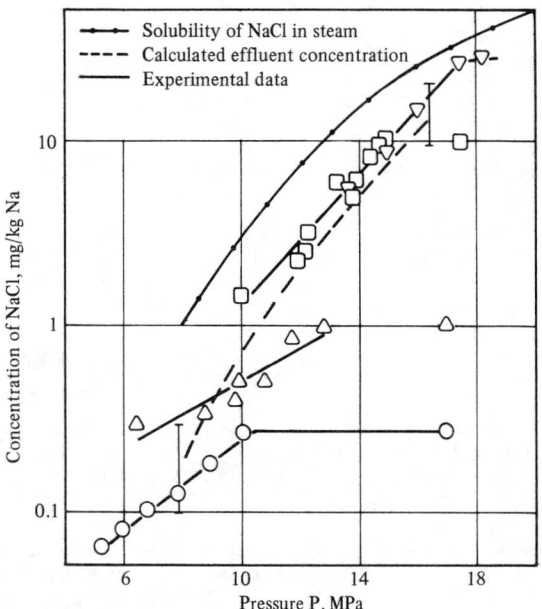

Fig. 7. Calculated and experimental data for NaCl concentration.

of low initial concentration of the contaminant in the flow. To obtain more reliable data on concentration degree, experiments under near to real operation conditions are necessary. Special consideration has to be given to the temperature regime in the post dry-out area, which will probably require experiments using the same heating methods as in real power plant equipment.

SYMBOLS

p	- pressure
ρW	- mass velocity
q	- heat flux
n	- concentration factor (degree)
K_d	- coefficient of distribution
c	- concentration
d	- diameter
ρ	- density
W	- velocity
i'	- enthalpy of saturated water

i'' — enthalpy of saturated steam

i_m — enthalpy of steam/water mixture

$x = \dfrac{i_m - i'}{i'' - i'}$ — steam (mixture) quality - by thermal balance

REFERENCES

1. Styrikovich, M. A., Polonsky, V. S. and Bezrukov, E. K. (1971), 'Experimental study of initial boundary of caso deposites in steam generating tube'. Teplofizika Vysokikh Temperature, v.9,N.1, p. 129-134.

2. Styrikovich, M. A., Polonsky, V. S. and Bezrukov, E. K. (1971), 'Mass transfer study in steam-generating channels by "Salt method".' Teplofizika Vysokikh Temperature, v.9,N.3, p. 583-590.

3. Styrikovich, M. A., Martynova, O. I., Miropol'sky, Z. L. et al. (1977), 'Analysis of porous deposit's influence on the concentration of impurities in steam-generating channels'. Teplofizika Vysokikh Temperature, v.15,N.1, p. 109-114.

4. Styrikovich, M. A., Martynova, O. I., Miropolsky, Z. L. et al. (1977), 'Experimental study of mass transfer in steam generating channels with porous deposites'. Teplofizika Vysokikh Temperature, v.15,N.2, p. 353-358.

5. Styrikovich, M. A., Leontiev, A. I., Polonsky, V. S. et al. (1977), 'Experimental study of mass transfer's conditions in steam generating channels with heat-realise distribution according to cosinus law'. Teplofizika Vysokikh Temperature, v.16,N.3, p. 548-555.

6. Polonsky, V. S., Zuikov, A. S., Leontiev, A. I. et al. (1978), 'A model of concentrating process at boiling in cappilary-porous structures'. Doklady Akademii Nauk SSSR, v.241,N.3, p. 579-582.

7. Mann, G. M. W. (1975), 'Distribution of sodium chloride and sodium hydroxide between steam and water at dry-out in an experimental once-through boiler'. Chem. Eng. Sci., Vol. 30, No. 2, pp. 249-260.

8. Bakker, N. A. and Hawtin, P. (1977), 'Solute concentration in highly rated high pressure steam generators, 11. A model for the concentration of salt in an evaporating film'. AERE-R7224, Harwell, 10 p.

9. Cohen, P. (1977), 'Chemical thermohydraulics in steam generating surfaces'. 17th National Heat Transfer Conference, USA, 8 p.

10. Pritchard, A. M., Peakall, K. A. and Smart, E. F. (1977), 'A miniature high-pressure loop to study corrosion and deposition processes in high-performance boilers'. AERE-R8669, Harwell, Oxfordshire, p. 66.

11. Styrikovich, N. (1978), 'The role of two-phase flows in nuclear power plants'. International Seminar, Momentum, heat and mass transfer in two-phase energy and chemical systems, Dubrovnik, Yugoslavia, 20 p.

12. Deddens, I. C. and Montgomery, D. W. (1978), 'Review of 1977 operations and product improvement programs'. Nuclear Operating Experience Seminar, Hershey, Pennsylvania, 9 p.

13. Sarver, L. W. and Rigdon, M. A. (1978), 'Tube damage once-through steam generators'. Corrosion Advisory Committee Meeting EPRI, February 7, 1978, The Babcock and Wilcox Company, Research and Development Division 1562 Beeson Street, Alliance, Ohio 44601, 12 p.

14. Report to Electric Power Research Institute, Corrosion Advisory Committee - Significant steam generator inspections relative to OTSG tube leaks - 1978, August, 1978.

Heat Transfer Apparatus of Nuclear Power Plants

E. D. FEDOROVICH, B. L. PASKAR, D. I. GREMILOV, and I. K. TERENTJEV
Central Boiler-Turbine Research Institute, USSR

ABSTRACT

Some aspects of the process in heat exchangers (heat transfer, hydrodynamics) and related phenomena (pulsating thermo-stresses) affecting safe operation of nuclear power plant heat exchangers are discussed.

Some aspects of heat transfer and hydrodynamics in the nuclear power plant heat exchangers (steam generators, intermediate liquid metal heat exchangers, intermediate steam reheaters) are discussed in the report.

It should be noted that the problem of developing a new heat exhcanger is a complex one. Thermohydraulic and heat transfer related parameters must be rationally correlated with many other factors, accounting for the apparatus actual performance, reliability, safety and active life. Complex approach is of particular importance when the operational process involves boiling and condensation processes. In these apparatus some local and usually unstable processes take place. Though these processes do not seriously affect the overall heat transport, very often they have to be taken into account as main factors for the reliability prediction.

Some examples are given below.

Impurities in coolants could be either introduced into the apparatus from outside or formed inside. In the areas of impurities accumulation special heat transfer and flow conditions arise.

The significant and varying differences between parallel paths of coolants, local velocities and temperature oscillations are also typical for the systems with the phase transformation (from liquid to vapour or vise versa). As a result, thermal or mechanical stresses in corrosive media significantly decrease the life of the heat exchanger.

1 STEAM GENERATORS

Nuclear power plant steam generator is the apparatus whose reliability determines the reliability of the plant to a great extent.

As heat exchangers, these steam generators have some distinctive features calling for additional requirements for the organization of the operational process:

- wide variations in the coolant temperatures and heat transfer intensivities along the heat transfer surface, resulting in a severe nonuniformity of heat fluxes along the surface

$$\left(\frac{q_{max}}{q_{min}} = 10 - 20\right):$$

- the existence of zones with the tube material temperature pulsations caused by the alternate flows of water and steam along the heated surface on the secondary loop side. Such zones could occur in the evaporator section and in places having two-phase flow stratification;

- low temperature differences between heating and evaporating media ($\Delta T_{mean} = 25 - 100°K$ and $\Delta T_{local, min} = 5 - 10°K$), especially typical of water-heated steam generators; low local ΔT lead to poor heat transfer intensity in the first part of the evaporation section;

- extremely high requirements for the integrity of tube walls in all the modes and periods of operation.

Thus, the investigations of the processes in the steam generator, in our opinion, should be treated as a complex task.

For instance, our studies of the wall temperature oscillations in the dry-out region of the sodium-heated steam generators were accompanied by the analysis of the pulsating thermal stresses and the prediction of the period of time before the first cracks in material would occur. It is important that the curve stress vs number of stressing cycles before cracking, necessary for this analysis must be registered in operational conditions, i.e. in steam - water media having operational parameters.

Now we shall describe the technique for using the measured data on the temperature pulsations, treated as a stochastic random process, to the stress and tube material life calculations.

The temperature pulsations can be found experimentally. The important feature of the technique lies in the natural way of the model heating (i.e. reactor coolant is used as a heating media). It gives correct boundary conditions automatically. A temperature control device (thermocouple) must be directly placed on the steam generating surface, at the points of maximum temperature pulsations.

The experiments show that the normal correlated function of the temperature pulsations can be expressed in the form:

$$R_T(\tau) = D_T \cdot (1 + \nu \cdot \tau) \cdot \exp(-\nu \tau), \qquad (1)$$

where D_T — dispersion, degrees;
ν — nondimensional interval of the correlation,
$$\nu = \frac{\nu_{dimens} \cdot h^2}{a};$$
τ — nondimensional time;
(a — thermal diffusivity; h - tube wall thickness)

Statistical treatment of the temperature pulsations data gives also the pulsation intensity (mean square divergence) S_T.

The similar value of the pulsating thermal stress intensity, S_σ, is expressed by equation:

$$S_\sigma = \frac{\alpha_T \cdot E}{1 - \nu} \cdot k \cdot S_T, \qquad (2)$$

where α_T - temperature coefficient of the thermal expansion, $\deg^{-1}$;
  E - Jung module, MPa;
  ν - Puasson coefficient;
  k - transmission coefficient, which generally can be found by plotting and integrating the pulsating stress spectral density $G_\sigma(\omega)$ for angular frequensies $\omega = 2\cdot\pi\cdot f$ (f - frequency, $\sec^{-1}$) from 0 to ∞. But the operational method presented in [1], makes it possible to determine approximate transfer functions and to plot curves $k = f(\nu)$ for various Bio numbers (fig.1).

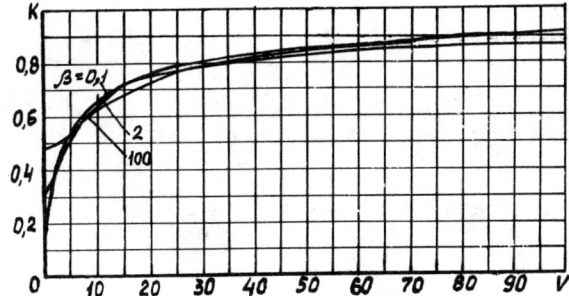

Fig.1. Intensity thermal stresses transmission coefficient.

If S_σ is known, one of the published life prediction methods can be used for parts subjected to cyclic stresses. Specifically, one can use the correlation, proposed in [2] for the general case of non-symmetrical cycles (more appropriate for practical problems):

$$\Theta_n = \frac{N_1 \cdot \Theta_e \cdot \mathcal{æ}_o^m}{\sum_{n=0}^{m} C_m^n \cdot 2^{\frac{m-n}{2}} \cdot \Gamma(\frac{m-n+2}{2}) \cdot \mathcal{æ}_1^n \cdot P\left[(\mathcal{æ}_o + \mathcal{æ}_1)^2, m-n+2\right]}, \qquad (3)$$

where Θ_n - operating time prior to fatigue cracking, hours;
  N_1^n - specimen fatigue test base;
  Θ_1 - effective period of pulsating stresses, hours;
  $\Theta_e = 2\cdot\pi\cdot k/k'$ (similar to K, K' is the gain factor of stress intensity derivative with respect to time; K' is plotted in fig 2);

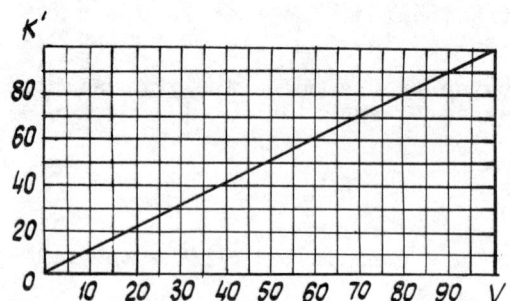

Fig.2 Rate changing intensity thermal stresses transmission coefficient.

$æ_o = \sigma_{-1}/S_\sigma$ (σ_{-1}-fatigue value with mean stresses below number of cycles before cracking tends to infinity, MPa);

m - fatigue curve parameter, $N = N_1 \left(\dfrac{\sigma_{-1}}{\sigma_A}\right)^m$;

N - number of cycles;
σ_A - pulsating stresses amplitude during fatigue test;
Γ - gamma-function;
P - Pearson distribution function (tabulated);

$æ_1 = \psi \cdot \dfrac{\bar\sigma}{S_\sigma}$ ($\bar\sigma$ - mean stress, MPa)

(coefficient ψ accounts for the difference of the fatigue value for symmetrical and non-symmetrical cases; its range is ψ = 0,05-0,3).

The complex approach is also applicable to the vibration analysis of tube bundles of a vertical steam generators for PWR nuclear power plants.

In this case, a non-stationary quasi-homogenious model of a two-phase flow is used for the analysis of flow induced dynamic effects on a tube bundle.

Life prediction for parts and components subjected to the combined action of temperature pulsations, corrosion, vibrations, impurities deposits, wear should eventually become a section of heat exchanger design handbooks.

As applied to once-through steam generators, various patterns of two-phase flows and heat transfer improvement were also studied, aimed at decreasing the post dryout region, having poor heat transfer intensity, and decreasing temperature pulsations in the dryout region.

Coiled tubes provide one of the most promising heat transfer surface for once-through steam generators. Compared to a straight tube, the coiled tube has better heat transfer performance due to secondary vortex flows inside the tube. The secondary flow pattern depend on the coil turning radius.

The heat transfer and pressure drop data analysis shows that coils should be divided into two groups: with $d_{tube}/D_{coil} < 0,05$ ("large" diameter of turning) and with $d_{tube}/D_{coil} > 0,05$ ("small" diameter).

The large amount of processed experimental data on laminar (but with secondary flows) and turbulent single phase flow resulted in following correlations:

$$\frac{Nu_{coil}}{Nu_{st.tube}} = 0,4 \cdot De^{0,37} \quad (\text{if } De = Re \cdot \sqrt{\frac{d_{tube}}{D_{coil}}} > 11,6; \quad Re > Re_{cr}) \tag{4}$$

$$\frac{Nu_{coil}}{Nu_{st.tube}} = 1 + 6,3 \left(1 - \frac{d_{tube}}{D_{coil}}\right)\left(\frac{d_{tube}}{D_{coil}}\right)^{1,15} \quad (\text{if } Re > Re_{cr}), \tag{5}$$

where $Re_{cr} = 2300 \cdot \left\{1 - \left[1 - \left(\frac{2000 \cdot d_{tube}}{D_{coil}}\right)^{-0,4}\right]^{2,2}\right\}^{-1}$ and $(Nu_{st.tube})_{lam}$ and $(Nu_{st.tube})_{turb}$ – Nusselt numbers for a straight tube.

It was found also that the nucleate boiling heat transfer coefficient for coiled tubes could be calculated using the same form of the correlation [3] as for straight tube [3], with the scale value of the heat transfer coefficient for the single-phase flow being taken from equations (4) or (5). That conclusion has been checked at any rate for the following range of the parameters:

$p = 0,1 - 10$ MPa; $\rho w = 80 - 800$ kg/m²s; $x = 0 - 1,0$;
$q = 6 \cdot 10^3 - 8 \cdot 10^5$ w/m²; $d_{tube}/D_{coil} = 0,002 - 0,145$.

The experimental data for the one phase pressure drop in coiled tubes can be expressed by the correlations:

$$\left(\frac{\xi_{coil}}{\xi_{st.tube}}\right)_{lam} = \left\{1 - \left[1 - \left(\frac{11,6}{De}\right)^{0,51}\right]^{2,9}\right\}^{-1} \quad (\text{if } De > 11,6; \; Re < Re_{cr}); \tag{6}$$

$$\left(\frac{\xi_{coil}}{\xi_{st.tube}}\right)_{turb} = 1 + 1,68 \left(\frac{d_{tube}}{D_{coil}}\right)^{0,65} \quad (\text{if } Re > Re_{cr}). \tag{7}$$

If $D_{coil}/d_{tube} > 860$, the difference in heat transfer and pressure drop for coiled and straight tubes disappears.

The two-phase pressure drop in coiled tubes could be expressed as:

$$\Delta P = \left\{ \text{single phase} \right\} \cdot \frac{1}{d_{tube}} \cdot \frac{\rho_1 \cdot W_o^2}{2} \cdot \psi \cdot \left[1 + x \cdot \left(\frac{\rho_1}{\rho_v} - 1 \right) \right], \tag{8}$$

where W_o - liquid phase velocity, related to the overall tube cross section;
ψ - correction coefficient, taking in account the non-homogenious properties of the flow;
x - quality.

The value of critical quality, corresponding to dryout in boiling flow in a high quality region, for coiled tubes is greater than for straight tubes (fig 3). But in a low (close to zero)

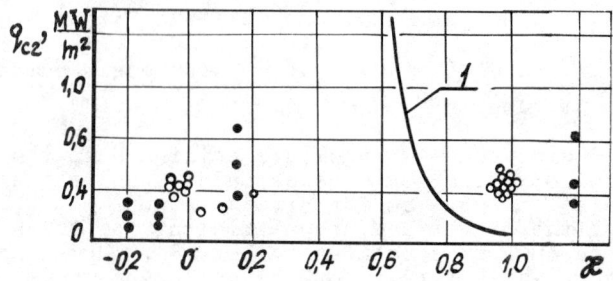

Fig 3 Critical steam water mixture qualities and heat fluxes for coils

1 - straight tube;
● - liquid metal heated coil;
○ - by electricity heated coil;
P = 10 MPa;
ρw = 300-700 kg/m²s;
d_{tube}/D_{coil} = 0,145.

quality region the boiling crisis in coiled tubes occurs for lower heat fluxes, as compared to straight tubes (fig 3). This phenomena could be related to the difficulties of removing the vapour phase from the perimeter part, which is internal relatively to the coil axis. It needs further studies.

2 INTERMEDIATE LIQUID METAL HEAT EXCHANGERS

In nuclear power plants high efficiency heat exchangers are used. This term denotes the apparatus, wherein the temperature changes of the heating (or cooling) media along the exchanger are much greater than the mean coolants temperature difference, ($T_{outlet} - T_{inlet} >> \Delta \overline{T}$). The typical example is an intermediate heat exchanger of nuclear power plant with liquid metal reactor cooling.

The flow distribution inside a real heat exchangers always differs from the ideal case, which is assumed in the course of a traditional designing. In a high efficiency heat exchanger the deviation of an actual distribution of the flow from the ideal uniform distribution is especially evident in the heat transfer. Therefore, the consideration of this phenomena must be incorporated in the thermohydraulic design.

It is well known, that the mean temperature difference across the heat transfer surface $\Delta \overline{T}$ can be expressed by the mean logariphmic ΔT_{log}, using correction factor $\psi \leq 1$:

$$\Delta \overline{T} = \psi \cdot \Delta T_{log}. \qquad (9)$$

Coefficient ψ depends in general on operating flow parameters, on the interacting coolants flow pattern and on specific features of heat transfer and flow distinguishing real cases from ideal ones. One of the specific features is the coolant distribution nonuniformity, which in many cases cannot be eliminated by design solutions In our opinion, the quantitative assessment of the hydraulic nonumiformity now has not been properly considered in present publications despite of its practical sighnificance

For the ideal design of a heat exchanger (for instance, for "pure" counter flow without nonumiformities in flow distribution) we assume $\psi = 1$. Then, if we take into account nonumiformities (also for a counter flow design), coefficient ψ will be less than unity and can be expressed from the analysis [5] as

$$\psi = \frac{m - \varepsilon}{m - 1} \cdot \frac{\ln \dfrac{m - \eta_2}{m(1 - \eta_2)}}{\ln \dfrac{\varepsilon \cdot (m - \eta_2)}{m(\varepsilon - \eta_2)}}, \qquad (10)$$

where ε - parameter of nonuniformity;

$$m = \frac{t_2^{outlet} - t_2^{inlet}}{t_1^{in} - t_1^{out}} \quad \text{and} \quad \eta_2 = \frac{t_2^{out} - t_2^{in}}{t_1^{in} - t_2^{in}} -$$

temperature parameters of coolants ("1" relates to "hot", and "2" - to "cold" liquid).

The parameters m and η_2 are khown from the heat transfer design practice for a flow pattern, other than counter flow.

The parameter ε can be found by both numerical and experimental methods. For sodium-sodium intermediate heat exchangers the experimental data yeilds as a high level $\varepsilon = 0,9$ and a lower level - $\varepsilon = 0,8$. It is necessary to note that the heat transfer decrease is caused mainly by hydraulic nonumiformities due to the deformation of a tube bundle during operation.

The coefficient ψ dependance on η_2 for equal coolant water equavalents (m = 1) is shown in fig 4. The shown curves bring about two important conclusions. Firstly (is we interprete the reciprocal value $1/\psi$ as a heat transfer surface margin), this margin for the high efficiency heat exchangers ($\eta_2 > 0,5$-0,6) must be very high, Secondly, specified outlet temperatures may not be reached with any values of the margin for heat exchangers with expected efficiencies near unity ($\eta_2 > 0,8$).

As an example the calculation is presented for the "sodium-sodium" heat exchanger with the heat capacity 750 MW and specified "hot" sodium temperatures 550/372 deg C; "cold" sodium temperature 508/330 deg C. The overall heat transfer coefficient (on the outside tube surface) is 8000 $W/m^2{}^\circ K$. We determine m = 1; $\eta_2 = 0,809$. Let us assume the probable value of nonumiformity parameter $\varepsilon = 0,85$.

Then ψ (see fig.4) is 0,461. The mean logariphmic temperature difference calculated from the specified temperatures, is $\Delta T_{lg} = 42$ deg; but the actual temperature difference is 42 x 0,461 = 19,4 deg. Thus, the required heat transfer surface is 3865 m^2.

3. INTERMEDIATE STEAM REHEATERS FOR TURBINES

For the saturated (wet) steam turbines of nuclear power plant the intermediate separation of moisture and steam reheating in separator - reheaters (MSR) are used. These apparatus as a rule have a separator and one - or two-stage reheater placed in a single vessel. There are various mutual arrangements of a separator and reheater. The reheater is made as an integral tube bundle or as an assembly of tubular sections. Several types of the heat transfer surfaces are used in reheaters for MSR: Straight tubes with various types of flow (cross flow, parallel flow); coils, finned tubes etc. The heating media (condensing steam) flows in the tubes; heated media (low pressure steam) on the shell side.

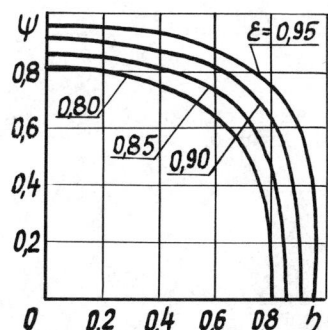

Fig.4. Correction coefficient vs heat exchanger effectiveness η_2 (m = 1)

Processes in MSR have some specific features:

1. One turbine has 2-4 MSR's with parallel lines of the heated and heating steam. So intermediate moisture separation and steam reheating system is the system of parallel paths with two-phase flows, with heat exchange between them. As the hydraulic resistance of these paths is usually small, it is necessary to determine correctly, to calculate and (if necessary) to eliminate heat and hydraulic descrepancies between parallel flows. Otherwise the heat transfer surface can be flooded (by condensate of heating or heated steam), resulting in significant temperature pulsation of the heat transfer surface elements.

2. When calculation of drainage lines with the saturated water flows has been made the influence of noncondensables, steam carry under and flashing of liquid must be taken into account. At present these calculations have not enough experimental background.

3. Nonsymmetrical conditions of the heated vapor inlet (with wetness about 10-15 %) may cause a significant nonuniformity in the wetness distribution in separator elements. So it is necessary to carry out the tests of inlets and another complicated nonsymmetrical elements of MSR using a two phase media.

4. The heated vapor has small wetness (about 0,5 %) after the separator. This condition must be considered when calculating first stage reheaters. The interaction on the moisture and heat surface may result in erosion and temperature pulsations.

Usually the net condensation of heating vapor takes place in MSR. The average heat transfer coefficient for the condensation of pure saturated vapor in vertical tubes may be determined as [6]:

$$Nu = \frac{\alpha}{\lambda_1} \cdot \frac{W_o^2}{g} = 0,1 \cdot \sqrt{A^{1,7} + 0,2 \cdot A^{2,8}}, \qquad (11)$$

where A - nondimensional coefficient,

$$A = F_{r_o} \cdot G_a^{1/3} \left(\frac{P_{r_l}}{P_{r_v}}\right) = \frac{W_o^2}{(g \cdot \nu_l)^{2/3}} \cdot \frac{P_{r_l}}{P_{r_v}},$$

W_o = velocity, defined as a ratio of mass velocity of mixture to density of liquid

$$W_o = \frac{4qL}{3600 \cdot r \, \rho_l \cdot d}$$

Subscripts "l" and "v" correspond to saturated liquid and vapor. Other symbols are well-known. Equation (11) is the correlation of the experimental data of many authors in regions of parameters:

$A = 0,6-4000$; $p = 0,1-9,0$ MPa; $q = 10-1000$ kWt/m^2;

$L = 1-7$ m; $d_{tube} = 10-20$ mm.

Average heat transfer coefficient for the condensation of pure saturated vapor in horizontal tubes may be defined as

$$Nu = \frac{\alpha \cdot d}{\lambda_l} = C \cdot Re_o^{0,8} \cdot Pr_l^{0,43} \frac{\sqrt{1 + x_{inlet}\left(\frac{\rho_L}{\rho_v} - 1\right)} + \sqrt{1 + x_{outlet}\left(\frac{\rho_L}{\rho_v} - 1\right)}}{2}, \quad (12)$$

where Re_o — Reynolds number for liquid; is calculated by velocity W_o and inner diameter of tube d;

x_{inlet}, x_{outlet} — steam quality for inlet and outlet of tube;

C — empirical coefficient; is equal 0,024 — for steel tubes and 0,032 — for copper and copper alloy tubes

Other symbols as in equation (11).

Equation (12) is valid for the follow parameter ranges:

$Pr = 0,86-0,96$; $Re_o = 5 \cdot 10^3 - 3 \cdot 10^5$; $p = 1,2-9,0$ MPa; $q = 160-1600$ kWt/m^2; $L = 2,2-12$ m; $d = 10-17$ mm;

$x_{inlet} = 0,25-1,0$; $x_{outlet} = 0-0,69$.

At present there are no reliable experimental data and correlation equations for the calculation of superheated vapor condensation heat transfer. Therefore numerical methods for the solution of conservation equations in differential form may be recommended. Heat transfer surface must be divided to three zones

1. Zone of the convective heat transfer from superheated steam to the wall. In this zone surface and steam temperature are higher than saturation temperature ($T_{wall} > T_{sat}$; $T_v > T_{sat}$)

2. Zone of the superheated steam condensation. In this zone two components of heat flux must be considered (convective q_c and condensation q_{cond}): while for temperature difference calculations saturation temperature is taken as a vapor temperature.

3. Zone of the wet vapor condensation. For this zone the correlations (11, 12) are used.

Effect of the noncondensable gases on condensation heat transfer may be allowed for by equation:

$$\frac{\alpha_{mix}}{\alpha_v} = 1 - B \cdot \varepsilon_r^m \qquad (13)$$

where α_{mix}, α_v - heat transfer coefficients for condensation vapor-gas mixture and pure vapor;

B, m - experimental coefficients depended from operation and construction parameters.

For example, for the case of steam condensation in a vertical tube $B = 0,25$, $m = 0,7$ and ε_r - volumetric gas concentration at the inlet of a tube.

REFERENCES

1. Sudakov, A.V.; Trofimow, A.S. Stresses as a result of temperature pulsations Ser "Nuclear Reactor Technology" n°8, Moscow, Atomizdat, 1980.

2. Vorobjev, V.A; Pilchenkov, A.H.; Remizov, O.J. The life of the steam generators tube evaluation: connection with the temperature pulsations in the beginning of the dryout region In book: "Boiling crisis in channels" Ed by Subbotin, V.I. Obninsk, 1974 Also: "Applied Mechnics Magazine", n°8, vol.10, p.90-97, 1974 (same authors)

3. Borishansky, V.M., at al. The generalized correlation for the heat transfer to the two-phase flow in tubes and channels. In collection of papers "Advances in two phase flow heat transfer and hydrodynamics in the power engineering" Leningrad, Nauka, 1973,

4. Grilihes, V.A.; Grishutin, M.M.; Krasnouhov, Ju.V.; Kudriavzev, I.S.; Paskar, B.L.; Fedorovich, E.D. About the choice of the correlation for the heat transfer to the two-phase flow in the coils "Proceedings of the higher education institutes. Power Engineering", n°9, 1980

5. Gotovski, M.A.; Lebedev, M.E.; Mizonov, N.N.; Firsova, E.V. Thermohydraulic irregularities influence on the effectiveness of shell and tube counter flow heat exchangers. Sixth international heat transfer conference. Ottawa, 1978, vol. 4 p. 213-217.

6. Andreev, P.A.; Paskar, B.L.; Fedorovich, E.D. The thermal and hydraulic design of the intermediate separators-reheaters for the saturated steam turbines of the nuclear power stations. Central-Boiler Turbine Research Institute, Leningrad, 1977, 131 pages.

On the Natural Circulating PWR U-Tube Steam Generators—Experiments and Analysis

S. P. KALRA
Electric Power Research Institute
3412 Hillview Avenue
Palo Alto, California 94304, USA

ABSTRACT

The integral simulation experiments and associated analysis are discussed in this paper. The flow and phase distribution due to asymmetric boiling (hot and cold legs) are analyzed using three-dimensional code (URSULA-2). The simple analytical model is described for evaluating circulation ratio. The sensitivity of the void-velocity distribution on the natural circulation process is presented.

1. INTRODUCTION

The steam generator provides a dynamic link between the reactor core and the turbine generator in PWR power plants and therefore it plays an important role for safe and reliable operation of the power plants. The physical process which determine the thermal performance of the steam generator under steady-state and transient conditions include coupled two-phase flow, natural circulation and heat transfer phenomena. A good understanding and prediction capability of the normal and off-normal behavior of a steam generator is essential for evaluating load-following mechanism, operating and accident transients of PWRs. It is necessary therefore to model steady-state and transient two-phase flow and void distribution in the steam generator in order to accurately predict the reactor response. The limitations of the system codes in predicting the steam generator behavior are discussed in reference [1].

The common theoretical approach to evaluate steam generator performance is by solving conservation equations with constitutive relations developed from simple tube experiments. However, the understanding of the physical phenomena in steam generator requires a knowledge of flow regime, flow distribution (including 3-D effects), and pressure loss for the momentum balance along the natural circulation loop in U-tube geometry. The two-phase flow in the U-tube steam generators develops under the following characteristic conditions: (i) phase separation in natural circulation, (ii) asymmetric boiling (hot/cold leg regions), (iii) separators (mechanical, centrifugal, etc.), and (iv) incomplete mixing due to carryover/carryunder. In view of the complexity of the flow problems, it is desirable to perform integral simulation experiments for model/code development with correct boundary conditions. Such studies have been performed. Also, some simple analytical models have been developed for evaluating the thermal-hydraulic performance of a U-tube steam generator.

2. EXPERIMENTAL APPROACH

We first briefly discuss U-tube steam generator simulation experiments and associated facility design rationale. To date limited data is in the open literature. The design specific vendor data is generally proprietary and therefore it is not available for general use. Hence a need exists for a general purpose facility to perform a broad spectrum of both steady-state and transient tests.

2.1 Facility Design Rationale

The general design principles were (i) to preserve integral geometric similarity, (ii) to model overall steam/water behavior of a prototypical steam generator, (iii) to visualize flow regime development within the steam generator, and (iv) to perform both steady-state and transient tests. Only by having appreciable number of tubes can we investigate all important phenomena.

As noted above, the geometric similarities have been preserved by incorporating all the principal features of a commercial steam generator (viz., downcomer, U-tube bundle, tube support plates, feedwater sparger, and separator) using a scaling rationale in the experimental design. A commercial PWR U-tube steam generator is shown in Fig. 1(a), and a schematic of the simulated steam generator interior is shown in Fig. 1(b). The model plainly has geometric similarity.

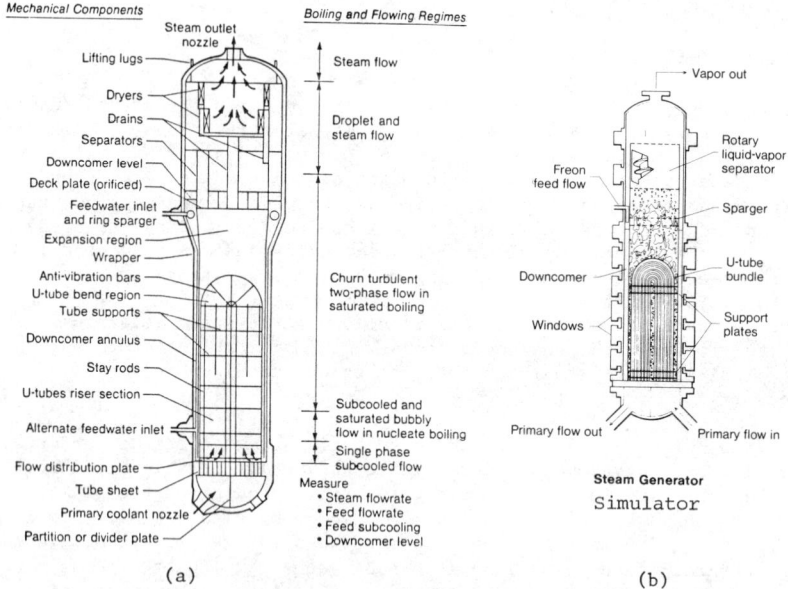

Fig. 1 (a). A Typical PWR U-Tube Steam Generator and Normal Two-Phase Boiling Flow Regimes
(b). A Scale (1/7) Model Used for Simulation Experiments

A prototypical* steam generator was chosen based on a compromise commercial design for scaling purposes. The question of how to properly scale (or model) a steam generator is very complex. In addition to the complexity of fluid-to-fluid modeling, it is very difficult to setdown rigorous sets for configuration scaling. A simple, but logical, approach was used to develop a design specifications for this experiment. The details are given in the earlier paper [2], which are typical of fluid-to-fluid modeling techniques used in critical heat flux experiments [3,4]. Basically retaining the liquid to vapor density ratio, we use a heavier vapor (flurocarbon, FREON-11) at lower pressure, but in a model with 1:1 tube diameter. The facility description is provided in the EPRI report [5].

2.2 Instrumentation

We need to characterize heat flux, pressure drop, circulation phenomena, local velocity and void distributions, fluid temperatures, etc. Preliminary code calculations indicate hot and cold leg assymetric and recirculating regions. Therefore we need to look for these phenomena.

The tube bundle is designed with a center row of seven tubes which can be removed from the tubesheet in order to facilitate instrumentation. In order to obtain temperature profiles, pressure drops, and local heat fluxes, this 7-tube row was thoroughly instrumented with thermocouples, thermisters, and pitot tubes. An extensive use of two separate techniques was made to measure the void fraction within the bundle. High-speed photography was used at selected tube bundle elevations during specific tests.

3. ANALYSIS

3.1 Steady State Code Calculations

A detail steam generator code, URSULA-2, has been used for calculating the local phase and velocity distributions for the 1/7th scale steam generator simulation experiments. The code is capable of calculating three-dimensional steady-state and trnasient flow field within a steam generator. A link between primary heat source and secondary heat sink is provided by coupling primary flow conditions via a heat conduction with the secondary flow conditions. In the development version available are three options for computing two-phase flow behavior on the secondary side. These are: (i) homogenous model-secondary side has mixture properties, (ii) algebraic slip model-a slip between vapor and liquid is accounted using empirical correlation, and (iii) two-fluid model-vapor and liquid are represented by separate equations, hence slip is calculated from conservation equations of mass momentum and energy of each phase. The most of the calculations reported here are based on two-fluid model; however, homogenous calculations have also been performed. The details of the URSULA-2 code have been reported in reference [6]. For code application to the Freon simulation experiments, the only change is Freon properties and users choice of an appropriate grid. One build in convergence criterion is to compare the magnitude of pressure correction.

---

*In view of the proprietary design details of the commercial steam generators (such as Westinghouse or Combustion Engineering), "prototypical" used here means representative of typical steam generators where, due to design differences, compromise in geometrical simulation is necessary.

In addition, the code requires flow regime information as an input which, in this case, is vapor bubbles in liquid continuum up to node 6 and liquid droplets in vapor continuum for the rest of the nodes. This information is used in computing interfacial resistance for two-fluid model calculations.

A 3 x 4 x 16 (r, θ, z) grid size was used in URSULA-2 for computing the secondary side and local conditions for the series of tests performed in the steam generator simulation test facility. The local liquid velocity vectors and corresponding void distribution for steady-state run SS19 and node structure used for computation are shown in Fig. 2(a), (b). This grid size provided enough spacial resolution. Although the grid size selection is arbitrary, however, care must be taken that each control volume cell has porosity for dynamic flow coupling between the neighboring cells.

<u>Local Velocity and Void Distributions</u>. In the present grid configuration, there are 3 radial (r) and 4 circumferential (θ) sectors at each of 16 axial cross-section locations [see Fig. 2(b)]. A symmetry is assumed; therefore, calculations are performed for only one half of the steam generator (180° sector).

The velocity arrow indicates the magnitude and the direction of flow as presented in Fig. 2(a). In this representation, the four circumferential nodes were merged into two to represent hot and cold legs, respectively. The fluid velocities show the effect of inlet fluid at the downcomer entrance

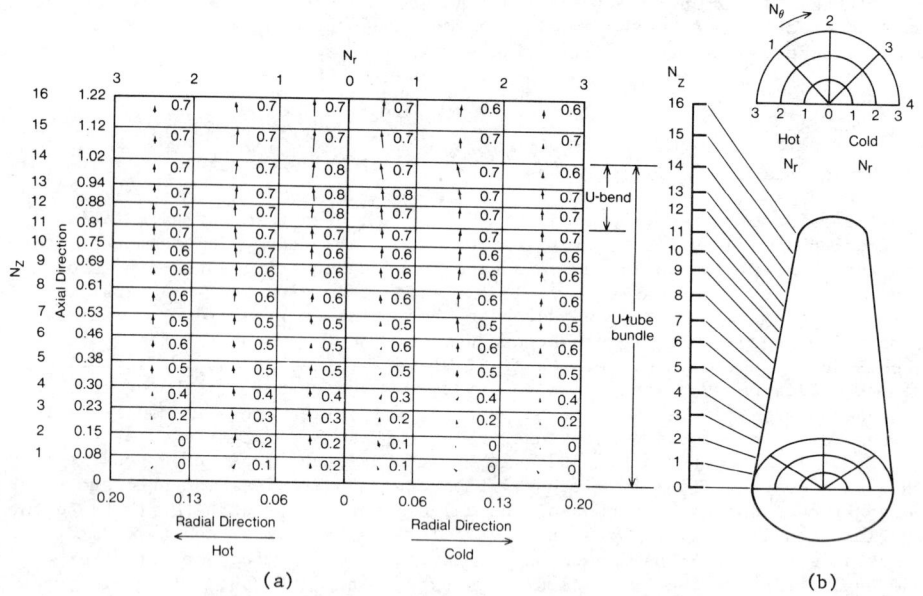

Fig. 2 (a). Local Liquid Velocity Vectors and Corresponding Void Distribution as Computed by 3-D URSULA-2 Code (Experiment Run #SS19)
(b). Node Structure Used for Computation

(angles are sharper near the wall). The velocity keeps on increasing as it moves upward. This is due to the combined effects of integrated heat input and density change. The similar trend is predicted for vapor velocity distribution. However, the difference in the hot and cold leg characteristics is pronounced for liquid compared to vapor due to slip difference between these two legs.

The axial velocity distribution for both phases in the hot leg control cells and in the cold leg cells are given in Figs. 3 for a steady-state run SS19. The 16 axial nodes, along with actual distance, are also shown in the figure. Note that the U-bend starts at axial node 11 and ends at axial node 14. The local phase velocities for corresponding hot and cold cells are superimposed in order to show the effect of hot and cold legs as shown in Fig. 3 (in the central zone). We observe reverse flow in nucleate boiling in cold leg circulating to hot side.

The axial velocities, V_g and V_ℓ, increase initially as downcomer head forces fluid into the heated riser. Also, area change, due to first support plate presence between nodes 1 and 2, enhances axial velocity component. The gravitational head (pressure drop) starts building up in the upward axial direction, the velocity component reduces in magnitude. However, this trend is governed by a delicate balance between the integrated heat input and the momentum balance (for example, it is different near the wall). Also high speed movies indicate a recirculating region at window #1. This trend changes at the axial node, $N_z = 6$, resulting in increase in velocity with axial

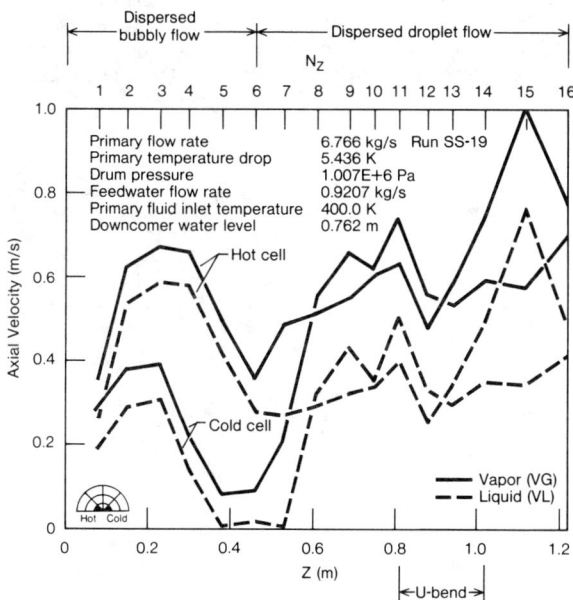

Fig. 3. Axial Variation of Vapor and Liquid Velocity (URSULA-2 Code Calculations)

distance. The vapor volume generation due to total heat input and enhanced
buoyancy effects lead to this increase in velocity component. Note that the
flow regime input to the URSULA-2 code is not a dynamic input but arbitrarily
selected such that the first six axial nodes have dispersed bubbles in liquid
continuum and the rest of the nodes have dispersed droplets in vapor continuum. However, the impact of flow regime selection on overall circulation
process is not significant as shown in the URSULA-2 sensitivity study [6].
The second support plate is between nodes 11 and 12. This, again, tends to
increase the velocity gradients. The U-bend region is between nodes 12 to
14. In this region, the flow sees larger flow areas in the axial direction,
thus causing an expected reduction in the z velocity component. This is
consistent with the cold leg side calculation, but the hot leg shows another
sharp increase in velocity between the middle of U-bend zone to the unheated
riser section. This is probably due to the change in void in the hot cell
(see vector plots) such that at approximately constant j_g, rsults in the
higher velocities ($V_g = j_g/\alpha$). For hot and cold cells near the walls, not
only that the corresponding velocities in general are lower, but also the
variations are much smoother. This behavior is due to the difference with
heat transfer characteristics between the central and outer locations of the
steam generator.

The void distributions (vapor volume fraction) in the corresponding cells
are shown in Fig. 4(a),(b). These plots indicate a significant variation in
the void distribution between hot and cold legs, especially near the lower end
of the steam generator. As the void moves upward, the mixing processes and
the approaching uniformity between the hot and cold legs primary/secondary
temperature differential reduces this difference. Near the lower end, a
significant temperature gradient may result in forming internal circulation
loops. As noted above such internal circulation loops have been observed in
the high-speed movies taken near the lowest window of the steam generator
simulation facility.

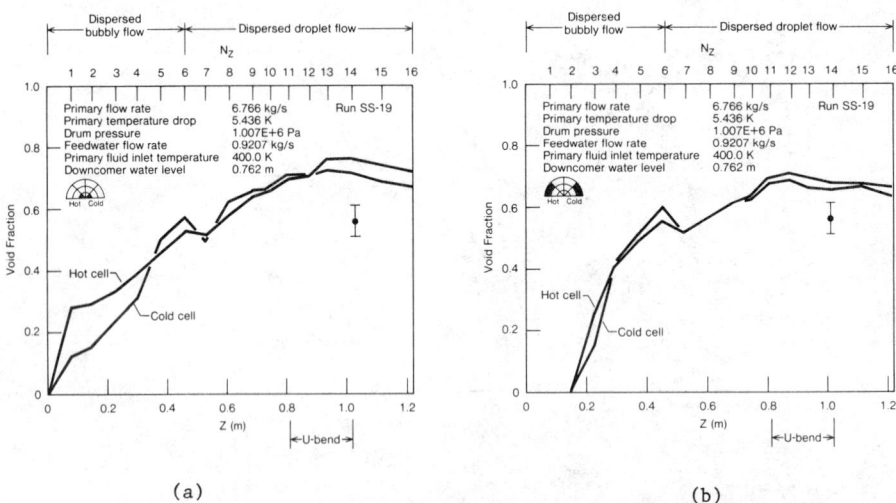

Fig. 4. URSULA Code Calculations for Axial Void Generation Rates (a) Center
Cells (b) Outer Cells. Cross-sectional Average Void Measurement Near
U-Bend Point is Also Shown

NATURAL CIRCULATING PWR U-TUBE STEAM GENERATORS

The void generation in the cold leg is offset in relation to the hot leg, as shown in Fig. 4(a),(b). The lower heat transfer in the cold leg (especially near the lower end) produces the difference. In addition, a comparison of Fig. 4(b) with Fig. 4(a) shows the effect of shroud wall; as a cell near the wall shows delay in void generation due to lower heat input. The local distribution of velocities and voids are sensitive to the primary to secondary heat transfer mechanism.

In the natural circulation U-tube steam generators, downcomer level is an important operating parameter. The downcomer level determines the driving head and hence velocity and void distributions. The velocity distribution for higher downcomer level shows larger values as compared to lower downcomer level. The impact of high downcomer level is to flow more fluid and therefore high fluid velocities.

<u>Circulation Ratio</u>. Under normal operation, circulation ratio indicates overall thermal-hydraulic behavior of a natural circulating U-tube steam generator. Therefore, it is an important performance indicator. A coupled momentum and energy processes define the (steady-state) circulation ratio. Hence, accurate prediction of this parameter by solving conservation equations indicates that the overall balance of mass, momentum and energy is preserved in the computing process. Table 1 shows the measured and URSULA-2 prediction of circulation ratio for a test series. A good agreement exists between the experimental results and the code predictions within experimental error.

3.2 Theoretical Model Development

<u>Circulation Ratio</u>. We proceed to devise an analytical solution to illustrate the interplay of the important parameters. Conservation of mass requires that, in the steady state, the recirculation flow rate is the difference between the downcomer rate (at the entry to the bundle) and the feed flow rate, i.e.,

$$\dot{m}_R = \dot{m}_D - \dot{m}_F = \dot{m}_F (R - 1) \tag{1}$$

since R is the ratio of downcomer to feed flow rate. Conservation of energy requires that the total power, $\dot{Q}_t$ is given by

$$\dot{Q}_t = \dot{m}_D (h_o - h_{in}) \tag{2}$$

where h_{in} is the mixed mean enthalpy at the bundle inlet, and h_o the mean enthalpy at the outlet. The momentum equation provides dynamic pressure drop as

Table 1

STEADY-STATE EXPERIMENTS USED FOR URSULA-2 CODE CALCULATIONS AND PREDICTION

RUN #	P_{dome} (MPa)	$\dot{M}_F$ (Kg/s)	T_F (°K)	L_{dc} (m)	EXPERI-MENTAL R	CALCU-LATED R	REMARK
SS18	1.013	1.183	355.8	0.762	6.6	6.31	Power effects
SS19	0.991	1.841	355.8	0.762	5.4	5.49	
SS20	0.998	1.892	356.3	0.965	8.0	8.56	Downcomer level effects
SS22	0.991	1.119	355.8	0.965	10.6	11.49	Primary inlet temperature effects
SS23	0.991	1.124	355.8	1.092	13.1	13.67	Effect of power (secondary feed-flow) and downcomer level
SS27	1.025	1.324	355.2	0.965	9.9	8.65	Downcomer level control effect including subcooling
SS29	1.032	1.334	357.4	1.092	12.0	12.00	

$$\Delta P_D = (\frac{K}{2\rho A_B^2}) \dot{m}_D^2 \qquad (3)$$

where K is the loss coefficient. K is a conventional engineering loss coefficient which is plainly the algebraic sum of losses due to bundle friction and support plate form losses, the accelerational losses being negligible in the steady state.

Now by definition, if the vapor (steam) is saturated, the vapor flowrate $\dot{m}_g$ is

$$\dot{m}_g = \frac{\dot{m}_D (h_o - h_f)}{h_{fg}} = \dot{m}_F, \text{ in the steady state} \qquad (4)$$

where h_{fg} is the enthalpy of vaporization and h_f the liquid enthalpy at the saturation (dome) pressure.

This expresses R in terms of measurable mixed mean quantities. Combining (2) and (4), we obtain the simple result

$$R = \frac{h_{fg} \dot{m}_D}{\dot{Q}_t} (1 + \frac{\Delta H_F}{h_{fg}}) \qquad (5)$$

Eliminating $\dot{m}_D$ between Eq. (5) and Eq. (3), we obtain

$$R = [\Delta P_D < \frac{2\rho}{K} >]^{1/2} \frac{A_B h_{fg}}{\dot{Q}_t} (1 + \frac{\Delta H_F}{h_{fg}}) \qquad (6)$$

as an initial estimate for the circulation ratio. From Eq. (6), it can be seen that the circulation ratio is only dependent on the system variables, and decreases as the bundle power and frictional losses increase.

The driving force, ΔP_D, can be expressed in terms of downcomer level, L_{dc}, collapsed two-phase level, L_c and therefore R can be written as

$$R = [\frac{(1 - L_c/L_{dc})^{0.5}}{<K \frac{\rho_f}{\rho}>}] (1 + \frac{\Delta H_F}{h_{fg}}) \frac{\rho_f A_B h_{fg} \sqrt{2g L_{dc}}}{\dot{Q}_t} \qquad (7)$$

The first term in eq. 7 depends upon the void distribution and the geometry of the steam generation and therefore shows the impact of these parameters on circulation, R. The second term shows linear dependence on the feed subcooling. The last term is a dimensionless power and its magnitude depends on the system operating conditions, viz., the downcomer level, the operating pressure, and the operating power.

<u>Evaluation of R</u>. If there are N identical support plates with flow area ratio of $\bar{A}$ in a steam generator and the losses at the downcomer entrance are neglected; the form loss term can be written as

$$<k \frac{\rho_f}{\rho}> = N (\bar{A} - 1)^2 \frac{\rho_f}{\rho_{2\phi}} \qquad (8)$$

where the secondary two phase fluid is assumed at the average density, $\rho_{2\phi}$. The density ratio $\rho_f/\rho_{2\phi}$ can be represented in terms of the ratio between the collapsed level and the two phase level which depends on the void distribution parameter <Co> [7-9], viz.,

$$\frac{\rho_f}{\rho_{2\phi}} = \frac{L_{2\phi}}{L_c} = \frac{<Co>}{<Co>-1} \qquad (9)$$

substituting eq. 8 and 9 in eq. 7

$$R = [\frac{1-L_c/L_{dc}}{N (\bar{A}-1)^2} \frac{(<Co> - 1)}{<Co>}]^{0.5} (1 + \frac{\Delta H_F}{h_{fg}}) \frac{\rho_f A_B h_{fg} \sqrt{2gL_{dc}}}{\dot{Q}_t} \qquad (10)$$

Equation 10 was used to perform the parametric study of the impact of operating conditions on the circulation ratio for a range of distribution parameters--the effect of power and subcooling in Figs. 5(a) and (b). As it is evident from these figures that (i) circulation process is very sensitive to the distribution parameters <Co> especially at low power and high subcooling, (ii) at a constant circulation, higher power and lower subcooling tend to enhance the nonuniformity in the phase distribution. This effect diminishes as one approaches low circulation, and (iii) distribution parameter, <Co>, approaches unity in the limit of R→1 (all vapor). The effect of pressure variation was found insignificant.

The test results from the scale simulation facility were also used to calculate the <Co> from measured R. The bundle pressure drops, ΔP_B, was used to compute collapsed level, as only gravitational pressure drop dominates. The range of experimental "R" values is shown in Figs. 5(a), (b) [see also Fig. 6]. The average <Co> of 1.23 was calculated from the range of tests performed in this series.

This model was also extended to evaluate the distribution parameter <Co> for given circulation ratio (or vice versa) at 100% power for Westinghouse System 51 and Combustion Engineering System 80. Under the assumptions that

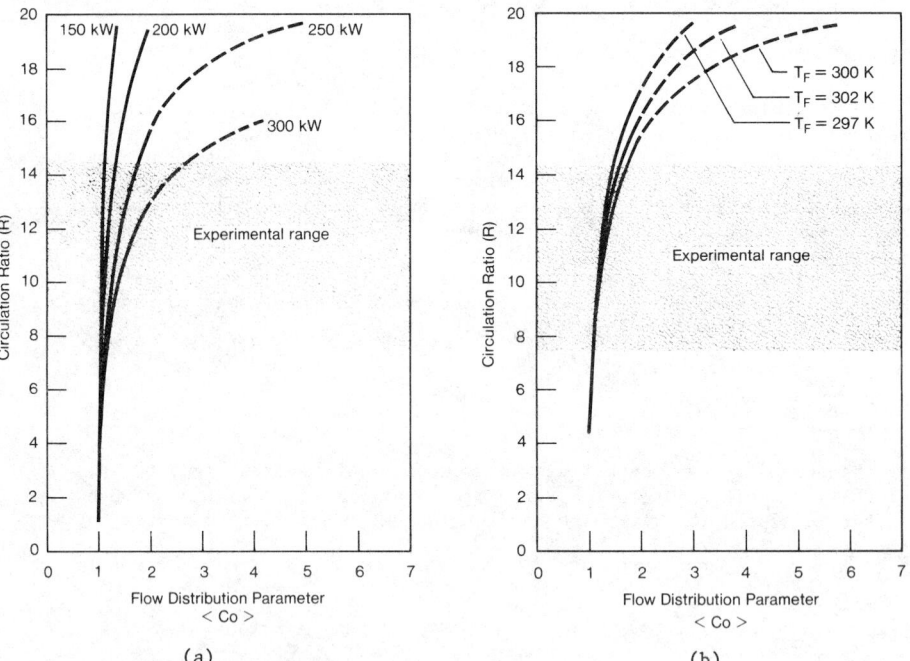

Fig. 5. The Effect of Flow Distribution Parameter on the Natural Circulation Process (a) at Different Power Levels (b) at Different Feed Temperatures (Subcooling).

(i) only support plates form losses are significant, (ii) secondary pressure drop (calculated using TRANSG-01 [10] code for W 51 and URSULA-2 for CE 80) is dominated by gravitational head; the calculated <Co> values are 1.02 (W 51) and 1.01 (CE 80), respectively. The average value of <Co> for the range of simulation experiments is ≃1.23. This difference may be due to the higher R values (R>7:5) in the experimental set compared to low values of R (R ≃ 3.2) in the commercial units. However, as indicated by the model (see Figs. 5(a) and (b)) the trend is correct; i.e., reduction in R will enhance more uniformity and therefore <Co> trend to 1.

4. DISCUSSIONS AND CONCLUDING REMARKS

PWR U-tube steam generators are used in Westinghouse and Combustion Enginering NSSS power plants. These are vertical tube-and-shell heat exchangers which provide two-phase natural circulation process on the

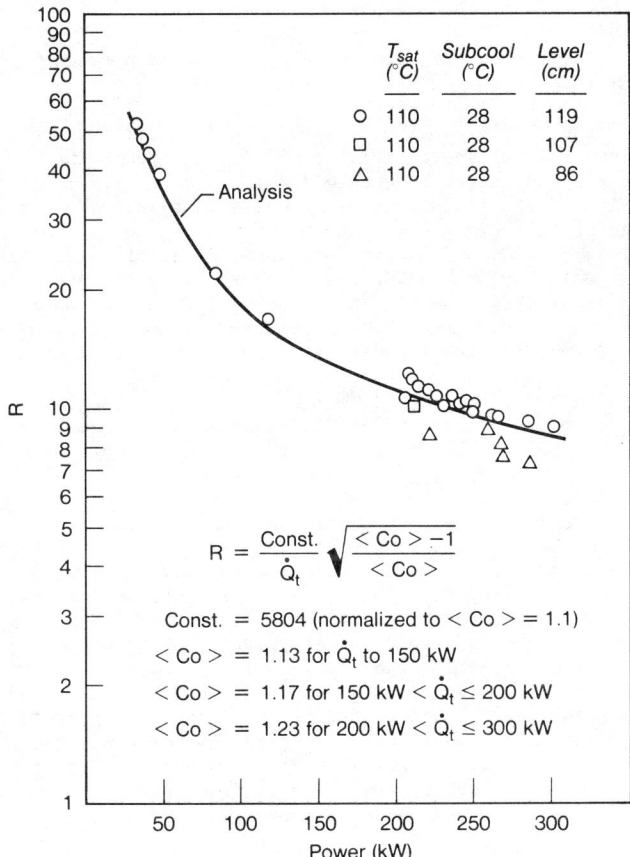

Fig. 6. The Comparison of Experiments and Analytical Model Predictions of Circulation Ratios

secondary side for generating steam. In the present study, the phenomena pertaining to the PWR U-tube steam generators under normal and off-normal conditions have been examined both experimentally and analytically using data generated in a Freon simulation test loop. In addition the flow regime visualization was accomplished using still and high-speed pictures.

The downcomer level is one of the major controlling parameters for the steam generator operation. The present series of experiments reveal the importance of this parameter in establishing the desired circulation, pressure drops and heat-transfer characteristics in a U-tube steam generator.

The circulation ratio determined by the driving force and hence the downcomer level for a given power (heat) to be transferred from primary to secondary. As this power level increases, this effect diminishes due to enhance boiling (two-phase) and corresponding pressure drop within the U-tube bundle. Also flow and phase distribution in the secondary side depends upon the downcomer level and this is strongly coupled with the local heat transfer chacteristics. The primary/secondary interaction is depicted through the primary inlet temperature and flow variations experiments.

On the analysis front, two parallel approaches have been used: (i) a detail 3-dimensional code, URSULA-2 and (ii) model development.

The code prediction of the local velocity and void distributions in the hot and cold legs have been presented (Figs. 2-4). Even though the predictions show possible physical trends; no detailed data is yet available to determine the validity of the calculations. Local void measurement at the top of the U-bend using optical void probe is superimposed on the figures showing an overprediction ($\simeq$15 to 20%). The downcomer level (control parameter) effects on the local velocity and void distributions have also been calculated. Again, this study shows sensitivity of such distribution on the downcomer height which also confirms the experimental observations. The prediction of overall system parameters (e.g., R) appears to be reasonably well [see Table 1].

Simple analytical models are desirable for understanding the complex physical process within the U-tube steam generator. Some efforts have been made in this direction. The knowledge of flow circulation for a given operating condition of a U-tube steam generator is essential in order to quickly evaluate its performance. Such model has been developed and its discussion is given in the analysis section of this paper. A drift flux approach [7] is used in defining void distribution for this model. A distribution parameter <Co>, defined in the text represents an integrated effect of both radial and axial void distribution within the steam generator. The coupling between flow circulation and <Co> is exhibited in terms of system conditions (Figs. 5a and b).

The analytical model developed for predicting circulation process (eq. 10) provides a tool for calculating R for given system conditions. The model's prediction capability is exhibited in Fig. 6 where a normalized version is shown to predict reasonably well R at various power levels. This comparison also shows the effect of void-velocity distribution (drift flux, <co>) on the circulation process.

The scale (1/7) simulation facility is designed to provide a complete data base for both normal and off-normal testing of the natural circulating U-tube steam generators. A detailed discussions of both steady state and transient thermal-hydraulics of U-tube steam generator is provided in reference [11]. This also includes a brief review of the state of the art on the steam generator analysis methods.

ACKNOWLEDGEMENT

The author wishes to thank Dr. R. B. Duffey for his valuable suggestions and comments which have been incorporated in this article.

REFERENCES

1. Lee, J. C., et al., "Transient Modelling of Steam Generator Units in Nuclear Power Plants," Electric Power Research Institute Report EPRI NP-1576, October 1980.

2. Kalra, S. P., et al., "Experimental Simulation Studies of PWR U-Tube Steam Generators," Boiler Dynamics and Control in Nuclear Power Stations, BNES London, 1979.

3. Ahmad, S. Y., "Fluid to Fluid Modelling of Critical Heat Flux: A Compensated Distortion Model," International Journal of Heat and Mass Transfer, Vol. 16, pp. 641-662, 1973.

4. Kalra, S. P. and Ahmad, S. Y., "Critical Heat Flux Management in a Vertical Multi-Element Segmented Cluster," Fluid Flow and Heat Transfer Over Rod in Tube Bundles, ASME Proc. pp. 171-180, 1979.

5. Adams, et al., "Integral Experiments for PWR U-Tube Steam Generator," EPRI Report NP-1088, Electric Power Research Institute Report EPRI NP-1088, 1979.

6. Singhal, A. K., et al., "The URSULA-2 Computer Program," Vol. 1 to 4, Electric Power Research Institute Report EPRI NP-1315, 1980.

7. Zuber, N. and Findley, J., "Average Volumetric Concentration in Two-Phase Systems," Trans. ASME J. Heat Transfer, Vol. 87, pp. 453-462, 1965.

8. Sun, K. H., et al., "As the Prediction of Two-Phase Mixture Level and Hydrodynamically Controlled Dryout Under Low Flow Conditioning," Electric Power Research Institute Report EPRI NP-1359SR, 1980.

9. Kalra, S. P., Duffey, R. B. and Adams, G., "Loss of Feed Water Transients in PWR U-Tube Steam Generators: Simulation Experiments and Analysis," AIChE Symposium Series 199, Vol. 76, pp. 36-44, 1980.

10. Lee, J. C., et al., "Transient Modeling of Steam Generator Units in Nuclear Power Plants: Computer Code, TRANSG-01," Electric Power Research Institute Report EPRI NP-1368, 1980.

11. Kalra, S. P., "Dynamic Thermal-Hydraulic Behavior of PWR U-Tube Steam Generators--Simulation Experiments and Analysis," Electric Power Research Institute Report EPRI NP-1837-SR, 1981.

A Model for Predicting the Steady-state Thermal-hydraulics of the Once-through Power Boilers

G. DEL TIN, M. MALANDRONE and B. PANELLA
Politecnico di Torino
Turin, Italy

G. PEDRELLI
ENEL-Ente Nazionale Energia Elettrica
Centro Ricerca Termica e Nucleare
Pisa, Italy

An analytical model to predict the steady-state water side thermal-hydraulic parameters for the furnace of the once-through forced flow power boilers at subcritical pressure is proposed. The model predicts, at a given heat flux distribution, the pressure drops, the flow-rate distribution in the tube-banks, and the boiling crisis in the ribbed and in the commercially rough tubes of the furnace panels. A set of the model predicted pressure drops shows a fairly good agreement between the calculated and the experimental operative conditions values.

1. INTRODUCTION

From the thermal-hydraulic point of view the once-through forced flow subcritical pressure power boilers main feature is that the working fluid is pumped into the unit as a liquid, passes sequentially through all the pressure part heating surfaces, where it is converted to steam as it absorbs heat and leaves the boiler section as saturated or slightly superheated steam; there is no recirculation of water within the unit. The furnace water-walls consist in a series of parallel membrane-wall connected panels, (except the top rear tube bank, to allow the combustion gases to flow towards the superheater, reheater and economizer). Fig. 1 shows the flow diagram of the once-through subcritical pressure 320MWe power boiler of the Babcock and Wilcox Company design. The furnace walls are subdivided in four passes; the subcooled water from the economizer flows through the first pass and enters in a tee header connected by feed pipes to the second pass panels. In every panel there are several groups of parallel tubes ending in a series of headers as fig. 1 shows schematically. The second and partly the first and the third pass panels are constituted by internally ribbed tubes.

The furnace thermal-hydraulics, and particularly the heat transfer inside the boiler tubes, has to be accurately analyzed because the heat fluxes are rather high and the tubes may approach the design temperature limits for carbon or low-alloy steel.

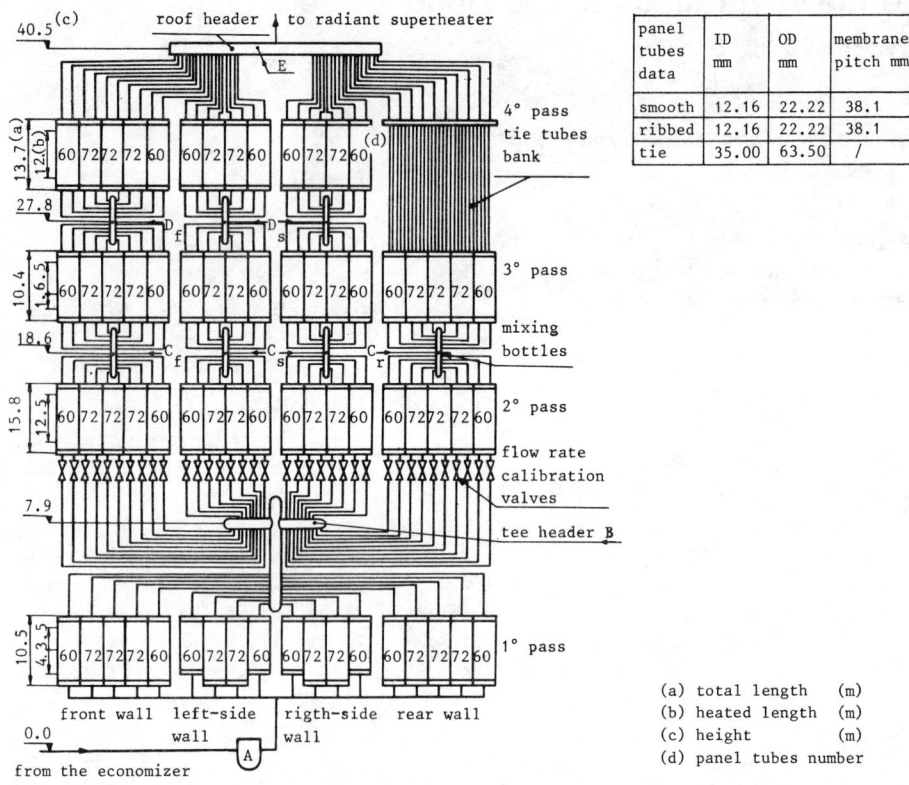

Fig. 1 - Furnace walls flow diagram

This paper presents an analytical model to predict the steady-state water-side thermal-hydraulic parameters for the furnace walls of the once-through subcritical boilers. The single and two-phase pressure drops for the ribbed tubes are investigated and a set of experimental results obtained at the Politecnico di Torino with air-water mixtures is reported. The boiling crisis in the internally grooved tubes is discussed and the parametric effects on the boiling transition for the ribbed tubes are analyzed. The derived computer code ENCO-3 predicts, at a given heat flux distribution, the flow rates, the pressure drops, the fluid enthalpy distributions, the wall temperatures and the transition boiling location in the tubes of the furnace panels. The code is particularly suitable for the parametric analysis and for the prediction of non usual operative conditions.

2. THERMAL-HYDRAULIC MODEL

In every panel evaporator tube the water or steam-water mixture enthalpy increases along the height depending on the distribution of the flux, and is evaluated by a simple heat balance. At the furnace bottom the coolant is subcooled liquid and the heat transfer coefficient is evaluated by the well known Dittus-Boelter correlation. In the ribbed tubes the heat transfer is enhanced owing to the intensive turbulence induced in the boundary layer by the grooves, but the effect seems relevant only at lower flow rates /1/; on the other hand the fouling on the inner tube walls decreases the heat transfer during the plant operation and the use of Dittus-Boelter equation is extended also to the ribbed tubes with good approximation.

The inner wall temperature is calculated by the heat transfer coefficient and the circumferentially variable heat flux up to the onset of the nucleate boiling which is evaluated by the Jens-Lottes correlation /2/, also in the ribbed tubes, in agreement with ref. /1/.

The effect of subcooled boiling on the liquid enthalpy evaluation is negligible. Bubbly flow and afterwards annular flow patterns are likely to occur along the tube depending on the pressure, flow rate, and quality; however there is a lack of data about the flow regimes of two-phase systems at high pressures slightly below the critical value. For the ribbed tubes there are very few data /3/, even at atmospheric pressure: the transition from bubbly flow to slug flow is harder to occur and the liquid film thickness in the annular flow is higher than for the smooth tubes; the flow seems to tend towards greater homogeneity owing to the boundary layer disturbance induced by the grooves. The more likely transfer regime is the nucleate boiling and the Jens - Lottes correlation can be adopted until a point is reached where the Boiling Transition (B.T.) (discussed later) takes place. A maximum wall temperature follows the B.T., since after the abrupt deterioration, the heat transfer is improved because of the mixture velocity increase as the flow quality increases, and, if dryout occurs, of the complex heat transfer mechanisms in the so called post-dryout regime. As a matter of fact, downstream of the dryout location the vapor is superheated and the liquid droplets, as dispersed spray, gradually evaporate so that the actual vapor temperature and quality are different from the thermal equilibrium ones. The heat transfer coefficient can be predicted by the Groeneveld-Delorme method /5/, which recommendes the Hadaller's superheated steam heat transfer correlation (modified for two-phase flow) and evaluates the actual quality and vapor enthalpy. The homogeneous flow assumption, adopted in /5/ is particularly true for the ribbed tubes. On the other hand at high pressure (p > 170 bar) and high flow rates the assumption of thermal equilibrium yields satisfactory results in most cases. However the Groeneveld-Delorme method has the advantage that is may be used to predict the wall temperature in the two-phase flow region as in the single phase superheated steam one. The ribbed tubes data /1/ show that when the boiling crisis occurs the wall temperature rise is lower than that of the smooth tubes; therefore the authors propose the assumption of the thermal equilibrium for the ribbed tubes post-dryout heat transfer owing to the higher turbulence induced by the ribs.

A great deal of studies about the boiling crisis phenomenon has been carried out in the past years (see ref. /6/), nevertheless the physical mechanisms are not completely understood, especially for the DNB, and there is a lack of data at high pressure (p > 200 bar) and at diameters larger than 10 mm.

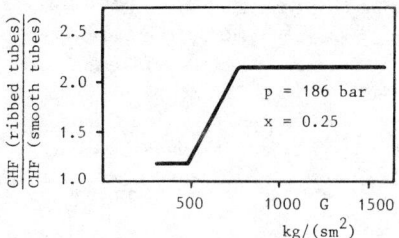
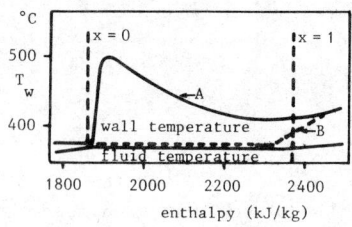

Fig. 2 - Ratio between critical heat flux in ribbed and smoooth tubes as a function of the mass velocity (vertical tube).

Fig. 3 - Typical wall teperature profiles for ribbed and smooth tubes at P= 207 bar, G=949 kg/sm^2, q=495 kw/m^2; A: smooth tube (ID=10.36 mm), B: ribbed tube (minimum ID=10.44 mm).

Regarding the once-through boilers B.T. (which is not likely to occur in the subcooled boiling region owing to the relatively low heat flux) there are few data obtained in facilities for conditions sometimes rather different as compared with the operative plant ones. Moreover the once-through boilers (like the Babcock-Wilcox ones) panels are connected by unheated large pipes (with horizontal sections, downflow sections and bends) and by headers in which the steam-water mixture coming up from the evaporator tubes homogeneizes before entering the next panel. Some of these conditions have been investigated separately. The circumferentially not uniform heat flux decreases the average CHF but the effect is less relevant as quality and pressure increase. The axially non uniform heat flux, like the cosine distribution, decreases the CHF but the effect decreases as pressure and flow rate increase /7/. An upstream unheated section, at high flow quality, may increase the CHF /8/ like the homogeneization in the headers. The liquid-vapor mixture at the inlet increases the CHF but the effect is relevant only at small length and it decreases with pressure /9/. As regards the effect of the grooves on the B.T. fig. 2 /11/ shows that the critical heat flux increase in comparison with the smooth tubes (that is the commercial rough tubes) is much more relevant at mass velocities higher than 500 kg/s m^2; this improvement may be attributed to the swirling of flow in proximity of the inside surface. Fig. 3 /5/ shows the inner wall temperatures as a function of the fluid enthalpy for smooth and ribbed tubes; a steep rise of wall temperature (curve A) was observed at about 3% steam quality whereas the ribbed tube prevented the boiling transition until about 90% quality and hence the wall temperature was kept low (curve B). In such conditions the flow pattern is likely bubbly flow and the heat transfer regime is nucleate boiling, so that the boiling transition is of DNB type for the smooth tubes and the ribs enhance the turbulence near the wall inhibiting the formation of the steam film, as a consequence preventing the film boiling.

The authors of ref. /1/ have experimentally compared the performance of the Babcock Wilcox ribbed tubes with smooth tubes, at pressure ranging from 167 to 206 bar, at heat flux ranging from 11.6 to 105 W/cm^2, at mass velocity

STEADY-STATE THERMAL-HYDRAULICS OF ONCE-THROUGH POWER BOILERS

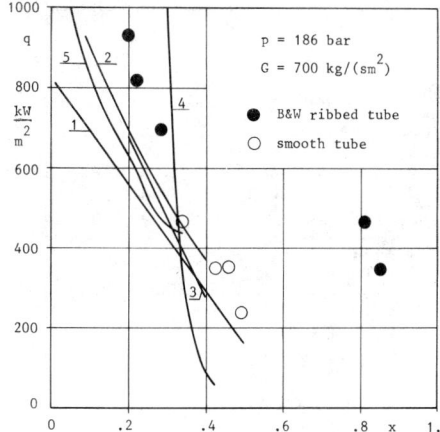
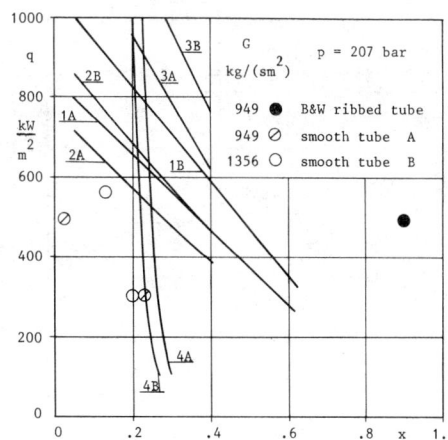

Fig. 4 - Critical heat flux as a function of the equilibrium quality: comparison between experimental values (ref. 1,7) and correlation predicted values (1-Becker, 2-Peskov, 3-Bailey, 4-Kon'Kov, 5-USSR Academy of Sciences).

in the 380 to 1270 kg/s m² range. In the smooth tube the boiling transition seems to the authors of DNB type as the steam quality is less than 50% and the bubbly flow pattern is estimated to occur from the extrapolations of the Baker's chart: the boiling transition is from nucleate to film boiling, especially at higher pressures. The ribbed tubes permit higher heat fluxes (twice or more) than the smooth tubes, especially with increase of flow rate; the DNB is inhibited except at low flow rate (mass velocity of about 400 kg/m² s at heat flux of about 46.0 W/cm²), but in such a case the wall temperature rise is much lower than with the smooth tube and the DNB steam quality is about the same as for the smooth tube. Fig. 4 presents the critical heat flux as a function of the steam quality for the smooth and ribbed tubes: data of ref. /1/ are compared with the prediction of several correlations. It is evident that such empirical correlations are not suitable for the ribbed tubes, and also their agreement with the smooth tubes data is sometimes poor, owing to the validity range of the experimental parameters.

In the present model, the ribbed tubes are considered able to prevent practically the DNB if the mass velocity is higher than a given value (say about 500 kg/m² s) and the heat flux is lower than a given value (say about 50 W/cm²): in such a case a value of 85% is assumed for the boiling transition steam quality.

Otherwise and for the smooth tubes of the panels Kon'kov correlation /15/ is adopted. The non uniformity of the heat flux effect is assumed negligible because of the high pressure and (for the ribbed tubes) of the homogeneization induced by the ribs. Also the effect of liquid-vapour mixture at the inlet seems to be negligible due to the high pressure and to the large length of the boiling tubes.

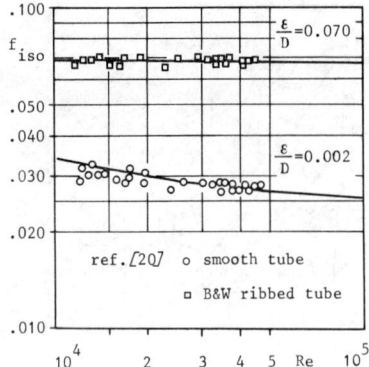

Fig. 5 - Single-phase friction factors as a function of Reynolds number; smooth tube, ID=12.15 mm; ribbed tube, max. ID=13.44 mm, ribs data: pitch =9 mm, width=5 mm, height=0.4 mm.

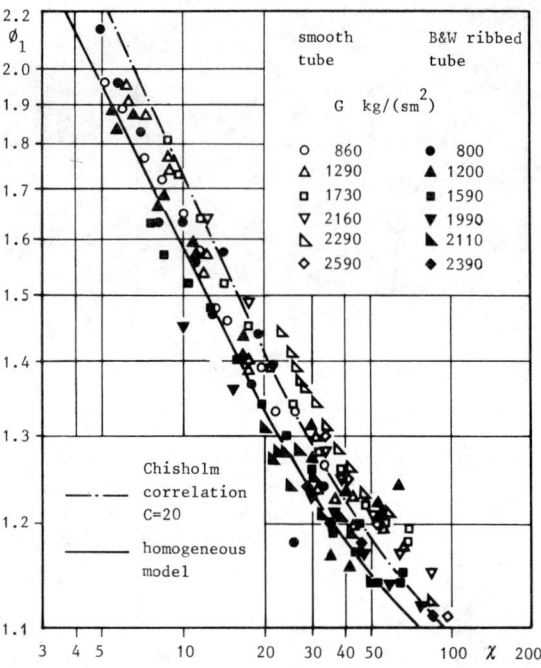

Fig. 6 - Two-phase friction local multiplier $\emptyset_1$ as a function of Martinelli parameter (tube data as in figure 5)

The pressure drops evaluation across the panels tubes of the furnace allows to predict the flow rate distribution between the furnace parallel tubes which in turn is an input to the thermal-hydraulic calculation. Besides, the pressure prediction along the channels is needed in order to evaluate the new fluid properties which also affect the termodynamics and the two-phase flow. The separated flow model is used in the smooth tubes upstream of the B.T., whereas the homogeneous model is adopted downstream of the B.T. point, according to ref. /17/; for the ribbed tubes, in agreement with the experimental results obtained by the authors /18/, the homogeneous model is adopted too. In the separated flow model the void fraction is evaluated by the Bankoff model /19/.

The pressure drop along the channels is calculated by summing the friction, elevation, acceleration and local pressure drop components terms. The isothermal single-phase friction factor, (if the tubes have no ribs) is evaluated by the correlation developed by Waggener /20/; in the two-phase region Lottes-Flinn correlation /21/ is used for smooth tubes and upstream of the B.T. point, whereas the homogeneous model is used downstream of the B.T. point.

If the tubes are ribbed the single-phase correlation with the commercial tubes roughness does not work, as the friction factor is much higher; in ref. /3/ the friction factor of Babcock-Wilcox ribbed tubes is more than twice the friction factor of smooth tube. Ribbed tubes friction factor experimental values have been obtained at Politecnico di Torino by an air-water loop /18/. Fig. 5 shows that the smooth tube single-phase friction factor data are in good

	Table 1a - Furnace water-wall calculated flow rates and pressure drops			Table 1b - Detailed analysis of the furnace front water-wall with non uniform heat flux distribution		
	front wall k = 137.8	side wall k = 158.7	rear wall k = 111.9	72 tubes k = 95.3	72 tubes k = 95.3	60 tubes k = 285.1
pass	Δp (a) (b)	Δp (a) (b)	Δp (a) (b)	Δp W	Δp W	Δp W
4°	5.52(0.1)(.2)	4.57(0.1)(-0.2)	3.43(0.0)(-1.0)	5.34 15.59	5.33 15.73	5.65 13.86
3°	3.48(1.1)(-4.4)	3.24(0.8)(-4.6)	4.23(0.9)(-4.7)	3.64 16.08	3.64 16.23	3.30 13.09
2°	5.07(-2.0)(-5.1)	4.83(-1.5)(-5.0)	6.15(-1.5)(-1.5)	5.73 17.05	5.60 17.26	3.87 11.60
1°	1.96	.84	1.88			
	W (a) (b)	W (a) (b)	W (a) (b)	Δp panel pressure drop (bar)		
2°-4°	74.76(.01)(-.06)	60.32(-.04)(.01)	87.61(.05)(0.0)	W flow rate (kg/s)		
1°	85.50	57.26	83.03	k calibration valves resistance coefficient		

(a) percent variation due to axially non uniform heat flux

(b) percent variation due to detailed geometrical simulation and non uniform heat flux across the water wall

Table 1 - Computational results for the maximum continuous power of 320 MWe (steam flow rate 1050 t/h; water inlet conditions: p=211.57 bar, (T=308.8°C).

agreement with the Waggener correlation if a relative roughness (ϵ/D) value of 0.002 is adopted; likewise the ribbed tube single-phase data are well fitted if $\frac{\epsilon}{D} = 0.07$ and this value is comparable to that obtained as the ratio between the ribs height and the average tube diameter in agreement with ref. /3/. Fig. 6 shows the two-phase local multiplier as a function of the Martinelli parameter at different flow rates for both smooth and ribbed tubes: data are compared with the Chisholm /22/ and homogeneous model. The homogeneous model appears more adequate for the ribbed tubes, due to the homogeneization induced by the grooves. Taking into account the experimental results and the boiler high pressure, the present model adopts the homogeneous correlation to predict the local two-phase multiplier for the ribbed tubes. The pressure drops across expansions, contractions, valves and bends are predicted by the usual correlations and in the case of the two-phase flow the homogeneous model is adopted.

3. COMPUTATIONAL MODEL AND RESULTS

The purpose of the <u>analytical model</u>, where the preceding correlations are used, is to perform a thermal-hydraulic analysis of the furnace water-walls consisting of parallel flow tubes. The lumped parameter technique is used and the parallel channels are handled with the flow convergence analysis. Input to the code includes system geometry, heat flux distribution within the furnace water-walls, inlet fluid temperature and pressure and the total inlet flow rate corresponding to the operating power level. Output yields the several water-walls pressure drops for a required inlet flow, flow distribution among the water-walls and among the parallel tubes of each panel, fluid enthalpy, quality

wall ΔP	front	rear	right side	left side
A-B 1° pass		2.64 (a) 2.61 (b) -1.1 (c)		
B-C 2° pass	12.70 12.86 1.3	14.30 14.94 4.5	13.20 13.76 4.2	14.60 13.76 -5.8
C-D 3° pass	4.43 4.35 -1.8		4.80 4.51 -6.0	4.05 4.51 11.4
B-D	17.13 17.21 0.5		18.00 18.27 1.5	18.65 18.27 -2.0
B-E		24.54 23.93 -2.5		

(a) experimental value (bar)
(b) calculated value (bar)

$c = \frac{b-a}{a} \cdot 100$

plant power 320 MWe

steam flow rate 1050 t/h

feedwater inlet conditions as in table 1

Table 2 - Comparison between measured and calculated pressure drops with reference to Figure 1.

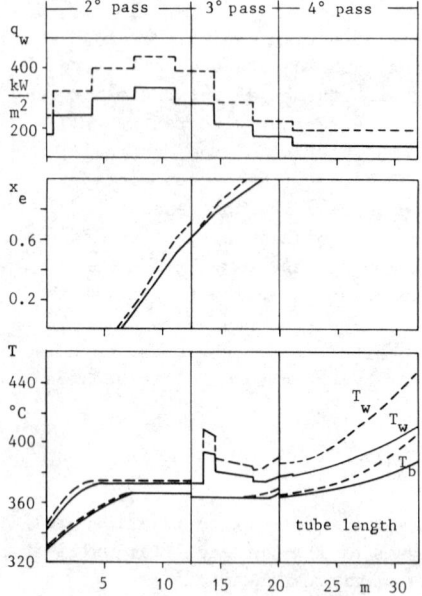

Fig. 7 - Furnace front water-wall: axial variation of the equilibrium quality, maximum wall temperature and fluid temperature with reference to two panels at different thermal loads.

and void fraction as a function of axial position in the tubes, transition boiling locations and inner tube wall temperature axial profile.

Two approaches have been made in order to simulate the complicated furnace water-walls geometry and heat flux distribution; in the first one, each water-wall is represented by an "average tube" with the same geometry characteristics as the real ones; the heat flux in each pass has been assumed axially uniform (like panel pass average value) or as predicted by the heat flux map. The second and more realistic approach is a detailed geometrical description of the furnace water-walls configuration. In this case the heat flux is the real one on the water walls.

Some computational results concerning the boiler maximum continuous power are summarized in the table 1, in the table 2 and in fig. 7. Table 1a shows the results relative to the "average tube" simulation with the heat flux in each pass either axially uniform or as predicted by the heat flux map. Owing to the low ratio between maximum and average heat flux values in each pass (ranging from 1.16 to 1.30) it can be observed that the calculated pressure drops are practically the same for both the heat flux distributions. The table 2 shows a comparison between calculated and measured pressure drops at various furnace water-walls locations. As it can be seen, the model predicted values are in satisfactory agreement with the measured ones. Results concerning the more detailed furnace water-walls simulation are also obtained; an example of them, with regard to furnace front water-wall, is summarized in the table 1 b. The comparison of the pressure drops shows that the detailed analysis of parallel tube banks does not affect considerably the predictions. Fig. 7 shows an example of the results, concerning the axial variations of the fluid temperature, maximum inner wall temperature and equilibrium quality with reference to two front water-wall panels; the heat flux distributions are reported too, corresponding respectively to the maximum and to the minimum thermal loads on the furnace front-wall panels.

4. CONCLUDING REMARKS

The proposed analytical model seems appropriate to predict the thermal-hydraulic behaviour of the once-through boiler furnace; the agreement between pressure drop calculated values and the measured ones seems a satisfactory achievement of the present work. Up to now there is no other available measured parameters to allow a complete check of the model. However the model improvement requires a better knowledge in the following areas:

- two-phase flow and heat transfer in ribbed tubes, particularly in the parameters range interesting once-through boiler applications;
- boiling transition in smooth and ribbed tubes subjected to circumferential and axial variations in heat flux, particularly at high pressures and for high tube lengths.

ACKNOWLEDGEMENT

This research was supported both financially and technically by ENEL (Italian National Electric Energy Agency) - Thermal and Nuclear Research Center Pisa).

REFERENCES

1. Nishikawa, K., Fujii, T., Yoshida, S., Ohno, M. 1974. Flow Boiling Crisis in Grooved Boiler Tubes, 5th Int. Heat Transfer Conf. Vol. 4, B6.7, p.270.
2. Jens, W.H. and Lottes, P.A. 1951. Analysis of Heat Transfer, Burnout, Pressure Drop and Density Data for High Pressure Water. ANL - 4627.
3. Nishikawa, K., Segokuchi, K., Nakasatomi, M. 1973. Two-phase Flow in Spirally Grooved Tubes. Bull. JSME, Vol. 16.102, p. 1918.
4. Ito, M. and Kimura, H. 1979. Boiling Heat Transfer and Pressure Drop in Internal Spiral-Grooved Tubes. Bull. JSME, Vol. 22, 171, p. 1251.
5. Groeneveld, D.C. and Delorme, G.G.J. 1976. Prediction of Thermal Non-equilibrium in the Post-dryout Regime. Nucl. Eng. and Design, vol. 36, p. 17.
6. Hewitt, G.F. 1978. Critical Heat Flux in Flow Boiling. 6th Int. Heat Transfer Conf., Vol. 6, KS-13, p. 143.

7. Swenson, H.S., Carver, J.R., Szoeke, G. 1962. Effects of Nuclear Boiling on Heat Transfer in Power Boiling Tubes. J. of Eng. for Power, vol. 84, p. 365.

8. Peskov, O.L., Remizof, O.V., Sudnitsyn, O.A. 1978. Some Feature of Heat Transfer Burnout in Tubes with Non-uniform Axial Heat Flux Distribution. 6th Int. Heat Transfer Conf., Vol. 5, NR-10, p. 53.

9. Hewitt, G.F. et al. 1966. Studies of Burnout in Boiling Heat Transfer to Water in Round Tubes with Non-uniform Heating. AERE-R5076.

10. Bennet, A.W., Collier, J.C., Kearsey, H.A. 1964. Heat Transfer to Mixtures of High Pressure Steam and Water in an Annulus. AERE-R3961.

11. Watson, G.B., Lee, R.A., Wiener, M. 1974. Critical Heat Flux in Inclined and Vertical Smooth and Ribbed Tubes. 5th Int. Heat Conf., Vol. 4, B6.8, p. 275.

12. Becker, K.M. 1971. Burnout Conditions for Round Tubes at Elevated Pressures, Symp. on Two-phase Systems, Haifa, Paper 1-9.

13. Peskov, O.L. et al. 1969. The Critical Heat Flux for The Flow of Steam Water Mixtures through Pipes. Pergamon Press, p. 48.

14. Lee, D.H. 1970. Studies of Heat Transfer and Pressure Drop Relevant to Subcritical Once-through Evaporators. IAEA SM 130-56.

15. Kon'kov, A.S. 1966. Experimental Study of The Conditions under which Heat Exchange Deteriorates when a Steam Water Mixtures Flows in Heated Tubes. Teploenergetika, Vol. 13, 12, p. 53.

16. USSR Boiling Water in Uniformely Heated Round Tubes. Teploenergetika, vol. 23, 9, p. 90.

17. Collier, J.C. 1972. Convective Boiling and Condensation. Mc Graw Hill Book Co. (UK), p. 54.

18. Impalà, F. et al. 1980. Cadute di pressione per attrito in tubi rigati per deflussi monofase e bifase. Proc. XXXV Congr. Naz. ATI, Vol. 1, p. 673.

19. Bankoff, S.G. 1960. A Variable Density Single-Fluid Model for Two-Phase Flow with Particular Reference to Steam-Water Flow. J. of Heat Transfer, Vol. 82, p. 265.

20. Waggener, J. 1961. Friction Factors for Pressure Drops Calculations. Nucleonics, vol. 19, 11, p. 145.

21. Lottes, P.A. and Flinn, W.S. 1956. A Method of Analysis of Natural Circulation Boiling Systems. Nucl. Sc. Eng., Vol. 1, p. 461.

22. Chisholm, D. 1967. Flow of Incompressible Two-Phase Mixture through Sharped Edged Orifices. J. of Mech. Eng., Vol. 9, 1.

Failures of the Turbines Condenser Tubes with the Direct Inflow of Make-up Water into Condenser

NIKOLA RUŽINSKI and MIJO MUSTAPIĆ
Faculty for Mechanical Engineering and
Naval Architecture
University of Zagreb
Zagreb, Yugoslavia

ABSTRACT

Higher ammonia concentrations in some parts of the condenser, and the higher content of oxygen introduced with make-up water, leads to condenser tube damage. The paper presents a calculation of possible ammonia concentrations in the condensate, and analyzes the possibility of reducing the oxygen content by thermal deaeration in the condenser.

1. INTRODUCTION

The direct inflow of make-up water into the condenser may cause substantial condenser tube damage. This occurs when the addition of ammonia - in order to provide and maintain a protective magnetite layer on steam generator tubes - and excess hydrazine increases the pH-value of feed water to approximately 9.2-9.5. An analysis of the causes underlying condenser tube damage in two Yugoslav power plants, rated 220 and 125 MW, established selective brass tube corrosion to have occurred precisely because of the higher feed water pH-value, and because of higher oxygen quantities introduced with make-up water into the condenser.

Severe damage of condenser brass tubes and of the tubes of the low-temperature feed water pre-heaters developed in operation, and higher quantities of corrosion products were found in the feed water tank and, in the form of sediment, in the feed water pumps. Chemical analysis of the sediment produced the following composition: SiO_2=0.10%; Fe_2O_3=5.23%; CuO=94.30%; CaO=0.10%; MgO=0.10%; P_2O_5=0.01%. It was obvious that the main component was copper oxide, while the other corrosion components were present in a lower degree. The turbine condensate had an average composition as follows: pH=9.0-9.2; NH_3=0.25-0.4 mg/kg; O_2=0.2-0.5 mg/kg. Such a low ammonia content cannot aggressively affect the brass condenser tubes, because it is only at pH-values in excess of 10.5, i.e., ammonia quantities of about 150 mg/kg, that brass changes its polarity, becomes a lower grade material than iron, and dissolves. Nevertheless, experience has shown that a quantity as low as 5 mg NH_3/kg, however with the concurrent presence of oxygen, can lead to corrosion in the case of copper alloys. Initially the process leads to the development of complex, water-solutable salt,

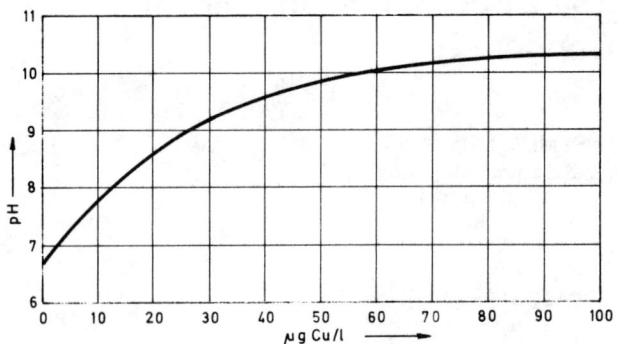

Fig.1. Brass dissolution vs pH curve

which decomposes in the boiler and is finally reduced to elementary copper. Copper may deposit on the surface of the steam generator steel tubes where it brings about - through local iron-boiler water-copper interaction - troublesome pitting. Formally, the mechanism of the reaction can be represented as follows:

$$2Cu + O_2 = 2CuO \tag{1}$$

$$CuO + 4NH_3 + H_2O \rightleftharpoons \left[Cu(NH_3)_4\right](OH)_2 \tag{2}$$

In the steam generator, equation (2) first runs from the right to the left, and the oxide is ultimately reduced to the pure elemnt.

The increase of oxygen concentration in the condensate above 0.02-0.05 mg/kg, which can be expected in normal operation, occured in both the studied cases because of the direct inflow of make-up water into the condenser. Increased quantities of make-up water - added because of the loss of part of the steam used in the process of a neighbouring plant - also increased the condensate oxygen content. Considering the low condensate ammonia content,

Fig.2. Charasteristic place of tube damage [9]

however, it was obvious that there occured locally higher ammonia concentrations in the condenser. This is borne out by the characteristic places of condenser tube damage. The tubes of the examined condensers are mainly damaged in the same spots, i.e., at the points where the tubes are fastened by partition plates (Fig.2.).

2. INCREASE OF CONDENSATE AMMONIA CONCENTRATION

Ammonia evaporates in high pressure steam generators, while excess hydrazine, which is not bound to oxygen, decomposes hydrothermally. Hydrazine - and its compounds - is frequently used in boiler plants as a chemical deaeration agent, i.e.,as a reducing agent for the complete removal of oxygen from feed water. The efficiency of oxygen-to-hydrazine binding, however, depends on a number of factors, e.g., pressure, temperature, concentration, etc. In order to obtain effective chemical deaeration, hydrazine is always added in quantities exceeding the actual requirement. The excess, which is not bound to oxygen, decomposes in the steam generator - at higher pressures and temperatures - into ammonia and nitrogen:

$$3N_2H_4 = 4NH_3 + N_2 \qquad (3)$$

The developed ammonia, and the ammonia used to increase feed water pH-value, flows through the turbine with the steam, and condenses in the condenser. These relatively low ammonia quantities in the condensate should not produce aggresive action on the brass condenser tubes. Nevertheless, as an equilibrium is established between the partial pressures of steam and ammonia vapors, abnormally high ammonia concentations occur in "dead" pockets of the condenser. This development can be explained by substantial increase of ammonia solubility through water temperature reduction.

Ammonia concentration in the condensate - at given parameters for the atmosfere in the condenser - can be calculated from the following:

$$m_a^f \cdot \delta_a = m_a^s \qquad (4)$$

$$m_a^s = 55.5 \frac{p_a}{p_w} \qquad (5)$$

where p_a : partial ammonia pressure
p_w : partial steam pressure
δ_a : distribution coefficient
m_a^f : ammonia concentration in the liquid
m_a^s : ammonia concentration in steam

Local pressure and temperature differences, and the very turbulent convective processes, preclude any calculations based on the observation of equilibrium states in the condenser. On the other hand, every condenser includes the so-called "dead pockets" beyond the reach of the steam flow, which can be considered as equilibrium systems. In such a dead pocket total pressure and temperature depend on pressure outside the pocket and on the temperature on the surface of condenser tubes. Hesselbrok [11] has studied the effect of subcooling, because it can be claimed withcertainity that the steam pressure on the tubes is lower than the total pressure upstream of the dead pocket. This pressure differential causes steam flow into the dead pocket, where the steam condenses and a slight part of ammonia is dissolved in the condensate. Partial ammonia pressure, therefore, increases continuously in the vapor phase, until - in the dead pocket $p_w + p_a = P$. Inflow stops at this point, and subsequent processes are based on diffusion. In this, partial ammonia pressure will depend on the degree of condensate subcooling on the condenser tube surface. Condenser design usually assumes the condensate to outflow at the temperature of steam saturation. As a matter of fact, the temperature of the condensate is lower by Δt_k. Subcooling occurs because the temperature of the water layer on the tube wall is equal to the temperature t_s of steam saturation, on its surface, and to the temperature t_w of the wall itself, next to the wall. The particles mix besause of outflow, and condensate temperature t_k is lower than saturation temperature by Δt_k. By using the Nusselt equation Kirschbaum [4] has calculated the average condensate temperature drop as follows:

$$\Delta t_k = 0.55(t_s - t_w) \tag{6}$$

For a steam saturation temperature in the condenser $t_s = 305.55$ K (which matches the pressure in the studied condensers), and wall temperature $t_w = 303$ K, the following value of condensate temperature drop is obtained:

$$\Delta t_k = 0.55(305.55 - 303) = 1.4 \text{ K}$$

Accordingly, condensate temperature is lower than saturation temperature by 1.4 K. This relates to film condensation, while the situa-

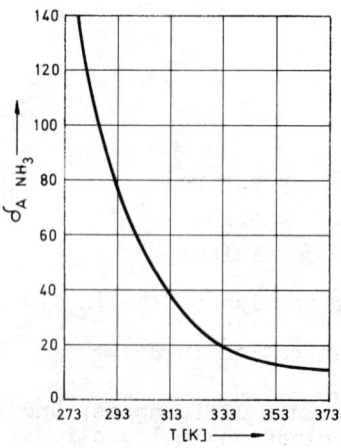

Fig.3. Ammonia distributin coefficient [5]

FAILURES OF THE TURBINES CONDENSER TUBES

tion - where ammonia dissolution is concerned - becomes even worse in cases of possible drip condensation. In our case, the total pressure in the condenser amounts to P=5 kPa, while condensate temperature on the cooling tubes is matched by p_w=4.62 kPa (for condensate temperature $t=t_s- t_k$=304.15 K). Ammonia pressure, then, amounts to:

$$p_a = P-p_w = 5-4.62 = 0.38 \text{ kPa}$$

This is the partial ammonia pressure after it has reached the steady state in the dead pocket of the condenser, and equation (5) can then be used to calculate ammonia concentration in the steam:

$$m_a^s = 55.5 \frac{0.38}{4.62} 17 = 77.6 \text{ g NH}_3/\text{kg}$$

On the basis of these values, and by using equation (4) and the value in Fig.3, the following ammonia concentration in the condensate is obtained:

$$m_a^f = \frac{77.6}{60} = 1.293 \text{ g/kg} = 1293 \text{ mg NH}_3/\text{kg of condensate}$$

Such enormous concentrations have actually been measured in practice [8,9] and, along with the presence of oxygen dissolved in the condensate, they fully explain the severe corrosion at specific points in the condenser.

3. POSSIBLE SOLUTIONS OF THE TUBE DAMAGE PROBLEM

Because higher ammonia concentrations do not cause brass tube damage unless oxygen is present at the same time, the possibility of ammonia concentration should be reduced to the least possible extent, while preventing oxygen introduction into condensate. Design modifications can help to reduce the influence of "dead" pockets and avoid places of enormously high ammonia concentrations. Condenser tubes should be fastened to the partition plates so that the tubes do not touch the plate along the entire circumference: this would permit continuous tube flushing in such spots, and reduce the likehood of occurence of high NH_3 concentrations [11]. A complete elimination of spots with major steam flow disturbances during condensation, however, is not feasible. Accordingly, make-up water with a high oxygen content should be fed by providing for oxygen removal in the condenser itself [8]. Vacuum ejectors would remove oxygen from the condenser, and O_2 concentration in the condensate would increase minimally.

3.1 Deaeration in the Condenser

In the studied condensers, make-up water is fed into the condenser by an ordinary tube. As the water flows in a jet, and in spite of its mixing with steam, the effect of water heating to the saturation state - the state where gases are no longer water soluble - is low, and the oxygen remains dissolved in make-up water which mixes with the condensate.

Our suggestion [8] was to feed the make-up water directly into the condenser by a system of nozzles, which would achieve, through water dispersion and mixing with steam, the effect of thermal deaeration. The deaeration process itself would involve two stages:

a) net material convection of gas (air) from the fluid in which it is dissolved (make-up water) to the fluid into which it is to be deaerated;

b) diffusion of the gaseous part from the fluid in which it is dissolved to the fluid into which it is to be deaerated along the phase interface.

The first reaction is naturally faster and, in adequate conditions, 90-95% of the dissolved gas can be removed by convection.

Bunsen equation [3], derived from Henry-Dalton's law on the proportionality of gas concentration in a solution and partial pressure in the gas phase, can be used to calculate the concentration of a gas in a dissolving fluid for given conditions:

$$c = \frac{\alpha_i}{\rho \cdot 22.4} p_i \tag{7}$$

where α_i : absorption coefficient
ρ : dissolving fluid density
c : concentration
p_i : partial gas pressure

Because liquid pressure p_S includes partial steam pressure p_W and partial gas pressure p_i, deaeration has to occur at boiling point

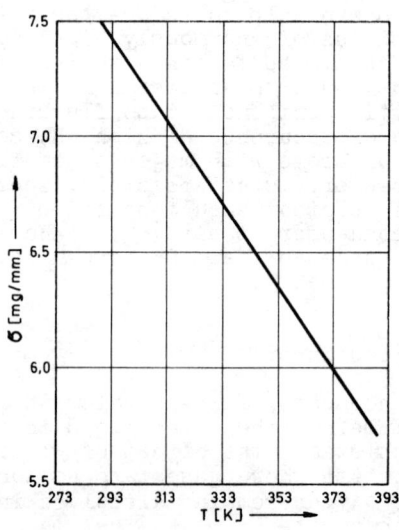

Fig.4. Surface tension of water against steam [1]

where $p_s=p_w$, i.e., the concentration of leftover gas should be zero. In the system of thermal deaeration, where liquid droplets are mixed with steam for the purpose of heating, we acutaly have $p_s-p_w=0$ on the surface of the droplet, and according to equation (7), because of $p_i=0$ there is no gas left in the solution. However, within the droplet, at the same temperature, $p_s-p_w>0$, and leaving gas can only be released by diffusion.

On the rounded phase interfaces, because of surface tension, partial pressure changes in the gas space. A thermodynamic equation for such partial pressure changes was developed by Thomson [1]; it covers the pressure increase within the droplet, due to surface tension:

$$\Delta p = \frac{2\sigma}{r} \qquad (8)$$

where σ : surface tension of the liquid
  r : radius of curvature of the spherical globule surface

For a droplet entering saturated steam, accordingly, pressure p_s within the droplet (water) is equal to saturation pressure plus Δp. Because saturation pressure is equal to p_w, equations (7) and (8) give:

$$c = \frac{2\cdot\sigma\cdot\alpha_i}{r\cdot\rho\cdot 22.4} \qquad (9)$$

In practical application, i.e., for cases of leftover oxygen content, the equation assumes the following form:

$$c_{O_2} = \frac{0.285\cdot\alpha_{O_2}}{r} \qquad mg\ O_2/kg \qquad (10)$$

This, the remaining share of gas can only be removed by diffusion. The rate of diffusion increases with the increase of contact area and of the concentration differential. The quantity of gas which crosses the interface of two phases in a unit of time can be calculated by Fick's law:

$$\frac{dQ_i}{dT} = F\cdot\alpha_i^g\cdot\Delta m_i^f \qquad (11)$$

where T : time
  α_i^g : diffusion coefficient
  Q_i : quantity of diffused particles
  m_i^f : concentration differential
  F : contact area

The dispersion of water into droplets for the purpose of enhanced deaeration is significant because it increases the specific contact area. Because the droplets fall freely in the steam space, or are even accelerated by pressure (dispersion by nozzles), the relative droplet-steam motion is increased, which has a favorable effect on gas transfer. Tietz has established [1] by measurement that the diffusion coeficient depends directly on the velocity of droplet movement. A safe assessment of the effect of deaeration by diffusion with liquid dispersion requires data on droplet size, their velocity and time of retention in the steam apace, i.e., data which are rather difficult to come by. The actual case, however, regards only deaeration by diffusion, and the conclusions do not apply to convective deaeration which, if the liquid is well-dispersed, is excelent and as a rule results in leftover oxygen quantities of the order of 0.1 mg/kg. Deaeration by diffusion, if the liquid is dispersed, plays a less important role throughout the process - diffusion is only responsible for subsequent deaeration. What is important is the convective, droplike deaeration through the development of gas bubbles, which occurs in the droplets because of good heat transfer. It is obvious, moreover, that thermal deaeration alone can never reduce the content of water-dissolved oxygen to zero.

4. CONCLUSION

The suggested solution can substantially reduce condenser damage. This implies - where make-up water is fed directly into the condenser - design modifications which would reduce the likelihood of ammonia concentration in specific spots, and provide a satisfactory deaeration of make-up water, thereby substantially decreasing the oxygen content in the condensate.

In both the studied cases - 220 MW and 125 MW power plants - the condensation pressure was 5 kPa (305.55 K), and the average oxygen content in make-up water about 10 mg O_2/kg, because of which the condensate contained up to 0.5 mg O_2/kg [8,9]. Equation (10) can be used to calculate the oxygen content left over in make-up water when deaeration is carried out in the condenser, i.e., if adequate dispersion nozzles are built into the water supply system:

$$c_{O_2} = \frac{0.285 \cdot 7.2 \cdot 0.0235}{0.2 \cdot 1} = 0.26 \text{ mg } O_2/\text{kg}$$

where σ : 7.2 from Fig.4
 ρ : 1 kg/l for water
 α : 0.0235 according to [1]
 r : 0.2 mm for average dispersed droplet radius

This water mixes with the condensate which, as a rule, has a much lower oxygen concentration - less than 0.05 mg/kg - so that feed water contains a mean oxygen concentration, depending on the respective feed water and condensate shares. Nevertheless, these concentrations are within permissible limits.

REFERENCES

1. E.Hömig. Phisikochemische Grundlagen der Speisewasserchemie. 2. Auflage, Vulkan-Verlag, 1963, pp 119-120, 299-320.

2. A. Splittgerber. Wasseraufbereitung im Dampfkraftbetrieb. Springer Verlag 1963, pp 168-188.

3. M. Salinger. Kraftwerkschemie. VEB Verlag 1964, pp 163-207

4. Z. Randt. Isparivanje i uparivanje. Tehnička knjiga, Zagreb 1965, pp 87-91.

5. G. Bonsack. Spezielle Korosivität aufgrund Verteilungsverheltnise im Zweiphasengebiet bei der Kondensation. VGB Kraftwerkstechnik 57/1978, pp 424-430

6. G. Resch-K. Zinke. Zuzammenhang zwichen Wasserqualität, Konstruktion und Korrosion in wasserberührten Anlagen. VGB Kraftwerkstechnik 57/1977, pp 373-378.

7. G. Bohnsak. Die Bedeutung des Sauerstoffs für die Wasser-Dampfkreislauf. VGB Kraftwerkstechnik 56/1976, pp 50-54.

8. I. Lončar, M. Mustapć, N. Ružinski. Izvještaj o ispitivanjima korozionih pojava u niskotemperaturnom dijelu bloka I TE-Sisak. FSB, Zagreb 1979.

9. I. Lončar, M. Mustapić, N. Ružinski. Studija o užrocima oštećenja kondenzatorskih cijevi termoenergetskog bloka TE-Plomin. FSB, Zagreb 1979.

10. N. Ružinski. Postupci termičke pripreme napojne vode - s posebnim osvrtom na otplinjavanje raspršenjem. Udruženje za tehnologiju vode, Beograd 1980.

11. H. Hesselbrock. Dampfseitige Korrosionschäden an Kondensatorrohren. VGB Krafrwerkstechnik 79/1962, pp 283-290.

12. P.H. Effertz, W. Fichte, P. Forschammer. Kühlwasserseitige Korrosion an Kondensatoren und Kühlern in Kraftwerken. Der Maschinenschaden 46/1973, pp 189-200.

13. Popplewel. Corrosion Performance of some Copper Alloy Condenser Tube Materials in Ammonical Condensate. Corrosion Vol 30/1974.

14. K. Eichhorn. Kondensatorrohre aus Kupferwerkstoffen. Werkstoffe und Korrosion, Heft 7/1970, pp 535-553.

15. R. Svoboda, G. Ziffermayer, H. Schmied. Verteilung von Konditionierungsmitteln und Verunruigungen im Dampf-Wasserkreislauf von Sattdampf-Turbinenanlagen. VGB Kraftwerkstechnik 2/1979, pp 150-157.

Direct-Contact Heat Exchangers for Large Steam Turbine Installations

G.I. EFIMOCHKIN
Cand.Sc.(Tech.)
All Union Heat Engineering Institute
Moscow, USSR

ABSTRACT

The results of investigations of the extreme conditions for industrial direct-contact heaters with nonramming spray-type water distribution at the 300 MW turbines are presented. Empirical formulas for estimation of the permissible limiting steam flow velocity in the lower stage of the heater and for determining the value of the relative subcooling with the rise of the steam air are given.

INTRODUCTION

In the USSR direct-contact heaters have become recently in wider use in the large steam turbine cycles. Heat transfer between water and steam is accomplished there by direct contact thereof. Elimination of the tube system makes such heat exchangers less expensive as compared with surface heat exchangers, and the water in them can practically be heated up to the heating steam saturation temperature.

In surface heat exchangers, the presence of the tube wall thermal friction and other factors affecting heat transfer, result in unavoidable subcooling of water, which for industrial low-pressure heaters of modern turbines is at the level of 5-7°C, reaching in some cases even higher values. Prevention of subcooling and of other heat losses, when two out of four 300 MW turbine surface low-pressure heaters at the USSR thermal power plants are substituted by direct-contact heaters, makes a gain in specific heat consumption of 0.3-0.5% /1/.

Of primary importance for normal operation of a direct-contact heater is proper selection of the design scheme. Water in the direct-contact heater steam space is either sprayed or admitted through a distribution system. In the former case, a head created by the pump is used, while in the latter one the water is distributed through perforated horizontal edge-type trays. Uniform free-stream flow of the heating steam contacting the falling water sprays, droplets or films must be ensured and adequate arrangements of vents for the air-steam mixture must preclude the origination of dead zones (air pockets).

The USSR 300 MW turbine installations employ both types of heaters. The spray-water distribution is utilized in some transfer pump based design schemes. Good operational behavior has been shown by a direct-contact heat exchanger with spray water film distribution and parallel-counter flow heat exchange. Besides, spray-type water distribution heaters have been field tested to show that the heating stage can in this case be located within the steam pipe.

With the cascade gravitational direct-contact heaters located at different levels where water gravitates from the high heater to the low one, it is a practice in the USSR to employ heaters with nonramming spray water distribution via perforated trays. Fig.1 shows the design scheme of two-stage horizontal direct-contact heaters using gravitation principle.

The heating steam to the heater is admitted into the lower part of the steam stage and moves as the cross-counter flow in respect to the falling water.

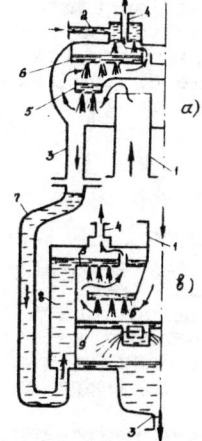

Fig.1. Design scheme of direct-contact low-pressure heater No.1(a) and No.2(b) with nonramming spray water distribution.
1 - heating steam input from 300 MW turbine bleed; 2 - full-flow condensate input to the l.p. heater No.1; 3 - heated condensate output; 4 - flash steam output from the heaters; 5 - bottom tray; 6 - upper tray; 7 - water distribution pipe; 8 - side hydroseal; 9 - partition.

The tests of such a heater installed at the 300 MW turbine have shown that the water heats mostly in the upper stage, where about 95% of the steam fed to heater is condensed /2/. The experimental data on the integral water heating in the spray stage are found to be in good agreement with the calculated values. The calculation was performed according to the formula /1/, which

HEAT EXCHANGERS FOR LARGE STEAM TURBINE INSTALLATIONS

is valid for the cross flow of the pure steam around the water sprays, 1 kg air-steam mixture flow per ton of the heated water and the following parameters:

$$\lg \frac{t_s - t_1}{t_s - t_2} = 0.053 \frac{1}{Pr^{0.62}} \sqrt[3]{\left(\frac{w''}{w'}\right)^2 \cdot \frac{\rho''}{\sigma' \cdot d}} \quad (1)$$

where $P'' = 1 - 130$ kPa : the heater pressure variation,

$W' = 0.8 - 1.7$ m/s : the initial water velocity in sprays,

$d = 2 - 15$ mm : the spray diameter,

$l = 0.2 - 0.7$ m : the spray length,

$\rho''\cdot(w'')^2 = 4 - 60$ kg/m.s^2 : the steam flow dynamic head,

t_s : heating steam saturation temperature at a pressure equal that before the sprays,

t_1 : the temperature of the water in the space before the sprays,

t_2 : the temperature of the water in the space past the sprays,

w'' : average steam velocity in sprays,

σ' : surface tension factor,

Pr : Prandtl criterion for water at the saturation line.

The higher the free-stream flow velocity of the steam contacting water sprays the more effective the heat exchange and the lower the cross section and, hence, the heater dimensions. However, with the increased velocity the deviation of the sprays from the vertical also increases, which is expressed by the following empirical formula:

$$\alpha = 0.625 \cdot \rho'' \cdot (w'')^2, \, °C \quad (2)$$

The heaters of two or of more stages feature a definite limiting free-stream flow velocity value of steam contacting water sprays in the lower stage, when the so-called "extreme regime" occurs. This is characterized by hydraulic shocks and noise inside the heater and sharp fluctuations of the water level on the tray as well as by increase of steam wetness in the air-steam mixture vented from the heater. Operation of the heater under such conditions is intolerable, because it is accompanied by sudden deterioration of the operational behavior and may result in destruction of the heater internals and depressurization of the shell.

The main cause of hydrodynamic instability with heaters employing spray nonramming water distribution is the breaking of water sprays by the steam flow and periodic ingress of the thus-formed droplets to the upper stage of the heater. In the points where the steam droplet flow turns, separation occurs and the mass of the hanging-up water is increased until its suddenly dropping down. When the water clogs the steam passages between the trays, the water level rises and the water may overflow the tray edge, which increases the mass of the hanging-up water.

The extreme regime arises readily when the steam entrains a certain amount of the water from the falling sprays. Therefore, much attention should be paid to control the uniform velocity of the steam moving across the falling sprays. The higher the non-uniformity, the lower the average velocity where such conditions are established. The nonuniformity of the steam flow velocity profile depends on the heater geometry, relative area of passages in the trays (i.e. the mouth) as well as the degree of steam condensation on the sprays in the heater lower stage.

The complex nature of the process and a great number of interdependent factors prevent calculating the flow rate of the supplied water and its heating where extreme regime in the heater occurs.

Presently, these parameters can be judged by the average value of the limiting permissible velocity which is connected therewith and can be obtained experimentally.

Since the increased velocity nonuniformity tends to decrease the average value of the limiting velocity, it should be minimized during heater designing.

The investigations have shown that to eliminate the effect of nonuniformity on the value of limiting velocity the water spray length in the heater and the mouth area with respect to the heater cross section should at least be 600 mm and 15-17%, respectively.

To obtain the experimental values of the average spray limiting velocity (and, with the low flow rate of the steam condensing in the heater stage, of the free-stream flow velocity of steam contacting the first sprays) the industrial direct-contact heaters have been tested for hydrodynamic stability. In this case, the extreme conditions were achieved by increasing the full-flow condensate rate at the heater inlet. Their occurrences were observed by hydraulic shocks.

These experiments have shown that with the pressure below the atmospheric the limiting values of the average free-stream flow velocity of steam and of the dynamic head increase sharply with lower vacuum, while the limiting value of the steam mass flow rate is kept practically constant (Fig.2).

HEAT EXCHANGERS FOR LARGE STEAM TURBINE INSTALLATIONS

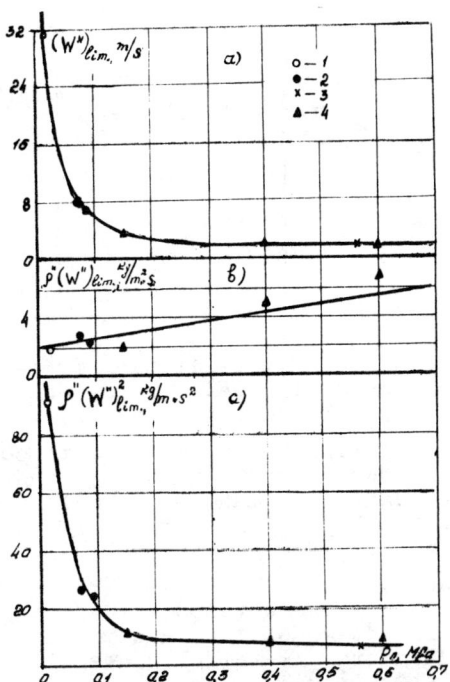

Fig.2. Variations of the limiting steam flow velocity before the sprays (a), the limiting mass flow rate of steam (b) and the limiting dynamic head of steam (c) with the pressure in the heater.

1 - tests of the 300-240 ХТГЗ turbine l.p. heater No.1; 2 - tests of the K-300-240 ЛМЗ turbine l.p. heater No.2; 3 - ДСП-400 deaerator tests; 4 - bench tests conducted at VTI.

To determine the limiting average steam flow velocity for sprays the following empirical correlation is suggested. This velocity happened to be calculated basically by the pressure magnitude in the heater (Fig.3):

$$(W'')_{lim.} = 5 \cdot 10^6 \cdot t_s^{-3}, \text{ m/s} \qquad (3)$$

where t_s is saturation temperature at the heater pressure, °C.

With the origination of the extreme regime during operation, the flow rate and heating of the water fed to the heater should be decreased first. Besides, other measures can also be of use to reduce the average free-stream flow velocity of steam by increasing the flow section of the heater lower stage or by increasing the flow rate of the condensing steam there. This can be obtained by providing higher edges of the low tray or by-passing the cold condensate through the upper stage to the lower one by the overflow pipes. However, it should be kept in mind

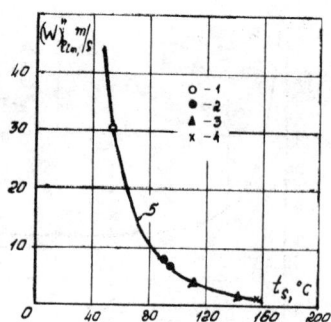

Fig.3. The dependence of the limiting free-stream flow velocity of steam contacting water sprays on saturation temperature with the heater under pressure.

1 - direct-contact l.p. heater No.1 tests conducted at the Troitzk thermal power plant; 2 - direct-contact l.p. heater No.2 tests conducted at the Karmanovo thermal power plant; 3 - bench tests conducted at VTI; 4 - deaerator ДСП-400 tests; 5 - according to the formula $(W'')_{lim} = 5 \cdot 10^6 \cdot t_s^{-3}$.

that the decreased average velocity of the steam through the sprays in the low stage of the heater can lead to a more subcooled heater condensate. Parallel connection of steam sprays is rather effective here.

As to the effect of the air on heat transfer in direct-contact heaters, it should be noted that with normal operation of a steam turbine installation its content in the turbine bleed is low and therefore this effect may be neglected. When the content of air or other insoluble gases in the heating steam due to certain causes is found to be in excess of 0.2-0.3%, the resultant water subcooling can be calculated from the following empririral correlation developed for this purpose, summarizing the results obtained for industrial heaters (Fig.4a):

$$\delta t = 7.7 \cdot \varepsilon_{air}^{1.45} \qquad (4a)$$

or, in relative quantities, as follows (Fig.4b):

$$\frac{\delta t}{\Delta t_o} = 0.355 \cdot \varepsilon_{air}^{1.45} \qquad (4b)$$

where $\delta t = t_s - t_2$: subcooling of the water to the heating steam saturation temperature,

$\Delta t_o = t_2 - t_1$: the heating of water in the heater stage in the absence of air in the heating steam,

$\varepsilon_{air} = \dfrac{G_{air}}{G'' + G_{air}} \cdot 100\%$: the relative content of air in the air-steam mixture.

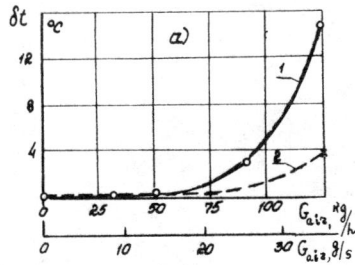 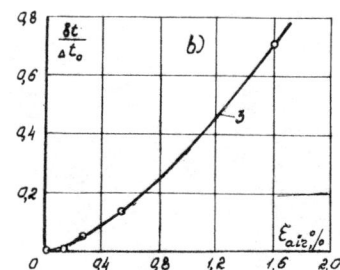

Fig. 4. Absolute (a) and relative (b) subcooling curves for the full-flow condensate in the K-300-240 ЛМЗ turbine direct-contact l.p. two-stage spray heater No.1.
G' = 170 kg/s; P" = 0.015 MPa, Δt_o = 21°C.

1 - according to formula (4.a); 2 - according to formula (1) with the factor $(1 - \frac{\mathcal{E}_{air}}{100})^7$; 3 - according to the formula (4.b.).

The comparative tests to determine the effects of air on the operational behaviour of 300 MW turbine direct-contact and surface heaters have shown different results. While with 0.4% air content in the heating steam the subcooling in direct-contact heaters increases by 1.5-2.0°C only, the surface heaters under such conditions fail to operate.

The proposed correlations are used for the prediction of spray heaters with nonramming spray water distribution.

REFERENCES

1. Technical instructions. PTM 108-038.01-76, Calculation and design of direct-contact low-pressure heaters and their connection. Official publication, Leningrad, 1976.

2. Efimochkin, G.I., Verbitzky, V.L., Shipilev, S.G. 1975. Static tests of 300 MW turbine regenerative system with direct-contact low-pressure heaters. Electrical stations, No.11, pp.36-38.

Calculation of Heat Transfer in Power Units with a Complex System of Boundary Surfaces

**O.G. MARTYNENKO, O.V. DIKHTIEVSKY,
and N.V. PAVLYUKEVICH**
Heat and Mass Transfer Institute
BSSR Academy of Sciences
Minsk, USSR

ABSTRACT

　　The paper deals with a method of calculation of heat exchangers represented by a class of elongated cylindrical structures with longitudinal, circular and asymmetrically situated cooling channels. Under sufficiently general conditions, this method allows reduction of a three-dimensional steady heat transfer problem to solution of a number of coupled plane problems. Calculation of a system of plane heat conduction equations in conjunction with the equations of coolant motion makes it possible to select appropriate resistances in order to considerably equalize the temperature field over the heater cross-section, or, which is equivalent, to decrease to some extent the total coolant flow rate.

1.　INTRODUCTION

　　Solution of three-dimensional heat transfer problems in multi-coupled regions with a large coupling number usually presents some specific difficulties. These difficulties originate from a limited computer memory and rapid realization of these problems by computers, which hampers, and sometimes makes impossible direct application of the generally known numerical, semi-analytical and analytical methods, the more so that in the general case conjugation of solutions at the boundary between the coolant and the heater is required.

　　The most widely used methods of solution of such problems are the following:

　　a) method of finite elements [1,2];
　　b) difference methods [3,4];
　　c) variational methods [5,6];
　　d) methods based on the Green equations [7].

　　Besides, auxiliary methods prove very useful, such as the alternating Schwarz method [8], which allows one to divide the region within which the solution is being sought into a number of subregions intercrossing in pairs, and to solve the initial problem successively in each region. The procedure goes on until convergence has been achieved, taking the solutions obtained in the neighbouring regions crossing the given one as the boundary condition.

Another auxiliary method is that of expansion given in /9/.

When solving unsteady problems, the Duhamel theorem may be used, which makes it possible to get rid of the temporal dependence in the boundary conditions and thus to simplify the problem.

The methods based on using the Green formulae /7/ have certain advantages, especially when the heat conduction equation solution must be found not for the entire region but at some of its points. The basic Green formula for elliptic-type equations is, as is known, of the form:

$$\Omega \cdot T(M) = \iint_\Sigma \left[\frac{1}{R_{MP}} \frac{\partial T}{\partial n_p} - T(P) \frac{\partial}{\partial n} \left(\frac{1}{R_{MP}} \right) \right] d\Sigma_p - \iiint_V \frac{\frac{\partial^2 T}{\partial x^2} + \frac{\partial^2 T}{\partial y^2} + \frac{\partial^2 T}{\partial z^2}}{R_{MP}} dV_p$$

where
$$\Omega = \begin{cases} 4\pi & \text{for the points M inside the region V} \\ 2\pi & \text{for the points M on the surface } \Sigma \\ 0 & \text{for the points M outside the region V} \end{cases}$$

Here, Σ, is the continuously differentiated surface; T, temperature; R_{MP}, distance between the points M and P.

Since in the boundary-value heat conduction problem either temperature or its derivative is assigned at the boundary point, while the presented formula yields the solution only with both values being known, an additional problem arises, consisting in determination of the temperature on the boundary with its derivative being assigned, or vice versa, in determination of the heat flux from the assigned boundary temperature based on the same formula. One of the methods of solution is to divide the surface Σ into a finite number of elementary platforms and to substitute sums for the integrals in the Green formula. In so doing, the values of both heat fluxes and surface and ambient temperatures together with heat transfer coefficients and the like are assumed constant at every division platform. This procedure leads to a finite system of algebraic equations with respect to lumped-constant temperatures or fluxes on elementary platforms. Solution of this system yields numerical solution of the initial problem at any point inside the region. Naturally, during computer realization of this method, an increase in the region complexity leads to an increase in the computer time and the memory volume.

Among the most widely used methods for solution of the heat transfer problems under consideration are the differential ones $\sqrt{3}\,\overline{/}$. The provide for reduction of solution of differential equations in specific derivatives to solution of algebraic equations whose order is determined by the number of the region division units. It is known that a matrix of differential equations is characterized by its special form, i.e. the presence of a large number of zero elements and its large orders, which, naturally, increase with an increase of the region complexity due to the necessity to decrease in this case the division step.

Variational methods are very useful, and deserve more attention than is usually paid to them. They are based on the problem of determination of the functional which is related to the initial differential problem. The main types of variational methods are: direct methods, counter-methods, projection methods. They allow reduction of the initial problem to solution of a system of either common differential equations or of linear algebraic equations. The order of the latter is determined by the number of coordinate functions and increases with the system complexity. Among these the structural method should be noted, which is effective only for relatively simple regions with continuously differentiated boundary. This method amounts to coordinate functions being written in the form which satisfies beforehand all boundary conditions (those of the fourth kind included). This is done on the basis of construction, according to certain rules, of a boundary function which becomes zero on the boundary itself and exceeds zero in the region V. The form of the coordinate functions being compiled becomes sharply more complex with an increase in the region coupling, thus imposing restrictions on the method applicability. A characteristic feature of variational methods should be noted. An analytical solution of the problem (if such exists) taken as a set of coordinate functions may replace the entire set. But if there is any information available apriori on the problem solution, it may be used in some way in the set of coordinate functions thus reducing their number and simplifying the solution.

In recent years the method of finite elements (MFE) became most widely used in computational mathematics $\sqrt{1,2,3}\,\overline{/}$. It is noteworthy that the method is closely related both to variational and to differential ones. Thus, in Ref.3 it is interpreted as a variety of variationa-differential methods. The same paper presents an example of its application for derivation of a differential scheme, emphasizing its being only one of many methods (nonstandard ones, as may be) of obtaining differential schemes. On the other hand, the MFE is often considered as a specific case of the Ritz-Galerkin method. According to the MFE, coordinate functions are selected in the form of splines, i.e. finite functions; matrices of equation systems are, just as in differential methods, of a band character, while Ritz matrices are completely filled and symmetric.

Thus, for calculation of heat transfer problems with complex systems of boundary surfaces the most appropriate methods are the differential, variational and MFE, which are roughly equivalent in the general case, although the priority should be given to the MFE. The situation is changed when some information on the solution is available apriori. In this case the MFE and variational methods are preferable, for the information available may be used both for construction of a network for MFE by its stretching, for example, in the places of the solution behaviour as a slowly changing function and thus decreasing the number of unit points, and for a set of basis functions for the variational method, which can considerably reduce the order of the set.

2. SOLUTION OF THREE-DIMENSIONAL HEAT TRANSFER PROBLEMS IN ELONGATED CYLINDRICAL STRUCTURES

Let us discuss the method of reduction of solution of three-dimensional steady heat transfer problems in elongated cylindrical bodies with longitudinal cooling channels to a series of coupled plane problems and solution of the latter by the Ritz method.

Heat conduction equation in the region V is given by

$$- \text{div}(\lambda \, \text{grad} \, T(x,y,z)) = F(x,y,z) \qquad (1)$$

with the boundary conditions on the longitudinal surfaces Σ_i:

$$\lambda \frac{\partial T}{\partial n} = A_i(x,y,z) \Big|_{\Sigma_i} \quad i = 1,2,\ldots,N$$
$$\lambda \frac{\partial T}{\partial n} = \alpha_i(\theta_i - T) \Big|_{\Sigma_i} \quad i = N+1, N+2, \ldots, M \qquad (2)$$

At the ends, the conditions of the type (2) or of the first kind may be assigned:

$$T = B_i(x,y,z) \Big|_{\Sigma_i} \quad i = M+1, M+2 \qquad (3)$$

Energy balance equations in channels necessary to determine the coolant temperature and averaged over the channels cross-sections are given by:

$$\dot{M}_i \frac{d}{dz}(c_p \theta_i) = \int_{S_i} q_i dS_i \qquad (4)$$

$$\theta_i(0) = \theta_{io} \qquad i = N+1, N+2, \ldots, M$$

The RHS of Eq.4 is the heat flux taken from the cross-section S_i of the surface Σ_i of the cooling channel and is related to the second boundary condition (2) by the following equation:

$$\int_{S_i} q_i dS_i = -\int_{S_i} \lambda \frac{\partial T}{\partial n} dS_i = S_i \alpha_i (\bar{T}_i(z) - \theta_i(z)) \qquad (5)$$

$$i = N+1, \ldots, M$$

Here, λ is the heat conduction coefficient; n, external normal, α_i, heat transfer coefficients; θ_i, coolant temperature in the i-th channel; $\dot{M}_i$, coolant mass flow rate; $\bar{T}$, solid body temperature averaged over the cross-section perimeter.

Further, a sufficiently general assumption is introduced, that the internal heat source, $F(x,y,z)$, and the assigned fluxes, $A_i(x,y,z)$ are representable in the form of series expansions over the longitudinal coordinate $(z-z_0)$ of the powers m and p, respectively. For the sake of definitness, let $m > p$. Then F and A_i can be written as

$$F(x,y,z) = \sum_{k=0}^{m} \varphi_k(x,y,z_0)(z-z_0)^k \qquad (6)$$

$$A_i(x,y,z) = \sum_{k=0}^{p} a_{ik}(x,y,z_0)(z-z_0)^k \qquad (7)$$

The expansions (6),(7) are conducted, generally speaking, in the region V_0, whose characteristic dimension, L_{zo}, is smaller than the characteristic longitudinal dimension of the region L_z (such L_{zo} is taken so that the internal source F can be well polynomial-approximated), but larger than the characteristic transverse dimension of the region L_{xy}, which, specifically, allows one to neglect circular effects. Thus, $L_{xy} \ll L_{zo} \ll L_z$. Taking into account the above, we can now represent temperatures of the coolant, θ_i, in the channels, and temperature of the structure, T, outside the immediate vicinity of the ends in the following form / 12 /:

$$T(x,y,z) = \sum_{k=0}^{m+1} t_k(x,y,z_0)(z-z_0)^k \qquad (8)$$

$$\theta_i(z) = \sum_{k=0}^{m+1} c_{ik}(z-z_0)^k \qquad (9)$$

The problem is solved under the assumption of a constant heat conduction coefficient, λ. The substitution of expansions (6),(7),(8),(9) into Eq.1 and boundary conditions (2) and separation of the terms with equal degrees, $(z-z_0)$, followed by their setting to zero yields a series of plane problems of the following form:

heat conduction equations:

$$-\lambda\left(\frac{\partial^2}{\partial x^2} + \frac{\partial^2}{\partial y^2}\right) t_k(x,y,z_0) = \ell_k(x,y,z_0) + (k+2)(k+1)t_{k+2}(x,y,z_0)$$
$$(x,y) \in D \qquad (10)$$

boundary conditions on the cooling channels perimeters:

$$\lambda \frac{\partial t_k(x,y,z_0)}{\partial n} = a_{ik}(x,y,z_0) \Big|_{S_i} \quad i = 1,2,\ldots,N \qquad (11)$$

$$\lambda \frac{\partial t_k(x,y,z_0)}{\partial n} = \alpha_i(c_{ik} - t_k) \Big|_{S_i} \quad i = N+1, N+2,\ldots,M \qquad (12)$$

heat transfer equations in the coolant, which are also the conjugation conditions:

$$c_p \dot{M}_i(k+1)c_{ik+1} = S_i \alpha_i \left(\frac{\int_{S_i} t_k(x,y,z_0)dS_i}{S_i} - c_{ik}\right) \qquad (13)$$

$$i = N+1, N+2,\ldots M$$
$$k = 0,1,2,\ldots m+1$$

with the boundary condition

$$\theta_i(0) = \sum_{k=0}^{m+1} c_{ik}(-z_0)^k \qquad (14)$$

Note that due to the above assumption of the finite expansion order of the internal source (6) of the heat conduction equation (1), the values of expansion coefficients φ_k in (10) for $k > m$ will be zeroes and the plane equations will be of the form:

$$-\lambda\left(\frac{\partial^2}{\partial x^2} + \frac{\partial^2}{\partial y^2}\right) t_m(x,y,z_o) = \varphi_m(x,y,z_o) \qquad (15)$$

$$-\lambda\left(\frac{\partial^2}{\partial x^2} + \frac{\partial^2}{\partial y^2}\right) t_k(x,y,z_o) = 0$$

for $k \geq m+1$

This makes it possible to confine ourselves to the $(m+1)$th expansion degree of the structure and coolant temperatures (8),(9).

Heat transfer coefficients, α_i, included into heat conduction equations (13) and boundary conditions (12) can be chosen in the form of this or that dependence on Re and Pr numbers. A sufficiently wide choice of such semi-empirical relationships can be found in Ref.11.

It is easily seen that the heat transfer equation in the coolant (13) represents a recurrent formula for the coefficients c_{ik} and can be written in a more convenient form, namely:

$$c_{ik+1} = \alpha_i \frac{\int_{S_i}(t_k(x,y,z_o) - c_{ik})dS_i}{kc_p m_i} \qquad (16)$$

$$i = N+1, N+2, \ldots, M$$

In the case when the internal source, $F(x,y,z)$, and the functions $A_i(x,y,z)$ are well approximated by the following successions

$$F(x,y,z) = \sum_i F_i(x,y) \Psi_i(z) \qquad (17)$$

$$A_i(x,y,z) = \sum_j A_{ij}(x,y) \mu_j(z) \qquad (18)$$

the solution for each z can be represented in the form of a linear combination of a fixed number of sets, $\{t_k, c_{ik}\}$, provided that the heat conduction and heat transfer coefficients are constant. This follows directly from the form of equations (10)-(13) and from the fact that in this case the expansion formulae (8),(9) and equations (10)-(13) have all coefficients, φ_k, and α_{ik}, as linear combinations of the coefficients of expansions (17)-(18), $F_i(x,y)$ and $A_{ij}(x,y)$, independently of z_o.

Turning back to solution of the initial three-dimensional problem reduced to a series of coupled plane ones (10)-(14), we may note that solution of the latter should begin with the last $(m+1)$th problem with a zero RHS (15). Expansion coefficient of the coolant temperature, c_{im+1}, in this case is determined from the recurrent relationship (16), allowing for $c_{im+2} = 0$:

$$c_{im+1} = \int_{S_i} t_{m+1} dS_i / S_i \qquad i = N+1,\ldots,M \qquad (19)$$

Solution of the problem (10)-(12) for k=m+1 under the assumption of closeness of channel surfaces temperatures with their definite perimeter values is then given by the indefinite constant:

$$c_{im+1} = \bar{t}_{m+1} = \text{const}$$

This constant will be determined when solving the next m-th problem (10)-(13), for it already requires fulfilment of an additional integral condition for the known values. This condition consists in obeying the conservation law imposed on the heat fluxes and source, when the boundary conditions are determined only by conditions of the second kind. Indeed, the condition (12) allowing for (19) and (13) at k=m assumes the form

$$\lambda \frac{\partial t_m}{\partial n} = \alpha_i \left(\frac{\int_{S_i} t_m dS_i - t_m S_i}{S_i} - \frac{c_p \dot{M}_i (m+1) c_{im+1}}{\alpha_i S_i} \right) \qquad (20)$$

$$i = N+1, N+2,\ldots,M$$

The RHS expression will be a constant value if we assume $\int t_m dS_i = t_m S_i$. Thus, plane problems take on the form of Neumann problems which require that the above conservation law be observed.

During successive solution of these plane equations from the m-th to the (m-1)th and so on, the difference between them is only in the RHS of the equations and the corresponding boundary conditions of the third kind, which makes it possible to compile a standard program for solution of such problems.

At the end of a successive transition to the zero problem (k=0) during solution of the latter an indefinite constant appears, which can no longer be determined using the above technique. Its determination requires satisfaction of some agreement condition, for example, we can assign the coolant temperature at the inlet, thus requiring fulfilment of the condition (14).

3. SOLUTION OF PLANE HEAT TRANSFER PROBLEMS BY THE RITZ VARIATIONAL METHOD

The procedure discussed in the previous sections makes possible the transition from a three-dimensional problem solution to

CALCULATION OF HEAT TRANSFER IN POWER UNITS

solution of a series of plane problems.

In the case when longitudinal heat overflows ($L_z \gg L_{xy}$) can be neglected, the form of the system (10)-(14) is simplified due to disappearance of the second term in Eq.10 and realization of connection between plane problems only via boundary conditions, i.e. via relationships (12),(13).

Heat conduction equation then assumes the form (let $z_0 = 0$ for simplicity):

$$-\lambda \left(\frac{\partial^2}{\partial x^2} + \frac{\partial^2}{\partial y^2} \right) t_k(x,y) = \varphi_k(x,y), \quad (x,y) \in D \qquad (21)$$

the boundary conditions of the third kind in a general form being written as

$$\lambda \frac{\partial t_k(x,y)}{\partial n} + \alpha_i t_k(x,y) = \delta_{ik} \bigg|_{S_i}, \quad i=1,2,\ldots,M \qquad (22)$$

At $\alpha_i = 0$, these conditions become the conditions of the second kind, in which $\delta_{ik} = a_{ik}$ and $i=1,2,\ldots,N$, while for $i=N+1, N+2, \ldots, M$; $\delta_{ik} = \alpha_i c_{ik}$. Further, the conjugation condition is maintained in its earlier form:

$$c_{ik+1} = \alpha_i \frac{\int_{S_i}(t_k(x,y) - c_{ik}) dS_i}{k c_p \dot{M}_i} \qquad i=N+1, N+2, \ldots, M \qquad (23)$$

and the coolant boundary condition is simplified:

$$\theta_i(0) = c_{i0} = \theta_0 \qquad (24)$$

The simplified system of equations (21)-(24) should be solved in succession reversed with respect to that used for (10)-(14), i.e. beginning with k=0 and so on. In so doing, the coefficients c_{ik} in (22) are determined directly from (23) allowing for (24).

As is known, solution of the problem (21),(22) has a corresponding solution of the problem of minimization of functionals of the form [5]:

$$J(t_k) = \int_D [\lambda(\text{grad } t_k)^2 - 2\varphi_k t_k] dD + \sum_{i=1}^{M} \int_{S_i} [\alpha_i t_k^2 - 2\delta_{ik} t_k] dS_i \qquad (25)$$

In this case, use of the Ritz method implies determination of the approximate solution in the form of a finite sum of coordinate functions satisfying the necessary requirements. Specifically, fulfilment of the boundary conditions by each coordinate function is necessary only for the conditions of the first kind, while for the natural boundary conditions, given by (22), this requirement is not necessary and the conditions are satisfied by the functional minimization (25). The approximation accuracy, as noted earlier, depends on the choice of the very succession of coordinate functions, which must be part of the complete system of functions (preferably orthogonal and normalized ones), and on the order of this succession. In this case, with an increase in the order not only the approximation accuracy increases, but also the time of numerical realization of the problem and computer memory volume.

For the problem of the functional minimization (25) using the Ritz method and providing for S_i being the round-shaped boundaries, a succession of coordinate functions is suggested, which allows representation of the approximate solution in the form:

$$t^{(n)}(x,y) = \sum_{i=0}^{M} A_i \ln \frac{R_i}{R_{oi}} + \sum_{i=M+1}^{n} A_i P_i(x,y) \qquad (26)$$

$P_i(x,y)$ are the Chebyshev and Legendre polynomials,
R_i, radius from the centre of the i-th cooling channel to the current point (x,y);
R_{oi}, radius of the i-th channel;
M, number of the cooling channels.

The choice of the succession of the type (26) is due to the following reasons. The analytical solution of the heat conduction problem in a cylindrical wall and of the problem of a pipeline in a semi-bounded massif is represented by a sum of the logarithmic function and polynomial. For example, for a cylindrical wall with constant heat generation, F_v, this well-known solution is given by

$$T = -\frac{F_v}{4} R^2 + A_1 \ln R + A_2 \qquad (27)$$

Evidently, the logarithmic function makes the largest contribution to the solution near the internal boundary, where it changes more rapidly. This suggests the idea to use the succession (26) when solving the variational problem (25). Thus, the information on the character of solution near the cooling channels, obtained from the analytical solution form for vicinity of these channels, is used to choose the type of the very succession of coordinate functions. It seems possible to carry out similar procedure for other surface geometries. In so doing, we isolate first a **recurrent** element of the geometry (vicinity of the channel of this or that cross-section etc.), then it should be used to obtain an analytical solution (if it exists), isolate from it the main part (special func-

CALCULATION OF HEAT TRANSFER IN POWER UNITS

tion) to be used as an addition to the system of functions from some complete system. The effect of the "special" functions obtained may be restricted if necessary by the isolated elements dimensions. It goes without saying that the number of special functions is determined by the number of recurrent elements, allowing us to expect the succession thus obtained to be a good approximation to the accurate solution.

The technique described above has been tested using a model problem for a cylindrical structure with two cooling channels. When solving plane problems, coordinate functions sets of different types have been used for the sake of comparison, namely: of the type of (26), Chebyshev polynomials and sets based on the structural method / 10 /. The calculation results obtained under the assumption of neglection of longitudinal heat overflows, demonstrated high efficiency of the method suggested and made it possible to choose from a number of variants of coordinate functions successions the succession of the type (26) as the best one, i.e. the one yielding good approximation at a minimum order and hence considerably reducing the volume of the computer memory required (almost by a factor of 25) and the computer time (by an order of magnitude for the same problem).

As noted before, solution of the three-dimensional problem will converge with the accurate one everywhere except in the end regions with the thicknesses on the order of the characteristic transverse dimension of the structure, L_{xy}. To satisfy the conditions at the ends (3) on the average, the Kantorovich's method may be applied if necessary / 9 /, using the three-dimensional problem solution obtained earlier and assuming that the temperatures of the structure, T, and the coolant, θ, are given by

$$T^{(m+1)}(x,y,z) = \sum_{k=0}^{m+1} t_k(x,y) z^k f_k(z) \qquad (28)$$

$$\theta_i(z) = \beta_i(z) \qquad (29)$$

in which $t_k(x,y)$ are the earlier obtained plane problems solutions, and $f_k(z)$ and $\beta_i(z)$ are the unknown functions. End conditions (3) should be replaced by the averaged ones:

$$\bar{T}\Big|_{\Sigma_{m+1}} = \frac{\int_{\Sigma_{M+1}} B_{M+1}(x,y,0) d\Sigma}{\Sigma_{M+1}} = T_o \Big|_{z=0} \qquad (30)$$

$$\bar{T}\Big|_{\Sigma_{m+2}} = \frac{\int_{\Sigma_{M+2}} B_{M+2}(x,y,L_z) d\Sigma}{\Sigma_{M+2}} = T_L \Big|_{z=L_z}$$

Solution of the problem (1),(2),(30),(4),(5) now is reduced to the problem of minimization of the functional of the form

$$J(T) = \int_V [\lambda(\text{grad } T)^2 - 2FT]\,dV + \sum_{i=1}^{M} \int_{\Sigma_i} [\alpha_i T^2 - 2\alpha_i \theta_i T]\,d\Sigma_i \quad (31)$$

In this case, for $i=1,2,\ldots N$, $\alpha_i = 0$, while $\alpha_i \theta_i = A_i(x,y,z)$. Substitution of the equations (28),(29) into the functional (31) and integration over the variables x and y yields the functional of the form:

$$J(T^{(i)}) = \int_0^{L_z} \varphi(z, f_1, f_2, \ldots, f_i, f_1', f_2', \ldots, f_i', \ldots)\,dz +$$
$$+ \int_0^{L_z} \varphi_1(z_1, f_1, f_2, \ldots, \beta_1, \beta_2, \ldots)\,dz \quad (32)$$

After introducing of a notation for the integrand function, $\varphi_0 = \varphi + \varphi_1$, the Eulerian equations for (32) yields a system of common differential equations:

$$\frac{d}{dz}\frac{\varphi_0}{f_k'} - \frac{\varphi_0}{f_k} = 0$$

with the boundary conditions:

$$\bar{T}(0) = T_0$$
$$\bar{T}(L_z) = T_L$$

This system of equations must be supplemented with differential equations of heat transfer in a gas (4) allowing for the relationship (5) and the boundary conditions $\theta_i(0) = \theta_0$. By this, a closed system of common differential equations is formed, with the corresponding edge conditions for determination of the unknown functions, f_k and β_k. Solution of this system yields the solution of the initial three-dimensional problem satisfying on the average the end conditions.

4. CALCULATION OF FLOW RATES REDISTRIBUTION BETWEEN THE COOLING CHANNELS AND POSSIBILITIES FOR REDUCING THE MAXIMUM TEMPERATURES IN THE STRUCTURES

When solving the heat transfer problems discussed in the

previous sections, it was assumed that the coolant mass flow rates, M_i, are assigned for each cooling channel. In reality, the cases are more frequent when the total coolant flow rate is assigned for all channels, while individual flow rates, M_i, assume the values depending on the pressure drop between the heat exchanger ends, which is the same for all channels, and also on hydraulic resistances and temperature factor.

With this approach the essence of the problem of flow rates redistribution is to find such a set of flow rates, $\{\dot{M}_i\}$, which provides for the constant pressure drop between the heat exchanger ends, i.e.:

$$P_i(L_z) - P_i(0) = \Delta P_0 = \text{const} \qquad (33)$$
$$i = 1, 2, \ldots, M$$

The above equation is fulfilled on condition that the total flow rate for all channels is an assigned value, $\dot{M}_0$, the temperatures θ_i being known:

$$\sum_{i=1}^{M} \dot{M}_i = \dot{M}_0 \qquad (34)$$

The equations used for solution of this problem, averaged over the cooling channels cross-sections, are:

equation of motion

$$\dot{M}_i \frac{d\rho u_i}{dz} = -F_i \frac{dP_i}{dz} - 2 \frac{\tau s_i}{D_{hi}} F_i \qquad (35)$$

equation of state

$$P_i = \rho_i \tilde{R} \theta_i \qquad (36)$$

continuity equation

$$(\rho u F)_i = \dot{M}_i \qquad (37)$$

and the boundary condition for pressure

$$P_i \Big|_{z=0} = P_0 \qquad i = 1, 2, \ldots M \qquad (38)$$

The coolant temperature in the channels, θ_i, necessary to calculate the written system, is assumed to be assigned, and can be, for example, the solution of the problem from the previous sections. The system (35)-(38) is easily solved by any numerical method for each channel, the coolant temperature, $\theta_i(z)$, and flow rate, $\dot{M}_i$, being known. In Eq.35, β is the averaging coefficient, and τ_{Si}, D_{hi}, tangential friction stresses and hydraulic diameters, respectively. Solution of (35)-(38) in the general case yields a different pressure value at the outlet of each channel, $P_i(L_z)$. Since it has been assumed that these values must coincide for all channels, such a flow rate distribution should be found, which yields the same pressure at the outlet of all channels, the cooler temperatures and total flow rate, $\dot{M}_O$, being assigned.

Equations (35)-(38) after appropriate substitutions of the pressure, ρ_i, velocity, u_i, and the equations for tangential stresses

$$\tau_S = \xi \frac{\rho u^2}{8}$$

are reduced to the following differential equation of the first kind:

$$\frac{dP_i}{dz} = \dot{M}_i^2 \, \tilde{R} P_i \, \frac{\xi_i \theta_i / 4 D_{hi} + \beta \frac{d\theta_i}{dz}}{\beta \tilde{R} \theta_i \dot{M}_i^2 - P_i^2 F_i^2} \tag{39}$$

Thus, the problem of flow rates redistribution is reduced to solution of equations (39) with the boundary conditions:

$$P_i(0) = P_o \tag{40}$$

and the additional conditions (33),(34).

The following procedure is suggested to solve the problem. An apriori pressure value is assigned at the outlet end, P_{L1}, determined using equation (39):

$$P_{L1} \cong P_o + \frac{\xi \theta / 4 D_h}{\beta \tilde{R} \theta \dot{M}_i^2 - P_o^2 F_i^2} \dot{M}_i^2 \tilde{R} L_z \tag{41}$$

Now the systems of the following usual first-order differential equations are solved:

$$\frac{dP_i}{dz} = \frac{\tilde{R}\left(\xi_i \frac{\theta_i}{4 D_{hi}} + \beta \frac{d\theta_i}{dz}\right)}{\beta \tilde{R} \theta_i \bar{\dot{M}}_i^2 - \bar{P}_i^2 F_i^2} \circ \frac{\bar{\dot{M}}_i^2 P_i + 2 \bar{P}_i \dot{M}_i \bar{\dot{M}}_i - \bar{P}_i \dot{M}_i^2}{}$$

$$\frac{d\dot{M}_i}{dz} = 0 \tag{42}$$

CALCULATION OF HEAT TRANSFER IN POWER UNITS

with the corresponding boundary conditions:

$$P_i(0) = P_0$$
$$P_i(L_z) = P_{L1} \quad \text{where } i = 1, 2, \ldots, M \tag{43}$$

The first equation of the system (42) is the linearized equation (39), the coolant temperatures, θ_i, are assigned and the values of $\bar{\dot{M}}_i$ and $\bar{P}_i$ are the iteration values of flow rates and pressure, obtained as a result of solving the problem (42,43) at the previous stage. For the first iteration, the flow rate values can be taken, for example, as a mean flow rate value in channels, and $\bar{P}_i(z)$, in the form of a linear function over z in the interval of values $[P_0 P_{L1}]$.

Iteration of the problem (42,43) with respect to $\bar{\dot{M}}_i$ and $\bar{P}_i$ goes on until the solution has been reduced to solution of the equation with a nonlinear RHS (39), i.e. until the following relationships have been satisfied:

$$\left|1 - \frac{\bar{\dot{M}}_i}{\dot{M}_i}\right| < \varepsilon_1 \quad \text{and} \quad \left|\frac{\bar{P}_i(z)}{P_i(z)} - 1\right| < \varepsilon_2 \tag{44}$$

Here, ε_1 and ε_2 are the pre-set small values determining the approximation accuracy.

Thus, sets of flow rates, $\{\dot{M}_i\}$, and pressures, $\{P_i(z)\}$, are obtained, which are the solutions of equations (39) with the boundary conditions (43). The next stage of the suggested procedure is verification of the condition imposed on the total flow rate (34). If the relationship (34) is not satisfied, the adopted pressure value on the end outlet, P_{L1}, must be changed for P_{L2}, and then the systems (42,43) must be solved with this new boundary condition. The pressure on the boundary in this case is chosen to correspond to the difference of the total flow rate:

$$\sum_{i=1}^{M} \dot{M}_i - \dot{M}_0 = \Delta \dot{M}_0 \tag{45}$$

It is understood that the increased pressure drop, $P_0 - P_{L2}$, has the corresponding increased flow rates, and vice versa, a decrease in the pressure drop results in decreased flow rates.

Several points, $P_{L1}, P_{L2}, \ldots$, and the corresponding defferences (45), $\Delta \dot{M}_{o1}, \Delta \dot{M}_{o2}, \ldots$, obtained when solving the problems (39,42,43) can be used to find, by means of simple linear or quadratic interpolation, the condition to be determined, (34), i.e. $\Delta \dot{M}_o \approx 0$. For numerical solution, three or four points $(P_{Lk}, \Delta \dot{M}_{0k})$ are usually sufficient to satisfy with the required accuracy through a successively linear interpolation the condition (34), and, hence, the inequality:

$$\left|\frac{\Delta \dot{M}_o}{\dot{M}_o}\right| < \varepsilon_3 \qquad (46)$$

Thus, the problem of flow rates redistribution is solved.

The method described in earlier sections (2,3) for calculation of heat transfer problems in power units with a complex system of boundary surfaces in conjunction with the represented method of calculation of flow rates redistribution problems provides a good possibility for numerical experiments which allow determination of the effect of various parameters of the problem on its solution.

This statement of the problem makes it possible, specifically, to study the effect of variation of the boundary conditions parameters, e.g. of heat transfer coefficients, or substitution of the second kind conditions (thermal insulation, maybe) for the third kind conditions at some surfaces. Moreover, this statement makes it possible to use a corresponding selection of parameters, e.g. channels resistances, to obtain more advantageous temperature field distributions, i.e. distributions with lowered values of pressure drops. Hydraulic resistance of the channel can be easily changed by erecting a throttle at its inlet. Pressure drop on the throttle in the general case is of the form:

$$\Delta P_d = \xi_d \frac{\rho_o u_o^2}{2} \qquad (47)$$

Here, the density, ρ_o, and velocity, u_o, of the coolant are related to the cross-section averaged values before the throttle. Expressing (47) through the coolant flow rate, $\dot{M}_i$, easily yields the relationship:

$$\Delta P_{di} = \xi_{di} \frac{\dot{M}_i^2}{2F_i^2 \rho_{oi}(\theta_o, P_o)} \qquad (48)$$

Now the pressure drop over the entire channel will be represented by the value ΔP_L, which is the sum of pressure drop on the throttle and pressure drop due to the channel hydraulic resistance:

$$\Delta P_L = \Delta P_d + \Delta P_h = P_0 - P_L \qquad (49)$$

Changing of resistances of some channels by erection of throttles at the inlet results in a decrease of coolant flow rate and hence of heat exhaust. This leads to an increase of temperature in the vicinity of these channels and its certain decrease in the remaining region due to an increase in flow rates caused by their redistribution.

Thus, solution of the problem of reducing the maximum temperature drop in the structure begins with assignment of throttle resistance coefficients, ξ_{di}. These coefficients are assigned for the channels situated in the "cold" regions of the structure and estimated roughly from the temperature field of the problem without throttles. After that the problem of flow rates redistribution among cooling channels of the type (39,42) is solved, although different boundary conditions are used, which account for throttle pressure drops:

$$P_i(0) + \xi_{di} \frac{1}{2F_i \rho_i(\Theta_o, P_o)} \overline{M}_i \dot{M}_i = P_o$$

$$P_i(L_z) = P_L \tag{50}$$

The linearized problem (42) is solved as before with the conditions (50), until convergence, allowing for the conditions (33, 34), while the set of flow rates obtained, $\{\dot{M}_i\}$, is used to determine the temperature field which is the solution of heat transfer problems discussed in sections 2,3. This field is analyzed and the necessary conditions are imposed on the throttle resistance coefficients. The procedure is repeated until satisfactory results have been obtained with respect to the maximum temperature drop. The calculations have shown that such a procedure makes it possible to considerably reduce this drop, thus providing for reduction in the total coolant flow rate.

5. APPLICATION OF THE SUGGESTED CALCULATION TECHNIQUE

The problems of heat transfer in bodies of complex form in three- or two-dimensional formulation are used for calculation of different heat exchangers with cooling channels and inner heat sources, of heat accumulators in solar power engineering, power mechanical engineering, and also for the analysis of durability of the elements of structures having complex configurations of boundaries and made of materials porous ones including with volumetric radiation absorption. It should be emphasized that the boundary surfaces may be of the most complex form.

Thus, the suggested calculation technique (sections 2,3) amounts to the following. Let, for the sake of generality, there be a heat source represented in the form of a finite expansion of the exponent "n" in the longitudinal coordinate "z". The heater and cooler temperatures are sought in the form of similar expansions of the by one larger exponent than that used for the heat source expansion. Flow rates, temperatures and pressures of the coolant at the inlet to the channels are assumed set. The energy equations for the coolant are taken averaged over the channels cross-sections. Under these assumptions the three-dimensional heat transfer problem is reduced to the n+2 plane problems. The obtained system of equations is solved by the variational technique, i.e. the Ritz method is used to determine the minimum of the corresponding functional. Then the order of coordinate functions should be chosen with account for the form of cooling channels. It has been shown (section 4) that the corresponding selection of local resistances can essetially equalize the temperature field over the body cross-section.

REFERENCES

1. Descloux, J. 1976. Method of finite elements. Izd.Mir, Moscow.

2. Zenkevich, O. 1975. Method of finite elements in engineering. Izd. Mir, Moscow.

3. Samarsky, A.A. 1977. Difference scheme theory. Izd. Nauka, Moscow.

4. Godunov, S.K. and Ryaben'ky, V.S. 1973. Difference schemes. Izd. Nauka, Moscow.

5. Bio, M. 1975. Variational principles in heat transfer theory. Izd.Energiya, Moscow.

6. Mikhlin, S.G. 1970. Variational methods in mathematical physics. Izd. Nauka, Moscow.

7. Tikhonov, A.N. and Samarsky, A.A. 1966. Mathematical physics equations. Izd. Nauka, Moscow.

8. Kantorovich, L.V. and Krylov, V.I. 1962. Approximate methods of higher-order analysis. GIFML, Moscow-Leningrad.

9. Rabinovich, N.R. and Koskinov, Yu.G. 1978. Expansion methods in heat conduction problems. Zh. Inzh. Phys., Vol.35, No.4, pp.728-733.

10. Rvachyov, V.L. and Slesarenko, A.P. 1978. Algebraic-logical and projection methods in heat transfer problems. Izd. Navukova Dumka, Kiev.

11. Mikheev, M.A. and Mikheeva, I.M. 1973. Heat transfer fundamentals. Izd. Energiya, Moscow.

12. Luikov, A.V., Perel'man, T.L. and Ryvkin, V.B. 1966. On determination of the heat transfer coefficient in simultaneous conductive and convective heat transfer. Proceedings of the Third International Heat Transfer Conference. Vol.2, Chicago, pp.12-24.

FOULING IN HEAT EXCHANGERS

Fouling of Heat Exchangers

NORMAN EPSTEIN
Department of Chemical Engineering
University of British Columbia
Vancouver, British Columbia V6T 1W5, Canada

ABSTRACT

The fouling of heat transfer surfaces, which gives rise to high economic penalties and is still dealt with by heat exchanger designers using the crude TEMA approach, is classified into six principal categories. The measurement of thermal fouling is critically described and this is followed by a discussion of the successive events, up to five in number, which characterize most fouling situations. The effect of fouling on enhanced surfaces is analytically distinguished from its effect on extended surfaces. Graphical procedures are presented for determining the optimum cleaning cycles for a heat exchanger when the cleaning time is composed of both a constant portion, θ_{cc}, and a variable portion, θ_{cv}, which is directly proportional to the cumulative throughput between cleanings.

1. INTRODUCTION

Fouling is a phenomenon which occurs with or without a temperature gradient in a great many natural, domestic and industrial processes. Some examples are arterial blood flow, membrane permeation, catalysis, fluid flow in conduits and heat exchange. A temperature gradient complicates, but is frequently not essential to, the phenomenon. If we define fouling as the accumulation of undesired solid material at phase interfaces and restrict ourselves to heat transfer surfaces, then five possible interfaces present themselves:

>   gas-liquid
>   liquid-liquid
>   gas-solid
>   liquid-solid
>   gas-liquid-solid

The first two involve direct-contact heat transfer between a fluid and a liquid phase, where the crud which develops at the interface is mobile and relatively easy to remove. The last, commonly referred to as the triple interface, occurs in change-of-phase operations such as condensation and boiling, the latter of which has received much specialized study in desalination evaporators and steam generators. Current interest in fouling is focussed on liquid-solid and, because of the revival of coal combustion as well as the demand for greater waste heat recovery from exhaust gases, increasingly also on gas-solid interfaces.

1.1 Cost of Fouling

The overall annual cost of heat exchanger surface fouling in industry has been estimated as being equivalent to about 0.3% of the U.K. Gross National Product in a recent British study [1]. This penalty for fouling can be attributed in roughly equal parts to

a. Higher capital expenditures through oversized plants
b. Energy losses through increased thermal inefficiencies and pressure drops
c. Maintenance, including cleaning of heat exchangers and use of anti-foulants
d. Loss of production during shutdown for cleaning or through reduced overall plant efficiency

The order of magnitude of these estimates is confirmed by an even more recent American study of petroleum refinery fouling costs [2].

1.2 TEMA Approach

The current practice in heat exchanger design for fouling is to select values of R_f from T.E.M.A. [3] for both the hot and cold sides of the exchanger and to add these to the total clean surface resistance (for a unit of surface area) in order to arrive at the required heat transfer surface area. This procedure has been frequently criticized on several grounds:

1. The TEMA tables of R_f, restricted as they are mainly to water and to hydrocarbon based process streams flowing through shell-and-tube heat exchangers, give incomplete coverage to the large variety of possible process fluids and heat exchanger configurations.

2. These tables barely give recognition to the variation of R_f with such process variables as fluid velocity, bulk temperature and composition, and surface temperature.

3. No indication is given in the tables as to how R_f was arrived at, e.g. whether it is meant to denote an asymptotic resistance or a resistance after some fixed operating time which is not specified.

4. Most importantly, this procedure treats fouling, which is a transient process, as if it were instantaneously at a steady state with a fixed value of R_f. Since initially the heat exchanger surfaces are clean, this means that the exchanger is over-designed at start-up, which may dictate the actual use of lower initial fluid velocities and give rise to higher initial surface temperatures than prescribed by the design, and thus result in more and faster fouling than would otherwise be the case.

The increased understanding of fouling achieved during the past decade has still made no significant dent in this procedure, although values of R_f other than those in TEMA have been proposed, e.g. for plate heat exchangers [4].

2.   CLASSIFICATION OF FOULING

Because of the wider variety of fouling mechanisms from liquids than from gases, thermal fouling is classified with the liquid-solid interface as the prototype. The classification scheme around which an increasing concensus has developed amongst workers in the field during the past ten years is one based on the key physical/chemical process essential to the particular fouling

FOULING OF HEAT EXCHANGERS

phenomenon. The six primary categories which have thus been identified are

1) *Precipitation Fouling* - the crystallization from solution of dissolved substances onto the heat transfer surface, sometimes called *scaling*. Normal solubility salts precipitate on subcooled surfaces, while the more troublesome inverse solubility salts precipitate on superheated surfaces.

2) *Particulate Fouling* - the accumulation of finely divided solids suspended in the process fluid onto the heat transfer surface. In a minority of instances settling by gravity prevails, and the process may then be referred to as *sedimentation fouling*.

3) *Chemical Reaction Fouling* - deposit formation at the heat transfer surface by chemical reactions in which the surface material itself is not a reactant (e.g. in petroleum refining, polymer production, food processing [43]).

4) *Corrosion Fouling* - the accumulation of indigenous corrosion products on the heat transfer surface.

5) *Biological Fouling* - the attachment of macro-organisms *(macro-biofouling)* and/or micro-organisms *(micro-biofouling* or *microbial fouling)* to a heat transfer surface, along with the adherent slims often generated by the latter.

6) *Solidification Fouling* - the *freezing* of a pure liquid or the higher melting constituents of a multi-component solution onto a subcooled surface.

Note that these categories do not denote the rate-governing process for the fouling (e.g. mass transfer, surface reaction, etc.), which can only be discovered by a detailed analysis of the effect which process variables such as fluid velocity and surface temperature have on fouling rate. Categories 1 and 6 both involve *crystallization fouling*, the first from solution and the last from the melt. The category Corrosion Fouling has been used by some authors to include deposition of corrosion products which originate from a source other than the heat transfer surface, but such fouling due to *ex situ* corrosion, which is more correctly categorized as either precipitation or particulate fouling depending on whether the corrosion products are soluble or insoluble at bulk conditions, should be excluded from the present fourth category, which strictly refers to *in situ* corrosion fouling. Even after its exclusion, however, some difficulties still exist in applying these categories. For example, it is at times not clear in the case of crystallization from solution whether precipitation is occurring directly onto the heat transfer surface, in which case we have an unambiguous precipitation fouling; or whether it is occurring in the bulk of the solution followed by deposition of the precipitated particles, in which case we have particulate fouling; or whether both processes are occurring simultaneously. Similarly, chemical reaction fouling at a heat transfer surface may sometimes be difficult to distinguish from chemical precipitation of a solid product in the bulk of the fluid which gives rise to particulate fouling. Despite such practical ambiguities, which can be analyzed on a case-by-case basis, this classification scheme has served to decompose the otherwise hopelessly complicated processes of real-life fouling into modes which can be studied in isolation from each other, thus allowing for a more fundamental understanding of their underlying behaviour.

Almost any pair of the above fouling modes are synergistic, i.e. mutually reinforcing [5]. This is particularly true of corrosion fouling in conjunction with each of the other modes. One exception is scaling accompanied by particle deposition, which tends to weaken an otherwise tenacious scale [6].

It will eventually be useful to undertake controlled experiments in which two or more fouling modes are *deliberately* allowed to occur together, but at the present state of knowledge, it is, in my view, more important to continue with single mode experiments (and initially, where feasible, a single species for any mode, e.g. one crystallizing salt, one chemical reactant, one microbial species) until better understanding of the individual categories is achieved. Even with such experiments, unintended effects such as those of corrosion or of inadvertent impurities may complicate the results and should certainly not be neglected.

3. MEASUREMENT OF FOULING

In-plant measurements, though useful, do not usually lend themselves to the degree of control necessary for acquiring reliable fouling data, and resort is therefore commonly made to small-scale fouling rigs. The variations in geometry, heating methods and monitoring techniques used in such rigs has been detailed elsewhere [5,7]. It should be noted that what is monitored with time is either the mass of deposit per unit heat transfer surface, m, or the deposit thickness, x, or the thermal fouling resistance, R_f. The relationship between these three terms is

$$dR_f = \frac{dx}{k_f} = \frac{dm}{\rho_f k_f} \qquad (1)$$

Equation (1) is written in differential form to allow for the possible variation of both deposit density ρ_f and deposit thermal conductivity k_f with distance from the heat transfer surface. Note that two deposits with the same incremental increase in m may give rise to very different corresponding increases in R_f, depending on the relative values of the product $\rho_f k_f$. A hard tenacious non-porous scale will typically have relatively high values of both density and thermal conductivity, while a soft non-tenacious porous scale will have considerably lower values of ρ_f and k_f. If monitored thermally, the latter (which typically may replace the former when an antiscaling additive is used) will thus show a higher resistance to heat transfer than the former after the same length of time, even though it is the more desirable deposit from the viewpoint of the ease with which it can be removed (e.g. by a velocity surge).

Fouling rigs which measure R_f commonly do so by subtracting a total thermal resistance at time zero from the corresponding thermal resistance at time θ. At constant heat flux, $\dot{q}$, with one thermocouple measuring the wall temperature T_w and another the constant fluid bulk temperature T_b, the initial thermal resistance under clean conditions is given by

$$\frac{1}{U_o} = \frac{T_{wo} - T_b}{\dot{q}} \qquad (2)$$

and the thermal resistance at time θ by

$$\frac{1}{U} = \frac{T_w - T_b}{\dot{q}} \qquad (3)$$

Then by simple subtraction,

$$R_f = \frac{1}{U} - \frac{1}{U_o} = \frac{T_w - T_{wo}}{\dot{q}} \qquad (4)$$

Deceptively low and even negative values of R_f may sometimes be recorded by this method, particularly at low values of θ in turbulent flow, where the deposit roughness gives rise to an increase in the convective heat transfer coefficient between the surface and the fluid, which counteracts the increase in resistance due to the deposit conduction barrier. This effect declines in relative importance as R_f gets large and, if desired, can be corrected for by estimating the roughness, e.g. from pressure drop measurements, and its effect on the heat transfer coefficient [8]. Blockage of the flow channel by the deposit at a fixed fluid throughput and hence rising velocity (and pressure drop) can also increase the convective heat transfer coefficient and thus mask the true rise in R_f, but this effect is more likely to be important at high rather than low values of x. In the case of a hot wire probe, added complications are the substantial increase in outer heat transfer surface area A_{out} as x increases and the partially compensating decrease in the convective heat transfer coefficient h between the deposit surface and the fluid flowing past it:

$$\frac{1}{U} - \frac{1}{U_o} = \frac{R_f A_w}{A_{l.m.}} + \frac{1}{h} \cdot \frac{A_w}{A_{out}} - \frac{1}{h_o} = \frac{(T_w - T_{wo})A_w}{q} \qquad (5)$$

where $U_o = h_o$, $R_f = x/k_f$, $A_{out} = A_w + 2\pi x$ and $A_{l.m.} = (A_{out} - A_w)/\ln(A_{out}/A_w)$.

A recent development of note, arising from the renewed interest in fouling from coal combustion gases onto steam generators [9], is the successful application of commercially available heat flux meters as sensors for detecting and monitoring ash deposits on boiler tubes [10]. Another interesting development is the use of a radial flow growth chamber [11], which in any one run displays a wide range of surface shear, to study the effect of shear stress on bacterial attachment to metallic surfaces [12], as well as on subsequent detachment of the adhering biofilm.

4. SEQUENTIAL EVENTS IN FOULING

For all the above categories of fouling, the successive events which commonly occur in most situations are up to five in number:

1. *INITIATION* (delay, nucleation, induction, incubation, surface conditioning)

2. *TRANSPORT* (mass transfer)

3. *ATTACHMENT* (surface integration, sticking, adhesion, bonding)

4. *REMOVAL* (release, re-entrainment, detachment, scouring, erosion, spalling, sloughing off)

5. *AGING*

The words in brackets are alternative terms often used to designate the given step in the fouling sequence, in some cases for particular categories of fouling (e.g. nucleation and surface integration are unique to crystallization) and in other cases more generally.

4.1 Initiation

Initiation is associated with the delay period, θ_D, so often (but certainly not always) observed before any appreciable fouling is recorded after starting an experiment or process with a clean heat transfer surface. For precipitation fouling it is closely associated with the crystal nucleation process, and thus θ_D tends to decrease as the degree of supersaturation is increased with respect to the heat transfer surface temperature, and as the general temperature level is increased for a given degree of supersaturation; the effect of operating velocity is, however, still in doubt [13]. For chemical reaction fouling, θ_D appears to decrease as the surface temperature is increased [14], presumably due to speeding up of the induction reactions. For all fouling modes, many investigators have reported that θ_D decreases as the surface roughness increases. The roughness projections provide additional sites for nucleation, adsorption and chemical surface-activity, while the grooves provide regions for deposition which are sheltered from the mainstream velocity. Surface roughness also decreases the thickness of the viscous sublayer and hence increases eddy transport to the wall.

In the case of biofouling, one of the interesting discoveries of the past decade, associated largely with the name of Baier [15], is that the initial events involve the surface adsorption of polymeric glycoproteins and proteoglycans, traces of which are always present in natural waters and which act as surface conditioning films, to which micro-organisms subsequently adhere. A key parameter in determining whether or not a given surface responds to this effect with high bio-adhesion is the *critical wetting tension*, σ_c, which is the surface tension of a hypothetical liquid which just wets (i.e. contact angle $\beta = 0$) the clean surface, and is obtained by linearly extrapolating to $\cos \beta = 1$ a plot of surface tension σ for various organic liquids vs. $\cos \beta$ on the clean surface. According to Baier, if σ_c falls between 20 and 30 dynes/cm (σ_c for metallic surfaces exceeds 30 dynes/cm while σ_c for fluorocarbons is less than 20 dynes/cm), then bio-adhesion is minimized. This gives a surface criterion to aim for in the case of biofouling, and might conceivably be relevant to other types of fouling.

4.2 Transport

Transport is the best understood of the fouling stages. A key component, such as the fouling species itself or oxygen or a crucial reactant, must be transported from the bulk of the fluid, where its concentration is C_b, to the heat transfer surface, where its concentration in the adjacent fluid is C_s. The local deposition flux, $\dot{m}_d$, is then given by

$$\dot{m}_d = k_t(C_b - C_s) \qquad (6)$$

where k_t is a transport coefficient. In the case of ions, molecules or submicron particles, the transport is diffusional in nature and k_t is the equivalent to the well known mass transfer coefficient, k_m, which can be obtained from the relevant empirical correlations or theoretical equations for forced convection mass transfer in the literature, provided the diffusivity of the key component can be determined. For a dilute suspension of spheres, the Brownian diffusivity D of the particles is given by the Stokes-Einstein [16] equation,

FOULING OF HEAT EXCHANGERS

$$D = \frac{k_B T}{3\pi\mu d_p} \tag{7}$$

which for 0.5μm spheres in water yields Sc = $\mu/\rho D \sim 10^6$. Providing we are not dealing with excessive mass fluxes of the key component towards the deposition surface, any solutions for the forced convection heat transfer coefficient in laminar or turbulent flow can be converted to a solution for forced convection mass transfer coefficient by substituting Sh for Nu and Sc for Pr (or St_m, Pe_m and Gz_m for St_H, Pe_H and Gz_H respectively).

If a momentum-mass transfer analogy for turbulent flow is used at the high values of Sc (i.e. low D) characteristic of colloidal particles, it must be one which allows for the fact that in the viscous sublayer the eddy diffusivity does not vanish, except at the wall, and is typically of the same order of magnitude as that of the Brownian diffusivity. Using the Reichardt analogy [17], which meets this criterion, Metzner and Friend [18] derived for turbulent flow of high Schmidt number binary solutions that

$$\frac{k_m}{u_b} = \frac{f/2}{1.20 + 11.8\sqrt{f/2}(Sc-1)Sc^{-2/3}} \tag{8}$$

For Sc in the order of 10^6, equation (8) simplifies to

$$\frac{k_m}{u_*} = \frac{1}{11.8 \, Sc^{2/3}} \tag{9}$$

where the friction velocity $u_* = \sqrt{\tau_s/\rho} = u_b\sqrt{f/2}$. Using an entirely different approach, based on a computation of the stagnation flow towards the wall, Cleaver and Yates [19] independently derived for the diffusion regime of particle deposition that

$$\frac{k_m}{u_*} = \frac{1}{11.9 \, Sc^{2/3}} \tag{10}$$

which is in excellent agreement with equation (9). This equation has recently been shown to correlate the initial deposition rates of fine magnetite particles from aqueous suspensions flowing turbulently through aluminum tubes [20].

If the key component consists of particles having a dimensionless relaxation time $t^+ (=\rho_p d_p^2 u_*^2 \rho/18\mu^2)$ in excess of about 0.1, inertial effects become important, but the transport coefficient for such particles too can now be estimated with some reliability from a large body of literature on particle-fluid mechanics developed during the past two decades [21]. For still larger particles in horizontal channels at relatively low fluid velocities, gravity may control the deposition process, especially if the particle density is high, and then

$$\dot{m}_d \propto u_t/u_b \tag{11}$$

One aspect of transport that tends to be neglected in the literature on particle fouling is *thermophoresis*. This is the phenomenon whereby a "thermal force" moves fine particles down a temperature gradient. Hence cold walls attract and hot walls repel colloidal particles. The thermophoretic velocity, v_t, of micron-size particles has been shown to be representable by [22]

$$v_t = -\alpha \cdot \frac{\nu}{T} \cdot \nabla T \qquad (12)$$

where, for continuum flow, the coefficient α is given by

$$\alpha = \frac{c}{(k_p/k) + 2} \qquad (13)$$

and the constant c is 1.8 for gases and 0.26 for liquids. Clearly, since the kinematic viscosity ν of a gas is typically at least an order of magnitude greater than ν of a non-viscous liquid at the same temperature (especially at high temperatures), and since conversely the thermal conductivity k for a gas is much smaller than k for a liquid so that at a given heat flux $\dot{q}$ (= $-k\nabla T$), $-\nabla T$ is again much larger for gases than for liquids, it follows from equation (12) that v_t for gases is considerably greater than v_t for liquids. Nevertheless, even for liquids, thermophoresis cannot be neglected in the presence of large heat fluxes. By assuming that the entire temperature difference between the bulk stream and the wall is confined to a thin laminar film near the wall, Whitmore and Meisen [22] have used equation (12) to develop the following expression for the fractional particle removal efficiency, ε_t, by thermophoresis from a suspension flowing through a duct with cold sticky walls (i.e. no particle rebound or re-entrainment):

$$\varepsilon_t = 1 - \left(\frac{T_{out}}{T_{in}}\right)^{\alpha \cdot Pr} \qquad (14)$$

If the fractional removals of particles on a cold wall by transport mechanisms other than thermophoresis e.g. Brownian diffusion, inertia, gravity, etc. are denoted by ε_d, ε_i, ε_g, etc., then it can be shown [22] that the overall fractional removal, ε, is given by

$$\varepsilon = 1 - (1-\varepsilon_t)(1-\varepsilon_d)(1-\varepsilon_i)(1-\varepsilon_g)\ldots \qquad (15)$$

The measurable *decrease* of particle removal from a heated as compared to an unheated wall has recently been demonstrated experimentally [23]. This effect of thermophoresis could explain why particle attachment data at hot walls do not always show an increase with wall temperature, as would be expected from electrokinetic theory, for example equation (20) below.

4.3 Attachment

Attachment of the fouling species to the wall follows transport of the key component to the wall region, where the solid which deposits is actually formed, except in the case of particulate fouling. For the latter case, particle attachment is sometimes treated by putting $C_s=0$ in equation (6) and multiplying the right-hand side by the sticking probability S_P, i.e. the probability that any particle which reaches a particle-free wall region will remain at the wall. The term S_P is usually characterized by an Arrhenius dependence on surface temperature [5]. If it also displays a stronger than linear dependence on the reciprocal of friction velocity [24], then the direct proportionality between k_m and u_* shown by equation (9) is more than counterbalanced in equation (6) by this inverse dependence of S_p on u_*. Thus for $S_p=1$, mass transfer controls and $\dot{m}_d$ is directly proportional to u_* while for $S_p<1$, surface attachment is important and $\dot{m}_d$ may decrease as u_* increases [24].

A less empirical approach to the particle attachment process may be arrived at by consideration of the dominant surface forces which come into play when a particle approaches a wall. Thus for a highly idealized mono-dispersed colloidal (submicron) particle suspension, the work of Bowen [25] has definitively shown that the initial stickiness of a non-corroding wall is virtually perfect when particles and wall have zeta potentials of opposite sign, in which case electrical double layer attraction enhances London-Van der Waals attraction to yield essentially a mass-transfer controlled initial deposition, given by equation (6) with $C_s=0$. The situation changes, however, as soon as a relatively few particles have accumulated at the wall, the zeta potential of which rapidly decreases (numerically) and changes its sign to that of the particles, while the deposition rate falls off sharply and starts to approach an asymptotic value. It is also very different when the zeta potentials of the initial wall and the particles are of like charge, particularly when the potentials are also large in magnitude, in which case a large repulsive energy barrier must be overcome before a particle can settle into the primary London-Van der Waals potential energy sink at the wall. Even the initial particle deposition rate is then much smaller than what would be predicted by equation (6) with $C_s=0$. Though electrokinetic theory has been able to qualitatively predict trends for this more common situation, its quantitative predictions of particle deposition are typically orders of magnitude smaller than what is measured, a deficiency which can be attributed primarily to its failure as yet to appropriately take account of heterogeneities in charge distribution and microscopic geometry of real surfaces [26].

For crystal growth, assuming nucleation has already occurred at the wall, attachment is by the process known as surface-integration, which, in the case of stoichiometric equality in solution of crystallizing cations and anions, is commonly represented by

$$\dot{m}_d = k_r(C_s - C_{sat})^n \tag{16}$$

where k_r is the attachment rate constant and C_{sat} is the saturation concentration of the crystallizing species at the temperature of the surface. Typically $n \sim 2$ for sparingly soluble salts [27], especially when their cations and anions have the same valence. Combination of Equations (6) and (16) with $k_t=k_m$ gives

$$\dot{m}_d = \frac{C_b - C_{sat}}{\dfrac{1}{k_m} + \dfrac{1}{k_r(C_s-C_{sat})^{n-1}}} \tag{17}$$

When mass transfer controls (e.g. crystal growth at sufficiently low fluid velocities), the first term in the denominator of equation (17) predominates over the second and therefore

$$\dot{m}_d = k_m(C_b - C_{sat}) \tag{18}$$

When surface attachment controls (e.g. crystal growth at sufficiently high fluid velocities), the second term predominates, $C_b \sim C_s$ and therefore

$$\dot{m}_d = k_r(C_b - C_{sat})^n \tag{19}$$

Note that Equations (16)-(19) with $C_{sat}=0$ apply also to colloidal particle deposition, chemical reaction fouling and commonly even pure corrosion fouling, to which must be added an additional term x/D_f in the denominator of equation (17) to take account of oxygen diffusion through the corrosion deposit, the effective oxygen diffusivity of which is D_f. For colloidal particle fouling, electrokinetic theory indicates additionally that $n=1$. For all fouling categories, an Arrhenius type equation relates the attachment rate constant to the surface temperature T_s:

$$k_r = A'e^{-E/R_g T_s} \tag{20}$$

4.4 Removal

Removal of the deposit may or may not begin right after deposition has started. That it does so is an assumption implicit in the removal model originally proposed by Kern and Seaton [28] and further developed by Taborek et al [6]:

$$\dot{m}_r = \frac{B\tau_s m}{\psi} \tag{21}$$

that is, the removal flux $\dot{m}_r$ is directly proportional to both the mass of deposit and the shear stress τ_s on the heat transfer surface, and inversely proportional to the deposit strength ψ. That removability increases linearly with deposit thickness and hence with m has recently been rationalized via a theory of deposit-shattering by thermal stresses [29]. Although the continuous co-existence of removal with deposition (especially particulate deposition) is more readily rationalized in turbulent [30] than in laminar flow [5], the fouling rate at any time θ according to this assumption is then given by

$$\frac{dm}{d\theta} = \dot{m}_d - \dot{m}_r = \dot{m}_d - \frac{B\tau_s m}{\psi} \tag{22}$$

Integration of equation (22) from the initial condition $\theta=0$, $m=0$, on the assumption that the only variables in equation (22) during the course of fouling are θ and m, yields the well known Kern-Seaton equation

$$m = m^*(1-e^{-\theta/\theta_c}) \tag{23}$$

where m^* is the asymptotic mass per unit surface area and the time constant θ_c is given by

$$\theta_c = \frac{m}{\dot{m}_r} = \frac{m^*}{\dot{m}_d} = \frac{\psi}{B\tau_s} \tag{24}$$

That θ_c decreases as a crystalline deposit acquires more impurities and hence loses strength, ψ, has been shown experimentally by Morse and Knudsen [31].

From equation (24) it is also seen that θ_c can be interpreted as the average residence time of an element of fouling deposit on the heat transfer surface, as well as the time it would take to accumulate the asymptotic fouling deposit, m*, if the fouling proceeded linearly at the initial deposition rate, $\dot{m}_d$. By putting $\theta=\theta_c$ in equation (23), m works out to be 0.632 m*, so that θ_c is also the actual time required to achieve 63.2% of the asymptotic fouling resistance.

Since $m^* = \dot{m}_d \theta_c$ and $\theta \alpha (1/\tau_s) \alpha (1/u_*^2)$, it follows that even if $\dot{m}_d$ is directly proportional to u_* as would be the case under conditions of turbulent mass transfer control at high values of Sc, m* and hence R_f^* would still decrease as the velocity increases. This generalization has commonly been found in practice, at least when deposit-removal occurs [24,21]. Only if deposit strength ψ is also directly proportional to u_*, as inferred by Gudmundsson [32] from the inverse proportionality of θ_c with fluid velocity for wax deposits solidifying from hydrocarbon streams, might this generalization falter. The evidence for a velocity dependence of ψ is still tenuous.

Asymptotic fouling can always be empirically fitted by equation (23), but it does not follow that it is always caused by deposit-removal. For example, in the asymptotic particle deposition results of Bowen [25], mentioned above, it was shown unambiguously by a radioactive tracer technique that, under the laminar flow conditions which prevailed in his experiments, no re-entrainment of particles occurred whatsoever. Autoretardation mechanisms other than increasing surface repulsion due to electrical double layer interactions (whereby particles stick to an oppositely charged wall more readily than they do to each other, or whereby they stick to each other more readily when their charges are neutralized by adsorbed corrosion ions than when they are not [33]) include the effect of deposit blockage in increasing the scouring velocity and hence suppressing attachment under constant mass flow rate conditions, the smoothing of a rough surface by a uniform deposit thus giving rise to a thicker viscous sublayer and reduced transport, an ever-weakening wall catalysis of chemical reaction fouling as the deposit builds up on the wall, a similarly ever-decreasing oxygen diffusion rate in corrosion fouling as the deposit thickens, and the drop in the temperature of the deposit-fluid interface as fouling proceeds under conditions of constant wall temperature (as opposed to constant heat flux). The last effect is the basis for the non-asymptotic falling-rate model of Hasson [34] and Reitzer [35] for inverse solubility scaling, while corrosion fouling controlled by oxygen diffusion has been modelled by Galloway [36] as a similar falling rate process.

Nevertheless, deposit-removal *has* been observed to occur simultaneously with deposition [5,6,20] in certain instances, and for those cases, θ_c can reasonably be interpreted by equation (24). According to Cleaver and Yates [30], it is not simple viscous shear that lifts (or is capable of lifting) particles from the deposit back to the mainstream, but the randomly periodic turbulent bursts which are randomly distributed over less than 0.5% of the surface at any instant. They have referred to these bursts as miniature tornadoes, and that this characterization is not just a metaphor has been vindicated by recent experiments [37] which have shown that there is a measureable wall suction associated with the turbulent bursting. For a *given* deposit and fluid, a minimum friction velocity u_* is required before the turbulent bursts can become effective in removing some of the deposit. By reference to equation (24) it is not unreasonable to generalize the criterion to be fulfilled by *any* deposit as

$$\theta_c < (\theta_c)_{crit} \tag{25}$$

or, since $\theta_c \alpha (\psi/\tau_s) \alpha (\psi/u_*^2)$,

$$\frac{u_*^2}{\psi} > \left(\frac{u_*^2}{\psi}\right)_{crit} \qquad (26)$$

where the subscript crit denotes some critical value for a given fluid. Note that the numerator in equation (26) represents hydrodynamic forces tending to disrupt the deposit while the denominator represents the adhesive or cohesive strength of the deposit, depending on which is weaker.

4.5 Aging

Aging of the deposit starts as soon as it has been laid down on the heat transfer surface. The aging processes may include changes in crystal or chemical structure, e.g. by dehydration or polymerization, respectively. Such changes, especially at constant heat flux and hence increasing deposit temperature, may strengthen the deposit with time. Alternately, changes in crystal structure, or chemical degradation e.g. of hydrocarbon gums to coke, or developing thermal stresses, or the slow poisoning of micro-organisms by corrosion cations released from the wall may result in gradually decreasing ψ with time. Thus a deposit which is not hydrodynamically removable at the beginning of a run may suddenly become so after some time when ψ has decreased sufficiently that the criterion of equation (26) is finally realized. Such behaviour can give rise to sawtooth-shaped fouling curves [5]. It should be noted that where the aging processes undermine the deposit by decreasing its strength, the final death blow to the deposit is usually provided by hydrodynamic removal, much as the shedding of leaves from trees which occurs in the autumn is primed by the weakening of their bonds to the twigs, but the final de-leafing is wrought by atmospheric turbulence.

5. AUGMENTED SURFACES

From the fouling viewpoint it is important to distinguish between *enhancement* or intensification of a plain surface in order to increase the heat transfer *coefficient*, and *extension* or enlargement of a plain surface in order to increase the heat transfer *area*. For example, turbulators enhance while conventional fins extend. In many cases both effects occur simultaneously, but they can still be distinguished conceptually. Thus the thermal resistance of an extended surface is $1/hA\eta$, where the total surface efficiency η is related to the fin efficiency η_f by

$$\eta = 1 - \frac{A_f}{A}(1 - \eta_f) \qquad (27)$$

The heat transfer improvement factor relative to the plain surface is then given by

$$\frac{(hA\eta)_{augmented}}{(hA)_{plain}} = \frac{h_{augmented}}{h_{plain}} \cdot \frac{(A\eta)_{augmented}}{A_{plain}} \qquad (28)$$

The first ratio on the right-hand side of equation (28) may be referred to as the enhancement factor, while the second ratio is the extension factor. An improvement in the enhancement factor is usually at the expense of some decline

FOULING OF HEAT EXCHANGERS

in the extension factor, since increasing the heat transfer coefficient on an extended surface results in decreasing the fin efficiency, and hence the total surface efficiency. Normally the latter loss is much smaller than the former gain, at least under non-fouling conditions.

Consider now a pure enhanced surface and a pure extended surface, and the effect of a uniform fouling deposit on each, as compared with its effect on the original plain surface. For a fixed value of R_f applied to both a plain surface and an enhanced surface, the latter, since by definition it starts with a smaller thermal resistance than the former, suffers a larger percentage increase in thermal resistance than the plain surface. Thus the enhancement, and hence the heat transfer improvement, factor decreases [38]. On the other hand, for a fixed value of R_f applied to both a plain surface and an extended surface, the fin efficiency and hence η of the latter is increased by the insulating effect of the fouling layer, so that the extended surface suffers a smaller percentage increase in thermal resistance than the plain surface. Thus the extension, and hence the heat transfer improvement, factor increases [39].

A given degree of uniform fouling is thus far more harmful thermally to an enhanced than to an extended surface [38]. The degree to which any given augmented surface actually fouls is, of course, another matter — to be investigated empirically rather than conceptually.

6. OPTIMUM CLEANING CYCLES

Periodic shutdown and cleaning of a heat exchanger is often required due its declining production rate, $dP/d\theta$, with time θ. Examples are an evaporator undergoing scaling which consequently delivers a steadily decreasing concentrated liquor rate of the required concentration, and a sensible heat exchanger undergoing fouling which delivers a steadily decreasing rate of product heated or cooled through the required temperature range. The problem of determining when to shut down for cleaning can be formulated in terms of either a maximum throughput cycle or a minimum cost (per kilogram product) cycle. In either case it is common to assume that the cleaning time θ_K is constant and independent of the cycle throughput, P_{cycle}, and simple graphical procedures for both cases have been developed on this basis [40].

In many instances it is more reasonable to assume that θ_K is a weak linear function of P_{cycle} than that it is independent of the cumulative throughput. Thus in general,

$$\theta_K = \theta_{cc} + aP \tag{29}$$

where θ_{cc} is the fixed unreducible portion of the cleaning time, e.g. at the very least the time required to dismantle and re-assemble the heat exchanger for cleaning, and aP is the variable portion, θ_{cv}, of the cleaning time which is directly proportional to the cumulative throughput between cleanings.

The criterion for the maximum throughput cycle is that

$$R = \frac{P}{\theta + \theta_K} \tag{30}$$

is a maximum. Substituting for θ_K according to equation (29) and differentiating equation (30) with respect to θ, setting $dR/d\theta=0$, the result is

$$\left.\frac{dP}{d\theta}\right|_{\theta=\theta_{opt}} = \frac{P_{opt}}{\theta_{opt} + \theta_{cc}} \qquad (31)$$

i.e. the maximum throughput cycle is one in which the tangent from $\theta = -\theta_{cc}$, P=0 to the curve of P vs. θ touches the latter at (θ_{opt}, P_{opt}). The graphical construction, Figure 1, is thus identical with that for constant θ_K [40], except that θ_{cc} replaces θ_K. The total cleaning time per cycle is $\theta_{cc} + aP_{opt}$.

The corresponding minimum cost problem can be formulated after Badger and Othmer [41] by first writing an expression for the total operating cost T_θ per unit time, taking into account both production and cleaning time:

$$T_\theta = H + \frac{F\theta}{\theta+\theta_K} + \frac{K\theta_K}{\theta+\theta_K} + SR \qquad (32)$$

The total cost T_m per unit mass of product is then given by

$$T_m = \frac{H}{R} + \frac{F\theta + K\theta_K}{R(\theta+\theta_K)} + S \qquad (33)$$

Substituting for R according to equation (30), for θ_K according to equation (29), and setting $dT_m/d\theta=0$ on the assumption that S is independent of θ, the result is

$$\left.\frac{dP}{d\theta}\right|_{\theta=\theta'_{opt}} = \frac{P'_{opt}}{\theta'_{opt} + \theta'_{cc}} \qquad (34)$$

where

$$\theta'_{cc} = \theta_{cc}\left(\frac{H+K}{H+F}\right) \qquad (35)$$

The term θ'_{cc} is a cost-modified value of the cleaning time constant, θ_{cc}, which becomes equal to θ_{cc} for the particular case where K=F, but is commonly greater than θ_{cc} because K is usually greater than F. Thus the optimum throughput for a minimum cost cycle is such that the tangent from $\theta = -\theta'_{cc}$, P=0 to the curve of P vs. θ touches the latter at (θ'_{opt}, P'_{opt}). Again the graphical construction, Figure 2, is identical with that for constant θ_K [40], except that θ'_{cc} now replaces θ'_K. The actual cleaning time is then $\theta_{cc} + aP'_{opt}$.

In both Figures 1 and 2 an alternative procedure is shown whereby the tangent is drawn to the curve of P vs. ($\theta+\theta_{cv}$) instead of P vs. θ. The virtue of these simple graphical methods is that they do not require a mathematical equation relating P and θ, the curve of which can be based directly on operating data. The minimum cost procedure can be modified for a variable rather than constant energy cost per unit mass of product [42].

FOULING OF HEAT EXCHANGERS

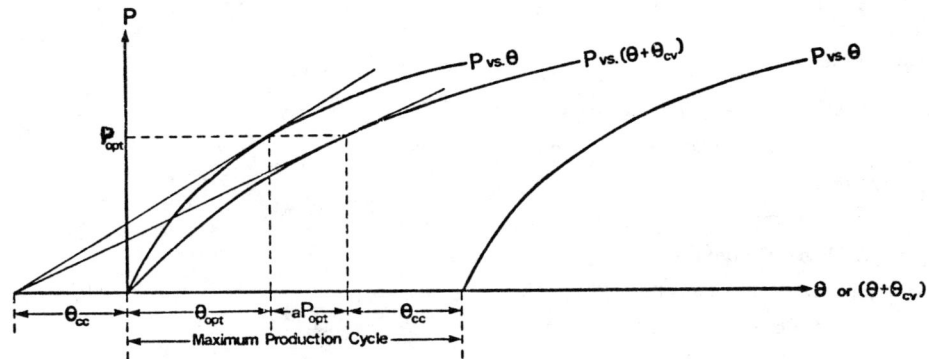

Figure 1. Maximum throughput cycle for cleaning time θ_K which is a linear function of P: $\theta_K = \theta_{cc} + \theta_{cv}$, $\theta_{cv} = aP$; from [42].

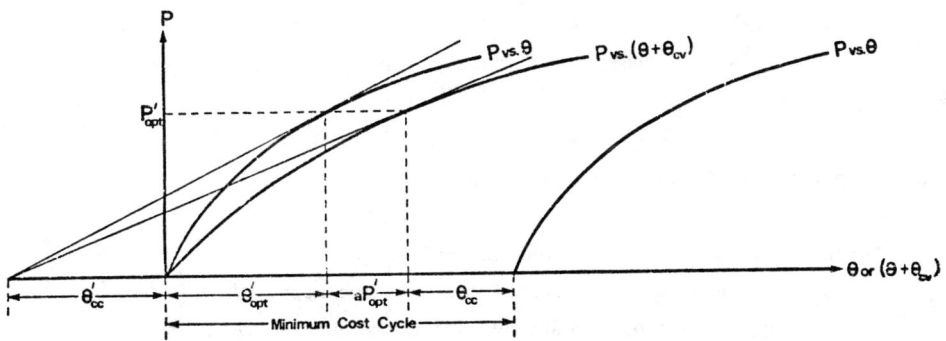

Figure 2. Minimum cost cycle for cleaning time θ_K which is a linear function of P: $\theta_K = \theta_{cc} + \theta_{cv}$, $\theta_{cv} = aP$; from [42].

ACKNOWLEDGMENT

I am grateful to the Natural Sciences and Engineering Research Council of Canada for continuing financial support, and to AERE Harwell for three months of telephone-free facilities.

NOMENCLATURE

A	heat transfer surface area
$A_{l.m.}$	logarithmic mean of A_{out} and A_w
A_{out}	surface area on outside of deposit
A_w	wire surface area
A'	Arrhenius coefficient
a	constant in equation (29), s/kg
B	constant in equation (21)
C_b	bulk concentration of key component, kg/m^3
C_s	concentration of key component adjacent to surface, kg/m^3
C_{sat}	saturation concentration of crystallizing species at surface temperature, kg/m^3
c	constant in equation (13), dimensionless
c_p	heat capacity of fluid at constant pressure, J/kgK
D	diffusivity of key component in fluid, m^2/s
D_f	effective diffusivity of oxygen through fouling deposit, m^2/s
d	differential
d_p	particle diameter, m
E	activation energy, J/mol
e	base of natural logarithms = 2.718
F	variable (e.g. labour) costs per unit time during production, \$/s
f	Fanning friction factor, $2\tau_s/\rho u_b^2$, dimensionless
G_{z_H}	heat transfer Graetz number, wc_p/kL, dimensionless
G_{z_M}	mass transfer Graetz number, $w/\rho DL$, dimensionless
H	fixed overhead costs per unit time, \$/s
h	convective heat transfer coefficient, W/m^{2}K
h_o	convective heat transfer coefficient at time zero, W/m^{2}K
K	variable (e.g. labour) costs per unit time during cleaning, \$/s
k	thermal conductivity of fluid, W/mK
k_B	Boltzmann constant = 1.38×10^{-23} J/K
k_f	thermal conductivity of fouling deposit, W/mK
k_m	convective mass transfer coefficient, m/s
k_p	particle thermal conductivity, W/mK
k_r	attachment rate constant

FOULING OF HEAT EXCHANGERS

k_t transport coefficient, m/s
L   duct length, m
l   characteristic linear dimension of duct (e.g. pipe diameter), m
m   mass of deposit per unit surface area, kg/m^2
m*   asymptotic value of m, kg/m^2
$\dot{m}_d$ deposition flux, kg/m^{2}s
$\dot{m}_r$ removal flux, kg/m^{2}s
n   order of surface reaction, dimensionless
Nu   Nusselt number, hl/k, dimensionless
P   cumulative product at time θ, kg
P_{cycle} cumulative product for one operating cycle, kg
P_{opt} optimum value of P per cycle for maximum daily production, kg
P'_{opt} optimum value of P per cycle for minimum cost per unit mass of product, kg
Pe_H heat transfer Peclet number, RePr, dimensionless
Pe_M mass transfer Peclet number, ReSc, dimensionless
Pr   Prandtl number, $c_p\mu/k$, dimensionless
q   heat transfer rate, W
$\dot{q}$ heat flux, W/m^2
R   overall production rate = $P_{cycle}/(\theta+\theta_K)$, kg/s
R_g universal gas constant = 8.314 J/molK
R_f thermal resistance of fouling deposit for a unit surface area, m^{2}K/W
R_f^* asymptotic value of R_f, m^{2}K/W
Re   Reynolds number, $lu_b\rho/\mu$, dimensionless
S   energy costs per unit mass of product, \$/kg
S_p sticking probability, dimensionless
Sc   Schmidt number, $\mu/\rho D$, dimensionless
Sh   Sherwood number, $k_m l/D$, dimensionless
St_H heat transfer Stanton number, Nu/Pe_H, dimensionless
St_M mass transfer Stanton number, Sh/Pe_M, dimensionless
T   fluid temperature, K
T_b bulk fluid temperature, K
T_{in} bulk fluid inlet temperature, K
T_m total operating cost per unit mass of product, \$/kg
T_{out} bulk fluid outlet temperature, K
T_s heat transfer surface temperature, K
T_θ total operating cost per unit time, \$/s
T_w wall or wire temperature, K
T_{wo} wall or wire temperature at time zero, K

812   FOULING IN HEAT EXCHANGERS

t^+	relaxation time, $\rho_p d_p^2 u_*^2 \rho / 18\mu^2$, dimensionless
U	overall heat transfer coefficient, $W/m^2 K$
U_o	overall heat transfer coefficient at time zero, $W/m^2 K$
u_b	bulk fluid velocity, m/s
u_t	terminal settling velocity of particles, m/s
u_*	friction velocity = $\sqrt{\tau_s/\rho}$ = $u_b \sqrt{f/2}$, m/s
v_t	thermophoretic velocity, m/s
w	mass flow rate, kg/s
x	deposit thickness, m

Greek

α	thermophoretic velocity coefficient, dimensionless
β	contact angle, dimensionless
ε	total fractional removal of particles from fluid by wall, dimensionless
ε_d	fractional removal of particles by Brownian diffusion, dimensionless
ε_g	fractional removal of particles by gravity, dimensionless
ε_i	fractional removal of particles by inertia, dimensionless
ε_t	fractional removal of particles by thermophoresis, dimensionless
η	total surface efficiency, dimensionless
η_f	fin efficiency, dimensionless
θ	time, production time, s
θ_c	time constant, s
θ_{cc}	constant portion of cleaning time, s
θ'_{cc}	$\theta_{cc}(H+K)/(H+F)$, s
θ_{cv}	variable portion of cleaning time = aP, s
θ_D	delay time, s
θ_K	cleaning time, s
θ'_K	$\theta_K(H+K)/(H+F)$, s
θ_{opt}	optimum production time per cycle for maximum daily production, s
θ'_{opt}	optimum production time per cycle for minimum cost per unit mass of product, s
μ	fluid viscosity, kg/ms
ν	fluid kinematic viscosity, μ/ρ, m^2/s
ρ	fluid density, kg/m^3
ρ_f	density of fouling deposit, kg/m^3
σ	surface tension, N/m
σ_c	critical wetting tension, N/m
τ_s	fluid shear stress at surface, N/m^2
ψ	deposit strength

Subscript

crit   critical

REFERENCES

1. Thackery, P.A. 1979. The cost of fouling in heat exchange plant. Inst. Corros. Sci Tech./I.Chem.E. Conference on Fouling - Science or Art? University of Surrey, Guildford, U.K., pp. 1-9.

2. Van Nostrand, W.L., Leach, S.H. and Haluska, J.L. 1981. Economic penalties associated with the fouling of refinery heat transfer equipment, in Fouling of Heat Transfer Equipment, E.F.C. Somerscales and J.G. Knudsen, eds. Hemisphere, New York, pp. 619-643.

3. Standards of the Tubular Exchangers Manufacturers Association (TEMA), 1978. New York, pp. 140-142.

4. Bond, M.P. April 1981. Plate heat exchangers for effective heat transfer. The Chemical Engineer, No. 367, pp. 162-167.

5. Epstein, N. 1981. Fouling in heat exchangers & Fouling: technical aspects, in Fouling of Heat Transfer Equipment, E.F.C. Somerscales and J.G. Knudsen, eds. Hemisphere, New York, pp. 701-734 & pp. 31-53.

6. Taborek, J., Aoki, T., Ritter, R.B., Palen, J.W. and Knudsen, J.G. 1972. Fouling — the major unresolved problem in heat transfer. Chem. Eng. Prog., Vol. 68, No. 2 pp. 59-67, No. 7 pp. 69-78.

7. Knudsen, J.G. 1981. Apparatus and techniques for measurement of fouling of heat transfer surfaces, in Fouling of Heat Transfer Equipment, E.F.C. Somerscales and J.G. Knudsen, eds. Hemisphere, New York, pp. 57-81.

8. Walker, R.A. and Bott, T.R. March 1973. Effect of roughness on heat transfer in exchanger tubes. The Chemical Engineer, pp. 151-156.

9. Bryers, R.W., ed. 1978. Ash Deposits and Corrosion due to Impurities in Combustion Gases, Hemisphere, 691 pp.

10. Chambers, A.K., Wynnyckyj, J.R. and Rhodes, E. 1981. Development of a monitoring system for ash deposits on boiler tube surfaces. Can. J. Chem. Eng., Vol. 59, pp. 230-235.

11. Fowler, H.W. and McKay, A.J. 1980. Chapter in Microbial Adhesion to Surfaces, R.C.W. Berkeley, J.M. Lynch, J. Melling, P.R. Rutter and B. Vincent, eds. Ellis Horwood Ltd., Chichester, pp. 143-161.

12. Duddridge, J.E., Kent, C.A. and Laws J.F. 1981. Bacterial adhesion to metallic surfaces. Inst. Corros. Sci. Tech. Conference on Progress in the Prevention of Fouling in Industrial Plant, Nottingham University, U.K., pp. 137-153.

13. Troup, D.H. and Richardson, J.A. 1978. Scale nucleation on a heat transfer surface and its prevention. Chemical Engineering Communications, Vol. 2 pp. 167-180.

14. Watkinson, A.P. and Epstein, N. 1969. Gas oil fouling in a sensible heat exchanger. Chem. Eng. Prog. Symp. Series, Vol. 65, No. 92, pp. 84-90.

15. Baier, R.E. 1981. Early events of micro-biofouling of all heat transfer equipment, in Fouling of Heat Transfer Equipment, E.F.C. Somerscales and J.G. Knudsen, eds. Hemisphere, New York, pp. 293-304.

16. Einstein, A. 1906. Ann. Physik., Vol. 19, p. 371.

17. Reichardt, H. 1951. Fundamentals of turbulent heat transfer. Translated from Archiv f. die Gesamte Warmetechnik No. 6/7. Nat. Adv. Comm. Aero. TM 1408 (1957) and N-41947 (1956).

18. Metzner, A.B. and Friend, W.L. 1958. Theoretical analogies between heat, mass and momentum transfer and modifications for fluids of high Prandtl or Schmidt number. Can. J. Chem. Eng., Vol. 36, pp. 235-240.

19. Cleaver, J.W. and Yates, B. 1975. A sub layer model for the deposition of particles from a turbulent flow. Chem. Eng. Sci., Vol. 30, pp. 983-992.

20. Newson, I.H., Bott, T.R. and Hussain, C.I. 1981. Studies of magnetite deposition from a flowing suspension. ASME HTD-Vol. 17, pp. 73-81.

21. Gudmundsson, J.S. 1981. Particulate fouling, in Fouling of Heat Transfer Equipment, E.F.C. Somerscales and J.G. Knudsen, eds. Hemisphere, New York, pp. 357-387.

22. Whitmore, P.J. and Meisen, A. 1977. Estimation of thermo- and diffusiophoretic particle deposition. Can. J. Chem. Eng., Vol. 55, pp. 279-285.

23. El-Shobokshy, M.S. 1981. A method for reducing the deposition of small particles from turbulent fluid by creating a thermal gradient at the surface. Can. J. Chem. Eng., Vol. 59, pp. 155-157.

24. Watkinson, A.P. and Epstein, N. 1970. Particulate fouling of sensible heat exchangers. Proc. 4th Int. Heat Transf. Conf., Vol. 1, Paper HE 1.6, 12 pp.

25. Bowen, B.D. 1978. Fine particle deposition in smooth channels. Ph.D. thesis, University of British Columbia, Vancouver, Canada.

26. Bowen, B.D. and Epstein, N. 1979. Fine particle deposition in smooth parallel-plate channels, J. Coll. Interf. Sci., Vol. 72, pp. 81-97.

27. Konak, A.R. 1974. A new model for surface reaction-controlled growth of crystals from solution. Chem. Eng. Sci., Vol. 29, pp. 1537-1543.

28. Kern, D.Q. and Seaton, R.E. 1959. A theoretical analysis of thermal surface fouling. Brit. Chem. Eng., Vol. 4, No. 5, pp. 258-262.

29. Loo, C.E. and Bridgwater, J. 1981. Theory of thermal stresses and deposit removal. Inst. Corr. Sci. Tech. Conference on Progress in the Prevention of Fouling in Industrial Plant, Nottingham University, U.K., pp. 137-153.

30. Cleaver, J.W. and Yates, B. 1976. The effect of re-entrainment on particle deposition. Chem. Eng. Sci., Vol. 31, pp. 147-151.

31. Morse, R.W. and Knudsen, J.G. 1977. Effect of alkalinity on the scaling of simulated cooling tower water. Can. J. Chem. Eng., Vol. 55, pp. 272-278.

32. Gudmundsson, J.S. 1977. Fouling of surfaces. Ph.D. thesis, University of Birmingham, U.K.

33. Rodliffe, R.S. 1978. Chemiphoresis as the controlling mechanism for particulate corrosion product deposition — an explanation for auto-retardation, CEGB Report No. RD/B/N4374, Berkeley Nuclear Laboratories, U.K.

34. Hasson, D. 1962. Rate of decrease of heat transfer due to scale deposition. Dechema-Monographien, Vol. 47, pp. 233-252.

35. Reitzer, B.J. 1964. Rate of scale formation in tubular heat exchangers. Ind. Eng. Chem. Proc. Des. Dev., Vol. 3, pp. 345-348.

36. Galloway, T.R. 1973. Heat transfer fouling through growth of calcareous film deposits. J. Heat Mass Transf., Vol. 16, pp. 443-460.

37. Dinkelacker, A. August 1979. Play tornado-like vortices a role in the generation of flow noise? Reprint from <u>Mechanics of Sound Generation in Flows</u>, E.-A. Müller, ed., IUTAM/ICA/AIAA Symposium, Max-Planck-Institut für Strömungsforschung, Göttingen, BRD, Springer-Verlag, Berlin Heidelberg.

38. Starner, K.E. May 1976. Effect of fouling factors on heat exchanger design. ASHRAE Journal, pp. 39-41.

39. Epstein, N. and Sandhu, K. 1978. Effect of uniform fouling deposit on total efficiency of extended heat transfer surfaces. Heat Transfer 1978 — Proc. 6th Int. Heat Transf. Conf., Vol. 4, pp. 397-402.

40. Epstein, N. 1979. Optimum evaporator cycles with scale formation. Can. J. Chem. Eng., Vol. 57, pp. 659-661.

41. Badger, W.L. and Othmer, D.F. 1924. Studies in evaporator design — VIII. Optimum cycle for liquids which form scale. Trans. AIChE, Vol. 16, Part 2, pp. 159-168.

42. Ma, R.S.T. and Epstein, N. 1981. Optimum cycles for falling rate processes. Can. J. Chem. Eng., in press.

43. Sandu, C. and Lund, D. 1981. Fouling of heat transfer equipment by food fluids: computational models. AIChE Symposium Series, in press.

Mechanisms of Furnace Fouling

JOHN R. WYNNYCKYJ and EDWARD RHODES
Department of Chemical Engineering
University of Waterloo
Waterloo, Ontario, Canada

ABSTRACT

Observations using a new device and technique for monitoring the onset and rate of growth of slag and ash as well as microscopic investigations of ash deposits have established a critical new fact. Deposits in coal fired furnaces do not grow by gradual, flat and even deposition of sticky fly ash particles onto the wall tubes. Rather, a very rough surface is produced by deposition of pyroclastic lumps. Then there is in-situ growth of these lumps. It seems likely that larger particles (Nodules), and not fly ash (Cenospheres), are the main contributors to in-situ growth. This is evident from the microstructure of the deposits as well as captured particles. It is, moreover, consistent with this that only the larger, heavier, particles would be able to penetrate the boundary layer between the stationary walls and the turbulent gas column in the furnace. It is when the occupation density of the Pyroclasts on the surface becomes large enough, then the dangerous consolidation and fusion begins.

1. INTRODUCTION.

In previous papers (1,2,3,4) the authors have presented the results of experimental work done on:

(a) the development and use of a monitoring system for measuring the amount of fouling on coal fired furnace walls (1,2),
(b) the dynamics of growth of fouling in coal-fired furnaces (3),
(c) and the microscopic structure of ash deposits found in pulverized-coal-fired boilers (4).

In this paper the results of the monitoring system results and the microscopic and micrographic results will be synthesized together with the general knowledge of furnace hydrodynamic behaviour to provide an interpretation of the mechanisms by which furnaces are fouled by ash deposits.

Because of the use of the more highly fouling coals there has been a dramatic increase in the size (as measured by the specific furnace release rate, W/m^2 of furnace area) and capital cost of new furnace installations between 1968 and 1978 in the U.S.A. and Canada. This trend is likely to continue as more Western coals come to be used. Fouling also increases operating costs. The better known effects of fouling are load cycling and shut-downs necessitated by catastrophic deposits.

Large-scale fouling of heat transfer surfaces has been studied mainly from the point of view of design of new installations (i.e., prediction of required heat transfer surface, and in what configuration to arrange it) (5,6), and evaluation of new coal sources (7,8). The major coal evaluation tests, such as coal ash softening and fusion temperatures, ash viscosity and sintering strength have been developed for this purpose. Some considerable success seems to have been achieved in correlating the tests with fouling behaviour, though pilot plant and full-scale testing seem still to be essential . The essential point is that these studies, though very important from the engineering point of view, are phenomenological. Their concern has been empirical correlation of measurable properties of coal ash with fouling. They have not addressed the question of mechanism. There has been relatively little experimental evidence reported on how one portion of the particles of mineral matter in the coal, rather than becoming fly ash, forms solid, large-scale deposits on either convection section or furnace wall tubes.

The literature reports that a property of "stickiness" (as yet ill-defined) is decisive in causing the fouling tendency of a coal (8,9). Stickiness, in turn, is associated with viscous, relatively low melting, liquid-phases which can be supercooled. Presence of sodium has been shown to be the principal factor (10,11) in Western coals. The mechanism of fouling proposed by Tuft and Beckering (9) envisages that stickiness arises as a result of condensation of compounds of vapour-phase sodium (and/or potassium) onto particles of "matrix parent". The matrix parent, in the case of fouling coals, is a complex silicate of a composition whose melting range is strongly influenced by sodium oxide. The fluxing envisaged would occur on the surface only and might be transient but present at the time of collision and deposit growth. This work, however, neither deals with the mechanistic aspects of collisions amongst the particles of mineral matter, nor does it deal with how the particles manage to impinge on the furnace wall.

In this paper an attempt is made to partly fill this gap. Experimental evidence is reported concerning growth of ash deposits on furnace walls of large steam generators. Of concern are the heavy, massive deposits, their formation, growth and consolidation on a macroscopic scale, the microstructure of such deposits, and the question of what are the building blocks of such deposits on the next smaller scale. Only when the mechanism of fouling is fully understood, will it be possible to design furnaces to eliminate the problem.

2. THE PARADOX OF FOULING IN FURNACES.

When the exit particles leaving furnaces are captured in electrostatic precipitators, it is found that, in the main, they are made up of tiny, hollow spheres about 5 to 20 microns in diameter (14). The problem posed by the fact that furnace walls do indeed become fouled, is that it is hard to see how such small and lightweight particles could penetrate a wall hydrodynamic boundary layer and impinge on the surface. It could be imagined that small numbers of such particles could diffuse to the surface, but it is hard to believe that the enormous amounts of deposits formed on furnace walls could be caused by such diffusion. Such a mechanism is, therefore, not very plausible and it is felt that another one, discussed below, satisfies the experimental furnace-fouling evidence far better.

MECHANISMS OF FURNACE FOULING

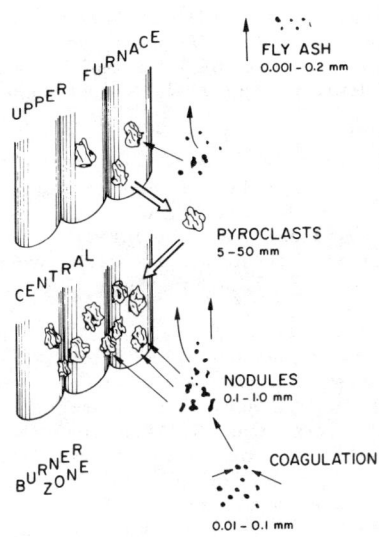

HYPOTHESIS OF DEPOSIT GROWTH

FIGURE 1.

3. HYPOTHESIS OF SCALE DEPOSITION AND GROWTH (see Figure 1).

Evidence will be discussed in this paper which will establish the fact that there are three main building blocks to the material which deposits on furnace walls. These are as follows:

"Cenospheres" (12,13,14). These are the basic ash particles which are in the order of 5-20 micrometers in diameter. These particles are, in the main, light-weight, hollow spheres. Just what proportion of the fly ash particles are Cenospheres is controversial. The most recent reviews (15,16) consider this number to be a few percent (based apparently on the original work of Raask (12) and Paulson and Ramsden (17)). A recent study, however, based on one Western lignite and one Eastern bituminous coal, concludes that most fly ash particles are Cenospheres (14). In fact it is shown that Cenospheres contain Cenospheres, contain Cenospheres, etc. and that this is the norm rather than an exception. (the name Plerosphere is proposed.) It seems to these authors that the wide density variations, without chemical compositional variations observed by Watt and Thorne (18) could be reinterpreted to mean void space within the spherical ash particles.

"Nodules". These are larger particles in the size range 0.1 to 1.0 mm in size. They appear to be made up of fused or agglomerated Cenospheres and are relatively dense. How they are first formed is not yet determined, however, it is suggested that they could be the result of particle-particle collisions in the very turbulent flame zone of the furnace. This kind of particle and inter-particle collision theory has been disputed in the literature on the grounds that only Cenospheres are noticed in the fly ash leaving the furnace in the exit gases. The hypothesis presented here is that the 0.1 to 1.0 mm particles have enough lateral momentum to penetrate the boundary layer adjacent to the furnace wall. Once in this layer, the vertical gas velocity is low enough so that the particles can no longer be fluidized, but rather they begin to fall downwards - some of them impinging on the slagging walls. Thus the furnace only elutriates the small, light particles and retains the heavier and larger ones.

"Pyroclasts". These are large lumps seen to be deposited in the early stages of wall fouling. It is believed these Pyroclasts (which are of the order of several centimeters in size) grow on the walls of the furnace by collecting more and more solid particles (Nodules). The Pyroclasts may become so large that they break from the surface and fall through the furnace, either to the very bottom, or to be deposited somewhere else on the furnace wall.

Thus the mechanism of wall fouling is postulated as follows: The basic Cenospheres are created during combustion. In the combustion zone, where enormous turbulence exists, a portion of the Cenospheres collide and agglomerate to form Nodules. The surviving Cenospheres are easily elutriated by the furnace gases, however, the Nodules penetrate the boundary layer adjacent to the furnace walls and are captured there either by collision with Pyroclasts (in the early stages of fouling) or by collision with the molten surface of the deposit (in the late stages of fouling). When Pyroclasts have become large and unstable they break away from the surface (or are blown off by the mechanical cleaning ("sootblowing") of the walls) to fall to the bottom or be redeposited elsewhere. In the remainder of this paper a wide variety of experimental evidence will be discussed which will be used to support the above hypothesis.

4. EXPERIMENTAL WORK

4.1 Generating Stations.

The studies were performed in three modern, large-scale, industrial boilers. One was at the Lakeview Generating Station of Ontario Hydro. This station was operated on a relatively non-fouling Eastern bituminous coal. Next was the Battle River G.S., owned by Alberta Power, Limited, burning the so-called Forestburg coal; a sub-bituminous, high sodium Western coal. Its feature was a high degree of variability in composition and ash-fusion characteristics. This variability; coupled with an inherently high fouling tendency, caused frequent and unpredictable fouling difficulties at this station. The third location was the Boundary Dam G.S. owned by Saskatchewan Power Corporation. The coal here was a Western lignite. From among the three, this station experienced most fouling under steady state operation, but was subject to less frequent catastrophic departures from normal than the Battle River G.S. Complete details of the furnaces and the coal analyses are given elsewhere (1,2,3,4).

4.2 Photography.

Still photographs were taken using a Pentax KM single reflex camera, a 200 mm focal length lens, and Kodachrome 64 film. In general illuminations as indicated by the camera's light meter were used. Later experience showed that use of a green filter gave improved results. The moving and time lapse pictures were taken with a N1Z0 S80, 8 mm camera, equipped with an 80 mm focal length lens using Kodachrome 40 cinecolor film.

Observation of furnace walls at the lower boiler levels was difficult. Not only was the heat intense but there was much interference from the turbulent flame with its high loading of incandescent particles. Time lapse photography demonstrated a wide random pattern of over and underexposure. The overexposed frames were also "washed out"; as if out of focus. The cause appeared to be flame bursts and light path interference through thermal convective effects. It was attempted to select the least "washed out" frames for reproduction but this was not always possible. For maximum information the reader is urged to view the photographs from as large a distance as possible. Detail integration is then improved and the three-dimensionality is brought out. The boiler wall tubes were 2.5 inches outside diameter at Battle River G.S. and Boundary Dam G.S. and 3.0 inches at Lakeview G.S. Remembering these dimensions will aid in visualizing the size of the ash pieces in the photographs. Because of the skewed angle of sight it is not possible to give a scale applicable to the whole area for these photographs.

4.3 Heat Flux Meters.

A system of measuring the amount of fouling deposit on the surface of a furnace wall was perfected during the course of this work. Basically a comparison between a heat flux meter mounted on the furnace wall and another meter mounted on a heat pipe and artificially kept clean by an air jet enabled a continuous output signal which represented the amount of wall fouling as a function of time.

The principle of the heat flux meters used and details of their installation in the three boilers are given elsewhere (1,2,3,4).

4.4 Material Sampling and Microscopy.

Samples of ash material were taken from both the flue gases and the ash pits in the bases of the furnaces. In addition lumps of slag were knocked from furnace walls (when the furnace was being repaired) and finally air-cooled metal probes were inserted into the furnaces to capture particles. All of the samples were observed with the naked eye and also by optical and scanning electron microscopes. Some of the samples were cut by diamond saws. Others were observed before and after they had been "cleaned" by a high velocity air jet. The results of these observations will be discussed below.

5. RESULTS AND DISCUSSION

5.1 Observed Characteristics of Large-Scale Deposition.

A characteristic feature of fouling in pulverized coal boiler furnaces is that it does not occur by deposition onto the tube surfaces of the fly ash particles to form an even flat layer of deposit. Rather there is initial deposition of relatively large ash aggregates which, once on the tube surfaces, grow substantially while the neighbouring tube surface retains a relatively clean condition. Thus, the term pyroclast or pyroclastic lump is introduced to describe these typical large aggregates. Figure 2 shows the macrospopic appearance of Pyroclasts.

A second feature of the deposition process is that it is highly dynamic. The growing Pyroclasts fall off or are torn off by the turbulent furnaces gases and reenter the furnace gas space. A significant fraction of them is again redeposited onto the tubes in another location. Others fall through the flame into the bottom ash hopper. It is these growing Pyroclasts superposed by others deposited in a secondary stage which eventually fuse over to form the semi-molten or slagging deposit often referred to in the literature.

Figure 3 shows three frames from a time-lapse photograph sequence taken at the Lakeview Station through an inspection port located at the corner of the furnace at elevation 95.2 meters. The viewing direction was from right to left at a very acute angle along the east wall so that any protruding wall deposits tended to be seen essentially in profile. The tubes in the right side of the photographs were close to the furnace corner and were relatively deposit free. Further to the left are the Pyroclastic deposits. The bright, washed out areas at the left of the photograph are due to flame radiation coming from that direction. The dark sides of the Pyroclasts were in the shadows of the flame. No magnification is indicated in the Figure. Because of the acute angle it is not constant, decreasing from right to left. The outside diameter of the tube was 3 inches (7.62 cm). This gives an approximate measure of the size of the individual Pyroclasts and shows them to be about the same as those in Figure 2.

The very uneven and spotty nature of the deposit is clearly evident in Figure 3. The surface coverage was quite incomplete. Other experience shows that, if the areas to the left of the photographs were viewed normally, they would show substantially less than 50 percent coverage. The times indicated in Figure 3 are from the beginning of filming and are not related to any significant operating parameter. It is evident that the

FIGURE 2. Typical appearance of pyroclastic lumps from boiler furnaces. (Boundary Dam G.S. Specimens)

MECHANISMS OF FURNACE FOULING

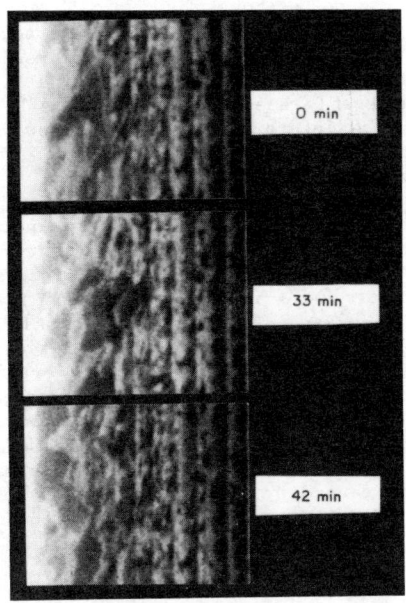

FIGURE 3.
Time lapse phogographs of a position on the wall of the Lakeview G.S. furnace.

deposit was of approximately the same severity throughout the 42 minute period but the individual details at a particular spot changed radically. Projecting this time-lapse film (which was taken at one frame every three seconds) at the normal speed of 18 frames per second, emphasises dramatically the three critical components of the deposition mechanism; impaction of previously formed Pyroclasts, growth of such Pyroclasts while attached to the walls (without there being much deposition on the neighbouring areas of the walls) and finally the frequent shedding of the Pyroclasts.

The Pyroclasts, as can be seen Figures 2 and 3, usually had a characteristic angular orientation to the tubes. In Figure 2 they are shown oriented in profile to further detail this point. When attached to the tube surface the latter was in contact with the straight edge at the Pyroclasts' left.

The conclusion from the macroscopic photographs is that Pyroclasts are a reality. They deposit from elsewhere in the furnace (they must break off or be blown off surfaces elsewhere). They grow while they are attached, presumably from the impactation of other smaller sized materials.

5.2 Slagging and Non-Slagging Deposits

Two of the Western furnaces, where our studies were carried out were subjected to extreme fouling and slagging of the furnace walls, causing costly operating problems. In the Eastern furnace, fouling was experienced to a significantly minor degree to be considered only a routine operating consideration, and not a problem. Figure 4 shows the typical degree to which heat fluxes were diminished by fouling. Even though fouling severity differed substantially among the three furnaces the fouling mechanism described above did not differ in principle, only in degree. In all three cases Pyroclasts deposited, grew, shed and redeposited. Thus, all three traces indicate random noise which can be interpreted to be the growing and shedding process. It is noticed that the noise is more frequent and has wider amplitude soon after a sootblow.

With the Eastern Station, the pyroclast population density increased only very slowly with time after a sootblow. The heat flux meter trace in Figure 4 shows nearly a steady-state condition and the fouling process was clearly slow. The individual Pyroclasts were never close enough to each other to consolidate into a coherent mass. The noise continued right through the run between sootblows. In the case of the Western Stations, on the other hand, the increase in the population density was very rapid as indicated by the rapid decrease in the heat flux after a sootblow. Moreover, about two hours after the sootblow, the deposit would show signs of consolidation. The latter is indicated in Figure 4 by the decrease in the random fluctuation in the heat flux traces, i.e., the smoothening out of the curve at long times after sootblow caused by the thermal inertia of the deposit. Clean heat flux meters showed (1,2) normal random flame noise in these periods of operation.

Further evidence to support this interpretation is given in Figure 5.

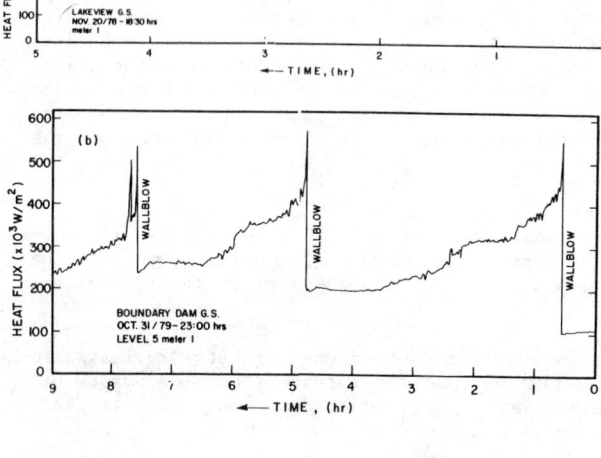

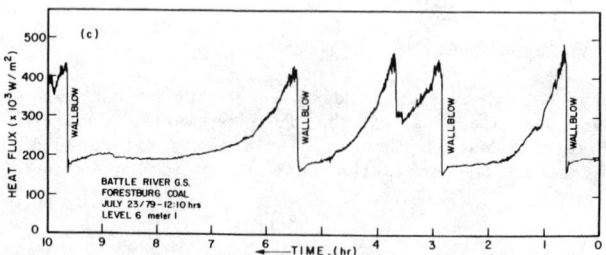

FIGURE 4. Comparison of typical heat flux meter outputs from the three generating stations studied

MECHANISMS OF FURNACE FOULING

The sharp peaks in this graph, as in Figure 4 before, correspond to deposit removal by sootblowing. The numerals along the rapidly decaying heat flux curve (between 1 and 4 hours) indicate the points in time when the respective photographs in the upper position of the Figure were taken. The photographs are of a position of the wall area containing the heat flux meter producing the trace and were taken by a still camera.

Picture 1 shows a relatively clean tube wall surface with several Pyroclasts already in evidence (the wall blower location is to the left of the pictures). Picture 2, taken about 7 minutes after the blow, shows that the tube surface is essentially fully covered, but the tubes are still fairly distinguishable. Significantly the coverage is by individual Pyroclasts of about the same size as shown in Figure 2 representing Boundary Dam Pyroclasts, as well as in Figure 3 representing Lakeview. After 19 minutes of operation (Picture 3) the deposit was quite heavy. Consolidation of the Pyroclasts into larger aggregates by local sintering accompanied by local fusion had begun. The latter had progressed significantly by 1 hour and 26 minutes of operation (Picture 4) leading to a dramatic coarsening of the deposit appearance. The fact that deposition started in the form of individual porous ash aggregates (the Pyroclasts) has resulted in this deposit having a strong relief. It is noteworthy, however, that the relief does not follow the tube pattern and it is shown elsewhere (3,4) that the tube pattern only reappears when the fusion and consolidation process has progressed much further.

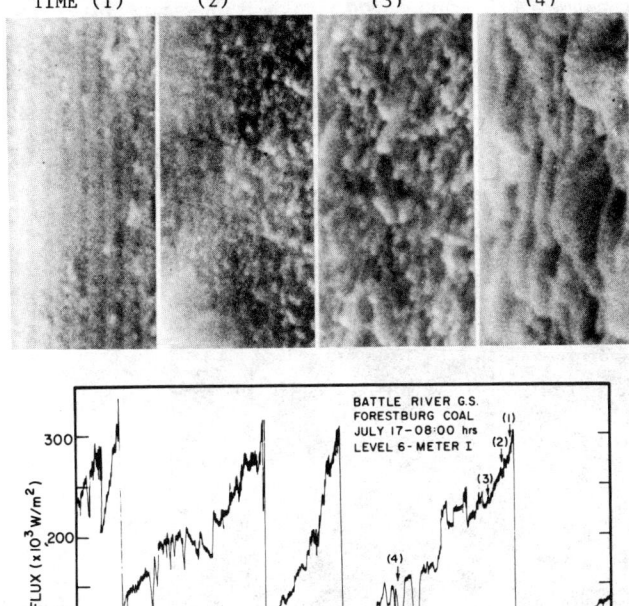

FIGURE 5. Time-lapse photograph and associated heat flux measurement

The conclusions from this work are that Pyroclasts are a reality. They can be seen with the eye and detected in the signals obtained from the heat flux meters. A dense covering of Pyroclasts leads to undesirable consolidation and fusion.

5.3 PYROCLAST MICROSTRUCTURE AND THE PROBLEM OF NODULES

The microstructure of the pyroclast described above has been investigated by optical and scanning electron microscopy. Figure 6a shows the typical microstructure of a sectioned specimen (cut with a diamond saw) and 6b shows the undisturbed pyroclast surface which had been exposed to the flame. The identity of the well-known typically spherical fly ash particles has been essentially lost during growth. The structure of the undisturbed surface is quite symptomatic of local consolidation or accretion of the individual particles into larger aggregates by local partial fusion and sintering. This consolidation process apparently stopped as soon as a new surface layer of particles was deposited and insulated the previous particles from the radiation from the flame. Accordingly the structure of the sectional cut in Figure 6a features intermeshed aggregates and large pores.

The appearance of the microstructure and the size of the pores and of the aggregates, typically 500 microns or more, suggests that the Pyroclasts were in the main built up from particles which were larger than the typically 5 to 20 micrometer Cenospheres. This conclusion was confirmed fully by an investigation of deposits in the convection section and of the furnace fly ash. One should note that at the higher levels in the boiler

FIGURE 6. Typical Pyroclast and its Microstructure.

the temperatures are not high enough to cause the sort of fusion that must have been present in the sintering consolidation observed in Figure 6b. In the deposition in the upper parts of the furnace or in the convection section the particles form local bridging bonds but there can be no general consolidation.

Figure 7 shows the various microscopic appearances of a sample obtained by impaction onto a cool sampling tube positioned just below the first convection pass. Figure 7a is of the deposit which had been in contact with the metal. Figure 7b is a cross-section and 7c is the deposit surface which faced the gas stream. Figure 7d is a SEM photomicrograph of the undisturbed surface of the deposit in Figure 7c at right. This shows some of the familiar spherical Cenosphere particles of the expected size range. There are also broken up aggregates with rough spongy appearing surfaces (e.g., in the top left corner) possibly products of sootblowing of wall ash. In addition, however, there are larger particles fully consolidated with smooth surfaces that are shaped in a way which suggests that they formed by aggregation of smaller fused or semi-fused spherical particles. They seem to form a very substantial portion of the total deposit. All these particles are joined to each other by local bonds.

These larger aggregate particles referred to above have been given the name Nodules. It was concluded that they play a major role in the growth of the Pyroclasts. Their microstructural characteristics are revealed well by blowing away the surface ash Cenospheres, shown in Figure 7d, by a strong compressed air jet or by careful fracturing or diamond saw sectioning of a deposit followed by an air jet treatment. The resulting structure is shown in Figure 7e and reveals clearly the basic Nodules. The right side of this photomicrograph shows the diamond saw cut through two neighbouring ones. The other fracture surfaces, the isolated mottled areas,

FIGURE 7. Appearance of a sample captured in the convection zone.

are the broken bonds between the Nodules and the smaller particles blown away by the air jet.

The existence of furnace ash in the form of a larger particle typically 0.1 to 1.0 mm in size and entrained in the furnace gases must be considered to be established. The orgin of the Nodules and the mechanism of their formation must, at this stage, remain a matter of some speculation but their role in deposition on heat transfer surfaces is clearly important.

6. THE RELEVANCE OF FLUID PARTICLE DYNAMICS TO THE FOULING PROBLEM.

6.1 Turbulence Conditions.

From the point of view of the dynamics of gas flow, observations show that a large pulverized coal fired furnace can be looked upon as a gas column noving relatively rapidly past the furnace walls, e.g., at 15 to 35 m/sec (19). Within this column there are two major sources of local turbulence patterns. The first is the flame or flames induced by the burners firing into a gas space. This is a source of very high intensity local turbulence. The second is the change in direction of the major flow by virtue of the fact that burners fire at right angles (or nearly so) to the direction of the rising gas column. The location of both the latter sources of turbulence is at the lower furnace levels. The intensity of turbulence, therefore, decreases with increasing level but, because of wall boundary layer effects, large scale turbulence fields of much lower intensity persist at the upper furnace levels as well.

If the furnace flow was purely potential flow, the chances of lightweight microscopic particles colliding and agglomerating would be negligible. However, the flow pattern in a furnace is not potential flow. Rather it is a complex mixture of highly turbulent, boundary layer and (to some small extent) potential type flow.

For deposits to grow there must be "successful" collisions of the relatively small ash particles which are produced from the accretion of mineral matter within the individual coal particles. Stickiness will determine the collision success. The following extreme, and somewhat idealized, types of these collisions are all possible in furnaces:

(a) between the moving particles and the stationary tubes set at sharp angles to the direction of bulk flow (e.g., lower convection section tube banks, but also critical areas of furnace walls in, and above, the burner zone). Relatively high probability of collision is expected
(b) between the moving particles and tube surfaces parallel to direction of bulk gas flow (middle and upper furnace levels). Very low probability of collision is expected
(c) collision between particles (of various sizes) suspended in the turbulent gas phase. Evidence exists to show probable high collision rates exist.

Particles suspended in a carrier gas will experience collisions only when they cross the streamlines established by local flow patterns. Collisions with (and deposition on) a surface parallel to the direction of bulk gas flow must overcome the resistance of the viscous sublayer and the buffer layer between the stationary wall and the turbulent bulk of the gas. Such deposition was successfully interpreted by Friedlander and Johnston

(20) as an eddy diffusion process, analogous to the eddy diffusion transport of one gas species within another. The model applies to particles of a size less than the scale of turbulence of the gas eddies. It envisages that the momentum imparted to a particle, which is moving within an individual eddy, is sufficient to propel the particle across the viscous layer and the buffer layer. The calculation proceeds by assuming a phenomenological particle eddy diffusivity which is numerically equal to the eddy diffusivity of the carrier gas. According to this model the collision rate is proportional to the fourth power of particle size.

Collisions with surfaces at sharp angles with the direction of bulk flow are a case of inertial impaction recognized in many particle collection applications. Particles of sufficient mass readily cross streamlines fixed by local changes in direction of carrier gas flow, including local eddies. According to Hedly et al (19) 30 μm particles will penetrate to cylindrical surfaces located at right angles in a carrier gas in potential flow at 8 m/sec.

The applicability of these ideas to the problem of collisions in boiler furnaces has been recognized (3) but there has been no research reported which was directly concerned with defining the properties of the large scale turbulence fields in the boiler furnace nor how they affect collisions of ash particles with each other or with the furnace walls. However, Nodule formation by particle particle collision seems quite possible in the turbulent conditions of a furnace.

6.2 Circumstantial Experimental Evidence for Gas Phase Collisions.

There is some indirect or circumstantial evidence to the effect that gas phase collision within the flame space may be frequent. The first is the fact already noted that burning of "fouling" coals produces a smaller ratio of the number of ash particles to coal particles. Although this could be due to less shattering of the char particles the possibility of coagulation is equally open. Another is the tracer test. Addition of cuprite, Cu_2O, melting point 1233°C, to the primary, fluidizing air in the combustion of a Western lignite resulted in the dispersion of copper to essentially all fly ash particles (as determined by electron microprobe analysis of the fly ash particle surfaces (21)). The third is the very tangible benefits in controlling fouling derivable from addition of powdered CaO or MgO to the coal stream. Characteristically the effects have been found quite insensitive to the degree of distribution during addition (22). All these observations suggest that there is extensive "communication" between particles of ash in the pulverized coal flame and in the boiler furnace in general. Such communication can only mean interparticle collision while still in the gas phase.

7. CONCLUSIONS

Visually observed, physically measured (heat flux probes) and circumstantial evidence all combine to support the hypothesis that deposit growth on furnace wall surfaces is dynamic and involves three basic particles, Cenospheres, Nodules and Pyroclasts. The Cenospheres combine to form Nodules which transport to the walls to enable the growth of Pyroclasts which in turn disengage and reattach and eventually fall to the bottom of the furnace.

It could be concluded that if any of the links in this dynamic process could be broken - e.g., prevention of Cenosphere collision or prevention of Nodule penetration of the wall boundary layer - then the tendency of the furnace to foul could be reduced.

8. ACKNOWLEDGEMENTS

The Canadian Electrical Association is gratefully acknowledged for funding and guiding this work.

Ontario Hydro, Alberta Power Limited and Saskatchewan Power Corporation are thanked for making the boiler experiments possible.

9. REFERENCES

1. Chambers, A.K., Wynnyckyj, J.R. and Rhodes, E. 1981. Development of a Monitoring System for Ash Deposits on Boiler Tube Surfaces. Can. J. Chem. Eng. 59, pp. 230-235.

2. Chambers, A.K., Wynnyckyj, J.R. and Rhodes, E. 1981. A Furnace Wall Ash Monitoring System for Coal Fired Boilers. Trans. A.S.M.E. J. Eng. Power, 103 (3) pp. 532-538.

3. Wynnyckyj, J.R., Chambers, A.K. and Rhodes, E. 1981. Fouling Dynamics in Furnaces of Large Pulverized Coal Fired Boilers. Accepted for publication in Can. J. Chem. Eng.

4. Wynnyckyj, J.R., Reiner, M.J. and Rhodes E. 1981. Structure of Ash Deposits on Large Pulverized Coal Fired Boilers. Accepted for publication in Can. J. Chem. Eng.

5. Smith, V.L. 1976. J. Eng. Power, 98 (3) pp. 297-304.

6. Reid, W.T. 1971. External Corrosion and Deposits - Boilers and Gas Turbines. American Elsevier, N.Y.

7. Sondreal, E.A., Selle, S.J., Tufte, P.H., Menze, V.H. and Laning, V.R. April 1977. Proceedings 30th Am. Power Conference, I.T.T., Chicago.

8. Boow, J. 1972. Fuel, 51, pp.170-177.

9. Tufte, P.H. and Beckering, W. 1975. J. Eng. Power, 97, pp. 407-412.

10. Sondreal, E.A., Gronhovd, G.H., Tufte, P.H. and Beckering, W. 1978. Ash Deposits and Corrosion due to Impurities in Combustion Gases. Ed. R.W. Byers, Hemisphere Publishing, pp. 85-111.

11. Babcock and Wilcox Staff. 1972. Steam/Its Generation and Use. Babcock and Wilcox Co., N.Y.

12. Raask, E. Sept. 1968. Cenospheres in Pulverized-Fuel Ash. J. of the Institute of Fuel, pp. 339-344.

13. Ramsden, A.R. 1969. A Microscopic Investigation into the Formation of Fly-Ash during the Combustion of a Pulverized Bituminous Coal. Fuel, 48, pp. 121-137.

14. Fisher, G.L., Chang, B.P.Y. and Brummer, M. 1976. Fly Ash Collected From Electrostatic Precipitators. Science, 192, pp. 553-555.

15. Wall T.F., et al. Pergamon Press 1979. Mineral Matter in Coal and the Thermal Performance of Large Boilers. Progress in Energy and Combustion Science, 5, pp. 1-29.

16. Flagan, R.C. and Friedlander, S.K. 1978. Particle Formation in Pulverized Coal Combustion - A Review. Recent Dev. Acrosol. Sci. Editor D.T. Shaw, Wylie, pp. 25-59.

17. Paulsen, C.A.J. and Ramsden, A.R. 1970. Some Microscopic Features of Fly Ash Particles and Their Significance in Relation to Electrostatic Precipitation. Atmos. Environ., 4, pp. 175-185.

18. Watt, J.D. and Thorne, D.J. 1965. J. Applied Chemistry, Vol. 15, pp. 585-594.

19. Reid, W.T. 1971. External Corrosion and Deposits - Boilers and Gas Turbines. American Elsevier, N.Y., pp. 16-18.

20. Friedlander, S.K. and Johnston, H.F. 1957. Ind. Eng. Chem., 49, p. 1151.

21. Sondreal, E.A., Granhovd, G.H., Tuft, P.H. and Beckering, W. 1974. Ash Fouling Studies of Low Rank Western U.S. Coals. ASME Symp., Hemisphere Press, pp. 85-111.

22. Anderson, B., Baker, B. and Gardiner, W. 1977. The Control of Slagging and Fouling When Burning Lignite Fuel in Pulverized Fuel Generators. Can. Electrical Association Meeting, Fall Session.

Some Aspects of Heat Exchangers Surface Fouling Due to Suspended Particles Deposition

M.A. STYRICOVICH
USSR Academy of Sciences

O.I. MARTYNOVA, V.S. PROTOPOPOV, and M.G. LYSKOV
Moscow Power Institute
Moscow, USSR

ABSTRACT

Deposition processes of suspended in water iron oxide particles on steam-generating tube surfaces were investigated. It was shown, that on a line with thermo-hydrodynamic conditions the influence of physico-chemical, especially electrochemical factors on these processes seems to be of significance.

As it is well known, under steam generating heat exchangers operation conditions, the tubes material being austenitic or carbon steel, two iron oxide layers are formed on the wall surface. The characteristics of these layers, their formation mechanisms, their structure, density, etc. differ quite significantly, resulting in different influence on heat and mass transfer. The inner layer (topotactic) results, as a rule, from direct oxidation of the metal surface due to its interaction with the heat medium. This layer consists mainly of magnetite, is very dense with good adhesion to the surface and is protecting the metal from further intensive corrosion. The thickness of this layer on austenitic steel surfaces is about 1-2 μm, the density - about 4 g/cm^3.

The outer -epitactic- layer of corrosion products on tube surfaces is formed by deposition of already existing in the fluid in various forms corrosion products and can, depending on operation conditions, consist of two sublayers: the first - formed by crystallization from supersaturated real solutions due to temperature or other parameters changes; the second - formed by deposition of suspended in the heat medium particles of various degree of dispersion, including colloidal size (mostly in the range from 10^{-3} μm to several μm).

The mechanism of crystallization from real solutions caused, for example, by temperature or pH values changes influence on solubility, is relatively simple and has been established rather well. At the same time the mechanism of suspended particles deposition under various thermo-physical and phasico-chemical conditions, though having been the subject of a number of experimental (as well as theoretical) investigations and theoretical hypothesis can not be considered to be so clear. Just this layer because of its high porosity, relatively low adherence to the underlaying surface, low density, etc. can be the source of, depending on operation conditions: a) disturbance of normal heat and mass transfer, in some cases very serious, leading to heavy consequences; b) transport of radioactivity, etc.

This is why one more experimental investigation of suspended particles of steel corrosion products on heat exchangers inner tube surfaces was carried out under various conditions: iron oxide concentrations, pH-values, steam quality, heat fluxes etc. with special attention to the electrochemical aspect of this phenomena. Most of experiments were carried out with austenitic steel tubes with diameter 8/6 mm at 7 MPa pressure. The inlet temperature was kept up at 210°C, the outlet equilibrium steam quality approached 0,3. The average mass velocities values made up 1100 kg/m^2.sec., heat flux density - 250 kW/m^2, (heating - by superheated steam with pressure 3,5 MPa and temperature about 400°C). Zones of subcooled non-boiling, surface boiling and nucleate boiling water were covered. Heat medium pH-values were kept at 4.5; 7.5 and 9.7; in most tests corrosion products concentrations were - artificially - kept on the level of about 1 ppm (though some tests were carried out with natural corrosion products of the installation's material of construction - about 40 ppb Fe) and dissolved in the water oxygen concentration was about 500 ppb. Some tests at 7 and 14 MPa were carried out with electrical heating of the experimental sections allowing to get higher heat flux densities up to 1000 kW/m^2 and more.

The experimental installation was a closed circuit formed by several heat exchangers. The main element of the circuit is an experimental steam generating tube divided under steam heating conditions in four, connected in consecutive order heated sections, with about 400 mm non-heated intervals (spaces) in between. This partition allowed to obtain practically constant heat flux densities by means of increasing the heating steam velocity to compensate decrease of steam temperature. Besides, the mentioned partition led to practically complete equilibrium at the inlet of each section. Some experimental sections were totally electroisolated (via the metal) from the other elements of the circuit, permitting to measure under steam heating conditions electrochemical values and to control their variation. The installation was equipped with instrumentations for the purpose to measure the electrophoretic mobility of corrosion products particles at high temperatures and pressures, as well as their degree of dispersion. The heat medium was high purity water with an electroconductivity about 0,1 μS/cm and pH_2 about 6.8; to get lower or higher pH values, HNO_3 or KOH were used. More detailed description of the used experimental installation is given in / I /.

Each test included three stages:

1) Depositions accumulation on the operating heating surfaces of the experimental installation under given operation parameters during a definite time interval (up to 300 h);

2) Cutting the experimental sections into samples and measuring deposition characteristics by metallographic and electronomicroscopic methods.

3) Handling and analysis of the experimental data.

The relative error (accuracy) of determination: the layer's thickness, the quantity and deposition grow rate, their volumetric density and porosity did not exceed 10%.

Under applied experimental conditions the mentioned above external corrosion products layer, loose incoherent, porous (porosity about 60-90%), relatively slightly linked with the surface, consisted of a mixture of hematite and magnetite (hematite was dominating); the layer was formed by particles with a size in the range of some thousandth to unities of μm /2/. Main attention during conducting the experiments was paid to generating data on structure, thermal properties, formation mechanism of the suspended particles layer with special regard to the electrochemical aspects.

The thickness of the outer iron oxides layer, the amount of deposition per unit of area and the average, during the test period, deposition rate in all tests were not uniformly along the steam generating tube. Maximal deposition rate was under conditions of surface boiling and at start of nucleate boiling, the most sharp change occurring at start of the surface boiling zone (Fig. 1). These results are in good accordance with results ob-

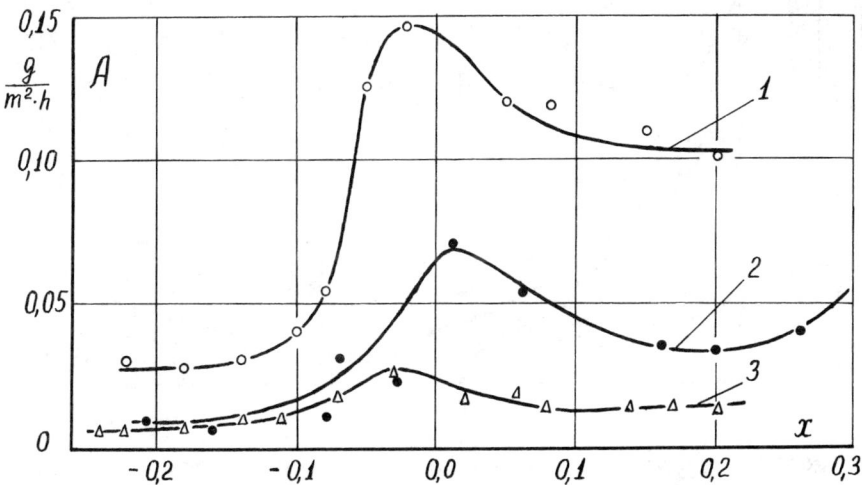

Fig.1. Influence of the heat medium relative enthalpy (x) on the suspended corrosion products layer formation rate(A).

$p= 7$ MPa; $w\rho = 1100$ kg/(m^2.s); $q=250$ kW/m^2;
$C_{Fe}=1,3$ ppm; $C_{O_2}=0.5$ ppm; 1.$pH_{25}=4.9$; 2.$pH_{25}= 9,6$;
3. $pH_{25} = 7.5$.

tained in the investigations / 3,4 /. In the boiling zone (especially surface boiling zone), depositions have an elevated porosity (up to 90%). Here a "chimney" structure, first mentioned by Macbeth / 5 / was observed. Typical for these conditions were high effective thermal conductivity values of the depositions equal to $\lambda_{ef} = 2,5$ W/m.K at $q = 250$ kW/m² and increasing with heat flux up to $\lambda_{ef} = 8-9$ W/m.K at $q = 1000$ kW/m² / 6 /.

Great influence on deposition formation rates and deposition thermophysical characteristics by the fluid's pH values was detected (Fig. 2).

When corrosion products concentration in the water are artificially equalized and under other equal conditions deposition formation rates were minimal at near to neutral pH, this minimum is most deeply expressed in zones of subcooled surface and nucleate boiling.

A similar conclusion about the most important determination of pH on the deposit weight of corrosion products on heated surfaces was obtained in investigations of some Canadian scientists / 7 /. But the optimal pH-value in their work providing minimal deposition rates, was much higher and at pH_{25} near 7, deposition rate was considerable.

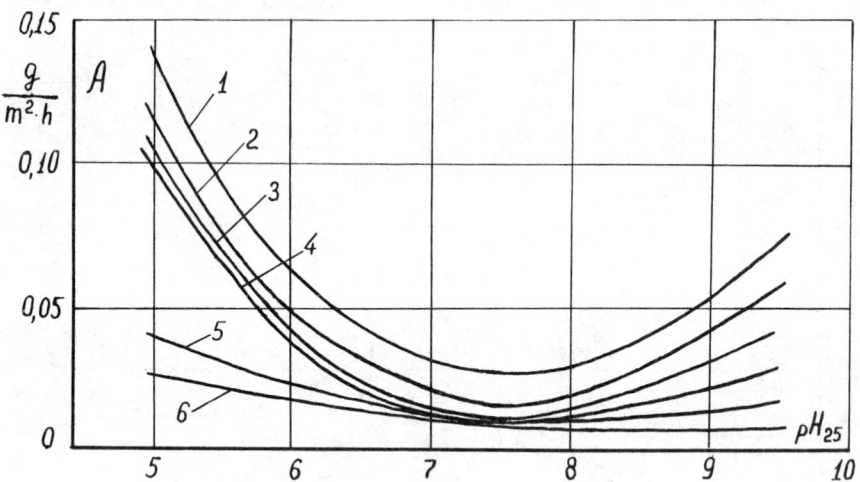

Fig.2. Medium pH_{25} values vs. suspended corrosion products layer formation rate (A).
p= 7 MPa; $W\rho$ = 1100 kg/m²s); q=250 kW/m²; C_{Fe} = 1,3 ppm; C_{O_2} =0,5 ppm; 1- x=0,0; 2- x=0,05; 3- x=0,1; 4 - x=0,15; 5 - x = -0,1; 6 - x = -0,2; x - relative enthalpy.

This discrepancy can, it seems, be explained the way that regardless of the broad range of all other variables in the Canadian experiments, they were carried out under conditions of molecular hydrogen excess in the water, whereas in our tests an excess of oxygen was maintained. It is interesting to mention that in the smaller part of the Canadian experiments, carried out under boiling conditions, a decrease of deposition rate by more than one order of magnitude was noticed at $pH_{25} = 7.2$, when hydrogen concentration was decreased.

An analysis of our investigation's results has shown the different deposition formation rates in zones with different thermal and hydrodynamic regimes to be determined by change of mass exchange processes intensity between the bulk flow and boundary layer. At the same time the obtained data point to the fact that a significant role in the deposition process falls to electrophoretic precipitation of charged particles. Electrochemical forces influence the particles transport in the fluid's boundary layer and their binding on the surface. Iron corrosion products particles, electrophoretic velocity measurements have shown that under conditions of the experiment their isoelectric points correspond to a pH_{25} value near to 7 (Fig.3). / 8 /.Measurements of potentials, originating between circuit sections, being in different heat conditions, have shown that the tube surface's potential magnitude and sign depend on a large number of factors. The most important of them are: temperature, heat fluxes, density, mass velocity, pH value, the medium's redox condition (excess of oxygen or hydrogen), surface material of construction etc. These measurements to a certain degree characterize the presence and magnitude of potential difference between the tube wall and the thermal boundary layer. The character of this difference, interdependence with pH values, according to several preliminary measurements, is shown on Fig. 4. At different combinations of water chemistry parameters, zero potential values in the thermal boundary layer can correspond to one or several pH_{25} data. Under given investigation parameters (oxygen concentration about 500 ppb), the potential in the thermal boundary layer at low pH value - is positive (positive charged particles move to

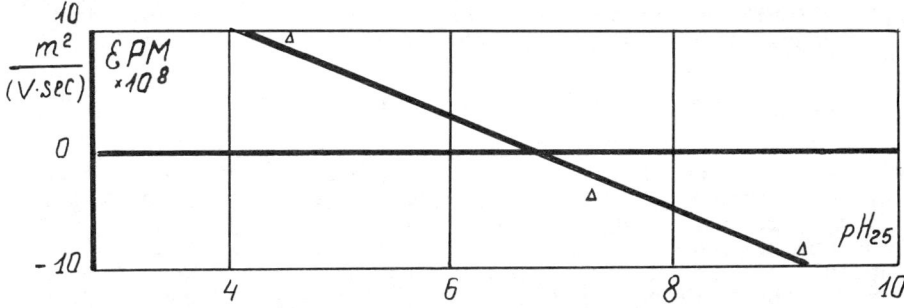

Fig.3. Electrophoretic mobility (EPM) of iron corrosion products particles.

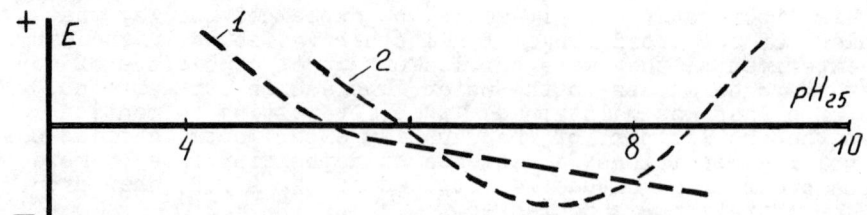

Fig.4. Qualitative characteristics of potential change in the heat medium thermal boundary layer. 1. C_{O_2} = 0.5 ppm; C_{O_2} < 0.06 ppm.

+E - positive charged particles move to the tube's wall;
-E - negative charged particles move to the tube's wall (surface boiling).

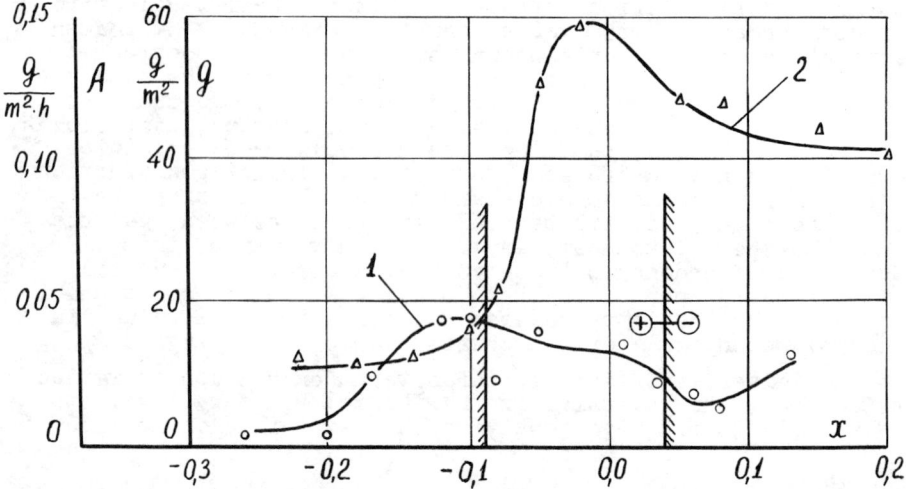

Fig.5. Comparison of iron oxide deposition formation rate in tests with external electric field put in (curve 1) and without electric field (curve 2).

pH_{25} = 4,5; C_{Fe} = 1,3 ppm; C_{O_2} = 0.5 ppm.

the tube wall), at high pH values - negative (negative charged particles move to the tube wall). These results are qualitative in good correlation with data on Fig.2. Depending from the magnitude and sign of the particles charge and the tube metal's potential, forces of electrical interaction either increase or decrease the number of particles, penetrating the boundary layer and approaching the surface up to a distance, where intermolecular attraction begins to act. At pH values equal to pH in the isoelectric point, the particle can overcome the boundary layer resistance only due to its kinetic energy after crossing the boundary from the fluid's bulk. This is why - so it seems - a dependence of deposition rate from pH value is observed.

To confirm such explanation of suspended corrosion products particles deposition mechanism, some special experiments were carried out. In so doing, on the surface of a electroisolated tube part some potential from an outer source was provided, approximately corresponding to the potential difference in the thermal boundary layer but reverse by sign. This led to a significant drop of the deposition formation rate as compared with tests, carried out when other conditions were equal but no external electric field was put on (Fig.5).

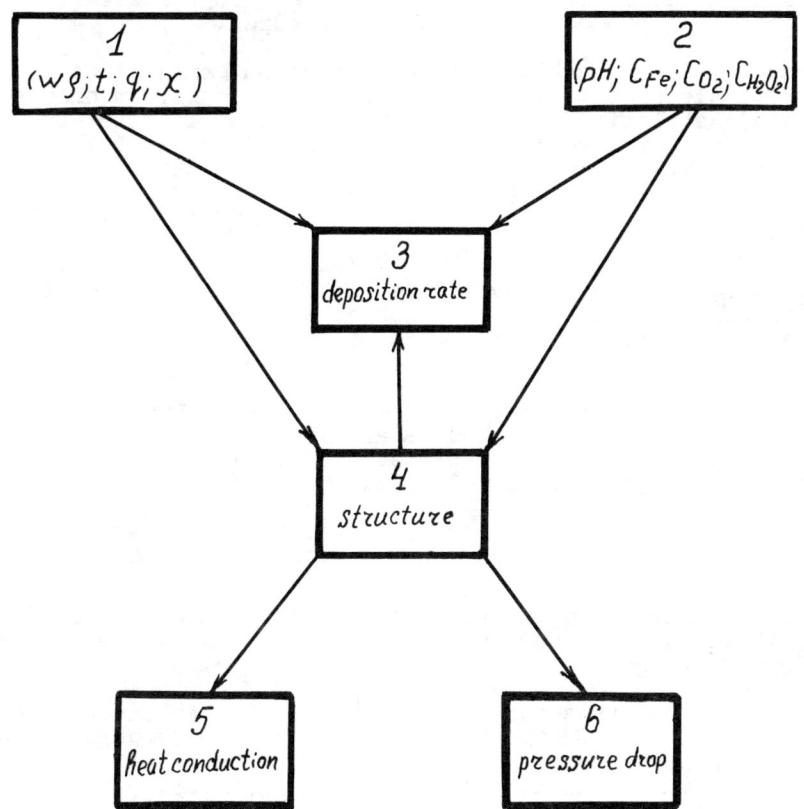

Fig.6. Interconnection of thermophysical and physico-chemical parameters in iron oxide deposition formation processes. 1.Thermo-hydrodynamic parameters ($w\rho$, t, q, x); 2.Water chemistry parameters (pH, C_{Fe}, C_{O_2}, $C_{H_2O_2}$). 3.Deposition rate; 4. Deposits structure; 5.Deposits effective thermal conductivity; 6. Deposition's influence on hydraulic pressure drop.

The accomplished investigation has shown the presence of a close interconnection between thermophysical and physico-chemical factors in iron-oxide deposition formation processes (Fig. 6). Received results can be considered to show that suspended particles deposition seems to be determined by turbulent diffusion of particles from the bulk into the boundary layer, as well as electrophoretic precipitation of these particles on the heated tube,s surface. The deposition rate can be reduced by optimization of parameters, determining each of the process stages.

When optimizing the water chemistry of steam generating equipment (heat exchanges), it is necessary to take into consideration not only the influence of water chemistry parameters on materials of construction, their corrosion characteristics, but also these parameter's influence on heat exchange surfaces fouling by corrosion products deposition formation.

Results of the carried out investigation confirm the at least principle possibility to create a method of iron oxides deposition restriction by putting on the thermoelectric effect. At present further investigations are carried out in this direction by using improved experimental methods.

REFERENCES.

1. Martynova O.I., Reznikov M.I. et al., 1974, Some regularities of corrosion products deposition on austenitic steel steam generating surfaces. Proc.Moscow Power Institute, vol.200, pp. 133-140.

2. Reznikov M.I., Menshikova V.L. et al., 1980, Fractional composition of corrosion products and its influence of iron oxide deposition process on steam generating surfaces. Proc. Moscow Power Institute, vol. 466, pp. 10-17.

3. Kabanov L.P. 1971. Heat and Mass transfer as related to corrosion product deposition. Energy Nuclear, v.18, N 5, pp.285-294.

4. Thomas D., Grigull U. 1974. Experimentelle Untersuchung über die Ablagerung von suspendiertem Magnetit bei Rohrströmungen in Dampferzeugern. B.W.K. 26, N 3, pp. 109-115.

5. Macbeth R.V., Trenbreth R., Wood R.W. 1971. An Investigation into the Effect of "Crud" Deposition Surface Temperature, Dry-Out and Pressure Drop with Forced Convection Boiling of water at 69 bar in an Annular Test Section, AEEF-705, p. 33.

6. Martynova O.I. et al., 1979. Thermal conductivity of iron-oxide depositions. Proc.Moscow Power Inst.,vol.405,pp.21-28.

7. Burril K.A., Shaddick A. 1980. Water chemistry and the deposition of corrosion products on nuclear reactor fuel: A summary of experimental results. Chalk River Nuclear Laboratories, Chalk River, Ontario.

8. Martynova O.I.,Gromoglasov A.A. et al.,1977.Investigation of temperature effect on the electrophoretic mobility of corrosion products particles,Teploenergetica,N2,ppp 70-73.

Fouling in Crude Oil Preheat Trains

GERALD A. LAMBOURN
Total C.F.R.
Paris, France

MARC DURRIEU
Total C.F.R.
Harfleur, France

ABSTRACT

The crude oil distillation unit is a major energy consumer in any refinery. The necessity to ensure efficient heat recovery from refluxes and product streams has been the subject of many articles (1). The efficient operation, monitoring and prevention of fouling in such systems have been sparsely reported ; most articles in the literature omit to compare different units or to suggest ways of improving existing exchangers or preheat trains so as to minimize fouling or its effects.

This paper addresses the topics of crude preheat train performance monitoring and compares actual fouling resistances with those normally recommended.

The different mechanisms of fouling are discussed and an innovative technique developed by TOTAL - UMP is described which has shown promising results in industrial trials.

By linking fouling rates to energy usage for a specific unit we are able to calculate an optimum interval for exchanger cleaning based on present day fuel, additive and cleaning costs.

Finally different exchanger and preheat train configurations are compared in an effort to point up useful criteria for their selection.

1. MONITORING PREHEAT EXCHANGER PERFORMANCE

1.1. The necessity for good hardware and software

Typically the preheat temperature at the inlet to the crude unit heater decreases by 30 - 50°C over a 12 month period (2) ; this loss of preheat temperature can be particularly serious for units which have elaborate heat recovery systems where preheat temperatures may approach 300°C.

As shown later some fouling phenomena are transient in nature, this fact together with the need to monitor preheat train performance efficiently convinced TOTAL - C.F.R. of the need to improve the frequency and accuracy of its procedures for calculating fouling resistances.

All exchangers downstream of the desalter are now equiped with thermocouples, replacing the previous thermowells; in some cases the additional thermocouples are directly linked to online process control computers. In order to provide a uniform basis of calculation HTRI' s ST-4 program was selected and used on an off-line basis; this type of advanced program is sufficiently accurate to check the efficiency of cleaning ($R_s \doteq 0$ after cleaning) which is not always the case of methods in the open literature.

Data input has now been simplified by using the ST - 4 program to generate simple power type relationships based on flow-rates and physical properties such as viscosity. These are then placed in the on-line computer for daily use by operating personnel.

1.2. Exchanger fouling rates measured by rigorous methods

Figure 1 below shows the measured fouling rates for the hottest crude/residue exchangers of two different crude distillation units of relatively modern design. The operating characteristics and constructional details of the two sets of exchangers are shown in Table 1. The fouling rates shown compare well with those reported by Haluska for similar exchangers.

Figure 2 shows the measured fouling rates for crude/hot reflux exchangers in the same two units, the evolution of fouling is here much more erratic being due to excessive break through of salt from the desalter.

Fouling curves of the same form as those shown were reported by Van der Wee and Tritsmans (3).

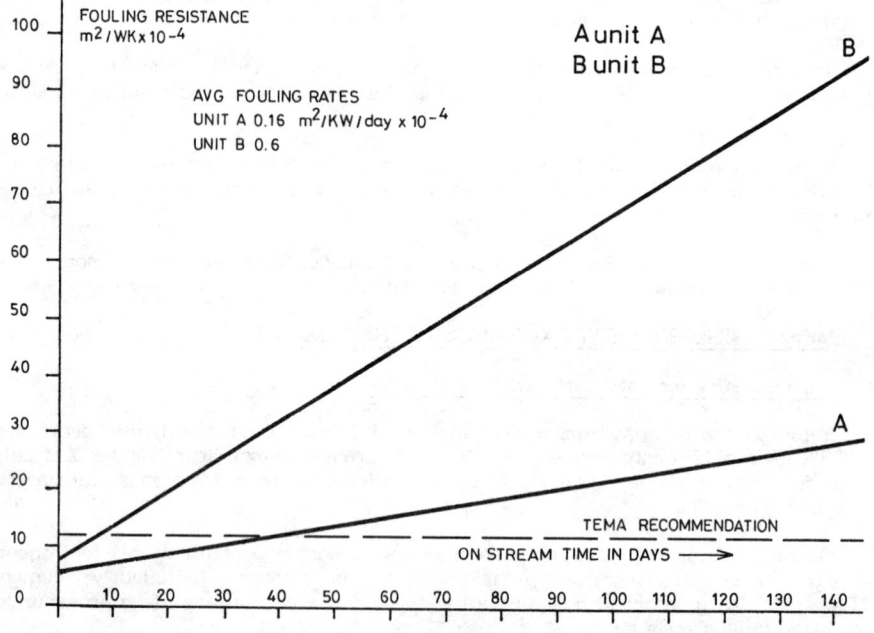

FIGURE 1 : EVOLUTION OF FOULING RESISTANCES FOR CRUDE/RESIDUE EXCHANGERS

FOULING IN CRUDE OIL PREHEAT TRAINS 843

TABLE 1
CRUDE/HOT RESIDUE EXCHANGER DESIGN CHARACTERISTICS
(BY HTRI ST - 4 PROGRAM)

	UNITS	UNIT A CRUDE 30 000 tonnes/d	UNIT A RESIDUE	UNIT B CRUDE 20 000	UNIT B RESIDUE
Flowrate	kg/s	323.5	117.1	217.7	75.7
Température IN	°C	229	374	199	361
OUT	°C	269	265	247	224
Duty	MW	37		29	
Shell type			AES		AES
Tube length	m	6.1		6.1	
Tube diam	mm	25.4		19.4	
N. of passes		2	1	2	1
Baffle spacing/cut	mm/%		250/15.0		229/10.0
Surface/Shell	m²	536		484	
N Shells in series		2	2	2	2
in parallel		2	2	2	2
Tube side velocity	m/s	1.34		1.32	
Reynolds No. in		30 000	13 800	25 420	6 870
out		37 000	6 420	35 000	1 720
Shell side velocity					
Bundle/Window	m/s		0.94/0.98		0.57/1.22
Clean heat trans coeff	W/m²K		490		468
Service " "			280		290
Fouling factor (Design)	m²/WK	.0005	.0009	.00035	.0009

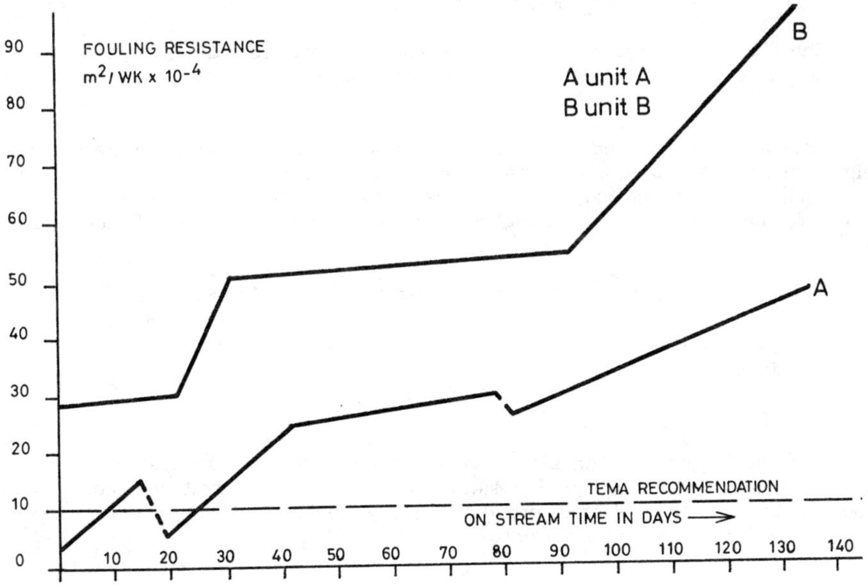

FIGURE 2 : EVOLUTION OF FOULING RESISTANCES FOR CRUDE/HOT REFLUX EXCHANGERS

In each case it is to be noted that the fouling resistances following 6 months of operation considerably exceed the recommended TEMA values.

In addition it is clear that the rate of fouling increase in the crude/residue exchangers of UNIT B is considerably greater than the same exchangers of UNIT A. A very careful review of all the pertinent operating conditions has failed to pin-point the exact reason for this difference ; however as will be explained later fouling on the residue side appears important for UNIT B. As can be seen from Table 1 UNIT B's exchangers suffer from excessively low velocities on the residue (shell) side.

1.3 Fouling resistances adopted for design purposes

As a result of these repeated observations TOTAL CFR now specifies a fouling resistance of 40×10^{-4} m^2/WK for average quality crude oils between 180 and 300°C, this increased fouling resistance is aimed to provide a cycle length of at least six months between cleaning.

The effect of this greater value on the required heat exchange area is not as large as might first be imagined as many preheat improvement projects are based either on recovering additional low level heat from products or by exporting heat from pumparound refluxes to other services.

2. MECHANISMS OF FOULING FOR CRUDE OILS

The reduction of the levels of fouling found in crude oil preheat trains requires first of all that valid mechanisms are available for the different types of fouling encountered.

The examination and analysis of the performance of different preheat trains enables the following conclusions to be drawn.

2.1. Exchangers up stream of desalter

The cold exchangers before the desalter are subject to fouling by salts which are easily eliminated by water injection to the charge pump, an injection of stripped process water (2 to 6 % vol) is sufficient to cure such a problem.

It is worthwhile to point out that although desalter temperatures are normally below 140°C the use of split-stream preheat trains before the desalter can excacerbate this problem. Such a configuration of preheat train has recently become popular as a means to improve low level heat recovery by a better match of heat capacity flowrates. (1)

It is relatively easy to allow the temperature in one branch of such a system to exceed 180°C while still maintaining the specified desalter temperature. Thus the monitoring and control of crude oil temperatures in each branch must be borne in mind when designing such a system.

2.2. Exchangers downstream of desalter

The exchangers situated after the desalter can suffer from deposits of calcium sulfate, of bicarbonate or magnesium salts from sea water. The first two deposits can be precipitated by a combination of crude oil salinity, caustic injection rate and a desalting water which is too hard and occur between 180 and 200°C.

The exchangers placed immediately after the desalter can in certain cases be subject to very rapid fouling by salt which is precipitated as the water dissolves in the crude as shown in figure 3 ; this phenomena occurs when the concentration of salt in the crude leaving the desalter exceeds 25 ppm Wt (6 pt b). Heavy crudes of less than 30 API exhibit unfavorable settling characteristics and thus such crudes can contain more than 250 ppm Wt (60 pt b) * before the desalter. Recourse must be had in such situations to fluxing or two stage desalting.

*   pt b : pounds per thousand barrels.

2.3. Crude/hot residue exchangers

The hottest exchangers (generally crude/residue exchangers) are subject to heavy fouling, generally on the tube side. The deposits are formed from a mixture of asphaltenes and iron sulfide strongly bound together, see Table 2.

The composition of these deposits varies with the crude oil treated ; in the case of light North African crudes (API gravity above 40°) it is composed essentially of iron salts (up to 80 %) and can be easily removed by acid washing. The treatment of heavier crudes (API gravity below 37°) results in the appearance of asphaltic deposits which cannot be easily removed by chemical treatment.

TABLE 2
AVERAGE COMPOSITION OF THE CRUDE SIDE DEPOSITS IN CRUDE/RESID EXCHANGERS

		Crudes processed	
		Various crudes API Gravity ≃ 34°	Light crudes API Gravity ⩾ 40°
Asphaltenes + Carbenes + Condensation Products	Wt %	60 - 75	3 - 10
Water soluble Salts	Wt %	1 - 5	1 - 5
Iron salts (Oxydes and/or sulfides)	Wt %	20 - 35	75 - 90

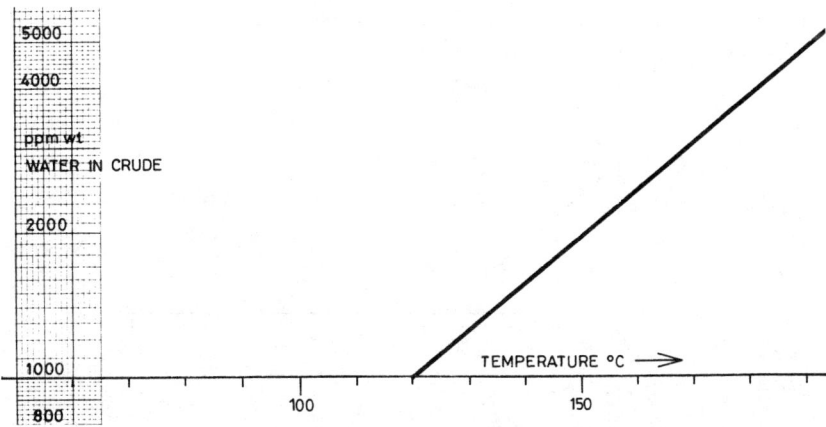

FIGURE 3 : SOLUBILITY OF WATER IN MIDDLE EAST CRUDES (30 API - 880 kg/m^3)

3. A NEW METHOD FOR CONTROLLING CRUDE OIL FOULING

3.1. Nature of stable desalter emulsions

About 4 years ago work was started by TOTAL - C.F.R. on a new approach to anti-fouling control which aims at limiting asphaltene - iron sulfide deposits in the hottest crude/residue exchangers and has formed the subject of a systematic study in our laboratories. (5) (6)

Van der Wee & Tritsmans emphasised some time ago the importance of the emulsion which is after formed at the water-interface level in a crude oil desalter. It is suprising that their pioneer work has not been further developed since desalter demulsifiers are selected and injected solely on the basis of maintaining an oil free desalter effluent water while anti-foulants are injected into the crude downstream of the desalter.

Evaluations of common crudes (density below 840 kg/m^3 or 37 API) show the following characteristics upstream of the desalter :

Water soluble salts (as NaCl) ppm Wt		50 - 300 (15 - 100 ptb)
water content	% Wt	0.2 - 0.8
Suspended solids content :	ppm Wt	100 - 530
Iron salts	ppm	5 - 30
Insoluble asphaltenes	"	100 - 500
Sand etc	"	5

Meanwhile we have found out that the insoluble asphaltene content varies significantly with temperature (Fig. 4) as suspended asphaltenes are partly redissolved near 130°C and then reprecipitated by thermal deasphalting above 200°C.

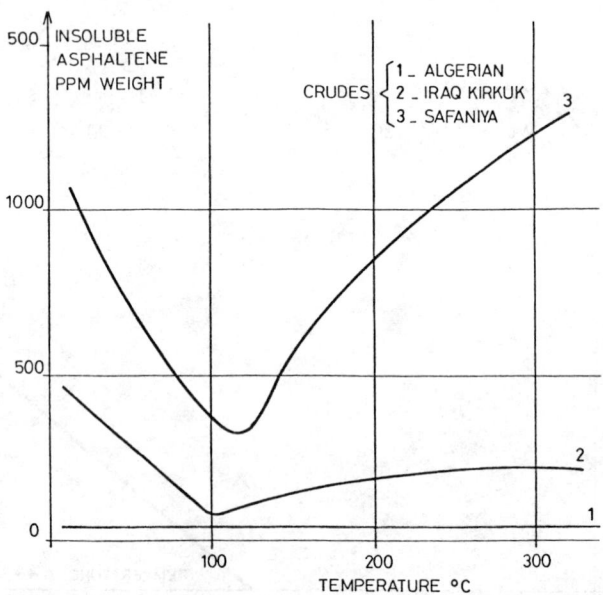

FIGURE 4 : THERMAL EVOLUTION OF INSOLUBLE ASPHALTENES IN CRUDE OILS

The usual water content of the desalted crude ranges from 0.2 to 0.6 % Wt and a typical single stage desalting efficiency is 90 %. Nevertheless certain situations can arise which result in the following composition :

Water soluble salts (as NaCl)	ppm Wt	10 - 20 (3 - 6 ptb)
Water	Wt %	0.7 - 1.0
Suspended solids content	ppm Wt	250 - 800
Iron salts	"	100 - 300
Insoluble asphaltenes	"	150 - 500

During such an upset a thick "stable emulsion" layer gradually forms above the normal oil/water interface, having the following composition :

Desalting water	% Wt on emulsion	50 - 75
Crude	" " "	50 - 25
Solidparticles	% Wt on crude	2 - 15
Viscosity	cPs 60°C	1000 - 10000
Waterdrop diameter	m	10 - 100

The "stable emulsion" is immiscible with water and oil at desalter temperatures.

In addition a microscopic examination shows a shell of colloidal asphaltenes and solid particles which coats the water droplets and effectively impedes any natural coalescence.

The solid particles have the following typical composition :

Iron oxides/sulfides	% Wt	50 - 70
Insoluble asphaltenes	"	50 - 30

This composition closely approximates that of the fouling deposits found in exchangers operating above 180°C on the crude side. The stability of the emulsion is promoted by the presence of solid particles that are equally well wetted by the oil and water phases.

PHOTO 1 : TUBE SIDE FOULING CRUDE HOT RESIDUE EXCHANGER UNIT A

The stable emulsion builds up and is then removed in an irregular fashion by the desalted crude. The severity of the fouling deposit thus caused can been seen from the thickness of the deposit shown in Photo 1.

It should be emphasised that this mechanism is applicable to crude oils having an asphaltene content above 1.3 % Wt, the practice of recycling recovered slop oils to the fresh crude can lower this threshold level to about 1.0 % wt.

3.2. Pilot plant study of fouling caused by stable emulsions

The mechanism proposed by Van der Wee & Tritsmans was confirmed in a series of pilot plant experiments previously described (4) in which a variety of clean crudes were "doped" with stable emulsion and heated to 320°C under laminar flow conditions (velocity 20 mm/s N_{Re} = 250).

The test rig was used to develop a dual purpose additive designed to:

Break the stable emulsion in the desalter
Prevent solids deposition by dispersing solids
and preferentially wetting exchanger tube walls.

In the pilot plant the additive formulation finally retained resulted in a six-fold reduction in the fouling rates measured under standardised conditions. (Fig. 5) (6)

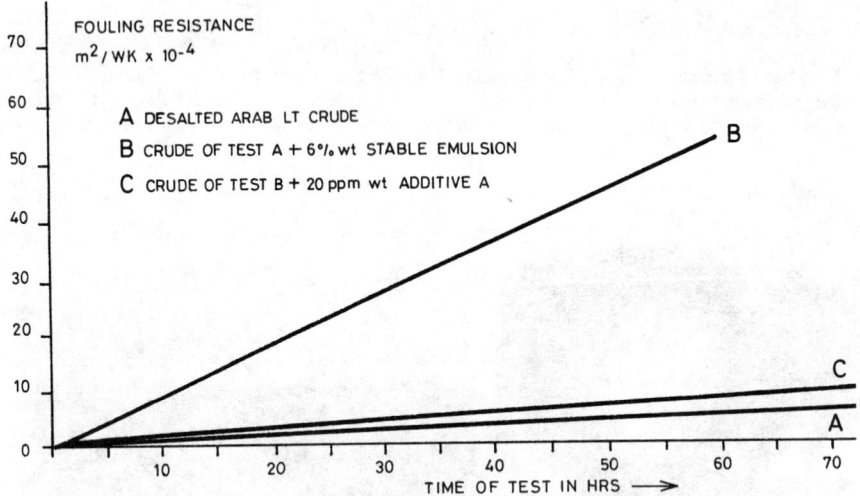

FIGURE 5 : FOULING RESISTANCE BUILD-UP MEASURED ON PILOT PLANT

3.3 An improved additive injection system

A novel injection system has also been developed for the injection of this additive by which the surfactant is injected directly into the "stable emulsion" layer thus increasing the effective concentration of the additive as compared to simple injection into the bulk of the crude. (5)

4. INDUSTRIAL RESULTS WITH ADDITIVE INJECTION

4.1. Fouling rates with additive injection

To date two industrial trials have been carried out using an additive developed by TOTAL and marketed by Universal Matthey Products.

The first trial was made on the crude/hot residue exchangers of UNIT A referred to in para. 1.2. ; injection rate was 13 ppm Wt. This trial was followed by a similar application on Unit B previously referred to.

The results are shown in Figure 6 from which it can be concluded that in each case the fouling rate has been halved as compared with the reference cases. As remarked on earlier it would appear that residue side fouling is a significant factor for the exchangers of Unit B.

Confirmation of the additives ' efficiency can be seen from Photos 2 and 3 which show the tube side conditions for Unit A's crude residue exchangers.

PHOTO 2 : CRUDE SIDE CRUDE/RESID EXCHANGER UNIT A WITHOUT ADDITIVE

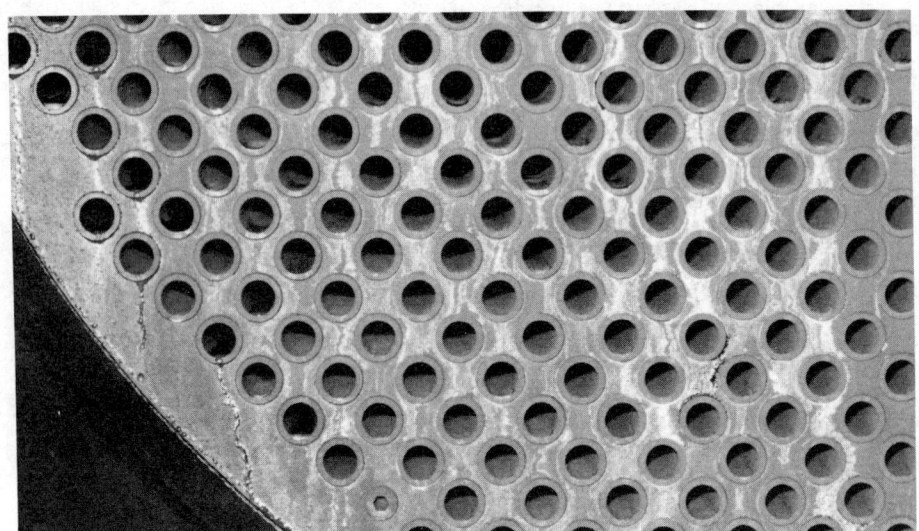

PHOTO 3 : SAME EXCHANGER AS PHOTO 2 WITH ADDITIVE

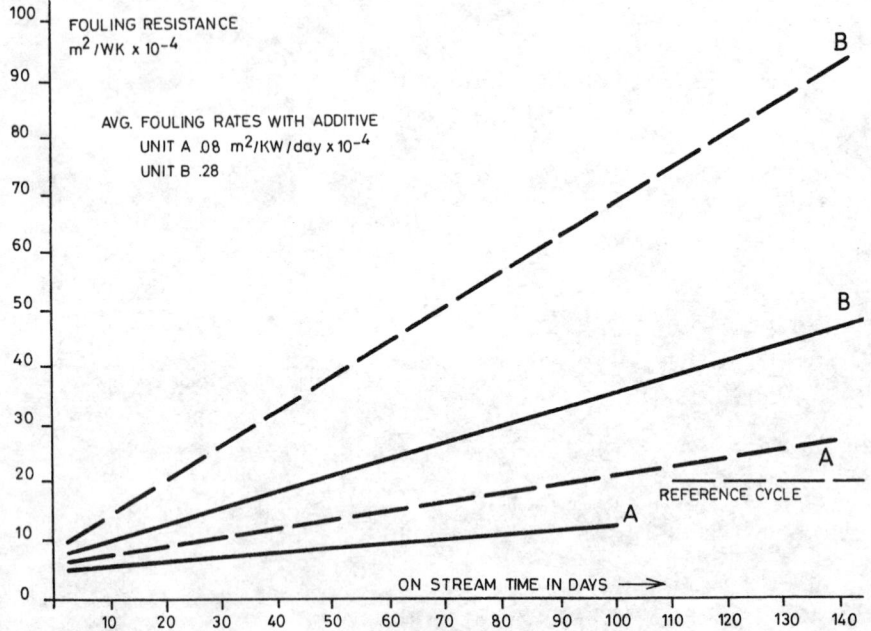

FIGURE 6 : EVOLUTION OF FOULING RESISTANCES IN CRUDE/RESIDUE EXCHANGERS USING TOTAL / UMP ADDITIVE INJECTION

4.2 Economies of additive injection

The efficiency of the additive has been defined with respect to a reference cycle as :

$$E = 100 \times \left(\frac{\Delta R - \Delta R'}{\Delta R}\right)$$

R : Evolution of the fouling resistance over a cycle without additive.

R' : Evolution of the fouling resistance over a cycle with additive.

for each of the cycles shown in Fig. 6 the efficiency is around 50 %.

In order to optimise the use of the additive we developed an empirical relationship for the cost of fouling as a function of the time since on stream cleaning can be accomplished on both of the units in question the cost of this operation is also taken into account.

$$C = K_1 K_2 t^{1.2} + K_3\left(\frac{12}{t} - 1\right) + K_4 + K_5$$

t = length of cycle in months
C = cost of fouling
K_1 = constant for the unit
K_2 = cost of fuel
K_3 = cost of one on stream cleaning
K_4 = cost of annual cleaning.
K_5 = annual cost of additive

For unit B (20 000 t/day) in our example the following cost table has been drawn up for end 1980 economics using an additive injection rate of 10 ppm Wt.

TABLE 3
ANNUAL COST OF FOULING K $ (1 $ = 5 F)

Annual cost	Annual cost with additive at an efficiency of :			Cost of fuel	140 $/tonne
without additive	75 %	50 %	20 %	Cost of additive	2 $/kg
4 month cycle				Cost of on stream	
1 040	580	780	1 000	cleaning	102 000 $
6 month cycle				Cost of off-line	52 000 $
1 440	600	920	1 300	cleaning	
12 month cycle					
3 000	920	1 680	2 540	On stream time	8 000 hours

The optimum cycle length is in fact of 4 months duration while the borderline efficiency which justifies use of the additive is an improvement of 20 % over the reference conditions.

Two other units are now being equipped with the necessary equipment to permit additive injection to the desalter.

5. CONCLUSIONS

This work which is still continuing has enabled a number of important conclusions to be drawn :

- The fouling of crude oil preheat trains is a complex phenomena which above all is not caused primarily by the crude oil itself but by some constituents of the crude, by various impurities and by synergistic action between different impurities and the crude.

- The values of fouling factor currently admitted by TEMA for crude above 180°C appear much too low if an economical cycle length between exchanger cleaning is to be attained.

- The use of dispersants within the desalter emulsion phase offers a major hope for decreasing the fouling of the hottest exchangers which recover the most valuable heat.

- Fouling on the shell side can be aggravated by inadequate turbulence and velocity.

- Care should be focussed on the temperature control of preheat trains using parallel circuits to increase heat recovery.

The authors wish to express their thanks to M. J. FEBVAY, Director of TOTAL TECHNIQUE, for permission to publish this paper and to all C.FR & UMP personnel associated with this project over the last three years.

REFERENCES

1 PLATT, G., OIL AND GAS JOURNAL 78, 159 - 173, Oct 13 1980
  HUANG F., ELSHOUT R., Chemical Engineering Progress, 72 (7), 68 - 74, July 1976.

2 HALUSKA, J.L., Hydrocarbon Processing, 55 (7) 153 - 156 (1976).

3 VAN DER WEE, P., TRITSMANS, P.A., Hydrocarbon Processing, 45 (8), 141 - 144 (1966).

4 SCHERRER, C., DURRIEU, M., RICHMOND, J.R., 1979 NPRA Meeting, Paper AM-79-50.

5 FRENCH PATENT   No. 2388037
  US. PATENT   No. 4200550

6 FRENCH PATENT   No. 2421958
  US. PATENT   No. 4222853

Water Quality Effects on Fouling from Hard Waters

A. PAUL WATKINSON
Department of Chemical Engineering
The University of British Columbia
Vancouver, British Columbia V6T 1W5, Canada

ABSTRACT

Artificial hard waters were recirculated through a steam-heated annular test section in which the fouling films deposited on the outside of the inner tube. Experiments were carried out in the constant heat flow mode using waters of pH 6.9-8.1, and total alkalinity between 90 and 700 ppm as $CaCO_3$. Fouling rates ranged from 0.1 to 100×10^{-6} m^2K/kJ. The ionic diffusion model of Hasson and co-workers fitted the data at fouling rates below 4×10^{-6} m^2K/kJ. The effects of suspended particulates and of the presence of magnesium were investigated.

1. INTRODUCTION

Precipitation fouling was recently reviewed by Hasson [1]. For hard waters the ionic diffusion model of Hasson, Sherman and Biton [2] is recommended to predict the rate of deposition of $CaCO_3$ given values for the pH, calcium concentration, total alkalinity of the water and the fluid flow parameters. If the product of the scale density and thermal conductivity, $\rho_s k_s$, is available or can be estimated, the thermal scaling rate can be calculated. This predictive model contrasts with the traditional approach of characterizing the scale potential of waters via the Langelier Saturation Index, or the Ryznar Stability Index.

In an experimental study [3] of scaling from artificial hard waters, it was found by tests on a series of dilutions of a stock hard water solution of $CaCl_2$ and $NaHCO_3$ that the scaling rate could be correlated with the calculated carbonate ion concentration. Less definite trends of scaling rate were found with the saturation index, stability index and alkalinity. The work was carried out under rapid scaling conditions and with about 300 ppm of suspended $CaCO_3$ present. Comparison of experimental scaling rates with predictions of ionic diffusion model showed a major discrepancy, with experimental fouling rates up to 3 to 4 times greater than predicted by the model. This deviation was attributed to the presence of particulates which would provide a second mechanism for deposition. Examination of the literature suggests that the ionic diffusion model has not been well verified for thermal scaling. In earlier work, Hasson et al. [4] did show scale deposition to be diffusion controlled, but the water chemistry effects were not elucidated. The model which was published subsequently did predict [2] scaling rates of the same order of magnitude as observed in three runs of Morse and Knudsen [5], however it was necessary to resort to literature values of the product $\rho_s k_s$ which resulted in an uncertainty of a factor of 4 in predicted scaling rates. Even so, measured values were a factor of 3 above predicted rates for one run of the three. At high pH, and in the absence of heat transfer, deposition tests show good agreement with the ionic diffusion model [1].

In this work a range of scaling conditions was studied and an attempt was made to determine conditions under which the scaling rate may be predicted by the ionic diffusion model.

2. EXPERIMENTAL

The test section consisted of a 133 cm. long, 1.91 cm. O.D. copper tube which was mounted in a glass annulus of 3.72 cm. I.D. Water passed through the annular section and was heated by steam condensing in the tube. Temperatures of the inlet and outlet water and steam were measured, as was the steam pressure. Upon leaving the heater the water was cooled in a second heat exchanger and returned to the feed tank. Water was made up in 140 liter batches by addition of $CaCl_2$ and $NaHCO_3$. In some cases NaOH, HCl or $MgCl_2$ were added.

Water was analyzed several times during a run for pH, total alkalinity, calcium hardness and dissolved solids by standard methods. Suspended solids were determined by filtration of a known volume through a Whatman GF/B glass filter which is claimed to remove particles down to 1 micron in size.

The test section was operated at a constant heat flow by making manual adjustments of the steam pressure as fouling occurred. The overall heat transfer coefficient based on the outside area of the copper tube, and the fouling resistance were calculated as shown below.

$$U_o = \dot{m} C_p \ln[(T_s - T_1)/(T_s - T_2)]/A_o \qquad (1)$$

$$R_f = (1/U_o - 1/U_{o_{\theta=0}}) \qquad (2)$$

3. RESULTS AND DISCUSSION

Tests were first done under conditions comparable to the previous study [3] with an inlet temperature of 56°C, and with hard water make up of 1.39 gpl $CaCl_2$ and 1.23 gpl $NaHCO_3$. Figure 1 shows the heat transfer coefficient, and several water quality parameters plotted versus time. The decrease in U_o with time is rapid, dropping from 2.8 to 1.8 kW/m^{2}K in twenty minutes. To maintain a constant heat flow of 10.6 kW during fouling the condensing steam temperature was raised from 107 to 130°C. The fouling process is accompanied by a rapid drop in total alkalinity and a rise in suspended solids content to 300 ppm in the first five minutes. A blank run with recirculation at 57°C but no heating in the exchanger resulted in suspended solids reaching 300 ppm in 15 minutes, and 380 ppm in 30 minutes. Thus, the suspended solids originate from precipitation at the bulk temperature, and not in the hotter film surrounding the steam heated tube, nor are they a removal product from the fouling deposit.

Lowering the inlet temperature had a marked effect on the scaling process. Figure 2 shows results of a typical experiment with an inlet temperature of 25°C. Compared to Figure 1, the scaling rate is much lower, the alkalinity remains essentially constant and the maximum suspended solids content is reduced to about 40 ppm. Some precipitation undoubtedly still occurs in the

WATER QUALITY EFFECTS ON FOULING FROM HARD WATERS

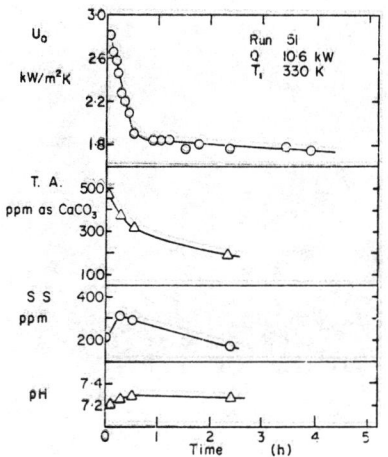

Figure 1. Scaling run at inlet temperature of 330K, velocity of 0.69 m/s and total alkalinity of 500.

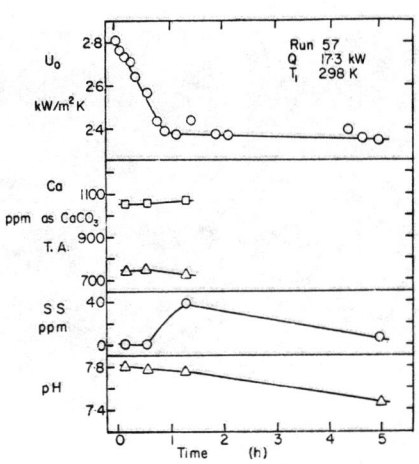

Figure 2. Scaling run at inlet temperature of 298K, velocity of 0.69 m/s and total alkalinity of 725.

hot film and in the bulk, giving rise to the suspended solids. Calculations show an expected drop in bulk precipitation rate of a factor of about 25 with the drop of 32°C in temperature.

A significant difference was also noted in the quality of the deposits. At higher temperature the deposits were fluffy on the top and consolidated but powdery underneath, whereas at lower temperatures the deposits were compact, although rough (Figure 3). Microscopic examination showed the deposits to be crystalline with single crystals of about 50-100 microns in size on the surface of the dense scale. There was no evidence of loosely trapped suspended particles. Solids filtered from the recirculating water were also crystalline, and 5-10 micron in size. The scale on the tube surface increased roughly linearly with distance from the fluid entrance, varying in thickness from less than 0.1 mm to almost 1 mm at the exit of the exchanger in some cases. The thicker parts of the deposits were readily cracked off by thermal shock, but the thinner films near the fluid entrance were tenacious and could only be removed by chemical cleaning. Deposit characteristics from a number of runs are given in Table 1. Thicknesses reported are average values over the tube length, and thermal conductivity values were calculated from the known thermal resistances at the beginning and end of the run. The thermal conductivity values must be viewed with some caution since they are averages over the tube length, and may be affected by roughness effects. Density values were determined by pycnometer. The average thickness of deposit was about 0.15-0.4 mm, and the density was close to 2600 kg/m^3 in all cases. Calculated thermal conductivity values varied from 2.0-4.5 W/mK. Chemical analysis showed the deposits to be primarily $CaCO_3$. Conductivity values compare with 2.9 W/mK reported by Hasson [4], and the average value of the product $\rho_s k_s$ of 5 kg kW/m^{4}K compares to the value of 5.62 quoted in [2] as an upper bound.

3.1 Comparison with Ionic Diffusion Model

With Run 57 in Figure 2 as a base case, various tests were done in which water chemistry and fluid mechanical parameters were varied to determine if the

a) Run 58   b) Run 59   c) Run 60   d) Run 56

Figure 3. Typical Scale Deposits (a,b,c) and Suspended Solids (d)
Magnification: ⊢―――⊣ 100 microns.

rate change was predictable by the low pH equation of Hasson, Sherman and Biton:

$$dR_f/d\theta = 0.5 k_D [Ca^{++}][(1+4ac/b^2)^{1/2}-1]b/a\rho_s k_s \qquad (3)$$

where $a = 1 - 4\, K_2' k_r [Ca^{++}]/k_D K_1'$

$b = [CO_2]/[Ca^{++}] + 4\, K_2' k_r [HCO_3^-]/K_1' k_D + K_{sp}' k_r /k_D [Ca^{++}]$

$c = K_2' k_r [HCO_3^-]^2/K_1' k_D [Ca^{++}] - K_{sp}' k_r [CO_2]/k_D [Ca^{++}]^2$

The equations for the various constants are given in Table 2, as they appear in references [2] and [6].

Changes were made from the base case of 25°C inlet temperature, 0.68 m/s velocity and initial T.A. of about 600 ppm as $CaCO_3$. The T.A. was varied by either acidifying the water with HCl, or reducing the amounts of $NaHCO_3$ and $CaCl_2$ added to the water. Figure 4 shows heat transfer coefficients versus time for several runs. Scaling took place over periods ranging from 0.8 to 68 hours, and the scaling rate, determined by fitting the initial linear portions of the U_o versus time plots, varied by over a factor of 1000. Both experiment and model predictions showed a drop in scaling rate with decreasing velocity, which is an indication that mass transfer controls the scaling rate. In the case of temperature and alkalinity changes the model correctly predicted the direction of the observed change in rate.

Figure 5 is a plot of measured scaling rate versus that predicted by the ionic diffusion model. Although virtually all the measured values lie above the predicted line, agreement can be considered very good at scaling rates up to 4×10^{-6} m^2K/kJ, above which a major discrepancy occurs. The highest rates measured were a factor of 70 larger than the ionic diffusion model predictions. The role of particulates was considered as a possible explanation for the

WATER QUALITY EFFECTS ON FOULING FROM HARD WATERS

deviation at high fouling rates.

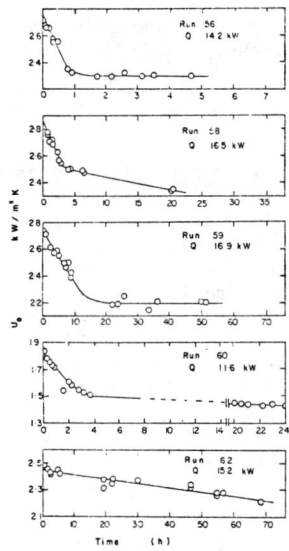

Figure 4. Heat transfer coefficient versus time for typical scaling runs.

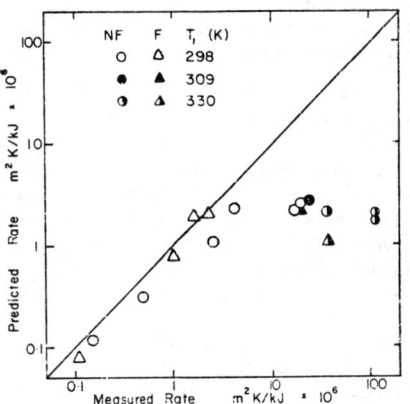

Figure 5. Comparison of measured scaling rate with rate predicted by the ionic diffusion model (F = in-line filter, N = no filter)

3.2 Role of Particulates

An in-line filter with an AMF Cuno Mikro-Klean 75 micron cartridge was installed downstream of the heat exchanger to remove particulates without affecting the water chemistry. This filter element reduced the suspended solids to levels below the detectable limit for the method described in the experimental section. Runs were attempted under conditions of previous tests without the filter, however identical waters were not produced. Results are illustrated in Table 3. In each case removal of particulates resulted in a decrease of about 50% in the fouling rate, however part of the decrease can be attributed to the lower alkalinity in the runs with the filter. At high inlet temperatures of Runs 50 and 51, where suspended solids reached 150-250 ppm, when the filter was present the rate still only decreased by about one-half. It appears that particulates also contribute significantly even at low concentrations where the corresponding alkalinity is low. The data from these runs are also shown in Figure 5, which illustrates that removal of solids for high temperature, high rate cases (Runs 70 and 74) does not bring that rate data into agreement with the model.

A calculation was carried out to estimate the magnitude of particulate deposition. Using the mass transfer equations recommended by Gudmundsson [7], for assumed particle sizes of 1 and 10 microns, the calculated particulate flux was less than 1% of the predicted ionic flux for the 250 ppm suspended solids, of Run 50 (Table 1), and about 10% and 4% respectively for the 10 ppm suspended solids of Run 62. Although the mass deposition fluxes appear small compared to ionic fluxes, they may be more important for thermal effects since particulate deposits are more porous and of lower conductivity than dense scales.

Explanations for the disagreement of the model and the data at fouling rates in excess of 4×10^{-6} m^{2}K/kJ remain speculative. However, these high scaling rates are far beyond the range encountered outside the laboratory, and the model seems reliable in the range of practical interest.

3.3 Effect of Magnesium

Magnesium may co-exist with calcium in hard waters. Thurston [1] suggested that for boiler waters low in silica when Mg^{++}/Ca^{++} exceeds 0.2, free flowing sludges rather than dense scales were produced. Magnesium has also been reported to lower the rate of scaling in certain concentration ratios [3]. In this study the effect of magnesium additions was studied both at high scaling rates where bulk precipitation was important, and at lower scaling rates. In Figure 6 the fouling resistance is plotted versus time for rapid scaling conditions at various Mg^{++}/Ca^{++} ratios for $Ca^{++}_i \approx 0.5$ gpl. As the Mg^{++}/Ca^{++} ratio increases the scaling rate and final thermal resistance go through a minimum. Figure 7 shows that the minimum occurs at Mg^{++}/Ca^{++} close to 0.2. At lower scaling rates corresponding to lower inlet temperatures, and much reduced bulk precipitation, a similar but weaker minimum is evident. Chemical analysis of deposits from waters containing $Mg^{++}/Ca^{++} > 0.2$ showed magnesium present at 1.4-1.8% by weight. Deposits in these runs were all powdery, and could not be removed by thermal shock.

4. SUMMARY

Thermal scaling rates from recirculating hard waters with pH 6.9-8.1 and total alkalinity between 100 and 600 can be predicted from the ionic diffusion model of Hasson and co-workers. Rough, compact calcium carbonate scales of average specific gravity 2650 kg/m^3, thermal conductivity about 2.2 W/mK and thickness up to 0.5 mm were produced. Calculated particulate mass fluxes were less than 10% of the ionic fluxes, however tests with an in-line filter suggested that particulates could account for about 25% of the thermal fouling.

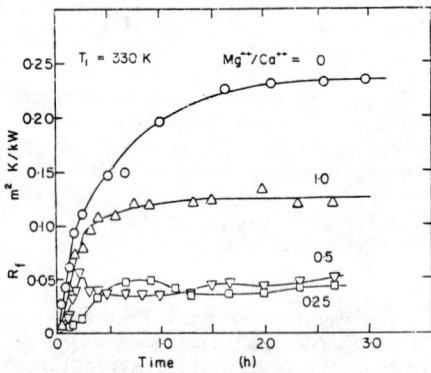

Figure 6. Fouling resistance versus time for different initial magnesium to calcium mass ratios. (Total alkalinity 520 ppm as CaCO$_3$ for T$_1$=330K, and 680 for T$_1$=298K)

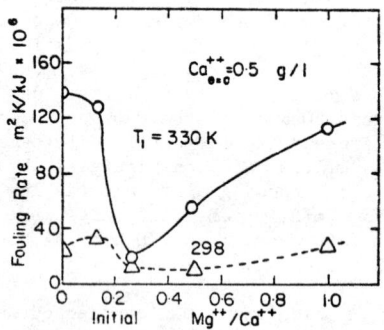

Figure 7. Effect of initial magnesium to calcium mass ratios on fouling rate at two inlet temperatures.

WATER QUALITY EFFECTS ON FOULING FROM HARD WATERS

ACKNOWLEDGEMENT

This work was carried out under a grant from the National Science and Engineering Research Council of Canada.

NOMENCLATURE

Ao	outside area of tube	m^2
a	activity	mol/m^3
C_p	specific heat of water	J/kgK
D_e	equivalent diameter for annulus	m
f_M, f_D	activity factors for monvalent and divalent ions	
I	ionic strength of solution	
k_D	mass transfer coefficient for ions	m/s
k_r, k_r'	crystallization rate constant m^2-s/kg $CaCO_3$ and m^2-s/mol	
k_s	thermal conductivity of scale	W/mK
$K_1 = K_1' f_M^2$	first dissociation constant of carbonic acid $[a_{H^+}] \cdot [a_{HCO_3^-}]/[a_{H_2CO_3}]$	mol/m^3
$K_2 = K_2' f_D$	second dissociation constant of carbonic acid $[a_{H^+}][a_{CO_3^=}]/[a_{HCO_3^-}]$	mol/m^3
$K_{sp} = K_{sp}' f_D^2$	solubility product of $CaCO_3$ $[a_{Ca^{++}}][a_{CO_3^=}]$	$(mol/m^3)^2$
$\dot{m}$	mass flowrate of water	kg/s
Q	heat flow	W
R_f	fouling resistance	m^2K/W
T	temperature	K
U_o	overall heat transfer coefficient	W/m^2K
V	fluid velocity	m/s
x	scale deposit thickness	m
Re	Reynolds number	
Sc	Schmidt number	
ρ	specific gravity	kg/m^3
θ	time	s
[ ]	concentration	mol/m^3

REFERENCES

1. Hasson, D. "Precipitation Fouling - A Review" in "Fouling of Heat Transfer Equipment", E.F.C. Somerscales and J.G. Knudsen (eds.) Hemisphere Publ. Co., New York, 1981 p. 527.

2. Hasson, D., Sherman, H. and Biton, M. "Prediction of $CaCO_3$ Scaling Rates", Proc. 6th Intern. Symp. Fresh Water from the Sea 2:193 (1978).

3. Watkinson, A.P. "Effects of Water Quality on Hard Water Scaling", 30th Canadian Chem. Eng. Conference, Edmonton, Oct. 1980 Preprints Vol. 2 pp. 616-623.

4. Hasson, D., Avriel, M., Resnick, W., Rozenman, T. and Windreich, S., "Mechanism of $CaCO_3$ Scale Deposition on Heat Transfer Surfaces", Ind. Eng. Chem. Fund. 7, 59 (1968).

5. Morse, R.W. and Knudsen, J.G. "Effect of Alkalinity on the Scaling of Simulated Cooling Tower Water", Can. J. Chem. Eng. 55, 272 (1977).

6. Wiechers, H.N.S., Sturrock, P. and Marais, G.V.R. "Calcium Carbonate Crystallization Kinetics", Water Research 9, 835 (1975).

7. Gudmundsson, J.S. "Particulate Fouling" in "Fouling of Heat Transfer Equipment", E.C.F. Somerscales and J.G. Knudsen (eds.), Hemisphere Publ. Co., New York, 1981 p. 357.

Table 1. Summary of Water, Deposit and Fouling Rate Data

Run No.	Inlet Temp. (K)	pH	T.A. (mg/l as $CaCO_3$)	Ca^{++} (mg/l as $CaCO_3$)	SS (p.p.m.)	DS (p.p.m.)	x (mm)	k_s W/m K	Scaling Rate (m^2K/kJ x 10^6) Meas.	Scaling Rate (m^2K/kJ x 10^6) Pred.
50	330	7.22	400		250	2200			113	2.0
51	330	7.23	430		250	2200			115	1.7
54	330	6.80	450	950	275	2100			37	2.
56	310	7.57	720	1130	85	1900			24	2.6
57	298	7.77	726	1055	10	2000			20	2.5
58	298	7.38	660	1230	8	2000	0.40	4.6	4	2.2
59	298	8.05	386	610	6	1000	0.28	2.6	2	1.0
60	298	7.55	680	867	20	2050	0.39	2.5	17	2.1
62	298	8.2	128	200	40	500	0.18	2.6	0.15	0.11
67	298	8.14	192	320	0*	483			0.48	1.8
68	298	7.57	483	1123	0	2475	0.15	2.1	1.6	1.8
69	298	7.62	497	1175	0	1700	0.20	2.1	2.2	1.9
70	310	7.80	578	1130	0	2612	0.15	1.7	14	2.2
71	298	8.05	337	635	0	575	0.14	3.4	1	0.76
72	298	8.10	96	207	0	100			0.11	0.075
74	332	7.30	350	835	0	1766			45	1.2

Note: Run 60 V = 0.343 m/s, otherwise V = 0.686 m/s   *Not detected
Runs after 62 used in-line filter
Deposit specific gravity 2650 ± 90 kg/m^3
S.S. = suspended solids
D.S. = dissolved solids
T_{steam} = 379K at time zero

Table 2. Equations for Model Calculations [2,6]

$\ln k_r$ = 41.04 − 10417.7/T
pK_1 = 17052/T + 215.21 log T − 0.12675 T − 545.56
pK_2 = 2902.39/T + 0.02379T − 6.498
pK_w = 4787.3/T + 7.1321 log T + 0.01037T − 22.801
pK_{sp} = 0.01183 (T − 273.2) + 8.03
pK = − log K
k_D = 0.023 V/Re$^{0.17}$ Sc$^{0.67}$
$-\log f_M$ = 0.51 [I$^{1/2}$/(1 + I$^{1/2}$) − .30 I]
$-\log f_D$ = 2.04 [I$^{1/2}$/(1 + I$^{1/2}$) − .30 I]
T.A. = [HCO$_3^-$] + 2[CO$_3^=$] + [OH$^-$] − [H$^+$]
[CO$_3^=$] = (T.A. + [H$^+$] − [OH$^-$])/2(1 + [H$^+$]/2K$_2'$)
[HCO$_3^-$] = (T.A. + [H$^+$] − [OH$^-$])/(1 + 2K$_2'$/[H$^+$])
[CO$_2$] = [H$^+$] (T.A. + [H$^+$] − [OH$^-$])/K$_1'$(1 + (2K$_2'$/[H$^+$])

Table 3. Effect of Filtration to Remove Particulates

Run	Filtration	T.A. (mg/l as CaCO$_3$)	S.S. (p.p.m.)	Scaling Rate (m^{2}K/kJ x 10^6) Measured	Predicted
50	No	400	250	113	2.
74	Yes	350	0	45	1.2
56	No	720	85	24	2.6
70	Yes	582	0	14	2.2
58	No	660	7.8	4	2.2
69	Yes	492	0	2.2	1.9
59	No	386	8	2	1.0
71	Yes	310	0	1	0.76

Precipitation Fouling of Cooling Water

L. LAHM, JR.
Fractionation Research Incorporated
Alhambra, California 91803, USA

J. G. KNUDSEN
Department of Chemical Engineering
Oregon State University
Corvallis, Oregon 97331, USA

ABSTRACT

This paper discusses the results of a systematic study of cooling water fouling. Effects of surface temperature, calcium hardness and velocity are discussed.

NOMENCLATURE

c_1, c_2 = constants
K_1, K_2 = constants defined after Equation (2)
k_f = thermal conductivity of deposited scale, w/mK
m = exponent determined by calibration of the heater rod
n = exponent
P_d = probability function
q/A = heat flux, w/m^2
R_f^* = asymptotic fouling resistance, m^{2}K/w
R_f = fouling resistance, m^{2}K/w
R_W = wall resistance, m^{2}K/w, determined by calibration of the heater rod
T_B = bulk temperature, K or C
T_{WC} = wall temperature when clean, K or C
T_{WF} = wall temperature when scaled, K or C
V = velocity, m/s

Greek Symbols

ϕ_d = scale deposition rate, m^{2}K/ws
ϕ_r = scale removal rate, m^{2}K/ws
ψ = factor representing structural strength of scale
θ = time, hr
τ = shear stress, J/m^2
Ω = water characterization factor

Subscripts

B = bulk fluid
C = clean condition
F = fouled condition

1. INTRODUCTION

Fouling has been described as the "major unresolved problem in heat transfer" (1)(2). The reference describes an extensive fouling research program undertaken by Heat Transfer Research Incorporated (HTRI), Alhambra, California. The form of a generalized predictive method for industrial cooling tower water in the absence of corrosion is presented based upon fouling data obtained on a wide variety of cooling tower water.

For several years, one of the authors (JGK) has been conducting a systematic study of the precipitation fouling characteristics of cooling tower water. The study has been directed toward determining the effect of the various parameters in the HTRI model (described below).

2. THEORETICAL BACKGROUND

The characteristic fouling resistance-time curve for industrial cooling tower water indicates a rather rapid increase initially, followed by an approach to an asymptotic resistance. The HTRI model describing this behavior is:

$$R_f = (K_1/K_2)\exp(-E/RT_s)[1 - \exp(-K_2 t)] \tag{1}$$

where
$$K_1 = c_1 P_d \Omega^n \quad \text{and} \quad K_2 = c_2 \tau k_f / \psi$$

If time becomes very large

$$R_f^* = \lim R_f = (K_1/K_2)\exp(-E/RT_s) \tag{2}$$

3. EXPERIMENTAL EQUIPMENT AND DATA REDUCTION

The main components of the test equipment are the cooling tower (6 m high x 0.61 m dia.), a holding tank, three test sections, a circulation pump, a heat exchanger and a hot water system. The test section has been described elsewhere (5),(6).

The following equation is used to calculate the fouling factor

$$R_f = \frac{T_{WF} - \frac{(q/A)_F}{(q/A)_C}(\frac{V_C}{V_F})^m (T_{WC} - R_W(q/A)_C - T_{BC})}{(q/A)_F} - R_W \tag{3}$$

4. STUDIES ON COOLING TOWER WATER

Table 1 shows a series of four waters that have been studied by Lee and Knudsen (5), Coates and Knudsen (6) and the authors. Waters 1, 2, and 3 are in a condition of near saturation. Water No. 4 is highly supersaturated. Table 2 gives the details of the tests that have been made on these waters.

4.1 Results for Water No. 1.

Lee and Knudsen (5) investigated the effect of surface temperature and velocity on the fouling characteristics of water No. 1. The results are shown in Fig. 1. The equations of the lines are

$$R_f^* = (1.5)(10^8)\exp(-9700/T_s) \tag{4}$$

and

$$R_f^* = (3.5 \times 10^{21})\exp(-20000/T_s) \tag{5}$$

Most data on Fig. 1 are obtained at 1.62 m/s. There appears to be some effect of velocity with the results from one high velocity run (2.56 m/s) being below the lines and one low velocity run (0.76 m/s) being above the lines.

PRECIPITATION FOULING OF COOLING WATER

Table 1. Water Studied by Knudsen and Co-workers (5)(6).

	No. 1	No. 2	No. 3	No. 4	City Water Make-Up
Total Hardness (ppm $CaCO_3$)	210	220	470	370*(800)**	40
Calcium Hardness (ppm $CaCO_3$)	150	150	400	270(640)	30
m-Alkalinity (ppm $CaCO_3$)	210	300	115	160(600)	42
Chloride (ppm)	300	400	600	500	40
Silica (ppm SiO_2)	105	110	115	150	20
pH	9	9	9	9	7

* As analyzed
** Effective concentrations based on mass balance.

4.2 Results for Waters Nos. 2 and 3.

Coates and Knudsen (6) investigated waters Nos. 2 and 3 at various velocities and heater surface temperatures. The results are shown in Fig. 2 on which is shown the same straight lines as appear in Fig. 1. Below an apparent threshold temperature of 60°C relatively low fouling resistances were obtained and they generally fell within the range of fouling factors for magnesium silicate.

The three solid symbols on Fig. 2 represent results for water No. 2 to which had been added about 25 ppm of Brazor River clay.

Velocity appears to have some effect at the lower surface temperatures but there appears to be no significant effect at surface temperatures above 60°C.

4.3 Results for Water No. 4.

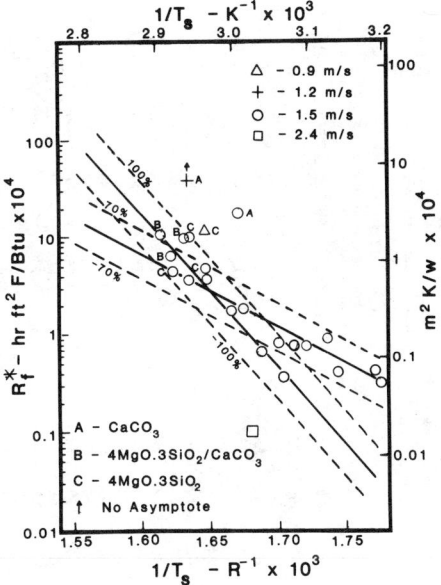

Fig. 1. Fouling Data for Water No. 1.

Knudsen and Coates (6) and the present authors investigated water No. 4 and the results are shown in Fig. 3. Again the lines from Fig. 1 are included. Three runs produced deposits in the non-asymptotic mode. All others showed asymptotic behavior and fall generally close to the line for magnesium silicate.

5. SUMMARY

Cooling tower water No. 1 produced magnesium silicate and showed an Arrhenius relationship with heater surface temperature. Cooling tower waters Nos. 2 and 3 showed minimal deposition at surface temperature below 60°C. Cooling tower water No. 4 showed minimal deposition over the whole temperature range studied and gave asymptotic fouling resistances similar in magnitude to water

TABLE 2
Experimental Data and Results on Fouling of Cooling Tower Water
Bulk Temperature-35°C

Run (Duration-Hrs)	Flow Velocity m/s	Heat Flux w/m² x10⁻³	Surface Temperature-°C TCA	TCB	TCC	TCD	Heater and Material[1]	Asymptotic Fouling Factor[2] m²K/w x10⁴	Average Surface Temp-°C	Scale[3]	Total Hardness[6]	Calcium Hardness	M-Alkalinity	Chloride	Silica	pH
1 (936)	1.62	300	71	64	65	72	167 ss	1.8	68	III	210	150	200	200		8.9-9.1
2 (1512)	1.62	47	40	40	41	41	169 cu	0.02	41	NS			DITTO			
3 (672)	1.62	300	68	61	61	68	167 ss	0.8	65	III	220	155	250	310	130	
4 (936)	0.76	132	67	64	64	65	164 ss	2.1	65	III	210	150	200	200		
5 (360)	1.62	47	41	39	40	40	167 ss	0.18	40	NS			DITTO			
6 (624)	1.62	287	68	66	71	72	169 cu	0.7	69	III	215	150	200	260	85	
7 (168)	1.62	48	41	41	40	40	111 cu	0.1	41	NS	210	150	200	200		
8 (216)	1.62	105	47	46	44	47	167 ss	0.18	46	NS			DITTO			
9 (192)	1.62	106	47	48	48	48	111 cu	0.18	48	NS			DITTO			
10 (192)	1.62	150	53	49	48	52	167 ss	0.18	51	NS	215	150	200	260	85	
11 (216)	1.62	148	53	54	54	58	111 cu	0.18	54	NS			DITTO			
12 (216)	1.62	196	59	56	54	58	167 ss	0.2	57	NS			DITTO			
13 (696)	1.62	192	59	58	60	59	111 cu	0.35	59	NS	210	158	200	300	90	
14 (696)	1.62	239	63	58	57	62	167 ss	0.35	60	NS			DITTO			
15 (1200)	1.62	288	67	67	68		164 ss	1.8	67	II	220	155	250	310	130	
16 (504)	1.62	237	68	63	65	64	111 cu	0.7	64	NS			DITTO			
17 (504)	1.62	298	69	64	62	69	167 ss	0.7	66	NS	210	150	200	300	90	
18 (792)	1.62	295	70	71	73	72	111 cu	1.8	72	II	220	155	250	310	130	
19 (792)	1.62	33	74	68	65	73	167 ss	1.1	70	II			DITTO			
20 (312)	1.62	158	52	52	54	54	169 cu	0.1	53	NS			DITTO			
21 (264)	1.62	205	55	54	57	57	169 cu	0.18	56	NS			DITTO			
22 (1392)	1.62	237	58	57	61	61	111 cu	3.1	59	I	230	150	260	320	140	
23 (648)	2.56	353	57	57	58	57	111 cu	0.02	57	NS			DITTO			
24 (528)	1.62	315	71	64	63	69	167 ss	7.4	67	I			DITTO			
25 (38)[4]	1.52	334	68	67	71	65	93 cs	2.5 (NA)	68	I	300	200	300	900	150	
26 (38)[4]	1.55	334	68	66	67	64	104 zn	0.5 (NA)	66	I			DITTO			
27 (38)	1.52	221	57	58	61	61	169 cu	1.1 (NA)	59	I			DITTO			
28 (1950)	1.68	356	67	66	65	64	104 zn	7.8 (NA)	66	II	215	151	213	232	103	
29 (1950)	1.62	365	71	67	68	68	93 cs	7.2 (NA)	68	II			DITTO			
30 (480)	1.68	214	58	59	62	61	169 ss	0.1	60	NS	215	150	215	400	105	
31 (1200)[5]	1.62	288	67	67	67	-	164 ss	1.8	67	II	215	151	213	232	103	
32 (300)[5]	1.65	47	28	28	28	29	111 cu	0.02	28	NS	210	149	290	274	105	
33 (600)	1.65	49	40	40	40	40	111 cu	0.1	40	NS	207	148	318	391	115	
34 (480)	1.55	37	40	40	39	39	118 cu	0.1	40	NS			DITTO			
35 (480)	1.65	96	46	46	46	46	169 cu	0.1	46	NS			DITTO			

[1] cs - carbon steel; zn - galvanized; cu - copper plated; ss - stainless steel.
[2] NA - no asymptote.
[3] I - calcium carbonate; II - calcium carbonate/magnesium silicate mixture; III - magnesium silicate; IV - copper oxide, silica and magnesium; NS - no sample.
[4] Water loss during Runs 25, 26, 27 caused data to be invalid.
[5] Bulk temperature - 24°C.
[6] Numbers in parentheses indicate effective concentration of water No. 4.
[7] 25 ppm Brazos River Clay in cooling tower water.

TABLE 2 (continued)
Experimental Data and Results on Fouling of Cooling Tower Water
Bulk Temperature-35°C

Run (Duration-Hrs)	Flow Velocity m/s	Heat Flux w/m² x10⁻³	Surface Temperature-°C TCA	TCB	TCC	TCD	Heater and Material[1]	Asymptotic Fouling Factor[2] m²K/w x10⁴	Average Surface Temp-°C	Scale[3]	Total Hardness[6]	Calcium Hardness	M-Alkalinity	Chloride	Silica	pH
36 (800)	1.62	123	52	53	50	49	118 cu	0.1	51	NS	230	142	347	524	169	8.9-9.1
37 (800)	1.65	183	57	56	57	57	169 cu	4.4 (NA)	57	I			DITTO			
38 (100)	1.65	242	63	62	62	63	111 cu	3.9 (NA)	62	I	212	146	298	359	117	
39 (270)	2.65	383	63	62	63	64	111 cu	2.8 (NA)	63	I	230	142	347	524	169	
40 (340)	1.62	158	56	57	53	52	118 cu	0.2	54	I			DITTO			
41 (264)[7]	1.40	51	41	42	43	41	169 cu	0.7	42	NS	250	160	310	650	160	
42 (264)[7]	1.50	148	50	52	56	54	118 cu	0.7	53	NS			DITTO			
43 (240)[7]	1.50	290	68	68	67	68	111 cu	2.5 (NA)	68	I			DITTO			
44 (390)	1.62	148	53	56	52	50	118 cu	1.1	53	I	231	188	191	530	91	
45 (390)	1.65	59	42	41	41	41	111 cu	0.2	41	I			DITTO			
46 (400)	1.62	259	63	63	63	64	118 cu	4.1 (NA)	63	I			DITTO			
47 (940)	1.65	147	53	55	52	50	111 cu	0.6	52	II	402	348	133	583	110	
48 (940)	1.65	60	42	41	42	42	169 cu	0.3	42	NS			DITTO			
49 (940)	1.65	243	62	63	63	64	111 cu	3.2 (NA)	63	I			DITTO			
50 (680)	1.52	177	57	61	57	54	118 cu	0.3	57	NS	490	418	118	760	128	
51 (540)	1.52	324	76	75	78	77	111 cu	5.1 (NA)	77	I			DITTO			
52 (680)	1.52	271	68	68	68	70	111 cu	2.5 (NA)	68	NS			DITTO			
53 (480)	2.26	49	39	41	39	39	118 cu	0.05	40	NS	448	385	98	548	134	
54 (580)	2.26	227	56	52	52	51	111 cu	0.1	53	NS			DITTO			
55 (280)	2.26	469	-	-	-	71	168 ss	1.1 (NA)	71	NS	502	420	106	735	146	
56 (720)	2.26	268	57	61	58	54	111 cu	0	57	I			DITTO			
57 (740)	2.26	437	63	67	-	66	118 cu	5.8 (NA)	64	I			DITTO			
58 (680)	1.19	66	-	-	-	46	168 ss	0.3	46	NS	463	426	115	586	119	
59 (680)	2.93	82	41	42	41	41	118 cu	0.02	41	NS			DITTO			
60 (680)	2.93	293	52	54	55	-	111 cu	0.02	54	NS			DITTO			
61 (680)	1.19	138	-	-	-	58	168 ss	2.1	58	I	462	418	118	623	114	
62 (680)	2.93	369	57	64	60	57	118 cu	0.4	59	NS			DITTO			
63 (500)	2.93	576	63	67	71	-	111 cu	5.8 (NA)	67	NS			DITTO			
64 (460)	3.14	353	59	61	60	54	118 cu	0.1	59	NS	454	419	124	555	117	
65 (430)	1.01	154	66	69	72	-	111 cu	4.9 (NA)	69	I	430	400	100	500	100	
66 (1450)	3.66	541	62	66	60	56	111 cu	3.2 (NA)	61	I	500	410	100	480	100	
											490 (590)	400 (500)	120 (400)	400	100	
67 (1450)	1.13	119	58	59	54	56	118 cu	3.52 (NA)	57	I	374 (726)	235 (560)	SAME AS RUN 66	360	185	
68 (540)	1.58	25	41	41	41	41	118 cu	0.1	41	NS			150 (477)			
69 (540)	2.41	43	40	40	40	39	168 ss	0.1	39	NS			DITTO			
70 (350)	0.94	24	40	40	40	-	150 cu	0.02	40	NS			DITTO			

[1] cs - carbon steel; zn - galvanized; cu - copper plated; ss - stainless steel.
[2] NA - no asymptote.
[3] I - calcium carbonate; II - calcium carbonate/magnesium silicate mixture; III - magnesium silicate; IV - copper oxide, silica and magnesium; NS - no sample.
[4] Water loss during Runs 25, 26, 27 caused data to be invalid.
[5] Bulk temperature -24°C.
[6] Numbers in parentheses indicate effective concentration of water No. 4.
[7] 25 ppm Brazos River Clay in cooling tower water.

867

TABLE 2 (continued)
Experimental Data and Results on Fouling of Cooling Tower Water
Bulk Temperature-35°C

Run (Duration-Hrs)	Flow Velocity m/s	Heat Flux w/m² x10⁻³	Surface Temperature-°C TCA	TCB	TCC	TCD	Heater and Material[1]	Asymptotic Fouling Factor[2] m²K/wx10⁴	Average Surface Temp.-°C	Scale[3]	Total Hardness[6]	Calcium Hardness	M-Alkalinity	Chloride	Silica	pH
71 (430)	1.58	72	48	49	47	46	118 cu	0.2	47	NS	366 (805)	273(641)	166(591)	500	142	8.9-9.1
72 (430)	0.88	67	48	48	48	--	150 cu	0.02	48	NS			DITTO			
73 (560)	1.58	97	52	54	51	49	118 cu	0.05	52	NS	368 (802)	273(641)	172(622)	560	134	
74 (560)	0.94	83	51	51	52	--	150 cu	0.1	51	NS			DITTO			
75 (560)	1.58	120	57	59	57	54	118 cu	0.05	58	NS	373 (773)	257(622)	172(620)	551	135	
76 (570)	2.41	315	56	59	57	60	109 cu	0.1	58	NS			DITTO			
77 (570)	0.91	110	57	58	58	--	150 cu	6.2(NA)	58	I			DITTO			
78 (1060)	1.46	204	64	67	63	58	118 cu	0.4	63	NS	377 (788)	264(627)	186(661)	569	141	
79 (1060)	2.26	382	63	62	68	65	109 cu	0.4	64	NS			DITTO			
80 (1060)	0.88	147	65	--	64	--	150 cu	0.4	64	NS			DITTO			
81 (840)	1.46	272	70	74	68	68	118 cu	0.7	69	NS			DITTO			
82 (840)	2.26	479	68	68	74	71	109 cu	0.4	70	NS			DITTO			
83 (840)	0.94	171	68	--	68	--	150 cu	0.7	68	NS			DITTO			
84 (1200)	1.59	305	74	79	73	70	118 cu	0.5	74	NS	430 (1045)	312(783)	169(750)	725	143	9.1
85 (1200)	2.41	539	72	73	76	76	109 cu	0.44	74	NS			DITTO			
86 (1200)	0.95	198	75	--	74	--	150 cu	0.44	75	NS			DITTO			
87 (648)	1.59	267	69	74	69	65	118 cu	3.5 (NA)	69	I	401	291	127	350	102	8.9
88 (648)	1.55	367	73	73	75	76	109 cu	3.5 (NA)	74	I			DITTO			
89 (564)	0.95	198	75	--	74	--	150 cu	5.2(NA)	75	I	369	271	124	329	95	8.9
90 (1224)	1.59	264	68	72	67	65	118 cu	1.6	68	NS	600	400	37	875	125	8.1
91 (1224)	1.55	353	72	71	73	74	109 cu	1.6	73	NS			DITTO			
92 (1224)	0.95	192	72	--	72	--	150 cu	2.6	72	NS			DITTO			
93 (1728)	1.95	279	69	75	69	65	118 cu	3.3	70	IV	575	380	9	707	87	6.9
94 (1728)	1.55	373	73	--	76	--	109 cu	3.3	75	IV			DITTO			
95 (720)	0.95	209	74	--	74	--	150 cu	2.6	74	IV	624	407	11	792	88	7.1
96 (840)	0.95	195	74	--	72	--	149	3.2	73	IV	536	359	7	636	86	6.8
97 (1488)	0.91	208	83	92	81	--	118 cu	8.8	85	IV	592	393	5	848	78	6.7

[1] cs - carbon steel; zn - galvanized; cu - copper plated; ss - stainless steel.
[2] NA - no asymptote.
[3] I - calcium carbonate; II - calcium carbonate/magnesium silicate mixture; III - magnesium silicate; IV - copper oxide, silica and magnesium; NS - no sample.
[4] Water loss during Runs 25, 26, 27 caused data to be invalid.
[5] Bulk temperature - 24°C.
[6] Numbers in parentheses indicate effective concentration of water No. 4.
[7] 25 ppm Brazos River Clay in cooling tower water.

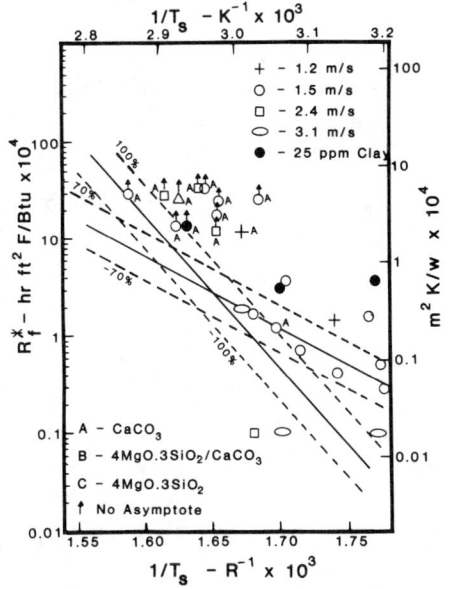
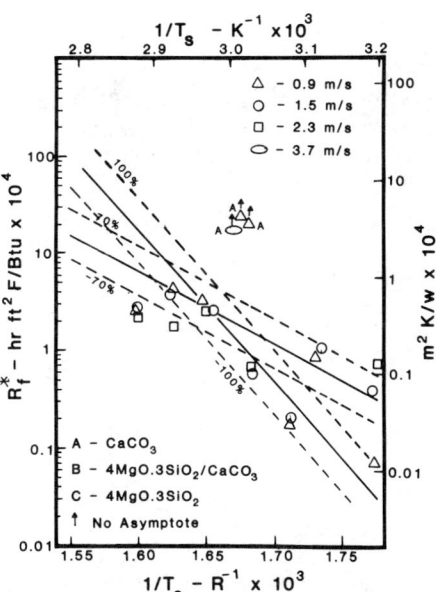

Fig. 2. Fouling Data for Waters Nos. 2 and 3

Fig. 3. Fouling Data for Water No. 4.

No. 1. Equation (4) which applies for magnesium silicate deposition for water No. 1 appears to be applicable to waters Nos. 2 and 3 below 60°C and to water No. 4 over the whole surface temperature range investigated. No quantitative effect of velocity has been determined but, in general, for the asymptotic mode of deposition, the higher the velocity the lower the asymptotic fouling resistance.

REFERENCES

1. Taborek, J., Knudsen, J.G., Aoki, T., Ritter, R.B. and Palen, J.W., 1972. Fouling: The Major Unresolved Problem in Heat Transfer. Chem. Eng. Prog., 68 (No. 2).
2. Taborek, J., Knudsen, J.G., Aoki, T., Ritter, R.B. and Palen, J.W., 1972. Predictive Methods for Fouling Behavior. Chem. Eng. Prog., 68 (No. 7).
3. Langlier, W.F., 1936. The Analytical Control of Anti-Corrosion Water Treatment. J. Am. Water Works Assoc., 28, 1500.
4. Ryznar, J.W., 1944. A New Index for Determining the Amount of Calcium Carbonate Formed by a Water. J. Am. Water Works Assoc., 36, 472.
5. Lee, S.H. and Knudsen, J.G., 1978. Scaling Characteristics of Cooling Tower Water. ASHRAE Transactions, Vol. 85, Part 1.
6. Coates, K.E. and Knudsen, J.G., 1980. Calcium Carbonate Scaling Characteristics of Cooling Tower Water. ASHRAE Transactions, Vol. 86, Part 2.

ACKNOWLEDGEMENT

Appreciation is expressed to the National Science Foundation for support of this research and to Heat Transfer Research Inc. for the donation of a Portable Fouling Unit.

Fouling in Plate Heat Exchangers and Its Reduction by Proper Design

LADISLAV NOVAK
Product Planning and Research
Alfa Laval AB
Box 1721
S-221 01 Lund 1, Sweden

1. ABSTRACT

This paper gives practical advice based on recent tests and many years' experience gained by the author's company in the design and manufacture of plate heat exchangers operating with fresh and saline waters as the cooling medium. The various design possibilities and arrangements to reduce the fouling are discussed in relation to the different fouling problems.

2. NOTATIONS

A – heat transfer area, m^2
d_p – particle diameter, m
G – mass flow rate, kg/s
K – overall heat transfer coefficient, W/m^2K
K_{Max} – overall heat transfer coefficient in design with zero fouling margins, W/m^2K
Q – transferred heat, W/s
R_f – fouling resistance, m^2K/W
$(R_f)_d$ – design fouling resistance, m^2K/W
S – distance between plates, m
t – time
v – linear flow velocity, m/s
τ – shear stress, Pascal

3. INTRODUCTION

A new heat exchanger is always clean both inside and outside and nicely finished but, in majority of applications, this attractive appearance and cleanliness are subject to more or less rapid change when the apparatus is taken into operation. This change depends on the simple fact that liquids (cooling water, process streams) entering the heat exchanger are, as a rule dirty and/or unstable, and cause more or less rapid build-up of fouling deposit.

Although it is very difficult to make a meaningful comparison of fouling in different exchanger designs, it is generally recognized that plate heat exchangers, due to their advanced design from the fluid dynamics point of view, foul less than corresponding units of other types of heat exchangers.

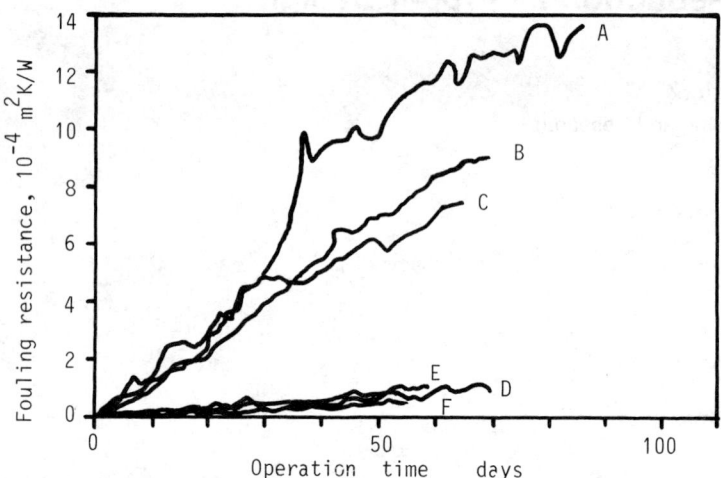

Fig. 1. Development of fouling resistance in correctly and wrongly designed plate heat exchangers.

Figure 1 shows the development of fouling deposit in a correctly designed plate heat exchanger - curves D,E and F, as well as in a poorly designed one - curves A,B and C. All curves show the development of fouling deposit consisting mainly of microbiological slime at a temperature of 30 - 45°C. Curves A and D show the development in river water, curves B and E the development in cooling tower water and curves C and F the development in seawater.

As vill be shown later in section 4.2 biological fouling (but not only biological fouling) is very strongly dependent on shear stresses between the flowing liquid and heat transferring surfaces. Since shear stresses in cases A,B and C were between 5 and 10 Pascal, thus approaching shear stresses normally found on tube side of a well designed shell and tube heat exchangers, this figure at the same time also indicates the difference between the fouling behavior of the plate type and of the shell and tube heat exchanger design in case of biological fouling up to about 50°C. Biological fouling is however not the only fouling mechanism which can be successfully reduced by using the plate heat exchangers. The cooling or heating of fine slurries containing up to 25 % of solids, or cooling by liquids tending to form a scale deposit (4) are other two examples.

Nevertheless, this advantage of plate heat exchanger design may be completely spoiled by incorrect dimensioning or unsuitable connection to the flow system

4. DESIGN CONSIDERATIONS

To achieve the optimum function of the plate heat exchanger the following objectives should be a part of the design procedure:
1) Examination of fouling margins
2) Prevention of coarse fouling
3) The use oj optimal flow rates
4) Estimation of scaling thresholds
5) Control and removal of biological fouling by chlorination

FOULING IN PLATE HEAT EXCHANGERS

6) The use of chemical cleaning

Although a number of other objectives affecting the design do exist, the ones above are the most important for the design discussions.

4.1 Examination of Fouling Margins

Fouling margins (fouling resistance, excess surface, cleanliness factor, additional surface) are introduced to the design in order to safeguard the operation of fouled heat exchanger. Introduction of fouling margins to the design of plate heat exchangers usually - but not always - leads to the increase of the number of plates and, therefore, also to the increase of the cross-sectional area for the flow of the two fluids. Since the flow rate of the two fluids is independent of fouling margins, and is constant, the channel velocity and shear stresses between the fluid and the heat surfaces , and consequently also K-values decrease with increased fouling margins (see Fig.2)

The fouling rate is dependent on flow rate - shear stresses, and in natural waters (sea, river, lake, cooling tower), as a rule, it increases with decreasing shear stresses. Experimental values of fouling rate as a function of shear stresses obtained on the Rhine water and the Öresund seawater are summarized in reference (1) and (2).

Consequently, heat exchangers designed with a high fouling resistance foul more rapidly than those designed with a low fouling resistance. This increase in fouling rate may be so extensive that exchangers designed with high fouling margins must be cleaned more frequently than those designed with low fouling margins.

Operation time (i.e. the time during which fouling resistance reaches the design value) of a plate heat exchanger, based on experimental results obtained on the Rhine water (1), was calculated for different, zero fouling, design K-values. The calculations summarized in Figure 3, were performed on following assumptions:

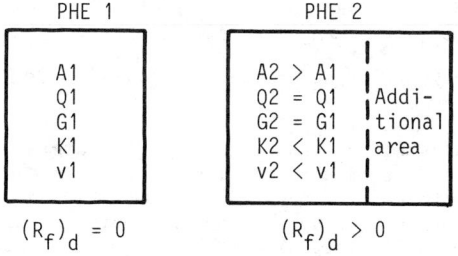

Fig. 2. Consequences of the Design with Fouling Margins

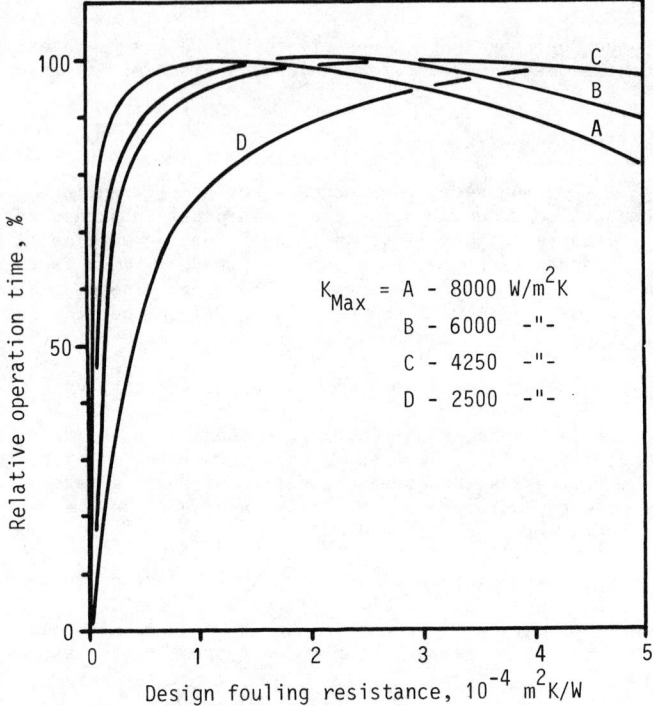

Fig. 3. The Effect of Design Fouling Margins on the Function of a Plate Heat Exchanger

a) Fouling margins are provided by additional plates and not by changed thermal length of plates (possible in ALFA-FLEX system)
b) Wall thermal resistance was $0.35 \times 10^{-4} m^2 K/W$
c) Flow rate and the coefficient of heat transmission (α) were the same for the cooled and cooling medium.

Figure 3 demonstrates clearly that when fouling margins of a plate heat exchanger having a clean "zero fouling design" K-value exceeding 4000 W/m²K (a typical value for plate heat exchangers is 4500-6000), are provided by a simple addition of plates and not by a change of thermal length of plates, then the use of design fouling margins exceeding approx. 25% or a value of $0.5-1.0 \times 10^{-4} m^2 K/W$ does not essentially improve the function of such plate heat exchanger. Figure 3 also demonstrates that the use of too high design fouling margins leads to shorter cleaning intervals.

In the mixed theta design (ALFA-FLEX)(3), the negative effect of design fouling margins on the fouling rate may be partly balanced by a substitution

FOULING IN PLATE HEAT EXCHANGERS

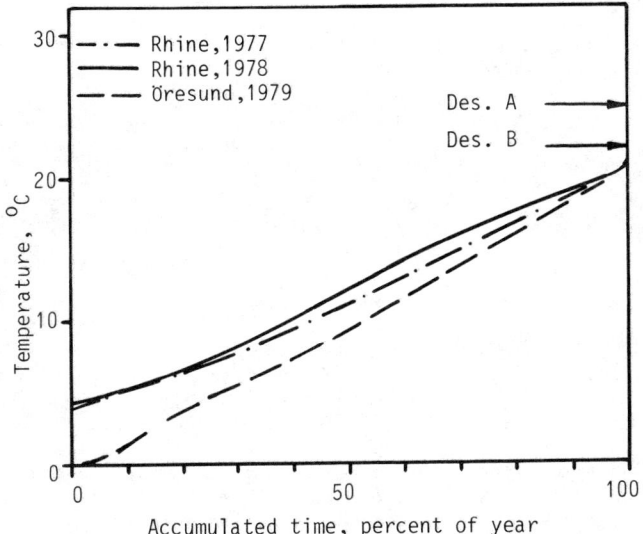

Fig.4. Seasonal Variations of the Rhine Water in Mannheim Region and in the Öresund Seawater

of low theta channels by medium or high theta ones. Therefore, the operation time (cleaning intervals) of one ALFA-FLEX plate heat exchanger is usually longer than the operation time of a conventional plate heat exchanger designed with the same fouling margins and having the same "zero design fouling" K-value.

The fouling margins introduced into the design should be considered not only on the basis of the fluid properties but also as a function of temperature program, temperature variations (seasonal variations of cooling water temperature), variations in heat load as well as the extent of auxiliary equipment.

Fig.4 shows seasonal variations of temperature of the Rhine river in Mannheim region (the years 1977 and 1978) and of the seawater of Öresund (1979). Similar temperature variations are certainly also found in other rivers and lakes in Europe as well as in the lakes and rivers within the tempered region of North America and Asia.

The effect of temperature program and seasonal variations of cooling water temperature on the operation of a heat exchanger is perhaps best illustrated by the following two examples of the existing heat exchangers:

Example A. Example A is a typical example of a central cooler (SECOOL). The exchanger is installed in the nuclear power station Philippsburg I, West Germany, and is designed as follows:

circulation water   38 → 30°C
cooling water (the Rhine) 25 → 32°C

LMTD   5.5°C
design, clean K-value   5000 W/m²K
design fouling resistance 0.6×10^{-4} m²K/W

<u>Example B</u>. Example B is nearly a typical process cooler. The exchanger is installed in a Nitric acid plant of Supra AB in Landskrona, Sweden and is designed as follows:

process water   86 → 51°C
cooling water (seawater)   22 → 63°C
LMTD   25.9°C
clean, design K-value   5800 W/m²K
design fouling resistance 0.6×10^{-4} m²K/W

Starting from the seasonal variations of the Rhine water and the seawater of Öresund, and also taking account of the increase in pressure drop in the fouled exchanger due to fouling, the max allowable fouling of the exchangers A and B has been calculated. Figure 5 shows the result of these calculations expressed in terms of fouling resistance and plotted against the accumulated running time expressed in percentage of one year's period.
Figure 6 shows then the result of these calculations in terms of the overcapacity of the clean exchangers throughout the year. The calculations were performed both for the design with fouling resistance of 0.6×10^{-4} m²K/W, and for design without any fouling resistance.

As seen, the overcapacity of the clean exchanger A was at least 95% during the summer, and up to 340% during the winter of the years 1977 and 1978. In other words, this exchanger which was designed with fouling resistance of 0.6×10^{-4} m²K/W could be 3 times during the summer months and 7 times during the winter months more fouled than it had been estimated in the design. The design of such heat exchanger with high fouling margins is a luxury, both unnecessary and undesirable, leading to heavy problems with regulation of such a system. On the contrary, the overcapacity of the clean exchanger B designed with zero fouling margins was only 12% during the summer and not more than

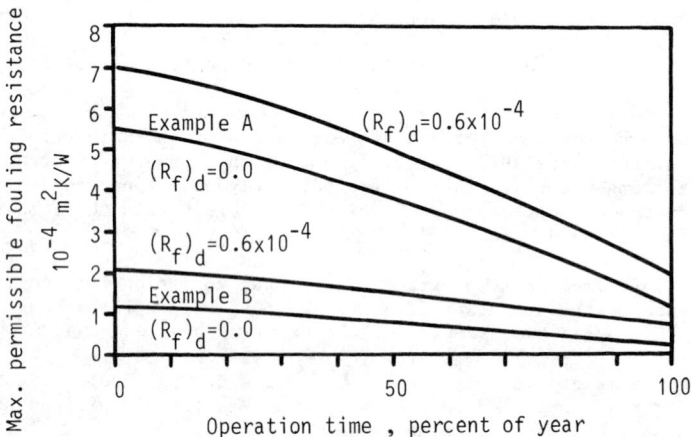

Fig. 5. The Maximum Admissible Fouling Resistance of Exchanger A and B as a Function of Season.

FOULING IN PLATE HEAT EXCHANGERS

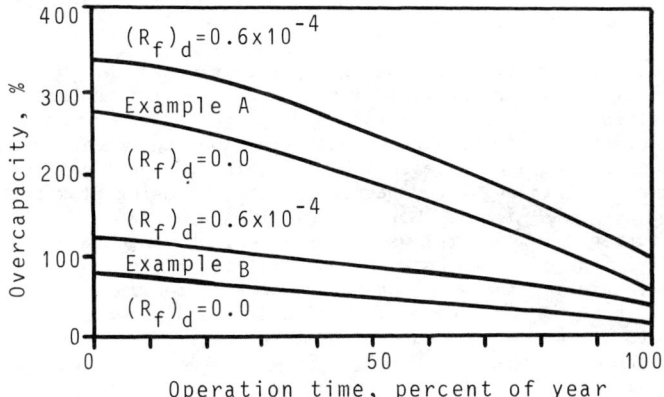

Fig. 6. The Overdesign of a Clean Exchanger A and B as a Function of Season.

80% during the winter 1979. Designing such an exchanger with zero fouling margins should surely jeopardize its functions during summer months. Therefore heat exchangers localized in tempered regions and having low logarithmic mean temperature difference can certainly be designed with lower fouling margins than exchangers having high logarithmic mean temperature difference.

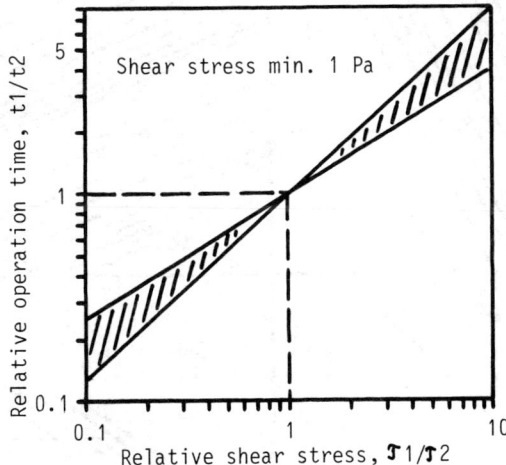

Fig. 7. Dependence of Operation Time on Shear Stresses in Case of Biological Fouling.

4.2 The Effect of Flow Rate

Generally, the formation of fouling deposit is dependent on flow rate. This dependence is, however, not identical for all fouling mechanisms. At operating conditions normally occurring in heat exchangers (i.e. shear stresses 2-200 Pa), the formation of fouling deposit from cooling water usually decreases with increasing shear stresses. However, in some applications such as milk sterilizers the fouling rate increases with increasing flow rate.

Cooling duties using natural waters as a cooling medium are the most frequent applications of plate heat exchangers. Microbiological fouling is usually the main fouling mechanism occurring on surfaces immersed in natural water of temperature up to 50°C. Fig.7 which is based on measurements carried out with seawater, river water and cooling tower water (1, 2), shows the probable change of the running time (cleaning intervals) of a heat exchanger as a function of change in shear stresses of the cooling water. Fig.7 presupposes constant fouling margins. For example, if the cleaning interval of one heat exchanger having shear stresses of cooling water 25 Pa, is two months, then we can expect that the cleaning interval of another heat exchanger designed with identical fouling margins but having shear stresses of cooling water increased to 75 Pa, will increase according to Fig.7 by a factor of 2.5, or to 5 months.

If no antifouling means are used for control of biological fouling, then for obtaining resonably long cleaning intervals, the plate heat exchangers should not be designed with shear stresses of the cooling water below 50 Pascal.

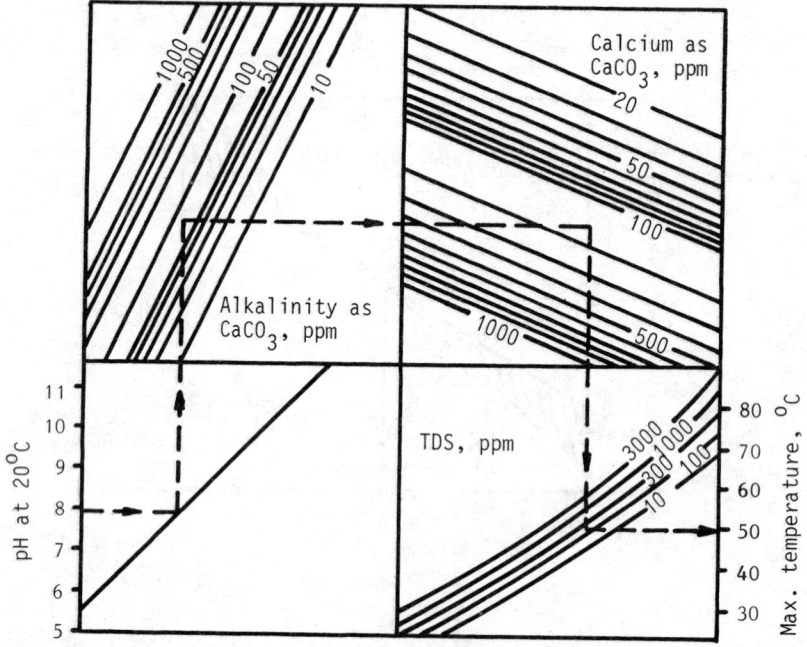

Fig. 8. Determination of Scaling Thresholds of Calcium Carbonate in Fresh water.

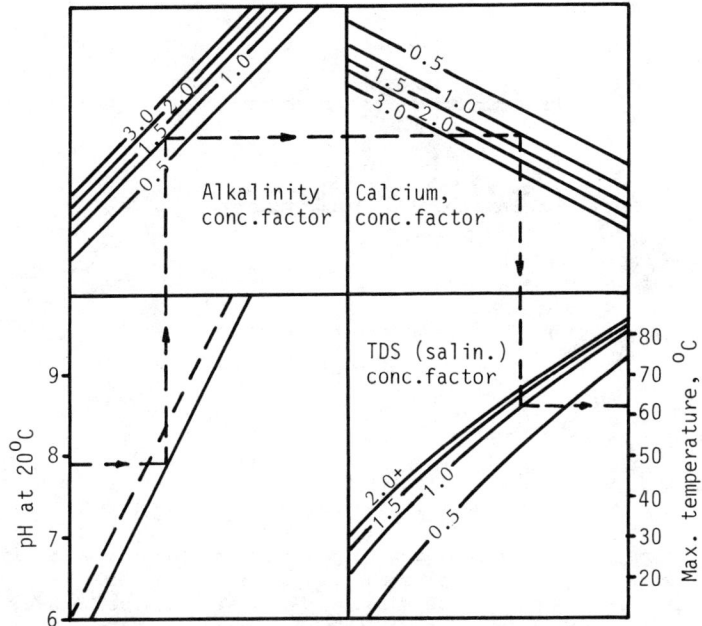

Fig. 9. Determination of Scaling Thresholds of Calcium Carbonate in Seawater

.3 Estimation of Scaling Thresholds

Calcium carbonate is the most common scaling species in natural waters at conditions occurring most frequently in plate heat exchangers (i.e. up to 90°C). Scaling thresholds of calcium carbonate is dependent on pH, calcium content, alcalinity and the ionic strength of the water. The exact estimate of calcium carbonate scaling thresholds is difficult and not always possible. Nevertheless with regard to uncertainty of the representativeness of a given water quality, the scaling thresholds may be, for the purpose of design, determined by Langelier and Ryznar indices. Scaling thresholds may be very quickly estimated for fresh water from Figure 8, and for seawater from Fig. 9. Both figures are based on available literature sources and are calculated for Ryznar stability index of approx. 6.5. Regulation of scaling thresholds in once-through systems by chemical additives is very expensive and is practically never used. Therefore, the designer should try to avoid the scaling region as far as possible. This may be done, for example, by cocurrent flow arrangement, by using intermediate circuits or simply by using high flow rates of cooling water and low flow rates of the cooled medium.

Similar diagrams may be used also for determining scaling thresholds of calcium sulphate, calcium phosphate and magnesium hydroxide.

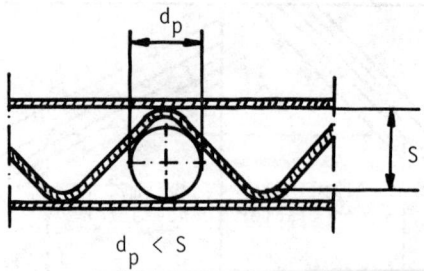

Fig. 10. The Narrowest Passage in a Modern Crosscorrugated Plate Heat Exchanger

.4 Prevention of Coarse Deposits

The plate heat exchangers are familiar by their outstanding heat transfer coefficients and also by their narrow passages for the heat transferring media. The space between plates in the plate heat exchangers is usually between 2.5 and 6 mm and one may easily imagine what happens if the cooling water does not pass through a strainer before passing through a heat exchanger. The max. size of spherical particles that can pass through the plate heat exchanger is always, as explained in Fig. 10, less than the distance between the plates. Therefore, the dimensions of screen openings should not exceed approx. 75% of the plates' spacing. The type and construction of the strainer itself is not important, but self-cleaning strainers should be preferred. Alfa-Laval has recently developed

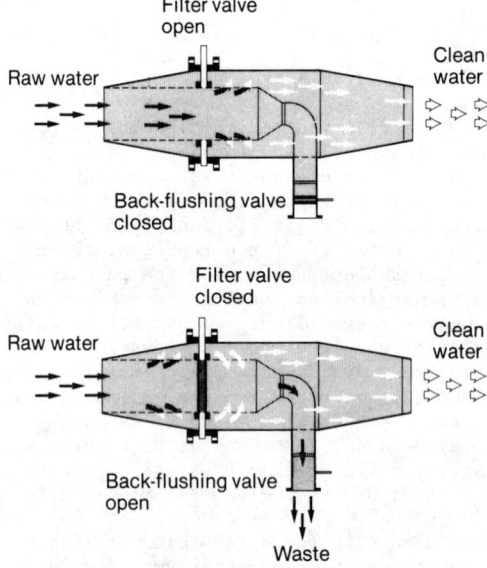

Fig. 11. The Principle of Alfa-Laval Secool Filter

FOULING IN PLATE HEAT EXCHANGERS

a very simple self-cleaning strainer having a minimum of moving parts. The principle of the filter is explained in Figure 11. During normal operation only a minor portion of raw water goes through the first filter basket which is kept clean due to the flushing effect of the water stream passing to the second filter basket. Debris accumulated in the second basket are then periodically flushed out through backflushing pipe. The filter has been found very effective. Central coolers of plate type on a Danish ferry 'Mette Mols' were especially exposed to clogging by sea algae. When the situation became very bad, the exchangers were opened and cleaned twice a day. After the strainer had been installed neither the plate heat exchanger nor the strainer have been opened during the last 6 months until this report has been prepared.

Another method aimed at reducing the coarse fouling is the installation of a back-flushing arrangement. Generally, the inlet of the cooling water should be arranged at the bottom of the plate heat exchanger, as any heavy debris will then settle on the bottom of the inlet part rather than clog the inlet of the channels.

4.5 Control and Removal of Biological Fouling by Chlorination

Chlorination is the least expensive chemical method for preventive control of biological fouling. However, recent tests (2) showed that chlorination may be used also for removal of already formed deposit. Figure 12 shows the reduction of fouling resistance of a small test heat exchanger due to chlorination. Chlorination was performed in normal operation conditions. Test result indicates that in situations where the heat exchanger may be on stand by for an hour or so, for example if a spare exchanger is available, it would be possible to soak the heat exchanger at regular intervals ranging from one day to approximately 2 weeks. In this manner, both the environmental pollution and operation costs are considerably reduced.

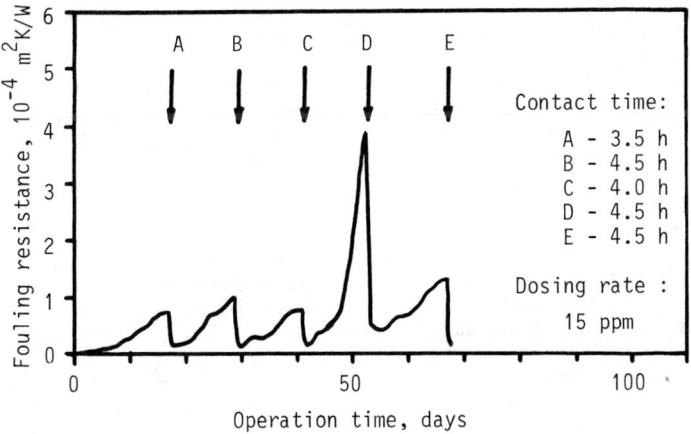

Fig. 12. Reduction of Fouling Resistance by Chlorination

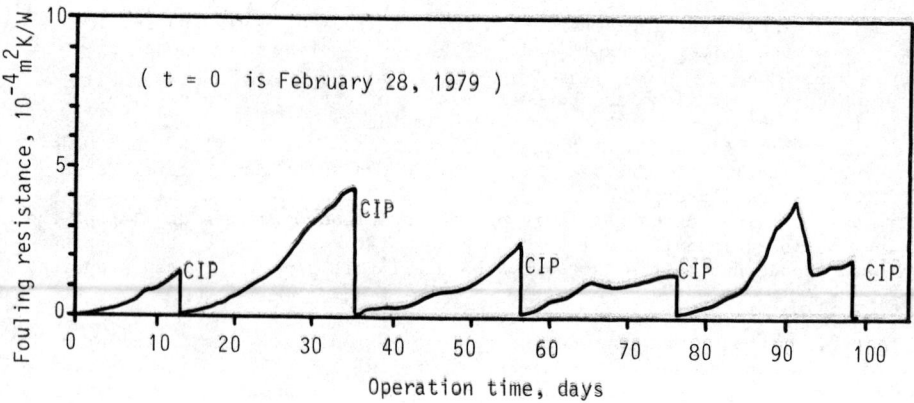

Fig. 13. Reduction of Fouling Resistance by Cleaning-in-Place Procedure.

4.6 The Use of Chemical Cleaning

In situations when no fouling preventive methods are used, heat exchangers must be cleaned afterwards. Chemical cleaning is one of possible cleaning methods.

Chemical, in-place cleaning (CIP) may be effectively used for removal of almost all kinds of fouling deposit, such as biological slime, rust, scale and organics. Common denominator of a CIP-procedure is the high cleaning efficiency as well as the use of elevated temperatures, high equipment requirements and last but not least, the handling of chemicals and chemical waste. In some applications, such as dairy industry, the CIP is a necessary part of the production procedure. An example of the efficiency of CIP procedure in cleaning of cooling water deposits is shown in Fig.13.
In this case biological fouling formed from the Rhine water in one small test heat exchanger was removed by slow circulation of 60°C hot alkaline solution (1). Similar cleaning efficiencies have been reached also in full scale procedures. Naturally, before any decision concerning CIP application is made, safe and, from environmental aspects, acceptable handling and discharge of used cleaning solution as well as the type and origin of fouling deposit with regard to proper choice of cleaning agent should be seriously discussed.

4.7 Other Considerations

From time to time the use of coatings, polished surfaces, ultrasonics, magnetic field etc. is also a subject of discussions. We have performed several tests with polished surfaces, coatings such as Teflon, Säkaphen and enamel. The common result obtained for these surfaces was that these sufaces only prolonged the initial time of fouling formation. Later on there was no difference between these surfaces and ordinary stainless steel or titanium surfaces. Taking account of the potential of all ionic, nonionic, polar and nonpolar as well as living species occurring in cooling water, which can be expected to settle on these surfaces, this is by no means a surprising observation.

Heat shocks are sometimes also considered for removal of biological fouling. Our tests were, however, not so successful. Microorganisms are, of course, sterilized, but the slime deposit is removed only in part. Some five years ago we have tested the use of ultrasonics in plate heat exchangers, but we have found that, due to the mechanical stability of the plate pack and the complicated flow pattern, the effect was very low and limited to only a few centimeters from the plate inlet.

5. CONCLUSIONS

The optimal design of a heat exchanger (and also of a plate heat exchanger) is by no means reduced to the thermal and strength calculations. The design becomes optimal only when all factors including operational factors such as fouling and cost factors are being considered and discussed in detail.

6. LITERATURE

1) Novak, L. 1979. Control of the Rhine Water Fouling
   International Conference on the Fouling of Heat Transfer Equipment, Troy, N.Y., August 1979.

2. Novak L. 1981. Comparison of the Rhine and the Öresund Seawater Fouling and its Removal by Chlorination.
   '20th National Heat Transfer Conference, Milwaukee, Wis., August 1981.

3) Marriott J. 1977. Performance of an Alfaflex Plate Heat Exchanger, CEP, February 1977, pp. 73-78.

4) Cooper A., Suitor J.W., Usher J.D., 1980. Cooling Water Fouling in Plate Heat Exchangers, Heat Transfer Engng. Vol.1, 1980, No 3, pp 50 - 55.

Fluid Bed Heat Exchanger: A Major Improvement in Severe Fouling Heat Transfer

DICK G. KLAREN
Esmil Research
Estel Hoogovens BV
IJmuiden, The Netherlands

ABSTRACT

The principle and operating characteristics of a fluid bed heat exchanger are discussed and it has been investigated to what extent a fluid bed heat exchanger functions more satisfactorily than a conventional heat exchanger under severe fouling conditions. A number of pilot plants based on the fluid bed principle have been operated in industry and their fouling behaviour has been studied and sometimes compared with conventional heat exchangers, operating parallel to the pilot plants.

1. INTRODUCTION

Large scale applications of fluid bed heat exchangers in severe fouling heat transfer, have been reported by Klaren [1] in 1975.

A fluid bed heat exchanger consists of a large number of parallel heat exchanger tubes, in which small solid particles are kept in a stationary fluidized condition by the liquid passing up the tubes. The solid particles regularly break through the boundary layer in the tubes, so that good heat transfer is achieved in spite of comparatively low liquid velocities in the tubes. Further, the solid particles have a slightly abrasive effect on the wall of the heat exchanger tubes, so that any deposit will be removed from the tube wall at an early stage. Particularly, the defouling capability of the fluidized particles is considered to be very valuable, as the effect of this property is that periodic cleaning of a fluid bed heat exchanger by chemicals or mechanical means can be dispensed with or at least be done less frequently.

Stable operation of all parallel heat exchanger tubes, in which the solid particles are fluidized, is the main concern with a fluid bed heat exchanger. It requires equal velocity distribution of the liquid in the tubes and also equal distribution of the solid particles over all the tubes. In order to achieve stable operation of a large number of parallel fluidized beds a throttling device was applied at each tube inlet. However, this method has the following disadvantages:
- sensitive to clogging;
- costly construction;
- more pumping power required.

In reference [1], the principle of individual throttling has been extensively explained.

The principle of the fluid bed heat exchanger based on more recent

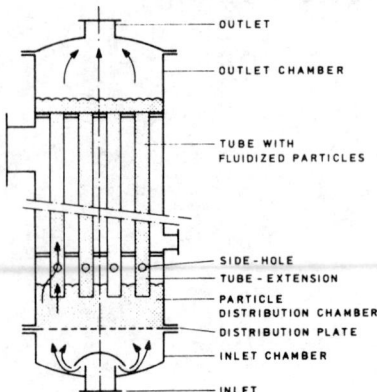

FIGURE 1 - Principle of fluid bed heat exchanger.

developments with respect to stabilisation of all parallel operating tubes is shown in figure 1. The heat exchanger consists of the tube bundle, an inlet section and the outlet chamber. The inlet section has been divided into two chambers by a distribution plate, viz.:
- the actual inlet chamber for the liquid;
- the particle distribution chamber from where the particles are equally distributed over all tubes.

The particle distribution chamber also contains the tube extensions, which are provided with a side hole. Design of the distribution plate, pressure drop across this plate and design of the tube extensions are of utmost importance to achieve satisfactory operation, which actually means stable operation or equal distribution of liquid and particles over all tubes.

2. HYDRAULICS OF THE FLUID BED HEAT EXCHANGER

From experiments, it has been determined that one of the conditions for stable operation is that the pressure drop across the distribution plate should satisfy the condition:

$$\Delta P_d > 0.1 \, \Delta P_t \tag{1}$$

where ΔP_d = pressure drop across distribution plate
ΔP_t = total pressure drop over the heat exchanger due to the weight of the solid particles.

The latter pressure drop follows from the equation:

$$\Delta P_t = g \, (\rho_s - \rho_\ell) \left\{ L_1 \, (1 - \varepsilon_1) + L_2 \, (1 - \varepsilon_2) + L_3 \, (1 - \varepsilon_3) \right\} \tag{2}$$

with
$\quad g$ = earth gravity
$\quad \rho_s$ = density of the particle material
$\quad \rho_\ell$ = density of the liquid
$\quad L_1, L_3$ = bed height in distribution chamber and outlet chamber respectively
$\quad L_2$ = tube length
$\quad \varepsilon_1, \varepsilon_2, \varepsilon_3$ = porosity of the bed in particle distribution chamber, tubes and outlet chamber respectively.

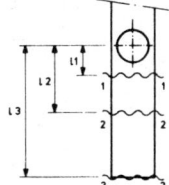

FIGURE 2 - Tube extension with side-hole.

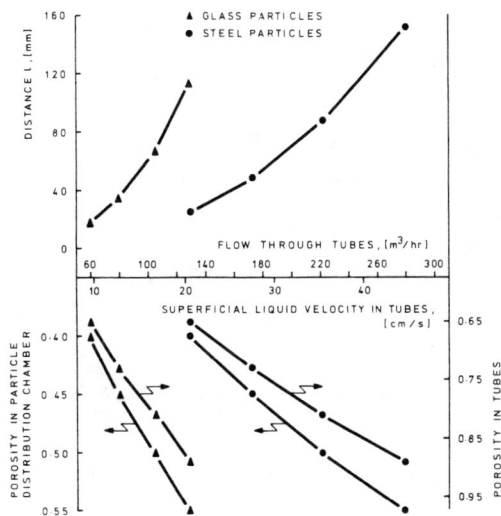

FIGURE 3 - Flow as function of length of tube extension and other process parameters.

The distribution plate can be designed as a perforated plate. Diameter and number of perforations should satisfy the pressure drop required for stable operation of the bundle. In practical applications, the diameter of the perforations in the distribution plate may approach half of the inner tube diameter of the actual heat exchanger tubes, which results in fairly large holes for 1½" and 2" tubes often used in severe fouling heat transfer.

The design of the tube extension with side hole determines the flow variation, which can be applied to the heat exchanger; see also figure 2. At minimum flow through the heat exchanger, the height of the fluidized bed around the tube extensions reaches the level 1 - 1 in the particle distribution chamber, which is located slightly below the side hole at a distance l_1 from the center of this hole; increasing the flow lowers this bed height to the level 2 - 2 at a distance l_2 from the center of the side hole. As long as the flow is not increased to a value which causes the bed height around the tube extension to sink below the level 3 - 3 and the pressure drop across the distribution plate satisfies equation (1), stable operation of the tube bundle is assured. The level 3 - 3 corresponds to a distance l_3 for the center of the side hole, which represents the actual length of the tube extension from the center of the side hole. Figure 3 shows the distance from the center of the side hole in the tube

TABLE 1 - Specification fluid bed heat exchanger.

Total number of tubes	100
Tube size	50 x 1.5 mm
Pitch	1.25 x OD
Particle size	2 mm
Side hole	18 mm
Liquid entering the tubes	water

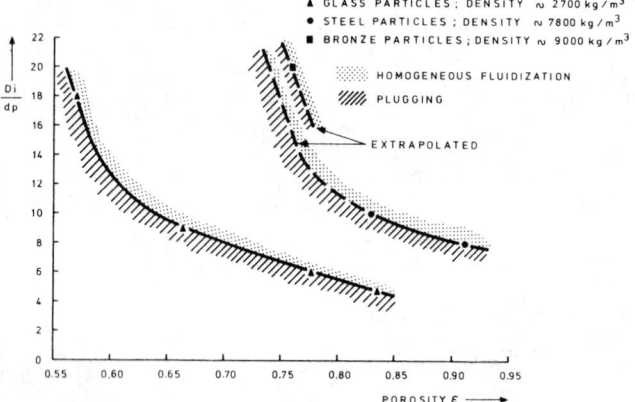

FIGURE 4 - Transition lines for homogeneous fluidization and plugging.

extension (referred to in figure 2 as "l") as a function of the flow and other parameters for the fluid bed heat exchanger as specified in table 1.

Minimum flow through the heat exchanger mostly corresponds to a nearly fixed bed in the particle distribution chamber and a porosity of this bed of approx. 0.4; minimum flow also requires the fluidized beds in the tubes to be extended over the full tube length. Increasing the flow through the heat exchanger means higher porosities of the beds and accommodating a large fraction of the particles in the outlet chamber, which requires an outlet chamber of sufficient volume. The length of the tube extension from the center of the side hole to the actual inlet of the tube in combination with the size of the side hole determines the flow variation which can be applied to the heat exchanger. Higher densities of the material of the particles make it possible to operate the heat exchanger at higher flows; this is also the case for larger particles. Diameter of the side hole in the tube extension is approx. 40 % of the internal diameter of the heat exchanger tubes and is proportional to the tube diameter for a given set of design parameters.

In the tubes, the porosities of the fluidized particles and the density of the particle material should satisfy the ratios of the inner tube diameter to the particle size, as shown in figure 4. Too low porosities and/or particles of high density material cause plugging of the particles in the tubes and reduces heat transfer between the liquid and the tube wall.

It has been explained that stable operation with expanded beds over the

FLUID BED HEAT EXCHANGERS

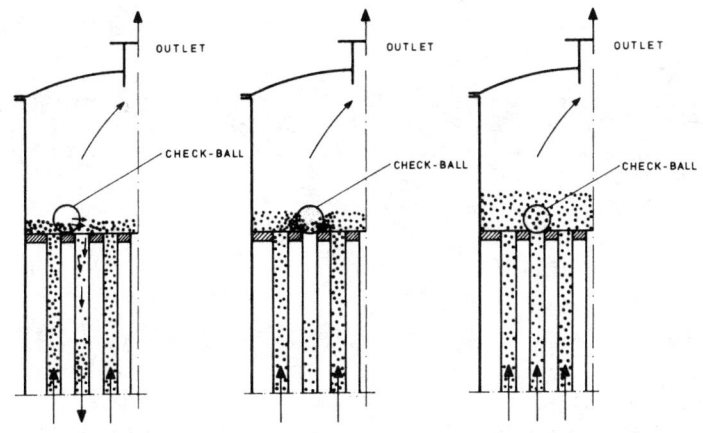

A, check-ball moves in place
B, check-ball stops circulation
C, normal operation restored

FIGURE 5 - Movement of a check-ball.

total tube length can be achieved between two flow limits. However, in the case of rough start-up procedures, the flow may overshoot its required value and also exceed the value which corresponds to the maximal value which can be applied to the heat exchanger. Then the level of the fluidized particles in the particle distribution chamber will sink below the level 3 - 3 as shown in figure 2, and circulation of particles and liquid from the outlet chamber to the particle distribution chamber and conversely, may occur. Actually, this means that the stable operation pattern is disturbed and that cold and hot end of the heat exchanger are short-circuited, which could result in a remarkable reduction of the performance of the heat exchanger.

Stable operation of all parallel fluidized beds can be restored in different ways: For example, the flow through the tubes is reduced slightly below its minimal value, which means that all beds are retracted in the tubes and circulation cannot occur, as there are no more particles in the outlet chamber. Then the flow is gradually increased up to its required value. According to another procedure, the flow is directly reduced to its required value. However, circulation of particles and liquid will continue, because, once induced, such a circulation pattern is stable by itself.

Adding so-called check-balls in the outlet chamber makes it possible to stop circulation. These check-balls are provided with apertures of sufficient size to allow the particles to pass through the ball at normal stable fluidization conditions. Further, the check-balls are made of such a material that they sink in the fluidized bed in the outlet chamber. Should circulation occur, causing for instance in the outlet chamber a migration of particles to the tubes in which the particles are moving downwards, the check-balls, which have a slightly larger diameter than the tubes, will also move towards these tubes and settle on top of the tube openings. Due to bridge formation in the particle mass against or in the check-ball, the flow of particles into the tubes is completely cut off. The result is that the pressure balance in the tubes with initial downward flow of particles and liquid, is redressed and the tubes can refill gradually with a stable fluidized bed along their entire length. The number of check-balls required may vary from a few percent of the total number of tubes to any other higher number. The attractive aspect of the check-balls is their free movement transversely through the outlet chamber which means, that they do not need to be mounted in cages. Figure 5 gives an impression of the movement of a check-ball.

3. HEAT TRANSFER PERFORMANCE OF A FLUID BED HEAT EXCHANGER

In the correlation used for the wall to liquid heat transfer coefficient of a fluidized bed, the superficial velocity of the liquid in the tubes is an important parameter. The way this velocity $U_{\ell,s}$ is calculated is presented in detail in reference [2]; this paper only summarizes the equations required to compute the superficial velocity:

$$U_{\ell,s} = U_i \, \varepsilon^n \qquad (3)$$

with n = empirical factor = 2.4

and
$$U_i = U_\infty \, 10^{-\frac{d_p}{D_i}} \qquad (4)$$

where d_p is the particle diameter and U_∞ is the terminal velocity of one single particle falling in a stagnant liquid with which the tube would normally be filled and which must behave as an infinite fluidum. For spherical particles, the terminal velocity can be determined from the equations:

$$c_{D_\infty} \, Re_\infty^2 = \frac{4}{3} \frac{d_p^3 \, \rho_\ell \, g \, (\rho_s - \rho_\ell)}{\eta_\ell^2} \qquad (5)$$

with
$$Re_\infty = \frac{\rho_\ell \, U_\infty \, d_p}{\eta_\ell} \qquad (6)$$

In the above equations the symbols have the following meaning:
c_{D_∞} = drag coefficient = 0.44 and η_ℓ = dynamic viscosity of the liquid.

For the wall to liquid heat transfer coefficient of a fluidized bed heat exchanger, it has been found by the author that the correlation of Ruckenstein [3] showed a good fit with his experimental results obtained over the past years. The relevant part of this correlation reads:

$$Nu = 0.067 \, Pr^{0.33} \, Re_{d_p}^{-0.237} \, Ar^{0.522} \qquad (7)$$

In the above equations the dimensionless numbers are composed as follows:

$$Nu = \frac{\alpha_\ell \, d_p}{\lambda_\ell} \qquad (8)$$

$$Pr = \frac{\eta_\ell \, c_\ell}{\lambda_\ell} \qquad (9)$$

$$Re_{d_p} = \frac{\rho_\ell \, U_{\ell,s} \, d_p}{\eta_\ell} \qquad (10)$$

$$Ar = \frac{g \, d_p^3 \, \rho_\ell \, (\rho_s - \rho_\ell)}{\eta_\ell^2} \qquad (11)$$

where α_ℓ = wall to liquid heat transfer coefficient
and λ_ℓ = heat conductivity of the liquid.

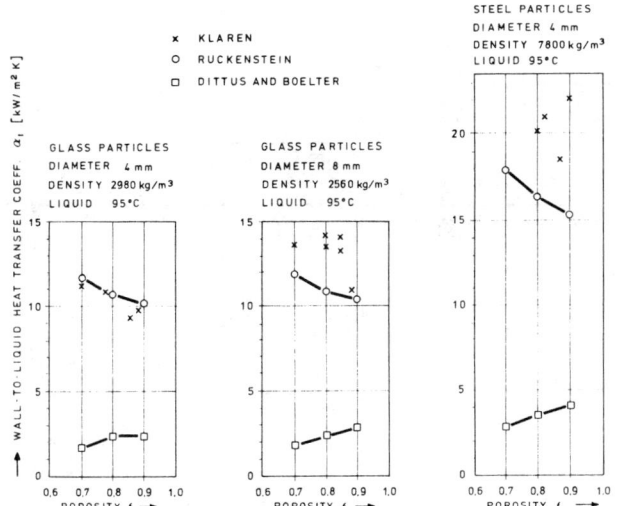

FIGURE 6 - Heat transfer coefficients in fluidized bed.

Figure 6 compares some results obtained experimentally by the author with the correlation of Ruckenstein. This figure also shows the wall to liquid heat transfer coefficient according to the well-known equation of Dittus and Boelter, that predicts these heat transfer coefficients when no fluidization is involved according to the equation:

$$\alpha_\ell = \frac{\lambda_\ell}{D_i} (0.023 \, Re_D^{0.8} \, Pr^{0.3}) \tag{12}$$

where $Re_D = \frac{\rho_\ell \, U \, D_i}{\eta_\ell}$ \hfill (13)

In the above equations the velocity U corresponds with the actual velocity in a fluidized bed, i.c.:

$$U = \frac{U_{\ell,s}}{\varepsilon} \tag{14}$$

From figure 6, it can easily be seen that the wall to liquid heat transfer coefficients for a fluidized bed determined by the correlation of Ruckenstein, exceed the values according to Dittus and Boelter five to six times; this must be caused by the break-down of the laminar-thermal boundary layer at the wall of the tube due to the action of the fluidized particles. Actually, the wall to liquid heat transfer coefficient of a fluidized bed heat exchanger requires only (superficial) liquid velocities in the tubes of 0.1 to 0.2 m/s depending on the size of the particles, the density of the particle material and the porosity of the bed. For a conventional heat exchanger a corresponding wall to liquid heat transfer coefficient is obtained for liquid velocities in the tubes of approx. 1.8 m/s.

From the above, it can therefore be concluded that a fluidized bed heat exchanger offers the possibility of obtaining heat transfer coefficients of the same magnitude as normally obtained in conventional heat exchangers at much lower liquid velocities in the heat exchanger tubes.

TABLE 2 - Application, location and specification of experimental fluid bed heat exchangers.

Case	Application	Location	Total operating time	Liquid at fluid bed side	Temperature range of the fouling liquid	Specification of the fluid bed heat transfer surface	Specification of the fluidized particles
1	1) 5-stage MSF/FBE; 50 tons/day	Delft University of Technology, The Netherlands	In operation since June 1975	Highly polluted canal water	20 - 100°C	A = 60 m^2 2) L = 5.5 m ODxt= 15.88x1.2 mm N = 288 MAT = AlBr	Glass beads; diameter 2 mm
2	26-stage MSF/FBE; 500 tons/day	Isle of Texel, The Netherlands	In operation since June 1978	Natural sea-water	20 - 115°C	A = 1000 m^2 L = 12.5 m ODxt= 19.05x0.9 mm N = 1557 MAT = AlBr	Glass beads; diameter 2 mm
3	Experimental heat exchanger (condenser)	Frigate of Royal Netherlands Navy	5000 hours	Natural sea-water	45 - 85°C	A = 0.12 m^2 L = 1.2 m ODxt= 31.75x1.2 mm N = 1 MAT = CuNi 90/10	Glass beads; diameter 2 and 3 mm
4	Experimental heat exchanger (condenser)	Paper mill, USA	In operation since December 1979	Highly polluted river water	25 - 45°C	A = 13.4 m^2 L = 4.5 m ODxt= 50.8x1.4 mm N = 19 MAT = SS 304	Glass beads; diameter 6 mm
5	Experimental heat exchanger (liquid/liquid)	Coking plant, The Netherlands	In operation since April 1979	Liquid contains naphthalene	20 - 70°C	A = 4.2 m^2 L = 5 m ODxt= 38 x 2 mm N = 7 MAT = C-steel	Glass beads; diameter 3 mm
6	Experimental heat exchanger (liquid/liquid)	Potato-starch plant, The Netherlands	In operation during October and November 1979	Liquid contains proteins	10 - 45°C	A = 0.5 m^2 L = 3 m ODxt= 50x1.5 mm N = 1 MAT = SS 304	Chopped stainless steel wire; diameter 1.8 mm, length 1.8 mm
7	Experimental heat exchanger (condenser)	Potato-starch plant, The Netherlands	In operation December 1979 and October and November 1980	Liquid contains proteins	45 - 110°C	A = 6.6 m^2 L = 6 m ODxt= 50x1.5 mm N = 7 MAT = SS 316	Chopped stainless steel wire; diameter 1.8 mm, length 1.8 mm
8	Experimental heat exchanger (condenser)	Pre-fried potato-chips manufacturer, The Netherlands	In operation since March 1980	Waste-water, which contains starch, fats and proteins	20 - 45°C	A = 1.72 m^2 L = 1.85 m ODxt= 42.4x1.6 mm N = 7 MAT = SS 304	Glass beads; diameter 5 mm
9	Experimental heat exchanger (hot flue gas/liquid)	Desalination plant, The Netherlands	In operation during September and October 1980	Natural sea-water	115 - 122°C	A = 2.35 m^2 L = 1.2 m ODxt= 16 x 1 mm N = 39 MAT = SS 304	Glass beads; diameter 2 mm
10	Experimental heat exchanger (condenser)	Alcohol production plant, The Netherlands	In operation since December 1980	Severe fouling cooling water	15 - 65°C	A = 11.5 m^2 L = 2.9 m ODxt= 33x1.5 mm N = 37 MAT = SS 316	Glass beads; diameter 2 mm
11	Experimental heat exchanger (chiller)	Major oil company, USA	In operation August 1980	Lubrication oil containing wax	35 - 0°C	A = 0.5 L = 3 m ODxt= 50x1.5 mm N = 1 MAT = SS 316	Chopped stainless steel wire; diameter 3 mm, length 3 mm

1) MSF/FBE = Multi-stage flash/fluidized bed evaporator; see also reference [1]

2) A = Area of heat exchanger
L = Tube length
OD = Outer diameter of tubes
t = Wall thickness
N = Total number of tubes
MAT = Material of the tubes

TABLE 3 – Fouling of conventional heat exchangers versus fluid bed heat exchangers.

Case	Fouling behaviour conventional heat exchanger	Fouling behaviour fluid bed heat exchanger
1	Drop of overall heat transfer coefficient (k-value) to approx. 50 % of its original clean value in only 4 weeks	Never fouling observed
2	Impossible to operate without chemical treatment	Never fouling observed
3	Drop in k-value from 3000 W/m^2 K to 1200 W/m^2 K in only 6 weeks	Never fouling observed; k-value remained constant during operation at 3100 W/m^2 K
4	Drop in k-value from 1950 W/m^2 K to 630 W/m^2 K in 3 months	Never fouling observed; k-value remained constant during operation at 2580 W/m^2 K
5	"Hot water wash" required once to twice a week to boost k-value from 250 W/m^2 K to 400 W/m^2 K	Never fouling observed; k-value remained constant at approx. 1000 W/m^2 K
6	Plate heat exchanger with "non stick" teflon coating clogged in a few hours	Over a period of 100 hours, k-value dropped from 2300 W/m^2 K to 1900 W/m^2 K
7	No heat exchanger applied; heating of severe fouling liquid is only possible by direct steam injection	Specially developed operating cycle which applies two fluid bed heat exchangers in series guarantees long operating times
8	Severe fouling of plate heat exchanger occured in a matter of days	Never fouling observed; even not after a period of 5 months of continuous operation
9	Impossible to operate without chemical treatment	No fouling observed, in spite of hot flue gas inlet temperatures above 600°C
10	K-value varied from its clean value of 300 W/m^2 K to 200 W/m^2 K, depending on the actual fouled condition	K-value remained constant at 960 W/m^2 K
11	Only scraped heat transfer surfaces can be applied	No fouling observed

893

4. EXPERIMENTAL RESULTS ON FOULING

Over the past years a large number of pilot plants and/or demonstration plants have been operated at various locations in industry. To avoid any lengthy explanation, the locations of the test units, the specifications of the heat exchangers and their operating results have been summarized in the tables 2 and 3.

The following comments should be added to the information listed in the above tables:
The fluid bed heat exchangers never used any chemicals to avoid fouling. The experiments referred to in the cases 1, 2, 8 and 10 have been carried out with real size equipment integrated in the actual production processes. The experiments referred to in the cases 3, 4 and 5 have been carried out parallel to conventional heat exchangers.

5. CONSEQUENCES AND LIMITATIONS OF THE FLUID BED HEAT EXCHANGER

Besides the advantages with respect to fouling in comparison with the conventional heat exchanger, the fluid bed heat exchanger also has some disadvantages.

Because of the low liquid velocities in the tubes, a fluid bed heat exchanger requires a large number of parallel tubes. As a consequence, more holes have to be drilled in the tube plates and support plates and a large number of tubes have to be aligned, rolled or welded into the tube plates. The accommodation of a large number of parallel tubes increases the diameter of the shell and the tube plates of the heat exchanger. Particularly at high operating pressures, a large shell diameter including large tube plates could become very expensive.

The shell diameter of a single-pass low-velocity fluid bed heat exchanger and that of a conventional heat exchanger may approach one another if the conventional heat exchanger applies more flow-passes. For example, a fluid bed heat exchanger with a superficial velocity in the tubes of 0.3 m/s requires the same number of parallel tubes as a conventional 6-pass heat exchanger with a velocity in the tubes of 1.8 m/s.

A fluid bed heat exchanger should always contain fluidized particles extended over the total tube length to assure maximum heat transfer performance; this makes it necessary to maintain a minimum velocity in the tubes. Increasing the flow could make it necessary that a considerable amount of particles would have to be accommodated in the outlet chamber, which could result in a quite voluminous outlet chamber. The total amount of particles depends very much on the design and the total heat transfer surface of the heat exchanger. The heat exchanger referred to in the tables 2 and 3 under case 2 requires more than 3,000 kg, i.e. approx. 3 kg per m^2 heat transfer surface.

The choice of the particle material is not only an important design parameter, as it influences the liquid velocity in the tubes; it also determines the abrasion exerted on the tube wall. Figure 7 shows the fouling behaviour in a fluid bed heat exchanger for different particle materials; according to this figure chopped stainless steel wire is much to be preferred to glass particles.

Wear of tube and particle material has always been a subject of much attention and has been investigated for different materials at various applications. For sea-water, it has been found that copper-alloys in combination with a fluid bed cannot be applied, if the sea-water contains oxygen; however, deaerated sea-water has demonstrated excellent wear resistance. Copper-alloys can be used in a fluid bed heat exchanger, even for water that is saturated with oxygen in

FLUID BED HEAT EXCHANGER

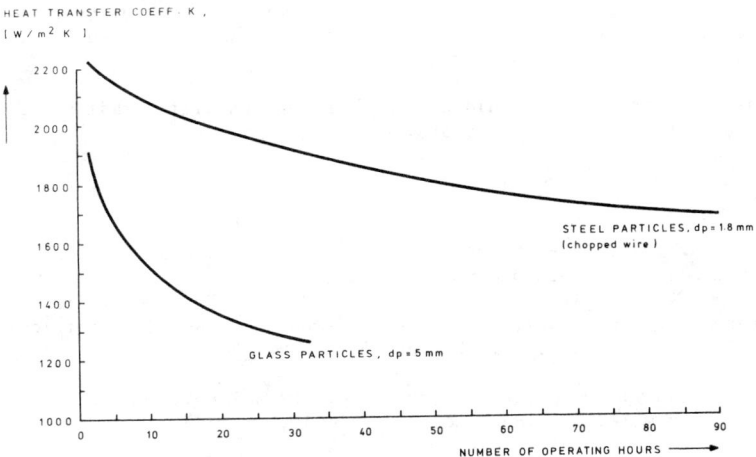

FIGURE 7 - Fouling in fluid bed heat exchanger for different particle materials.

the case that the water contains only fairly low chloride ion concentrations, i.c. for example 200 mg/l.
Stainless steel 304 and 316 have demonstrated excellent wear resistance, even in the case that the particles were made of highly abrasive chopped metal wire.
As far as the particles are concerned, neither glass nor chopped metal wire have shown any signs of excessive wear.

For a fluidized bed heat exchanger containing glass material, pumping power requirements are generally lower than for conventional heat exchangers. In the case that chopped metal wire is used, these requirements may become slightly higher than for conventional heat exchangers.

6. FUTURE DEVELOPMENTS

The following goals are set for further improvements:
- amount of particles will be reduced by a factor 5 to 10;
- velocities and turn-down ratios for the liquid flow in the tubes will be increased.

Above improvements can be achieved by circulating the particles between inlet chamber and outlet chamber, without short-circuiting the cold and hot liquid at both ends of the heat exchanger.

Other developments will lead to a fluid bed liquid/liquid shell and tube heat exchanger of new design and heat transfer characteristics as normally achieved by plate heat exchangers.

7. CONCLUSIONS

The fluid bed heat exchanger seems to have the potential of becoming an extremely useful piece of equipment in heat exchange applications, which suffer from severe fouling.

It is to be expected, that introduction of this new type of heat exchanger into the market will begin in the area of scraped heat transfer surfaces, direct heating and plate type heat exchangers.

Further improvements of fluid bed heat exchangers will steadily widen its field of application in the next years.

REFERENCES

1. Klaren, D.G. 1975. Development of vertical flash evaporator. Ph.D. Thesis. Delft University of Technology.

2. Richardson, J.F. and Zaki, W.N. 1954. Sedimentation and fluidization: Part 1. Trans. Inst. Chem. Eng., Vol. 28, p. 35.

3. Ruckenstein, E. 1959. On heat transfer between a liquid-fluidized bed and the container wall. Rep. Pop. Române, Vol. 10, pp. 235-246.

PERFORMANCE
OF ENHANCEMENT DEVICES

Performance Optimization of Internally Finned Tubes for Laminar Flow Heat Exchangers

A.C. TRUPP and H.M. SOLIMAN
Department of Mechanical Engineering
University of Manitoba
Winnipeg, Manitoba R3T 2N2, Canada

ABSTRACT

A feasibility study was conducted in which the optimized performances of internally finned tubes were compared to smooth tubes for laminar flow heat exchangers. Both straight and spiral fins were examined in four scenarios at the two extreme thermal boundary conditions. In all cases, internally finned tubes were found capable of superior performance, often by a wide margin. Almost consistently, the best finning configurations amounted to a small number of long fins. Results are also included which indicate that substantial exergy savings can be achieved by using finned tubes.

NOMENCLATURE

A	wetted perimeter (actual surface area per unit length)	P	pressure
A_f	flow cross-sectional area	Pr	Prandtl number, $\mu c_p / k_f$
f	friction factor, $(\pi^2 \rho r^5 / \dot{m}^2)(-dP/dx)$	Q	total heat transfer rate
G	mass velocity, ρU	Q'	total heat transfer rate per unit tube length
h	convective heat transfer coefficient	r	inside radius of tube
H	dimensionless fin height, ℓ/r	Re	Reynolds number, $2\dot{m}/(\pi r \mu)$
k_f	thermal conductivity of fluid	S'	net entropy generation rate per unit tube length
k_s	thermal conductivity of fin material	T	temperature
K	fin conductance parameter, $\beta k_s / k_f$	T_b	bulk temperature of fluid
ℓ	fin height	T_w	tube wall temperature
LMTD	log-mean temperature difference	U	cross-sectional average velocity
$\dot{m}$	mass flow rate	V	total volume of heat exchanger tube material
M	number of fins		
N	number of tubes (in parallel)	V'	tube material volume per unit tube length
N_s	augmentation entropy generation number	$\dot{W}$	pumping power
Nu	Nusselt number, $Q'/[\pi k_f (T_w - T_b)]$	x	axial direction

X	axial distance for a half turn of the fin along the periphery	Subscripts	
Y	$1 - M\beta(2H - H^2)/\pi$	e	exit
Z	$1 + M(2H - H^2)/[4\pi(1+\beta)]$	i	inlet
α	twist ratio, $\pi r/X$	max	maximum
β	half the angle subtended by one fin (see Fig. 1)	min	minimum
		o	smooth tube
ϕ	irreversibility distribution ratio	opt	optimum
		t	total

1. INTRODUCTION

The thermal effectiveness of tubular heat exchangers in which the convective resistance on the inner surface of the tube constitutes the main barrier to heat flow, can be substantially increased by a number of augmentative techniques such as twisted tape or wire coil inserts, internal surface roughness or fins, etc. Unfortunately for such modifications, the enhanced heat transfer over smooth tube conditions is typically accompanied by extra cost and maintenance and by increases in weight, pressure drop and pumping power. Collectively these disadvantages (especially increased capital and operating costs) may more than offset the thermal advantage in certain applications. Thus there is no assurance that an augmented heat exchanger is automatically superior to the conventional heat exchanger, and each case must be examined on its own merits. Ideal selection of a heat exchanger for a given application would require the smooth tube configuration and each enhancement candidate to be optimized individually, with final selection made by comparison of the competing alternatives. The optimization criterion is invariably of an economic nature (e.g. initial cost plus lifetime operating and maintenance costs), although for mobile applications, weight and size are likely to be major factors. Since relative costs vary with time and location, complete a priori optimization is impractical. It is therefore necessary in a general treatment to select objective functions of a physical nature. In the case of tube volume, the objective function is directly related to components of weight and first cost. A second complication arises from the fact that each application is somewhat unique in that it has its own special set of constraints. Even so it is possible to identify certain scenarios which are expected to occur frequently. Performance evaluations for these cases should therefore be of use to designers as a starting point.

This paper is concerned with only one augmentative technique, viz internally finned tubes. Their performance under single-phase fully developed, laminar flow conditions in heat exchanger applications are compared to conventional smooth tube heat exchangers. Several scenarios are examined in which a specific objective function is optimized subject to several system constraints. The selection of scenarios was guided by the comprehensive performance evaluation criteria proposed by Bergles et al [1], and by the studies of Webb and Eckert [2] and by Webb and Scott [3]. The latter authors have recently examined the performance of internally finned tube heat exchangers under turbulent flow conditions. Since fully developed laminar flow data are now available (as discussed below), the present study is timely and should serve as a basis for the optimization of thermal systems in this category.

2. DATA BASE

Internally finned tubes are now available commercially in both straight and helical patterns. Typically, these fins are not quite rectangular in cross-section, but rather trapezoidal with the fin thickness decreasing somewhat from base to tip. In fact the geometry selected for analysis by Soliman and Feingold [4] is quite similar to real cross-sections when β is taken equal to $\pi/60$ radians (3 degrees). This characteristic geometry is illustrated as an inset to Fig. 1. Theoretical solutions (constant fluid properties) for fully developed laminar flow for straight fins of this geometry ($\beta = \pi/60$) are now available for friction factor and axial velocity distribution [4], and for temperature distribution and Nusselt numbers for constant temperatures axially and circumferentially [5] (hereafter referred to as B.C. 1), and for constant heat input per unit axial length and uniform temperature circumferentially [6] (B.C. 2). The heat transfer results include the influence of fin conductance. The data for $K = \infty$ are shown in Fig. 1[1] for $0.2 \leq H \leq 0.8$ and $M \leq 32$. In addition to these data, Ivanovic [7] has recently reported theoretical results for laminar friction factor and heat transfer for B.C. 2 for helical fins of zero thickness ($\beta = 0$). These ([4-7]) constitute the data base for the present study. This data base has been reviewed in considerable detail in our original report [8]. Due to space limitations, no attempt will be made here to summarize the assessment, however two points require elaboration before proceeding with the performance evaluation.

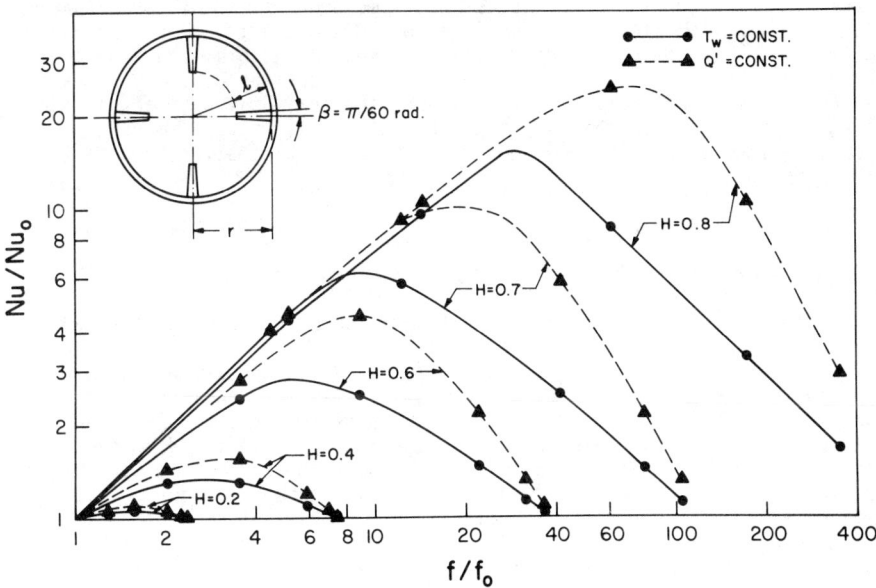

Fig. 1. Predicted heat transfer and friction factor enhancement over smooth tube conditions for straight internally finned tubes ($K = \infty$).

---

[1] The five data points shown for each curve correspond, from left to right, $M = 4, 8, 16, 24$ and 32. The curve shapes (especially near the maxima) reflect data additional to those shown.

The present study is confined to trapezoidal fins with $\beta = \pi/60$. Hence for straight fins, fRe = fRe (M, H). For spiral fins, fRe = fRe (M, H, α, Re), but Ivanovic [7] has shown that the dependence on α and Re as separate parameters, can be replaced by a single dependence on the product αRe when $\alpha < \sim 0.3$, and his predictions were made on this basis. In order to maintain β fixed at $\pi/60$, fRe data for spiral fins were obtained by multiplying the results of [4] by Ivanovic's ratio of twisted-to-straight fin friction factor. This makes the not unreasonable assumption that the ratios for $\beta = \pi/60$ are the same as for $\beta = 0$. In addition, rather than treating twist ratio as a separate parameter, the ratios were taken at αRe = 300 to provide a single set of data. For fixed Re (and hence fixed α), this data set reflects the proper relative effect of twisting on the M, H spectrum. However as applied in this study, Re may vary and hence α is a floating parameter which is constrained implicitly by αRe = 300. Accordingly, the present results for spiral fins should be treated as approximate. These results are for B.C. 2 only. In this regard it is noted for spiral fins that Nu depends on αRe and Pr in addition to M, H and K. For the purpose of this study, a set of Nu data for $\beta = \pi/60$ was (parallel to the procedure adopted for fRe) obtained by multiplying the values of Nu from [6] by Ivanovic's ratio which was taken at αRe = 300 and Pr = 0.7. To form a self-consistent set of data (e.g. gas flow-copper tube), the data from [6] was taken at $K = \infty$, and the resultant set was treated as corresponding to $K = \infty$.

The second point concerns the fact that the present results correspond to pure laminar forced convection and constant fluid properties. In practise, Nusselt numbers will differ from these due to fluid property variations and free convection. But these effects will be present in both smooth tube and finned tube heat exchangers. Hence (hopefully) the optimum configurations deduced here will remain optimum (or at least near optimum) under actual operating conditions. Nonetheless, since actual "merit figures" may be different, designers are cautioned accordingly to use discretion in applying the present results.

3. OPTIMIZATION SCENARIOS

This study involves laminar flow heat exchangers and examines the prospects for using internally finned tubes as a design alternative to smooth tubes. Four cases (referred to as Case A, B, C and D) are examined with the object of determining (for each case) the optimum finning configuration for straight fins with B.C. 1 and 2 and for spiral fins for B.C. 2. Each case consists of a scenario involving certain operating conditions and a specific objective. The objective function is written so as to express the performance of the finned tube exchanger relative to the smooth tube exchanger. This relative objective function is to be minimized or maximized subject to a set of operating constraints. The global optimum is determined by a search method whereby the objective function is optimized for each of 25 finning configurations defined by the intersections of H = 0.2, 0.4, 0.6, 0.7 and 0.8 and M = 4, 8, 16, 24 and 32.

The groundrules for the study are itemized as follows:

a) The fins are trapezoidal with $\beta = \pi/60$ (Fig. 1).

b) The tube wall thickness is $2\beta r$ for finned tubes (i.e. wall thickness equal to fin base thickness) and $2\beta r_o$ for finless tubes.

INTERNALLY FINNED TUBES FOR LAMINAR FLOW HEAT EXCHANGERS

c) Each heat exchanger involves the same incompressible fluid (constant properties ρ, μ, c_p and k_f), has the same inlet temperature (T_i), and consists of N (or N_o) tubes in parallel, each of length L (or L_o), giving a total length of tubing equal to NL (or $N_o L_o$). In some cases the total mass flow rates are also the same.

d) The thermal resistance of each heat exchanger is vested exclusively on the tube inside. This is precise for B.C. 2 and typical for condensers/evaporators (B.C. 1).

e) In general, the radius ratio (r/r_o), the tube numbers ratio (N/N_o) and the tube length ratio (L/L_o) were all left as floating parameters within limits. These limits are:

$$0.5 \leq (r/r_o) \leq 2 , \tag{1a}$$

$$(L/L_o) \geq 0.25 , \tag{1b}$$

$$(N_o/N) \geq 0.45, \text{ i.e. } (N/N_o)_{max} = 2.22 . \tag{1c}$$

It is emphasized first that these limits are arbitrary, but the choices made do influence the results. Concerning (1a), usually $r > r_o$ and the upper limit simply attempts to face reality. The need for a lower limit initially stems from the fact that $(V'/V'_o) \to o$ as $(r/r_o) \to o$. The value 0.5 was initially selected on purely physical grounds (finned tubes from half as small to twice as large), however for Case A (for example), limits (1b) and (1c) govern, hence this lower limit could be raised to about 0.7 before becoming effective as a constraint. For (1b), usually $L < L_o$ and the lower bound is meant to reflect a physical limit. Finned tube lenghts can not be ridiculously small, but there is a natural reluctance to setting the boundary much higher than was done since internally finned tubes are expected to provide "compact" heat exchangers. Regarding (1c), usually $N > N_o$ and this means a larger frontal area. But shell size exerts a sizeable influence on the cost of a heat exchanger. The limit chosen corresponds (for $r \approx r_o$) approximately to a 50% increase in shell diameter.

In our original report [8], the general equations were derived first and then additional constraints were introduced to form common equations for the two thermal boundary conditions. Each case was then developed in detail. Due to space limitations here, it is not possible to include the derivations. Accordingly, the objective function, constraints and relations for the geometric parameters are simply summarized in Table 1. Also included in Table 1 are certain auxiliary relations usually involving the ratio of mass velocities. It is suggested that the reader carefully examine the information contained in the last two columns in order to better understand the various design cases. For Case A, for example, it should be noted that the relative mass velocity is independent of radius ratio, tube length ratio and tube number ratio. In fact, for a given heat transfer boundary condition and fin conductance ratio, G_o/G is unique for each internally finned tube. But at the same time, the constraints interrelate the three geometric parameters as shown in Table 1, hence choosing any one, fixes the other two.

//
Table 1
Summary of Cases

Case	Objective Fcn	Primary Constraints	Addit. Constraints B.C.	Constraint	Relations for Geometric Parameters	Auxiliary Relations
A	$\dfrac{V}{V_o} = \left[\dfrac{Nu_o}{Nu}\right]\left[\dfrac{r}{r_o}\right]^2 (Z)$ (minimize)	$\dfrac{\dot{W}}{\dot{W}_o} = 1 = \dfrac{Q}{Q_o} = \dfrac{\dot{m}_t}{\dot{m}_{to}}$	1 2	$T_w = T_{wo}$ $\dfrac{(T_w - T_b)}{(T_w - T_b)_o} = 1$	$\dfrac{N}{N_o} = \left[\dfrac{Nu_o}{Nu}\right]\left[\dfrac{L_o}{L}\right] = \left[\dfrac{fReNu_o}{16Nu}\right]^{1/2}\left[\dfrac{r_o}{r}\right]^2$	$\dfrac{G_o}{G} = \left[\dfrac{fReNu_o}{16Nu}\right]^{1/2} (Y)$ $\dfrac{Q'}{Q'_o} = \dfrac{Nu}{Nu_o}$ for B.C.2
B1	$\dfrac{Q}{Q_o} = \left[\dfrac{Nu}{Nu_o}\right]\left[\dfrac{N}{N_o}\right]\left[\dfrac{L}{L_o}\right]$ (maximize)	$\dfrac{\dot{W}}{\dot{W}_o} = 1 = \dfrac{V}{V_o}$	1 2	$T_w = T_{wo}$ $\dfrac{(LMTD)}{(LMTD)_o} = 1$ $T_e = T_{eo}$ $\dfrac{T_w - T_b}{(T_w - T_b)_o} = 1$	$\dfrac{L}{L_o} = \left[\dfrac{Nu}{NuZ}\right]^{0.4}\left[\dfrac{16}{fRe}\right]^{0.2}\left[\dfrac{N_o}{N}\right]^{0.6}$ $\dfrac{r}{r_o} = \left[\dfrac{NLZ}{N_o}\right]^{1/2}$	$\left[\dfrac{\dot{m}_t}{\dot{m}_{to}}\right]\left[\dfrac{N_o}{N}\right] = \left[\dfrac{Nu}{Nu_o}\right]\left[\dfrac{L}{L_o}\right] = \left[\dfrac{G}{G_o}\right]\left[\dfrac{A_f}{A_{fo}}\right]$
B2	$\dfrac{hA}{h_o A_o} = \dfrac{Nu}{Nu_o}$ (maximize)	$\dfrac{\dot{W}}{\dot{W}_o} = 1 = \dfrac{V}{V_o} \dfrac{\dot{m}_t}{\dot{m}_{to}}$	None		$\dfrac{N}{N_o} = \left[\dfrac{fRe}{16Z}\right]^{1/2}\left[\dfrac{r_o}{r}\right]^3$ $\dfrac{L_o}{L} = \left[\dfrac{fReZ}{16}\right]^{1/2}\left[\dfrac{r_o}{r}\right]$	$\dfrac{G_o}{G} = \left[\dfrac{fRe}{16Z}\right]^{1/2}\left[\dfrac{r_o}{r}\right] (Y)$
C	$\dfrac{\dot{W}}{\dot{W}_o} = \left[\dfrac{fRe}{16}\right]\left[\dfrac{Nu_o^2}{Nu}\right]\left[\dfrac{L}{L_o}\right]\left[\dfrac{N}{N_o}\right]$ (minimize)	$\dfrac{Q}{Q_o} = 1 = \dfrac{V}{V_o} \dfrac{\dot{m}_t}{\dot{m}_{to}}$	Same as for Case A		$\dfrac{r}{r_o} = \left[\dfrac{Nu}{Nu_o Z}\right]^{1/2}$ $\dfrac{N}{N_o} = \left[\dfrac{Nu_o}{Nu}\right]\left[\dfrac{L_o}{L}\right]$	$\dfrac{G_o}{G} = 2.22\left[\dfrac{Y}{Z}\right]\left[\dfrac{Nu}{Nu_o}\right]$ if $Nu<1.8Nu_o$ $\dfrac{G_o}{G} = 4\left[\dfrac{Y}{Z}\right]$ if $Nu \geq 1.8Nu_o$
D	$\dfrac{(\Delta T)}{(\Delta T)_o} = \left[\dfrac{Nu_o}{Nu}\right]\left[\dfrac{N_o}{N}\right]\left[\dfrac{L_o}{L}\right]$ (minimize)	$\dfrac{Q}{Q_o} = 1 = \dfrac{V}{V_o} = \dfrac{\dot{W}}{\dot{W}_o} = \dfrac{\dot{m}_t}{\dot{m}_{to}}$	None		$\dfrac{r}{r_o} = \left[\dfrac{fRe}{16Z}\left[\dfrac{N_o}{N}\right]^2\right]^{1/6}$ $\dfrac{L_o}{L} = \left[\dfrac{N}{N_o}\right]\left[\dfrac{r}{r_o}\right]^2 Z$	$\dfrac{G_o}{G} = \left[\dfrac{N}{N_o}\right]\left[\dfrac{r}{r_o}\right]^2 \; Y = \left[\dfrac{L_o}{L}\right]\left[\dfrac{Y}{Z}\right]$

4. RESULTS AND DISCUSSION

4.1 Case A: Reduced tube material volume for equal heat duty, equal pumping power and the same total mass flow rate.

Before addressing the general case, the special case of $r = r_o$ is examined for the purpose of comparison to Webb and Scott [3]. A sample result (B.C. 2, $K = \infty$) is shown in Fig. 2. These results do not include the limits embodied in (1). The best straight fin is $M = 16$, $H = 0.8$ (which was also the case for B.C. 1). In general, the best performers are configurations involving a relatively small number of long fins. Concerning trends, for low M ($M \leq 8$), V/V_o decreases with increasing fin height. For higher M, V/V_o first increases with H and then finally drops rapidly beyond about $H = 0.6$. These trends are basically opposite to the behaviour under turbulent flow conditions [3], although it must be conceded that both the present results and those in [3] are quite sensitive to the fin thickness-to-height relationship. In general, Webb and Scott [3] concluded that the best straight fins under turbulent flow conditions are short fins ($H < 0.2$) with M in the range of 16 to 25, however material savings are small with $(V/V_o)_{opt} \approx 0.88$. In constrast, for laminar flow conditions, the best fins are long fins ($H \approx 0.6$ to 0.8) with $M < \sim 20$, and material savings appear to be substantial. But actually some of the attractive configurations appearing in Fig. 2 are not practical in the sense that $(N/N_o) \gg 1$ or $(L/L_o) \ll 1$. Solutions for the general case subject to the limits prescribed in (1) are considered next, and this provides a

INTERNALLY FINNED TUBES FOR LAMINAR FLOW HEAT EXCHANGERS

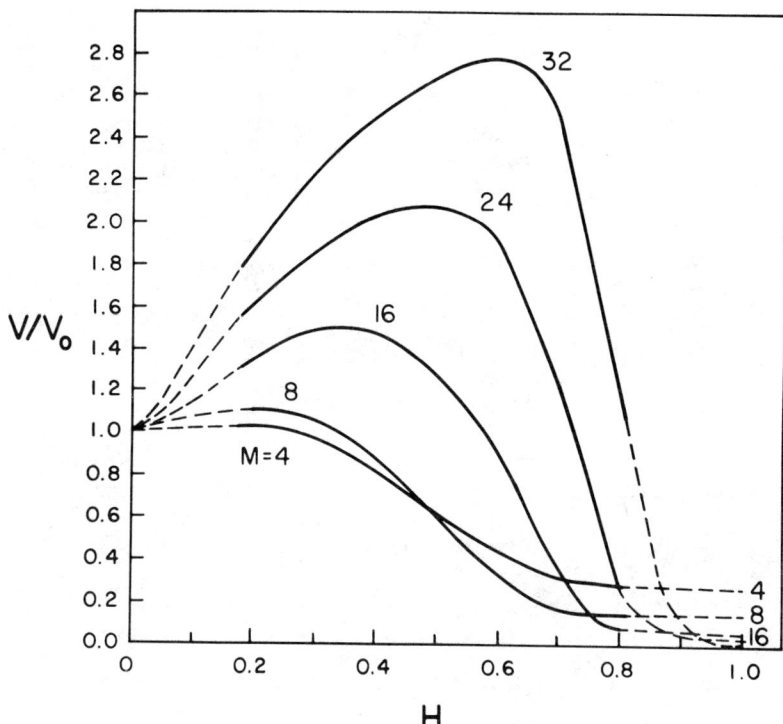

Fig. 2. Relative Tube Material Volumes for Case A with $r = r_o$, B.C. 2, $K = \infty$ (Str. Fins).

truer indication of the potential for material savings under practical heat exchanger conditions.

It can be shown [8] that minimization of the objective function is governed by $(L/L_o)_{min}$ when $Nu \geq 1.8\ Nu_o$, and by $(N_o/N)_{min}$ when $Nu < 1.8\ Nu_o$. According to Fig. 1 (straight fins), each governs for roughly half of the 25 data points. In general, material savings are limited by tube length when the fins are long, and by tube number when either the fins are short or, for longer fins, when M is large. The results for Case A are shown in Fig. 3. The plots do not include the results of configurations for which $(V/V_o) > 1$ (not competitive) or $r > 2r_o$ (impractical). The tic mark on each curve indicates the approximate location where $r = r_o$. Points to the left of this mark have $\sim 0.70 < (r/r_o) \leq 1$, whereas points to the right have $r > r_o$. (Note, for M = 4 and K = 1, $r < r_o$ always). The results for straight fins are discussed first. For M = 4 and both boundary conditions, the best finned tubes are $H \approx 0.7$ for $K = \infty$ ($V/V_o \approx 0.34$) and $H \approx 0.6$ for $K = 1$ ($V/V_o \approx 0.44$). For M = 8, the best fins are $H \approx 0.8$ (0.7) for $K = \infty$ for B.C. 1 (2), but for K = 1, H = 0.2 outperforms longer fins for both boundary conditions. For M = 16, only short fins provide $(V/V_o) < 1$, but (as with turbulent flow) material savings are marginal. In general, the results presented in Fig. 3 show that internally finned tubes offer excellent potential for light-weight laminar flow heat exchangers.

Regarding spiral fins, the results shown in Fig. 3 for B.C. 2 are quite

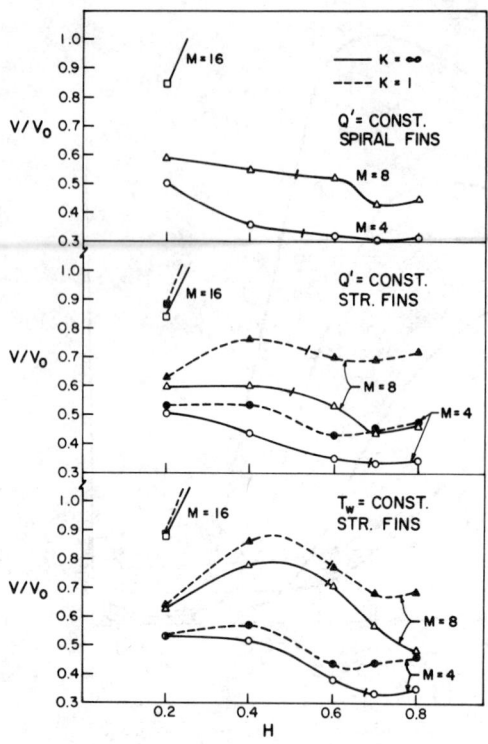

similar to the results for straight fins which suggests that spiralling offers no additional benefit in saving material. However in view of the reservations expressed earlier, no firm conclusion should be drawn here. Additional study (with α as an independent parameter and also using data which is valid for $\alpha > 0.3$) is needed to properly determine the effect of spiralling for Case A. In this connection, for turbulent flow, Webb and Scott [3] found that increasing α provided substantially increased material savings.

Fig. 3. Relative Tube Material Volumes for Case A for Practical Heat Exchangers.

4.2 Case B: Increased heat duty for equal pumping power and the same tube material volume.

Two versions of this case are considered. The results for Case B1 subject to the limits of (1) are shown in Fig. 4. The plots do not include the results of configurations for which $(Q/Q_0) < 1$. Again a tic mark is used on each curve to indicate the approximate location where $r = r_0$. Points to the left of this mark are in the range $\sim 0.83 < (r/r_0) \leq 1$, i.e. the limit $(r/r_0)_{min} = 0.5$ is never invoked. On the other hand, the limit $(r/r_0)_{max} = 2$ becomes operative (for the data shown) only for spiral fins (B.C. 2) for $M = 8$ at $K = \infty$. It is reached between $H = 0.6$ and 0.7, and is effective at $H = 0.7$ and especially at $H = 0.8$. This causes the large downswing in the curve compared to the corresponding straight fin case. Note, the results for Case B1 (Fig. 4) are qualitatively similar to the results for Case A (Fig. 3).

For Case B2, instead of increased heat duty, per se, the objective function is switched to hA/h_0A_0 which is to be maximized. As a preliminary step, the constraint $V = V_0$ is set aside temporarily in order to examine the special case of $r = r_0$ and $NL = N_0L_0$. This is done mainly to allow direct comparison to the results of Webb and Scott [3]. It is noted for this special case, that the constraints require $L_0/L = N/N_0 = (fRe/16)^{\frac{1}{2}}$ whereas $G_0/G = (fRe/16)^{\frac{1}{2}}$ (Y). The latter is unique for any given finned tube and is independent of the thermal boundary condition. The objective function for this special case remains as listed in Table 1, hence optimum configurations

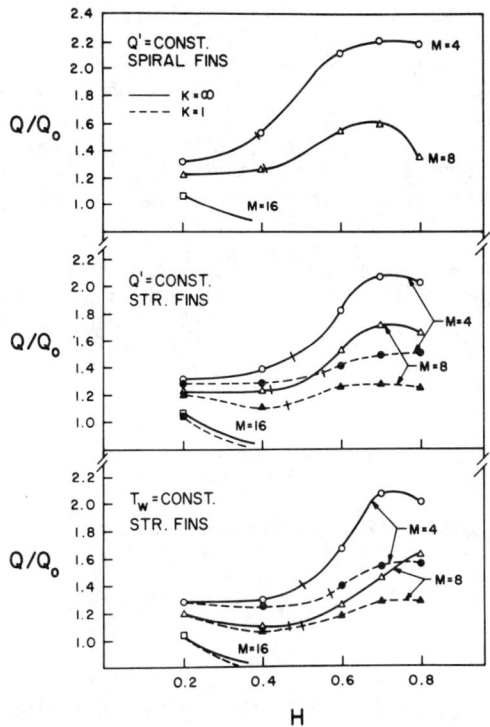

Fig. 4. Total Heat Transfer Ratios for Case B1 for Practical Heat Exchangers.

can be read directly from Fig. 1. The limits imposed by (1) can be shown to amount to:

$$fRe/16 = 4.938, \text{ or } fRe \leq 79.0, \qquad (2)$$

which is a limit due to number of tubes.

As shown in Fig. 1, for fixed H, $hA/h_o A_o$ first increases with M and then decreases. These patterns are very sensitive to H. Ignoring (2), the best straight fins ($K = \infty$) are $H = 0.8$, $M \approx 11$ for B.C. 1 ($hA/h_o A_o \approx 15$) and $H = 0.8$, $M \approx 16$ for B.C. 2 ($hA/h_o A_o \approx 25$). With (2), the best practical straight fins ($K = \infty$) are $H = 0.7$, $M = 4$ ($fRe = 71.8$) with $hA/h_o A_o = 4.0$ for both boundary conditions. The trends exhibited in Fig. 1 differ from behaviour under turbulent flow conditions. Webb and Scott [3] considered only relatively short fins ($H < \sim 0.23$) and for this region, found for fixed H that $hA/h_o A_o$ increased monotonically with M. These increases were only weakly proportional to H. Consequently the best finning arrangements amounted to a large number of short fins. The calculations were then extended to include the effect of tube material volume, and the general conclusion was reached that internally finned tubes offer little advantage for Case B2 under turbulent flow conditions. As shown in the following, prospects are much better under laminar flow conditions.

The analysis for Case B2 is now continued with the constraint $V = V_o$ operative. The objective function is represented by Nu/Nu_o which is a constant for a given straight finned tube, hence individual optimization is redundant, and only a global optimum exists for each boundary condition. In the absence of constraints on the geometric parameters, optimum finning configurations may again be taken directly from Fig. 1 for straight fins. But of course geometric constraints have been imposed, hence the investigation reduces to determining whether or not the primary constraints for Case B2 can be met with the geometric parameters within the bounds of (1). These constraints may be combined in various ways to obtain links between any two geometric paramaters (see Table 1 for examples). If, for a given finned tube, a combination complying with (1) is possible, the design is deemed practical and the results are shown in Fig. 5. In general, for each configuration appearing in Fig. 5, there exists a range of solutions for the geometric parameters. For example, for H = 0.2, M = 8, a designer might select $(r = r_o, N = 1.135 N_o, L = 0.7235 L_o)$ or $(r = 1.2 r_o, N = 0.6568 N_o, L = 0.8682 L_o)$ or $(r = 0.80 r_o, N = 2.217 N_o, L = 0.5788 L_o)$, etc. On the other hand, certain configurations (e.g. H = 0.8, M = 16 and all others not appearing in Fig. 8) are impractical. For these cases, even with $r = 2r_o$ (i.e. maximum) finned tube lengths (meeting the primary constraints) are too short (i.e. $L < 0.25 L_o$) but tube numbers are satisfactory $(N < 2.22 N_o)$.

The results shown in Fig. 5 correspond to $K = \infty$. For spiral fins, only two fin heights (0.7 and 0.8) are shown in order to retain clarity. The best practical fins are H = 0.8 and M in the range of about 11 to 13. Enhancements in $hA/h_o A_o$ range from about 15 to 20 depending on B.C. and α. It follows that finned tubes offer large enhancements vis-a-vis smooth tubes for Case B2 under laminar flow conditions.

4.3 Case C: Reduced pumping power for equal heat duty, the same total mass flow rate and the same tube material volume.

It is noted first that the equal heat duty constraint amounts to:

$$Q/Q_o = 1 = (Nu/Nu_o)(N/N_o)(L/L_o) . \quad (3)$$

It is also noted that Webb and Scott [3] examined Case C for turbulent flow but they replaced $V = V_o$ by $r = r_o$ and $NL = N_o L_o$. They found that internally finned tubes offered a significant potential for reduced friction power designs. But as may be seen from (3), for laminar flow it is impossible to meet the equal heat duty constraint with $NL = N_o L_o$ since $Nu > Nu_o$ always. Hence the present results for Case C (with $V = V_o$) are not directly comparable to [3].

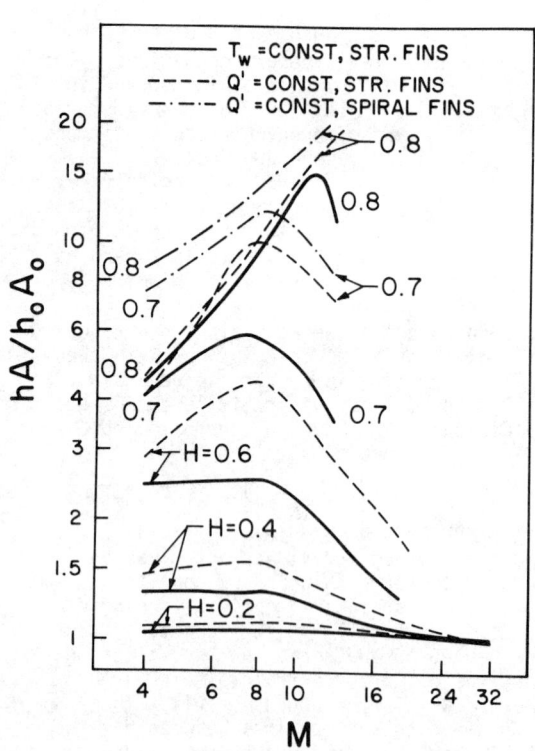

Fig. 5. Thermal Conductance Ratios for Case B2 for Practical Heat Exchangers $(K = \infty)$.

For Case C, due to the constraints, it turns out that the radius ratio is given by:

$$r/r_0 = (Nu/Nu_0)^{1/2} (Z)^{-1/2}, \qquad (4)$$

which is unique for each finning configuration. It may also be shown that the minimization of $\mathring{W}/\mathring{W}_0$ is governed by tube length when $Nu \geq 1.8\ Nu_0$, and by tube number when $Nu < 1.8\ Nu_0$. The results for Case C are shown in Fig. 6 for practical heat exchangers. The abscissa spans the r/r_0 range given by (1a). Finning configurations with $M > 16$ (also $M = 16$, $H > 0.2$) yield $(\mathring{W}/\mathring{W}_0) > 1$[2] and hence are not shown in the figure. Also not shown are certain cases of $(\mathring{W}/\mathring{W}_0) < 1$ but $(r/r_0) > 2$; notably for spiral fins (B.C. 2, $K = \infty$) pumping power ratios reached as low as 0.092 for $M = 4$, $H = 0.7$ but the radius ratio here was 2.42. As may be seen from Fig. 6, the best practical straight fin for $K = \infty$[3] is $M = 4$, $H = 0.7$ for both boundary conditions with $(\mathring{W}/\mathring{W}_0) = 0.113$ for each B.C. For spiral fins ($K = \infty$), the best practical fin is $M = 4$, $H = 0.6$ with $(\mathring{W}/\mathring{W}_0) = 0.104$.

The results shown in Fig. 6 display some rather unusual features. For fixed H with $H \leq 0.4$, r/r_0 decreases with M; even where Nu increases with M (see Fig. 1), the effect of M in the second term of (4) is dominant. As a consequence of the reducing flow cross-sections, pumping power rises rapidly with M. For fixed H with $H \geq 0.7$, r/r_0 first increases with M and then decreases, but $\mathring{W}/\mathring{W}_0$ still continues to increase with M due now mainly to the relatively high fRe values. These patterns are of

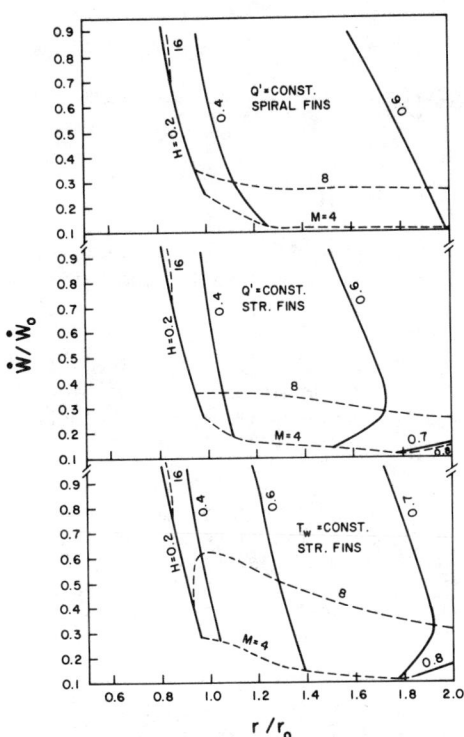

Fig. 6. Pumping Power Ratios for Case C ($K = \infty$).

---

[2] For large M and H, typically $(\mathring{W}/\mathring{W}_0) \gg 1$ but radius ratios comply with (1a).

[3] In comparison for $K = 1$: B.C. 1, best is $M = 4$, $H = 0.6 - 0.7$, $(\mathring{W}/\mathring{W}_0) = 0.192 - 0.194$ (flat); B.C. 2, best if $M = 4$, $H = 0.6$, $(\mathring{W}/\mathring{W}_0) = 0.188$.

course due to the combined constraints which dictate a unique r/r_o value for each finned tube. Superficially, this would seem to act disadvantageously on fin number. To explore this, the feasibility of replacing $(V/V_o) = 1$ by $(V/V_o) \leq 1$ was examined. The only change is that the equality sign in (4) becomes a 'less than or equal' situation, which only permits still smaller radius ratios and still larger pumping power ratios. Only cases of $(\dot{W}/\dot{W}_o) < 1$ with $(r/r_o) > 2$ (as noted earlier) can possibly benefit from such a trade-off. The opposite direction, i.e. $(V/V_o) > 1$, permits larger radius ratios and hence smaller pumping power ratios. But this approach is highly subjective (a tube volume penalty for extra pumping power savings), and is best left for consideration at the detailed design level. On the basis of Fig. 6 and the preceding discussion, it is concluded that internally finned tubes offer excellent potential for pumping power reductions.

4.4 Case D: Reduced temperature difference for equal heat duty, equal pumping power, the same total mass flow rate and the same tube material volume.

The objective function is:

$$\frac{(\Delta T)}{(\Delta T)_o} = \left(\frac{Nu_o}{Nu}\right)\left(\frac{N_o}{N}\right)\left(\frac{L_o}{L}\right) \to \text{minimize} , \qquad (5)$$

where $(\Delta T)/(\Delta T)_o \equiv (LMTD)/(LMTD)_o$ for B.C. 1,
$(\Delta T)/(\Delta T)_o \equiv (T_w - T_b)/(T_w - T_b)_o$ for B.C. 2.

For B.C. 1, since the two heat exchangers have the same inlet temperature (T_i) and the same total mass flow rate, they will also have the same exit temperatures ($T_e = T_{eo}$) for the same heat duty. Furthermore, if $(\Delta T) = (\Delta T)_o$, they will also have the same condensing/evaporating temperature, i.e. $T_w = T_{wo}$. But if $(\Delta T) < (\Delta T)_o$, this means $T_w < T_{wo}$; or equivalently if $T_w = T_{wo}$[4], the tube-side fluid in the finned tube heat exchanger would operate at a higher (lower) temperature level than in the conventional condenser (evaporator). Similarly for B.C. 2, $(\Delta T) < (\Delta T)_o$ means the finned tube heat exchanger (relative to smooth tubes) can operate either with the fluid at higher temperature levels (e.g. to ultimately increase cycle thermal efficiency) or with the tube wall at lower temperature levels (e.g. to meet some temperature limit). More generally, for both boundary conditions, since pumping powers are equal, $(\Delta T) < (\Delta T)_o$ means reduced thermodynamic irreversibility for the steady flow heat transfer process involving finned tubes, and hence there is more effective energy utilization. Analyses of heat exchanger processes on an exergy basis have been made by several authors, e.g. [9-12].

The results of (5) subject to the constraints and to the limits of (1) are shown in Fig. 7 which includes all the results for which $(\Delta T)/(\Delta T)_o \leq 1$. Regarding radius ratios, for straight fins with M = 4, r/r_o ranges from 0.785 at H = 0.2 to 0.966 at H = 0.8. Similarly for M = 8, the range is 0.799 to 1.199. For spiral fins, the radius ratios are slightly large (due to higher fRe) but the trends are similar. As may be seen from Fig. 7, the best straight fins for K = ∞ are M = 8, H = 0.8, (LMTD) = 0.235 (LMTD)$_o$ for

---

[4] Inlet temperatures not equal now.

INTERNALLY FINNED TUBES FOR LAMINAR FLOW HEAT EXCHANGERS

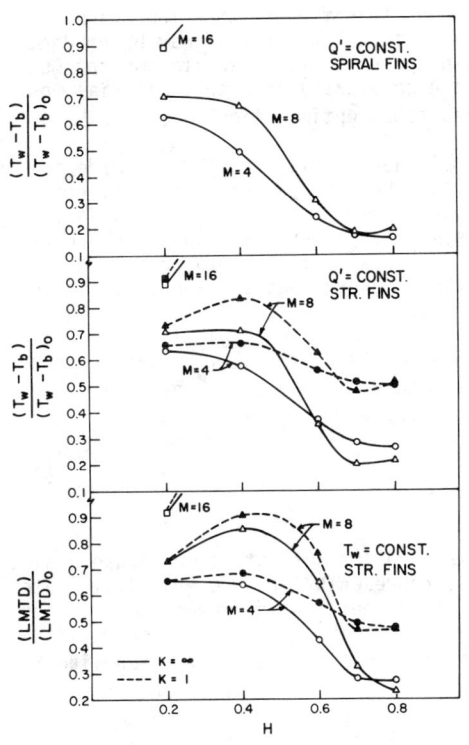

B.C. 2. For K = 1, the best finning configurations for both boundary conditions are 4 to 8 straight fins with H ≈ 0.7 to 0.8 which give temperature difference ratios of about 0.48. For spiral fins with K = ∞ and B.C. 2, the best fin is M = 4, H = 0.8, $(T_w - T_b) = 0.162 (T_w - T_b)_o$, however 4 to 8 long fins all perform much the same.

Fig. 7. Temperature Difference Ratios for Case D for Practical Heat Exchangers.

5. FURTHER CONSIDERATIONS

The results of the preceding analyses are expected to be of some use to both designers of thermal systems and to manufacturers of internal finned tubing. The key findings are summarized in the conclusions section of this paper. The results for Cases A to D should serve as a guide to designers mainly for occasions where internally finned tube heat exchangers are to be considered as replacement equipment. The results are also pertinent to new designs, however for such occasions it is more appropriate to approach the thermal system design in a manner so as to maximize the effectiveness of energy management. It is well known for heat exchangers that thermodynamic irreversibilities are due to two factors, viz fluid friction (i.e. pressure drop/pumping power) and heat transfer across a finite temperature difference. The two contribute additively to the total irreversibility. But the two losses are coupled (oppositively) in that increasing the heat transfer area (so as to reduce temperature differences) invariably causes increases in pumping power. Hence whereas one contribution to the total irreversibility is reduced, the other is increased. In total, increasing surface area is only effective if the increased operating and capital costs are more than offset by savings accrued by reduced total irreversibility. There is obviously scope for optimization. Unfortunately the optimum balance between the two factors is strongly dependent on local operating conditions, i.e. each individual design has its unique optimum, and hence the exergy approach is not completely amenable to a general treatment. Nonetheless, it is apparent from the above that even for a particular design, the optimum balance for a conventional heat exchanger may be quite different than the optimum balance for a finned tube heat exchanger. It follows that it may be prudent for Case D (for example) to allow pumping power to float to its own optimum with $(\Delta T)/(\Delta T)_o$. One might also (for the same case) free tube material volume, and attempt to attain a less constrained total optimization.

In view of the preceding, an attempt (necessarily limited) is made in the following to examine internally finned tube heat exchangers vis-a-viz conventional exchangers from the contemporary viewpoint of conservation of exergy. This should be of interest to thermal system designers who undoubtedly have been finding it necessary (due to high fuel costs) to analyze processes on an exergy basis in addition to the conventional energy basis. The analysis is restricted to B.C. 2.

It can be shown [10] under conditions of $(T_w - T_b) \ll T_w$, that the rate of net entropy generation per unit tube length for any B.C. 2 heat exchanger tube, is given by:

$$S' = \frac{Q'(T_w - T_b)}{T_w^2} + \frac{\dot{m}}{\rho T_w}\left(-\frac{dP}{dx}\right), \qquad (6)$$

where T_w is the wall temperature expressed on the absolute temperature scale. The first term (hereafter designated S_T') represents the irreversibility rate due to heat transfer across the wall-fluid temperature difference, while the second term (S_p') is the irreversibility due to fluid friction. The ratio (S_p'/S_T') forms [11] the irreversibility distribution ratio, ϕ, which like S_p' and S_T' is a local parameter. For example, for smooth tubes under fully developed laminar flow conditions, although Nu_o and $f_o Re_o$ are constants, $S_{To}' = Nu_o \pi k_f [(T_w - T_b)/T_w]$ whereas $S_{po}' = 16 \dot{m}_o^3/(\pi\rho)^2 R_o^5 T_w Re_o$, hence ϕ_o is a complex function of several design parameters. For a smooth tube heat exchanger consisting of N_o tubes each of length L_o, the total rate of net entropy generation is $S_o \triangleq S_o' N_o L_o$ where (as an approximation in (6)), T_w is taken to be the average absolute temperature over L_o. It is assumed in the following that ϕ_o has been optimized so as to minimize S_o for a given application[5].

In order to compare an internally finned heat exchanger to a smooth tube heat exchanger, an augmentation entropy generation number (N_s) is defined (similar to Ouellette and Bejan [11]) as:

$$N_s = \frac{S'NL}{S_o' N_o L_o}, \qquad (7)$$

which gives the ratio of the total rates of net entropy generation in the two heat exchangers. $N_s < 1$ means the internally finned tube heat exchanger produces less total irreversibility and is therefore thermodynamically superior.

Consider the case where $Q = Q_o$ and $\dot{m}_t = \dot{m}_{to}$. For simplicity, assume $r = r_o$ and $NL = N_o L_o$. The latter in conjunction with $Q = Q_o$ yields $Q' = Q_o'$.

---

[5] Strictly speaking it is the overall system objective function (total cost formula) which is to be minimized.

INTERNALLY FINNED TUBES FOR LAMINAR FLOW HEAT EXCHANGERS

From (7) with $NL = N_0 L_0$, the objective function is:

$$N_s = \frac{S'}{S'_0} = \frac{S'_T + S'_P}{S'_{T_0} + S'_{P_0}} = \frac{(S'_T/S'_{T_0}) + \phi_0 (S'_P/S'_{P_0})}{1 + \phi_0} \to \text{minimize}. \quad (8)$$

It can be shown [8] (for the case under consideration) that $S'_T/S'_{T_0} = Nu_0/Nu$ and $S'_P/S'_{P_0} = (fRe/16)(N_0/N)^2$, hence (8) becomes:

$$N_s = \frac{(Nu_0/Nu) + \phi_0 (fRe/16) (N_0/N)^2}{1 + \phi_0}. \quad (9)$$

For a given finned tube, Nu and fRe are constants, therefore for a given application for which ϕ_0 = constant, the minimum N_s corresponds to $(N_0/N) = 0.45$[6] via (1c), which incidentially makes $L = 0.45 L_0$.

For a given application, the constant value for the irreversibility distribution ratio will (theoretically) lie somewhere in the range $0 < \phi_0 < \infty$. For example, for M = 8, H = 0.7 and $(N_0/N) = 0.45$, depending on the attendant value for ϕ_0, the augmentation entropy generation number (N_s) will lie between 0.107 ($\phi_0 = 0$) and 2.50 ($\phi_0 = \infty$). It is only if $\phi_0 < 0.594^S$ (or $\phi_0 < 0.0786$ for $N = N_0$) that $N_s < 1$ and hence thermodynamically advantageous. With this in mind, Fig. 8 has been prepared to show the ϕ_0 values for which $N_s = 1$.

Exergy savings occur only if the finned heat exchanger operates in the domain lying below its curve position. Again, as an example, for M = 8, H = 0.7 and $N = N_0$, if the operation is at $\phi_0 < 0.0786$, irreversibilities are reduced ($N_s < 1$) compared to the smooth tube exchanger. But if $\phi_0 > 0.0786$, using this

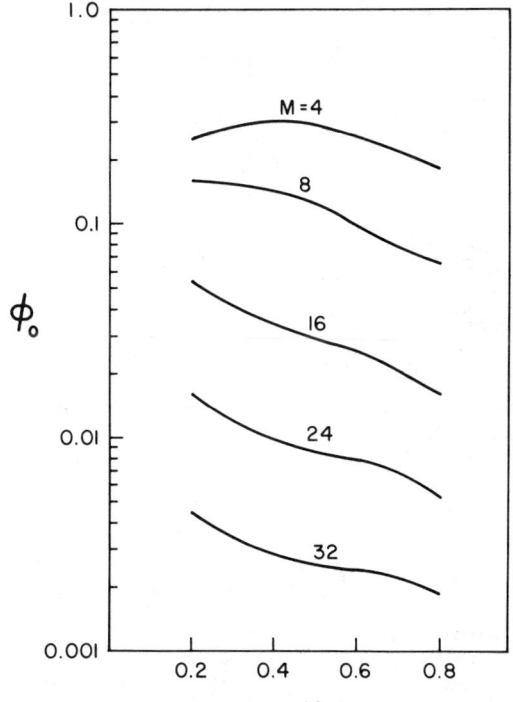

Fig. 8. Values of ϕ_0 for $N_s = 1$; B.C. 2, $K = \infty$, $N = N_0$.

---

[6]This approach is slightly artificial in that as $H \to 0$, $Nu \to Nu_0$ and $fRe \to 16$, hence $N_s \to 1$ (as it should) only if $N \to N_0$.

finned tube heat exchanger will actually result in an increase in the rate of exergy waste ($N_s > 1$). As shown by Bejan [12], the situation is similar under turbulent flow conditions except that the domain boundary (for $N_s = 1$) is determined by both ϕ_o and Reynolds number.

As a further illustration of the use of Fig. 8, if $\phi_o = 0.06$ (for example), finning configurations with M > 16 can be immediately dismissed from consideration. Of course the designer must bear in mind that this corresponds to $N = N_o$, and furthermore, the entire scenario is for $Q = Q_o$, $\mathring{m}_t = \mathring{m}_{t_o}$, $r = r_o$ and $NL = N_o L_o$. Under these circumstances $V > V_o$ and in general $\mathring{W} \neq \mathring{W}_o$. Contending finning configurations would now be examined on the basis of their N_s value; the best finning configuration being the one having the lowest N_s value. To illustrate this step, Fig. 9 has been prepared for $\phi_o = 0.10$, but this time with $(N_o/N) = 0.45$. For completeness, all 25 cases have been included. As shown in Fig. 9, the best fin is M = 4, H = 0.8 ($N_s = 0.296$), but finning arrangements involving M = 4 to 8 and H ≈ 0.6 to 0.8 all perform much the same with substantial exergy savings ($N_s \approx 0.34$). Despite the rather simple scenario involved, this suggests that internally finned tubes are feasible as a superior alternative to smooth tubes for laminar flow heat exchangers.

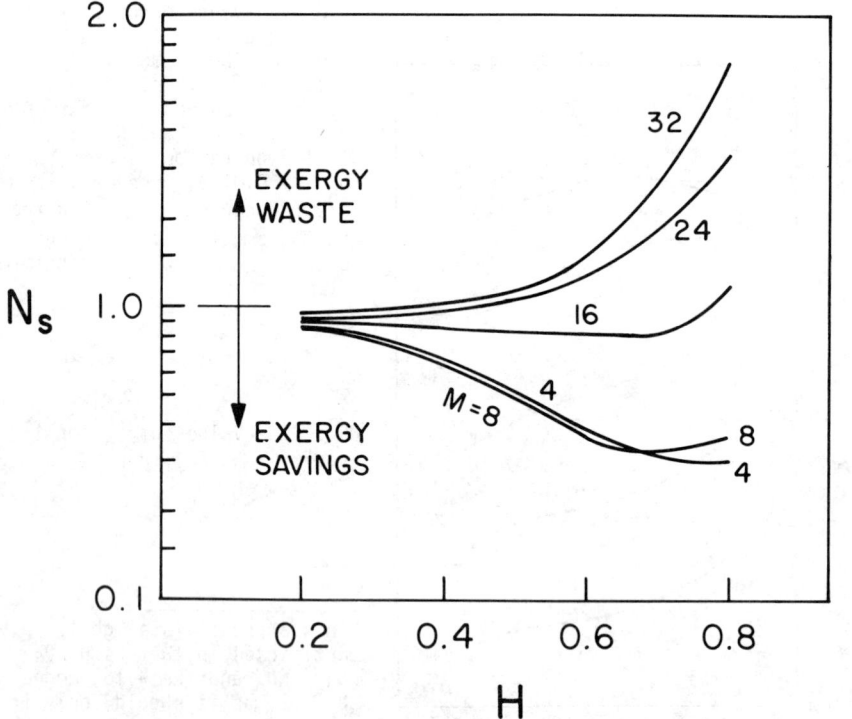

Fig. 9. Values of N_s for $\phi_o = 0.10$; B.C.2, $K = \infty$, $(N_o/N) = 0.45$.

6. CONCLUSIONS

A feasibility study was conducted concerning the optimized use of internally finned tubes as a replacement for smooth tubes for laminar flow heat exchangers. Four cases were examined in detail for the two extreme thermal boundary conditions. For brevity, only the results for $K = \infty$ are summarized as follows:

a) Case A: Reduced tube material volume.
   Reductions in tube material volume (weight) by as much as a factor of 3 are possible (for the same heat duty and pumping power) by using a small number of long fins (Fig. 3).

b) Case B: Increased heat duty.
   Again by using a small number of long fins, total heat exchange rates can be increased by at least a factor of 2 for the same pumping power and tube material volume (Fig. 4). Alternatively, thermal conductances can be increased by a factor of 15 to 20 by using 11 to 13 fins with $H = 0.8$ (Fig. 5).

c) Case C: Reduced pumping power.
   Reductions in pumping power (for the same heat duty and tube material volume) by a factor of about 9 can be achieved by (again) using a small number of long fins (Fig. 6).

d) Case D: Reduced temperature differences.
   Temperature differences can be reduced (same heat duty, pumping power and tube material volume) by a factor of about 4 by using 4 to 8 fins with $H = 0.8$ (Fig. 7).

It is concluded that internally finned tubes offer significant improvements over smooth tubes for laminar flow heat exchangers. Substantial exergy savings appear possible. Thermal system designers are therefore encouraged to consider internally finned tubes as a design alternative for laminar flow heat exchangers. Since thermal boundary conditions will most often lie between the two extremes considered here, designers are reminded that the present results correspond to the limit where external thermal resistances are negligible. Also the present results are influenced by the choices made in (1).

It is also concluded, assuming a viable market, that the incentive exists for manufacturers to develop production techniques for tubes having a small number of long fins. Concerning spiralling, it appears that spiral fins perform slightly better than straight fins, however, further study is required to determine if spiralling produces a significant additional benefit under laminar flow conditions.

Finally, it is noted that prospects for internally finned tubes are much better under laminar flow conditions than under turbulent flow conditions. However, at the same time, it must be recognized that Webb and Scott [3] dealt only with relatively short fins. Their analysis was also restricted to finned tubes having the same inside diameter as smooth tubes.

7. ACKNOWLEDGEMENT

The authors gratefully acknowledge the support provided for this research by the Natural Sciences and Engineering Research Council Canada.

8. REFERENCES

1. Bergles, A.E., Blumenkrantz, A.R., and Taborek, J., "Performance Evaluation Criteria for Enhanced Heat Transfer Surfaces", Proc. Fifth Int. Heat Tr. Conf., Tokyo, Sep 3-7, 1974, "Heat Transfer 1974", Vol. II, pp. 239-243.

2. Webb, R.L., and Eckert, E.R.G., "Application of Rough Surface to Heat Exchanger Design", Int. J. Heat Mass Transfer, Vol. 15, 1972, pp. 1647-1658.

3. Webb, R.L., and Scott, M.J., "A Parametric Analysis of the Performance of Internally Finned Tubes for Heat Exchanger Applications", ASME Journal of Heat Transfer, Vol. 102, 1980, pp. 38-43.

4. Soliman, H.M., and Feingold, A., "Analysis of Fully Developed Laminar Flow in Longitudinal Internally Finned Tubes", The Chemical Engineering Journal, Vol. 14, 1977, pp. 119-128.

5. Soliman, H.M., Chau, T.S., and Trupp, A.C., "Analysis of Laminar Heat Transfer in Internally Finned Tubes with Uniform Outside Wall Temperature", ASME Journal of Heat Transfer, Vol. 102, 1980, pp. 598-604.

6. Soliman, H.M., "The Effect of Fin Conductance on Laminar Heat Transfer Characteristics of Internally Finned Tubes", To appear in Can. J. of Chem. Eng., Apr, 1981.

7. Ivanovic, M., "Prediction of Flow and Heat Transfer in Internally Finned Tubes", Ph.D. Thesis, University of Minnesota, July 1978.

8. Trupp, A.C., and Soliman, H.M., "Performance Optimization of Internally Finned Tubes for Laminar Flow Heat Exchangers", University of Manitoba, Dept. of Mech. Eng. Rep. ER25.29, Nov, 1980.

9. Rogers, J.T., "The Thermodynamic Basis for Effective Energy Utilization", Paper presented at the Third New Zealand Energy Conference, Victoria University of Wellington, N.Z., May 12-14, 1977.

10. Bejan, A., "General Criterion for Rating Heat-Exchanger Performance", Int. J. Heat Mass Transfer, Vol. 21, 1978, pp. 655-658.

11. Ouellette, W.R., and Bejan, A., "Conservation of Available Work (Exergy) by Using Promoters of Swirl Flow in Forced Convection Heat Transfer", Energy, Vol. 5, 1980, pp. 587-596.

12. Bejan, A., Discussion on Reference [3], ASME Journal of Heat Transfer, Vol. 102, 1980, pp. 586-587.

Two-dimensional Heat Flow through Fin Assemblies

P.J. HEGGS
Chemical Engineering Department
Leeds University
Leeds LS2 9JT, UK

P.R. STONES
BNFL, Windscale and Calder Works
Sellafield, Seascale
Cumbria CA20 1PG, UK

ABSTRACT

A theoretical investigation of the effects of two-dimensional temperature distributions on the heat flow through rectangular fins is presented. The two-dimensional effects can be significant and current design method can over-predict by 16% and under-predict by 10%. The fin temperature distribution is mainly uni-dimensional, however severe two-dimensional temperature distributions occur in the supporting wall.

1. INTRODUCTION

Extended or finned assemblies are formed by attaching longitudinal strips, annuli, spines or wire loops to the surfaces of heat transfer equipment. Similar assemblies are formed by extruding or pressing these shapes into the surfaces. The extension of the surface in this way is to increase the rate of heating or cooling of the surrounding fluids. In many industrial situations, heat transfer between two fluids is limited by the controlling thermal conductance on one side of the interface separating the fluids. Often the rate of heat transfer in these situations is significantly increased by the attachment of fins to the side of the interface with the controlling resistance. The finned assembly provides a greater heat transfer area; however, the inherent temperature distribution along the fins results in a heat transfer rate which decreases from the fin base to the fin tip. The temperature distribution in the interface also changes with the addition of the fins. The total heat transfer through the assembly is a combination of that from the fin surface and that from the remaining interface surface not covered by the fins. Thus a thorough knowledge of the temperature distribution in both the interface and the attached fins is necessary in order to design the fin dimensions and its spacing to provide the largest heat transfer rate.

2. CURRENT DESIGN METHODS

The transfer of heat between two fluids separated by a partition, Figure 1a, is best envisaged by the summation of resistances and is simply given by

$$Q = (T_W - T_A)/(RE1 + RE2 + RE3) \tag{1}$$

where $RE1 (=1/h_W(S+W))$ and $RE3 (=1/h_A(S+W))$ are the respective convective resistances and $RE2 (= WT/k_1(S+W))$ is the conductive resistance. In many instances one of the convective resistances is controlling, say, RE3 and one method of increasing the heat flow is to reduce RE3 by attaching fins to that particular side. Hence a useful criterion for deciding whether to attach fins to a surface is

$$\frac{h_A}{h_W} + \frac{h_A WT}{k_1} < 1. \tag{2}$$

The heat flow with fins attached to one side, Figure 1b, is now given by

$$Q1 = (T_W - T_A)/\left(RE1 + RE2 + (RE4^{-1} + RE5^{-1})^{-1}\right) \tag{3}$$

where RE4 $(=1/h_A S)$ the resistance at the surface not covered by fins and RE5 $(= 1/h_A \eta (2H+W))$ the resistance of the fin. RE5 includes the fin efficiency (η): defined as the ratio of the heat dissipated by the fin to that which would be dissipated by the fin all at its base temperature, which lies between zero and one, and takes into account the temperature distribution along the fin height [1]. Values for the fin efficiency can be easily read from the charts produced by Gardner [2] for various fin geometries.

This method incorporates several simplifying assumptions, the overriding one being that the temperature distribution within the assembly is uni-dimensional in the direction of heat flow. In addition, the expressions for evaluating the fin efficiency ignore the existence of wall and the other resistances affecting the heat flow. However, the authors have verified [3] that the heat flow through the whole assembly assuming uni-dimensional temperature distribution is given by equation (3). The assumption of uni-directional temperature distribution is most questionable and has been investigated for single isolated fins, single fins attached to an infinite mass and arrays of fins attached to a surface of finite thickness.

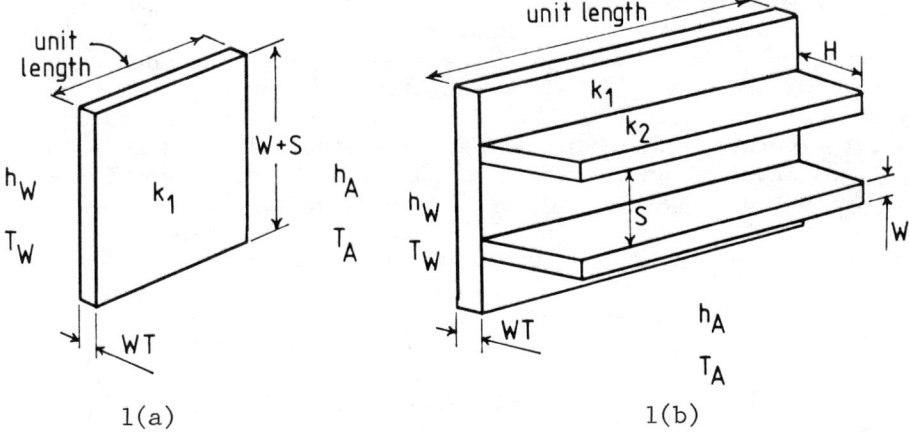

1(a) 1(b)

Figure 1. Schematics of partition and fin assembly.

3. TWO-DIMENSIONAL HEAT FLOW EFFECTS

3.1 Previous Investigations

Harper and Brown [1] showed that the efficiency of a single longitudinal fin assuming a two-dimensional temperature distribution was only 0.6 percent different to that assuming uni-dimensional distribution. Keller and Somers [4] concluded that a two-dimensional solution was necessary for an annular fin of rectangular profile, if the fin height/thickness was less than 10, Irey [5] illustrated the value of the fin Biot number (=$h_A W/k_2$) as a criterion for determining dimensional effects for a pin fin, and Lau and Tan [6] confirmed this for longitudinal and annular fins. If this Biot number is less than 0.2, then a uni-dimensional analysis was adequate. All these investigations ignored the wall and other parameters and assumed the fin base temperature was uniform.

Sparrow and Hennecke [7], and Klett and McCulloch [8] clearly showed that temperature depressions exist at the base of a fin attached to an infinite mass. Two-dimensional numerical solutions were obtained for the arrangements, and show larger errors in the total heat flow than Irey for similar Biot numbers. Klett and McCulloch [8] obtain a 26% overestimation of the total heat flow for $h_A W/k_2 = 0.06$, whereas Irey [5] found only 1.0%.

Sparrow and Lee [9] and Suryanarayana [10] considered

assemblies of fins and there is general agreement that dimensional effects increase as the heat transfer coefficients on either side of the assembly become equal. Suryanarayana [10] concludes that a uni-dimensional analysis gives satisfactory values for fin-height equal to fin half thickness, but otherwise overestimates the heat flowrate. The error increases as the wall thickness and fin height increase and as the fin spacing and ratio of coefficients decreases. Sparrow and Lee [5] conclude that increasing the spacing between longitudinal fins on a tube increases the error, which is contrary to Suryanarayana's results.

The above two investigations have not always used examples which satisfy the criterion, equation (2). Sparrow and Lee [9] found a difference of 40% between the uni-and two-dimensional representations when the left hand side of equation (2) was 45. Suryanarayana [10] found differences of 80%, when the left hand side of equation (2) was 1.5 which is obviously a borderline case.

Hence it is apparent that two-dimensional temperature distribution can affect the overall heat flow, but no definitive conclusions can be drawn from previous work.

3.2 Rectangular fins attached to a wall or tube

The mathematical equations describing the temperature distribution and resultant heat flowrate (Q2) for two-dimensional effects are given in the appendix. Inspection of Figure 1b shows that the heat flow through the assembly for both uni- and two-dimensional representations is a function of eight variables, i.e.

$$Q1 \ \& \ Q2 = f(h_W, k_1, h_A, k_2, W, WT, S, H) \qquad (4)$$

This is more succinctly expressed by considering three Biot numbers and three aspect ratios: $B11 = h_W W/k_1$, $B21 = h_A W/k_1$, $B22 = h_A W/k_2$, $X2 = H/W$, $YSW = S/W$ and $XW = WT/W$. In addition the tube radius must be specified for annular configurations.

The heat flowrate through the assembly is finally expressed as an autmentation factor, AUGF, defined as the ratio of the heat flowrate through the finned assembly to that through a plain partition under the same conditions. Thus the augmentation factor is a direct measure of the increase in the heat transfer brought about by the addition of fins [11].

To investigate all combinations of the six dimensionless parameters is beyond the scope of this paper, although a number of charts of the two-dimensional augmentation factor, AUGF2, have been produced for the following ranges B11/B21 (=h_W/h_A): 2-20000, B21/B22 (= k_1/k_2): 1-25, B11 : 2.0 and 0.8, X2: 1-160, YSW:1-50 and XW:1-100 [11, 12].

The dimensional effects for these ranges have been investigated by comparing the uni-dimensional augmentation factor, AUGF1, with AUGF2, in Figures 2 and 3.

Each figure shows how the percentage difference between the one- and two-dimensional augmentation factors (AUGF1 and AUGF2 respectively) varies over the chosen range of Biot numbers and aspect ratios.

Figure 2 is for a fixed unfinned side Biot number of B11 = 0.8 and figure 3 for B11 = 2.0. These two figures show similar trends, for example, in both figures the difference between AUGF1 and AUGF2 decreases as XW increases also AUGF1 is greater than AUGF2 provided B21/B22 = 8.0 or 25.0 and AUGF1 is only less than AUGF2 for fins and wall of the same material (B21/B22 = 1.0). However, figure 2 with B11 = 2.0 shows greater extreme values of (AUGF1 - AUGF2)/AUGF1. For example, comparing curves B11/B21 = 200 at XW = 1.0 : the percentage difference between AUGF1 and AUGF2 is 15.0 in Figure 3 when B11 = 2.0 but is reduced to 7.0 in Figure 2 when B11 = 0.8.

The following generalisations can be drawn from figures 2 and 3 concerning the dimensional effects within longitudinal fin assemblies:

i) The effects are greatest (up to 16.0 percent difference) for small values of B11/B21 (e.g. B11/B21 = 20) corresponding to a large fin side heat transfer coefficient similar in magnitude to that on the unfinned side. Conversely the difference is generally small (less than 4.0 percent) for large values of B11/B21 (e.g. B11/B21 = 2000).

ii) The effects tend to be greater when the fin metal has a greater thermal conductivity than the wall. For example, in figure 2 for YSW = 30 and B11/B21 = 200, the effects are 2.0 per cent when the fin and wall thermal conductivities are equal (B21/B22 = 1.0); however, increasing the fin thermal conductivity by a factor of 8.0 (B21/B22 = 8.0) increases the effects to 8.0 per cent.

iii) The left hand chart in both figures shows that the effects are greater for a thin wall (1.0 < XW < 40.0) than a thick wall (XW > 40.0).

iv) The middle chart in both figures shows that for B21/B22 = 8.0 or 25.0 there is a value of YSW the fin spacing which causes the largest effect. For example, in figure 2 with B11/B21 = 200 and B21/B22 = 8.0 the curve for YSW shows a maximum of 8.0 percent at

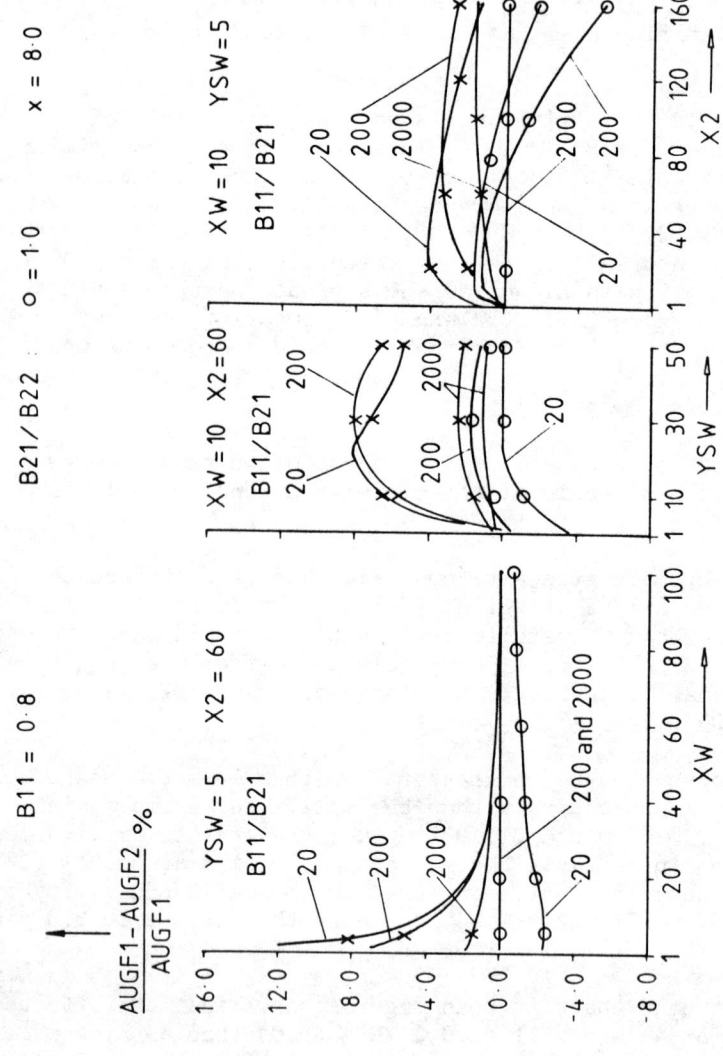

Figure 2. Dimensional Effects in longitudinal fin assemblies, for B11 = 0.8

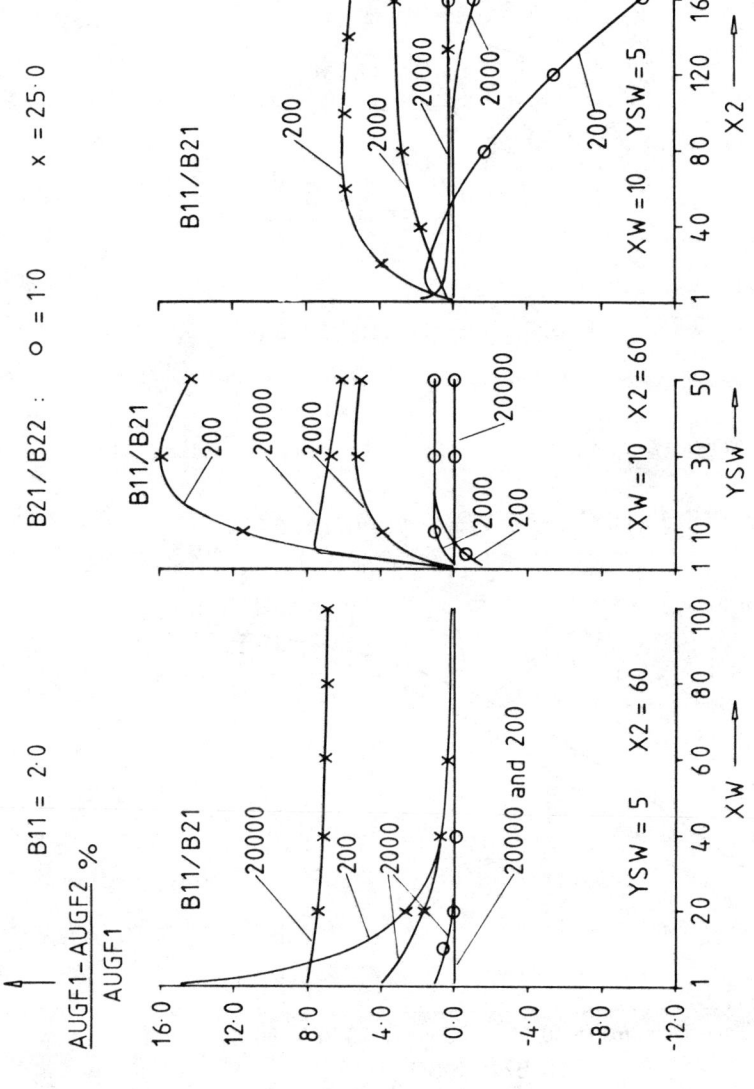

Figure 3. Dimensional Effects in longitudinal fin assemblies, for B11 = 2.0

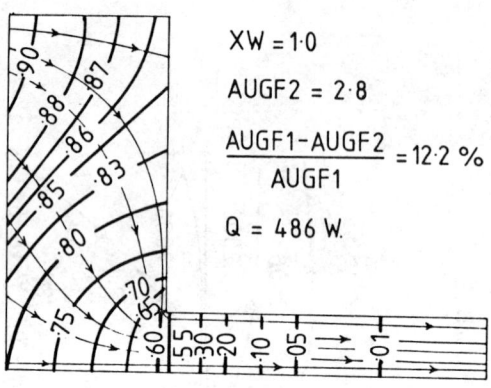

Figure 4. Temperature distributions in a longitudinal fin assembly

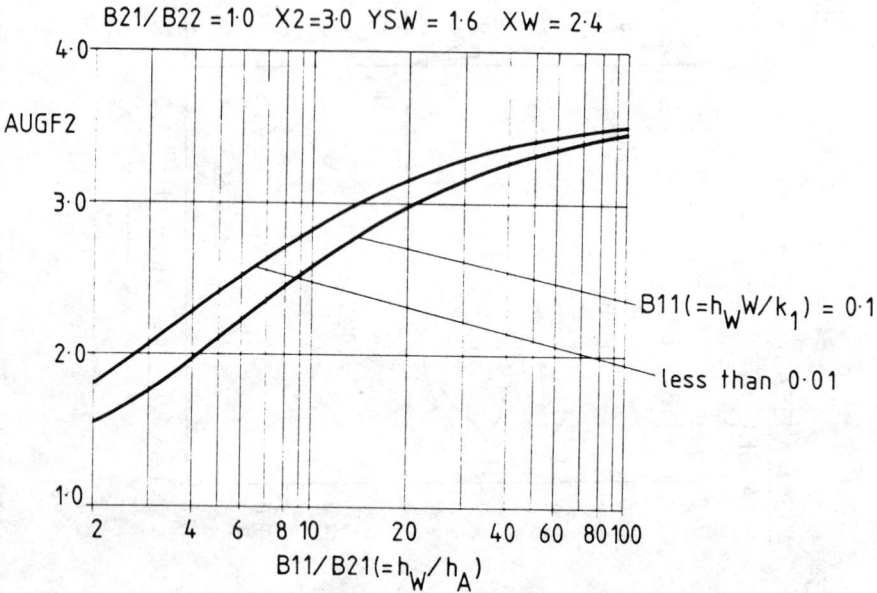

Figure 5. Augmentation factors for one type of Integron Low Fin tubing

YSW = 30.0. The YSW curves for B21/B22 = 1.0 do not show maxima, rather they tend to be asymptotic to zero effects as YSW increases.

v) The right hand chart in both figures shows that for B21/B22 = 8.0 or 25.0 the effects tend to increase as X2 the fin height increases. For B21/B22 = 1.0 (e.g. integral assemblies) and B11/B21 = 200 the effects first increase to a maximum at X2 = 10.0 with AUGF2 < AUGF1, then decrease becoming zero at X2 = 8.0 in figure 2 and X2 = 60.0 in figure 3. The effects then increase as X2 increases but now AUGF2 > AUGF1.

Hence the dimensional effects cannot be predicted by a single parameter, e.g. B22, as was proposed by Irey [5] and although the effects can be significant, the regions of severe two-dimensional temperature distributions are not apparent from this discussion.

4. TEMPERATURE DISTRIBUTIONS WITHIN THE LONGITUDINAL FIN ASSEMBLIES

The need to consider the complete assembly of fins and supporting surface is further illustrated by plotting isotherms within the fin and wall of the assembly. Figure 4 is for a longitudinal assembly taken from figure 2 in which B11 = 0.8, B11/B21 = 20 and B21/B22 = 8.0 in which the dimensional effects are relatively large (12.2%). The high curvature of the isotherms in the wall region is characteristic of a two-dimensional problem. The curvature of the isotherms causes the heat flow lines leaving the unfinned side to diverge from their linear one-dimensional path and converge at the fin-wall junction, as shown in figure 4.

It is significant that it is within the wall in figure 4 that the isotherms exhibit curvature, whereas within the fin the isotherms are straight and parallel (typical of a one-dimensional problem). Thus investigations restricted to a single unattached fin, such as those by Harper and Brown [1], Keller and Somers [4] and Irey [5], would not show up these two-dimensional effects within the wall. In addition, the two-dimensional temperature distribution illustrated in Figure 4 reveals the presence of a "hot spot" in the wall mid-way between the fins, this phenomenon is not shown by the one-dimensional analysis. The one-dimensional analysis predicts a maximum wall temperature of 0.84 at the unfinned side as opposed to 0.71 for the two-dimensional analysis. Phenomena like this may be significant if the finned equipment is operating close to the ultimate tensile strength of its material of construction or under some critical corrosive conditions such as can occur in certain applications in the nuclear industries.

5. INTEGRON LOW FIN TUBING

It was mentioned earlier that Keller and Somers [4] had shown that

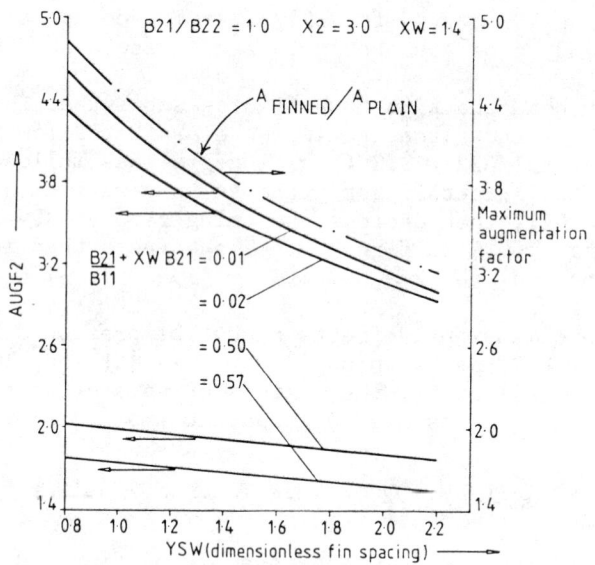

Figure 6. Maximum augmentation factors for Integron Low Fin tubing

dimensional effects in annular fins were greatest when the fin height to fin thickness ratio was less than 10.0, i.e. $X2 < 5$ of the commercially used fins, Integron Low Fin [13] tubing is the closest to this short stubby configuration with $X2 = 3.0$. In addition, once the dimensions of a particular configuration are fixed, then the usefulness of the augmentation factor in design calculations becomes very apparent. The augmentation factor is now a function of only two variables,

$$AUGF2 = f(B11, B11/B21) \tag{5}$$

because X2, YSW and XW are fixed and $B21/B22$ $(k_1/k_2) = 1.0$.

Figure 5 shows the augmentation factor plotted against $B11/B21$ (h_W/h_A) for a range of B11 values from an upper limiting value of $B11 = 0.1$ corresponding to an upper limit of the tube side coefficient of 6000 W/m^2k and a low metal thermal conductivity (monel) down to B11 0.01. The curve for $B11 = 0.01$ is the bound for all values of B11 less than 0.01. The augmentation is always greater than 1 and as B11/B21 becomes greater than 100 then an asymptotic augmentation factor of around 3.5 is approached. If a designer wished to use this particular configuration, then the reduction in the number of tubes can be directly found by evaluating the film coefficients and finding the augmentation factor from Figure 5. The inverse of the augmentation factor is the reduction in tube numbers.

The maximum possible augmentation factor for a given fin height, spacing and wall thickness is limited by the ratio of the outer surface area of the finned tube to that of a plain tube with the same outside diameter. The maximum possible augmentation factor for a particular Integron Low Fin configuration is plotted in Figure 6 against the fin spacing for different combinations of the inner and outer Biot numbers. The maximum augmentation would only be possible if the conductive resistance of the metal tube and fin, and the tube-side convective resistance were zero. Each curve in Figure 6 has a particular value of the left hand side of the inequality of equation (2), i.e. $Z = B21/B11 + XW\ B21$. When this is 0.0, we have the maximum augmentation, when it is 0.01, the curve is 4% below the maximum, whereas when it is 0.57 it is over 60% less than the maximum. Also it is very obvious that the augmentation falls with increasing fin spacing. The effect of the fin spacing is less significant the larger the value of Z.

For all the configurations investigated, the difference between AUGF1 and AUGF2 never exceeds 1.0%. This is contrary to the results of Keller and Somers [4], who only considered a single fin and thus the effects of neighbouring fins was neglected. Figures 7 and 8 show one- and two-dimensional temperature distributions respectively within one size of Integron Low Fin tube. The curvature of the isotherms in Figure 8 show that the two-dimensional effects extend deep into the tube wall and are not restricted to the fin region. However, the total heat flow for both representations differ by less than 1.0%. This apparent contradiction can be explained because although the fin-side surface (AB) in Figure 8 is hotter than that in Figure 7, the fin surface (BCD) is cooler in Figure 7. The heat dissipated from the wall surface is Q_{AB} = 4.877 W for the two-dimensional analysis and = 4.771 W for the one-dimensional analysis. The heat dissipated from the fin surface is Q_{BCD} = 11.796 W in Figure 8 and = 11.965 in Figure 7. Thus the increased flow from the wall surface in Figure 7 is counteracted by a decreased heat flow from the fin surface, the result being that the total heat dissipated by the fin and wall assembly is almost identical in both cases, i.e. 0.4%.

Hence assemblies of stubby fins can be confidently designed using the sum of resistances approach although two-dimensional temperature distributions occur both in the tube wall and the fins.

6. CONCLUSIONS

The effects of multi-dimensional temperature distributions on the overall heat flow through a fin assembly cannot be predicted by a single parameter. The current design techniques which

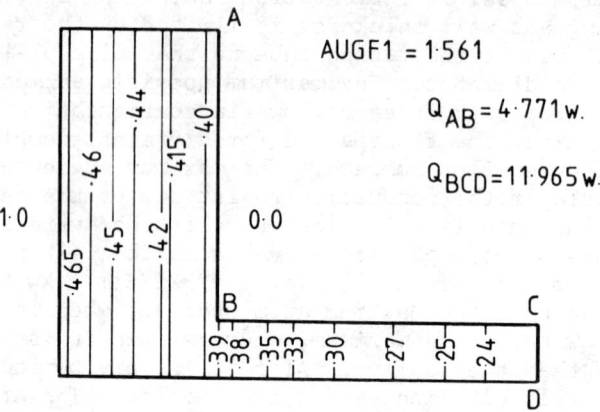

Figure 7. Uni-dimensional temperature distribution within Integron Low Fin tubing

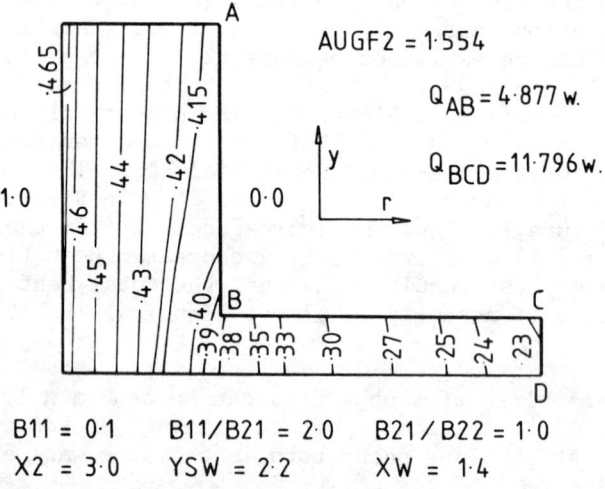

Figure 8. Two-dimensional temperature distribution within Integron Low Fin tubing

ignore multi-dimensional effects can over- and under-predict the heat flow rate through the fin assembly. For rectangular fins the over-prediction can be as high as 16% and the under-prediction can be as low as 10%. The multi-dimensional temperature distribution occurs mainly in the assembly wall and the temperature distribution in the fin is almost uni-dimensional. Plots of the augmentation factor for the two-dimensional representation of particular fin assemblies will provide good predictions of the heat flow. However, for stubby fin arrangements, such as Low Integron Tubes, the current design methods provide good estimates of the heat flow.

NOMENCLATURE

AUGF1, AUGF2	one- and two-dimensional augmentation factors respectively.
B11, B21, B22	Biot numbers = $h_W W/k_1$, $h_A W/k_1$ and $h_A W/k_2$ respectively.
h_A, h_W	finned and unfinned side heat transfer coefficients, $W/m^2 K$.
H	fin height, m.
k_1, k_2	wall and fin conductivities, W/mK.
Q, Q1	heat flow with and without fins, W.
RE1, RE2, RE3, RE4, RE5	thermal resistances of an assembly, K/W.
T_A, T_W	finned and unfinned side fluid temperature, K.
W	fin thickness, m.
WT	wall thickness, m.
XW	WT/W.
X2	H/W
YSW	S/W.
η	fin efficiency.
θ	$(T-T_W)/(T_A-T_W)$

REFERENCES

1. Harper, D.R. and Brown, W.B. Mathematical Equations for Heat Conduction in the Fins of Air-Cooled Engines. NACA Report 158, 1922.

2. Gardner, K.A. Efficiency of Extended Surface. Trans. ASME. Vol. 67, No. 8, pp. 621-631, 1945.

3. Heggs, P.J. and Stones, P.R., The Effects of Dimensions on the Heat Flowrate through Extended Surfaces, J. Heat Transfer, Trans. ASME. Vol. 102, pp. 180-182, 1980.

4. Keller, H.H. and Somers, E.V, Heat Transfer from an Annular Fin of Constant Thickness. J. Heat Transfer, Trans. ASME. Fol. 81, series C, No. 2, p.151, 1959.

5. Irey, R.K. Errors in the One-Dimensional Fin solution. J.Heat Transfer, Trans. ASME. Vol. 90, series C, No. 1, p.175, 1968.

6. Lau,W. and Tan, C.W., Errors in One-Dimensional Heat Transfer Analysis in Straight and Annular Fins, J.Heat Transfer, Trans. ASME, Vol. 95, p.549, 1973.

7. Sparrow, E.M. and Hennecke, D.K., Temperature Depression at the Base of a Fin. J. Heat Transfer, Trans. ASME. Vol. 92, series C, No. 1, pp. 204-206, 1970.

8. Klett, D.E. and McCulloch, J.W., The Effect of Thermal Conductivity and Base-Temperature Depression on Fin Effectiveness. J.Heat Transfer, Trans. ASME. Vol. 94, series C, No. 3, pp. 333-334, 1972.

9. Sparrow, E.M. and Lee, L., Effects of Fin Base-Temperature Depression in a Multifin Array. J.Heat Transfer, Trans. ASME. Vol. 97, series C, No. 3, pp. 463-465, 1975.

10. Suryanarayana, N.V., Two-Dimensional Effects on Heat Transfer Rates from an Array of Straight Fins. J. Heat Transfer, Trans. ASME. Vol. 99, series C, No. 1, pp. 129-132, 1977.

11. Heggs, P.J. and Stones, P.R., Improved Design Methods for Finned Tube Heat Exchangers, Trans. Instn. Chem.Engrs., Vol. 58, pp. 147-154, 1980.

12. Stones, P.R., The Effects of Dimensions on the Heat Flowrate through Extended Surfaces, Ph.D. Thesis, Leeds University,1980.

13. Integron Tube for Refrigeration and Air Conditioning, Yorkshire, Imperial Metals Ltd., PO Box 166,Leeds LS1 1RD,England.

APPENDIX The Mathematical equations describing the two-dimensional distribution and heat flowrate

The Laplace equation describes the temperature distribution in both regions A B C E and D E F G in figure 9.

$$\frac{\delta^2 \theta}{\delta N^2} + p \frac{\delta \theta}{\delta N} + \frac{\delta^2 \theta}{\delta M^2} = 0 \qquad (7)$$

with $N = x/W$, $p = 0$ and $M = y/W$ for the longitudinal case and $N = r/W$, $p = W/r$ and $M = y/W$ for the annular situation.

The boundary conditions are:

$$\text{AB} \qquad \text{B11} \; (\theta_i - 1) = \frac{\delta \theta_i}{\delta N} \qquad (8)$$

TWO-DIMENSIONAL HEAT FLOW THROUGH FIN ASSEMBLIES

BC, AE and EF $\quad\quad\quad \dfrac{\delta\theta_i}{\delta M} = 0 \quad\quad\quad (9)$

CD $\quad\quad -B21 \quad \theta_i = \dfrac{\delta\theta_i}{\delta N} \quad\quad (10)$

DE $\quad\quad\quad \dfrac{\delta\theta_i}{\delta N} = \dfrac{k_2}{k_1}\dfrac{\delta\theta_{ii}}{\delta N} \quad\quad (11)$

and $\quad\quad\quad \theta_i = \theta_{ii} \quad\quad\quad (12)$

DG $\quad\quad -B22 \quad \theta_{ii} = \dfrac{\delta\theta_{ii}}{\delta M} \quad\quad (13)$

GF $\quad\quad -B22 \quad \theta_{ii} = \dfrac{\delta\theta_{ii}}{\delta N} \quad\quad (14)$

The heat flowrate by convection into the unfinned side of the assembly is

$$Q2 = \int_{y=0}^{y=(W+S)/2} h_W \left(T_W - T_i \Big|_{x=0} \right) dy \quad (15)$$

The solutions to equations (7) to (14) were found numerically either by finite difference or finite elements methods [12] and equation (15) was solved numerically by Simpson's Rule.

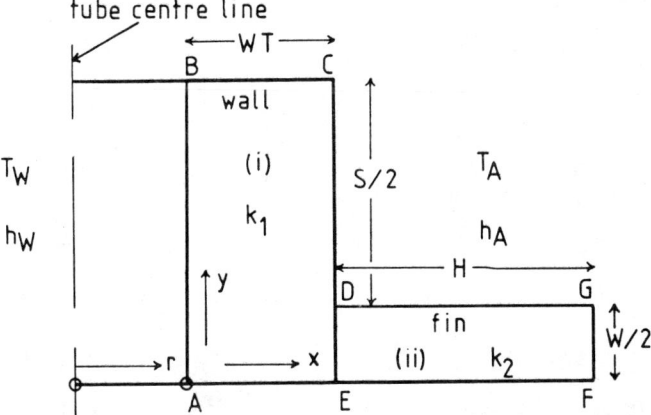

Figure 9. Nomenclature for the two-dimensional representation of longitudinal and annular fin assemblies

Effect of Swirl Angle and Geometry on Heat Transfer in Turbulent Pipe Flow

M.S. ABDEL-SALAM, M.M. HILAL, E.E. KHALIL, and A.M.A. MOSTAFA
Mechanical Power Department
Faculty of Engineering
Cairo University
Cairo, Egypt

ABSTRACT

This research work was carried out to study the effect of swirler vane angles on heat transfer characteristics in turbulent pipe flow. Vane swirlers were inserted at the entrance of the main tube to swirl the air flowing in the test section. Also, the effect of vane swirlers with sudden tube enlargement was studied. The flowing air was heated by means of electric heaters wound round the main tube so as to give a uniform heat flux.

The augmentation in heat transfer due to different vane angles and different enlargement ratios was represented by the increase in Nusselt number. Relations for this increase, pressure drop and thermal entry length for various entry conditions were obtained.

NOMENCLATURE

A	area of cross-section of tube
D	inner diameter of tube = 2R
H	rate of heat transfer by convection
H_x	local rate of heat transfer at distance x from tube inlet
h	average heat-transfer coefficient
h_x	local heat-transfer coefficient at distance x from tube inlet
h_∞	asymptotic heat-transfer coefficient
I.M.H.	power input to the main heater
k	thermal conductivity of air
L	length of tube
L_{th}	thermal entry length
ΔP	pressure drop of air
ΔP_o	pressure drop of air in the plain tube
Q	volumetric rate of air discharge
r_1	inner radius of swirler
r_2	outer radius of swirler
S	area of heat transfer surface
t_a	mean temperature of air at the working section
t_s	temperature of the heating surface
Z	hub ratio = r_1/r_2
Nu	Nusselt number, Nu_o for the plain tube
Re	Reynolds number

ρ density of air, mean at working section
μ viscosity of air, mean at working section
θ swirler vane angle, in degrees

1. INTRODUCTION

The heat transfer problem of simultaneous development of velocity and temperature fields in a heated straight tube in a heat exchanger has been studied by many investigators. The main aim of the present study is to augment the heat transfer from the tube inner surface to the flowing fluid. One of the methods used to achieve this target was the insertion of different turbulence promoters in the entry section of the tube. Many workers arrived at good results(e.g. references 1-8) by using helical, tape and spiral swirlers. The study of the fully developed swirling pipe flow due to a vane swirler placed at the tube entrance is the concern of this work.

In the present work the effect of swirler vane angle, with and without sudden tube enlargement(at swirler exit), on the heat transfer characteristics in a fully developed pipe flow is investigated. The effects of these parameters on the average heat-transfer coefficient, pressure drop and thermal entry length are studied. Various Reynolds numbers, for the air heated with uniform heat flux, are used in all experiments. General relations for the effect of swirler vane angle and sudden tube enlargement on heat transfer characteristics are obtained. Also, relations for pressure drop and thermal entry length for various entry conditions are obtained. A comparison with the results of previous workers is made.

2. EXPERIMENTAL INVESTIGATION

2.1. Apparatus

The apparatus used here is similar to that used by Abdel-Salam and coworkers[1], with some modifications. It consists mainly of the air passage, vane swirlers, heat transfer tube and the measuring instruments.

Air is sucked from the laboratory atmosphere and blown into the main heat-transfer tube through a conical section, a calming section with honey combs as shown in figure 1. Near the end of the air passage an orifice-meter is placed in order to determine the rate of air discharge. This discharge could be controlled by varying the air-blower speed, and using a by-pass valve before the conical section.

The swirl angles investigated were $0°, 15°, 30°$ and $45°$. The enlargement ratios were 1:1, 1:0.85 and 1:0.7, classified as groups I, II and III respectively. The hub ratio was taken 0.4. Twelve different swirlers were manufactured. Each swirler has five copper vanes whose thickness is 0.6 mm. The dimensions of the vane swirler are shown in Fig.2 . The design parameters of the different swirlers are indicated in table 1.

The main heat-transfer tube was made of brass, and had inner and outer diameters of 0.038 and 0.043 m respectively. Its length was 4 m. Two electric heaters were used; a main heater to supply

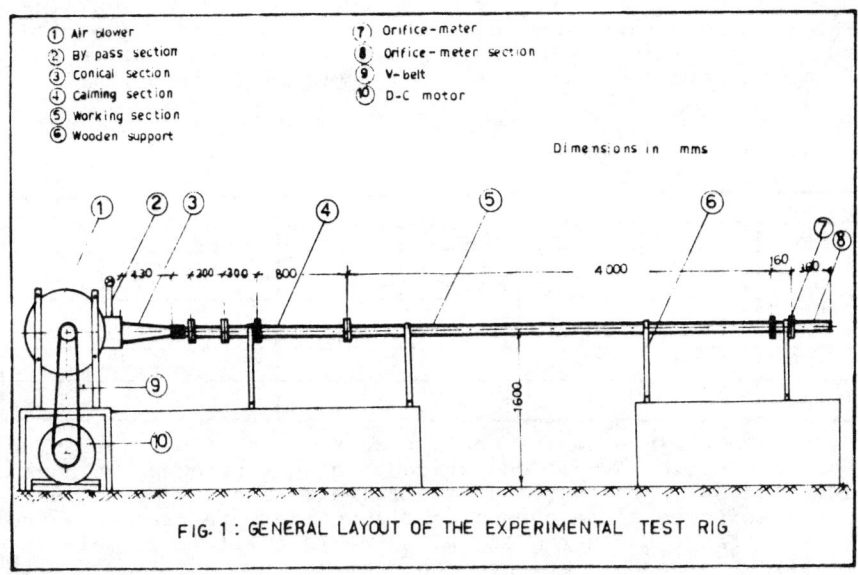

FIG. 1: GENERAL LAYOUT OF THE EXPERIMENTAL TEST RIG

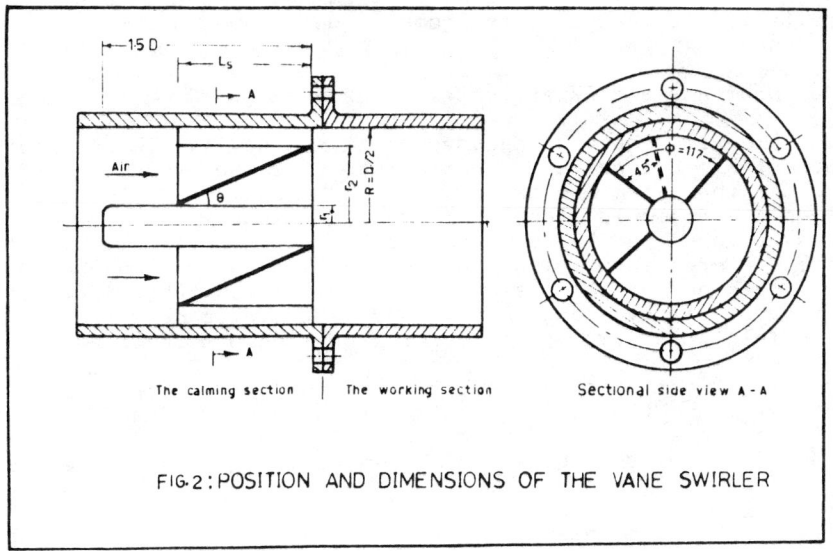

FIG. 2: POSITION AND DIMENSIONS OF THE VANE SWIRLER

heat to the heating surface, and a guard heater to prevent the heat flow from the outer side of the main heater. The two heaters were separated by an asbestos layer with five pairs of thermocouples to assure that there is no heat flow in the radial direction. These two heaters were surrounded by electrical insulation. Then, the guard heater was surrounded by asbestos at the outer surface. Also precautions were made to prevent axial heat flow from the main tube.

Table 1 : Swirler Parameters

Group No	I				II				III			
angle θ,°	0	15	30	45	0	15	30	45	0	15	30	45
Swirler length, mm	39.6	39.6	26.9	15.5	33.9	33.9	23.0	13.3	28.7	28.7	19.6	11.2
enlargement ratio	1:1				1:0.85				1:0.7			
$2r_1$, mm	15.2				13.0				10.6			
$2r_2$, mm	38.0				32.3				26.6			

The temperature of the heating surface was always measured at 16 points by using 0.25mm copper-constantan thermocouples. Special calibration was carried out on the thermocouples. A potentiometer, selector switches and a common cold junction were used to measure the e.m.f. for each thermocouple. Suitable ammeter, voltmeter and wattmeter were used to measure and control the heat input to the main and guard heaters. Also, a micromanometer was used to measure the pressure difference across the orifice, which was calibrated in sito to define the air flow rate across the working section. The apparatus was supported on four wooden brackets mounted on a wooden bench. The main heat-transfer tube was placed in a horizontal position. The measuring instruments were mounted on the wooden bench.

2.2. Measurements and Experimental Procedure

The blower speed was adjusted to give steadily the required discharge. This discharge was determined from the pressure difference across the orifice-meter, and was kept constant during the experiment. A suitable electrical input was supplied to the main heater. The input to the guard heater was adjusted to give identical temperatures across the asbestos insulation between the two heaters. Readings of all thermocouples were taken every 15 minutes until thermal steady state was reached. The static pressure at both the inlet and exit of the working section was recorded.

In all experiments the heat flux was always uniform around all the heating surface. The air-flow rates ranged from 31 to 111 m^3/hr for different swirlers. The input to the main heater ranged from 180 to 360 watts. The sum of 94 experiments were carried out for the twelve swirlers.

2.3. Method of Calculation

The net rate of heat transfer by convection to the air stream can be calculated from the following relation:

$$H = I.M.H. - \text{Heat losses} \tag{1}$$

The heat losses include the axial, radial and radiation losses. The sum of these losses has been reduced to about 0.3% from the input to the main heater. Hence, equation (2) may be used:

$$H = I.M.H. \tag{2}$$

The local heat-transfer coefficient has been calculated as:

$$h_x = H_x / (S(t_s - t_a)_x) \tag{3}$$

The average heat-transfer coefficient has been determined from the formula:

$$h = \frac{1}{L} \int_0^L h_x \, dx \tag{4}$$

where dx = element of the length along the tube.
To calculate the dimensionless groups, the air properties were calculated at the mean air temperature. The characteristic length is the inner tube diameter. Thus,

$$Nu = hD/k \quad \text{and} \quad Re = 4\rho Q / (D \pi \mu)$$

3. RESULTS

The results consist of two parts; the heat transfer results and the pressure drop results.

3.1. Heat Transfer Results

The heat transfer results of the present work could be represented by the following equation:

$$Nu = K \cdot Re^n \tag{5}$$

The constant n takes a value of 0.8 for all experiments, while the constant K is given in table 2.

Table 2 Values of constant K in equation (5)

Group No	I				II				III			
angle, θ,°	0	15	30	45	0	15	30	45	0	15	30	45
K	.0193	.0217	.0249	.0284	.0208	.0235	.02668	.03092	.02230	.02533	.02859	.03246

The experimental results of groups I, II and III are shown in Fig 3a, b and c respectively. In all these experiments, the average surface temperature varied between 30 and 54 °C. The inlet air temperature varied between 18 and 37 °C.

3.2. Pressure-Drop Results

The pressure drop (ΔP) in the working section, including the vane swirler, was measured in the different experiments. This was done by measuring the gauge pressure before the swirler and at the working-section exit. The variation of ($\Delta P / \Delta P_o$) with Re for the three groups is shown in figures 4a, b and c.

The relation between the pressure drop and the Reynolds number could be expressed in equation (6) and table 3 as follows:

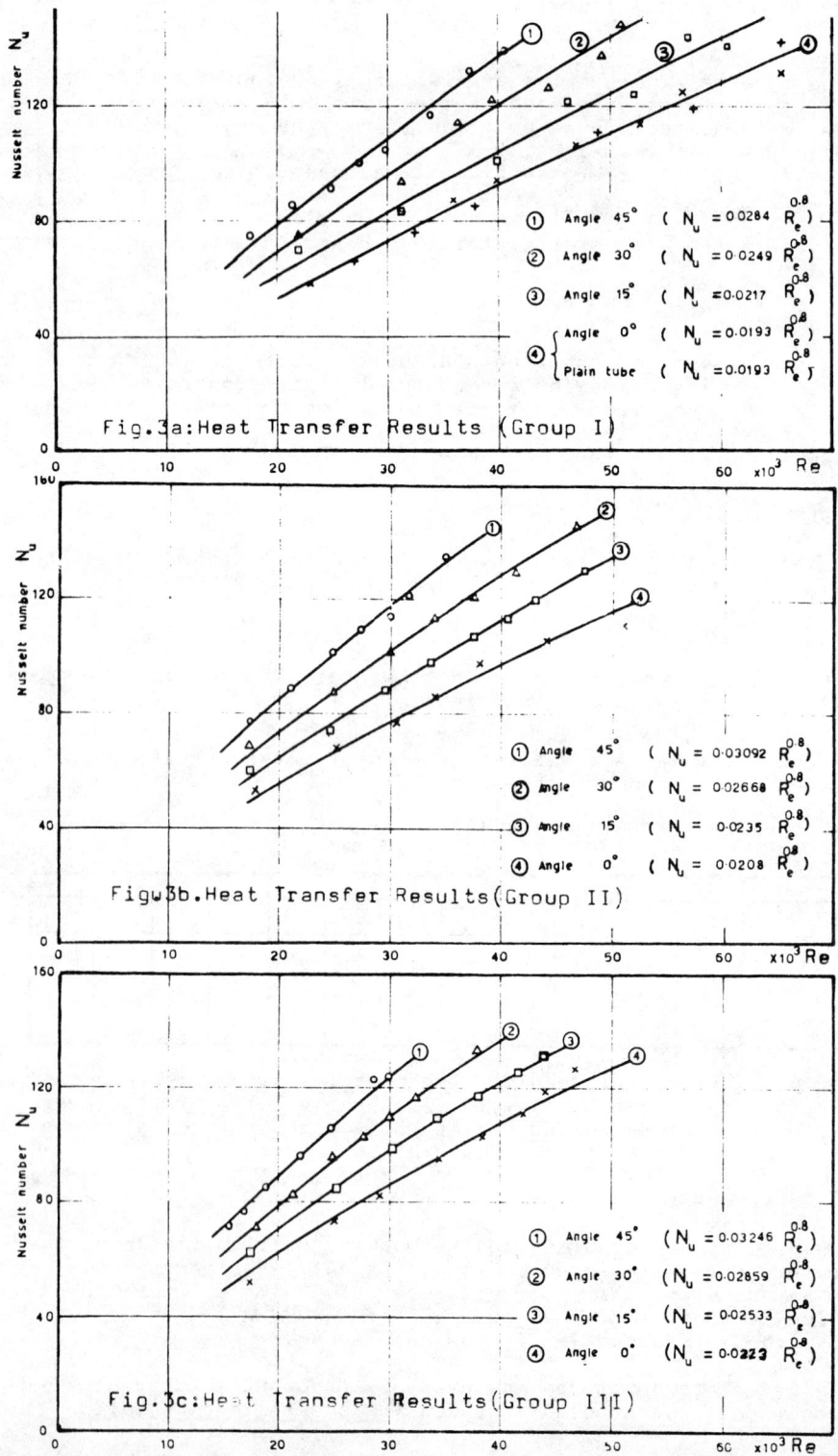

Fig.3a: Heat Transfer Results (Group I)
① Angle 45° ($N_u = 0.0284\ R_e^{0.8}$)
② Angle 30° ($N_u = 0.0249\ R_e^{0.8}$)
③ Angle 15° ($N_u = 0.0217\ R_e^{0.8}$)
④ Angle 0° ($N_u = 0.0193\ R_e^{0.8}$)
  Plain tube ($N_u = 0.0193\ R_e^{0.8}$)

Fig.3b. Heat Transfer Results (Group II)
① Angle 45° ($N_u = 0.03092\ R_e^{0.8}$)
② Angle 30° ($N_u = 0.02668\ R_e^{0.8}$)
③ Angle 15° ($N_u = 0.0235\ R_e^{0.8}$)
④ Angle 0° ($N_u = 0.0208\ R_e^{0.8}$)

Fig.3c: Heat Transfer Results (Group III)
① Angle 45° ($N_u = 0.03246\ R_e^{0.8}$)
② Angle 30° ($N_u = 0.02859\ R_e^{0.8}$)
③ Angle 15° ($N_u = 0.02533\ R_e^{0.8}$)
④ Angle 0° ($N_u = 0.0223\ R_e^{0.8}$)

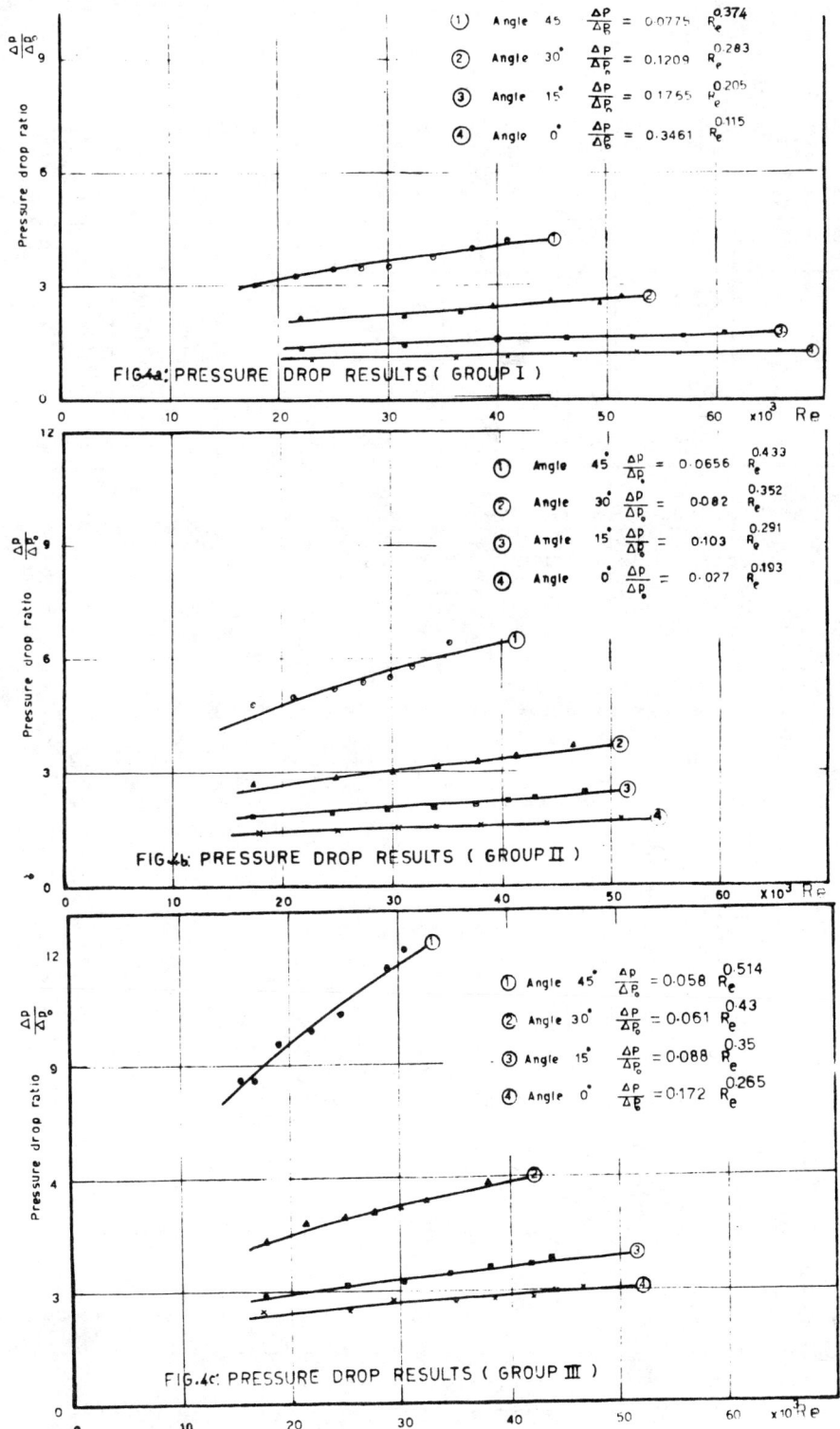

FIG.4a: PRESSURE DROP RESULTS (GROUP I)

FIG.4b: PRESSURE DROP RESULTS (GROUP II)

FIG.4c: PRESSURE DROP RESULTS (GROUP III)

$$\Delta P/\Delta P_o = K_1 \cdot Re^{K_2} \tag{6}$$

where K_1 and K_2 are constants depending on the swirler vane angle and enlargement ratio.

Table 3: Values of K_1 and K_2 in equation (6)

Group No	I				II				III			
angle, θ,°	0	15	30	45	0	15	30	45	0	15	30	45
K_1	0.346	0.176	0.121	0.078	0.217	0.103	0.082	0.066	0.172	0.088	0.061	0.058
K_2	0.115	0.205	0.283	0.374	0.193	0.291	0.352	0.433	0.265	0.350	0.430	0.514

4. ANALYSIS AND DISCUSSION OF RESULTS

4.1. Analysis of Heat Transfer Results

The heat-transfer results obtained for different types of swirlers located at the tube entrance may be expressed by the following dimensionless relation:

$$Nu/Nu_o = 1 + a + b \tag{7}$$

where a = constant depending on the hub ratio and swirl angle.
b = constant depending on the enlargement ratio and swirl number; in this work, the swirl number = 0.743 tan θ.
The values of the constants (a) and (b) were obtained from the plotting of Nu/Nu_o against the swirl angle and for different enlargement ratios, Fig 5. Thus, the heat-transfer correlation for flow

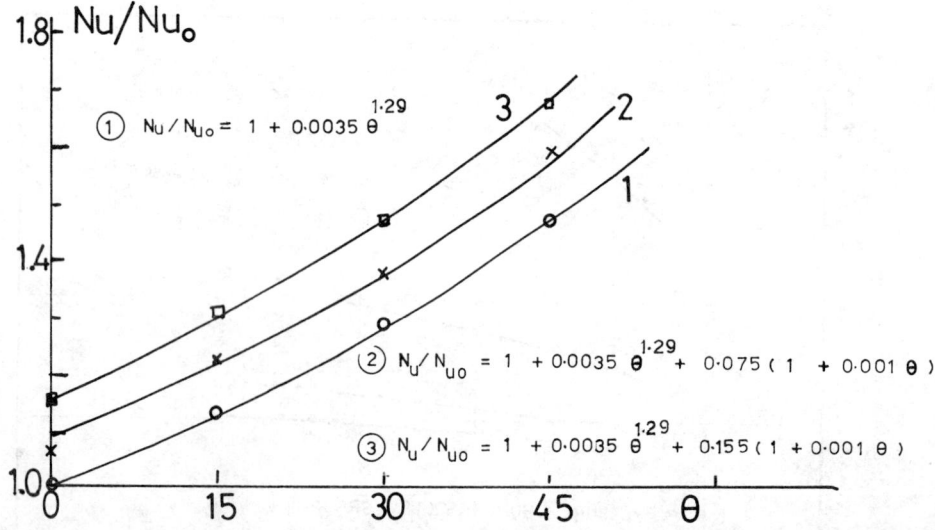

Fig.5: Relation between Nusselt number ratio and swirl angle

inside tubes with inserted vane swirlers(of different enlargement ratios and different vane angles) becomes:

$$Nu = 0.0193 \, Re^{0.8} \cdot f \tag{8}$$

where
$$f = 1 + 0.0035 \, \theta^{1.29} + 0.51 \cdot (1 - r_2/R)(1 + 0.001 \, \theta) \tag{9}$$

This formula is valid for $Re = 15000$ to 65000.

4.2. Effect of Turbulence on Thermal Entry Length

The thermal entry length is defined as the tube length at which the flow becomes fully developed, and the local heat-transfer coefficient attains its asymptotic value. This length was obtained from the variation of the local heat-transfer coefficient with tube length. The variation of the thermal entry length with Re for different swirl angles and different enlargement ratios were plotted, and could be represented by the following equation:

$$L_{th}/D = a_1 \cdot Re^{a_2} \tag{10}$$

where a_1 and a_2 are constants which depend on the swirl angle and the enlargement ratio. In table 4, the values of a_1 and a_2 are given for all experiments.

Table 4: Values of a_1 and a_2 in equation (10)

Group No	I				II				III			
angle, θ,°	0	15	30	45	0	15	30	45	0	15	30	45
a_1	0.86	.42	.23	.15	.40	.26	.18	.14	.32	.22	.16	.14
a_2	.37	.45	.52	.58	.44	.50	.55	.59	.47	.52	.57	.61

4.3. Relation between Heat Transfer and Pressure Drop

The relation between the augmentation of heat transfer, represented by the ratio Nu/Nu_o, and the pressure drop ratio $\Delta P/\Delta P_o$ for the different groups of experiments is plotted in figures 6a, b and c. The comparison between the effects of swirl angle and sudden tube enlargement on the heat transfer is shown in Fig.7. From these figures it can be seen that the increase in the swirl angle increased the heat transfer by a significant amount. However, the sudden enlargement of the tube had smaller influence on heat transfer. On the other hand, and as noticed from figures 6a, b and c, the sudden tube enlargement resulted in larger pressure drops than those observed using vane swirlers only.

4.4. Comparison with Previous Work

The results of the present work are compared with those of Kovalnogov[2], Migay[7] and ElMahallawy[9]. The comparison is shown in Fig 8; it illustrates the augmentation of heat transfer with the increase in the swirl angle. The present results are in reasonable agreement with those of Migay, and differ by less than 8% from those of ElMahallawy.

5. SUMMARY AND CONCLUSIONS

In this work, the heat transfer characteristics of an air stream flowing in a tube and heated with uniform heat flux under

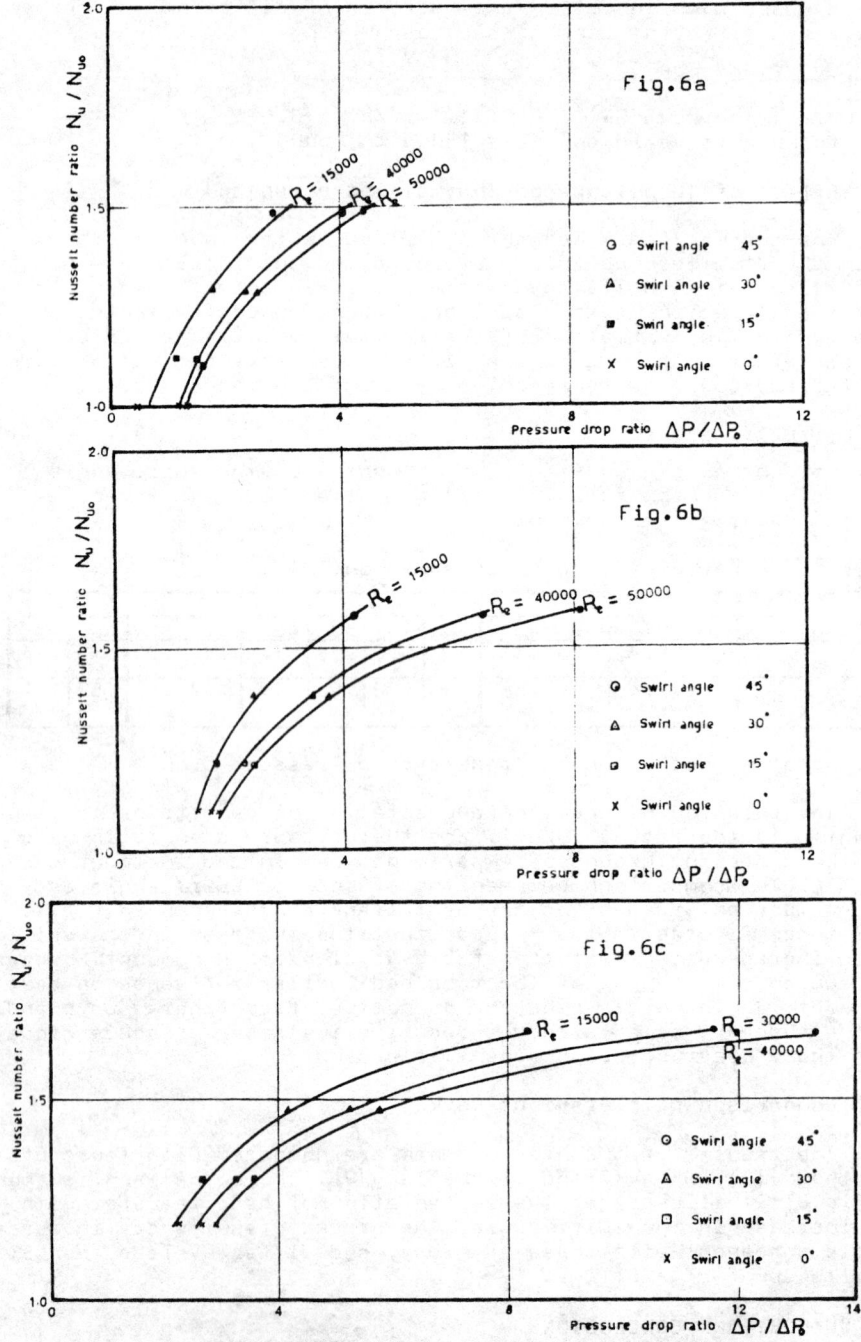

FIG.6 : RELATION BETWEEN NUSSELT NUMBER RATIO AND PRESSURE DROP RATIO FOR DIFFERENT R_e & DIFFERENT SWIRL ANGLES

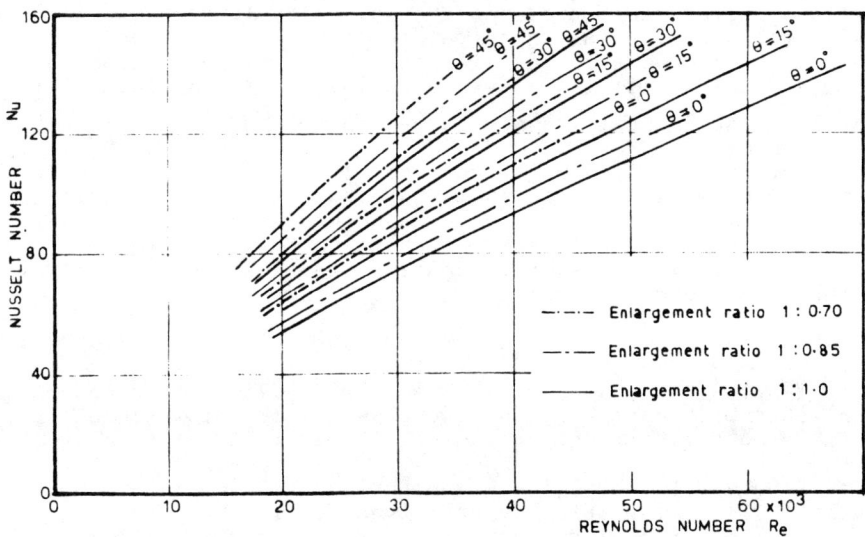

FIG. 7: COMPARISON BETWEEN THE EFFECTS OF VANE SWIRLER ANGLES AND ENLARGEMENT RATIOS

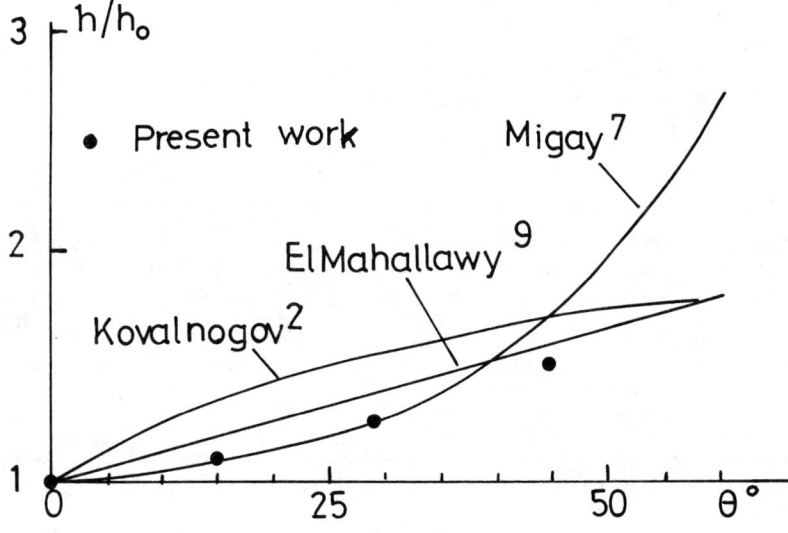

Fig. 8: Effect of Swirl on Wall Heat Transfer

swirling flow conditions were investigated. The swirling flow was generated by the insertion of a vane swirler with and without sudden tube enlargement at the main-tube inlet. The variations of Nusselt number,pressure drop and thermal entry length with Reynolds number were investigated. A general relation that related the augmentation of heat transfer with the vane angle and enlargement ratio was obtained. It is,therefore,worthwhile to conclude that:

1. The increase in heat transfer depended on the vane angle and enlargement ratio used.
2. The increase of vane angle increased the heat transfer appreciably. However,its effect on the pressure drop was not great, except at large angles and high Reynolds numbers.
3. The effect of sudden tube enlargement on heat transfer was relatively small. However,it caused a large increase in the value of the pressure drop.

REFERENCES

1. Abdel-Salam,M.S.,Hilal,M.M.,Khalil,E.E. 1976. Forced Convection in tubes with partially opened valves at tube inlet. Bulletin of the Faculty of Engineering,Cairo University,paper 15.

2. Kovalnogov,A.F.,Shchukinv,K.and Gortyshev,Yu.F. 1974. The effect of local flow swirl on heat transfer in the initial section on a cylindrical pipe. Heat Transfer,Soviet Research.Vol.6,pp. 109-114.

3. Raw , K.V .and Jyotirmoy,D. 1978. A note on turbulent swirling flows . AIAA Journal, Vol. 16,pp. 409-411.

4. Hay,N.and West,P.D. 1975. Heat transfer in free swirling flow in a pipe. J.Heat Transfer, Vol. 97,pp. 411-416.

5. Huang,F. and Tsou,F.K. 1979. Friction and heat transfer in turbulent free swirling flow in pipes. ASME 18th National heat transfer conference,paper 79-HT-39.

6. Mathur,M.L.and Maccallum,N.R.L.1967. Swirling air jets issuing from vane swirlers. Part I. Free jets . J.Inst.of Fuel, Vol. 40,pp. 214-224.

7. Migay,V.K.and Golubevl.K.1970. Friction and heat transfer in turbulent swirl flow with a variable swirl generator in a pipe. Heat Transfer-Soviet Research, Vol. 2,pp. 68-73.

8. Smithberg.E.and Landis,F.1974. Friction and forced convection heat transfer characteristics in tubes with twisted tape swirl generators. J.Heat Transfer, Vol. 86,pp. 39-49.

9. ElMahallawy,F.M.1978. Effect of the air swirl on the heat transfer by convection in furnaces. in Flow,Mixing and Heat transfer in furnaces,Edt.Khalil,HMT.2,Pergamon Press,pp. 169-179.

Spirally Fluted Tubing for Enhanced Heat Transfer

J.S. YAMPOLSKY
General Atomic Company
San Diego, California 92138, USA

1. INTRODUCTION

   A type of spirally fluted tubing is under development at General Atomic Company. The results to date indicate three advantages for this tubing concept:

   1. The fabrication technique of rolling flutes on strip and subsequently spiralling and simultaneously welding the strip to form tubing results in low fabrication costs, approximately equal to those of commercially welded tubing.

   2. The heat transfer coefficient is increased without a concomitant increase of the friction coefficient on the inside of the tube. The physical process is that in single-phase axial flow of water, the helical flutes continuously induce rotation on the flow both within and without the tube as a result of the effect of curvature. This rotation enhances turbulent exchange both on the inside and outside of the tube in the immediate vicinity of the wall. When heat is transferred from a high-temperature fluid external to the tube through the tube wall to a cold fluid within the tube, the density gradients are such as to enhance turbulent exchange in both flows. The improvement in the heat transfer coefficient results from the enhancement of the turbulent exchange.

      An additional effect from rotation of the fluid inside of the tube has been observed for a helix angle of 30°. The friction coefficient does not increase with the increased heat transfer; in fact, it is somewhat lower than that of a smooth round tube at equivalent Reynolds numbers.

      The combination of these two effects results in the achievement of values considerably in excess of one for the ratio of the turbulent exchange coefficients of heat and momentum (the reciprocal of the turbulent Prandtl number). This ratio is equivalent to the ratio of twice the Stanton number to the friction coefficient multiplied by the Prandtl number raised to the two-thirds power.

Work supported by the U.S. Department of Energy under Contract DE-AC-22-79ET15210

3. An increase in condensation heat transfer on the outside of the tube is achieved. The reason is that in a vertical orientation with fluid condensing on the outside of the helically fluted tube, the flutes provide a channel for draining the condensed fluid. The surface tension forces draw the condensate film from the crests of the flutes into the troughs. This results in the major portion of the crest having a very thin film, which greatly reduces the resistance to heat flow through the crest area. The result of providing the drainage in the troughs and decreasing the resistance to heat flow over the crests is a substantial improvement in heat transfer performance over conventional smooth tubes. This effect, which was first proposed by Gregorig [1], is not novel to this tubing concept. In fact it is the basis for a number of different proposed enhanced tubes, some of which have straight flutes and some spiralled. However, the combination of this enhancement on the outside of the tube in condensation with the enhancement on the inside of the tube in single-phase water flow can lead to significant improvements in heat transfer performance.

2. HEAT TRANSFER AND FRICTION

2.1. Performance Measurements of Helically Fluted Tubes in Single-Phase Water Flow

The heat transfer performance on the inside of helically fluted tubes as indicated by the Nusselt modulus, $Nu/Pr^{0.4}$, as a function of the Reynolds number is shown in Fig. 1. The Nusselt number is defined as $Nu = hd_{hyd}/k$, where h is the conductance coefficient (Btu/hr-ft^2-°F), d_{hyd} is the hydraulic diameter (ft) and is equal to 4 x cross-sectional area (ft^2)/perimeter (ft),

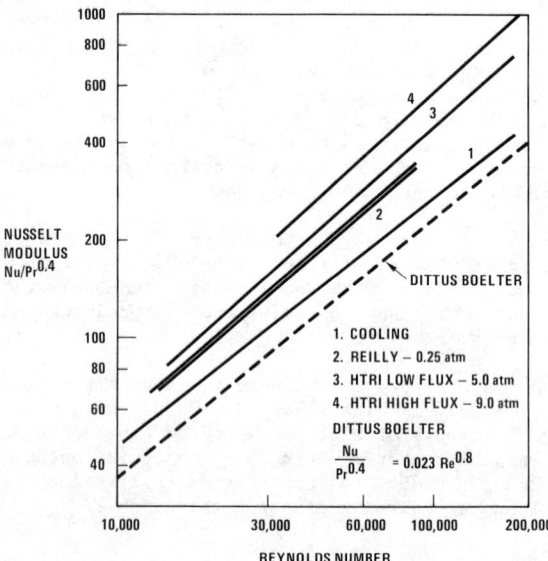

Fig. 1. Enhancement of 30° Helix for Varied Heat Flux (does not include area extension)

and k is the thermal conductivity of the water (Btu/hr-°F-ft). The Reynolds number is defined as vd_{hyd}/ν, where v is the mean axial velocity (ft/sec) and ν is the kinematic viscosity (ft^2/sec). The Prandtl number is Pr = $c_p\mu/k$, where c_p is the specific heat at constant pressure and μ is the absolute viscosity. The Dittus Boelter correlation, $Nu/Pr^{0.4} = 0.023\ Re^{0.8}$, for a smooth tube is shown in comparison.

The ratio of the ordinates for the four curves shown in Fig. 1 to that of the smooth tube curve indicates the amount of heat transfer enhancement achieved on the inside of the helically fluted tubes relative to a smooth tube. These curves do not include the effect of the extended area of the helically fluted tubes relative to that of a smooth tube of the same cross-sectional area. This is an additional factor of 1.33 for the tube reported here; in some of the tubes under development the extended area is 1.6 greater than the area of a smooth tube of equal cross-sectional area. Curves 1, 3, and 4 are the results from tests conducted by Heat Transfer Research, Inc. (HTRI), on a 1-in.-o.d. tube that had flutes with a 30° helix angle and was approximately 7 ft long. (The helix angle is defined as the angle the helix makes with the tube axis.) Curve 1 is for cooling; that is, the direction of heat flow is radially outward. Hot water flowing on the inside of the tube was cooled by colder water flowing on the outside. Although enhancement is achieved in this mode of operation, it is less than in the other curves since the density gradient is stabilizing.

Curves 3 and 4, however, show significant enhancement. These data were obtained with the direction of the heat flow radially inward, which leads to density gradients that are destabilizing and hence increases turbulent exchange. Steam was condensed on the outside of the tube by colder water flowing on the inside. The condensing pressure was approximately 9 atm for curve 4 and 5 atm for curve 3 in these tests. The steam flow was parallel to the tube axis, which was horizontal. Data from two tests conducted by D. J. Reilly at the Naval Postgraduate School [2] on a 5/8-in. 30° helix angle spirally fluted tube are shown by curve 2. In these tests steam at 0.25 atm flowed transverse to the tube axis, which was horizontal. Curves 2, 3, and 4 indicate that as the condensing pressure and hence temperature were increased relative to the tube-side coolant, heat transfer increased. The higher condensing temperatures indicate higher heat fluxes and result in larger temperature gradients on the inside of the tube.

The frictional performance of the 30° helix angle tube can be seen in Fig. 2, where the isothermal friction coefficient (Moody definition) is plotted against Reynolds number for both the HTRI tests (indicated by the line)

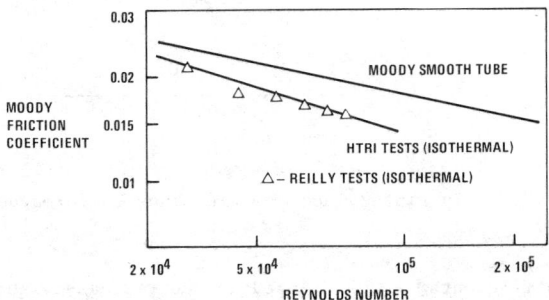

Fig. 2. Moody Friction Coefficient Versus Reynolds Number for Helically Fluted Tube

and the Reilly tests (indicated by the triangles). The hydraulic diameter is used in the calculation for the frictional coefficient.

As a decrease in the frictional coefficient in a tube in which enhancement of the heat transfer is achieved was surprising, a series of additional tests were undertaken. These experiments were designed to eliminate the uncertainties in the measurement of frictional coefficients in test rigs that were designed for heat transfer measurements. The preliminary data from the tests underway at the present time confirm the frictional coefficients depicted in Fig. 2 over a Reynolds number range between 40,000 and 80,000 with air as the fluid. Tests with water as the working fluid are scheduled which will also extend the Reynolds number range of the data. The cross-sectional area of the tubes varied from 0.127 in.^{2} in the Reilly tests to 0.885 in.^{2} in the tests underway at present. The cross-sectional area of the tube in the HTRI tests was 0.576 in.^{2}.

2.2. Combined Heat Transfer and Friction

While the analogy between momentum and heat transfer suggested by Reynolds and modified by Colburn is valid only under rather restrictive conditions, it is almost by definition an index of the degree of heat transfer enhancement achieved relative to the frictional penalty. It is directly related to the ratio of the exchange coefficients of heat to momentum. Therefore, it is of interest to examine the heat transfer and friction data in the context of the Colburn analogy. The Colburn analogy is defined as $2j/f^*$ and is equal to unity for a straight smooth tube, where $j = N_S Pr^{2/3}$. It is shown in Fig. 3 for the high flux case and in Fig. 4 for the low flux case for the 30° spiral tube.

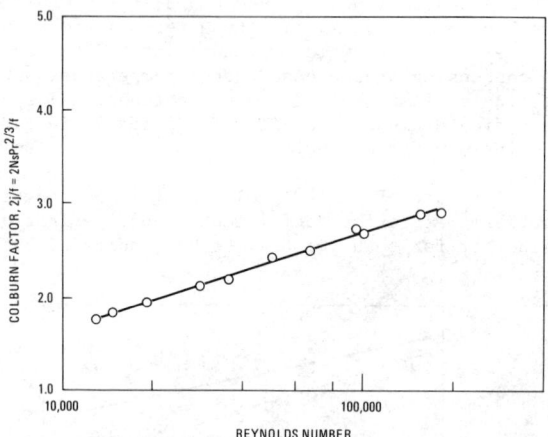

Fig. 3. 30° Spirally Fluted Tube Low Flux Tube Performance Factor

---

*The frictional coefficients, λ and f, used in the text and figures are respectively based on the Moody and Fanning definitions, which differ by a factor of 4. The Moody coefficient is larger than the Fanning coefficient by this factor.

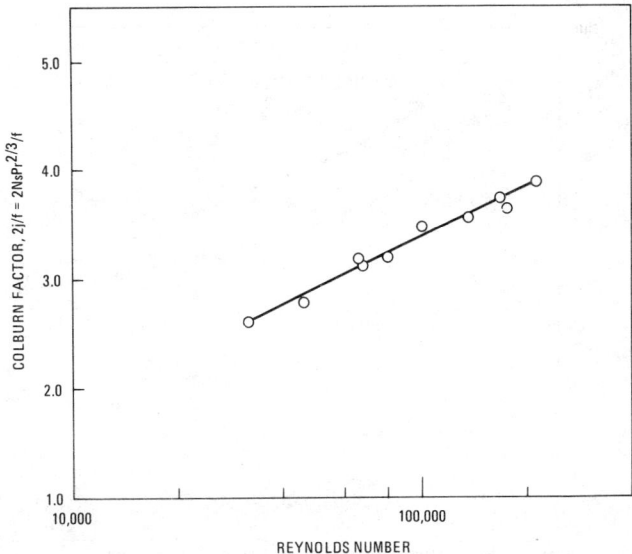

Fig. 4. 30° Spirally Fluted Tube High Flux Tube Performance Factor

There are two surprises. First, and least significant, is the positive slope that results from the negative slope of the friction coefficient/Reynolds number characteristics of Fig. 1, which is greater than that of a smooth tube. More important is the level of the Colburn factor, $2j/f$, which achieves a value of 2.3 at low flux and a value of 2.75 at high flux, when a value of one is the case for a smooth tube and less than one for most other enhancement techniques.

2.3. Condensation of Helically Fluted Tubes in a Vertical Configuration

Two tubes with different helical angles (one a 30° helix and the other a 45° helix) were tested at Oak Ridge National Laboratory with Refrigerant 11 condensing on the outside of the tube, which was cooled by water on the inside. The test data depicting the performance of these tubes are shown in Figs. 5 and 6. In Fig. 5, the condensing film coefficient is the ordinate and the heat flux is the abscissa. The increase in performance over the smooth tube decreases with heat flux but remains quite substantial. Figure 6 shows the same data with the heat load as the ordinate and the condensing temperature difference as the abscissa. Again, the gain over the smooth tube is larger at the lower heat loads. The 45° helix appears to have better performance at the lower heat flux. It should be pointed out that the 45° tube had fewer flutes and hence troughs, so the troughs would fill sooner. The liquid capacity of each of the flutes in both of these tubes was approximately the same. Therefore, a larger number of flutes would drain more liquid. Another fact that should be kept in mind is that the latent heat of Refrigerant 11 is quite low, so a considerably larger amount of liquid must be drained from the tubes than would be the case for steam or ammonia.

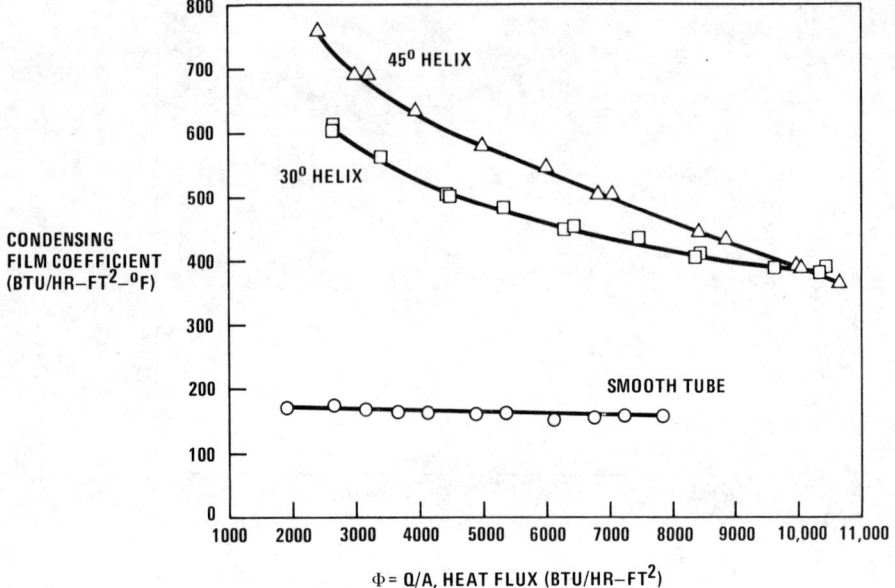

Fig. 5. Condensing Film Coefficient Versus Heat Flux for Refrigerant 11

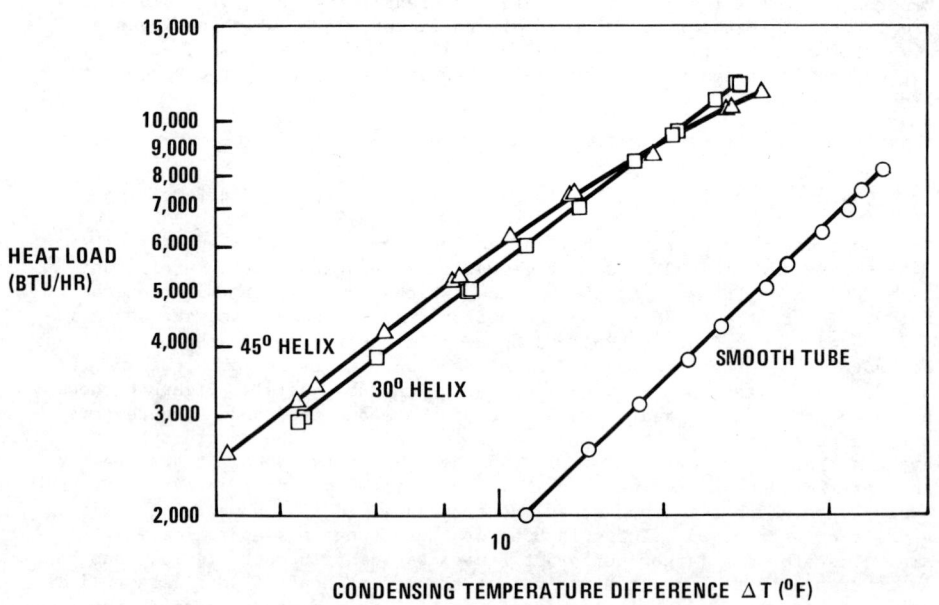

Fig. 6. Heat Load Versus Condensing Temperature Difference for Refrigerant 11

3. DISCUSSION

The heat transfer performance enhancement increased with heat flux and disappeared when the direction of heat flow was reversed, supporting the contention that the density gradient combined with the rotational flow induced by the flutes resulted in an instability in the vicinity of the flutes. The result of this instability is to increase the turbulent exchange coefficient for heat relative to that of momentum, similar to the effect reported by Mizushina, et al. [3] for a rectangular open channel where gravity supplied the body force. However, their experiments showed that both exchange coefficients individually increased relative to the neutral case (no density gradient), but that the increase in the exchange coefficient for heat was much greater. In the stable case, both decreased relative to the neutral case. The effect was limited to a relatively narrow range of Richardson numbers.

In the measurements reported herein, the reduction in the frictional coefficient is therefore surprising, although it has also been shown by Reilly for tubing of the same helical angle but a smaller diameter. Furthermore, the measured friction coefficient is very nearly independent of heat flow and its direction; it is apparently not affected by the density gradient imposed by the heat flux through the wall. This implies that while the momentum exchange coefficient may increase with the density gradient created by the heat flux radially inward through the wall, the overall momentum loss does not increase.

An explanation for this may be offered by examining an experiment on an axially rotating tube. White [4] showed by means of flow visualization that rotation suppressed the turbulence which existed over most of the cross section with no rotation at the same Reynolds number. Furthermore, his measurements of the pressure drop showed a reduction of pressure loss with rotation, which at the highest rotational speeds of his tests amounted to a reduction of the frictional coefficient to 60% of its value for the same tube without rotation. In this case, the fluid in the rotating tube is in solid body rotation, which is stable. Evidence that bears directly on this, cited by Bradshaw, is that the effect of surface curvature on the spreading of a wall jet implies that the Reynolds stresses are decreased on the concave surface and increased on the convex surface. Bradshaw [5] showed a photograph of a trailing vortex behind a C-47 aircraft, where it can be seen that the inner core of the vortex is laminar, being stabilized by streamline curvature.

It is now possible to qualitatively draw an image of the physics of the flow in the helically fluted tube. The axial pressure gradient along the tube causes a flow through the tube and in the flutes. The helical angle of the flutes induces rotation of the flow within the flutes and of the bulk flow as a result of the curvature of the flutes. The core flow is primarily in solid body rotation, has no strain, and is stable. In the region between the core flow and the flute flow, there is an interchange of angular momentum from the individual flutes to the core flow, resulting in a decrease of the angular momentum in the flutes. This is the cause of instability, since the decrease of the peripheral velocity is destabilizing. The instability increases with radially inward heat flow through the wall and decreases with the direction of heat flow outward. Instability enhances the turbulent exchange near the wall, leading to improved heat transfer since most of the resistance to heat flow is in the laminar sublayer. The rotation of the core flow reduces the axial momentum loss as in White's experiment, so the friction coefficient does not increase with the increased heat transfer coefficient.

REFERENCES

1. Gregorig, R. 1954. Zeit, Agnew. Math. Phys., Vol. 5, pp. 36-49.

2. Reilly, D. J. 1978. An experimental investigation of enhanced heat transfer on horizontal condenser tubes. M. S. Thesis, Naval Postgraduate School.

3. Mizushina, T., et al. 1979. Buoyancy effect on any diffusivities in thermally stratified flow in an open channel. Proceedings of the Sixth International Heat Transfer Conference, Vol. 1, pp. 91-96 (Paper MC16).

4. White, A. 1964. Flow of a fluid in an axially rotating pipe. J. Mech. Eng. Sci., Vol. 6, pp. 47-52.

5. Bradshaw, P. 1973. Effects of streamline curvature on turbulent flow. AGARDo-graph No. 169.

Heat Transfer Enhancement by Static Mixers for Very Viscous Fluids

TH.H. VAN DER MEER and C.J. HOOGENDOORN
Delft University of Technology
Department of Applied Physics
Lorentzweg 1, Delft, The Netherlands

ABSTRACT

Experimental results are given in this paper for heat transfer at constant heat flux and constant temperature of a straight pipe filled with Sulzer mixing elements. A semi-empirical model is presented. Heat transfer enhancement by a factor of 5 could be gained with a pipe filled with elements compared to an empty pipe, mainly due to the short thermal entrance length.

NOMENCLATURE

a	thermal diffusivity, m^2/s	α	coefficient of heat transfer, $W/m^2\ ^\circ C$
C_p	specific heat, $J/kg\ ^\circ C$		
D	diameter of the pipe, m	η_b	dynamic bulk viscosity, kg/ms
d_h	hydraulic diameter, m	η_w	dynamic viscosity of fluid at the wall kg/ms
F	surface of the pipewall, m^2		
L	pipelength, m	λ	coefficient of heat conduction, $W/m\ ^\circ C$
L_m	length of a mixing section, m		
L_r	refreshment length, m	ν	kinematic viscosity, m^2/s
Δp	pressure drop, N/m^2	ρ	density kg/m^3
R	electric resistance of the pipe, Ω	τ_w	wall shear stress
T_b	bulk temperature, $^\circ C$	$Gz = Re\ Pr\ \frac{D}{x}$, Graez number
T_i	inlet temperature, $^\circ C$	$Nu = \frac{\alpha D}{\lambda}$, Nusselt number
T_o	outlet temperature, $^\circ C$	$Re = \frac{v_m D}{\nu}$, Reynolds number
T_w	wall temperature, $^\circ C$	
T_{wg}	average wall temperature, $^\circ C$	$Pe = Re\ Pr$, Péclet number
ΔT	$T_o - T_i$, $^\circ C$	$Pr = \frac{\nu}{a}$, Prandtl number
v	axial velocity, m/s	
v_m	averaged axial velocity, m/s	
V	voltage, V	
x	axial coordinate	
y	normal coordinate	

1. INTRODUCTION

For heating or cooling of very viscous liquids often compact heat exchangers are needed. Especially when thermal degradation of the fluid requires a short residence time in the heat exchanger. Also when in view of a chemical reaction the residence time distribution should be narrow, static mixers are beneficial. However not much experimental work has been done to describe the heat transfer in static mixers.

A static mixer acts as an in line mixer for fluids by dividing the flow in separated flows and then combining these with other parts, dividing once more, etc. Two commercial types of mixers are well known: the Kenics Static Mixer and the Sulzer Static Mixer. The flow performances of these mixers are described in references [1] and [2]. Some of our experimental results on the heat transfer of a Sulzer Static Mixer type SMV are reviewed in this paper and compared to the heat transfer characteristics of the Koch [3] and Kenics Static Mixer, and the new type of Sulzer Static Mixer (SMXL)[4,5].

2. DESCRIPTION OF MIXER

The static mixer elements of the SMV are composed of corrugated sheets of metal which are arranged in layers on top of each other forming open crossing ducts (see Fig. 1). The fluid flow is divided in flows through the open intersecting ducts which make an angle of $45°$ with the axis of the pipe wherein the elements are inserted. After one element the fluid is united and divided again. The "overall" velocity profile in the pipe with inserts is that of a plug flow.

Next to the good mixing features in laminar flow of very viscous fluids the mixing elements cause an increase of the velocity gradient at the wall and a refreshment of the liquid at the wall by liquid from the bulk, both improving the heat transfer.

Fig. 1. The working principle of the SMV-inserts

HEAT TRANSFER ENHANCEMENT BY STATIC MIXERS

3. THE SEMI-EMPIRICAL MODEL

In a semi-empirical model, discussed in [6], we assume that the hydrodynamic entrance length is very short compared to the thermal entrance length. This is true for our type of flow with high Prandtl numbers ($Pr > 1000$) and low Reynolds numbers ($Re < 3$). Then the relations for heat transfer in the thermal entrance of a straight circular pipe (Levêque solutions) can be used. Heat transfer coefficients for constant heat flux at the wall and constant wall temperature are [7,8]:

$$\alpha = C \cdot \lambda \frac{\left(\left(\frac{\partial v}{\partial y}\right)_{y=0}\right)^{1/3}}{ax} \qquad \begin{array}{l} C = 0.65 \; (q_w = \text{constant}) \\ C = 0.54 \; (T_w = \text{constant}) \end{array} \qquad (1)$$

with the velocity gradient at the wall from a Poiseuille flow

$$\left(\frac{\partial v}{\partial y}\right)_{y=0} = \frac{\rho \tau_w}{\nu} = \frac{8 v_m}{D} \qquad (2)$$

For a pipe with SMV - inserts the hydraulic diameter has to be used in (2) leading to a higher velocity gradient at the wall and to

$$\alpha = C\lambda \left(\frac{D}{d_h}\right)^{1/3} \left(\frac{8 v_m}{aDx}\right)^{1/3} \qquad (3)$$

$$Nu_x = 2C \left(\frac{D}{d_h}\right)^{1/3} \left(Re \; Pr \; \frac{D}{x}\right) \qquad (4)$$

The hydraulic diameter was determined by measuring the flow volume and the wetted surface of the mixing elements and pipe.

Equation (4) now gives the local Nusselt number in the thermal entrance region of a pipe filled with elements. Frequently the liquid at the wall is refreshed by liquid from the bulk. The model assumes that all liquid along the wall is refreshed after a length L_r (refreshment length).
The averaged Nusselt over this length is:

$$\overline{Nu}_{L_r} = C_1 \left(\frac{D^2}{d_h L_r}\right)^{1/3} (Re \; Pr)^{1/3} \qquad (5)$$

with $C_1 = 1.95$ for constant heat flux and $C_1 = 1.62$ for constant wall temperature. Due to the repeated refreshment the Nusselt number is independent of the pipelength for $L > L_r$:

$$Nu_L = \overline{Nu}_{L_r} = C_1 \left(\frac{D^2}{d_h L_r}\right)^{1/3} Pe^{1/3} \qquad (6)$$

Clearly the refreshment length will depend on the geometry of the mixing elements. For one geometry type of mixing elements we will have

$$\left(\frac{L_r}{D}\right) \sim \left(\frac{d_h}{D}\right)^n \qquad (7)$$

Dimensional analysis gives $n = 1$.

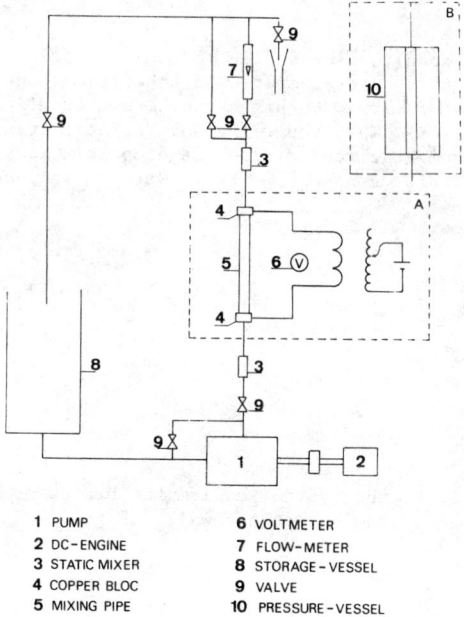

```
1 PUMP              6 VOLTMETER
2 DC-ENGINE         7 FLOW-METER
3 STATIC MIXER      8 STORAGE-VESSEL
4 COPPER BLOC       9 VALVE
5 MIXING PIPE      10 PRESSURE-VESSEL
```

Fig. 2. The experimental set-up

To reduce the pressure drop over a pipe filled with mixing elements one can use free parts of pipe between the mixing sections. A model for the heat transfer in this case also is discussed in [6].

4. THE EXPERIMENTAL SET UP

4.1 Constant heat flux.

Fig. 2 shows the experimental set up for measuring local heat transfer coefficients at constant heat flux. A mono-pump driven by a d-c engine with a continuously variable speed pumps the liquid from a storage vessel through the electrically heated mixing pipe. The liquid (silicone oil) in the storage vessel is thermostated at about $20°C$. Before and after the mixing pipe the liquid flows through separated mixing elements to eliminate possible temperature differences. The liquid temperatures at the in- and outlet cross section are then measured each by three thermocouples.

The stainless steel mixing pipe has an inner diameter equal to the diameter of the mixing elements (37.2 mm) a pipewall thickness of 1 mm and a length of 0.8 m. At both ends of the steel pipe copper blocks are soldered which are connected to a transformer. Electric energy is dissipated in the pipewall causing a constant heat flux from the pipewall to the liquid, which is precisely known when the pipe is thermally isolated from the environment. Measurements with plastic mixing elements show that radial conduction through the material of the elements can be neglected. Pipewall temperatures are measured along the circumference as well as in axial direction by thermocouples. When the steady state is reached the heat balance over the mixing pipe results in:

HEAT TRANSFER ENHANCEMENT BY STATIC MIXERS

$$Nu = \frac{\alpha D}{\lambda} = \frac{V^2}{R} \frac{D}{\lambda F(T_w - T_b)} \quad (8)$$

or

$$Nu = (\phi_v \rho c_p \Delta T - \phi_v \Delta p) \frac{D}{\lambda F(T_w - T_b)} \quad (9)$$

With T_b (the bulktemperature) equal to

$$T_b = T_i + \frac{x}{L}(T_o - T_i) \quad (10)$$

This method allows us to measure local Nu-values around the circumference and in the axial direction.

4.2 Constant wall temperature.

Figure 2(b) shows the experimental set up for measurements of overall heat transfer ceofficients at constant wall temperature.

The mixing pipe is surrounded by a pressure vessel which is supplied with steam. This steam is condensating on the mixing pipewall causing a constant wall temperature. Using the logarithmic temperature difference the steady state heat transfer can be described by:

$$Nu = \frac{\phi_v}{\pi \lambda L}(\rho c_p - \frac{\Delta p}{T_o - T_i}) \ln(\frac{T_{wg} - T_i}{T_{wg} - T_o}) \quad (11)$$

4.3 Experimental conditions

Two mixing elements were tested with hydraulic diameters d_h = 5.2 mm and d_h = 7.7 mm.

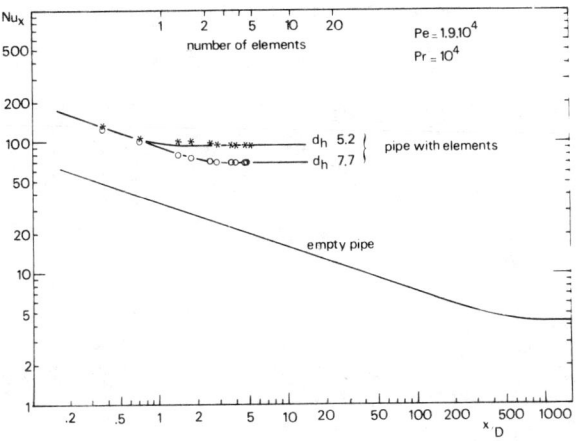

Fig. 3. Local heat transfer, q_w = constant

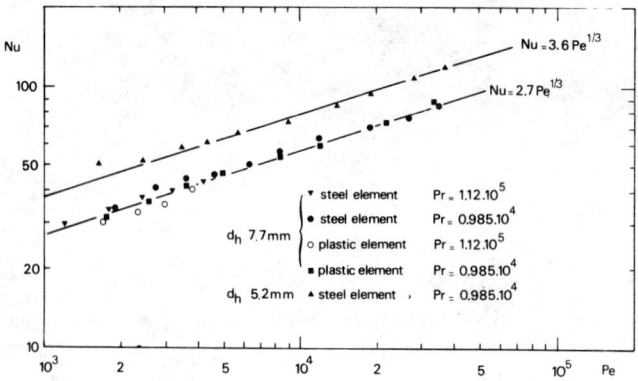

Fig. 4. Nusselt thermally developed, q_w = constant

Silicone oils were used as fluids ($0.6 \cdot 10^{-3}$ m^2/s $< \nu < 1.1 \cdot 10^{-2}$ m^2/s). Thus the Prandtl numbers were $0.6 \cdot 10^4 < Pr < 1.1 \cdot 10^5$. The range of Reynolds numbers tested was $10^{-2} < Re < 3$.

5. RESULTS AND DISCUSSION.

Fig. 3 gives the local Nusselt number at constant heat flux for the two mixing element types as a function of pipelength at constant Pé. In the thermal entrance region the Nusselt number is given by:

$$Nu_x = 3.4 \, (Pe \, \frac{D}{x})^{1/3} \qquad (12)$$

The thermal entrance length appears to be for the two mixing elements with d_h = 5.2 mm and d_h = 7.7 mm respectively about 1D and 2D, which is very much shorter than the thermal entrance length of an empty pipe for this Pe-number. Fig. 3 also shows the heat transfer enhancement relative to the empty pipe for a specific case.

For the developed region ($x > L_r$) the Nusselt number is independent of x/D. We form the Nusselt-Péclet - relationship for this region as shown in Fig. 4. With equation (6) the value of L_r can be calculated resulting in:

$$d_h = 5.2 \text{ mm} \quad Nu = 3.6 \, Pe^{1/3} \, : \, L_r = 44 \text{ mm} \qquad (13)$$

$$d_h = 7.7 \text{ mm} \quad Nu = 2.7 \, Pe^{1/3} \, : \, L_r = 68 \text{ mm} \qquad (14)$$

For both elements we have found $L_r/d_h \approx 8.5$, which leads to n = 1 in eqn. (7). We observe that L_r is almost equal to the thermal entrance length as found above. In fact it is the same mechanism of the refreshment of liquid at the wall which takes one refreshment length.

Both elements tested had about 4 corrugations over their refreshment length. Because of the asymmetrical shape of the mixing elements there are circumferential differences in the local heat transfer coefficient.

HEAT TRANSFER ENHANCEMENT BY STATIC MIXERS

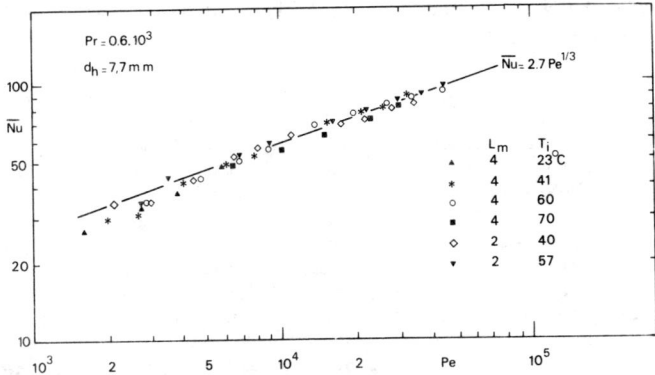

Fig. 5. Overall Nusselt number, T_w = constant

For the mixing elements with d_h = 5.2 mm the averaged difference between maximal and minimal heat transfer was about 20%. For the second type of mixing element this was less than 5%. From Fig. 4 we also can see that there is no difference in heat transfer for plastic and stainless steel mixing elements in the range of our experiments.

Some results of measurements at constant wall temperature are given in Fig. 5. It shows that the inlet temperature does not influence the Nusselt number which means that there is no significant η_w/η_b dependence. We explain this by the good refreshment at the wall which prevents the development of a boundary layer. From Fig. 5 also can be concluded that the length of the mixing elements L_m (40 mm or 20 mm) hardly has an influence.

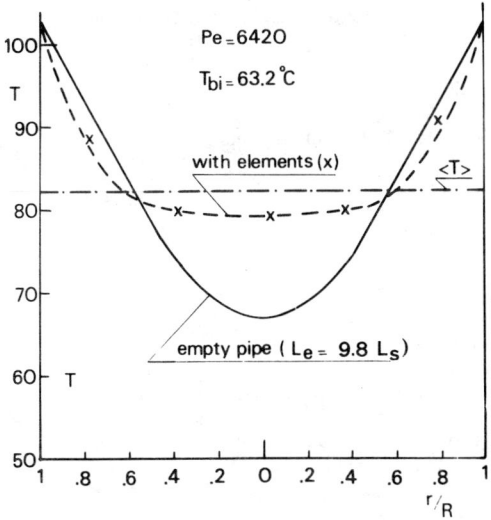

Fig. 6. Radial temperature distributions

For the averaged Nusselt-number at constant wall temperature for a 80 cm long pipe (L/D = 21.5) filled with mixing elements we find:

$$Nu = 2.7 \, Pe^{1/3} \qquad (15)$$

This is 8% higher than would be expected from the model with L_r from eqn. (14) and taking into account a thermal entrance length similar to that of the measurements at constant heat flux.

At constant wall temperature some measurements were done of the radial temperature distribution at the outlet of the mixing pipe. The results are given in Fig. 6 and compared to a calculated temperature profile of a 9.8 times longer empty pipe. For both cases the total heat exchange and the inlet temperature (63.2°C) are the same. The figure clearly shows the good mixing performance of the static mixer giving a more uniform temperature profile in the fluid.

For small Pe (e.q at low velocities), if $Gz = Pe^{-1} L_r/D > 0.1$, the refreshment of the liquid at the wall is no longer effective and the conduction in the radial direction over full D/2 will dominate before x equals L_r. This allows us to use the solutions for thermally fully developed flows. Insertion of mixing elements is still beneficial because it causes a plug flow instead of a parabolic flow in the pipe. Also the heat conduction in radial direction is enhanced by the stainless steel inserts. However heat conduction through the mixing element material did not effect heat transfer at high Peclet numbers, as measurements with plastic mixing elements proved, for Pe → 0 it may have a noticeable influence. The volume percentage of steal in the pipe is 10% (d_h = 7.7 mm) and 16% (d_h = 5.2 mm). Measurements on heat conduction in paraffins (λ = 0.13 W/mK) containing 10% of aluminium (λ = 230 W/mK) showed an effective coefficient of heat conduction λ_{eff} = 2.2. [9]. Interpreting this result for Silicone-oil (λ = 0.157 W/mK) and stainless steal (λ = 25 W/mK) would give λ_{eff} = 0.38 meaning an improvement by a factor 2.4. Considering the Nusselt number for plug flow in the thermally fully developed region this would give for a pipe filled wity mixing elements (d_h = 7.7 mm, steel elements, Pé → 0):

$$Nu = 19.2 \quad (q = \text{constant})$$

$$Nu = 13.9 \quad (T_w = \text{constant})$$

The total enhancement compared to the empty pipe (Nu = 4.36, q = constant) then is still a factor of 4.4.

6. PRESSURE DROP CONSIDERATIONS

Heat exchangers with extended surfaces pay the penalty for the achieved enhanced heat transfer by an increased pressure drop. Evidently also the pressure drop over a pipe filled with static mixer elements will be rather high. Some rough pressure drop measurements were done on both mixing elements tested. Fig. 7 gives the result on the pressure drop as a function of flowrate. The accuracy of the measurements is about $0.1 \cdot 10^5$ N/m². From these maesurements it can be calculated that for the two types of elements tested:

$$\Delta p = 4.1 \cdot 10^9 \, L \, \phi_v \quad N/m^2 \quad (d_h = 5.2 \text{ mm}) \qquad (16)$$

$$\Delta p = 1.8 \cdot 10^9 \, L \, \phi_v \quad N/m^2 \quad (d_h = 7.7 \text{ mm}) \qquad (17)$$

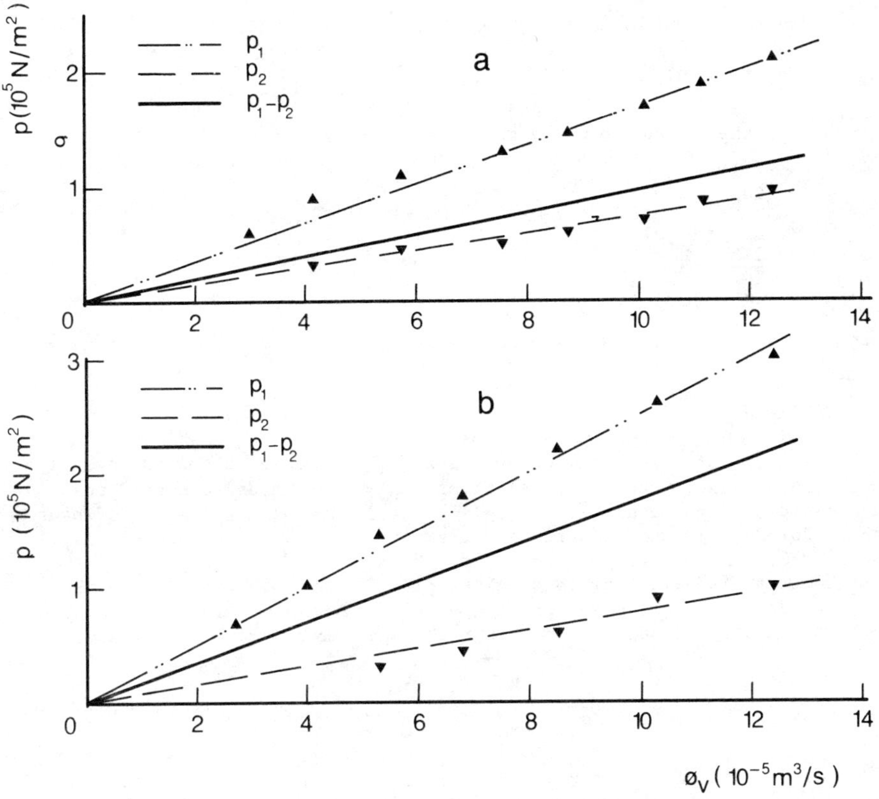

Fig. 7. Pressure drop ($p_1 - p_2$) of a SMV-packed pipe.
a) d_h = 7.7 mm., L_s = .56 m.   b) d_h = 5.2 mm., L_s = .4 m.

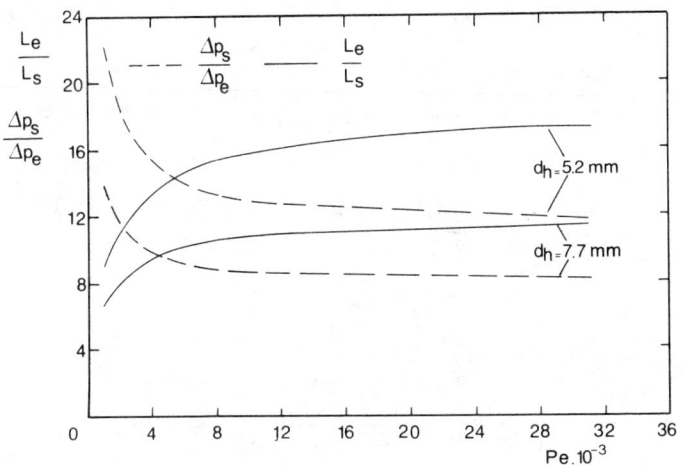

Fig. 8. Comparison empty and SMV-packed pipe with equal heat exchange

with the correlation for laminar flow

$$\Delta p = 4f \cdot \tfrac{1}{2} \rho v_m^2 \cdot \frac{L}{d_h} \tag{18}$$

this gives with the Re-number based on d_h:

$$4f = \frac{251}{Re_h} \quad (d_h = 5.2 \text{ mm}) \tag{19}$$

$$4f = \frac{242}{Re_h} \quad (d_h = 7.7 \text{ mm}) \tag{20}$$

These results are in agreement with data given by Tauscher [2].

7. COMPARISONS

The heat transfer enhancement by inserting SMV mixing elements in a pipe is considerable. Fig. 8 shows the length of empty pipe needed compared to a SMV packed pipe of one meter (25 mixing elements) for the same amount of heat exchanged. For $Pe > 10^4$ the length of the empty pipe has to be larger than 11 meters ($d_h = 7.7$ mm) or 16.5 meters ($d_h = 5.2$ mm).

Since for Poiseulle flow in an empty pipe it is known that $4f = 64/Re$, the pressure drop increase of a pipe filled with SMV mixing elements compared to the pressure drop of an empty pipe for the same amount of heat exchange is:

$$\frac{\Delta p_s}{\Delta p_e} = 4 \cdot \left(\frac{D}{d_h}\right)^2 \cdot \frac{L_s}{L_e} \tag{21}$$

Fig. 8 gives this comparison for a mixing pipe of 1 meter. For $Pe > 10^4$ pressure drop increases by a factor of about 13 ($d_h = 5.2$ mm) and 8 ($d_h = 7.7$ mm).

If these high pressure drops are too large a disadvantage in applications it is possible to use the mixing elements with free intersections. Measurements on a pipe with half the number of mixing elements halving the pressure drop showed a decrease in heat transfer by about 33%.

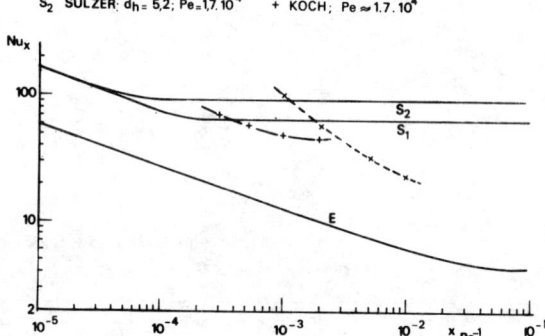

Fig. 9. Local Nu_x for q_w = constant

In Fig. 9 some results of the SMV inserts with d_h = 5.2 mm. are compared with results from Marner and Bergles [10] on the Kenics and the Koch Static Mixer. From the comparison it is clear that the heat transfer enhancement of the SMV-inserts is very high mainly due to the short thermal entrance length, which is longer for the other types of mixing elements.

Results for the average heat transfer at constant wall temperature for various types of mixing elements and a rather long mixing length (L/D = 100) are gathered in Fig. 10. For the SMV-elements the results from chapter 5 are used to extrapolate to this case. The heat transfer correlations in Fig.10 are (q_w = constant)

$$\overline{Nu} = 3.24 \, Pe^{1/3} \qquad SMV - d_h = 5.2 \text{ mm}$$
$$\overline{Nu} = 2.43 \, Pe^{1/3} \qquad SMV - d_h = 7.7 \text{ mm}$$
$$\overline{Nu} = 0.98 \, Pe^{0.38} \qquad SMXL \ [6]$$
$$\overline{Nu} = 3.65 + 3.8 \, (Pe\frac{D}{L})^{1/3} \qquad Kenics \ [11] \ (\frac{L}{D} = 100)$$
$$\overline{Nu} = (3.66^3 + 1.61^3 \, Pe\frac{D}{L})^{1/3} \qquad empty \ pipe \ (\frac{L}{D} = 100)$$

7. CONCLUSIONS

The refreshment length concept appears to give a good description of the heat transfer of a pipe containing mixing elements, which was concluded from the length of the thermal entrance region. The described model is a useful help to predict and extend the experimental results of convective heat transfer in a pipe filled with SMV-elements. A main feature of the inserts is the D/L-independance of the local Nusselt number after a short entrance region.

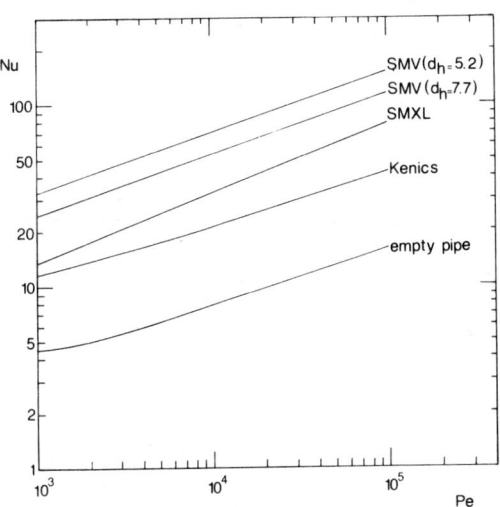

Fig. 10 $\overline{Nu}$ for L/D = 100, T_w = constant

Although the pressure drop of a pipe with SMV-inserts is considarably higher than the pressure drop of other mixers (a factor of 10 between SMV (d_h = 7.7) and SMXL), SMV-inserts are very useful for applications where a very short residence time with a narrow distribution is needed in combination with a uniform radial temperature distribution in heating up very viscous fluids that show thermal degradation.

[1] Sununu, J.H., Heat Transfer with Static Mixer Systems, Technical Report 1002, Kenics-Corp. Danvers, Mass. 1970.

[2] Tauscher, Internal Report, Sulzer Brothers Ltd., Winterthur.

[3] Koch Engineering Company, Inc. Bulletin KSM - 2

[4] Grosz- Röll, F., Wärmeübertrager für erzwungene, Laminare Strömung, Chemie-anlagen + verfahren, Heft 11/1979, p. 50 - 54

[5] Heïerle. A., Wärmeaustauscher für temperaturempfindliche, viskose Flüssigkeiten, Chemie-Technik, 9, 1980, p. 83 - 85

[6] Van der Meer, Th.H., Hoogendoorn, C.J., Heat Transfer Coefficients for viscous fluids in a static mixer, Chem. Eng. Science, 33, 1978, p. 1277

[7] Bird, R.B., Stewart, W.E. and Lightfoot, E.N., Transport Phenomena, Wiley, New York, 1960

[8] Lévêque, M.A., Ann. Mines, 13, 1928, p. 201

[9] De Jong, A.G., Hoogendoorn, C.J., Improvement of heat transport in paraffins for latent heat storage systems. Proc. TNO-symposium on thermal Storage of Solar Energy. To be published by Martinus Nijhoff Publishers B.V.

[10] Marner, W.J., Bergles, A.E., Augmentation of tubeside laminar flow heat transfer by means of twisted-tape inserts, static mixer inserts, and internally finned tubes,6th Int. Heat Transfer Conf., Toronto 1978, Vol. 2, p. 583

[11] Grace, C.D., 'Static Mixing' and Heat Transfer, Chemical and Process Engineering, Vol. 52, 1971, p. 57

Index

Absorbents, 556
Adam-Moulton predictor-corrector method, 88–89
Additive injection, 851
Adiabatic systems, flooding in, 55
Adsorption, 556, 805
Aging, in fouling, 799
Air conditioners, use of fins in, 457
Air-cooled exchangers, 357, 363–383
  (See also Coolers)
Air separation plants, 470, 495
Air-to-air regenerator, 608
Alkalinity, 853
Alloy recuperators, 508
Alternative energy sources, 571
Aluminosilicates, 509
Aluminum melting furnace, 563
Aluminum tubes, 458
Ammonia, 647
  condensation of, 13, 761
  and tube damage, 757
Anemometer measurements, 337–338
Annealing, prevention of, 650
Annular flow model, 68–70
Annulus, condensation heat transfer in, 102
Argon, drift phenomenon, 247
Argonne National Laboratory, 560
Arrhenius equation, 804
Artificial turbulence (see Turbulence, artificial)
Ash, 647, 817
Aspect ratio, of channels, 439
Augmentation, factors in, 920–927
  (See also Efficiency)
Autocorrelation function, 326, 334
Axial conduction, 485

Backflushing, 638
  and fouling, 881
Baffles:
  arrangements of, 188
  clearances for, 617
  and pressure drop, 620
  rod-baffle exchanger, 637
  and sound damping, 376
Bankoff variable density model, 139
Baroczy multiplier, 697
Barometric evaporators, 84
Baskakov formula, 539
Batch processes, 586

Batch regenerators, 640
Bayonet tube bundle, 637
B-coefficient, 112
Beds (see Fluidized beds)
Bessel function, 524
Biery heat-pipe recuperator, 642
Biological fouling, 797, 800
Biot number, 587, 919
Blasius equation, 405
Blowdown facilities, 586
Bluff bodies, 287
Boilers:
  counterflow boilers, 557
  economizers for, 456
  on tankers, 559
  thermal storage boilers, 572
  waste heat boilers, 644
Boiling:
  of argon, 233–251
  asymmetrical, 733
  boiling crisis, 242, 251, 748
  and enchanced surfaces, 175–191, 193–203
  film boiling, 750
  flow boiling, 19–34
  forced convective, 20, 143
  helium boiling, 251–262
  in horizontal tubes, 19–34
  in hot and cold leg, 733
  kettle reboilers, 7–8
  Ledinegg forms, 483
  at low temperatures, 234
  of multi-component liquids, 9–10
  of nitrogen, 233–251
  nucleate boiling, 24, 29, 197, 474–475
  in plate-fin exchangers, 473
  pool boiling, 153
  saturated flow boiling, 19, 143–156
  Stephan's equation, 240
  subcooled boiling, 749
  in tube bundles, 195
  in vertical tubes, 19–34
Boundary conditions, 435
Boundary layer effects, 160–161, 268, 287–298
Boundary surfaces, 775–792
Boussinesq equation, 161
Brazing, of exchangers, 427
Break-away of droplets, 714
Brownian diffusivity, 800
Bubbles:
  bubbly flow, 111, 749

965

Bubbles (*Cont.*):
  and flooding, 136
  in fluidized beds, 551
  in pool boiling, 242
  separation bubbles, 271
Bundles:
  flow resistance in, 299, 310
  peak heat flux in, 8
  recirculation in, 183
  (*See also* Tubes)
Bunsen equation, 762
Burnout, 242
Burrs, on fins, 443
BWR plants, 131–141
By-pass efficiency, 498–499

Calcium carbonate, 379
  (*See also* Deposits; Scaling)
Calibration curves, 226
Carbide, for ceramic exchanges, 517
Carry-over, 702
Cellular recuperators, 509
Cement industry, clinker units for, 650
Cenospheres, 819
Ceramics, 507–519
  in checkerworks, 577
  for recuperators, 643
Channels, 435
  flow rate calculation for, 786–788
Chawla correlation, 139
Chebyshev polynomials, 784
Check-balls, 889
Checkerworks, 577
Chen's F-function, 22–23
Chevron dryers, 703
Chillers, design of, 176–190
Chilton-Colburn analogue, 615
Chlorides, 881
  and denting, 698
Circulation ratio, 733, 739, 741
Circumferential heat conduction, 600, 610, 712
Claude OTEC system, 84
Cleaning cycles, 807
  (*See also* Corrosion; Fouling)
Clearances, of baffles, 617
Coal:
  conversion of, 631–657
  fouling from, 799
  gasification of, 535
  hydrogenation plants, 651
Co-current exchangers, 502
Coefficient correlations, 652–653

Coils, 447
  convection in, 489
Coke formation, 631
Coke gasifier, 546
Colburn analogy, 948
Colburn factor, 431, 475
Colburn-Hougen method, 45–47
Collectors (*see* Solar energy; Storage of heat)
Colorimetric methods, 521–522
Compactness, 265–285, 397, 425–468
Computer modeling (*see* Models; Numerical solutions)
Concentration factors, 707–712
Condensation, 413–422, 478
  annulus condensation, 102
  condensation rates, 45
  design methods, 35–55
  differential condensation, 51
  drip condensation, 12, 761
  and enhanced surfaces, 11
  in honeycomb matrix, 599
  inside horizontal tubes, 67–83
  in hydrocarbon services, 175
  integral condensation, 51
  reflux condensation, 55–66
  and surface tension, 478
  and two-phase pressure drop, 67–83
  (*See also* Condensers)
Condensers:
  dead packets in, 760
  deaeration in, 761
  direct contact design, 15
  flow angles for, 10
  make-up water inflow, 757–765
  for OTEC applications, 83–95
  side condenser, 55–66
  tube damage in, 757
  two-phase flow in, 108
Conduction coefficient, 606
Conductivity, of storage material, 587
Conservation equations, 88
Contact angle, in nuclear boiling, 25
Convection:
  augmentation of, 269
  and forced boiling, 234–278
  free, 225
  gas convection, 539
  particle convection, 535
  and scraping of surface, 278
  transfer coefficients for, 582
  transfer equations, 605
Convergence criterion, 504

INDEX

Coolants:
  impurities in, 721
  temperature calculation for, 787
Coolers:
  airflow rates in, 357
  direct contact design, 650
  quench coolers, 650
Cooling towers, 117-118, 361, 363
Copper, deposits in tubes, 758
Corrosion, 631, 797, 833
  denting type, 698
  and deposition, 716
  in gasifiers, 648
  (*See also* Fouling)
Corrugated tubes, 101
Costs, energy-cost diagram, 659-672
Counter-current exchanges, 502
Counterflow, 117-129, 413-422, 581
Coupling, of heat and mass transfer, 601
Coupling number, 775
Cowper stove, 577
Cracking, of hydrocarbons, 470
Cracks, in tubes, 704
Creep rupture, of tubes, 647
Critical heat flux:
  and boiling crisis, 254
  and wall thickness, 152
Critical velocity, and turbulence, 326
Cross flow:
  in bundles, 300, 618
  and entrainment, 112
  and fluidelastic stability, 325-337
  and liquid entrainment, 197
  and multipassing, 428
  and permeable fins, 482
  power relation, 307-308
  and pressure drop, 109
  in reactors, 287
  and roughness, 311-322
  in tube banks, 265-285
  (*See also* Yawed bundles)
Cross-permeability, 482
Cryogenic systems, 4-5, 246, 251, 470, 483, 633, 673
Crystallization, 833
Cyclones, 625
Cylinders (*see* Tubes)

Damping factor, 343
Dana mineral classifications, 562
Dead pockets, in condensers, 760
Deaeration, 90, 757
Dedusting equipment, for gases, 625

Defrosting, of surfaces, 413
Degrees of freedom, 341
Demulsifiers, 846
Density-wave phenomena, 483
Denting, 689, 698, 700
Department of Energy (DOE), 561
Dephlegmators, 472
Deposits, 823-824, 833-840
  (*See also* Corrosion; Fouling; Scaling)
Desaltation, 844
Design, and TEMA method, 795
Desuperheating, 41
Dew point, 601
Diabatic flows, 259
Diehl correlation, 488
Diffusion:
  and deaeration, 764
  and fan systems, 387
  ionic diffusion, 853
  multi-component, 37
  one-way, 122
Dimensionless parameters, 580
Direct contact condensers, 15, 767
Dissipation, of turbulence, 314
Distillation, 55, 841
Distribution coefficient, for films, 714
Dittus-Boelter Correlation, 891, 947
DOE (*see* Department of Energy)
Double-tube evaporator, 104
Downcomers, 131, 138, 734, 744
Drag, 313, 444, 447
Drift velocity, 247
Drenching, in tube fields, 10
Drop-off, and Reynolds number, 451
Drop-wise condensation, 12
Droplets, 477, 763
  break-away of, 714
  carry-over, 702
  in heaters, 770
  in two-phase flow, 260-261
Dry cooling towers, 361
Dryers, 703
Dryout, 255
  in horizontal flow, 96
  post-dryout region, 712
  and shellside boiling, 182
  in steam generators, 722
  in tubes, 646

Economizers, 747
Eddies:
  eddy diffusion, 829

Eddies (*Cont.*):
   eddy viscosity, 160
   and turbulence, 326
Efficiency:
   by-pass efficiency, 495
   and compactness, 282
   in conversion processes, 631
   and crossflow, 265-285
   energy-cost diagram, 659-672
   fin efficiency, 475
   high efficiency exchangers, 726-728
   and reliability, 707-719
   waste heat boilers, 644
   (*See also* Augmentation; Waste heat, recovery of)
Electric power plants, 707
Electrochemical model, 616
Electrolytic experiments, 163
Electron microscopy, 826
Electrostatic precipitators, 625
Elliptic-type equations, 776
Energy Cost Characteristic Diagrams (CAREC), 662
Enhanced surfaces, 11, 175-191, 193-203, 426
Enthalpy, temperature profiles, 504
Entrainment, 107-116
   in annular flow, 71-72
   in crossflow, 112
   and deposition, 260
   entrainment parameter, 111
   from fins, 477
Entrance effects, 297-299
   and pressure drops, 435
Entropy, augmentation number, 912
Erosion, and impingement plates, 636
Evaporation:
   in annulus, 97-99
   countercurrent, 83
   deaeration in, 90
   double-tube systems, 95-107
   with enhancing tubing, 175
   of falling films, 83-95, 206-208
   flash evaporation, 84
   and liquefied gas, 221
   pre-evaporator, use of, 95
   steam-heated systems, 221-233
   thermo-syphon methods, 133
   tube-and-shell systems, 131-141
   in vertical tubes, 157-174
   water bath evaporators, 221-233
Exchangers:
   air-cooled, 357-362, 363-383

   BEM design, 636
   ceramic, 507-519
   coils in, 447
   compact (*see* Compactness)
   with condensation, 413-422
   design of, 615-629
   direct-contact types, 767-773
   double-tube exchangers, 95-107
   effectiveness of, 87
   efficiency of (*see* Efficiency)
   energy-cost diagram, 659-672
   extended surface exchangers, 427
   falling cloud type, 564
   fluid bed (*see* Fluidized beds)
   Hampson design, 485
   and heat storage, 572
   helium/helium type, 299
   High Flux exchangers, 177
   liquid metal exchangers, 726-727
   modeling of (*see* Models)
   multi-stream, 495-506
   with phase change, 3-18
   plate-type, 495-506, 638
   prestressed tubes in, 339-353
   regenerators, 413, 599-611
   rod-baffle exchanger, 637
   rotary air-to-air systems, 599
   safety of, 639
   shell-and-tube types, 131-141, 265-285, 287-298, 427, 743
   wound-coil exchangers, 470, 485
   yawed tube, 299-310
Exergy, conservation of, 909-912
Exhausts, 413, 591
   (*See also* Pollution; Waste heat, recovery of)
Extreme regime, 769

Fanning friction factor, 166, 433, 948
Fans:
   in air-cooled exchangers, 386, 394
   noise from, 362
Fatigue:
   fatigue curve, 724
   structural, 352
F-chart method, 591
Feolite, 573
F-function of Chen, 22
Fick's law, 763
Films:
   coefficients for, 181, 651
   condensation of, 181, 206, 760

INDEX **969**

Films (Cont.):
  disturbance waves in, 56
  draining film, 14
  falling films, 83-95, 122, 206-208
  film boiling, 242, 750
  liquid micro-films, 254
  mass-transfer models for, 51
  matrix film model, 37
  and Nusselt number, 477
  and Reynolds number, 157, 161
  and rivulet model, 716
  thin film model, 552
  turbulent films, 140
  and vapor shear, 11
  in vertical tubes, 157-174
Finite conductivity model, 591
Finite difference methods (see Numerical solutions)
Fins, 274, 397
  in air conditioners, 457
  back-to-back design, 481
  with burrs, 443
  circular fins, 460
  and Colburn factor, 472
  continuous fins, 454
  efficiency of, 475, 496
  and entrainment, 477
  and friction factor, 449, 472
  louver fins, 444
  offset strip fins, 440
  perforated fins, 446
  plain fins, 439, 449
  plate-fin assembly, 427, 428, 495-519
  segmented fins, 456
  serrated fins, 480
  and Sherwood number, 439
  slotted fins, 457
  spacing correlation for, 449
  spine fins, 456
  strip fins, 407
  in tubes, 269, 899-916, 917-931
  types of, diagrams, 429-430
Firebricks, materials in, 573
Fixed bed regenerator, 578
Flash hydropyrolysis, 646
Flat slab units, 592
Flooded chiller (see Kettle reboilers)
Flooding:
  in adiabatic systems, 55
  and bubble formation, 136
  in reflux condenser, 56
Flow analysis, in plate exchangers, 521-531
Flow boiling (see Boiling)

Flow-induced vibrations, 339
Flow rates redistribution, 788
Flow resistance coefficient, 303
Fluid stream equation, 579
Fluidelastic stability, 325-337
Fluidization, of magnetic particles, 564
Fluidized beds, 535-548, 650, 885-896
  bed expansion equation, 543
  fouling in, 562
  magnetic particles in, 564
  materials used in, 555
  for melting furnace, 563
  and pollution, 550
  pressure drop in, 557
  and waste heat recovery, 549-568
Fluted tubes (see Tubes, fluted)
Flux meters, 799, 821
Fog formation, 13
Fouling, 283, 632, 634, 795-815
  auto-retardation in, 805
  chemical cleaning, 882
  in coal conversion, 653
  of cooling water, 863-871
  by deposits, 804
  in fluidized beds, 562
  of furnaces, 817-831
  from hard waters, 853-861
  in plate exchangers, 871-883
  by precipitation, 853, 863-871
  in preheat trains, 841
  by suspended particles, 833-840
  and throttling, 885
Fourier analysis, 146, 340
Free stream turbulence, 325
Freons (see Halocarbons)
Fretting, in OTSG, 705
Friction:
  Blasius solution, 405
  coefficient of, 126, 315
  Fanning friction factor, 404, 433
  from fins, 449
  and heat transfer, 946-947
  and louvred surfaces, 397
  and Reynolds number, 126
  skin friction, 438
  (See also Pressure drop; Surfaces)
Friedel multiplier, 697
Frosting, 413-422
  in air-to-air system, 608
  in honeycomb matrix, 599
  and nonlinear behavior, 601
Froude number, 110, 230
Functional design description, 696

Furnaces, 747
  fluidized bed for, 563
  fouling of, 817-831

Gas-solid system, 535
Gas turbines, 507
Gases:
  and convection, 539
  and cracking process, 650
  dedusting equipment for, 625
  in fluidized beds, 564
  friction factor for, 73
  gas convection, 539
  gas flow, correlations for, 436
  gasification, of coal, 535, 632
  gasifiers, Koppers-Totzek, 647
  heat transfer coefficient for, 426
  kinetic theory of, 537
  liquefaction of, 221-233, 469, 495
  non-condensing, 35-55
  partial oxidation systems, 647
  quenching of, 645-646
  recovery from wastes, 469
Geothermal systems, 708
Geysering, 483
Glass plants, 593, 639-640
Glauber's salt, 574
Glycol, injection of, 67
Gradient problem, 674
Green's equations, 775
Gregorig surface (see Surfaces, fluted)
Grids:
  and turbulence, 325-337
  in wind tunnels, 329
Groeneveld-Delorme method, 749

Half-exchanger, costs for, 660
Halocarbons, 175
Heat exchangers (see Exchangers)
Heat pipe arrays, 642
Heat shocks, 882
Heat wheels, 507-519
Heaviside function, 524
Heliostats, 514
Helium-cooled reactors, 287, 339
Helium, two-phase flow of, 251-262
Heller-cycle power plant, 16
Helmholtz resonator, 65
Henry-Dalton law, 762
Henry's number, 87
High Temperature Gas Cooled Reactor (HTRG), 299

High Temperature Institute, 708
Homes (see Residential systems)
Honeycomb matrix, 599
Honeycomb recuperators, 509
HTRG (see High Temperature Gas Cooled Reactor)
Humidity ratio, 414, 601
Hydraulic resistance coefficient, 303
Hydraulic shocks, 769
Hydrazine, 757, 759
Hydrocarbon processes, 175, 181, 194, 469, 841-852
Hydrogen, preheating of, 640
Hydrogenation plants, 651
Hydropyrolysis reactor, 646

Icing, 413
Immiscible liquid phase, 35-55
Impingement plates, 616, 636
Independence principle, 300
Infrared thermography, 522
Inhibitors, 700
In-line bundles, 287-298
Instability:
  in cryogenic systems, 483
  density-wave form, 483
  fluidelastic, 333
  and transfer equations, 503
  and turbulence, 951
Interfacial resistance, 736
Intergranular attack (IGA), 689
Integron fin configuration, 926-927
Ionic diffusion model, 853, 855

Jens-Lottes correlation, 749
Jung module, 723

Kantorovich method, 785
Kapitza number, 163
Kerosene, as transfer fluid, 648
Kettle reboilers, 6-8, 179
Knudsen number, 538-539
Kolerator, of Hewlett Packard, 124
Kon'kov correlation, 751
Kutateladze equation, 153, 256
Kutateladze-Zuber equation, 182

Laminar boundary layer, 270
Laminar flow, 117, 436, 652

INDEX

Lancing, of sludge, 702
Langelier saturation index, 853
Laplace equation, 930
Ledinegg forms, in boiling, 483
Legendre polynomials, 784
Leidenfrost point, 225, 254
Leveque solutions, 955
Lignite converters, 631
Liquefaction plants, 632
Liquid metal heat exchangers, 726
Liquid sheets, 16
Load coefficient, definition of, 384
Lockhart-Martinelli parameter, 23, 72, 96
Lottes-Flinn correlation, 752
Louvred surfaces, 397-412
Lumped heat transfer coefficient, 582

Magnesium, and scaling, 858
Magnetic particles, in fluid beds, 564
Magnetic systems, superconducting, 251
Martinelli parameter, 487
Matrix film model, 37
Melting furnace, 563
Methanol conversion, 631
Microstreams, 279
Mineral classifications, 562
Miscible liquid phase, 35-55
Mixing:
  inertial losses in, 442
  mixing length theory, 120
  in PWR systems, 733
  static, 953-964
Models:
  annular flow model, 68-70
  Bankoff variable density model, 139
  Colburn and Hougen method, 37
  for complex networks, 673-686
  for counter flow, 117-129
  for eddy diffusivity, 829
  electrochemical modeling method, 615
  for evaporation, 85-87
  falling-rate model, 805
  film models, 22, 35-55, 157-174, 552
  finite thermal conductivity model, 591
  for heat storage, 586, 591
  homogeneous models, 735, 753
  ionic diffusion model, 853
  Levich model, 157
  Levich-van Driest model, 164
  of louvred surfaces, 399
  lumped parameter model, 55
  Mann model, 716

  mass-transfer models, 51
  matrix models, 37, 47
  moving particle model, 552
  network simulation, 673-686
  of once-through boilers, 747-765
  for packing beds, 589
  of plate exchangers, 522
  for precipitation fouling, 864
  for prestressed tubes, 339-353
  of regenerators, 579
  rivulet model, 716
  Rohsenow's model, 162
  scaling of, 735
  separated flow model, 111-112
  shear model, zero interface, 109
  Silver method, 13
  slip model, 735
  of steam generator, 734
  thermal-hydraulic model, 748-749
  thin film model, 552
  transient conduction model, 552
  tube models, 345-346
  for turbulence, 157-174, 314
  for U-tube generator, 744
  of vapor condensation, 417-420
  of vapor shear, 11
  wall models, 162, 535-548
  (*See also* Numerical solutions)
Momentum equation, 63
Moody friction coefficient, 947
Moussalli correlation, 139
Multi-component liquids, 9-10
Multipassing effects, 428, 511
Multi-stream exchangers, 495-506

Naphtha stabilizer, 184
National Engineering Laboratory, 8
National Research Council, 571
Natural gas, liquefaction of, 469-471, 495
Navier-Stokes equations, 314
Network analysis, 92
Neumann problems, 782
Newton-Raphson method, 163
Nikuradse formula, 120
Nitrogen:
  flow boiling of, 233-251
  liquefaction of, 469
  and nucleate boiling, 29
Nodules, 827
  in deposition processes, 819
Noise:
  in air-cooled exchangers, 363-383

Noise (Cont.):
  in heaters, 769
  noise level limitation, 357
  and vortex shedding, 447
Noncondensable gas effects, 84
Nonlinear behavior, 600
Nuclear technology, 721-732
  crisis conditions, for plants, 714
  and flow boiling, 233
  helium cooling in, 708
  HTRG reactors, 287, 339
  NSSS system, 689-692
  PWR plants, 733
  steam generators, 689-705
  waste heat recovery, 550
  wound-coil exchangers, 485
Nucleate boiling, 29, 197, 474-475, 803
  bubble nucleation, 85
  and forced convective boiling, 20
  in single tube, 193-194
  stable, 154
  and transfer coefficient, 24
  in vertical tubes, 25
Numerical solutions, 120, 215, 261, 599-611, 674, 775, 931
  Adam-Moulton method, 88
  Euler's method, 604
  URSULA code, 733
  (See also Models)
Nusselt number, 162, 293, 301, 433
  and channel shapes, 433-434
  definition of, 946-947
  in entrance region, 439, 955
  and horizontal tubes, 11
  and laminar flow, 162
  Nusselt equation, 760
  Nusselt-Peclet relationship, 958
  for particle convection, 538
  and Reynolds number, 164, 486
  and vapor shear, 11

Oblique flow, 299
Ocean Thermal Energy Conversion (OTEC), 5, 83-95
  open cycle system, 84
Offset strip fins, 440
Oil:
  embargo of, 507
  fouling from, 180
  oil coolers, 446
  preheat trains, 841
  in transformers, 282
  (See also Petroleum refining)
Olefins, 469
  processing plants, 176
Once-through power boilers, 747-765
One-way diffusion, 122
Open cycle OTEC system, 84
Operability, of exchangers, 4
Organics, vaporization of, 175
Orifice baffles, 623
Oscillations, equations for, 63
OTEC (see Ocean Thermal Energy Conversion)
Oxygen:
  in fouling, 637
  and hydrazine, 759
  liquefaction of, 469
  in make-up water, 757
Ozaki quench cooler, 650

Packed beds:
  gravel in, 573
  model for, 589
Packing, of cooling towers, 117-118
Parabolic equations, 120
Particles:
  collision of, 829
  in fluid beds, 539-541
  magnetic, 564
  moving particle model, 552
  particle convection, 537-538
  scouring action of, 552
  suspended, 833-840
Partition factors, 307
Peak heat flux, in bundles, 8
Pearson distribution function, 724
Perforated fins, 446
Perimeter angle, and transfer coefficient, 154
Perimeter averaging, of transfer coefficient, 20-21
Petrochemical operations, 634
Petroleum refining, 175
Phase change, 601
  in heat exchangers, 3-18
  phase change materials, 573, 590
Phase equilibrium, 9
Photography, 820
  high-speed, 738
  of reboiler circulation, 7
Pin fins, 466
  and vortex shedding, 447

INDEX

Pipes, 642
  pipe dust, 180
  turbulent pipe flow, 120
  (*See also* Tubes)
Pitting, 758
  in OTSG, 705
Plain tube bundles, 194, 197
Planar jets, 84
Plate exchangers, fluid flow in, 521
Plate-fin exchangers, 427, 470, 495-519
  in air separation, 495
Plug flow, in reboilers, 132
Pohlhausen expression, 405
Pollution:
  adsorbents, 556
  and fluid beds, 550
Polynomial curve-fitting, 194
Polytropic state change, 63
Pool boiling, 153
  in chillers, 180
  equations for, 24
  and kettle reboilers, 7
  Kutateladze equation for, 257
  Stephan's equation, 240
Porosity, of regenerator, 584
Post-dryout region, 712
Power boilers, 747-765
Power plants, 707
  flow analysis of, 10
  Heller-cycle type, 16
  pre-heating in, 576
  tube damage in, 757
Power spectral density, 326
Power units, transfer calculation for, 775-792
Prandtl mixing length theory, 120
Prandtl number, 121, 162, 955
  turbulent, 121
Precipitation fouling, 797, 853
Pre-condenser, 100
Predictor-corrector method, 88
Preheating, 635, 665
  heat wheel system, 640
  oil preheat trains, 841-852
  in power plants, 576-577
  and recirculation, 158
  rotary systems, 641
Pressure:
  and crossflow, 108
  and nucleate boiling, 238
  pressure transducers, 58, 74
  and shear stress, 67
  and transfer coefficient, 24, 27

and vapor flow, 11
and void fraction, 62
(*See also* Pressure drop)
Pressure drop:
  in air-cooled exchangers, 383
  across baffle clearance, 618
  Chenoweth-Martin correlation, 488
  in diabatic flow, 251
  across distribution plate, 886
  effects of, 266
  and entrainment, 112
  in exchanger networks, 674
  at exit distributor, 472
  and extended surfaces, 960
  and Fanning factor, 438
  and flow reversal, 62
  in fluidized beds, 557
  frictional pressure drop, 70-72, 257
  along furnace channels, 752
  gradients, frictional, 74
  and heat storage, 575
  in hydrocarbon reboilers, 188
  hydrostatic, 8-9
  isothermal, 455
  measurements of, 616
  in parallel tube banks, 755
  partition factor, 308
  in PWR system, 735
  in reboiler, 137
  reduction of, 188
  in regenerators, 582
  and Reynolds number, 287-298
  and roughness, 277
  and saturation temperature, 84
  of SMV-packed pipe, 961
  and swirlers, 937-939
  on throttle, 790
  two-phase pressure drop, 67-83, 726
  two-phase multipliers, 697
  and yaw angle, 306
  (*See also* Friction)
Prestressed tubes, 339-353
Process heaters, 550
Propylene, 181
Pseudo-homogeneous equation, 109
Pumping power, 909-910
Purification processes, 4
Pyroclasts, 817, 820

Quench cooler, 650

Radiation, 539, 544
Radioactive tracer technique, 805
Raining packed bed, 564
Rankine cycle, 633, 649
  and OTEC system, 84
Reactors (see Nuclear technology)
Reattachment, 289
Reboilers, 9-10, 131-141, 178
  flooding, of vapor space, 133
  for hydrocarbons, 194
  and plug flow, 132
  sealing steam reboiler, 132
  shell-tube, 175
  thermosyphon, 470
  vertical, 186
Recirculation:
  of air flow, 120
  in bundles, 183
  and preheating, 158
  region of, 317
Recovery, of waste heat, 456
Recuperators, 299, 508, 593
  Biery system, 642
  cage type, 639-640
Recycling, of waste heat, 456, 549-568
Re-entrainment, 802
Re-entrant cavities, 5
  and high flow velocity, 197
Refineries, 841
Reflux condensers, 14
  complete, 57
  in vertical tubes, 55-66
Refreshment length, 955
Refrigerants:
  refrigerant R-12, 28, 47
  refrigerant R-22, 95
  test use of, 95
  and wall friction, 24
Regenerators, 280, 428, 507-519, 573
  batch type, 640
  counterflow, 581
  finite-difference approximation of, 604
  industrial use of, 593
  nonlinear behavior of, 600
  and recuperators, 593
  regenerative wheel, 413
  rotary, 577, 599-611
  and thermal storage, 571-597
Regression analysis, 164
Reheaters, 714, 747
  in nuclear plants, 728
Reliability:
  of exchangers, 707-719
  of nuclear plants, 721
Residential systems, ventilation, 599
Resistance, in mass transfer, 13
Resistance, in yawed tubes, 299-310
Resistance thermometers, 208
Reynolds number, 121, 287-298, 622
  and electrochemical model, 616
  and Fanning friction factor, 433
  and film friction factor, 70
  for films, 161
  and fluted tubes, 213
  and friction coefficient, 126
  and in-line tube bank, 299
  and laminar flow, 121, 432
  and Nusselt number, 486
  and organics, 181
  and pressure drop, 288, 297
  resultant Re number, 279
  and rollover, 451
  and Stanton number, 404
  vapor phase number, 84
  and vapor shear, 11
  and yaw angle, 299
Richardson-Zake equation, 543
Ritz method, 784, 792
Rivulet model, 716
Rod-baffle exchanger, 637
Rollover, 483
Rotary exchangers, 599
Rotary preheaters, 641
Rotary regenerators, 599
Rothemuhle air heater, 641
Rotor, equations for, 605-606
Roughness, 275
  and annular flow model, 67
  and augmentation, 311-322
  and corrugation, 101, 103
  interfacial, 67
  and pressure drop, 277, 287, 291, 320
  and Reynolds number, 291
  and Stanton number, 284
  and transfer coefficient, 27
  in tube banks, 311-322
  and Wallis correlation, 77
Row effects, in tube banks, 460
Ryznar stability index, 853

Safety:
  plate exchangers, 639
  of PWR plants, 733
  in reboiler design, 179

INDEX

Safety (*Cont.*):
  shielding, use of, 639
  of steam injection system, 231
Salts, 637
  hydrates, 574
  tube deposits, 709-712
  (*See also* Corrosion; Fouling)
Sand particles, in fluidized bed, 545
Saturated flow:
  and boiling, 19, 143-156
  and plug flow, 132
  saturation curves, 40
  saturation pressure, 11, 84, 763
  saturation temperature, 30
  vapor structure curve, 38
Scaling:
  by calcium carbonate, 879
  from hard waters, 853
  (*See also* Deposits)
Schlunder formula, 538
Schmidt number, 87, 121, 801
  for dissolved air, 87
  Schmidt correlation, 487
  turbulent, 121
Schwarz method, 775
Scrubbers, 625
Sealing steam reboiler, 132
Seide formula, 345
Segmented fins, 456
Separate cylinders model, 69
Separated flow, 315
  annular flow model, 68
  and entrainment, 115
  separated flow model, 108, 111
Separation bubble, 271
Separation point, 294
Separator-reheaters, 778
Shape drag, 274
Shared surface model, 44
Shear stress, 160
  interfacial, 70
  and pressure gradient, 67
  vapor shear, 4, 11, 158
  zero interface model, 109
Shell, integration of downcomers in, 131
Shell-and-tube exchangers, 265-285, 592, 615, 669
  costs for, 636
Shell-side coefficient, 139
Shell-side, of exchangers, 107-116
Shell-tube reboilers, 175
Sherwood number, 127
Shielding, of plate exchangers, 639

Ships, waste heat boiler for, 559
Shutdown corrosion, 135
Side condenser, 55-66
Sieve trays, and Weber number, 230
Silicon, in storage systems, 573
Silver approximation method, 13, 35-55
Simpson's rule, 931
Simulation methods, 734
  (*See also* Models)
Simultaneous equations, matrix inversion, 14
Single blow problems, 586
Sintered layer, 105
Sintered surfaces, 195
Skin friction, 293, 438
  Chien-Ibele correlation, 86
Slag, 817
Slip ratios, and enhanced bundles, 198
Slot type distributor, 93
Slotted fins, 457
Slug flow, 139, 749
  and condensation, 68
Slurries, 631
  erosion effects, 652
  saltation of, 637
SMV mixing elements, 962
Soaking, and inhibitors, 700
Solar energy, 516, 571-572, 591
  solar collectors, 593
Solidification fouling, 797
Sootblowers, 647, 824
Sound, 357-383
  in air-cooled exchangers, 383
  damping with baffles, 376
  power level, 385
  propagation law, 368
Spectral analysis, 340, 345
Spine fins, 456
Spiny tubes, 275
Spiral exchangers, 638
Spiral fins, 905
Splitters, 176
Spodumene, 509
Spray distribution system, 93
Spray flow, 132
Spray heaters, 769-773
Spray-water distribution, 768
Stabilizer reboilers, 178
Staggered exchangers, 287-298, 317, 459
Stagnant film theory, 87
Stagnation flow, 801
  and baffle spacing, 627
Stagnation point, 289
Stanton number, 284, 404

INDEX

Static mixers, 953–964
Static pressure drop, 138
Static pressure, and frictional drag, 483
Stationary mass regenerators, 507–519
Steady-state model, 748
Steam:
    mass flow rate, 89
    non-radioactive, 131
    supply system for, 226
    void fraction for, 139
Steam generators:
    double tube, 645
    gland sealing, 131
    in nuclear plants, 689–705, 721
    sodium heated, 722
    steam enthalpy, 674
    U-tube generators, 733
    wound-coil systems, 485
    (See also Gases; Vapor)
Steam injection system, 230
Steam void fraction, 139
Steam-water-mixture, pressure drops for, 113
Steel plants, regenerators for, 593
Steel soaking pit:
    furnaces for, 507
    recuperator for, 511
Stefan-Maxwell equations, 14
Stephan's equation, 240
Stickiness, in fouling, 818
Stirling engines, 428, 641
Stochastic processes, 722
Stokes-Einstein equation, 800
Storage of heat, 571–597
    flat slab units, 586
    materials used in, 573
    and pressure drops, 575
    for solar systems, 591
Stratified flow, 31, 109
    and evaporation, 102
    and nucleate boiling, 31
    in two-phase flow, 722
Stresses, thermal, 645
Strip fins, 443
    louvred, 399
Structural method, 785
Studded fins, 457
Subcooling, 95, 132, 742, 760
    in direct contact condenser, 15
    and downcomers, 133
    of liquid film, 86, 166
    prevention of, 767
    thermal entry effect, 166
Sublimation, of nitrogen, 247

Submerged bundles, 107–116, 221
Sulfates, in denting, 700
Sulfidation, of walls, 648
Superconduction, 251
Supercooling, 251
Supercritical flow regime, 272
Superheating, 474, 692, 731, 747
    wet-wall desuperheating, 37
    (See also Desuperheating)
Super position techniques, 591
Suppression factors, for enchanced boiling models, 198
Surfaces:
    and bubble nucleation, 197
    burnout of, 242
    of collectors, 591
    compact surfaces, 425
    defrosting of, 413
    with dimples, 398
    enhanced, 5, 175–191, 193–203, 426
    in fluidized beds, 535–548
    fluted, 12
    freezing on, 413
    High Flux surface, 176
    interrupted surfaces, 438
    iron oxide formation, 833
    louvred, 397–412
    non-wetting, 12
    offset strip fins, 440
    perforated, 446
    porus, 178
    power unit boundaries, 775–792
    roughness of (see Roughness)
    shared surface model, 44
    sintered, 195
    subcooling of, 11
    surface factor, 194, 197
    surface tension, 12, 157, 162
    tube-fin surfaces, 447–449
    waves on, 168, 278
    wetting of, 245
Surface tension:
    and condensation, 478
    of cryogenic fluids, 470
    and eddy damping, 162
    and turbulence, 157
Swirl angle, 933–945
Swirlers, 269
Symmetric flow, model for, 127

Tangential conduction, 29
Tankers, 559
Teflon spacers, 57

INDEX

TEMA exchangers, 221, 636
  design methods, 795
Temperature/enthalpy relationship, 38
Tenemax, 573
Thermal acoustics, 483
Thermal boundary layer (*see* Boundary layer)
Thermal ratio, for regenerators, 584
Thermal stresses, 645
Thermocouples, 74
  placement of, 119
Thermodynamic ratio, 63
Thermometer, Pt-resistance, 160
Thermophoresis, 801
Thermographic analysis, 521-531
Thermo-syphon reboiler, 131-141, 178, 185, 470
  High Flux type, 185
  instability in, 483
  model for, 8
Throttle pressure drops, 790-791
Throttling, and fouling, 885
Time-averaged variables, 120
Top flooding, 56
Toulene, 46, 176
Towers:
  control systems for, 178
  packing of, 117
Tracer measurements, of wetness, 135
Transcritical flow conditions, 293-296
Transfer coefficients, correlations for, 86
Transfer constraint, 602
Transformer oil, 282, 311
Transient analysis, 591
Transient conduction model, 552
Transition points, 48
Transmission coefficient, 723
Triangular relationship, in annular flow, 69
Triple interface, 795
Tubes:
  ACHE bundle design, 389
  aluminum, 458
  baffle holes in, 179
  bayonet tube bundle, 637
  boiling in, 19-34
  bundle geometry factor, 196
  check-balls in, 889
  coiled tubes, 230, 724
  with continuous fins, 447, 454
  copper deposits in, 758
  corrugated, 101, 103
  with cracks, 704
  creep rupture of, 647
  and crossflow, 265-285

double-tube evaporator, 104
double-tube exchangers, 95-107
dry-out in, 646
dynamic equations for, 341-342
and eddy currents, 704
enhanced evaporator tubing, 175
and enhanced surfaces, 193-203
enlargement, and pressure drop, 943-944
evaporation in, 97-101
finned, 269, 455, 558, 899-916
and fluidelastic stability, 325-337
fluted, 186, 205-219
fretting of, 702
Gewa-T tubes, 194
half-tubes, 113
heat transfer enhancement in, 214-217
helices in, 281
High Flux tubing, 176
impacting of, 339
inclined bundles, 300
inline, 459
low, finned, 11
and nitrogen boiling, 233-251
nucleate boiling in, 29
partition plates in, 759
pitches of, 313
pressure variation along, 76
prestressed, 339-353
removal of, 621
resistance coefficients, 303
ribbed tubes, 748, 751
roughness of surfaces, 311-322
salt deposits in, 711
shell-and-tube exchangers, 265-285, 743
spiny tubes, 275
spirally fluted, 945-952
staggered banks, 272, 293, 317, 454, 459
submergence of, 107-116, 221
sulfidation, of walls, 648
surface factor for, 195
swirled flow, 281
transfer along perimeter of, 143-156
tube/baffle clearances, 617
tube damage, 761
tube-fin exchangers, 427
tubeside condensation, 67
turbine condenser tubes, 757-765
turbulence in, 325-337
and two-phase flow in, 108
two-phase pressure drop in, 67-83
U-tube bundle, 637
vertical, condensation in, 55-66
vibration of, 269, 341, 689, 702
wall thickness, 152

Tubes (Cont.):
　yawed, 266
　(See also Fouling; Pipes; Surfaces; Walls)
Turbine systems, 514
　Brayton system, 516
　and compact exchangers, 397
　condensor tubes, 757-765
　design of, 703
　in direct-contact exchangers, 767-773
　gland sealing steam generation, 131-141
　in PWR plants, 733
　steam reheaters for, 728
Turbulators, 446
Turbulence, 907
　artificial, 272
　and autocorrelation, 334
　of boundary layer, 268
　falling turbulent films, 84
　and fluidelastic stability, 325
　in fluidized beds, 565
　and fouling, 828
　free stream turbulence, 325
　grid generated, 325-337
　and instability, 951
　k-e model of, 314
　kinetic energy of, 124, 317
　and mass transfer, 9
　in pipe flow, 120
　Prandtl mixing length theory, 128
　and Prandtl number, 162
　pseudo-turbulent flow, 475
　reattachment, 289
　and surface tension, 157
　and swirl angle, 933-945
　and thermal entry length, 941
　turbulence grids, 329
　turbulent films, 157
　turbulent Schmidt number, 121
　and vapor shear, 11
　and viscosity, 268, 314
Turbulent film flow, 157-174
Turbulent flow, and laminar flow, 120
Turbulizers, 269
Twisted tapes, in evaporators, 226
Two-phase applications, 470
Two-phase flows, 95-107
　countercurrent flow, 4
　diabatic, 261
　of helium, 251-262
　Kesper method, 140
　in steam generators, 724
Two-phase pressure drop, 487
Two-phase processes, 480

Universal velocity profile, 162
U-tubes:
　bundles of, 637
　regenerators, 733-747
　steam generator, 693
　for TEMA exchanger, 221

Valve leakoff steam, 131
Van Driest equation, 121
Van Driest profile, 163
Vane swirlers, 933-945
Vapor, 3-18
　binary vapors, 13
　Colburn-Hongen method, 37
　condensation of, 35-55, 413-422
　and drenching, 10
　enthalpy of, 740
　integral condensation of, 51
　molar flux, 38
　in MSR systems, 729
　and multicomponent boiling, 9
　multicomponent diffusion, 14
　organics, vaporization of, 175
　pressure drop in, 14
　reboiler vapor space, 133
　saturated vapor, 158
　superimposed flow of, 10, 157-174
　vapor blanketing, 179
　vapor core, 22
　vapor drag, 181
　vapor-liquid interface, 14
　vapor phase Reynolds number, 11, 84
　vapor shear, 4
　vaporizers, 3-18, 177
　volume fraction (see Void distributions)
　(See also Gases)
Variable density model, of void fraction, 139
Variational methods, 775, 777, 791-792
Velocity field equations, 120
Velocity profile of Prandtl-Nikuradse, 162
Ventilation system, 599
　and heat recovery, 413
Venting, optimal design of, 13
Vertical tubes, 55-66
　evaporation inside, 157-174
　and film condensation, 157-174
　nucleate boiling in, 25, 29
Vertical reboilers, 186
Vibrations, 278-279, 446-447, 689, 724
　flow induced, 339-353
　and fluidelastic stability, 325
　and fretting, 702
　nonlinear, 339-353

INDEX

Viscosity, 652
   eddy-viscosity distribution, 168
   and mean free path, 539
   turbulent viscosity, 160, 314
   and turbulence, 268
   very viscous fluids, 953-964
   viscous sublayer, 121
Viscous dissipation, 65
Void distributions, 736
Void fraction, 139, 542
   calculation of, 69
   capacitance probe for, 58
   measurement of, 55, 735
   variable density model, 139
Void-velocity distribution, 733
Volatility, 9
   of isopropanol, 47
   spectrum of, 9
   (See also Vaporization)
Von Karman method, 326
Vortex shedding, 272, 447

Waggener correlation, 752
Wallis correlation, 56, 79
   for adiabatic systems, 56
Walls:
   during boiling crisis, 243
   at constant temperature, 957
   deposits on (see Deposits; Fouling)
   desuperheating of, 37
   and dry-out region, 722
   and frictional pressure drop, 257
   furnace walls, 747, 754
   and Lockhart-Martinelli equation, 23
   and pressure drop, 266
   sulfidation of, 648
   temperature of, 148, 149-152, 314, 484, 500, 749
   thickness of, 151
   wall friction, 24
   wall models, 162
   wall-to-bed transfer, 546
   wetted surfaces, 26
   (See also Corrosion; Fouling; Surfaces; Tubes)
Waste heat, recovery of, 456, 507, 549-568
   fluidized bed system, 560
   Rankine cycles for, 649
   waste heat boilers, 644
Water film, 127
Water hammer, 703
Water vapor condensation, 413-422
   (See also Steam; Vapor)
Water-walls, in furnace, 754
Watertube boilers, 647
Waves, on surfaces, 168, 278
Wavy annular flow, 57
Wavy fins, 439, 453
Weber number, 230
Weeping, and sieve trays, 230
Westinghouse steam generator, 694
Wet-wall desuperheating, 37, 41
Wetness, steam wetness, 135
Wetting, effect of flow direction, 20
Whalley-Hewitt correlation, 79
White noise, 347-348
Wind tunnels, 325-337, 403
   grids in, 329
Windings, in superconduction, 251
Window baffles, 188
Window zones, in bundles, 618
Wire loop fins, 457

Yawed bundles, 299-310

Zeroth Law, of equilibrium, 673